Aeronautics and Astronautics
AIDAA XXVII International Congress

AIDAA XXVII International Congress, 4th-7th September 2023, Padova, Italy

Editors

Sergio De Rosa[1], Marco Petrolo[2], Mirco Zaccariotto[3]

[1]Università di Napoli Federico II, Italy

[2]Politecnico di Torino, Italy

[3]Università di Padova, Italy

Peer review statement

All papers published in this volume of "Materials Research Proceedings" have been peer reviewed. The process of peer review was initiated and overseen by the above proceedings editors. All reviews were conducted by expert referees in accordance to Materials Research Forum LLC high standards.

Published under License by **Materials Research Forum LLC**
Millersville, PA 17551, USA

Published as part of the proceedings series
Materials Research Proceedings
Volume 37 (2023)

ISSN 2474-3941 (Print)
ISSN 2474-395X (Online)

ISBN 978-1-64490-280-6 (Print)
ISBN 978-1-64490-281-3 (eBook)

Distributed worldwide by

Materials Research Forum LLC
105 Springdale Lane
Millersville, PA 17551
USA
https://www.mrforum.com

Manufactured in the United State of America
10 9 8 7 6 5 4 3 2 1

Table of Contents

Aeroacoustics

Aeroelasticity

Aeronautical Propulsion

Aeronautical Systems

Air Traffic Control, Aircraft Operations and Navigation

Aircraft Design and Aeronautical Flight Mechanics

Artificial Intelligence Application

Fluid-Dynamics

General Session

Materials and Aerospace Structures

MOST Project

Satellite and Space Systems

Space Flight Mechanics

Space Propulsion

Special Session in Memory of Professor Debei

Vibroacoustics

XR and Human Factors for Future Air Mobility

Preface

AIDAA (www.aidaa.it), founded in 1920, was one of the first Aerospace Associations worldwide and aims to promote the development and diffusion of aeronautical and space science. AIDAA organizes an international congress every other year, sharing the most recent progress in Aerospace Science and Technology.

AIDAA is the hosting society of three world congresses in 2024: the International Astronautical Congress (IAC 2024, Milano); the Congress of the International Council of Aeronautical Sciences (ICAS, Firenze); the AIAA/CEAS Aeroacoustics Congress (Roma). For the first time, these congresses are hosted by the same country just a few weeks apart. To pave the way to 2024, AIDAA launched the Aerospace Italy 2024 Initiative, www.aidaa.it/aerospaceitaly2024/, to promote and support the organization of events in Italy and define the Road to 2024.

The 27th AIDAA Congress (https://www.aidaa.it/aidaa2023/) was held in the Beato Pellegrino Complex of the University of Padua from the 4th to the 7th of September 2023. About 176 papers were presented in all scientific and aerospace engineering fields, including aerodynamics and fluid dynamics, propulsion, materials and structures, aerospace systems, flight mechanics and control, space systems, and missions. Two hundred fifteen attendees participated, and eight plenary talks were delivered: the Vice-President of the Azerbaijan Space Agency, Azercosmos, Dunay Badirkhanov; Professor Gianluca Iaccarino from Stanford; Franco Malerba, the first Italian astronaut; Professors Giorgio Guglieri and Erasmo Carrera from the Politecnico di Torino; Dr Ian Carnelli of the European Space Agency; the director of the ZAL Centre for Applied Aeronautical Research, Roland Gerhards; and Professor Daniele Ragni from the Delft University of Technology.

Scientific Committee

Aeroacoustics

Aeroacoustic assessment of blended wing body configuration with low noise technologies

Francesco P. Adamo[a]*, Mattia Barbarino[2,b] et al.
[1]CIRA, Italian Aerospace Centre, Capua (Italy)
[2] CIRA, Italian Aerospace Centre, Capua (Italy)
[a]f.adamo@cira.it, [b]m.barbarino@cira.it

Keywords: Aeroacoustics, Low Noise Technologies, BWB, Distributed Electric Propulsion, Regional Aircraft, Shielding effect, Airport Noise, Footprint

Abstract. An aeroacoustic assessment of promising novel aircraft concepts (BOLT and REBEL, two Blended wing bodies respectively with conventional and hybrid engines) devoted to fly in 2035-2050 scenario coupled with Low Noise Technologies (LNT) developed inside the framework of ARTEM project (H2020)[1] has shown. The noise assessment of each noise source of the AAC (Advanced Air Concepts) has been provided including the attenuation due to the masking effects due to the fuselage. The results are then used for the noise impact on the ground, through a ray-tracing method and taking into account the installative effects, with a comparison with standard/similar aircraft. Finally, the noise assessment of a AAC&Standard fleet on a reference airport has been provided.

Introduction

The main noise sources of novel BWB configuration BOLT and REBEL have been analyzed though some of the most used semi-empirical or numerical models well known in literature, then noise predictions have been coupled with Low Noise Technologies in order to satisfy the community requests. The effects of the installation of these LNTs, together with the shielding effect due to the particular shape of the Blended-Wing-Body, have been analyzed, following the most common noise metrics, in terms of single aircraft and fleet simulation on a reference airport.

Aeroacoustic Approach

The methodology proposed for the aeroacoustic analysis of the noise impact of BOLT and REBEL follows four steps: i)Aerodynamic modelling, focused on REBEL and the effects of a set of DEP; ii)Aeroacoustic modelling for each noise sources, generating noise hemispheres; iii) Scattering modeling where the fuselage effects of the BWB have been evaluated; iv) Ground propagation modelling where the noise assessment at ground level for a single fly-over event and then a fleet simulation at two certification points have been performed.

Validation of noise models

Reference data for an aircraft with similar characteristic to one of two disruptive concepts analyzed – BOLT – have been collected in order to compare them with the results predicted with the noise source models.

The comparison has been made using the NPD (Noise Power Design) curves provided by ANP database[1] for a range of altitudes that varies from 630 to 12000 ft and the airspace has been defined to correctly predict the SEL (*Sound Exposure Level*) as a time integrated noise index derived from the SPL (*Sound Pressure Level*) as shown, for several thrusts, in *Figure 1* . The results show good agreement between experimental data and theoretical prediction.

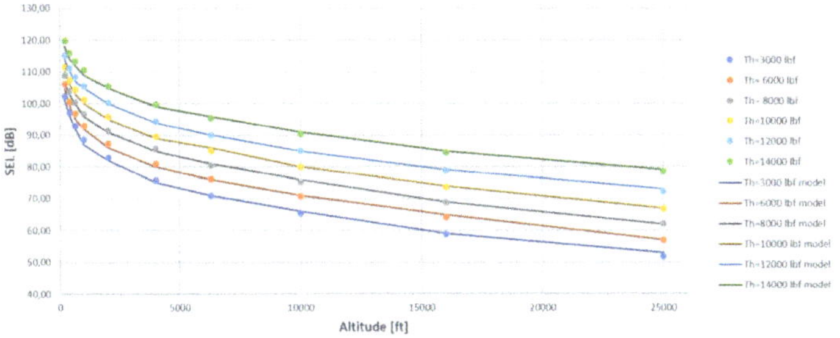

Figure 1: SEL comparison calculated for a reference aircraft similar to BOLT, for different thrusts. Circles represent exp data (SEL, ANP); Lines represent model (SEL predicted).

Fuselage shielding

The shielding effects by the BWB fuselage have been evaluated on hemispheres surrounding the aircraft, with a radius of 50m and with a twofold approach assuming the absence of mean flow: the first one is based on discontinuous Galerkin method developed by ACTRAN[2] while the second one is based on the fast shielding approach (method of Maekawa[3]), more suitable for medium-high frequencies. The results show a good agreement among two approaches and in particular show that, for the first two blade passing frequencies (the most annoying part of the overall noise), the attenuation due to the configuration can reach 20 dB (in front of the aircraft): the fuselage shielding plays a crucial role for the noise reduction of such aircrafts (*Figure 2*).

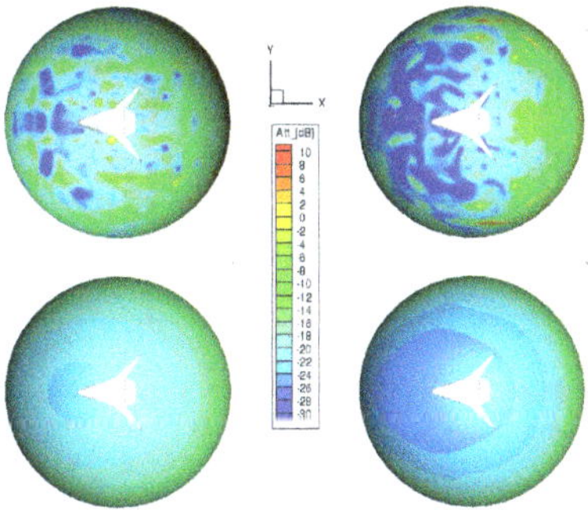

Figure 2: Attenuation effects at the BPF and comparison between DGM Method (on the top) and fast shielding approach (bottom figure). On the left 1st BPF, on the right 2nd.

BOLT and REBEL assessment

The attention is then focused on the noise analysis of BOLT and REBEL with and w/o low noise technologies (LNT) developed inside the framework of ARTEM project. Their coupling was done taking in account the feasibility of the single LNT analyzed, the degree of risk and the scaling to BOLT and REBEL. The main LNT analyzed and the related noise source involved can be summarized in the following list: i) Innovative liners on slanted septum core (fan noise); ii) High Lift Devices; iii) Flap porous treatments for Jet Interactions; iv) Jet Installation Effects; v) Landing Gear Effects. The contribution of each LNT has been considered in terms of Insertion Loss (IL). The noise predictions have been done considering firstly the noise results got with the noise models for each noise source of the aircraft, then applying the LNT related to each noise source and finally summing energetically the new noise results and projecting them on takeoff/flyover and landing dedicated trajectories. The final assessment of this procedure is shown, only for BOLT, into *Figure 3*:

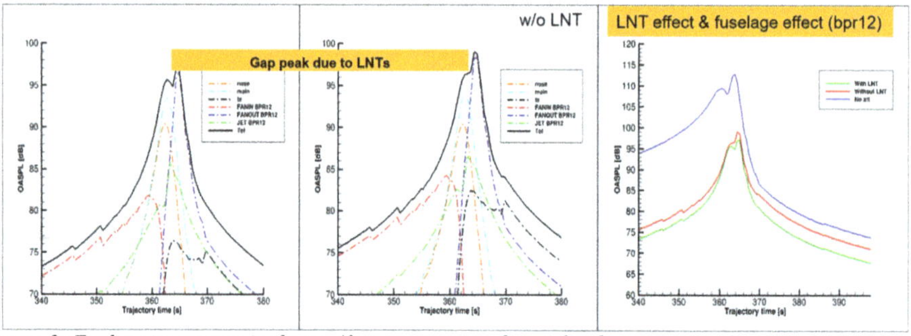

Figure 3: Each noise source is shown (fan, jet, main and nose landing gear, trailing edge, total). First two pictures on the left shows the gap due to LNTs; third on right the gap due to the fuselage effect. BOLT, takeoff, mic at x=2300 m.

Then BOLT and REBEL SELs are predicted for each noise sources as shown in figure below:

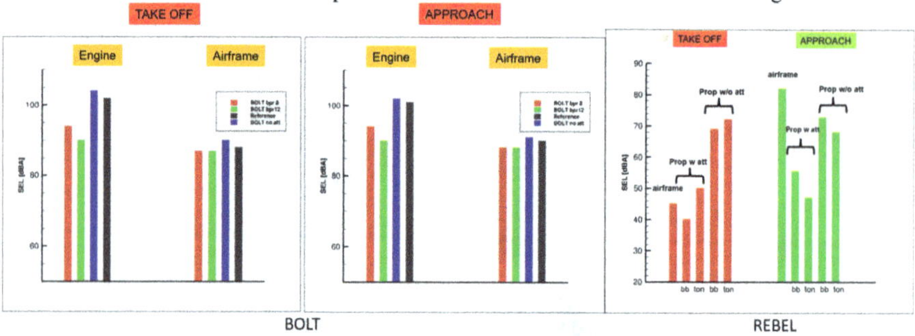

Figure 4: first two figures on the left: SEL of each noise source in BOLT; third figure on the right: REBEL SELs of each noise source (propeller noise, tonal and broadband component).

Acoustic impact on a reference airport

The acoustic impact on a selected airport (Naples, Italy) of BOLT and REBEL has been simulated. *Figure 5* shows the results derived from comparing BOLT with a reference aircraft. The results clearly show a noise reduction due to BOLT configuration. Same conclusions arise simulating a fleet simulation of AACs and comparing with standard aircraft. (*Table 1*).

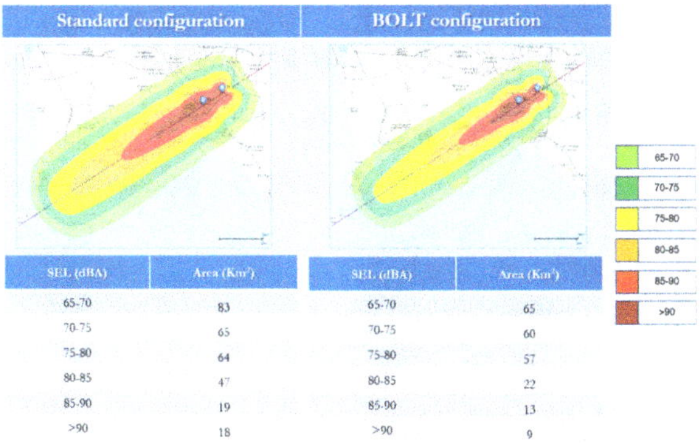

Figure 5: Acoustic impact comparison: Single event contours (take off).

L_{DEN} [dBA]	Area (standard – novel) [Km^2]
60-65	-3
65-70	-1
70-75	-1
75-80	0
80-85	0
>85	-0,3

Table 1: Equal loudness LDEN contours: standard scenario - 2050 scenario.

Conclusions

AAC as BOLT and REBEL satisfy the community requests of noise reduction. This is mainly due to the fuselage shielding effect of BWB configuration and secondary to the application of Low Noise Technologies. Further investigations will have to pursuit this twice way in order to reach the main target of the noise reduction.

References

[1] https://cordis.europa.eu/project/id/769350

[2] https://doi.org/https://www.acare4europe.org/

[3] Chevaugeon N. et al., Efficient discontinuous Galerkin methods for solving Aeroacoustics problems, 11th AIAA/CEAS Aeroacoustic Conference, 23-25 May 2005, Monterey, California.

[4] Maekawa Z., Noise Reduction by Screens. Applied Acoustics. 1968. 1,157-173. https://doi.org/10.1016/0003-682X(68)90020-0

Leonardo I4N research program – design of novel acoustic liners for aero engine nacelles

Giuseppe Dilillo[1,a], Paul Murray[1,b], Nicola Gravagnone[2,c] and Massimiliano Di Giulio[2,d]

[1]Leonardo S.p.A., Aircraft Division, Venegono Superiore, 21040, Italy

[2]Leonardo S.p.A., Aircraft Division, Pomigliano d'Arco, 80038, Italy

[a]giuseppe.dilillo@leonardo.com, [b]paul.murray.ext@leonardo.com, [c]nicola.gravagnone@leonardo.com, [d]massimiliano.digiulio@leonardo.com

Keywords: Novel, Aero Engine, Inlet, Acoustic, Liner, Broadband, Attenuation

Abstract. The minimisation of aircraft noise remains a major challenge for the aerospace industry. Noise certification limits continue to be driven lower over time to counter the impact of the steadily increasing number of noise events in the vicinity of airports, along with an increased sensitivity of the public to aircraft noise. This, in turn, ensures that considerable engineering time and effort is used to target all of the major contributors to aircraft noise. Aircraft noise sources include engine fan inlet, fan bypass, jet, and airframe noise, among others. Fan inlet, bypass, and core, ducts are typically lined with absorbing acoustic panels in order to minimise radiated fan noise. This paper is focussed on the latter, describing work to address the design and optimisation of novel acoustic liners.

Introduction

In recent years, the design and manufacture of more efficient aero engine duct acoustic panels has been receiving increased attention. However, the design process cannot be performed in isolation, as the requirement for ever lower aircraft noise levels may be impeded somewhat by the drive to reduce Specific Fuel Consumption (SFC). In particular, modern aero engine ducts have progressively larger fan diameters, and higher bypass ratios. They also have reduced duct lengths in order to minimise weight. While both of these modifications help to reduce fuel consumption, from a noise perspective the ensuing reduction in duct lined length-to-height ratio reduces the attenuation for a given liner design. Hence, for a fixed source level, liner efficiency must be improved just to maintain the status quo. Further challenges are also introduced as modern engine fans generally rotate less quickly, and have fewer blades, than their predecessors. This leads to a tonal content shifted towards lower frequencies, while higher frequencies remain important for broadband noise sources, thus broadening the target frequency range for the acoustic linings. This scenario is particularly challenging for traditional liner designs, as their maximum efficiency is realized over a relatively narrow bandwidth. Hence, if possible, novel designs must be introduced with larger attenuation bandwidths adapted to the source of modern engines.

The Leonardo Innovation for Nacelles (I4N) programme has a work stream dedicated to the design of novel aero engine duct liners. This paper summarises the progress realised to date in liner design. The studies have led to the development of a number of low weight novel liner configurations of varying complexity which are predicted to show improved broadband attenuation when compared to that realised for traditional designs. Work has included the development and validation of acoustic liner impedance models [1], liners with porous cell walls and cells which are considerably wider than traditional designs [2], and the design and manufacture of novel liners with complex cavities (replacing traditional straight cavities) [3,4]. The attenuation potential of the novel designs is improved further by increasing the integrity of duct liner attenuation modelling significantly, in particular by using a more representative source content and by including the influence of boundary layer refraction at the duct walls [5]. Each of the above subjects are now addressed.

Development and validation of a single degree of freedom perforate impedance model under high SPL and grazing flow

Normal incidence impedance, grazing flow in-situ impedance, and insertion loss testing of a range of Single Degree-Of-Freedom (SDOF) perforates was performed, along with insertion loss tests on a complex cavity panel with varying cell widths and path lengths [1]. Grazing flow measurements, performed in the Santa Catarina grazing flow facility in Brazil, acquired liner impedance and attenuation data at Mach numbers up to 0.6, and at SPLs up to 155dB. The semi-empirical impedance models developed were validated by employing them in calculations of lined duct insertion loss (Figure 1). The studies also confirmed the potential for complex cavity designs to provide improved broadband attenuation (Figure 1).

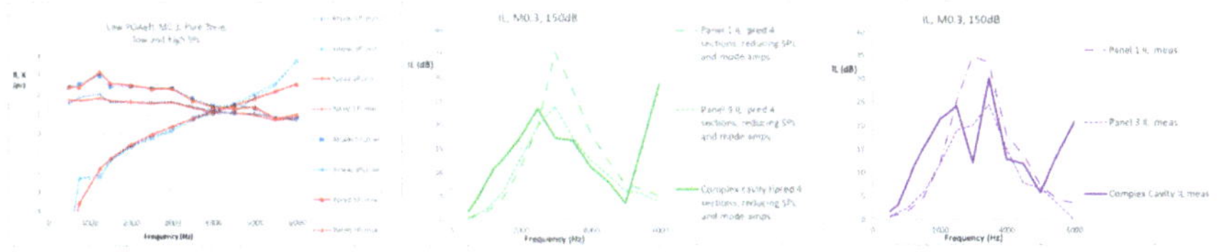

Figure 1. SDOF Perforate Impedance Prediction vs Measurement (left), plus predicted (centre) and measured (right) insertion loss showing bandwidth benefits of complex cavities

Optimisation of non-locally reacting liners for improved duct attenuation

Traditional aero engine liner configurations have narrow cells (approx. 10mm wide) which allow only a plane wave to propagate within them at the frequencies of interest for community noise. The liner response is then independent of the angle of incidence of the impinging sound, and it is considered “locally reacting”. As such, the acoustic response may be characterized uniquely by an impedance, defined as the complex ratio of the acoustic pressure and velocity at the liner surface. Should the cell width increase, additional modes may propagate within the cells, and the liner response becomes a function of the incident mode content.

It is well known that non-locally reacting porous materials show good broadband behaviour. However, they may not be used in aero engines given their propensity to retain fluids. In this work [2], a set of parametric studies were also performed to look at the acoustic attenuation of non-locally reacting liners which are also suitable to fly within aero engine ducts.

The COMSOL Multiphysics® simulation software was used to model the lined propagation for a number of designs under uniform flow. The FEM model was validated initially against insertion loss measurements performed for traditional liners in the NLR rectangular Flow Duct Facility (FDF) in Holland. Thereafter, a parametric study was performed to look at the impact of adjusting these designs to incorporate varying cell widths and varying cell wall resistances. The study demonstrated potential gains in insertion loss for the non-locally reacting designs over that predicted for traditional designs (Figure 2).

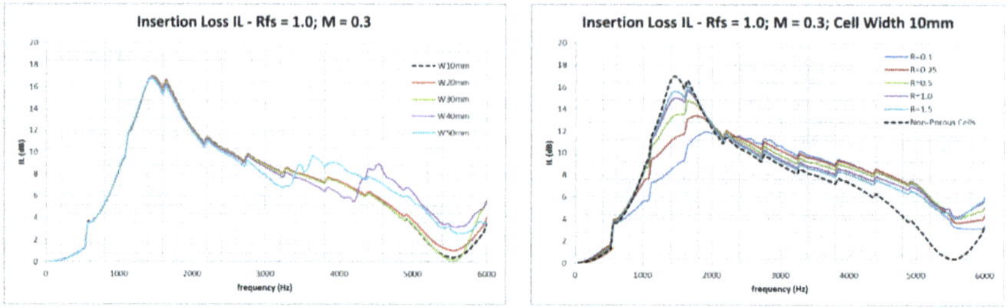

Figure 2. Downstream propagation at Mach 0.3 – Wide cell performance (left) and porous cell wall performance (right)

Improved aero engine inlet attenuation from novel broadband liners

The work presented in [4] continues the research activity initiated in [3], where a novel broadband liner, which uses a core of complex cavities, was designed and optimised to maximise the normal incidence sound absorption over a wide frequency range, including low frequencies. In [4], the novel broadband liner concept was optimised to improve the AneCom inlet attenuation over a wide bandwidth. The liner optimisation incorporated the measured AneCom fan noise circumferential modal content and it also included the impact of wall boundary layer refraction. The geometry of the broadband liner was optimised for two different overall panel depths, 24mm and 40mm, with the deeper panel allowing lower frequencies to be targeted via the inclusion of longer folded cavities. The predicted performance (Figure 3) of the two broadband liner geometries was compared with that predicted for the traditional liners tested by Leonardo previously at AneCom, along with their re-optimised designs, which also account for the impact of the measured source and the wall boundary layer refraction. The novel designs were able to provide significant improvements in attenuation across the frequency range where the AneCom source was loudest in the far-field.

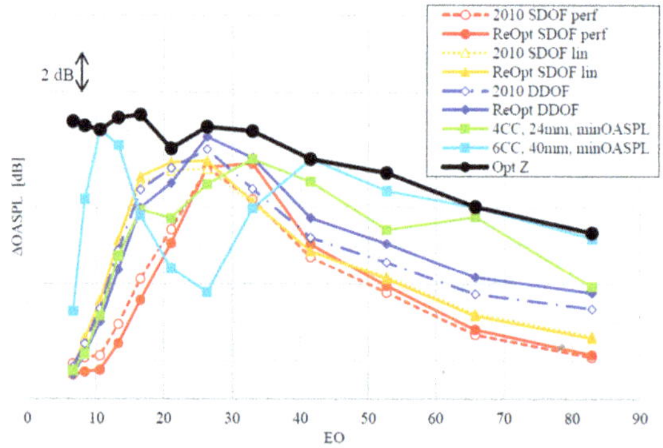

Figure 3. Predicted far-field attenuation at 18.5m from the AneCom inlet, between 40° and 90°, using the measured fan noise source and assuming shear flow in the inlet duct.

Impact of the engine fan source and wall boundary layer on inlet liner design

This work [5] sought to validate far-field predictions of engine liner attenuation. An improved version of the Leonardo cylindrical duct propagation and radiation code (NextGen Liner Multiphysics Code, NLMC) was developed to allow the inclusion of the measured modal source content along with the impact of refraction when sound propagates through the shear layer at the inlet wall. These aspects are generally not included by the industry. Application of the NLMC code was shown to provide excellent agreement with in-duct and far-field measurements, providing confidence in the acceptability of the assumed simplifications in the modelling. The predictions for an equal energy per mode source and no boundary layer were also found to be significantly different from those using a more representative source (measured circumferential amplitudes and equal energy per associated radial modes) and including boundary layer refraction (Figure 4).

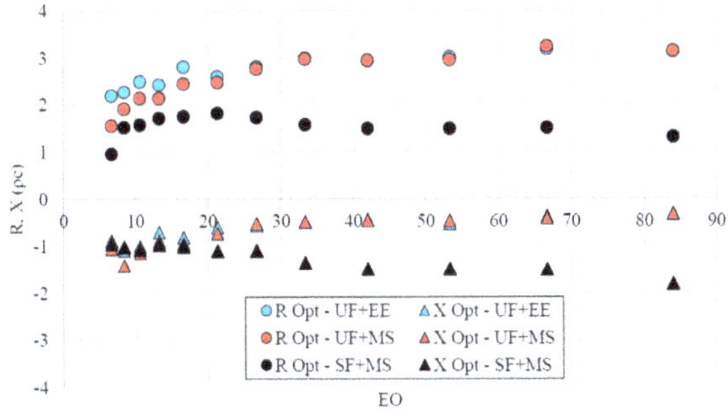

Figure 4. Impact of modelling integrity on the engine inlet optimum resistance (R) and reactance (X) at Approach. Uniform flow with equal energy per mode (NoBL+EE) or measured source (NoBL+MS), and shear flow measured source (BL+MS)

Conclusion

This paper has provided a brief overview of the aero engine acoustic design studies performed by Leonardo under the auspices of the I4N research programme. A series of coordinated activities has led to a significantly improved understanding and capability in the design of advanced aero engine nacelle liners. The novel designs can provide improved attenuation levels, accompanied by an increased bandwidth of attenuation, when compared to that of traditional designs currently flying. The design activity has led to the manufacture of four novel inlet liners which were tested successfully in the AneCom facility in May 2023.

References

[1] P. B. Murray and M. Di Giulio, "Development and validation of a single degree-of-freedom perforate impedance model under high SPL and grazing flow", *AIAA 2022-2929*, Southampton, UK, 2022. https://doi.org/10.2514/6.2022-2929

[2] N. Gravagnone, P. Murray and M. Di Giulio, "Optimisation of non locally reacting liners for improved duct attenuation", *AIAA 2022-2931*, Southampton, UK, 2022. https://doi.org/10.2514/6.2022-2931

[3] G. Dilillo, Design and Optimisation of Acoustic Liners with Complex Cavities for Improved Broadband Noise Absorption, Politecnico di Milano, Tesi di Laurea Magistrale in Ingegneria Aeronautica, 2020.

[4] G. Dilillo, P. Murray and M. Di Giulio, "Improved aero engine inlet attenuation from novel broadband liners", *AIAA 2022-2965*, Southampton, UK, 2022. https://doi.org/10.2514/6.2022-2965

[5] G. Dilillo, P. Murray and M. Di Giulio, "Impact of the engine fan source and wall boundary layer on inlet liner design", *AIAA 2022-2964*, Southampton, UK, 2022. https://doi.org/10.2514/6.2022-2964

Aeroacustics computation based on harmonic balance solution

Luca Abergo[1,a*] and Alberto Guardone[1,b]

[1]Department of Aerospace Science and Technology, Politecnico di Milano, Building B12, Via La Masa 34, Milano, MI 20156, Italy

[a]luca.abergo@polimi.it, [b]alberto.guardone@polimi.it

Keywords: Aeroacoustics, Harmonic Balance, ROM, FWH

Abstract. This paper presents a new open-source framework to compute the noise emitted by aerodynamic bodies whose motion is dominated by a specific frequency. This flow behavior is typical of propellers, pitching blades and wind turbines. The reduced order model called harmonic balance is used to compute the unsteady flow solution reducing the computational cost. K frequencies are solved by obtaining the conservative variables at N-time instances inside one period, where N = 2k+1. The time history of the surface flow solution is reconstructed with a Fourier integration. The Kirchhoff Ffowcs Williams Hawkings integral formulation, implemented in SU2, is used to compute the sound pressure level perceived by farfield observers. The integral formulation propagates the acoustic solution with a computational cost independent of observer distance. The noise emittance of a pitching wing is computed with the proposed framework and compared with a fully time accurate solution showing a very good agreement.

Introduction

Noise emittance due to the airframe is gaining importance in many aeronautical areas, in particular concerning wind turbines, pitching blades and propellers for urban air mobility. Both need to interact with residential neighbors and comply with strict regulations and noise certification requirements. It is necessary to develop reliable and verified noise emission models that can be integrated in design processes. Obtaining an accurate prediction of the noise emmittance is challenging and computationally expensive due to its unsteadiness and turbulent nature. An accurate time-resolved flow solution is required. In a modular style, a CFD-CAA aeroacoustic solver had been implemented in the open-source high fidelity software SU2 [1]. Although high-fidelity models are essential, approaches with a modest level of computational complexity are required to include noise into a design environment. The Kirchhoff Ffowcs Williams Hawkings integral formulation (FWH), implemented in SU2 in the Di Francescantonio version [2], is used to compute the sound pressure level (SPL) perceived by farfield observers. The integral formulation propagates the acoustic solution with a computational cost independent of observer distance. In this work, only the solid surface version is used therefore the surface flow unsteady solution is needed as sound source.

Typically, a fully Unsteady Reynolds Averaged Navier Stokes (URANS) is solved with a dual time stepping method and the flow solution at every time step is stored. The novelty of this work is the introduction of a reduced order model to compute the flow solution and use it as noise sources, reducing the computational effort needed. Quasi-periodic flows, like wind turbines or propellers, which are dominated by a set of frequencies can be efficiently solved with a harmonic balance method. It avoids the time-consuming transient calculation that is typical in time accurate CFD. With respect to other frequency methods, the harmonic balance is a time-spectral method where the k frequencies solved are not necessarily integral multiples of one another. Moreover, it is highly parallelized. The harmonic balance method [3] can well capture the time history of the aerodynamic coefficient with a relatively low number of frequencies solved. The HB matrix and the N-instance, where N = 2k + 1 conservative variable vector are combined to generate a matrix-vector product corresponding to the time derivative of the state variables. To sum up, the problem is formulated and solved as N steady-state problems, advancing in parallel in the pseudo-time and linked by a low order representation of the time derivative. Once N time solutions are obtained a more time accurate flow history inside a single period is reconstructed through a Fourier interpolation. Since for the tonal acoustic only the variables on the solid surface are needed, the interpolation is limited to this portion of the CFD domain.

This paper is structured as follows. In Sections 2 and 3, we present the aeroacustics framework and the numerical tools employed. In Section 4, the results obtained on a pitching wing are presented and the difference between the fully time accurate and the solution obtained with the reduced order method are analyzed. In the end, in Section 5 we summarize the findings and comment on future perspectives.

Harmonic Balance

This section introduces the governing flow equations and the reduced order model used to solve them. After temporal discretization and spatial integration across a control volume, the unsteady compressible Navier-Stoker equations are provided using the finite volume approach and are written as:

$$D_t U|\Omega| + R(U) = 0 \tag{1}$$

Where Ω is the control volume and U the conservative variables. The residual R(U) contains the convective and viscous fluxes integrated over the control volume's interfaces. D_t is the derivative operator with respect to the control volume and to time. A dual time-stepping integration methos is used in SU2, at each physical time instance a steady-state problem is solved in the pseudo-time τ:

$$|\Omega|\frac{\partial U_n}{\partial \tau} + D_t|\Omega|U_n + R(U_n) = 0 \tag{2}$$

When dealing with high computational demanding time accurate method, it becomes interesting to introduce reduced-order models and study the range of applicability. Harmonic balance (HB) can be used to solve quasi-periodic flows, dominated by a set of frequencies that are not necessarily integral multiples of one another. The theory about HB, here briefly presented, was implemented in SU2 software by [3]. The harmonic balance time operator $\mathcal{D}_t$ must be introduced.

Given $\bar{\omega}$ the vector of K frequency to be solved, the N flow solution U will be obtained at the t_n time instance in the period T, where $N = 2K + 1$ and $t_n = (n - 1)\, T/N$. Being E the Discrete Fourier Transform (DFT) matrix, defined as:

$$E_{k,n} = \frac{1}{N} e^{-i\omega_k t_n} \tag{3}$$

The spectral operator matrix H is found to be:

$$\mathrm{H} = \mathrm{E}^{-1}\,\mathrm{DE} \text{ where } \mathrm{D} = \mathrm{diag}(\bar{\omega}) \tag{4}$$

Defined $\bar{U}$ as the vector containing the conservative variables for all the N time instance, the harmonica balance operator is given by:

$$\mathcal{D}_t(\bar{U}) = \mathrm{H}\,\bar{U} \tag{5}$$

Replacing the harmonic balance operator in equation (2) and defining q as the pseudo time step it is obtained that each time instance is solved in a steady-state manner, with all the time instances marching with the corresponding local pseudo-time step:

$$|\Omega|\frac{\partial U_n}{\partial \tau} + |\Omega|\mathrm{H}\ \bar{U} + R\left(U_n^{q+1}\right) = 0 \tag{6}$$

In conclusion with harmonic balance, we obtain a discrete flow solution in a flow period with a time step $\Delta t = T/(N-1)$ without computing the typical transitory of the unsteady flows. However, concerning acoustic propagation a more time accurate surface flow solution is needed. To obtain it, a Fourier interpolation is performed for each conservative variable ϕ at each surface node of the grid. An arbitrary time resolution Δt^* related to a larger number of time instances inside N^* the period can be selected and the relative ϕ^* are obtained with:

$$\phi^* = E^{*-1}(E\phi) \tag{7}$$

where E^{*-1} is the bigger IDFT rectangular matrix of dimension $N^* \times N$.

Acoustic Formulation

The Ffowcs Williams-Hawkings (FW-H) equation is solved in this work to transmit pressure changes on the emission surface to observers in the farfield. The so-called "wind tunnel configuration" is taken into consideration, in which the source and observer move simultaneously while maintaining a constant distance between them. In a modular style, the FWH formulation implemented in C++ [4] takes as input the interpolated flow solution obtained with HB. Concerning SU2, the version of the integral formulation proposed by Di Francescantonio is implemented, which is an extension of the Farassat work [5] to general moving surfaces in which combines the positive aspect of the Kirchhoff and FWH equations:

$$4\pi p' = \int_S \left[\frac{\rho_0\left(\dot{U}_i n_i + U_i \dot{n}_i\right)}{r|1-M_r|^2}\right]_{ret} dS + \int_S \left[\frac{\rho_0 U_i n_i K}{r^2|1-M_r|^3}\right]_{ret} dS$$
$$+ \frac{1}{c}\int_S \left[\frac{\dot{F}_i \hat{r}_i}{r|1-M_r|^2}\right]_{ret} dS + \int_S \left[\frac{F_i\hat{r}_i - F_i M_i}{r^2|1-M_r|^2}\right]_{ret} dS + \frac{1}{c}\int_S \left[\frac{F_i \hat{r}_i K}{r^2|1-M_r|^3}\right]_{ret} dS$$

where

$$K = \dot{M}_i \hat{r}_i r + M_r c - M^2 c,$$
$$U_i = u_i + [(\rho/\rho_0) - 1](u_i - v_i),$$
$$\hat{r}_i = r_i/r.$$

M stands for the Mach number, r for Euclidean observer distance from the source node, u is the flow velocity, v represents the grid velocity and P_{ij} stands for the perturbation stress tensor. The derivatives are calculated in the time that the observer hears the noise signal, which is known as the retarded time. This formulation can handle both permeable and solid surfaces; a solid surface is one that prevents mass transit through it, such as the aerodynamic body itself. The loading noise contribution is connected to the final three components, whereas the first two terms are related to thickness noise contribution.

Numerical Simulation

To validate the capability of the proposed framework to predict tonal noise emitted by quasi period flow a pitching straight wing has been simulated. The wing has a chord of 1m and a span of 3m, the section is a NACA6410 airfoil. In standard conditions of pressure and temperature, the freestream Mach number is M = 0.796, resulting in a Reynold number of 12.56 E6. Jameson-Schmidt-Turkel convective scheme is used with the $2th$ order artificial dissipation term set at 0.02 and the $4th = 0.02$. SA one equation turbulence model is applied. A hybrid grid of around 1 million elements is generated with Pointwise, with a first cell high low enough to obtain y+<1 on the entire surface. The rigid motion is imposed by a prescribed time-varying angle of attack, calculated with:

$$\alpha(t) = 1.06\sin(\omega t) + \alpha_0$$

where $\alpha_0 = 0°$ and $\omega = 109.339$.

The test case is simulated both with a standard fully time accurate URANS with 100 time steps per period and with the Harmonic Balance method. Concerning HB, the solution is obtained both for 3- and 5-time instances corresponding to a frequency vector of $\overline{\omega} = [0; \pm\omega_1]$ in the first case and $\overline{\omega} = [0; \pm\omega_1; \pm\omega_2]$ in the second one, with $\omega_1 = \omega = 109.339\ and\ \omega_2 = 2\omega_1$. After, with the Fourier interpolation the surface flow time solution with the same Δt of the URANS is obtained. For every time instance in the HB simulation, the residual density must be reduced by eight orders of magnitude.

The next figures show the comparison of the lift and drag coefficient. Concerning the C_l a very good matching is obtained already with only three frequencies, instead for the C_d it is necessary to increase the length of $\overline{\omega}$ to reduce the relative error.

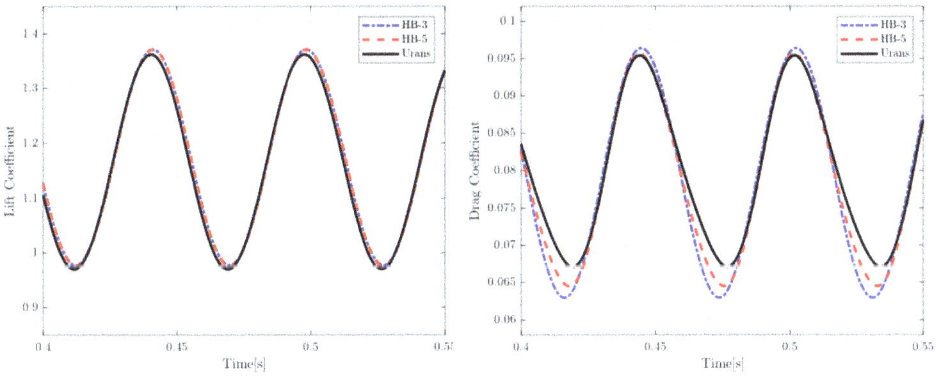

Figure 2 Lift (left) and drag (right) coefficient comparison between HB and Urans for one period of oscillation.

The sound pressure level (SPL) in dB perceived by 17 microphones is computed with the FWH module. The observers are equally distributed on an arch with radius 10m, centered in the middle of the span and the chord, on a plane perpendicular to the wing.

Materials Research Proceedings 37 (2023) 10-16 https://doi.org/10.21741/9781644902813-3

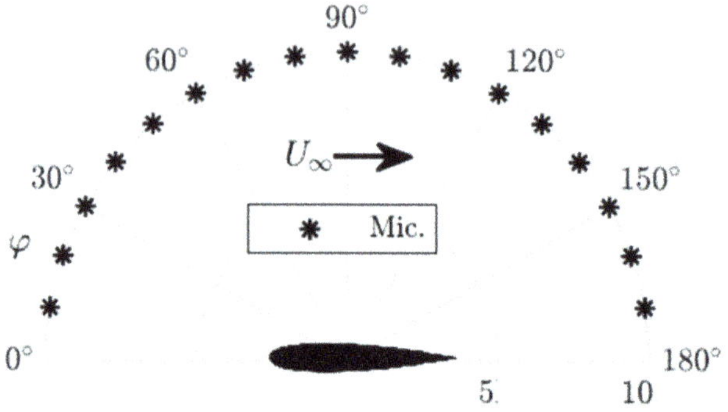

Figure 3 Farfield observer position and reference of system.

First it is reported the pressure perturbation perceived by two observers. The harmonic balance reported is the one obtained with 5 frequencies. It can be observed that there is perfect matching for the microphone positioned at $\varphi = 90°$ instead a higher relative error is found in the downstream direction $\varphi = 170°$, Fig(4). The same trend is reflected for the SPL computed for the entire arch, Fig(5). The relative error is everywhere lower than 1%, which is a remarkable result, Fig(5). There is not only a good matching in the integral value, but also the spectrum directivity plots obtained with the two solutions show are almost identical, see Fig(6).

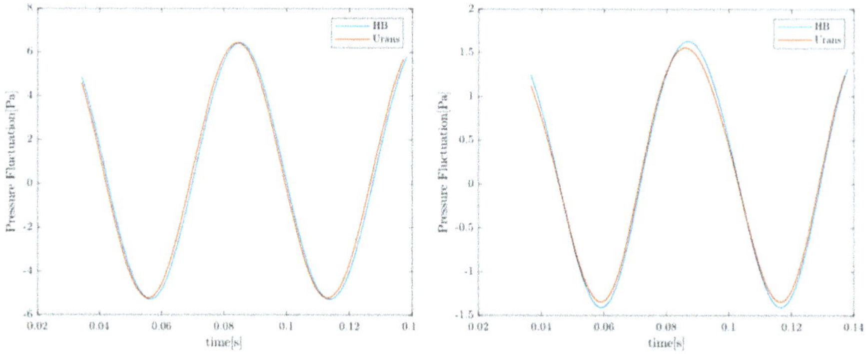

Figure 4 Pressure perturbation perceived by the microphone at $\boldsymbol{\varphi} = \mathbf{90°}$*(left) and* $\boldsymbol{\varphi} = \mathbf{170°}$ *(right).*

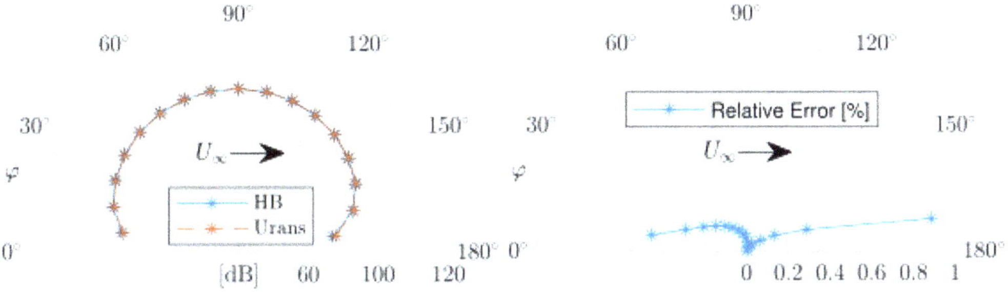

Figure 5 SPL perceived by 17 microphones (left) and relative error between the HB and Urans solution (right)

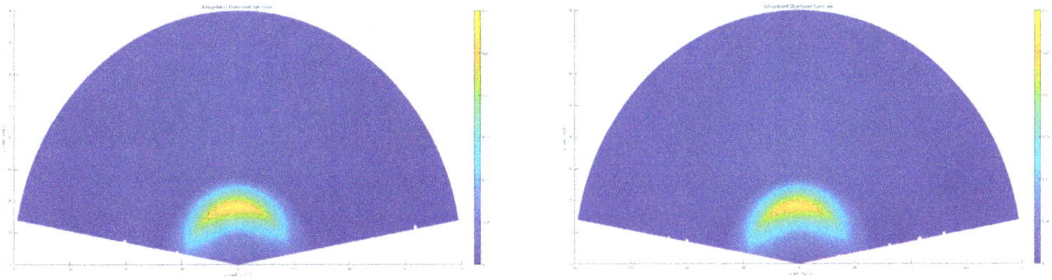

Figure 6 Directivity spectrum obtained with HB solution(left) and Urans solution (right).

Conclusions

In this work, a new open-source framework for computing the sound pressure level perceived by farfield observers is presented. The framework is specific for quasi-periodic flows. The reduced order model Harmonic Balance method is used to compute the flow solution. With this time-spectral method it is not necessary to simulate the transient period typical of an unsteady CFD but the converged solution is directly obtained advancing in the pseudo time. Following, a more time accurate solution is obtained with a Fourier interpolation only for the surface nodes. The acoustic module, where the integral FWH formulation is implemented, takes the interpolated solution as noise sources, and propagates to farfield observers. The proposed framework has been tested on a pitching wing and the results compared with fully time accurate solution obtained with a standard Urans. Five frequencies are sufficient to obtain a good approximation of the integral aerodynamic coefficients. Concerning the acoustics, a good matching is found between the two solutions not only in terms of SPL but also comparing the directivity spectrum. The promising results obtained suggest the application of the proposed framework to more complex cases like propellers or helicopter blades.

References

[1] T. D. Economon, F. Palacios, S. R. Copeland, T. W. Lukaczyk and J. J. Alonso, "SU2 An open-source suite for multiphysics simulation and design," *Aiaa Journal,* vol. 54, no. 3, pp. 828-846, 2016. https://doi.org/10.2514/1.J053813

[2] P. Di Francestantonio, "A new boundary integral formulation for the prediction of sound radiation," *Journal of Sound and Vibration,* vol. 202, no. 4, pp. 491-509, 1997. https://doi.org/10.1006/jsvi.1996.0843

[3] A. Rubino, M. Pini, P. Colonna, T. Albring, S. Nimmagadda, T. Economon and J. Alonso, "Adjoint-based fluid dynamic design optimization in quasi-periodic unsteady flow problems using a harmonic balance method," *Journal of Computational Physics,* vol. 372, pp. 220-235, 2018. https://doi.org/10.1016/j.jcp.2018.06.023

[4] L. Abergo, M. Morelli, S. F. Pullin, B. Y. Zhou and A. Guardone, "Adjoint-Based Aeroacoustic Optimization of Propeller Blades in Rotating Reference Frame," in *AIAA Aviation*, San Diego, 2023. https://doi.org/10.2514/6.2023-3836

[5] F. Farassat, "Linear Acoustic Formulas for Calculation of Rotating Blade Noise," *AIAA journal*, vol. 19, no. 9, pp. 1122-1130, 1981. https://doi.org/10.2514/3.60051

[6] P. Spalart and S. Allmaras, "A one-equation turbulence model for aerodynamic flows," in *30th aerospace sciences meeting and exhibit*, 1992. https://doi.org/10.2514/6.1992-439

Experimental investigation of the noise emitted by two different propellers ingesting a planar boundary layer

Michele Falsi[1,a *], Ismaeel Zaman[2,b], Matteo Mancinelli[1,c], Stefano Meloni[3,d], Roberto Camussi[1,e], Bin Zang[2,f] and Mahdi Azarpeyvand[2,g]

[1]Department of Civil, Computer Science and Aeronautical Technologies Engineering, Roma Tre University, Via Vito Volterra, 62, 00146 Roma RM (IT)

[2]Faculty of Engineering, University of Bristol, Bristol, BS8 1TR (UK)

[3]Department of Economics, Engineering, Society and Business, Organization, University of Tuscia, 01100 Viterbo (IT)

[a]michele.falsi@uniroma3.it, [b]i.zaman@bristol.ac.uk, [c]matteo.mancinelli@uniroma3.it, [d]stefano.meloni@unitus.it, [e]roberto.camussi@uniroma3.it, [f]nick.zang@bristol.ac.uk, [g]m.azarpeyvand@bristol.ac.uk

Keywords: Aeroacoustics, Rotor Noise, Boundary Layer Ingestion

Abstract. Novel-aircraft concepts consider the possibility of placing the propulsor very close to the fuselage to ingest the incoming airframe boundary layer. In this configuration, the engine takes in flow at a reduced velocity, thus consuming less fuel in the combustion process. However, this induces a series of noise consequences that alter the noise perceived by an observer. The present work reports an experimental investigation to compare the far-field noise directivity emitted by two different propellers ingesting a boundary layer at two different states. The experiments have been performed in the anechoic wind tunnel at the University of Bristol. The experimental setup consists of a propeller placed in the proximity of a tangential flat plate, which represents a simplified model of a fuselage. Two tripping devices placed 1 m (6.5 rotor radii) upstream of the propeller have been used to generate distinct boundary layer thicknesses. Results from two distinct propellers with three and five blades have been compared, varying the advance ratio J from 0.56 to 0.98. Far-field noise has been acquired using a microphone array positioned in the plate plane. The data have been analysed in the frequency domain, providing an extensive characterization of the far-field directivity. Results show a general increase in noise when the propeller ingests a thicker boundary layer. Furthermore, a change in directivity pattern is observed varying the advance ratio, suggesting a variation of the underlying physics. Finally, considering different J, the overall noise emission appears to be dependent on the number of blades.

Introduction

In the pursuit of more sustainable and efficient air transportation, researchers and engineers are continuously seeking innovative solutions to enhance aircraft performance. An approach that has garnered considerable attention is the concept of Boundary Layer Ingestion (BLI), which involves the ingestion of the slow-moving boundary layer air into the propulsion system of an aircraft. In fact, the boundary layer experiences lower velocities compared to the free stream airflow. This low-velocity air has significant kinetic energy remaining, which can potentially be harnessed to improve the overall aerodynamic efficiency of the aircraft. BLI aims to capture this underutilized energy by ingesting the boundary layer air into the propulsion system, thereby reducing drag. As a consequence, it has the potential to increase the propulsive efficiency of an aircraft by capitalizing on the kinetic energy in the boundary layer air [1]. By ingesting and re-energizing this slower-moving air, the propulsion system can produce more thrust for the same amount of fuel, resulting in improved fuel economy and reduced emissions [2,3]. However, one of the potential disadvantages of BLI is the increase in the aircraft noise emitted in the far-field [4,5].

Experimental setup

The experiments herein presented were carried out at the University of Bristol's Lawson anechoic wind tunnel. This facility is a closed-circuit, temperature-controlled wind tunnel that is 16.6 m long, 6.8 m wide, and 4.6 m high. The wind tunnel uses a nozzle with a contraction ratio of 8.4 and exit dimensions of 775 mm in height and 500 mm in width, which can achieve freestream velocities of up to 40 m/s and has a high flow uniformity across its exit plane. The anechoic chamber is acoustically lined with acoustic foam wedges and it allows for anechoic measurements down to 160 Hz, according to the ISO 3745 standardized testing procedure [6].

A general overview of the experimental setup is shown in Fig. 1 in which an array of 21 GRAS 40 PL microphones with a radius of 1.75 m (11.5 rotor radii) was positioned parallel to the plate. The array covers angles from $\theta = 35°$ to $\theta = 135°$ from upstream to downstream with $\Delta\theta = 5°$ between every microphone. Two propellers, one with three blades and the other with five blades, both featuring a common radius of R = 0.152 m, are mounted on a steel rig positioned 1 m (6.5 rotor radii) downstream of the wind tunnel contraction.

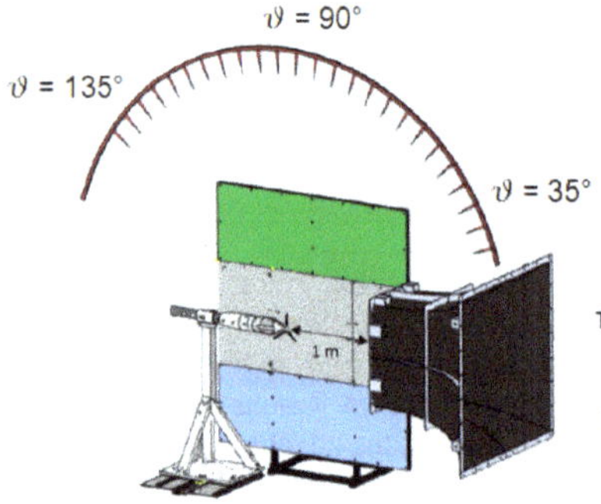

Fig. 1: Experimental setup.

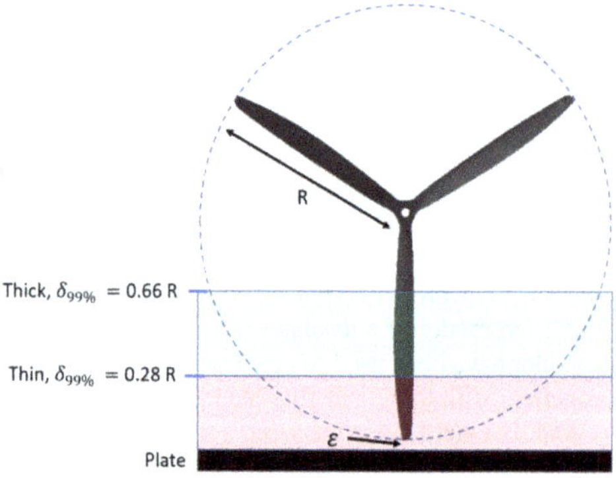

Fig. 2: Boundary layer thicknesses at the propeller location.

Both propellers share the same airfoil shape and geometry, ensuring consistent baseline characteristics for comparison (see [4]). The wind tunnel velocity is fixed at $U_\infty = 20\ m/s$, while the propeller RPMs are varied to achieve a range of advance ratios spanning from J = [0.56, 0.98]. Two Aluminium metal foam porous material tripping devices, referred to as 'Thick' and 'Thin', were strategically placed after the contraction exit to manipulate the boundary layer thickness at the propeller location. The Thick tripping device had a thickness of 10 mm, while the Thin tripping device had a thickness of 5 mm. Prior to the experimental measurements, a preliminary hot-wire test campaign was conducted to assess the boundary layer thickness and the turbulence intensity (TI) at the propeller position. The results showed that the boundary layer thickness was δ = 0.66 R with TI = 7.76% for the Thick trip and δ = 0.28 R with TI = 3.89% for the Thin trip, as shown in Fig. 2. These measurements were obtained without the propeller installed and served as an estimation of the boundary layer ingested by the propeller during subsequent experiments. In all presented results, a tip gap of $\epsilon = 5\,mm$ ($\epsilon/R = 0.03$) was maintained between the propeller tip and the plate.

Results

Four representative advance ratios were chosen to compare the three and five-bladed propellers: J = 0.56, J = 0.65, J = 0.75, J = 0.98, where J = U_∞/nD, with $n = RPM/60$ and $D = 2R$. For each microphone the Overall Sound Pressure Level (OASPL) was calculated as:

$$OASPL = 10log_{10}\int_{f_1}^{f_2}\frac{PSD}{P_{ref}^2}df\ , \quad [dB]$$

where $f_1 = 160\ Hz$ which is the wind tunnel cut - off frequency, $f_2 = 10000\ Hz$, $P_{ref} = 20\cdot10^{-6}\ Pa$ and PSD indicates the Power Spectral Density of the signal evaluated with the Welch's method. Fig. 3 displays the OASPL results as a function of polar angles for both the three-bladed and five-bladed configurations, considering various values of J. The figure distinguishes between the thicker boundary layer cases, represented by dashed lines with cross markers, and the thinner boundary layer cases, represented by solid lines with dots.

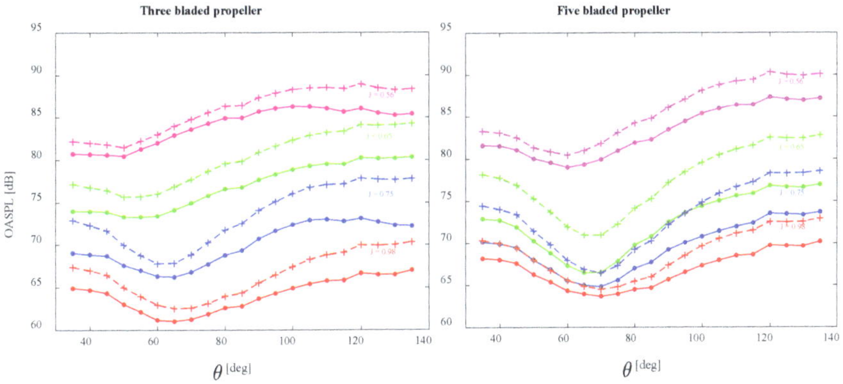

Fig. 3: OASPL for three-bladed propeller (left) and five bladed propeller (right). Dashed lines with cross markers refer to the thicker boundary layer (Thick) whereas solid lines with dots refer to the thinner boundary layer (Thin).

In both configurations, the ingestion of a thicker boundary layer by the propeller leads to an amplified level of noise across all polar angles. This increase in noise is attributed to heightened pressure fluctuations on the propeller blades, induced by turbulence within the thicker boundary layer. The effect is particularly pronounced at higher polar angles (wake side), specifically for J values of 0.65 and 0.75. Additionally, a noticeable change in noise emission directivity is observed at high J, characterized by a distinct dip around $\theta = 70°$. This change is more pronounced in the five-bladed configuration. As a result, comparing the two plots in Fig. 3, a general trend of increased noise is evident in the three-bladed case compared to the five-bladed case, considering equal J values.

Conclusions

This study aimed to investigate the far-field noise characteristics of propellers operating in close proximity to a flat plate, ingesting a boundary layer with two different thicknesses. We compared the experimental results obtained from two propellers with three and five blades, with the advance ratio varying from J = 0.56 to J = 0.98. Far-field noise data were acquired using a microphone array positioned in the plate plane. The results of the study revealed a general increase in noise levels when the propeller ingested a thicker boundary layer. This observation underscores the influence of boundary layer thickness and turbulence intensity on noise generation. Additionally, variations in the advance ratio led to changes in the noise directivity pattern, suggesting alterations in the underlying physics of noise emission. Finally, considering different advance ratios, the overall noise emission appeared to exhibit a trend based on the number of propeller blades with a general increase considering the three-bladed case. In a potential future development, varying the thickness of two boundary layers while maintaining the same turbulence intensity could help differentiate their individual impacts on far-field noise.

Acknowledgments

This work has been supported by the European Union's Horizon 2020 research and innovation program under project ENODISE (Enabling optimized disruptive airframe-propulsion integration concepts) grant agreement No. 860103.

References

[1] Smith L, 1993, Wake Ingestion Propulsion Benefit, Journal of Propulsion and Power 9 74–82. https://doi.org/10.2514/3.11487

[2] Ahuja J and Mavris D, 2021, A Method for Modeling the Aero-Propulsive Coupling Characteristics of BLI Aircraft in Conceptual Design, AIAA Scitech Forum 2021. https://doi.org/10.2514/6.2021-0112

[3] Yildirim A, Gray J, Mader C and Martins J, 2021, Performance Analysis of Optimized STARC-ABL Designs Across the Entire Mission Profile, AIAA Scitech Forum 2021. https://doi.org/10.2514/6.2021-0891

[4] Zaman I, Falsi M, Zang B, Azarpeyvand M and Camussi R, 2023, Experimental Parametric Investigation of the Haystacking Phenomenon for Propeller Boundary Layer Ingestion, AIAA AVIATION 2023 Forum. https://doi.org/10.2514/6.2023-4054

[5] Zaman I, Falsi M, Zang B, Azarpeyvand M and Camussi R, 2023, Effect of Tip Gap on Nearfield and Farfield Acoustics of Propeller Boundary Layer Ingestion, AIAA AVIATION 2023 Forum. https://doi.org/10.2514/6.2023-4055

[6] Mayer Y, Kamliya Jawahar H, Szoke M, Showkat Ali S and Azarpeyvand M, 2019, Design and Performance of an Aeroacoustic Wind Tunnel Facility at the University of Bristol, Applied Acoustics 155 358–370. https://doi.org/10.1016/j.apacoust.2019.06.005

Predicting noise spectrum of a small drone rotor in a confined environment: a lattice Boltzmann Vles analysis

Riccardo Colombo[1,a], Lorenzo Maria Pii[2,b], Gianluca Romani[2,c], Maurizio Boffadossi[1,d]

[1]Dipartimento di Scienze e Tecnologie Aerospaziali - Politecnico di Milano, Via La Masa, 34 - 20156 Milano - Italy

[2]Dassault Systèmes Italia Srl, Viale dell'Innovazione, 3 - 20125, Milano Italy

[a]riccardo15.colombo@mail.polimi.it, [b]lorenzomaria.pii@3ds.com, [c]gianluca.romani@3ds.com, [d]maurizio.boffadossi@polimi.it

Keywords: Propeller Aerodynamics, Laminar Separation Bubble, Aerodynamic Noise, Lattice-Boltzmann Method, Very-Large-Eddy-Simulation, Confined Aeroacoustics

Abstract. The objective of this paper is to study the predictive capabilities of a Very-Large-Eddy-Simulation CFD solver for the simulation of the flow past a small drone propeller blade. The solver is based on a Lattice-Boltzmann Method coupled with an FW-H acoustic analogy to compute the far field noise. The method is able to cope with the anechoic test chamber to predict the complex flow-field and the characteristic boundary layer phenomena (such as laminar separation, transition and reattachment). The acoustic hybrid formulation provides tonal and broadband noise radiation in agreement with the experimental data. Frequency spectrum prediction exhibited a strong low-frequency tonal contribution at multiples of the blade passing frequency related to the interaction with coherent vortical structures in hover, and a high-frequency broadband hump due to the laminar separation bubble at high advance ratio.

Introduction

The design of small rotors, aerodynamically and aeroacoustically efficient, is a research field of great interest. According to a recent experiment of propeller at low Reynolds number (Re < 70000) Grande et al. [5] at TU Delft have observed a very complex flow, involving laminar separation, transition, and reattachment, where the size and the position of the Laminar Separation Bubble (LSB) greatly affect the propeller broadband noise spectrum.

The prediction of such non-linear phenomena can be extremely challenging even for high-fidelity (Computational Fluid Dynamics) CFD solvers. Typically, CFD solvers are based on a hybrid turbulence model and use an artificial trip to enforce boundary-layer transition. Previous studies performed by Casalino et al. [2, 3] have been focused on predicting far field noise using Dassault Systèmes PowerFLOW® software, which is based on Lattice Boltzmann Method (LBM) and Very Large Eddy Simulation (VLES). The Lattice-Boltzmann method [3] is based on statistical advection and collision of fluid particles by a number of distribution functions aligned with pre-defined discrete lattice velocity directions. Flow variables such as density and velocity are computed by taking the appropriate moments of the distribution function. The relaxation time and other parameters of the distribution function are computed by considering the turbulent motion computed using a two-equation transport model based on the $k-\epsilon$ re-normalization group theory. Conversely to RANS models, Reynolds stresses are not explicitly added to the flow governing equations but are a consequence of an alternation of the gas relaxation properties that lead the flow towards a state of dynamic equilibrium. LBM/VLES model can be interpreted as an extension of the kinetic theory from a gas of particles to a gas of eddies [4]. It can be demonstrated that the effective Reynolds stresses have a nonlinear structure and are better suited to represent turbulence in a state far from equilibrium, such as in the presence of shear and rotation. Being the LBM low dissipative, compressible, and intrinsically unsteady, it constitutes a well-suited model for aeroacoustics simulations.

In this work, the aerodynamic noise generated by the propeller is evaluated by means of an hybrid FW-H calculation [1] based on a time-advanced solution of Farassat's formulation of the Ffowcs Williams & Hawkings' (FW-H) equation applied to the propeller blades, hub and nacelle surface pressure LBM solution. Although a favorable agreement with experiments was obtained by Casalino et al. in previous studies [2, 3], some inconsistencies were found and possibly attributed to the absence of the experimental test section in the simulation [5] and grid resolution. Therefore, the aim of the present study is to further validate the aeroacoustics and aerodynamic predictive capabilities of the LBM/VLES approach, including the simulation of the actual anechoic test section by means of an Acoustic Porous Medium. Moreover, no artificial trip is used to trigger transition, and the effect of grid resolution refinement on the prediction of Laminar Separation Bubble (LSB) and related phenomena is further assessed.

Computational setup

The propeller geometry considered in TU-Delft experiments [5] is based on 2-bladed APC 9x6 propeller, with 0.30 m of diameter. The angular velocity considered is 4000 RPM, corresponding a Blade Passing Frequency (BPF) of 133.3 Hz; the blade-tip Mach number is 0.19. Two different advance ratios were considered: $J=V_\infty/nD = 0.0$ and 0.6.

The unconfined computational fluid domain is a spherical volume with a radius of 80m, centered around the propeller hub. The mesh (Fig. 1) increases the resolution as the distance from the propeller decreases, using different volumes for each resolution level. The overall mesh size is of 3.8 million voxels. The anechoic TU-Delft A-Tunnel is modelled by means of an Acoustic Porous Medium, using an equivalent acoustic porosity and tortuosity. The geometry of the inner volume of the test section, as well as the nozzle's geometry were simulated with the actual dimensions of the chamber (6.4m×6.4m×3.2m for the test section, with a 1m wide opening on the ceiling and a 1.4m walls thickness).

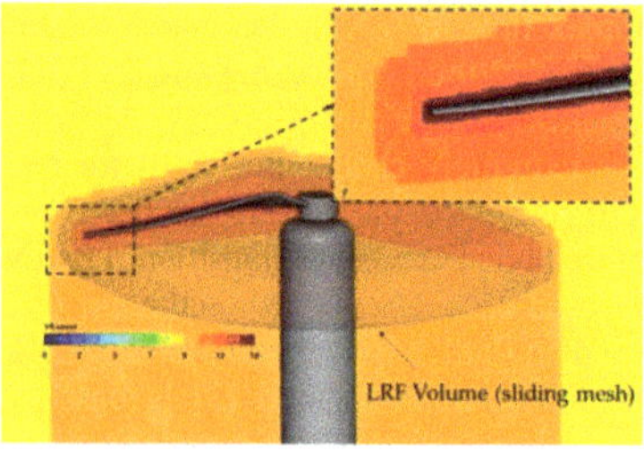

Fig.1 Mesh with different VR level near the blades

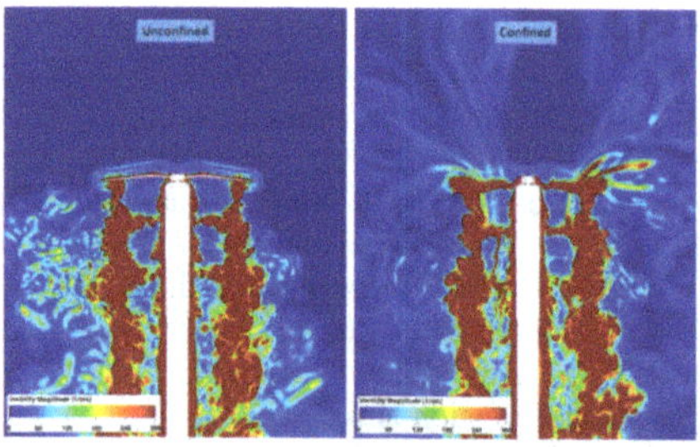

Fig. 2 Istantaneous vorticy field for the unconfined (left) and confined (right) simulation

Results

The results firstly presented are relative to the hover case in which confinement effects are more significant. To investigate the impact of confinement effects on the turbulence ingested and generated by the propeller, an instantaneous vorticity snapshot can be considered as shown in Fig.2. The confined case shows the expected presence of large vortical structures that are stretched along the flux tube streamlines and ingested by the rotor.

For the aeroacoustics analysis in Fig.3 the comparison between numerical and experimental far field noise for both unconfined and confined simulations is shown. The noise spectra are reported in terms of Power Spectral Density (PSD) versus the frequency normalized by the Blade Passing Frequency (BPF). The unloaded electric motor noise and wind tunnel background noise are included in the plots. It is worth noting that the primary sources of experimental uncertainty are: (i) the background noise, which is responsible for high levels of broadband noise at low frequency; (ii) the loaded electric motor noise and non-perfect balance of blade loading, leading to low-frequency tone increases at harmonics of the shaft frequency (BPF-0.5, 1, 1.5, etc.); and (iii) the unloaded electric motor noise, adding mid-frequency tonal contributions approximately between BPF-5 and BPF-25 and between BPF-50 and BPF-70. It can be observed that BPF-1 is accurately predicted by the numerical simulations for both cases. In the hovering confined case, all tonal peaks from BPF-1 to BPF-10 exhibit an additional tonal contribution resulting from unsteady loading compared due to the ingestion of recirculating vortical structures.

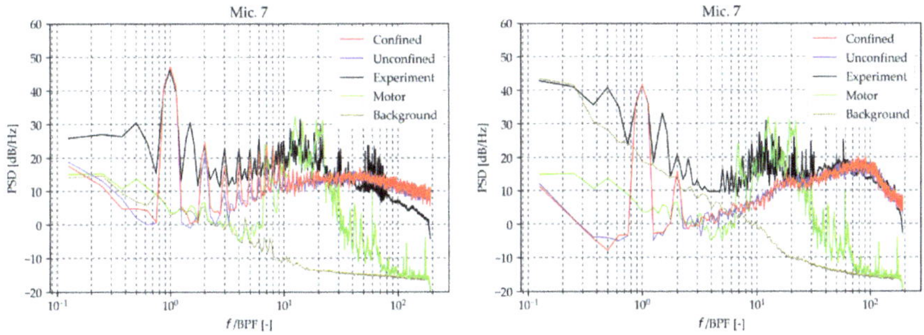

Fig. 3 Far field noise prediction for J=0 (left) and J=0.6(right)

With regards to the broadband noise component, while the case J=0.6 excellently matches experimental results, there appears to be an overestimation of noise levels above BPF-50 in hover for both the confined and unconfined simulations.

Aiming to address the discrepancies of the broadband noise between simulations and experiments and to obtain a better simulation of LSB, additional simulations were considered for the hover case. The mesh resolution was increased with an additional refinement (one level higher than previous simulations, with voxel size half of the previous one). An improvement of the broadband contribution at frequencies above BPF-50 can observed in Fig. 4. The enhanced prediction can be attributed to a better simulation of reattachment points of the LSB. Time-averaged surface streamlines of Fig. 4 clearly show the presence of the Laminar Separation Bubble. The most significant improvement can be observed in the last 25% of the blade span, close to the blades tip where the reattachment point is better aligned with the experiments.

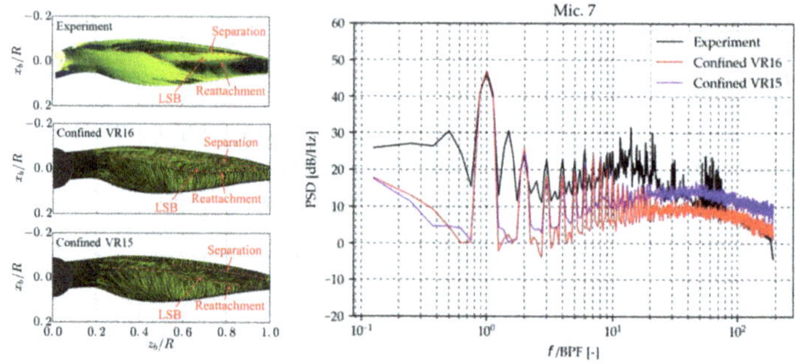

Fig. 4 Comparison of averaged streamlines for the refined simulation along with experimental oil-flow viz (left) and far field noise prediction (right).

Conclusion

This study focused on predicting far field noise spectrum, aerodynamic characteristics and boundary layer evolution (laminar separation, transition and reattachment) for a small drone rotor. Numerical simulations are performed, in both free-field and confined environments, for two different advance ratios by means of the industrial LBM-based CFD solver PowerFLOW®. The simulation aims to replicate experimental studies conducted in TU-Delft. Flow confinement simulation have a significant impact on recirculating turbulence ingested by the propeller, resulting in enhanced tonal noise prediction during hovering. An analysis of the LSB indicates that the observed broadband misprediction in hover was likely caused by an incorrect prediction of the reattachment point of LSB, resulting in an overestimation of the broadband noise. Extra simulations, performed for the hovering case with increased resolution mesh in the proximity of the expected LSB give a better accuracy in predicting trailing-edge noise, with agreement between the simulated broadband component and experimental data. Confined simulations also reveal an enhancement in the mid frequency spectrum due to turbulence impingement noise. This study demonstrates the capabilities of the industrial LBM/VLES software PowerFLOW® in accurately predicting complex flow-field, noise levels and natural boundary layer transition phenomena without the use of artificial triggering systems.

References

[1] D. Casalino; An advanced time approach for acoustic analogy predictions. Journal of Sound and Vibration, Vol. 261, n. 4, 583-612, 2003. https://doi.org/10.1016/S0022-460X(02)00986-0

[2] D. Casalino, G. Romani, E. Grande, D. Ragni, and F. Avallone; Definition of a benchmark for low reynolds number propeller aeroacoustics. Aerospace Science and Technology, 2020. https://doi.org/10.1016/j.ast.2021.106707

[3] D. Casalino, G. Romani, R. Zhang, and H Chen; Lattice-boltzmann calculations of rotor aeroacoustics in transitional boundary layer regime. AIAA 2022-2862, 2022. https://doi.org/10.2514/6.2022-2862

[4] Hudong Chen, Steven A Orszag, Ilya Staroselsky, and Sauro Succi; Expanded analogy between Boltzmann kinetic theory of fluids and turbulence. Journal of Fluid Mechanics, 519:301–314, 2004. https://doi.org/10.1017/S0022112004001211

[5] E. Grande, G. Romani, D. Ragni, F. Avallone, and D. Casalino; Aeroacoustic investigation of a propeller operating at low Reynolds numbers. AIAA Journal, Vol. 60, No. 2, 2022. https://doi.org/10.2514/1.J060611

Aeroelasticity

On the influence of airframe flexibility on rotorcraft pilot couplings

Carmen Talamo[1,a *], Andrea Zanoni[1,b], Davide Marchesoli[1,c], Pierangelo Masarati[1,d]

[1]Politecnico di Milano, Dipartimento di Scienze e Tecnologie Aerospaziali, Campus Bovisa, Via La Masa 34, 20156 Milano, Italy

[a]carmen.talamo@polimi.it, [b]andrea.zanoni@polimi.it, [c]davide.marchesoli@polimi.it, [d] pierangelo.masarati@polimi.it

Keywords: Rotorcraft-Pilot Coupling, Multibody Modelling, Biodynamic Feedthrough

Abstract. A set of numerical simulations of the interactional dynamics of the pilot-rotorcraft system is performed. The aim of the numerical analysis is to evaluate the stability of the closed-loop system in hovering conditions, focusing on the influence of the first airframe flexible mode, the one with the most participation in relative motion between the main rotor and the pilot's seat and closest in frequency to pilot-vehicle interaction. Two approaches are employed. First, a linear analysis, in which the modal representation of the airframe flexible mode is added to a linearized model of the helicopter vertical motion and the helicopter dynamics is coupled with a single degree of freedom linearized model of the pilot biomechanics. Subsequently, a multibody model of the helicopter is coupled with the same linear model of the pilot, trimmed in hover and perturbed by a vertical gust. A sensitivity analysis shows that such mode has a significant influence on the stability of the closed loop system, especially if its frequency is close to the natural frequency of the pilot's biomechanics, as one might expect.

Introduction

The interaction between pilot and vehicle dynamics represents a challenging issue for rotorcraft. Rotorcraft-Pilot-Couplings (RPC) can be at the root of different kinds of unwanted, adverse feedback loops: PIO (Pilot-Induced Oscillations) and PAO (Pilot-Assisted Oscillations). The latter, the object of this research, are characterized by the involuntary participation of the pilot. Human-machine coupling can be described by a feedback loop that connects the rotorcraft and the pilot. While aeroelastic effects on PAOs due to structural flexibility have been the subject of previous research [1], also addressing tiltrotor aeroelasticity [2], no extensive sensitivity analysis has been performed so far. This work aims at filling this gap through a numerical investigation.

Problem description

The goal of this work is to perform a closed loop stability analysis for the coupled rotorcraft-pilot system. For both parts of this work the pilot is described by a state space representation of the BDFT (Biodynamic feedthrough). The BDFT is defined as the transfer function between the control inceptor rotation and the airframe acceleration input. The pilot model is coupled with another state space system that describes the helicopter dynamics. The vertical acceleration of the airframe is fed through the pilot biomechanics to the collective control deflection, which in turn produces a vertical acceleration of the airframe.

Methods

The work has been divided into two parts. In the first part, the analysis is performed using Linear Time Invariant (LTI) models. To represent the selected rotorcraft, a simple analytical model consisting of the helicopter heave motion and the main rotor coning motion as proposed in [3] has been used, considering data representative of a light helicopter of the class of the BO105. Modal data available from previous work supports the modal representation of an analytical model including the dynamics of the airframe. A sensitivity analysis has been performed, changing the frequency of the first airframe flexible mode, the one closest to the frequency of the pilot's biomechanics and the most involved in the relative vertical motion between the main rotor mast and the pilot' seat, to understand how this parameter can affect the stability of the closed loop system composed by helicopter and pilot. In the second part of this work the rotorcraft is represented through a flexible multibody model, implemented in the free, general-purpose multibody solver MBDyn. The results are used to investigate how well such a simple analytical model can predict the results obtained from a full flexible multibody model.

Analytical model

Elastic airframe addition

The analytical airframe representation proposed in [3] has been enhanced by adding airframe flexibility, described using a modal model obtained from a finite element analysis. The first four mode frequencies are listed in Table 1.

Table 1: Airframe modes

Mode	**1**	**2**	**3**	**4**
Frequency [Hz]	5.8	7.7	11.4	12.6

The first and the third modes present the most participation of hub and pilot seat vertical relative displacement. It is worth noticing that the third mode is outside the frequency range typical of PAO events (3-7 Hz): *therefore, only the effect of the first mode is analyzed in detail.*

Closed loop sensitivity analysis

The loop transfer function of the pilot-vehicle system can be written as:

$$H_{LTF}(s) = -G\, H_{BDFT}(s) H_{HELI}(s)$$

Equation 1

where G is the gear ratio between the collective inceptor deflection and blade pitch, $H_{HELI}(s)$ is the transfer function between the collective pitch and the pilot seat vertical acceleration.

Figure 1 shows how the 1st airframe mode (at 5.8 Hz) affects the loop transfer function.

After adding the airframe elasticity, a sensitivity analysis has been performed moving the 1st mode frequency in the range [f ± 40%]. The gain and phase margins on the system's closed-loop transfer function have been chosen as indices to evaluate the system's stability.

The sensitivity analysis results are presented in Figure 2. The mode frequency increases from red to blue. The gain margin rises when the mode frequency increases. When the frequency is set to the lower end the gain margin is negative, therefore the system is unstable. Moving towards the upper bound the gain margin increase shows a non-linear trend. The most significant improvements are visible up to 5.5 Hz; the value tends to settle for higher frequencies. The explanation behind this behavior is related to that of the pilot. The typical frequencies that describe the human response in these conditions are in the range [1.5 Hz–4.5 Hz]; departing from this range the mode frequency has less influence on the system's stability.

Materials Research Forum LLC
https://doi.org/10.21741/9781644902813-6

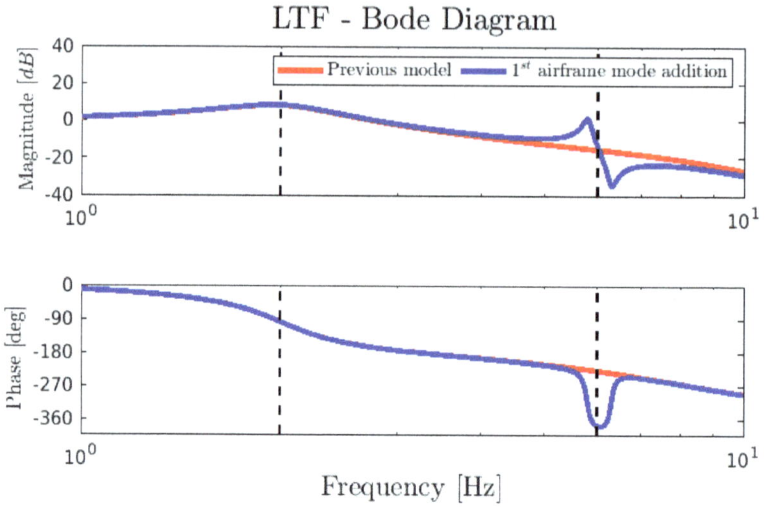

Figure 1: Loop transfer function, with and without 1st airframe mode

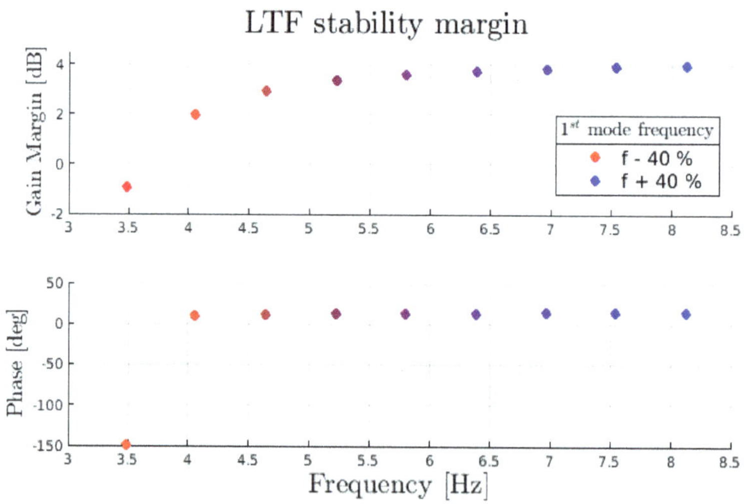

Figure 2: LTF stability margin changing 1st mode airframe frequency

Multibody model

After studying the collective bounce using a simple analytical model, a more detailed, fully flexible multibody model has been used to describe the dynamics of the helicopter.

For fairness of comparison the same pilot model is used and, as in the analytical study, only the first airframe mode has been enabled for the airframe dynamics.

The 1st airframe frequency has been moved to the value corresponding to zero gain margin. The model had been trimmed in hover, after which a perturbation in the form of a gust in the vertical direction has been introduced, after 5 seconds of simulation time, and the system's response has been evaluated. (Figure 3)

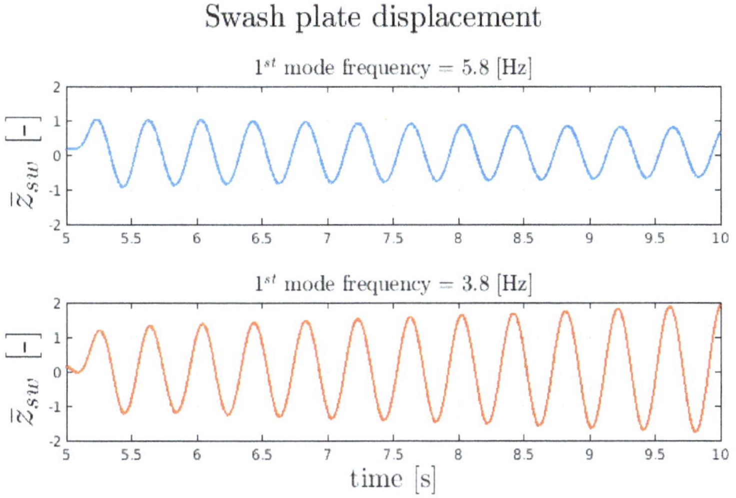

Figure 3: Closed-loop system vertical gust response

In Figure 3, $\bar{z}_{sw}$ (the normalized swashplate displacement with respect to its value at 5 s) is shown. After 5 seconds, the perturbation starts. When the 1st frequency is set to the nominal value listed in Table 1, the amplitude of the oscillations slowly decreases (damping $\xi = 0.005$). If the 1st airframe frequency is reduced to 3.8 Hz, i.e., the value corresponding to marginal stability in the LTI system, the system is not able to absorb the disturbance and the oscillations increase: the system is unstable (damping $\xi = -0.007$). It is worth stressing that the results in Figure 3 have been obtained increasing the pilot gain transfer function by 35%. The increased stability margin is probably due to increased damping introduced by the flexible dynamics of the system and by the nonlinearities included in the multibody simulations.

Conclusions

This work analyzes how the airframe mode can play a key role in rotor pilot coupling. It shows that the reduction of stability margins, and the possible development of instability, is also related to the airframe elasticity. More analyses and investigations are nonetheless needed to enhance the comprehension of the phenomenon.

References

[1] Dieterich O., et al. "Adverse rotorcraft-pilot coupling: Recent research activities in Europe," 34th European Rotorcraft Forum 2008

[2] Muscarello V., et al. 2019, "Aeroelastic rotorcraft-pilot couplings in tiltrotor aircraft," Journal of Guidance, Control, and Dynamics. https://doi.org/10.2514/1.G003922

[3] Muscarello V., et al. "The Role of Rotor Coning in Helicopter Proneness to Collective Bounce," Aerospace Science and Technology

Tiltrotor whirl-flutter stability analysis using the maximum Lyapunov characteristic exponent estimated from time series

Gianni Cassoni[1,a,*], Alessandro Cocco[1,2,b], Aykut Tamer[3,c], Andrea Zanoni[1,d], Pierangelo Masarati[1,e]

[1] Politecnico di Milano, Milano - Italy

[2] University of Maryland, College Park, MD - USA

[3] University of Bath, Bath - UK

[a]gianni.cassoni@polimi.it, [b]acocco1@umd.edu, [c]at2849@bath.ac.uk, [d]andrea.zanoni@polimi.it, [e]pierangelo.masarati@polimi.it

Keywords: Lyapunov Characteristic Exponents, Whirl-Flutter, Stability, Multibody Dynamics, Jacobian-Less Methods

Abstract. Stability analysis and assessment are fundamental in the analysis and design of dynamical systems. Particularly in rotorcraft dynamics, problems often exhibit time-periodic behavior, and modern designs consider nonlinearities to achieve a more accurate representation of the system dynamics. Nonlinearities in rotorcraft may arise from factors such as nonlinear damper constitutive laws or the influence of fluid-structure interaction, among others. Regardless of their origin, quantifying the stability of nonlinear systems typically relies on calculating their Jacobian matrix. However, accessing the Jacobian matrix of a system is often challenging or impractical, calling for the use of data-driven methods. This introduces additional complexity in capturing the characteristic dynamics of the system. Hence, a data-driven approach is proposed that utilizes the Largest Lyapunov Characteristic Exponent, obtained by analyzing the system's time series.

Introduction

When faced with a nonlinear problem in its general form,

$$\dot{x} = f(x,t)$$

stability is a local property of a specific solution, $x(t)$, resulting from a corresponding set of initial conditions, $x(t_0) = x_0$, i.e., of a Cauchy problem. One commonly encountered instance is the Linear, Time-Invariant (LTI) scenario. In this context, Lyapunov Characteristic Exponents (LCEs) quantify the growth or decay rate of disturbances from a typical solution in the nonlinear differential problem across distinct directions within the state space, providing insight into the stability of the reference solution in relation to these directions. Consider a solution $x(t)$ for $t \geq t_0$ (some call it the 'fiducial trajectory'), and a solution ${}_ix(t)$ of the problem.

$$ {}_i\dot{x} = f_{/x}|_{x(t),t}\ {}_ix, \qquad {}_ix(t_0) = {}_ix_0$$

for a perturbation ${}_ix_0$ of arbitrary magnitude and direction. LCEs are defined as

$$\lambda_i = \lim_{t\to\infty} \frac{1}{t} \| {}_ix(t) \|$$

When all the Lyapunov Characteristic Exponents (LCEs) are negative, the solution exhibits exponential stability. Conversely, if at least one LCE is positive, the reference solution is unstable or may lead to a chaotic attractor. When the largest LCE – or LCEs – are zero, a limit cycle oscillation (LCO) can be expected. This means that in the state space there exists one or multiple independent directions along which the solution neither expands nor contracts, converging instead to a self-sustained periodic motion. When multiple largest LCEs are equal to zero, a higher-order periodic or quasi-periodic attractor arises, such as a multi-dimensional torus. It is important to exercise caution when interpreting the LCEs as eigenvalues of Linear, Time-Invariant (LTI) problems, as demonstrated in [5], as they are not always equivalent.

Jacobian-Less Methods: Max LCE from Time Series

The MLCE is the LCE associated with the least damped principal direction of the problem, which represents the most critical stability indicator. Among the algorithms proposed in the literature, the one proposed by Rosenstein et al. [1] is used in this work. It is defined by the following steps. By utilizing the trajectory matrix, $\boldsymbol{X} \in \mathbb{R}^{M \times m}$, the full phase-space can be reconstructed using the time delay method, if needed, along with estimating the embedding dimension, m (estimated following Takens' theorem), and the reconstruction delay, J, where $M = N - (m-1)J$ and N is the length of the time series. In this context, each column of matrix $\boldsymbol{X}$ is a phase-space vector.

$$\boldsymbol{X} = [\boldsymbol{X}_1 \, \boldsymbol{X}_2 \dots \boldsymbol{X}_m]$$

After constructing the trajectory matrix, the algorithm locates the nearest neighbor, $\boldsymbol{X}_{\hat{j}}$, of each point on the trajectory, which is found by searching the point that minimizes the distance from each reference point, $\boldsymbol{X}_j$. The distance is expressed as

$$d_j(0) = \min_{\boldsymbol{X}_{\hat{j}}} \left\| \boldsymbol{X}_j - \boldsymbol{X}_{\hat{j}} \right\|$$

where $d_j(0)$ is the initial distance from the jth point to its nearest neighbor, and $\|\cdot\|$ denotes the Euclidean norm. Nearest neighbors must have a temporal separation greater than the mean period ($\bar{T}$, the reciprocal of the mean frequency of the power spectrum, although it can be expected that any comparable estimate, e.g., using the median frequency of the magnitude spectrum, yields equivalent results) of the time series, $|j - \hat{j}| > \bar{T}$. The largest Lyapunov exponent is then estimated as the mean rate of separation of the nearest neighbors. The jth pair of nearest neighbors diverge approximately at a rate given by the largest Lyapunov exponent:

$$d_j(l) \approx C_j e^{\lambda_1 (l \, \Delta t)}$$

where C_j is the initial separation. By taking the logarithm of both sides which represents a set of approximately parallel lines, for $j = 1, \dots, M$, each with a slope roughly proportional to λ_1. The largest Lyapunov exponent is calculated using a least-squares fit to the "average" line defined by

$$y(l) = \frac{1}{\Delta t} \langle \log d_j(l) \rangle$$

where $\langle \cdot \rangle$ denotes the average over all values of j.

XV-15 Model Whirl Flutter

Building upon the previous research conducted in [2], the aeroelastic simulation of the XV-15 tiltrotor is now considered. For this analysis, an aeroservoelastic model is employed, encompassing all significant structural components. The airframe model comprises various elements, such as the flexible wing, rigid fuselage, empennages, control surfaces (elevator, rudder, flaps, and flaperons), and nacelle tilt mechanisms. Additionally, the model, originally developed in [4], incorporates crucial cockpit elements (a seat and control inceptors - collective and cyclic) to explore rotorcraft-pilot couplings. To develop this model, the MBDyn multibody solver (https://mbdyn.org/) is utilized, to represent the fundamental frequencies and mode shapes of the complete aircraft, with specific emphasis on the wing-nacelle section. The proprotor, featuring a three-bladed stiff-in-plane rotor with a gimballed hub, comprises the blades, flexible yoke, and pitch control chain. The flexibility of the wing, rotor blades, and yokes of the two rotors is modeled using an original

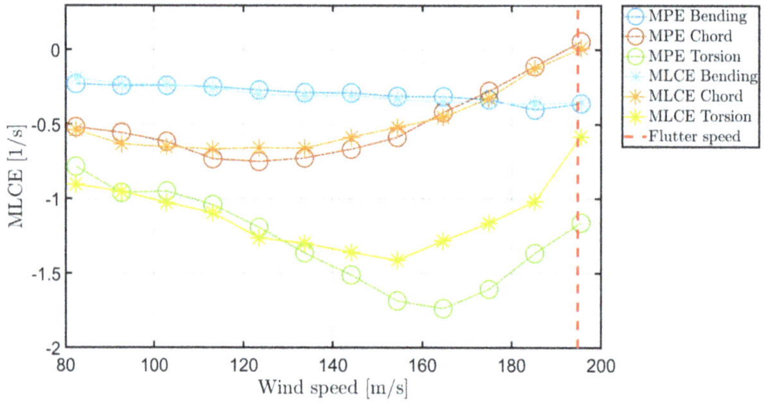

Figure 1: MLCE of the XV-15 in the standard operating condition for the airplane mode.

geometrically exact composite-ready beam finite element model known as “finite volume” [3], which is well-suited for multibody dynamics. The Whirl-Flutter phenomenon was observed through a two-phase approach involving excitation with a sinusoidal input through the swashplate, after reaching the desired trim configuration, followed by a free response phase. By linear interpolation, the instability region under the standard operating condition for the airplane mode was estimated to be U_∞ = 195.5 m/s (indicated by the dotted red line in Fig. 1). Notably, the proposed method successfully recovered the accurate estimation of the Whirl-Flutter instability previously documented in [4], validating its reliability. However, it is important to note that time series analysis methods are sensitive to changes in the observation window. Due to the system's rapid convergence compared to other directions, evaluating it accurately poses challenges, particularly in the case of torsion. To verify the obtained results, a comparison was made with those obtained using the Matrix Pencil Estimation method (MPE). In the chord direction (with a configuration in airplane mode and idle engines), a time series depicting a slow limit cycle was obtained at U_∞ = 185.2 m/s (Fig. 2). Interestingly, when solely considering the linear component, the system appears to exhibit growth. However, the system converges to a limit cycle. The amplitude, A, in Fig. 2 used to compare the solution obtained by the MPE method is determined by selecting the maximum value present in the time series.

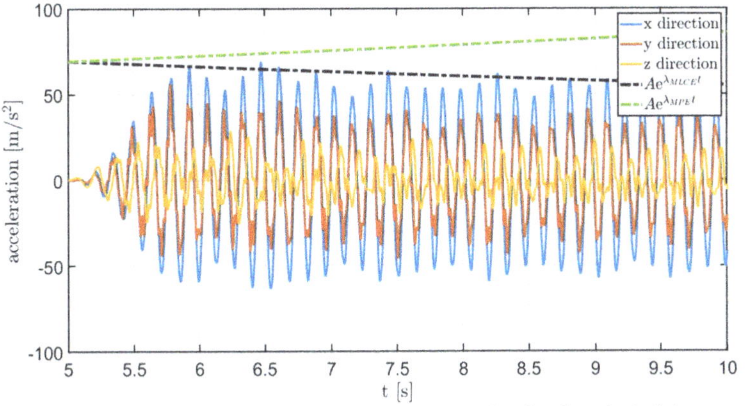

Figure 2: Time history with an input perturbation in the chord, at the hub location.

Conclusion

The presented approach is employed for investigating the whirl flutter instability of a tiltrotor. By extensively exploring its nonlinear dynamics, it becomes feasible to expand the understanding of various instabilities, particularly in cases where linear methods fall short in capturing the entirety of the response domain. Future advancements will involve coupling the method with the vortex particle solver DUST (https://www.dust-project.org/) to have a more detailed account for the aerodynamic interactions.

References

[1] M.T. Rosenstein, J.J. Collins, C.J. De Luca. A practical method for calculating largest Lyapunov exponents from small data sets. Physica D: Nonlinear Phenomena, 65(1–2):117–134, May 1993. https://doi.org/10.1016/0167-2789(93)90009-P

[2] A. Cocco, A. Savino, P. Masarati. Flexible multibody model of a complete tiltrotor for aeroservoelastic analysis. 18th International Conference on Multibody Systems, Nonlinear Dynamics, and Control (MSNDC) of ASME IDETC/CIE Conference, 08 2022. V009T09A026. https://doi.org/10.1115/DETC2022-89734

[3] G.L. Ghiringhelli, P. Masarati, P. Mantegazza. A multi-body implementation of finite volume C^0 beams. AIAA-Journal, 38(1):131–138, January 2000. https://doi.org/10.2514/2.933

[4] A. Cocco, A. Savino, A. Zanoni, P. Masarati. Comprehensive simulation of a complete tiltrotor with pilot-in-the-loop for whirl-flutter stability analysis. In 48th European Rotorcraft Forum, Winterthur, Switzerland, September 6–8 2022. https://hdl.handle.net/11311/1220824

[5] G.A. Leonov, N.V. Kuznetsov. Time-varying linearization and the perron effects. International Journal of Bifurcation and Chaos_Vol. 17, No. 04, pp. 1079-1107 (2007). https://doi.org/10.1142/S0218127407017732

Wind tunnel flutter tests of a strut-braced high aspect ratio wing

Luca Marchetti[1,a *], Stephan Adden[2,b], Michael Meheut[3,c], and Sergio Ricci[1,d]

[1] Politecnico di Milano, Department of Aerospace Science and Technology, via La Masa 34, 20156, Milano, Italy

[2] IBK Innovation GmbH & Co. KG, 21129 Hamburg, Germany

[3] The French Aerospace Lab ONERA, Location Palaiseau, France

[a]luca.marchetti@polimi.it, [b]stephan.adden@ibk-innovation.de, [c]michael.meheut@onera.fr, [d]sergio.ricci@polimi.it

Keywords: Strut-Braced Wing, High Aspect Ratio Wing, Nonlinear Aeroelasticity, Flutter

Abstract. Increasing the wing aspect ratio is one way to improve aircraft aerodynamic efficiency. This reduces the induced drag term but, at the same time, produces an increment of the wing loads, hence an increase of the structural weight. One design solution to reduce the wing root bending moment, which is the main driver of the weight of the wing, is the addition of a strut. This work deals with the experimental identification of the flutter behavior of an ultra-high aspect ratio (19) strut-braced wing in a wind tunnel. The inherent non-linear behavior of such a structure that has two different effects on the wing when loaded in compression and in tension is coupled with large deformations due to its extreme flexibility. From here derives the extreme importance of experimental tests to understand how different parameters of such a design can impact its aeroelastic behavior.

Introduction

New aircraft configurations such as the strut-braced wing require experimental studies, as results coming from simulations are less trustworthy than those regarding traditional and well-tested configurations. Data should be collected to assess both static and dynamic aeroelastic behavior, with the ultimate goal of validating numerical models and techniques. After a scaling process of a typical strut-braced wing reference aircraft provided by ONERA, a 1:10 scale model for wind tunnel testing is designed by Polimi and IBK allowing for some variations in the geometry and kinematics of the strut. A pneumatic system is designed for the excitation of the model inside the wind tunnel when turbulence is not enough. The dynamic response is measured both in terms of accelerations and displacements, using accelerometers and an optical system to track the motion of some markers on the model.

Reference aircraft

Reference aircraft is an Onera concept [1,2]. Typical mission and fuselage use an Airbus A321 as a baseline. It's a strut-braced configuration with an aspect ratio of 19. This results in a 55 m wingspan. Two engines are mounted on the rear of the cabin, the empennage layout is a T-tail. The connection between the strut and the main wing is not straight, but the strut is bent in a way that allows it to enter the wing perpendicularly, improving its aerodynamic behavior at transonic speeds.

Description of the model

The test model consists of a scaled left-wing clamped at the root and mounted in a vertical position. A constant Froude scaling approach was adopted with a geometric scale factor of 1:10.

Figure1 Wind tunnel model

The design of the structure is decoupled from the aerodynamics. For the main wing, a single glass fiber spar is used. It is obtained by milling a thick glass fiber plate with full rectangular cross-sections defined at 9 spanwise locations to correctly match out-of-plane and in-plane bending stiffnesses. The resulting torsional stiffness is representative of the correct one. The aerodynamic shape is obtained with 8 3D printed sectors attached to the spar at just one point spanwise so that they can transfer the aerodynamic loads to the structure without contributing to the overall stiffness.

For the strut, two options were designed: one stiffer than the reference with a full rectangular section aluminum spar and 3D printed aerodynamic sectors; and a second one with the correct stiffness obtained with a steel wire, but no aerodynamics attached. From here on we will refer to these two configurations as the "Aero" configuration and the "Wire" configuration. Both struts are hinged to the ground, the axis of rotation being parallel to the fuselage direction. A parallel hinge is also used to connect the strut to the wing. A rail on the floor and a plate at the connection with the wing allow 4 different chordwise positions for both strut configurations. This is done with the goal to obtain a sensitivity regarding this design parameter.

Description of measurement system

Two quantities are measured during the tests: accelerations and positions: 17 accelerometers are installed inside the main wing, all reading accelerations in the out-of-plane direction. 2 accelerometers are installed on the strut in the Aero configuration. All the wires pass inside the model and through a hole in the floor of the test chamber, where signals are acquired by a SCADAS acquisition system. Acquisitions and postprocessing are performed using Simcenter Testlab. As a main source of excitation for the structure, the turbulence inside the wind tunnel is used, which is near to a white noise excitation. In addition, a compressed air pulse is also available to get an impact-like effect. The high-pressure line of the wind tunnel (8 bar) is linked to a small tube passing inside the model to reach the 7th aerodynamic sector (near the tip of the wing). A hole in the lower skin allows the air pulse to act perpendicular to the chord exciting the out-of-plane bending of the wing. Since the hole is positioned towards the trailing edge, a partial excitation of the torsion is obtained, too. A button in the control room opens a valve allowing for remote control of the pulse excitation. This system is used both to excite the structure in the no-wind conditions and to tune the optical system. This consists of 6 infrared cameras which track the position of 53 optical markers placed on the model (only 32 in the wire configuration).

Testing procedure

Wind tunnel tests were carried out at the large wind tunnel facility at Politecnico di Milano (GVPM). The test section is a 4 m by 4 m square, top speed is 54 m/s (near the scaled cruise dynamic pressure).

Static aeroelastic assessment was performed changing the angle of attack and measuring the deformed shape with the optical system. This was done at low speeds (10 m/s and 20 m/s) with angles ranging from -3° to +5° with 1° increments.

Dynamic characteristics of the structure were assessed with modal analysis at different airspeeds. Operational Modal Analysis was used to identify frequency, damping and shape of the normal modes of the structure for wind speeds between 10 m/s and 54 m/s (maximum speed of the wind tunnel) with 5 m/s increments. Air pulse was used to identify normal modes at 0 m/s.

The first test consisted in changing the angle of attack to evaluate the influence of the prestress in the strut on the aeroelastic response. Three angles were tested: 0°, 1° and 2°. The chordwise position of the strut was fixed, and its configuration was too (aero). The second test consisted in changing the chordwise position of the strut and its configuration. 8 tests were performed at a fixed angle of attack (1°) to investigate all available options (4 chordwise positions and 2 strut configurations). As a third and last test, the aero configuration was modified to block the hinge connecting the strut and the wing. Data were collected for just one chordwise position at the usual angle of attack (1°) to be able to compare the aeroelastic response in three different conditions for the strut.

Results: Static Aeroelasticity

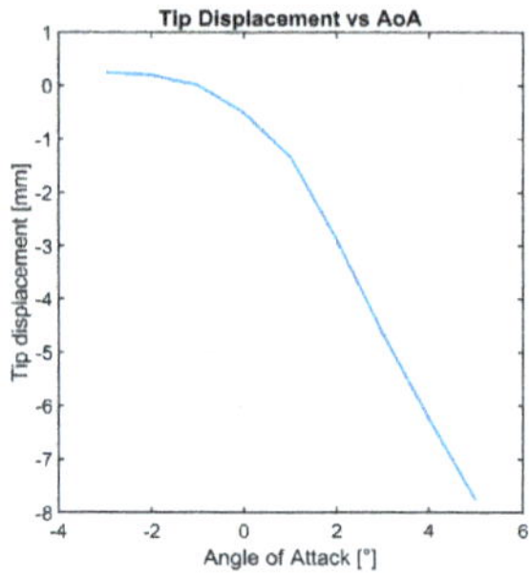

Figure 2 Tip displacement VS AoA

The measured displacement of the tip of the wing is reported in Fig.2

This is obtained at a fixed airspeed of 20 m/s and a range of angles of attack from -3° and +5°. The behavior is linear just in a small range of angles of attack. When the strut is compressed, the nonlinear trend is evident, but also at the upper bound of the tested range, the concavity of the curve starts to change.

Results: Dynamics and Flutter

As a first way to interpret the results, all chordwise positions for the same strut configuration are plotted overlapped, just to see the general trend of the curves and the dispersion of the data. In both the strut configurations, we can see a convergence of the frequency of the 3rd bending mode and the first torsional mode. The damping of the bending mode changes concavity and tends to the zero-damping axis. We can't conclude that this will result in a flutter condition, but this behavior is typical of a pre-flutter condition. Results from the wire configuration are less scattered than those coming from the aero configuration, both for the absence of the aerodynamics on the strut and for the absence of hybrid modes coming from the coupling between the wing and strut bending with different phases. Lower frequency modes are far less consistent, especially around 30 m/s where results become very scattered. At some speeds, modes

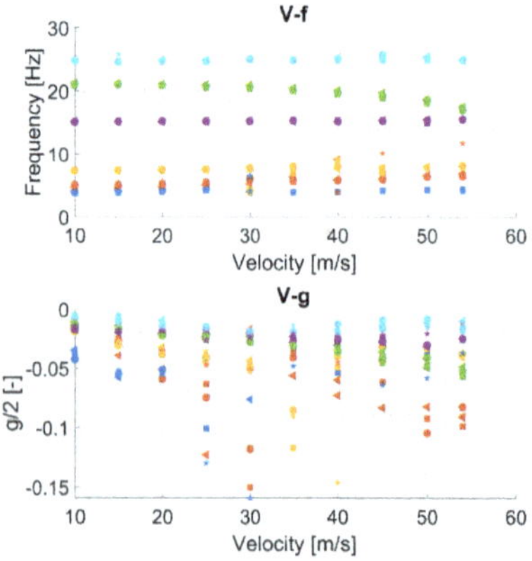

Figure 3 Wire configuration Vf-Vg

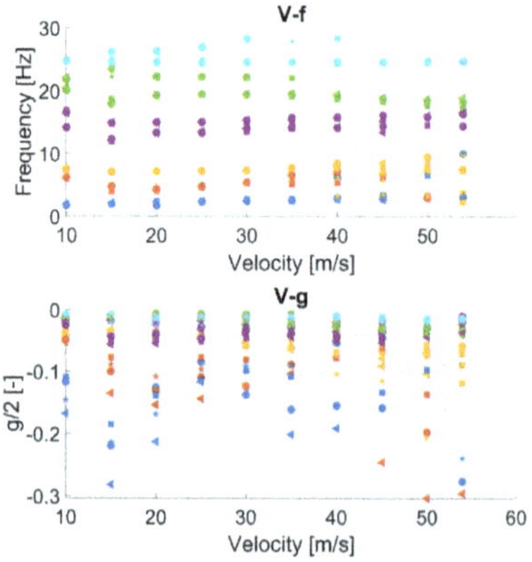

Figure 4 Aero configuration Vf-Vg

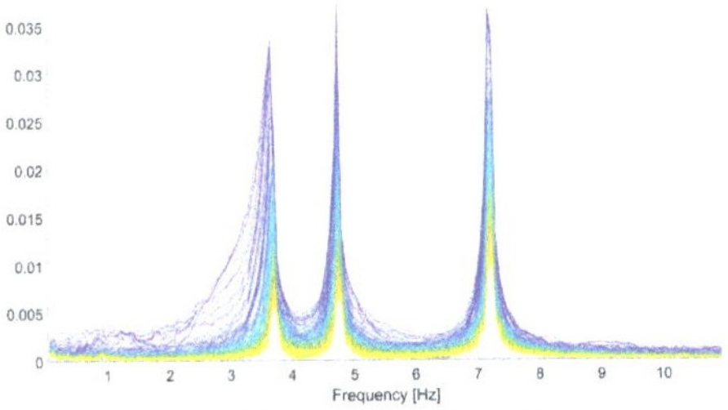

Figure 5 Zeroed Early-Time Fast Fourier Transform

with the same shape as the second bending mode appear both at their expected frequency and at the frequency of the first bending mode. This is a behavior found in nonlinear modes, where frequency and damping depend on vibration amplitude and modal shapes are not unique anymore. A test is performed to check if the modes of this kind of structure can be nonlinear. It is the Zeroed Early-Time Fast Fourier Transform [3]. At least the first mode appears to be nonlinear.

Conclusions

Data was collected for different configurations of the strut at different airspeeds. Initial analyses of the static deformations show a clear nonlinear behavior. Flutter analyses show a potential flutter condition inside the flight envelope of the reference aircraft. A nonlinearity of the normal modes was detected.

Summary

A clamped wing in a strut-braced configuration was tested at Politecnico di Milano in its large wind tunnel facility to gain insights into its static and dynamic aeroelastic behavior. This was achieved using accelerometers for the dynamics and an optical system for the statics. Both static and dynamic characteristics appear to be nonlinear.

Acknowledgement

This project U-HARWARD has received funding from the Clean Sky 2 Joint Undertaking (JU) under grant agreement No 886552. The JU receives support from the European Union's Horizon 2020 research and innovation programme and the Clean Sky 2 JU members other than the Union.

References

[1] CARRIER, Gerald G., et al. Multidisciplinary analysis and design of strut-braced wing concept for medium range aircraft. In: *AIAA SCITECH 2022 Forum*. 2022. p. 0726 https://doi.org/10.2514/6.2022-0726

[2] Delavenne, M., et al. "Multi-fidelity weight analyses for high aspect ratio strut-braced wings preliminary design." *IOP Conference Series: Materials Science and Engineering*. Vol. 1226. No. 1. IOP Publishing, 2022. https://doi.org/10.1088/1757-899X/1226/1/012009

[3] ALLEN, Matthew S.; MAYES, Randall L. Estimating the degree of nonlinearity in transient responses with zeroed early-time fast Fourier transforms. *Mechanical Systems and Signal Processing*, 2010, 24.7: 2049-2064. https://doi.org/10.1016/j.ymssp.2010.02.012

New insights on limit cycle oscillations due to control surface freeplay

Nicola Fonzi[1,a *], Sergio Ricci[1,b] and Eli Livne[3,c]

[1]Politecnico di Milano, Milano, Italy

[2]University of Washington, Department of Aeronautics and Astronautics, Seattle, USA

[a]nicola.fonzi@polimi.it, [b]sergio.ricci@polimi.it, [c]eli@aa.washington.edu

Keywords: Aeroelasticity, Flutter, Limit Cycle Oscillation

Abstract. A new experimental wind tunnel test-bed has been developed for the study of limit cycle oscillations induced by control surface freeplay. Studies of the effects of a single nonlinearity, made possible by the new horizontal tail plane, are described here. Several effects are considered, starting from a reference configuration: the effect of changes in inertia and stiffness, a time-varying gap size, and an aerodynamic preload due to an angle of attack. Both time marching simulations and describing functions analytical methods have been used to understand the experimental measurements and study the capability of the methods to capture the physical behavior. Good agreement was found in all cases and physical insights are gained from the mathematical models. Limitations of the analytical tools are also addressed, focusing on the important difference between the self-excited dynamics of the nonlinear system and its forced response to external excitations.

Introduction

Research on limit cycle oscillations (LCO) and other nonlinear aeroelastic mechanisms due to control surface hinge nonlinearities has gained significant traction since the 1990's. The drivers include the prevalence of the problem in many aeroelastic flight vehicle systems, the growing number of large dynamically actuated control surfaces where keeping tight tolerances on the freeplay over time can be demanding, the growing power of simulation capabilities (in computing hardware and the theory involved), and major developments over the last forty years or so in the area of nonlinear dynamic systems. The challenge has been known and tackled for years in analysis, simulation, as well as wind tunnel and flight tests. Recent reviews include [1], [2] and the technically thorough [3]. Those three sources cite many, if not most, of the works on control surface freeplay and on aeroelastic nonlinear behavior in the years prior to their publication, providing a state of the art view of the field.

An examination of the work done in this area so far reveals needs for more work in a few particular areas. First, the effects of time-dependent freeplay gap variation, the effects of interacting multiple local nonlinearities, and the effects of control / actuator freeplay on the active control of aeroelastic systems, including gust load alleviation and flutter suppression. Some research on interacting multiple structural nonlinearities has been reported over the years (see [4]–[9]). This area, however, still lacks sufficient experimental work.

Very little work has been reported on the time varying freeplay gap problem, and never in the context of an actuation failure [10].

Driven to cover by analytical and experimental work areas in which not enough research has been done to date, a new project was launched by the Politecnico di Milano (POLIMI) and the University of Washington (UW) to study realistic aeroelastic systems, representing real aircraft, and investigate the effects of multiple control surface hinge nonlinearities, time-varying freeplay gaps, a wide range of freeplay gaps from the very small to the large, and the effects of control surface freeplay on active flutter suppression.

A series of studies, using systems of increasing complexity, tackled first a system with a single nonlinearity and a constant gap size [11], then a system where the gap size was dynamically changing with time [12], and finally the same system subject to preload. Meanwhile, the development of a wind tunnel model of a full aircraft configuration, designed to have multiple nonlinearities, was carried out, and wind tunnel tests were performed in February of 2023. The work with the full-configuration wind tunnel model will be described in future papers. In the present paper, a review of the results obtained with the single nonlinearity system is presented. Building on the results previously presented previously by the authors, the effect of aerodynamic and mechanical preload on the LCO is investigated. Recent advances in the Describing Function (DF) technique are used to shed light on the phenomena observed in the simulations and in the test. The effect of gravity is discussed as a source of natural quenching of the nonlinear phenomena. Finally, the sensitivity of different systems to perturbations is established by studying the effects of different angles of attack of the lifting surface.

The results presented here for the case of single nonlinearity will be the base for the upcoming analysis and test results considering multiple nonlinearities.

Figure 1: Photo of the test model, installed in the wind tunnel

Test models

The test model is the right horizontal tail of the modified X-DIA, with a nonlinearity in the elevator's hinge attachment rotation. The test model is shown above. At the root of the model a hinge nonlinearity-generating mechanism can be seen. A close up view of this mechanism is also presented. The model was designed to create an accurate freeplay gap, taking into account production and mounting tolerances, which can also be dynamically or statically varied. The gap-generating mechanism also allows for the accurate measurement of the hinge movement itself. The entire mechanism is held by two structural ribs, 18 millimetres apart. Both ribs host a radial bearing, which sustains the same shaft, glued to the elevator aerodynamic surface.

Selected results

As mentioned in the introduction, several studies have been performed to understand the effect of different parameters on the LCO. These include the effect of different gap sizes, different inertial and elastic configurations, time varying gap sizes, and different preloads. Here, as an example, the experimental measurements related to the effect of the angle of attack are reported. In figure below, the experimental LCO amplitude, computed as the RMS rotation, normalized by the gap size, is reported for various gap sizes and for two angles of attack (0.5° and 1.0°).

The angle of attack has the immediate effect of quenching the LCO in some conditions. By looking at the freeplay values for which LCO is completely quenched, it can be noticed that in a practical application, with a realistic gap size, the control surfaces that are usually at some angle of attack will often show no LCO due to this quenching effect. However, during maneuvers when the aerodynamic moment (and thus aerodynamic preload) disappears, LCO reappears, leading to vibrations that can range from the uncomfortable to dangerous.

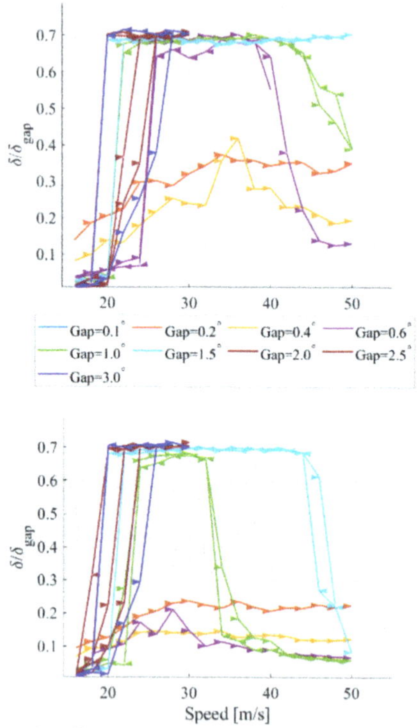

Figure 2: Effect of angle of attack on the LCO amplitude

With an angle of attack as small as half of a degree, LCO due to freeplay is quenched for freeplay gaps smaller than 0.4°, thus well within the limits imposed by regulatory agencies. In the middle of the plot it can be seen how the green and red curves are not overlapping with the others and are placed at a significantly smaller amplitude. This is signifying the disappearance of LCO and the start of a forced response movement of smaller amplitude due to external excitation.
An angle of attack of a degree quenches LCOs with a gap smaller than 0.6° and significantly limits the range of speed for which some larger gaps are creating the limit cycle oscillations. This is an important consideration for practical applications, but it must be approached with care. As previously mentioned, LCO can indeed develop with small gaps, including those within the regulations, if no angle of attack is present.

Relatively large gap values were also tested, up to a value of 3.0°. Due to the important vibrations transmitted to the main structure in those cases, the wind speed was limited to preserve the model integrity.

Summary

In this presentation, a comprehensive overview of the physics of limit cycle oscillations is provided. Several effects affecting the LCO amplitude and frequency are explored. The obtained conclusions will serve as a foundation for further work, exploiting a more complex system, with multiple nonlinearities.

Acknowledgements

Support by the Federal Aviation Administration as well as contributions by Wael Nour and Sohrob Mottaghi from the FAA are gratefully appreciated.

References

[1] J. Panchal and H. Benaroya, "Review of control surface freeplay," *Prog. Aerosp. Sci.*, vol. 127, p. 100729, Nov. 2021. https://doi.org/10.1016/j.paerosci.2021.100729

[2] D. D. Bueno, L. D. Wayhs-Lopes, and E. H. Dowell, "Control-surface structural nonlinearities in aeroelasticity: A state of the art review," *Aiaa J.*, vol. 60, no. 6, pp. 3364–3376, 2022. https://doi.org/10.2514/1.J060714

[3] G. Dimitriadis, *Introduction to Nonlinear Aeroelasticity*. Chichester, UK: John Wiley & Sons, Ltd, 2017. doi: 10.1002/9781118756478

[4] E. J. Breitbach, "Flutter Analysis of an Airplane With Multiple Structural Nonlinearities in the Control System," NASA Technical Paper 1620, 1980.

[5] R. M. Laurenson and R. M. Trn, "Flutter Analysis of Missile Control Surfaces Containing Structural Nonlinearities," *AIAA J.*, vol. 18, no. 10, pp. 1245–1251, 1980. https://doi.org/10.2514/3.50876

[6] C. L. Lee, "An Iterative Procedure for Nonlinear Flutter Analysis," *AIAA J.*, p. 8, 1986. https://doi.org/10.2514/3.9352

[7] B. H. K. Lee and A. Tron, "Effects of structural nonlinearities on flutter characteristics of the CF-18 aircraft," *J. Aircr.*, vol. 26, no. 8, Art. no. 8, Aug. 1989. https://doi.org/10.2514/3.45839

[8] M. Manetti, G. Quaranta, and P. Mantegazza, "Numerical Evaluation of Limit Cycles of Aeroelastic Systems," *J. Aircr.*, vol. 46, no. 5, Art. no. 5, Sep. 2009. https://doi.org/10.2514/1.42928

[9] Y.-J. Seo, S.-J. Lee, J.-S. Bae, and I. Lee, "Effects of multiple structural nonlinearities on limit cycle oscillation of missile control fin," *J. Fluids Struct.*, vol. 27, no. 4, Art. no. 4, May 2011. https://doi.org/10.1016/j.jfluidstructs.2011.02.009

[10] M. A. Padmanabhan, "Sliding wear and freeplay growth due to control surface limit cycle oscillations," *J. Aircr.*, vol. 56, no. 5, Art. no. 5, 2019. https://doi.org/10.2514/1.C035438

[11] N. Fonzi, H. Curasi, S. Ricci, and E. Livne, "Experimental Studies on Dynamic Freeplay Nonlinearity," in *19th International Forum on Aeroelasticity and Structural Dynamics (IFASD 2022)*, 2022, pp. 1–20.

[12] N. Fonzi, S. Ricci, and E. Livne, "Numerical and experimental investigations of freeplay-based LCO phenomena on a T-Tail model," in *AIAA Scitech 2022 Forum*, San Diego,CA,U.S.A., May 2022. https://doi.org/10.2514/6.2022-1346

Aeroelastic design and optimization of strut-braced high aspect ratio wings

Francesco Toffol[1,a *], Sergio Ricci[1,b]

[1] Department of Aerospace Science and Technology, Politecnico di Milano, via la Masa 34, 20126, Milano, Italy

[a]francesco.toffol@polimi.it, [b]sergio.ricci@polimi.it

Keywords: Strut-Braced Wing, High Aspect Ratio Wing, Conceptual Design

Abstract. To improve aircraft aerodynamic efficiency, a possible solution is to increase the wing aspect ratio to reduce the induced drag term. As a drawback, the span increase introduces an increment of the wing loads, specially of the wing root bending moment that drives the sizing of the wing. Structural mass must be added to withstand higher loads, reducing the aerodynamic advantage from a fuel consumption point of view, as it can be instinctively seen in the Breguet's range equation. To limit the load increment due to the increased span, a possible solution is the usage of a strut: this kind of structure modifies the load path spanwise, diminishing the wing internal forces and reducing the wing penalty mass. In this framework, a lot of research is done studying Ultra-High Aspect Ratio Strut-Braced Wing, where the aspect ratio of such configuration is exasperated above 15, and the resulting wing is extremely flexible and may experience large deformation under loading. Moreover, the over determined structure realized by the fuselage-wing-strut connections deserves particular attention to fully characterize the aeroelastic interaction among the structural elements. For this reason, a two-step design approach that exploits NeoCASS (GUESS + NeOPT) is used to provide a sizing of the wing and of the strut considering several structural and aeroelastic constraints (e.g. flutter and ailerons effectiveness).

Introduction

In the framework of the Clean Aviation program, the HERWINGT project studies new wing configurations for a future regional aircraft carrying 80 passengers. In the project, alongside with innovative powertrain solutions, Ultra High Aspect Ratio Wings (UHARWs) are studied to improve the aerodynamic performance thanks to the reduction of the induced drag. As a drawback, the increase of the aspect ratio introduces higher bending moment for the wing sizing, that starts a snowball effect on the structural mass needed to withstand the loads. Therefore, the design becomes multidisciplinary and it's a trade-off between the aerodynamic performance advantages and the structural mass penalty. A possible solution to limit the structural mass is the Strut-Braced Wing (SBW) configuration, that redistributes the loads between the wing and the strut. This solution is becoming popular in the aviation industry, and it is proved by many project and programs focused on such configuration, for example [1]-[5]. This new configuration introduces new challenges since the conceptual design phases, where statistical approaches like [6][7] are no more valid for the wing structural mass estimation due to the lack of statistical sample for such configuration. Moreover, to fully exploit the benefits of using composite materials and their higher strength to weight ratio, novel physical based approaches must be adopted since the early design phases.

For this reason, in the WP1 of the HERWINGT project, Polimi is involved in the identification of the best wing configuration to be developed in the following of the project, using its in-house developed code NeoCASS [8][11] and its optimization module NeOPT [13][14] (Figure 1).

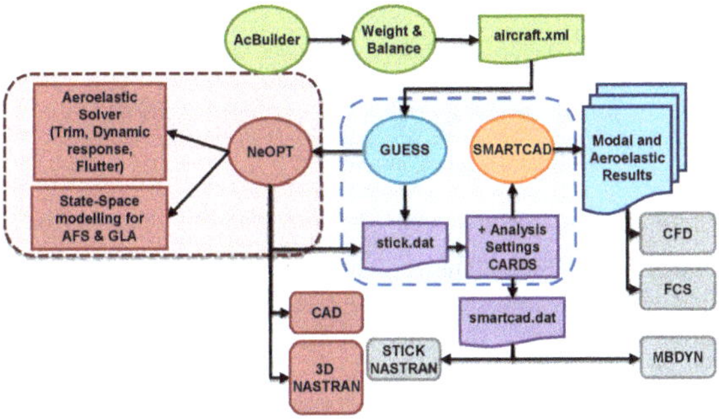

Figure 1: NeoCASS environment from [15]

When dealing with UHARWs, classical fully-stressed approaches may miss some important aeroelastic aspect of such configurations, for example the wing deformation, the divergence and control surfaces effectiveness. For this reason, an approach like the one adopted by NeOPT is convenient: in addition to classical stress and buckling criteria, whichever aeroelastic response can be accounted during the optimization, e.g. flutter or control surface efficiency.

Wing conceptual design in the HERWING project

In the framework of the HERWING project, a reference aircraft carrying 80 passengers is studied. The wing is high-mounted and equipped with two propellers. An initial wing layout was used as starting point for the study of the most promising solution, and its main characteristics are reported in Table 1. It is a trapezoidal wing that is rectangular up to the kink, then a constant tapering is applied. The wing is completely flat, i.e. no dihedral is present. From the material point of view, the wing is required to be in composite material with a symmetric and balanced layup, with at least 10% of ply oriented in the three main directions (0°-45°-90°).

Table 1: Reference wing geometry

Area [m2]	Span [m]	Root Chord [m]	Tip Chord [m]	Aspect Ratio [-]	Engine Span [m]	Kink [m]	Outboard Sweep [°]
73.3	29.65	2.92	1.18	12	5.555	5.555	2.62

Starting from this configuration, the wing was stretched increasing the aspect ratio and keeping the same wing surface, to have the same wetted area, associate to the viscous drag, but increasing the aspect ratio so reducing the induced drag term. The constraints accounted in the wing stretching are: the maximum span achievable remaining in the same Aerodrome Reference Code (ARC= C), that is 36m; the wing tip minimum chord that must be guaranteed to fit the actuation systems, that is 1.4m; the kink position is kept fixed as well as the LE sweep angle.

This approach resulted in a reduction of the root and tip chords, while the LE position was kept constant.

Performing a full geometry optimization would take too much time in this design phases, where it is way more important to understand the sensitivity of the design to the macro parameters rather than finding the optimal solution. For this reason, a parametric study increasing the aspect ratio was performed, considering AR=12,13,15,17. The wings considered for the analysis are illustrated in Figure 2. Despite the maximum span allowed is 36m, the wing is stretched to a maximum of 35m to leave 0.5m for each side to eventually install wing tip device like winglets.

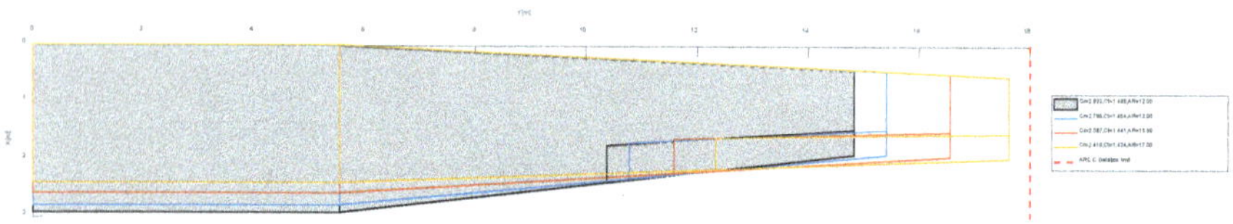

Figure 2: Wing layout selected for the conceptual parametric study.

The design strategy adopted was to perform an initial screening of the solution with the GUESS module, the module of NeoCASS dedicated to the quick and simplified structural sizing, and then to perform a refined optimization of some configuration with NeOPT able to consider a 3D wingbox and different aeroelastic constraints.

Initially, the wing was studied for all the aspect ratio in both cantilevered (CNT) and strut braced (SBW) configurations. The SBW with AR17 immediately showed a considerable weight reduction w.r.t. the CNT wing, for this reason only the AR17 was considered for the SBW configuration.

All the wings were designed with a maneuver envelope compliant with EASA CS25 regulations [16], that results in 45 maneuvers (pull-up, push-down, high lift, roll, sideslip, etc,...), in different flight points (VA,VS,VMO,VD,...). The structural masses of the wingbox and strut-wingbox (spar, stringers, caps, skins) are reported in Table 2 and plotted in Figure 3. Wing group it is intended as the sum of the wing and the strut items.

Table 2: GUESS sizing results for the considered configuration, wing and strut structural masses

Item	CNT AR=12	CNT AR=13	CNT AR=15	CNT AR=17	SBW AR=17
Wing [kg]	1145	1255	1516	1821	828
Strut [kg]	0	0	0	0	615
Wing + Strut [kg]	1145	1255	1516	1821	1211

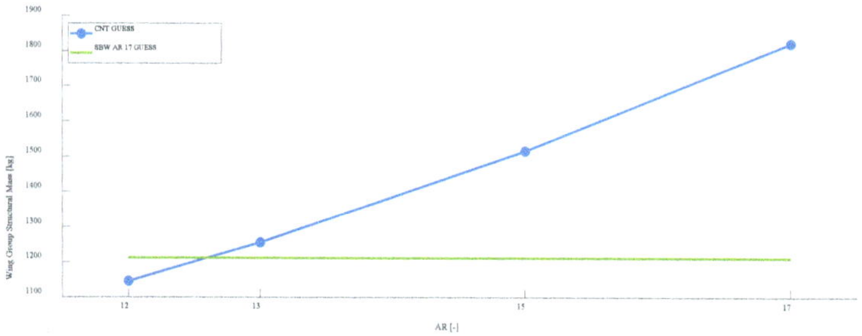

Figure 3: Wing group structural mass evolution with AR

The results obtained show that the SBW wing group with AR=17 has the same weight of the CNT with AR = 12.5. Despite CNT AR = 12 is lighter than the SBW AR=17 solution, the induced drag is term is around 40% higher for the CNT solution. For this reason, the focus is moved on the SBW configuration only, which design was refined with NeOPT considering different structural layout.

The refinement performed with NeOPT considered the same load conditions (45 maneuvers). In addition to the structural constraints, a minimum aileron efficiency must be guaranteed to preserve the aircraft handling qualities. The aileron efficiency is expressed as the ration between the flexible and rigid roll moment derivative w.r.t. the aileron deflection $C_{L,\delta}$ (Eq.(1)).

$$\frac{C_{L,\delta\, flexible}}{C_{L,\delta\, rigid}} > 0.5 \qquad (1)$$

The internal structure layouts considered are listed in Table 3.

Table 3: Structural configuration considered for the NeOPT refinement.

SBW ID	η, Wing Strut Connection [Span %]	Rib Pitch [m]	Stringer Pitch [m]	Spar Number
1	50	0.5	0.16	2
2	50	0.5	0.16	3
3	50	1.0	0.16	3
4	50	0.5	No stringer	3
5	50	1.0	0.16	4
6	66	0.5	0.16	2
7	31	0.5	0.16	2

The results of the sizing are reported in Table 4

*Table 4: NeOPT Sizing results. * Not converged*

SBW ID	1	2	3	4	5	6	7
Wing Mass [kg]	1108	1181	3686*	1700	3355	992	1499
Strut Mass [kg]	383	383	383	383	383	536	386
Wing Group Mass [kg]	1491	1564	4069	2083	3738	1528	1885

From the optimization results, the most convenient layout is a conventional 2 spar with 0.5m rib pitch. Three and four spar configurations were considered to increase the rib pitch, but this solution was not effective for the SBW configuration. Indeed, the strut introduces relevant compression load on the inboard portion of the wing and the increased rib pitch makes the buckling of the stiffened panels critical.

The sensitivity w.r.t. the connection point between the wing and the strut is studied for the two-spar configuration in three different section spanwise, which are 31% 50% and 66%. The firs value is studied to evaluate a possible configuration were the engine and the strut are connected to the wing in the same span location, having a single reinforced wing rib for both the items. The second point is exactly the mid wing, while the last one is the optimal point found in previous studies [17]. The obtained data were fitted with a 2nd order polynomial function and a minimum in the structural mass was found for an η=0.56, as shown in Figure 4.

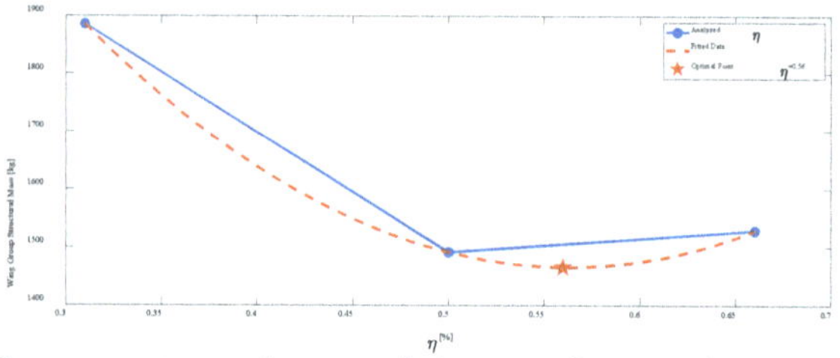

Figure 4: Wing group structural mass evolution w.r.t. the strut-wing connection point.

Conclusion

In the HERWINGT project different SBW configuration were sized, investigating the different structural layout options and connection point for the strut. The most promising solution in terms of wing structural mass was not directly considered, but it was identified with a parametric study: it is a 2 spar layout with 0.5m rib pitch and a connection between the wing and the strut placed at 56% of the span.

Acknowledgements

This project has received funding from the Clean Aviation Joint Undertaking under grant agreement No. 101102010. The granting authority receives support from the European Union's Horizon Europe research and innovation programme and the Clean Aviation Joint Undertaking members other than the Union.

CLEAN AVIATION

Hybrid Electric Regional Wing Integration Novel Green Technologies
HERWINGT

Co-funded by the European Union

Disclaimer

Funded by the European Union. Views and opinions expressed are however those of the author(s) only and do not necessarily reflect those of the European Union or Clean Aviation Joint Undertaking. Neither the European Union nor the granting authority can be held responsible for them.

References

[1] Gur, O.; Bhatia, M.; Schetz, J.A.; Mason, W.H.; Kapania, R.K.; Mavris, D.N. Design Optimization of a Truss-Braced-Wing Transonic Transport Aircraft. J. Aircr. 2010, 47, 1907–1917. https://doi.org/10.2514/1.47546

[2] Bradley, M.K.; Droney, C.K.; Allen, T.J. Subsonic Ultra-Green Aircraft Research (No. NF1676L-19776). Available online: http://web.archive.org/web/20220303155711/https://ntrs.nasa.gov/api/citations/20120009038/downloads/20120009038.pdf (accessed on 19 January 2023).

[3] Hosseini, S.; Ali Vaziri-Zanjani, M.; Reza Ovesy, H. Conceptual Design and Analysis of an Affordable Truss-Braced Wing Regional Jet Aircraft. Proc. Inst. Mech. Eng. Part G: J. Aerosp. Eng. 2020. https://doi.org/10.1177/0954410020923060

[4] Harrison, N.A.; Gatlin, G.M.; Viken, S.A.; Beyar, M.; Dickey, E.D.; Hoffman, K.; Reichenbach, E.Y. Development of an Efficient m= 0.80 Transonic Truss-Braced Wing Aircraft. In Proceedings of the AIAA Scitech 2020 Forum, Orlando, FL, USA, 6 January–10 January 2020. https://doi.org/10.2514/6.2020-0011

[5] Ricci, S., Paletta, N., Defoort, S., Benard, E., Cooper, J. E., & Barabinot, P. (2022). U-HARWARD: a CS2 EU funded project aiming at the Design of Ultra High Aspect Ratio Wings Aircraft. In AIAA Scitech 2022 Forum (p. 0168). https://doi.org/10.2514/6.2022-0168

[6] D. Raymer, Aircraft design: a conceptual approach. American Institute of Aeronautics and Astronautics, Inc., 2012. https://doi.org/10.2514/4.869112

[7] E. Torenbeek, Synthesis of subsonic airplane design: an introduction to the preliminary design of subsonic general aviation and transport aircraft, with emphasis on layout, aerodynamic design, propulsion and performance. Springer Science & Business Media, 2013.

[8] L. Cavagna, S. Ricci, and L. Travaglini, "Neocass: an integrated tool for structural sizing, aeroelastic analysis and mdo at conceptual design level," Progress in Aerospace Sciences, vol. 47, no. 8, pp. 621–635, 2011. https://doi.org/10.1016/j.paerosci.2011.08.006

[9] L. Cavagna, S. Ricci, and L. Travaglini, "Structural sizing and aeroelastic optimization in aircraft conceptual design using neocass suite," in 13th AIAA/ISSMO Multidisciplinary Analysis Optimization Conference, p. 9076, 2010. https://doi.org/10.2514/6.2010-9076

[10] L. Cavagna, S. Ricci, and L. Riccobene, "A fast tool for structural sizing, aeroelastic analysis and optimization in aircraft conceptual design," in 50th AIAA/ASME/ASCE/AHS/ASC Structures, Structural Dynamics, and Materials Conference 17th AIAA/ASME/AHS Adaptive Structures Conference 11th AIAA No, p. 2571, 2009. https://doi.org/10.2514/6.2009-2571

[11] L. Cavagna, S. Ricci, and L. Riccobene, "Structural sizing, aeroelastic analysis, and optimization in aircraft conceptual design," Journal of Aircraft, vol. 48, no. 6, pp. 1840–1855, 2011. https://doi.org/10.2514/1.C031072

[12] Fonte, F., & Ricci, S. (2019). Recent developments of NeoCASS the open source suite for structural sizing and aeroelastic analysis. In 18th International Forum on Aeroelasticity and Structural Dynamics (IFASD 2019) (pp. 1-22)

[13] Toffol, F., & Ricci, S. (2023). A Meta-Model for composite wingbox sizing in aircraft conceptual design. Composite Structures, 306, 116557. https://doi.org/10.1016/j.compstruct.2022.116557

[14] Toffol, F., & Ricci, S. (2023). Preliminary Aero-Elastic Optimization of a Twin-Aisle Long-Haul Aircraft with Increased Aspect Ratio. Aerospace, 10(4), 374. https://doi.org/10.3390/aerospace10040374

[15] Toffol, F. (2021). Aero-servo-elastic optimization in conceptual and preliminary design.

[16] EASA Easy Access Rules for Large Aeroplanes. Available online: http://web.archive.org/web/20230119132207/https://www.easa.europa.eu/en/downloads/136694/en. (accessed on 19 January 2023).

[17] Carrier, G. G., Arnoult, G., Fabbiane, N., Schotte, J. S., David, C., Defoort, S., ... & Delavenne, M. (2022). Multidisciplinary analysis and design of strut-braced wing concept for medium range aircraft. In AIAA SCITECH 2022 Forum (p. 0726). https://doi.org/10.2514/6.2022-0726

Ten years of aero-servo-elastic tests at large POLIMI's wind tunnel for active flutter control and loads alleviation

Sergio Ricci

Politecnico di Milano, Department of Aerospace Science and Engineering, Via La Masa 34, Milano, Italy

sergio.ricci@polimi.it

Keywords: Aeroelastic Wind Tunnel Testing, Gust Loads Alleviation, Flutter Suppression

Abstract. Born in the autumn of 2001, but fully operational since 2023, the large wind tunnel of Politecnico di Mlano turns 20 this year and today it is one of the 4 large research infrastructures of the University. Designed, from the point of view of fluid dynamics, entirely within Politecnico, it is close circuit wind tunnel characterized by a particular design that includes two test sections. The boundary layer section (14m x 4m x 38m), located in the return circuit, is particularly suitable for testing objects subjected to the action of the wind such as bridges, skyscrapers, stadiums and large roofing systems; this section is widely used for research in the wind energy sector. The low turbulence section (4m x 4m x 6m), located as usual between the convergent and divergent parts of the tunnel, is mainly used for aeronautical tests, such as airplanes and helicopters, but not only. Suffice it to recall the numerous tests carried out in the field of sport aerodynamics, because the dimensions allow the equipment to be tested directly with the athletes. The paper briefly describes the most relevant aero-servo-elastic tests carried out during the last 10 years of activity.

Introduction

In the late 1990s Politecnico di Milano decided to organize itself over a network of different campus, two in Milano, and other outside Milano in other cities such as Lecco, Como, Mantova and Piacenza. The new Bovisa campus, north side of the city, allowed to create large new laboratories for the Departments that decided to move there, at first Aerospace Department, followed by Mechanical and Energy. The design of the new spaces and new buildings has made it possible to create the large new wind tunnel, I would say unique in the university context.

The new wind tunnel, designed by taking advantage of the internal competences, is a traditional close circuit plant, but with unusual space organization. Indeed, the return circuit is used as a large chamber for wind engineering applications, with a maximum speed of 16 m/s, while two interchangeable test rooms of 4x4x6 m and a maximum speed of 55 m/s are used for typical aeronautical applications. Two characteristics make this wind tunnel especially useful for aeroelastic testing. At first, the large test rooms allow for testing large scale models, in some cases components at full scale. Second, the flow generation system based on 14 fans for a total of 1.5 MW power is fully protected by a steel grid, so that it cannot be damaged by possible brake of aeroelastic models. During years of activity the wind tunnel has been equipped with dedicated equipment for aeroelastic tests that will be presented in the following. Due to the limited space, mainly two kinds of aeroelastic tests will be briefly described: gust loads alleviation, and active flutter suppression tests.

Materials Research Proceedings 37 (2023) 48-51 https://doi.org/10.21741/9781644902813-11

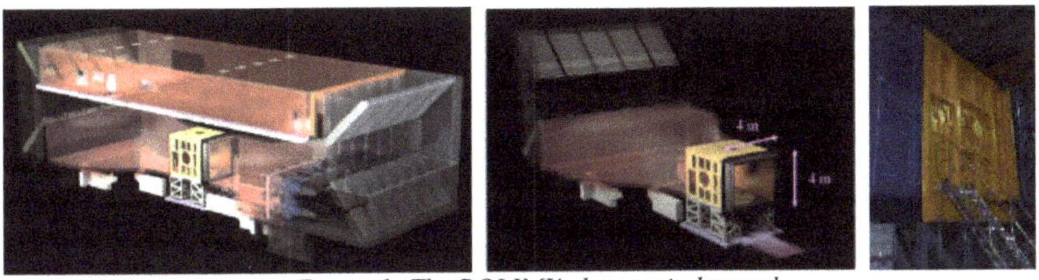

Figure 1: The POLIMI's large wind tunnel

Gust Load Alleviation Tests

During recent years a lot of studies have been dedicated to the implementation of maneuver and gust load alleviation techniques (MLA and GLA), as part of a more general strategy to decrease the environmental impact of future transport aircraft. Indeed, these techniques are believed capable for a further structural weight reduction of about 20%. Two are typically the challenges in testing in wind tunnel GLA control systems. The biggest part of the gust response is due to rigid body motion, so at least plunge and pitch motions must be reproduced. Second, due to the scale factor the bandwidth of actuators requested demands for sufficient space to install inside the model the actuators. These challenges have been addressed at POLIMI by means of the following strategies.

At first, a half model is usually adopted, and due to installation and accessibility problem, the half model is vertically mounted, the connection between the half fuselage and the wind tunnel floor is realized with a pivot, that preserves the pitch free body motion, mounted on a sledge that preserves the plunge free body motion. Since the gravity acts perpendicularly to the lift, a dedicated Weight Augmentation System (WAS) is used to artificially generate the weight force. The WAS consists in a linear electric actuator that acts on the moving sledge in correspondence of the center of gravity of the aircraft, reproducing the weight force indeed. An important ratio which cannot be preserved is the one between the inertial and gravity forces, represented by the Froude number. Thanks to the WAS, it is possible to impose the force that counteracts the lift force, and in the case of unit load factor n=1 it is the weight force. In this way the Froude scaling is preserved only for the plunge motion. For what concerns the model scaling, an iso-frequency approach is adopted, so that the control system designed for the full-scale aircraft can be plugged and played for the wind tunnel test. The different control surfaces are driven by Harmonic drive electric motors that guarantee large bandwidth, typically 15 Hz, and zero free play due to a patented gear box. A dedicated six vanes gust generator has been designed and manufactured able to produce typical 1-cos discrete gust as well as continuous gust excitation with prescribed PSD.

The equipment described above has been successfully tested during different European projects, such as GLAMOUR "*Gust Load Alleviation techniques assessment on wind tUnnel MOdel of advanced Regional aircraft*" (JTI-CS-2013-1-GRA-02-022), as well as the Clean Sky 2 AIRGREEN2 project. Figure 2, left shows the WTT3 1:8 scaled aeroelastic model representative of the future twin prop regional aircraft designed, manufactured, and tested by POLIMI in AIRGREEN project while in the same figure, on the right, a summary of results comparing the root bending moment reduction capabilities for different flight envelope velocities as well different control laws.

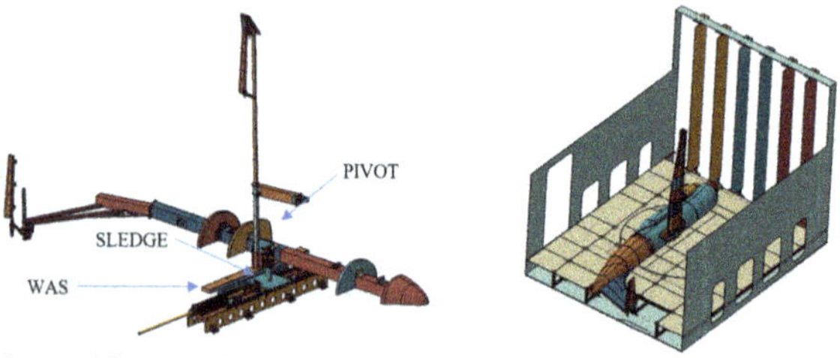

Figure 2: Photo of the aeroelastic half model installed on the WAS system (left) and the gust generator (right)

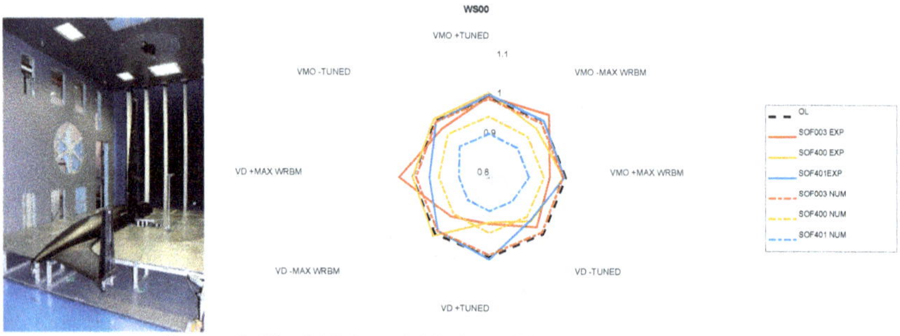

Figure 3: The WTT3 model (left) and the typical test results (right)

Active Flutter Suppression Tests

During the last 10 years POLIMI also acquired a deep knowledge in the design and manufacturing of aeroelastic scaled wind tunnel models used for validation of new active control strategies. One of them is for sure the so-called X-DIA model, three surfaces fully dynamically aircraft 1:10 scaled with respect the reference aircraft developed originally in the 3AS EU project and since then used for many wind tunnel test campaigns for multi surface aeroelastic control. The model has been updated in 2017 to a new, conventional configuration in the framework of the Active Flutter Suppression project in cooperation with the University of Washington in Seattle, USA, and FAA to investigate the robustness of active flutter suppression technologies. Different control laws have been developed and successfully tested, from the simple ones such as Static Output Feedback and ILAF, up to the more sophisticated and robust, such as H infinity. The ailerons, driven by Harmonic drive motors have been used for control. Dedicated safety devices installed on the tip based on a small moving mass from the rear to the forward position were adopted to stop the flutter in an automatic way in case of failure of the control system. The tests have been conducted with the model installed in the testing room by means of cable so to simulate the free-free conditions. During this study, more than 50 flutter conditions have been tested without any damage to the model, showing a very high reliability for the entire system.

During last two year the project focused on the effect of LCOs due to the free play in the control surfaces. Special devices have been designed and manufactured so to be able to apply accurate and small enough gaps to the elevators of X-DIA model. They have been tested separately and last February 2023 the new tailplanes have been installed on the X-DIA model to repeat the active flutter suppression tests in presence of LCOs. Due to the need to accurately set the angle of attack and side du to their strong effect on the preload in presence of gap on control surfaces, the model has been installed on a pylon, equipped with accurate alfa and beta positioning system. Due to the presence of the flexible pylon, the model shows now two flutter mechanisms: the original one, i.e. a symmetric bending torsion at 41.5 m/s, and a new one, an antisymmetric bending torsion at 47.5 m/s. This situation creates a challenging goal since, to identify the second flutter, it is necessary to control the first one. This has been done successfully, then also the second flutter has been actively controlled.

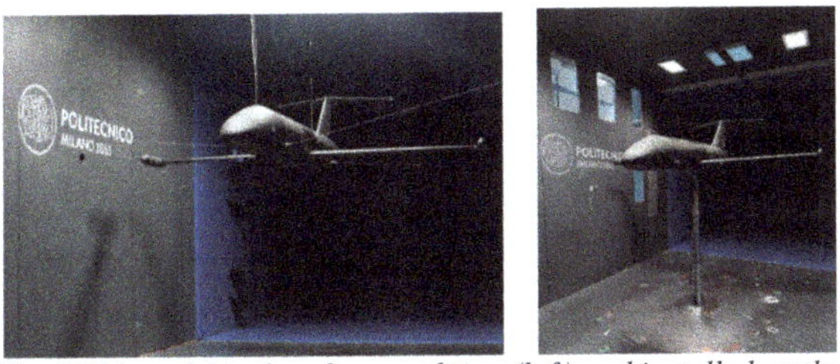

Figure 4: The XDIA model in free-free condition (left) and installed on the pylon (right)

Summary

In this paper a summary of the aeroelastic wind tunnel testing performed at POLIMI's large wind tunnel has been reported.

Acknowledgement

The large set of aeroelastic activities in the framework of different international projects has been possible thanks to the passion, dedication, and competence of many colleagues and students at different level involved. I would like to thank first Prof. Paolo Mantegazza that about 35 years ago founded the aeroelasticity branch at DAER-POLIMI. Then, the contribution of L. Riccobene, A. De Gaspari, F. Fonte, F. Toffol, L. Marchetti, N. Fonzi, V. Cavalieri and G. Bindolino is kindly acknowledged.

References

[1] Ricci, S., Toffol, F., De Gaspari, A., Marchetti, L., Fonte, F., Riccobene, L., Mantegazza, P., Berg, J., Livne, E., Morgansen, K., Wind Tunnel System for Active Flutter Suppression Research: Overview and Insights, AIAA Journal, Vol. 60, N. 12, 2022, p. 6692-6714. https://doi.org/10.2514/1.J061985

[2] N. Fonzi, S. Ricci, and E. Livne, "Numerical and experimental investigations of freeplay-based LCO phenomena on a T-Tail model," in AIAA Scitech 2022 Forum, San Diego, CA,U.S.A. May 2022. https://doi.org/10.2514/6.2022-1346

[3] Ricci, S., De Gaspari, A., Fonte, F., Riccobene, L., Toffol, F., Mantegazza, P., Karpel, M., Roizner, F., Wiberman, R., Weiss, M., Cooper, J.E., Howcroft, C., Calderon, D., Adden, S., Design and Wind Tunnel Test Validation of Gust Load Alleviation Systems, 58th AIAA/ASCE/AHS/ASC Structures, Structural Dynamics, and Materials Conference AIAA SciTech 2017, ISBN: 9781624104534, p. 1-12, Grapevine, TX, USA, 9-13 Jan. 2017. https://doi.org/10.2514/6.2017-1818

[4] Toffol, F., Marchetti, L., Ricci, S., Fonte, F., Capello, E., Malisani, S., Gust and Manouvre Loads Alleviation Technologies: Overview, Results and Lesson Learned in the Framework of the Cs2 Airgreen2 Project, 19th International Forum on Aeroelasticity and Structural Dynamics (IFASD 2022), 2023, ISSN: p. 1-24, Madrid, Spain, 13-17 June 2022.

Aeronautical Propulsion

Studies in hydrogen micromix combustor technologies for aircraft applications

Ainslie D. French[1,a*], Giuseppe Mingione[1,b], Antonio Schettino[1,c], Luigi Cutrone[1,d], Pietro Roncioni[1,e]

[1]CIRA (Italian Aerospace Research Centre), Aerothermodynamics & Fluid Mechanics Dept., Via Maiorise, 81043 Capua (CE), Italy

[a]a.french@cira.it, [b]g.mingione@cira.it, [c]a.schettino@cira.it, [d]l.cutrone@cira.it, [e]p.roncioni@cira.it

Keywords: Hydrogen, Micromix Combustion, Flamelets

Abstract. Within the context of future aircraft turbine engine development technology, hydrogen has fast become one of the most favored candidates as an alternative fuel due to the possibility of producing extremely low levels of pollutants in particular NOx. The principal component of such an engine technology is the hydrogen micro-mix combustor which provides a solution to safe hydrogen combustion avoiding auto-ignition and flashbacks by addressing novel methods of hydrogen-air mixing. The current design concepts include a typical injection manifold composed of multiple concentric arrays of micro-mix combustors which produce hundreds of miniature low temperature diffusion flames having very low NOx levels.

Introduction

Air transport requires much energy and therefore it is necessary to have a fuel with a high energy density. Conventionally, jet kerosene produced from fossil fuels has been the main propellant used in aviation. The main problem with using kerosene as an aviation fuel is the environmental impact due to the production of the greenhouse gas, CO_2, which accounts for about 2,5% of global emissions as well as CO, SO_x.

A switch to hydrogen propelled aircraft would produce zero CO_2, CO, SO_x emissions and, when combined with a suitable technology may substantially reduce the levels of NO_x emitted into the atmosphere and potentially eliminate other pollutants. Additionally, hydrogen has an energy density almost three times that of kerosene but requires heavier storage facilities to kerosene. The main disadvantage of hydrogen is that it requires about five times the volume of conventional fuel to carry the same amount of energy and consequently new concepts for aircraft design need to be addressed to enable safe, light storage of liquid hydrogen.

Currently two lines of research are being pursued towards the application of combusted hydrogen as a propellant both of which aim to reduce the levels of NO_x produced: Lean Direct Injection (LDI), which tends to limit the levels of NOx emissions to those of current kerosene fuelled engines and the Micromix combustor technology which aims to lower the levels of NOx produced compared to kerosene. This latter technology is investigated in this paper.

Preliminary investigations and experiments using hydrogen were first conducted by Funke-et-al [1] using a hydrogen air mixture using an experimental test rig constructed for the GTCP 36-300 APU which was modified to incorporate a hydrogen combustor mechanism consisting of concentric arrays of numerous miniature micro-mix combustors as shown in Fig. 1.

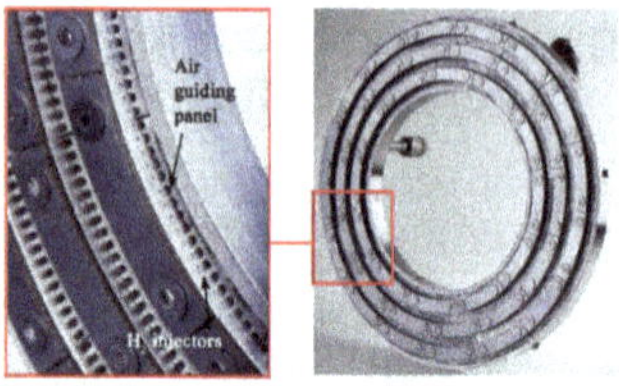

Fig. 1 Prototype micro-mix combustor for gas turbine Honeywell/Garrett APU GTCP 36-300 [2]

Numerical method

The commercial program ANSYS-FLUENT [3] was used as the main software used to obtain flow solutions for all test cases. Calculations using the FGM method with partially premixed combustion were conducted and compared with results from simulations using the eddy dissipation method also performed for a baseline test case hypothetical APU micromixer design at a relatively low pressure described by Ghali [4]. The chemical kinetic mechanism for the first series of FGM simulations used the Naik model [5] consisting of 9 species and 21 reactions. Turbulence was modelled using the k-ω SST turbulence model by Menter [6].

This turbulence model is particularly suited to these kinds of simulations where the recirculation and mixing of hydrogen is of primary importance. The model is capable of accurately capturing the detail of the turbulent flow both through the viscous sublayer down to the solid wall as well the large-scale eddies in the recirculation region and the small-scale eddies in the vicinity of the hydrogen jet.

To calculate the concentration of NO in ANSYS-FLUENT the extended Zeldovich chemical kinetic scheme was used, the details of which are given in [3] and where the calculation is restricted to thermal NOx formation.

Computational domain and numerical results

A structured grid having almost 580,000 cells was created using the ANSYS-ICEMCFD grid generator which satisfied the grid quality requirements where the air inlet is over the left vertical face and the hydrogen inlet is the cylindrical vertical tube shown in Fig. 2.

Mass flow rate boundary conditions are set with an air inlet temperature of 422K at a pressure of 2.5 atm., a hydrogen inlet temperature of 300K and the mass flow rates listed in Table 1 at a pressure of 1 atm. and pressure outlet boundary conditions of 2.4 atm. Adiabatic no slip boundary conditions were applied at the walls.

Table 1: Equivalence ratios and hydrogen mass flow rates for combustor

Equivalence ratio ϕ	H_2 mass flow rate ($kg{\cdot}s^{-1}$)
0.3	4.335×10^{-6}
0.4	5.75×10^{-6}
0.5	7.225×10^{-6}
0.6	8.67×10^{-6}

The results are presented from Fig. 3 to Fig. 9. Fig. 3 shows the flame anchored between two relatively large recirculation regions on the symmetry plane, the upper region being slightly larger than the lower one. The upper inner vortex originates from the main air jet whilst the lower vortex originates from the recirculated combusted gas downstream of the hydrogen jet. The velocity at the air channel exit is approximately 100 $m{\cdot}s^{-1}$ which is also consistent with that associated with the low pressure drop of a typical combustor. The velocity at the downstream edge of the H_2 channel at the entrance to the chamber is about 250 $m \cdot s^{-1}$.

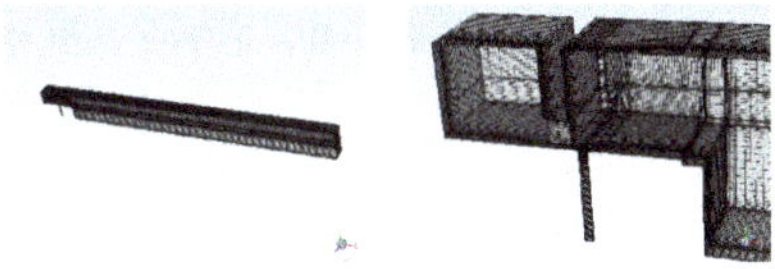

Fig. 2: Computational domain used in calculations

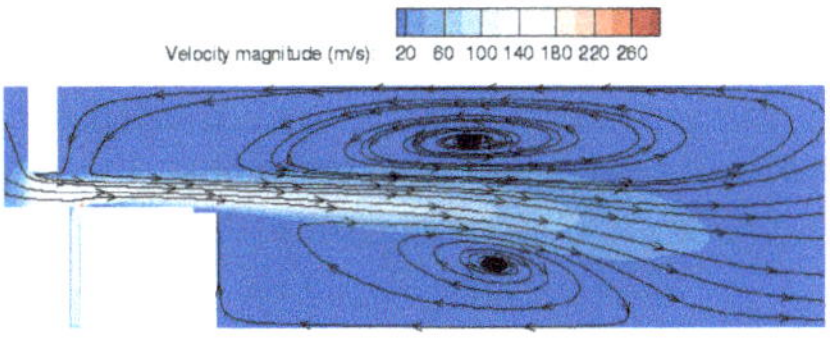

Fig. 3 Calculated flow field and stabilising vortices with EDM model for ϕ = 0.4

Fig. *4* and Fig. 5 show the temperature and NO distributions for different equivalence ratios on the symmetry plane obtained with the EDM model. The temperatures are consistent with those in [4], where a maximum flame temperature of about 2600 K is predicted for the highest equivalence ratio of $\phi = 0.6$ associated with a larger more extended flame.

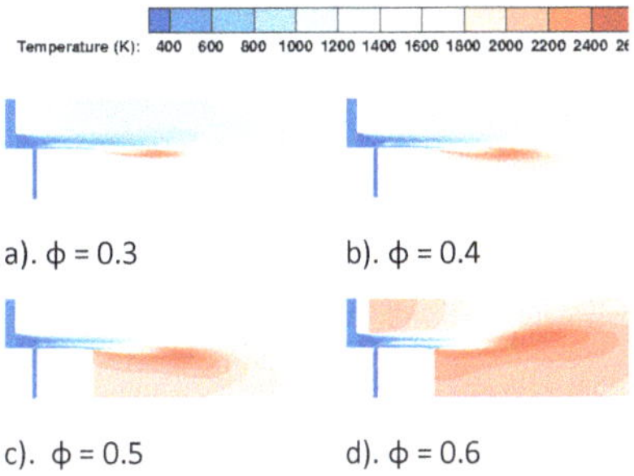

Fig. 4: Flame temperature for different equivalence ratios with EDM

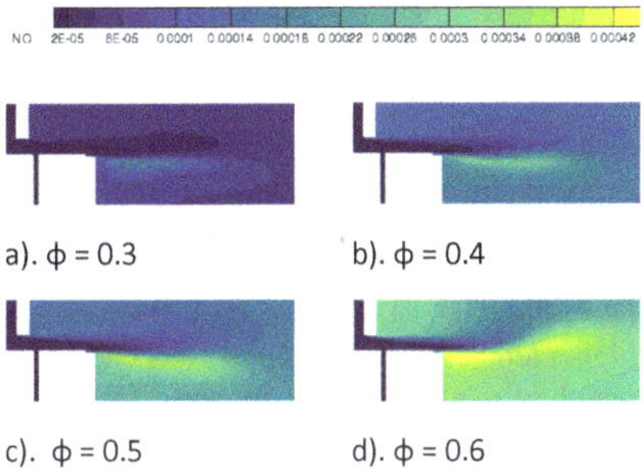

Fig. 5: Mass fraction of pollutant for different equivalence ratios with EDM

Similar distributions using the FGM method are shown in Fig. 6 and Fig. 7

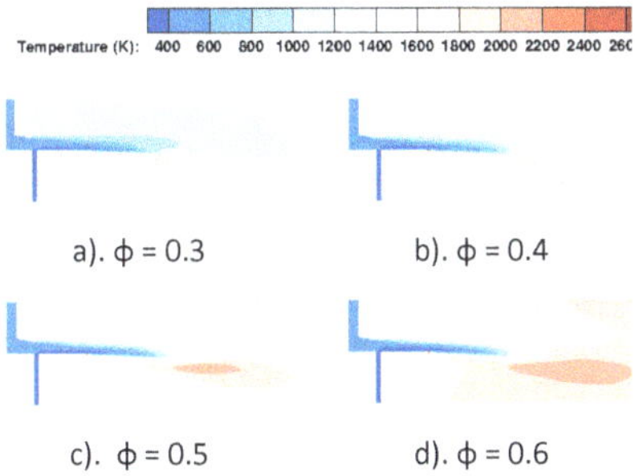

Fig. 6: Flame temperature for different equivalence ratios with FGM - Naik

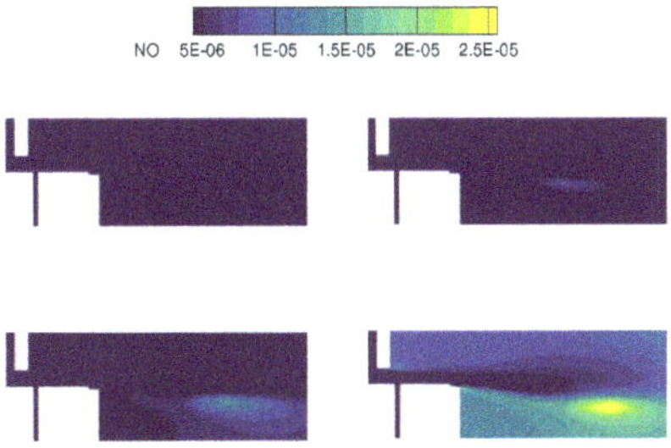

Fig. 7: Mass fraction of pollutant for different equivalence ratios with FGM - Naik

The flame temperature is about 200K lower than with the EDM due to the energy consumed in the chemical reactions in the FGM model compared to the single step reaction in EDM calculation with a corresponding reduction in NO concentrations.

The results for mass-averaged temperature and NO emissions corrected to 15% O_2 are shown in Fig. 8 and Fig. 9. The flame temperature and NO emissions are a direct consequence of equivalence ratio, the lower the equivalence ratio, the shorter and cooler the flames with the associated reductions in NO emissions.

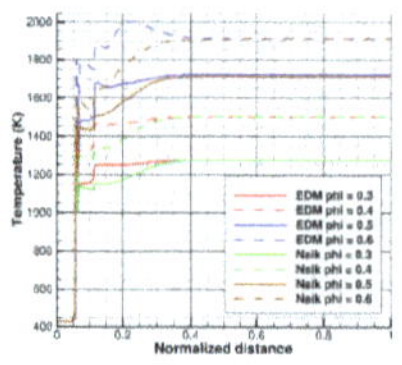

Fig. 8: Mass weighted average temperature along combustor length for EDM and FGM -Naik for different equivalence ratios

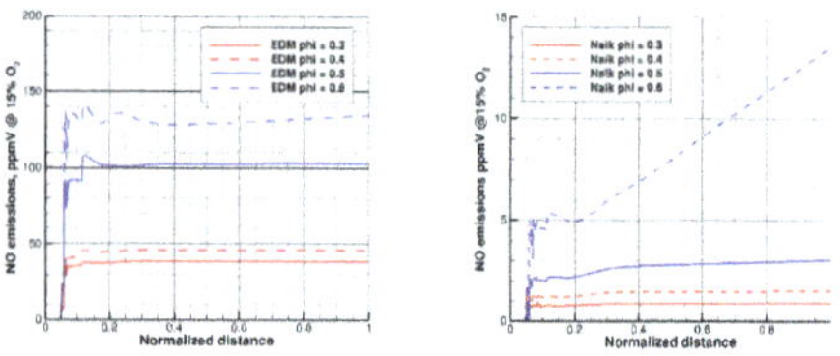

Fig. 9: NO distributions along combustor length for EDM and FGM-Naik models for different equivalence ratios

Summary

Comparison of results between the EDM and the FGM-Naik model demonstrated that the temperature distributions were in good agreement but much lower NO levels, of about three orders of magnitude were obtained with the latter model. Qualitative comparison of the temperature and NO emission levels indicated satisfactory agreement with the trends expected in the physics where higher temperatures and higher NO concentrations were obtained for increasing equivalence ratios which were also in broad agreement with similar results available in the literature [4].

References

[1] Funke, H. H.-W., Boerner, S., Keinz, J., Kusterer, K., Kroniger, D., Kitajima, J., Kazari, M., Horikawa, A., "Numerical and Experimental Characterization of Low NOX Micromix Combustion Principle for Industrial Gas Turbine Applications", GT2012-69421, Proceedings of ASME Turbo Expo 2012, June 11-15, 2012. https://doi.org/10.1115/GT2012-69421

[2] Haj Ayed, A., Kusterer, K., Funke, H.H.-W., Keinz, J., Striegan, C., Bohn, D., "Experimental and numerical investigations of the dry-low. NOx hydrogen micromix combustion chamber of an industrial gas turbine", Propulsion and Power Research 2015; 4(3):123-131. https://doi.org/10.1016/j.jppr.2015.07.005

[3] ANSYS, ANSYS Fluent Theory Guide Release 15.0, ANSYS Inc., 2013.

[4] Ghali, P.F., Khandelwal, B., "Design and Simulation of a Hydrogen Micromix Combustor", AIAA Sci-Tech Forum, 19-21 January 2021., M. Rodriguez, J.A. Voigt and C.S. Ashley, U.S. Patent 6,231,666. (2001).

[5] Reaction Design: San Diego, 2015. Ansys Chemkin Theory Manual 17.0 (15151).

[6] Menter, F.R., "Two-equation eddy-viscosity turbulence models for engineering applications", AIAA Journal, Vol. 32, No. 8, 1994, pp. 1598-1605. https://doi.org/10.2514/3.12149

Performance assessment of low-by-pass turbofan engines for low-boom civil supersonic aircraft

Francesco Piccionello[1,a *], Grazia Piccirillo[1,b] and Nicole Viola[1,c]

[1]Politecnico di Torino, Department of Mechanical and Aerospace Engineering, Corso Duca degli Abruzzi 24, 10129, Torino, Italy

[a]s301549@studenti.polito.it, [b]grazia.piccirillo@polito.it, [c]nicole.viola@polito.it

Keywords: Low-By-Pass Turbofan, Supersonic Aircraft, Conceptual Design, MORE&LESS

Abstract. This paper presents an approach to evaluate the performance of low-bypass turbofan engines without afterburner for a low-boom supersonic aircraft operating at Mach 1.5. The proposed method focuses on optimizing the propulsive performance by minimizing fuel consumption while meeting mission profile requirements. The study contributes to the MORE&LESS project, providing methods for rapidly designing novel supersonic propulsion concepts with improved environmental performance. The research conducts a thermodynamic analysis for on-design engine conditions based on the Modified Specific Heat (MSH) gas model. Specific non-installed thrust and fuel consumption are estimated for cruise phase. Then, the engine cycle analysis is also performed to study off-design performance, including simplified models to account for engine drag and calculate installed thrust and fuel consumption. MATLAB simulations are employed to determine thrust and consumption based on the specific mission profile of the Mach 1.5 case-study, allowing for comparison of different engine types. Ongoing work involves the optimization of engine parameters such as compression ratio, bypass ratio, and turbine inlet temperature, targeting further fuel consumption reduction and pollutant emission estimations.

Introduction

The next generation of supersonic aircraft is shifting towards low-boom configurations with slender structures and cruising Mach numbers of about 1.5. Extensive research is underway to identify the most suitable propulsion system that can effectively fulfil the specific mission requirements of these aircraft. Low-bypass turbofan engines are currently considered the most promising solution. Therefore, this paper aims at analysing the use of low-by-pass turbofan for low-boom civil supersonic aircraft cruising at Mach 1.5. The proposed approach relies on a series of previous studies. In [1] the performances of the engine were computed performing on-design and off-design analyses. Similar considerations could be done for [2], which focused on the cycle and engine layout of similar boom jet. The aim of this paper is to develop the approach proposed in [4] and [6], in which the engine is designed considering some mission phases estimating fuel consumption and expanding off-design analysis to the entire mission profile.

The initial analysis focuses on the engine's parametric cycle, or on-design cycle, considering the aircraft high-level requirements related to the propulsion system, such as the desired cruise thrust, and geometrical constraints. Subsequently, the sea-level thrust is calculated solving the off-design equations using the Newton-Rapson algorithm. During this phase, limitations to the operation of the afterburner are also considered, in order to minimize the acoustic impact of the engine. Once the engine meets all the requirements and constraints, the main parameters are fixed. The final analysis involves evaluating the Thrust-Specific Fuel Consumption (TSFC) during each phase of ASTOS (Aerospace Trajectory Optimization Software) mission profile. Multiple iterations can be performed to assess the mission average TSFC reduction, which serves as figure of merit for performance comparison. The analyses were conducted using MATLAB routines, which were developed to contribute to the EU-funded MORE&LESS project, providing methods for rapidly designing novel supersonic propulsion concepts with better environmental performance. The high-level requirements of the considered aircraft are in Table 1.

Table 1. Low-boom supersonic jet high-level requirements

Low-boom supersonic jet high-level requirements	
Cruise Mach Number	1.5
Cruise altitude	16 km
Payload	8-12 passengers
Range	6500 km
Propellant	biofuel

The Methodology section specifies how these requirements were used as inputs to develop an *ad hoc* propulsive system. Outputs in terms of fuel consumption and fuel mass flow rate for different turbofan engine designs are presented in the Results section. Considerations and final remarks are drawn in the Conclusion section.

Methodology

The workflow adopted is depicted in Fig. 1, while the inputs required to start the analysis are listed in Table 2.

Table 2. Inputs: Engine requirements

Cruise Thrust	30 kN
SL Thrust	143.1 kN
Inlet Diameter	1.10 m
Number of Engines	2

First, the required thrust during cruise and geometrical constraints (inlet and nacelle size) are considered. On-design analysis is carried out, using altitude and flight Mach number as main inputs. According to [7], two turbofan mixed flows engines are the appropriate propulsion systems configuration for this type of jet. It also possible to consider separated flow turbofans [2], however the increase of fan diameter lead by by-pass must be taken in consideration due to aerodynamics implications. Then, sea-level thrust is computed setting Mach number and altitude equal to zero (sea level static thrust). The mission duration is around 16000 seconds, dived into 486 points, so each step is around 32 seconds. At each step time, a flight Mach number, altitude and required thrust is given, it is possible to calculate the engine thrust from off-design tuning the throttle, to reach required thrust with a residual less 10 N. The inlet temperature is chosen as throttle parameters, varying it until convergence with a step of 2 Kelvin. Additionally, Fig.1 reports the output of engine/mission analysis. It also possible to consider the installation losses due to nacelles.

However simplified model in this paper is adopted for the inlet (Eq. 1) and an average value of 0.03 of uninstalled thrust for the nozzle.

$$T = F - \phi_{inlet} \cdot F - \phi_{nozzle} \cdot F \approx 0.95 \cdot F \tag{1}$$

It is possible to summarize the uninstalled thrust calculations using off-design relationship, reported in the following equation (Eq.2):

$$F = F(M_0(t_k), Alt(t_k), \tau(t_k)) \tag{2}$$

Known the installed thrust, it is possible to calculate the installed thrust fuel consumption directly, using Eq. 3:

$$TSFC = \frac{\dot{m}_f}{T} \tag{3}$$

Mass fuel rate $\dot{m}_f$ is computed from off-design algorithm using Newton-Raphson method to solve non-linear system of equations. TSFC varies during the mission and the thrust fuel consumption during mission profile is known at each point, so it is possible to optimize the engine achieving the fuel consumed reduction, changing the on-design engine parameters such as by-pass (BPR), overall pressure ratio (OPR) and fan pressure ratio (FPR). For on-design inlet turbine temperature, it is possible to consider the level of technology of the evaluated engine, assign the desired value.

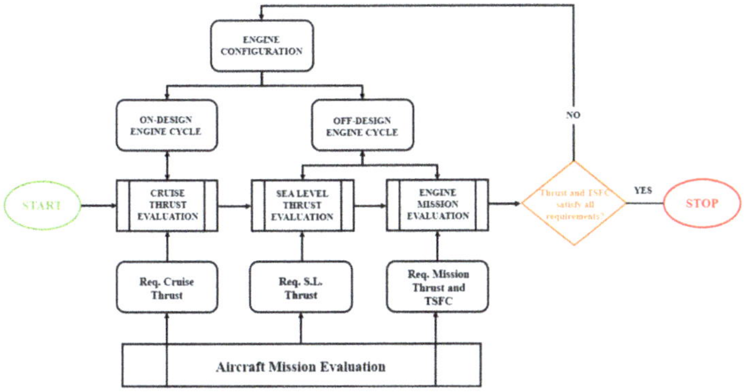

Figure 1. Methodology workflow

Results

The methodology previously described has been used to optimize a turbofan engine for supersonic low-boom jet. Letter B indicates the by-pass ratio and letter O the engine OPR. The compression ratio of the fan is calculated so that the ratio of total pressures inside the mixer is equal to 1 [4]. It is assumed that the power and flow rate spills to the engine to be zero according to a technological level equal to 4 [6] as the efficiencies of the various components that make it up. The starting engine has a bypass ratio of 0.20, and an OPR of 20 (TURB MIX B020-O20). The engine delivers sufficient thrust throughout the mission while consuming about 11121.40 kg of fuel. An attempt can be made to increase the bypass ratio in order to reduce fuel consumption. It is observed that excessive increase in the bypass ratio (0.600) results in insufficient thrust output from the engine. Consequently, the bypass ratio is adjusted to 0.400, and the process is iterated. As can be seen in Table 3, there is a reduction in engine consumption. Another way to reduce the TSFC of the engine is to increase the total compression ratio of the engine, up to a limit value equal to 30.

Table 3. Engine fuel used.

	Single Engine	Two Engine	[%] of fuel saved
TURB MIX B020-O20	11121.40 kg	22242.79 kg	0%
TURB MIX B040-O20	8919.91 kg	17839.81 kg	-20%
TURB MIX B040-O24	9036.41 kg	18072.83 kg	-19%
TURB MIX B040-O26	8625.68 kg	17251.36 kg	-22%

Keeping the engine by-pass constant, we can reach up to 26. Note that the TURB configuration MIX B040 O24 has higher fuel consumption than TURB MIX B040 O24. The explanation for these counterintuitive results lies in the numerical error related to the throttle setting during the mission. Fig. 2 and 3 show the fuel flow rate and TSFC values of the four engines.

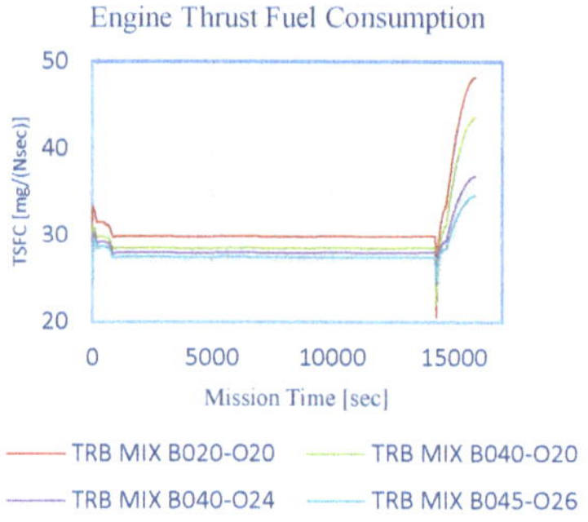

Figure 2: Engine TSFC

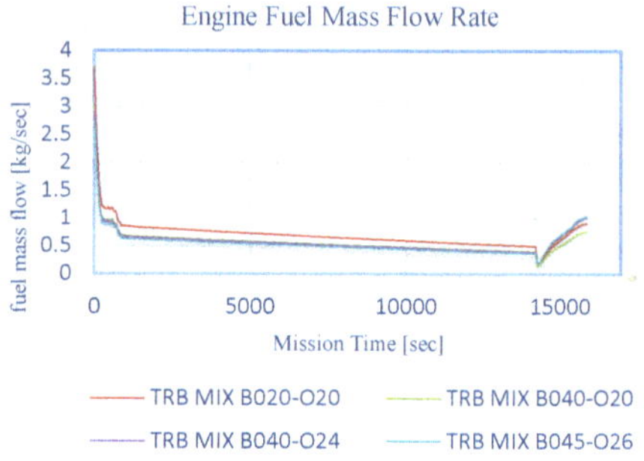

Figure 3: Fuel Mass Flow Rate

Conclusions

The method presented here calculates the performance of a mixed flow turbofan engine for a Mach 1.5 low boom supersonic aircraft, supporting the activities of the MORE&LESS project. Simple equations implemented in MATLAB were used for on-design and off-design thermodynamic cycle calculations. It was found that the model has limitations in throttle adjustment but can still yield results in agreement with literature using minimal input. The method is suitable for preliminary sizing in early project phases, with potential for further improvement in component mapping, throttle adjustment, and emissions modeling.

References

[1] Aria Tahmasebi, Linköpings. (n.d.). *Turbofan Engine Modeling - For the The fighter Aircraft of The Future.* Aeronautical Engineering Spring 2022.

[2] Debiasi, M. (2003). Conceptual Development of Quiet Turbofan Engines for Supersonic Aircraft. *Journal of Propulsion and Power.*

[3] Fronseca, V. F., & Lacava, R. F. (2018). Turbofan engine performance optimization on aircraft cruise thrust level. *Journal of the brazilian Society of Mechanical Sciences and Engineering.*

[4] J. Mattingly and W. Heiser United States Air Force Academy, Colaro Springs. (1986). Performance Estimation of the Mixed Flow, Afterburning, Cooled, Two-spool Turbofan Engine with Bleed and Power Extraction, 1986. *AIAA.* https://doi.org/10.2514/6.1986-1757

[5] Mattingly, J. D., & Jr., J. D. (1996). *Elements of Turbine Propulsion.* McGraw-Hill Series in Aeronautical and Aerospace Engineering.

[6] Mattingly, J. D., Heiser, W. H., & Pratt, D. T. (2002). *Aircraft Engine Design Second Edition.* AIAA Education Series. https://doi.org/10.2514/4.861444

[7] Melker Nordqvist, J. k. (2017). Conceptual Design of a Turbofan Engine for Supersonic Business Jet (ISABE-2021-22635).

Numerical and experimental studies on BLI propulsor architectures

A. Battiston[1*], A. Magrini[1,2], R. Ponza[1], E. Benini[2], J. Alderman[3]

[1]HIT09 S.r.l., Padova (Italy)

[2]University of Padova, Padova (Italy)

[3]Aircraft Research Association ARA, Bedford (United Kingdom)

Keywords: Boundary Layer Ingestion, Jet Propulsion, Aerodynamics, Optimization, Computational Fluid Dynamics

Abstract. An increasing awareness about the impact of civil air transportation emissions is currently driving a low-carbon technology transition, towards more sustainable propulsion strategies. Boundary layer ingesting systems are one of the most promising solutions, as a closer integration between fuselage and propulsors is considered a key in the achievement of more sustainable architectures. Such architecture is characterized by a high level of integration between the airframe and propulsors, making the design process become a major challenge. The present work deals with a complete CFD based design and optimization of a propulsive fuselage concept, both in terms of airframe shape and fan design.

Introduction

The present work deals with the activities carried out in the Clean Sky 2 project SUBLIME. The aim of this project was to advance the state of art in BLI studies by means of wind tunnel activities supported by high fidelity CFD simulations to consistently predict the behavior of BLI architectures that minimize inlet flow distortion and maximize the power saving. The design activity has been separately carried out for the wind tunnel test model (WTT) – subjected to geometrical constraints – and an unconstrained full scale (FS) model, using for both cases a momentum-based approach for the definition of a proper performance metric, a surrogate of the propulsive power defined starting from the aircraft net assembly force (NAF): starting from the Net Assembly Force, a surrogate of the net propulsive thrust is introduced by subtracting to the NAF a reference airframe drag:

$$\Delta NAF = NAF - D_{ref} \qquad (1)$$

ΔNAF takes into account the installation effects of the BLI nacelle and the variations in the fuselage shape with reference to a non-BLI configuration. The surrogate of the propulsive efficiency is then defined as:

$$\eta_{\Delta NAF} = \frac{-\Delta NAF\ V_{\infty}}{W_{fan}} \qquad (2)$$

This performance metric has been used during the CFD-based design optimization of the propulsors. For both the WTT and the FS cases, a sequential approach has been followed : (a) geometric parametrization of the propulsor and its connection to the airframe ; (b) design space investigation aimed at determining the most influential design variables and the research space ; (c) shape optimization carried out using GeDEA-II[2-8], a proprietary genetic algorithm. A CFD approach has been considered at each step for the BLI propulsor performance evaluation.

BLI360 Design Optimization

The BLI360 design has been carried out for both the WTT and FS cases. In particular, the WTT configuration was subject to a high degree of geometrical constraints, thus reducing the design variables space and therefore limiting the optimization process. As stated in the previous section, the design process followed a similar approach for each case, starting with a 2D axisymmetric design space investigation. The results of the WTT design space investigation suggested that geometries with low hub radii and short nacelles feature higher values of the ΔNAF efficiency. Furthermore, this metric has a maximum for a certain range of fan pressure ratio, between 1.3 and 1.5, as seen in Fig. 1. The FS Design of Experiments led to an analogue result.

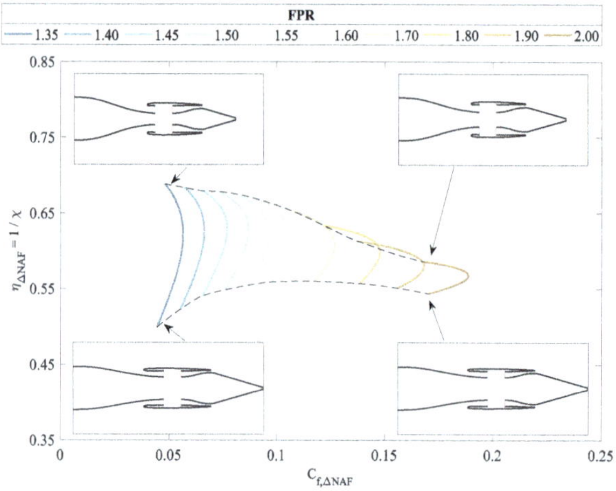

Figure 1: WTT design space investigation results.

A 2D optimization has been conducted to obtain a starting geometry for the 3D design. The objective was the maximization of the ΔNAF efficiency for an arbitrary set of NAF values. The three-dimensional design has therefore been set up by considering three different azimuth profiles of the propulsor: such profiles have been parametrized in terms of highlight height and axial displacement of the cowl maximum radius point (points "A"). Furthermore, interpolation laws between the highlight points and the points "A" of each profile have been defined. Given the high set of constraints of the WTT model, a design space investigation between the design variables defining the interpolation laws and profiles shapes has been conducted, leading to a population of individual with slight variations in terms of performance. Being the FS model geometrically unconstrained, an optimization has been produced. Fig. 2 shows the optimization results in the objectives space. The lack of a Pareto front has been justified by having fixed the fan pressure ratio and the nozzle contraction ratio to the 2D optimum values. The previous design space investigations suggested that these parameters were the main performance drivers. Therefore, any increase in the efficiency is due to a decrease in the drag-components of the axial force.

Materials Research Proceedings 37 (2023) 61-64 https://doi.org/10.21741/9781644902813-14

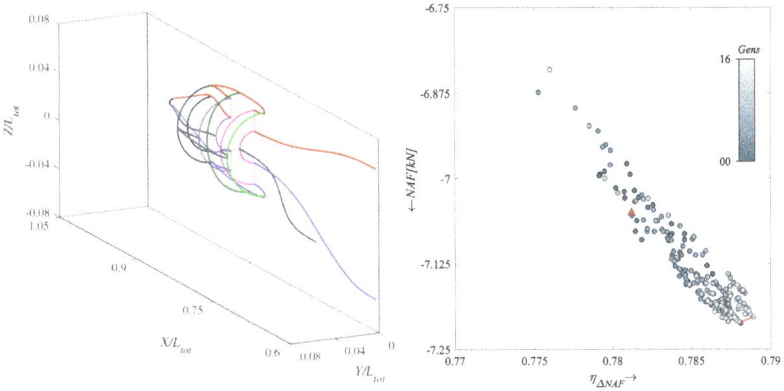

Figure 2: BLI360 FS Model Parametrization and Optimization Results.

SUBLIME Experimental campaigns

The SUBLIME project involved two wind tunnel test campaigns. The first was un-powered where three main configurations were tested which included a reference (without a propulsor), BLI360 and BLI180 boundary layer ingesting propulsor configurations which included exhaust plug changes for intake mass flow variation. The second test campaign concerned only the BLI360 configuration but the nacelle pressurized with an ejector system to simulate the exhaust flow. The test campaigns were carried out in the 9'x8' ARA Transonic wind tunnel at Mach = 0.2 to 0.8 with variation in α and β. Figure 3 shows the model in the wind tunnel & with center cover removed showing the high pressure air manifold for the ejector.

The model instrumentation included a rear fuselage balance, fuselage static pressures and provision for boundary layer rakes. The configurations with propulsors were equipped with AIP rakes for assessment of intake mass flow & distortion characteristics. The wake of the model was measured via a motorized traversing rake system mounted between the support booms to provide high-resolution wake plane data for all configurations.

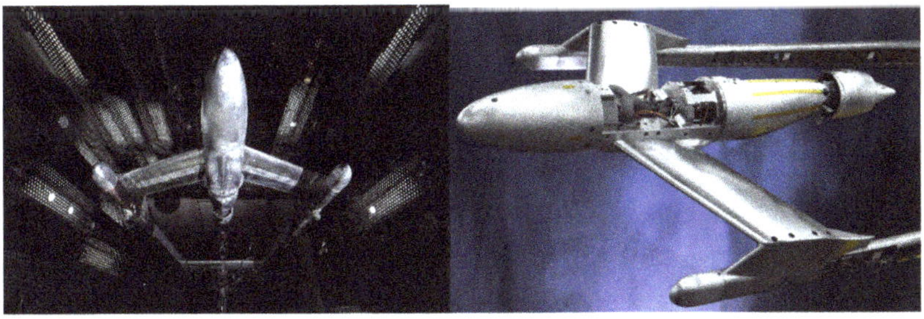

Figure 3: SUBLIME BLI360 configuration in wind tunnel

Acknowledgments

This study is financed by the Clean Sky 2 project SUBLIME (Supporting Understanding of Boundary Layer Ingestion Model Experiment). The project has received funding from the

European Union's Horizon 2020 research and innovation program under grant agreement number 864803.

References

[1] Lee, D., Fahey, D., Skowron, A., Allen, M., Burkhardt, U., Chen, Q., Doherty, S., Freeman, S., Forster, P., Fuglestvedt, J., Gettelman, A., De León, R., Lim, L., Lund, M., Millar, R., Owen, B., Penner, J., Pitari, G., Prather, M., Sausen, R., and Wilcox, L., "The contribution of global aviation to anthropogenic climate forcing for 2000 to 2018," *Atmospheric Environment*, Vol. 244, 2021, p. 117834. https://doi.org/https://doi.org/10.1016/j.atmosenv.2020.117834

[2] Toffolo, A., and Benini, E., "Genetic Diversity as an Objective in Multi-Objective Evolutionary Algorithms," *Evol. Comput.*, Vol. 11, No. 2, 2003, p. 151–167. https://doi.org/10.1162/106365603766646816

[3] Ronco, C., and Benini, E., "GeDEA-II: A Simplex Crossover Based Evolutionary Algorithm Including the Genetic Diversity as Objective," *Applied Soft Computing*, Vol. 13, 2013, pp. 2104–2123. https://doi.org/10.1016/j.asoc.2012.11.003.

[4] Toffolo, A., and Benini, E., "A new pareto-like evaluation method for finding multiple global optima in evolutionary algorithms," *Late Breaking Papers at the 2000 Genetic and Evolutionary Computation Conference*, 2000, pp. 405–410.

[5] Comis Da Ronco, C., Ponza, R., and Benini, E., "Aerodynamic shape optimization of aircraft components using an advanced multi-objective evolutionary approach," *Computer Methods in Applied Mechanics and Engineering*, Vol. 285, 2015, pp. 255–290. https://doi.org/https://doi.org/10.1016/j.cma.2014.10.024.

[6] Benini, E., Ronco, C., and Ponza, R., "Aerodynamic Shape Optimization in Aeronautics: A Fast and Effective Multi-Objective Approach," *Archives of Computational Methods in Engineering*, Vol. 21, 2014, pp. 189–271. https://doi.org/10.1007/s11831-014-9123-y.

[7] Massaro, A., and Benini, E., "A Surrogate-Assisted Evolutionary Algorithm Based on the Genetic Diversity Objective," *Applied Soft Computing*, Vol. 36, 2015, pp. 87–100. https://doi.org/10.1016/j.asoc.2015.06.026.

[8] Benini, E., Venturelli, G., and Łaniewski, W., "Comparison between pure and surrogate assisted evolutionary algorithms for multiobjective optimization," *Front. Artif. Intell. Appl*, Vol. 281, 2016, pp. 229–242.

Aeronautical Systems

Neural networks for the identification of degraded components of aircraft fuel quantity system

Rosario Arcuri[1,a*], Roberta Masciullo[2,b] and Roberto Bertola[3,c]

[1]Aircraft Technologies and Systems, AT&S Virtual Engineering, Leonardo Aircraft Division, Corso Francia 426, 10146, Turin (TO), Italy

[2]Aircraft Systems, Hydraulic & Fuel System, Leonardo Aircraft Division, Corso Francia 426, 10146, Turin (TO), Italy

[3]Aircraft Systems, Eurofighter Project Integrator, Leonardo Aircraft Division, Corso Francia 426, 10146, Turin (TO), Italy

[a]rosario.arcuri@leonardo.com, [b]roberta.masciullo@leonardo.com, [c]roberto.bertola@leonardo.com

Keywords: Aircraft Systems, Fuel system, Fuel Gauging System, Machine Learning

Abstract. The physical and software architecture design of the fuel system of high-performance aircrafts is very complex and represents a challenging topic for aeronautic engineers. Among the main functionalities, there is the calculation of the on board fuel quantity, consisting in data computed by the fuel gauging sub-system, shown to the pilot on the cockpit display and used by the flight control system for aircraft controllability. Due to the large number of sensors and to a strongly ramified calculation code, faults and performance degradation of components, are difficult to be detected/isolated, since the operation would require an invasive investigation on the aircraft. To reduce the impact of the fault detection and isolation process in terms of time and costs, a digital twin of the fuel system has been developed and coupled with a condition-monitoring algorithm based on machine-learning methods. It is thus possible to quickly replicate the mission profiles during which the faults can occur, to calculate the residual fuel mass in parallel with the fuel computer and to precisely identify which component caused the system failure. A neural network, trained on experimental flight data, has been developed to provide reliable data. Once validated , the neural network is coupled with a 0D model that simulates the movement of the fuel inside the tank. In this way, it is possible to obtain the mass of fuel, simulating any flight mission profile. This approach optimizes the analysis of system malfunctions in terms of time and costs, highlighting unexpected mass values, otherwise undetectable. The reliability of the neural network can clearly be increased by training the algorithm with additional flight data, which can be derived from experimental or virtual flights simulated using the 0D model. The versatility of this process makes it applicable for different aircrafts as well as for further developments.

Introduction

During last years, more and more studies and applications are focusing on Artificial Intelligence (AI). AI allows to solve complex problems by processing large amounts of data. In industrial context, investing in the development of new products that use artificial intelligence recommends optimizing time/costs thanks to its versatility and taking advantage of predictive analysis with autonomous machine learning [1,2].

In particular, in the aeronautical industry, AI, supported by digital models, can be a valid support in predicting behavior problems from complex aircraft systems which are constituted of different subsystems with dedicated functionality. With data acquired during the operational life of the aircraft, it is possible to train a machine learning algorithm for debugging failures that have occurred on one or more functionalities, or for predicting new ones. In this context, the digital twin of the aircraft can be a valid support for testing the system recursively, avoiding experimental tests. The knowledge enhanced with this type of activity would allow the deploy of more accurate future system already in the design phase.

This paper presents the application of machine learning, supported by a digital twin, for the identification of the degradation of the fuel gauging system components. For this purpose, two models have been developed that can run alone or interact with each other:

1. Behavioral System Model: tank volume discretization in AMESsim starting from 3D CAD data. The model describes the whole tank geometry and connect each tank compartment domain with the respective rib and spar holes. Imposing the motion due to the aircraft manoeuvre, it is possible to simulate the fuel dynamics inside the tank. Furthermore, the engine feeding, refueling, defueling and venting subsystem has been modelled.

2. Neural Network (NN) for Fuel Gauging: Machine-learning based model developed in Matlab/Simulink environment, which, starting from probes and load factors data, determines the total quantity of fuel. This model exploits the Neural Networks capability to interpolate and extrapolate complex multi-dimension data with very high accuracy without the use of fuel gauging design algorithm.

Fuel Gauging System

The main function of fuel management and gauging subsystem is to provide an accurate measure of on-board fuel quantity [3]. The probes and compensators of gauging system relate to fuel computer, that elaborate the signals derived by gauging component and calculate total fuel mass and aircraft lateral and longitudinal CG position. The fuel tanks quantity data are displayed on cockpit to be available for pilot during flight and are used by the flight control system for aircraft controllability.

Fuel System Behavioral Model

This kind of model is necessary to be used to train a neural network by replacing data retrieved from experimental flights, allowing save time/cost to run tests on the system that would otherwise not be possible in terms of number and variety.

The model has been developed in two phases: the first is the extraction of geometry from CAD data of fuel tank in Catia V5 and then 1D sketching in Simcenter AMESim 2021.1. The CAD contains details (presence of flanges or grooves) and elements (presence of rivets, gaskets) which are not of interest for this application. For this reason, each compartment within the tank has been approximated as a closed cuboid following ribs, spars, top surface and bottom surface of the tank. The tank compartments geometry has been discretized through Amesim Aircraft Fuel System library, and then the whole tank reconstructed with holes, pumps and probes using Amesim 1D sketching (Fig. 1). The model has been validated by matching flight tests: probe signals and fuel mass outputs by Amesim model has been correlated with the same data available from flight recorder, refining the simplifications made on tank geometry. This model can be considered the digital twin of fuel system: it makes possible to simulate the behavior of the fuel inside the tank and thus the center of gravity displacement used by the flight control system for aircraft controllability. In particular, it is possible to have as output the wetted height of fuel probes starting from aircraft load factors and initial fuel quantity. These data can be used to monitor life cycle of the probes and their eventual degradation comparing simulation and real flights data.

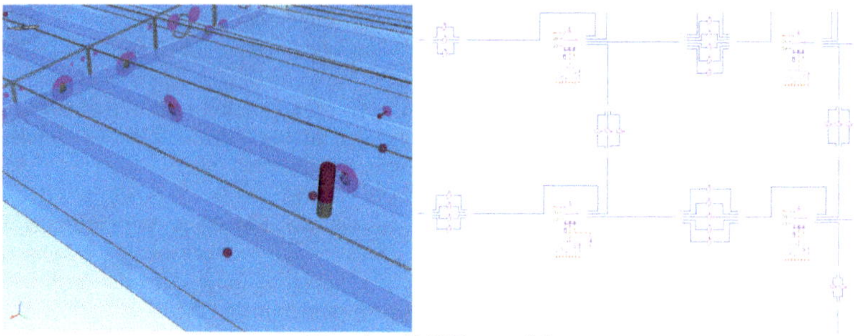

Fig. 1 – AMESim model

NN for Fuel Gauging

For the development of the NN it is necessary to have a training dataset correlating NN inputs with NN outputs that must be as large as possible. This approach is useful when one or more output variables depend on several input variables: in this case it is not always easy or possible to find an analytical correlation using more traditional methods. The resulting model will be a black box with the possibility of being stimulated in different ways.

In this application the main goal of the model is to correlate the signals of the fuel probes and aircraft load factors with the quantity of fuel inside the tank. This correlation can derive either from experimental flights or from virtual flights simulated using the digital twin model, either way the reliability of the neural network increases automatically by increasing the number of flight data provided to the algorithm. Different NN configurations have been studied: the most performant NN as trade-off between accuracy and training time required, resulted in a two-layer feedforward network with sigmoid hidden neurons and linear output neuron exploited with Bayesian regularization training algorithm.

The main application of this NN is the investigation of degraded component of fuel system: a test has been reported in Fig. 2 and Fig. 3. The signal supplied by a specific probe during the test flight is taken, and in one case it is fictitiously amplified (Test #1) and in the other fictitiously reduced (Test #2) for a limited amount of time, as shown in Fig. 2. This determines an altered trend of the computed mass of fuel contained in the tank, clearly departing from the actual trend. In particular, in Fig. 3 it is possible to observe that an amplification on the signal of this specific probe determines a reduction (almost a rigid translation) of fuel; opposite effect on the total mass for a reduction of the signal. This is an expected trend: indeed, the relationship between probe output signal and fuel mass is not univocal, each probe concurs with different weight to fuel quantity computation. Using the described NN algorithm, it is possible to identify which probe is degraded comparing NN fuel mass data with flight fuel quantity.

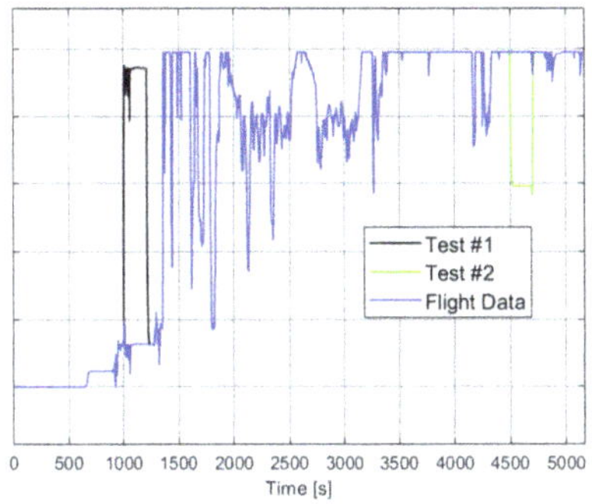

Fig. 2 - Probe Signal Output

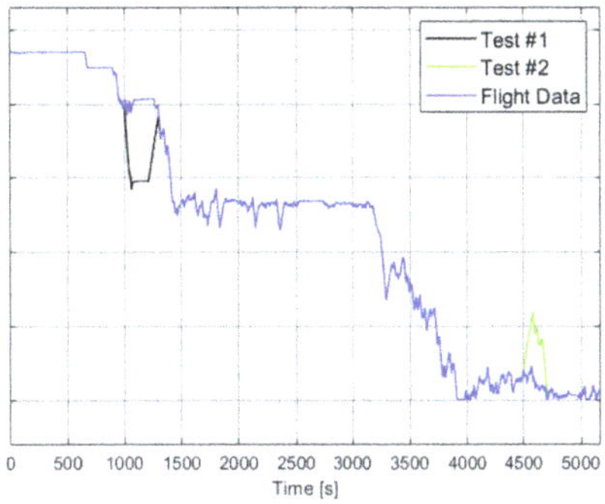

Fig. 3 - Total Fuel Mass

Conclusion

The paper presents the modelling activity carried out for on-service aircraft fuel system: a digital twin has been developed coupled with a machine learning algorithm. The results provide high accuracy with respect to flight test data, by permitting to predict faults and performance degradations of fuel system components with a robust approach as well as by avoiding invasive investigation of the aircraft. Therefore, this approach, highlighting unexpected mass values, optimizes in terms of time and costs the analysis of system component malfunction cases which would otherwise be more time demanding and invasive. The generality of this process makes it applicable for different on-service aircraft as well as for further developments.

References

[1] G. Silvestri, F. Bini Verona, M. Innocenti, M. Napolitano, Fault detection using neural networks, IEEE World Congress on Computational Intelligence, 1994. https://doi.org/10.1109/ICNN.1994.374815

[2] M. T. Hagan, H. B. Demuth, M. Hudson Beale, O. De Jesus, Neuron Model and Networks Architectures, in "Neural Network Design", 2nd Edition.

[3] R. Langton, C. Clark, M. Hewitt, L. Richards, Aircraft Fuel Systems, United Kingdom, 2009. https://doi.org/10.2514/4.479632

Hardware-in-the-loop validation of a sense and avoid system leveraging data fusion between radar and optical sensors for a mini UAV

Marco Fiorio[1,a *], Roberto Galatolo[1,b] and Gianpietro Di Rito[1,c]

[1]Largo Lucio Lazzarino 2, 56122, Pisa, Italy

[a]marco.fiorio@dici.unipi.it, [b]roberto.galatolo@unipi.it, [c]gianpietro.di.rito@unipi.it

Keywords: Sense And Avoid, Unmanned Aerial Vehicles, Hardware in the Loop, Data Fusion

Abstract. The present work illustrates the results obtained at the conclusion of the three-year project TERSA (Tecnologie Elettriche e Radar per Sapr Autonomi), involving the aerospace section of the Dept. of Civil and Industrial Engineering (DICI) of the University of Pisa and its industrial partners. The project aimed at the design and development of a fully autonomous Sense and Avoid (SAA) prototype system, based on data fusion between optical and radar sensors data, for a tactical lightweight surveillance UAV (MTOW<25Kg). The problem of non-cooperative collision avoidance is well known in literature and is currently a central theme of investigation within the aeronautical industry, considering the growing UAV traffic and the consequent need to employ autonomous self-separation technologies in the market. Several past works have investigated the most varied solutions for the Sense problem utilizing optical, acoustic, electro-magnetic signals or a combination of the previous. Likewise, the Avoidance problem has been successfully tackled in literature by means of a wide variety of different approaches ranging from rule-based methods, strategies based on game theory, force field methods, optimization frameworks leveraging genetic algorithms and nonlinear programming techniques and geometric methods. Yet, to the knowledge of the authors, no previous work found in literature has successfully demonstrated and validated the real-time simultaneous interaction of both sense and avoid functionalities within a highly integrated simulation environment. The present work describes the implementation of a complex, nonlinear, simulation environment conceived in order to perform real-time, Hardware-in-the-Loop (HIL), testing of the effective cooperation between sense and the avoid algorithms constituting the core of the SAA system developed within the context of the project. The system effectiveness has been validated by means of complex dynamic simulations, comprising an accurate, fully nonlinear, flight mechanic model of the aircraft, a graphic rendering engine of the scene, proper video capture and transmission pipelines, computer vision algorithms and collision avoidance logics running on the target hardware (Nvidia Jetson Nano) and tailored noise resilient data fusion algorithms. Results show the effectiveness of the system in detecting impending collisions and performing last-resort resolution manoeuvres with high computational efficiency and update frequencies compatible with real world applications in Unmanned Aircraft Systems (UAS).

Introduction

Unmanned aircraft systems (UAS) undoubtedly represent the future of the aeronautic world. The growth in popularity of these highly autonomous aircraft system is driven by the numerous potential applications, their inherent flexibility, lower maintenance cost w.r.t to manned aircraft and the current general willingness to achieve a greener and more sustainable aviation industry. UAS, due to their inherent lower weights, represent indeed the ideal field of application to test innovative electric propulsion systems, characterized by lower emissions. Furthermore, the disruptive military potential of UAS proved itself beyond any doubts in the ongoing Ukrainian conflict, where, for the first time in history, UAS and small drones have been massively used in order to carry out hundreds of thousands of sorties on both sides since the start of the hostilities. For civil applications however, the non-trivial problem of integrating autonomous aircraft systems into non segregated airspace still presents to date a combination of technical and regulatory challenges to solve. Under the general and widely accepted principles that a UAS shall comply with existing regulations and procedures, its operations shall not increase the risk to other users and its integration shall not force other users to carry additional equipment, the development of a suitable Sense and Avoid (SAA) is considered a fundamental milestone towards the future integration of UAS into non segregated airspaces. Said system, should allow the UAS to perceive and avoid obstacles as a human pilot does, effectively empowering the aircraft with separation provision and collision avoidance capabilities. When it comes to the physical implementation of a SAA system, taking into account the dimensional and ponderal constraints of mini-UAVs which often hinder the applicability of more conventional systems (such as ADB-S or TCAS), multisensory architectures seem to be the most prominent solution widely investigated in literature. In particular, SAA system based on data fusion between RADAR and optical sensors have been the subject of several studies [1, 2, 3] which have outlined the benefits of a solution that allows to compensate for the shortcomings of each sensor type, reducing noise and increasing at the same time system robustness and accuracy. During the design phase of such a complex system, simulation models are helpful to validate system requirements and predict system response before actual flight tests can take place. Unfortunately, the simulation of the interaction between the avoidance and the sensing algorithms of a multisensory SAA system is no simple matter. The nontrivial dependency between the source data fed as input to the sensing algorithms (optical data from the camera and range information from the radar) and the attitude and position states of the aircraft (which in turn depend on the output of the avoidance algorithms) give rise to an interconnected problem, to address which, a dedicated HIL simulation environment was developed in the laboratories of the aerospace section of the Dept. of Civil and Industrial Engineering of the University of Pisa. The present work shortly describes the architecture of the SAA system developed within the context of TERSA project and focuses on the description of the nonlinear, HIL, simulation framework specifically created to test the effectiveness of the system utilizing an accurate dynamic model of the aircraft, with aerodynamic data provided by courtesy of Sky-Eye-Systems (SES). The last section briefly describes the simulation conducted and comments on the results.

System Overview

The SAA system has been developed following a multisensory architecture approach that leverages data fusion between a radar, which provides target distance and velocity reading, and Electro Optical (EO) sensor which allows for an accurate angular reconstruction of the target within the field of view of the camera [4, 5]. An High level diagram of the system architecture is presented in Fig. 1; Fig. 2 shows the prototype of the system built by the aerospace section of DICI, in partnership with Echoes [6] within the scope of TERSA project, that was successfully tested at the airfield of Tassignano (Lucca) using a Rapier X-25 aircraft as a moving target [7].

Materials Research Proceedings 37 (2023) 70-75 https://doi.org/10.21741/9781644902813-16

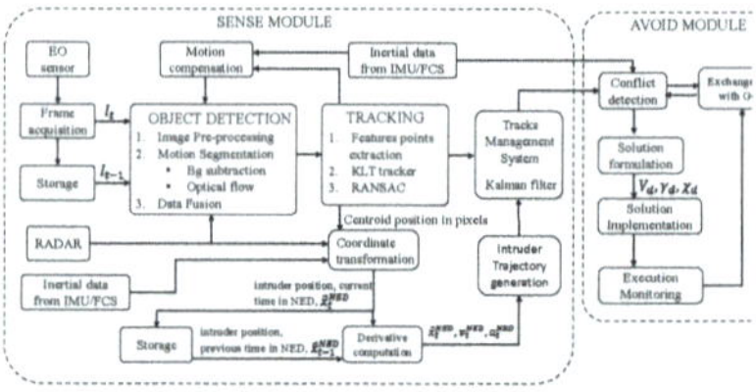

Fig. 1: SAA system architecture

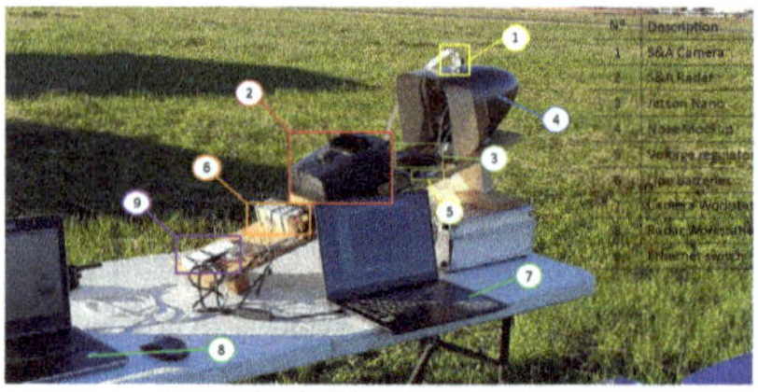

Fig. 2: SAA prototype

In a nutshell, both radar and the optical sensor, continuously scan the air space in front of the aircraft in search of a potential non cooperative intruder. The detection algorithm for the optical sensor is based on an implementation of Single Gaussian Models (SGM) background subtraction algorithm. Motion compensation schemes are employed in order to greatly reduce the noise caused by the ego motion of the camera. Specifically, at each iteration step, the background model is continuously warped (updated) according to an homography matrix which is iteratively obtained via a RANSAC procedure, where matching points between current frame and background model are fed as input. Such correspondences between points are computed by an efficient matching algorithm (i.e. pyramidal implementation of the KLT tracker), while feature points are extracted by means notorious Harris Corner algorithm. Various blobs obtained from the detection stage are then collapsed into single tracks using an implementation of Connected Component Analysis (CCA). Validation based on statistical criteria are then applied to the track readings, in order to make sure that such tracks exhibit a certain regularity in terms of position within the field of view (FOV) and dimension of the visual track itself to further suppress background noise. Once tracks are validated, the pixel position of the intruder aircraft centroid within the FOV is transformed into elevation and azimuth angles in a body fixed reference frame using geometric relationships involving intrinsic parameters of the camera. Such angles, together with the distance reading provided by the radar and the attitude, position/velocity states of the UAV are fed as input to an Extended Kalman Filter (EKF) which performs data fusion and outputs the position and velocity states of the target aircraft in an inertial reference system. These states in turn, constitutes the input to the avoidance routines which continuously check for impending violations of the self-separation condition. When a collision is predicted (meaning that at a certain future time the relative distance between UAV and intruder aircraft will be less than a certain safety threshold), a geometric conflict resolution approach based on the strategies discussed in [8, 9], transforms the dynamic problem into a relative static kinematic problem and computes the tangency solution (i.e. a trajectory which results in the UAV being tangent to the safety bubble at the point of closes approach). The system has been extensively tested in a simulation campaign detailed in the following sections.

Simulation Framework

Initially, a simplified simulation model was devised in order to test avoidance algorithms independently from computer vision part. A Simulink model has been created featuring a realistic 6 DOF, nonlinear, data base driven model of the reference UAV, where the aerodynamic database and the FCS laws has been provided by courtesy of SES. The model features an automatic trim routine leveraging gradient descent optimization, that computes the initial inputs in terms of control surface deflections and engine settings in order to trim the aircraft at various conditions. Avoidance algorithms have been implemented by means of tailored "m-functions", together with the EKF and a waypoint follower block for the trajectory recovery after the end of the avoidance manoeuvre. The position states of the intruder aircraft are considered known in advance, albeit injected with gaussian noise in order to simulate sensor noise. System effectiveness has been tested for a variety of initial conditions in terms of relative position w.r.t to the intruder aircraft ($\Psi_0 =$

$0^{\circ} \div 45^{\circ}$) and intruder speed ($V_0^{Intr} = 22 \div 42\ m/s$), which is assumed to be constant during the simulation, see Fig. 3; the initial heading angle of the intruder aircraft Ψ_{Intr} is precomputed taking into account each aircraft relative initial position and speed magnitude to ensure that a collision will always take place. The avoidance manoeuvre is accomplished by means of a coordinated turn in the horizontal plane. The effectiveness of the manoeuvre is evaluated according to a series of performance metrics (Fig. 4) together with the respect of safe envelope boundaries in terms of load factor limits, maximum bank angle and control saturations.

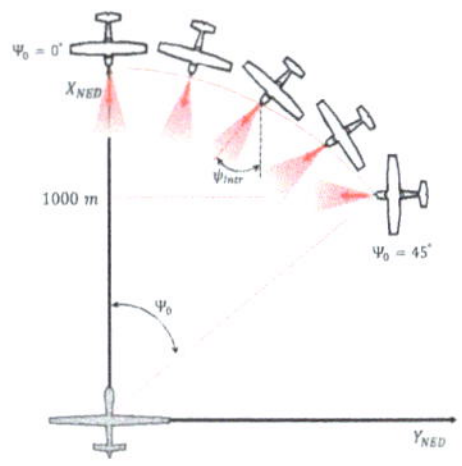

Fig. 3: Simulation initial conditions

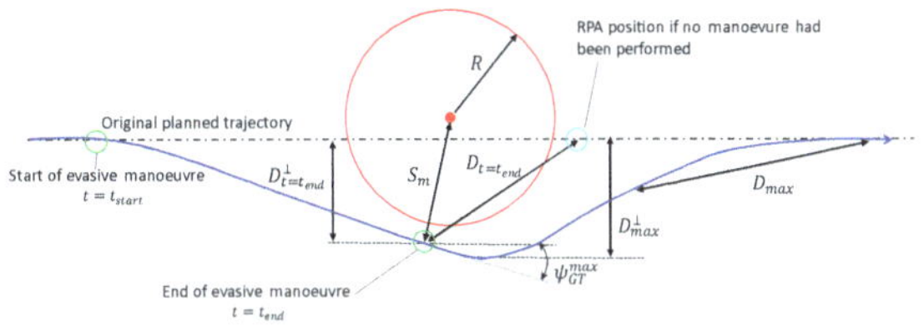

Fig. 4: Evaluation metrics

After the verification of the correct behaviour of the avoidance algorithms and the correct operation of the EKF, more complex simulations have been conducted inserting computer vision algorithms in the close loop, with the aim of validating the simultaneous interaction between sense and avoid modules. This time, intruder aircraft states are no longer assumed known beforehand but only the relative distance is known in a body fixed reference frame. Once again, this measure is injected with gaussian noise. Position states are reconstructed in real time based on the output of the computer vision algorithms that process a realistic rendering of the scene produced by the Open-source flight simulator *Flight Gear (FG)*.

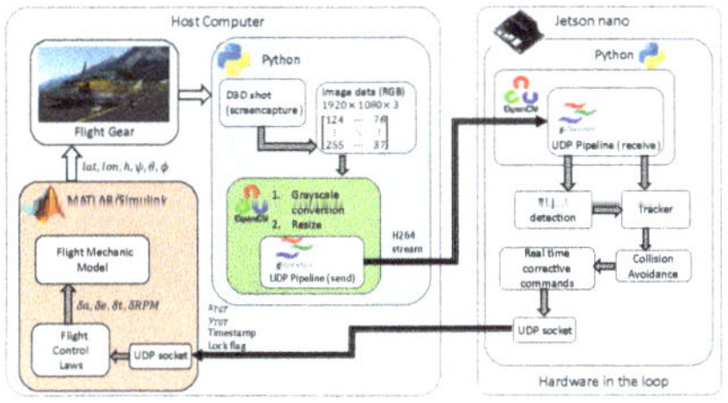

Fig. 5: Scheme of the HIL simulation framework

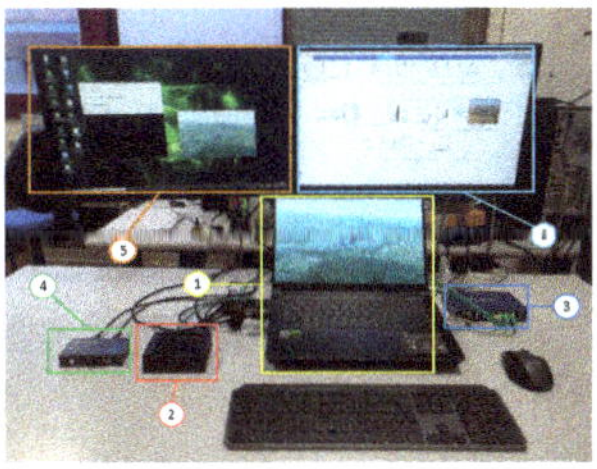

N°	Description
1	Workstation
2	Jetson nano
3	Ethernet switch
4	KVM switch
5	Jetson Display
6	Workstation 2nd Display

Fig. 6: HIL Simulation setup of the SAA system

Basically, as shown in Fig. 5, the Simulink model outlined above, drives an instance of the FG software, running on the workstation which in turn generates a realistic 3D scene comprising accurate weather, lightning conditions, background settings etc. Being an open-source software, tailored modifications have been implemented in order to setup a camera view positioned in front of the aircraft (as to replicate the real EO sensor of the SAA system). Likewise, FG is also configured to display an intruder aircraft model, following a pre-cooked trajectory. Proper synchronization functions make sure that both the Simulink model, the intruder aircraft and FG rendering output are in sync. The workstation screen where FG is opened is captured at a high rate by a dedicated Python module; the obtained frames are then packed into a video stream leveraging the Gstreamer framework and sent to the Video Processing Unit (VPU), Jetson Nano embedded platform, where, thanks to extensive optimisation efforts in the design of the algorithms, object detection and tracking algorithms run in real time (~25Hz) together with avoidance routines. When a collision is predicted and a real time corrective manoeuvre is computed, the prescribed heading angle is fed back in real time to the UAV's FCS in the Simulink model via a dedicated UDP port. This allows for the generation of a realistic synthetic imagery representative of the scene, which evolves according to the dynamic manoeuvre of the aircraft. The simulation setup implemented at the laboratories of the DICI is shown in Fig. 6.

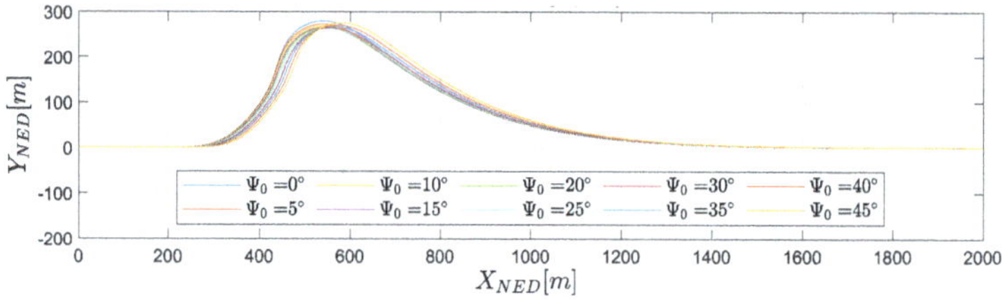

Fig. 7: Avoidance trajectory at various initial intruder positions

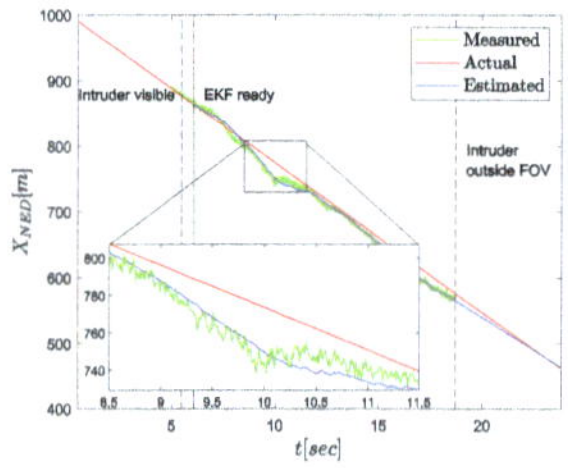

Fig. 8: Intruder position state X

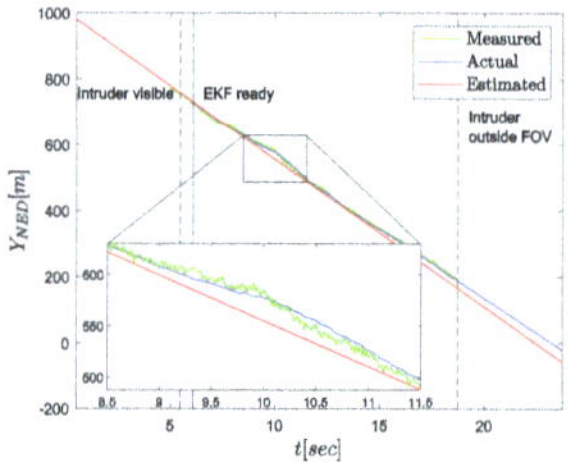

Fig. 9: Intruder position state Y

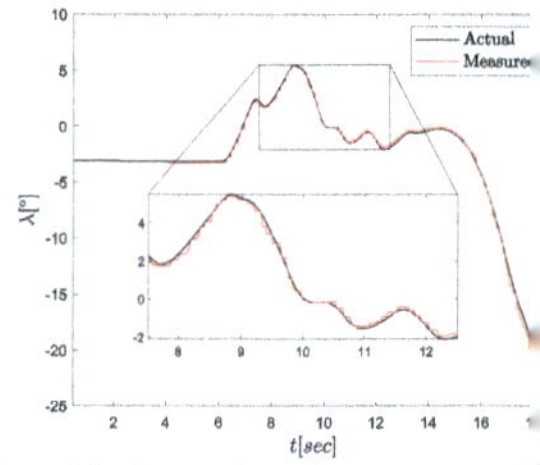

Fig. 10: Intruder elevation angl

Fig. 11: Rendering # 1

Fig. 12: Rendering # 2

Fig. 13: Rendering # 3

Simulation Results

Fig. 7 shows the evasive manoeuvre (seen from above) for various values of Ψ_0 for the highest intruder speed ($V_0^{Intr} = 42\ m/s$) obtained with the simplified simulation model. Fig. 8-9 instead show the position state X, Y of the intruder aircraft (Cessna 172p), reconstructed by the EKF in the complete simulation with the Jetson Nano as hardware in the loop (intruder approaching from port side, @ 52 m/s). Fig. 10 shows the intruder elevation angle reconstructed by the computer vision algorithms plotted against the actual one, while Fig. 11-13 depicts the actual rendering of the scene generated by FG. Plots of the performance parameters of the evasive manoeuvre are omitted for brevity. However, simulation have shown that the system satisfies the requirements for all the tested initial conditions, without invalidating the safe boundary limits. Minimum distance to the intruder is never violated and the evasive manoeuvre is conducted correctly.

Conclusions

The present work has illustrated the complex simulation environment implemented at the Fly-by-wire laboratories of the DICI of the University of Pisa, within the context of the TERSA project which has proved a crucial and invaluable tool for the preliminary validation of a multisensorial Sense and Avoid system. The framework has allowed the verification of the correct simultaneous interaction between avoid and sense algorithms as well as the verification of the requirements of the chosen hardware in terms of available computational power.

Funding & Acknowledgements

This research was co-funded by the Italian Government (Ministero delle Imprese e del Made in Italy, MIMIT) and by the Tuscany Regional Government, in the context of the R&D project "Tecnologie Elettriche e Radar per SAPR Autonomi (TERSA)", Grant number: F/130088/01-05/X38.

The authors wish to thank Sky Eye Systems (Italy) for the support to the research activity and for providing the UAV aero-mechanical data and FCS laws that allowed the creation of an accurate dynamic simulation model.

References

[1] G. Fasano, D. Accardo and A. Moccia, "Sense and Avoid for Unmanned Aircraft Systems," *IEEE A&E SYSTEM MAGAZINE,* pp. 82-110, November 2016. https://doi.org/10.1109/MAES.2016.160116

[2] G. Fasano, D. Accardo, A. E. Tirri and A. Moccia, "Radar/electro-optical data fusion for non-cooperative UAS sense and avoid," *Aerospace Science and Technology,* vol. 46, pp. 436-450, 2015. https://doi.org/10.1016/j.ast.2015.08.010

[3] S. Ramasmy and R. sabatini, "A Unified Approach to Cooperative and Non-Cooperative Sense-and-Avoid," in *International Conference on Intelligent robots and Systems (IROS)*, Vancouver, BC, Canada, 2017., 2017.

[4] M. Fiorio, R. Galatolo and G. D. Rito, "Sense and Avoid system for a mini UAV based on data fusion between electro-optical and radar sensors," in *AIDAA XVI International Congress*, Pisa, 2021.

[5] M. Fiorio, R. Galatolo and G. D. Rito, "Object Detection and Tracking Algorithms Based on KLT Feature Tracker for a Sense and Avoid System," in *AIDAA XXVI International Congress*, Pisa, 2021.

[6] "Echoes Tech," 2022. [Online]. Available: https://www.echoes-tech.it/

[7] "Sky Eye Systems," [Online]. Available: https://www.skyeyesystems.it/

[8] C. Carbone, U. Ciniglio, F. Corraro and S. Luongo, "A Novel 3D Geometric Algorithm for Aircraft Autonomous Collision Avoidance," in *Proceedings of the 45th IEEE Conference on Decision and Control*, San Diego, CA, USA, 2006. https://doi.org/10.1109/CDC.2006.376742

[9] K. D. Bilimora, "A GEometric Optimization Approach to Aircraft Conflict Resolution," in *AIAA Guidance, Navigation, and Control Conference and Exhibit* , Denver, CO, USA, 2000.

A parametric model for thermal management system for more electric and hybrid aircraft

Sofia Caggese[1,a*], Marco Fioriti[1,b] and Flavio Di Fede[2,c]

[1]Department of Mechanical and Aerospace Engineering, Politecnico di Torino, C.so Duca degli Abruzzi n.24, Turin, Italy

[2]Leonardo Labs, Future Aircraft Technologies, Leonardo S.p.A., c.so Castelfidardo 22, 10138 Turin, Italy

[a]sofia.caggese@polito.it, [b]marco.fioriti@polito.it, [c]flavio.difede.ext@leonardo.com

Keywords: Thermal Management System, Hybrid Aircraft, Liquid-to-Liquid Heat Exchanger, Nanofluid

Abstract. In the last decade, the hybrid and electric propulsive systems have been gaining increasing interest to cut down greenhouse gas emissions and thus reduce the environment impact of the aerospace sector. The paper reports the development of a parametric model to design and simulate the Thermal Management System (TMS) of an hybrid electric regional aircraft. Considering the need for a compact design and avoiding the generation of additional drag, a liquid-to-liquid offset strip fin surface heat exchanger is selected. Analysis and modelling of the system are performed for both traditional and innovative coolant, namely, as nanofluids. Five different thermal load conditions are analyzed, which correspond to five different levels of hybridization defined in terms of reduction of CO_2 emission. The most demanding one entails a reduction up to 50% of CO_2 and a thermal load of 67.2 kW to be dissipated. The paper also aims to investigate the most challenging conditions for TMS design and whether the suitability of nanofluids as superior heat carriers. In fact, using nanofluids it is possible to reduce the size of heat exchanger, thanks to the higher thermal conductivity compared to conventional coolant.

Introduction

The primary concerns for climate change are emission of CO_2, NO_x and non-volatile particulate matter (nvPM). In 1983 ICAO Council established the Committee on Aviation Environmental Protection (CAEP), as technical assistant to formulate new strategies and to adopt new Standards and Recommended Practices (SARPs) related to international civil aviation noise and emissions [1] . In 2019, the European Parliament declared a climate emergency and established the European Green Deal. The main goals are to reduce emissions by 55% by 2030 and to achieve net carbon neutrality in Europe by 2050, in order to limit global warming below the 1,5 °C rise [2] . The aviation target is to halve emissions from 2005 levels by 2050. To reach this task new technologies and policies are improved. Electrified systems, biofuels, synfuels, hydrogen, fuel cells, electric motors and batteries are potential sustainable solutions[1] . One of the main challenges is the thermal power control and dissipation of new adopted propulsive system. The Thermal Management System (TMS) is the appropriate system to accomplish these tasks. The TMS acquires, through heat acquisition techniques, the generated thermal power from the heat sources and transfers it, through heat rejection mechanisms, to the terminal heat sinks. The new components needed to implement the electrified propulsion systems, such as the electric motor, the PEM fuel cell, and the battery, are sources of "low grade" heat [2] . For the traditional thermal engine, the heat load is higher than the new heat sources, due to the lower efficiency, but it is directly dissipated from combustion gases. Consequently, for traditional aircraft, little attention is paid to the TMS design. Therefore, the waste heat temperature of low-grade heat source is lower than traditional propulsive system. The absence of exhaust gases is a limitation for the new adopted heat sources and the thermal load to dissipate could exceed one megawatt. Traditionally, aircraft TMS adopts ram air cooling system or liquid cooling system using water, oil or fuel. Heat exchanger is the major implemented device in both conventional and innovative propulsive system. Especially compact heat exchangers are used into aerospace applications [4] The reference aircraft is a regional aircraft with parallel hybrid propulsion system [5] .

Architecture and design

The devised concept features a liquid cooling system involving a coolant, such as water-ethylene glycol mixture, in the main loop, and fuel as the working fluid in the second loop. The main loop transfers the thermal load from the heat source to the liquid-to-liquid heat exchanger, while the second loop transfer it from the heat exchanger to the heat sink. Using fuel as the working fluid, wing fuel tanks are used as terminal heat sinks. The fuel is taken from the hub of the wing tanks and, after the heat exchange with the coolant, it is then returned to the tip of the wing tanks. A schema of TMS concept is reported in Fig. 1.

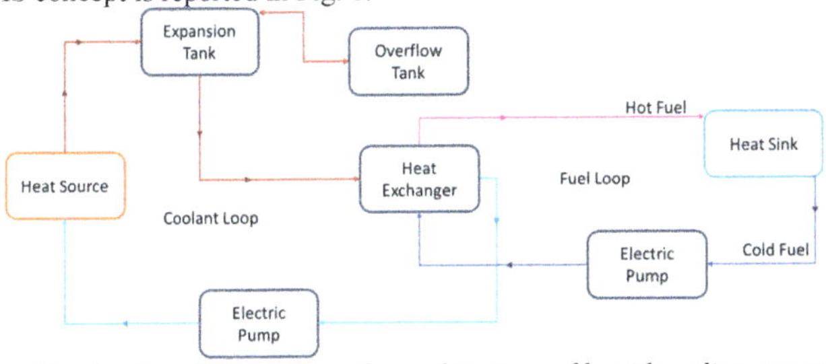

Fig. 1 – Simplified schema for architecture of liquid cooling system

The work aims to design a parametric model of a liquid-to-liquid heat exchanger and to integrate it into the TMS. The heat exchanger is a device that allow heat exchange between two fluids operating at different temperatures. The design differs according to the field of application. In the aerospace domain, the use of compact heat exchangers, characterized by small volumes due to lamellar surfaces that expand the heat exchange surface area, is favoured. The developed device is an offset strip fin surface heat exchanger, considering their typical compactness, with unmixed crossflow arrangement. Sizing is implemented through an iterative python code. At each iteration, the heat exchanger grows in the three dimensions to reach the size to satisfy the design temperature convergence. Type of internal geometry, types of fluids involved in, inlet temperature of both fluids and relative mass flow rates are input required by the parametric model. As output, the python script provides the spatial dimensions, effectiveness, volume and mass of the heat exchanger, core pressure drop, outlet temperature and regimes established of the two fluids involved, and also addresses the system electric consumption, distribution pipe's characteristics and overall TMS mass. The effectiveness-NTU method is used as design approach for heat exchanger. The model is based on the formulation presented by [4] . Effectiveness, ε, is defined as

$$\varepsilon = \frac{\dot{Q}}{\dot{Q}_{max}} = \frac{C_h(T_{h,i} - T_{h,o})}{C_{min}(T_{h,i} - T_{c,i})} = \frac{C_c(T_{c,o} - T_{c,i})}{C_{min}(T_{h,i} - T_{c,i})}, \tag{1}$$

where C_c and C_h are respectively the cold and hot stream heat capacity rates, $T_{h,i}$ and $T_{h,o}$ indicate inlet and outlet hot stream temperatures, $T_{c,i}$ and $T_{c,o}$ indicate inlet and outlet cold stream temperatures. Defining NTU, the number of thermal units, ε-NTU relationship (Eq. 2) can be used.

$$\varepsilon = f(\mathrm{NTU}, C^*) = f\left({}^{UA}/_{C_{min}}, {}^{C_{min}}/_{C_{max}} \right). \tag{2}$$

ε-NTU relationship differs for each configuration, for crossflow arrangement configuration is:

$$\varepsilon = 1 - \exp\left\{\frac{\mathrm{NTU}^{-0.22}}{C^*}\left[\exp(-C^*\mathrm{NTU}^{0.78}) - 1\right]\right\}. \tag{3}$$

Test case

Analysis and modelling of the system are performed for the electric motor and the inverter as heat sources. The selected heat source sets the design temperature for coolant convergence loop. The outlet coolant temperature from the heat source is 90 °C and the inlet coolant temperature in the heat source is expected to reach 65 °C. The analyses are performed in hot day condition for both ground and ceiling conditions (FL250), considering both traditional and innovative coolant, respectively 40% water-ethylenic glycol mixture and nanofluids. Both heat source and heat exchanger are located into the engine nacelle, and the heat sinks are represented by wing tanks, so 80% of wingspan is considered as distance to be covered by fuel. The five test cases represent the requirements for five different levels of hybridization, corresponding to a progressive reduction in carbon dioxide emissions due to the increased propulsive power generated from electric motor at the expense of conventional propulsion. They are proposed to evaluate the operation of liquid cooling system with several levels of thermal load, as indicated by Table 1.

Table 1 – Test Case Input

Test case	ΔCO_2	Thermal power EM+PE [kW]		TMS coolant mass flow [kg/s]	
		Take Off	Ceiling	Take Off	Ceiling
Test case 1	-10%	14.3	14.3	0.136	0.136
Test case 2	-20%	28.3	28.3	0.270	0.270
Test case 3	-30%	41.2	41.2	0.393	0.393
Test case 4	-40%	54.2	54.2	0.517	0.517
Test case 5	-50%	67.2	67.2	0.641	0.641

Fig. 2 reports the sizing script outputs: ground condition in blue line, ceiling condition in green line. Each dot depicts a test case. The dimensioning condition is identified in Test case 5, -50% CO2 emission, in the ground condition.

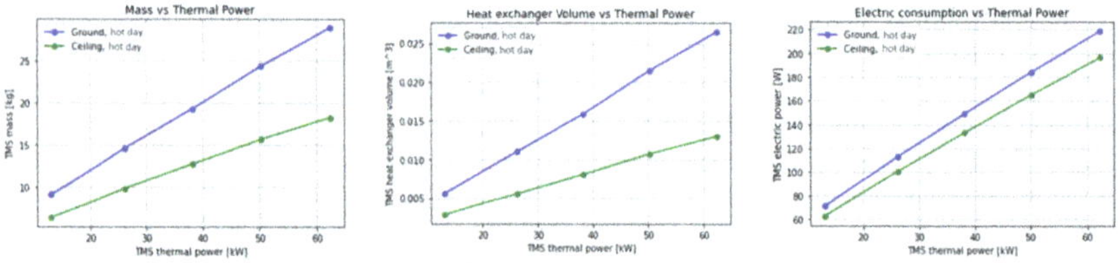

Fig. 2 – Test Case Output

The mass of heat exchanger is 6.6 kg, that is 22.7% of overall TMS weight, 28.9 kg and its volume is 0.026 m^3. The electric consumption amounts to 219 W, distributed in 135 W required by coolant loop dedicated electric pump and 84 W required by fuel loop dedicated electric pump. Table 2 reports significant output of heat exchanger model for test case 5 at ground level.

Table 2 - Test Case 5 output: ground condition

Output Test Case 5 - ground					
Efficiency	[-]	0.450	Volume	[m^3]	0.026
Pressure Drop - coolant	[Pa]	244.307	Length side 1	[m]	0.325
Pressure Drop - fuel	[Pa]	311.548	Length side 1	[m]	0.325
Weight	[kg]	6.56	Height	[m]	0.251

A sensitivity analysis is performed for three values of fuel mass flow rate, a design parameter. The increase in fuel flow rate achieves a reduction on heat exchanger mass of 15.8% and a 15.5% reduction on volume, but with an 11.4% increase on electric consumption. Despite the lower weight of the heat exchanger, due to the other components of the TMS, the TMS overall mass does not achieve the same benefit.

Nanofluids

Nanofluids are a new class of nanotechnology. Escalating interest of the last decade is based on the higher thermal conductivity compared to conventional coolant. They are engineered by suspending nanoparticles (1-100 nm) in a base fluid, such as water, oil, mixture of water and ethylenic glycol or traditional heat transfer fluids. The goal is to reach a significant improvement in thermal properties by a uniform and stable suspension of nanoparticles, thanks to a small concentration, less than 1% by volume. A lot of experiments are reported in the state of the art [5] of nanofluids coolant, with ceramic oxides nanoparticles, metallic carbides, metals, carbon nanotubes and others and for each of them experimental equations are presented. In this work, theorical thermophysical equations are adopted, as defined in Chapter 5 of [5] Thanks to 0.502% of Al_2O_3, TMS obtains a mass reduction of 10% and a smaller heat exchanger. Different types of nanoparticles at different concentrations are tested. The best case resulted in a 12% overall weight loss and 15% heat exchanger volume reduction.

Conclusion

The developed parametric model provides the design of a liquid-to-liquid heat exchanger integrated into TMS concept. For the most demanding test case, over 67 kW to be dissipated, the overall TMS weight is 29 kg and the heat exchanger volume is 0.026 m^3. The sensitivity analysis reports, by increasing the fuel flow rate, an increase in the electrical power required by the electric pumps, a marked reduction in the heat exchanger dimensions and a slight reduction of TMS mass. If the purpose of design is volume optimization, the increasing fuel flow rate is promising, without achieving significant mass benefits. The use of nanofluids offers a good potential, but presently, not so great as promise in state of the art. Further investigation is needed to better understand the interactions between nanoparticles and base fluids and to predict the actual behavior.

References

[1] ICAO, 2019 Environmental Report, Aviation and Environment, https://www.icao.int/environmental-protection/Documents/ICAO-ENV-Report2019-F1-WEB%20(1).pdf

[2] https://eur-lex.europa.eu/legal-content/EN/TXT/HTML/?uri=CELEX:52019DC0640

[3] A.S.J. van Heerden, D.M. Judt, S. Jafari, C.P. Lawson, T. Nikolaidis, D. Bosak, Aircraft thermal management: Practices, technology, system architectures, future challenges, and opportunities, (2021). https://doi.org/10.1016/j.paerosci.2021.100767

[4] J. E. Hesselgreaves, R. Law, D. Reay, Compact Heat Exchangers. Selection, Design and Operation, 2nd ed., Elsevier Ltd, 2017. https://doi.org/10.1016/B978-0-08-100305-3.00002-1

[5] https://www.clean-aviation.eu/about-us/who-we-are

[6] S. K. Das, S. U. S. Choi, W. Yu, T. Pradeep, NANOFLUIDS, Science and technology, Wiley, 2008. https://doi.org/10.1002/9780470180693

Preliminary design of an electromechanical actuator for eVTOL aircrafts in an urban air mobility context

Roberto Guida[1,a*], Antonio Carlo Bertolino[1,b], Andrea De Martin[1,c], Giovanni Jacazio[1,d] and Massimo Sorli[1,e]

[1]Department of Mechanical and Aerospace Engineering, Politecnico di Torino, Corso Duca degli Abruzzi, 24, Torino, 10129, Piemonte, Italy

[a]roberto.guida@polito.it, [b]antonio.bertolino@polito.it, [c]andrea.demartin@polito.it, [d]giovanni.jacazio@polito.it, [e]massimo.sorli@polito.it

Keywords: Preliminary Design, Preliminary Sizing, EMA, e VTOL, Urban Air Mobility

Abstract. Urban areas face issues like traffic congestion, noise, and pollution. In this context Urban Air Mobility (UAM) offers a solution by utilizing the urban airspace for transportation, in particular, the exploiting of Electric Vertical Take-Off and Landing (eVTOL) vehicles is promising in terms of noise and environmental pollution, despite challenges like safety issues and financial constraints. To overcome these issues, Prognostic and Health Management (PHM) plays a vital role in ensuring safety and reliability. The proposed case study focuses on a compact Electro-Mechanical Actuator (cEMA) for flap control surface. A preliminary design is proposed with the aim of reducing dimensions and weight while maintaining performance and reliability. This work represents one of the first steps in the creation of a digital twin for the design, sizing, and application of PHM logics.

Introduction

The issue of overcrowding in cities and the resulting environmental and noise pollution, has spurred the search for alternative mobility solutions. One potential solution is Urban Air Mobility (UAM), which aims to establish a safe, and sustainable air transportation system within cities [1]. UAM should be exploited for various purposes, including passenger transport, package delivery, and emergency services [2]. It is part of the broader concept of Advanced Air Mobility (AAM) [3], which encompasses emerging aviation markets. UAM primarily utilizes electrically powered Vertical Take-Off and Landing (VTOL) aircrafts. The development of all-electric aircraft with compact Electro-Mechanical Actuators (cEMAs) has been driven by the need to reduce environmental pollution in cities [4]. However, progress in UAM has been hindered by a few fatal accidents, noise restrictions, and financial challenges [5]. In this context, safety, airworthiness, propulsion efficiency, and performance are crucial aspects in UAM research [6]. A way to improve the safety and the reliability of these systems is the application of Prognostic and Health Management (PHM) techniques. About that the development of High-fidelity (HF) model allows the study of performance variations and the identification of potential defects to obtain the definition of a Digital Twin (DT) to be used as a virtual test bench of the real system.

Requirements of the cEMA

The first step in the cEMA design and sizing is the definition of the control surface chosen as case study and then the related specific requirements. Despite variations in configuration [7], eVTOL aircrafts share similar actuation systems. The case study chosen is the flap actuator responsible for the thrust vectoring in a tiltrotor aircraft.

Performance requirements. The steady-state performance requirements involve accuracy, resolution, and hysteresis. The first one must be achievable under all operating conditions including actuator faults that do not result in the complete system failure. In the present work, a minimal accuracy equal to 0.3° is considered [8]. The resolution is due to the sensors system and must be smaller than the required accuracy to properly control the surface. The hysteresis represents the difference between the input command required for the two different directions moving of the surface to achieve the same angle movement. At a first approximation this requirement can be negligible and calculated after the overall EMA sizing. The dynamic performance requirements address the analysis of the actuator response to commands and disturbances in both time and frequency domains, and it aims at studying the stability, the tracking, the impedance, and the damping characteristics of the system. These requirements are mostly linked to the definition of the control algorithm and parameters. Another technical requirements regard the maximum load the servo actuator must bear: it depends on factors such as overall inertia, maximum external load, and friction losses. Finally, the maximum and effective strokes as well as the rated speed of the actuator need to be determined. For this application a minimum stroke of 90° is needed, in addition a minimum-security margin of about 20% in both directions is considered and the total operational actuator stroke shall be from -20° to 110°. The rated speed is defined for assuring that the actuator, without external load and with the power supply at the minimum guaranteed level, reaches its maximum stroke within a specified time. For this application a minimum speed has been set to 360°/s [9,10].

Safety and reliability requirements. To define safety and reliability requirements, it is essential to determine the expected operational lifespan of the system. During this time the system should ensure the requirements without the need for structural component replacement. The flap actuator system reliability is commonly associated with the failure rate, which represents the frequency of failure occurring within a specific period of time. The loss of control, due to a single failure or combination of failures, shall be less than 1×10^{-6} per flight hour [11,12].

Proposed architecture design

To meet the demanding requirements of UAM EMAs in terms of compactness, simplex solutions often fall short of airworthiness standards. Previous literature has presented various fault-tolerant architectures based on redundancy for EMAs [13]. The conventional EMA architecture for flight control actuators is typically composed of an electric motor, power electronics and control electronics (including sensors), mechanical transmission (with gearboxes, reducers or ball screws), and fail-safe devices (e.g., clutches, brakes) [14].

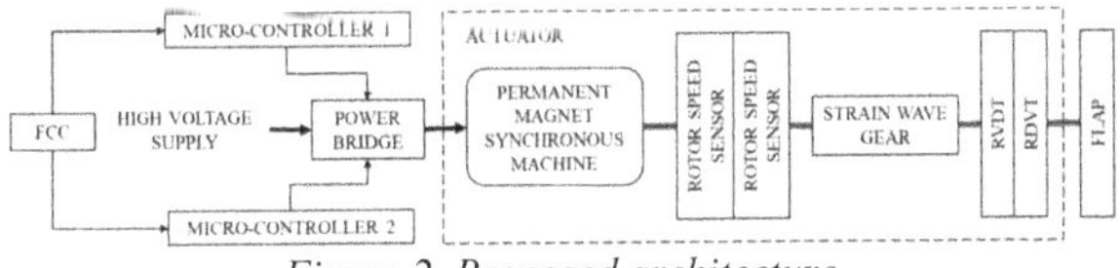

Figure 2. Proposed architecture.

In the case study, a compact rotary-output actuator architecture is chosen to fulfill the size and weight requirements for UAM applications. This architecture comprises a six-phase permanent magnet synchronous machine, fault-tolerant control electronics, a strain wave gear, and several sensors (Fig. 2).

Electric motor. In the aerospace industry, various types of electric motors have been considered for converting electrical energy into mechanical energy. The selection process involves evaluating factors such as losses, cooling, weight, volume, and other distinctive features, advantages, and disadvantages [11]. For aviation purposes, the chosen electric motor must exhibit thermal robustness and high efficiency to minimize power losses and associated cooling requirements. Four types of electric motors are considered: Permanent Magnet Synchronous Machine (PMSM), Electrically excited Synchronous Machine (ESM), Switched Reluctance Machine (SRM), and Induction Machine (IM). Among these options, the PMSM is identified as the most suitable for the present application. It offers higher efficiency and power density, lower heat production and is capable of sensor-less control, while it presents drawback as the relatively higher cost of magnets and its total loss of operation in case of one phase loss.

Mechanical transmission. Certain criteria must be considered, including a compact structure, high efficiency, and characteristics such as high loading capabilities, low kinematic error, and minimal backlash. The reducer plays a critical role in ensuring system reliability [17]. Among the available options, the strain wave gear (SWG), stands out as the preferred choice due to its superior performance in terms of compactness and reliability, keeping a good efficiency [15].

Sensors. To perform position control loop a Rotary Variable Differential Transformer (RVDT) will be exploited. For the speed loop, a rotational sensor is required and positioned on the high-speed shaft of the gearbox to enhance measurement accuracy.

Control electronics. The Electronic Control Unit (ECU) typically manages the position and speed control loops, while the current control is handled by the Motor Drive Electronics (MDE). To achieve the reliability requirements, a fault-tolerant architecture with two control channels is implemented. The architecture utilizes a dual-control/single-drive setup, and each control channel communicates with the Flight Control Computer (FCC) via separate lines.

Preliminary Fault Tree Analysis

The Preliminary System Safety Assessment (PSSA) is a top-down process that assigns reliability and safety requirements from systems to components. In Fig. 3 a preliminary example of a qualitative FTA for this specific case study is proposed [11,16,17].

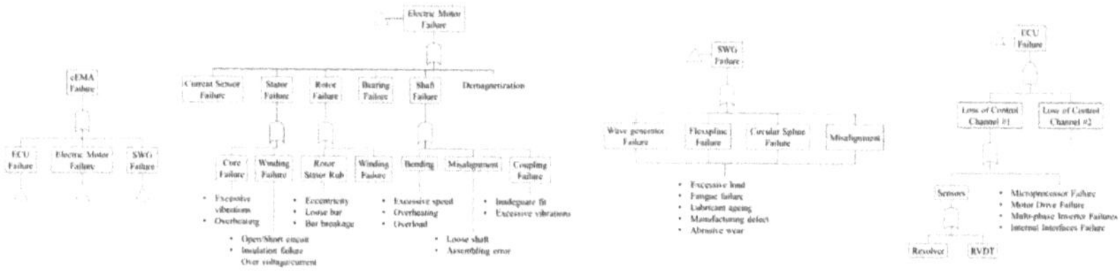

Figure 3. Preliminary FTA of an EMA for eVTOL flap surface in an UAM context.

Conclusion

UAM offers a solution to novelty urban transport challenges and eVTOL vehicles show promise in reducing noise and environmental pollution. The presented case study focuses on a cEMA for flap surface. A preliminary design sizing approach is proposed to reduce dimensions and weight while maintaining performance and efficiency. This work represents a first step in the development of a digital twin, enabling design, sizing, and application of PHM for the cEMA.

Acknowledgement

This study was carried out within the MOST – Sustainable Mobility National Research Center and received funding from the European Union Next-GenerationEU (PIANO NAZIONALE DI RIPRESA E RESILIENZA (PNRR) – MISSIONE 4 COMPONENTE 2, INVESTIMENTO 1.4 – D.D. 1033 17/06/2022, CN00000023). This manuscript reflects only the authors' views and opinions, neither the European Union nor the European Commission can be considered responsible for them.

References

[1] "What is UAM." https://www.easa.europa.eu/en/what-is-uam.

[2] R. Goyal and A. Cohen, "Advanced Air Mobility: Opportunities and Challenges Deploying eVTOLs for Air Ambulance Service," Applied Sciences (Switzerland), vol. 12, no. 3, Feb. 2022. https://doi.org/10.3390/app12031183

[3] R. Goyal, C. Reiche, C. Fernando, and A. Cohen, "Advanced air mobility: Demand analysis and market potential of the airport shuttle and air taxi markets," Sustainability (Switzerland), vol. 13, no. 13, Jul. 2021. https://doi.org/10.3390/su13137421

[4] D. P. Rubertus, L. D. Hunter, and G. J. Cecere, "Electromechanical Actuation Technology for the All-Electric Aircraft," IEEE Trans Aerosp Electron Syst, vol. AES-20, no. 3, pp. 243–249, 1984. https://doi.org/10.1109/TAES.1984.310506

[5] A. P. Cohen, S. A. Shaheen, and E. M. Farrar, "Urban Air Mobility: History, Ecosystem, Market Potential, and Challenges," IEEE Transactions on Intelligent Transportation Systems, vol. 22, no. 9, pp. 6074–6087, Sep. 2021. https://doi.org/10.1109/TITS.2021.3082767

[6] W. Johnson, C. Silva, and E. Solis, "Concept Vehicles for VTOL Air Taxi Operations."

[7] A. Bacchini and E. Cestino, "Electric VTOL configurations comparison," Aerospace, vol. 6, no. 3, Mar. 2019. https://doi.org/10.3390/aerospace6030026

[8] S. E. Lyshevski, "Electromechanical Flight Actuators for Advanced Flight Vehicles."

[9] I. Chakraborty, D. N. Mavris, M. Emeneth, and A. Schneegans, "A methodology for vehicle and mission level comparison of More Electric Aircraft subsystem solutions: Application to the flight control actuation system," Proc Inst Mech Eng G J Aerosp Eng, vol. 229, no. 6, pp. 1088–1102, May 2015. https://doi.org/10.1177/0954410014544303

[10] J. W. Bennett, B. C. Mecrow, A. G. Jack, and D. J. Atkinson, "A prototype electrical actuator for aircraft flaps," in IEEE Transactions on Industry Applications, May 2010, pp. 915–921. doi: 10.1109/TIA.2010.2046278

[11] M. Mazzoleni, · G. Di Rito, and F. Previdi, "Electro-Mechanical Actuators for the More Electric Aircraft." [Online]. Available: https://www.springer.com/gp/authors-editors/journal-author/journal-author-helpdesk/

[12] J. D. Booker, C. Patel, and P. Mellor, "Modelling green vtol concept designs for reliability and efficiency," Designs (Basel), vol. 5, no. 4, Dec. 2021. https://doi.org/10.3390/designs5040068

[13] M. A. A. Ismail, S. Wiedemann, C. Bosch, and C. Stuckmann, "Design and evaluation of fault-tolerant electro-mechanical actuators for flight controls of unmanned aerial vehicles," Actuators, vol. 10, no. 8, Aug. 2021. https://doi.org/10.3390/act10080175

[14] M. Budinger, A. Reysset, E. Halabi, C. Vasiliu, and J.-C. Maré, "Optimal preliminary design of electromechanical actuators, Electro-Mechanical Actuators for the More Electric Aircraft," Proceedings of the Institution of Mechanical Engineers, vol. 228, no. 9, 2013. https://doi.org/10.1177/0954410013497171ï

[15] I. Schäfer, "Improving the Reliability of EMA by using Harmonic Drive Gears."

[16] Y. Cao, J. Wang, X. Rong, and X. Wang, "Fault Tree Analysis of Electro-mechanical Actuators."

[17] A. Raviola, A. De Martin, R. Guida, G. Jacazio, S. Mauro, and M. Sorli, Harmonic Drive Gear Failures in Industrial Robots Applications: An Overview.

Landing gear shock absorbers guidelines

Michele Guida[1,a], Giovanni Marulo[1,b] and Francesco Marulo[1,c*]

[1]Department of Industrial Engineering, Via Claudio, 21 – 80125 Napoli – Italy

[a]michele.guida@unina.it , [b]nanni.marulo@hotmail.it , [c]francesco.marulo@unina.it

Keywords: Aeronautical History, Landing Gear, Shock Absorber, Numerical-Experimental Correlation

Abstract. This paper is based on an old paper presented by Eng. Ermanno Bazzocchi almost seventy years ago [1] and is here presented again in his memory, to take the opportunity for showing to young generation of engineers how applied research was performed and presented when the computer age was not still born. The topic is on landing gear shock absorber design guidelines, and it has been selected because of the importance of such device for airplanes, which represents a very important system for the efficiency of the entire aircraft. The original paper [1] has represented a milestone for the design and dimensioning of landing gear shock absorbers highlighting parameters which often have not been discussed so clearly in papers which came later in the scientific literature and therefore the idea of collecting those information and revisiting them in a modern framework has been particularly exciting. It is in the idea of the authors that revisiting fundamental classic papers and projecting them in a modern scenario could be beneficial for the real understanding of the physical aspects of aircraft systems, covering calculations which could not being performed because of low computing power. Classic papers were forced to strongly rely on physical understanding from which creating simple models to correlate experimental data and theoretical calculations. Such physical background should not be lost, but hopefully improved by the actual computer power and this paper is an attempt proposed to the scientific community for discussing on the validity of such an approach.

Introduction

The aircraft landing gear is one of the most critical part of an aircraft, being this system responsible for ensuring the safety of the payload during take-off, landing and for the taxiing procedures, too. The general arrangement of the landing gear consists of the shock absorbers, retraction mechanism, steering, shimmy control, tires, wheels and brakes. It represents about 3.5 to 7 percent of the gross weight and from 2 to 4 percent of the aircraft sales price.

The design of landing gears has been considered one of the more challenging and technically satisfying engineering task, since it demands expertise in mechanical and structural engineering, hydraulics, kinematics, materials and a very good understanding of detail design.

For these reasons, but even more to bring back memories, we remember, in the year of the first centenary of the Italian Air Force, Eng. Ermanno Bazzocchi, the designer of one of the most iconic aircraft, the MB-339, which still successfully equips the national aerobatic team, the PAN.

In 1955 Eng. Bazzocchi publishes in the magazine L'Aerotecnica a work on the design of one of the main elements of the landing gear of an aircraft, the shock absorber, based on a presentation held the previous year at a scientific congress in Paris.

This work has been taken up again in this article as evidence of its technical-scientific validity, after almost seventy years, and above all for the organization and setting given by the author to this work. An attentive technician who reads this work with dedication comes out capable of designing and realizing the structural component under discussion and with the modern tools of calculation and assisted design, can enhance concepts that at the time of its first draft could not be optimized due to lack of calculation tools.

The main wish behind recalling this work, underlining its importance and highlighting its merits, is to show young researchers how significant it can be to revisit past experiences, when the low computing power was overcome by knowledge of the physics of the problem as well as by a thorough ability in the use of engineering tools. From the authors of this article viewpoint, there is the hope to have been able to convey the pleasure of having revisited an important scientific document, [1], that still represents a milestone in the design of aircraft landing gear, together with other "well-seasoned" papers, [2] and [3].

Mechanics of the landing gear

One of the possible schemes of the landing gear considers that it is attached to a rigid mass which has a degree of freedom for the vertical displacement, only. Both the systems, mass of the airplane and landing gear constitute a 2-dof system, fig. 1(a). To obtain energy dissipation, the hydraulic fluid is forced to flow at high velocity because of the telescoping strut. To maximize such dissipation, the passage of the hydraulic fluid through the orifice should be properly designed by defining a variation of the orifice area by introducing a metering strut which controls the size of the orifice and governs the performance of the shock absorber, fig. 1(b). The balance of the forces is reported in fig. 1(c) which shows the mutual interaction between the tire and the strut.

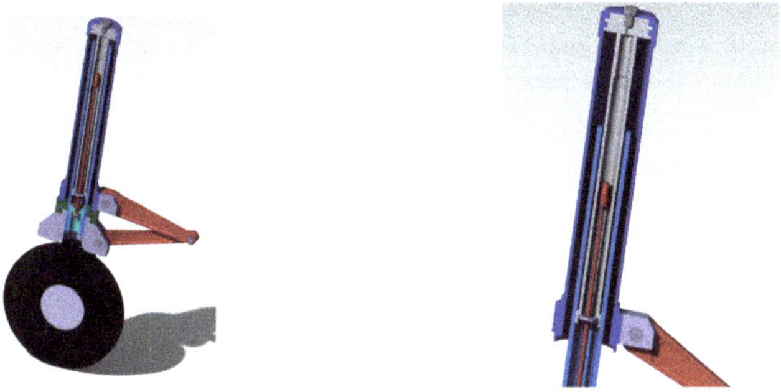

Figure 1 – Schematic of a typical landing gear (left) and shock absorber (right)

One of the most interesting aspects emerging by comparing the results of [1] and [2] is the choice of the plane for their representation. Ref. [1] uses a force versus displacement representation for the behavior of the landing gear. Ref. [2], on the other hand, presents the results in force, or displacement, versus time plots. This latter could appear more intuitive, as generally is for a time behavior of parameters, while the force versus displacement may appear a bit more difficult, but it is result (and efficiency) related, regarding the design of the landing gear because it gives an easy representation of its efficiency and opens to opportunities for improvements, if any.

Equations of motion

The approach for writing the equations of motion follows a common path which initially studies each component separately, then they come together for assessing the full behavior of the landing gear. This general approach may be split in modelling a passage from one to two degree of freedom system or, which is equivalent, to an initial motion of the tire only and, reaching a certain force, activating the shock absorber, too. In [1] this latter approach is followed along with an engineering procedure which assumes, due to the short time, a static behavior of the tire, before defining the complex behavior of the shock absorber. Ref. [2] instead, employs the passage from 1 to 2-dof system in a chronological sequence which computes also the initial conditions for the beginning of the second phase, when the inertia, weight and lift forces become sufficiently large to overcome the preloading force in the shock strut due to the initial air pressure and internal friction.

The overall dynamic non-linear equilibrium equation for a 2-dof system is given by the following relationship:

$$\frac{W_1}{g}\ddot{z}_1 + \frac{W_2}{g}\ddot{z}_2 + F_{V_g}(z_2) + L = W_1 + W_2$$

where the subscripts 1 and 2 refer, respectively, to the upper mass (typically the partial mass of the airplane on the shock absorber) and the lower mass (typically the tire with its systems mass), L is the lift force and F_{V_g} is the vertical force on tire.

The previous equation uses physical degrees of freedom for representing the equilibrium. Clearly it needs to be manipulated for explicitly showing the main parameters before being integrated and becoming effectively useful for designing the shock absorber.

A different approach, more energy based, is used in [1], ending with one scalar non-linear equation which is stepwise integrated according to the time evolution of the physical phenomenon. Such equation, appeared at the time of Ref. [1] as impossible to solve, is written as

$$v v_a^2 \frac{dv_a}{dC_a} + v_a(A + B) - A = 0$$

where $v = v_p + v_a$ is the sum of the velocity of the tire and the shock absorber, and A and B are variables grouping several parameters as the stiffness of the tire and the shock absorber.

Results

Based on the previous equations, mainly those of Ref. [2], computer codes, [8], have been developed, tested and used for correlating laboratory measurements and numerical results. Fig. 2 presents the results obtained from the discussed methodologies and applied in comparing predictions with experimental measurements for a general aviation landing gear.

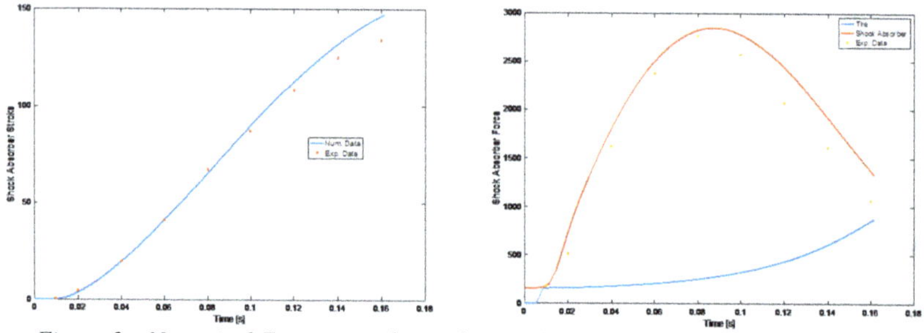

Figure 2 – Numerical-Experimental correlation of shock absorber stroke (left) and force (right) versus time, according to Ref. [2]

Figure 3 shows the results which can be obtained during the design phase of the main elements of a landing gear equipped with a shock absorber, together with the definition, for example, of the geometry of the metering pin, fig. 4, for reaching a specified value of the absorbing efficiency in some specific landing condition.

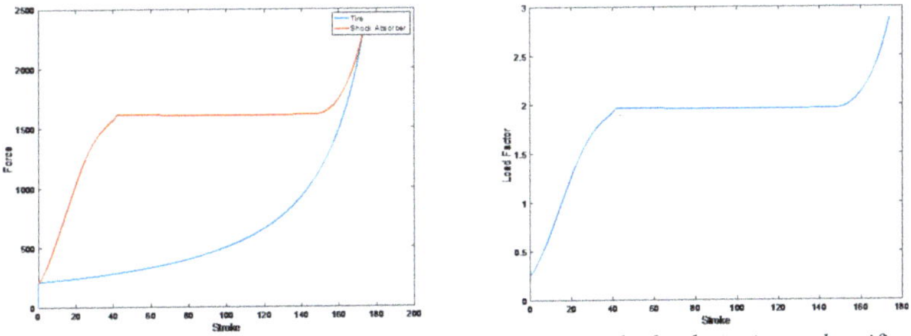

Figure 3 – Force and load factor versus landing gear stroke for designing and verification purposes, according to Ref. [1]

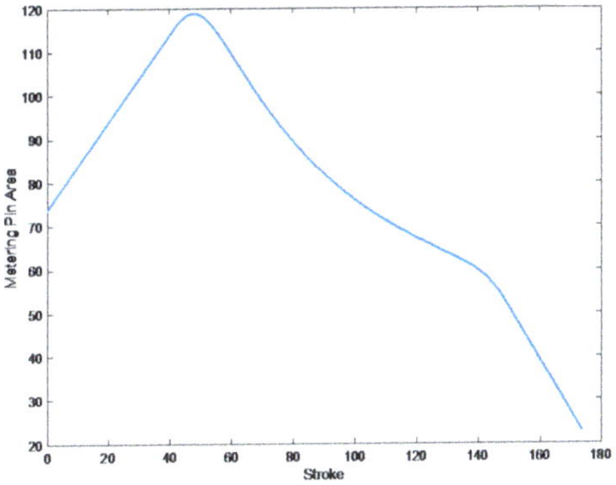

Figure 4 – Metering pin area versus landing gear stroke

Conclusions

The main purpose of this article has been to resume some old papers on the design and analysis of landing gear and to review considering modern computer technologies, trying to evidence the importance of their content and how they are still useful in the actual environment.

Acknowledgements

The authors are indebted with Prof. Fabrizio Ricci for having shared the basic idea of this work, for having encouraged its writing, for the interesting resulting discussions and for allowing the use of experimental data and Matlab scripts which are not reported here for sake of brevity, but very useful for summing up the main approaches described throughout the paper.

References

[1] Ermanno Bazzocchi, "Metodo di calcolo degli ammortizzatori oleopneumatici e confronti con i risultati ottenuti alle prove", L'Aerotecnica, Vol. XXXV, fasc. 3°, 1955

[2] Benjamin Milwitzky, Francis E. Cook, "Analysis of Landing-Gear Behavior", NACA Report 1154, Langley Field, Va., USA, 1953

[3] Ladislao Pazmany, Landing Gear Design for Light Aircraft Vol. I, Pazmany Aircraft Corporation, San Diego, Ca, USA, 1986

[4] Norman S. Currey, Aircraft Landing Gear Design: Principles and Practices, AIAA Education Series, Washington D.C., USA, 1988. https://doi.org/10.2514/4.861468

[5] Benjamin Milwitzky, Francis E. Cook, "Analysis of Landing-Gear Behavior", NACA Report 1154, Langley Field, Va., USA, 1953

[6] AGARD Landing Gear Design Loads, CP-484, Portugal, 1990

[7] Military Specifications, Landing Gear Systems, MIL-L-87139, July 1979

[8] Rinaldi F., "Metodologie di progetto di un ammortizzatore oleopneumatico", Tesi di Laurea, Dipartimento di Progettazione Aeronautica, a.a. 1994-'95

Experiments and simulations for the development of a dual-stator PMSM for lightweight fixed-wing UAV propulsion

Aleksander Suti[1,a] *, Gianpietro Di Rito[1,b], Roberto Galatolo[1,c],
Luca Sani[2,d], Giuseppe Mattei[3,e]

[1]Università di Pisa, DICI, Largo Lucio Lazzarino 2, 56122, Pisa, Italy

[2]Università di Pisa, DESTEC, Largo Lucio Lazzarino 2, 56122, Pisa, Italy

[3]Sky Eye Systems, Via Grecia 52, 56021, Cascina, Italy

[a] aleksander.suti@dici.unipi.it , [b] gianpietro.di.rito@unipi.it , [c] roberto.galatolo@unipi.it,
[d] luca.sani@unipi.it, [e] giuseppe.mattei@skyeyesystems.it

Keywords: UAV, Full-Electric Propulsion, Axial-Flux PMSM, Simulation, Testing

Abstract. The work summarizes the experimental and simulation studies carried out on the full-electric propulsion system of a lightweight fixed-wing UAV developed in the research program TERSA. The electric motor, specifically designed for the project, is a double-stator axial-flux PMSM with single output shaft, directly connected to a twin-blade fixed-pitch propeller. The system dynamics is firstly addressed by nonlinear simulation, modelling the motor and related sensors, the propeller and the digital controllers, to design the closed-loop control architecture. Successively, an experimental campaign is defined to identify/substantiate the main motor parameters (resistance and inductance of phases, torque and speed constants, back-electromotive force waveforms) and to validate the closed-loop control design.

Introduction

In the air transport sector, Full-Electric Propulsion Systems (FEPSs) are expected to obtain large investments in forthcoming years, aiming to replace/support the operation of conventional internal combustion engines, especially in small-size power applications [1]. Even if immature nowadays in terms of reliability and energy density (e.g., gasoline energy density is about 100 times higher than lithium-ion battery packs [2], typically ranging about 300 kJ/kg), the design of the next-generation long-endurance UAVs is moving toward the use of FEPSs, pulled by the wider objectives of aerospace electrification. In this context, the Italian Government and the Tuscany Regional Government funded the research program TERSA[1] [3], led by Sky Eye Systems (Italy) in collaboration with University of Pisa and other Italian industries. The TERSA project aims to develop a Unmanned Aerial System (UAS) based on a lightweight fixed-wing UAV (Fig. 1), having the following main characteristics:

- Main performance data: MTOW from 35 to 50 kg; Endurance >6 h; Range >3 km;
- Take-off/landing systems: pneumatic launcher and parachute/airbags;
- Propulsion system: FEPS with a twin-blade fixed-pitch propeller;
- Payloads: SAR (Synthetic Aperture Radar) and SAAS (Sense-And-Avoid System).

[1] Tecnologie Elettriche e Radar per Sistemi aeromobili a pilotaggio remoto Autonomi, Fund ID F/130088/01-05/X38

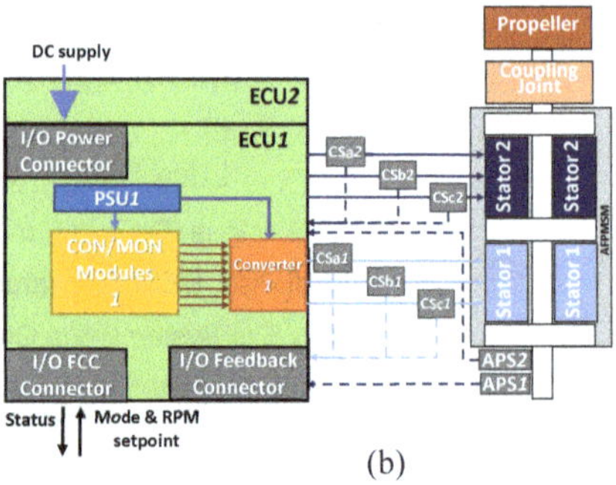

(a) (b)

Figure 1. (a) TERSA UAV layout; (b) FEPS architecture.

The FEPS of TERSA UAV is based on Axial-Flux PMSM (AFPMSM) technology [4], which, though characterized by lower technology readiness level, is preferable to conventional radial flux PMSMs in terms of weight (core material is reduced), torque-to-weight ratio (magnets are thinner), efficiency (losses are minimized), and versatility (air gaps are easily adjustable) [5], [6]. Another key aspect for application in UAV flying into unsegregated airspaces is clearly reliability and safety. Due to the weight limitations, hardware redundancy in FEPSs is often a drawback, and reliability is enhanced via motor phase redundancy [7], or unconventional converters [8]. Although the use of unconventional converters leads to a more compact solutions, it requires an ad-hoc design of motor and power electronics [8],[9], so PMSMs with multiple three-phase arrangements using conventional converters driven by standard techniques are in some extent convenient [10]. Starting from these considerations, the reference FEPS has been equipped with a dual-stator AFPMSM that can operate in both active/active or active/stand-by configurations, providing the system with fault tolerant potentialities [11] This paper summarizes the activities carried out for the modelling, the simulation, the control design and the experimental characterization of the TERSA UAV FEPS. The work is articulated in two parts: the first one dedicated to the system description and to the nonlinear modelling, and the second to testing and simulations for model validation.

System description

The TERSA UAV FEPS, Fig. 1(b), is composed of a AFPMSM mechanically connected to a twin-blade APC22×10E fixed-pitch propeller and electrically connected to two Electronic Control Units (ECUs). Each ECU includes a three leg converter, a Power Supply Unit (PSU) and a I/O connector interface. The ECUs receive commands from the Flight Control Computer (FCC) and feedbacks from Current Sensors (CSa, CSb, CSc) and an Angular Position Sensor (APS). The closed-loop control applies two nested loops, on the propeller speed and motor currents (via Field-Oriented Control, FOC) respectively. The cascade regulators implement proportional/integral actions on tracking error signals, plus anti windup function with back-calculation to compensate for commands saturation.

Experiments and simulations

Two test campaigns with different experimental setup have been carried out to identify/substantiate the main parameters of the propulsion system model and to validate the closed-loop regulators. In the first campaign, since the motor model identification requires three parameters (the motor speed constant k_m, the resistance R and the inductance L of the phases) and three BEMF waveforms as functions of the electrical angle ($k_{ex}(\theta_e) = e_x/k_m\dot{\theta}_e$, where e_x is the BEMF of the phase x ($= a, b, c$) and θ_e is the electrical angle), two tests have been designed. The

Materials Research Proceedings 37 (2023) 89-93 https://doi.org/10.21741/9781644902813-20

first one is done with blocked rotor to identify R and L, and the second one with the rotor dragged by the bench motor to identify k_m and k_{ex}. In the second campaign, a first test has been done by blocking the rotor while commanding the motor current with step inputs of different amplitudes, and a second test (Fig. 2) has been performed by connecting the propeller and by commanding the angular speed with tracking signals of different amplitudes. Figure 3 reports the motor speed constant and the BEMF waveforms for both stators. Figure 3(a) highlights that the dependency of the speed constant on the angular speed is more relevant in the stator 1 (S1), even if the maximum variation is about 1%. In addition, Fig. 3(b) shows an electrical angle phase misalignment (about 15 deg) in the BEMF waveforms. Figure 4 summarizes the simulation results. Figure 4(a) points out the validation of the current controller, accomplished by requesting step commands of quadrature current (40 A and 60 A, respectively) while the output shaft is blocked. The model exhibits a slight delay (1 ms) when the command signal is applied, but the prediction error rapidly diminishes (within 5ms). The validation of the speed controller is finally reported in Fig. 4(b), and it is performed by requesting two ramped-step speed commands, with 2 and 4 krpm amplitudes, both characterised by a 1 krpm/s slope. In both cases, the test ends by removing the electrical power and letting the motor passively decelerate (thus permitting a more direct identification/substantiation of system inertias). Apart from a low frequency (about 8 Hz) harmonic disturbance due to imperfect rig grounding, the model behaves satisfactorily during both controlled and uncontrolled phases.

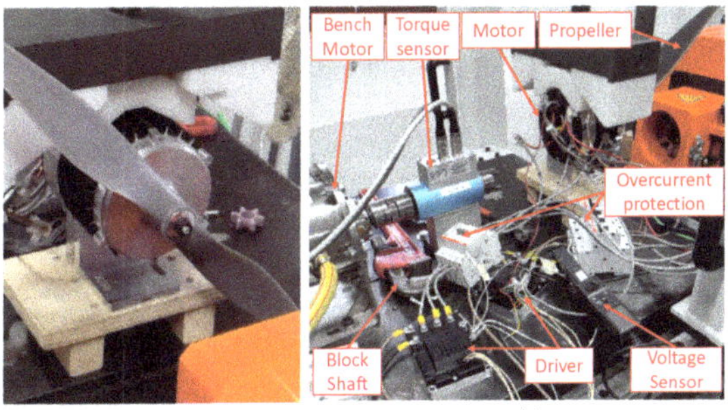

(a) (b)

Figure 2. Experimental setup for speed tracking tests: (a) connected propeller (b) rig layout.

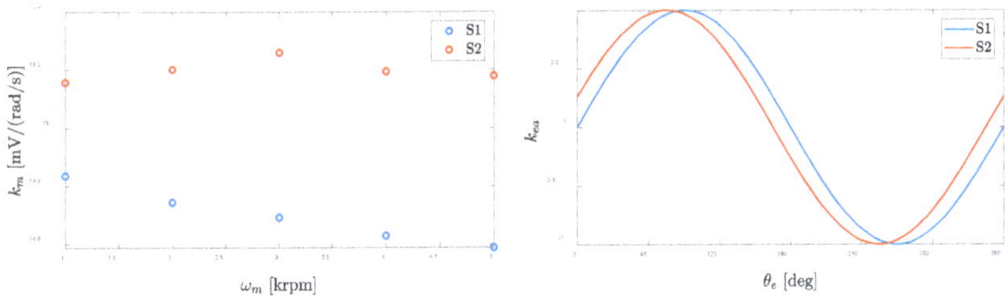

(b)

Figure 3. BEMF of motor modules: (a) speed constant; (b) BEMF wrt electrical angle.

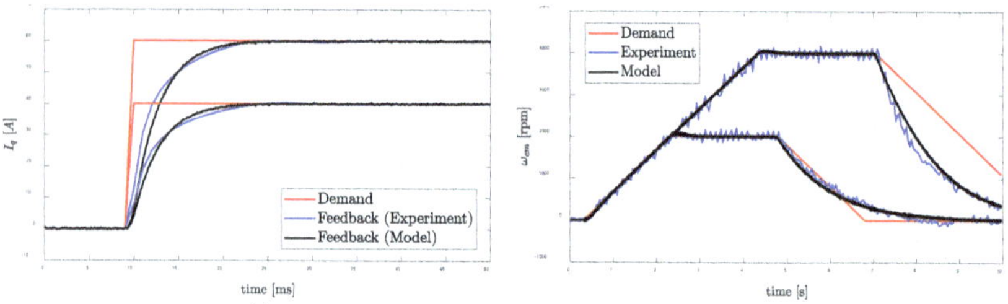

Figure 4. Closed-loop control validation: (a) current (b) angular speed.

Summary

The main parameters of the nonlinear model of the TERSA UAV propulsion system are identified/substantiated through experiments and the closed-loop control performances are validated. Due to manufacturing imperfections, the BEMF of the stator modules are misaligned wrt the motor angle and they have different torque constants. The asymmetry, though partially compensated by the control actions, can cause a reduction of the propulsion system efficiency and high-frequency torque ripples that, affecting bearings and seals, can potentially lead to premature failure. The FEPS model can be used to develop advanced model-based monitoring techniques, for the enhancement of the system diagnostics and prognostics.

References

[1] A. Suti, G. Di Rito, and R. Galatolo, "Climbing performance enhancement of small fixed-wing UAVs via hybrid electric propulsion," in *2021 IEEE Workshop on Electrical Machines Design, Control and Diagnosis (WEMDCD)*, IEEE, Apr. 2021, pp. 305–310. https://doi.org/10.1109/WEMDCD51469.2021.9425638

[2] C. C. Chan, "The state of the art of electric and hybrid vehicles," *Proceedings of the IEEE*, vol. 90, no. 2, pp. 247–275, 2002. https://doi.org/10.1109/5.989873

[3] A. Suti, G. Di Rito, and R. Galatolo, "Fault-Tolerant Control of a Dual-Stator PMSM for the Full-Electric Propulsion of a Lightweight Fixed-Wing UAV," *Aerospace*, vol. 9, no. 7, p. 337, Jun. 2022. https://doi.org/10.3390/aerospace9070337

[4] R. Huang, C. Liu, Z. Song, and H. Zhao, "Design and Analysis of a Novel Axial-Radial Flux Permanent Magnet Machine with Halbach-Array Permanent Magnets," *Energies (Basel)*, vol. 14, no. 12, p. 3639, Jun. 2021. https://doi.org/10.3390/en14123639

[5] A. ; R. N. , A. ; H. W. , P. Mahmoudi, "Axial-flux permanent-magnet machine modeling, design, simulation and analysis," *Scientific Reseach and Eassays*, vol. 6, no. 12, pp. 2525–2549, 2011.

[6] Z. Wang, J. Chen, and M. Cheng, "Fault tolerant control of double-stator-winding PMSM for open phase operation based on asymmetric current injection," in *2014 17th International Conference on Electrical Machines and Systems (ICEMS)*, IEEE, Oct. 2014, pp. 3424–3430. doi: 10.1109/ICEMS.2014.7014114

[7] G. Liu, Z. Lin, W. Zhao, Q. Chen, and G. Xu, "Third Harmonic Current Injection in Fault-Tolerant Five-Phase Permanent-Magnet Motor Drive," *IEEE Trans Power Electron*, vol. 33, no. 8, pp. 6970–6979, Aug. 2018. https://doi.org/10.1109/TPEL.2017.2762320

[8] A. Suti, G. Di Rito, and R. Galatolo, "Fault-Tolerant Control of a Three-Phase Permanent Magnet Synchronous Motor for Lightweight UAV Propellers via Central Point Drive," *Actuators*, vol. 10, no. 10, p. 253, Sep. 2021. https://doi.org/10.3390/act10100253

[9] R. Zhang, V. H. Prasad, D. Boroyevich, and F. C. Lee, "Three-dimensional space vector modulation for four-leg voltage-source converters," *IEEE Trans Power Electron*, vol. 17, no. 3, pp. 314–326, May 2002. https://doi.org/10.1109/TPEL.2002.1004239

[10] M. Mazzoleni, G. Di Rito, and F. Previdi, *Electro-Mechanical Actuators for the More Electric Aircraft*. Cham: Springer International Publishing, 2021. doi: 10.1007/978-3-030-61799-8

[11] Z. Wang, J. Chen, and M. Cheng, "Fault tolerant control of double-stator-winding PMSM for open phase operation based on asymmetric current injection," in *2014 17th International Conference on Electrical Machines and Systems (ICEMS)*, IEEE, Oct. 2014, pp. 3424–3430. doi: 10.1109/ICEMS.2014.7014114

Target localization with a distributed Kalman filter over a network of UAVs

Salvatore Bassolillo[1,a], Egidio D'Amato[2,b], Massimiliano Mattei[3,c], Immacolata Notaro[1,d] *

[1]Department of Engineering, University of Campania "L. Vanvitelli", 81031 Aversa, Italy

[2]Department of Science and Technology, University of Naples "Parthenope", Napoli, Italy

[3]Department of Electrical Engineering and Information Technology, University of Naples "Federico II", Napoli, Italy

[a]salvatorerosario.bassolillo@unicampania.it, [b]egidio.damato@uniparthenope.it, [c]massimiliano.mattei@unina.it, [d]immacolata.notaro@unicampania.it

Keywords: Distributed Kalman Filter, Target Localization, UAV Formation

Abstract. Unmanned Aerial Vehicles (UAVs) have gained significant usage in various kinds of missions, including reconnaissance, search and rescue, and military operations. In rescue missions, timely detection of missing persons after avalanches is crucial for increasing the chances of saving lives. Using UAVs in such scenarios offers benefits such as reducing risks for rescuers and accelerating search efforts. Employing a formation of multiple drones can effectively cover a larger area and expedite the process. However, the challenge lies in achieving autonomous and scalable systems, as drones are typically operated on a one-to-one basis, requiring a large team of rescuers. To enhance situational awareness and distribute communication load, this paper proposes a decentralized Kalman filtering algorithm that exploits sensor data from multiple drones to estimate target positions and support guidance and control algorithms. The algorithm combines Consensus on Information and Consensus on Measurements techniques. Preliminary validation is conducted through numerical simulations in a sample scenario.

Introduction

In recent decades, the use of UAVs (Unmanned Aircraft Vehicles) has experienced exponential growth in both civil and military sectors. The widespread adoption of these aircraft can be attributed to their ease of use and versatility in various missions. However, in certain scenarios, a single UAV may not suffice, and a collaborative team of UAVs is preferred. Such formations offer comparable or even greater mission capabilities, along with improved flexibility and robustness [1]. According to the Center for Research on the Epidemiology of Disasters (CRED) [2], the issue of fatalities and disappearances due to natural events remains a significant concern globally. To mitigate risks for human rescuers, UAVs should be prioritized for initial rescue operations in hazardous environments. Collaborative efforts among multiple drones have garnered attention from researchers in recent years, as they provide an effective solution to expedite the search process [3]. The coordination and collaboration of UAVs are crucial to function as a unified entity, maintaining formation and directing their flight toward the target.

In this paper, a distributed estimation algorithm for a formation of UAVs is proposed to locate a possible missing skier under the snow. The situational awareness is obtained by equipping UAVs with on-board heterogeneous sensors. The distributed algorithm is based on a decentralized Kalman filtering technique that involves a set of local Kalman filters, one for each UAV. By considering the formation as a network of vehicles, every node provides onboard sensors data and contributes to the estimation of the overall state [4]. With respect to a centralized architecture, running on a leader, that receives information from the other nodes, the decentralized scheme decreases the computational burden on the central node of the formation and it is less vulnerable against system failure, being not depending on a single aircraft.

The proposed sensor fusion algorithm receives measurements from a Global Positioning System (GPS) receiver and an Inertial Navigation System (INS), that represent the most widely used sensors for navigation purposes. To enhance system capabilities in hazardous environments and the robustness to sensors fault, each aircraft is equipped with a radio transponder to measure the relative distance between vehicles. Such device can be based on time-of-flight measurements over ultra-wideband radio signals or on the Received Signal Strength Indication (RSSI) of the same communication standard used to create the formation network. In order to find targets, each aircraft is equipped with a modern avalanche transceiver (ARTVA), typically used by back country skiers, whose signal power measurements, from multiple drones, can be used to perform triangulation and locate missed people.

Problem Statement

Let us consider a heterogeneous swarm of N UAVs, composed of aircraft in multi-rotor or fixed wing configurations. The swarm must survey a specific area in order to identify the presence of a snow-covered skier after an avalanche event. For first rescue mission in the search of a missing skier and for navigation purposes, UAVs must be able to estimate the position of the skier and their position in the airspace.

The discrete-time model of each UAV can be described in the inertial reference frame by the following equations:

$$\begin{cases} \mathbf{S}_i(k) = \mathbf{S}_i(k-1) + \mathbf{V}_i(k-1)T_s + \boldsymbol{\omega}_i^S(k) \\ \mathbf{V}_i(k) = \mathbf{V}_i(k-1) + \boldsymbol{\omega}_i^V(k) \end{cases} \quad \forall i = 1,2,\dots,N \tag{1}$$

where k is the time step, $\mathbf{S}_i(k) = [X_i(k), Y_i(k), Z_i(k)]^T$ is the vector of the spatial coordinates, $\mathbf{V}_i(k) = \left[V_{X_i}(k), V_{Y_i}(k), V_{Z_i}(k)\right]^T$ is the velocity vector of the i-th UAV, $\left[\boldsymbol{\omega}_i^{S^T}(k), \boldsymbol{\omega}_i^{S^T}(k)\right]^T$ is a process noise vector, and T_s is the sample time.

The global state vector $\mathbf{x}(k) \in \mathbb{R}^m$ is defined as:

$$\mathbf{x}(k) = [\mathbf{S}_1^T(k), \mathbf{V}_1^T(k), \dots, \mathbf{S}_N^T(k), \mathbf{V}_N^T(k), \mathbf{S}_t^T(k)]^T \tag{2}$$

where the position of the missing skier is identified by $\mathbf{S}_t(k) = [X_t(k), Y_t(k), Z_t(k)]^T$.

Each UAV is equipped with a set of sensors composed of:

- a GPS, to measure its position in the airspace;
- $N-1$ transponders, to evaluate the relative distance to each UAV in the swarm;
- an avalanche receiver antenna (ARTVA), to detect the signal from the skier avalanche transceiver.

Consider the measure provided by the GPS

$$\mathbf{z}_i^{GPS}(k) = [X_i(k), Y_i(k), Z_i(k)]^T + \mathbf{v}_i^{GPS}(k) \tag{3}$$

the measurement provided by the transponders

$$\mathbf{z}_i^{Tr}(k) = [d_{i1}(k), \dots, d_{iN}(k)]^T + \mathbf{v}_i^{Tr}(k) \tag{4}$$

and the measurement provided the ARTVA

$$\mathbf{z}_i^{AT}(k) = d_{it}(k) + \mathbf{v}_i^{AT}(k) \tag{5}$$

where $\mathbf{v}_i^{GPS}(k)$, $\mathbf{v}_i^{Tr}(k)$, and $\mathbf{v}_i^{AT}(k)$ are the corresponding sensor noises. The term $d_{ij}(k) = \left\|\mathbf{S}_j(k) - \mathbf{S}_i(k)\right\|_2$ represents the Euclidean between two aircraft, i and j, whereas $d_{it}(k) = \left\|\mathbf{S}_t(k) - \mathbf{S}_i(k)\right\|_2$ is the distance of the $i-th$ aircraft from the target skier, with $\|\cdot\|_2$ denoting the Euclidean norm.

The measurement vector of the overall set of sensors is represented by

$$\mathbf{z}_i(k) = \left[\mathbf{z}_i^{GPS^T}(k), \mathbf{z}_i^{Tr^T}(k), \mathbf{z}_i^{AT}(k)\right]^T + \mathbf{v}_i^T(k) \tag{6}$$

with $\mathbf{v}_i(k) = \left[\mathbf{v}^{GPS}{}_i^T(k), \mathbf{v}^{Tr}{}_i^T(k), \mathbf{v}_i^{AT}(k)\right]^T$ as the overall measurement noise vector.

Consensus Estimation

At any time k, the communication structure of the formation can be represented by an undirected graph $\mathbb{G}(k) = \{\mathcal{V}, \mathcal{A}(k)\}$, where $\mathcal{V} = \{1,2,\dots,N\}$ is the set of UAVs, and $\mathcal{A}(k) \subseteq \mathcal{V} \times \mathcal{V}$ is the set of edges describing the communication link between the aircraft i and j. The i-th UAV can receive data from the j-th vehicle if the arc $(i,j) \in \mathcal{A}(k)$. For each aircraft i, $\mathcal{M}_i(k) = \{j: (i,j) \in \mathcal{A}(k)\}$ is the set of its neighbors, including the i-th vehicle, and $D_i(k) = card(\mathcal{M}_i(k)) - 1$ represents its degree, with $card(\cdot)$ representing the cardinality of a generic set.

Let $\xi_i(k)$ be the generic information associated with the $i-th$ agent. We assume that $\xi_i(k)$ is updated according to the Average Consensus Protocol [8]:

$$\xi_i^{(l+1)}(k) = \sum_{j \in A} W_{ij}(k) \xi_j^{(l)}(k) \qquad l = 0,1,\dots,L-1 \tag{7}$$

performed on L consensus steps for each time instant k.

Coefficients $W_{ij}(k)$ can be computed using the Metropolis formula [9]:

$$W_{ij}(k) = \begin{cases} \frac{1}{1+max\{D_i(k), D_j(k)\}} & \text{if } j \in \mathcal{M}_i(k) \\ 1 - \sum_{j \in \mathcal{M}_i(k)} W_{ij}(k) & \text{if } i = j \\ 0 & \text{otherwise} \end{cases} \tag{8}$$

So that, to determine the weight, each agent i does not need any knowledge of the communication graph but only the degrees of its neighboring nodes, $D_j(k)$, with $j \in \mathcal{M}_i$.

The UAVs asymptotically reach the consensus if, for any initial condition $\xi_i^0(k)$, $\lim_{l\to\infty} \|\xi_i^l(k) - \xi_j^l(k)\| \to 0$, for each $(i,j) \in \mathcal{A}(k)$, where l is the generic consensus step.

Results

The performance of the proposed algorithm was evaluated by means of numerical simulations. As reference, a centralized version of the KF was considered, in order to verify the performance cost of the decentralized technique. In this section, to resume results, only an example scenario is considered, with a formation of 9 UAVs, flying together and surveying an area of interest to look for a target. Each aircraft flies along the same direction, at a constant speed of 5 m/s, with an altitude of 100 m above the terrain. Every agent is able to communicate with two neighbors, in order to form a cycle graph and it is equipped with an avalanche receiver to sense the signal from the target skier. Such receiver has a detection range of $d_{sens} = 200$ m.

At the beginning of the simulation, each aircraft knows only its position, initializing the overall state estimate with random values. Such assumption was useful to evaluate the convergence of the algorithm to the correct estimate and the ability to reach the consensus.

Fig. 1 show the estimated components of the target position vector, comparing the real value, the centralized IKF and the proposed decentralized KF. As shown in Figures, at the beginning of the simulation, the target position is still unknown being outside UAVs sensor range.

At $t = 5.5s$, the target enters in the sensor range of UAV_1, but it is not sufficient to be localized. Indeed, for localization problems using fusion algorithms, the measurements of at least three aircraft are required to effectively locate the target [5].

From $t = 17s$, the target is in the sensor range of UAV_1, UAV_2 and UAV_3 and it becomes localizable. The skier estimated position definitively converge at $t = 26s$.

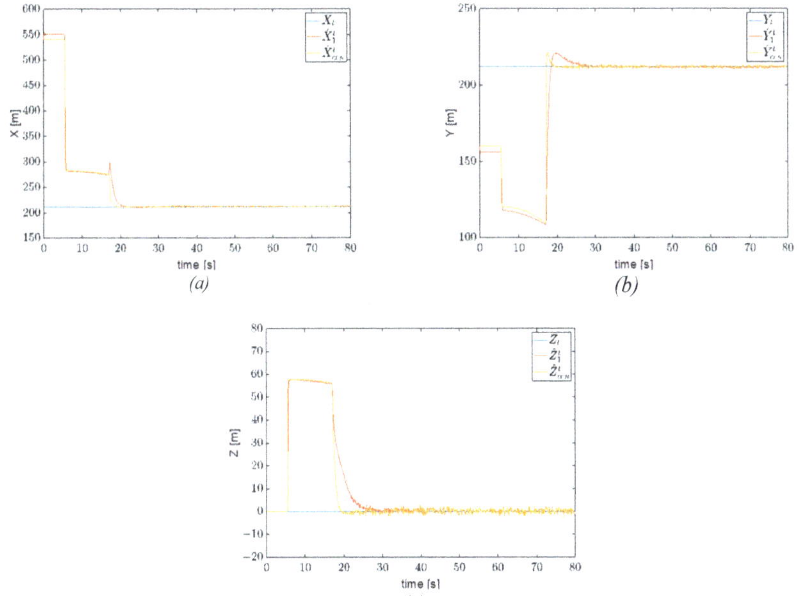

(a) (b)

(c)

Figure 1- Comparison of the estimates of the skier's coordinates: (a) x-component, (b) y-component, (c) z-component .

References

[1] F. Y. Hadaegh, S.-J. Chung, and H. M. Manohara, On Development of 100-Gram-Class Spacecraft for Swarm Applications, IEEE Systems Journal, vol. 10, no. 2, pp. 673–684, Jun. 2016. https://doi.org/10.1109/JSYST.2014.2327972

[2] D. Guha-Sapir, F. Vos, and R. Below, Annual Disaster Statistical Review 2011. 2012.

[3] L. Ruetten, P. A. Regis, D. Feil-Seifer, and S. Sengupta, Area-Optimized UAV Swarm Network for Search and Rescue Operations, in 2020 10th Annual Computing and Communication Workshop and Conference (CCWC), Jan. 2020, pp. 0613–0618. https://doi.org/10.1109/CCWC47524.2020.9031197

[4] M. Cicala, E. D'Amato, I. Notaro, and M. Mattei, Scalable Distributed State Estimation in UTM Context, Sensors, vol. 20, no. 9, Art. no. 9, Jan. 2020. https://doi.org/10.3390/s20092682

[5] M. A. Azam, S. Dey, H. D. Mittelmann, and S. Ragi, Average Consensus-Based Data Fusion in Networked Sensor Systems for Target Tracking, in 2020 10th Annual Computing and Communication Workshop and Conference (CCWC), Jan. 2020, pp. 0964–0969. https://doi.org/10.1109/CCWC47524.2020.9031250

Air Traffic Control, Aircraft Operations and Navigation

Integration in controlled airspace: definition and validation of link loss contingency procedures for RPAS in terminal manoeuvring areas

G. D'Angelo[1,a]*, G. Pompei[2,b], A. Manzo[3,c], G. Riccardi[3,d], and U. Ciniglio[4,e]

[1]Leonardo, C.so Francia, 426 – Torino 10146, Italy

[2]Leonardo, Via Tiburtina km 12,400 – Roma 00131, Italy

[3]ENAV, Viale Fulco Ruffo. di Calabria – Napoli 80144, Italy

[4]CIRA, Via Maiorise – Capua 81043, Italy

[a]giuseppe.dangelo@leonardo.com, [b]gino.pompei@leonardo.com, [c]alessandro.manzo.1@enav.it, [d]giovanni.riccardi.1@enav.it, [e]u.ciniglio@cira.it

Keywords: RPAS Integration, Terminal Manoeuvring Area (TMA), Contingency

Abstract. Remotely Piloted Aircraft Systems (RPAS) are increasingly becoming a part of our day-to-day lives. The vast range of possible applications in both military and civil contexts (e.g. border surveillance, search and rescue, civil protection) is creating a new industry with a large economic potential that, consequently, is pushing the rulemaking and standardization authorities to define and publish the rules, standards and procedures the industry has to comply with in order to integrate its products with the Air Traffic Management System. From an operational perspective, the challenge lies in integrating the worlds of manned and unmanned aircraft in a safe and efficient way allowing both types of aircraft to share the same airspace. An aspect of this challenge consists in defining and validating operational procedures and technical capabilities that allow to manage safely the RPAS even when a (Command and Control and/or ATC radio) link loss occurs. This paper is aimed at describing the procedures for the management of link loss events that affect RPAS flying under Instrument Flight Rules (IFR) in a Terminal Manoeuvring Area (TMA). This paper also describes the distributed facility used to validate such procedures by means of real time simulations in an Italian operational scenario with pilots and Air Traffic Controller (ATCO) in the loop. Positive feedback was provided by both ATCO and Remote Pilots (RPs) on the overall acceptance of the proposed operational procedures which were considered satisfactory even in non-normal conditions (e.g.: degradation of the communication channel quality). The insertion of RPAS in TMA was considered feasible, even in case of single or multiple Command and Control Link Loss (C2LL) contingencies. The experiment described in this paper was part of the SESAR 2020 Programme, PJ13 ERICA project.

Background

In 2012, experts in the RPAS field were called upon by the European Commission to develop a European roadmap for the integration of civil RPAS. In 2013, in response to the 'Roadmap for the integration of civil RPAS into the European aviation system', the SESAR Joint Undertaking launched a set of projects within the SESAR1 framework, then continued in the frame of SESAR 2020 in 2019 [4]. Bringing partners together from across ATM and Europe, the projects aimed at validating emerging RPAS technologies and operational procedures in non-segregated airspace and supporting the update of the related avionic standards and flight rules ([2][3][6]). Overall, these projects perceived no significant difference between the behaviour of an RPAS and a general aviation aircraft of the same (small or medium) category, when operating in the Air Traffic Control (ATC) environment. However, the following threads needed to be addressed before integration could be considered:

- Updated and well-established civil regulation and certification system by the required certification authorities [7];
- Policies and procedures on how ATC should interact with RPAS to ensure efficient operations and to meet safety-level requirements [1];
- A detect & avoid (D&A) capability and compliance with European aircraft equipage requirements;
- A reliable Command & Control (C2) link as well as voice link together with contingency procedures in case of link failures.

Operational context and user needs

Accepting the integration of RPAS into the ATM system poses many challenges and the need to address operational, performance and safety concerns for each of the flight phases of the RPA and considered ATM scenarios. The operational context considered in this paper is the integration of civil RPAS in the 'certified' category flying under IFR in low/medium density non-segregated airspaces from class A to C. The considered ATM scenario was a TMA wherein RPAS and manned aircraft operate simultaneously (mixed traffic), flying Standard Initial Departure and Standard Arrival Route procedures (SID/STAR) and responding to ATCOs clearances and instructions (vectoring included).

In such context two main user needs can be identified:

- Ensure that the RPA are able to flow the same SIDs and STARs designed for manned aviation without penalizing the traffic operations (e.g. by introducing delays or degrading the overall flight safety) in the involved ATC sector[1];
- Ensure that a C2 link-loss or voice loss condition affecting an RPA, has an acceptable impact on RP and ATC in terms of workload and procedures.

Simulation objectives

This paper describes a specific Real Time Simulation (RTS) campaign executed in the frame of SESAR 2020 W2 PJ13 ERICA project, Solution 117 [5][8], aimed at defining and validating the long-term operational, procedural and technical capabilities required to allow the integration of certified MALE and Tactical RPAS into the Italian Area of the Brindisi Control Centre (Brindisi ACC), by using the BARI-PALESE (LIBD) as main airport and BRINDISI-CASALE (LIBR) as alternate one.

The validation experiment described in this paper is the last step of an extensive validation activity which comprised two other propaedeutic RTS campaigns. The final RTS campaign, based on a set of four defined validation scenarios, considered both nominal and link loss contingency situations [5][8]. The experiment aimed at proving the feasibility of mixed operations in nominal conditions, as well as at the acceptance by RPs and Traffic Controllers of the implemented Human System Interfaces and automatic C2LL contingency procedures. It was also used to define a specific ATC phraseology for handling such C2LL contingencies. The simulation framework as well as the functions and procedures used, the outcomes and the resulting recommendations, main scope of this paper, are reported in the next sections.

Distributed simulation framework

To get significative results against the validation objectives, several complex scenarios were built and executed by using the Capua-Roma-Torino Air Ground Operation (CARTAGO) simulation framework depicted in Figure 1.

[1] A defined airspace region for which the *associated* controller(s) has *ATC* responsibility.

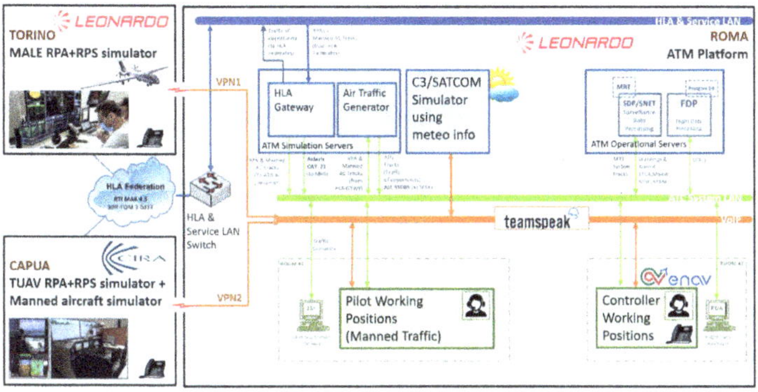

Figure 1 - The CARTAGO Real Time, geographically distributed, simulation facility

CARTAGO consisted of a Real Time geographically distributed simulation facility including:

- A Leonardo real ATC platform equipped with a Short-Term Conflict Alert (STCA) tool and upgraded to manage C2LL contingencies.
- A Leonardo MALE RPAS (10 tons of MTOW, 325 kt of maximum speed) full simulator hosting a Flight Management System model and a C2LL contingency function.
- A CIRA fixed wing Tactical RPAS (550 Kg of MTOW, 90 kt of maximum speed) simulator.
- A CIRA General Aviation manned aircraft simulator.
- A SATCOM model used by RPS to exchange both C2 messages with RPA and voice messages with ATCO. The model allowed to simulate the effects of different weather conditions on the link in terms of delay and degradation.

The geographically distributed simulation platforms were integrated by using the High Level Architecture (HLA) standard [9], specifically customized for ATM data run time exchange so as to keep under neglectable limits the related site-to-site transmission latency. In addition, as shown in Figure 1, the voice communication between RPs and ATCO was implemented by using the open-source TeamSpeak IP-based framework, adequately modified to emulate realistic disruptions over VHF radio link.

Technical functions, operational procedures and scenarios

An automatic C2LL contingency function was developed and tested together with a set of contingency procedures designed to allow the handling of a link loss condition in a wide range of possible situations.

Figure 2 shows the most complex operational scenario that was simulated where the MALE RPA lost the C2 link via SATCOM and the Tactical RPA lost the VHF radio. It can be noted the presence of a ground-ground voice link between the RPS and the ATC. Indeed, it is assumed that the integration of the RPAS into the ATM also requires a ground-ground voice link as essential enabler in order to ensure a back-up link in case of a VHF radio loss. It is still under discussion if this ground-ground link will be implemented by upgrading the existing telephone lines (quick solution) or by creating a dedicated ATM VoIP infrastructure (medium-term solution but with the advantage to avoid the communication latencies introduced by the SATCOM link).

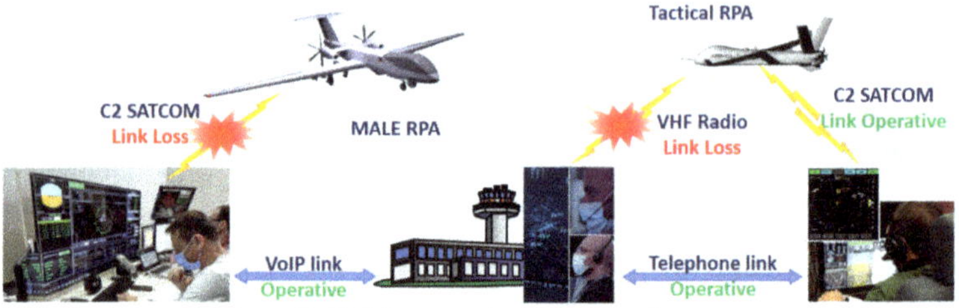

Figure 2. Link loss continency - Complex scenario

Figure 3 shows one of the C2LL contingency procedures automatically executed by the RPA under the control of the on-board C2LL contingency function. The light blue arrow represents the direction from where the RPA could arrive. If the RPA loses the C2 Link (C2L) before the BAR Initial Approach Fix (IAF), it shall continue to fly in automatic way to BAR waypoint by maintaining the last authorized speed and Flight Level (FL). Once BAR IAF is reached, the RPA turns right (yellow path) direct to APFIB fix, then the RPA enters the holding pattern (red path) and holds it for 7 minutes. If the C2L is not recovered during the holding, the RPA shall go direct to BAR IAF (brown arrow) continuing to maintain speed and flight level. Once BAR IAF is reached, the RPA enters a second holding pattern to recover the IAF altitude if needed (green path). As soon as the IAF altitude is reached, the RPA flies to BD554 and starts an automatic RNP approach (purple path) and auto-landing procedure.

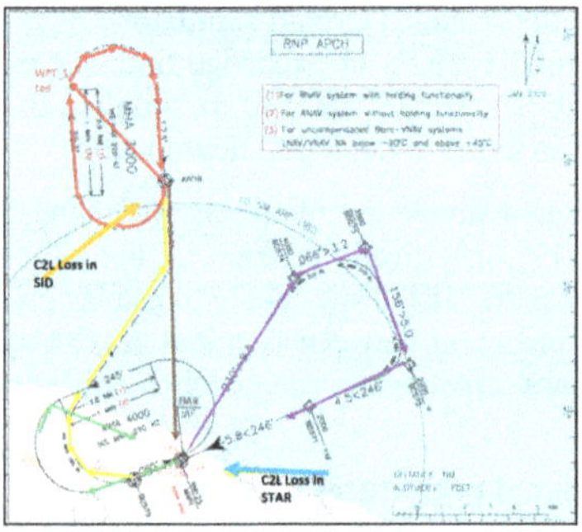

Figure 3. C2LL Procedure for SID and STAR, RWY25

The orange arrow represents, instead, the direction from where the RPA could come if the RPA lost the C2L before the Final SID Fix (FSF). In fact, if the C2L is lost during the departure phase, the C2LL contingency function will take the control of the RPA and, by maintaining the last authorized speed and FL, will complete the climb phase. Once the FSF is reached, the RPA flies automatically direct to APFIB (orange arrow) then enters the holding pattern (red path) and continue the procedure as already described above.

The C2LL contingency function is working properly even when the C2L is lost during the execution of a vectoring instruction. In that case, the C2LL contingency function maintains the last assigned instruction for two minutes then it commands the RPA to rejoin the original flight plan and reach the FSF or IAF as applicable, by maintaining the last authorized speed and FL. Once the FSF or IAF is reached, the RPA flies automatically direct to APFIB and continue the procedure as already described above.

Validation results and recommendations

Throughout the exercise, human performance and safety related aspects were investigated using a range of qualitative and quantitative assessment techniques including post run and post simulation questionnaires, debriefings sessions and operational expert observations annotated during the runs.

Operational acceptance. Positive feedback was provided by both ATCO and RPs on the overall acceptance of the concept and the procedures, even in case of single or multiple C2LL contingencies.

Human performance and perceived level of safety. Both ATCO and RPs stated that the overall workload was tolerable. Single and multiple C2LL contingencies generated a moderate workload but no concerns were raised about potential increase in human error linked to management of RPAS traffic. Positive feedback was provided by the ATCO in terms of impact on safety in both nominal and non-nominal conditions regarding the nominal and C2LL operational procedures, indeed the safety levels were not degraded.

Recommendation and future studies. As additional outcome of the exercise, the following set of recommendations was collected and reported by the ATCO and the RPs, suggested to be addressed in future studies and validations:

- in case a first C2LL procedure is already in place and another RPA is arriving at the same TMA, it was proposed that the ATCO (if needed) instructs this second RPA (with the C2 link still working) to update the pre-programmed C2LL trajectory in order to avoid potential overlapping on the same contingency path.
- to implement the automatic execution of the "open loop" clearances[2] with a flight level limit. In fact, only automatic execution of open loop clearances with time limit were validated, but the ATCO reported that this kind of clearance is rarely used, while flight level limit is more commonly used.
- to implement the automatic execution of "closed loop" clearances. In fact, despite the implementation of "open loop" clearance with limit was acceptable, ATCOs would prefer this implementation to allow them to specify the re-join waypoint together with a vectoring instruction.
- to implement an on-board emergency function to handle an FMS failure under a C2LL condition.

References

[1] CANSO, ANSP CONSIDERATIONS FOR UNMANNED AIRCRAFT SYSTEMS (UAS) OPERATIONS, 2019 https://canso.org
[2] ICAO, Doc 10019 Manual on Remotely Piloted Aircraft Systems (RPAS), February 2019.
[3] SESAR 2020, PJ.10-05-W1 V2 OSED SPR/INTEROP v01.00.00, July 2020. https://doi.org/10.5005/pid-2-1-iv
[4] EUROCONTROL, RPAS ATM CONOPS V4.0, February 2017.
[5] Alessandro Manzo et al 2023 J. Phys.: Conf. Ser. 2526 012103. https://doi.org/10.1088/1742-6596/2526/1/012103
[6] EASA, Advance Notice of Proposed Amendment, 2015-10.
[7] EASA, Commission Implementing Regulation (EU) No 923/2012. https://www.easa.europa.eu
[8] SESAR 2020, PJ13-W2-117 Validation Report (VALR) for V2, December 2022.
[9] Richard M. Fujimoto (Georgia tech) & Richard M. Weatherly (MITRE), HLA TIME MANAGEMENT AND DIS, 1999.

[2] An ATC clearance that does not include a specified or implied point where the restriction on the trajectory ends.

Hybrid graph-clothoid based path planning for a fixed wing aircraft

Luciano Blasi[1,a] *, Egidio D'Amato[2,b], Immacolata Notaro[1,c], Gennaro Raspaolo[1,d]

[1]Department of Engineering, University of Campania "L.Vanvitelli", Via Roma, 29, Aversa, 81031, Italy

[2]Department of Science and Technology, University of Naples "Parthenope", Centro Direzionale di Napoli, isola C4, Napoli, 80143, Italy

[a]luciano.blasi@unicampania.it, [b]egidio.damato@uniparthenope.it, [c]Immacolata.notaro@unicampania.it, [d]gennaro.raspaolo@studenti.unicampania.it

Keywords: Path Planning, Clothoids, RRT, UAVs

Abstract. Planning of safe and efficient trajectories is a critical task in the operation of unmanned aerial vehicles (UAVs), especially in urban or complex environments. With the increasing use of UAVs for various applications, such as surveillance, delivery, and inspection, it is becoming more important to automatically generate collision-free paths that also consider aircraft dynamics. This paper proposes an algorithmic approach based on the Rapidly exploring random tree (RRT) algorithm combined with a clothoid-based smoothing procedure to account for aircraft performance.

Introduction

The Rapidly exploring random tree (RRT) is a path planning algorithm commonly used in robotics and other applications where complex path planning is necessary. The algorithm constructs a tree of randomly generated exploration locations, iteratively adding nodes connected to existing ones if they do not cause collisions with their surrounding environment. RRT is particularly useful for trajectories planning in high-dimensional spaces or with complex obstacle spaces as it can quickly find approximate solutions to such problems. Typically, the solution produced by this algorithm is a piecewise linear path that overlooks the aircraft inability to follow instantaneous heading changes between segments. In literature, RRT is often hybridized with smoothing techniques based on Dubins circles or optimal control methods. The proposed method follows this idea, using clothoid curves to smooth the path. Clothoids, also known as Euler spirals, are mathematical curves that have a constant rate of curvature change along their length. They are widely used in various fields, including engineering, robotics, and transportation, due to their unique properties. Clothoids are particularly useful in path planning problems because they allow for smooth and continuous changes in curvature, enabling vehicles or robots to follow complex trajectories without abrupt changes in direction, which can be detrimental to performance and efficiency. The aim is to provide UAVs with the ability to navigate challenging environments while optimizing a flyable path that minimizes fuel consumption and avoids no-fly zones. In our approach, clothoid construction is integrated into an enhanced RRT procedure, with a reduction algorithm to avoid too many heading changes during flight.

Single aircraft clothoid-based path planning

To follow a sequence of waypoints, a flight trajectory can use straight and circular paths [1]. However, switching from one to the other is hard for fixed wing aircrafts because of the sudden change in yaw rate and bank angle. To track paths with continuous curvature and limits to maximum curvature and sharpness, clothoids can be used [2, 3]. These curves can achieve the desired direction with a linear relationship between curvature and arc length.
Spatial position coordinates x and y can be expressed as a function of the arc length s as follows [4]:

$$x(s) = x_0 + \int_0^s \cos\left(\frac{1}{2}\sigma\zeta^2 + \kappa_0\zeta + \psi_0\right) d\zeta \tag{1}$$

$$y(s) = y_0 + \int_0^s \sin\left(\frac{1}{2}\sigma\zeta^2 + \kappa_0\zeta + \psi_0\right) d\zeta \tag{2}$$

where σ is the sharpness, κ_0 represents the initial curvature, ψ_0 defines the initial heading and ζ is the integration variable. Multiple clothoids can be combined into a spline to create a continuous curvature path [5, 6, 7, 8] by matching the curvature at the tips of clothoid segments.

The maximum bank angle ϕ_{max} and the maximum bank angle rate $\dot{\phi}_{max}$ are important factors for setting the curvature and sharpness boundaries. The maximum path curvature depends on the aircraft speed v, the gravity acceleration g and the maximum bank angle ϕ_{max} [9]. The maximum sharpness can be obtained by taking the derivative of the curvature function with respect to time.

$$\kappa_{max} = \frac{g}{v^2}\tan\left(\phi_{max}\right) \tag{3}$$

$$\sigma_{max} = \frac{g}{v^2}\dot{\phi}_{max}\sec^2(\phi_{max}) \tag{4}$$

Being curvature a linear function with the arc length, the lowest and highest curvature values are found at the endpoints of a clothoid segment.

Rapidly exploring random tree (RRT)

In this chapter a RRT based path planning procedure is presented, combined with a reduction algorithm and clothoid smoothing curves to make the final paths more compliant with aircraft dynamics. RRT represents a wide spectrum of search algorithm extensively used in path planning [10]. Each RRT is based on a graph whose shape resembles a tree, built incrementally by adding random samples chosen from the search space. RRT is often build so that the graph tends to expand towards less explored areas; however, it is possible to properly guide the search towards specific areas of interest.

Let C be the search space, α an infinite sequence of samples in C and α(i) the i-th sample. The first vertex of the graph is q_0 and, starting from it, the tree is generated for k iterations. At every iteration a random sample is picked and linked to the nearest node of the graph, thus becoming a node itself.

To avoid too many changes of flight attitude a reduction algorithm is implemented inside the RRT. In a traditional RRT each new sample, out of the obstacle region, is linked only with the nearest node of the graph. With a reduction algorithm for each new point the most convenient path is searched; the starting point is the nearest node, then the search proceeds backwards by assigning a cost to each link between the new point and the nodes of the graph, provided the link does not enter an obstacle zone.

Only the link with the lesser cost is added to the graph and become one of its edges. During this phase the clothoids are inserted. Each edge between two nodes is composed by a linear path and two clothoids, one for each node. This is done to obtain a feasible trajectory without the need of a post processing phase.

Test case

The algorithm has been tested taking a real-world scenario. In particular, the city of Padova is considered and two of its squares, piazza Cavour and piazza degli Eremitani are chosen, respectively, as the starting point and the ending point, with a minimum height considered for the obstacles of 10 meters.

Since RRT is based on random samples, it is reasonable to expect some minor differences between different run of the algorithm, given that a solution (i.e. a path between the starting and ending points) is always found. A test was done by running the algorithm 100 times. The mean path length found was 456.62m while 200.24s was the mean computational time. On the other hand, the minimum and maximum path length were, respectively, 417.52m and 606.34m, whereas the minimum and maximum time were, respectively, 44.04s and 806.32s. Figure 1 (Left) shows the effect of the reduction procedure. Without reduction the path (blue dashed line) would have been too long and convoluted, whereas the path obtained applying the reduction algorithm (red solid line) has just a few direction changes. In Figure 1 (Right) a detail of the smooth transition between different directions obtained with clothoid curves is shown.

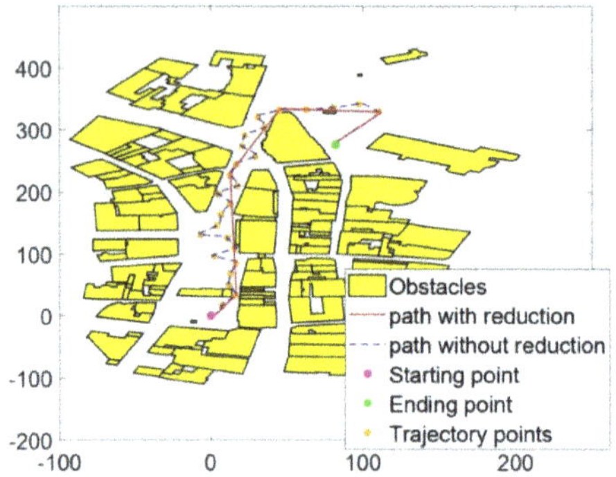

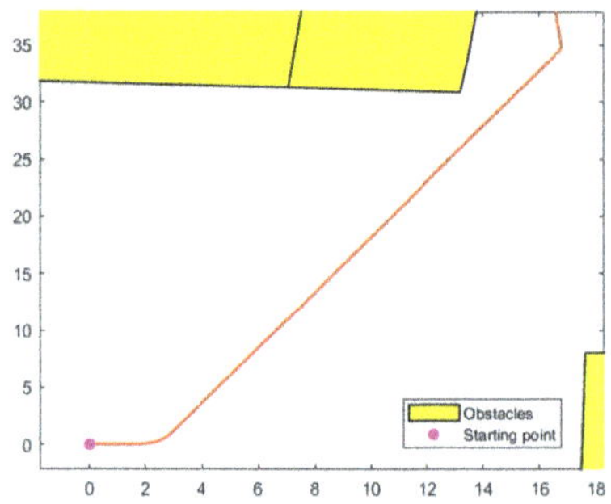

Figure 1 - (Left): comparison between paths obtained with (red solid line) and without (blue dashed line) the reduction algorithm; (Right): detail of the smooth transition between different directions obtained with clothoid curves.

Conclusions

This paper proposes an air vehicle path planner based on the Rapidly exploring random tree (RRT) algorithm combined with a clothoid-based smoothing procedure to better account for aircraft performance. The analysis carried out in the paper proved the effectiveness of the proposed procedure also in complex real-world scenarios, where the use of clothoids ensures the path compliance with aircraft dynamics. Moreover, the application of a reduction algorithm allows to avoid an excessive number of direction changes, although it increases the computational burden.

Future developments will be focused on the edge-adding procedure, being the basis of the algorithm, to guide the creation of new nodes more efficiently and minimize the length of the edges.

References

[1] L. Blasi, E. D'Amato, M. Mattei e I. Notaro, «UAV Path Planning in 3D Constrained Environments Based on Layered Essential Visibility Graphs,» IEEE Transactions on Aerospace and Electronic Systems, pp. 1-30, 2022. https://doi.org/10.1109/TAES.2022.3213230

[2] M. Al Nuaimi, Analysis and comparison of clothoid and Dubins algorithms for UAV trajectory generation, 2014.

[3] T. Tuttle e J. P. Wilhelm, «Minimal length multi-segment clothoid return paths for vehicles with turn rate constraints,» Frontiers in Aerospace Engineering, vol. 1, 2022. https://doi.org/10.3389/fpace.2022.982808

[4] E. Bertolazzi e M. Frego, «Interpolating clothoid splines with curvature continuity,» Mathematical Methods in the Applied Sciences, vol. 41, p. 1723–1737, 2018. https://doi.org/10.1002/mma.4700

[5] D. S. Meek e D. J. Walton, «Clothoid spline transition spirals,» Mathematics of computation, vol. 59, p. 117–133, 1992. https://doi.org/10.1090/S0025-5718-1992-1134736-8

[6] T. Fraichard e A. Scheuer, «From Reeds and Shepp's to continuous-curvature paths,» IEEE Transactions on Robotics, vol. 20, p. 1025–1035, 2004. https://doi.org/10.1109/TRO.2004.833789

[7] D. K. Wilde, «Computing clothoid segments for trajectory generation,» in 2009 IEEE/RSJ International Conference on Intelligent Robots and Systems, 2009. https://doi.org/10.1109/IROS.2009.5354700

[8] S. Gim, L. Adouane, S. Lee e J.-P. Derutin, «Clothoids composition method for smooth path generation of car-like vehicle navigation,» Journal of Intelligent & Robotic Systems, vol. 88, p. 129–146, 2017. https://doi.org/10.1007/s10846-017-0531-8

[9] T. McLain, R. W. Beard e M. Owen, «Implementing dubins airplane paths on fixed-wing uavs,» 2014.

[10] S. M. LaValle, «Rapidly-exploring random trees : a new tool for path planning,» The annual research report, 1998.

Navigation services from large constellations in low earth orbit

Giovanni B. Palmerini[1,a *] and Prakriti Kapilavai[2,b]

[1]Scuola di Ingegneria Aerospaziale, Sapienza Università di Roma, via Salaria 851 – 00138 Roma, Italy

[2]Dipartimento di Ingegneria Astronautica, Elettrica ed Energetica, Sapienza Università di Roma, via Salaria 851 – 00138 Roma, Italy

[a] giovanni.palmerini@uniroma1.it, [b] prakriti.kapilavai@uniroma1.it

Keywords: Global Navigation Satellite Systems, Large Constellation Design, Dilution of Precision (DOP), Doppler Observable

Abstract. Very large satellite constellations in Low Earth Orbits (LEO) devoted to data broadcast could also help in providing navigation services. Lacking a specific payload onboard, the downlink can be exploited as a signal of opportunity, as an example looking at the carrier's Doppler shift. The number of sources and the short distance to users, enabling indoor positioning, are significant advantages of this option. However, recent studies confirmed that commercially-oriented designs partly miss the advantage on the number of sources by directing just one or two beams at a given time to any area on the Earth: it is enough for communication services, it is not for navigation when several signals need to be received by the user at the same time. Looking at a possible service combining downlinks from more than one system to achieve the requested minimum of four signals, this work focusses on the dilution of precision proper to the novel concept. Therefore, the paper updates previous studies - concerning the effects of the orbital configuration of a single LEO system - extending the results to the new scenario.

Introduction

Satellite-based navigation services are a technological asset everyday more present in our life. Starting from GPS, continuing with GLONASS, Galileo and Beidou, these systems, continuously updated and increasingly completed by regional additions (e.g. EGNOS, QZSS) are quickly becoming an essential infrastructure. All of these Global Navigation Satellite Systems (GNSS) work on the time-of-arrival (TOA) principle, measuring the distances from the observer to at least four signal sources. The TOA principle calls for a specialized nature of the payload, including atomic time sources of exceptional stability onboard. Furthermore, a rich control segment is requested to achieve great accuracy in position and timing [1]. GNSS constellations are mainly deployed in Medium Earth Orbits (MEO) to reduce the number of spacecraft and the effect of perturbations: their large orbital radius, between 25000 and 30000 km, makes signals reaching users quite low in power, limiting the capability to receive in certain environments.

The recent appearance of extremely large constellation, with hundreds or thousands of platforms in LEO devoted to data broadcasting, suggested alternatives to traditional systems. In fact, the widespread distribution of these satellites offers a huge number of signal sources at a far shorter distance, enabling indoor service (see [2] and references therein). The interest for LEO systems is proofed also by some emerging commercial venture explicitly devoted to navigation and timing [3]; however, the huge number of satellites requested to provide an effective service suggested the idea of using instead the rich set of sources already in orbit for successful and large data broadcast systems, so called big-LEOs. Work has been done on the way to exploit the relevant signals from big-LEOs [4] as well as on the quality of the service attainable as function of their orbital geometry [5], [6]. These studied considered that Big-LEOs, not devoted – until now – to navigation, should be better seen as sources of signals of opportunity and the service completed by tracking and timing functions adds-on. More recent studies [7] changed the scenario making it far more complex. In fact, it has been observed that the more developed big-LEO, i.e. Starlink, provides at most two beams for any location on the Earth, with a dynamic distribution of resources focusing on current clients' demand. Indeed, the expected rich coverage is definitely apparent, ending up to be similar or even worse with respect to first generation large telecommunication constellations as Iridium or Globalstar. Notwithstanding this strong limitation, the large number of satellites into orbit still calls for the analysis of services combining signals of opportunity from different systems. This paper aims indeed to update the results referred to orbital aspects of the navigations solution presented in [6] by considering the integrated contribution of the two nowadays larger systems, i.e. Starlink and OneWeb. As the same idea could be applied with proper caution to other systems too, it is expected that findings will be useful independently on the current, still limited, development status of data broadcasting constellations. Focusing on the orbital aspects only, this approach acknowledges that several other, more important technical issues ([6], [7]) should be also considered.

Doppler as observable

Basic goal while navigating a standard terrestrial user (i.e. where high performance should not be requested) is the definition of a 4x1 set of unknowns given by the coordinates and the time

$$X = [\, x_u \;\; y_u \;\; z_u \;\; t]^T \tag{1}$$

(notice that this reduced kinematic state, missing velocity variables, perfectly fits stationary or very low-dynamics users). As recalled, traditional TOA technique is not exploitable form current LEO platforms due to the need of a specialized payload. Instead, it would be possible to extract the carrier of the (complex) modulated data broadcast signal [4]. Neglecting errors and noise, the carrier from source *i* recovered from a user *u* will be Doppler-shifted in frequency:

$$\Delta f_i = \frac{f_i}{c}(\vec{v}_i - \vec{v}_u) \cdot \vec{e}_i \; = \dot{\rho}_i \tag{2}$$

leading to the following relation between the range rate (expressed in Hz) and the variables:

$$\dot{\rho}_i = \frac{f_i}{c\,d}[(\dot{x}_i - \dot{x}_u)\cdot(x_i - x_u)+(\dot{y}_i - \dot{y}_u)\cdot(y_i - y_u) + (\dot{z}_i - \dot{z}_u)\cdot(z_i - z_u)] + \Delta f_u + \varepsilon \tag{3}$$

where $d = \sqrt{(x_i - x_u)^2 + (y_i - y_u)^2 + (z_i - z_u)^2}$ is the distance between the source and the user, Δf_u the frequency error at the receiver and ε the other errors affecting the measurement. Considering a linearization, it is possible to focus on the corrections with respect to the reference condition, obtaining a relation as

$$\Delta\dot{\rho} = H\Delta X \;, \qquad \Delta X = [\, \Delta x_u \;\; \Delta y_u \;\; \Delta z_u \;\; \Delta f_u \,]^T \tag{5}$$

where the time variable has been substituted by the drift of the receiver's oscillator with respect to the nominal frequency, from which the time is computed. According to Eq.3, the terms in H will clearly differ from the ones proper of the TOA scheme (i.e., the direction cosines of the line of sight of the sources with respect to the receiver, with the last column given by unity terms). Specifically, also following [8], it is possible to write

$$H_{i1} = -\frac{f_i}{c d_i^3}(x_i - x_u)[(\dot{x}_i - \dot{x}_u)(x_i - x_u) + (\dot{y}_i - \dot{y}_u)(y_i - y_u) + (\dot{z}_i - \dot{z}_u)(z_i - z_u)] - \frac{f_i}{c d_i}(\dot{x}_i - \dot{x}_u)$$

$$H_{i2} = -\frac{f_i}{c d_i^3}(y_i - y_u)[(\dot{x}_i - \dot{x}_u)(x_i - x_u) + (\dot{y}_i - \dot{y}_u)(y_i - y_u) + (\dot{z}_i - \dot{z}_u)(z_i - z_u)] - \frac{f_i}{c d_i}(\dot{y}_i - \dot{y}_u)$$

$$H_{i3} = -\frac{f_i}{cd_i^3}(z_i - z_u)[(\dot{x}_i - \dot{x}_u)(x_i - x_u) + +(\dot{y}_i - \dot{y}_u)(y_i - y_u) + (\dot{z}_i - \dot{z}_u)(z_i - z_u)] - \frac{f_i}{cd_i}(\dot{z}_i - \dot{z}_u)$$

$$H_{i4} = 1 \quad (6)$$

The relation for Doppler observables (Eq. 5) can be processed as done for pseudorange observables in TOA systems [1], to obtain an approximated error evaluation as the product of the error in measurements (σ_{UERE}) and a geometric factor known as dilution of precision (GDOP):

$$\varepsilon_{GNSS} = GDOP\ \sigma_{UERE}\ . \quad (7)$$

GDOP includes the effects of the relative geometry between satellites and receiver, indeed collecting the properties of the orbital configuration of the sources. Assuming the hypotheses of a stationary geometry during the measurements, of an uncorrelated behaviour in terms of errors among satellites and of a common error statistics among them, it can be computed as:

$$GDOP = \sqrt{tr(H^T H)^{-1}} \quad (8)$$

The correctness of the hypotheses leading to GDOP definition should be re-evaluated in the case of the Big-LEO. Specifically, the first hypothesis is certainly more relevant due to the higher velocity of the lower-altitude platforms, while the second hypothesis is in some way relaxed as platforms belong to different systems. The third hypothesis should be fully re-considered as the sources could even have different possible causes of errors (as an example, they can adopt different frequencies and very different hardware), and the condition of an equal value of the overall standard deviation of the error is really difficult to judge. If, pending some specific analysis, the hypotheses can be preliminarily accepted, it is possible to numerically evaluate the GDOP in the frame of an orbital propagation for some selected BigLEOs. In such a way, GDOP can be evaluated in time for any site, defining indeed the amplification factor on the error due to the specific, considered orbital configurations.

Simulations and discussion

Figures 1-6 refer to the configurations of Starlink and OneWeb constellations, currently the two largest big-LEOs, as per Fall 2022, when Starlink included 3049 satellites in various inclinations and altitudes, and OneWeb featured 424 satellites mostly at 1200 km altitude and 87° inclination. More recent data (July 2023) provide 4347 Starlink's and 631 OneWeb's spacecraft, not really changing the rationale of the analysis. The ephemerides requested for the orbital propagation have been obtained from Two Lines Elements (TLE) sets available online [9]. It can be first confirmed that large LEO have the possibility to provide navigation services. Fig. 1 shows the coverage in Padua area (45.41N, 11.88E) limited to sources above 15° elevation. As expected, Starlink is offering most of the coverage, even if, due to the nature of the constellation, in a way which is definitely non uniform in time. Notice that the coverage from OneWeb has a significant periodic behavior, with repetitive gaps: it reflects the more ordered nature of the OneWeb architecture, with higher, less perturbed orbits and a more regular spacing of the orbital planes.

Materials Research Proceedings 37 (2023) 108-112 https://doi.org/10.21741/9781644902813-24

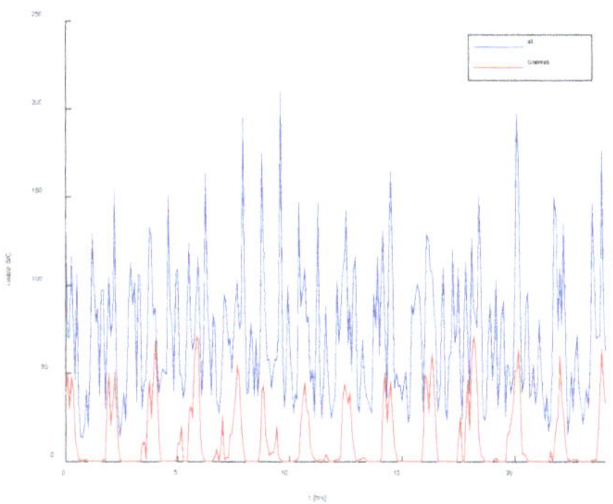

Fig. 1 – S/C above 15° from Padua (cumulative Starlink/OneWeb and OneWeb only)

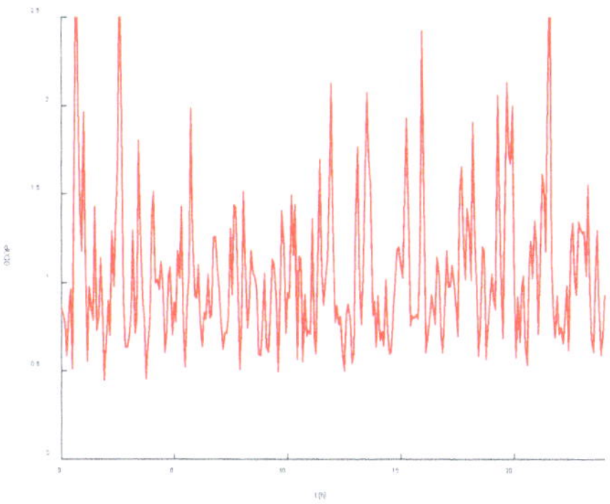

Fig. 2 – GDOP(Time of Arrival observables), full set of visible S/C

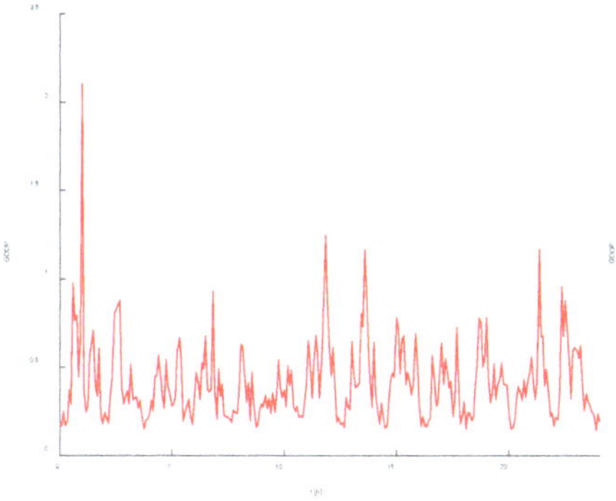

Fig. 3 – GDOP(Doppler), full set of visible S/C

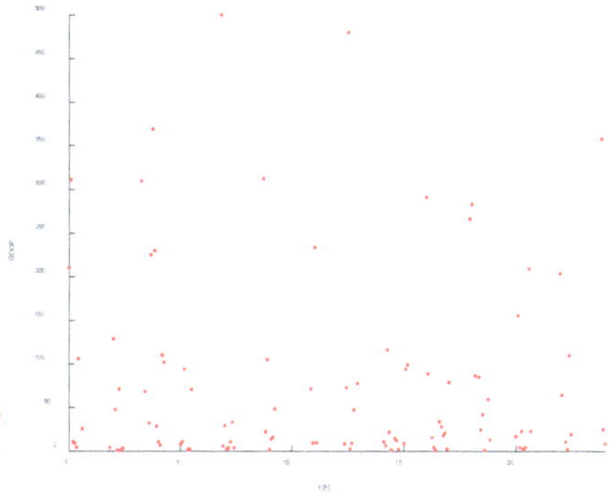

Fig. 4 – GDOP(Doppler), 4 satellites case

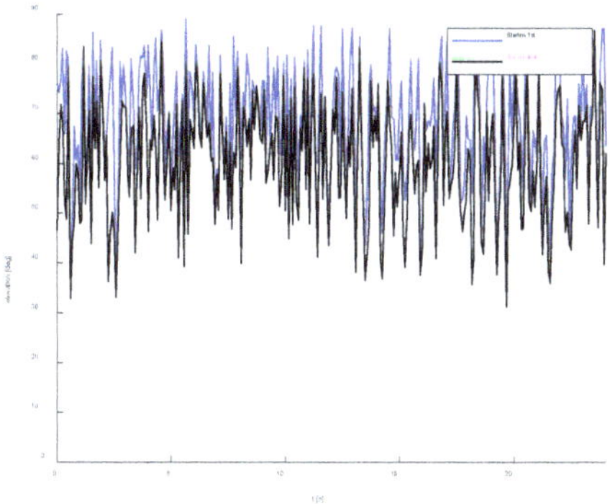

Fig. 5 – Elevation of the best and second best visible S/C of the Starlink constellation.

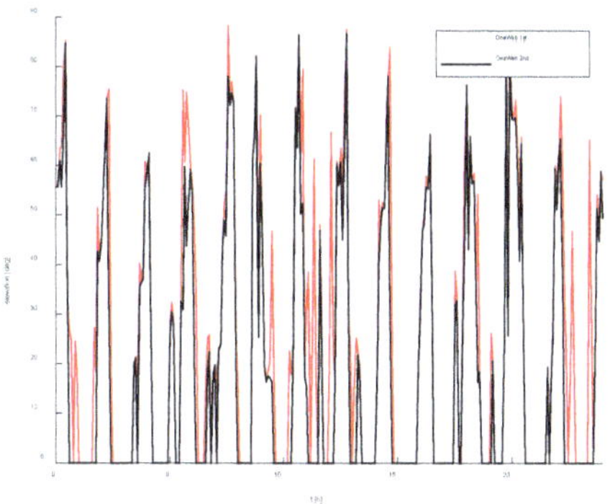

Fig. 6 – Elevation of the best and second best visible S/C of the OneWeb constellation.

Comparable plots referred to July 2023 constellation would show these gaps disappearing with at least one satellite visible above the horizon (elevation threshold equal to 0°). Fig. 2 and Fig. 3 show extremely low, and indeed good, values of GDOPs, for both time-of-arrival and Doppler observable cases, if all sources are considered: and obvious result of the rich coverage offered. If, looking at the new findings about signal distribution [7], the focus is instead on the case of 2-sources only from each constellation, Fig. 4 clarifies that spotted GDOP is higher and the coverage is non continuous. Only the Doppler observable case is here reported, but the behavior does not change for the traditional time-of-arrival case except for the magnitude of values, which is by the way not too significant as observable themselves do have a different magnitude. In fact, active satellites would be the same, i.e. the two from each constellation seen by the receiver with the higher elevation (indeed the closest one at the site). Fig. 5 clarifies that the rich, yet non regular, distribution of Starlink perfectly satisfy the coverage request, with even the second best always above 30°, and generally 40° elevation. Instead, Fig. 6 reports the spaced-out distribution of OneWeb, with the gaps where not even a satellite was available (as previously reported, OneWeb current configuration would offer a single coverage in these intervals, still not enough for a navigation service). Notice that, notwithstanding the (far from trivial) issues in capturing and elaborating different signals, the same approach could be extended to more than two constellations (as an example by adding Orbcomm or Iridium), achieving in such a way the requested coverage without gaps.

Conclusion

Big-LEOs systems can be helpful in providing navigation services in addition, or as a back-up, of existing GNSS, if considered as sources of signals of opportunity. The paper reports the evaluation of the geometry matrix for the Doppler shift observable and the simulations for the GDOP in case of two large (Starlink and OneWeb) systems, considering their current configuration and their specific limited irradiation characteristics.

References

[1] E.D. Kaplan, C.J. Hegarthy, Understanding GPS Principles and Applications, 2nd ed., Artech House, Norwood (MA, USA), 2006.

[2] T.G.R. Reid et al. (part 1), and Z.M. Kassas (part 2), Navigation from Low Earth Orbit, in: J. Morton, F. van Diggelen, J. Spilker, Jr., B. Parkinson (Eds.), Position, Navigation, and Timing Technologies in the 21st Century: Integrated Satellite Navigation, Sensor Systems, and Civil Applications, Volume 2, Wiley–IEEE, 2021, pp. 1359 ff.

[3] T. Reid, Commercial Satnav from LEO, Inside GNSS, May 2022.

[4] J.J. Khalife, Z.M. Kassas, Receiver Design for Doppler Positioning with Leo Satellites, IEEE International Conference on Acoustics, Speech and Signal Processing (ICASSP), 2019. https://doi.org/10.1109/ICASSP.2019.8682554

[5] G.B. Palmerini, P. Kapilavai, Navigation Services from LEO Constellations, paper IAC-22-B2.7.8, 73rd International Astronautical Congress (IAC), Paris, France, 18-22 September 2022.

[6] G.B. Palmerini, P. Kapilavai, Orbital Configurations for Large LEO Constellations Providing Navigation Services, IEEE Aerospace Conference Proceedings, 2023. https://doi.org/10.1109/AERO55745.2023.10115563

[7] P.A. Iannucci, T.E. Humphreys, Fused Low-Earth-Orbit GNSS, IEEE Transactions on Aerospace and Electronic Systems (2020, *in press*). http://doi.org/10.1109/TAES.2022.3180000. https://doi.org/10.1109/TAES.2022.3180000

[8] M. Di Mauro, Doppler Positioning Satellite System, EWP1959, ESA ESTEC 1997.

[9] https://celestrak.org/NORAD/elements/

Aircraft Design and Aeronautical Flight Mechanics

Aeronautics and Astronautics - AIDAA XXVII International Congress Materials Research Forum LLC
Materials Research Proceedings 37 (2023) 114-117 https://doi.org/10.21741/9781644902813-25

A tool for risk assessment after a catastrophic event during suborbital flight operations

Giulio Avanzini[1a,*], Giovanni Curiazio[2b], Lorenzo Vampo[2c]

[1] Department of Engineering for Innovation, University of Salento, Campus Ecotekne (Building "O"), Via per Monteroni, 73100 Lecce (Italy)

[2] Course in Engineering of Aerospace Systems, Politecnico di Bari (Centro Magna Grecia), Via del Turismo, 74123 Taranto (Italy)

[a]giulio.avanzini@unisalento.it, [b]g.curiazio@studenti.poliba.it, [c]l.vampo@studenti.poliba.it

Keywords: Suborbital Flight; Risk Analysis; Space Flight Boundaries

Abstract. Suborbital flights represent a new frontier for the aerospace industry. Together with technological challenges and legal aspects, suitable tools for risk analysis are required to evaluate the potential damage produced by a catastrophic event during a suborbital flight. In particular, an explosion during the powered acceleration phase can cause dispersion of a large number of debris over a wide area, potentially harming the population living close to the launch site. A tool for the determination of the impact footprint of debris after an explosion is proposed, with the objective of supporting the definition of suitable ascent trajectories which reduce the risk for third parties below a publicly acceptable threshold. Legal aspects are also discussed.

Introduction

Suborbital flight is being envisaged as a new market for space tourism and as a means for low-cost accessto microgravity environment (although for time intervals limited to a few minutes). Safety issues are a primary concern for full commercial development of this novel class of activities at the threshold between atmospheric and space flight. This is relevant not only for people on board of the suborbital vehicle, butalso for third parties on the ground. Vehicle reliability should be high enough for commercial operations, with a risk level adequate for public acceptance. Simultaneously, tools are needed for evaluating the riskof third parties exposed to the passage of this novel class of vehicles in case of a (hopefully unlikely, but not impossible) catastrophic event. In this respect, also legal aspects require to be taken into due consideration, possibly requiring *ad hoc* regulations defined at a national as well as international level. During the descent phase, suborbital ballistic flight is less critical than conventional re-entry trajecto- ries. Assume as a reference the configuration of Space Ship II, developed by Virgin Galctic: after release from its mother-plane, the vehicle accelerates by means of a solid-fuel rocket, reaching its apogee in proximity of the Kármán line at 100 km with a velocity close to zero. Thermal loads and peak values ofdeceleration during descent remain within bounds which does not require a heat shield, nor it producesextreme structural loads. This makes the possibility of a major catastrophic event with vehicle fragmentation less likely during this phase. Conversely, failure of the solid rocket during the ascent may result into an explosion, with vehicle fragments impacting the ground on a large area. This implies that a risk analysis for third parties on the ground requires evaluating the impact footprint of debris produced by an explosion at different points along the trajectory.

The objective of the present paper is focused on this latter issue. Several potential explosion pointsare evaluated along the trajectory. A cloud of fragments and velocity increments along tangential, normal to the trajectory, in the vertical plane, and transverse directions are randomly generated. A correction is applied, in order to enforce that the total linear momentum after the explosion equals the momentumof the vehicle at the instant before the explosion. The magnitude of the increments is then scaled in order to match an estimate of the kinetic energy increase due to the energy released by the explosion. This energy is higher, at early stages of the ascent trajectory, when more unburned fuel is present in the rocket, decreasing close to zero at rocket burnout. Stemming from previous experience with risk analysisfor remotely piloted vehicle operations over inhabited areas [1], statistical properties of impact footprintsin terms of number of fragments per unit area and kilograms of debris per unit area are determined, together with the distance of the centroid of the footprint. Combining this information with population density in the areas possibly interested by the fallout allows one to evaluate the risk for communities and individuals in the region, making it possible to design the ascent trajectory in such a way that the probability of damage to people on the ground remains within acceptable levels.

Legal aspects

Before the end of World War II, technologies developed for long-range bomber aircraft were paving theway towards the blossomong of commercial flight. At the same time it was clear that a supranational setof regulations was required for the sake of harmonization of flight procedures, aircraft certification andcrew licensing, together with other activities essential for civil commercial flight. The Convention on International Civil Aviation, usually referred to as the Chicago Convention, signed in 1944, features asmany as 19 Annexes, covering issues from meteorology to accident investigation, from aircraft noise endengine emissions to safety and secutiry aspects. Although other Conventions followed, such as the Con-vention on Offences and Certain Other Acts Committed on Board Aircraft in 1963 (known as the TokyoConvention), The Hague Hijacking Convention (formally the Convention for the Suppression of Unlaw-ful Seizure of Aircraft), signed in 1970, and The Montreal Convention (formally, the Convention for theUnification of Certain Rules for International Carriage by Air) signed in 1999, the Chicago Convention,updated in 2006, is still the backbone of international regulations for Civil Aviation.

Unfortunately, such an effort for providing an internationally recognized set of regulations for spaceactivities has yet to be undertaken. The Outer Space Treaty, signed in 1967 [3], is an early attempt, formally accepted by all Nations with relevant space activities, which states only basic principles, such as the freedom for all Nations to access space or the impossibility to claim portion of space under a single Nation sovereignty. Coming to more specificand technically relevant issues, there is no set of space rules, which can be the counterpart of the Rules of the Air listed in Annex 2 of the Chicago Convention. As an example, the United Nations delivered guidelines for the mitigation of the danger related to the increasing number of space debris [4], but these guidelines only provide a set of non-binding recommendations, without any actualconstraints (let alone, sanctions) for potentially dangerous space activities of sovereign states.

There are two major aspects that pose a serious obstacle to the development of a supranational spacelaw. First of all, the concept of airspace extends the sovereignty of a state to the volume where aeronautical activities are carried out above its territory and an airplane can follow a trajectory which avoids theairspace of war zones, as it is currently happening over Ukraine. This is not possible in space, where or-bits follow a prescribed pattern due to gravity and perturbing forces, and a continuous trajectory controlis not available. Hence it is impossible to prescribe boundaries in space which follow in any form thosepresent on the Earth surface and extended vertically for conventional air operations. A second issue is represented by the definition of an unambiguous threshold for separating the domains of atmospheric and space flight. Conventionally, the Kármán line, placed at 100 km, is often adopted as the boundary that marks the entry into space flight, but air traffic never gets even close to those altitudes, most air activities being limited to altitudes well below 30 km. Conversely, spacecraft orbit the Earth at an altitude higher than 250 km for avoiding a fast orbit decay. Conventional space vehicles rapidly cross the region between 30 and 250 km during launch and, much less frequently, reentry. The development ofsuborbital flight operation will soon require to adequately regulate activities taking place in the region of space between those used for conventional air and space operations.

Model for explosion, fragmentation and fallout

Among many other aspects, tools for risk analysis are required in order to define safe operations which do not result into a hazard for communities living in the neighborhood of the spaceport. This requires the identification of an impact footprint of debris in case of a catastrophic event that causes the fragmentation of the vehicle during the powered ascent trajectory. Flight data for Virgin Galactic SpaceShip 2 were used as a reference, which allow for the determination of suitable initial conditions in terms of vehicle speed and climb rate at different altitudes.

At the time of the catastrophic event, t_0, the velocity components of the vehicle are equal to $V_{x,0} = V_0 \cos\gamma_0$, $V_{y,0} = 0$, and $V_{z,0} = V_0 \sin\gamma_0$, with $\sin\gamma_0 = \dot{h}/V_0$. We assume that the explosion generates a cloud of N fragments. A uniform distribution of N random numbers r_k between 0 and 1 is generated. Assuming a vehicle mass of approximately M = 4535 kg and letting $R = \sum_{k=1}^{N} r_k$ and $f = M/R$, the mass of the k-th fragment is $m_k = f\, r_k$. Three sets of N Gaussian distributed velocity increments $\Delta v_{x,k}$, $\Delta v_{y,k}$, and $\Delta v_{z,k}$, are also generated. Provided that the momentum of fragments after t_0 must equate vehicle momentum right before t_0, three corrections, $\Delta v_{x,C}$, $\Delta v_{y,C}$, and $\Delta v_{z,C}$ are introduced, which satisfy the relation

$$\sum_{k=1}^{N} m_k(\Delta v_{x.k} + \Delta v_{x,C}) = 0\,;\ \sum_{k=1}^{N} m_k(\Delta v_{y.k} + \Delta v_{y,C}) = 0\,;\ \sum_{k=1}^{N} m_k(\Delta v_{z.k} + \Delta v_{z,C}) = 0$$

Velocity increments are then multiplied by a factor K_E, related to the intensity of the explosion, higher, when more unburned fuel is present in the rocket, and smaller at engine shut off. Assuming anexplosion intensity proportional to kinetic energy increase, K_E is obtained solving

$$\sum_{k=1}^{N} \mathcal{E}_k - \mathcal{E}_0 = \Delta\mathcal{E}_{expl}$$

where $\Delta\mathsf{E}_{expl}$ is the increment of kinetic energy due to the explosion, $\mathsf{E}_0 = \frac{1}{2} M V_0^2$ is the kinetic energy of the vehicle just before t_0 and the kinetic energy of the k-th fragment after t_0 is

$$\mathcal{E}_k = \frac{1}{2} m_k \left\{ [V_{x,0} + K_{\mathcal{E}}(\Delta v_{x.k} + \Delta v_{x,C})]^2 + [K_{\mathcal{E}}(\Delta v_{y.k} + \Delta v_{y,C})]^2 + [V_{z,0} + K_{\mathcal{E}}(\Delta v_{z.k} + \Delta v_{z,C})]^2 \right\}$$

A simple ballistic trajectory is assumed after the explosion. This allows to analytically determine thetime of impact on the ground of the k-th fragment, $t_{k,F}$, from the equation

$$h_0 + [V_{z,0} + K_{\mathcal{E}}(\Delta v_{z.k} + \Delta v_{z,C})]\, t_{k,F} - \frac{1}{2} g t_{k,F}^2 = 0$$

The distance flown in the along track and cross track directions are thus respectively equal to

$$x = [V_{x,0} + K_{\mathcal{E}}(\Delta v_{x.k} + \Delta v_{x,C})]\, t_{k,F}\,;\ \ y = [V_{y,0} + K_{\mathcal{E}}(\Delta v_{y.k} + \Delta v_{y,C})]\, t_{k,F}$$

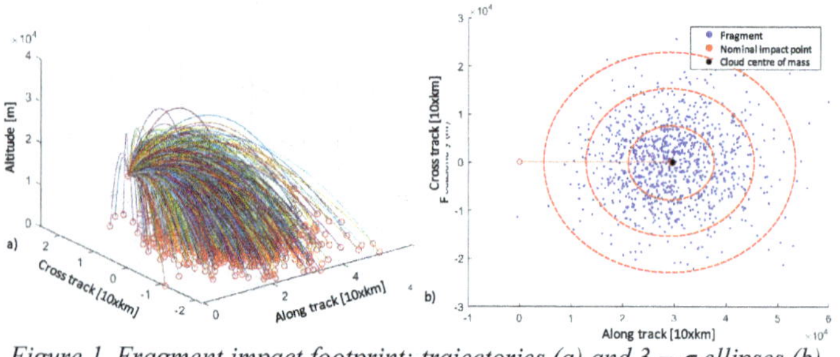

Figure 1. Fragment impact footprint: trajectories (a) and $3-\sigma$ ellipses (b).

Results

The last equations are used for the determination of the impact footprint of the N fragments, when they hit the ground. Figures 1.a and b represent the parabolic trajectory and the impact footprint after a fragmentation due to an explosion at an altitude of 21 640 m, after 20 s from rocket engine ignition. In this preliminary analysis only the position of the fragments at impact is considered, neglecting the effects of aerodynamic drag on the resulting trajectory (hence also wind and turbulence). Regardless of these simplifying assumptions, it is possible to determine the standard deviation σ_x and σ_y in the along-track and cross-track directions of the positions of impact points with respect to the nominal point, represented by the impact point of the center of mass of the fragment cloud. The relevant data for a bivariate Gaussian distribution become thus available, which represents the possibility of the impact of a fragment in a given area. As it was done in [1], the three ellipses with 1σ, 2σ, and 3σ semiaxes contain 39%, 87%, and 99% of the fragments, respectively. These allows one to estimate the number of fragments per unit area expected to fall in each circular region. By matching these data with population density it is possible to derive the probability of hitting somebody on the ground. Future studies will address the effect of aerodynamic drag, wind and turbulence on the dispersion of the fragments. Moreover, together with the number of fragments per unit area, other risk parameters will be evaluated, such as the mass of debris per unit area and the kinetic energy of the fragments at impact, which are significantly affected by the deceleration due to drag.

Conclusions

A procedure for the determination of the impact footprint of fragments generated by the explosion of a suborbital vehicle during its powered ascent phase is outlined and some preliminary results proposed. Once the altitude and velocity profile of the mission are known, it is possible to perform a statistical analysis of the expected impact points of a cloud of debris, thus identifying the number of fragments per unit area expected to fall on the ground after a catastrophic event at a given altitude.

References

[1] G. Avanzini, D.S. Martí nez, Risk assessment in mission planning of uninhabited aerial vehicles, Proc. I.Mech.E. Part G: J. Aerosp. Eng., 233:10 (2019) 3499-3518. https://doi.org/10.1177/0954410018811196

[2] Various Authors, Convention on International Civil Aviation, Doc. 7300/9, 9th edition, Interna- tional Civil Aviation Organization, 2006

[3] Various Authors, Treaty on Principles Governing the Activities of States in the Exploration and Use of Outer Space, including the Moon and Other Celestial Bodies, United Nations Office for Disarmament Affairs, 1967

[4] Various Authors, Space Debris Mitigation Guidelines of the Committee on the Peaceful Uses of Outer Space, United Nations Office for Outer Space Affairs, 2010

Cruising by air and sea: brief history, status and outlook for a submersible aircraft

Sergio De Rosa[1,2,*], Marco Cinque[1], Giuseppe Petrone[1,2] and Leonardo Lecce[3]

[1] Department of Industrial Engineering, Università di Napoli Federico II, Italy

[2] WaveSet S.r.l., Napoli, Italy

[3] Novotech S.r.l., Casoria (NA), Italy

* sergio.derosa@unina.it

Keywords: Submersible Aircraft, Innovative Design, Transitional Configuration

Abstract. Even in a post-modern era, some possible new configurations for aircrafts and spacecrafts remain to be explored. Normally, the main attention was given detaching from ground and by looking at the sky, but the sea should not be forgotten and thus the possibility to go from/to the sky to/from the sea should have to be considered. This paper is just centered around the historical first developments for the submersible aircraft and the possible future developments. Market segments for manned aircraft seems to be largely unexplored.

Introduction and Historical References

In the field of aeronautics, there exists an aircraft configuration that remains unrealized, only existing as a concept in works of fiction such as books and movies. This refers to the idea of a craft capable of cruising through both air and sea. As far back as 1904, Jules Verne, a visionary who shaped the modern era through his fictional writings, had already imagined a machine called *l'Épouvante*, which could transform and adapt for terrestrial, aerial, or marine motion [1].

Not surprisingly, there is a long-standing tradition of attempting to tackle this challenge. The emergence of unmanned vehicles has reinvigorated the research in this fascinating area [2-12]. It is worth noting that in 2008, DARPA made a statement regarding the potential design of a manned version, as shown in Figure 1, [13].

Looking at the historical context, the initial concepts for these configurations emerged in the military sectors of both the United States and Russia, which is a common trend in the history of aeronautics and aerospace [14-16].

Crouse [17] delves into the conceptual design of a submersible airplane, and notably, Longobardi's patent [18] is explicitly mentioned for proposing the incorporation of foldable wings in such a configuration. This patent, dating back to 1918, also envisioned the vehicle functioning as a car.

Manned configurations and a Possible New Market

Living in Napoli, the idea was to have a mission profile really related to the land, sky and sea in order to give a full range of opportunities for tourists. Let's image a small aircraft which can

(i) take-off and land over a small field;
(ii) take-off and land over a sea area and
(iii) submerge and cruise underwater the fantastic natural and archeological scenarios in Baia or the blue sea around Capri.

The task at hand is highly demanding due to the evident conflict between the requirements for air and sea domains. However, recent indications suggest that the design issues may have been overlooked or insufficiently addressed, along with the associated opportunities for funding in the market.

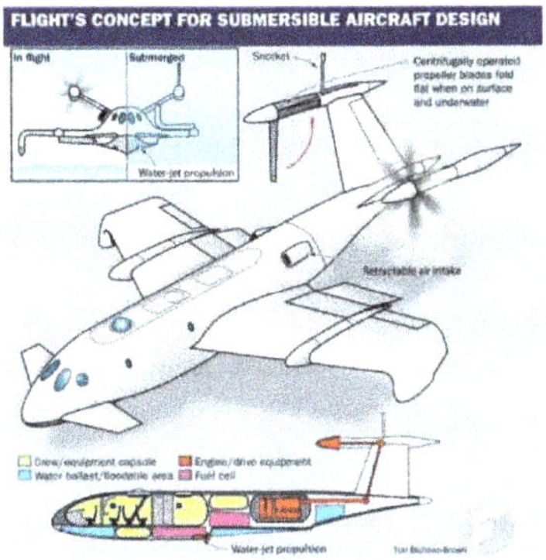

Fig. 1. DARPA Concept (https://www.flightglobal.com/picture-darpa-seeks-submersible-aircraft-concepts/83329.article)

Effectively managing the path of this novel submersible aircraft will necessitate a wholly innovative approach. This entails establishing designated areas for landing and submergence over the sea, implementing signals to indicate the presence of the submerged aircraft, and addressing other related considerations.

Nonetheless, the primary challenge lies in devising and determining the feasibility of potential new configurations for the aircraft.

The Seagull

A suitable starting point for analysis could be the Seagull aircraft, recently developed by Novotech (Italy). This aircraft features a main hull, two auxiliary side floats (sponsons), and a V-tail, as depicted in Figure 2.

The Seagull is a hybrid-electric, two-seater (side by side) amphibian aircraft with a pusher propeller. It stands out for its automated folding wing system and incorporates extensive use of composite structures manufactured through eco-compatible production processes.

The Seagull program commenced in January 2018, and the production of the first prototype was completed by December 2020. As shown in Figure 2, the Seagull aircraft underwent initial water tests in the early part of 2021 to evaluate maneuverability and floating capabilities. Currently, the aircraft is in the process of obtaining flight permits from the Italian Civil Aviation Authority, while the second prototype is nearing completion.

Designed to facilitate communication between individuals and overcome existing barriers in public and private transport, the Seagull can autonomously moor at common seaports thanks to its folding wing system. The high-wing configuration enables panoramic views both in the air and over the sea, while the sponsons can be adapted and transformed into water tanks for loading and unloading. Indeed, thanks to the folding wing system, it can be moored in a common seaport.

The high-wing configuration allows the possibility to look around in both air and over sea, while the sponsons can be adapted and transformed in two tanks able to load and unload water for diving and emerging from water. Now, this vehicle in air

- can accommodate 1 passenger and 1 pilot;
- has foldable wings for increasing maneuverability in the water taxiing;
- can fly 220 km at 4000 ft altitude;

- has an estimated cruise speed of 185 km/hr;
- can accommodate up to 700 kg;
- can be equipped with hybrid electric or I.C. engine with a power up to 100 HP @5800 RPM.

It should be modified so to allow a transit under surface for about 45 minutes at a cruise speed of 4 knots (about 7 km/hr). This goal will be achieved if these design problems will be faced and solved:

- immersion and emersion phases
- additional engine/propeller for underwater cruise
- estimation of the increased weight and installed power
- management of the breathable air for the pilot and the passenger
- sealing of the interiors and the air systems (engine, electrical wiring, etc.)

Fig. 2. Seagull: (top) sketch and (bottom) picture of the prototype in marine configuration

Managing these challenges is undoubtedly complex, but it is seeming emerging from the literature that utilizing foldable wings represents the most promising starting point for exploring the possibilities.

It is also important to keep in mind that the market segment associated with this submersible aircraft is highly specific, well-defined, and somewhat limited in scope.

Concluding Remarks

As humankind sets its sights on returning to the moon and embarking on the first manned mission to Mars, along with the advent of super- and hypersonic transportation, it seems that the time is ripe for exploring underwater cruising as well, despite the design challenges that remain largely theoretical at this point.

References

[1] Jules Verne, https://en.wikisource.org/wiki/The_Master_of_the_World, 1904

[2] P. Marks. *A sub takes to the skies*. New Scientist, 3 July 2010: 32-35. https://doi.org/10.101 6/S0262-4079(10)61629-6
[3] P. Drews-Jr, et alii, *A Survey on Aerial Submersible Vehicles*. (2009). https://www.researc hgate.net/publication/263314909
[4] Q. Wang, et alii, *Submersible Unmanned Aerial Vehicle: Configuration Design and Analys is Based on Computational Fluid Dynamics*, MATEC Web Conf., 95 (2017) 07023. https:// doi.org/10.1051/matecconf/20179507023
[5] Y. Xu, W. Son, *Conceptual design of a submersible aircraft with morphing technology*, As ian Workshop on Aircraft Design Education, AWADE 2016, Nanjing University of Aerona utics and Astronautics (NUAA).
[6] L. Qiu, W. Song, *Efficient Decoupled Hydrodynamic and Aerodynamic Analysis of Amphi bious Aircraft Water Takeoff Process*. Journal of Aircraft, 2015, 50(5): p. 1369-1379. https: //doi.org/10.2514/1.C031846
[7] M. Maia, et alii, *Demonstration of an Aerial and Submersible Vehicle Capable of Flight an d Underwater Navigation with Seamless Air-Water Transition*, https://arxiv.org/pdf/1507.0 1932.pdf, 2015.
[8] X. Yang, et al., *Survey on the novel hybrid aquatic–aerial amphibious aircraft: Aquatic un manned aerial vehicle (AquaUAV)*, Progress in Aerospace Sciences. Beihang University, C hina, 2014. https://doi.org/10.1016/j.paerosci.2014.12.005
[9] T. A. Weisshaar, *Morphing Aircraft Systems: Historical Perspectives and Future Challeng es*, Journal of Aircraft, Vol. 50, No. 2 (2013), pp. 337-353. https://doi.org/10.2514/1.C0314 56
[10] H.X. Liu. *Investigation on the mechanism of a bionic aquatic–aerial aircraft and prototype aircraft project*, Beihang University, Beijing, 2009
[11] A. Gao, A. H. Techet, 2011. *Design considerations for a robotic flying fish*. Proceedings of Oceans 2011, Washington D.C., IEEE, pp. 1–8. https://doi.org/10.23919/OCEANS.2011.61 07039
[12] L. Dong et alii, *Numerical Study on the Water Entry of a Freely Falling Unmanned Aerial-Underwater Vehicle*. *J. Mar. Sci. Eng.*, 2023. https://doi.org/10.3390/jmse11030552
[13] DARPA. *Submersible Aircraft*, DARPA-BAA-09-06. October 3, 2008.
[14] G. Petrov, *Flying submarine*, Journal of fleet, 1995 (http://www.airforce.ru/ aircraft/miscell aneous/flying submarine/index.htm)
[15] B. Reid. *The Flying Submarine: The Story of the Invention of the Reid Flying Submarine, R FS-1*. Heritage Books, 2004
[16] N. Polmar and K. J. Moore. *Cold War Submarines: The Design and Construction of U.S. a nd Soviet Submarines*. Potomac Books Inc., 2003
[17] G. L. Jr. Crouse. *Conceptual Design of a Submersible Airplane*. Auburn University, 2010, 4 8th AIAA Aerospace Sciences Meeting. https://doi.org/10.2514/6.2010-1012
[18] F. Longobardi. Combination Vehicle. US Patent #1,286,679, Issued Dec. 3, 1918. https://pa tentimages.storage.googleapis.com/17/4a/c4/bbd85eb1cbb025/US1286679.pdf

Low-boom supersonic business jet: aerodynamic analysis and mission simulation towards a CO2 emission standard

Giacomo Richiardi[1,a*], Samuele Graziani[1,b], Oscar Gori[1,c] and Nicole Viola[1,d]

[1]Politecnico di Torino, Torino, Italy

[a] s296139@studenti.polito.it, [b] samuele.graziani@polito.it, [c] oscar.gori@polito.it, [d]nicole.viola@polito.it

Keywords: Supersonic Aircraft, Environmental Sustainability, CO_2 Emission Standard

Abstract. This study aims at investigating the aerodynamic characteristics and mission performance of a supersonic business jet at a conceptual design stage. Moreover, the environmental impact of such concept is analyzed to support the development of a potential CO_2 emissions standard for supersonic transport aircraft. The case study considered for the analysis is a supersonic business jet.

Introduction

High-speed transport has gained a renewed interest within the aerospace community during the past few decades. However, concerns about environmental impact, specifically CO_2 emissions, require a thorough analysis of aerodynamics and mission performance. This study focuses on a Mach 1.5 low-boom supersonic business jet, analyzing its aerodynamic characteristics and mission simulation to support the development of a CO_2 emissions standard for supersonic transport aircraft. CFD simulations are exploited to examine lift, drag, and pitching moment coefficients. Mission simulation is used to evaluate the performance in terms of fuel consumption and maximum range. The study also assesses CO_2 emissions standards, which are compared to subsonic limits and other supersonic concepts. The findings are expected to contribute to the design and regulation of environmentally sustainable future supersonic aircrafts.

Case Study

The case study is a Mach 1.5 low-boom supersonic business jet, 100% SAF-powered. An isometric view of the aircraft is presented in Fig. 1, while the main data are reported in Table 1.

The vehicle's configuration is derived from the *Nasa X-59 QueSST* [1], which is a configuration specifically studied for minimizing the sonic boom signature, ensuring a modest far-field pressure distribution and a reduced time distance between the two peaks of the N-wave [2].

However, to accommodate up to twelve passengers and three members of the crew, the central part of the fuselage has been enlarged, assuming a seat pitch of *1.4m*. This change in geometry generated a gap between the leading edge of the root chord of the wing and the cockpit, that were originally at the same longitudinal coordinate from the front of the vehicle, allowing the placement of the passengers' entrance door. Moreover, due to the necessity of having two thrusters for range and safety-related reasons, the two state of the art turbofan engines have been moved from the tail, under the vertical stabilizer, to the wing of the plane.

Fig. 1 Isometric view of the airplane

Table 1 Aircraft main data

Payload [kg]	1500
MTOW [kg]	39283
Empty weight [kg]	19048
Fuel mass [kg]	18434
Wing surface [m^2]	112
Wingspan [m]	14
Fuselage diameter [m]	2.2
Length [m]	44
Range [km]	3800
Mach cruise	1.5

Aerodynamic analysis

To investigate the aerodynamic characteristics of the case study, inviscid and steady *CFD* simulations are performed. Two different mesh grids are generated, one for the subsonic domain (about 5.2 million elements) and the second one for the supersonic one (about 2.8 million elements). *ANSYS ICEMCFD* [3] is used to generate the mesh grids, while *ANSYS FLUENT 2022R2* [4] is used as pre-processor and solver. An overview of the mesh grid is shown in Fig. 2.

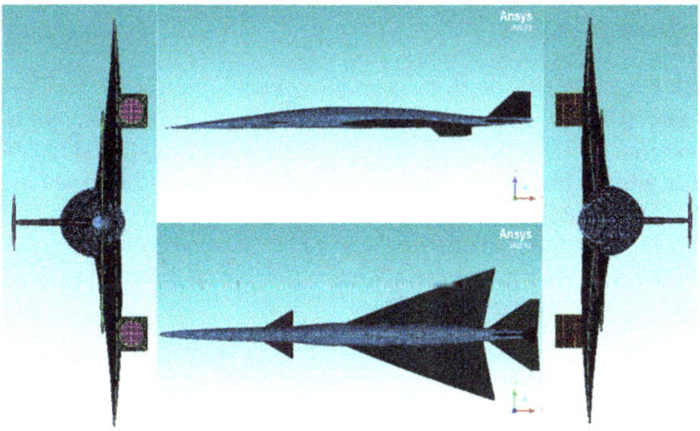

Fig. 2 Mesh grid

The resulting lift and drag coefficients as a function of the angle of attack and for different Mach numbers are reported in Fig. 3 and Fig. 4, respectively. The drag polar is also reported in Fig. 5, while the pitching moment coefficient trend for different angles of attack is shown in Fig. 6.

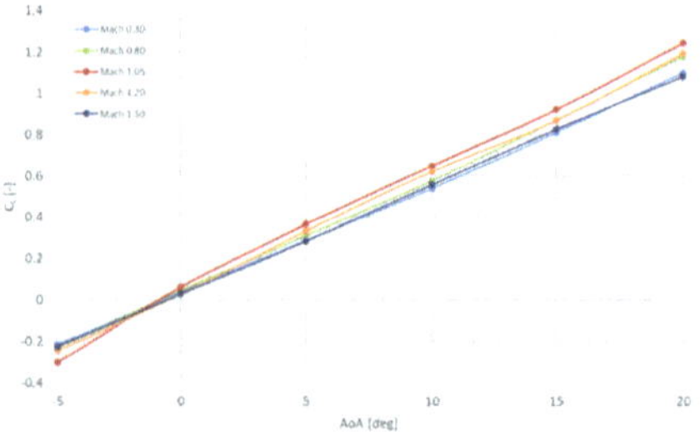

Fig. 3 Lift coefficient vs Angle of Attack for different Mach numbers

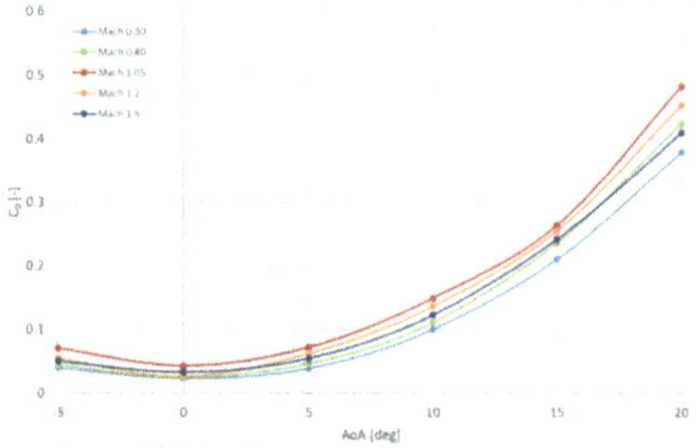

Fig. 4 Drag coefficient vs Angle of Attack for different Mach numbers

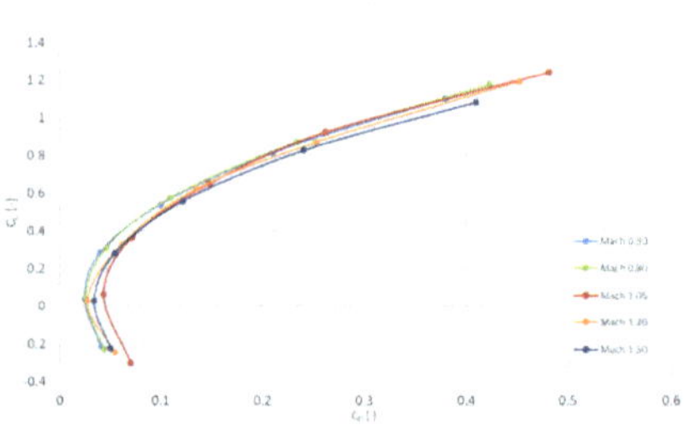

Fig. 5 Lift coefficient vs Drag coefficient for different Mach numbers

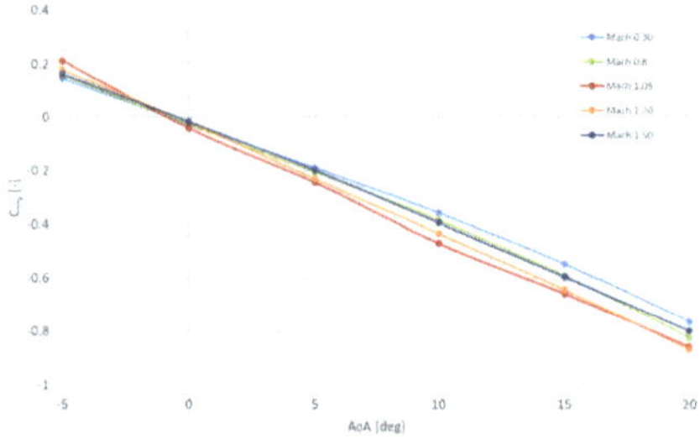

Fig. 6 Pitching moment coefficient vs Angle of Attack for different Mach numbers

Mission simulation

Once the aerodynamic database is available, the aircraft's performance along the reference mission can be studied using the ASTOS software. The main results of the mission simulation are presented in this section. The altitude and Mach profiles during the mission are reported in Fig. 7, while the total and propellant mass variation over time is shown in Fig. 8. The aircraft performs the cruise at Mach = 1.5, while the altitude varies from 14 to 17 km. The propellant on-board is sufficient to cover a range of 3800 km. The angle of attack variation during the mission is reported in Fig. 9, while the L/D ratio is shown in Fig. 10.

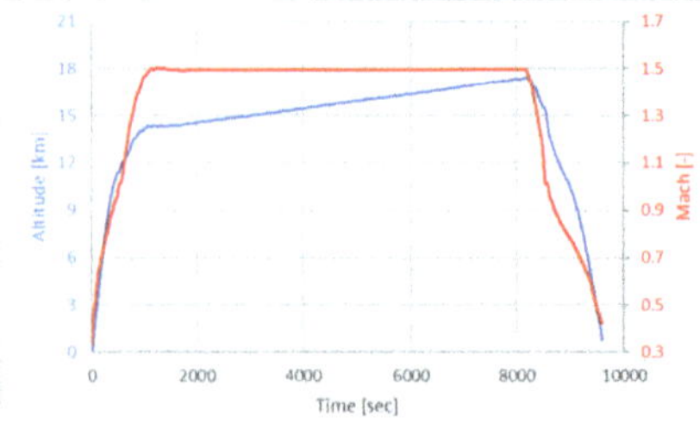

Fig. 7 Altitude and Mach profile during the mission

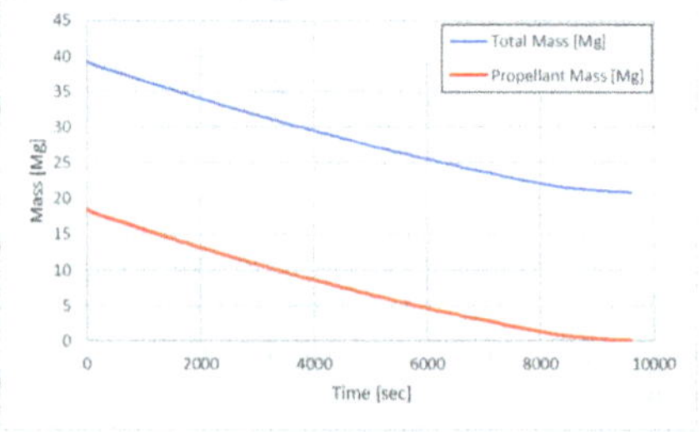

Fig. 8 Total and propellant mass variation during the mission

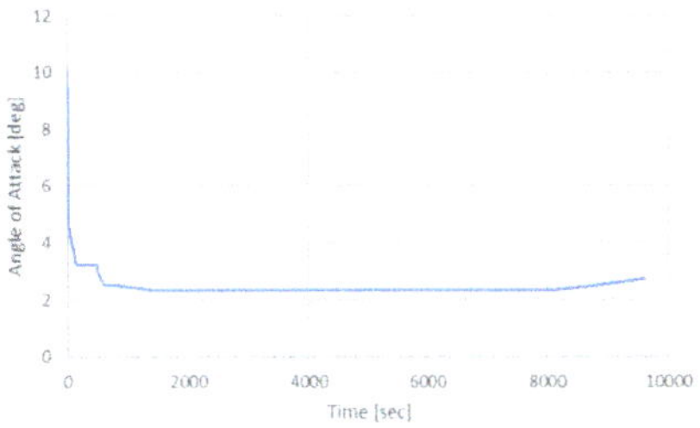

Fig. 9 Angle of attack vs time

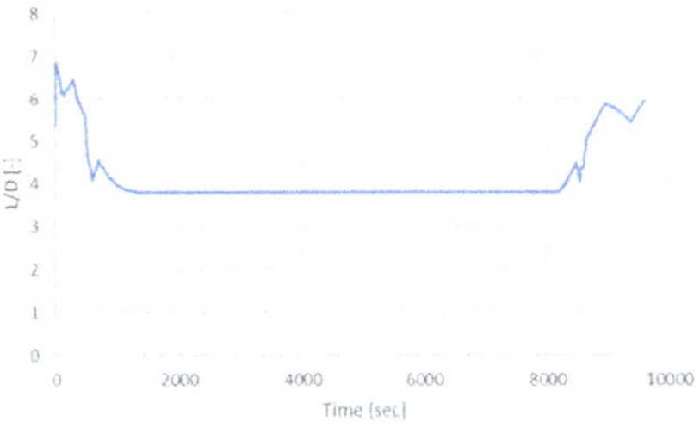

Fig. 10 Lift to Drag ratio vs time

CO_2 emission standard

The CO_2 emission standard is based on *Specific Air Range (SAR)* in cruise flight and *Reference Geometric Factor (RGF)* as presented in the following equation [5], [6]:

$$CO_2\ Metric\ Value = \frac{\left(\frac{1}{SAR}\right)_{avg}}{RGF^{0.24}}$$

Where $SAR = kilometer\ range/unit\ of\ kg\ fuel$ is a cruise point fuel burn performance while *RGF* is just a measure of cabin size. In line with requirements for subsonic airplanes, *SAR* values were computed for 3 specific reference points, which are function of Maximum Take-Off Mass (MTOM) and are presented in Fig. 11:

1. $High\ mass\ point = 0.92 \cdot MTOM$
2. $Low\ mass\ point = (0.45 \cdot MTOM) + (0.63 \cdot MTOM^{0.924})$
3. $Mid\ mass\ point =$ average of high and low

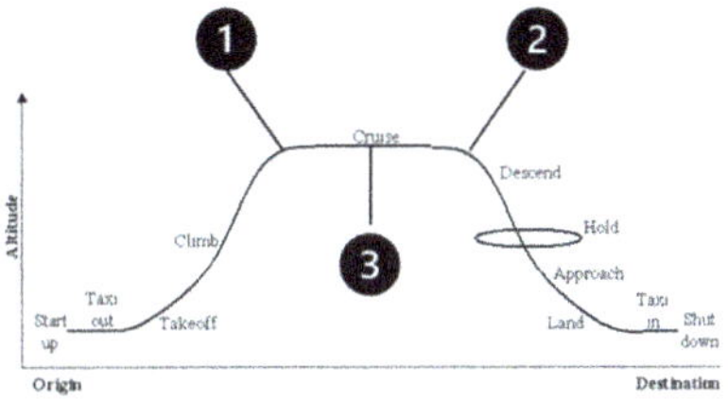

Fig. 11 Mass points for a subsonic mission

Since for supersonic aircrafts these reference points may not be representative of cruise conditions, *SAR* was evaluated at modified points, so that the high and low mass points coincided with the actual cruise start and end conditions. The evaluated CO_2 metric value for both the subsonic reference mass points and the modified ones is reported in Fig. 12. These results are compared to the CO_2 limits for subsonic aircrafts (reported with continuous lines) and to other supersonic concepts, such as a Mach 2 passenger aircraft, a Mach 1.4 and a Mach 1.6 business jet concepts.

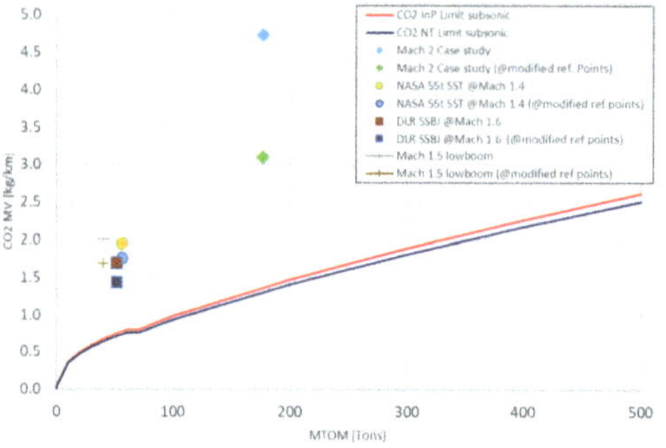

Fig. 12 CO_2 Metric value

Conclusion

A Mach 1.5 low-boom supersonic business jet concept has been analyzed in this study. The aerodynamic characteristics and mission performance of such aircraft have been computed at a preliminary design level. The capability to cover a range up to 3800 km was verified. Moreover, the Co2 metric value has been computed according to present regulations. Eventually, a comparison with other supersonic concepts has shown that the evaluated metric value has some similarities with the values computed for those aircrafts.

References

[1] D. Durston, "NTRS - NASA Technical Reports Server," 12 June 2022. [Online]. Available: https://ntrs.nasa.gov/api/citations/20220009235/downloads/Supersonics%20talk%20to%20AirVenture%2C%20Durston%20July%202022%20no%20vid.pdf.

[2] Y. Sun and H. Smith, "Low-boom low-drag solutions through the evaluation of different supersonic business jet concepts," *The Aeronautical Journal,* vol. 124, no. 127, pp. 76-95. https://doi.org/10.1017/aer.2019.131

[3] I. ANSYS, ANSYS ICEM CFD Tutorial Manual, Canonsburg, PA 15317, 2012.

[4] I. ANSYS, ANSYS FLUENT User's Guide, Canonsburg, PA 15317, 2010.

[5] ICAO, ICAO Annex 16 Vol III.

[6] ICAO, "ICAO Aircraft CO2 Emissions Standard," [Online]. Available: https://www.icao.int/environmentalprotection/Documents/CO2%20Metric%20System%20-%20Information%20Sheet.pdf. [Accessed 30 06 2023].

Effects of different drag laws on ice crystals impingement on probes mounted on a fuselage

A. Carozza[1*], P.L. Vitagliano[1], G. Mingione[1]

[1]Centro Italiano Ricerche Aerospaziali, Dipartimento di Meccanica dei Fluidi, via Maiorise snc, Capua, Italia

*a.carozza@cira.it

Abstract. In this work the effects of different drag laws regarding the ice crystals impingement on the fuselage of a regional aircraft are investigated. Different probes are considered on the surface of interest. Along each of these instrumentations the collection efficiency has been calculated by using a RANS structured solver named UZEN and an eulerian impingement code IMP3D, both developed internally at CIRA. The solvers are parallelized and well assessed. The computational grid has been generated with ICEM CFD. Results show the strong influence of the shape considered for the ice crystal particles. Results are shown in terms of collection efficiency and total ice mass collected.

Keywords: Impingement, Probes, Ice Crystals, Eulerian Scheme

Nomenclature

Latin symbols

A	surface area of the particle	[m^2]
d	particle diameter	[μm]
MVD	mean volume diameter	[μm]
X	x axis - coordinate of probe	[m]
Y	y axis - coordinate of probe	[m]
Z	z axis - coordinate of probe	[m]
V	ice crystals velocity	[m/s]

Greek symbols

β_{inst}	installation coefficient, ratio of the local particle concentration to the upstream particle concentration, is calculated along the direction normal to the fuselage	
y+	dimensionless wall distance	[m]
Φ	$\frac{\pi d_p^{2}}{A}$	[0,1]

Subscripts

p	particle

Introduction

Icing is a major hazard for aviation safety. In flight icing is caused by water droplets that froze after impacting on aircraft surface. Ice crystals were assumed to not be an hazard for in flight aircraft operations, nevertheless in the last decades an additional risk has been identified when flying in clouds with high concentrations of ice-crystals where ice accretion may occur on warm parts of the engine core, resulting in engine incidents such as loss of engine thrust, strong vibrations, blade damage, or even the inability to restart engines. Ice crystals can also accumulate on probe: on July 2029 an A320 from Paris to Rio de Janeiro had a fatal incident, investigators demonstrated that the main cause of the incident was ice formation on pitot tube due to a very strong concentration of ice crystals, that caused provision of false information on pilots instruments that caused the incident. Performing physical engine tests in icing wind tunnels is extremely challenging, expensive and currently limited to partial tests for engine components and in addition very few facilities are able to simulate ice crystals.

The need for the European aeronautics industry to use numerical simulation tools able to accurately predict ICI (Ice Crystal Icing) is therefore urgent and paramount, especially regarding the development of the new generation engines (UHBR, CROR, ATP) which are expected to be even more sensitive to the ICI threat than current in-service engines and for which comparative analysis methods will not be applicable any more.

MUSIC-HAIC research project is devoted to complete the development of ICI models, implement them in existing industrial 3D multidisciplinary tools, and perform extensive validation of the new ICI numerical capability through comparison of numerical results with both academic and industrial experimental data.

The resulting capability will allow the replacement of physical tests by cheaper virtual tests, which would be easier to configure and run, allowing substantial gains in development costs and more design choices to be explored and de-risked.

Most importantly, MUSIC-HAIC will provide the aeronautical sector with the confidence to move away from a step-by-step incremental evolution of engine design to a more radical breakthrough approach, because the ability to simulate the behaviour of ICI on these designs with a high degree of confidence will be available. This will reinforce the competitiveness of the European aircraft and engine manufacturers. MUSIC-HAIC will also enhance the expertise of the scientific and research community on ICI.

The scope of this work is to show the effects of the sphericity and deformation of a particle on the collection efficiency when ice crystals with high MVD are considered. At this aim different drag laws have been implemented in the Eulerian impingement code developed at CIRA IMP3D.

Methods

Since the end of the 1990s CIRA has invested in the study, development and application of numerical methods for the analysis of in flight icing conditions. Over the years, the presence of the in-house IWT facility has promoted the development of tools for supporting experiments as well as offering a viable and low-cost means to industrial customers. Besides, the participation in EU-funded projects has implied a continuous upgrade of in-house competences, allowing at the same time the monitoring of the most promising finds. Currently, almost all the project's proposals on the icing topic include, in addition to the experimental measurements, work packages dedicated to numerical analyses and/or the study of new simulation techniques.

For 2D ice accretion a very fast low-order model, was developed, the Multi-Ice tool, which solves two-dimensional potential flow. It is a classic panel method often used for the aerodynamic analysis of single- and multi-element airfoils, capable to compute impingement and ice accretion on each component. In case of 3D complex geometries two software have been developed at CIRA: UZEN and SIMBA [7][8]. They solve the compressible RANS equations on two- and three-dimensional block-structured (UZEN), and Cartesian meshes (SIMBA).

Water droplets trajectories can be calculated by using IMP3D and SIMBICE-ICE, respectively coupled to UZEN and SIMBA. They both solve the water phase by using a Eulerian approach.

For the objective of this study the in-house Eulerian solver named IMP3D is available to estimate the ice crystal impingement around 2D and 3D geometries. Several models can be used for the drag coefficient C_d of the ice crystals as a function of the particle sphericity [2] and the Reynolds number based on the particle equivalent diameter. It is crucial for the trajectories of ice crystals.

In this work the following drag correlations have been used:

1. Clift and Gauvin [1][3], with no sphericity (Φ=1)
2. Ganser et al. [4],
3. Haider and Levenspiel [5],
4. Hölzer and Sommerfeld [6],
5. Nakayama et al. with the effect of the surface tension (EXTICE) on Φ.

On the other hand, evaporation and melting processes have been simulated within an Eulerian formulation by means of source terms for mass and energy transfer.

Results and Discussion

A research configuration of fuselage has been considered with five probe locations. Indeed, the probe blockage represents a risk pilots face when they travel in icing conditions. In order to have a representative overview of what recent numerical tools can predict on these instrumentations in terms of ice impingement the effects of several drag correlations are shown.

Four probes were considered: a pitot probe, a total temperature probe and two angle of attack probes. In addition, it was decided to add a fifth location for post-processing at the nose of the fuselage.

The probes locations on the aircraft nose are provided in Table 1.

Table 1: Coordinates of the probes on the fuselage configuration for post-processing purposes

	X [m]	Y [m]	Z [m]
PT2	8.9195	1.2950	-1.1660
TAT2	9.5832	0.6389	-1.9203
AoA1	13.7037	2.6183	-0.6364
AoA2	13.6996	2.5187	-0.9685
Nose	6.3825	0	0.2925

The aircraft is assumed to be flying at Mach number 0.78 at altitude 34000 ft. corresponding to an atmospheric pressure of 25000 Pa. The angle of attack is 2.05 degree (Table 2).

Table 2: Flight conditions

Altitude [ft]	Angle of attack [°]	Mach	Pressure [Pa]	Temperature [K]
34000	2.05	0.78	25000.	233.15

The impingement has been computed considering the MMD and the IWC indicated in Table 3 without considering droplets but only ice crystals.

Table 3: Ice crystal characteristics

Mean Mass Diameter [μm]	Ice Water Content [g/m^3]
336.5	1.15

Only half of the configuration is meshed and a symmetry plane condition is used. The geometry shown in the Figure 1 is meshed using ANSYS/ICEMCFD. The limits of the computational domain surrounding the geometry are located at a distance of 30 times the length of the nose. The ratio between two adjacent cells is equal or lower than 1.2. The height of the first cell at the vicinity of the wall is 0.006 mm. 50 cells have been used to capture the boundary layer characteristics near the wall and in direction perpendicular to it. A Navier-Stokes multiblock structured mesh of about 18 Million of cells and 26 blocks has been generated.

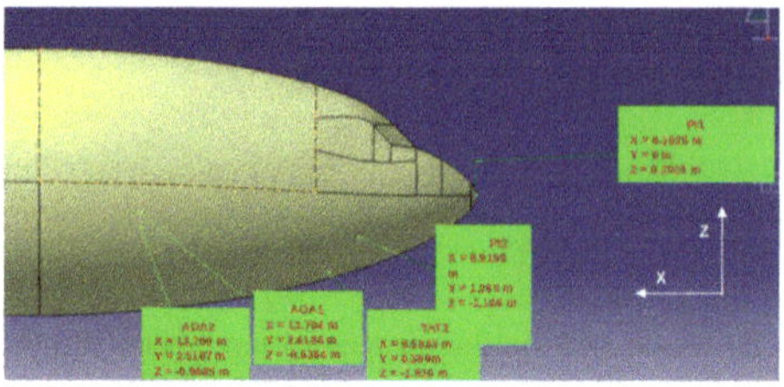

Figure 1: nose fuselage configuration

Results show that the wall y+ lies between 0.5 and 1.0, Figure 2. Pressure coefficient is shown in Figure 3 while a collection efficiency distribution is plotted in Figure 4.

The computation of the aerodynamic flow field around the geometry made of nose and fuselage is realized using the in house UZEN Multi-block CFD simulation software. The Reynold number based on the length of the geometry nose is 5.9×10^6. The aerodynamic flow field is computed using RANS modelling with TNT k-ω turbulence model in fully turbulent approach.

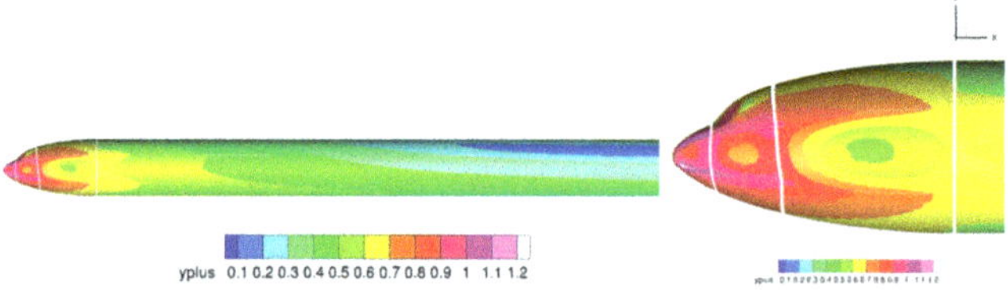

Figure 2 y+ contours

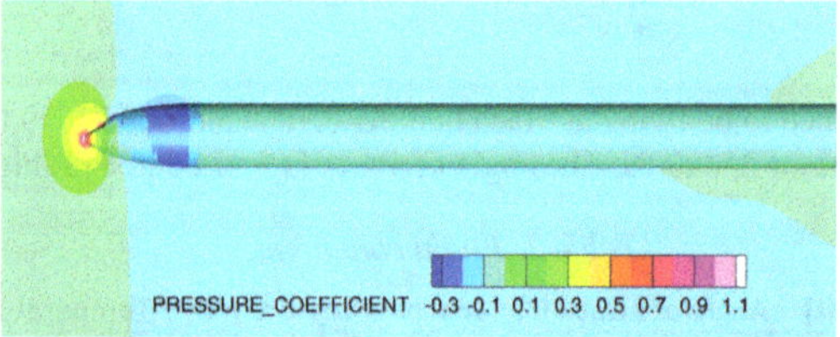

Figure 3 Pressure coefficient C_p contours

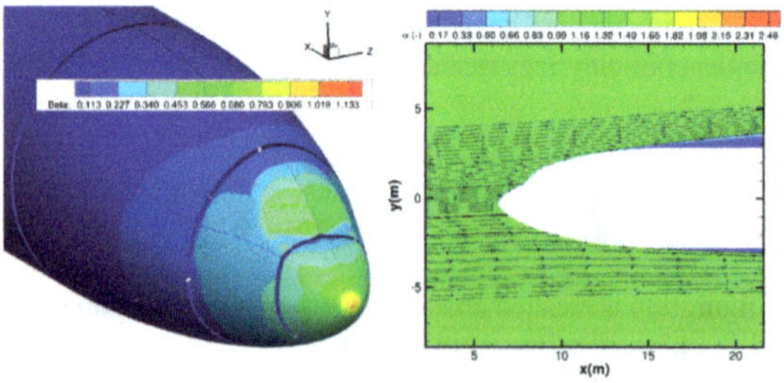

Bucknell (with fragmentation threshold)

Figure 4 Ice crystals collection efficiency using the Bucknell impact model

Observing the collection efficiency contours, see Figure 4, it is possible to state that the two zones mainly affected by the problem of ice crystals impact are the nose and the glass in front of the cabin. The maximum value of the collection efficiency does not go over 1.15 also for this simulation, while the total mass flux deposited on the surface is about 2.12 kg/s on the whole surface. In the Figure 5 the installation coefficient β_{inst}, ratio of the local particle concentration to the upstream particle concentration, IWC, is calculated along the direction normal to the fuselage, calculated by using the Bucknell model [9] including the fragmentation without re-emission are shown.

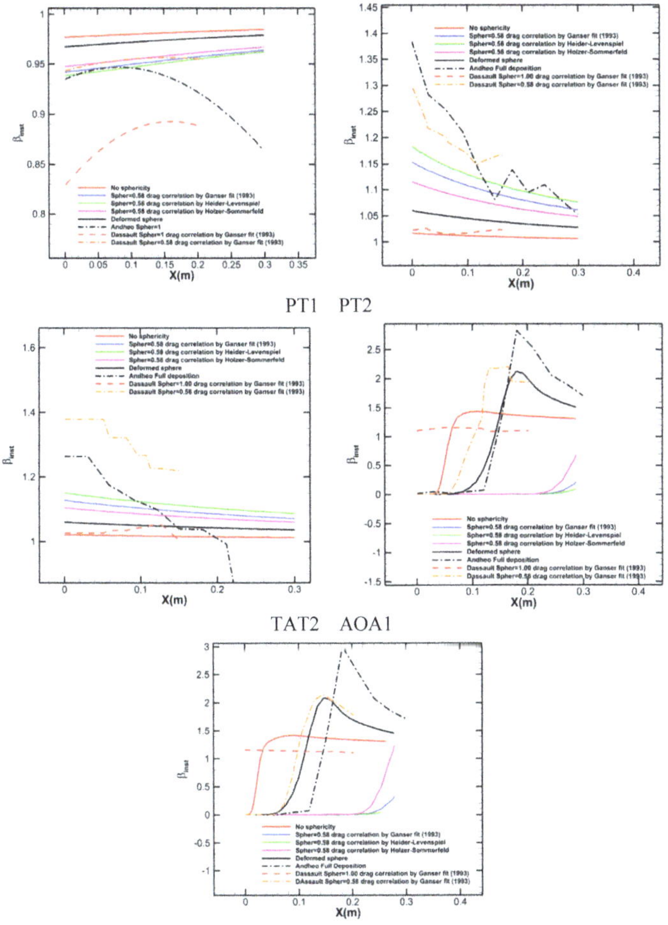

PT1 PT2

TAT2 AOA1

AOA2

Figure 5 Installation coefficient for different shapes of ice crystals along the probes axis

The installation coefficients can be looked at. Probes AoA1 and AoA2, that are the furthest ones from the nose, show the more relevant impact of the ice crystals shapes and related correlations on the installation coefficient and mass deposition along the probes. The ice crystals current is subjected to fragmentation with a larger mass loss when compared to the case with ice crystals modelled like simple spheres. Moreover, it has to be noticed that there are strong over concentrations of ice crystal particles on the probes AoA1 and AoA2 without considering the effect of particle shape. This behaviour is inverted when probes Nose, PT2 and TAT2 are considered.

Some comparisons have also been added in the installation coefficient plots, considering data coming from Andheo or Dassault companies. They are in agreement with data computed by the authors.

Conclusions

An impingement study has been carried out on some probes installed on a fuselage. At this aim a RANS finite difference code and a Eulerian scheme impingement 3D code have been used on the same computational grid generated around a fuselage. Different drag laws related to the ice crystals injected in the domain have been adopted in order to assess the sphericity importance in the collection efficiency of a specific geometry of aeronautical interest. The main conclusions are:

1. Sphericity has a relevant rule into determining the impingement values and the shadow zones
2. It generates a variable water concentration of ice crystals along the probes axis

Acknowledgments

Authors would thank Andheo and Dassault for their support and permission to use the data collected in the Eu project Music-haic.

References

[1] Clift R, Grace JR, Weber ME. Bubbles, drops, and particles. New York: Academic Press 1978.

[2] E. Loth, Drag of non-spherical solid particles of regular and irregular shape, Powder Technology 182 (2008) 342–353

[3] Clift, R., and Gauvin, W.H., The motion of particles in turbulent gas streams, Proceedings CHEMECA 1970, Butterworth, Melbourne, 1, 14-28, 1970.

[4] G.H. Ganser, A rational approach to drag prediction of spherical and nonspherical particles, Powder Technol. 77 (1993) 143.

[5] A. Haider, O. Levenspiel, Drag coefficient and terminal velocity of spherical and non-spherical particles, Powder Technol. 58 (1989) 63–70.

[6] A. Hölzer, M. Sommerfeld, New simple correlation formula for the drag coefficient of non-spherical particles, Powder Technology 184 (2008) 361–365

[7] C. Marongiu, P. Catalano, M. Amato, G. Iaccarino, U-ZEN : A computational tool solving U-RANS equations for industrial unsteady applications, 34th AIAA FluidDynamics Conference, Portland (Or), June 28 -July 1 2004, AIAA Paper 2004–2345.

[8] Capizzano et al., CIRA contribution to the first AIAA Ice Prediction Workshop, AIAA AVIATION 2022 Forum, https://doi.org/10.2514/6.2022-3400

[9] Bucknell, A., McGilvray, M., Gillespie, D., Yang, X. et al., "ICICLE: A Model for Glaciated & Mixed Phase Icing for Application to Aircraft Engines," SAE Technical Paper 2019-01-1969, 2019, https://doi.org/10.4271/2019-01-1969.

Morphing technology for gust alleviation: an UAS application

Fernando Montano[1,a*], Vincenzo Gulizzi[1,b] and Ivano Benedetti[1,c]

[1]Department of Engineering, University of Palermo, Viale delle Scienze, Edificio 8, 90128, Palermo, Italy

[a]fernando.montano@unipa.it, [b]vincenzo.gulizzi@unipa.it, [c]ivano.benedetti@unipa.it

Keywords: Gust Alleviation, Morphing Wings, UAVs

Abstract. Atmospheric turbulence can significantly affect aircraft missions in terms of aerodynamic loads and vibration. These effects are particularly meaningful for MALE-HALE UAS because of their high aspect ratios and because of their low speed, sometimes comparable with that of the gust itself. Many studies have been conducted to reach the goal of efficient gust alleviation. A viable solution appears the application of morphing technology. However, the design of morphing aircraft is a strongly multidisciplinary effort involving different expertise from structures to aerodynamics and flight control. In this study, a multidisciplinary wing-and-tail morphing strategy is proposed for attaining gust attenuation in UAVs. The strategy is based on the combined use of: i) an automatic detection system that identifies gust direction and entity and ii) an aeroelastic model stemming from the coupling between a high-order structural model that is able to resolve the motion and the strain and stress distributions of wings with complex internal structures and a Vortex Lattice Method (VLM) model that accounts for the aerodynamics of the wing-tail system. The gust alleviation strategy employs the information from the detection system and the aeroelastic model to determine the modifications of the wing and the tail surfaces aimed at contrasting wind effects, reducing induced loads and flight path errors. Numerical results are presented to assess the capability of the framework.

Introduction

UAS flight is widely affected by atmospheric turbulence for two principal motivations: small dimensions and slow speed. This is more pregnant especially for small UAS because they flight at low altitude and have speed components comparable with atmospheric ones. For the above-mentioned reasons, it is very important to devise an efficient gust alleviation strategy to achieve safe flight conditions. This requires a two-step process: first, wind components need to be accurately identified. Second, suitable actions must be taken on the lifting surfaces to counteract the effect of the gust. Among the various approaches proposed in the literature, morphing appears a viable option to implement the latter step [1, 2].

Morphing covers all those technologies that result in a continuous shape variation of one or more elements of the aircraft during flight, aimed at obtaining maximum performance in multiple flight phases. For example, morphing technologies are increasingly applied in micro unmanned aircraft (MAVs) [3].

Airfoil modifications are morphing technologies that allow to change the camber or the thickness of the wing profile during flight.

This approach could be efficiently applied to gust alleviation modifying aerodynamic surfaces to generate aerodynamic coefficient modification to contrast wind induced ones.

Wind identification algorithm

An accurate non-linear mathematical model based on the classical rigid body equations of motion in body axes has been used [4]. The identification algorithm is based on an Extended Kalman Filter (EKF) [5, 6, 7], in which the corrector employs a set of measurements gathered in turbulent air.

The measurement vector is composed of the airspeed (V), pitch rate (q), elevation angle (θ) and the spatial coordinate of the center of mass (x; h).

As it is well known, the wind components modify the airspeed (V), the angle of attack (α) and the pitch rate (q) as follows:

$$V = \sqrt{(u + u_g)^2 + (w + w_g)^2} \qquad (1)$$

$$\alpha = atan\frac{w+w_g}{u+u_g} \qquad (2)$$

$$q = q + q_q \qquad (3)$$

where (u_g, w_g, q_g) are the unknown wind components in body axes.

To tune the EKF, an optimization procedure, based on the control of prediction errors, has been used. In previous paper [6, 7] authors demonstrated the robustness of the tuning procedure, which allows to identify various kind of atmospheric disturbances with appreciable accuracy. In fact, it has been demonstrated that the algorithm is able to identify either infinite step gusts or finite ones.

Vortex Lattice Method

The aerodynamic model is based on the Vortex Lattice Method (VLM). As it is well known, VLM is a numerical method used in computational fluid dynamics based on the following assumptions:

- the flow field is incompressible, inviscid and irrotational;
- the lifting surfaces are thin, the influence of thickness on aerodynamic forces are neglected;
- the angle of attack and the angle of sideslip are both small, small angle approximation;
- the wing (or tail) is replaced by a lifting surface.

Morphing technology

In this research a combined wing and tail morphing technique is proposed. The morphing approach allows to modify wing and tail camber either simultaneously or not, in terms of both position of maximum camber and camber itself, depending on the gust entity and loads.

In Fig.°1 and Fig.°2 examples of camber modification of wing airfoil are reported.

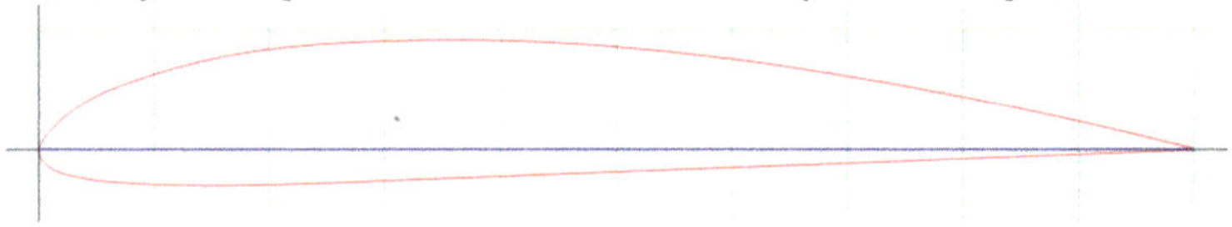

Figure 1: Clark Y wing airfoil

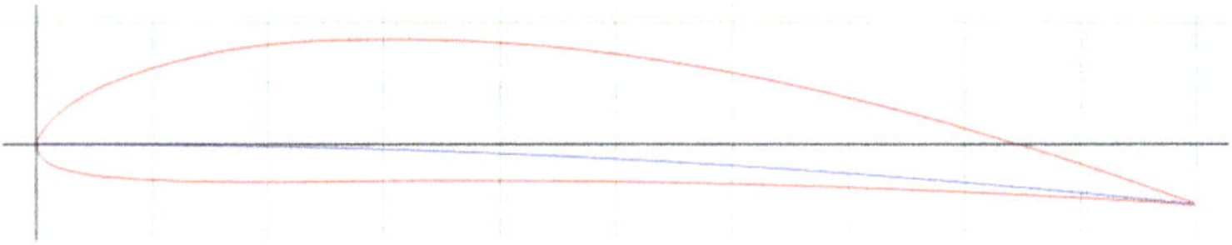

Figure 2: Clark Y wing airfoil with maximum camber variation

Materials Research Proceedings 37 (2023) 133-136 https://doi.org/10.21741/9781644902813-29

Simulations and results

The proposed procedure has been applied to a UAS that is a 1:5 scale model of a real airplane. Its geometrical and weight features are well known as it is the same model studied in previous applications [5, 6, 7].

Using VLM, various wind-tail geometric configurations have been modeled. These ones can be applied to the UAS during flight with morphing techniques. All obtained data has been analyzed and a polynomial relationship between the C_L and camber has been extracted.

The obtained relationship has been inserted in an algorithm developed in MATLAB environment. Such an algorithm identifies the configuration that UAS needs to reject the external disturbance.

Various simulations have been performed and results show that, when the wind identification procedure identify the external disturbance (in few time steps), the gust alleviation system comes into operation modifying aerodynamic surfaces to reach the goal of the reduction of gust induced variation of angle of attack.

Fig.°3 shows the desired flight path and the real flight path for a landing procedure in turbulent air, while Fig.°4 shows the tracking error.

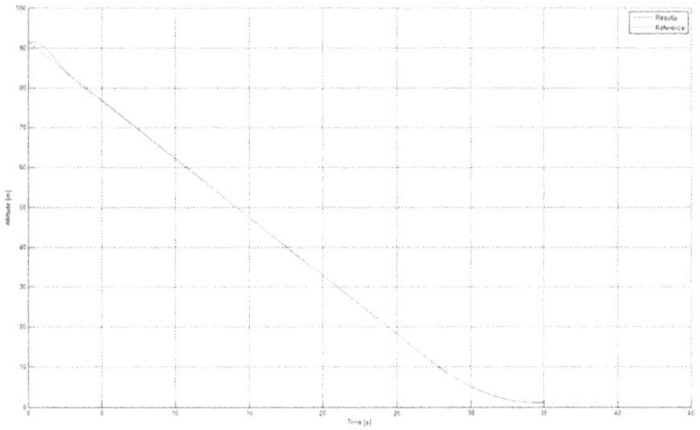

Figure 3: Desired and controlled flight path

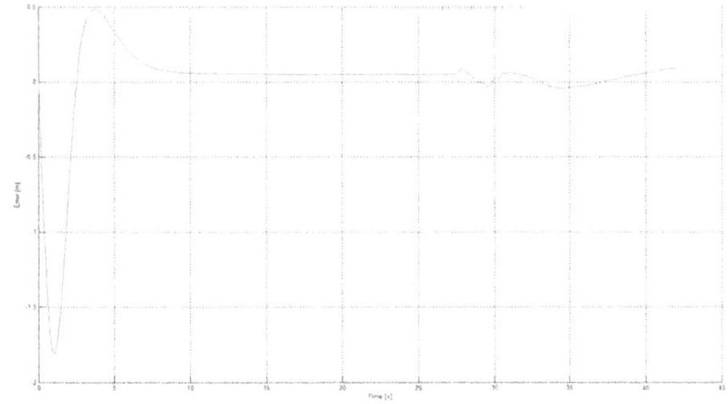

Figure 4: Tracking error

Acknowledgements
FM and IB acknowledge the support of the European Union through the FESR o FSE, PON Ricerca e Innovazione 2014-2020 - DM 1062/2021 co-funding scheme.

References

[1] D. Li, S. Zhao, A. Da Ronch, J. Xiang, J. Drofelnik, Y. Li, L. Zhang, Y. Wu, M. Kintscher, H. P. Monner, A. Rudenko, S. Guo, W. Yin, J. Kirn, S. Storm, R. De Breuker, A review of modelling and analysis of morphing wings, Progress in Aerospace Sciences, 100 (2018), 46-62. https://doi.org/10.1016/j.paerosci.2018.06.002

[2] R. M. Ajaj, M. S. Parancheerivilakkathil, M. Amoozgar, M. I. Friswell, W. J. Cantwell, Recent developments in the aeroelasticity of morphing aircraft, Progress in Aerospace Sciences, 120 (2021), 100682. https://doi.org/10.1016/j.paerosci.2020.100682

[3] R. Bardera, A. Rodríguez-Sevillano, A. García-Magariño, Aerodynamic investigation of a morphing wing for micro air vehicle by means of PIV, Fluids, Volume 5, issue 4 191 (2020). https://doi.org/10.3390/fluids5040191

[4] B. Etkin, Dynamics of atmospheric flight, J. Wiley & Sons, New York, 1972

[5] C. Grillo, F. Montano, F., Automatic EKF tuning for UAS path following in turbulent air, International Review of Aerospace Engineering, 11(6) (2018), 241-246. https://doi.org/10.15866/irease.v11i6.15122

[6] C. Grillo, F. Montano, Wind component estimation for UAS flying in turbulent air, Aerospace Science and Technology, Volume 93 (2019), 105317. https://doi.org/10.1016/j.ast.2019.105317

[7] F. Montano, I. Benedetti, Automatic wind identification for UAS: a case study, Modelling Progress in Aerospace Science (MPAS 2022) in ICNAAM 2022, Heraklion, Greece, 19th-25th September 2022

Materials Research Proceedings 37 (2023) 137-139 https://doi.org/10.21741/9781644902813-30

Multidisciplinary design, analysis and optimization of fixed-wing airborne wind energy systems

Filippo Trevisi[1,a *], Alessandro Croce[1,b] and Carlo Emanuele Dionigi Riboldi[1,c]

[1]Department of Aerospace Science and Technology, Politecnico di Milano, Via La Masa 34, 20156 Milan, Italy

[a]filippo.trevisi@polimi.it, [b]alessandro.croce@polimi.it, [c]carlo.riboldi@polimi.it

Keywords: AWE, MDAO, Wind Energy, Flight Mechanics

Abstract. Airborne wind energy (AWE) is the second generation of wind energy systems, an innovative technology which accesses the large untapped wind resource potential at high altitudes. It enables to harvest wind power at lower carbon intensity and, eventually, at lower costs compared to conventional wind technologies. The design of such systems is still uncertain and companies and research institutions are focusing on multiple concepts. To explore the design space, a new multidisciplinary design, analysis and optimization framework for fixed-wing airborne wind energy systems (T-GliDe) is being developed. In this work, the framework of T-GliDe and its problem formulation are introduced.

Introduction

Airborne Wind Energy (AWE) refers to the field of wind energy in which tethered airborne systems are used to harvest wind power at high altitudes. Compared to conventional wind energy, AWE opens new areas for energy from wind, offers increased energy generated per square kilometer, has the potential to provide energy at lower cost and has lower environmental impact [1].

Airborne Wind Energy Systems (AWESs) are typically classified based on their flight operations, which can be crosswind, tether-aligned and rotational as described by Vermillion at al. [2]. Electric power is generated with onboard wind turbines and transferred to the ground through the tether (Fly-Gen) or generated directly on the ground by a moving or fixed ground station (Ground-Gen). This work focuses on crosswind AWESs and the results are applicable to both Ground-Gen and Fly-Gen systems which are characterized by a single fixed wing.

The design of such systems is still uncertain and companies and research institutions are focusing on multiple concepts. To explore the design space and perform a robust design, the usage of MDAO techniques is crucial. This paper aims at introducing the underdevelopment MDAO framework T-GliDe (Tethered Gliding systems Design) [3] and the related problem formulation.

T-GliDe architecture and problem formulation

T-GliDe features an optimization module and an uncertainty quantification module, allowing for a number of algorithm-based design techniques (Fig. 1). The disciplines currently involved are related to the flight dynamics [4], to the optimal control [5], to the structural design, to the aerodynamics [6] and to the economics [7].

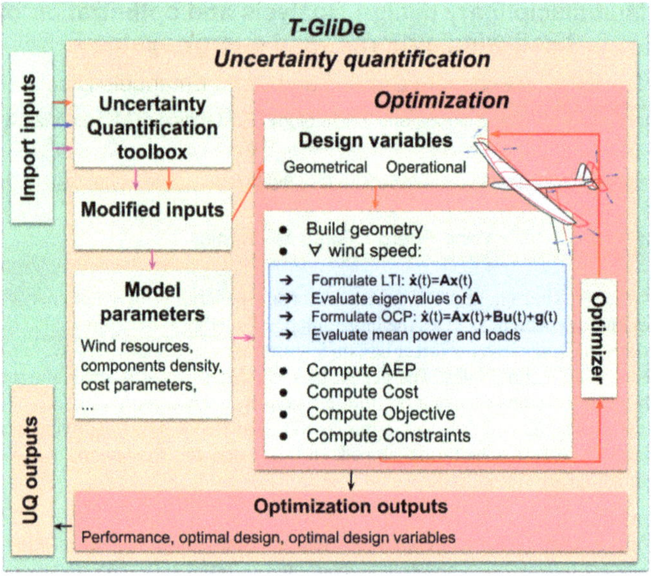

Figure 1: T-GliDe architecture.

The flight dynamics is modelled by linearizing the 6 d.o.f. equations of motion about a fictitious steady state on the circular trajectory where the fluctuating terms $\boldsymbol{g}(t)$ (e.g. gravity) are treated as disturbances. In this way, the dynamics is reduced to a linear time-invariant system

$$\dot{\boldsymbol{x}}(t) = \boldsymbol{A}\boldsymbol{x}(t) \quad (1)$$

and the eigenvalues of $\boldsymbol{A}$ are designed to be stable by the optimizer.

The fluctuating terms and the control inputs $\boldsymbol{u}(t)$ are added to formulate the optimal control problem (OCP)

$$\dot{\boldsymbol{x}}(t) = \boldsymbol{A}\boldsymbol{x}(t) + \boldsymbol{B}\boldsymbol{u}(t) + \boldsymbol{g}(t). \quad (2)$$

Since the OCP is periodic for a steady inflow, it is solved in the frequency domain, reducing the problem size and enhancing the physical understanding of optimal trajectories. Small circular-shaped trajectories, obtained by decreasing the airborne mass, are found to be beneficial for the dynamics, as they decrease the gravitational potential energy exchange over the loop.

Tight trajectories, however, increase the aerodynamic induction and thus decrease the aerodynamic power potential. To model this, an analytical vortex-based aerodynamic model, validated with the lifting line free vortex wake method implemented in QBlade [8], is included. These disciplines are coupled to an underdevelopment economic model based on companies' data.

The optimization problem is being built with a monolithic architecture in a "all-at-once" fashion [9], employing a gradient-based algorithm with algorithm differentiation for the derivative calculation. The uncertainty quantification module allows to study how the optimization problem is influenced globally by uncertainties in the model parameters and in the modelling, to achieve a robust design.

Both Ground-Gen and Fly-Gen Airborne Wind Energy Systems (AWESs), with fixed wing, can be designed with T-GliDe. The optimal designs are expected to perform well in the energy market, while having good dynamic qualities which could enable "soft" trajectories and relieve the control system in presence of turbulence and gusts.

Summary

In this work, the underdevelopment MDAO framework T-GliDe (Tethered Gliding systems Design) and the related problem formulation are introduced. T-GliDe features an optimization module and an uncertainty quantification module, allowing for a number of algorithm-based design techniques. The disciplines currently involved are related to the flight dynamics, to the optimal control, to the structural design, to the aerodynamics and to the economics. T-GliDe will be used to explore the design space and achieve robust conceptual designs of fixed-wing crosswind AWESs.

References

[1] Getting airborne – the need to realise the benefits of airborne wind energy for net zero. Tech. rep. BVG Associates on behalf of Airborne Wind Europe, Sept. 2022. url: https://airbornewindeurope.org/wp-content/uploads/2023/03/BVGA-Getting-Airborne-White-Paper-220929.pdf

[2] Vermillion, C., Cobb, M., Fagiano, L., Leuthold, R., Diehl, M., Smith, R. S., Wood, T. A., Rapp, S., Schmehl, R., Olinger, D., and Demetriou, M.: Electricity in the air: Insights from two decades of advanced control research and experimental flight testing of airborne wind energy systems, Annual Reviews in Control, 52, 330–357, 2021. https://doi.org/10.1016/j.arcontrol.2021.03.002

[3] Trevisi, F., Riboldi, C. E. D., and Croce, A.: Sensitivity analysis of a Ground-Gen Airborne Wind Energy System design., Journal of Physics: Conference Series, 2265, 042 067, 2022. https://doi.org/10.1088/1742-6596/2265/4/042067

[4] Trevisi, F., Croce, A., and Riboldi, C. E. D.: Flight Stability of Rigid Wing Airborne Wind Energy Systems, Energies, 14, 2021. https://doi.org/10.3390/en14227704

[5] Trevisi, F., Castro-Fernández, I., Pasquinelli, G., Riboldi, C. E. D., and Croce, A.: Flight trajectory optimization of Fly-Gen airborne wind energy systems through a harmonic balance method, Wind Energy Science, 7, 2039–2058, 2022. https://doi.org/10.5194/wes-7-2039-2022

[6] Trevisi, F., Croce, A. and Riboldi, C. E. D.: Vortex model of the aerodynamic wake of airborne wind energy systems, Wind Energy Science, 8, 999-1016, 2023. https://doi.org/10.5194/wes-8-999-2023

[7] R. Joshi, R., Trevisi, F., Schmehl, R., Croce, A., and Riboldi, C.: A Reference Economic Model for Airborne Wind Energy Systems, Presented at the Airborne Wind Energy Conference (AWEC 2021), Milan, 22-25 June, 2022. http://resolver.tudelft.nl/uuid:3e9a9b47-da91-451b-b0af-2c26c7ff9612

[8] Marten, D., Lennie, M., Pechlivanoglou, G., Nayeri, C. N., and Paschereit, C. O.: Implementation, Optimization and Validation of a Nonlinear Lifting Line Free Vortex Wake Module Within the Wind Turbine Simulation Code QBlade, Volume 9: Oil and Gas Applications; Supercritical CO2 Power Cycles; Wind Energy, 2015. https://doi.org/10.1115/GT2015-43265

[9] Martins, J. R. R. A. and Lambe, B.: Multidisciplinary Design Optimization: A Survey of Architectures, AIAA Journal 2013 51:9, 2049-2075

Artificial Intelligence Application

Refinement of structural theories for composite shells through convolutional neural networks

Marco Petrolo[1,a *], Pierluigi Iannotti[1,b], Mattia Trombini[1,c] and Mattia Melis[d]

[1]MUL2 Lab, Department of Mechanical and Aerospace Engineering, Politecnico di Torino, Corso Duca degli Abruzzi 24, 10129, Torino, Italy

[a]marco.petrolo@polito.it, [b]pierluigi.iannotti@polito.it, [c]mattia.trombini@polito.it, [d]mattia.melis@studenti.polito.it

Keywords: CUF, Structural Theories, Neural Networks, FEM

Abstract. This study examines the use of Convolutional Neural Networks (CNN) to determine the optimal structural theories to adopt for the modeling of composite shells, to combine accuracy and computational efficiency. The use of the Axiomatic/Asymptotic Method (AAM) on higher-order theories (HOT) based on polynomial expansions can be cumbersome due to the amount of Finite Element Models (FEM) virtually available and the problem-dependency of a theory's performance. Adopting the Carrera Unified Formulation (CUF) can mitigate this obstacle through its procedural and lean derivation of the required structural results. At the same time, the CNN can act as a surrogate model to guide the selection process. The network can inform on the convenience of a specific set of generalized variables after being trained with just a small percentage of the results typically required by the AAM. The CNN capabilities are compared to the AAM through the Best Theory Diagram (BTD) obtained using different selection criteria: errors over natural frequencies or failure indexes.

Introduction

The modeling of composite structures involves the balancing between accuracy and computational costs. Focusing on 2D models, using refined theories based on higher-order polynomial thickness expansions [1,2] is particularly useful in describing crucial aspects such as transverse anisotropy and shear deformability. However, their accuracy is strictly problem-dependent, and their variety is virtually unlimited.

The Axiomatic/Asymptotic Method [3-6] can be used to identify the best models for different levels of numerical complexity. The AAM starts with selecting a maximum order for the polynomial expansion and then gradually suppresses its terms. The resulting models are then compared to a reference using a control parameter, e.g., displacements or frequencies. Different theories can emerge as optimal for the same structural problem depending on the control parameter chosen. Implementing the AAM may be cumbersome due to the vast number of results required, becoming even less manageable for complex structures. The numerical efficiency of AAM can be augmented by Machine Learning (ML), more specifically through Convolutional Neural Networks [7]. By exploiting feature extraction capabilities, CNN can create surrogate models that identify the best theories at a fraction of the cost required by the AAM. This result is achieved by drastically reducing the FEM analyses needed to train the network successfully.

The new methodology presented in this paper is based on the Carrera Unified Formulation (CUF) [6] for deriving the finite element results used to train the network and to perform the comparison with the outcome expected from the AAM. For the structural case presented here, the best models were selected based on the accuracy in estimating different natural frequencies and failure indexes, with the results summarized through the Best Theory Diagram (BTD).

CUF and FEM Formulation
CUF efficiently obtains the governing equations and the finite element formulations for virtually any higher-order theory. For shells, the displacement field is expressed as

$$u(\alpha, \beta, z) = F_\tau(z)u_\tau(\alpha, \beta) \qquad \tau = 1, \dots, M \tag{1}$$

$F_\tau(z)$ are the expansion functions adopted along the thickness, $u_\tau(\alpha, \beta)$ is the vector of the generalized unknown displacements, and M is the total number of expansion terms. The Einstein notation is used on τ. As an example of displacement field formulation stemming from a higher-order theory, a complete fourth-order model (E4) is reported herein extended format,

$$\begin{aligned} u_\alpha &= u_{\alpha_1} + zu_{\alpha_2} + z^2u_{\alpha_3} + z^3u_{\alpha_4} + z^4u_{\alpha_5} \\ u_\beta &= u_{\beta_1} + zu_{\beta_2} + z^2u_{\beta_3} + z^3u_{\beta_4} + z^4u_{\beta_5} \\ u_z &= u_{z_1} + zu_{z_2} + z^2u_{z_3} + z^3u_{z_4} + z^4u_{z_5} \end{aligned} \tag{2}$$

From the geometrical and constitutive relations described in [8], and by applying the Principle of Virtual Displacements (PVD), the governing equation for the free-vibration problem can be derived for the k-th layer:

$$m^k_{\tau isj}\ddot{u}^k_{\tau i} + k^k_{\tau sij}u^k_{\tau i} = 0 \tag{3}$$

$k^k_{\tau sij}$ and $m^k_{\tau isj}$ are 3x3 matrices known as fundamental nuclei of the stiffness and mass matrices, respectively. By assembling all nodes and elements and introducing the harmonic solution, the complete formulation of the eigenvalue problem can be obtained,

$$(-\omega_n^2 M + K)U_n = 0 \tag{4}$$

Using the same approach for the static case, the governing equation reads:

$$k^k_{\tau sij}u^k_{\tau i} = p^k_{sj} \tag{5}$$

Here, p^k_{sj} is the fundamental nucleus for the external mechanical load. Similarly , the well-known static problem formulation is derived through the assembly procedure.

$$KU_n = P \tag{6}$$

For a more in-depth description of the assembly procedure and other mathematical details, the reader can refer to [8].

Axiomatic/Asymptotic Method
AAM selects the optimal set of expansion terms, or generalized variables, to adopt for a specific problem configuration. The aim is to find the most convenient structural theory to provide the best accuracy at the lowest computational cost. Dealing with polynomial expansions, this procedure's first step is defining the maximum order allowed. For the work presented in this paper, a maximum order of four was considered, leading to a total amount of possible theories equal to 2^{15}. This number was reduced to 2^{12} by always considering the constant terms for each of the three displacement components. The accuracy of each model can be evaluated by choosing a reference solution, e.g., the full fourth-order expansion, E4, and a control parameter. The first one

considered in this paper was the percentage error over the single natural frequency, defined as follows:

$$\%E_{f_i} = 100 \times \frac{|f_i - f_i^{E4}|}{f_i^{E4}} \tag{7}$$

where f_i^{E4} is the i-th frequency evaluated using the reference full fourth-order Taylor expansion. The second indicator adopted was the percentage error over a failure index evaluated at a specific location in the structure. The 3D Hashin criterion [9] was selected for this purpose. For example, the percentage error over the index for the matrix tension (MT) mode is

$$\%E_{MT} = 100 \times \frac{|MT - MT^{E4}|}{MT^{E4}} \tag{8}$$

The outcome of this selection procedure is summarized by the BTD, a graphical representation of the accuracy achievable by varying the number of generalized variables adopted. Each point of the BTD corresponds to the best theory, given the number of active expansion terms.

Convolutional Neural Network

A CNN able to handle multi-dimensional inputs and outputs was used. The input is represented by the set of active generalized variables identifying a specific structural theory. This information is first encoded into a series of 0 and 1, corresponding to a deactivated and active term, respectively. An example of this procedure is presented here,

$$\begin{aligned} u_\alpha &= u_{\alpha_1} + z u_{\alpha_2} + z^4 u_{\alpha_5} \\ u_\beta &= u_{\beta_1} + z u_{\beta_2} + z^3 u_{\beta_4} \\ u_z &= u_{z_1} + z u_{z_2} + z^2 u_{z_3} \end{aligned} \quad \Rightarrow \quad [111001010100] \tag{9}$$

Note that the three constant terms were not included in the sequence because they were always considered active. This sequence is then re-shaped into a 3x4 matrix, constituting the actual input to the network. The output consists of the percentage errors over the first ten natural frequencies or the two evaluated failure indexes. The complete architecture is presented in Table 1. The training of the network was performed only using 10% of all possible theories and related errors.

Table 1. Parameters and architecture of the adopted CNN.

Layer Type	Filters (Size) / Neurons	Activation Function
Convolutional	128 (3x3)	ReLU
Convolutional	128 (3x3)	ReLU
Convolutional	128 (3x3)	ReLU
Flatten	-	-
Dense	128	ReLU
Dense	128	ReLU
Output	10 or 2	Sigmoid

Numerical Results

A simply-supported shell with [0°/90°/0°] stacking sequence was considered. The curvature radii were kept equal along the two curvilinear coordinates α and β, imposing R/a=5. A thickness ratio a/h=10 was used, and the following material properties were employed: $E_{11}/E_{22}=25$, $G_{12}/E_{22}=G_{13}/E_{22}=0.5$, $G_{23}/E_{22}=0.2$, $\upsilon_{12}=\upsilon_{13}=\upsilon_{23}=0.25$. Only a quarter of the shell was modeled to reduce computational costs further. This choice required the use of symmetry boundary conditions, thus allowing to consider symmetrical vibration modes exclusively. A 4x4 mesh of Q9 elements was adopted.

Starting with the free-vibration problem, Figs. 1 and 2 show the BTDs for the first and third natural frequencies, respectively. In each of them, the results obtained from the direct application of the AAM are compared to those obtained by the CNN. Table 2 shows the best theories provided by CNN for the first five frequencies having eight active degrees of freedom; an active term is indicated by a black triangle, with their order increasing from left to right, up to the fourth.

In the case of failure indexes as control parameters, a static analysis was performed on the same structure. A bi-sinusoidal pressure of unit amplitude was applied to the top surface. Failure indexes were evaluated at the center of the top and bottom edges, for the compressive and tensile modes, respectively. Figure 3 shows the resulting BTD. In each diagram, some of the resulting best models are presented.

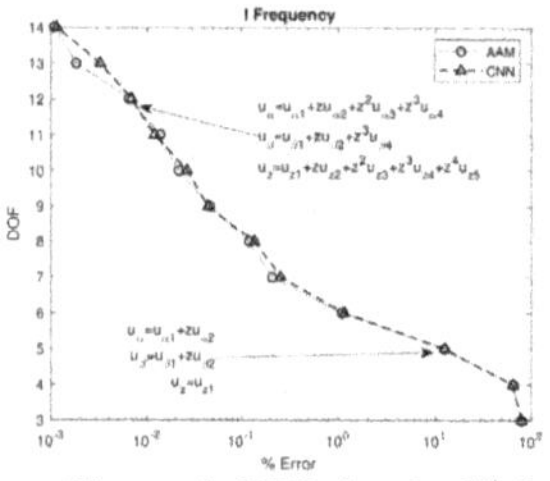

Figure 1. BTD for the 1st frequency.

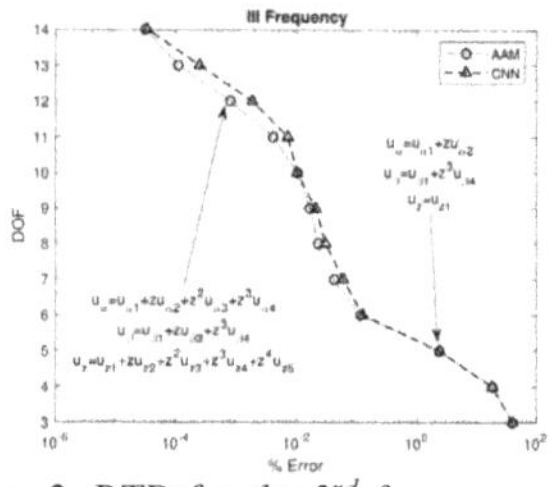

Figure 2. BTD for the 3rd frequency.

For the free-vibration case, the CNN accurately reproduced the BTD for different frequencies while also correctly providing indications regarding the relevance of specific terms. The results obtained over the failure indexes show similar levels of accuracy, with just a slight reduction when dealing with larger amounts of degrees of freedom. This behavior is related to the influence of each expansion term on the various stress components involved in the failure index estimation.

Table 2. Best models with eight active terms for the first five frequencies.

I						II						III					
u_α	▲	▲	△	▲	△	u_α	▲	▲	△	▲	△	u_α	▲	▲	△	△	△
u_β	▲	▲	△	▲	△	u_β	▲	▲	△	▲	△	u_β	▲	▲	△	▲	△
u_z	▲	△	▲	△	△	u_z	▲	△	△	△	▲	u_z	▲	▲	△	△	▲

IV						V					
u_α	▲	▲	△	▲	△	u_α	▲	▲	△	▲	△
u_β	▲	▲	△	▲	△	u_β	▲	▲	△	▲	△
u_z	▲	△	△	△	▲	u_z	▲	△	△	△	▲

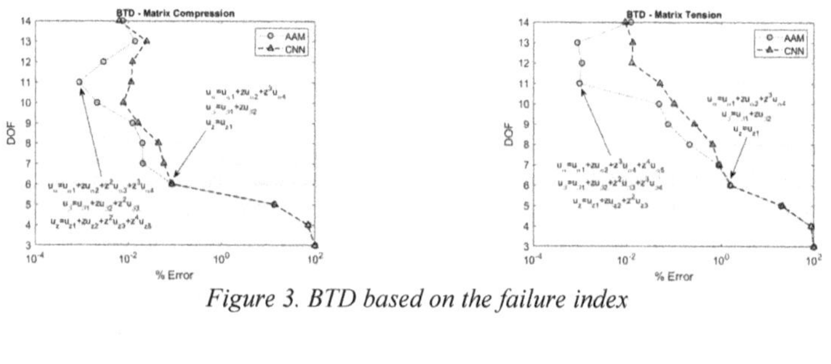

Figure 3. BTD based on the failure index

Summary

This paper explores the use of Convolutional Neural Networks in the analysis of composite shells. Focusing on higher-order theories obtained through polynomial expansions, CNN can identify the best models for various structural configurations with a fraction of the computational overhead required by the Axiomatic/Asymptotic Method. This new efficient approach can be extended to different families of problems maintaining consistent precision levels in providing the optimal modeling strategy.

References

[1] F.B. Hildebrand, E. Reissner, G. B. Thomas, Notes on the foundations of the theory of small displacements of orthotropic shells, Technical Report, Massachusetts Institute of Technology, 1949.

[2] J. N. Reddy, A simple higher-order theory for laminated composite plates, Journal of Applied Mechanics, December 1984, 51(4) 745–752. https://doi.org/10.1115/1.3167719

[3] D. S. Mashat, E. Carrera, A. M. Zenkour, S. A. Al Khateeb, Axiomatic/asymptotic evaluation of multilayered plate theories by using single and multi-points error criteria, Composite Structures, 106 (2013) 393–406. https://doi.org/10.1016/j.compstruct.2013.05.047

[4] M. Petrolo, M. Cinefra, A. Lamberti, E. Carrera, Evaluation of mixed theories for laminated plates through the axiomatic/asymptotic method, Composites Part B: Engineering, 76 (2015) 260–272. https://doi.org/10.1016/j.compositesb.2015.02.027

[5] M. Petrolo, P. Iannotti, Best Theory Diagrams for Laminated Composite Shells Based on Failure Indexes, Aerotecnica Missili & Spazio, In Press. https://doi.org/10.1007/s42496-023-00158-5

[6] M. Petrolo, E. Carrera, Best Spatial Distributions of Shell Kinematics Over 2D Meshes for Free Vibration Analyses, Aerotecnica Missili & Spazio, 99 (2020) 217-232. https://doi.org/10.1007/s42496-020-00045-3

[7] S. Albawi, T. A. Mohammed, S. Al-Zawi, Understanding of a convolutional neural network, 2017 International Conference on Engineering and Technology (ICET), Antalya, Turkey, 2017. https://doi.org/10.1109/ICEngTechnol.2017.8308186

[8] E. Carrera, M. Cinefra, M. Petrolo, E. Zappino, Finite element analysis of structures through unified formulation, Wiley, Chichester, 2014. https://doi.org/10.1002/9781118536643

[9] Z. Hashin, Failure criteria for unidirectional fiber composites, Journal of Applied Mechanics, 47 (1980) 329–334. https://doi.org/10.1115/1.3153664

Aeronautics and Astronautics - AIDAA XXVII International Congress Materials Research Forum LLC
Materials Research Proceedings 37 (2023) 146-149 https://doi.org/10.21741/9781644902813-32

Decision trees-based methods for the identification of damages in strongly damped plates for aerospace applications

Alessandro Casaburo[1,a*], Cyril Zwick[1,b], Pascal Fossat[1,c], Mohsen Ardabilian[2,d], Olivier Bareille[1,e], Franck Sosson[3,f]

[1] Laboratoire de Tribologie et Dynamique des Systèmes (LTDS), École Centrale de Lyon, 36 Av. Guy de Collongue, 69134 Écully, France

[2] Laboratoire d'Informatique en Image et Systèmes d'Information (LIRIS), École Centrale de Lyon, 36 Av. Guy de Collongue, 69134 Écully, France

[3] Materials Departement, SMAC S.A.S., 66 Impasse Edouard Branly, 83079 Toulon, France

[a]alessandro.casaburo@ec-lyon.fr, [b]cyril.zwick@ec-lyon.fr, [c]pascal.fossat@ec-lyon.fr, [d]mohsen.ardabilian@ec-lyon.fr, [e]olivier.bareille@ec-lyon.fr, [f]franck.sosson@smac.fr

Keywords: Composite Plate, Vibration Test, Damage Identification, Machine Learning

Abstract. Damage identification and localization is fundamental in industrial engineering, since it helps perform corrective actions in time to reduce as much as possible system downtime, operational costs, perform quick maintenance and avoid failure. Recently, structural health monitoring has found in machine learning an extremely useful tool, making the monitoring of complex systems more manageable. In this work, composite plates manufactured with the purpose of damping vibrations in aerospace structures are experimentally tested; the strong damping suddenly reduces the vibrations, leading to responses very similar to one another, without noticeable or structured differences between undamaged and damaged plates. To overcome this issue, machine learning methods are applied. Decision trees-based methods are chosen since they provide a combination of feature selection capabilities and robust classification performances. The used methods are decision trees themselves and two boosting methods: AdaBoost and RUSBoost. All three methods perform well in identifying damaged plates, the type (thickness damage and debonding) and sub-type of damage (thickness/debonding of types A and B).

Introduction

Damage is defined as an intrinsic change in geometrical or material characteristics of an engineering system that negatively affects its operational life, safety, reliability, and performance [1]. The detection, diagnosis, and prognosis of failure can be performed through Structural Health Monitoring (SHM). The most challenging step in SHM is damage detection, interpreted as the systematic and automatic process of finding the existence of a damage. As Yuan et al. [2] report, damage detection has been performed in SHM with two approaches up to now: physics-based and data-driven. The former becomes unreliable as the system complexity increases. Improvements in computational power and advances in information and sensing technologies allow monitoring of many parameters, which opens the path to data-driven approaches. As reported by Avci et al. [3], during the last decades, Machine Learning (ML) has been widely applied to SHM, with the objective of generating models mapping input patterns in measured sensor data to output targets for damage assessment.

This work is executed in the framework of IDEFISC (IDEntification de FISsures dans les Composites) project, aimed at the identification and quantification of damages in composite structures with the aid of machine learning methods. The test articles under investigation are composite plates consisting of three layers: a metallic, an elastomeric and a composite one, aiming to damp the vibrations in aerospace vehicles. Such strong damping leads to a sudden reduction of vibration amplitude, making the responses remarkably similar between healthy and damaged plates, impairing the damage identification task. The need of fast and reliable methods paves the way to the powerful classification capabilities of machine learning methods, exploited here to distinguish which plate is damaged and which not, but also to identify the types (reduction or debonding) and sub-types (changes in position) of damages.

Theoretical framework

Three machine learning methods are used in this work. The first method is decision trees. They have a flowchart-like structure, characterized by nodes, in which a test on an attribute is executed, and branches representing the outcome of the test [4]. Each split is executed to maximize the information gain, so that the most informative feature is used to determine the status of the sample. Decision trees are attractive models because of interpretability, they allow mixing feature types, and automatic selection of the optimal feature. However, they tend to overfit when trees are too deep. A typical approach to fight overfitting is to build a more robust model through ensemble methods: they combine several weak classifiers into a meta-classifier having better generalization performances than an individual classifier alone. There are different types of ensemble learning; the one used herein is AdaBoost (Adaptive Boosting), in which several decision trees are trained in series so that, at each iteration, the training examples are re-weighted to build a more robust classifier which learns from the errors of the previous classifier in the ensemble. The final prediction combines the outputs of all the weak learners and is taken by majority voting. A modification of AdaBoost is used as third method: RUSBoost (Random UnderSampling Boosting). It is very effective when the classes are not evenly distributed in the dataset. Instead of involving the entire training set (like AdaBoost), RUSBoost takes the basic unit for sampling equal to the number of members N in the class with the smaller number of instances in the training data.

Experimental tests and application of machine learning

The plates may present two types of purposely made damages: thickness damage, and debonding between the aluminum and elastomeric layers. In turn, each damage type appears in two sub-types, labeled A and B, in which size and location of the damages change. Eleven plates are tested, in total. Three plates are undamaged, while the remaining ones have one sub-type of damage.

Global vibration tests are executed on the plates to obtain their response in both the time and frequency domains. The test articles are clamped on the short side, they are excited with a 1.5 second sweep sine signal provided through an electrodynamic shaker on the aluminum part, ranging from 5 to 5000 Hz. Velocity measurements are performed with a Polytec LDV (Laser Doppler Vibrometer) on the composite part in 187 points. The frequency response of one point is displayed in Fig. 1: there are no noticeable differences among the several damage conditions.

The dataset is made of 10 plates and 25 features, extracted from both time and frequency measurements, such as summary statistics, frequency centroid, roll-off frequency, time of flight, etc. The features are estimated for all the measured points, thus the entire training set contains 1870 observations. Three identification tasks are performed: damage identification, damage type identification, and damage sub-type identification. Thus, a label is assigned to each observation, corresponding to the status of the plate. The generalization capabilities of the machine learning methods are checked considering one plate at time for testing. Hence, all the observations of one plate are extracted from the dataset to generate the test set, and all the remaining observations, belonging to all the other plates, are shuffled and constitute the training set. This procedure is performed several times, one for each plate.

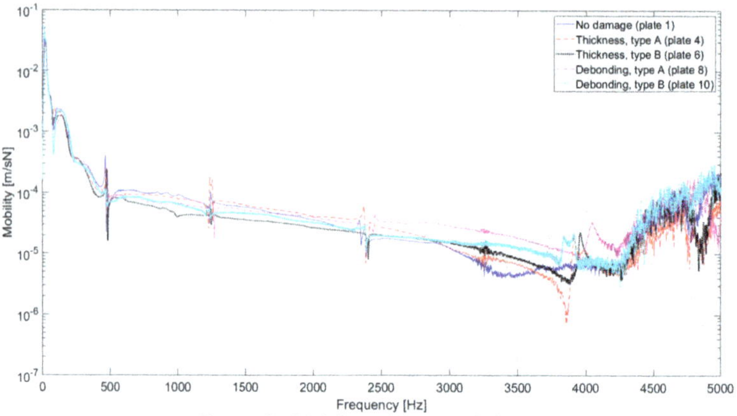

Figure 1 - Mobility of the tested plates.

For sake of brevity, the results of all the tasks are summarized in Tables 1-3, where the acronyms DT, AB, and RUSB stand for Decision Trees, AdaBoost, and RUSBoost, respectively. Plate 1, without damage, is used as baseline for the estimation of those features requiring the comparison of the test plate with an undamaged plate, thus it cannot be used for machine learning. The test plates are reported along the rows, the columns refer to the methods; each cell provides the proportion of correctly classified measurements with respect to the total number of measurements performed for each plate. The results show that not all the labels are correctly predicted: Plate 8 always provide misclassifications, and the ratio of wrong predictions increases as the complexity of the task increases. However, the overall performance capabilities are very good, since the majority of the measurements are associated with a correct prediction of the status of the plates.

Plate ID	DT	AB	RUSB
Plate 2	187/187	187/187	187/187
Plate 3	187/187	0/187	187/187
Plate 4	187/187	187/187	187/187
Plate 5	187/187	187/187	187/187
Plate 6	187/187	187/187	187/187
Plate 7	187/187	187/187	187/187
Plate 8	0/187	0/187	0/187
Plate 9	187/187	187/187	187/187
Plate 10	187/187	187/187	187/187
Plate 11	187/187	187/187	187/187

Table 1 - Classification performances of DTs, AB, and RUSB, undamaged-damaged classification task.

Plate ID	DT	AB	RUSB
Plate 2	187/187	187/187	187/187
Plate 3	187/187	0/187	187/187
Plate 4	187/187	187/187	187/187
Plate 5	187/187	187/187	187/187
Plate 6	187/187	187/187	187/187
Plate 7	187/187	187/187	187/187
Plate 8	0/187	0/187	0/187
Plate 9	187/187	187/187	187/187
Plate 10	187/187	0/187	187/187
Plate 11	0/187	187/187	187/187

Table 2 - Classification performances of DTs, AB, and RUSB, damage type identification task.

Table 3 - Classification performances of DTs, AB, and RUSB, damage subtype identification task.

Plate ID	DT	AB	RUSB
Plate 2	187/187	187/187	187/187
Plate 3	187/187	0/187	187/187
Plate 4	187/187	187/187	187/187
Plate 5	187/187	187/187	187/187
Plate 6	187/187	187/187	187/187
Plate 7	187/187	187/187	187/187
Plate 8	0/187	0/187	0/187
Plate 9	187/187	187/187	187/187
Plate 10	187/187	0/187	187/187
Plate 11	0/187	0/187	187/187

Conclusions

This work's main aim is to classify the health status of plates characterized by strong damping. Such a damping explains the responses of experimental vibration tests, remarkably similar among plates with and without damages. However, what is undistinguishable for the human eye provides valuable information for machine learning techniques. In fact, the application of three methods, namely decision trees, AdaBoost, and RUSBoost, proves that data-driven methods have excellent classification capabilities: the presence of damage itself, damage type and sub-type are correctly predicted with high accuracy. This opens the way to new, more advanced types of tasks, such as identification of smaller damages, as well as their localization.

Acknowledgments

Regarding the here-reported works, the authors would like to gratefully acknowledge for the financial support to the Region Auvergne Rhône Alpes through the scientific program RDI BOOSTER 2019 - RRA 20 010276 01 – IDEFISC.

References

[1] D. Frangopol, J. Curley, Effects of damage and redundancy of structural reliability, journal of Structural Engineering 113(7) (1987) 1533-1549. https://doi.org/10.1061/(ASCE)0733-9445(1987)113:7(1533)

[2] F. Yuan, S. Zargar, Q. Chen, S. Wang, Machine learning for structural health monitoring: challenges and opportunities, in: H. Huang (Ed.), Sensors and Smart Structures Technologies for Civil, Mechanical, and Aerospace Systems 2020, International Society for Optics and Photonics (SPIE), 2020, pp. 1-23. https://doi.org/10.1117/12.2561610

[3] O. Avci, O. Abdeljaber, S. Kiranyaz, M. Hussein, M. Gabbouij, D. Inman, A review of vibration-based damage detection in civil structures: From traditional methods to machine learning and deep learning applications, Mechanical Systems and Signal Processing 147 (2021). https://doi.org/10.1016/j.ymssp.2020.107077

[4] T. Mitchell, Machine Learning, McGraw-Hill Education, New York, 1997.

[5] S. Raschka, Y. Liu, V. Mirjalili, Machine Learning with PyTorch and Scikit-Learn, Packt, Birmingham, 2022.

SHM implementation on a RPV airplane model based on machine learning for impact detection

G. Scarselli[1,a *], F. Dipietrangelo[1,b], F. Nicassio[1,c]

[1]Department of Engineering for Innovation, University of Salento, Building O, Ekotekne, Via per Monteroni, Lecce 73100, Italy

[a]gennaro.scarselli@unisalento.it, [b]flavio.dipietrangelo@unisalento.it, [c]francesco.nicassio@unisalento.it

Keywords: Lamb Waves, RC Airplane, Impact Detection, Machine Learning

Abstract. In this work an on-working Structural Health Monitoring system for impact detection on RC airplane is proposed. The method is based on the propagation of Lamb waves in a metallic structure on which PZT sensors are bonded for receiving the corresponding signals. After the detection, Machine Learning algorithms (polynomial regression and neural networks) are applied to the data obtained by the processing of the acquired ultrasounds in order to characterize the impacts. Furthermore, this work presents the development of a mini-equipment for acquisition and data processing based on a Raspberry Pi micro-computer.

Introduction

The localisation of impacts on aerospace structures is one of the main goals of Structural Health Monitoring (SHM) systems [1]. Even a small impact at low speed can cause a crack that can become serious damage in the long run. Therefore, SHM has reached a certain level of maturity for what concerns the choice of the best sensor network for the impact detection [2],[3]. Furthermore, other works focus on the application of machine learning (ML) algorithms for the elaboration of acquired data and the prediction of the impact localisation and of the damage [4]-[8]. In the present work a ML model is built in order to characterize the real impacts on a fixed specimen in laboratory and then it is tested in the presence of vibrations due to the engine of a balsa wood RC model of the Piper J3 CUB airplane.

Experimental Setup

In a first activity, low speed impacts were performed on a 25×25 cm specimen made of aluminium alloy with density 2700 kg/m3, elastic modulus E = 72 GPa, Poisson's ratio v = 0.33 and thickness = 1.2 mm. Four piezoelectric ceramic PZT $\mathrm{Pb}[Z_{r_x}\mathrm{Ti}_{1-x}]\mathrm{O}_3$ sensors [9] (diameter equal to 10 mm) were bonded on the surface, at the four vertices of a 12.5×12.5 cm square area. The impacts were performed inside the area above by dropping a steel ball from the top of a "drop tower" built up in the AeroSpace Structural Engineering Lab (AS.S.E. Lab – University of Salento) [10]. The waves generated by the impacts were processed by a Picoscope 6402D oscilloscope connected to an Intel CPU workstation running Picoscope6 software. The processed data became the input for a ML model implemented in MATLAB, as described below. In a second activity, the vibration due to the engine were detected by the sensors bonded on the fuselage and the wings of a balsa wood RC model of the Piper J3 CUB airplane (Fig. 1). The vibrations were processed by a Pimoroni HAT Explorer Pro connected to an ARM CPU Raspberry Pi minicomputer running Python software. The processed data became the input for a ML model implemented in C++ through MATLAB Coder, run on the same Raspberry Pi.

Fig. 1. Experimental setup: acquisition of the vibrations via Raspberry Pi

Machine Learning Application

In the first activity, L impacts were performed and, for each impact, four ToFs (Time of Flight) were calculated on the basis of the Lamb waves detected by the four PZT sensors. The ToF was defined as the arrival time of A0 mode of the Lamb wave to a sensor, in the range 0-40 kHz, where the A0 mode is dominant, while the S0 can be considered neglectable [11-12]. The ToF was calculated by evaluating the Short Time Fourier Transform (STFT) of the signals in the range 0-40 kHz. Because of the absence of an absolute clock signal, the differences t_1, t_2, t_3 between the ToFs at three sensors and the ToF at one reference sensor were chosen as the features for the ML application. After the evaluation of the ToFs, a dataset was built, made of L rows corresponding to the samples (impacts points) and five columns containing, for each sample, the actual coordinates (x,y) and the three ToF differences t_1, t_2, t_3. Two supervised ML algorithms, polynomial regression (PR) and artificial neural network (ANN), were applied to this dataset, in order to build models able to predict the position of an unknown impact, and to identify the best one. Considering a PR algorithm, the coordinates (x,y) of the impact are polynomial functions of the three ToF differences t_1, t_2, t_3. The building of the model consists of identifying the degree d and the coefficients θ in order to best fit a set of given data:

$$x = \theta_{x_0} + \sum_{i=1}^{3} \theta_{x_i} t_i + \sum_{j=1}^{3} \sum_{k=1}^{3} \theta_{x_{jk}} t_j t_k + \cdots \tag{1}$$

$$y = \theta_{y_0} + \sum_{i=1}^{3} \theta_{y_i} t_i + \sum_{j=1}^{3} \sum_{k=1}^{3} \theta_{y_{jk}} t_j t_k + \cdots \tag{2}$$

Fixing the polynomial degree d and extending the above equations to L impacts, it is possible to use the matrix form:

$$\boldsymbol{U} = \boldsymbol{TB} \tag{3}$$

where: $\boldsymbol{U}$ is $L\times 2$ matrix, in which the columns contain the coordinates for the L impacts respectively; $\boldsymbol{T}$ is $L\times p$ matrix, in which the p columns contain the so-called polynomial features (ToF differences, their power and relative cross multiplication); $\boldsymbol{B}$ is $p\times 2$ Design Matrix of weight coefficients θ. A subset of M impacts was used as training data, in order to calculate the design matrix $\boldsymbol{B}$ as specified in [13]. The model was validated by estimating the Mean Radial Error (MRE) over the M training data, the N test and total L data, where the Radial Error (RE) was defined as the Euclidean distance between the actual coordinates (x_i,y_i) and the coordinates $(\bar{x}_i,\bar{y}_i)$ calculated by the algorithm for the ith impact. Considering an ANN, the coordinates (x,y) of the impact are calculated in the basis of the three ToF differences t_1, t_2, t_3 by computations in succession through connected nodes called neurons. Each neuron performs the computation of an intermediate output

z using the input vector t, the vector of weights w and the vector of biases b and the computation of the output a as an activation function g (a linear function in a regression problem, like a *sigmoid* one) of z:

$$z = w^T t + b \tag{4}$$

$$a = g(z) \tag{5}$$

The neurons are aggregated into layers and a shallow neural network (SNN) was chosen, consisting of only one hidden layer. As for PR, a subset of M impacts was used for the training phase of the network, that consists of an iterative procedure in order to set the weights w and the biases b of each neuron. For this procedure, three learning algorithms were compared: Levenberg-Marquardt, Bayesian Regularization and Scaled Conjugate Gradient [14]-[16]. As for PR, the model was validated by estimating the Mean Radial Error (*MRE*).

Analysis and Results

About the PR, in order to generalise the model, the calculation of ***B*** and the corresponding *MRE* was performed by the mean in a *K-Fold* cross validation procedure, considering the polynomial degrees from 1 to 7 and, for each degree, 5 different combinations of training/test sets with an 80/20 ratio. In this calculation, L (total number of impacts) was equal to 167, M (number of training impacts) was equal to 134, N (number of test impacts) was equal to 33. The best model, in terms of generalising, was chosen considering both minimum total *MRE* and minimum gap between training and test *MRE*s: this condition occurred with degree equal to 3. Moreover, the threshold 110 appeared to be the best training size: for bigger values of size there was no littler *MRE*. After the evaluation of the best degree and training size, 50 test cases were implemented, with a main result in terms of *MRE* on the entire dataset equal to **1.50** mm. About the SNN, the performances were evaluated considering all the three learning algorithms above, increasing the complexity of the model in terms of number of neurons in the hidden layer (10, 20, 30, 40, 50). The increasing in neurons number did not lead to a significant improvement of *MRE*, while the best training algorithm was the Bayesian Regularization. After the evaluation of the best neurons number and training algorithm, 50 test cases were implemented in MATLAB [17], with a training/test ratio equal to 70/30 ($L = 167$, $M = 117$, $N = 50$) and a main result in terms of *MRE* on the entire dataset equal to **1.20** mm. In a second activity, the vibrations due to the engine of a balsa wood RC model were acquired and processed by a Pimoroni HAT Explorer Pro connected to an ARM CPU Raspberry Pi minicomputer. The processed waveforms were used to reproduce the same vibrations on the 25×25 cm specimen using a LMS Test Lab shaker. The best ML model (based on SNN algorithm) trained and built during the first activity was then tested in presence of vibrations, obtaining a similar *MRE*. The ML model was implemented in C++ through MATLAB Coder, run on the same Raspberry Pi.

Conclusions

This work focused on the implementation of a SHM system on a balsa wood RC model of the Piper J3 CUB airplane. A machine learning model able to predict the location of low-speed impacts on aluminium plate was built and it was tested in the presence of the vibrations due to the engine of the airplane model. The best algorithm was found to be a shallow neural network trained with Bayesian Regularization procedure. The best *MRE* value was equal to 1.20 mm, configuring the model with a 70% training sample ratio and 10 hidden neurons, and a similar *MRE* was calculated in the presence of the vibrations. The results can be considered excellent, because the mean radial error falls in an acceptable range if compared to the size of the plates. Furthermore, this work presents the development of two innovative mini-equipment: (i) impact detection and wave acquisition system via Pimoroni HAT Explorer Pro; (ii) data processing and prediction via ML learning software running on a Raspberry Pi micro-computer. These two mini-systems can be considered very efficient because of their performance in terms of precision and for their little size, that allows to install it on unmanned aerial vehicles.

References

[1] V. Giurgiutiu, "Structural health monitoring (SHM) of aerospace composites," Polymer Composites in the Aerospace Industry, pp. 491–558, Jan. 2020. https://doi.org/10.1016/B978-0-08-102679-3.00017-4.

[2] S. Carrino, A. Maffezzoli, and G. Scarselli, "Active SHM for composite pipes using piezoelectric sensors," Mater Today Proc, vol. 34, pp. 1–9, 2019. https://doi.org/10.1016/J.MATPR.2019.12.048.

[3] F. Nicassio, S. Carrino, and G. Scarselli, "Elastic waves interference for the analysis of disbonds in single lap joints," Mech Syst Signal Process, vol. 128, pp. 340–351, Aug. 2019. https://doi.org/10.1016/J.YMSSP.2019.04.011.

[4] L. Capineri and A. Bulletti, "Ultrasonic Guided-Waves Sensors and Integrated Structural Health Monitoring Systems for Impact Detection and Localization: A Review," Sensors 2021, Vol. 21, Page 2929, vol. 21, no. 9, p. 2929, Apr. 2021. https://doi.org/10.3390/S21092929.

[5] N. D. Boffa, M. Arena, E. Monaco, M. Viscardi, F. Ricci, and T. Kundu, "About the combination of high and low frequency methods for impact detection on aerospace components," Progress in Aerospace Sciences, vol. 129, p. 100789, Feb. 2022. https://doi.org/10.1016/J.PAEROSCI.2021.100789.

[6] D. F. Hesser, G. K. Kocur, and B. Markert, "Active source localization in wave guides based on machine learning," Ultrasonics, vol. 106, Aug. 2020. https://doi.org/10.1016/j.ultras.2020.106144.

[7] S. Carrino, F. Nicassio, and G. Scarselli, "Development and application of an in-flight structural health monitoring system," in Proceedings of Meetings on Acoustics, 2019, vol. 38, no. 1. https://doi.org/10.1121/2.0001177.

[8] R. Gorgin, Z. Wang, Z. Wu, and Y. Yang, "Probability based impact localization in plate structures using an error index," Mech Syst Signal Process, vol. 157, Aug. 2021. https://doi.org/10.1016/j.ymssp.2021.107724.

[9] "Piezoelectric Discs." https://www.physikinstrumente.com/en/products/piezoelectric-transducers-actuators/disks-rods-and-cylinders/piezoelectric-discs-1206710/#downloads (accessed Mar. 01, 2022).

[10] "https://asselab.unisalento.it/en/."

[11] G. B. Santoni, L. Yu, B. Xu, and V. Giurgiutiu, "Lamb Wave-Mode Tuning of Piezoelectric Wafer Active Sensors for Structural Health Monitoring," J Vib Acoust, vol. 129, no. 6, pp. 752–762, Dec. 2007. https://doi.org/10.1115/1.2748469.

[12] P. M. Schindler, R. G. May, R. O. Claus, and J. K. Shaw, "Location of impacts on composite panels by embedded fiber optic sensors and neural network processing," Smart Structures and Materials 1995: Smart Sensing, Processing, and Instrumentation, vol. 2444, pp. 481–489, Apr. 1995. https://doi.org/10.1117/12.207698.

[13] C. M. Bishop, "Pattern Recognition And Machine Learning - Springer 2006," Antimicrob Agents Chemother, vol. 58, no. 12, pp. 7250–7, 2014.

[14] X. Glorot and Y. Bengio, "Understanding the difficulty of training deep feedforward neural networks", Accessed: Feb. 24, 2022. [Online]. Available: http://www.iro.umontreal.

[15] M. Kayri, "Predictive abilities of Bayesian regularization and levenberg-marquardt algorithms in artificial neural networks: A comparative empirical study on social data," Mathematical and Computational Applications, vol. 21, no. 2, May 2016. https://doi.org/10.3390/MCA21020020.

[16] M. F. Møller, "A scaled conjugate gradient algorithm for fast supervised learning," Neural Networks, vol. 6, no. 4, pp. 525–533, Jan. 1993. https://doi.org/10.1016/S0893-6080(05)80056-5.

[17] M. Paluszek and S. Thomas, MATLAB machine learning recipes: A problem-solution approach, Second edition. Apress Media LLC, 2019. https://doi.org/10.1007/978-1-4842-3916-2.

Fluid-Dynamics

Assessment and optimization of dynamic stall semi-empirical model for pitching aerofoils

Enrico Galli[1,a], Gregorio Frassoldati[2,3,b], Davide Prederi[2,c], Giuseppe Quaranta[1,d] *

[1]Politecnico di Milano, Dipartimento di Scienze e Tecnologie Aerospaziali, via La Masa, 34, 20156, Milano, Italy

[2]Leonardo S.p.A., Helicopter Division, via Giovanni Agusta, 520, Cascina Costa (VA), Italy

[3]Università di Roma Tre, Department of Engineering, via della Vasca Navale, 79, 00146, Roma, Italy

[a]enrico5.galli@mail.polimi.it,[b]gregorio.frassoldati@leonardo.com, [c]davide.prederi@leonardo.com, [d]giuseppe.quaranta@polimi.it

Keywords: Dynamic Stall, Helicopter Rotors, Optimization

Abstract. Dynamic stall is a phenomenon affecting aerofoils in unsteady flows which is particularly relevant in the helicopter field. Semi-empirical models are reliable tools to simulate this phenomenon, especially during preliminary design phases and for aeroelastic assessments. However, they need a large number of tuning parameters to provide reliable estimations of unsteady airloads. To face this problem, a parameter identification procedure based on sequential resolutions of optimization problems by means of a Genetic Algorithm is developed and it is applied to the state-space formulation of a modified version of the so-called "Second Generation" Leishman-Beddoes model. The effects of the optimal parameters on the model prediction capabilities are discussed and the variability of the parameters with reduced frequency is studied. The estimations of the unsteady airloads obtained by applying the optimization of parameters show a great improvement in the correlation of the experimental data if compared to the predictions obtained by using the parameters provided in literature, especially for pitching moments where the negative peaks are very well described. These improvements justify the need for optimization to set the parameters.

Introduction

Dynamic stall is a very impacting phenomenon in the helicopter field because it is the main cause of the occurrence of large blade torsional airloads and rotor in-plane vibrations which are usually limiting factors both for design purposes and for performance evaluation. So, a reliable prediction of unsteady airloads is paramount.

Mathematical tools, aiming at modelling the effects of dynamic stall, range from semi-empirical models to sophisticated CFD methods. Despite the continuously increasing applications of the CFD methods for research purposes and their capability of providing physically realistic simulation of the rotor flow field, their daily use during the design process is still impractical nowadays due to the huge computational costs, the large memory requirements and the numerical issues.

It is here that semi-empirical models, developed since the eighties and the nineties of the last century, return as competitive and trustworthy prediction tools. Indeed, although they are based on simplified nonlinear equations, they can provide very quickly reliable predictions of unsteady airloads if properly tuned. Therefore, their application could be fundamental, especially in those cases where many predictions of unsteady airloads are needed in a short time, as is the case of aeroelastic assessments.

Among all the semi-empirical models available inthe literature, the Leishman-Beddoes model is chosen for the present investigation because it is one of the most used by the industries and it is one of those which has known a great number of modifications and improvements over the years. Moreover, differently to the models developed starting from the ONERA model [1] or from the Peters model [2], the Leishman-Beddoes is more physically basedsince it is easier giving a physical interpretation to the involved parameters.The state-space formulation of the modified version of the "Second Generation" of the LB model which is considered in the present investigation is extensively explained in [3].

One of the main drawbacks of semi-empirical models is the necessity to identify many parameters that have to be extracted from experimental measurements or very reliable numerical simulations. Unfortunately, there are a couple of issues about the available parameters:

- the parameters can be applied with confidence only for the combinations of aerofoil shapes, Mach and Reynolds numbers used for their identification, limiting the extrapolation capability of semi-empirical models.
- the ranges of the parameters reported in the literature have a very limited extension in terms of aerofoil shapes and aerodynamic conditions. So, the number of experimental test cases where they can be used decreases.

To solve this issue, it has been decided to focus on the development of a parameter identification procedure based on optimization problems that aims at identifying the proper tuning parameters of the considered version of the Leishman-Beddoes model. The goal of this investigation is to provide a general methodology which could be applied routinely to aerodynamic conditions and aerofoil shapes different from those already assessed in the past, allowing to enlarge the range of effective applicability of the Leishman-Beddoes model.

Optimization of tuning parameters for the Leishman-Beddoes model

The considered version of the model needs sixteen parameters depending on aerofoil shapes, Mach and, Reynolds numbers which can be extracted from the static curves of normal force and pitching moment coefficients and from the unsteady airloads of a few selected pitching motions performed at different combinations of mean angle, oscillation amplitude and reduced frequency. The developed parameter identification procedure has some characteristics:

- it is a sequential procedure becausethe parameters are not identified simultaneously by means of aunique optimization, instead, they are divided into smaller groups and then they are identified performing successive optimizations. Indeed:
  - the slope of the static normal force coefficient ($C_{N/\alpha}$), the maximum of the normal force coefficient ($C_{N_{max}}$), the angle of attack at zero lift (α_{zL}) and the pitching moment coefficient at null incidence (C_{M_0}) are easily extracted from the static curves of the coefficients and therefore they are not identified by means of any optimization,
  - the first optimization allows to identify parameters α_1, S_1, S_2 which are used for the approximation of the curve of the static trailing edge separation point,
  - then, the second optimization allows to identify parameters m, K_0, K_1, K_2 which approximates the static curve of the pitching moment coefficient,
  - finally, the last optimizations are performed to extract parameters T_p, T_f, T_v, T_{vl}, δ_{α_1} which account for all the delays occurring during dynamic stall due to: evolution of the unsteady pressure distribution (T_p), evolution of the unsteady boundary layer (T_f), vortex formation (T_v), vortex travel over aerofoil upper surface (T_{vl}) and flow reattachment (δ_{α_1}).

- The optimization problems are bounded problems,and the design spaces of the variables are defined by exploiting the physical interpretation of the parameters. The parameters reported in [4] are used as guess values around which the design spaces are set.
- The objective functions of the optimization problems are error functions computed between the static curves of the aerodynamic coefficients or the unsteady airloads from pitching motions and their numerical approximations.
- The bounded optimization problems are solved by means of the Genetic Algorithm Toolbox provided by Matlab [5]. A genetic algorithm is chosen to solve the problems due to the high non-linearity and the complexity of the objective functions.

Approximation static trailing edge separation point curve. The LB model uses the Kirchhoff-Helmholtz theory to account for the nonlinear airloads due to flow separation at the trailing edge. In particular, the static normal force coefficient is related to angle of attack α and to the effective trailing edge separation point, f:

$$C_N = C_{N/\alpha}(\alpha - \alpha_{zL})\left(\frac{1+\sqrt{f}}{2}\right)^2 \quad (1)$$

By measuring the experimental static behavior of the normal force coefficient and by manipulating Eq.(1), it is possible to obtain an experimentally derived distribution for the static trailing edge separation point. Then, this distribution can be fitted by the following exponential approximation:

$$f(\alpha) = \begin{cases} 1 - 0.3e^{\frac{\alpha-\alpha_1}{S_1}}, & \alpha \leq \alpha_1 \\ 0.04 + 0.66e^{\frac{\alpha_1-\alpha}{S_2}}, & \alpha > \alpha_1 \end{cases} \quad (2)$$

To extract α_1, S_1 and S_2, the fitting problem of Eq.(2) is considered as an optimization problem where the optimal set of parameters is the one which minimizes the fitting error against the distribution of the separation point f previously obtained by inverting Eq.(1).

Approximation pitching moment static curve. The LB model suggests the following expression for the approximation of the static curve of the pitching moment where the non-linearity of the airloads is accounted by means of the effective separation point, f:

$$\frac{C_M - C_{M_0}}{C_N} = K_0 + K_1(1-f) + K_2\sin(\pi f^m) \quad (3)$$

Since the parameters for the exponential approximation of Eq. (2) and the static curves of the aerodynamic coefficients are available from experimental measurements, then the parameters for the pitching moment approximation can be extracted by considering the problem of Eq. (3) as an optimization problem where the optimal set of parameters is the best one over many optimization problems. The total error to be minimized is the sum of a fitting error between static pitching moment curve and its numerical approximation, the error between the maxima and the minima of the experimental curve and its numerical approximation.

Time constants computation. To identify the non-dimensional time constants, the bounded optimization problems are performed by considering the time history of the unsteady airloads coming from pitching oscillations.The objective function to be minimized is complex and it is the sum of a total error on the lift coefficient and a total error on the pitching moment coefficient. These total errors are themselves the sum of a global error between the time history of the coefficient and its numerical approximation and an error between the maximum and an error between the minimum over the pitching cycle.To limit the interdependent effects between the variables, the identification is further split in two successive steps:

- At first, a unique optimization is performed by considering the unsteady airloads of a pitching motion in a condition of moderate stall with mean angle lower than the static stall one and a large variation of incidence both under and above static stall angle ($\pm$ 10°)at high reduced frequency ($k = 0.1$). This specific condition is assumed to be appropriate because all the stages of dynamic stall can have enough time to develop, all the involved delays can occur and the vortex shedding effects are not yet too much strong. This first optimization is performed by using all the five variables but only T_p, T_f and δ_{α_1} are correctly computed.
- Then, to identify parameters T_v and T_{vl}, which describe vortex-induced effects, many optimizations are performed by considering the unsteady airloads from pitching motions at different conditions of deep stall characterized by a mean angle greater than the static stall one, a large variation of incidence both under and above static stall angle ($\pm$ 10°) and different values of reduced frequency smaller than the previous one. The condition of deep stall is considered suitable to compute vortex-related parameters because in this case the vortex is very strong. These computations are performed by keeping fixed the values of T_p, T_f and δ_{α_1} obtained by previous resolution and allowing only parameters T_v and T_{vl} to vary. It is interesting to note that performing the optimizations considering the unsteady airloads at different reduced frequencies allows to study the variability of T_v and T_{vl} with this parameter which is defined as $k = \frac{\omega c}{2V}$ (where ω is the frequency of the pitching oscillation of the aerofoil, c is the chord length and V is the constant free stream velocity). So, by modifying k, it is possible to explore the widest possible range of conditions that a blade section could encounter on the rotor disk during operative conditions.

Results

The previously described identification procedure is applied to NACA 0012 at conditions of $Ma = 0.3$ and $Re = 3.8 \cdot 10^6$. All the static curves of the aerodynamic coefficients and the experimental unsteady airloads come from the database of [6].

The predictions of the unsteady airloads obtained by applying the optimization of parameters are shown in Fig. 1 and Fig. 2. It is possible to see how the predictions obtained by the optimal set of parameters (red curves) allow a great improvement in the correlation of the experimental data (black dashed lines), if compared to the predictions obtained by using the parameters provided in literature (blue curves). The improvement applies especially for pitching moments where the negative peaks are very well described and for small values of the reduced frequency. These improvements justify the need for optimization in the parameters setting.

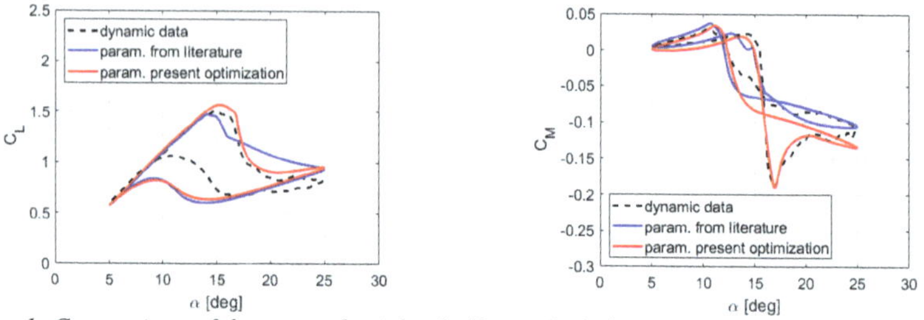

Figure 1: Comparison of the unsteady airloads (C_L on the left, C_M on the right by means of LB model for NACA 0012 oscillating in pitch with $\alpha = 15° + 10° \, sin(\omega t)$ *and* $k = 0.025$.

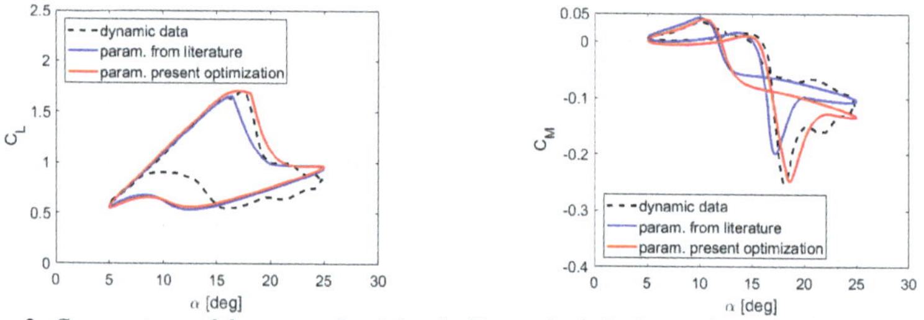

Figure 2: Comparison of the unsteady airloads (C_L on the left, C_M on the right) by means of LB model for NACA 0012 oscillating in pitch with $\alpha = 15° + 10° \, sin(\omega t)$ *and* $k = 0.05$.

References

[1] C. T. Tran, D. Petot, Semi-empirical model for the dynamic stall of airfoils in view of the application to the calculation of the responses of a helicopter blade in forward flight, Vertica 5 (1981) 35-53.

[2] D. A. Peters, Toward a unified lift model for use in rotor blade stability analysis, Journal of the American Helicopter Society 30 (1985) 32-42. https://doi.org/10.4050/JAHS.30.3.32

[3] G. Dimitriadis, Introduction to nonlinear aeroelasticity, first edition, John Wiley & Sons Ltd, Chichester, West Sussex, UK, 2017.

[4] J. G. Leishman, G. L. Crouse, State-space model for unsteady airfoilbehavior and dynamic stall, Technical Report Paper 89-1219, AIAA, 1989. https://doi.org/10.2514/6.1989-1319

[5] Matlab, Genetic Algorithm Toolbox, https://it.mathworks.com/help/gads/genetic-algorithm.html.

[6] W. J. McCroskey, K. W. McAlister, L. W. Carr, S. L. Pucci, An experimental study of dynamic stall on advanced airfoil sections, Volume 1, 2, 3, Technical Memorandum TM-84245, NASA, 1982.

Pattern recognition of the flow around a pitching NACA 0012 airfoil in dynamic stall conditions

Giacomo Baldan[1,a *], Alberto Guardone[1,b]

[1]Department of Aerospace Science and Technology, Politecnico di Milano, Via La Masa 34, Milano 20156, Italy

[a]giacomo.baldan@polimi.it, [b]alberto.guardone@polimi.it

Keywords: Dynamic Stall, Proper Orthogonal Decomposition, Pitching Airfoil, Pattern Recognition, Helicopters, Wind Turbines

Abstract. The present work numerically investigates the flow evolution of a pitching NACA 0012 airfoil incurring in deep dynamic stall phenomena. The experimental data at Reynolds number Re $= 1.35 \cdot 10^5$ and reduced frequency $k = 0.1$, provided by Lee et al., are compared to numerical simulation using different methods. Firstly, 2D URANS with different turbulence models are explored highlighting the advantages and the drawbacks of each strategy. On the one hand, simulations are able to describe most characteristic flow features of dynamic stall. On the other hand, numerical models still struggle in describing the inherent complexity of instability and transition from laminar to turbulence, resulting in a misprediction of the angle of attack at which the dynamic stall vortex (DSV) is generated and convected rearward. Finally, a Proper Orthogonal Decomposition (POD) is proposed to analyze the main flow features and to recognize flow patterns. The decomposition of both velocity magnitude and pressure fields shows a high frequency content requiring a large portion of the modes to recover most of the flow energy.

Introduction

Dynamic stall is a complex unsteady aerodynamic phenomenon that is present in a large range of engineering applications such as retreating blades of helicopter rotors, wind turbines, turbomachinery and maneuvering fixed wing aircraft [1]. Compared to static stall, it is characterized by a temporary delay of the boundary layer separation followed by a large flow detachment on the suction side of the profile. This results in a loss of lift, a strong pitch down moment, and an increase in drag. Before the flow reattachment, the profile underlays varying loads due to the chaotic nature of the involved phenomena.

Especially during the last decade, many experiments and computational simulations have been performed to further investigate dynamic stall phenomena and study the flow around pitching and plunging airfoils [2]. Low-fidelity tools are still unable to capture the main features of the flow and computational fluid dynamics should be leveraged [3]. Most difficulties rely in the modelling of the laminar separation bubble (LSB) and the subsequent creation of the dynamic stall vortex (DSV). Laminar to turbulent transition plays a major role in the formation and the convection of the DSV that influences a significant portion of the flow evolution.

In this paper, the flow around a pitching NACA 0012 profile underlying deep dynamic stall regime is numerically investigated. Firstly, a space and time convergence study is conducted using 2D RANS equations. Then, three different turbulent models are compared focusing on the LSB and DSV description. Finally, a Proper Orthogonal Decomposition (POD) is performed using the pressure and velocity magnitude fields.

Numerical setup

The numerical setup aims at reproducing the experimental campaign presented in Lee et al. [4]. A NACA 0012 profile with c=0.15 m chord underlying a sinusoidal pitching motion, at reduced frequency k=0.1 ($k=\omega \cdot c/2 / V_\infty$) and Reynolds number $Re = 1.35 \cdot 10^5$, is investigated. The free-stream velocity is $V_\infty = 14$ m/s with a turbulent intensity equal to 0.08%, pressure is $P_\infty = 1$ atm, and the pitching frequency ω is set equal to 18.67 Hz. Three different O-grids have been generated as reported in Tab. 1, where N_x is the number of points around the profile and N_y in the normal direction. All grids respect the $y^+ < 1$ requirement at the wall.

Table 1: NACA 0012 O-grid specifications

O-Grids	N_x	N_y	$\Delta x_{le}\,(\cdot 10^{-3}c)$	$\Delta x_{te}\,(\cdot 10^{-4}c)$
G1	384	96	3.3	8.4
G2	512	128	2.0	5.0
G3	1024	256	0.67	3.6

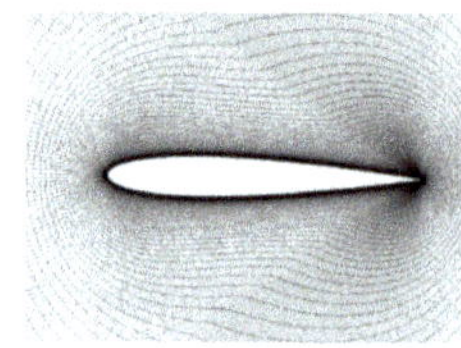

Figure 1: O-grid detail

Numerical simulations are performed using ANSYS Fluent 2023R1. The unsteady incompressible RANS equations are solved using second-order upwind discretization and second-order implicit time integration scheme. Gradients are retrieved through a least square cell-based method and fluxes are obtained with the Rhie-Chow momentum-based formulation. Pressure-velocity equations are solved using the SIMPLE method. A rigid motion of the entire grid is prescribed to allow the airfoil pitching, $\alpha(t) = 10° + 15° \sin(\omega t)$.

The POD decomposition is chosen, in this work, because it is the linear decomposition that, on average, minimize the energy loss considering a subset of *k* modes and it also grants the orthogonality of the modes [5]. The POD is computed using the Singular Value Decomposition (SVD) method.

Results

As a first step, time and grid independence study is performed. For each mesh three different simulations have been computed using 900, 1800, and 3600 time steps per pitching cycle. SST model with intermittency equation close the RANS equations system. All simulations are run until the solution of subsequent cycles overlaps, usually requiring three or four cycles. In Fig. 2 the last converged cycle is reported. The convergence has to be mainly verified in the upstroke part and on the angle at which the DSV generates the peak in lift and drag. All three grids show a good agreement in the aforementioned points but G1 should not be used since the low resolution does not satisfactory describe the DSV evolution, resulting in a larger mismatch in the downstroke part.

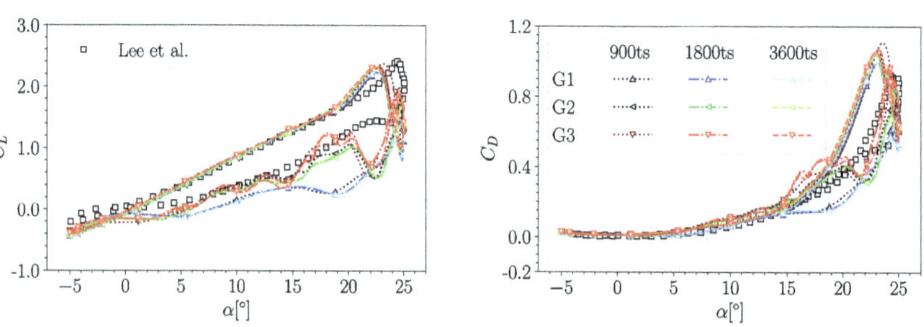

Figure 2: Space and time convergence study. Experimental data [4].

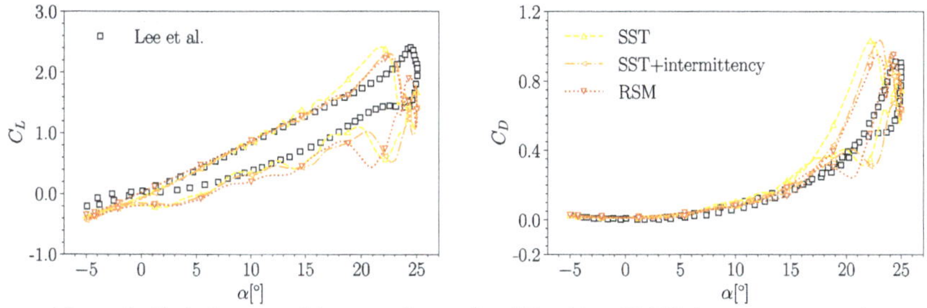

Figure 3: Turbulence model comparison using G2 grid and 3600 time steps per cycle. Experimental data [4].

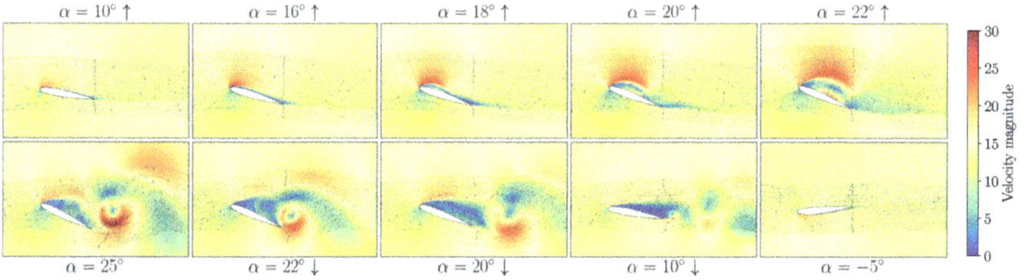

Figure 4: Velocity magnitude contours and streamlines at different pitch angles.

Another critical aspect in dynamic stall simulations is turbulence modelling. SST, SST with intermittency equation, and RSM models are tested using the G2 grid and 3600 time steps per cycle. In Fig. 3, it can be noted that standard SST model fails to describe the LSB and results in oscillation in the DSV formation. The same behavior is present in the RSM even if the effect is weaker since the individual components of the Reynolds stress tensor are computed. The SST model coupled with the intermittency equation describes in more detail the transition from laminar to turbulent giving more accurate results. For this last case, the velocity magnitude contour and streamlines are reported in Fig. 4 to represent the dynamic stall phenomena.

Finally, the POD decomposition is computed for pressure and velocity magnitude fields in the airfoil reference frame during an entire cycle. In Fig. 5 the eigenvalues are shown while in Fig. 6 the first ten modes are represented. The first mode, which is the most energetic one, represent the mean flow over the entire cycle. The following modes confirm the crucial role of the DSV evolution. As the number of the mode increases higher frequency content is represented. It is visible how the DSV that detaches from the leading edge interacts with the stall that originates from the trailing edge. The high frequency content of the flow is confirmed by the cumulative eigenvalue spectrum in which the number of modes required to represent 99% of the energy are 155 and 503 for pressure and velocity, respectively.

Conclusion

In this work, a pitching NACA 0012 underlying dynamic stall phenomena is investigated using 2D URANS with three turbulence models. Numerical simulations are able to describe the LSB and the DSV, showing a good agreement with experiments in the upstroke phase but are not able to capture the angle at which the DSV is convected rearwards anticipating the drag and lift picks. The LSB is shown to play a crucial role in the flow evolution requiring a turbulence model capable of describing the transition from laminar to turbulent, as in the case of SST with intermittency equation. In addition, a POD is provided for both pressure and velocity magnitude fields to extract the main flow features and to emphasize the interactions of leading and trailing edge stall.

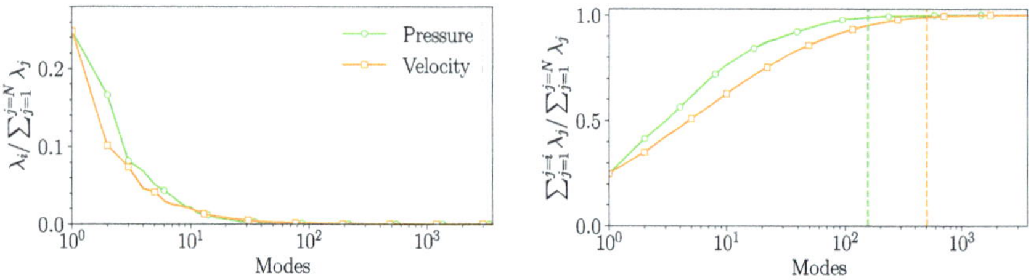

Figure 5: Pressure and velocity magnitude POD eigenvalues and cumulative eigenvalue spectrum. The dashed lines show the number of modes required to represent 99% of the energy.

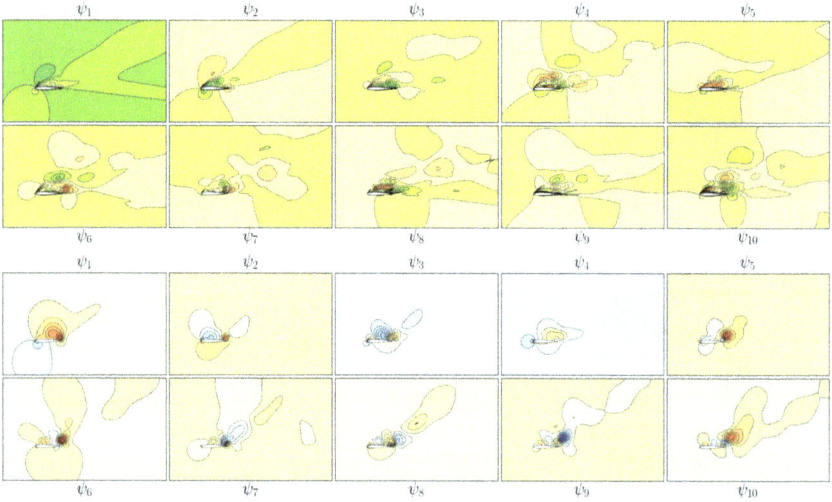

Figure 6: First ten POD modes of velocity magnitude (above) and pressure (below) fields.

References

[1] M. Visbal, D. Garmann, "Analysis of Dynamic Stall on a Pitching Airfoil Using High-Fidelity Large-Eddy Simulations". AIAA Journal 2018; 56(1):46-63. https://doi.org/10.2514/1.J056108

[2] F. Avanzi, F. de Vanna, Y. Ruan and E. Benini, "Enhanced Identification of Coherent Structures in the Flow Evolution of a Pitching Wing". AIAA 2022-0182. AIAA SCITECH 2022 Forum. https://doi.org/10.2514/6.2022-0182

[3] A. D. Gardner, A. R. Jones, K. Mulleners, J. W. Naughton, M. J. Smith. "Review of rotating wing dynamic stall: Experiments and flow control". Progress in Aerospace Sciences 2023; 137:100887. https://doi.org/10.1016/j.paerosci.2023.100887

[4] T. Lee, P. Gerontakos, "Investigation of flow over an oscillating airfoil". Journal of Fluid Mechanics 2004; 512:313–341. https://doi.org/10.1017/S0022112004009851

[5] H. Eivazi, S. Le Clainche, S. Hoyas, R. Vinuesa. "Towards extraction of orthogonal and parsimonious non-linear modes from turbulent flows". Expert Systems with Applications 2022; 202:117038. https://doi.org/10.1016/j.eswa.2022.117038

RANS transition model predictions on hypersonic three-dimensional forebody configuration

Luigi Cutrone[1,a*], Antonio Schettino[1,b]

[1] CIRA (Centro Italiano Ricerche Aerospaziali), Unità di Aerotermodinamica, Via Maiorise, 81043 Capua (CE), Italy.

[a]l.cutrone@cira.it, [b]a.schettino@cira.it

Keywords: Laminar-to-Turbulent Transition, CFD, Hypersonic Flows

Abstract. Future space transportation systems will heavily rely on predicting and understanding Boundary Layer Transition (BLT) during atmospheric entry, especially in the hypersonic phase. Several models, compatible with RANS solvers, have yet been proposed, but not validated in the hypersonic regime. This paper focuses on evaluating prediction capabilities for such models on complex 3D geometries, using the International Boundary Layer Transition (BOLT) Flight Experiment as a test case.

Introduction

The success of future "Apollo" or "Shuttle"-like spacecraft programs and other concepts based on air-breathing propulsion will require an accurate prediction of Boundary Layer Transition (BLT) given that a boundary layer turbulence can amplify surface heating by a factor in excess of five with respect to laminar conditions. The Reynolds-averaged Navier–Stokes equations (RANS equations) are widely used for modeling turbulent flows, but they can't predict laminar-to-turbulent transition. This limitation arises from the RANS averaging procedure itself, which effectively removes the influence of linear disturbance growth—an essential factor in the transition process. A common approach is then to combine the turbulence model with a transition criterion based on experimental correlations. Correlation-based models are frequently linked to an intermittency transport equation, such as that developed by Steelant and Dick [1] or more complex formulations as proposed by Suzen et al. [2], even though these models require nonlocal information to trigger the transition. More recently, several local correlation-based transition modeling (LCTM) methods have been developed and implemented into in modern parallel RANS code. Examples include models developed by Menter [3][4], based on solving one or more differential equations or even based on fully algebraic frameworks [5]. Furthermore, based on the LCTM framework, extensions targeting the hypersonic flow regime have been developed [6]. In all cases, these models have only been partially explored for hypersonic flows, and rely on a large number of constants to tune the results.

Expanding on the prior research conducted in a previous paper [9], we will now apply the 2015 and 2021 Menter's models to a three-dimensional configuration. The primary goal is to validate their predictive abilities when multiple transition mechanisms such as Mack waves, crossflow instabilities, Goertler vortices, and others may simultaneously occur. The configuration selected is the one proposed in the International BOundary Layer Transition (BOLT) Flight Experiment, which was specifically designed to have multiple mechanisms to transition that interact with each other. The geometry was extensively tested in several wind tunnels, including full-scale tests at the CUBRC LENS II wind tunnel, [7].

Models description

The 2015 Menter γ transition model[4], from now on referred as Model-1, is based on the solution of the $k-\omega$ equations (accordingly to the SST turbulence model [8]) and an additional transport equation for intermittency:

$$\frac{\partial(\rho\gamma)}{\partial t}+\frac{\partial(\rho U_j\gamma)}{\partial x_j} = P_\gamma - E_\gamma + \frac{\partial}{\partial x_j}\left[\left(\mu+\frac{\mu_t}{\sigma_f}\right)\frac{\partial\gamma}{\partial x_j}\right]. \quad (1)$$

The transition to turbulence is controlled by the F_{onset} factor in the intermittency production term, $P_\gamma = F_{lenght}\rho S\gamma(1-\gamma)F_{onset}$, which can be triggered from both the streamwise and crossflow transition modes:

$$F_{onset} = \max(F_{onset,sw}, F_{onset,cf}). \quad (2)$$

The model can effectively incorporate information from freestream turbulence and the streamwise pressure gradient using only local variables, and, by activating the cross-flow transition term, it can also account for local variations in the flow direction. The labels Model-1A and Model-1B will refer to the version without and with the crossflow transition term, respectively.

Menter recently introduced a new algebraic correlation for intermittency in the 2021 version of his γ transition model [5], thus avoiding the solution of an independent transport equation for it. Similar to Model-1, this new fully algebraic γ transition model, referred to as Model-2 from now on, can consider the streamwise pressure gradient, but currently not the crossflow transition.

Application to 3D geometry

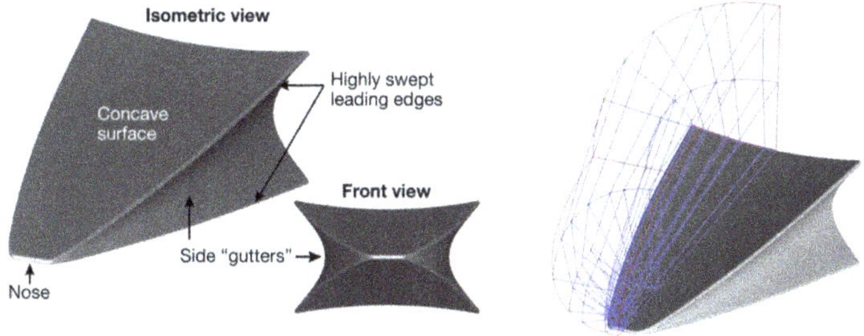

Fig. 1 (left) BOLT geometry, from [7]; (right) domain decomposition for a quarter of the body.

Test case consists in the numerical rebuilding of some of the ground experiments carried out on the geometry of the BOLT (BOundary Layer Transition) project which was designed to investigate the hypersonic boundary layer transition on a low-curvature concave surface with highly swept leading edges, Fig. 1. A full-scale model of the BOLT geometry underwent extensive ground test experimentation in the LENS-II hypervelocity reflected shock tunnel at CUBRC, and here the conditions of RUN-03 in [7] will be used as reference. These conditions are here briefly summarized: $M = 5.17$, $Re_L = 3.92 \cdot 10^6 [m^{-1}]$, with stagnation pressure and temperature equal to 1.5[MPa] and 1130[K], respectively. A wall temperature of 294.4[K] was imposed. Four inlet values of the turbulent intensity level, $Tu_\infty = [0.1,0.3,0.5,1.0]\%$, were selected to study the effect of freestream turbulence intensity on the transition onset, provided that no specific information on tunnel noise or specific freestream turbulence measures are available. A quarter of the forebody was meshed (exploiting both symmetries). Two shock-fitted structured multi-blocks meshes with 3.5M and 24M cells were used. All the BOLT simulations presented in this paper are based on the 24M cell mesh. All the simulations have been carried out using Ansys Fluent 2021R2.

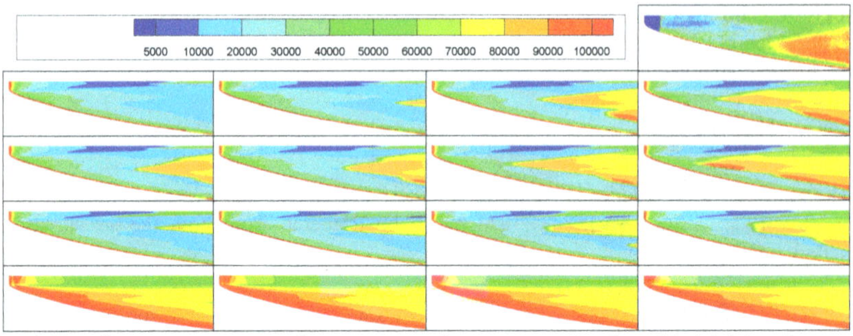

Fig. 2 Heat flux contour map on the primary surface, progressing from Model-1A at the top to Model-2 at the bottom. It showcases various levels of freestream turbulent intensity, ranging from 0.1% leftmost to 1.0% rightmost. Fully turbulent results are also reported for comparisons.

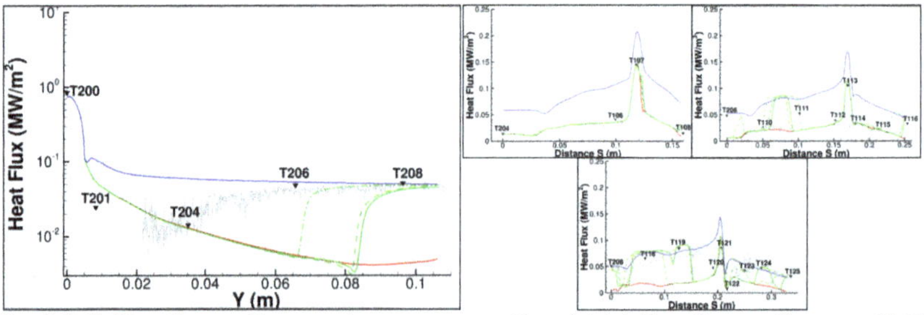

Fig. 3 (left) Heat flux profiles on primary symmetry line; (right) Heat flux profiles at x=10,20, and 30 inches sections. —: TSP heat flux, ▼ thin-films heat flux, —: Fully laminar, —: Fully turbulent, —: Model-1A, – –: Model-1B, – · –: Model-2. Numerical results for Tu $= 0.5\%$.

Results from computations using different transition models can be seen in Fig. 2. Here, four set of results per freestream turbulence intensity level are available, three sets employing a different transition model and one fully turbulent. An experimental heat flux map obtained with TSP for this run is also shown.

Figure 2 shows quite clearly the dependence of the transition front position on the transition model choice. It is also evident that Model-1A and Model-2 show a significant dependence of their transition front position on Tu_∞, with a trend towards earlier transition for increasing Tu_∞ levels as expected, [10][9]: at extremely low Tu_∞ levels, both transition models tend to converge towards the laminar solution (not reported here for the sake of brevity). On the other hand, when the correction for cross-flow transition is activated by using Model-1B (and BOLT geometry was specifically designed to experience this mode of transition) the solutions seem to be less sensitive to the Tu_∞ levels, suggesting that crossflow term is entirely responsible for predicting a transition front even at very low Tu_∞ levels: only at the highest value considered, $Tu_\infty = 1\%$, the streamwise onset terms became predominant and shift significantly forward the transition front.

With respect to the TSP contour map reported in the top right corner of Fig. 2, it seems that all the three transitional results provide the best alignment with the experiment when the Tu_∞ is set at 0.5%: notably, the transition front obtained with Model-1B demonstrates a tendency to be closer to the lateral leading edge, similar to what observed in the experimental map. Then, a significant difference between the models appears in central part of the primary surface (a symmetry plane in the simulations), where Model-2 systematically predicts an earlier transition.

The aforementioned tendency becomes more apparent in Fig. 3(left), which provides a quantitative comparison between experimental and numerical heat flux profiles. Here, the heat flux at the stagnation point (T200) is accurately predicted, but at the second point (T201) location, where the flow is still laminar, all the numerical results consistently overestimate the experimental value. The authors intend to investigate the effects of a non-homogeneous wall temperature distribution, closer to the experimental conditions. Additionally, when observing the temperature-sensitive paint (TSP) data on the same graph, it becomes evident that the transition needs approximately 0.04 m to fully take place, being extensively distributed across the surface. Conversely, the simulations predict a rapid but delayed transition to turbulence onset, with only Model-2 displaying slightly less delay. Finally, the cross-cut sections in Fig. 3(right) clearly show that Model-1B is everywhere capable of predicting a wider transition zone with respect to all the other models.

Summary

Two transition models have been investigated, one of which equipped with a specific term for crossflow transition. This latter proves to be especially valuable at lower Tu_∞ values, although an accurate turbulence characterization of the experiment is mandatory for reducing uncertainties.

References

[1] Steelant, J., Dick, E., Modelling of Bypass Transition with Conditioned Navier–Stokes Equations Coupled to An Intermittency Transport Equation. J. Num. Meth. Fluids, 23:193-220. https://doi.org/10.1002/(SICI)1097-0363(19960815)23:3%3C193::AID-FLD415%3E3.0.CO;2-2
[2] Suzen, Y. B., and Huang, P. G., Modeling of Flow Transition Using an Intermittency Transport Equation. ASME. J. Fluids Eng. June 2000; 122(2): 273–284. https://doi.org/10.1115/1.483255
[3] R. B. Langtry and F. R. Menter. Correlation-Based Transition Modeling for Unstructured Parallelized Computational Fluid Dynamics Codes. AIAA Journal. 47(12), 2009, pp. 2894–2906. https://doi.org/10.2514/1.42362
[4] Menter, F. R., Smirnov, P. E., Liu, T., and Avancha, R., A One-Equation Local Correlation-Based Transition Model, Flow, Turbulence and Combustion, Vol. 95, No. 4, 2015, pp. 583–619. https://doi.org/10.1007/s10494-015-9622-4
[5] Menter, F. R., Matyushenko, A., Lechner, R., Stabnikov, A., and Garbaruk, A., An Algebraic LCTM Model for Laminar–Turbulent Transition Prediction, Flow, Turb. Comb., Vol. 109, No. 4, 2022, pp. 841–869. https://doi.org/10.1007/s10494-022-00336-8
[6] Liu, Z., Lu, Y., Li, J., and Yan, C. Local correlation-based transition model for high-speed flows. AIAA Journal, 60(3):1365–1381, 2022. https://doi.org/10.2514/1.J060994
[7] Berridge, D. C., McKiernan, G., Wadhams, T. P., Holden, M., Wheaton, B. M., Wolf, T. D., and Schneider, S. P., Hypersonic Ground Tests in Support of the Boundary Layer Transition (BOLT) Flight Experiment, Fluid Dynamics Conference, 2018, p. 2893. https://doi.org/10.2514/6.2018-2893
[8] Menter, F. R., "Two-Equation Eddy-Viscosity Turbulence Models for Engineering Applications," AIAA Journal, Vol. 32, No. 8, August 1994, pp. 1598-1605. https://doi.org/10.2514/3.12149
[9] Infante, G. M., Cutrone, L., Schettino, A., Numerical methods for laminar/turbulent transition prediction in Hypersonic regime, ESA FAR conference, 2020.
[10] J. I. Cardesa, G. Delattre Comparison of RANS transition model predictions on hypersonic three-dimensional forebody configurations 57th 3AF International Conference on Applied Aerodynamics 29-31 March 2023, Bordeaux – France

A finite-volume hybrid WENO/central-difference shock capturing approach with detailed state-to-state kinetics for high-enthalpy flows

Francesco Bonelli[1,a *], Davide Ninni[1,b], Gianpiero Colonna[2,c] and Giuseppe Pascazio[1,d]

[1]DMMM & CEMeC, Politecnico di Bari, Bari, 70125, Italy

[2]CNR-ISTP, Via Amendola 122/D, 70126 Bari, Italy

[a]francesco.bonelli@poliba.it, [b]davide.ninni@poliba.it, [c]gianpiero.colonna@cnr.it, [d]giuseppe.pascazio@poliba.it

Keywords: WENO, Hybrid Schemes, Hypersonics, Thermochemical Non-Equilibrium

Abstract. This work shows novel space discretization capabilities of an innovative fluid dynamics solver able to deal with thermochemical non-equilibrium by using a detailed state-to-state model. The implementation of a WENO hybrid scheme is verified and thermochemical non-equilibrium effects are investigated by considering a high temperature shock tube test case. The work represents a first step to enable the solver to perform LES and DNS simulations of turbulent hypersonic flows.

Introduction

The investigation of high enthalpy flows is important in several fields, e.g., hypersonic flows in the context of entry objects (space capsules, reusable space vehicles, debris, meteoroids, etc.), laser applications (laser induced breakdown spectroscopy, etc.), high-enthalpy wind tunnels, rocket engines, etc. [1]. Such flows can involve different phenomena (e.g., strong shock waves, thermochemical non-equilibrium, turbulence, etc.) each of which requires a specific expertise and adequate modeling. In the last years, the authors have developed a finite-volume solver of the Navier-Stokes equations for the simulation of hypersonic flows in thermochemical non-equilibrium, see, e.g. [2,3]. From a numerical point of view the solver is based on conventional approaches that have demonstrated to be robust and affordable. A third order Runge-Kutta scheme and the flux vector splitting of Steger and Warming, with a second order MUSCL reconstruction, were employed for time and space discretization, respectively. On the other hand, to deal with thermochemical non-equilibrium an accurate state-to-state (StS) approach, in addition to the classical multi-temperature Park's model, makes the software a unique tool in the present scene. In order to extend code capabilities to problems that involve both shocks and turbulence, a hybrid WENO/central-difference scheme has been implemented. Indeed, this approach has shown to be among the most convincing when dealing with problems involving shocks and turbulence interacting dynamically [4]. In the present implementation, the popular fifth-order WENO scheme is coupled with a sixth-order central scheme with a shock sensor that limits the use of the shock capturing scheme to region of strong gradients, thus reducing numerical dissipation in smooth regions. To verify the algorithm implementation and to show code capabilities, the one dimensional Sod [5] and Shu–Osher [4] problems, and the two dimensional double Mach reflection (DMR) [5] have been analyzed. Finally, a high temperature shock tube has been investigated by using the hybrid scheme along with the StS approach.

Materials Research Proceedings 37 (2023) 170-173 https://doi.org/10.21741/9781644902813-37

Governing Equations and Numerical Method

The system of Euler equations is solved for a calorically perfect gas ($\gamma = cp/cv = 1.4$) or, to consider high temperature effects, for a reacting mixture in thermochemical non-equilibrium.
A cell centered finite-volume (FV) approach is employed and the method of lines is used to separate space and time discretization. Upwind space discretization can be performed either by the Steger and Warming (SW) or by the Lax-Friedrichs (LF) Flux Vector Splitting (FVS). The accuracy of the upwind scheme is increased by a fifth order WENO reconstruction following a characteristic-wise FV method [6]. Then, to reduce numerical dissipation, the WENO scheme is hybridized with a numerical flux built from a sixth order central scheme (CD6). The hybrid approach switches from CD6 to WENO when a shock sensor (χ) [7] is larger than a threshold (χ_L). Time integration is performed by a third order explicit Runge-Kutta scheme [5]. In the case of reacting mixtures an operator splitting approach is applied to deal with stiff source terms [2,3]. Finally, thermochemical non-equilibrium is handled by a StS approach for a N_2-N mixture [2].

Results

Code verification has been performed by considering 1D and 2D test cases. All problems have been solved by using the SW-FVS except for the double Mach reflection for which the LF-FVS has been employed.

The first test is represented by Sod's shock tube [5]. In Fig. 1 (left) the exact solution is compared with the results obtained by the WENO scheme with three different resolutions: a very good agreement is obtained. Figure 1 (right) shows the results of the hybrid scheme with three different χ_L using 100 cells. Oscillations due to dispersive error increase with χ_L .

The Shu-Osher problem [4] is a benchmark for studying shock turbulence interactions being the propagation of a shock in a perturbed density field. A reference solution has been obtained by using the WENO scheme with mesh size $\Delta x = 6.25 \cdot 10^{-3}$ (1600 cells). Figure 2 shows that with the same resolution ($\Delta x = 5 \cdot 10^{-2}$, 200 cells) the hybrid scheme provides better results with respect to the WENO approach in terms of density profiles, whereas no difference emerges from velocity profiles.

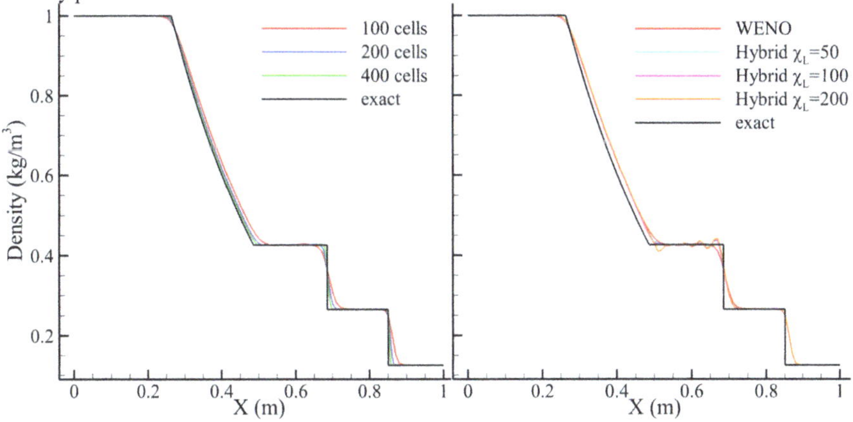

Fig. 1 Sod's shock tube at t=0.2: (left) WENO scheme; (right) WENO/CD6 hybrid scheme.

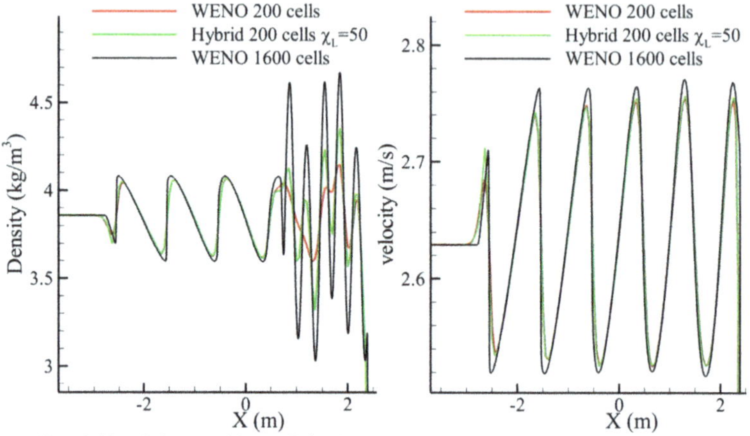

Fig. 2 Shu-Osher problem: (left) density profiles; (right) velocity profiles.

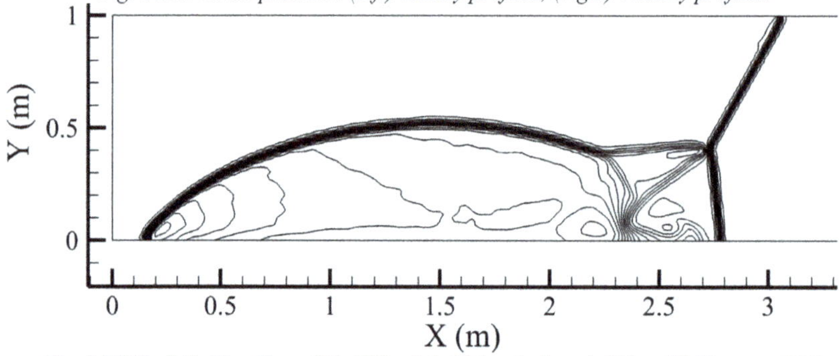

Fig. 3 DMR: 240x59 cells, t=0.2, CFL=0.6, 30 levels from 1.731 to 20.92, $\chi_L = \mathbf{200}$.

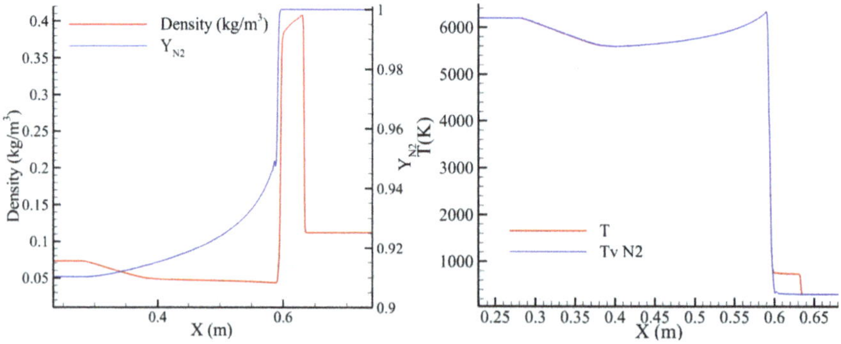

Fig. 4 High temperature shock tube at $\mathrm{t} = 1.3 \cdot 10^{-4}\mathrm{s}$*: (left) density and* Y_{N2}*; (right) T and* Tv_{N2}

Figure 3 shows the density isolines for the double Mach reflection problem (hybrid scheme). A very good agreement is obtained with Jiang and Shu [5]. Finally, a high temperature shock tube has been simulated by using the StS approach. The initial conditions are those given in Grossmann and Cinnella [8] except for the initial mixture composition that here is a pure N_2 mixture (mass fraction Y_{N2}=1). The hybrid scheme with χ_L=100 has been employed. Figure 4 shows an important N_2 dissociation which causes a temperature reduction in the upstream region. The translational (T) and the vibrational (Tv_{N2}) temperature differ in the region between the shock wave and the contact discontinuity thus showing a small thermal non-equilibrium.

Conclusions

In this work a WENO/CD6 hybrid scheme was implemented in a fluid dynamics solver able to deal with thermochemical non-equilibrium by using a StS approach. The solver was verified by considering benchmark test cases. Finally, a high temperature shock tube was analyzed showing the ability of the scheme to deal with high temperature gases in thermochemical non-equilibrium.

Acknowledgement

D. N., F. B. and G. P. were partially supported by the Italian Ministry of Education, University and Research under the Program Department of Excellence Legge 232/2016 (Grant No. CUP - D93C23000100001). F. B. was supported by PON "Ricerca e Innovazione" 2014-2020 Azione I.2 "Mobilità dei Ricercatori" - Avviso di cui al D.M. n. 407 del 27/02/2018 AIM "Attraction and International Mobility" Linea 1 Grant No. AIM1895471 - CUP D94I18000210007

References

[1] J.D. Anderson Jr., Hypersonic and High-Temperature Gas Dynamics, second ed., American Institute of Aeronautics and Astronautics, Inc., Reston, Virginia, 2006.

[2] G. Pascazio, D. Ninni, F. Bonelli, G. Colonna, Hypersonic flows with detailed state-to-state kinetics using a GPU cluster. In Plasma Modeling (Second Edition): Methods and applications. IOP Publishing. Bristol, UK, 2022, pp. 10-1–10-41. https://doi.org/10.1088/978-0-7503-3559-1

[3] D. Ninni, F. Bonelli, G. Colonna, G. Pascazio, On the influence of non equilibrium in the free stream conditions of high enthalpy oxygen flows around a double-cone. Acta Astronaut., 201 (2022) 247-258. https://doi.org/10.1016/j.actaastro.2022.09.017

[4] E. Johnsen et al., Assessment of high-resolution methods for numerical simulations of compressible turbulence with shock waves, J. Comput. Phys. 229.4 (2010) 1213-1237. https://doi.org/10.1016/j.jcp.2009.10.028

[5] G.S. Jiang, C.W. Shu, Efficient implementation of weighted ENO schemes. J. Comput. Phys., 126.1 (1996) 202-228. https://doi.org/10.1006/jcph.1996.0130

[6] CW. Shu, Essentially non-oscillatory and weighted essentially non-oscillatory schemes for hyperbolic conservation laws. in: Quarteroni, A. (eds) Advanced Numerical Approximation of Nonlinear Hyperbolic Equations. Lecture Notes in Mathematics, vol 1697. Springer, Berlin, Heidelberg, 1998, pp. 325-432 https://doi.org/10.1007/BFb0096355

[7] D.J. Hill, D.I. Pullin, Hybrid tuned center-difference-WENO method for large eddy simulations in the presence of strong shocks, J. Comput. Phys. 194 (2004) 435–450. https://doi.org/10.1016/j.jcp.2003.07.032

[8] B. Grossman, P. Cinnella, Flux-split algorithms for flows with non-equilibrium chemistry and vibrational relaxation. J. Comput. Phys. 88.1 (1990) 131-168. https://doi.org/10.1016/0021-9991(90)90245-V

Quantum computing CFD simulations: state of the art

Giulio Malinverno[1,a *], Javier Blasco Alberto[2,b] and Jon Lecumberri SanMartin[2,c]

[1]Engineering Department, FIMAC S.p.A., via Piemonte 19, 20030 Senago (MI) - Italy

[2]Departamento de Ciencia y Tecnología de Materiales y Fluidos, Universidad de Zaragoza, calle Marie de Luna 3, 50018 Zaragoza, Spain

[a]g.malinverno@fimac.aero, [b]jablasal@unizar.es, [c]jonlecum@gmail.com

Keywords: Quantum Computing, CFD, Scientific Machine Learning, Lattice Boltzmann Method, Hydrodynamic Schrödinger Equation, Navier Stokes Equations

Abstract. This document is meant to review and discuss the possible applications of Quantum computing in the area of computational fluid dynamics (CFD). A review of the current state-of-the-art of quantum computing applied to computational fluid dynamics has been carried out, highlining how the technology is promising but still in an early stage of development. Furthermore, within the approaches developed to solve CFD problems with the use of quantum algorithms and / or quantum computers, this article discusses a quantum algorithm approach, based on the Lattice Boltzmann Method and developed to the study of 2D flow around a cylinder, a model which can be related to several industrial problems and, in the future, modified to simulate the refrigeration cycle used in aeronautical environmental control systems (ECS). This preliminary code helped to highlight the inherent difficulties to implement a quantum algorithm but helped also to demonstrate the applicability of quantum computing.

Introduction

Quantum computing is a type of computation that harnesses the collective properties of quantum states, such as superposition, interference, and entanglement, to perform calculations [1]. The devices that perform quantum computations are known as quantum computers.

Quantum Computing is currently a very active field of study and development, with expected applications in various fields ranging from the classical computer science problem, like cryptography and search problem, to engineering application, like structural optimization [2] and mechanical dynamics [3], but additional applications in different fields have been identified and developed [4], also for pure financial investments [5].

Indeed, several approaches have been proposed for the solutions of CFD problems with the use of quantum algorithms and / or quantum computers, as summarized in [6], being the rationale of the development the estimated scalability of quantum computing [7] and the analogy between Navier Stokes equations (NSE) with the Schrödinger equation through the Madelung transform, i.e., the Hydrodynamic Schrödinger equation (HSE) [8].

Known methodologies review and related works.

Basically, three approaches can be envisioned to solve fluid mechanics problems, namely the algorithmic approach, in which is focused on the development of numerical algorithms to be run on quantum computers, the analog approach, which can be described as the "design" of a quantum mechanical system able to "mimic" the fluid mechanic problem, and, last and not least, the development of machine learning codes for quantum computer.

An example of the "algorithmic" or "circuital" approach, implicitly demonstrating the applicability of quantum computing to the classical computational fluid dynamics equations despite the inherent complexity of these equations [7], is the resolution of classical Navier-Stokes equations for a typical De Laval nozzle, described by Gaitan [9].

Another examples of the algorithmic approach are the solution of the vortex-in-the-cell method in a parallel environment by Steijl and Barakos [10] or the solution of Collisionless Boltzmann Equations by Steijl and Todorova [11], which highlights how the streaming operations can be effectively implemented in a quantum circuit by the use of one of elementary quantum gate, i.e., the controlled NOT (shortly CNOT) logic gates, under the assumed periodic boundaries conditions.

It is worth to noting that the problem associated with a (classical) lattice gas can be equivalently solved with a quantum computer implementing the "analog" approach, i.e., the use of a lattice gas quantum computer in which quantum bits replace classical bit and are arranged in a lattice-based array [12], whereas the "streaming" operations are carried out by the quantum system evolution.

Contributions of present work.
To verify the applicability of quantum computer to industrial problem, a relatively simple fluid dynamic problem has been considered, i.e., the flow around a cylinder in a 2D domain, solved with a quantum algorithm based on a two dimensional approach on nine variables (streaming/collision directions), or D2Q9, configuration of the Lattice Boltzmann Method (see the following figure for the calculation mesh), extending the previous works with a non-rarified fluid and consequently including collisions.

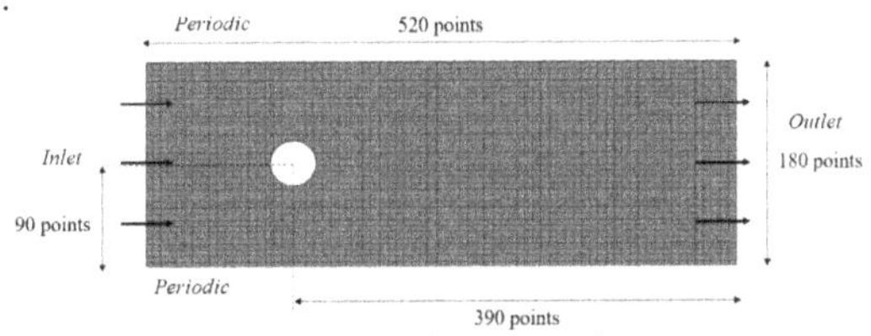

Figure 1 – Lattice mesh

The rationale of this selection is that flow around a cylinder is routinely used as benchmark problem for CFD algorithms because of the phenomena it exhibits but it is representative, in its simplicity, of many applications such as heat exchangers design or subsea pipeline assessment [13], as well as the modelling of porous media [14].

Indeed, the Lattice Boltzmann method has been considered due to the opportunity of extension of this study to porous media (as the filtering mesh used in aeronautical refrigerant system can be modelled) or to the refrigeration cycle itself (due to the presence of multiphase flow).

The flow is considered unsteady, incompressible, laminar, and with constant fluid properties. Inlet has been modelled imposing the flow velocity, whereas the outlet is a simple opening, with periodic boundary conditions on the upper and lower edge. The cylinder wall has been modeled with the bounce back technique to model the non-slip condition.

The equations regulating the flow can be summarized by the classical Lattice Boltzmann Equations with the BGK (Bhatnagar–Gross–Krook) approximation [15]:

$$f_k(\boldsymbol{x} + \Delta\boldsymbol{x}\,, t + \Delta t) = f_k(\boldsymbol{x}, t) \cdot (1 - \omega) + \omega \cdot f_k^{eq}(\boldsymbol{x}, t)$$

where **x** is the position vector, ω is the relaxation time, f_k is the particle distribution function, and f_k^{eq} is the local equilibrium distribution function defined as:

$$f_k^{eq}(x,t) \doteq w_k \cdot \rho(x,t) \cdot \left(1 + 3 \cdot \frac{\boldsymbol{c_k} \cdot \boldsymbol{u}}{c_s^2} + \frac{9}{2} \cdot \frac{(\boldsymbol{c_k} \cdot \boldsymbol{u})^2}{c_s^4} - \frac{3}{2} \cdot \frac{\boldsymbol{u}^2}{c_s^2}\right)$$

Where **u** is the particle velocity vector, w_k a weighting factor, while c_s is the isothermal speed of sound and $\mathbf{c_k}$ is the unitary velocity vector along the streamlines [15].

In the quantum algorithm, the equilibrium distribution function expression has been truncated to the first order terms respect the velocity to avoid non linearities for this first algorithm and rely only on linear operations, introducing an averaged expected velocity vector $\mathbf{u_0}$:

$$f_k^{eq}(x,t) \simeq w_k \cdot \rho(x,t) \cdot \left(1 + 3 \cdot \frac{c_k \cdot u}{c_s^2} + \frac{9}{2} \cdot \frac{(c_k \cdot u) \cdot (c_k \cdot u_0)}{c_s^4} - \frac{3}{2} \cdot \frac{u \cdot u_0}{c_s^2}\right)$$

The numerical expected error is quite small due to the range of velocity considered. Please note that the classical code has been implemented considering the full expression of the distribution function.

The full state of the quantum system is given by:

$$|\psi\rangle \equiv |a\rangle|F\rangle|y\rangle|x\rangle = |a_0 a_1 a_2\rangle|F_0 F_1 F_2 F_3 F_4\rangle|y_0 y_1 y_2\rangle|x_0 x_1 x_2 x_3\rangle$$

where $|a\rangle$, $|F\rangle$, $|y\rangle$, and $|x\rangle$ represent respectively the state vector relative to the ancilla qubits, the distribution function qubits, the macroscopic variables, and the spatial (*x,y*) coordinates.

The quantum algorithm has been developed and tested using the IBM's quantum computing software development framework *Qiskit* [16], and it can be generally discretized in five major steps [17]: initialization, collision, propagation, boundary condition implementation and, calculation of macroscopic quantities.

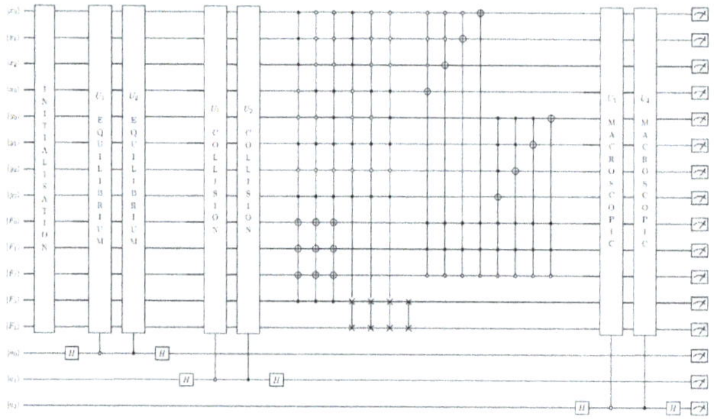

Figure 2 – Quantum algorithm scheme for the single iteration (simplified)

It is worth noting that, after one time step simulation is finished, it is necessary to re-initialize the quantum state for the next step. Indeed, classical programming tools like for-cycles or variables overwriting cannot be implemented into a quantum system.

Results, comparison, and validation

To assess the quantum algorithm, a verification of the implementation of the Lattice Boltzmann equations on classical computer has been carried for several Reynolds numbers, to verify the correctness of the underlying theoretical algorithm, retrieving not only the velocity and vorticity distribution but also evaluating the drag coefficient on the cylinder. Results are in accordance with available data from bibliography and results obtained through the solution of the classical Navier Stokes equations.

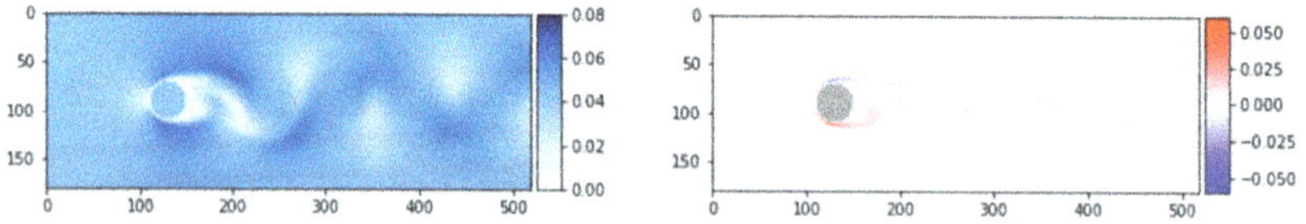

Figure 3 –Classical LBM algorithm results: velocity and vorticity profiles for R_E=1000

The quantum algorithm is still under investigation, but on-going preliminary results seem to be (at least qualitatively) in accordance with the results obtained at the previous step, even considering the strong simplification introduced by the linearization of equilibrium function.

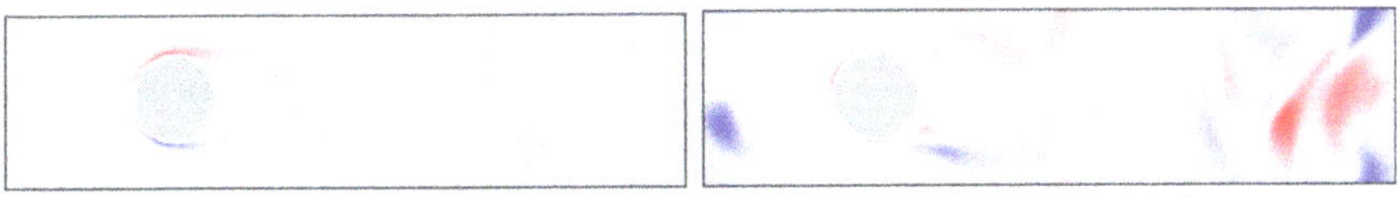

Figure 4 –Classical vs. quantum LBM algorithms results (vorticity)

Unfortunately, a significant drawback (beside the numerical error introduced by the linearization) is the time required to run the above-described algorithm, i.e., execute the algorithm on a quantum computer simulator, mainly due to the initialization and measurement process (roughly 2 minutes for iteration instead of few tenths of seconds for the classical algorithm) [18].

Conclusions

This paper summarized the three approaches that can be envisioned to solve fluid mechanics problems, namely the algorithmic o circuital approach, the analog or annealing approach and, finally, the machine learning applied to quantum computers.

A quantum algorithm for solving a classical two-dimensional fluid mechanic problem is introduced in the present work, based on the "translation" to a quantum computing framework of the numerical procedure known as Lattice Boltzmann method. The rationale of this choice is that the method shows similarities with the quantum operations themselves and it can be extended to multiphase flows and to complex geometrical domains, typical conditions of interesting industrial problems such as the modeling of aeronautical heat exchangers and in general, the refrigeration cycle of aircraft environmental control system.

Very preliminary results show that the quantum algorithm is able to achieve (even if under heavy mathematical simplifications) a result which is comparable with the results obtained with classically implemented Lattice Boltzmann codes, but the implementation shows the inherent difficulties to translate a classical code into a quantum framework (e.g., linearization, time required to simulate the quantum system on classical hardware) and the discrepancies due to the linearization.

Future steps include the implementation of nonlinear distribution function as well as the extension of algorithm to more sophisticated problem, including heat exchanges, also to confirm that the algorithm works fine even outside the range of unitary velocities. Furthermore, another aspect that requires a strong improvement is the numerical implementation in order to avoid the re-initialization of the code at each iteration, which is a major bottleneck in the current implementation.

References

[1] Wikipedia, "Quantum Computing", https://en.wikipedia.org/wiki/Quantum_computing (Last visited: February 23rd, 2022)

[2] Wils, K.A., Quantum Computing for Structural Optimization, M.Sc. thesis, TU Delft, (2020).

[3] Smaili A. & Alt., Application of the flexible link model (FLM) and quantum computing-based algorithm for the optimum synthesis of partially compliant mechanism, ESDA2012-82467, (2012). https://doi.org/10.1115/ESDA2012-82467

[4] Palmer J., Quantum technology is beginning to come into its own", The Economist, https://www.economist.com/news/essays/21717782-quantum-technology-beginning-come-its-own (Last visited: February 23rd, 2022)), (2017).

[5] Walters R., Early quantum computing investors see benefits, Financial Times, https://www.ft.com/content/b2f1c0ea-e4ff-11e7-a685-5634466a6915, (2018).

[6] Bharadway S.S. et Alt., Quantum Computation of Fluid Dynamics, Pramana–Journal of Physics, (2020), DOI 12.3456/s78910-011-012-3.

[7] Chen et alt., "Quantum Finite Volume Method for Computational Fluid Dynamics with Classical Input and Output", arXiv:2102.03557 (2021) [quant-ph].

[8] Meng and Yang, Quantum computing of fluid dynamics using the hydrodynamic Schrödinger equation., arXiv:2302.09741 (2023) [physics.flu-dyn]. https://doi.org/10.1103/PhysRevResearch.5.033182

[9] Gaitan F., Finding flows of a Navier -Stokes fluid through quantum computing, npj Quantum Information, (2020), s41534-020-00291-0. https://doi.org/10.1038/s41534-020-00291-0

[10] Steijl, R. and Barakos, G. N., Parallel evaluation of quantum algorithms for computational fluid dynamics. Computers and Fluids, 173, (2018), pp. 22-28. https://doi.org/10.1016/j.compfluid.2018.03.080

[11] Todorova B. and Steijl R. (2020), "Quantum Algorithm for the Collisionless Boltzmann Equation", Journal of Computational Physics, 409, (2020), 109347. https://doi.org/10.1016/j.jcp.2020.109347

[12] Yepez J. (1998), "Quantum Computation of Fluid Dynamics", Quantum Computing and Quantum Communications Lecture Notes in Computer Science, Colin P. Williams (Ed.) Vol. 1509 Springer-Verlag Berlin (1999). https://doi.org/10.1007/3-540-49208-9_3

[13] Doreti, L. and Dineshkumar, L. Control techniques in flow past a cylinder- A Review. IOP Conference Series: Materials Science and Engineering. 377, 012144 (2018,6). https://doi.org/10.1088/1757-899X/377/1/012144

[14] Fattahi et alt., Lattice Boltzmann methods in porous media simulations: From laminar to turbulent flow. Computers Fluids. 140 pp. 247-259 (2016). https://doi.org/10.1016/j.compfluid.2016.10.007

[15] Mohamad A.A., Lattice Boltzmann method - Fundamentals and Engineering Applications with Computer Codes, Springer-Verlag (London), 2011. https://doi.org/10.1007/978-0-85729-455-5

[16] https://qiskit.org/

[17] Budinski L., Quantum algorithm for the Navier Stokes equations by using the stream function vorticity formulation and the lattice Boltzmann method, Quantum Physics, 2022, https://doi.org/10.48550/arXiv.2103.03804

[18] Lecumberri J., Quantum computing implementation of LBM, thesis, Universidad de Zaragoza, B.Sc. Physics degree, 2023

Analysis of plasma formation during hypersonic flight in the earth atmosphere

Salvatore Esposito[1,a*], Domenic D' Ambrosio[2,b]

[1]Dipartimento di Elettronica e Telecomunicazioni, Politecnico di Torino, Torino, Italy

[2]Dipartimento di Ingegneria Meccanica e Aerospaziale, Politecnico di Torino, Torino, Italy

[a]salvatore_esposito@polito.it, [b]domenic.dambrosio@polito.it

Keywords: Hypersonic Aerodynamics, Plasma, Non-Equilibrium, Thermo-Chemistry

Abstract. In this study we investigate the formation of plasma in hypersonic flight and its impact on radio communications and radar tracking. The transfer of kinetic energy from the vehicle to the surrounding gas in the hypersonic regime leads to the formation of plasma, which can cause interference with electromagnetic waves. By conducting a numerical simulation campaign using Computational Fluid Dynamics (CFD), we are determining the critical Mach number and altitude conditions that lead to plasma formation. The plasma generated at the nose of the vehicle and its subsequent convection along the body and in the wake are the main subjects of our investigation. The simulations include physical models that account for chemical, vibrational and electron-electron energy non-equilibria, using a two-temperature approach. The results indicate the Mach numbers and altitudes at which plasma formation can significantly affect the propagation of electromagnetic waves.

Introduction

Hypersonic flight presents a significant challenge to aircraft design and operation due to the highly complex gas flow characteristics that occur when objects travel at speeds well in excess of the speed of sound. In hypersonic flight, the transfer of kinetic energy from the object to the surrounding gas creates a region of high temperature around the body, leading to the formation of plasma [1], which can greatly affect the propagation of electromagnetic waves. If the charge density in the plasma is high enough, the wave can be completely reflected. Specifically, if the collision frequency tends to zero and the frequency of the radio link is less than or equal to the plasma frequency (satisfying the cut-off condition), the real part of the permittivity tends to zero or becomes negative. As a result, the electromagnetic wave becomes evanescent, resulting in an exponential decay of its intensity as it traverses that region of space [2]. The plasma surface thus replaces the surface of the body, distorting the reflected radiation and altering the radar trace. Even if the plasma frequency is below the cut-off values, refraction and absorption can still occur, causing a redistribution of electromagnetic waves and a reduction in re-radiation.

Understanding and predicting plasma formation around hypersonic vehicles is therefore critical for accurate tracking and evaluation of radio communication capabilities.

We present a numerical simulation campaign based on CFD tools focused on the prediction of plasma formation in suborbital hypersonic flight. The aim of the research is to determine the Mach number and altitude conditions that could produce regions around the vehicle where the plasma frequency, collision frequency (and hence permittivity) reach critical levels. For this purpose, we consider a test matrix with Mach numbers between 8 and 16 and altitudes between 20 and 70 km. We analyze the plasma formation at the nose of the vehicle and its subsequent convection along the conical nose and in the wake. We use this data to show the flight regimes in which plasma formation can interfere with the propagation of electromagnetic waves, and we superimpose this with surface temperature data to eliminate flight conditions that are not feasible for thermal protection reasons.

Physical Model and Numerical Method

The adopted physical model is based on the Navier-Stokes equations and includes non-equilibrium phenomena involving vibrational and electronic energy relaxation, as well as chemical and ionization reactions. We consider air as a gas mixture potentially composed of 7 chemical species, namely monoatomic oxygen and nitrogen, O and N, nitric oxide, NO, diatomic oxygen and nitrogen, O_2, N_2, the positive ion NO^+, and electrons, e^-. The effects of non-equilibrium energy on air chemistry are represented by a two-temperature model. The rates of the chemical reactions are derived from [4,5], while the thermodynamic properties of each chemical species are taken from [6].

Due to the flight conditions, a non-magnetized, inhomogeneous, collisional cold plasma model is assumed. Under this premise, the relative permittivity, ε_r, which determines the wave transmission in a medium, can be approximatively described by the Drude model [2]:

$$n^2 = \varepsilon_r = 1 - \frac{(f_pe^2)}{(f(f + if_c))} = 1 - \frac{(f_{pe}^2)}{(f^2 + f_c^2)} + \frac{i(f_{pe}^2 f_c)}{\left(f(f^2 + f_c^2)\right)}. \quad (1)$$

In Eq. (1), n is the refractive index at the frequency of the electromagnetic wave, f. The electron number density, n_e , determines the electron plasma frequency as in Eq. (2):

$$f_{pe} = \frac{1}{2\pi}\sqrt{e^2 n_e/\varepsilon_0 m_e}. \quad (2)$$

where e is the electric charge, m_e is the effective mass of the electron, and ϵ_0 is the permittivity of free space.

The term f_c is the collision frequency between electrons and neutral particles, defined as:

$$f_c = \frac{1}{2\pi}\sum_i n_i \sigma_{i,e}\sqrt{\frac{8k_bT}{\pi m_e}}. \quad (3)$$

where n_i and $\sigma_{i,e}$ are respectively the number density and the neutral-electron scattering cross section for the neutral species i, T is the temperature and k_b is the Boltzmann constant.

The real part of ϵ_r governs the wave propagation, while the imaginary part controls the collisional absorption, i.e., the transfer of energy from electrons to neutral species.

The mathematical formulation of the physical model is solved numerically using the CFD software ICFD++ by Metacomp Technologies. It utilizes a finite volume discretization approach, employing a Harten-Lax-van Leer-Contact (HLLC) approximate Riemann solver with Total Variation Diminishing (TVD) limited second-order reconstruction for the convective fluxes. The diffusive fluxes are computed using a naturally second-order centered scheme. The computational mesh is unstructured and consists primarily of polyhedral cells, except for the wall region where a structured, stretched quadrilateral grid is employed to accurately capture the boundary layer. The mesh undergoes local refinement through an Adaptive Mesh Refinement technique based on the magnitude of the Mach number gradient. The total number of cells in the mesh varies between 350,000 and 400,000, depending on the specific test case. We apply the model to an axi-symmetric blunt-nosed cone exposed to hypersonic flow. The body has an overall length of 1.125 m. The nose of the cone is an ellipsoid with minor and major radii of 2.5 cm and 5.5 cm, respectively. This is followed by a cylindrical section that is 2.5 cm long, and finally, a cone with a semi-opening angle of approximately 8.2047°, as described in [3]. Despite its simplicity, this geometric configuration serves as a suitable starting point for representing plasma formation around a slender body in hypersonic atmospheric flight. To capture the plasma in the wake, the computational domain extends up to 2.875 m behind the nose.

Results

The results were obtained considering non-catalytic and radiation-adiabatic wall conditions, assuming a zero angle of attack to ensure axisymmetric flow. Applying a radiation-adiabatic wall boundary condition implies that the wall is considered adiabatic, but it can radiate heat received from the flow. Such a condition is known as 'radiative equilibrium.' It is assumed that the gas is fully transparent to the radiation flowing away from the wall, while the wall benefits from radiative cooling. Previous research has demonstrated that, in many scenarios, this condition provides a reasonably accurate estimate of surface temperature compared to flight data [7]. The radiative equilibrium condition states that the heat transfer into the wall is determined by the approximate relation:

$$q_w = \varepsilon\sigma T_w^4 \tag{4}$$

implying that the wall will reject all heat from the flow, except for what it can radiate away. In Eq. (4), σ=5.67×10^{-8} W/m^2/K^4 is the Stefan-Boltzmann constant, while ϵ is the surface emissivity, which we assumed to be equal to 0.8 in this work.

The permittivity was calculated for an electromagnetic wave frequency of 1 GHz, which is typical for land-based long-range surveillance radars [8]. Setting the maximum possible temperature at the vehicle surface to 3500 K, Table 1 shows that the Mach 14 and 16 conditions at 20 km altitude and the Mach 16 condition at 30 km altitude are not feasible. The results also show that no thermochemical phenomena are observed at an altitude of 70 km for Mach numbers between 8 and 10, and therefore these observations have not been reported.

Table 1 – Maximum wall temperature in Kelvin degrees.

z[Km]\M	8	9	10	12	14	16
70	/	/	/	1.47E+03	1.68E+03	1841
50	1.56E+03	1.74E+03	1.91E+03	2.05E+03	2.27E+03	2.51E+03
30	2041	2.30E+03	2.43E+03	2.82E+03	3.29E+03	3761
20	2.25E+03	2.43E+03	2.83E+03	3.34E+03	3845	4.44E+03

Figures 1 to 4 indicate that along the symmetry axis in the region of the cone nose and at four different stations in the wake, the plasma frequency and the collision frequency reach their maximum values, and the real part of the permittivity reaches its minimum values. For the highest Mach numbers (at least M=12), the permittivity decreases below unity even in the wake, indicating refraction of electromagnetic waves in this region. However, significant permittivity values are only observed at altitudes of 30 and 50 km.

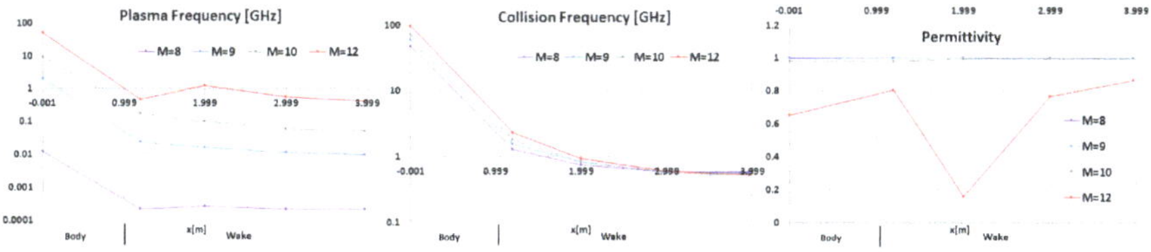

Figure 1. Maximum plasma and collision frequencies, and minimum real part of permittivity at 20 km altitude.

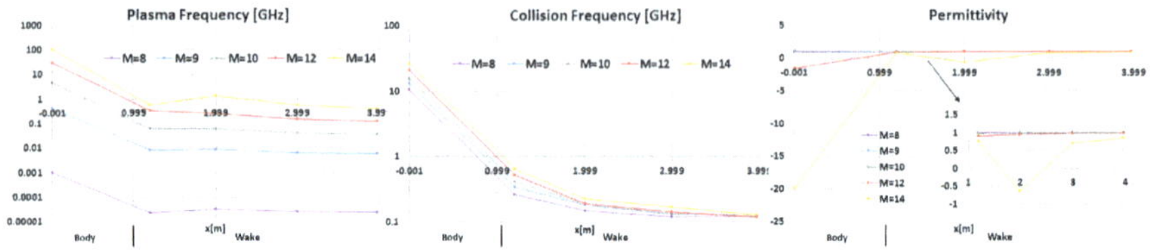

Figure 2. Maximum plasma and collision frequencies, and minimum real part of permittivity at 30 km altitude.

Specifically, the results presented in Fig. 1 indicate that the conditions at 20 km altitude correspond to highly collisional regimes with relevant collision frequencies. Significant permittivity values are observed only at Mach 12. Regarding the results in Fig. 4, the very low pressure at 70 km altitude limits the thermodynamic activity, resulting in relevant conditions only for Mach 16 and at the nose, while plasma formation in the wake remains negligible.

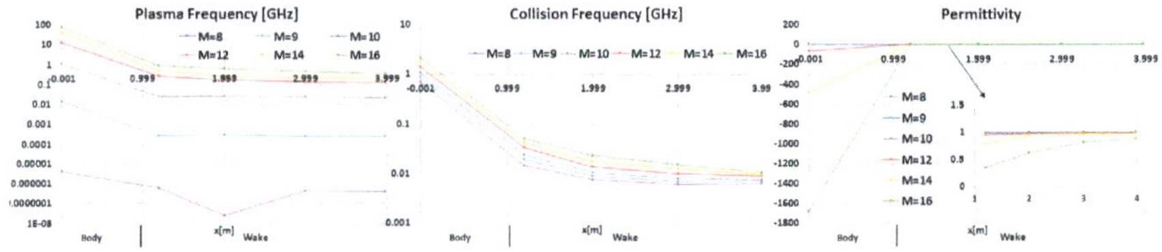

Figure 3. Maximum plasma and collision frequencies, and minimum real part of permittivity at 50 km altitude.

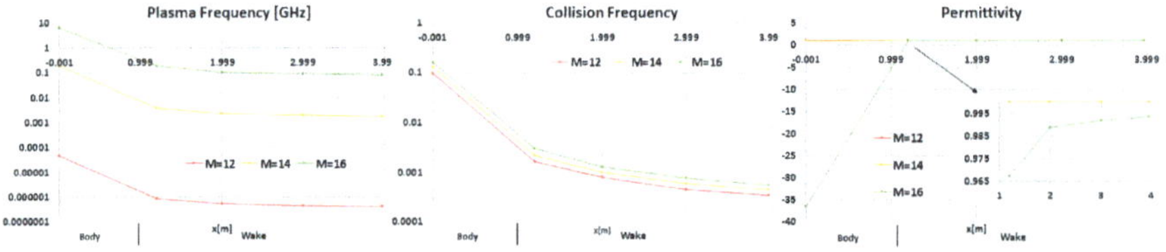

Figure 4. Maximum plasma and collision frequencies, and minimum real part of permittivity at 70 km altitude.

Conclusions

The objective of this study was to understand plasma formation during hypersonic flight and its impact on radio communications and radar tracking using CFD tools and non-equilibrium thermochemical models. The results showed that at altitudes between 30 and 50 kilometers and Mach numbers above 12, the plasma sheath can have a significant effect on the propagation of electromagnetic waves. Subsequent research will include an analysis of the scattering of electromagnetic waves induced by the plasma, with focus on the study of the radar cross section.

References

[1] Anderson, John David. Hypersonic and high temperature gas dynamics. AIAA, 1989.

[2] Stix, Thomas H. Waves in plasmas. Springer Science & Business Media, 1992.

[3] Qian, Ji-Wei, Hai-Li Zhang, and Ming-Yao Xia. "Modelling of Electromagnetic Scattering by a Hypersonic Cone-Like Body in Near Space." International Journal of Antennas and Propagation 2017 (2017), Article ID 3049532. https://doi.org/10.1155/2017/3049532

[4] Park, Chul. "Review of chemical-kinetic problems of future NASA missions. I - Earth entries." Journal of Thermophysics and Heat transfer 7(3), 1993: 385-398. https://doi.org/10.2514/3.431

[5] Park, Chul, Richard L. Jaffe, and Harry Partridge. "Chemical-Kinetic Parameters of Hyperbolic Earth Entry." Journal of Thermophysics and Heat transfer 15(1), 2001: 76-90. https://doi.org/10.2514/2.6582

[6] Gupta, Roop N., et al. "A review of reaction rates and thermodynamic and transport properties for an 11-species air model for chemical and thermal nonequilibrium calculations to 30000 K." NASA-RP-1232, 1990.

[7] Gnoffo, Peter A., Johnston, Christopher O., and Thompson, Richard A. "Implementation of Radiation, Ablation, and Free Energy Minimization in Hypersonic Simulations." Journal of Spacecraft and Rockets. 47(2), 2010: 251–257. https://doi.org/10.2514/1.44916

[8] Skolnik, Merrill I. Radar handbook. McGraw-Hill Education, 2008.

Materials Research Proceedings 37 (2023) 184-188 | https://doi.org/10.21741/9781644902813-40

Multi-step ice accretion on complex three-dimensional geometries

Alessandro Donizetti[1,a*], Tommaso Bellosta[1,b], Mariachiara Gallia[1,c], Andrea Rausa[1,d], Alberto Guardone[1,e]

[1]Politecnico di Milano, Department of Aerospace Science and Technology, Via Privata Giuseppe La Masa, 34, 20156 Milano, Italy

[a]alessandro.donizettti@polimi.it, [b]tommaso.bellosta@polimi.it, [c]mariachiaria.gallia@polimi.it, [d]andrea.rausa@polimi.it, [e]alberto.guardone@polimi.it

Keywords: In-Flight Icing, Multi-Step, Level-Set Method, Unstructured Grids, Mesh Adaptation

Abstract. This work presents the Politecnico di Milano Icing Research Group's contribution to developing new numerical tools and methodologies for simulating long-term in-flight icing over complex three-dimensional geometries. PoliMIce is an in-house ice accretion software that includes state-of-the-art solvers for the dispersed phase to compute the droplets' impact on the aircraft, and ice accretion models, including the exact local solution of the unsteady Stefan problem. PoliMIce has also been extensively developed for the simulation and robust design optimization of thermal ice protection systems. A crucial aspect that characterizes and makes numerical simulations challenging is the formation, and evolution in time of complex ice geometries, resulting from the ice accretion over the body surface and/or previously formed ice. A multi-step procedure is implemented since the aerodynamic flow field is coupled with ice accretion. The total icing exposure time is subdivided into smaller time steps. At each time step, a three-dimensional body-fitted mesh suitable for the computation of the aerodynamic flow field around the updated geometry is generated automatically. The novel remeshing procedure is based on an implicit domain representation of the ice-air interface through a level-set method and Delaunay triangulation to generate a new conformal body-fitted mesh. In this work, the unique capabilities of the PoliMIce suite are employed to perform automatic multi-step ice accretion simulations over a swept wing in glaze ice conditions. Numerical simulations are hence compared with the available experimental data.

Introduction

In-flight icing is a complex problem that involves various disciplines, including aerodynamics, multi-phase flows, thermodynamics, and meshing capabilities.

It is a critical safety issue in aeronautics since it disrupts the aircraft's aerodynamics. Computational fluid dynamics techniques are valuable for simulating different and potentially extreme conditions that complement experimental and in-flight campaigns, better understanding how ice accretion affects aerodynamic performances, and designing optimal ice protection systems [1].

For the numerical simulation of in-flight ice accretion, most of the icing tools, such as LEWICE [2], FENSAP-ICE [3], SIMBA-ICE [4], rely on a standard and well-established segregated approach as multi-step approach. Under a quasi-steady assumption, the total icing exposure time is divided into smaller time steps. The aerodynamic flow field, the amount and distribution of the cloud water droplets impinging on the selected surfaces, and the ice growth rate are computed sequentially at each step. Then, the new geometry and the corresponding surrounding mesh must be updated. This operation is usually the most critical phase of the multi-step loop, especially if a fully automated procedure is desired.

The following sections describe how the in-house code PoliMIce [5] computes the aerodynamic flow field, the amount and distribution of the cloud water droplets impinging on the selected surfaces, the ice growth rate, and the updating of the geometry and corresponding surrounding mesh.

Flow solver

Aerodynamic simulations of the external airflow are performed here using SU2 [6]. A node-centered finite volume method (FVM) is applied on arbitrary unstructured meshes using a standard edge-based data structure on a dual grid with median-dual control volumes. Convective fluxes are discretized at each edge mid-point using either centered or upwind schemes. The aerodynamic field is computed as a solution to the Reynolds-averaged Navier-Stokes (RANS) equations, used in this study in tandem with the Spalart–Allmaras (SA) turbulence model.

Collection efficiency solver

Due to the scales at play in ice accretion and the concentration of water droplets, their effects on the solution of the airflow field can usually be neglected so that the computation of the aerodynamics can be performed independently of the water droplets. This assumption leads to the so-called one-way coupled approach; only the airflow field can affect the motion of water droplets. The in-house particle tracking code is based on a Lagrangian framework and simulates clouds containing supercooled water droplets [7, 8].

The Lagrangian framework allows straightforward modeling of supercooled water droplets' effects, such as splashing, aerodynamic breakup, and deformation, and can deal with secondary particles. As the result depends on the particle resolution, a strategy was developed to automatically refine the seeding region by adding new particles where needed. Elements are incrementally split at each iteration, evolving the current cloud front, and computing the collection efficiency on the surface. The simulation stops when the difference in the L_2 norm between two consecutive collection efficiency calculations is below a user-supplied threshold.

Thermodynamic solver

The in-house code PoliMIce [5] is used for computing the ice accretion. Computing the thickness of the forming ice layer amounts to solving a phase change problem over the body surface. Typically surfaces are first discretized in computational cells, and a one-dimensional Stefan problem is solved for each control volume. Early icing tools rely on the approximate solution of the Stefan problem proposed by Messinger in 1953 [9] for aeronautical applications. In 2001 Myers [10] proposed an improved version of Messinger's model, obtaining a better representation of the transition between the rime and the glaze ice. A further modification to Myers' model, based on the exact local solution of the unsteady Stefan problem, is implemented in the current version of PoliMIce [11]. PoliMIce can also perform numerical simulation of Electro-Thermal Ice Protection Systems (ETIPS) [12], both in the anti-icing and de-icing mode, identifying six different icing conditions and the transition between one and the other, including accretion, melting of ice, and water evaporation.

The surface roughness is estimated a-priori with the empirical formula of Shin, as a function of liquid water content (LWC), static air temperature, and freestream velocity. It provides an equivalent sand-grain roughness, k_s, which is needed by the Spalart-Allmaras turbulence model.

In the multi-step procedure, the water film distribution and all the icing data are interpolated from the previous step through a nearest-neighbor search algorithm.

Updating geometry module

To avoid mesh entanglement and grid intersections typical of standard mesh deformation techniques, PoliMIce updating geometry module [13, 14] is based on an implicit domain remeshing strategy to obtain a robust and automatic remeshing procedure that permits long-term in-flight icing simulations without the user intervention.

At each step, the new ice-air interface is represented as the zero-contour level of a level-set function defined over the computational volume, such that $\varphi < 0$ in the portion of the domain occupied by material, $\varphi > 0$ in the portion of the domain not occupied by material, and $\varphi = 0$ at the interface. Keeping the clean body elements untouched, only the ice-air interface ones are then isotropically remeshed to obtain a body-fitted surface mesh with good-quality elements, which is finally used as a base for generating the volume grid required for the next time step computations.

Results

This section presents a 10 steps ice accretion simulation over a 30° degree swept NACA0012 wing in glaze ice conditions, taken from the 1st Ice Prediction Workshop. Numerical results are compared with the experimental results obtained at NASA Glenn Icing Research Tunnel (IRT).

Test conditions are reported in Table 1. The inputs to the problem are the angle of attack (AoA), the freestream velocity, temperature and pressure, the liquid water content (LWC) of the cloud, i.e. the grams of water contained in a cubic meter of air, the droplet median volume diameter (MVD) and the total icing exposure time.

Table 1: Icing test conditions.

Case	Sweep Angle [deg]	AoA [deg]	Vel [m/s]	Temp [K]	Pres [Kpa]	LWC [g/m^3]	MVD [μm]	Time [s]
362	30	0	103	266	92.32	0.5	34.7	1200

The simulated ice shape is characterized by high temperatures near the stagnation point, which result in thin ice layers at the leading edge, and runback water driven by shear stresses.

The length and the angle of the horns are well predicted, although there is an underprediction in the overall mass of ice accreted, particularly near the stagnation point.

The final ice shape and the evolution of the geometry represented through a slice perpendicular to the leading edge are represented in Figure 1 and Figure 2, respectively. Comparing the solution obtained with the single-step approach with the multi-step one, it is evident how the latter is fundamental to correctly capture the glaze ice shape, accounting for the progressive modification of the flow field around the wing.

Conclusions and future work

This paper briefly presented the PoliMIce toolkit, which can perform multi-step simulations of in-flight ice accretion over three-dimensional geometries, avoiding mesh entanglement and allowing for long-term simulations necessary to simulate full aircraft configurations and bringing the possibility of certification by simulation in the near future.

In the future, PoliMIce will be employed for multi-step simulations around other swept wings with different sweep angles, testing new roughness and density models to better understand ice accretion's physics and consequently improve its modeling. Finally, complex three-dimensional ice shapes, easily managed by the proposed methodology, will be adopted to evaluate the aerodynamic losses due to ice formation on commercial and military aircraft.

Figure 1: Glaze ice shape at t= 1200 s.

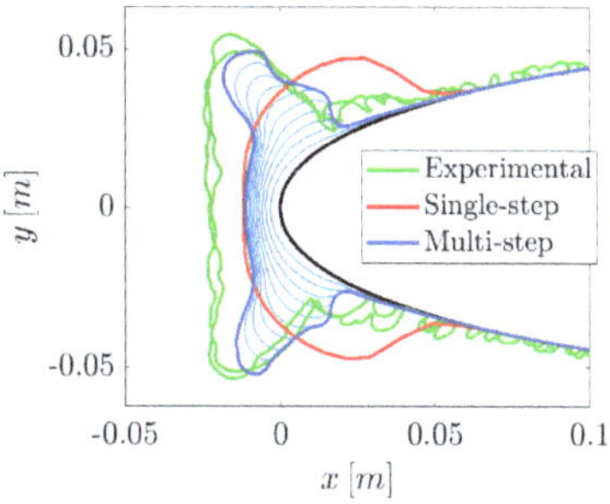

Figure 2: Ice shape evolution.

References

[1] M. Gallia, B. Arizmendi Gutierrez, G. Gori, A. Guardone et P. M. Congedo, «Robust Optimization of a Thermal Anti-Ice Protection System in Uncertain Cloud Conditions,» *Journal of Aircraft,* pp. 1-15, 2023. https://doi.org/10.2514/1.C037223

[2] W. B. Wright, «User Manual for the NASA Glenn Ice Accretion Code LEWICE,» 2002.

[3] H. Beaugendre, W. Habashi et F. Morency, «FENSAP-ICE's three-dimensional in-flight ice accretion module: ICE3D,» *Journal of Aircraft,* vol. 40, pp. 239-247, 2003. https://doi.org/10.2514/2.3113

[4] D. de Rosa, F. Capizzano et D. Cinquegrana, «Multi-step Ice Accretion by Immersed Boundaries,» chez *International Conference on Icing of Aircraft, Engines, and Structures*, 2023. https://doi.org/10.4271/2023-01-1484

[5] M. Morelli, B. Tommaso, A. Donizetti et A. Guardone, «Assessment of the PoliMIce toolkit from the 1st AIAA Ice Prediction Workshop,» chez *AIAA AVIATION 2022 Forum*, 2022. https://doi.org/10.2514/6.2022-3307

[6] T. D. Economon, F. Palacios, S. Copeland, T. Lukacyzk et J. Alonso, «SU2: An Open-Source Suite for Multiphysics Simulation and Design,» *AIAA Journal,* vol. 54, pp. 828-846, 2015. https://doi.org/10.2514/1.J053813

[7] T. Bellosta, G. Baldan, G. Sirianni et A. Guardone, «Lagrangian and Eulerian algorithms for water droplets in in-flight ice accretion,» *Journal of Computational and Applied Mathematics,* vol. 429, 2023. https://doi.org/10.1016/j.cam.2023.115230

[8] G. Baldan, T. Bellosta et A. Guardone, «Efficient Lagrangian particle tracking algorithms for distributed-memory architectures,» *Computers & Fluids,* vol. 256, 2023. https://doi.org/10.1016/j.compfluid.2023.105856

[9] B. Messinger, «Equilibrium Temperature of an Unheated Icing Surface as a Function of Air Speed,» *Journal of the Aeronautical Sciences,* vol. 20, pp. 29-42, 1953. https://doi.org/10.2514/8.2520

[10] T. Myers, «Extension to the Messinger model for aircraft icing,» *AIAA Journal,* vol. 39, pp. 211-218, 2001. https://doi.org/10.2514/3.14720

[11] G. Gori, G. Parma, M. Zocca et A. Guardone, «Local Solution to the Unsteady Stefan Problem for In-Flight Ice Accretion,» *Journal of Aircraft,* vol. 55, pp. 251-262, 2018. https://doi.org/10.2514/1.C034412

[12] M. Gallia, A. Rausa, A. Martuffo et A. Guardone, «A Comprehensive Numerical Model for Numerical Simulation of Ice Accretion and Electro-Thermal Ice Protection System in Anti-icing and De-icing Mode, with an Ice Shedding Analysis,» chez *International Conference on Icing of Aircraft, Engines, and Structures*, 2023. https://doi.org/10.4271/2023-01-1463

[13] A. Donizetti, T. Bellosta, A. Rausa, B. Re et A. Guardone, «Level-Set Mass-Conservative Front-Tracking Technique for Multistep Simulations of In-Flight Ice Accretion,» *Journal of Aircraft,* pp. 1-11, 2023. https://doi.org/10.4271/2023-01-1467

[14] A. Donizetti, A. Rausa, T. Bellosta, B. Re et A. Guardone, «A Three-Dimensional Level-Set Front Tracking Technique for Automatic Multi-Step Simulations of In-Flight Ice Accretion,» chez *International Conference on Icing of Aircraft, Engines, and Structures*, 2023. https://doi.org/10.4271/2023-01-1467

Large Eddy simulations and Reynolds-averaged Navier-Stokes simulations of separation-induced transition using an unstructured finite volume solver

Manuel Carreño Ruiz[1,a] *and Domenic D'Ambrosio[1,b]

[1] Department of Mechanical and Aerospace Engineering, Politecnico di Torino, C.so Duca degli Abruzzi, 24, 10124 Torino, Italy

[a]manuel.carreno@polito.it and [b]domenic.dambrosio@polito.it

Keywords: Laminar Separation Bubbles (LSB), Large Eddy Simulation (LES), Transition Modelling, Reynolds Averaged Navier Stokes (RANS)

Abstract. The study aims to assess the capability of different methodologies in capturing the separation-induced transition phenomenon. This transition mechanism occurs when the flow separates from the airfoil surface, and transitions from a laminar to a turbulent state due to the amplification of the Kelvin-Helmholtz instability developed in the separated shear layer. The simulations employ high-order numerical methods for solving the Navier-Stokes equations, while the transition modeling for RANS is based on the $\gamma - Re_\theta$ transition model. LES enables prediction of the onset and location of transition and provides turbulent flow statistics.

Introduction

Predicting flow transition on airfoils is crucial for designing unmanned aerial systems, as it directly impacts their aerodynamic performance, stability, and control. However, accurately predicting flow transition remains challenging due to the complex nature of flow physics and the limited availability of high-fidelity experimental data, especially in the very-low Reynolds number regime. One extensively studied benchmark case in this regard is the SD7003 airfoil. Galbraith et al. [1] carried out Implicit Large Eddy Simulations (ILES) using a Discontinuous Galerkin method to study flow transition on the SD7003 airfoil. Uranga et al. [2] corroborated the good performance of ILES in computing separation-induced transition, testing Reynolds numbers as low as 22,000. Catalano et al. [3] presented LES results using a second-order scheme for the chordwise and wall-normal directions, employing Fourier colocations in the spanwise direction. They contested the conclusions of [1], which claimed the necessity of a high-order scheme to capture laminar separation bubbles. In fact, second-order schemes can adequately capture separation-induced transition, albeit with extremely refined grids. RANS approaches have also been employed for studying the SD7003 airfoil. Windte et al. [4] coupled a k-ω model with an e^N transition model to predict transition around the airfoil. Catalano et al. [5,6] proposed modifications to the k-ω SST turbulence model to better capture lower Reynolds number flows, applying it to this specific airfoil. De Santis et al. [7] recently introduced a modification of the γ transition model [8] to enhance turbulent kinetic energy production within separation bubbles. Carreño et al. [9] also attempted to increase turbulent kinetic energy production in separation bubbles by tuning the s_1 parameter in the $\gamma - Re_\theta$ transition model [10].

This study uses LES to predict flow transition on the SD7003 airfoil at Reynolds number 60,000 and Mach number of 0.2. We examine different numerical settings for LES, focusing on the impact of selecting higher-order and less dissipative schemes. We compare the velocity field created by the separation bubble using the RANS model described earlier and LES. Additionally, we present a comparison between the WALE and dynamic Smagorinsky subgrid-scale models.

Numerical Methods

Numerical simulations in this paper were conducted using the commercial software STAR-CCM+. The SD7003 airfoil geometry at an angle of attack of 4 degrees was employed for this analysis. To minimize the influence of the far-field boundary, the computational domain was extended to approximately 100 chords. The RANS equations were solved on a two-dimensional grid. A time-accurate implicit second-order integration was utilized to capture the vortex shedding that occurs behind the trailing edge of low Reynolds number airfoils at low angles of attack prior to transition. The time-step was set to 0.01 convective turnovers. Further details regarding the numerical setup of the RANS simulations and the implementation of the $\gamma - Re_\theta$ transition model can be found in [9]. For LES, the SD7003 geometry was extruded in the spanwise direction for 0.1 chords, which has been determined by [1] as sufficient for computing flow statistics at low and moderate angles of attack. The lateral boundaries were meshed conformally and assigned periodic boundary conditions. Several grids were tested to evaluate the influence of resolution on capturing small-scale structures, with the finest grid consisting of approximately 15 million cells. The grid ensures y_+ values below 0.2 and x_+ and z_+ values below 5 near the airfoil. Furthermore, we verified that the grid resolves at least 80% of the turbulent kinetic energy in the spectrum. Simulations were performed using 128 cores of 4 Intel Xeon Scalable Processors Gold 6130 2.10 GHz. The time step employed in our simulation was 0.002 convective turnovers. The inner solver executed 10 iterations per time step, resulting in residuals dropping between 2 and 3 orders of magnitude. Simulations were conducted for 20 turnovers, and statistics were computed within the last 10. The Wall Adaptive Local Eddy-viscosity (WALE) sub-grid turbulence model and the Dynamic Smagorinsky approach were compared. The recommended spatial discretization in STAR-CCM+ is the second-order Bounded Central Difference (BCD) scheme. This scheme was compared with a third-order Central Difference (CD3) scheme, a third-order Monotonic Upwind Scheme for Conservation Laws (MUSCL3), and a hybrid scheme that combines third-order central difference and upwind schemes (CD/MUSCL3). Third-order schemes demonstrated no significant overhead compared to the bounded central difference scheme implemented in STAR-CCM+. Compared to the standard second-order upwind scheme, the overhead was approximately 10%. Despite these small overheads, the enhanced accuracy of third-order schemes, as demonstrated by Ricci et al. [11], could outweigh the additional computational costs.

Results and Discussion

Figure 1 illustrates the Q-criterion iso-surface obtained using the CD3 scheme, revealing an artificial structure outside the boundary layer resulting from solver instability. Conversely, the hybrid CD/MUSCL3 scheme with a blending factor of 0.15 proved to be stable and robust. In Figure 2, the friction coefficient for our simulations is compared to the results presented by Galbraith & Visbal [1]. It is important to note that their simulations were conducted at a Mach number of 0.1. However, we believe this discrepancy is likely insufficient to account for the variations observed in Figure 2. Our simulations exhibit a delayed separation transition and reattachment, with separation bubbles of the same length but shifted toward the trailing edge. These differences indicate that dissipation remains a significant factor, necessitating further grid refinements to ensure that grid-related issues do not influence this phenomenon. Nevertheless, some intriguing observations can still be made. Firstly, reducing the blending factor clearly decreases dissipation, which is associated with an artificially rapid decay of turbulent kinetic energy when using upwind schemes. Additionally, we can observe that the Dynamic Smagorinsky approach demonstrates a higher level of turbulent kinetic energy production, anticipating the reattachment of the boundary layer.

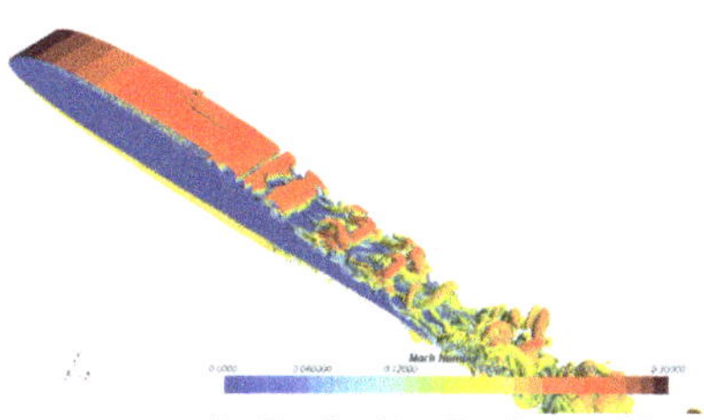

A) Upwing blending factor of 0.

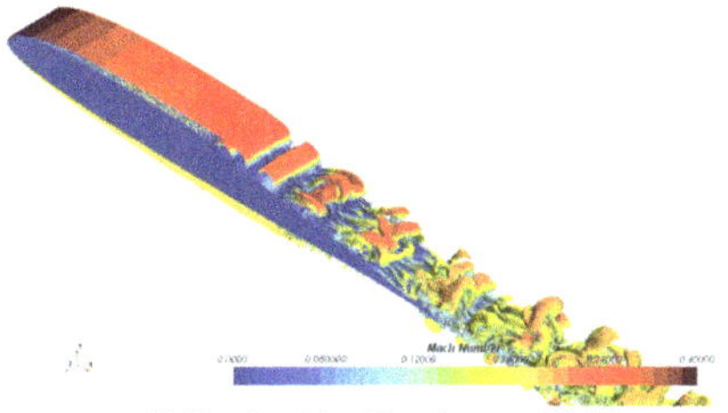

B) Upwing blending factor of 0.15.

Figure 1. Q-criterion=500 Isosurface coloured with Mach number. CD3/MUSCL3 Scheme.

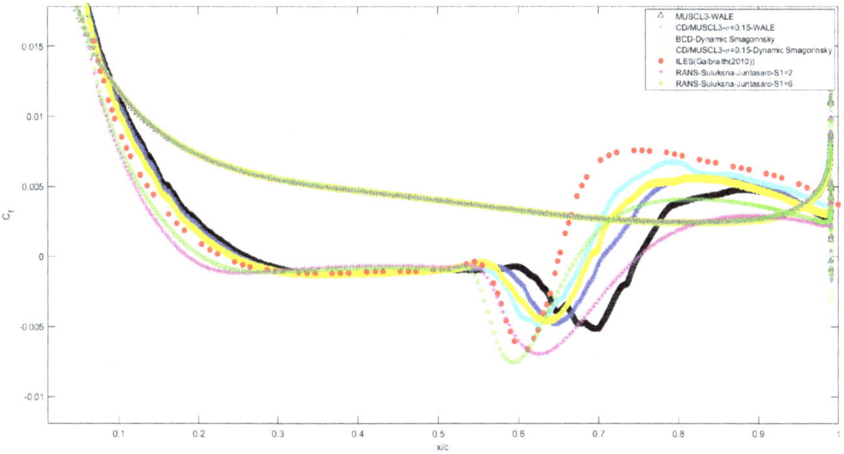

Figure 2. Time-averaged friction coefficient at the mid-span plane.

When comparing our RANS and LES approaches, we observe a noticeable shift of the separation bubble towards the trailing edge. This discrepancy arises because the RANS model follows the recalibration performed in reference [9], which utilized the Galbraith ILES results [1] as a reference. Nevertheless, the overall agreement is satisfactory. Figure 3 depicts the averaged velocity fields, revealing the presence of a separation bubble in both cases. Once again, we observe a slight shift of the bubble towards the trailing edge in the LES simulation. Notably, the LES simulation predicts a much sharper closure of the separation bubble, leading to a higher friction coefficient after reattachment. This discrepancy in skin friction under-prediction becomes more pronounced at lower Reynolds numbers [9] and is associated with excessive damping of turbulent kinetic energy production near the airfoil wall in the RANS model. Despite these discrepancies, Figure 3 highlights the usefulness of a well-tuned transition model, enabling accurate performance predictions with two-dimensional simulations that require a computational cost of approximately 1 CPU-hour, compared to around 40,000 CPU-hours for the 3D LES simulations. Furthermore, it opens up the possibility of enhancing the accuracy of RANS simulations in complex three-dimensional scenarios, such as flow over a rotor, where LES remains currently unaffordable.

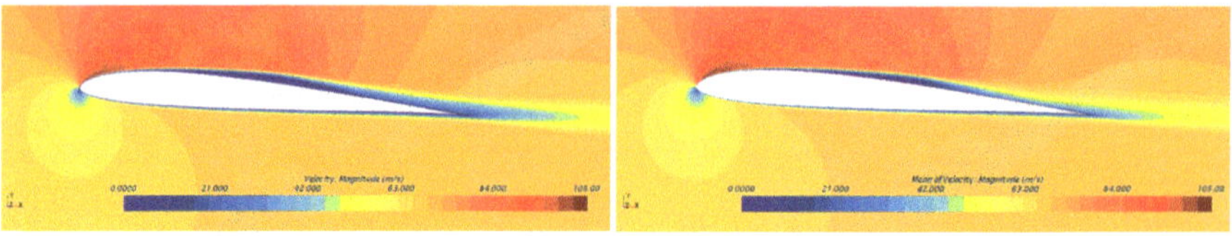

A) RANS. $S_1 = 6$. **B) LES. Upwing blending factor of 0.15.**

Figure 3. Time-averaged velocity magnitude fields at the mid-span plane.

References

[1] Galbraith, M., & Visbal, M, Implicit large eddy simulation of low-Reynolds-number transitional flow past the SD7003 airfoil. In 40th fluid dynamics conference and exhibit (2010) AIAA-4737. https://doi.org/10.2514/6.2010-4737

[2] Uranga, A., Persson, P. O., Drela, M., & Peraire, J. Implicit large eddy simulation of transitional flows over airfoils and wings. In 19th AIAA Computational Fluid Dynamics (2009) AIAA- 4131. https://doi.org/10.2514/6.2009-4131

[3] Catalano, P., & Tognaccini, R. Large eddy simulations of the flow around the SD7003 airfoil. In AIMETA Conference (2011) pp. 1-10.

[4] Windte, J., Scholz, U., and Radespiel, R., "Validation of RANS Simulation of Laminar Separation Bubbles on Airfoils," Aerospace Science and Technology Journal, Vol. 10, No. 7, 2006, pp. 484–494. https://doi.org/10.1016/j.ast.2006.03.008

[5] Catalano, P. and Tognaccini, R., "Turbulence Modelling for Low Reynolds Number Flows," AIAA Journal, Vol. 48, No. 8, 2010, pp. 1673–1685. https://doi.org/10.2514/1.J050067

[6] Catalano, P. and Tognaccini, R., "RANS analysis of the low-Reynolds number flow around the SD7003 airfoil," Aerospace Science and Technology Journal, 2011. https://doi.org/10.1016/j.ast.2010.12.006

[7] De Santis, C., Catalano, P., & Tognaccini, R. (2022). Model for enhancing turbulent production in laminar separation bubbles. AIAA Journal, 60(1), 473-487. https://doi.org/10.2514/1.J060883

[8] Menter, F. R., Smirnov, P. E., Liu, T., & Avancha, R. (2015). A one-equation local correlation-based transition model. Flow, Turbulence and Combustion, 95, 583-619. https://doi.org/10.1007/s10494-015-9622-4

[9] Carreño Ruiz, M., D'Ambrosio, D. Validation of the γ-Re θ Transition Model for Airfoils Operating in the Very Low Reynolds Number Regime. Flow Turbulence Combust 109, 279–308 (2022). https://doi.org/10.1007/s10494-022-00331-z

[10] Langtry, R. B., & Menter, F. R. (2009). Correlation-based transition modeling for unstructured parallelized computational fluid dynamics codes. AIAA journal, 47(12), 2894-2906. https://doi.org/10.2514/1.42362

[11] Ricci, F. & Strobel, P. & Tsoutsanis, P. & Antoniadis, A. Hovering rotor solutions by high-order methods on unstructured grids. Aerospace Science and Technology. (2019). https://doi.org/10.1016/j.ast.2019.105648

High-fidelity simulation of the interaction between the wake of a descent capsule and a supersonic parachute

Luca Placco[1*], Giulio Soldati[3], Alessio Aboudan[1], Francesca Ferri[1], Matteo Bernardini[3], Federico Dalla Barba[2] and Francesco Picano[2]

[1] Centro di Ateneo di Studi e Attività Spaziali 'Giuseppe Colombo' (CISAS), Università degli Studi di Padova, via Venezia 15, 35131 Padua, Italy

[2] Department of Industrial Engineering, Università degli Studi di Padova, via Venezia 1, 35131 Padua, Italy

[3] Department of Mechanical and Aerospace Engineering, Sapienza Università di Roma, via Eudossiana 18, 00184 Rome, Italy

*luca.placco@unipd.it

Keywords: Supersonic Parachute, Supersonic Flows, Large Eddy Simulation, Unsteady

Abstract. The objective of the project is to analyze the unsteady dynamics of the parachute-capsule system in a supersonic airflow while descending during planetary entry. Currently, a combination of Large-Eddy Simulation and an Immersed-Boundary Method is being utilized to examine the evolving flow of a rigid supersonic parachute trailing behind a reentry capsule as it descends through the atmosphere of Mars. The flow is simulated at $Ma = 2$ and $Re = 10^6$. A massive GPU parallelization is employed to allow a very high fidelity solution of the multiscale turbulent structures present in the flow that characterize its dynamics. We show how strong unsteady dynamics are induced by the interaction of the wake turbulent structures and the bow shock which forms in front of the supersonic decelerator. This unsteady phenomenon called 'breathing instability' is strictly related to the ingestion of turbulence by the parachute's canopy and is responsible of drag variations and structure oscillations observed during previous missions and experimental campaigns. A tentative one-dimensional model of the flow time-evolving dynamics inside the canopy is proposed.

Introduction and case approach

The recent unsuccessful European missions (i.e. ExoMars 2016) proved how the prediction and the understanding of the dynamics of the descent capsule under the effect of a supersonic decelerator is still an open question in the active research scene that revolves around space exploration. The failure of Schiaparelli EDM landing indeed was ultimately caused by an improper evaluation of the coupled oscillatory motions existing between the descent module and the deployed parachute. The models and the experimental evaluations that were employed to predict the general behaviour of the capsule under the effect of a supersonic decelerator proved to be insufficient, triggering the premature end of the mission [1]. In this context, the main aim proposed by this research activity is to develop a novel technique to study effectively how compressible and turbulent flows interact with non-rigid structures, to properly evaluate and predict their non-steady behaviour. The description of this phenomenon is very elaborate, being affected by several uncertainties such as atmosphere fluctuations, unsteady flow dynamics and structure oscillations [2],[3]. The proposed approach involves Large-Eddy Simulations (LES) [4] for solving the multi-scale flow dynamics and Immersed Boundary Methods to deal effectively with moving solid boundaries [5]. A novel technique to deal with the fluid-structure interaction of compressible flows and thin membranes is in the process of development, starting from the existing IBM strategies. As a starting point for the implementation of the final configuration, a Large-Eddy simulation of a rigid mock-up parachute trailing behind a reentry capsule has been performed, showing both the potential of the LES approach and the primary dependence of the breathing phenomenon to the interaction of the turbulent wake of the module with the bow shock produced by the parachute.

Computational approach and simulation setup

Compressible Navier-Stokes equations are solved with the high-order finite difference solver STREAmS [4]. Turbulent structures are ultimately identified using the implicit large eddy simulation (ILES) approach; in this way, conventional LES turbulence modeling has been omitted, using instead the numerical dissipation given by the numerical discretization as artificial viscosity acting at small scales. Thus, the 3D Navier-Stokes equations solved are the following:

$$\frac{\partial \rho}{\partial t} + \frac{\partial(\rho u_j)}{\partial x_j} = 0$$

$$\frac{\partial(\rho u_i)}{\partial t} + \frac{\partial(\rho u_j u_i)}{\partial x_j} + \frac{\partial p}{\partial x_i} - \frac{\partial}{\partial x_j}\left(\mu\left(\frac{\partial u_i}{\partial x_i} + \frac{\partial u_j}{\partial x_i} - \frac{2}{3}\frac{\partial u_k}{\partial x_k}\delta_{ij}\right)\right) = 0 \tag{1}$$

$$\frac{\partial(\rho E)}{\partial t} + \frac{\partial(\rho E u_j + p u_j)}{\partial x_j} + \frac{\partial}{\partial x_j}\left(\lambda\frac{\partial T}{\partial x_j}\right) + \frac{\partial}{\partial x_j}\left(\mu\left(\frac{\partial u_i}{\partial x_i} + \frac{\partial u_j}{\partial x_i} - \frac{2}{3}\frac{\partial u_k}{\partial x_k}\delta_{ij}\right)u_i\right) = 0$$

where ρ is the density, u_i denotes the velocity component in the i Cartesian direction ($i = 1,2,3$) and p is the thermodynamic pressure. With the intent of reproducing the effect of Mars' atmosphere, the fluid is considered as an ideal gas of CO_2; the ratio between the specific heat at constant pressure C_p and the specific heat at constant volume C_v is set to 1.3 while Prandtl number is 0.72. $E = C_vT + u_i^2/2$ represents the total energy per unit mass and the dynamic viscosity μ is assumed to follow the generalized fluid power-law. The thermal conductivity λ is related to μ via the Prandtl number with the following expression: $\lambda = C_p\mu/Pr$. Convective and viscous terms are discretized using a sixth-order finite difference central scheme while flow discontinuities are accounted through a fifth-order WENO scheme. Time advancement of the ODE system is given by a third-order explicit Runge-Kutta/Wray algorithm. No-slip and no-penetration wall boundary conditions on the body are enforced through an Immersed-Boundary Method (IBM) algorithm.

The simulation was performed at $Ma = 2$ and $Re = 10^6$ to simulate the condition at which the parachute deploys. The reference fluid properties associated to the free-stream condition correspond to an altitude of about 9 km from the planet surface and have been obtained using a simulated entry and descent trajectory through the Mars atmosphere of a generic reentry probe [2].

The flow domain selected to perform this first simulation has a size of $L_x = 20D$, $L_y = 5D$, $L_z = 5D$, where D=3.8 m is the maximum diameter of the descent module; parachute diameter is set to $2.57D$. the mesh is a rectilinear structured grid that consists of $N_x \cdot N_y \cdot N_z = 2560 \cdot 840 \cdot 840$ nodes. The grid density changes in both axial and transverse directions, gaining resolution in the central portion of the domain; the position of the capsule nose is set at [1D, 0, 0] while the parachute center lies at [10D, 0, 0]. Computations have been carried out on CINECA Marconi100 cluster, allowing the domain parallel computing on a total of 64 GPUs.

Results

In figure 1 we observe the two-dimensional instantaneous flow field obtained by isolating the y=0 slice from the full 3D domain; Mach number contours are shown. Subsonic flow regions (in red), sonic regions (in white) and supersonic areas (in blue) can be identified. We observe the generation of two bow shocks ahead of the capsule and the canopy and the wake produced by the two bodies. The flow at the vent section is sonic. Pushed by the high pressure within the canopy and finding a larger passage section, it rapidly accelerates to the highest Mach number of the flow field. The breathing motion involves inhomogeneous fluctuations in pressure and density, resulting in substantial variations in drag, even though the canopy area remains constant. The primary cause of the breathing cycle appears to be the aerodynamic interaction between the wake of the capsule and the bow shock created by the parachute canopy.

Figure 2 shows density ratio contours in the area around the canopy. Different phases of the cycle that surrounds the periodic motion of the front bow shock along the flow direction: an increasing density inside the canopy pushes the shockwave away, allowing a larger flux to escape from the canopy (from [1] to [2]). Thus, this creates a decrease in the density that in turn draws back in the shockwave ([2] to [3]) and restarts the cycle ([3] to [4]). The

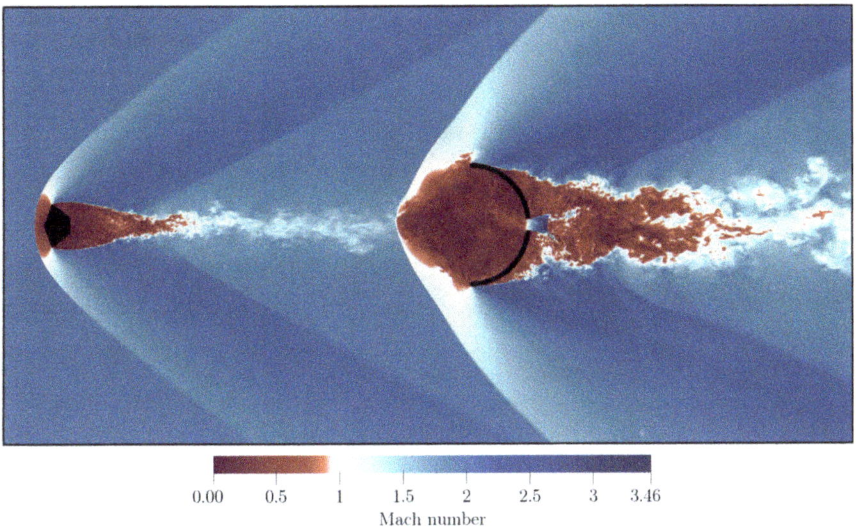

Figure 1: Instantaneous Mach contours (y = 0 cross section) of the simulated flow domain.

Conclusions

The present work proposes a high-fidelity time-evolving simulation of the interaction between the turbulent wake of a supersonic descent module and a generic rigid mock-up thick decelerator. We show how the critical 'breathing' instability associated to supersonic parachutes is intrinsically connected to the interaction of the turbulent wake flow of the descent module and the front bow shock produced by the decelerator. To overcome the limitation of the current setup and further extend the representation of its dynamics, the implementation of a novel immersed boundary method technique is in progress. This will require the solution of fluid-structure interaction of compressible supersonic flows and flexible thin membranes. The new framework will involve an extension of the current IBM module and a finite element method model to deal with flexible moving boundaries (zero-thickness), representing the very thin structure of the simulated decelerator. In this way, the approach in development will allow to represent properly both the entire deployment sequence and the system unsteadiness in all its components, thus providing the full representation of the 'breathing' phenomenon. The oscillation cycles align with the dynamics of the wake, as observed in previous experimental studies [3]. These cycles exhibit a frequency that is consistently around 0.16 in terms of the Strouhal number.

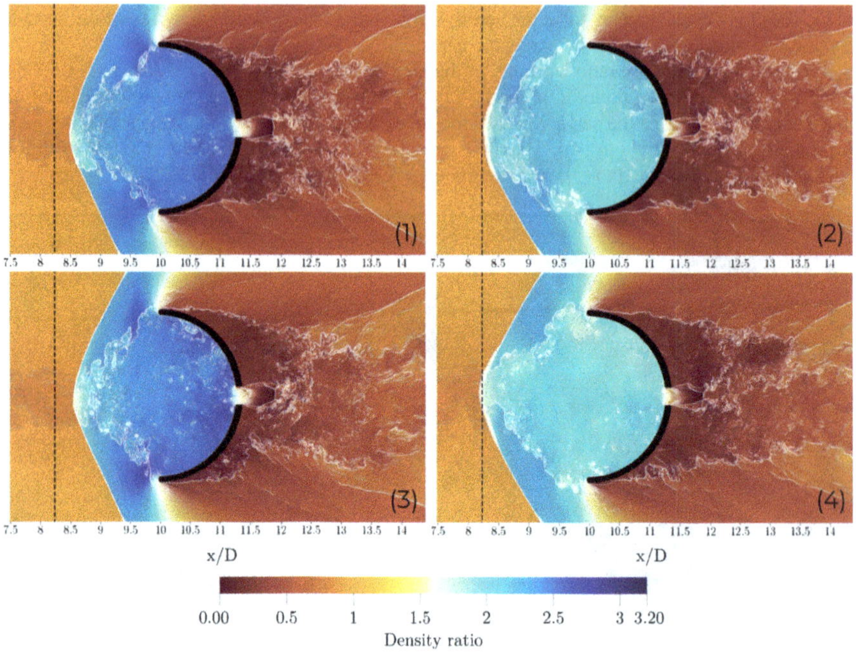

Figure 2: Instantaneous density ratio contours (y = 0 cross section) at different progressive timestep around the parachute canopy.

References

[1] T. Tolker-Nielsen. EXOMARS 2016 - Schiaparelli Anomaly Inquiry, 2017.

[2] A. Aboudan et al. ExoMars 2016 Schiaparelli module trajectory and atmospheric profiles reconstruction. Space Science Reviews, 214: 97, 08 2018. https://doi.org/10.1007/s11214-018-0532-3

[3] X. Xue and Chih-Yung Wen. Review of unsteady aerodynamics of supersonic parachutes. Progress in Aerospace Sciences, 125:100728, 2021. https://doi.org/10.1016/j.paerosci.2021.100728

[4] M. Bernardini, D. Modesti, F. Salvadore, and S. Pirozzoli. Streams: a high-fidelity accelerated solver for direct numerical simulation of compressible turbulent flows. Co.Ph.Co., 263, 2021. https://doi.org/10.1016/j.cpc.2021.107906

[5] H. Yu and C. Pantano. An immersed boundary method with implicit body force for compressible viscous flow. Journal of Computational Physics, 459:111125, 2022. https://doi.org/10.1016/j.jcp.2022.111125

Impact of a wedge in water: assessment of the modeling keyword, presence of cavitation and choice of the filter most suitable for the case study

D. Guagliardo[1,a], E. Cestino[1,b*], G. Nicolosi[1,a], E. Guarino[1,a], A. Virdis[1,a], A. Alfero[1,a], D. Pittalis[1,a] and M.L. Sabella[1,a]

[1]Politecnico di Torino (DIMEAS), Corso Duca degli Abruzzi, 24 10129 Torino, Italy

[a] teams55.polito@gmail.com, [b] enrico.cestino@polito.it

Keywords: Fluid-Structure Interaction, SPH, Cavitation, Pressure Filter

Abstract. The purpose of this paper is to compare the results obtained from a rigid wedge impacting water that is modelled using different techniques based on the SPH (Smoothed Particle Hydrodynamics) method. The study aims to evaluate the quality of the results, optimizing the computational time, which is obtained when the wedge is discretized as a section or as a half of it. The comparison of the results obtained considers the different materials that the ANSYS LS-DYNA software allows to assign to water through different keywords. The effect of cavitation on the pressures reached during the vertical impact was evaluated as a function of ambient temperature. Finally, given the high noise recorded in the pressure files, the study uses a filter created in MATLAB. The latter involves a double pass through the Kalman filter first and the Gauss filter later. All results obtained through the numerical method are compared with Von Karman and Wagner analytical theories.

Introductio

The student team "TEAM S55" of Politecnico of Turin was born in 2017 to rebuild the SIAI-Marchetti S55 seaplane on a 1:8 scale [1-3]. The following paper is created by the FSI section which studies the interaction between the aircraft and the water at ditching. The purpose of this paper is to describe through ANSYS LS-DYNA the water impact of a wedge, at a vertical speed of 5.8 m/s, to evaluate the effect of various parameters in wedge and water modelling. Due to the high computational requirements inherent in the SPH method, the influence of air is neglected to reduce the computational time, since it doesn't have an influence on the accuracy of stress prediction [4]. The analysis results are purified from numerical noise through a filter implemented in MathWorks MATLAB.

Model description

Full model. The model under study is composed of two parts, here under described. The first one is the wedge, it has a width of 254.8 mm, a height of 105 mm, a thickness of 2 mm and a dihedral angle of 30°, as shown in *Fig. 1*. The item is discretized in 1320 shell elements, mainly concentrated in the impacting wall. The material used is an infinitely rigid steel, modelled through the keyword *020 MAT_RIGID*. The boundary conditions forced the wedge to translate only along the z-direction and lock any rotation.

The second one is the water. The water box has dimensions x = 1200 mm, y = 40 mm, z = 300 mm, and it got an SPH number of 56000. It has got **BOUNDARY_SPH_NON_REFLECTING* plans in the bottom and perpendicular to the x-direction, and **BOUNDARY_SPH_SYMMETRY-PLANE* perpendicular to the y-direction. It's possible to use three different materials to model the water [5]: *009 MAT_NULL*, *001 MAT_FLUID-ELASTIC_FLUID* and *010 MAT_ELASTIC_PLASTIC_HYDRO*. For each material, it's possible to define the cavitation pressure. The phenomenon of cavitation is a function of the ambient temperature through vapor pressure, and it is defined differently for each material. For materials 009 and 010 cavitation is indicated by the pressure cut-off, while for materials 001 it is defined by the cavitation pressure parameter.

Figure 1. Wedge and water

Figure 2. Half wedge with symmetry plane

Model description

Half model. The geometry in this case turns out to be the exact half as shown in *Fig. 2*; both the wedge and the water box were halved along the x direction and a symmetry wall was then added via the keyword **BOUNDARY_SPH_SYMMETRY_PLANE*, green in *Fig. 2*, as used by [4].

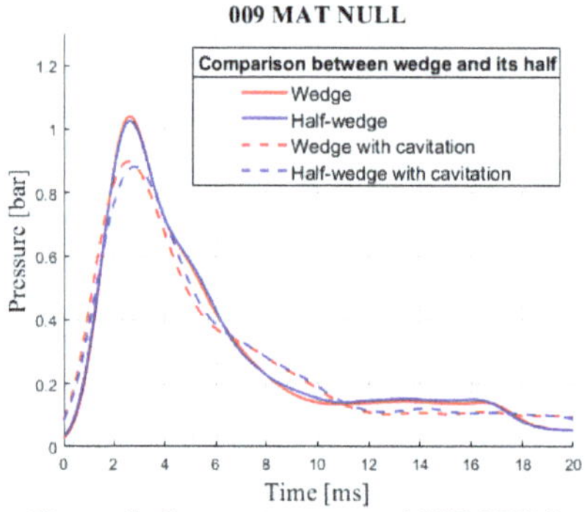

Figure 3. Comparison using MAT_NULL

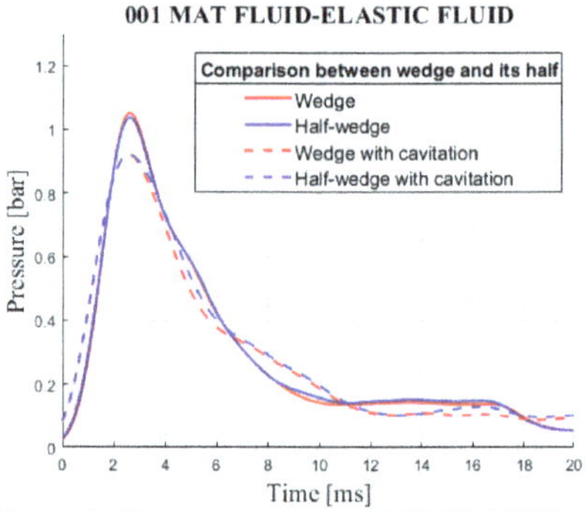

Figure 4. Comparison using MAT_ELASTIC

Between a complete model and a halved one there is a time saving of 48% with a discrepancy in the results, as regards the peak pressure in *001 MAT_FLUID-ELASTIC_FLUID*, of 1% for the results in which cavitation is neglected and almost nothing as regards the case in which cavitation is considered. The pressure peaks for *009 MAT_NULL* differ by 1.34% when cavitation is ignored and by 1.84% when cavitation is considered. The pressure detection sensor, in all analyses, is positioned at 13.93 mm along the x direction; this position will be the same one considered for the analytic theories of von Karman [6] and Wagner [7].

The pressure results for the cases just mentioned are shown in *Fig. 3* and *Fig. 4* and are related respectively to the materials *009 MAT_NULL* and *001 MAT_FLUID-ELASTIC_FLUID*. Furthermore, the vertical impact velocity of the wedge is 5.8 m/s and the vapor pressure considered for cavitation is referred to standard conditions (25 °C).

Water keyword comparison

As studied by Q.W. Ma and D.J. Andrews [7] the water can be defined in three different ways. *009 MAT_NULL* and 0*10 MAT_ELASTIC_PLASTIC_HYDRO* need an EOS (Equation Of State), it was chosen the **EOS_GRUNEISEN* for both materials. In modelling a fluid, it can be observed that some keywords can be advantageous due to computational cost, the *001 MAT_FLUID-ELASTIC_FLUID* appears to be the fastest in computation, because it's the only one without an EOS. The *009 MAT_NULL* is 6% slower, and the 0*10 MAT_ELASTIC_PLASTIC_HYDRO* is even 26% slower. *Fig. 5* shows the results obtained for the three different keywords, in the absence of cavitation, compared with the maximum values predicted by the analytical theories of von Karman [6] and Wagner [7]. It's noteworthy that in the peak pressure zone, which is the most important in terms of structural strength, the values are consistent among the various materials, with an error lower than 2%.

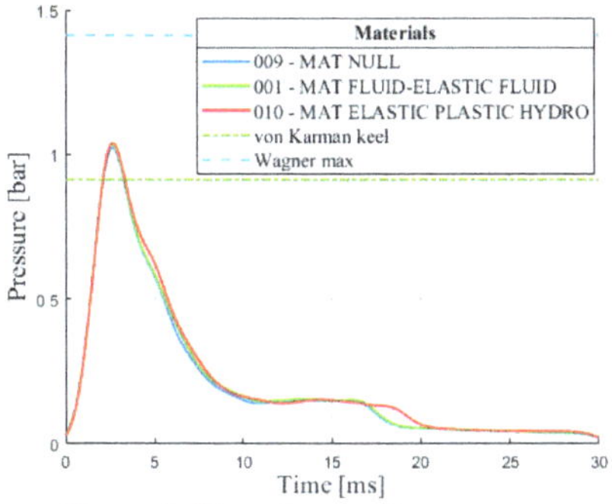

Figure 5. Water keyword comparison

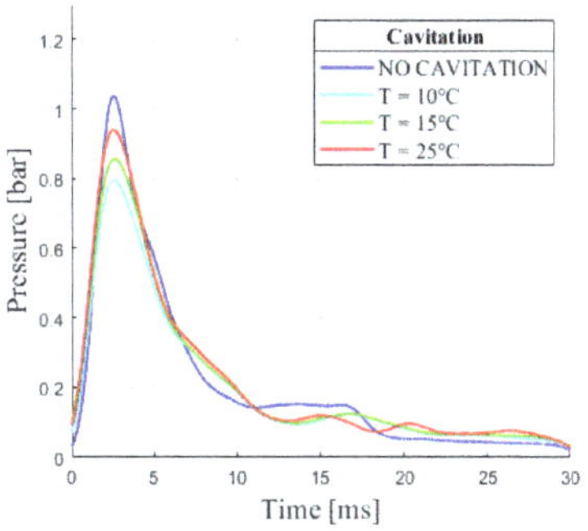

Figure 6. Pressure as function of temperature

Effect of cavitation

Cavitation is a phenomenon that can occur when an object impacts a liquid, causing its phase change to gas due to the variation in fluid pressure, as predicted by Wagner [8]. That condition is verified in the study, but it was found a similar phenomenon even with velocities that don't verify the relation. This phenomenon is defined by Korobkin [9] as interface cavitation. The study shows that in a zone close to the first impacting object part, the hydrodynamic pressure is smaller than the atmospheric pressure. This leads to the formation of a cavity that, however, does not extend to the peripheral zones. The formation of this cavity, whose shape and size are described analytically in [9], changes the liquid flow and the pressure distribution. On the LS-DYNA software [10], pressure cut-off (PC) is defined to allow for a material to "numerically" cavitate. In other words, when a material undergoes dilatation above a certain magnitude, it should no longer be able to resist this dilatation. Since dilatation stress or pressure is negative, setting PC as the vapor pressure value, for the desired temperature, would allow for the material to cavitate once the pressure in the material goes below this negative value. Cavitation pressure (CP) is also defined in *001 MAT_FLUID-ELASTIC_FLUID*, but in this case the chosen value is positive. Analysis have been carried out by comparing the pressure trend in the case where the cavitation is not considered and if the cavitation is considered at the temperatures of 10°C, 15°C and 25°C. For the comparison the material 001 MAT_FLUID-ELASTIC_FLUID is used. As can be seen in *Fig. 6* the pressure peak without cavitation is greater with respect to the cases where cavitation is considered, but after the peak, signal noise is lower. Considering that the vapor pressure decreases as the temperature decreases, it was found that the peak pressure decreases with decreasing water temperature and the effect of cavitation becomes more significant.

Filter choice and effects

The pressure results contained in the outputs provided by LS-DYNA are very noisy due to the unstable nature of the SPH. From previous studies [11], the importance of data filtering in this type of experiment was understood; so, the new goal was to search for the best filter that approximated data accurately with a lower error correlation. The choice was the Kalman filter following the indications given by [12] as it also reconstructs the trend without adding delays. From experimental results, Kalman filter is superior in terms of filtering accuracy, it's implemented in two iterative repeated phases: prediction and update. The *prediction* part consists in predicting the estimated state variable and the predicted state variance variable; the update part includes the Kalman gain and inside it the filtered data corresponds to the *updated* estimated state variable and is shown as the output of the algorithm. However, filtering performance depend on the parameters R, measurement constant, and Q, process variance constant. As reported by [13], the parameters $R = 1$ and $Q = 0.01$ experimentally provide the best filter results because noise is reduced but the original characteristics of the data are preserved. In general, the ratio of R to Q should not exceed 100 to avoid excessively filtered results. After the application of the Kalman filter, the graph is still partially affected by noise, for this reason a Gaussian filter is also applied to further smooth the curves (*Fig. 7*).

A Gaussian filter attenuates the high-frequency components in the signal and passes the low-frequency components. It is based on input signal convolution with a Gaussian function. The input signal is convoluted with the Gaussian filter kernel. This involves point-by-point multiplication

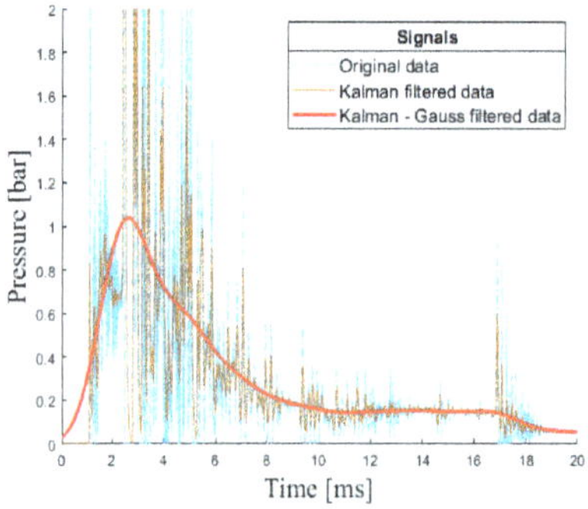

Figure 7. Kalman – Gauss filter

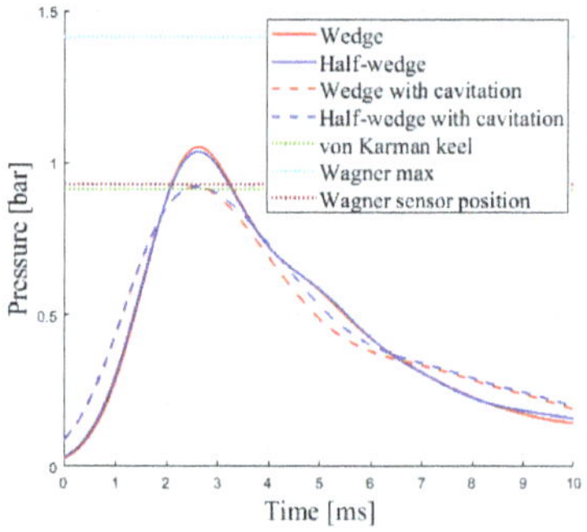

Figure 8. Numerical-analytical comparison

between kernel samples and input signal samples, followed by the sum of the results. The smoothing effect produced by a Gaussian filter will depend on the width of the kernel, which is controlled by the sigma parameter, standard deviation, automatically calculated by MATLAB.

Conclusions

Using a half geometry, it's worth choice to maintain the precision in the results and to reduce the computational cost. The cavitation effect or the cavity formed under the wedge reduce the pressure peak, but it can produce undesired vibrations. It's highlighted the importance of the cavitation as function of temperature variation in the range from 25 °C to 10 °C. A lower water temperature can produce a lower pressure due to the reduction of vapor pressure and bulk modulus. The reduction of these two values prevails over the increase of density and viscosity coefficient. Furthermore, the peak pressure, obtained in the analyses considering cavitation with the water temperature at 25 °C, well fit with the pressure values predicted by the analytical theories of Wagner and von Karman for the sensor position (*Fig. 8*), that were based, especially the von Karman theory, on practical experiments in which is naturally included the cavitation effect.

Acknowledgement
The authors thank the Politecnico di Torino and the DIMEAS department for the provision of resources and facilities. We would thank ANSYS and EnginSoft SpA for the support and for providing us with the software licenses.

References

[1] Cestino, E.; Frulla, G.; Sapienza, V.; Pinto, P.; Rizzi, F.; Zaramella, F.; Banfi, D. (2018) Replica 55 Project: A Wood Seaplane in The Era Of Composite Materials, In: Proc of 31st ICAS 2018 Congress, 9-14 September 2018, Belo Horizonte (Brasil)

[2] Nicolosi G., Valpiani F., Grilli G., Saponaro Piacente A., Di Ianni L., Cestino E., Sapienza V., Polla A., Piana P. Design Of A Vertical Ditching Test. Proc. 32nd ICAS Congress 6-10 September 2021 - Shanghai, China

[3] Cestino, E., Frulla, G., Polla, A., Nicolosi, G. (2023). Equivalent Material Identification in Composite Scaled Hulls Vertical Impact Tests. In: Lopresto, V., Papa, I., Langella, A. (eds) Dynamic Response and Failure of Composite Materials. DRAF 2022. Lecture Notes in Mechanical Engineering. Springer, Cham. https://doi.org/10.1007/978-3-031-28547-9_6

[4] Fragassa C, Topalovic M, Pavlovic A, Vulovic S. Dealing with the Effect of Air in Fluid Structure Interaction by Coupled SPH-FEM Methods. *Materials*. 2019; 12(7):1162. https://doi.org/10.3390/ma12071162

[5] Q.W. Ma and D.J. Andrews. On techniques for simulating effects of cavitation associated with the interaction between structures and underwater explosions using LS-DYNA. 3rd European LS-DYNA Conference, Paris, 2001.

[6] von Kármán T. The impact on seaplane floats during landing. NACA Technical Notes N.321, 1929.

[7] Wagner H. Über Stoß- und Gleitvorgänge an der Oberfläche von Flüssigkeiten. Zeitschrift Für Angewandte Mathematik Und Mechanik, Vol. 12, No. 4, 1932. https://doi.org/10.1002/zamm.19320120402

[8] Panciroli R, Pagliaroli T, Minak G. On Air-Cavity Formation during Water Entry of Flexible Wedges. *Journal of Marine Science and Engineering*. 2018; 6(4):155. https://doi.org/10.3390/jmse6040155

[9] "Korobkin, A. Cavitation in liquid impact problems. In Proceedings of the Fifth International Symposium on Cavitation (CAV2003), Osaka, Japan, 1 January 2003; Volume 2, pp. 1–7."

[10] Keyword Manual, 1999, "LS-DYNA keyword user's manual", Livermore Software Technology Corporation.

[11] F. Valpiani, P. Cicolini, D. Esposto, A. Galletti & D. Guagliardo. Numerical modeling of Fluid-Structure Interactione of a 3D wedge during water impact with variation of velocity and pitch angle. 33rd ICAS Congress 4-9 September 2022 – Stockholm, Sweden

[12] V, Kutz J N, Brunton B W. Numerical differentiation of noisy data: A unifying multiobjective optimization framework. IEEE Access, Vol. 8, 2020. https://doi.org/10.1109/ACCESS.2020.3034077

[13] Ma'arif, Alfian & Iswanto, & Nuryono, Aninditya & Alfian, Rio. (2020). Kalman Filter for Noise Reducer on Sensor Readings. Signal and Image Processing Letters. 1. 11-22. https://doi.org/10.31763/simple.v1i2.2

Thermal fluid-structure interaction by discontinuous Galerkin methods

Vincenzo Gulizzi[1,a*]

[1]Department of Engineering, University of Palermo, Viale delle Scienze, 90128, Palermo

[a]vincenzo.gulizzi@unipa.it

Keywords: Thermal Fluid-Structure Interaction, Discontinuous Galerkin Methods, High-Order Accuracy

Abstract. This research study presents a novel high-order accurate computational framework for thermal fluid-structure interaction problems. The framework is based on the use of block-structured Cartesian grids where level set functions are employed to define both the fluid and the solid regions. This leads to a mesh that consists of a collection of standard *d*-dimensional rectangular elements and a relatively smaller number of irregular elements at the fluid-solid interface. The embedded boundaries are resolved with high-order accuracy thanks to the use of high-order accurate quadrature rules for implicitly-defined regions. The fluid is assumed compressible and governed by the inviscid Navier-Stokes equations, whilst the solid region obeys the equations of thermo-elasticity within the small-strain regime. Numerical examples are provided to assess the capability of the proposed approach.

Introduction

The interest in developing reliable, sustainable and reusable transportation systems that are capable of flying at Mach numbers ranging from 0 to 12 is continuously growing. It is well-known that, within such a wide flight regime, the aircraft structure must endure extreme conditions in terms of pressure and temperature loads. These loads induce a complex thermo-elastic interaction that is generally resolved via the aid of numerical methods as analytical solutions exist for very special combinations of boundary conditions and material properties.

In the context of computational methods, the Finite Volume (FV) method is the industry-standard numerical approach to fluid mechanics problems and is found in many open-source and commercial software libraries; on the other hand, thermo-mechanical problems are very often tackled by the Finite Element (FE) method. Both the FV and the FE methods are extremely robust and widely employed in science and engineering; however, their coupling may become involved and may represent the bottleneck for fluid-structure interaction simulations.

Among the various alternatives to FV- or FE-based approaches, the discontinuous Galerkin (DG) method has proved a powerful numerical technique for both fluid- and solid-mechanics; see, e.g., [1], among several recent contributions. With respect to other techniques, DG-based formulations use spaces of discontinuous basis functions to approximate the solution fields; this naturally enables high-order accuracy with generally shaped mesh elements, block-structured mass matrices and massive parallelization. Additionally, as DG methods enforce both boundary and interface conditions in a weak sense, the coupling between different formulations for the same or for different sets of partial differential equations is significantly simplified. This includes the coupling between different DG formulations or between a DG formulation and a FVM scheme, see Ref.[2].

This study introduces a novel formulation for unsteady thermal fluid-structure interaction problems coupling a shock-capturing FV scheme and a high-order DG scheme. Numerical tests are presented for a thermo-elastic cylinder moving at supersonic speed in an inviscid gas.

Materials Research Proceedings 37 (2023) 202-205 https://doi.org/10.21741/9781644902813-44

Geometry representation and discretization

The coupled thermal fluid-structure interaction problem involves the modeling of two regions consisting of a fluid domain and a solid domain. Here, the geometry is represented via a level set function φ defined in a rectangular domain $\mathcal{R} \subset \mathbb{R}^d$, such that the fluid domain $\mathcal{D}^g$ and the solid domain $\mathcal{D}^s$ are identified by the points belonging to $\mathcal{R}$ where φ is negative and where φ is positive, respectively. It follows that the interface $\mathcal{J}$ between the fluid and the solid domains is identified by $\{\boldsymbol{x} \in \mathcal{R} : \varphi(x) = 0\}$. To illustrate, Fig.(1a) shows a level set function defining a circle in a square domain, whilst Fig.(1b) shows the corresponding fluid and solid regions.

The fluid and the solid domains are eventually discretized. Here, we use the implicitly-defined mesh approach developed in Refs.[2,3,4], which is based on intersecting a structured grid with the zero-contour of the level set functions and allows resolving the curved boundaries with high-order accuracy. Fig.(1c) shows the implicitly defined mesh for the geometry shown in Fig.(1b); in the figure, the darker elements represent the extended elements that prevent the presence of overly small elements in the mesh. See Refs.[2,3,4] for further detail on this meshing strategy.

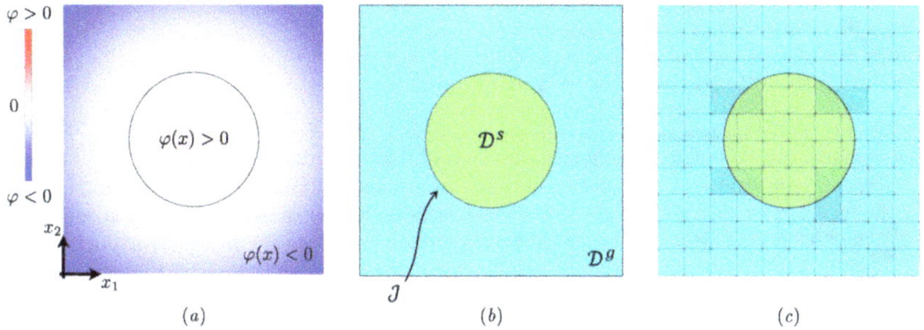

Fig. 1: (a). Level set function defined in a two-dimensional square and (b) corresponding fluid and solid regions identified by the sign of the level set function. (c) Implicitly defined mesh.

Fluid dynamics model

The considered fluid is a compressible gas assumed to obey the Euler equations, which are expressed as the following conservation law:

$$\frac{\partial \boldsymbol{U}^g}{\partial t} + \frac{\partial \boldsymbol{F}_k^g}{\partial x_k} = 0, \tag{1}$$

where t is time, x_k is the k-th spatial component, and $\boldsymbol{U}^g$ and $\boldsymbol{F}_k^g$ denote the $(d+2)$-dimensional vectors of the conserved variables and the flux in the k-th direction, respectively; these are

$$\boldsymbol{U}^g \equiv \begin{pmatrix} \rho^g \\ \rho^g \boldsymbol{v}^g \\ \rho^g e_0 \end{pmatrix} \quad \text{and} \quad \boldsymbol{F}_k^g \equiv \begin{pmatrix} \rho^g v_k^g \\ \rho^g v_k^g \boldsymbol{v}^g + p\boldsymbol{\delta}_k \\ (\rho^g e_0 + p) v_k^g \end{pmatrix}, \tag{2}$$

where ρ^g is the fluid density, $\boldsymbol{v}^g \equiv \left(v_1^g, \dots, v_d^g\right)^{\mathrm{T}}$ is the gas velocity vector, e_0 is the gas total energy and p is the gas pressure. The governing equations are closed by the ideal-gas equation of state with ratio γ of specific heats. In Eq.(1) and in the remainder of the paper, Latin indices will take value in $\{1,\dots,d\}$ and, when repeated, imply summation.

Thermo-mechanical model

The thermo-mechanical model considered here is based on the theory of linear elasticity and Fourier's law of heat conduction. In absence of external sources, it is possible to show that the governing equations of coupled, unsteady thermo-elasticity may be written as:

$$\frac{\partial \boldsymbol{U}^s}{\partial t} - \frac{\partial}{\partial x_k}\left(\boldsymbol{Q}_{kl}\frac{\partial \boldsymbol{U}^s}{\partial x_l} + \boldsymbol{R}_k \boldsymbol{U}^s\right) + \boldsymbol{R}_k^* \frac{\partial \boldsymbol{U}^s}{\partial x_k} + \boldsymbol{S}\boldsymbol{U}^s = \boldsymbol{0}, \tag{3}$$

where

$$\boldsymbol{U}^s \equiv \begin{pmatrix} \boldsymbol{u}^s \\ \boldsymbol{v}^s \\ \vartheta \end{pmatrix}, \quad \boldsymbol{Q}_{kl} \equiv \begin{pmatrix} \boldsymbol{0} & \boldsymbol{0} & \boldsymbol{0} \\ \frac{\boldsymbol{c}_{kl}}{\rho^s} & \boldsymbol{0} & \boldsymbol{0} \\ \boldsymbol{0} & \boldsymbol{0} & \frac{\kappa_{kl}}{c^s} \end{pmatrix}, \quad \boldsymbol{R}_k \equiv \begin{pmatrix} \boldsymbol{0} & \boldsymbol{0} & \boldsymbol{0} \\ \boldsymbol{0} & \boldsymbol{0} & -\frac{\boldsymbol{m}_k}{\rho^s} \\ \boldsymbol{0} & \boldsymbol{0} & 0 \end{pmatrix}, \quad \boldsymbol{R}_k^* \equiv \begin{pmatrix} \boldsymbol{0} & \boldsymbol{0} & \boldsymbol{0} \\ \boldsymbol{0} & \boldsymbol{0} & \boldsymbol{0} \\ \boldsymbol{0} & \frac{T_0 \boldsymbol{m}_k^{\mathrm{T}}}{c^s} & 0 \end{pmatrix} \tag{4a}$$

and

$$\boldsymbol{S} \equiv \begin{pmatrix} \boldsymbol{0} & -\mathbf{I}_d & \boldsymbol{0} \\ \boldsymbol{0} & \boldsymbol{0} & \boldsymbol{0} \\ \boldsymbol{0} & \boldsymbol{0} & 0 \end{pmatrix}. \tag{4b}$$

In Eq.(4), $\boldsymbol{u}^s \equiv (u_1^s, \dots, u_d^s)^{\mathrm{T}}$ is the solid displacement vector, $\boldsymbol{v}^s \equiv (v_1^s, \dots, v_d^s)^{\mathrm{T}}$ is the solid velocity vector, $\vartheta \equiv T^s - T_0$ represents the variation of the solid temperature field T^s with respect to a reference temperature T_0, ρ^s and c^s are the density and the heat capacity per unit volume, respectively, of the solid domain, $\boldsymbol{c}_{kl}$ is a $d \times d$ matrix collecting subsets of elastic coefficients, see, e.g., Refs.[5], κ_{kl} is the kl-th entry of thermal conductivity tensor, $\boldsymbol{m}_k$ is the d-dimensional vector containing components of the thermo-elasticity tensor, and $\mathbf{I}_d$ is the $d \times d$ identity matrix. It is noted that the thermo-elastic properties of the solid are assumed temperature independent.

Thermal fluid-structure coupling

The coupling between the gas region and the solid region occurs at the interface between the two domains, i.e. at $\mathcal{J}$ shown in Fig.(1b). Recalling that the solid is assumed to undergo small deformations, its interface with the gas do not change with time and, as such, behaves like a fixed wall for the gas dynamics equations. Additionally, as the gas is assumed inviscid and non-conducting, its temperature T^g is determined by the equation of state.

The thermal fluid-structure coupling problem is then solved as follows: the conserved variables of Eq.(1) are updated from the time instant t to the time instant $t + \mathrm{d}t$ using an explicit time-integration algorithm; then, at the time $t + \mathrm{d}t$, the computed values of the gas pressure and temperature provide the required boundary conditions at the gas-solid interface to solve the unsteady thermo-elastic problem.

Discontinuous Galerkin formulation

The governing equations of the gas domain, i.e. Eq.(1), are numerically solved via the time-explicit Runge-Kutta discontinuous Galerkin formulation coupled to a shock-capturing second-order FV scheme [2,4]. On the other hand, the equations governing the thermo-elastic solid are solved by extending the DG formulation for elliptic PDEs developed in Refs.[5] with suitably-defined terms accounting for the temporal derivatives in Eq.(3). See Ref.[6] for further detail.

Results

Numerical results are presented for a cylinder with radius $\boldsymbol{r} = \mathbf{0.2}$ m moving at a Mach number $\mathbf{M}_\infty = \mathbf{2}$ at an altitude $\boldsymbol{h} = \mathbf{10}$ km; the geometry and the boundary conditions of the problem are sketched in Fig.(2a). The final time of simulation is $\boldsymbol{I}_t = \mathbf{3}$ ms. The gas is assumed perfect with $\boldsymbol{\gamma} = \mathbf{1.4}$, while the solid is assumed isotropic with properties: density 2700 kg/m^3, Young's modulus 70 GPa, Poisson's ratio 0.33, thermal conductivity coefficient 210 W/(m K), thermal expansion coefficient $\mathbf{24} \times \mathbf{10^{-6}}$ 1/K and volumetric heat capacity $\mathbf{2.43} \times \mathbf{10^{6}}$ J/(m^3 K).

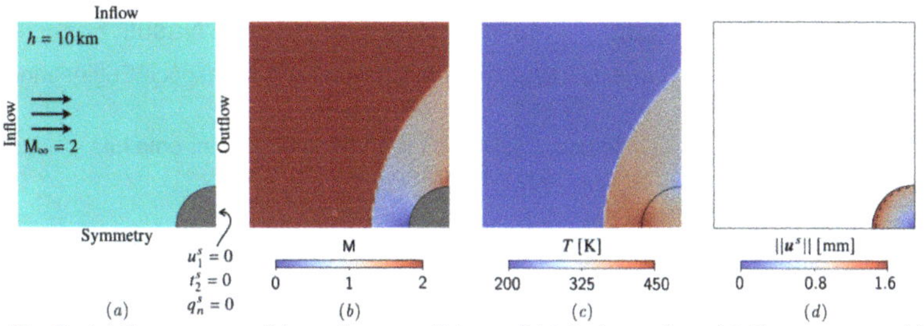

Fig. 2: (a) Geometry and boundary conditions. (b) Mach number. (c) Temperature. (d) Displacement magnitude (the dashed line denotes the undeformed shape).

Fig.(2b) shows the distribution of the Mach number, Fig.(2c) shows the temperature distribution within both the gas and the solid, while Fig.(2d) shows the displacement magnitude of the solid. The figures confirm the ability of the formulation to capture the shock wave, the thermal loads induced by the fluid flow and the deformation of the body.

Conclusions

A novel formulation for unsteady thermal fluid-structure interaction problems has been presented. The formulation uses a high-order accurate represented of embedded geometries, a shock-capturing FV scheme to resolve flow discontinuities, and a high-order accurate DG scheme for the thermo-elastic problem. Numerical results have been presented for a thermo-elastic cylinder moving at M = 2 and have showed the capability of the proposed approach.

References

[1] Z.Cai and B. Thornber. A high-order discontinuous Galerkin method for simulating incompressible fluid-thermal-structural interaction problems. International Journal of Heat and Fluid Flow, 83 (2020), p.108572. https://doi.org/10.1016/j.ijheatfluidflow.2020.108572

[2] V. Gulizzi, A.S. Almgren and J.B. Bell. A coupled discontinuous Galerkin-Finite Volume framework for solving gas dynamics over embedded geometries. Journal of Computational Physics, 450 (2022), p.110861. https://doi.org/10.1016/j.jcp.2021.110861

[3] R. Saye. Implicit mesh discontinuous Galerkin methods and interfacial gauge methods for high-order accurate interface dynamics, with applications to surface tension dynamics, rigid body fluid–structure interaction, and free surface flow: Part I. Journal of Computational Physics, 344 (2017), pp.647-682. https://doi.org/10.1016/j.jcp.2017.04.076

[4] V. Gulizzi and R. Saye. Modeling wave propagation in elastic solids via high-order accurate implicit-mesh discontinuous Galerkin methods. Computer Methods in Applied Mechanics and Engineering, 395 (2022), 114971. https://doi.org/10.1016/j.cma.2022.114971

[5] V. Gulizzi, I. Benedetti and A. Milazzo. A high-resolution layer-wise discontinuous Galerkin formulation for multilayered composite plates. Composite Structures, 242 (2020), 112137. https://doi.org/10.1016/j.compstruct.2020.112137

[6] V. Gulizzi, I. Benedetti and A. Milazzo. Discontinuous Galerkin Methods for Solids and Structures. in: M.H. Ferri Aliabadi, W.O. Soboyejo (Eds.), Comprehensive Structural Integrity 2nd edition, Oxford: Elsevier, 2023, pp. 348-377. https://doi.org/10.1016/B978-0-12-822944-6.00024-4

Conjugate heat transfer applied to transitory analysis for rocket engine cooling systems design

Vincenzo Barbato[1,a*], Matteo Fiore[1,b], Francesco Nasuti[1,c]

[1]University La Sapienza of Rome, Department of Mechanical and Aerospace Engineering, Via Eudossiana 18, Rome, Italy

[a]barbato.1739632@studenti.uniroma1.it, [b]matteo.fiore@uniroma1.it, [c]francesco.nasuti@uniroma1.it

Keywords: Conjugate Heat Transfer, Liquid Rocket Engine, Regenerative Cooling Systems

Abstract. This study investigates the use of an in-house Conjugate Heat Transfer (CHT) numerical solver for the modelling of transient phenomena in liquid rocket engines active cooling systems. Heat transfer considerations place great limitations in the development of rocket engines and transient operative conditions are amongst the most critical. The current lack of models and numerical tools capable of accounting for the complexities of this time-dependent multi-physics problem, results in oversized cooling systems, long development times and increased risk of failure. The fine modelling of all the involved phenomena and their interaction with each other is crucial to achieve a correct prediction of the thermal fluxes and wall temperatures involved. Hence, CHT simulations are the state-of-the-art for this application. The CHT solver proposed in this work utilizes a partitioned coupling strategy where two extensively validated single-physics solvers exchange information through their interfaces at discrete time steps. A simplified version of the RL-10A-3-3A regenerative cooling jacket is considered as reference to test the strengths and the limits of this approach. Both a complete chilldown of the engine and part of the start-up transient have been simulated. The analyses performed show the ability of the solver proposed to deal with transient phenomena where fluid-structure interaction occurs. In addition, they provide a complete overview of the numerical issues related to the partitioned coupling approach. These preliminary results pave the way for further developments aimed at increasing the reliability of the solutions and extending the application field of the software developed.

Introduction

Transient heat transfer conditions can be encountered during thrust build-up (engine start-up) and engine shutdown in all rocket propulsion systems. These phases of the rocket engine operational life are amongst the most critical and thus, must be carefully analysed to avoid failures: a cooling system sized on the engine nominal conditions is not sufficient to guarantee overall safety and integrity. During the start-up or shut-down transients, the chamber or nozzle walls may reach temperatures higher than those recorded at steady-state, as well as experience large temperature gradients. Furthermore, transient analysis is particularly important for reusable engines, where the knowledge of the thermal loads involved, can be crucial to guarantee safety through the engine thermal cycles. The aim of the present work is to present a numerical solver for the CHT problem completely developed by the authors, to study transient phenomena occurring in rocket engine regenerative cooling jackets. This multi-physics solver is based on a partitioned coupling approach, where two separate single-physics solvers exchange information through the boundary conditions at their interfaces. Two transient conditions have been analysed, engine chilldown and engine start-up, assuming a simplified version of the cooling jacket of the RL-10A liquid rocket engine as a reference for both geometry and operative conditions.

Conjugate Heat Transfer

All active cooling techniques, such as regenerative cooling, are characterized by the interaction between the fluid coolant and the solid chamber walls. To model such interaction and to predict accurate values of temperature and heat flux, one must focus on conjugating the boundary conditions at the fluid-solid interface through coupled heat transfer analysis. Such a coupled field of study is termed Conjugate Heat Transfer (CHT) analysis [1]. Over the years, CHT has evolved as the most effective method of heat transfer study. A review of the coupling techniques currently adopted in the CHT community together with a thorough analysis of the stability of each method, can be found in the work of Verstraete and Scholl [2]. Amongst the various alternatives, a *partitioned approach with a serial coupling and a Dirichlet-Neumann boundary exchange method* has been selected for the present work. As suggested by M.B. Giles [3], this coupling strategy guarantees overall stability through the coupling iterations, thus resulting in the best choice for the first implementation of a new CHT solver.

Reference Case

The CHT approach for transient analyses developed in the present work has been tested on a simplified version of the cooling jacket of the RL-10A LOX/LH_2 liquid rocket engine [4]. Two transient operative conditions have been simulated: the complete chilldown of the engine and the start-up transient. The RL-10A has been chosen as reference because a great number of details are available for this engine. However, since the scope of the present work is to investigate the limits of the CHT approach proposed, a simplified version of the RL-10A cooling jacket has been considered to neglect multiple phenomenology arising from a complex geometry (i.e., curvature effects, variable cross section, etc..). As a result, a rectified version of the cooling jacket is considered, assuming a constant rectangular cross section for each of the 180 tubes that make up the cooling jacket, made of Type-347 Stainless Steel. Channel dimensions have been chosen equal to the dimensions of the real channel section at the nozzle throat [5].

Numerical Setup

A conjugate heat transfer model based on the coupled numerical integration of the Navier-Stokes equations for the coolant flow and the Fourier's law of conduction for the heat transfer within the solid has been adopted [6], exploiting in-house validated solvers [7]. The flow solver integrates the Reynolds Averaged Navier Stokes Equations, written in the conservation form, by a Godunov-type finite volume scheme, which is second order accurate in space. Turbulence is computed with the one-equation model of Spalart-Allmaras [8].

The fundamental hypothesis on which the coupling strategy is based, is that the heat capacity of the solid is much greater than that of the fluid. This translates in the assumption that when the flow field reaches equilibrium, the structural thermal conduction has not yet begun. In terms of the coupling logic, this implies that at each time step, the steady flow analysis and the unsteady thermal analysis are implemented in turn, as shown in Fig. 1. Steady flow field computed by an isothermal boundary condition and uniform temperature field in the solid are specified as initial conditions for the coupled flow-thermal analysis. Thereafter, the thermal boundary condition for the flow analysis is updated after every thermal conduction calculation for the solid. A maximum temperature variation ΔT is considered as stopping criteria for the transient solid simulation: when at least one cell in the solid domain changes its temperature by an amount greater than the ΔT selected, the unsteady thermal analysis is stopped and the solution is evaluated. From the analysis carried out, it results that this method is first order accurate in time. This is probably a consequence of keeping the heat flux constant during the unsteady thermal analysis in the solid.

Given the symmetry of the problem, to reduce the computational cost of the simulations, only half of the channel has been simulated. The solid is divided in five 3D blocks whereas for the fluid, a single 3D component is sufficient to characterize the entire domain. The number of cells adopted for the fluid domain and its cell-clustering parameters have been chosen to guarantee that $y_+ \leq 1$. Instead, for the solid domain, the discretization has been chosen to include a sufficiently high number of cells to reduce the mapping error in the coupling phase caused by non-matching grids.

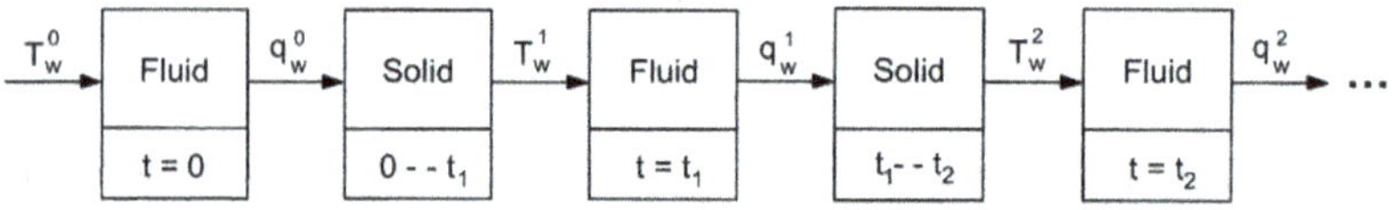

Figure 1 - Schematic representation of the coupling procedure adopted.

Results

To investigate the capabilities of the proposed CHT approach, a complete chilldown of the engine cooling jacket has been simulated. The analysis has been conducted assuming the following constant operative conditions for the fluid: inlet temperature $T_{in} = 25$ K, inlet mass flow rate $m_{in} = 0.01$ kg/s and outlet pressure $p_{out} = 40$ bar. Moreover, all external walls of the solid domain have been assumed adiabatic and the initial wall temperature is set to $T_w = 240$ K. With this study, the sensitivity of the solution to the variation of the stopping criteria (i.e., ΔT) has been investigated. Four different simulations have been realized with the same initial and boundary conditions, but with different simulation parameters. The results obtained showed that, for a simple problem as chilldown where the heat transfer process is monotonous, the variation of ΔT only slightly affects the overall time needed to simulate the same phenomenon. One of the major non-physical phenomena introduced by the discretization of the problem is the possible occurrence of *heat flux inversion*. In the case under analysis, heat flux inversion may occur towards the end of the chilldown process, when the temperature difference between the solid and the fluid approaches zero. When this happens, in the same time in which any cell of the solid changes its temperature of the ΔT required by the stopping criteria, the temperature of one or more of the wall cells, may become lower than the local fluid temperature. Thus, in the successive iteration, the heat flux locally changes its sign and rather than having the fluid cooling the solid, the inverse happens. Once heat flux inversion is triggered, solutions lose their validity and start oscillating because the

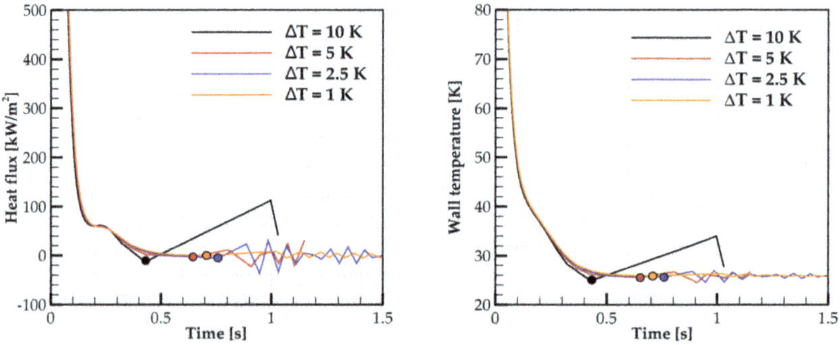

Figure 2 - Oscillations induced by the heat flux inversion phenomenon.

heat flux reverses its sign alternatively from one coupling iteration to the other (Fig. 2). Furthermore, the use of decreasing values of ΔT suggested the possibility of a convergence analysis analogous to the ones usually carried out for the spatial convergence. It has been found that the pseudo-order of convergence in time of this coupling strategy is approximately one.

Transient operative conditions in liquid rocket engines are generally much more complex than those characterizing a simple engine chilldown. During start-up and shut-down, cooling channels are subject to time-varying inflow and outflow conditions, as well as asymmetrical heating caused by the burning of fuel and oxidizer in the combustion chamber. In this work, part of the start-up transient of the RL-10A has been simulated. Several assumptions have been made to adapt the real start-up transient of the engine to the simplified geometry utilized. Not all the relevant quantities could be found in literature [4,5], thus a best-guess approach has been adopted to account for the missing or partial data. In particular, the heat flux value enforced on the hot-gas-side wall at each time $q_w(t)$ has been approximated by scaling the steady-state heat flux $q_{w,ss}$, through a function of the chamber pressure (Eq. 1) utilizing the information available in literature [5].

$$q_w(t) = q_{w,ss} \left(p_c(t) / p_{c,ss}\right)^{0.8} \quad (1)$$

It is crucial to emphasize that the scope of the present work is not to represent the real conditions of the start-up transient of the RL-10A. Instead, the aim of this further transient analysis is to show the ability of the solver to correctly represent the physical phenomena occurring in a much more complex scenario. Because of the wide temperature range involved in this simulation, an intermediate value of $\Delta T = 5K$ has been considered as stopping criteria, to reduce the number of couplings performed and the solutions evaluated. To show the consistency of the results obtained, in Fig. 3, the final solution of the transient simulation has been compared to a steady-state solution obtained with a parallel validated steady-state CHT solver [9]. In particular, the temperature field is shown along the channel symmetry plane. The presence of an intense monotonically increasing external heat flux, leads to a major modelling problem: the external heat flux is always underestimated. Indeed, the temperature distribution shows the same pattern suggesting that the solver is heading in the right direction. However, by constantly underestimating the real heat flux, lower temperatures are obtained with respect to the actual steady-state solution. To reduce this modelling error, a smaller ΔT should be adopted. However, for a simulation demanding as the one presented, this choice implies a non-negligible increase in the computational cost. Thus, to obtain a better trade-off between solution accuracy and computational time, a logic that automatically modifies the ΔT depending on the current solution of the problem, could be introduced. Given the multiple problems related to the constancy of the heat flux, such logic should be based on the percentage change, from one iteration to the next, of the magnitude of the heat flux itself.

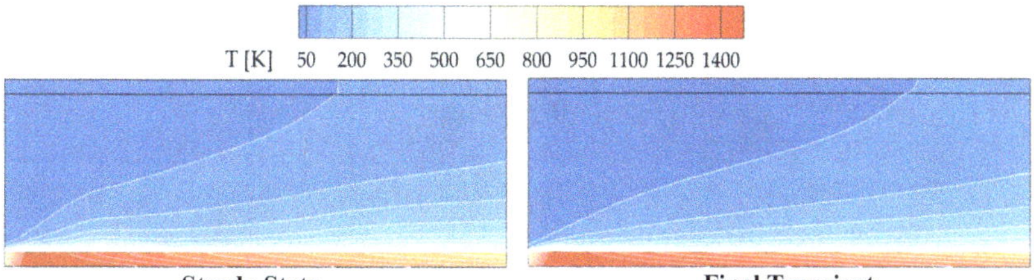

Figure 1 – Temperature field comparison for steady-state and final transient solutions.

Conclusions

This study presented a thorough analysis of the main limits of the in-house Conjugate Heat Transfer numerical solver developed by the authors. Overall, the partitioned approach proved to be a viable way of developing a new multi-physics solver, allowing for reuse of already existing validated tools. The results obtained show the capability of this CHT solver of dealing with the modelling of transient phenomena in liquid rocket engines active cooling systems. Despite the reported results being only preliminary, the methodology here proposed is quite general and can be easily extended to a wide range of problems where fluid-solid interaction occurs.

References

[1] John B., Senthilkumar P. and Sadasivan S., "Applied and theoretical aspects of conjugate heat transfer analysis: A review," Archives of Computational Methods in Engineering, Vol. 26, 2019, pp. 475–489. https://doi.org/10.1007/s11831-018-9252-9

[2] Verstraete, T., and Scholl, S., "Stability analysis of partitioned methods for predicting conjugate heat transfer," International Journal of Heat and Mass Transfer, Vol. 101, 2016, pp. 852–869. https://doi.org/10.1016/j.ijheatmasstransfer.2016.05.041

[3] Giles, M. B., "Stability analysis of numerical interface conditions in fluid–structure thermal analysis," International journal for numerical methods in fluids, Vol. 25, No. 4, 1997, pp. 421–436. https://doi.org/10.1002/(SICI)1097-0363(19970830)25:4%3C421::AID-FLD557%3E3.0.CO;2-J

[4] Binder, M., Tomsik, T., and Veres, J. P., "RL10A-3-3A rocket engine modeling project," Tech. rep., 1997.

[5] Binder, M., "A transient model of the RL10A-3-3A rocket engine," 31st Joint Propulsion Conference and Exhibit, 1995, p. 2968. https://doi.org/10.2514/6.1995-2968

[6] Pizzarelli, M., Nasuti, F., and Onofri, M., "Coupled wall heat conduction and coolant flow analysis for liquid rocket engines," Journal of Propulsion and Power, Vol. 29, No. 1, 2013, pp. 34–41. https://doi.org/10.2514/1.B34533

[7] Pizzarelli, M., Urbano, A., and Nasuti, F., "Numerical analysis of deterioration in heat transfer to near-critical rocket propellants," Numerical Heat Transfer, Part A: Applications, Vol. 57, No. 5, 2010, pp. 297–314. https://doi.org/10.1080/10407780903583016

[8] Spalart, P., and Allmaras, S., "A one-equation turbulence model for aerodynamic flows," 30th aerospace sciences meeting and exhibit, 1992, p. 439. https://doi.org/10.2514/6.1992-439

[9] Nasuti, F., Torricelli, A., and Pirozzoli, S., "Conjugate heat transfer analysis of rectangular cooling channels using modeled and direct numerical simulation of turbulence," International Journal of Heat and Mass Transfer, Vol. 181, 2021, p. 121849. https://doi.org/10.1016/j.ijheatmasstransfer.2021.121849

Assessment of aerodynamics of low Martian atmosphere within the CIRA program TEDS

Francesco Antonio D'Aniello[1,a *], Pietro Catalano[1,b] and Nunzia Favaloro[1,c]

[1]Cira, Via Maiorise, Capua (CE), Italy

[a]f.daniello@cira.it, [b]p.catalano@cira.it, [c]n.favaloro@cira.it

Keywords: RANS, Low Reynolds Number, Compressible, Martian Atmosphere

Abstract. The space exploration and colonization Roadmap, and the growing interest of the international scientific community toward the Mars colonization, are highlighting the need to develop, or improve to higher TRLs, those technologies enabling human exploration and colonization. In this framework, the CIRA - PRORA research program TEDS has identified some of the most promising technologies and research areas for human/robotic exploration, and human survivability in hostile environments in future space missions. In this contest, Aerodynamics in the Martian low atmosphere, characterized by low Reynolds number and high Mach number, is one of the research areas to investigate. The compressible aerodynamics of low Reynolds number flow (Reynolds number of orders of magnitude 10^4-10^5 and Mach number of 0.2-0.7) characterizes the low altitude Martian atmosphere. The need in the exploration of the Martian surface has increased the interest in this "particular" aerodynamic regime, currently, scarcely investigated, and an assessment of the numerical methods is necessary. Three suitable airfoils for compressible low-Reynolds aerodynamics in Martian atmosphere have been selected through a bibliographic study: Triangular, NACA 0012-34 and Ishii airfoils. The experiments in the low-density CO_2 facility of the Mars Wind Tunnel, at Tohoku University, over a Triangular, NACA0012-34 and Ishii airfoils with global forces and local PSP measurements, have been considered. The CIRA in-house developed flow solver UZEN (Unsteady Zonal Euler Navier-Stokes) code has been applied by employing several turbulence models. The flow over the Triangular airfoil has been simulated inside the wind tunnel and the free air flow over the NACA 0012-34 and Ishii airfoils have been simulated, and in this paper many results are reported.

Introduction

During the last years, the interest of the international scientific community in space exploration and colonization is growing up, particularly toward Mars. For this scope the CIRA-PRORA research program TEDS [1] has identified some of the most promising technologies and research areas for human/robotic exploration, and human survivability in hostile environments in future space missions. Within this program, focusing on Mars exploration and colonization, the study of Aerodynamics in the Martian low atmosphere, characterized by low Reynolds number (10^4 - 10^5) and high Mach number in the compressible or transonic range (0.2 - 0.7), is one of the research areas to investigate. The task concerns the evaluation of the aerodynamic characteristics, of airfoils and wings, in the low Martian atmosphere. Numerical results will also support the feasibility study of a Martian rotorcraft for future survey and exploration missions.

Unfortunately, the pair of compressible/transonic Mach and low Reynolds does not occur in the Earth's atmosphere which is characterized by the incompressible regime. For this reason, low-Reynolds number compressible flows were scarcely investigated so far, and thus the experimental data of airfoils, flying in this Aerodynamics, are very limited. Only NASA and JAXA agencies have experimentally studied this aerodynamic regime as in ***Errore. L'origine riferimento non è stata trovata.***, therefore, an assessment of the numerical methods has been necessary.

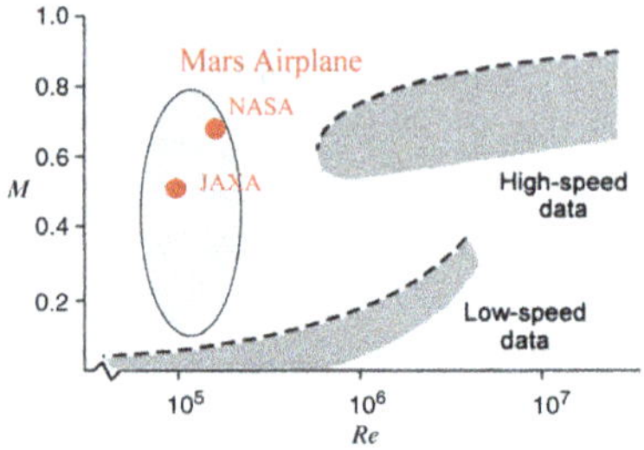

Figure 1 Martian atmospheric flight [2,3].

RANS numerical simulations through the CIRA built in-house UZEN code has been conducted on three suitable airfoils, for compressible low Reynolds aerodynamics in Martian atmosphere, that have been selected through a bibliographic study: Triangular, NACA 0012-34 and Ishii airfoils. The experiments in the low-density CO_2 facility of the Mars Wind Tunnel of Tohoku University over a NACA0012-34, Triangular and Ishii airfoils (sketched in Fig. 2, considering [2,4,5], respectively) with global forces and local PSP measurements, have been considered to first assess the UZEN code.

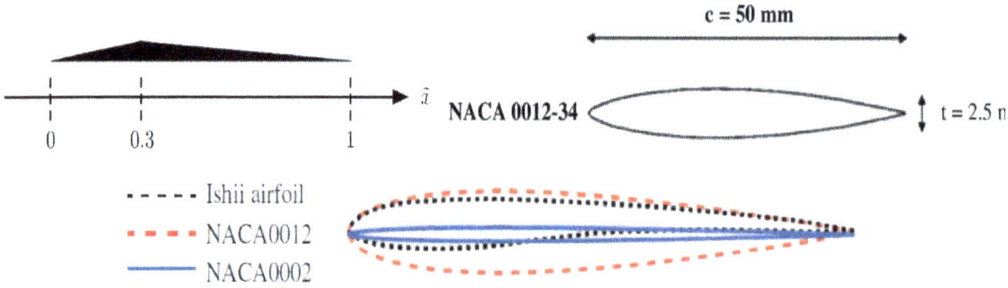

Figure 2 Triangular (top left), NACA0012-34 (top right) and Ishii airfoils (bottom).

Mars Low atmosphere and Aerodynamics

Martian low atmosphere is more rarefied than Earth's one, in fact it mostly consists of carbon dioxide CO_2 (95%) and it is characterized by low pressure (p~0.0075 × 101.3 kPa) low density (ρ~0.017 kg/m^3) and low temperature, at the surface, respect to the Earth surface. In the low Reynolds number compressible regime, the flow on the upper wing surface is prone to separate forming the laminar separation bubble (LSB), and after the transition zone the flow reattaches in turbulent manner (as shown in Fig. 3). This complicated flow field scenario strongly affects aerodynamic performances of airfoils and wings. Because of these unusual flow characteristics, the airfoil shape largely impacts on the aerodynamic characteristics, in fact in [3,6] it has suggested three shape characteristics involving high aerodynamic performances in low Reynolds number compressible regime:

- sharp leading edge to fix the separation point at the edge and can improve its Reynolds-number dependence on the aerodynamic performance;
- flat upper surface to reduce the separation region;
- cambered airfoil to gain a higher lift than a symmetric airfoil.

Following the trace of prescribed suggestions, Triangular, NACA 0012-34 and Ishii airfoils have been selected to be numerically simulated.

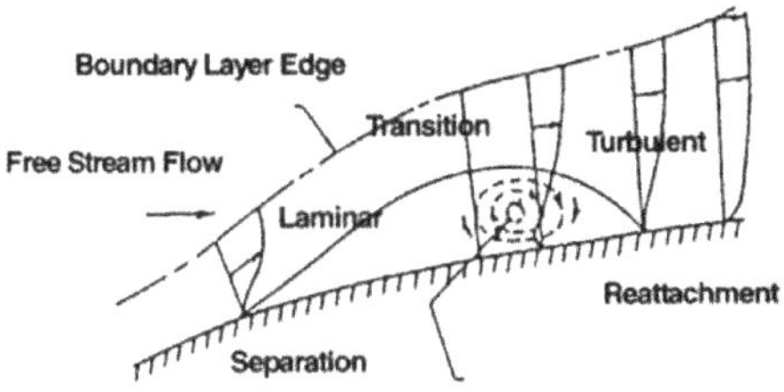

Figure 3 Flow field development when separation bubble occurs [7,8].

Numerical method

The numerical analysis is conducted by using the CIRA in-house code UZEN. The code UZEN [9] solves the compressible 3D steady and unsteady RANS equations on block-structured meshes. The spatial discretization adopted is a central finite volume formulation with explicit blended 2nd and 4th order artificial dissipation. The dual-time stepping technique is employed for time accurate simulations [10,11]. The pseudo-time integration is carried out by an explicit hybrid multistage Runge-Kutta scheme. Classical convergence acceleration techniques, such as local time stepping and implicit residual smoothing, are available together with multigrid algorithms. Turbulence is modelled by either algebraic or transport equation models [12]. Structured multi-block grids were built by using ICEM CFD© commercial code for all the selected airfoils. Both RANS and URANS numerical simulations were conducted to reproduce the MWT experimental data collected in the selected bibliography for all three airfoils, with large interest in the aerodynamic performance coefficients, i.e. lift coefficient (C_L) and drag coefficient (C_D), to better understand how the aerodynamic performance change in the Mars atmosphere respect to the Earth one.

Triangular wing has been simulated within the MWT, while NACA 0012-34 and Ishii airfoils has been exclusively simulated in bi-dimensional domains. In Table 1, the level of grid mesh and the number of cells, for each level and for the three considered airfoils, are listed. Where not specified, the number of cells in spanwise direction is equal to 1 (i.e. nk = 1); otherwise the choice of nk is due to the tri-dimensional grid construction.

Table 1 Grid levels and number of cells (ni × nj × nk) for Triangular, NACA 0012-34 and Ishii airfoils.

	1st lev	2nd lev	3rd lev	4th lev
Triangular	2×10^6 (nk = 64)	15×10^6 (nk = 130)	/	/
NACA 0012-34	200×25 = 5000	400×50 = 20000	800×100 = 80000	1600×200 = 320000
Ishii	64×32 = 2048	128×64 = 8192	256×128 = 32768	512×256 = 131072

Numerical results and discussion

In Fig. 4, the dimensionless wall distances (y^+) on Triangular wing (M=0.5, Re=1.0×10^4, α=5°), and over the upper region of NACA 0012-34 (M=0.2, Re=1.1×10^4, α=0°) and Ishii (M=0.2, Re=2.3×10^4, α=0°) airfoils, obtained by RANS calculations, are reported.

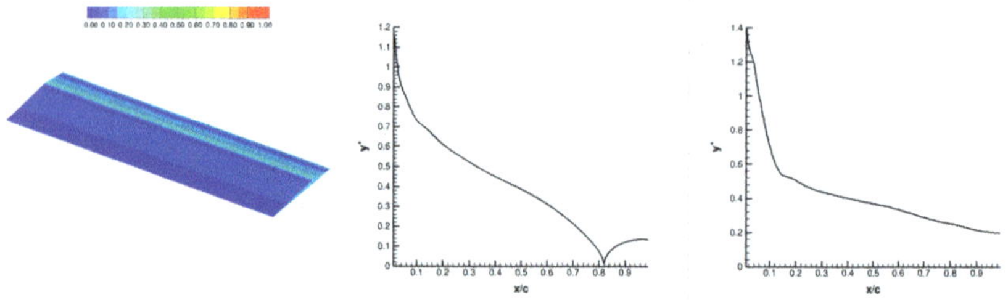

Figure 4 Dimensionless wall distance (y^+) on Triangular wing (left), over the upper region of NACA 0012-34 airfoil (centre) and over the upper region of Ishii airfoil (right).

Through the comparison with the available experimental data, numerical results for the three selected airfoils, globally confirm experimental reports, highlighting that the Mars aerodynamic regime strongly affects aerodynamic performances of airfoils. In fact, looking at Fig. 5, Fig. 6 and Fig. 7, convergence simulations have been discretely reproduced the global trend of polar curves or C_D curves for all the Reynolds and Mach numbers considered. The differences between numerical results and MWT experimental data are due to the used computational grids (bi-dimensional in the case of NACA 0012-34 and Ishii) because their coarsening or design (Ishii trailing edge is open). Turbulence models were not always be able to reproduce exactly the very complex flow field (vortex structures, laminar separation bubble, etc) generated around the wing or on the airfoils. For the Triangular wing (refer to Fig. 5), at Reynolds number equal to 1.0×10^4, numerical results are in good agreement with the experiments at $\alpha = 5^{\circ}$, at all the three Mach numbers considered. As the incidence increases, the comparison gets worse. For NACA 0012-34 airfoil (refer to Fig. 6), polar curves are partially reproduced due to the lack of numerical convergence. Numerically reproduced polars are slightly underestimated in terms of C_D than those experimentally recorded. The comparison between numerical results and experimental data slightly improves at high Mach numbers, perhaps as a result of the compressibility effect introduced. At Mach number equal to 0.61, a good agreement with experimental data is reached through the $\kappa - \omega$ SST. For Ishii airfoil (refer to Fig. 7), a globally slight underestimation is detectable in terms of C_D with respect to the experimental data. The $\kappa - \omega$ turbulence models used does not allow to reproduce perfectly the complex flow field that occurs on this airfoil, already at low angles of attack. The $\kappa - \omega$ SST model gave a good agreement with experimental data, and also respect to the LES numerical simulations reported in literature, for almost all the considered incidences.

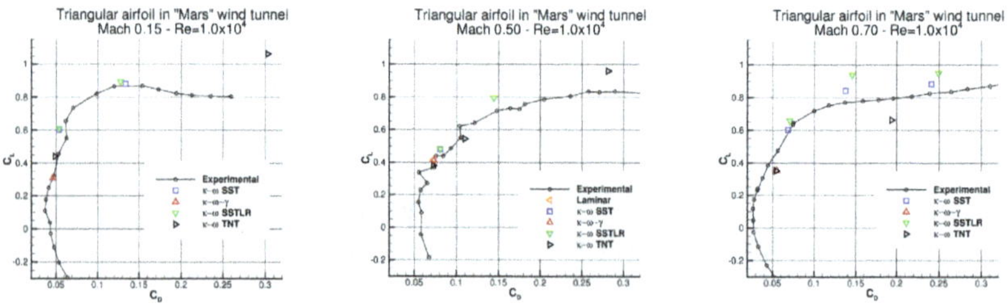

Figure 5 Triangular wing: Drag Polars at Reynolds number 1.0×10^4, at $M = 0.15$ (left), $M=0.50$ (centre) and $M=0.70$ (right), calculated with different $\kappa - \omega$ turbulence models.

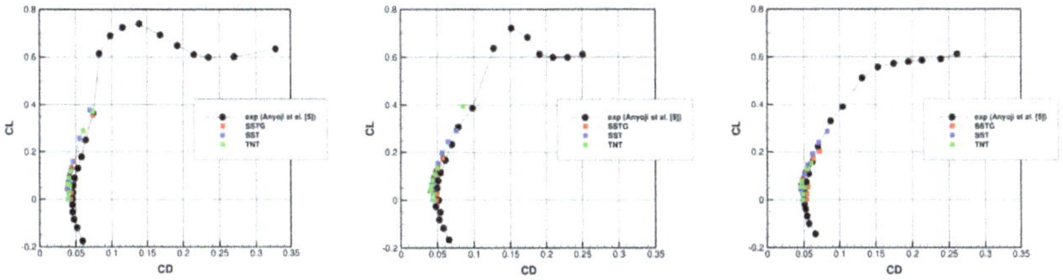

Figure 6 NACA 0012-34 airfoil: Drag Polars at Re = 1.1 × 10⁴, at M = 0.20 (left), M=0.48 (centre) and M=0.61 (right), calculated with different κ – ω turbulence models.

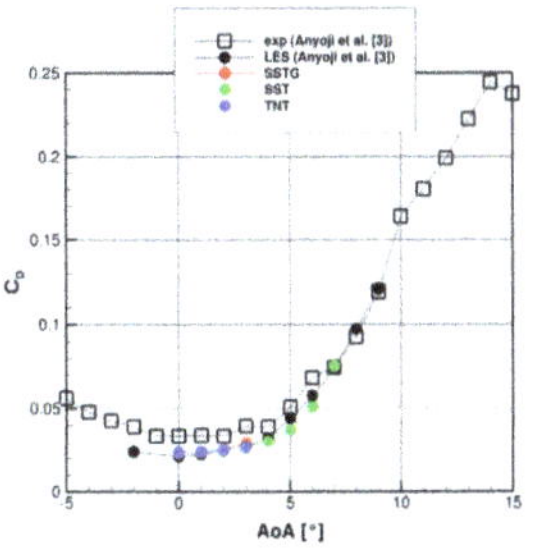

Figure 7 Ishii airfoil: Drag Coefficient versus angle of attack at Reynolds number 2.3 × 10⁴ and at Mach number 0.20, calculated with different κ – ω turbulence models.

Focusing on Triangular airfoil, the flow field around the wing, inside the wind tunnel, provided by the numerical simulations is actually very complex, above all at high incidence. Flow structures forming at the side-ends of the body and developing in the wake can be also appreciated in Fig. 8 that, following the Q-criterion, reports an iso-surface of $Q = \frac{1}{2}(\Omega_{i,j}\Omega_{i,j} - S_{i,j}S_{i,j})$, where $\Omega_{i,j}$ is the vorticity magnitude and $S_{i,j}$ is the rate of strain (Q-criterion defines vortices as areas where the vorticity magnitude is greater than the magnitude of the rate of strain). The vortex regions in the central part are visible. It can be also noted as the side-end structures tend to disappear at the highest Reynolds number.

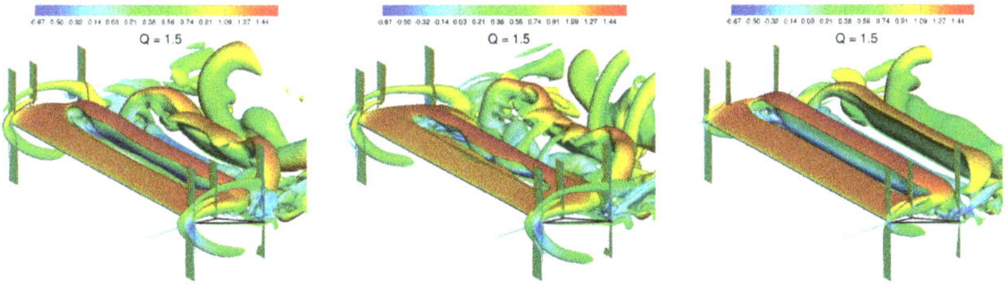

Figure 8 Iso-surface of Q at α = 15°, M=0.15, Re=3.0 × 103 (left), M=0.50, Re=3.0 × 103 (centre), M=0.15, Re=1.0 × 104 (right).

Summary

Numerical simulations conducted with CIRA built in-house UZEN code through all the employed κ-ω turbulence models, have been partially reproduced the flow field around the wing in the wind tunnel (Triangular airfoil) and around the airfoils in free air (NACA0012-34, Ishii) and with a global good agreement with the experimental data in terms of aerodynamic performances (i.e. lift and drag coefficients), especially at low and medium incidences.

Although the scarce 3D simulations conducted on the wing based on the Triangular airfoil, they allow to underline the strongly 3D structure of the flow field that influences the aerodynamic performance degradation, above all at high incidences.

NACA 0012-34 and Ishii are more feasible to fly airfoils than the Triangular one. Their design has been thought to fly at high incidences in a low-Reynolds environment and, at the same time, to allow the flow to reattach downstream the formation of the laminar separation bubble.

The influence of the Mach number has been also investigated. The numerical results have shown that the increase in Mach numbers seems to energize the flow, that is more able to withstand the adverse pressure gradients and becomes less prone to the separation.

References

[1] N. Favaloro, G. Saccone, F. Piscitelli, R. Volponi, P. Leoncini, P. Catalano, A. Visingardi and M. C. Noviello, "Enabling Technologies for Space Exploration Missions: The CIRA-TEDS Program Roadmap Perspectives," Aerotecnica Missili & Spazio, 2023. https://doi.org/10.1007/s42496-023-00159-4

[2] Anyoji, Numata, Nagai and Asai, "Effects of Mach Number and Specific Heat Ratio on Low-Reynolds-Number Airfoil Flows," AIAA Journal, vol. 53, no. 6, 2015. https://doi.org/10.2514/1.J053468

[3] F. W. Schmitz, "Aerodynamics of the Model Airplane Part1," NASA, Technical Rept., NASA-TM-X-60976, 1967.

[4] Munday, Taira, Suwa, Numata and Asai, "Nonlinear Lift on a Triangular Airfoil in Low-Reynolds-Number Compressible Flow," Journal of Aircraft, 2015. https://doi.org/10.2514/1.C032983

[5] Anyoji, Nonomura, Aono, Oyama, Fujii, Nagai and Asai, "Computational and Experimental Analysis of a High-Performance Airfoil Under Low-Reynolds-Number Flow Condition," Journal of Aircraft, vol. 51, no. 6, 2014. https://doi.org/10.2514/1.C032553

[6] F. W. Schmitz, "The Aerodynamics of Small Reynolds number," NASA TM, 1980.

[7] Tsukamoto, Yonemoto, Makizono, Sasaki, Tanaka, Ikeda and Ochi, "Variable-Pressure Wind Tunnel Test of Airfoils at Low Reynolds Numbers Designed for Mars Exploration Aircraft," Trans. JSASS Aerospace Tech. Japan, vol. 14, no. 30, pp. 29-34, 2016. https://doi.org/10.2322/tastj.14.Pk_29

[8] Rinoie and Kamiya, "Laminar Separation Bubbles Formed on Airfoils," Journal of Japan Society of Fluid Mechanics, no. 22, pp. 15-22, 2003.

[9] Catalano and Amato, "An evaluation of RANS turbulence modelling for aerodynamic applications," Aerospace Science and Technology, vol. 7, no. 7, pp. 493-509, 2003. https://doi.org/10.1016/S1270-9638(03)00061-0

[10] Marongiu, Catalano, Amato and Iaccarino, "U-ZEN: A computational tool solving U-RANS equations for industrial unsteady applications," in 34th AIAA FluidDynamics Conference and Exhibit, 2004. https://doi.org/10.2514/6.2004-2345

[11] Capizzano, Catalano, Marongiu and Vitagliano, "URANS modelling of turbulent flows controlled by synthetic jets," in 35th AIAA FluidDynamics Conference and Exhibit, 2005. https://doi.org/10.2514/6.2005-5015

[12] Catalano, Mele and Tognaccini, "On the implementation of a turbulence model for low Reynolds number flows," Computers and Fluids, vol. 109, pp. 67-71, 2015. https://doi.org/10.1016/j.compfluid.2014.12.009

Aeronautics and Astronautics - AIDAA XXVII International Congress | Materials Research Forum LLC
Materials Research Proceedings 37 (2023) 218-221 | https://doi.org/10.21741/9781644902813-47

A combustion-driven facility to study phenomenologies related to hypersonic sustained flight

Antonio Esposito

Dept. of Industrial Engineering, University of Naples Federico II, Via Claudio 21, 80125 Napoli (Italy)

antespos@unina.it

Keywords: Combustion-Driven Facility, Oxy-Fueled Guns, Hypersonics

Abstract. This paper reports on the development of a new Blowdown-Induction Facility driven by two different Oxy-Fueled Guns. The facility is conceived and realized to simulate different phenomenologies and flow conditions related to hypersonic sustained flight.

Introduction

Plasma Guns used in the context of Thermal Spray are often engineered versions of similar devices originally designed for aerospace research; for these applications the plasma gun is particularly relevant because it couples the reliability of an industrial device with the desired operating conditions, i.e. very high temperatures and very low pressures of the considered gas (heated by the electric arc). Nowadays however, new applications emerging in the general fields of Aerothermodynamics and Propulsion require (a) lower temperatures and higher pressures, (b) different process gases (Methane or Hydrogen) and (c) different operational modes (combustion instead of electric arc) respect to those generated by plasma torches. All these requirements call for the design and development of facilities with combustion-driven Thermal Spray Guns, and the present study may be regarded as a relevant effort along these lines; in particular, here we describe the development of a new facility based on two different Oxygen-Fueled guns (DJ2700 and 6P-II) that exhausts into a low pressure ambient. Unique properties of this facility are its ability to simulate different phenomenologies and flow conditions related to hypersonic sustained flight.

Experimental apparatus

Figure 1 shows the overall Vacuum Oxy-fueled Facility (VOF).

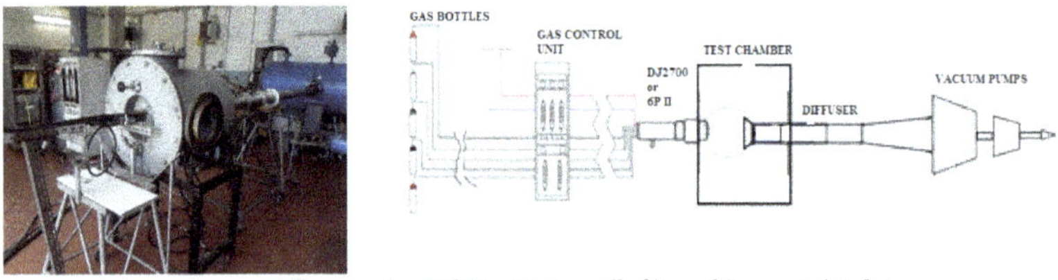

Figure 1 – VOF –Picture (left) and Layout (right).

Main facility components can be listed as:

1. Gas Supply Systems
2. Gas Control Units
3. Hybrid HVOF gun (DJ2700) / LVOF gun (6P-II)
4. Test Chamber and Diffuser

Materials Research Proceedings 37 (2023) 218-221 https://doi.org/10.21741/9781644902813-47

5. Vacuum Tanks and Vacuum Pump System

Gas Supply System

The details of the gas supply systems are shown in Fig. 2. In both systems Nitrogen is used only to prevent melting of the powder injectors in the guns.

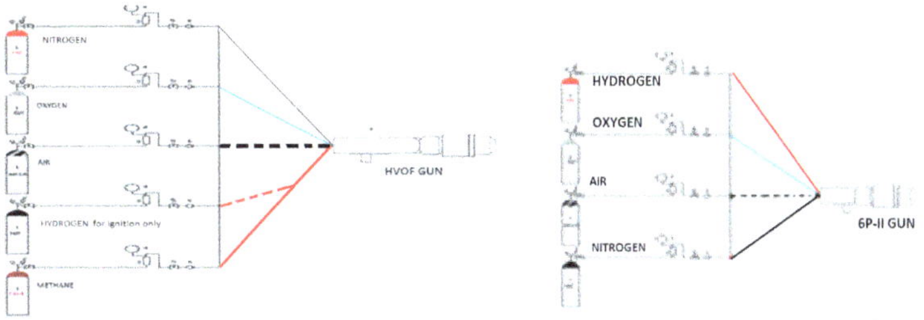

Figure 2 – Layout of the gas supply system for DJ2700 (left) and 6P II (right).

For the DJ2700 gun the gas flow requirements are: Oxygen 340 NLPM at 12 bar - Methane 200 NLPM at 7 bar - Air 439 NLPM at 7 bar - Nitrogen 18 NLPM at 12 bar - Hydrogen 8 NLPM at 10 bar. For the 6P-II gun the gas flow requirements are: Oxygen 45 NLPM at 2.6 bar - Hydrogen 170 NLPM at 2.4 bar - Air 50 NLPM at 6 bar - Nitrogen 15 NLPM at 5 bar.

Gas Control Units

A sketch of the Gas Control Units is shown in Fig. 3, with some relevant details being made evident through the associated legend. Both the units are realized in the laboratory.

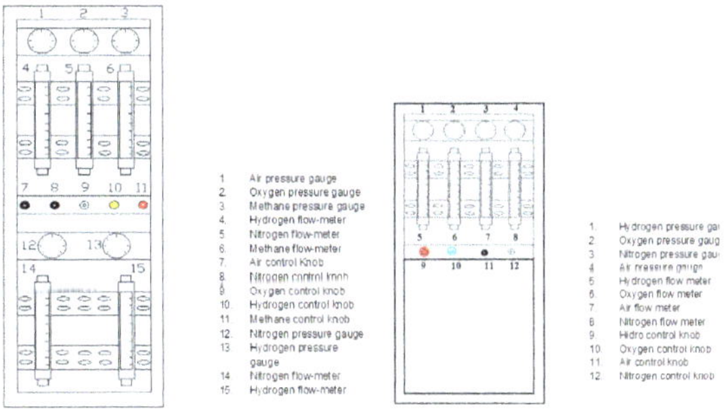

Figure 3 – Gas Control Units Front Panels for DJ2700 (left) and 6P-II (right)

HVOF gun (DJ2700) and LVOF gun (6P-II)

The Sulzer-Metco Diamond Jet (DJ) 2700 has been chosen as HVOF gun, the reader being referred to Figure 4 for a detailed sectional drawing of the gun. The DJ gun relies on a combination of oxygen, fuel and air to produce a high pressure annular flame, which is characterized by a uniform temperature distribution. The exhaust gases, together with the air injected from the annular inlet orifice, expand through the nozzle to reach a supersonic state. The air cap is cooled by both water and air to prevent it from melting.

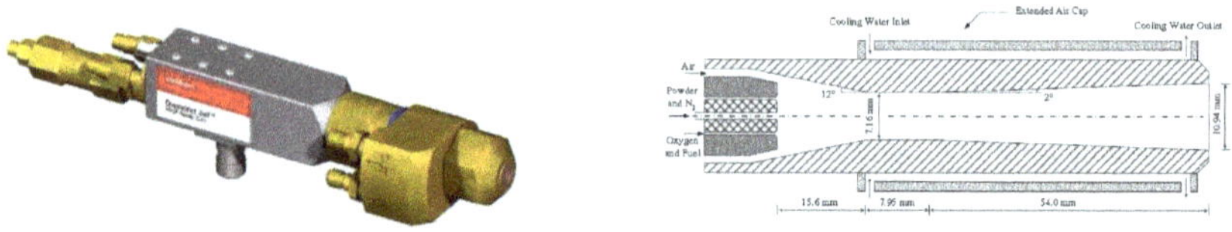

Figure 4 – DJ 2700 Hybrid Thermal Spray torch.

The Flame Spray Technology 6P-II has been chosen as LVOF gun, the reader being referred to Figure 5 for a detailed sectional drawing of the gun. The 6P-II can be used with Hydrogen as fuel gas. A siphon plug system mixes fuel and oxygen in precise volumetric proportions at the gun to provide consistent operation and prevents the possibility of backfire. A reversible air cap is used to create a parallel air flow to cool the gun or as a convergent pinch air flow.

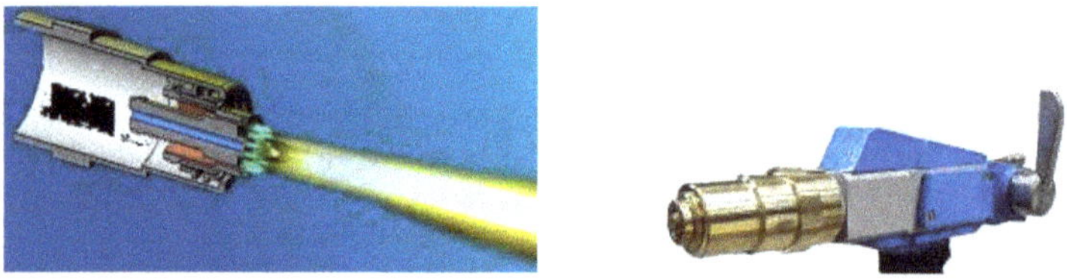

Figure 5 – 6P II Flame Spray torch.

Test Chamber and Diffuser
The test chamber is a iron cylinder with a diameter of 600 mm (Fig. 8), flanged at the ends and hosted inside the first section of the supersonic diffuser, Figs. 6(left) and (right).

Figure 6 – VOF Test Chamber (left) and Diffuser (right).

Vacuum Tank and Vacuum Pumps system
Vacuum pumps are a Stokes Microvac 212H (220 m^3/h) and a Edwards E2M275 (292 m^3/h); they evacuate the facility and the tank in a relatively brief time -5 minutes about - up to100 (Pa) ultimate pressure.

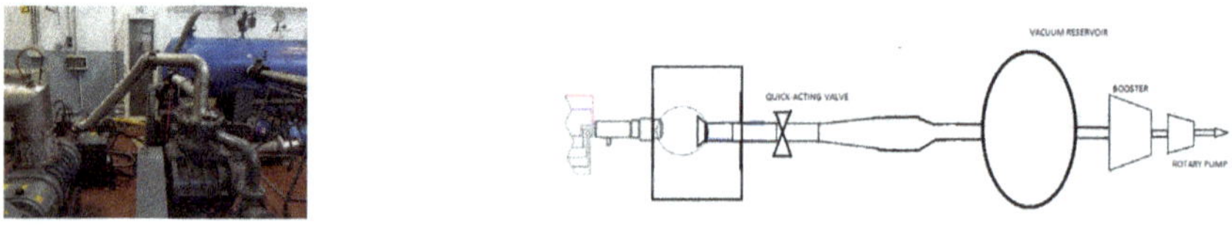

Figure 7 – Vacuum System: Rotary pumps (left) and system layout (right).

Applications and preliminary results

DJ2700 – HVOF

The HVOF gun (DJ700) produces a supersonic, overexpanded flow at atmospheric pressure (p ~ 0.7 (bar) at nozzle exit) of residual exhaust gases from Methane-Oxygen combustion; this flow is suitable for simulating the jet at the exit of a supersonic combustor (e.g. ramjet) if added with air and brought to the correct expansion conditions by means of a particular starting procedure of the facility set up in [1], see Figure 8.

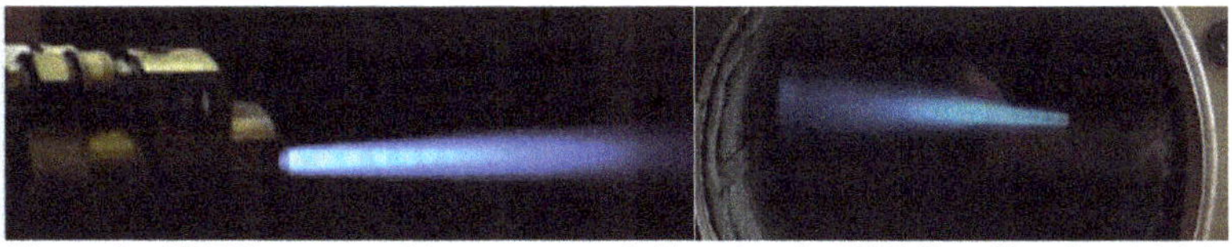

Figure 8 – HVOF gun at ambient pressure (left) and in the VOF Test Chamber (right).

6P – II LVOF

As for the 6P-II gun, this device produces a subsonic, high velocity flow in which Hydrogen and Oxygen burn generating thermal energy (and a residual water vapour), which heats Air (flowing in the cooling cap) and Nitrogen (flowing in the injector); by correcting the composition with Oxygen in the right percentage, Air at high temperature and atmospheric pressure is therefore obtained which can be used as a source for a hypersonic tunnel of the blowdown-induction type (see figure 7, right side). At the time of writing this paper ignition tests with hydrogen have begun and great precaution is needed.

References

[1] A. Esposito, C. Allouis, M. Lappa "A new facility for hypersonic flow simulation driven by a high velocity oxygen fuel gun" - ICAS 2022, 33rd Congress of the International Council of the Aeronautical Sciences, Stockholm Sweden 4 -9 September 2022.

Development of a DNS solver for compressible flows in generalized curvilinear coordinates

Giulio Soldati[1,a *], Alessandro Ceci[1,b] and Sergio Pirozzoli[1,c]

[1] Via Eudossiana 18, Rome 00184, Italy

[a]giulio.soldati@uniroma1.it, [b]alessandro.ceci@uniroma1.it, [c]sergio.pirozzoli@uniroma1.it

Keywords: Generalized Curvilinear Coordinates, Shock-Wave/Boundary Layer Interaction, Turbulent Compression Ramp, Curved Channel Flow

Abstract. We present a solver for DNS of turbulent compressible flows over arbitrary shaped geometries. The code solves the compressible Navier-Stokes equations in a generalized curvilinear coordinates system, using high-order central finite-difference schemes combined with WENO reconstruction for shock-waves treatment. An innovative stabilization strategy for central schemes based on skew-symmetric-like splitting of convective derivatives is used. The code is oriented to modern HPC platforms thanks to MPI parallelization and the ability to run on GPU architectures. The robustness and accuracy of the present code is assessed both in the low-subsonic case and in the supersonic case. We show here the results of a turbulent curved channel flow and a turbulent supersonic compression ramp, which proved to be in excellent match with previous studies.

Introduction

Compressible flows over curved surfaces are ubiquitous in the aerospace field, both in external (e.g., aircraft wings) and internal (e.g., turbomachinery) configurations. The dynamics of these flows involves complex phenomena, such as shock-wave boundary layer interactions, which still need further investigation, and make it challenging to accurately estimate skin friction and wall heat transfer. A key tool in this context is the DNS, which demands high-order numerical schemes and good shock-capturing capabilities to accurately describe high-speed flows. However, due to the numerical complications arising from geometric complexities, the application of the DNS is often limited to simple Cartesian cases [1,2]. Dealing with complex geometries leads to the use of local mesh refinement or unstructured meshes, which in turns causes high computational cost or decreased accuracy, respectively [3]. Another approach to deal with such geometries is the use of body fitted mesh in a generalized curvilinear coordinates system, which guarantee an accurate simulation of the fluid dynamics near the wall at a competitive computational cost. In this work, we developed a solver for DNS of turbulent compressible flows over complex geometries, using high-order central finite-difference schemes in a generalized curvilinear coordinate system. Central approximations of the Navier-Stokes equations, being non-dissipative, exhibit numerical instability when used at small viscosity. To address this problem, we implemented energy-preserving schemes [4], allowing to accurately simulate the wide range of turbulent scales without relying on artificial diffusivity [5] or filtering of physical variables [6]. Moreover, those schemes can be efficiently combined with modern shock-capturing methods as WENO reconstructions, yielding hybrid schemes that currently represent an optimal strategy for the computation of shocked flows [7].

Methodology

The code solves the compressible Navier-Stokes equations for a perfect gas in a generalized curvilinear coordinate system:

$$\frac{1}{J}\frac{\partial \mathbf{Q}}{\partial t}+\frac{\partial \mathbf{F}_j}{\partial \xi_j}=\frac{\partial \mathbf{F}_j^v}{\partial \xi_j}$$

where **Q** is the vector of conservative variables, $\mathbf{F}_j$ is the vector of convective fluxes, $\mathbf{F}^v_j$ is the vector of viscous fluxes and J is the Jacobian of the coordinate transformation. Here, we consider stationary grids. A full description of the fluxes vectors can be found in [8]. A curvilinear body-fitted mesh is first represented in the physical space, x_i. Through the transformation $x_i(\xi_j)$ it is then mapped to the computational space, ξ_j, where it can be seen as a regular hexahedron. Non-uniform skewed input cells of the mesh are thus re-stretched into uniform cubical cells. Finite-difference schemes are applied in the computational space to approximate spatial derivatives, which must be reconstructed in the physical space by using the metrics, $\partial\xi_j/\partial x_i$. Since the mesh is directly described in the physical space, first we compute the inverse of the metrics, $\partial x_i/\partial\xi_j$, by numerically deriving the physical mesh coordinates; then the metrics are obtained with a matrix inversion. To guarantee free-stream preservation, we use the same approximation for both metric and convective derivatives [9].

In smooth regions of the flow, a skew-symmetric-like splitting of the convective derivatives is employed. This approach guarantees preservation of kinetic energy in the semi-discrete, inviscid low-Mach-number limit. A computationally effective implementation of convective derivatives cast in split form was proposed by Pirozzoli [10]. The locally conservative formulation allows straightforward hybridization of central schemes with classical shock-capturing reconstructions. In our case, shock-capturing capabilities rely on WENO reconstruction of the numerical flux in the proximity of discontinuities. To judge on the local smoothness of the numerical solution and switch between the energy preserving and the shock capturing discretization, our code relies on a modified version of the Ducros shock sensor [11]. As for the viscous terms, they are expanded to Laplacian form to avoid odd-even decoupling phenomena. Spatial derivatives are approximated in the computational space with central formulas and reconstructed in the physical space by applying the chain-rule. The accuracy order of each scheme can be selected by the user and goes up to eight in the case of central schemes, up to seventh for WENO ones. The system is advanced in time using a three-stage, third order Runge-Kutta scheme [12].

The code is designed to efficiently work on the most common HPC architectures operating today. MPI parallelization and CUDA-Fortran porting enable the use of multi-GPUs architectures, while retaining the possibility to compile and use the code on standard CPU-based systems.

Results

We simulated low-Reynolds-number, mildly curved, turbulent channel flow. The computational domain is bounded by sectors of concentric cylinders, with a centreline radius of curvature $r_c = 79h$, where h is the channel semi-height. An imposed mean-pressure gradient in the azimuthal direction drives the flow. Periodic boundary conditions are imposed in the streamwise and spanwise directions, so that the simulated flow is fully evolved. The resolution is $216x72x144$ in the streamwise, wall-normal and spanwise directions, respectively. Mach number and Reynolds number based on the bulk velocity are set to $M_b = 0.1$ and $Re_b = 2600$. The Reynolds numbers based on the inner and outer friction velocities resulted in $Re^i_\tau = 156$ and $Re^o_\tau = 178$, respectively. Fig. 1 illustrates an instantaneous field of the velocity magnitude U normalized with respect to bulk velocity U_b. Fig. 2 shows the root-mean-square (rms) of the velocity fluctuations of the present DNS (STREAmS) together with the data from Moser and Moin [13] and Brethouwer [14]. The excellent agreement of the results can be noted.

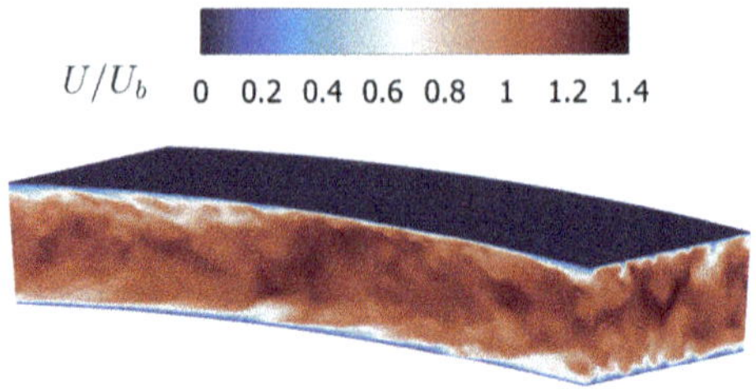

Fig. 1 Instantaneous field of the velocity magnitude.

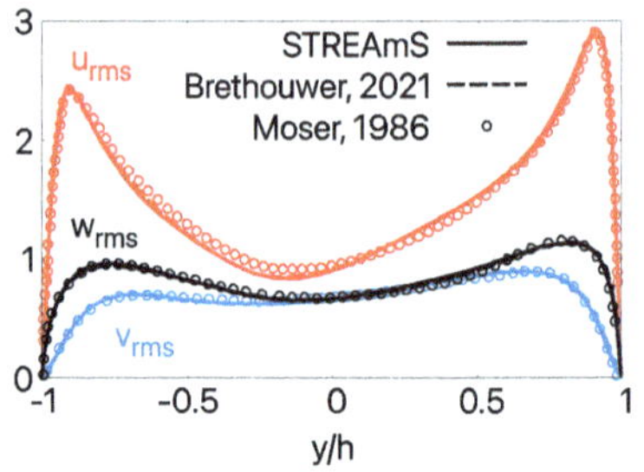

Fig. 2 Rms of the velocity fluctuations.

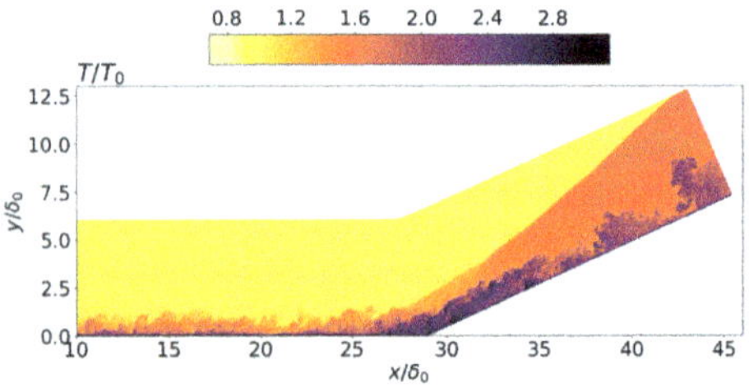

Fig. 3 Instantaneous field of the temperature.

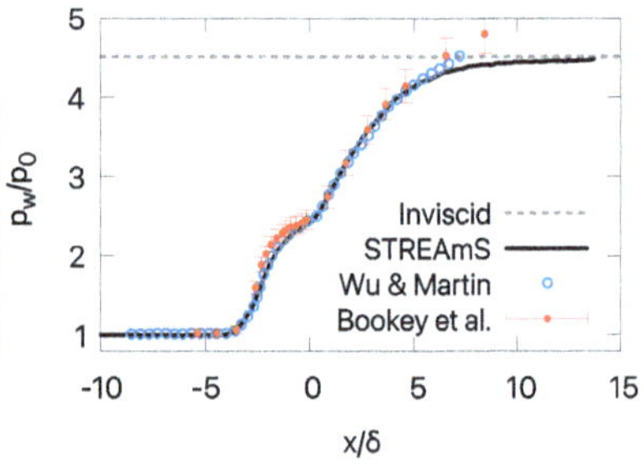

Fig. 4 Mean wall-pressure distribution.

Shock wave/turbulent boundary layer interaction is investigated by simulating a supersonic flow over a 24° compression ramp. Incoming flow conditions are Mach number $M = 2.9$ and Reynolds number $Re_\theta = 2300$ based on the inflow boundary layer momentum thickness. Synthetic turbulence is imposed at the inflow using a recycling and rescaling method. Supersonic outflow boundary conditions with non-reflecting treatment are used at the outlet and the top boundary. Non-slip condition is imposed at the wall, which is isothermal. The number of grid points used is $2432x256x160$ in the streamwise, spanwise, and wall-normal directions, respectively. Fig. 3 illustrates an instantaneous field of the normalized temperature. Fig. 4 shows the mean wall-pressure distribution of the present DNS (STREAmS) together with the data from Wu and Martin [15] and Bookey et al. [16]. Again, results are in excellent agreement.

Conclusions

We developed an in-house solver for DNS of turbulent compressible flows over complex geometries in the framework of the generalized curvilinear coordinates. The code is based on high-order central finite-difference schemes hybridized with WENO reconstruction for shock capturing. The high-fidelity of the solver is ensured by the innovative stabilization technique of central schemes, which relies on physical principles related to the preservation of the total kinetic energy from convection, without the addition of any artificial viscosity. The optimization to run on multi-GPUs architectures makes our solver capable to perform large-scale DNS of a wide range of different flow configurations. The validation of the code was carried out by simulating a subsonic turbulent curved channel flow and a supersonic turbulent compression ramp flow. In both cases, the results proved to be in excellent agreement with previous numerical and experimental studies.

References

[1] Bernardini, M., Modesti, D., Salvadore, F., & Pirozzoli, S. (2021). STREAmS: a high-fidelity accelerated solver for direct numerical simulation of compressible turbulent flows. *Computer Physics Communications*, *263*, 107906. https://doi.org/10.1016/j.cpc.2021.107906
[2] Li, Y., Fu, L., & Adams, N. A. (2021). A low-dissipation shock-capturing framework with flexible nonlinear dissipation control. *Journal of Computational Physics*, *428*, 109960. https://doi.org/10.1016/j.jcp.2020.109960
[3] Piquet, A., Zebiri, B., Hadjadj, A., & Safdari Shadloo, M. (2020). A parallel high-order compressible flows solver with domain decomposition method in the generalized curvilinear coordinates system. *International Journal of Numerical Methods for Heat & Fluid Flow*, *30*(1), 2-38. https://doi.org/10.1108/HFF-01-2019-0048
[4] Pirozzoli, S. (2010). Generalized conservative approximations of split convective derivative operators. *Journal of Computational Physics*, *229*(19), 7180-7190. https://doi.org/10.1016/j.jcp.2010.06.006
[5] Kawai, S., & Lele, S. K. (2008). Localized artificial diffusivity scheme for discontinuity capturing on curvilinear meshes. *Journal of Computational Physics*, *227*(22), 9498-9526. https://doi.org/10.1016/j.jcp.2008.06.034
[6] Visbal, M. R., & Gaitonde, D. V. (2002). On the use of higher-order finite-difference schemes on curvilinear and deforming meshes. *Journal of Computational Physics*, *181*(1), 155-185. https://doi.org/10.1006/jcph.2002.7117
[7] Pirozzoli, S. (2011). Numerical methods for high-speed flows. *Annual review of fluid mechanics*, *43*, 163-194. https://doi.org/10.1146/annurev-fluid-122109-160718
[8] Chandravamsi, H., Chamarthi, A. S., Hoffmann, N., & Frankel, S. H. (2023). On the application of gradient based reconstruction for flow simulations on generalized curvilinear and dynamic mesh domains. *Computers & Fluids*, *258*, 105859. https://doi.org/10.1016/j.compfluid.2023.105859
[9] Gaitonde, D., & Visbal, M. (1999, January). Further development of a Navier-Stokes solution procedure based on higher-order formulas. In *37th Aerospace Sciences Meeting and Exhibit* (p. 557). https://doi.org/10.2514/6.1999-557
[10] Pirozzoli, S. (2011). Stabilized non-dissipative approximations of Euler equations in generalized curvilinear coordinates. *Journal of Computational Physics*, *230*(8), 2997-3014. https://doi.org/10.1016/j.jcp.2011.01.001
[11] Ducros, F., Ferrand, V., Nicoud, F., Weber, C., Darracq, D., Gacherieu, C., & Poinsot, T. (1999). Large-eddy simulation of the shock/turbulence interaction. *Journal of Computational Physics*, *152*(2), 517-549. https://doi.org/10.1006/jcph.1999.6238
[12] Spalart, P. R., Moser, R. D., & Rogers, M. M. (1991). Spectral methods for the Navier-Stokes equations with one infinite and two periodic directions. *Journal of Computational Physics*, *96*(2), 297-324. https://doi.org/10.1016/0021-9991(91)90238-G
[13] Moser, R. D., & Moin, P. (1987). The effects of curvature in wall-bounded turbulent flows. *Journal of Fluid Mechanics*, *175*, 479-510. https://doi.org/10.1017/S0022112087000491
[14] Brethouwer, G. (2022). Turbulent flow in curved channels. *Journal of Fluid Mechanics*, *931*, A21. https://doi.org/10.1017/jfm.2021.953
[15] Wu, M., & Martin, M. P. (2007). Direct numerical simulation of supersonic turbulent boundary layer over a compression ramp. *AIAA journal*, *45*(4), 879-889. https://doi.org/10.2514/1.27021
[16] Bookey, P., Wyckham, C., Smits, A., & Martin, P. (2005). New experimental data of STBLI at DNS/LES accessible Reynolds numbers. In *43rd AIAA Aerospace Sciences Meeting and Exhibit* (p. 309). https://doi.org/10.2514/6.2005-309

Numerical tank self-pressurization analyses in reduced gravity conditions

Francesca Rossetti[1,a *], Marco Pizzarelli[2,b], Rocco Pellegrini[2,c], Enrico Cavallini[2,d], and Matteo Bernardini[1,e]

[1]Department of Mechanical and Aerospace Engineering, "Sapienza" University of Rome, Via Eudossiana 18, Rome, 00184, Italy

[2] Italian Space Agency (ASI), Via del Politecnico s.n.c., Rome, 00133, Italy

[a]f.rossetti@uniroma1.it, [b]marco.pizzarelli@asi.it, [c]rocco.pellegrini@asi.it, [d]enrico.cavallini@asi.it, [e]matteo.bernardini@uniroma1.it

Keywords: VOF, Tank Self-Pressurization, Cryogenic, Reduced Gravity

Abstract. In this study, a suitable numerical methodology to study the self-pressurization phenomenon inside a cryogenic tank, in a reduced gravity environment is proposed. This methodology is validated with the results of a benchmark self-pressurization experiment, carried out in the liquid hydrogen tank of the second stage of the Saturn IB AS-203 vehicle. The time-varying acceleration and heat flux due to solar radiation to which the tank was exposed during the experiment, have been modeled in our analysis. Finally, the numerical results show that the proposed methodology allows to reproduce the experimental data with a reasonably good accuracy.

Introduction

Studying the thermo-fluid-dynamics behavior of cryogenic propellant in reduced gravity is crucial for the design of upper stage cryogenic storage tanks. Cryogenic propellant, having a low boiling point, is very sensitive to the heat leaks to which the tank is unavoidably subjected. These heat leaks in a closed, no-venting tank, cause propellant boil-off, self-pressurization, and thermal stratification. The gravity level, g, influences the boiling process, in particular, according to Fritz expression [1], the bubble diameter at departure varies as $g^{-1/2}$ on an upward-facing horizontal surface. Not only the boiling process, but also the dynamics of the free-surface is influenced by the gravity level. Indeed, in case of low Bond number, the free-surface tends to go up along the tank walls. This phenomenon must be countered, through settling strategies, in cases when in-orbit engine start-up and operation is necessary. Important results on propellant behavior in tanks under reduced gravity conditions have been provided by the experiments, nevertheless, the uncertainty of the experimental data and the cost and complexity associated to experiments carried out in a reduced gravity environment highlight the potentialities of CFD simulations. The first self-pressurization experiment in reduced gravity, with data sufficiently detailed for validating storage tank models, was carried out in the liquid hydrogen (H_2) tank of the second stage of the Saturn IB AS-203 vehicle [2]. In the present study, a state-of-the-art numerical methodology [3], which allows to describe the main thermo-fluid-dynamics phenomena occurring in cryogenic tanks during self-pressurization in reduced gravity, is presented. This methodology is validated with the experimental data of Ward et al. [2].

Mathematical formulation and thermophysical properties

Mathematical formulation. The Volume-of-Fluid (VOF) method [4] is used to track the two-phase fluid interface. Moreover, to compute the mass transfer due to phase change, the Lee model [5] is selected. In addition, the Continuum Surface Force (CSF) model [6] is used to address the effects of surface tension force. The flow is modeled as turbulent, being the modified Rayleigh number (based on the liquid height) of liquid H_2 of the order of 10^{14}, so highly above the critical value of 10^{11}. In particular, the SST k-ω model of Menter [7] is selected.

Thermophysical properties. For the liquid H_2, the Boussinesq approximation is selected for the density, and constant thermophysical and transport properties, taken from the NIST [8] database, at the average pressure and liquid temperature, are used. The gaseous H_2 is modeled as an ideal gas, with temperature dependent specific heat, thermal conductivity, and viscosity. The latter are modeled as a piecewise linear fit of NIST data, at the average pressure. The H_2 saturation curve is approximated with a piecewise linear fit of NIST data.

Computational setup and preliminary results

Test case description. The test case under consideration is a self-pressurization experiment carried out in the liquid H_2 tank of the second stage of the Saturn IB AS-203 vehicle [2]. This tank has a height, H, of 11.3 m, a radius, R, of 3.3 m. The initial liquid mass is 7103.3 kg, corresponding to an initial liquid height, h_l, of 4.1 m, and the ullage contains only evaporated H_2. The tank characteristic dimensions are schematized in Fig. 1 (a). The tank shares a common bulkhead with the liquid oxygen (O_2) tank, which is placed below it. The test tank was in a circular low Earth orbit, and, thus, absorbed a time-varying heat flux due to solar radiation. Moreover, it was subjected to a time-varying axial acceleration resulting from the balance between an axial thrust and the drag force.

Flow solver. The pressure-based solver of the commercial CFD software Ansys Fluent® [9] is used to simulate the transient self-pressurization phenomenon. The flow is modeled as 2D axisymmetric, a minimum time step of 0.01 s is used in order to keep the residuals below a proper limit. Second order upwind schemes are used for spatial discretization of the convective terms in the governing equations. Ansys Fluent®'s Compressive scheme [9] is used for the volume fraction equation. The chosen pressure-velocity coupling scheme is SIMPLE. An interpolation of the experimental time-varying heat flux is imposed, as boundary condition, on each part of the tank wall, instead the walls of the baffles are treated as adiabatic. The different parts of the tank wall are indicated in Fig. 1 (a). Fig. 1 (b) represents the evolution of both the experimental and the interpolated heat transfer rate on the different parts of the tank wall. The tank axial acceleration level is modeled as an interpolation of its experimental values. This interpolation, together with the experimental acceleration, are represented in Fig. 1 (c). The tank is initialized at a pressure of 85495 Pa, the initial liquid temperature is 19.72 K, the ullage is initialized using a linear stratification between the liquid temperature and a temperature of 22.5 K [2].

Grid independence study. A grid independence study has been carried out, comparing the results obtained with two grids, a coarse one, having 23401 cells, and a medium one, having 34017 cells. For both grids the height of the first cell at the wall and at the interface is of 4 mm. Fig. 2 *(a)* shows the experimental and the numerical pressure evolutions obtained with the two grid levels. The pressure rise rate obtained during the first 500 s is in good agreement with the experimental one. Unfortunately, experimental data are lacking in the period between 500 s and 4350 s, so it is not possible to verify the accuracy of the numerical prediction in that time

Materials Research Proceedings 37 (2023) 226-229 https://doi.org/10.21741/9781644902813-49

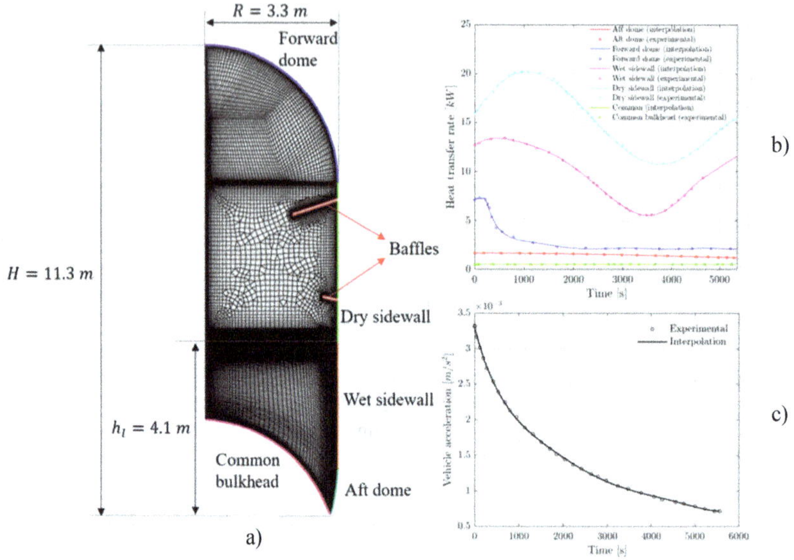

Figure 1: *(a) Selected computational grid with the indication of the tank characteristic dimensions and of the names of the different parts of the tank wall, (b) Experimental [2] and interpolated heat transfer rates, (c) Experimental [2] and interpolated vehicle acceleration.*

interval, but, certainly, some discrepancies with the experimental data arise, as it is confirmed by the numerical underestimation of the pressurization rate, in the last part of the experiment. The difference between the self-pressurization rates obtained with the two grids is negligible, but the finest grid is considered more appropriate because it allows a better representation of the vapor bubbles generating at the wall, as it will be clear in the following paragraph.

Preliminary results. Fig. 2 *(b)* shows the contours of the liquid H_2 volume fraction at 330 s, obtained with the coarse grid (left) and with the medium grid (right). Bubble nucleation is visible at the wall for both grid levels, and the resolution of the finest grid is able to better represent the shape of the bubbles. Nucleate boiling develops due to the high heat flux imposed at the wall. The liquid-ullage interface remains settled during the whole simulation, in agreement with the experiment [2]. Indeed, even if the tank was in orbital flight, the low acceleration imposed to it was sufficient to keep the propellant stabilized. This behavior is explained by the estimation of the Bond number based on the liquid level, which is of the order of 10^3, evidencing the prevalence of the gravity force over the surface tension force.

Conclusions

In this work, a numerical methodology to study the self-pressurization of cryogenic propellant in tanks characterized by reduced gravity conditions is proposed, and validated with experimental data from [2]. The agreement between the numerical and the experimental data is reasonably good, but further analyses are necessary in order to refine the methodology, and assess its applicability to test cases with different propellants, dimensions, and operating conditions.

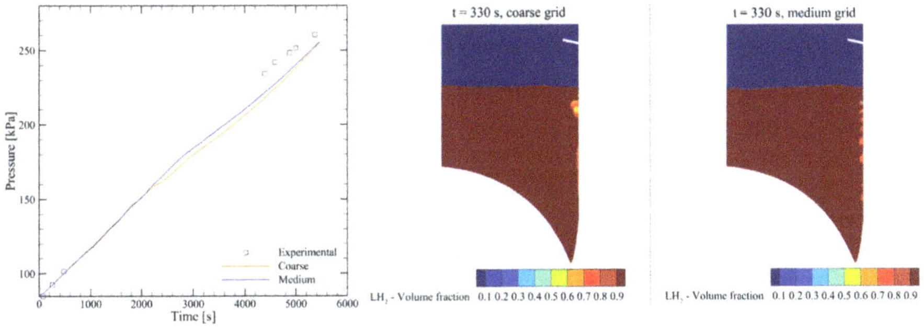

(a) *(b)*

Figure 2: (a) Experimental [2] and numerical pressure evolutions obtained with the two grid levels used for grid independence study. (b) Contours of the liquid H_2 volume fraction at 330 s, obtained with the coarse grid (left) and with the medium grid (right).

Acknowledgments

This research was jointly funded by Sapienza University and the Italian Space Agency - Agenzia Spaziale Italiana (ASI) as part of the research project: "Technical Assistance for launchers and propulsion" N.2019-4-HH.0 carried out under a framework research agreement CUP:F86C17000080005.

References

[1] W. Friz, Maximum volume of vapor bubbles, J. Physic. Zeitschz. 36 (1935) 354-379.

[2] W. D. Ward, et al., Evaluation of AS-203 low-gravity orbital experiment, NASA CR 94045 (1967) 13.

[3] F. Rossetti, et al., Setup of a numerical methodology for the study of self-pressurization of cryogenic tanks, submitted to Journal Cryogenics (2023).

[4] C. Hirt, B. Nichols, Volume of fluid (VOF) method for the dynamics of free boundaries, J. Comput. Phys. 39 (1981) 201–225. https://doi.org/10.1016/0021-9991(81)90145-5

[5] W. H. Lee, A pressure iteration scheme for two-phase flow modeling, Computational Methods for Two-Phase Flow and Particle Transport (2013) 61–82. https://doi.org/10.1142/9789814460286_0004

[6] J. Brackbill, et al. A continuum method for modeling surface tension, J. Comput. Phys. 100 1992) 335–354. https://doi.org/10.1016/0021-9991(92)90240-Y

[7] F. R. Menter, Two-equation eddy-viscosity turbulence models for engineering applications, AIAA J. 32 (1994) 1598–1605. https://doi.org/10.2514/3.12149

[8] M. M. Lemmon, E.W., D. Friend, Thermophysical properties of fluid systems, NIST Chemistry WebBook, NIST Standard Reference Database Number 69, Eds. Linstrom, P.J., and Mallard, W.G., National Institute of Standards and Technology, Gaithersburg MD, 20899. URL: https://webbook.nist.gov

[9] Ansys Fluent documentation, release 2022 R1, 2022.

Aerodynamic analysis of a high-speed aircraft from hypersonic down to subsonic speeds

Giuseppe Pezzella[1,a] and Antonio Viviani[1,b *]

[1] Università della Campania "Luigi Vanvitelli", Dipartimento di Ingegneria, via Roma 29, 81031 Aversa (CE), Italy

[a]giuseppe.pezzella@unicampania.it, [b]antonio.viviani@unicampania.it

Keywords: Subsonic, Hypersonic, CFD, Aerodynamics, Flying Test Bed

Abstract. Unmanned flying-test bed aircraft are fundamental to experimentally prove and validate next generation high-speed technologies, such as aeroshape design, thermal protection material and strategy; flight mechanics and guidance-navigation and control. During the test, the aircraft will encounter realistic flight conditions to assess accuracy of new design choices and solutions. In this framework, the paper focuses on the longitudinal aerodynamic analysis of an experimental aircraft, with a spatuled body aeroshape, from subsonic up to hypersonic speeds. Computational flowfield analyses are carried out at several angles of attack ranging from 0 to 15 deg and for Mach numbers from 0.1 to 7. Results are detailed reported and discussed in the paper.

Introduction

Advancements in high-speed technologies strongly rely on the development of flying-test beds [1].

In fact, performing flight and Wind Tunnel (WT) test campaigns represent the only and ultimate proof to demonstrate the technical feasibility of next generation high-speed aircraft (HAS) concepts and technologies [2], [3]. In this framework, the paper reports on the longitudinal aerodynamics of the Vanvitelli-one (V-one) flying test bed, shown in Figure 1 [4],[5].

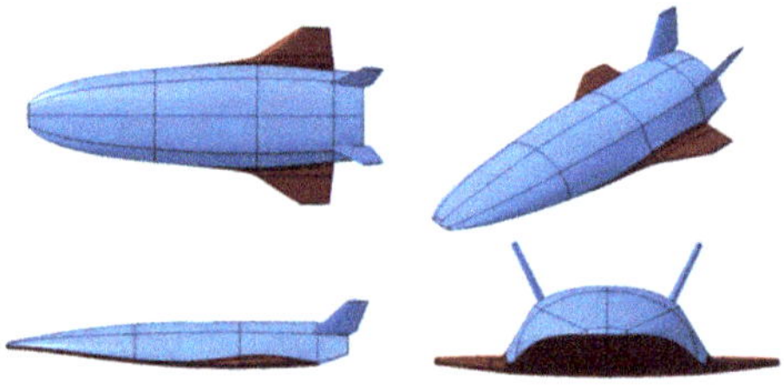

Figure 1 – Four views of the concept aeroshape [4], [5].

The V-one aircraft aims to provide a research platform suitable for a step-by-step increase of the readiness level of several enabling hypersonic technologies by means of both WT and in-flight experimentations. The V-one aeroshape features a classical lifting-body aeroshape which embodies all the features of an operational HSA, such as a low aspect ratio double-delta wing, two full movable vertical stabilizers in butterfly configuration, a spatuled fuselage forebody, characterized by a rounded off two-dimensional leading edge, mated on top of the wing. The wing flap, which must be actuated as elevon and aileron, is shown in purple in Figure 2 along with the complete moving fins (ruddervators) in green. The ruddervators pivot point is located at 50% of the root chord length.

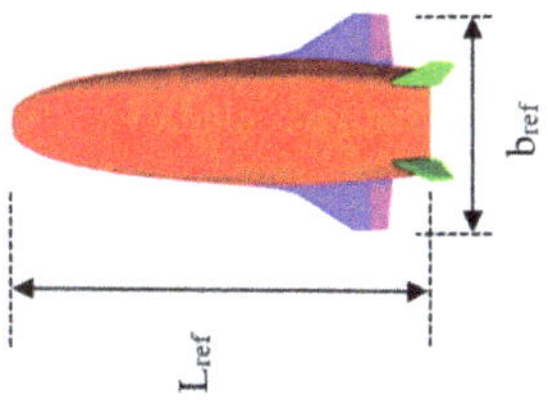

Figure 2 – Elevons and ruddervators of the aeroshape [4],[5].

The spatuled-body architecture allows to validate hypersonic aerothermodynamic design databases and passenger experiments, including thermal shield and hot structures, suitable for the successfully development of full-scale HSA [6],[7]. Within a typical mission scenario, in fact, HSA will encounter free-stream flows ranging from hypersonic to low subsonic speed.

The assessment of V-one longitudinal aerodynamics is undertaken with a Computational Fluid Dynamics (CFD) analysis with the goal to provide aerodynamic database (AEDB) to feed Flight Mechanics analyses. Both low speed and high-speed conditions were investigated at several angles of attack, α. Indeed, CFD analyses, carried out with ANSYS FLUENT® tool, address the flowfield past the aircraft for α ranging from 0 to 15 deg, for Mach from 0.1 to 7 and for clean configuration (i.e., no aerodynamic surfaces deflected) only.

Aerodynamic Flowfield Analysis

Concept longitudinal aerodynamics is addressed in terms of force and moment coefficients, according to the ISO-1151 standard. Lift (C_L), drag (C_D), and pitching moment (C_{Mx}) coefficients are calculated according to the following equations.

$$\begin{cases} C_i = \dfrac{F_i}{q_\infty S_{ref}} & i = L, D \\ C_{M_x} = \dfrac{M_x}{q_\infty L_{ref} S_{ref}} \end{cases} \quad (1)$$

The moment coefficient refers to the aircraft moment reference centre (MRC), and the reference length, L_{ref}, coincides with the fuselage length, see Fig. 2. Then, the vehicle planform area is considered as reference surface, S_{ref}. Aerodynamic coefficients are obtained by means of steady state Reynolds Averaged Navier-Stokes (RANS) flowfield computations. The pressure-based coupled solver was used for all CFD analyses carried out at free-stream Mach number lower than M_∞=0.3; for Mach higher than this threshold (i.e., M_∞=0.3) the density-based solver was considered. In this case, the Flux Difference Splitting (FDS) second-order upwind scheme (least square cell based) has been used for the spatial reconstruction of convective terms, while for the diffusive fluxes a cell-centred scheme has been applied. An implicit scheme has been considered for time integration. For both uncompressible and compressible simulations, the k-ω SST turbulence model was used for Reynolds stress closure due to its ability to model separated flows and regions of flow circulation. Further, the ideal gas model was assumed for air. Recall that, even though CFD simulations have been carried out at hypersonic flow conditions (up to M_∞=7), the ideal gas assumption was still valid. The reason is that the aeroshape features a very slender configuration and shall fly at rather low AoA (i.e., weak attached shock waves). A temperature-dependent formulation was considered for the specific heat at constant pressure, c_p, to accommodate the rather high flow energy at hypersonic speed [6]. Both unstructured and multi-block structured grids with an overall number of about 10M cells (half body) were considered for the flowfield computations. For each mesh, a body of influence surrounding the vehicle for grid refinement was considered, with the condition of $Y^+= O(1)$ at wall. Adiabatic wall is assumed at subsonic flow conditions, while the radiative cooled wall (ε=0.8) was assumed in the other cases. Sixty-four fully three-dimensional flowfield simulations were carried out and results, in terms of lift and drag coefficients, are summarized in Fig. 3.

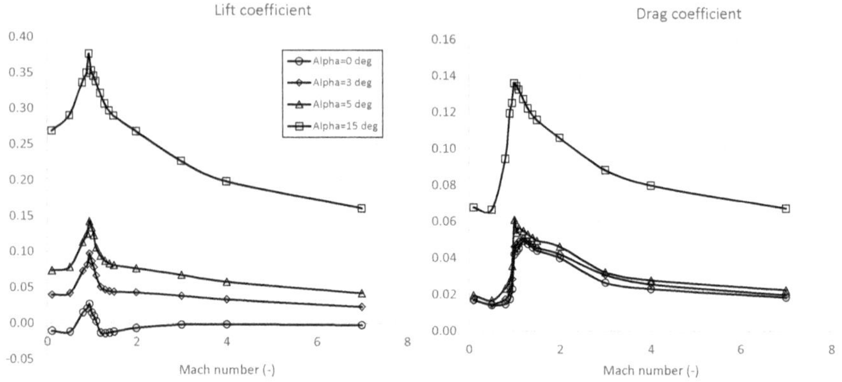

Figure 3 – Lift and drag coefficients versus Mach at α=0°, 3°, 5°, and 15°.

Figure 3 provides aircraft C_L and C_D versus Mach at four angles of attack, namely α=0°, 3°, 5°, and 15°. As shown, lift and drag coefficients rise in the transonic region and as expected increase as α increases. In particular, shock waves take place in the flowfield and determine a large increase of aerodynamic drag due to both wave drag and base drag which, in this speed regime, reach their maximum values. On the contrary, when Mach number further increases up to hypersonic flow conditions, aerodynamic coefficients decrease and reach a limit value, according to the Mach number independence principle.

Summary

Flying test bed vehicles are an efficient way to experimentally validate next generation high-speed technologies. In this framework, the paper focused attention on the appraisal of the longitudinal aerodynamic performance of a streamlined flying test bed aircraft with a spatuled-forebody aeroshape. Several computational flowfield analyses are carried out at angles of attack ranging from 0 to 15 deg and for Mach numbers from 0.1 to 7.

Initial findings for the hypersonic speed range pointed out that the high streamlined sharp leading edges of the V-one aeroshape led to a rather low wave drag component. In fact, up to 5 deg angle of attack the drag coefficient is equal to about 0.02; while aerodynamic lift at this attitude is close to 0.05.

Low speed aerodynamic analyses show that the lift coefficient features a mostly linear lift curve slope as expected for this flow regime and wing profile. The rounded leading edges of the double-delta planform prevent the formation of strong vortices, which would otherwise result in sharp changes in lift curve slope at high AoA. Vortices were instead observed forming along the forebody of the aircraft, which would contribute to aerodynamic forces and stability. A comparison of a structured domain with an unstructured showed little difference in extracted coefficients.

References

[1] McClinton, C. R., Rausch, V. L., Nguyen, L. T., Sitz, J. R., "Preliminary X-43 flight test results", Acta Astronautica, Volume 57, Issues 2–8, 2005, Pages 266-276, ISSN 0094-5765, https://doi.org/10.1016/j.actaastro.2005.03.060

[2] Jeyaratnam, J., Bykerk, T., Verstraete, D., "Low speed stability analysis of a hypersonic vehicle design using CFD and wind tunnel testing" (2017) 21st AIAA International Space Planes and Hypersonics Technologies Conference, Hypersonics 2017, 10 p. https://doi.org/10.2514/6.2017-2223

[3] Bykerk, T., Verstraete, D., Steelant, J., Low speed lateral-directional dynamic stability analysis of a hypersonic waverider using unsteady Reynolds averaged Navier Stokes forced oscillation simulations (2020) Aerospace Science and Technology, 106, art. no. 106228. https://doi.org/10.1016/j.ast.2020.106228

[4] Bykerk, T., Pezzella, G., Verstraete, D., Viviani, A., "High and Low Speed Analysis of a Re-usable Unmanned Re-entry Vehicle". HISST. International Conference on High-Speed Vehicle Science and Technology. Moscow. Russia. November 25-29, 2018. hisst-2018_1620897.

[5] Bykerk, T., Pezzella, G., Verstraete, D., Viviani, A., "Lateral-Directional Aerodynamics of a Re-Usable Re-Entry Vehicle". 8th European Conference for Aeronautics and Space Sciences (Eucass-2019). July 2019. Madrid. Spain.

[6] Scigliano, R., Pezzella, G., Di Benedetto, S., Marini, M., Steelant, J., "Hexafly-Int Experimental Flight Test Vehicle (EFTV) Aero-Thermal Design". Proceedings of the ASME 2017 International Mechanical Engineering Congress & Exposition IMECE 2017. November 3-9, 2017, Tampa, Florida, USA. ASME 2017 International Mechanical Engineering Congress and Exposition Volume 1: Advances in Aerospace Technology Tampa, Florida, USA, November 3–9, 2017.Conference Sponsors: ASME. ISBN: 978-0-7918-5834-9. Paper No. IMECE2017-70392, pp. V001T03A022; 14 pages. https://doi.org/10.1115/IMECE2017-70392

[7] Schettino, A., Pezzella, G., Marini, M., Di Benedetto, S., Villace, V.F., Steelant, J., Choudhury, R., Gubanov, A., Voevodenko, N. "Aerodynamic database of the HEXAFLY-INT hypersonic glider". (2020) CEAS Space Journal, 12 (2), pp. 295-311. https://doi.org/10.1007/s12567-020-00299-4

General Session

Virtual testing application to ESA micro vibrations measurement system

Lorenzo Dozio[1,a *], Leonardo Peri[2,b], Michele Pagano[3,c] and Pietro Nali[4,d]

[1]Politecnico di Milano, Department of Aerospace Science and Technology, via La Masa, 34, 20156, Milano, Italy

[2]Politecnico di Milano, Department of Aerospace Science and Technology, via La Masa, 34, 20156, Milano, Italy (currently employed as PhD student at KU Leuven, Dept. Mech. Eng.)

[3]Politecnico di Milano, Department of Aerospace Science and Technology, via La Masa, 34, 20156, Milano, Italy (currently employed as Mechanical Environment Engineer at CNES)

[4]Thales Alenia Space, Domain Observation & Navigation Italy, Strada Antica di Collegno, 253, 10146, Torino, Italy

[a]lorenzo.dozio@polimi.it, [b]leonardo.peri@kuleuven.be, [c]michelepagano919@gmail.com, [d]pietro.nali@thalesaleniaspace.com

Keywords: Virtual Testing, Microvibrations, MOR Techniques, Enhanced Craig-Bampton, Balanced Truncation

Abstract. The challenging application of Virtual Testing (VT) to ESA's six-degree-of-freedom Micro Vibrations Measurement System (MVMS) is described in this work. The digital replicate of MVMS is first obtained from a high-fidelity finite element model, whose order is later appropriately reduced. A state-space model representative of the dynamic behaviour of the MVMS is finally obtained. MVMS VT simulations are thus exploited as a key enabling technology to perform the ad-hoc design of MVMS control system design. This work focuses on different model-order reduction techniques applied to MVMS, which were evaluated and compared in terms of performance and computational issues. Classical and more recent approaches belonging to the family of Component Mode Synthesis (CMS) methods are addressed. State-space based techniques are considered as well, also in two-stage combination with CMS methods. Challenges and advantages of VT are lastly discussed.

Introduction to MVMS and virtual testing methodology

MVMS is a novel 6DOFs microvibration facility developed for the European Space Agency (ESA) by the UK's National Physical Laboratory. It is designed to measure/impose microvibration accelerations, forces, and moments in the frequency range from 0.03 Hz to 100 Hz, thus allowing both the characterisation of potential microvibration source and the assessment of an item's performance subjected to a microvibration environment [1]. Figure 1 shows MVMS, which is composed of three main components: a base support (BS), the VIBration ISOlation platform (VIBISO), and the vibration Measurement PLatform (MPLAT). The base support interfaces with the ground, and it is mechanically connected to VIBISO through a MINUS-K device which acts as a passive mechanical low-pass filter. The function of BS is to sustain the upper MVMS components and to hold the seismometers and the fixed parts of the voice coil actuators, which form the set of sensors/actuators used by VIBISO to complement and improve the passive isolation provided by the MINUS-K with an active control action. Indeed, the aim of VIBISO is to actively isolate the upper part of the facility from vibrations transferred via the ground such as seismic disturbance. The function of MPLAT is to measure/impose the microvibration environment from/to the test specimen. It interfaces VIBISO through a second MINUS-K device and an additional set of voice coil actuators.

The digital replicate of MVMS is obtained by a properly correlated high-fidelity finite element model of the facility. A set of reduced-order models are later derived by retaining only the most relevant dynamics information, with the final aim of using the resulting state-space representation for the control system design. Figure 2 provides an overview of the computational methodology put in place for MVMS Virtual Testing in line with [2].

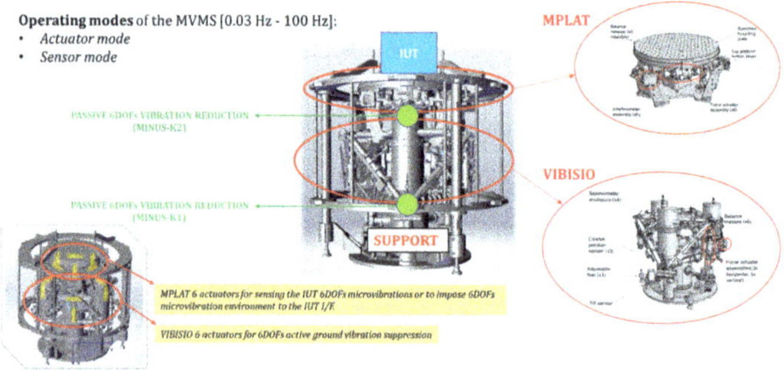

Figure 1. General description of the MVMS facility.

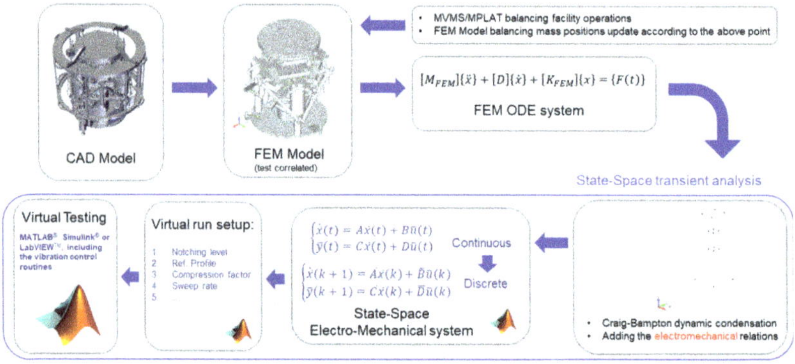

Figure 2. Overview of the computational methodology of MVMS Virtual Testing.

Assessment of model-order reduction techniques

In this work, several different model-order reduction techniques are considered, evaluated, and compared in terms of performance and computational issues. Classical and more recent approaches belonging to the family of Component Mode Synthesis (CMS) methods are applied to MVMS, as well as state-space based techniques such as the simple modal truncation (MT) and the powerful balanced truncation (BT). A hybrid technique consisting in a two-stage reduction combining CMS methods with BT or MT is also developed to overcome the numerical difficulties associated with the direct application of BT to very high-order problems [3]. A very restricted selection of results is presented below to give a concise comparison of the various methodologies, along with advantages or disadvantages of the adopted techniques.

Concerning CMS methods, in addition to the classical Craig-Bampton (CB) method [4], the Rubin (RU) method [5] and the Enhanced Craig-Bampton (ECB) [6] technique are evaluated. The reductions are carried out by considering as substructures the three main components of MVMS. Both ECB and RU aim to improve CB reduction performance by taking into account a contribution coming from discarded modes in the transformation matrix. While in RU this is achieved by approximating the dynamic behaviour via free-interface normal modes, ECB considers fixed-interface modes as in the original CB formulation. The selection of component modes to be retained for the three substructures is carried out by resorting to the Effective Interface Mass criterion [7], resulting in 15, 37 and 4 modes, respectively, for BS, VIBISO and MPLAT, along with 60 boundary DOFs.

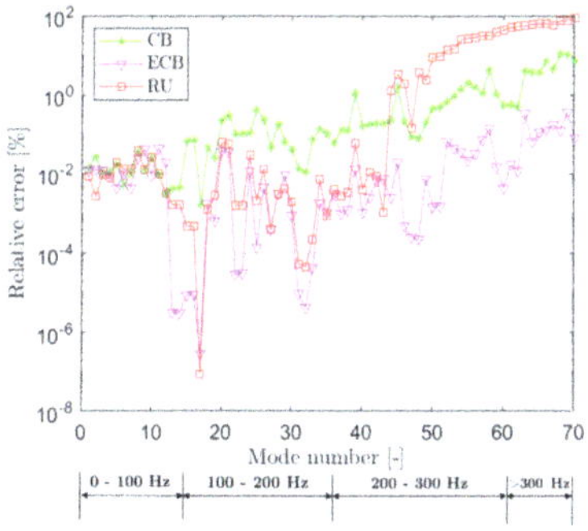

Figure 3. Relative error on natural frequencies among CMS methods.

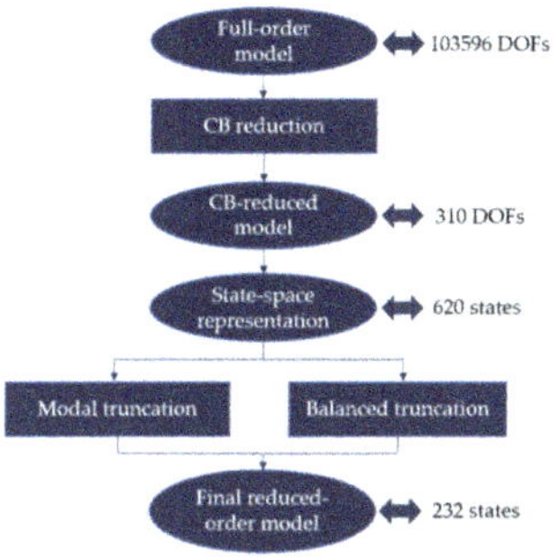

Figure 4. Two-stage reduction workflow.

Figure 3 reports the relative error between the natural frequencies computed by the high-fidelity FE model and those obtained from the reduced models. Both RU and ECB provide higher accuracy with respect to the classical CB reduction. In particular, ECB maintains superior performance in approximating both low and higher frequency modes, while a significant worsening is experienced with RU above approximately 250 Hz.

A similar analysis was carried out by employing MT and BT methods [8]. The latter is particularly appealing in the present VT approach, as it allows the input-output behaviour of the system to be preserved as much as possible. However, the high computational effort required by BT prevents its direct application to large-scale systems. This issue is overcome in this work by resorting to a hybrid two-stage reduction approach [9], envisaging a preliminary reduction via one of the CMS methods (CB is considered here), followed by either MT or BT, as reported in Fig. 4. The performance of the various reduction methods is assessed by comparing the frequency response in terms of the transmissibility from a longitudinal seismic disturbance to the item under test as shown in Fig. 5. All the methods, except for the hybrid MT, provide a high-performance approximation of the system response. The lack of a notable difference among the methods is explained by the peculiar dynamic behaviour of MVMS. Indeed, the low-pass filtering action of the MINUS-K devices strongly affects the dynamic response, which is dominated by a few low-frequency structural modes. With the aim of providing a deeper insight into the performance of the reduction techniques, the same analysis is carried out by retaining a significantly lower amount of DOFs. Figure 6 reports the relative error in the transmissibility function between the full-order and the reduced models. The largest improvements with respect to the classical CB method performance are provided by ECB and hybrid BT techniques.

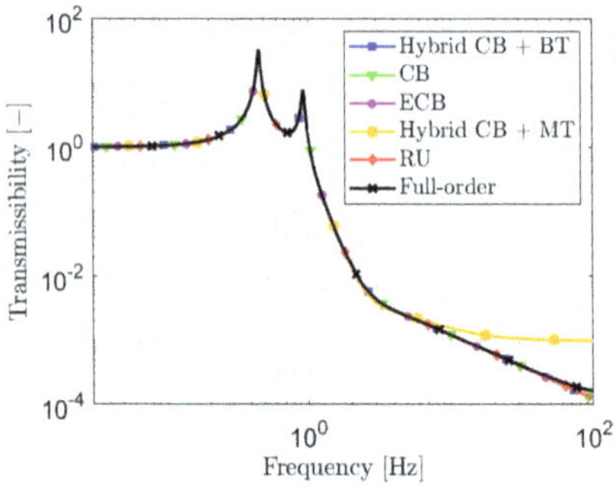

Figure 5. Comparison of the frequency responses.

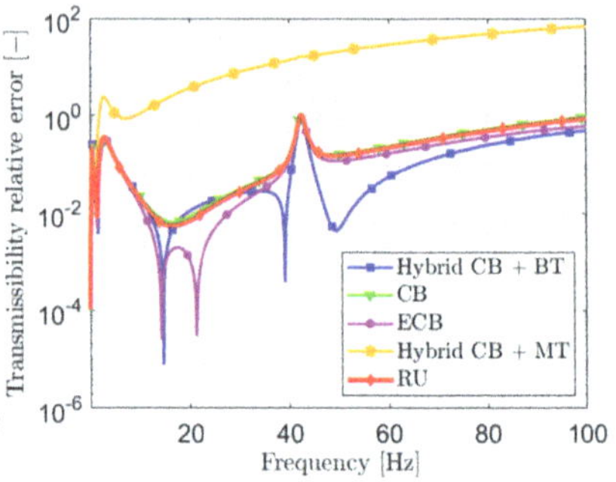

Figure 6. Relative error on the transmissibility.

In conclusion, referring to the MVMS application, hybrid BT and ECB appear as the most promising methods in terms of performance improvements with respect to classical reduction techniques typically adopted in the VT framework.

References

[1] Statement of Work ESA Express Procurement – EXPRO SOW 6DOF MVMS Simulation Model, SA-TEC-MXE-MECH-SOW-00147.

[2] P. Nali, et al., A virtual shaker testing experience: modeling, computational methodology and preliminary results, ECSSMET 2016 (European Conference on Spacecraft Structures, Materials & Environmental Testing), Toulouse, France, 27-30 September 2016.

[3] B. Besselink, et al., A comparison of model reduction techniques from structural dynamics, numerical mathematics and systems and control, J Sound Vib., 332 (2013), 4403–4422. https://doi.org/10.1016/j.jsv.2013.03.025

[4] R.R. Craig, M.C.C. Bampton, Coupling of Substructures for Dynamic Analyses, AIAA Journal, 6 (1968), 1313–1319. https://doi.org/10.2514/3.4741

[5] S. Rubin, Improved Component-Mode Representation for Structural Dynamic Analysis, AIAA Journal, 13 (1975), 995–1006. https://doi.org/10.2514/3.60497

[6] J. Kim and P. Lee. An enhanced Craig–Bampton method, International Journal for Numerical Methods in Engineering, 103 (2015), 79–93. https://doi.org/10.1002/nme.4880

[7] D. Kammer, M. Triller, Selection of component modes for Craig-Bampton substructure representations, ASME J Vib. Acoust., 118 (1996), 264–270. https://doi.org/10.1115/1.2889657

[8] A. Antoulas. Approximation of Large-Scale Dynamical Systems, SIAM, Philadelphia, USA, 1 edition, 2005. ISBN 978-0-89871-529-3.

[9] J. Spanos and W. Tsuha. Selection of Component Modes for Flexible Multibody Simulation, J Sound Vib., 14 (1991), 278–286. https://doi.org/10.2514/3.20638

Nonlinear transient analyses of composite and sandwich structures via high-fidelity beam models

Matteo Filippi[1,a *], Rodolfo Azzara[1,b] and Erasmo Carrera[1,c]

[1]MUL2 Group, Department of Mechanical and Aerospace Engineering, Politecnico di Torino, Corso Duca degli Abruzzi 24, 10129, Torino, Italy

[a]matteo.filippi@polito.it, [b]rodolfo.azzara@polito.it, [c]erasmo.carrera@polito.it

Keywords: Finite Element Method, Transient Nonlinear Analyses, One-Dimensional Formulations, Carrera Unified Formulation

Abstract. In this study, we employ low and high-fidelity finite beam elements to conduct geometrical nonlinear transient analyses of composite and sandwich structures. The equations of motion for various structural theories are derived in a total Lagrangian scenario using the Carrera Unified Formulation. The unified formalism's three-dimensional nature enables us to include all components of the Green-Lagrange strain tensor. To solve the equations, we utilize the Hilber-Hughes-Taylor (HHT)-α algorithm in conjunction with a Newton-Raphson procedure. We present the dynamic response of a sandwich stubby beam subjected to a step load, calculated using both equivalent-single layer and layer-wise approaches. Additionally, we discuss the effects of geometrical nonlinearity.

Introduction

In recent decades, the aerospace, automotive, and other engineering fields have faced new challenges that necessitate the adoption of sophisticated and lightweight components. These highly flexible structures are extensively utilized in various engineering applications as they can exhibit large displacements and rotations without undergoing plastic deformations. Furthermore, many of these components comprise sandwich structures made of composite materials to ensure a significant strength-to-weight ratio. As highlighted in numerous scientific papers [1,2], analyzing these structural configurations requires refined kinematic theories that can overcome the well-known limitations of the Euler-Bernoulli and the first-order-shear deformation theories.

This research aims to analyze the nonlinear dynamic behavior of composite and sandwich structures using variable-fidelity one-dimensional finite elements. The mathematical models are derived using the Carrera Unified Formulation (CUF). This hierarchical formalism enables the selection of the order of the structural model as an input of the analysis. Therefore, any theory can be obtained by arbitrarily expanding the generalized variables. Specifically, this work employs Lagrange (LE) and Taylor (TE) polynomials to develop kinematic expansions. According to the layer-wise concept, the LE models allow for the independent discretization of each lamina. In contrast, the Taylor-based models homogenize the cross-section properties with polynomials of arbitrary orders. Regardless of which theory is adopted, the governing equations and the related finite element arrays are formulated in terms of *Fundamental Nuclei* (FNs), the invariants of the methodology. The nonlinear equations are formulated using the total Lagrangian approach and solved using a suitable Newton-Raphson method. We utilize the Hilber-Hughes-Taylor (HHT)-α algorithm as the implicit time integration scheme to evaluate the nonlinear dynamic response. This algorithm proves particularly effective in stabilizing the time integration process under highly nonlinear effects.

To highlight the relevant discrepancies between low- and high-fidelity solutions in studying the response of laminated beams, we consider a stubby sandwich structure characterized by a significant degree of anisotropy between the core and external layers. Moreover, we have compared geometrical linear and nonlinear solutions for two different load magnitudes.

Unified formulation of geometrical nonlinear beam theory

The theoretical basis required for solving transient analyses in geometrically nonlinear regimes can be found in [3]. However, to ensure the self-contained nature of this work, we provide some basic equations here. Based on the one-dimensional finite element unified formulation, we express the displacement vector $\boldsymbol{u}^T(x,y,z,t) = (u_x, u_y, u_z)$ as a sum of products between cross-sectional (defined over the x-z plane) functions $F_\tau(x,z)$, finite element shape functions $N_i(y)$ (defined over the y-axis) and the nodal unknown vector $\boldsymbol{q}_{\tau i}(t)$

$$\boldsymbol{u}(x,y,z,t) = F_\tau(x,z)N_i(y)\boldsymbol{q}_{\tau i}(t) \qquad \tau = 1,\dots,M \text{ and } i = 1,\dots,nn_{el} \tag{1}$$

In Eq. 1, the subscripts indicate summation, while M and nn_{el} represent the number of functions included in the structural model and the number of nodes belonging to a single finite beam element, respectively. As previously mentioned, the geometrically nonlinear FE governing equations are obtained according to a total Lagrangian formulation by including all Green–Lagrangian strain tensor components. The strain–displacement relation and the constitutive law reported in Eq. 2 are obtained using the linear and nonlinear differential operators, $\boldsymbol{b}_l$ and $\boldsymbol{b}_{nl}$, respectively, and the stiffness matrix for linear elastic materials, $\boldsymbol{C}$.

$$\boldsymbol{\varepsilon} = (\boldsymbol{b}_l + \boldsymbol{b}_{nl})\boldsymbol{u} \qquad \boldsymbol{\sigma} = \boldsymbol{C}\,\boldsymbol{\varepsilon} \tag{2}$$

By substituting Eq. 1 and Eq. 2 into the principle of virtual work, it becomes possible to express the virtual variations of both strain energy (δL_{int}) and the work done by inertial forces (δL_{ine}) and external loads (δL_{ext}) in the CUF formalism.

$$\begin{aligned} \delta L_{int} &= \delta \boldsymbol{q}_{sj}^T \boldsymbol{K}_S^{ij\tau s} \boldsymbol{q}_{\tau i} \\ \delta L_{ine} &= \delta \boldsymbol{q}_{sj}^T \boldsymbol{M}^{ij\tau s} \ddot{\boldsymbol{q}}_{\tau i} \\ \delta L_{ext} &= \delta \boldsymbol{q}_{sj}^T \boldsymbol{F}^{sj} \end{aligned} \tag{3}$$

The FNs of the secant stiffness and mass matrix are denoted as $\boldsymbol{K}_S^{ij\tau s}$ and $\boldsymbol{M}^{ij\tau s}$, respectively, while $\boldsymbol{F}^{sj}$ represents the fundamental nucleus of the loading vector. Here, the indexes s and j are adopted for the virtual variations of the displacements and have the same bounds as τ and i. The assembled matrices and vectors associated with any arbitrary structural model are constructed by permuting these four indexes. Based on this notation, the equations of motion are

$$\boldsymbol{M}\ddot{\boldsymbol{q}}(t) + \boldsymbol{K}_s(\boldsymbol{q})\boldsymbol{q}(\boldsymbol{t}) = \boldsymbol{F}(t) \tag{4}$$

Equation 4 is solved by using the Newton-Raphson method and the HHT-α implicit time integration scheme. For a dynamic conservative problem, the linearization of the residual nodal forces leads to

$$\begin{aligned} \delta(\delta L_{int} + \delta L_{ine} - \delta L_{ext}) &= \delta \boldsymbol{q}_{sj}^T\left(\boldsymbol{K}_0^{ij\tau s} + \boldsymbol{K}_{T1}^{ij\tau s} + \boldsymbol{K}_\sigma^{ij\tau s}\right)\delta \boldsymbol{q}_{\tau i} + \delta \boldsymbol{q}_{sj}^T \boldsymbol{M}^{ij\tau s}\delta\ddot{\boldsymbol{q}}_{\tau i} = \\ &= \delta \boldsymbol{q}_{sj}^T \boldsymbol{K}_T^{ij\tau s}\delta \boldsymbol{q}_{\tau i} + \delta \boldsymbol{q}_{sj}^T \boldsymbol{M}^{ij\tau s}\delta\ddot{\boldsymbol{q}}_{\tau i} \end{aligned} \tag{5}$$

In Eq. 5, $\boldsymbol{K}_T^{ij\tau s}$ represents the FN of the tangent stiffness matrix (see [4]).

Results

Figure 1 and Table 1 present the dimensions, material properties, boundary conditions, and loading conditions of the sandwich beam. The structure was subjected to a pressure load that was constant in time but variable in magnitude.

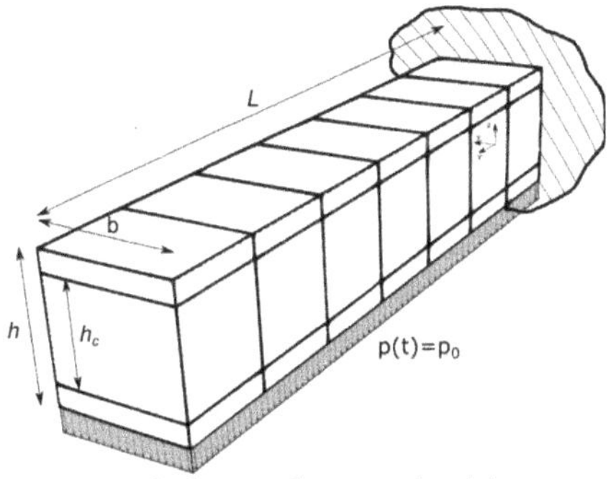

Figure 1: the cantilever sandwich beam.

Table 1: dimensions and material data of the sandwich beam.

Geometrical data	Material data	Face	Core
L = 0.1 [m]	Young's modulus, E [GPa]	200	0.66
h = b = 0.02 [m]	Poisson's ratio	0.27	0.3
h_c = 0.014 [m]	Density [kg/m^3]	7800	60

The finite element models utilized in this study consisted of seven 4-node beam elements placed along the longitudinal axis. Transient analyses were performed using the Taylor-based expansions of second (TE2) and third order (TE3) - as well as a layer-wise model consisting of three (one per layer) bi-cubic Lagrange elements (3-LE16) placed over the cross-section. The degrees of freedom (d.o.f.) corresponding to the TE2, TE3 and 3LE16 solutions were 396, 660, and 2640, respectively. In Figure 2, the transverse deflection of the sandwich beam is depicted for two different values of the applied pressure. As expected, the equivalent-single-layer solutions, although improved compared to classical beam models, significantly underestimate the deformation of the structure when compared to the layer-wise prediction. Additionally, it is noteworthy that both geometrically linear and nonlinear approaches provide almost indistinguishable results for relatively small values of p_0, irrespective of the structural theory employed. However, it can be observed that with the increase of the load value, the linear and nonlinear curves obtained using the 3-LE16 model differ significantly. This effect can be attributed to the superior ability of the layer-wise model to accurately describe the deformation and stress fields compared to the equivalent-single-layer kinematics.

Summary

This work presented preliminary results concerning composite and sandwich structures' geometrical nonlinear transient responses calculated with one-dimensional finite element models based on various kinematic assumptions. The comparisons between low- and high-fidelity solutions demonstrated the importance of accurately describing the cross-section deformations, especially in the nonlinear regime. Further results will be provided on composite laminated structures subjected to different loads.

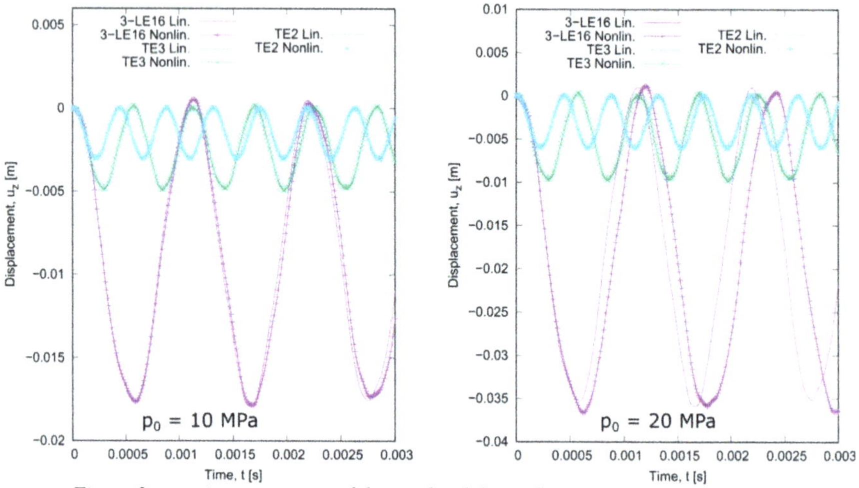

Figure 2: transient responses of the sandwich beam for two pressure values.

References

[1] T. Kant, C.P. Arora, J.H. Varaiya, Finite element transient analysis of composite and sandwich plates based on a refined theory and a mode superposition method, Composite Structures. 22 (1992) 109-120. https://doi.org/10.1016/0263-8223(92)90071-J

[2] J.R. Kommineni, T. kant, Large Deflection Elastic and Inelastic Transient Analyses of Composite and Sandwich Plates with a Refined Theory, Journal of Reinforced Plastics and Composites. 12 (1993) 1150-1170. https://doi.org/10.1177/073168449301201102

[3] R. Azzara, M. Filippi, A. Pagani, Variable-kinematic finite beam elements for geometrically nonlinear dynamic analyses, Mech. Adv. Mater. Struct. (2022). https://doi.org/10.1080/15376494.2022.2091185

[4] A. Pagani, E. Carrera, Unified formulation of geometrically nonlinear refined beam theories, Mech. Adv. Mater. Struct. 25 (2018) 15-31. https://doi.org/10.1080/15376494.2016.1232458

BEA: Overview of a multi-unmanned vehicle system for diver assistance

Leonardo Barilaro[1*], Jason Gauci[2], Marlon Galea[2], Andrea Filippozzi[3], David Vella[2], Robert Camilleri[2]

[1] The Malta College of Arts, Science & Technology, Aviation Department, Paola PLA 9032, Malta

[2] Institute of Aerospace Technologies, University of Malta, Msida, MSD 2080, Malta

[3]Divers Code Ltd., St Paul's Bay, SPB 2080, Malta

* leonardo.barilaro@mcast.edu.mt

Keywords: Swarm Unmanned Vehicles, UAV, USV, UUV, Safety at Sea

Abstract. This paper presents an overview of a solution to address the issue of marine traffic endangering scuba diving and free diving. Diving is a popular recreational activity, and it is estimated that there are around six million active scuba divers worldwide. When diving, it is essential to signal one's presence with universal markers, however, boat drivers do not always recognize them and can speed too close to dive zones, posing a risk to divers. To mitigate these risks, a multi-unmanned vehicle system consisting of an Unmanned Aerial Vehicle (UAV), an Unmanned Surface Vehicle (USV), and an Unmanned Underwater Vehicle (UUV) has been developed. The proposed system works in synergy to monitor and protect divers. The UAV monitors the surface of the sea near the dive zone for any traffic, while the USV tracks the UUV, communicates with the other unmanned vehicles, and provides a takeoff/landing surface for the UAV. The USV can also be used to tow divers and equipment to/from the shore. Finally, the UUV tracks the diver and warns them if it is unsafe to surface. The paper provides an overview of the design and system's architecture, algorithms for boat detection, precision landing and UUV tracking, as well as preliminary tests carried out on the prototype. The proposed system is found to be suitable for the intended application. The BEA (Buoy Eau Air) system is the first in the world to use a multi-drone system to create a geo-fence around the diver and monitor the area within it. The paper also highlights the potential benefits of such a system for the touristic sector, especially for countries where diving is a popular recreational activity.

Introduction

The project presented in this paper develops the first system in the world composed of a multi-drone system: BEA. Before the global pandemic due to Covid-19, the touristic sector in Malta contributed to approximately 25% of GDP and to 28% of fulltime employment. Around 5% of inbound tourists have engaged in diving activities such as snorkelling, scuba diving and freediving [1]. However, such activities have an element of risk. With the recovery of the economic situation worldwide it is important to investigate ways to reboost the touristic sector with new solution to improve the safety at sea. To mitigate the risks associated, BEA proposes a system of drones, one operating in the air, one on the water surface and one underwater to support and monitor the safety of divers. While the hovering UAV creates a geo-fence around the diver, therefore offering protection against boat incursions [2, 3]. The self-propelled Buoy acts as a resting platform while hosting critical support. It also acts as the communication link between the hovering drone and the underwater ROV. Finally, the ROV is able to follow the diver emulating the "buddy philosophy", while providing reassurance to the diver. The multi-drone system works in synergy, with the sub-systems being in communication with each other, and have the ability to relay a message to the diver. The buddy system is set of safety procedures to improve the chances of avoiding accidents in or under water by having divers in a group of at least two. The group dive together and co-operate, so that they can help or rescue each other in the event of an emergency. The key point is to respond in time, which is related with the experience of the divers. The multi-drone system enhances this, working in synergy, with the sub-systems being in communication with each other, and having the ability to relay a message to the diver.

The use of a group (or swarm) of unmanned vehicles to perform a task is advantageous because it makes it possible to overcome the limitations of individual vehicles and improves overall performance, flexibility and fault-tolerance, amongst other things. Various instances of collaboration between unmanned aerial, ground, surface and underwater vehicles are available in the literature. For instance, collaborations are reported between Unmanned Ground Vehicles (UGVs) and UAVs [4-7]; between USVs and UAVs [8-10]; and between UAVs, USVs and UUVs [11, 12]. The applications for such collaborations are endless and include: maritime patrol [10], power pylon inspection [7], search and rescue [11], object transportation [5] and terrain navigation [6]. BEA will contribute towards maintaining the safety track record of the local Maltese industry while offering a novelty which boosts the diving experience.

The rest of the paper is organized as follows. Section 2 describes the BEA concept overview while Sections 3 detail the architecture of the proposed system and, finally, Section 4 concludes the paper and describes future work.

BEA concept overview

The BEA system is composed of 3 main modules, the Buoy, developed from scratch, a drone monitoring the geo-fence around the diver and an underwater drone.

More in detail, the primary objectives of BEA are to develop a functional prototype of the multi-drone system and to conduct comprehensive testing under various ambient conditions, including different sea and diving scenarios. By capturing data from test campaigns, divers, and engineers, the secondary objective of BEA is to facilitate design improvements and industrialization while creating a valuable database of average diving safety conditions.

The engineering and divers' requirements have been captured to ensure the system's effectiveness and usability. Figure 1 illustrates the conceptual architecture of the system, outlining the roles and interactions of each component.

The UAV plays a critical role in detecting boat traffic through an onboard camera. It takes off from the Unmanned Surface Vehicle (USV) and hovers at a specified altitude. The UAV can initiate its flight either at the beginning of a dive or a few minutes prior to the expected end, during the divers' safety stop. Upon detecting a boat or vessel, the UAV transmits a warning signal to the USV, which subsequently activates a light indicator located on its underside. The warning signal is also relayed to the Underwater Unmanned Vehicle (UUV), triggering a light indicator for the divers. This system ensures that divers are visually alerted to the presence of boats, enabling them to exercise caution before surfacing.

Throughout the dive, the UUV closely follows the divers, detecting their movements and adapting its speed and direction accordingly, while avoiding obstacles such as rocks. Simultaneously, the USV tracks the UUV's position, maintaining a fixed distance from it by maneuvering and propelling itself. Furthermore, the UAV continuously tracks the current position of the USV. These three vehicles maintain mutual tracking during the dive, ensuring comprehensive surveillance and support.

At the conclusion of the dive, as the divers surface, the UAV descends and lands on the USV by accurately detecting and tracking fiducial markers. Additionally, the UUV docks with the USV, facilitating seamless retrieval. Prior to or after a dive, the USV can be manually operated by the divers, allowing them to tow themselves and their equipment to or from the shore or a boat. In rescue situations, the USV can also serve as a towing mechanism for divers requiring assistance.

Figure 2 shows the 3D model of the finalized conceptual design.

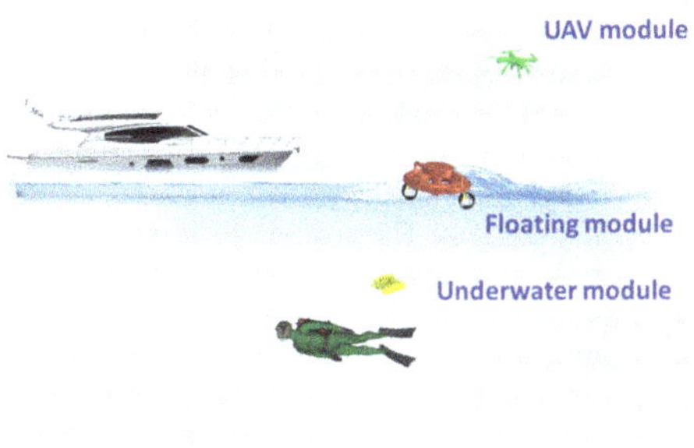

Fig 1: BEA system concept

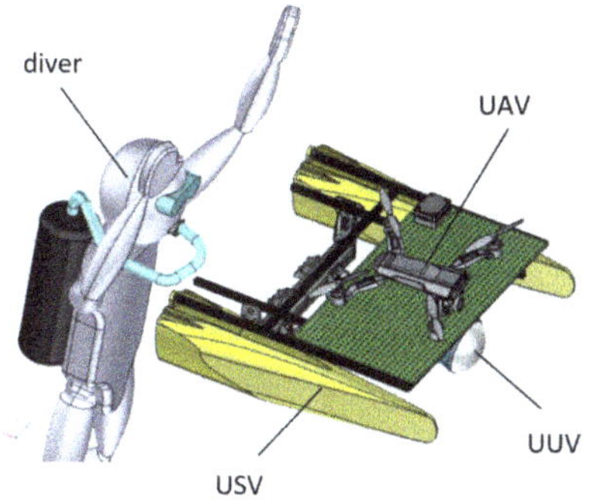

Fig 2: BEA conceptual design

System architecture

This paragraph will describe in detail the system architecture.

UAV: The UAV is the Cuta-Copter EX-1 drone. This quadcopter has an IP67 rating, a payload capacity of 2 kg and an endurance of 23 minutes (without a payload). It is equipped with a Pixhawk 2.4.8 flight controller (running ArduPilot autopilot software) and supports manual and automatic flight. For this application, additional components were added to the Cuta-Copter. The first of these is a Raspberry Pi 4 (Model B) companion computer with 8 GB of RAM. This computer exchanges MAVLink messages with the flight controller via a Universal Asynchronous Receiver-Transmitter (UART) communication interface. It also exchanges data with the USV via a wireless, half-duplex, serial communication link. Two sensors are connected to the Raspberry Pi: a visible light camera module and a distance sensor. The camera is a Raspberry Pi camera module (v1) with a weight of 3 g, a sensor resolution of 2592 x 1944 pixels, a horizontal Field of View (FoV) of 53.5 degrees and a vertical FoV of 41.4 degrees. The camera is mounted beneath the Cuta-Copter and points downwards. It is used both for boat detection and precision landing. The distance sensor is a Benewake TF02-Pro IP65 LiDAR with an outdoor range of 13.5 m and a weight of 50 g. This sensor is mounted next to the camera and measures the height of the UAV above the USV. This sensor is used for precision landing.

USV: The USV was custom-built for this work (Fig. 3). It is made of fiberglass and consists of two hulls; a takeoff/landing pad measuring 1 x 1 m; a waterproof compartment for the electronics; and a handle bar for manual control. Its overall size is 110 x 80 x 15 cm. The USV is controlled by an Arduino Mega 2560 microcontroller and is powered by a 14.8 V 46 Ah LiPo battery pack, complete with a Battery Management System (BMS). Propulsion is provided by four BlueRobotics T200 thrusters at the bottom of the USV – two in front and two at the back. The speed and direction of the USV is controlled by varying the Pulse Width Modulation (PWM) signals to these thrusters. The GPS position of the USV is determined by an Adafruit Ultimate GPS module which provides the controller with NMEA-0183 messages. The position of the USV is transmitted to the UAV via the wireless communication link, and this position is sent to the UAV's autopilot as a new waypoint. Thus, the UAV can follow the USV and remain overhead. Two BlueRobotics ping sonar altimeters and echosounders are mounted in front of the USV, 70 cm apart. Each sensor operates at 115 kHz and has a beam width of 30 degrees and can detect objects up to 50 m underwater. The output of each sensor consists of a distance measurement and a confidence value. These sensors are used to estimate the position of the UUV. An underwater Light Emitting Diode (LED) – whose brightness can be controlled by means of a PWM signal – is mounted beneath the USV. This LED can be turned on by the controller to warn the divers in the event of boat traffic. The USV and UUV communicate via Water Linked M64 underwater modems which provide a wireless half-duplex acoustic communication link.

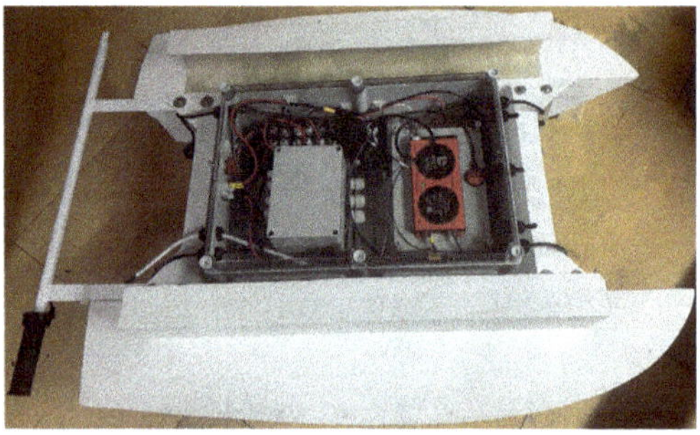

Fig 3: USV

UUV: The UUV is the iBubble Evo. This is an untethered drone with seven thrusters, an endurance of one hour, and a maximum operating depth of 60 m. The iBubble has a proprietary sonar positioning system. This is capable of tracking a diver by using four hydrophones to pick up the acoustic signal transmitted by a remote controller worn by the diver. The iBubble can also detect and avoid obstacles by means of an integrated visible light camera module. As in the case of the UAV, additional electronic components were added to the iBubble. A Water Linked M64 acoustic modem relays warning messages from the USV to an Arduino Mega 2560 microcontroller which, in turn, turns on an LED indicator to warn the diver in the event of marine traffic.

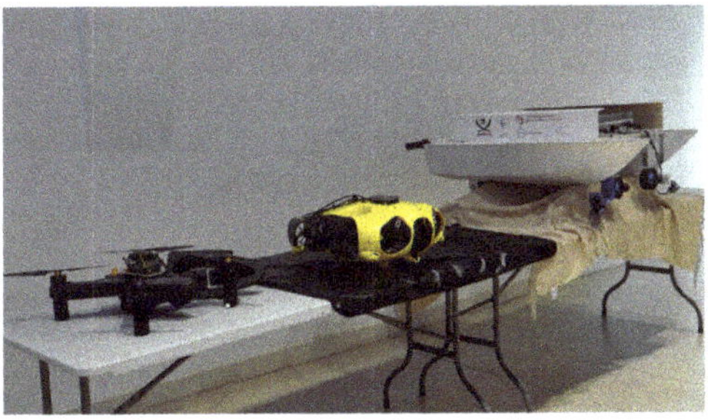

Fig 4: BEA system with Cutacopter, iBubble EVO, Buoy (from left to right)

Conclusions

This paper presented a multi-unmanned vehicle system consisting of a UAV, a USV and a UUV, to assist divers and warn them of boat traffic in their proximity. By integrating the Buoy, UAV, and ROV modules, this multi-drone system, Figure 4, effectively tracks and supports divers throughout their underwater activities. The requirements capture process, functional architecture and versatile surface operation capabilities collectively position BEA as a promising solution for improving diving safety conditions. The initial simulations and real-world test results have showcased the system's suitability and effectiveness for the intended application. Future developments will focus on the full integration of the three vehicles, enabling further real-world testing with divers to evaluate the system's performance in a wider range of diving scenarios.

Acknowledgement

The work described in this paper was carried out as part of the BEA (R&I-2018-005T) project which was financed by the Malta Council for Science & Technology, for and on behalf of the Foundation for Science and Technology, through the FUSION: R&I Technology Development Programme.

References

[1] "How Many Divers Are There? 8 Reasons Make Diving Is So Popular." Blue Calmness. https://bluecalmness.com/how-many-divers-are-there-8-reasons-make-diving-is-so-popular/ (accessed Jan. 31, 2023).

[2] "Safe Boating Guidelines." DAN World. https://world.dan.org/safety-prevention/diver-safety/psa/safe-boating-guidelines/ (accessed Jan. 31, 2023).

[3] M. Agius. "Vessels caught speeding next to diving flags." Newsbook. https://newsbook.com.mt/en/vessels-caught-speeding-next-to-diving-flags/ (accessed Jan. 31, 2023).

[4] F. Cocchioni et al., "Unmanned Ground and Aerial Vehicles in extended range indoor and outdoor missions," 2014 International Conference on Unmanned Aircraft Systems (ICUAS), Orlando, FL, USA, 2014, pp. 374-382. https://doi.org/10.1109/ICUAS.2014.6842276.

[5] E. H. C. Harik, F. Guérin, F. Guinand, J. -F. Brethé and H. Pelvillain, "UAV-UGV cooperation for objects transportation in an industrial area," 2015 IEEE International Conference on Industrial Technology (ICIT), Seville, Spain, 2015, pp. 547-552. https://doi.org/10.1109/ICIT.2015.7125156.

[6] T. Miki, P. Khrapchenkov and K. Hori, "UAV/UGV Autonomous Cooperation: UAV assists UGV to climb a cliff by attaching a tether," 2019 International Conference on Robotics and Automation (ICRA), Montreal, QC, Canada, 2019, pp. 8041-8047. https://doi.org/10.1109/ICRA.2019.8794265.

[7] A. Cantieri et al., "Cooperative UAV–UGV Autonomous Power Pylon Inspection: An Investigation of Cooperative Outdoor Vehicle Positioning Architecture," Sensors, vol. 20, no. 21, p. 6384, Nov. 2020. https://doi.org/10.3390/s20216384.
[8] M. Zhu and Y. Wen, "Design and Analysis of Collaborative Unmanned Surface-Aerial Vehicle Cruise Systems," Journal of Advanced Transportation, vol. 2019, Jan. 2019. https://doi.org/10.1155/2019/1323105.
[9] G. Shao, Y. Ma, R. Malekian, X. Yan and Z. Li, "A Novel Cooperative Platform Design for Coupled USV–UAV Systems," in IEEE Transactions on Industrial Informatics, vol. 15, no. 9, pp. 4913-4922, Sept. 2019. https://doi.org/10.1109/TII.2019.2912024.
[10] W. Li, Y. Ge, Z. Guan, and G. Ye, "Synchronized Motion-Based UAV–USV Cooperative Autonomous Landing," Journal of Marine Science and Engineering, vol. 10, no. 9, p. 1214, Aug. 2022. https://doi.org/10.3390/jmse10091214.
[11] C. Ke and H. Chen, "Cooperative path planning for air–sea heterogeneous unmanned vehicles using search-and-tracking mission," Ocean Engineering, vol. 262, Oct. 2022. https://doi.org/10.1016/j.oceaneng.2022.112020.
[12] J. Ross, J. Lindsay, E. Gregson, A. Moore, J. Patel and M. Seto, "Collaboration of multi-domain marine robots towards above and below-water characterization of floating targets," 2019 IEEE International Symposium on Robotic and Sensors Environments (ROSE), Ottawa, ON, Canada, 2019, pp. 1-7. https://doi.org/10.1109/ROSE.2019.8790419.

Requirements definition in support of digital twin platform development

Castrese Di Marino[1,a], Valeria Vercella[1,b*], Rocco Gentile[1,c], Giacomo Nasi[1,d] and Stefano Centomo[1,e]

[1]Leonardo Labs, Leonardo S.p.A, C.so Castelfidardo 22, 10138 Torino , Italy

[a]castrese.dimarino.ext@leonardo.com, [b]valeria.vercella.ext@leonardo.com, [c]rocco.gentile.ext@leonardo.com, [d]giacomo.nasi.ext@leonardo.com, [e]stefano.centomo.ext@leonardo.com

Keywords: Digital Twin Platform, User Experience, Aeronautics, Requirements

Abstract. This paper discusses the exploitation of a System Engineering approach and, specifically, of Requirements Engineering to derive a set of desired features based on stakeholders' needs to be implemented into a Digital Twin (DT) platform. Key focus is on the development of a collaborative and highly integrated simulation environment tailored for the design of breakthrough aeronautical products and able, in principle, to cover all the phases of product lifecycle. Specifically, a preliminary list of platform requirements is elicited and from them a set of desired features to be implemented is derived. Then, basing on these features, a Kano questionnaire is set up and used to question a pool of engineers and experts in the aeronautical field. Eventually, by analysing the questionnaire results, the list of desired characteristics is prioritized and used to provide guidelines for the development of the front-end interface of the collaborative platform.

Introduction

The high competitiveness within the aeronautical sector drives the ambitions for shortening the time to market of advanced aircraft concepts. In a context of increasing digitalization, the emerging Digital Twin (DT) technology can play a key role in supporting the introduction of cost-effective innovative solutions since the early design phases thanks to collaborative DT platforms able to facilitate the exchanges among involved stakeholders all along the aircraft life cycle. The DT concept is based on the idea that a physical system can be represented as a digital information or a "twin" of the information originally embedded into the physical system itself [1, 2]. Applications of the DT span all the phases of product lifecycle, starting from design, through manufacturing and service, up to retire phase [3]. In the aviation industry framework, all major aviation companies have been developing DT platforms [4] to be made available internally. A noteworthy example is the Digital Design, Manufacturing & Services (DDMS) platform by Airbus, providing a complete end-to-end process from preliminary design to final assembly to "reduce costs and time to market for our products, while meeting our customers' expectations for quality, safety and environmental performance" [5]. Supposing to be in charge of developing a custom DT environment to be used to assess the feasibility of highly innovative aeronautical products, the substantial lack of manufacturing and in-service data for such breakthrough solutions can hamper the development of a comprehensive DT platform like DDMS. However, this limitation can be turned into a chance to focus on the establishment of the design component of the DT (i.e. the Design DT), paving the way to potential cost savings in the subsequent phases of "real" product development. Indeed, it is well established that the cost to introduce a change into design direction increases along the product lifecycle phases [6]. Therefore, the possibility to fully explore all feasible design alternatives at early design stages can avoid the introduction of an abrupt and costly design change during manufacturing.

On this basis, with the goal to support the development of a DT platform for the aviation sector, Section 2 will discuss the derivation of main platform requirements basing on a possible set of stakeholders' needs. The proposed list of requirements, grouped in five macro-groups, is exploited by means of brainstorming and a Use Case Analysis to obtain a list of features to be implemented into the platform. Basing on these features, Section 3 describes the exploitation of Kano questionnaire to investigate a platform user experience and extract a set of basic information (attractive, must be, indifferent, etc.) useful for defining the features to be implemented and leading the development of the front-end Graphical User Interface (GUI) of the platform. Eventually, main conclusions will be drawn and ideas for future works will be elicited.

Digital Twin Platform Requirements

The System Engineering approach can support the definition of key DT platform characteristics in a structured way [7]. In particular, Requirements Engineering can be used to list high-level platform requirements basing on stakeholders' needs and expectations. For the current application, after internal discussion with involved stakeholders, the set of requirements is derived and represented through SysML (Fig. 1).

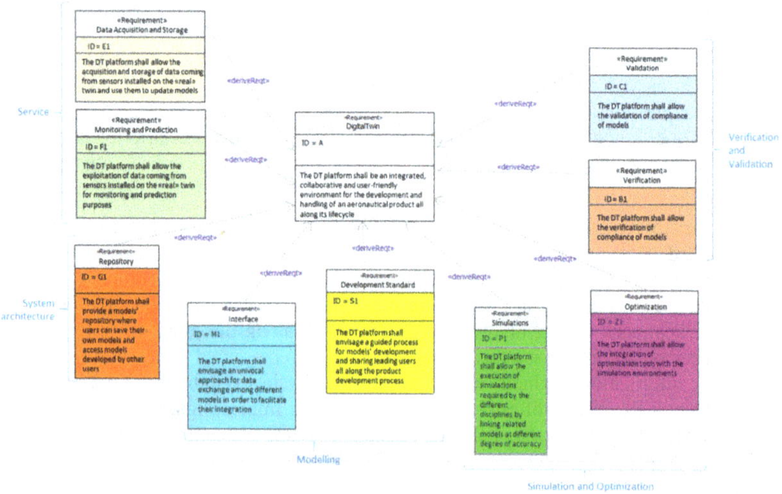

Fig. 1 Requirements and Pillars related to DT platform

According to Fig. 1 and considering the current focus on Design DT, the platform shall be a collaborative simulation environment allowing the user to build the aircraft DT by exploiting the simulation models loaded by himself or by other users. For sake of clarity, the two requirements connected to the in-service component of the DT (i.e. "Data Acquisition and Storage" and "Monitoring and Prediction") are here reported just to strengthen the link between the DT and the "real" twin, basing on the idea to exploit real in-service data to improve the simulation models used during design activities. The requirements under analysis can be grouped into five macro-categories herein considered as the pillars of DT platform development:

1. *System Architecture*, describing how the platform should work;
2. *Modelling*, including all modelling processes, protocols and rules to be followed within the platform;

3. *Simulation and Optimization*, connected to the need of performing simulation and optimization from the platform;
4. *Verification and Validation*, related to the possibility to verify and validate the models loaded into the platform;
5. *Service*, representing the need to link the DT with the «real» twin.

Basing on the pillars and related requirements, it is then possible to derive a set of desired features to be implemented into the DT platform (Fig. 2).

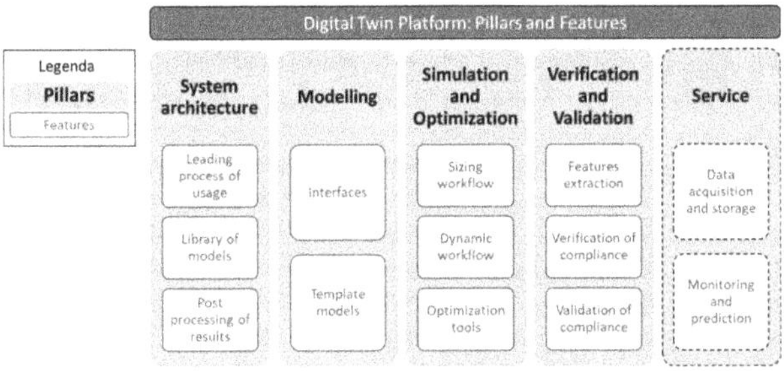

Fig. 2 Pillar and Features connected to DT platform

For sake of clarity, the leading process of usage is intended as a guided sequence of steps that leads the user alongside the whole process. This feature, like the one linked to the capability to provide a proper post-processing of results, is strictly connected to the need to deliver a user-friendly environment. As a result, the DT platform should embed the following features:
1) The user should be guided through the platform usage,
2) Access to the library of simulation models and
3) Easy visualization and manipulation of simulation results within the platform.

Dealing with the "Modelling" pillar, the platform should be able to:
1) Manage the interfaces and signals (inputs and outputs) of the models and their position and value within the database and
2) Guide the users by providing a set of predefined coding rules to be used while developing models (i.e. to provide coding templates for models development).

Eventually, for "Simulation and Optimization" pillar, by means of the DT platform, the user should be able to:
1) Analyse simulation results and perform feature extraction activities [8],
2) Automatically verify models compliance to System Requirements (i.e. the model "works" within the platform),
3) Validate models compliance to Stakeholders' Requirements (i.e. the model "works well" within the platform).

User Experience Investigation Through Kano Questionnaire

Since the current work points to develop an innovative platform for designing and exploring complex systems, it shall be characterized by collaboration among developers and usability. The first analysis conducted on Stakeholders' needs led to a set of requirements consequently exploded into a list of features as reported in the previous section. To select the most valuable features, a very powerful tool reported in literature for requirements analysis and prioritization is adopted.

Materials Research Proceedings 37 (2023) 249-253 https://doi.org/10.21741/9781644902813-54

The Kano Model [8] is a framework that allows exploring customer satisfaction with respect to product functions. A feature is presented in positive (functional) and negative (dysfunctional) form to measure the degree of satisfaction from the user in order to extract an opinion that can fall into a category: A (Attractive), M (Must-have), O (One-dimensional), R (Reverse), Q (Questionable), or I (Indifferent). The developers shall focus on Attractive to improve satisfaction of the customer and Must-have to avoid dissatisfaction. Concurrently, the features classified as One-dimensional (also named Performance) are characterized by a linear trend between satisfaction and functional and are the valuable to implement. On another hand, it may turn out that a feature considered valuable results indifferent for customer. If it happens, the developers can decide to invest time according to team needs. An example of a questions couple is reported in the following:

- If to develop a model you are asked to follow predefined coding rules, how would you feel?
- If to develop a model you are not asked to follow predefined coding rules, how would you feel?

The acceptable answers are: I like it!; I expect it; I'm neutral; I can live with it; or I dislike it.

The Kano Matrix [8] compares the answers of a single feature per customer. The sum of all answers provide the main category where a feature falls. The categories with second and third frequency can be considered for a detailed analysis. In Table 1 the answers of a pool of engineers is presented reporting not only the category, but also a Satisfaction Coefficient (S) and Dissatisfaction Coefficient (DS). The combination of S and DS provides the Overall Satisfaction Coefficient (OS) that indicates how much the satisfaction increases and decreases if the requirement is implemented or not. Some results are commented as sample of this preliminary analysis.

- *Feature extraction*: it results the only M category, that means it does not improve customer satisfaction but it must be implemented in order not to reduce the degree of satisfaction since the customer expects to find it.
- *Leading process of usage*: it results to be A on the first frequency, O and I on the second and third. This indicates that designers are available to work on this platform, appreciating the potential benefits even though usability is still not ready.
- *Sizing Workflow/Dynamic Workflow*: the majority of customers consider that features as O. Since the high importance in a DT for design as well as for simulation, it will be further implemented through innovative methods and agnostic tools.
- *Template of models*: although this feature results indifferent for customers, it is an outcome of an analysis of possible compliance within the DT platform, therefore, in future application it can be explored and implemented without producing dissatisfaction in customers.

The majority of the features are considered O, so they are well suited to be constantly improved during development.

Table 1 Results of Kano questionnaire (A: Attractive, M: Must-have, O: One-dimensional, R: Reverse, Q: Questionable, I: Indifferent, S: Satisfaction, DS: Dissatisfaction, OS: Overall satisfaction)

FEATURE	A	M	O	R	Q	I	CATEGORY	S	DS	OS
LEADING PROCESS OF USAGE	39%	17%	22%	0%	0%	22%	A	0,61	-0,39	0,22
LIBRARY OF MODELS	17%	33%	44%	0%	0%	6%	O	0,61	-0,78	-0,17
POST PROCESSING OF RESULTS	28%	17%	50%	0%	0%	6%	O	0,78	-0,67	0,11
INTERFACES	17%	17%	33%	0%	0%	33%	O	0,50	-0,50	0,00
TEMPLATE MODELS	17%	6%	17%	6%	0%	56%	I	0,35	-0,24	0,12
SIZING WORKFLOW	33%	11%	44%	0%	0%	11%	O	0,78	-0,56	0,22
DYNAMIC WORKFLOW	22%	11%	67%	0%	0%	0%	O	0,89	-0,78	0,11
OPTIMIZATION TOOLS	44%	17%	28%	0%	0%	11%	A	0,72	-0,44	0,28
FEATURES EXTRACTION	22%	39%	28%	0%	0%	11%	M	0,50	-0,67	-0,17
VERIFICATION OF COMPLIANCE	22%	11%	50%	0%	0%	17%	O	0,72	-0,61	0,11
VALIDATION OF COMPLIANCE	22%	17%	33%	0%	0%	28%	O	0,56	-0,50	0,06
DATA ACQUISITION AND STORAGE	33%	11%	39%	0%	0%	17%	O	0,72	-0,50	0,22
MONITORING AND PREDICTION	17%	11%	33%	0%	0%	39%	I	0,50	-0,44	0,06

Conclusions and Future Works

In conclusion, this research work presents a preliminary analysis towards the development of a collaborative and integrated environment for design and simulation of a DT. One of the main goals is to provide an environment where users can collaborate alongside all the design and production steps and where the integration and validation of models can be easily carried out. Currently, a set of requirements and features are identified and classified according to Kano Model. In the next step, a more complex questionnaire may support the decision making process considering a demographic section where roles of users are highlighted and functions and features of different application and phases of development are reported in order to extract more precise information about the real needs of involved people.

References

[1] L. Li, S. Aslam, A. Wileman and S. Perinpanayagam, Digital twin in aerospace industry: a gentle introduction, *IEEE Access,* 10 (2021) 9543-9562. https://doi.org/10.1109/ACCESS.2021.3136458

[2] H. Aydemir, U. Zengin and U. Durak, The digital twin paradigm for aircraft review and outlook, AIAA Scitech 2020 Forum (2020). https://doi.org/10.2514/6.2020-0553

[3] M. Liu, S. Fang, H. Dong and C. Xu, Review of digital twin about concepts, technologies, and industrial applications, Journal of Manufacturing Systems, 58 (2021) 346-361 https://doi.org/10.1016/j.jmsy.2020.06.017

[4] H. Meyer, J. Zimdahl, A. Kamtsiuris, R. Meissner, F. Raddatz, S. Haufe and M. Bassler, Development of a digital twin for aviation research, Deutscher Luft- und Raumfahrt Kongress, Hamburg, 2020

[5] Information on http://www.airbus.com/en/innovation/disruptive-concepts/digital-design-manufacturing-services

[6] S. J.Kapurch, NASA systems engineering handbook, Diane Publishing, 2010.

[7] B. Vogel-Heuser, D. Schütz, T. Frank. and C. Legat, Model-driven engineering of Manufacturing Automation Software Projects – A SysML-based approach, Mechatronics, 24, n. 7 (2014) 883-897. https://doi.org/10.1016/j.mechatronics.2014.05.003

[8] K. Samina, T. Shehryar and S. Nasreen, A survey of feature selection and feature extraction techniques in machine learning, in Proceedings of 2014 Science and Information Conference, SAI 2014, 2014

[9] N. Kano, Attractive quality and must-be quality, Journal of the Japanese society for quality control, 31, n. 4 (1984) 147-156

Coupling effect of acoustic resonators for low-frequency sound suppression

G. Catapane[1,a*], L.M. Cardone[1,b], G. Petrone[1,c], O. Robin[2,d], F. Franco[1,e]

[1] PASTA-Lab (Laboratory for Promoting experiences in Aeronautical STructures and Acoustics), Università degli Studi di Napoli "Federico II", Via Claudio 21, Napoli, 80125, Italy

[2] Centre de Recherche Acoustique-Signal-Humain, Université de Sherbrooke, 2500 boulevard de l'Université, Sherbrooke, J1K 2R1, Quebec, Canada

[a]giuseppe.catapane@unina.it, [b]luigimaria.cardone@unina.it, [c]giuseppe.petrone@unina.it, [d]olivier.robin@USherbrooke.ca, [e]francof@unina.it

Keywords: Sound Absorption, Acoustic Resonators, Noise Suppression

Abstract. Acoustic resonators like Helmholtz resonators, micro-perforated panels and quarter wavelength tubes are employed to suppress tonal noise for several industry application. The issue related to the design of these resonators is their bulkiness for low-frequency application and their narrow band behaviour. In this paper, microperforated panels and coiled quarter wavelength tubes are coupled in series and in parallel, tested inside impedance tube for sound absorption. The experimental samples are 3D printed with filament (PLA) additive manufacturing technique. The two acoustic devices are coupled and tested to reach broadband low-frequency noise suppression just by positioning one respect to the other in series or in parallel. The reported results demonstrate that the tonal behaviour of the acoustic devices can lead to enlarged absorption if they are tuned at similar frequencies. The disposition of the acoustic resonators and their frequency tuning hardly impact absorption: indeed, anti-resonance and filtering effect are experienced for series configuration, while parallel configuration is the sum of the two acoustic devices standalone absorption behaviour.

Introduction

Transport engineering has to develop vehicles, planes and ships with limited CO_2 and noise emissions. With regards to noise, sound absorbing structures are designed to suppress disturbance produced by acoustic sources and vibrating devices. Typical solutions include porous materials, Micro-Perforated Panels (MPPs), Helmholtz Resonators (HRs) and Quarter Wavelength Tubes (QWTs). Porous materials like fibers and foams are mostly used in transportation and building applications; MPPs are used for the design of acoustic liners that find several applications in the reduction of aircraft engine or automotive noise, but also in building acoustics. MPPs are composed by a thin plate with perforations followed by a backing cavity. Although a hole diameter of the order of 1 mm guarantees a large acoustic resistance and a reduced acoustic mass reactance, which results in a wide-band sound absorption [1], this cannot cover more than one or two octaves at low frequency [2]. Their micro-perforated structure can also be beneficial to suppress the sound disturbance without affecting the source performance: for instance, the small drag produced by a micro-perforated panel in the acoustic liners is the best compromise to reduce the engine noise without affecting the flow-path.

The main aim of this communication is to investigate the effect of the coupling of an MPP with a QWT in series and in parallel. Quarter wavelength tubes exhibit multiple resonance frequency when the length of the tube is an odd-multiple integer of the quarter of the wavelength (λ). Multiple resonance frequencies are exploited to extend the low-frequency region of influence of a hybrid model made by MPP and the tube. To cope with their excessive length requirement for low-frequency application, their channel is stretched into a spiral. In this paper, spiral-coiled quarter wavelength tubes (spiral resonator) are coupled with different MPPs to highlight interesting coupling properties like filtering, anti-resonance behavior and general low-frequency wide sound absorption. Generally, MPPs are made in titanium or steel; for this testing phase, they are herein 3D printed with PLA, and post-drilled to avoid manufacturing imperfection effects.

Definition of The Problem

An absorber based on the micro perforated panel has a perforated plate followed by a cavity (Figure 1a). An MPP, a perforated plate followed by a cavity (see Figure 1(a)) with holes diameter d and plate thickness t, has an impedance equal to $Z_{plate} = R_{plate} + j\omega M_{plate}$, with the resistance $R_{plate} = (32\eta t/\sigma d^2)([1+(k^2/32)]^{1/2} + (\sqrt{2}/32)\,k(d/t))$ and $M_{plate} = (\rho_0 \omega t/\sigma)(1 + [1+(k^2/2)]^{-1/2} + 0.85(d/t))$, with ρ_0 and c_0 density and speed of sound of air, η is the dynamic viscosity of air, σ is the perforation ratio and the generic coefficient $k = d\sqrt{\omega\rho_0/4\eta}$. The backing cavity is function of its dept D, with impedance: $Z_{cavity} = -jZ_0 \cot(\omega D/c_0)$ [1].

An Archimedean-spiral quarter wavelength tube (Figure 1b) is modeled as a perforated plate modelled with Johnson-Champoux-Allard (JCA) approach [3], followed by a QWT, defined following the Low Reduced Frequency (LRF) model [4]. LRF theory studies the sound propagation inside the QWT with a lossy Helmholtz equation, where the density ρ_{eff} and speed of sound c_{eff} are modelled taking into account visco-thermal dissipation inside the narrow tube, and so the effective impedance $Z_{eff} = \rho_{eff} c_{eff}$ and the effective wavenumber $k_{eff} = \omega/c_{eff}$. The impedance of the QWT of length L writes $Z_{QWT} = -jZ_{eff}\cot(k_{eff}L)$. The impedance of the spiral resonator writes $Z_{SR} = 1/\phi_{inlet}\left[Z_{d_{in},JCA}\frac{-jZ_{QWT}\cot(k_{d_{in},JCA}t_{in}) + Z_{d_{in},JCA}}{Z_{QWT} - jZ_{d_{in},JCA}\cot(k_{d_{in},JCA}t_{in})}\right]$, with $Z_{d_{in},JCA} = \rho_{JCA} c_{JCA}$ and $k_{d_{in},JCA} = \omega/c_{JCA}$ impedance and complex wavenumber of a perforated plated of thickness t_{in} and a circular inlet hole of diameter d_{in}, with $c_{JCA} = \sqrt{K_{JCA}/\rho_{JCA}}$, ρ_{JCA} and K_{JCA} effective speed of sound, density and bulk modulus. $\phi_{inlet} = A_{hole}/A_{plate}$ is the perforation ratio between the hole and the plate areas. The MPP is coupled with the spiral resonator in series (Figure 1c) and in parallel (Figure 1d), with the impedance of both systems evaluated through electro-acoustical analogy:

$$Z_{series} = Z_{plate} + \left[\frac{1}{Z_{cavity}} + \frac{1}{Z_{SR}}\right]^{-1}, \qquad Z_{parallel} = \frac{(Z_{plate} + Z_{cavity}) \cdot Z_{SR}}{(Z_{plate} + Z_{cavity}) + Z_{SR}}. \qquad (1)$$

MPP and QWT coupling is evaluated through analytical, numerical and experimental viewpoints. Experimental tests are made to measure sound absorption coefficient α_{exp} inside a 100 mm diameter impedance tube: a speaker placed at one end of the tube excites the tube with a normal plane wave radiation; the sample is placed at the opposite end, backed by a rigid cavity. The dimension of the tube D_{tube} is a design parameter for any samples, that must have circular cross section with a 100 mm diameter. Therefore, the perforation ratio of the MPPs can be written as $\sigma = N_{holes} d/D_{tube}$, with N_{holes} number of the MPP holes. The sound absorption coefficient is estimated following the ISO 10534-2 1998 standard. Acoustic simulations mimic this experimental measurement with the impedance tube and the wall of both the MPP and the spiral resonator are considered rigid. The analyses are made with COMSOL Multiphysics, *Pressure Acoustics Module*. The geometrical properties of the four tested configurations are listed in Table 1.

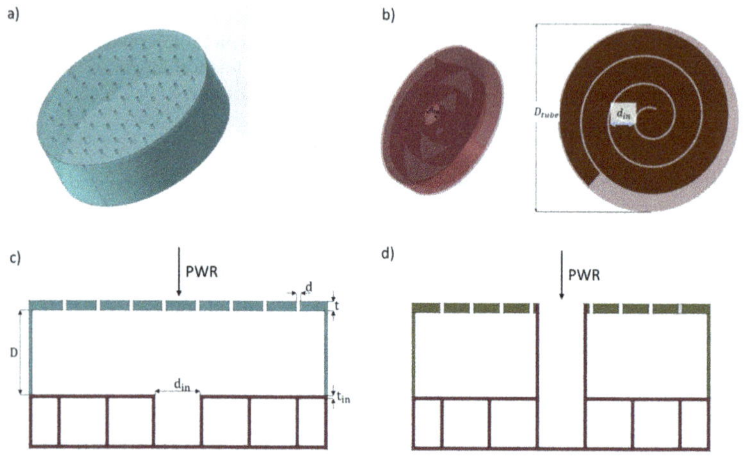

Figure 1: coupling scheme of the MPP and coiled QWT in a) series and in b) parallel.

Table 1: geometrical properties of the objects of study.

	d [mm]	N_{holes}	D [mm]	t [mm]		d [mm]	N_{holes}	D [mm]	t [mm]
$MPP_{1,series}$	1.0	33	27	3.0	$MPP_{2,series}$	1.4	65	27	3.0
$MPP_{1,parallel}$	1.0	32	27	3.0	$MPP_{2,parallel}$	1.4	64	27	3.0
	d_{in} [mm]				t_{in} [mm]	L [mm]			
SR_{series}	15				1.0	451			
$SR_{parallel}$	15				1.0	481			

Samples are 3D printed with PLA, and post-drilled to maximize the accuracy of holes diameter. Spiral resonators are different for series and parallel configuration: while for the first one the length L represents the length of the spiral path, for the parallel configuration is the sum of the length of the spiral path and the elongation to have the inlet at the top of the perforated plate, as it is possible to see in Figure 1b. Respective lengths L for SR in series and in parallel are reported in Table 1. The SR_{series} has first two harmonics at 190 and 570 Hz; $SR_{parallel}$ has first two harmonics at 178 and 535 Hz. MPP_1 has resonance peak around 283 Hz, hence between the first and the second SR harmonics. MPP_2 has resonance peak around 560 Hz, and it is designed to see resonance interaction effects between the second SR harmonics and the MPP characteristic frequency.

Results

Experimental sound absorption of the aforementioned samples is plotted in Figure 2 and each test shows a good match with theoretical and numerical predictions, which are nearly superimposed. The series coupling shows interesting properties: when the spiral and the MPP has no superimposed resonances, any spiral harmonics after the MPP resonance is filtered, and poor absorption peaks are visible (Figure 2a); on the other hand, when the MPP resonance is tuned at a similar frequency of the SR second harmonics, their interaction implies an anti-resonance effect, with the peak split in two and with a drop down of the sound absorption in correspondence of the resonance (Figure 2a). Filtering and antiresonance effect completely disappear in the parallel configurations (Figure 2c-d), with each peak preserved by their coupling.

Conclusions

Coupling an MPP and a QWT can extend the bandwidth of the acoustic efficiency. The parallel configuration induces properties that are deemed more useful with respect to the series

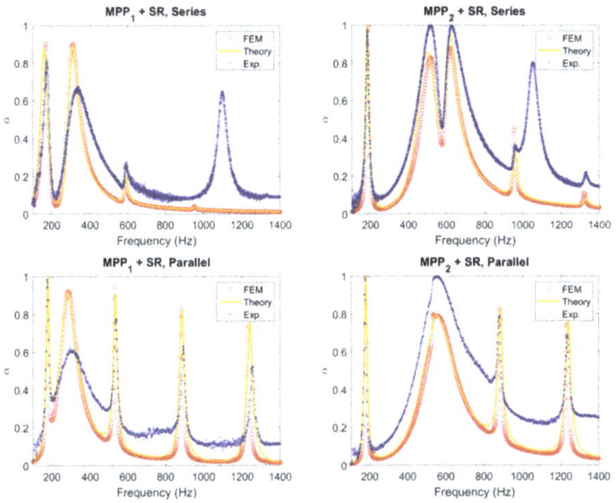

Figure 2: Sound absorption plot of the series and parallel coupling of two $\boldsymbol{MPP_s}$ coupled with 190Hz SR.

configuration. The larger inlet dimension that is required for the parallel configuration nevertheless limits its application. For instance, an engine acoustic liner cannot accept a 15 mm opening, because it would be unacceptable for aerodynamic and engine performance reasons. The series configuration could be easier implemented for an acoustic liner, but the relative position of each absorption peak brought by the MPP and the QWT is crucial. Future developments will involve different materials and manufacturing considerations

References

[1] D.-Y. Maa, "Potential of microperforated panel absorber," *J Acoust Soc Am*, vol. 104, no. 5, pp. 2861–2866, 1998. https://doi.org/10.1121/1.423870.

[2] C. Wang and L. Huang, "On the acoustic properties of parallel arrangement of multiple micro-perforated panel absorbers with different cavity depths," *J Acoust Soc Am*, vol. 130, no. 1, pp. 208–218, 2011. https://doi.org/10.1121/1.3596459.

[3] N. Atalla and F. Sgard, "Modeling of perforated plates and screens using rigid frame porous models," *J Sound Vib*, vol. 303, no. 1–2, pp. 195–208, 2007. https://doi.org/10.1016/j.jsv.2007.01.012.

[4] A. Magnani, C. Marescotti, and F. Pompoli, "Acoustic absorption modeling of single and multiple coiled-up resonators," *Applied Acoustics*, vol. 186, 2022. https://doi.org/10.1016/j.apacoust.2021.108504.

New UAV ice tunnel characterization

Arrigo Avi[1,a *], Giuseppe Quaranta[1,b] and Riccardo Parin[2,c*]

[1]Politecnico di Milano, Milano, Italy

[2]Eurac Research – terraXcube , Bolzano, Italy

[a]arrigo.avi@polimi.it, [b]giuseppe.quaranta@polimi.it, [c]riccardo.parin@eurac.edu

Keywords: UAV, Ice Tunnel, Ice Accretion, System Characterization, Climatic Chamber, Hypobaric Chamber

Abstract. In the recent years, the field of unmanned aircraft vehicles (UAVs) has shown great technological progresses and many new applications have born. To assess the potential of this technology and to improve the availability and reliability of the rising services it is critical to overcome operational limitations. One key operational hazard is atmospheric in-flight icing, resulting in large aerodynamic penalties, unbalances and other detrimental phenomena that sometimes can lead to catastrophic consequences. In this paper, a new ice tunnel developed in the large hypobaric and climatic chamber of the terraXcube facility of Eurac will be presented. A preliminary characterization and calibration of the system has been also performed following the EASA regulation reported in the Easy Access Rules for Large Rotorcraft (CS-29) (Amendment 6).

Introduction

While UAV systems offer numerous features and performance capabilities, there are areas in need of improvement, particularly in terms of reliability. Unmanned aircraft are often deployed on high-risk missions, operating in marginal weather conditions for purposes such as search and rescue or military operations. Therefore, the development of an effective tool to assess performance in these challenging conditions would be highly valuable. Currently, only a few unmanned aircraft are specifically designed to address the challenges of icing. These aircraft are primarily large military UAVs developed for long-range missions in harsh environments. While smaller UAVs may have some level of rainproof or snowproof certification, there is generally no specific consideration for icing (EU 2019/945 EU 2019/947). In the aviation industry, there are well-known and established facilities dedicated to icing testing, however, these facilities are both complex and expensive. Consequently, integrating an icing testing program during the development of a small or mini-UAV can be highly challenging. For example, the estimated cost of regional transport aircraft icing tests, as mentioned in the IPHWG Task 2 WG Report - Appendix J, is approximately 700,000 € [1]. Here we propose a more cost-effective alternative, a home-made ice tunnel developed in our climatic chamber. This new facility could allow for the testing of a full-scale UAV under realistic icing conditions.

Experimental setup and Methods

Facility

TerraXcube is an infrastructure for research and testing that leverages a multi-dimensional approach for environmental simulation part of Eurac Research Institution. With a useable volume of 360 m^3, the climatic chamber Large Cube can simulate the most extreme environmental conditions on the Earth's surface, allowing the synchronous control of multiple complex environmental parameters for long duration analysis.

Wind tunnel

Inside the climatic chamber an open-loop wind tunnel was developed. The chamber was hypothetically divided into two sections as showed in Fig. 1: testing section on the right and production section on the left. The testing section is the volume where the UAV is placed and the subcooled cloud is fully developed. The production section is composed by the wind tunnel, fan and nozzles which contributes to the formation of the ice conditions.

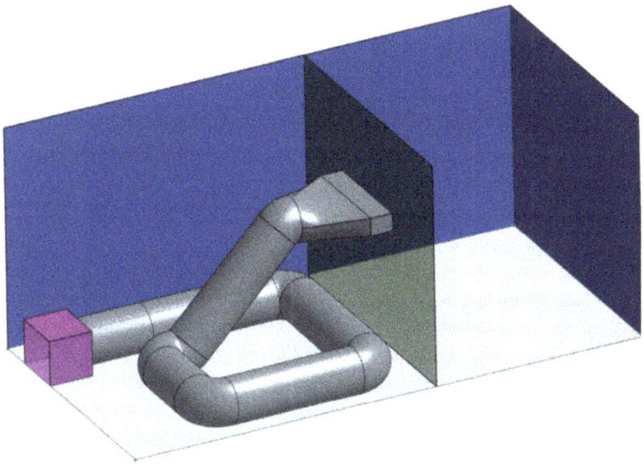

Fig. 1 - Layout

Fig. 2 - Setup

The climatic chamber works as a closed loop system. The air is extracted from the wall on the right side of Fig. 1 (North side), it is treated by heat exchangers and it returns from the left side (South side) inside the chamber. Almost all the flow treated by the chamber is streamed to the axial fan, depicted in purple on Fig. 1. The fan (volume purple in Fig 1) controls the air speed inside the wind tunnel. Immediately after the fan there are 25 nozzles which can be controlled in terms of air and water pressure leading to different subcooled cloud properties described in the next section. The ice tunnel allows to test UAV platforms up to 2 m^3 (Fig. 2).

The main characteristics of the open-loop wind tunnel are briefly reported in the following Table 1.

Table 1 Technical data of wind tunnel

Fan	Konz Lufttechnik
Speed [m/s]	0 ÷ 30
Diameter [m]	1
Length [m]	About 18 m

Icing system

The icing system is composed by 25 nozzles supplied by pressurized air and water and managed as prescribed in [2-3] to produce different icing conditions. The main aspects of subcooled clouds are the mean volume droplets (MVD), which refers to the average size or diameter of water droplets, and the LWC, liquid water content, which refers to the amount or concentration of liquid water present in a given volume of air or cloud. Modifying the number of nozzles involved, the air pressure and the water pressure, MVD and LWC can be varied as desired. The water supplied is ultrapure water (18 MΩ·cm) which allows to the water to remain in the liquid state until the impact with the UAV surface. The technical data are as follows:

Table 2 Technical data of icing system

LWC [g/m^3]	0 ÷ 3
MVD [μm]	15 ÷ 40
Operating temperature [°C]	-40 ÷ +0
Nozzles	Spraying System SUJ-12

Droplets measurements

The real time droplets measurements system consists in a low-cost optical system [4,5]. Based on shadowgraph, it is composed by a telecentric objective coupled with an industrial CMOS camera. After a proper calibration, the system takes pictures of the stream, an OpenCV algorithm detects droplets and measures their diameter. With such method, it is possible to have a real time measurement of MVD and LWC (indirect estimate) of the stream.

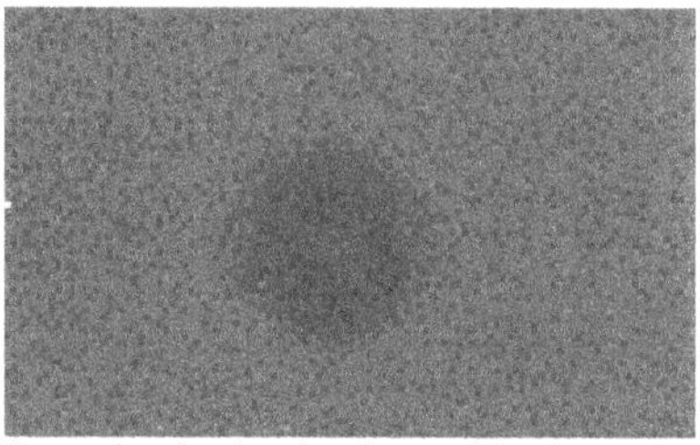

Fig. 3 - Example of a droplet detected by the optical system.

Results

The characterization and calibration of the system has been performed following the standard for calibration and acceptance of icing wind tunnel described by EASA [6] with the cylinder collection method. Three set of parameters were analyzed, the duration of each test was fixed at 10 minutes.

Table 3 System parameters

	Test #A	Test #B	Test #C
Temperature [°C]	-10 ± 2	-10 ± 2	-10 ± 2
Air pressure [bar]	3.1 ± 0.3	2.3 ± 0.3	1.1 ± 0.3
Water pressure [bar]	1.5 ± 0.3	1.5 ± 0.3	0.7 ± 0.3
Number of nozzles	4	4	4

Mean Volume Diameter. MVD data, collected with the optical measurement system, were compared to calibration chart as in [3].

Table 4 MVD measurement comparison

	Test #A	Test #B	Test #C
MVD theoretical [μm]	15	25	35
MVD measured [μm]	20 ± 4	31 ± 4	38 ± 4

Liquid Water Content. LWC data, obtained by the ice cylinder method, showed a good agreement between the technical datasheet of nozzles and [6].

Table 5 LWC measurement comparison

	Test #A	Test #B	Test #C
LWC datasheet [g/m^3]	0.16	0.23	0.18
LWC cylinder [g/m^3]	0.11	0.19	0.25

Conclusion

The preliminary characterization of the icing wind tunnel in the terraXcube facility yielded promising outcomes. The primary advantage of this new system lies in its potential to conduct full-scale UAV trials within the real operational environment. However, further refinements and investigations are required to ensure a complete understanding of its capabilities. Additionally, an important area for exploration involves the simultaneous utilization of the icing system and hypobaric capabilities of the chamber. This integration holds the potential to significantly impact UAV testing within icing conditions at high altitudes (up to 9000 m).

References

[1] Ice Protection Harmonization Working Group, Task, 2, appendix J (2005).

[2] EASA CS-25 Appendix O

[3] Ludovico Vecchione et al., Cloud calibration update of the CIRA Icing Wind Tunnel, (2003). https://doi.org/10.2514/6.2003-900

[4] Arrigo Avi, "Development of icing testing tools for large climatic chamber", MSc thesis, Politecnico di Milano, Milano, Italy (2019)

[5] Staffan Rydblom. "Measuring Water Droplets to Detect Atmospheric Icing". PhD thesis, Mid Sweden University, Sundsvall, Sweden, (2017)

[6] SAE (Society of automotive engineers) Aerospace Recommended Practice (ARP), (2015)

Remote sensing validation with in-situ measurements for efficient crop irrigation management

I. Terlizzi [2,3,a*], F. Morbidini [2,b], C. Maucieri [2,c], C. Bettanini [1,3,d], G. Colombatti [1,3,e], S. Chiodini [1,3,f], F. Toson [1,g], M. Borin [2,h]

[1] CISAS G. Colombo, University of Padova, Via Venezia, 15, Padova (PD), Italy

[2] DAFNAE, University of Padova, Viale dell'Università, 16, Legnaro (PD), Italy

[3] DII, University of Padova, Via Venezia, 1, Padova (PD), Italy

[a]irene.terlizzi@unipd.it, [b]francesco.morbidini@unipd.it, [c]carmelo.maucieri@unipd.it, [d]carlo.bettanini@unipd.it, [e]giacomo.colombatti@unipd.it, [f]sebastiano.chiodini@unipd.it, [g]federico.toson@phd.unipd.it, [h]maurizio.borin@unipd.it

Keywords: Vegetation Index, Satellite, Multi-Spectral, Remote Sensing

Abstract. The multi-spectral data acquired with either satellite imagery, UAV or tethered and stratospheric balloons can be used to calculate vegetation indices directly related to the well-being of the crops providing a quantitative information about its health and growth. The vegetation indices are calculated combining measurements from different parts of the electromagnetic spectrum, typically in the visible and near-infrared ranges. The aim of this work is to integrate the remote sensing data with in-situ collected measurements in order to validate remote observations for monitoring soybean water status and requirements. The study is conducted in Italy on a field of 160 x 40 m^2, divided into four plots of 40 x 40 m^2; two of them are irrigated at 100% of the CRW (Crop Water Requirement) and two irrigated at 70% of CWR. In each plot tensiometers and capacitive probes directly measure the soil moisture, along with a climate station used to monitor environmental parameters. The in-situ data are correlated with multi band satellite images by the PlanetScope constellation providing a ground resolution of 3 m. The use of UAV or balloons is needed to monitor the diurnal variation of the indices, as the satellite revisit time is once per day around 9:00 and 10:00 UTC on the site. The balloon payload is equipped with commercial cameras and dedicated filters to acquire images in the same spectral bands as satellites. The importance of this study lies in the possibility of managing the fields irrigation basing on the actual physiological need of the crop rather than relying on a predefined timetable, resulting in a more efficient and environmentally responsible irrigation. The article will present the methodology and the instruments applied, together with the results obtained.

Introduction

Soybean is one of the most important crops worldwide as it is used not only for livestock feed and human nutrition, but also assumes it has key role in the biofuel sector [1]. Moreover, the need to expand and increase irrigated agriculture appears to be extremely necessary in view of the increased demand for food in the coming decades [2].

For correct irrigation management, it is necessary to know the crop evapotranspiration index (ET_c) and the right value of the crop coefficient (K_c).

$$K_c = \frac{ET_c}{ET_0} \tag{1}$$

where ET_c is the evapotranspiration of the crop under optimal conditions and ET_0 is the potential evapotranspiration of the reference crop calculated with the Pennam-Withman equations.

Materials Research Proceedings 37 (2023) 262-265 https://doi.org/10.21741/9781644902813-57

A linear relationship between K_c values and some vegetative indices, first of all NDVI, has been observed from several studies [3]. Agronomical indices can be obtained from remote-sensing images from cameras aboard instrumentation such as tethered balloon and satellites. Remote sensing can provide useful information with high accuracy regarding optimal irrigation management allowing to obtain information not only of the crop but also of the soil and environmental factors [4].

Study area

The study was conducted in Castelfranco Veneto (IT) on a field constituted of 4 plots of 40 x 40 m^2 each. Two different irrigation regimes were applied: full irrigation (application of 100% of the CWR) and regulated deficit irrigation (application of 70% of the CWR with 100% restoration between the beginning of flowering and the beginning of pod formation) as can be seen in *Figure 1*.

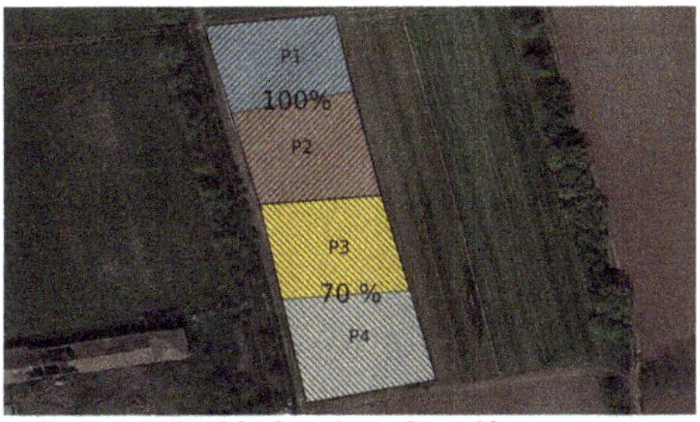

Figure 1: Field of study in Castelfranco veneto

Irrigation was managed by referring to probes capable of measuring soil moisture at three different depths (at 20, 40 and 60 cm) and with the data collected was calculated the water balance. Then consumption by evapotranspiration was obtained with the equation:

$$ET = I + P \pm \Delta S - R - D \tag{2}$$

Where ET is the evapotranspiration (in mm), I is the irrigation water (in mm), P is precipitation (in mm), ΔS is the change in soil water storage (in mm), R is surface runoff and D is deep percolation. Then with (1) was obtained K_c.

Satellite data and vegetation index calculation using multiband images

Images from Planetscope constellation have been used to calculate the indices. Planetscope is a constellation of 130+ Cubesat in sun-synchronous orbits which permits a revisit time of 24h on the field.

The satellites' camera operates in eight bands each one with a resolution of 3 m: Coastal Blue (431 - 452 nm), Blue (465 – 515 nm), Green I (513 - 549 nm), Green (547 – 583 nm), Yellow (600 - 620 nm), Red (650 – 680 nm), Red-Edge (RE) (697 – 713 nm), Near-Infrared (NIR) (845 – 885 nm).

The bands given by the camera allow to calculate numerous indices, each of these utilizes specific combination of spectral bands to capture different aspects of the crop health and vigour:

- Normalized Difference Vegetation Index (NDVI): used to assess the health and vitality of plants, allowing the recognition of areas with development issues. [5]

$$NDVI = \frac{NIR - Red}{NIR + Red} \tag{3}$$

- Green Normalized Difference Vegetation Index (GNDVI): GNDVI is a variant of NDVI that uses reflectance in the green band instead of the red band. This index is more sensitive to chlorophyll concentration and is used to determine the presence of water and humidity. [5]

$$GNDVI = \frac{NIR - Green}{NIR + Green} \tag{4}$$

- Normalized Difference Red Edge (NDRE): NDRE is an index combines reflectance in the Red Edge band with reflectance in the NIR band. This index is particularly sensitive to the presence of chlorophyll and is useful for assessing crop health and detecting water stress. [5]

$$NDRE = \frac{NIR - RE}{NIR + RE} \tag{5}$$

Figure 2: NDVI index of 15 July 2022

The purpose was to identify which index exhibited the strongest correlation with Kc.

The correlations aimed to determine which index could provide the most accurate estimation of K_c to study and compare the health of the two subplot.

Correlation of satellite based vegetation indexes and in situ based

The linear correlation was performed by considering the pixel mean value calculated for each of the two plots and the K_c data obtained from the water balance for each day when satellite data was available. The correlation coefficients were very close to each other, as can be seen in *Table 1*, so further analysis need to be conducted. In the meantime, this study helped defining the payload for the tethered balloon which will fly the next months to collect data on the variation of the indices over one day.

Table 1: Correlation coefficients between K_c and NDVI, GNDVI, NDRE for the two subplot

	70%CWR	100%CWR
NDVI	0,231445	0,23948
GNDVI	0,251871	0,23865
NDRE	0,272279	0,24143

Integration of balloon based multispectral images

The tethered balloon will fly over the field approximately once a month throughout the crop development phase. It will acquire data for at least one day at a height of 50 meters, providing a field of view of about 100 x 100 m^2, it will be positioned at the center of the field, acquiring with half of his Field Of View the field at 70% CWR and with the other half the field at 100% CWR.

It is equipped with a thermal camera, to measure the temperature of the crop and of the soil, and with two other cameras: a monochrome camera with a Red-Edge filter and a colored camera with a triple band filter in the bands of Blue, Green and NIR, to obtain data on the bands respectively of 735 nm, 475 nm, 550 nm and 850 nm.

In this way, the bands will be acquired to measure GNDVI, NDRE, and ENDVI (Enhanced Normalized Difference Vegetation Index) as substitutes for NDVI.

$$ENDVI = \frac{((NIR + GREEN) - 2BLUE}{((NIR + GREEN) + 2BLUE} \tag{6}$$

Conclusions

In conclusion, after acquiring data from the balloon and conducting a more detailed analysis of the correlations, it will be possible to effectively utilize the indices to assess the health status of the crop and optimize irrigation practices. This is especially crucial during a time when water resources are more critical than ever. By leveraging the information provided by the indices, farmers and irrigation can make informed decisions to ensure efficient water usage and maximize crop productivity while considering the limited water availability.

Acknowledgments

Image data services were provided by Planet team (2017), Planet Application Program Interface: in Space for Life on Earth, San Francisco, Ca. https://api.planet.com

References

[1] Aydinsakir K. "Yield and quality characteristics of drip-irrigated soy bean under different irrigation levels". In: Agronomy Journal 110 (2018),pp. 1473–1481. https://doi.org/10.2134/agronj2017.12.0748

[2] Massari C et al. "A review of irrigation information retrievals from space and their utility for users." In: Remote Sensing 13 (20 2021), p. 4112. https://doi.org/10.3390/rs13204112

[3] Campos I et al. "Estimation of total available water in the soil layer by integrating actual evapotranspiration data in a remote sensing-driven soil water balance." In: Hydrology 534 (2016), pp. 427–439. https://doi.org/10.1016/j.jhydrol.2016.01.023

[4] Ahmad U, Alvino A, and Marino S. "A review of crop water stress assessment using remote sensing. Remote Sensing". In: Agronomy Journal 13 (2021), p. 4155. https://doi.org/10.3390/rs13204155

[5] Information on https://leonardolaureti.wordpress.com/2021/01/08/interpretazione-degli-indici/

Aeronautics and Astronautics - AIDAA XXVII International Congress Materials Research Forum LLC
Materials Research Proceedings 37 (2023) 266-269 https://doi.org/10.21741/9781644902813-58

DUST mitigation technology for lunar exploration and colonization: existing and future perspectives

Guido Saccone[1,a*], Nunzia Favaloro[1,b]

[1]Italian Aerospace Research Centre (CIRA), Space Exploration Technologies Lab, Via Maiorise snc, 81043, Capua (CE), Italy

[a]g.saccone@cira.it, [b]n.favaloro@cira.it

Keywords: Moon, Dust Mitigation, Space Exploration, High-Performance Polymers

Abstract. Micrometric dust particles of lunar regolith represent one of the most serious issues of the harsh moon environment. Indeed, the extremely high vacuum conditions expose the lunar soil minerals to intense ultraviolet and galactic cosmic rays' bombardment during the Moon's daylight producing photoionization of the constituent's atoms and electron release. Moreover, Moon periodically interacts with the surrounding solar wind which generates a continuous flux of charged particles is generated accompanied by electric fields around the terminator region able to lift off the lunar regolith dust up to ~100 km above the geometrical surface. In this way, micrometric granular matter forms a subtle veil of contaminants. This electrically charged and extremely adhering dust environment can cause various critical drawbacks not only to several robotic parts e.g., mechanical components, electronic devices, solar panels, thermal radiators, rovers seals and bearings, etc. but also can dramatically damage the respiratory systems of humans if accidentally inhaled. For these reasons, lunar dust was recognised, by several agencies including NASA and ESA, as one of the main potential showstoppers for the ongoing robotic and manned exploration and colonization of our natural satellite. For overcoming or at least mitigating these issues, several technologies were developed and assessed ranging from the active ones requiring a source of energy e.g., mechanical, fluidal and, above all, electric devices, to the passive technologies involving suitable material design and development. In the work here reported, the design and development of innovative high-performance polymers simultaneously exhibiting outstanding thermo-mechanical properties and superior non-sticking capacity i.e., *abhesion* to be applied for structural purposes on the Moon is presented. Further improvement of these suitable designed materials with the addition of appropriate electric properties will make them ideal candidates as dielectric substrates of a combined passive and electroactive system able to repel micrometric regolith particles i.e., lunar dust shield.

Introduction

Dust in space environments, especially Moon and Mars has been recognized by several Space Agencies, including NASA and ESA, as a major concern for the successful robotic and manned exploration and colonization of extra-terrestrial bodies.

Indeed, the Lunar surface is covered by regolith i.e., a loose granular material, consisting of a broad range of shapes, sizes, and types of sediments [1]. It mainly consists of silicate minerals e.g., olivine, pyroxene, plagioclase, and non-silicate minerals such as ilmenite [2], crushed by meteoric bombardment causing the fracturing, scattering and agglutination of lunar regolith in the form of micrometric (approximately 50% of the lunar regolith consists of granular, dust particle with diameters less than 60 μm [3]), sharp, abrasive, porous, chemically reactive dust particles. These granulometric particles are electrostatically charged by the interactions with the local plasma environment and the solar ultraviolet (UV) radiations as well as the solar wind due to the photoelectric effect i.e., the UV and X-ray induced photoemission of electrons of the grain-materials [4],5]. Even without any mechanical activities, the dust grains are levitated by the electrostatic fields and transported away from the surface in the near vacuum environment of the Moon. Above all, the lunar, regolith dust particles can adhere strongly to the exposed surfaces.

As verified during the Apollo missions, these particles were an unforeseen problem and caused several issues [6] including vision obscuration during landing [7]], abrasion of visors, gloves and boots of astronaut's suits [9], degradation of radiators and thermal control surfaces, deterioration of solar panels efficiency, deposition on optical surfaces, vision obscuration, creation of wrong responses from measuring devices and false instrument reading, interference with the operation of Extra-Vehicular Activity (EVA) systems including suits, airlocks, suitports tools, etc. [8].

In order to efficiently overcome the serious issues represented by the challenging lunar and Martian dust environments (or small bodies like asteroids, comets, and Near-Earth Objects), innovative technologies/approaches have to be investigated and employed.

They can be generally subdivided into passive and active approaches:

- **Passive dust mitigation technologies:** they consist in preventing the sticking and adherence of the loose granular fraction of lunar or Martian regolith i.e., micrometric dust (on the Moon the term dust usually refers to particles with a Sauter mean diameter < 25 μm). This ambitious and challenging goal is achieved without the application of any kind of external source of energy, by means of suitable design and development of special materials with the lowest possible surface energy, thanks to a tailored modification of the chemical structure further improved by a suitable alteration of the topographical/morphological features of the material's external layers due to application of several possible chemical and physical strategies.
- **Active dust mitigation technologies**: they foresee the direct cleaning of the examined surfaces or the protection of them from dust deposition and contamination. Active technologies utilize external forces: electrodynamic, electrostatic, fluidal, mechanical, etc. But the most effective ones are based on regolith particle charge. In detail, solar radiation and the solar wind produce a "plasma scabbard" near the lunar surface and lunar grains acquire charge in this environment and can exhibit unusual behaviour, including levitation and transport across the surface because of electric fields in the "plasma scabbard".

Results and Discussion

In order to solve or mitigate the dust issue, CIRA aims to pursue a hybrid strategy consisting of the simultaneous application of passive and active methods as depicted in Figure 1.

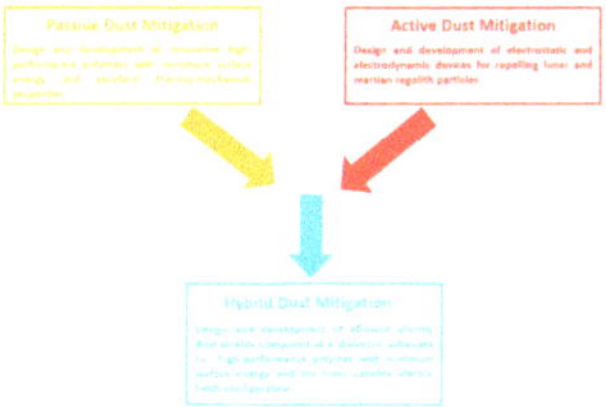

Figure 1. Scheme of the CIRA dust mitigation approach.

In particular, it can be accomplished by the design and development of a suitable material exhibiting at the same time minimum surface energy, i.e., *abhesion* capacity consisting of non-sticking ability against micrometric lunar dust particles, and excellent dielectric properties making it a good candidate as an insulating substrate of electrodynamic or electrostatic dust mitigation devices.

The harsh Moon environment requires also the design and development of a material endowed with outstanding thermo-mechanical properties in order to resist the strong temperature fluctuations, UV and galactic cosmic rays' irradiation, micro-meteoritic bombardment and ultra-vacuum conditions. Moreover, low density is another appealing characteristic for every material, which has to be transported from Earth and fly as payload within a space launcher. Considering also these constraints, the class of high-performance polymers has been recognized to be the most appropriate for lunar applications. In particular, aromatic polyimides (PI) have been selected as starting materials for achieving hybrid dust mitigation capacity.

According to this approach, CIRA has designed innovative aromatic PIs consisting of monomers endowed with a greater amount of trifluoromethyl groups in order to successfully employ the dipole-dipole repulsion forces exerted by the C-F groups, due to the highest electronegativity of the fluorine atom among all the elements. Furthermore, monomers containing a limited number of benzene rings have been identified since they guarantee the achievement of superior thermo-mechanical properties. Moreover, specially designed additives, Surface Migrating Agents (SMA) based on silicon moieties, have been formulated and they are able to spontaneously migrate toward the outermost layers of the material drawn by thermodynamic forces and react with the dianhydrides, used for the PI synthesis, forming a co-polyimide as shown in Figure 2.

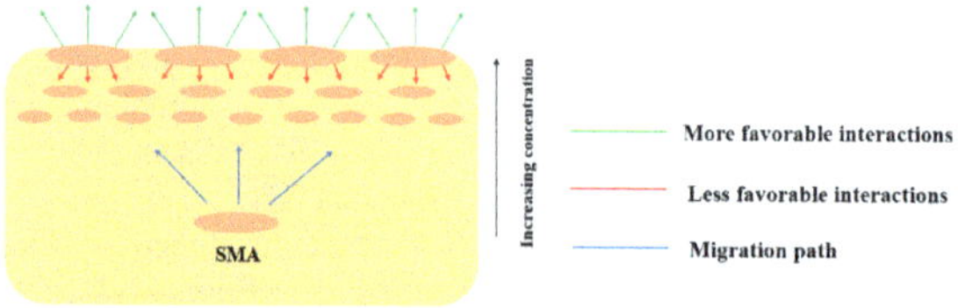

Figure 2. Schematic representation of the SMA migration mechanism.

The SMAs, reaching the external layers of the material, contribute to further lowering the surface energy, already minimized by the presence of elevated -CF_3 groups. This behaviour has been experimentally observed by the increase of the contact angles these coupons form with micrometric water droplets.

In order to realize a device endowed with passive and active dust mitigation capacity, this innovative PI has to be combined with a suitable electrodynamic system consisting of an array of thin conducting electrodes (Figure 3), embedded in the surface, and separated among them by an insulating material, designed to prevent electrical breakdown between the independent electrodes [7].

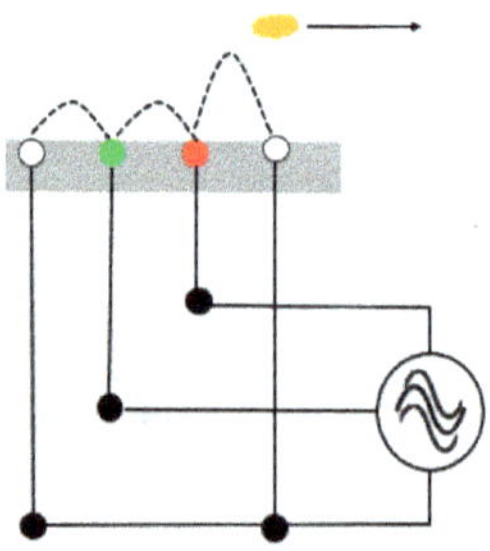

Figure 3: Schematic depicting dust particle motion on an electrodynamic system.

Electrodes are connected to a multi-phase low-frequency AC power supply, able to generate pulsed standing waves to shift, lift and transport far away charged particles, approaching the protected surface. Electrostatically charged particles, such as those encountered on the Moon, Mars, or on asteroids, are carried along by the traveling field due to the actions of Coulomb and dielectrophoretic forces.

Innovative materials developed by CIRA can be optimal candidates as insulating materials for electrodynamic devices for hybrid dust mitigation technology due to their inherent low dielectric strength.

References

[1] Wohl, C.J., Atkins, B.M., Belcher, M. A., Connell, J.W., Copoly(imide siloxane) Abhesive Materials with Varied Siloxane Oligomer Length, International SAMPE Symposium and Exhibition (Proceedings), (2010).

[2] Fateri, M., Gebhardt, A., Gabrielli, R.A., Herdrich, G., Fasoulas, S., Groβmann, A., Schnauffer, P., Middendorf, P., Additive Manufacturing of Lunar Regolith for Extra-terrestrial Industry Plant, 30th ISTS Conference, Kobe, Japan, (2015).

[3] Taylor, L.A., Schmitt, H.H., Carrier, W.D., Nakagawa, M., The Lunar Dust Problem: From Liability to Asset, in 1st Space Exploration Conference: Continuing the Voyage of Discovery, American Institute of Aeronautics and Astronautics, Orlando, FL., (2005). https://doi.org/10.2514/6.2005-2510

[4] Abbas, M., Tankosic, D., Craven, P.D., Spann, J.F., LeClair, A., West, E.A., Lunar dust charging by photoelectric emissions, Planetary and Space Science 55, 953-965, (2007). https://doi.org/10.1016/j.pss.2006.12.007

[5] Stubbs, T.J., Vondrak, R.R., Farrell, V.M., A dynamic fountain model for lunar dust, Advanced in Space Research 37, 59-66, (2006). https://doi.org/10.1016/j.asr.2005.04.048

[6] Afshar-Mohajer, N., Wu, C.-Y., Curtis, J., Gaier, J. R., Review of Dust Transport and Mitigation Technologies in Lunar and Martian Atmospheres, Advances in Space Research, (2015). https://doi.org/10.1016/j.asr.2015.06.007

[7] Calle, C.I., Buhler, C.R., McFall, J.L., Snyder, S.J., Particle removal by electrostatic and dielectrophoretic forces for dust control during lunar exploration missions, Journal of Electrostatics 67, 89-92, (2009). https://doi.org/10.1016/j.elstat.2009.02.012

[8] Gaier, J.R., The effects of lunar dust on EVA systems during the Apollo missions, NASA/TM 213610: (2005).

[9] Goodwin, R., 2002. Apollo 17: The NASA Mission Reports, vol. 1. Apogee Books, Ontario, Canada.

A static, refractive and monolithic Fourier spectrometer for an HEMERA balloon flight

Fabio Frassetto[a1*], Lorenzo Cocola[b1], Paola Zuppella[1], Vania Da Deppo[1], Riccardo Claudi[2] and Luca Poletto[1]

[1] CNR-IFN Via Trasea, 7 – 35131 Padova, Italy

[2] Astronomical Observatory, INAF, Padova, Italy

[a]fabio.frassetto@cnr.it, [b]lorenzo.cocola@cnr.it

Keywords: Fourier Transform Spectroscopy, Space Instruments, Interferometry

Abstract. In this work we present the characteristics of a static Fourier transform spectrometer designed for a balloon flight and meant to test the instrument in an actual working environment. The flight campaign has been provided within the HEMERA Research Infrastructure. The interferometric assembly, the core of the instrument, is made with only refractive and reflective surfaces, without diffraction gratings. It is realized with four glass elements that are glued in a single monolithic structure with no spacers or framed supports. There are no hollow spaces, thus improving the mechanical stiffness. The spectral band accessible to the presented optical design is 450 – 850 nm and the working spectral region can be chosen to be either 550 – 850 nm or 450 – 550 nm. The spectral resolution is variable as it is driven by the refractive index of the used glasses in the considered spectral region. In this design, the resolution varies from 1500 at 550 nm to 200 at 800 nm and to 4000 at 450 nm. The total field of view of the instrument is 3 degrees. All the optical elements used to operate the spectrometer are installed on a 3D printed structure. This is possible due to the low sensitivity of the instrument to vibrations. The interferometric assembly has undergone thermal test to quantify the temperature range it can tolerate. We will present the instrument design and the first results of the flight campaign.

Introduction

Static Fourier transform spectroscopy is nowadays a well-documented technique [1-3], the interested reader is invited to start from the references for a description of the technique. Here is sufficient to recall that the technique is based on the recording, via a 2D detector, of an image signal, called interferogram, whose Fourier transform returns the spectrum of the light that has produced the interferogram itself. What we have tested in this flight campaign is the possibility to realize the interferometer as a monolithic optical assembly in which the beam-splitter and the retro-dispersive elements, two Littrow's prisms [4-6], are glued together forming a compact structure (hereafter the interferometric assembly, IA) with no hollow volumes in the optical path.

Being the IA realized by gluing together glasses with different coefficients of thermal expansion (BK7 and SF57), the proposed optical concept has been tested to evaluate the possible occurrence and effect of thermal stresses in the interferometric assembly. The spectrometer, with the indication of all its major parts, is shown in Fig. 1a; Fig. 1b shows the instrument, closed with its cover, during the pre-flight tests. The supporting base is realized via SLS 3D printing technology in Nylon 12.

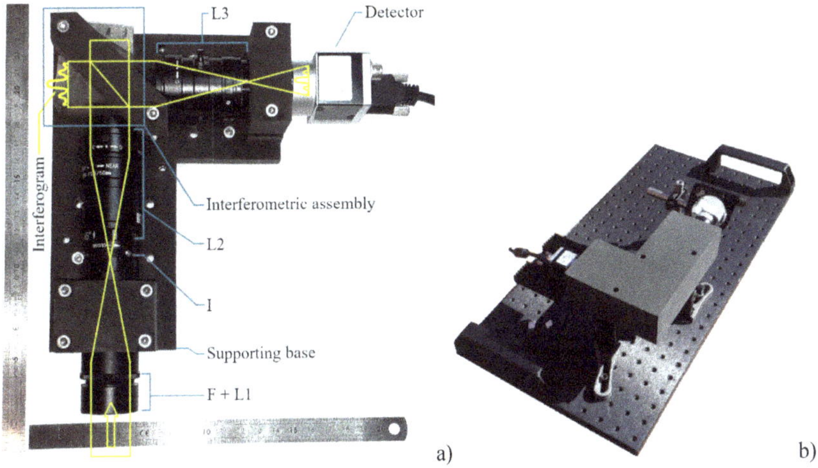

Fig. 1. Image of the realized instrument. a) The spectrometer without the cover and with all the main parts indicated. The yellow overlay mimics the light path and illustrates the imaging action of L3, which projects the interferogram on the detector focal plane. b) The spectrometer installed on a breadboard, during the pre-flight tests.

The flight

As in the first flight [7], the spectrometer, prepared for the flight campaign, has been installed inside a polystyrene box with 1 cm thick walls. The box contains also all the electronics and the power source, see Fig. 2 for details on the internal sub-assembly arrangement. The flight took place the 11th of May 2023 in the morning at the Aire-Sur-L'Adour "Centre d'Opérations Ballons" - CNES. The ascending phase lasted about 1 hour and 30 minutes and the descending one 50 minutes, for a total flight duration of about 2 hours and 20 minutes. The maximum reached altitude, as inferred from the internal sensor (Bosch BME680) with its default calibration curve, was 26 km. Pressure altitude and temperature as measured from the internal transducers are plotted in Fig. 3.

An interferogram has been acquired every second. Fig. 4a presents a 2D map of the acquired spectra during the entire ascending and descending phase. To obtain a spectrum, the interferogram is line by line multiplied with a "Blackman" window, then for each line the squared absolute value of the 1-D Fourier transform is calculated. The results are then averaged on the full frame. For each acquisition the spectrum is normalized to its maximum value. Fig. 4b shows in detail three spectra acquired after 1031 s, 3007 s and 5504 s.

a) b)

Fig. 2. The instrument assembled for the flight campaign. a) Internal view: the arrow indicates the entering direction of the light. A) the instrument, partially hidden behind the two PCBs, B) the single board computer, C) the power and environmental sensing unit, D) the batteries, E) one of the four connecting pylons between the instrument and the external box in polystyrene. b) The instrument, landed in an open field, after the flight.

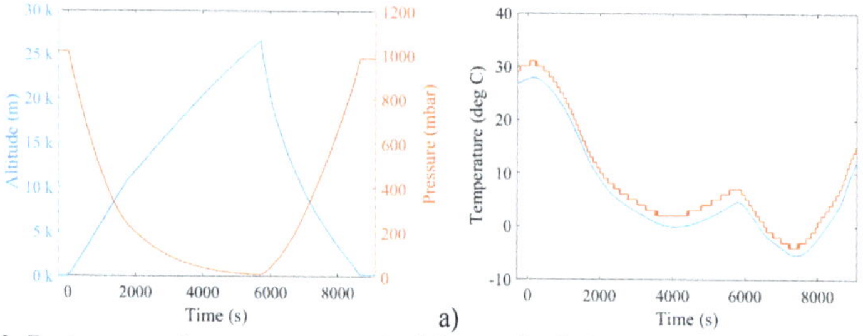

a) b)

Fig. 3. Environmental parameters acquired during the flight. a) Altitude and pressure from Bosch BME680. b) Temperature inside the polystyrene box measured by two distinct transducers.

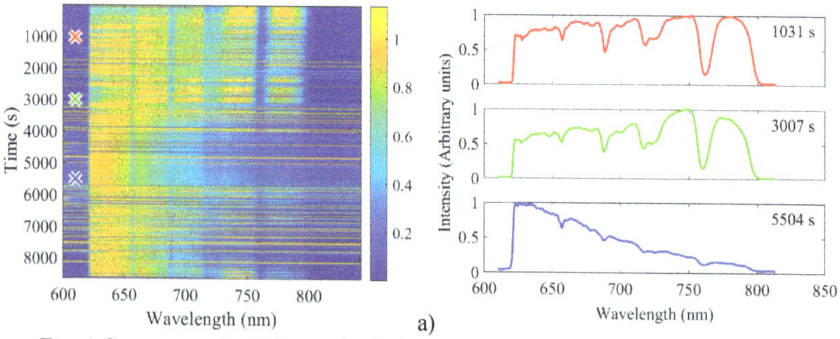

a) b)

Fig. 4. Spectra acquired during the flight. A) 2D map: each spectrum is normalized to its maximum value. B) for the three horizontal cuts corresponding to the crosses in figure a), the corresponding spectra are plotted. The spectra are acquired after 1031, 3007, 5504 seconds from the beginning of the flight.

Conclusions

The proposed optical design proved to operate in a real flight environment showing negligible thermal drift effects. Moreover, compared to the previous spectrometer design [7] which featured separate prisms and beam splitter on an aluminum optical bench, this new layout showed an improved mechanical stability. Despite the lightweight 3D printed supporting base, the new design enabled operation with longer integration times without any spectral degradation due to loss of contrast in the interferogram.

Acknowledgments

We desire to thank all the people from CNES, ASI and INAF that supported us during the realization of the instrument and the flight campaign. In particular, and knowing we are mentioning just a few, André Vargas, Cruzel Serge, all the CNES operation team, Angela Volpe, Lorenzo Natalucci, Pietro Ubertini and all the HEMERA community.

Fundings

This work has been supported by ASI, Agenzia Spaziale Italiana, Agreement n. 2019-33- HH.0, for the payload realization and the flight opportunity has been provided by the European Commission in the frame of the INFRAIA grant 730790-HEMERA.

References

[1] J. M. Harlander, Spatial Heterodyne Spectroscopy: Interferometric Performance at any Wavelength Without Scanning, Thesis (Ph.D.) University of Wisconsin, Madison, 1991, https://ui.adsabs.harvard.edu/abs/1991PhDT........62H

[2] J. Harlander, F. Roesler, J. Cardon, C. Englert and R. Conway, Shimmer: a spatial heterodyne spectrometer for remote sensing of Earth' middle atmosphere, Appl. Opt. 41, 1343-1352 (2002). https://doi.org/10.1364/AO.41.001343

[3] M. Kaufmann, F. Olschewski, K. Mantel, B. Solheim, G. Shepherd, M. Deiml, J. Liu, R. Song, Q. Chen, O. Wroblowski, D. Wei, Y. Zhu, F. Wagner, F. Loosen, D. Froehlich, T. Neubert, H. Rongen, P. Knieling, P. Toumpas, J. Shan, G. Tang, R. Koppmann and M. Riese, A highly miniaturized satellite payload based on a spatial heterodyne spectrometer for atmospheric temperature measurements in the mesosphere and lower thermosphere, Atmos. Meas. Tech., 11, 3861–3870. https://doi.org/10.5194/amt-11-3861-2018

[4] F. Frassetto, L. Cocola, V. Da Deppo, L. Poletto and P. Zuppella, High signal-to-noise ratio spectrometer based on static Fourier transform interferometer (for hemera balloon flight), Italian Association of Aeronautics and Astronautics, XXV International Congress, 9-12 September 2019, Rome, Italy

[5] F. Frassetto, L. Cocola, R. Claudi, V. Da Deppo, P. Zuppella and L. Poletto, Low noise Fourier transform spectrometer for a balloon borne platform, Italian Association of Aeronautics and Astronautics, XXVI International Congress, 31 August -3 September 2021, Pisa, Italy. https://doi.org/10.5194/egusphere-egu22-11428

[6] F. Frassetto, L. Cocola, P. Zuppella, V. Da Deppo and L. Poletto, High sensitivity static Fourier transform spectrometer, Opt. Express 29, 15906-15917 (2021). https://doi.org/10.1364/OE.422645

[7] F. Frassetto, L. Cocola, R. Claudi, V. Da Deppo, P. Zuppella and L. Poletto, Report on a refractive static Fourier transform spectrometer flown on a stratospheric balloon for the HEMERA program, ICSO International Conference on Space Optics 3-7 October 2022, Dubrovnik, Croatia. https://doi.org/10.5194/egusphere-egu22-11428

Ascent trajectory of sounding balloons: dynamical models and mission data reconstruction

C. Bettanini[1,2] *, M. Bartolomei[2], A. Aboudan[2] and L. Olivieri[2]

[1] Department of Industrial Engineering, University of Padova, via Venezia 1, Padova (Italy)

[2] CISAS - Center for Studies and Activities for Space "Giuseppe Colombo", University of Padova, via Venezia 15, Padova (Italy)

carlo.bettanini@unipd.it

Keywords: Sounding Balloon Dynamics; Trajectory Reconstruction, Linear Quadratic Estimation Method

Abstract. Small sounding balloons are a fast and cost-effective transport system to lift up scientific payloads up to stratospheric burst altitudes below 40 kilometres; during ascent and descent phase dedicated instruments may be operated to monitor atmospheric parameters and optical payloads may be used for remote observation. This work will focus on the reconstruction of the trajectory of the ascent phase, which is the longest and dynamically less perturbed part of the flight; in this section the dynamics of the flight system is determined by the lift of the balloon guiding the vertical motion and the local winds controlling the horizontal motion. The presented reconstruction algorithm is based on a linear quadratic estimation predictor corrector using the standard equations of motions in ECEF system to propagate the simulation and the measurement of the on-board sensors (triaxial accelerometer, GPS, pressure and temperature sensors) to correct the estimation and reduce the uncertainty in the reconstruction, which is mainly related to the value of balloon canopy drag coefficient Cd, the lifting gas volume and local wind perturbations. Two different balloon flights, both launched within a joint effort between teams by University of Padova and University of Pisa, are considered: one conducted during daytime, the other in night time. The different environmental conditions and in particular the different temperature evolution within the lifting balloon in the day flight due to Sun heating provide a good proving ground to investigate sensitivity of algorithm to environmental conditions. The prediction of flight dynamic models implementing horizontal and vertical equations of motion are compared with real mission data acquired by on board systems, highlighting the influence of local perturbations on the foreseen ascent trajectory.

Introduction

Stratospheric sounding balloons are small high-altitude balloons that can reach altitudes below 40 kilometres before burst; to date thanks to the availability of powerful low-cost command and data acquisition systems it is possible to carry scientific instruments, experiments or complex autonomous systems with an overall payloads mass below 3 kilograms to high altitudes and conduct in situ research and remote observation of ground targets. [1] [2].

One of the key challenges associated with stratospheric balloon flights is the accurate prediction and reconstruction of the balloon's trajectory; an accurate prediction is needed since launch decisions must be based on the likelihood of landing in a desirable location for safety issues and recovery; post flight reconstruction is both used to tune the parameters of the prediction algorithms and to provide along with attitude data georeferentiation for remote sensing data.

The simplest dynamic model utilised in the simulation of balloon trajectory is a lumped mass 3 DOF system and is the one applied in several freely available trajectory prediction software tools [3][4] or in the more complex Monte Carlo based simulation codes[5].

Several factors may affect the accuracy of these prediction algorithms: environmental conditions and atmosphere layering above the launch site may differ from the nominal standard atmosphere profile and the ascent profile depends strongly on the temperature reversal near the tropopause (about 15 km) [6]. Wind variations and atmospheric turbulence, along with cloud presence and changing ground albedo can affect also the accuracy of the prediction; the last two contributions affect lifting gas equilibrium temperature inside the canopy [7] [8] and ,as a consequence, the temperature inside the balloon may differ from surrounding atmosphere of more than 30 K due to balloon radiative heating both from the Sun and from the Earth's surface[5].

During daytime launches it is therefore critical to develop a thermal model for the balloon to investigate internal temperature variation and consequently calculate balloon volume and shape along ascent trajectory. [9] For the typical size of latex balloons temperature and pressure gradients can be disregarded and density considered constant over the volume of the balloon.

Furthermore, although helium gas mass at launch may be inferred by measuring the lift at the hook, the value is usually not accurate since is calculated by reaching the floating condition using a ballast mass; this requires to reconstruct the correct ascent velocity by mission data. Finally flight Reynolds numbers span within the so called "drag crisis" zone and accurate modelling of the balloon's ascent rate requires that the Cd be treated as a function of Reynolds number and therefore iteratively calculated depending on altitude and ascent velocity [10]. Adjustments and corrections are therefore needed to improve the accuracy of the dynamical model and due to the variability of the environmental conditions a reconstruction process is usually performed relying on mission data from a variety of sensors and instruments, including GPS receivers, accelerometers, magnetometers and external temperature sensors. Such exercise is particularly important when optical systems are part of the payload and no active control is present to guarantee platform's stability; an accurate trajectory reconstruction along with the elaboration of inertial attitude may minimise the error on pointing and therefore on the data extracted for targeted areas on ground.

Equations of motion for balloon ascent phase

The balloon ascent dynamic is a result of the lift, drag and gravitational forces on the flight train. The balloon's motion is mainly driven by the atmosphere winds governing the horizontal motion and the net lift force of the balloon determining the variation of altitude versus time and so the vertical motion. Usually the horizontal velocity of the balloon can be assumed as equal to the atmospheric wind so the drag created by relative motion between the atmosphere and the balloon system may be limited to the vertical motion into the atmosphere.

This assumption is essentially correct for lightly loaded balloons, especially at lower altitudes, but becomes less accurate at higher altitudes where inertia of the accelerated balloon must be taken into consideration. [11]

During the motion into the atmosphere, surrounding air is assumed to behave like a perfect gas in hydrostatic equilibrium. This assumption in the Earth's atmosphere is justified by the fact that the deviation from the perfect gas equation of state is negligible within the range of density and temperature conditions between the ground and a 40 km maximum altitude

The free lift force , also called net buoyancy of the balloon, is a function of the density of the lifting gas and the volume of the balloon and can be expressed as

$$\overrightarrow{F_{FL}} = \left(\rho_A V - m_{tot}\right)\vec{g} \quad (1)$$

where ρ_A is ambient air density, V is the volume of the balloon, g is the gravitational acceleration vector and m_{tot} is the total mass for lift. Total mass m_{tot} is the sum of the flight train mass (payload mass, harness mass and balloon fabric mass), helium mass and the added mass which is the inertia of the mass of fluid displaced by the body, since the balloon must shift some volume of surrounding fluid as it moves through it

$$m_{ftrain} = m_{pay} + m_{harness} + m_{ball} \ ; \quad m_{tot} = m_{ftrain} + m_{gas} + m_{add} \quad (2)$$

The added mass for the sounding balloon may is typically approximated as 40% of the mass of the displaced air volume of the balloon [12][13]

$$m_{add} = 0.4\, \rho_A V \quad (3)$$

It can be noted that due to added mass contribution, effective mass of the balloon system varies with altitude and must be properly updated especially if high volume balloons are considered.
In zero pressure balloons (like the sounding balloons) the elastic latex canopy does slightly compress the lifting gas, but internal and external pressure remain within 1% of each other on average over the course of the flight, so treating the interior and exterior pressures as equal is reasonable and treating the lifting gas as an ideal gas, the density may be solved for by using the ideal gas law for balloon volume calculation.

The drag force of the aerostat is a function of the relative velocity of the balloon and atmosphere, the ambient density, the size of the balloon, and the drag coefficient. It can be calculated as F_D using:

$$\overrightarrow{F_D} = \frac{1}{2} \rho_A S\, v_{rel}^2\, C_D \overrightarrow{u_{vrel}} \quad (4)$$

where S is the top projected area of the balloon, and C_D is the drag coefficient representative of the balloon flight train and can be considered as the one related to balloon canopy.

The C_D may be then assumed to be close to the one of a sphere and is a function of Reynolds number; Reynolds number must be therefore calculated using balloon diameter as reference dimension following the classical expression

$$Re = \frac{\rho_A v_{rel} D_{ball}}{\mu_{dyn}} \quad (5)$$

It may be underlined than velocity is given relative to the atmosphere, so a balloon's horizontal movement which is supposed to be similar to the horizontal wind does not affect the Reynolds number. As already mentioned balloon flight Reynolds numbers span the region where the coefficient of drag of the sphere decreases from 0.4-0.5 to 0.1; for the present work Conner's model was selected, as it is based on experimental data from latex balloon flights.[5]

Linear quadratic estimation model for Ascent trajectory reconstruction

The implemented model simulates the evolution of the balloon system considered as a point mass using a linear dynamic system of 12 variables (state) and considers the measures of on-board sensors as corrector building 7 observation parameters. The list of the variables is shown in listed the following equations (6)

The State is constituted by position (Rx, Ry, Rz) and velocity (Vx, Vy, Vz) of balloon system in ECEF Earth reference; Wind velocity (Wx, Wy, Wz) also expressed in ECEF is needed to calculate the vector expression for velocity relative to atmosphere as the difference of balloon velocity and local wind velocity and afterwards used in drag force evaluation. Atmospheric pressure and temperature are the last two parameters of the state and used to calculate balloon pressure and to estimate balloon temperature using a simplified thermal model for balloon canopy energy balance. [11]

The prediction of the system state at every time step performs the evaluation of total force on the balloon system (equations (2) and (4)) and calculates acceleration components (velocity time derivatives) for the trajectory.

$$State = (Rx|Ry|Rz\,|Vx|Vy|Vz|Wx|Wy|Wz|Pa|Ta)$$

$$Observation = \left(Rx_{GPS}\middle|Ry_{GPS}\middle|Rz_{GPS}\,\middle|Alt_{GPS/BARO}\middle|Ext\_T\middle|Ext\_P\middle|Acc\_tot\right) \quad (6)$$

Standard Atmosphere models for atmospheric pressure (Pa) and temperature (Ta) are used to predict evolution depending on vertical velocity. The correction of the prediction is performed using GPS data acquired at 1 Hz sampling rate during the flight and transforming latitude and longitude data in cartesian coordinates in ECEF reference. Depending on flight altitude, the altitude measurement from GPS or inferred from IMU barometer is used as observation. At altitudes near 25 km ambient pressure falls close to barometer resolution so increases variance in the model is assigned to barometer readings forcing the algorithm to disregard the data. Atmospheric temperature and pressure observation are obtained by data from external sensors although as previously described pressure data need to be disregarded above a certain altitude. The module of total acceleration measured by IMU Triaxial accelerometer is used to estimate vertical accelerations linked to vertical turbulence. [14]

Flight data analysis

The data of two different sounding balloon flight have been analysed. The MINLU flight was launched on July 7th 2020 in an astronomical night condition with no Moon to investigate light polluting sources on ground. AREO flight was launched in day time on October 20th 2021 to provide remote observation with high resolution of crop fields providing calculation of vegetation indexes.

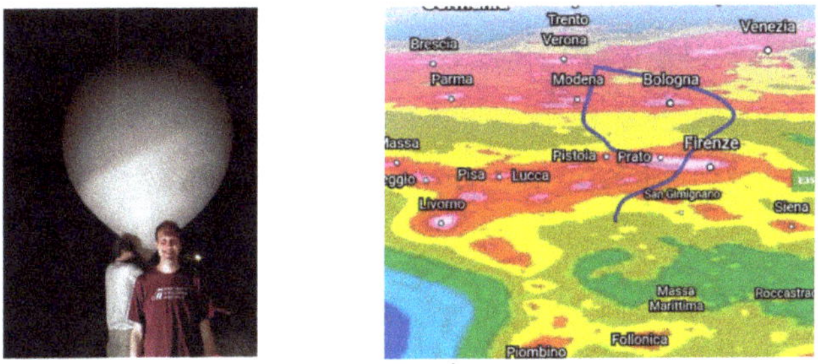

Figure 1 – MINLU night flight in the night of July 7th 2020 - Burst altitude 32104 m (3D trajectory reconstructed on satellite-based ground light pollution map)

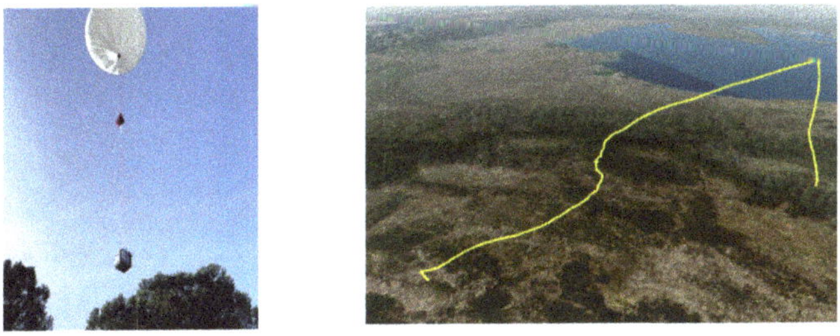

Figure 2 – AREO flight on October 20th 2021 - Burst altitude 36907.9 m (3D trajectory reconstructed on Google Earth)

The elaboration of trajectory profiles and vertical velocities are reported in following figures 3 and 4 showing the ability of LQE model to predict system dynamic to fit real mission data.

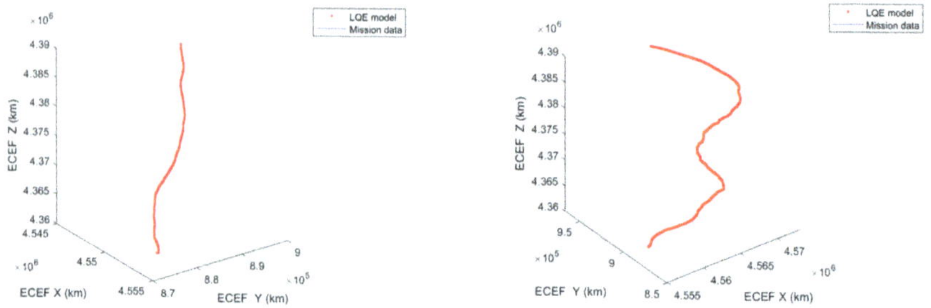

Figure 3 – Reconstructed trajectory from LQE model compared with GPS vertical velocity from mission data (left MINLU flight, right AREO flight)

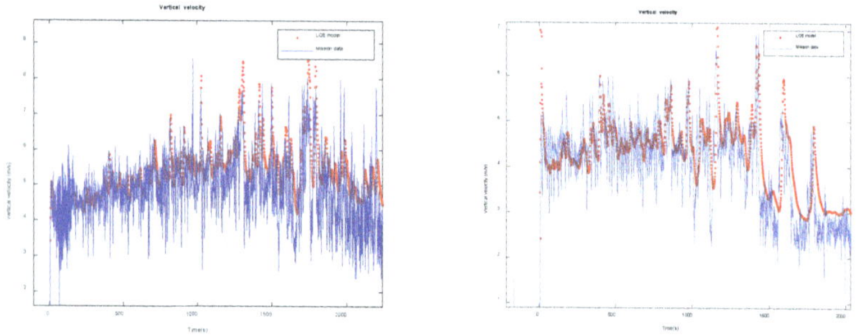

Figure 4 – Vertical ascent velocity from LQE model compared with GPS vertical velocity from mission data (left MINLU flight, right AREO flight)

LQE model is designed to adjust at every time step the parameters with higher uncertainty and no measured mission data (local Wind vector and Balloon drag coefficient Cd) allowing to estimate their time evolution

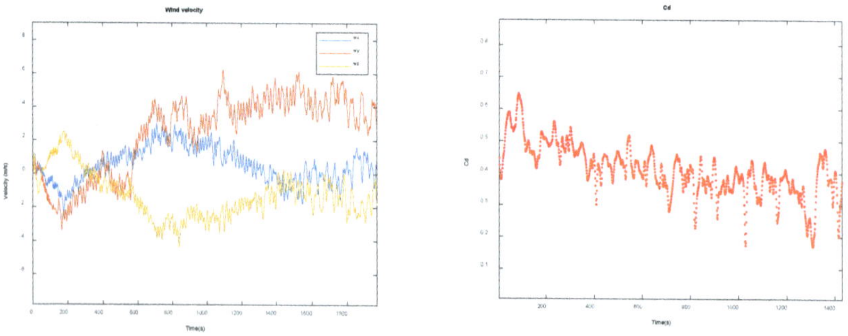

Figure 5 –LQE model estimation of local wind velocity and balloon Cd values (AREO flight)

Conclusions

LQE algorithms for trajectory reconstruction of ascent phase of high-altitude balloons have been developed to increase the accuracy of mission predictors and tested using real mission data. Post flight analysis has been conducted on two balloon-launched Earth observation payloads, one dedicated to the determination of ground sources of Artificial Light at Night, the other to the analysis of vegetation Indexes. Although the presented results are preliminary and further validtions shall be conducted , the algorithm allows to limit sudden changes in trajectory due to temperature variation and wind turbulences and improves the correct correlation of position and pointing direction needed to provide georeferentiation of images acquired during the balloon flights.

References

[1] C. Bettanini, M. Bartolomei, P. Fiorentin, A. Aboudan and S. Cavazzani, "Evaluation of Sources of Artificial Light at Night With an Autonomous Payload in a Sounding Balloon Flight," *IEEE JOURNAL OF SELECTED TOPICS IN APPLIED EARTH OBSERVATIONS AND REMOTE SENSING,* vol. 16, pp. 2318-2326, 2023. https://doi.org/10.1109/JSTARS.2023.3245190

[2] F. Toson, M. Pulice, M. Furiato, M. Pavan, S. Sandon, D. Sandu and R. Giovanni, "Launch of an Innovative Air Pollutant Sampler up to 27,000 Metres Using a Stratospheric Balloon," *Aerotecnica Missili & Spazio,* vol. 102, no. 2, pp. 127-138, June 2023. https://doi.org/10.1007/s42496-023-00151-y

[3] L. W. Renegar, "A Survey of Current Balloon Trajectory Prediction Technology". https://via.library.depaul.edu/cgi/viewcontent.cgi?referer=&httpsredir=1&article=1125&context=ahac

[4] A. Gallice, F.G. Wienhold, C.R. Hoyle, F. Immler and T. Peter "Modeling the ascent of sounding balloons: derivation of the vertical air motion", Atmos. Meas. Tech., 4, 2235–2253, 201. https://doi.org/10.5194/amt-4-2235-2011

[5] Luke W. Renegar "Development of a Probabilistic Trajectory Model for High-Altitude Scientific Balloons", University of Maryland, College Park

[6] H.D. Voss, N.A. Ramm and J. Dailey "Understanding High-Altitude Balloon Flight Fundamentals" , Proceedings of 3rd Annual Academic High-Altitude Conference

[7] C. Olsson,"Sensitivity Analysis for Ascending Zero Pressure Balloons", Master thesis ,2019, Luleå tekniska universitet

[8] Q. Dai, X. Fang , X. Li, L. Tian, "Performance simulation of high altitude scientific balloons", Advances in Space Research 49 (2012), 1045–1052. https://doi.org/10.1016/j.asr.2011.12.026

[9] F. Kreith , J.F. Kreider "Numerical Prediction of the Performance of High Altitude Balloons", National Center for Atmospheric Research, Technical Note NCAR-TN/STR-65

[10] J. Söder, M. Gerding, A. Schneider, A. Dörnbrack, H. Wilms, J. Wagner and F. Lübken "Evaluation of wake influence on high-resolution balloon-sonde measurements", Atmos. Meas. Tech., 12, 4191–4210, 2019. https://doi.org/10.5194/amt-12-4191-2019

[11] F.P. Camara, "Flight Dynamics of a High-Altitude Balloon," Master Thesis , 2018, Universidad Carlos III, Madrid

[12] M. Tuveri, A. Ceruti, "Added masses computation for unconventional airships and aerostats through geometric shape evaluation and meshing"

[13] W.J. Anderson, G.N. Shah, J. Park "Added Mass of High-Altitude Balloons", JOURNAL OF AIRCRAFT Vol. 32, No. 2, March-April 1995. https://doi.org/10.2514/3.46714

[14] G. J. Marlton "On the development, characterisation and applications of a balloon-borne atmospheric turbulence sensor" , Thesis Work, University of Reading, Department of Meteorology

Italian space agency space transportation activities and programs

Marta Albano[1]*, Rocco Carmine Pellegrini[1], Roberto Bertacin[1], Simone Ciabuschi[1], Simone Illiano[1], Rocco Maria Grillo[1], Enrico Cavallini[1]

[1]Italian Space Agency (ASI), Italy

marta.albano@asi.it

keywords: Launchers, Stratospheric Balloon, Hypersonic Vehicles, Suborbital Flight

Abstract Space transportation systems are the key elements for the space exploitation thought space-based services (for telecommunication, navigation and earth observation) and space exploration. As more government and commercial players show interest in effective and sustainable space transportation systems and services, affordable, regular and resilient transportation systems have become increasingly important for sustainable space services. Italy, thought the activities of the Italian Space Agency, is increasingly investing in the space sector and it confirms to be one of the top players on the international scenario. The paper presents an overview of the activities of the Italian Space Agency in this sector, both through national and European framework.

Introduction

The Italian Space Agency has the role to promote and coordinate the space activities at national and international level. The main space transportation activities and programs will be presented in this paper showing the complexity and the variety of the Italian investments in this sector. The activities cover three main areas. Launchers: Italy thought ASI, as National Space Agency with the aim of maintaining the national leadership in the sector participates to ESA launchers activities, such as VEGA family and develops, thorough ASI national programs, critical subsystems such as liquid and hybrid engines, avionics systems etc. Re-entry vehicles: Italy, throught ASI, is the main contributor in Europe for the development of the first European re-entry and re-usable vehicle that will enable new type of services for institutional and commercial customers through Space Rider, based on the experience gained in IXV program. Sub-orbital, hypersonic and stratospheric flights: with the aim to perform missions, develop a national capacity in the specific sectors promotes collaboration with European entities for balloon borne launch campaigns, participates to European working groups dedicated to suborbital law and activities, works on the realization of a national hypersonic demonstrator.

Launchers

Italy, throught the Italian Space Agency supports the development and exploitation of European launchers through its participation in ESA Programmes. The main support is to the VEGA family launchers (VEGA, VEGA-C and VEGA-E) where the leadership is Italian and the prime contractor is AVIO S.p.A. In particular, based on the successful development and exploitation of VEGA launch system entered into service in 2012, Italy has supported the development of the VEGA-C launcher[13] up to its maiden flight occurred in July 2022 with the successful release in orbit of LARES2 satellite (developed by ASI itself) and six CubeSats. After the successful maiden flight, Italy is now supporting the return to flight of VEGA-C after its failure on its second flight in December 2022, and its ramp-up in the stabilized exploitation phase, through a dedicated program aiming also at providing improved robustness to the Launch System in view of its commercial exploitation. Italy has also supported since long time the future evolution of VEGA, starting with ASI national Programs such as Lyra, fostering the introduction of a larger liquid propulsion upper stage exploiting LOX-CH4 propellant combination substituting Z9 and AVUM. In the frame of Lyra program, the basis for mastering this technology have been set up through a collaboration with Russia leading to the firing test of a 10 ton class engine demonstrator called MIRA [7]. Those achievements have been the basis for introducing within the ESA programmes the VEGA Evolution Preparatory Program aimed at developing VEGA-E that, starting form VEGA-C configuration, will substitute Z9 and AVUM+ with a new upper stage making use of M10 engine, developed in the frame of the same program. M10 [12] first development model (DM1) has been already successfully fire tested at engine level in the SPTF test facility located in Sardinia in 2022, while the second development model (DM2) will undergo a complete firing test campaign in the coming weeks. Italy has also a leading role in the P120C Solid Rocket Motor development program which constitutes a common element for the future VEGA and ARIANE launcher families, guaranteeing the synergy between the two launchers. P120C, indeed is used as first stage on VEGA-C and as strap-on booster on Ariane 6 in its version with 2 or 4 boosters. This will also allow higher production volume and cadence for the P120C production line, contributing to lowering the exploitation costs though an economy of scale and building blocks approach. In order to improve ARIANE 6 performances to allow the deployment of large constellations a further development of this SRM has been started in 2021 and then confirmed at the last Ministerial Conference 2022, for a more powerful version of the rocket called P120C+ (or P160C) with an extended length allowing to load a higher mass of solid propellant, increasing the overall total impulse of the motor.

Figure 1 First VEGA C -Lares 2 mission -courtesy of ESA.

In the logic of the P120C, the P120C+ (or P160) will maintain its characteristic of common building block of Ariane and Vega European family of launchers, increasing also the performance of Vega launchers, VEGA-C and VEGA-E. ASI conceived within the National PNRR and today supports the implementation of the so-called PNRR-STS project, funded by EU in the framework of the Italian PNRR. The project will be dedicated to the technology pushing towards the new generation of liquid propellant LOX-CH4 launchers, through the in flight testing and demonstration of the innovative LoX-methane propulsion, based on M10 engine developed in the frame of VEGA-E Program, and the development of critical and enabling technologies. Those critical technologies consist of composite material cryogenic tanks, non-pyrotechnic separation systems and integrated avionics systems based on low-cost hardware. In addition, in the frame of the National PNRR funds, the development of a larger thrust class liquid rocket engine with respect to M10 has been started in the frame of an optional ESA Programme fully funded by Italy, so called Hight-Thrust Engine. This High thrust Engine (HTE), in the class of thrust and performance complementary and synergic with M10 and with Prometheus will constitute the base for future building blocks to be used in the next generation of European launchers fully based on Lox-Methane propulsion.

Avionic systems for Launchers

The development of future generation launchers is strongly characterized by the efficiency of the service offered both in terms of performance and competitiveness. This requires technological innovations aimed at improve the services provided on the market. The growing amount of data to be managed on board requires new concepts of avionic architectures. To innovate these systems, worldwide, various actions are undertaken such as the improvement of the hardware with the development of new on-board computers, the creation of new architectures such as for example the modular approaches or the implementation of new procedures aimed at streamline integration and ground testing activities with the use, for example, of wireless technologies or real-time simulations of environmental stimuli (rotation of the earth, gravity, sloshing of propellant tanks, temperature and thrust, weather conditions, etc.) . It should also be considered that avionics systems account for approximately 20-25% of the cost of the launch service and they are a fundamental element of innovation for the management of on-board data, launcher safety and state recovery. Therefore, avionics systems constitute a key element in order to place the launch service on the market in a strategic way. In order to promote the innovation and to consolidate the role of Italy in the launcher sector, avionic systems are in development through national programs. Wireless systems will allow to facilitate the ground procedures and the launcher design. Different GNC systems are in phase of study both for the re-entry of the launcher stages and for optimization of GNC systems itself by the means of data fusion. Neutralization systems are also in development phase in order to assure the safety of the flights. Communication systems based on Flexible TTEthernet are in early stage of development in order to lower the costs guaranteeing high reliability of those systems.[1]

Green Innovative propulsion and Technology Developements

Funded by EU in the framework of the Italian PNRR, ASI has contracted in April 2023 a major project to design, develop and on-ground qualify an innovative Multi-Purpose Green Engine (MPGE) to be exploited as "*building-block*" for the future Space Logistics scenarios, as element shared between the Space Transportation roadmap and the In-orbit servicing roadmap whose MPGE is part of. Thanks to an intrinsic versatility, mainly related to its throttability, MPGE will be able either to operate as propulsive module of orbiting platforms for next generation in-orbit services, also in reusable configurations (e.g. Space Rider), and to power the upper stages of VEGA-class launchers.

This kilo-Newton class engine will be designed to operate with an innovative combination of green liquid propellants, characterized by long-term storability. Main drivers of the project will be also the wide use of additive manufacturing processes, to optimize the layout and minimize the overall mass, and of a fast prototyping approach, by use of a large number of test campaign and development/qualification models.

Figure 2 First draft of MPGE.

In the field of Hybrid Propulsion, ASI has set up the PHAEDRA Program (Paraffinic Hybrid Advanced Engine Demonstrator for Rocket Application) [5] with the aim to gather the main Italian competences in the field of hybrid propulsion both in industry and in Universities and Research Centers in order to develop a technological demonstrator in relevant scale to be fire tested at the beginning of next year. The Demonstrator will devote particular attention to the re-ignition characteristics and throttling capabilities, adopting paraffin-based fuel in solid form as a single port grain and oxygen peroxide (HTP) as liquid oxidizer. ASI has also funded research activities on solar sails propulsion with the aim to create highly specialized skills in various technological areas relating to the main issues of solar photonic propulsion such as mission analysis, materials development for membrane substrate and reflecting coating depositions, deployment structures (telescopic booms or memory form material booms) and attitude control system relying on photochromic materials. The activities on solar sails technologies are carried out in collaboration with some university departments and with specialized laboratories at Italian research institutes[3]. As a result of the activities carried out, material production processes have been set up, several samples at laboratory scale have been produced and characterized, both for the membrane substrate, reflecting coating optical properties, and boom subscale prototypes [6]. ASI participated also to project called “MAGIC”, funded by the Regione Lazio, which has developed enabling technologies aimed at the full industrialization of additive manufacturing processes (Additive Layer Manufacturing - ALM) of alloys and superalloys of nickel and copper for the production of aerospace components, complementary joining technologies, with a particular focus on laser welding, and certification and qualification protocols through Non-Destructive Testing (NDT). In particular components for the propulsion systems have been developed. ASI has supported the development of technologies for propulsion systems such as the study carried out for segmented solid rocket motor cases using composite materials culminated with the mechanical characterization tests of a sub-scale model applying this technology that can be of interest also for the application on inter-stage joints manufactured in composite materials with the aim to guarantee a more reliable and lightweight overall launch system.

Re-entry vehicles and technologies

Re-entry environment is the upmost challenging condition for a space vehicle. ASI has promoted over the years the development of technologies and system competences for re-entry. The national programs allowed Italy to have one of the most relevant European competence in this sector. Italy, thought ASI, funded also ESA programs such as IXV, which successfully flew in 2015 and now is supporting with a primary role Space Rider.

The overarching objective of the Space Rider program is to develop a reusable, fully autonomous European space transportation system service, providing a European independent capability to routinely access and return from orbit with precise landing capability and a space vehicle serving as a platform for experimentation, in-orbit demonstration and validation with application missions in Low Earth Orbit. The Space Rider program is capitalizing the IXV experience and technological baseline qualified in flight. The Space Rider system is composed of two modules: the Re-entry Module (RM) and the Orbital Module (OM). The RM, being developed under the Design Authority of TAS-I, is a lifting body and it will be designed to embark experiments and payload inside a cargo bay and to return to Earth for landing and re-flight. The OM, under the Design Authority of AVIO, consists of the ALEK (AVUM Life Extension Kit) and acts as service module during the orbital phase of the system. The overall system is completed by a ground segment under the responsibility of ALTEC and TELESPAZIO and a landing site. The system has in orbit lifetime for a minimum duration of 2 months with the following applications: Micro-gravity experimentation laboratory, In-Orbit Demonstration and Validation for a wide range of technologies (e.g. Earth monitoring, satellite inspection, etc.). The vehicle shall be able to perform a high precision soft landing on ground on the landing site. The Space Rider system shall be designed to fly for 6 times. [10][11]

Figure 3 A possible configuration of Space Rider [2]

Suborbital, hypersonic and stratospheric flights

Complementary to classical space transportation topics, the Italian Space Agency funds and promotes national technological and research projects in the field of sub-orbital flights, meant as non-orbital, including support of their mid-stream component i.e. the spaceports, stratospheric platforms and suborbital-hypersonic vehicles. Stratospheric flights are operated by the means of platforms such as balloons at variable altitudes up to about 40km. ASI promoted these activities by its participation to the European project Hemera which developed technologies for the flight train and offered flights to the scientific community. ASI is continuing to keep in contact with the main European actors which participated also to the Hemera project[9]. Agreements and possible future European projects are ongoing. Suborbital flights have become over the years an interesting opportunity for both new space transportation services and research and development activities and a viable option for access to Space of nanosatellite systems. A suborbital flight usually does not exceed 100 km in height and will not orbit the Earth. In this sector ASI participate to discussions on the regulation of these flights. together with ENAC and to the European Commercial Spaceport Forum (ECSF), which brings together several European countries active in light spacecraft launch and suborbital flights. Hypersonic flights are performed by vehicles able to fly at Mach over 5. These flights conditions are typical of re-entry vehicles (winged such as the space shuttle or the aforementioned Space Rider or not winged such as Carina, ARD etc.) and of airbreathing cruise and acceleration vehicles (such as Sanger [4]). The last category, named also as CAV, is objective of many European studies in the recent years. Based on the high experience cumulated on these studies, the Italian Space Agency promoted a dedicated national agreement with CIRA in order to develop a small-scale propelled demonstrator. The project will develop the system and the necessary technologies focused on the propulsion[8].

Conclusions

Space transportation vehicles and technologies are of increasing interests at European and international level. ASI supports the development of the national capacities in order to maintain the strategic position of Italy in this field and to support though its national and ESA programs the European strategic independent access to space for Europe Space Program, both with its role in the frame of the ESA European launchers family and their exploitation and evolution in the European cooperation framework, and through R&D activities through its national programs. In this paper, a general overview of the main programs carried out at national and ESA level has been provided for the main lines of developments in the fields of Space Transportation, including Launchers, re-entry vehicles, suborbital, hypersonic and stratospheric flight and related technologies.

References

[1] Flexible Time Triggered Ethernet: A Cost Efficient COTS-Based Technology for the Development of Launcher Networks "2023 IEEE INTERNATIONAL WORKSHOP ON Metrology for AeroSpace, Milano (Italy)

[2] https://www.esa.int/Enabling_Support/Space_Transportation/Space_Rider_overview

[3] Piano Triennale delle Attività dell'Agenzia Spaziale Italiana 2022-2024

[4] Sanger II, A hypersonic flight and space transportation system, ICAS, 1988

[5] A. Reina, M.L. Frezzotti, G. Mangioni, A. Cretella, F. Battista, C. Paravan, F. Nasuti, D. Pavarin, R. C. Pellegrini, E. Cavallini, Hybrid Propulsion System for future rocket applications, Space Propulsion Conference 2022,

[6] Giovanni Vulpetti, Christian Circi, Rocco Pellegrini, Enrico Cavallini, Sailcraft Helianthus: a Solar-Photon Sail for Geostorm Early Warning, 6th International Symposium on Space Sailing (ISSS 2023)

[7] M. Rudnykh, S. Carapellese, D. Liuzzi, L. Arione, G. Caggiano, P. Bellomi, E. D'Aversa, R. Pellegrini, S. D. Lobov, A. A. Gurtovoy, V. S. Rachuk Development of LM10-MIRA LOX/LNG expander cycle demonstrator engine, Acta Astronautica 126(2016)364–374. https://doi.org/10.1016/j.actaastro.2016.04.018

[8] Sara Di Benedetto, Marco Marini, Pietro Roncioni, Antonio Vitale, P. Vernillo, Salvatore Cardone, Marta Albano, Roberto Bertacin, "Design of the Scramjet Hypersonic Experimental Vehicle", Eucass 2023

[9] HEMERA: a European Stratospheric Balloon Research Infrastructure, Raizonville P., S. Payan, K. Dannenberg, D. Hagsved, L. Stephane, P. Ubertini, K. Pfeilsticker, F.Vallon, M. Albano, IAC-21,A7,1,5,x66353, 25-29 October 2021, Dubai

[10] Italian contribution to the esa ministerial conference 2016: next generation of the european vega launcher for new green and reusable space missions, IAC-17,D2,IP,22,x39093, A.Gabrielli; Mr. A. Cramarossa; E. D'Aversa; S. Ianelli; R.C. Pellegrini; M. Albano; IAC 2017

[11] A. Fedele, G. Guidotti, G. Rufolo, G.Tumino et al, The Space Rider Programme: End user's needs and payload applications survey as driver for mission and system definition, Acta Astronautica, volume 152, nov 2018 pp. 534-541. https://doi.org/10.1016/j.actaastro.2018.08.042

[12] D. Kajon, D. Liuzzi, C. Boffa, N.Ierardo et al., Development of the liquid oxygen and methane M10 rocket engine for the Vega-E upper stage, Eucass 2019

[13] https://www.arianespace.com/vega-c/

Materials and Aerospace Structures

Nonlinear mechanical analysis of aerospace shell structures through the discontinuous Galerkin method

Giuliano Guarino[1,a *] and Alberto Milazzo[1,b,*]

[1]Department of Engineering, University of Palermo, Viale delle Scienze, Bld. 8, 90128, Palermo, Italy

[a]giuliano.guarino@unipa.it, [b]alberto.milazzo@unipa.it

Keywords: Multilayered Shells, Nonlinear Structural Behavior, Discontinuous Galerkin Method, High-Order Modelling

Abstract. The geometrically non-linear mechanical response of multilayer composite shells is addressed via an innovative discontinuous Galerkin formulation. In the framework of the Carrera Unified formulation, equivalent single layer kinematics with different through-the-thickness accuracy is adopted. The variational statement governing the shell nonlinear behavior is derived. The corresponding governing equations are solved via a discontinuous Galerkin approach, which employs the pure penalty method to weakly enforce the connection between the mesh elements. Numerical tests are presented to show the capabilities of the proposed approach.

Introduction

Multilayered composite shells are extensively employed as high-performance lightweight components in aerospace engineering. In advanced applications they may undergo large displacements, requiring non-linear analysis to characterize accurately their behavior. In this framework a fundamental role is played by the modelling and analysis of these structure that needs to be carried out with appropriate fidelity and cost effectiveness.

From the modelling point of view, besides fully three-dimensional models, shell structures are studied within the context of the so-called shell theories which can be classified into equivalent single layer (ESL) theories and layer-wise (LW) theories. Due to their complexity, the solution of shell theories models commonly requires the employment of numerical methods. The most common approach in the literature is the finite element method (e.g. Ref. [1]); the Ritz method [2], mesh-less solutions [3, 4] and the isogeometric analysis approach [5] have been also proposed.

Recently, the discontinuous Galerkin (dG) method emerged as a viable alternative showing interesting advantages in the use of nonstandard element and shape functions, in the application of non conformal meshes as well as high-order elements, in the implementation of meshing strategies such as hierarchical refinement and adaptativity and in scalable implementations [6, 7]. These features can underlie a robust treatment of complex problems as those involved in the analysis of multilayered shells [8, 9]. This motivate the present work in which high-order, equivalent single layer shell theories are solved through a pure penalty discontinuous Galerkin approach formulated accounting for geometrical nonlinear behavior [10].

Shell model and governing equations

In the context of a total Lagrangian approach , the shell deformation is described in terms of the displacement components expanded as [8, 9]

$$u_{\xi_i}(\xi_1,\xi_2,\xi_3) = \sum_{k=0}^{N_i} Z_k^i(\xi_3) U_{ik}(\xi_1,\xi_2) \tag{1}$$

where ξ_m are the curvilinear coordinates used to describe the shell geometry, N_i is the order of the assumed expansion, $Z_k^i(\xi_3)$ is the prescribed k-th function of the expansion and $U_{ik}(\xi_1,\xi_2)$ are the unknown generalized displacements. In Eq. (1) the N_i are considered as parameters whose different values allow to build different order ESL shell theories. The shell theory corresponding to the expansion orders N_1 , N_2 and N_3 is denoted as $ED_{N_1N_2N_3}$.

To develop the proposed dG formulation, the shell reference domain Ω_ξ is partitioned into N_e elements Ω_ξ^e over which the generalized displacements $U_{ik}(\xi_1, \xi_2)$ are approximated via polynomial basis function.

Following Ref. [10] one obtains the following variational statement:

$$\sum_{e=1}^{N_e} \int_{\Omega_\xi^e} \left[\frac{\partial \boldsymbol{V}^T}{\partial \xi_\alpha} \left(\boldsymbol{Q}_{\alpha\beta} \frac{\partial \boldsymbol{U}_h}{\partial \xi_\beta} + \boldsymbol{R}_{\alpha 3} \boldsymbol{U}_h \right) + \boldsymbol{V}^T \left(\boldsymbol{R}_{3\alpha} \frac{\partial \boldsymbol{U}_h}{\partial \xi_\alpha} + \boldsymbol{S}_{33} \boldsymbol{U}_h \right) \right] d\,\Omega_\xi$$
$$+ \sum_{i=1}^{N_i} \int_{\partial\Omega_{\xi I}^i} \mu [\![\boldsymbol{V}]\!]_\alpha^T [\![\boldsymbol{U}_h]\!]_\alpha d\partial\Omega_\xi + \sum_{e=1}^{N_e} \int_{\partial\Omega_{\xi D}^e} \boldsymbol{V}^T \boldsymbol{U}_h d\,\partial\Omega_\xi \tag{2}$$
$$= \sum_{e=1}^{N_e} \int_{\Omega_\xi^e} \boldsymbol{V}^T \overline{\boldsymbol{B}} d\,\Omega_\xi + \sum_{e=1}^{N_e} \int_{\partial\Omega_{\xi N}^e} \boldsymbol{V}^T \overline{\boldsymbol{T}} d\,\partial\Omega_\xi + \sum_{e=1}^{N_e} \int_{\Omega_{\xi D}^e} \mu \boldsymbol{V}^T \overline{\boldsymbol{U}} d\,\partial\Omega_\xi$$

where the Einstein's summation is assumed for $\alpha, \beta = 1, 2$, $\boldsymbol{U}_h$ is the vector containing the dG approximation of the generalized displacements field, $\boldsymbol{V}$ is the vector of the test functions chosen of the same form of the generalized displacements. In Eq (2), $\overline{\boldsymbol{B}}, \overline{\boldsymbol{T}}$ and $\overline{\boldsymbol{U}}$ are the generalized domain forces, the generalized boundary forces and the boundary prescribed generalized displacements, respectively, $\boldsymbol{Q}_{\alpha\beta}, \boldsymbol{R}_{\alpha 3}, \boldsymbol{R}_{3\alpha}$ and $\boldsymbol{S}_{33}$ are the generalized stiffness matrices and $[\![\cdot]\!]$ is the interelement jump operator. Finally, $\partial\Omega_{\xi I}^i$, $\partial\Omega_{\xi D}^e$ *and* $\partial\Omega_{\xi N}^e$ are the inter-element interfaces and the elements portions of boundaries where the essential and natural boundary conditions are enforced, respectively. The definitions of the above-mentioned quantities can be found in Ref. [10], to which the reader is referred for the formulation details.

In the primal form of the proposed pure penalty dG method, namely Eq. (2), μ is the penalty parameter used to enforce the inter-elements continuity of the solution and the essential boundary conditions. The choice of μ is crucial for the method to be efficient as discussed in Refs. [7,10]

Applying the variational calculus procedures, the final nonlinear algebraic system is inferred, which is solved via a Newton-Raphson arc-length scheme. It is remarked that the use of a pure penalty formulation enables to compute the elements interface boundary integrals only once with the consequent computational time savings during the iterative solution scheme.

Numerical results

To illustrate the capabilities of the proposed method some results relative to the nonlinear analysis of cylindrical shells are proposed. The present results are representative of the proposed approach effectiveness that was proved by a lot of numerical tests whose results are not reported here for the sake of brevity.

Fig. 1a shows the geometry, boundary conditions and loads of the cylindrical shell having dimensions L = 254 mm, R = 2540 mm and θ = 0.1 rad and subjected to a central pinch load $\boldsymbol{F}$. Only a quarter of the structure is modelled for symmetry conditions. Three different shell sections have been considered: *i*) a single-layer section of isotropic material (see Table 1) and thickness τ = 6.35mm, which is labeled as C1 case (thin shell); ii) a single-layer section of isotropic material M1 (see Table 1) and thickness τ = 12.7mm, which is labeled as C2 case (moderately thick shell); iii) a three-layer section with [0/90/0] layup of 4.233 mm thick plies having orthotropic properties as M2 material in Table 1, which is labeled as C3 case.

Table 1: Material properties.

	Isotropic M1	Orthotropic M2
Young's modulus E_1	75000 [MPa]	3300 [MPa]
Young's moduli $E_2 = E_3$	75000 [MPa]	1100 [MPa]
Shear Moduli $G_{12} = G_{13} = G_{23}$	28846 [MPa]	660 [MPa]
Poisson's coefficients $\nu_{12} = \nu_{13} = \nu_{23}$	0.3	0.25

Fig. 1b show the transverse displacement of the load application point $u_3$versus the load amplitude *F*. These Equilibrium paths are computed using both the ED_{222} and ED_{333} shell theories and a 2 × 2 mesh grid of elements with polynomial trial and test function order p = 5. The results show that, as expected, snap-back or snap-through behavior occurs depending on the shell thickness ratio. The comparison of the present results with those of Refs [1], [11] and [12] shows very good agreement for both isotropic and multilayered shells, that proves the proposed approach effectiveness.

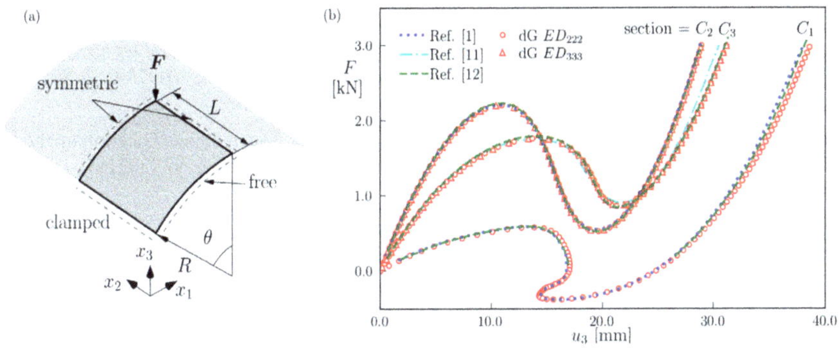

Figure 1: Cylindrical shells. (a) Geometry, boundary conditions and applied load. (b) Nonlinear equilibrium path for different shell sections.

Conclusions

A novel pure penalty discontinuous Galerkin method has been presented for the geometrically nonlinear static analysis of multilayered plates and shells described by refined equivalent single-layer kinematics with generality of the through-the-thickness resolution.

The numerical tests prove the ability of the method to deal with complex nonlinear behavior of shells and evidence very good agreement with literature results.

References

[1] B.Wu, A. Pagani, W. Chen, E. Carrera, Geometrically nonlinear refined shell theories by Carrera Unified Formulation, Mechanics of Advanced Materials and Structures. 28 (2021) 1721–1741. https://doi.org/10.1080/15376494.2019.1702237

[2] A. Milazzo, G. Guarino, V. Gulizzi, Buckling and post-buckling of variable stiffness plates with cutouts by a single-domain Ritz method, Thin-Walled Structures. 182 (2023) 110282. https://doi.org/10.1016/j.tws.2022.110282

[3] M. Fouaidi, A. Hamdaoui, M. Jamal, B. Braikat, A high order mesh-free method for buckling and post-buckling analysis of shells, Engineering Analysis with Boundary Elements. 99 (2019) 89–99. https://doi.org/10.1016/j.enganabound.2018.11.014

[4] S. Hosseini, G. Rahimi, D. Shahgholian-Ghahfarokhi, A meshless collocation method on nonlinear analysis of functionally graded hyperelastic plates using radial basis function, ZAMM - Journal of Applied Mathematics and Mechanics. 102 (2022) 202100216. https://doi.org/10.1002/zamm.202100216

[5] Y. Guo, Z. Zou, M. Ruess, Isogeometric multi-patch analyses for mixed thin shells in the framework of non-linear elasticity, Computer Methods in Applied Mechanics and Engineering, 380 (2021) 113771. https://doi.org/10.1016/j.cma.2021.11377.

[6] D. N. Arnold, F. Brezzi, B. Cockburn, L. D. Marini, Unified analysis of discontinuous Galerkin methods for elliptic problems, SIAM Journal on Numerical Analysis. 39 (2002) 1749–1779. https://doi.org/10.1137/S0036142901384162

[7] V. Gulizzi, I. Benedetti, A. Milazzo, Discontinuous Galerkin methods for solids and structures, in: F. M. H. Aliabadi, W. Soboyejo (Eds.), Comprehensive Structural Integrity, 2nd Edition, vol. 3, Elsevier, Oxford, 2023, pp. 348–377. https://doi.org/10.1016/B978-0-12-822944-6.00024-4

[8] G. Guarino, A. Milazzo, V. Gulizzi, Equivalent-single-layer discontinuous Galerkin methods for static analysis of multilayered shells, Applied Mathematical Modelling, 98 (2021) 701–721. https://doi.org/10.1016/j.apm.2021.05.024

[9] G. Guarino, V. Gulizzi, A. Milazzo, High-fidelity analysis of multilayered shells with cut-outs via the discontinuous Galerkin method, Composite Structures, 276 (2021), 114499. https://doi.org/10.1016/j.compstruct.2021.114499

[10] G. Guarino, A. Milazzo, A discontinuous Galerkin formulation for nonlinear analysis of multilayered shells refined theories, International Journal of Mechanical Sciences, 255 (2023), 108426. https://doi.org/10.1016/j.ijmecsci.2023.108426

[11] E. Carrera, A. Pagani, R. Azzara, R. Augello, Vibration of metallic and composite shells in geometrical nonlinear equilibrium states, Thin-Walled Structures. 157 (2020) 107131. https://doi.org/10.1016/j.tws.2020.107131

[12] K. Sze, X. Liu, S. Lo, Popular benchmark problems for geometric nonlinear analysis of shells, Finite Elements in Analysis and Design. 40 (2004) 1551–1569. https://doi.org/10.1016/j.finel.2003.11.001

Exploring aerospace advancements and global collaborations: a comprehensive analysis of MCAST's aerospace program in Malta

Leonardo Barilaro[1,a*], Lorenzo Olivieri[2,b], Mark Wylie[3,c], Joseph Borg[4,d]

[1] The Malta College of Arts, Science & Technology, Aviation Department, Paola PLA 9032, Malta

[2] CISAS "G.Colombo", University of Padova, Via Venezia 15, 35131 Padova (PD, Italy)

[3]South East Technological University (SETU), Aerospace, Mechanical and Electronic Engineering Department, Carlow campus, Kilkenny Rd, Moanacurragh, Co. Carlow, Ireland

[4] Department of Applied Biomedical Science, Faculty of Health Sciences, University of Malta, Msida MSD, 2080, Malta

[a]leonardo.barilaro@mcast.edu.mt, [b]lorenzo.olivieri@unipd.it, [c]mark.wylie@setu.ie, [d]joseph.j.borg@um.edu.mt

Keywords: Space Debris, Aerospace Structures, Satellites Shields, International Space Station

Abstract. This paper provides an overview of the Aerospace Program at the Malta College of Arts, Science and Technology (MCAST). The program comprises of four main projects that aim to address different challenges in the aerospace industry, in particular in the field of protection of aerospace structures and systems from space debris impacts. The first project focuses on the development of 3D printed Kevlar shields for aerospace applications. The outcome of this project is the development of repair strategies for inflatable manned modules in space and efficient small satellite shields. The second project describes the use of cold-welding phenomenon for spacecraft repair, in collaboration with South East Technological University (SETU), Ireland. The project aims to develop an experimental test rig to apply custom repair patches of different materials to pre-damaged metallic structures and monitor the performance of the adhered joint in low orbit and during re-entry. The third project presents a collaboration between MCAST and the University of Padova to develop a single stage Light-Gas Gun (LGG) impact facility in Malta. Finally, the paper discusses MCAST's participation in Malta's third space bioscience experiment launched to the International Space Station led by the University of Malta. The experiment aims to investigate how microgravity affects the behaviour of foot ulcer microbiomes in Type 2 Diabetes Mellitus patients. The project marks a significant milestone for both MCAST and the University of Malta. The projects presented in this paper reflect MCAST's commitment to contribute to the advancement of the aerospace industry and offer new opportunities for research, development, and commercialization.

Introduction

Over the past years, Malta has emerged as a promising player in the Space Sector, positioning itself as one of the up-and-coming nations in this field. This progress is underscored by Malta's Cooperation Agreement with the European Space Agency (ESA) on in 2012, aimed at establishing a framework for enhanced collaboration in ESA projects [1]. With its strategic location, access to the European Union (EU) market, administration and human capital, Malta presents a very good opportunity for the commercialization of space exploration. Guided by the central government, the Malta Council for Science and Technology (MCST) has taken on the crucial role of coordinating and governing space-related matters in the country since the signing of the Cooperation Agreement. MCST's Space Directorate actively fosters connections with foreign space agencies, the Maltese government, businesses, and educational institutions. Recognizing the potential for space exploration to become a pillar of Malta's economy, the government has outlined its commitment through the National Space Strategy 2022. Drawing inspiration from the success achieved in the aviation sector and incorporating best practices from other countries, this strategy encompasses a wide range of initiatives undertaken in the Maltese space sector. It focuses on upstream and downstream activities and examines the economic possibilities for Malta. Furthermore, it is noteworthy that Malta is set to join the ESA Plan for European Cooperating States (PECS) in 2024, further solidifying its commitment to and integration with the European space community. In light of these developments, this paper aims to present a comprehensive analysis of the aerospace activities carried out at the Malta College of Arts, Science and Technology (MCAST) in Malta, in particular in the field of protection of aerospace structures and systems from space debris impacts. Section 2 will present the development of 3D printed Continuous Aramid Fibers (Kevlar^{R}) based shields for aviation and aerospace applications, Section 3 the use of cold-welding adhesion for spacecraft repairing, Section 4 presents a collaboration to develop a single stage Light-Gas Gun (LGG) impact facility in Malta and Section 5 provides an overview of Maleth 3 mission that was onboard the International Space Station.

Smart ballistic optimization for repairing of aerospace structures using 3d printed Kevlar
The SBORAEK project introduces an innovative concept by proposing the development of shields for aviation and aerospace applications using 3D printed Continuous Aramid Fibers (Kevlar^{R}). In the aviation industry, maintaining the structural integrity of aircraft is crucial for ensuring satisfactory performance [2]. The availability of repair materials and time often plays a significant role in deciding whether a part should be repaired or replaced. Repairable damage by patching refers to damage exceeding the limits that can be fixed by installing splice members to bridge the affected area of a structural component. Regarding aerospace applications and protection against impacts and hull damages, regulations and strategies for mitigation are currently under scrutiny. Active and passive debris removal technologies, as well as post-mission disposal techniques, are being developed. The current state and projected evolution of the debris environment necessitate the exploration and development of shields to protect active spacecraft. In crewed International Space Station (ISS) modules, mitigation techniques include reinforced hulls and shields to reduce the risk of hull perforation and subsequent depressurization [3]. Avoidance maneuvers are also performed for detected objects above a certain size threshold that may collide with the ISS. In contrast, unmanned systems utilize simpler and more cost-effective structures such as sandwich panels to mitigate the effects of space debris impacts. However, there is ongoing research to determine the best strategy for repairing damaged spacecraft hulls and shields while in orbit. In recent years, innovative processes like additive manufacturing have found applications in the satellite industry, enabling solutions that would have been challenging or expensive to implement using traditional manufacturing techniques. For instance, within the scope of the EU H2020 ReDSHIFT project [4], new 3D printed shields were developed for the protection of microsatellites. The SBORAEK will allow to go further, this will be the first time 3D printed Kevlar structures with optimized geometry will be applied to aviation and aerospace. There is currently no system targeting the same objectives of this project. SBORAEK, Figure 1, is however made from a number of subsystems, whose technology levels are mature enough so that they can be used and integrated together. The technology used for the SBORAEK project is divided into three segments: 3D printing, optimization algorithms and design techniques for high-energy impact conditions.

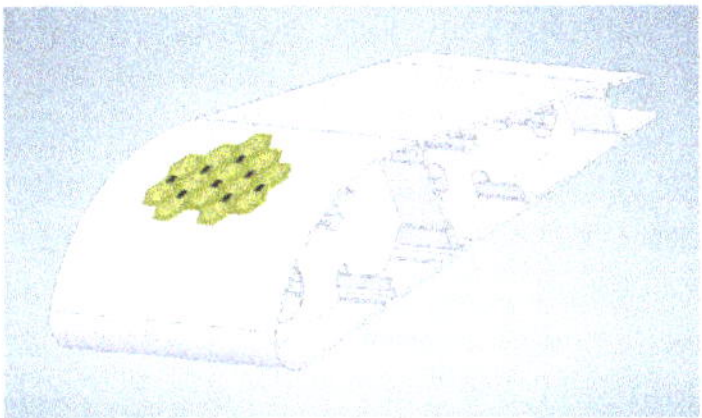

Fig 1: SBORAEK applications - Aviation structures (left), prototype spacecraft shields (right)

Cold welding adhesion for spacecraft repair

This research, in collaboration with South East Technological University (SETU), aims to explore the intentional cold-welding of metals as a method for spacecraft hull repair. This approach involves applying a repair patch to a perforated metal sample using an apparatus capable of applying combined axial and tangential forces [5]. Experimental evidence has shown that, in a near vacuum environment, cold-welding using this technique can be achieved with limited loads below 100 N. To validate this technique experimentally, it is necessary to recreate low orbit gravity and pressure conditions. The research plan consists of two phases: terrestrial experimentation and the subsequent development of an experimental payload to be tested in space. The terrestrial experimentation phase, Figure 2, establishes the primary system requirements, including perforation diameter and crater specifications, selection of material candidates for repair patches, mechanical interface design, actuation method (rotational, translational, or a combination), and applied forces. Secondary requirements encompass power needs and sensor specifications such as pressure transducers, load cells, and cameras. Currently, three non-ferrous materials are being considered as candidates: high ductility Indium foil (1-3 mm) shaped into a step tapered plug rivet, austenitic steels with a high Nickel content (Stainless Steel SS17-7PH plate), and a commonly used aerospace-grade Aluminum Alloy (AL AA-2024).

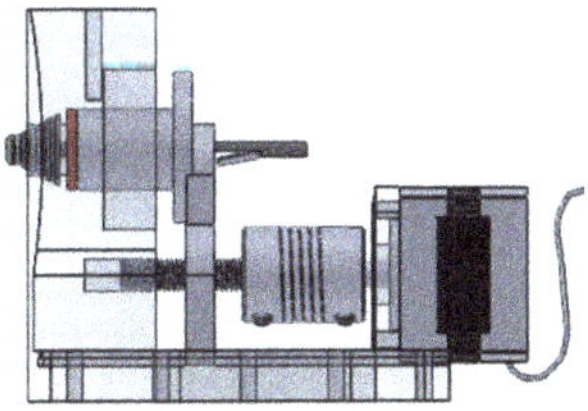

Fig 2: Test rig concept (left) and test rig prototype (right)

Single-stage Light-Gas Gun for high-velocity impacts

MCAST started a collaboration with the Center of Studies and Activities for Space (CISAS) at the University of Padova in Italy for the design of a hypervelocity research facility [6]. Considerable attention is directed towards essential components, such as the launch package and sabot-stopping system, in the construction of a hypervelocity impact facility. The launch package safeguards the projectile during acceleration, while the sabot-stopping system is responsible for terminating the sabot. Typically, Light Gas Guns (LGGs) employ an expandable sabot system to encase the projectile. Subsequently, following the acceleration phase, the sabot separates from the projectile, fragmenting to prevent contamination of the target. However, this approach can inflict damage on the sabot-stopping system, necessitating either complete replacement or time-consuming maintenance. In order to establish a test facility capable of conducting high-velocity impact tests for the aviation and space sectors, a range of projectile velocities has been selected. This range is suitable for testing the impact of aircraft components as well as simulating low-speed impacts akin to those experienced in GEO orbit. The LGG is designed to primarily replicate impacts caused by metal impactors, particularly aluminum alloys. Its modular configuration can also be adapted to accommodate other types of projectiles, simulating space debris such as plastics and silica materials.

Maleth 3

The first-time collaboration for a space project between MCAST and the University of Malta, took place for the launch of the space bioscience experiment 'Maleth 3' on March 15th, 2023. The experiment was part of a commercial re-supply mission (CRS27) by SpaceX to the International Space Station (ISS). The objective of the experiment, titled "Microgravity effects on microbiome studies of Diabetic Patients", was to explore how microgravity influences the behavior of foot ulcer microbiomes, a major complication prevalent among patients with Type 2 Diabetes Mellitus, often requiring amputation surgeries [7]. The project entails a collaborative effort between also other institutions facilitated by Spaceomix in conjunction with Space Applications Services from Brussels, Belgium; Weill Cornell Medicine, New York, USA; King Faisal Specialist Hospital & Research Centre, Jeddah, Saudi Arabia; Metavisionaries based in Oxford, UK; and the Mohammed Bin Rashid Space Centre (MBRSC), whose astronaut is part of NASA's Crew-6 and handled Maleth 3 in space, Figure 3. The outcomes of this experiment have the potential to yield breakthroughs in life science research and treatment methods.

Fig 3: Maleth 3 experiment

Conclusions

This paper presented the Aerospace Program at MCAST, focusing on the protection of aerospace structures and systems from space debris impacts. The program consists of four main projects: 3D printed Kevlar shields, cold-welding for spacecraft repair, a single-stage Light-Gas Gun impact facility, and MCAST's participation in the Maleth 3 space bioscience experiment. The development of 3D printed Kevlar shields enables repair strategies for inflatable manned modules and efficient small satellite shields. Cold-welding adhesion shows promise as a method for spacecraft hull repair, with experiments conducted under near vacuum conditions. Collaboration with the University of Padova is designing a single-stage Light-Gas Gun facility for high-velocity impact testing in the aviation and space sectors. MCAST's collaboration with the University of Malta resulted in the successful launch of the Maleth 3 experiment, investigating the effects of microgravity on foot ulcer microbiomes in Type 2 Diabetes Mellitus patients. These projects reflect MCAST's commitment to advancing the aerospace industry, contributing to research, development, and commercialization.

Acknowledgments
The work described in this paper was partially carried out as part of the SBORAEK (R&I-2022-002L) project which was financed by the Malta Council for Science & Technology, for and on behalf of the Foundation for Science and Technology, through the FUSION – Technology Development Programme LITE.

References

[1] Malta Space strategy: https://maltaspace.com, last visited: 18th May 2023

[2] Roach D., Rackow K. "Development and Validation of Bonded Composite Doubler Repairs for Commercial Aircraft". Aircraft Sustainment and Repair (2018) pp.545-743. https://doi.org/10.1016/B978-0-08-100540-8.00011-X

[3] Barilaro L., Francesconi A. et al. "Impact damage and ballistic limit equations for flexible multilayer meteoroid and debris protection shields. ARA Congress (2010)

[4] Rossi A. et al. "The H2020 ReDSHIFT project: a successful European effort towards space debris mitigation". 70th International Astronautical Congress (2019)

[5] L. Barilaro, M. Wylie, L. Olivieri, Cold welding adhesion for spacecraft repair: Experiment design and roadmap, Acta Astronautica, Vol. 210 (2023), pag 511-517. https://doi.org/10.1016/j.actaastro.2023.04.011

[6] L. Barilaro, L. Olivieri, R. Tiscio, A. Francesconi, Evaluation of a single-stage Light-Gas Gun facility in Malta: Business analysis and preliminary design, Aerotecnica Missili & Spazio 101(3) (2022). https://doi.org/10.1007/s42496-022-00113-w

[7] C. Gatt et al., The Maleth Program: Malta's first space mission discoveries on the microbiome of diabetic foot ulcers, Heliyon 8(639396) (2022). https://doi.org/10.1016/j.heliyon.2022.e12075

Comparison of lattice core topologies in sandwich structures

G. Mantegna*[1], C.R. Vindigni[1], D. Tumino[1], C. Orlando[1], A. Alaimo[1]

[1]Kore University of Enna, Faculty of Engineering and Architecture, 94100 Enna, Italy

*giuseppe.mantegna@unikore.it

Keywords: Additive Manufacturing, Design for Manufacturing, Lattice Structures, Hybrid Structures

Abstract. Hybrid sandwich structures are often used in the aviation industry thanks to their high strength-to-weight ratio and resistance to bending and buckling. Today, through Additive Manufacturing technologies, it is possible to use different materials to create topology-optimized structures with complex shapes using lattice structures. In this work, a numerical approach is proposed to study the behaviour of a hybrid sandwich structure which can be used as a reinforcement for a control surface of a lightweight aircraft. A comparative analysis is conducted between a conventional honeycomb lattice core and lattice truss core structures.

Introduction

The use of sandwich structures in the aeronautics field has gained significant attention due to their high stiffness-to-weight ratio and buckling loads [1]. However, the reliability of the bond between the core and face-sheets remains a challenge, as adhesive bonding can limit the strength of the sandwich panel [2]. Additive Manufacturing (AM) allows a direct connection between the core structure and face sheets. Furthermore, thanks to its free-form tailoring ability, it is possible to create a topology-optimized core through the use of lattice cells. Multiple studies have demonstrated the structural advantages of truss core sandwich panels, which, owing to their open geometric configuration, can also be employed for multifunctional purposes. These types of structures have shown remarkable structural performance against bending and compression loads, as demonstrated by Wicks and Hutchinson [3]. Additionally, truss core sandwich panels can be used in heat transfer, such as anti-icing systems [4], and have proven effective in damping vibrations [5] and for impact absorption.

In this paper, the mechanical properties of three different lattice structures, namely lattice honeycomb, truss lattice Body Centred Cubic BCC and sine-Waved truss lattice Body Centred Cubic WBCC [6] are investigated. These lattice structures are evaluated in the context of an asymmetric sandwich core, which is intended for application in a new-generation tilt rotor control surface.

Lattice core homogenisation

In order to reduce the computational costs on the macro scale numerical analyses, the equivalent homogenised properties of the three selected lattice structures, shown in Figure 1, have been determined. All three specimens are manufactured through a standard Aluminium alloy, $E_{Al} = 71.0\ GPa$, $\nu = 0.33$ and have the same specific density $\rho^* = Volume_{cell}/Volume_{box} = 0.094$. The solid size length $L = 10\ mm$ is chosen, while the other geometric parameters are accordingly selected in order to guarantee the desired density.

Materials Research Proceedings 37 (2023) 296-299 https://doi.org/10.21741/9781644902813-64

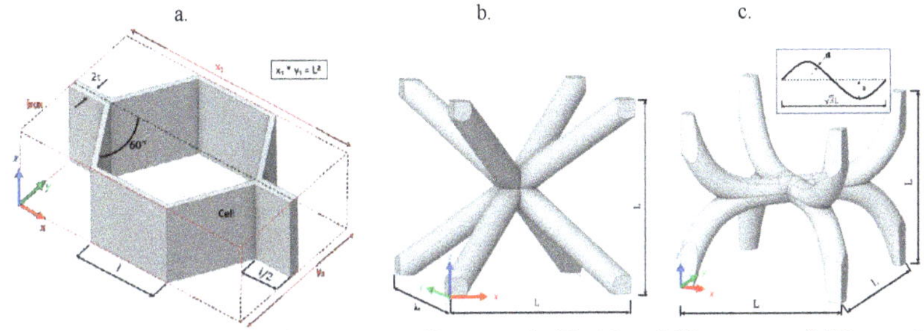

Figure 1 – Lattice cells configuration. a. Honeycomb (Hex) $l = 4.38\ mm, t = 0.29\ mm$, *b. Body Centred Cubic (BCC)* $d = 1.4\ mm, a = 0\ mm$, *c. Waved Body Centred Cubic (WBCC)* $d = 1.5\ mm, a = 2.0\ mm$.

The homogenised properties of the Hex cell are retrieved analytically from the work of Kumar et. al. [7], while an ad-hoc routine on Ansys Parametric Design Language (APDL) was developed for the BCC and WBCC lattice cells. More specifically, each node on the lateral faces is paired with its respective node on the opposite face through constraint equations:

$$u_{i^-} - u_{i^+} = \Delta u_p \tag{1}$$

with: u displacement along the x, y and z directions, i^- and i^+ nodes on two opposite faces sharing the same relative position. Δu_p is the difference in displacement of two pilot points chosen on the two faces.

The homogenised mechanical properties of each cell are reported in Table 1. It can be noted that the Hex and WBCC, whose material direction 3 is aligned with axis z in Figure 1, have an orthotropic behaviour.

Table 1 - Lattice cells homogenised mechanical properties.

	E_1 [Pa]	E_2 [Pa]	E_3 [Pa]	v_{12}	v_{23}	v_{13}	G_{12} [Pa]	G_{23} [Pa]	G_{13} [Pa]
Hex	4.48E+07	5.10E+07	6.70E+09	9.24E-01	2.51E-03	2.21E-03	3.43E+07	1.05E+09	1.43E+09
BCC	1.11E+08	1.11E+08	1.11E+08	4.82E-01	4.82E-01	4.82E-01	9.08E+08	9.08E+08	9.08E+08
WBCC	1.61E+08	1.61E+08	1.11E+09	8.24E-01	7.46E-02	7.46E-02	8.51E+08	9.26E+07	9.26E+07

Finite element analysis comparison

To compare the mechanical properties of the three chosen lattice cores in asymmetric sandwich panels, a control surface of a new-generation Tiltrotor, illustrated in Figure 2.a, is selected. A schematic representation of the asymmetric sandwich panels placed on both the upper and lower skins is also depicted. The whole structure is modelled with two-dimensional surface elements using Ansys Composite PrepPost (ACP) module. The composite stabilizing and working skins use Epoxy-Carbon woven prepreg plies with the following properties: $E_{1,2} = 61.3\ GPa, E_3 = 6.9\ GPa, v_{12} = 0.04, v_{23,13} = 0.3$ and $G_{12} = 3.3\ GPa, G_{23,13} = 2.7\ GPa$.

In this preliminary study stage, a uniform pressure is applied to both the upper and lower skins, while displacement constraints are imposed as additional boundary conditions in accordance with the real aircraft model. The contour map deformation along the z-direction for the control surface with a honeycomb homogenised core is reported in Figure 2.b highlighting the most deformed region on the lower skin surface in accordance with the superimposed uniform load.

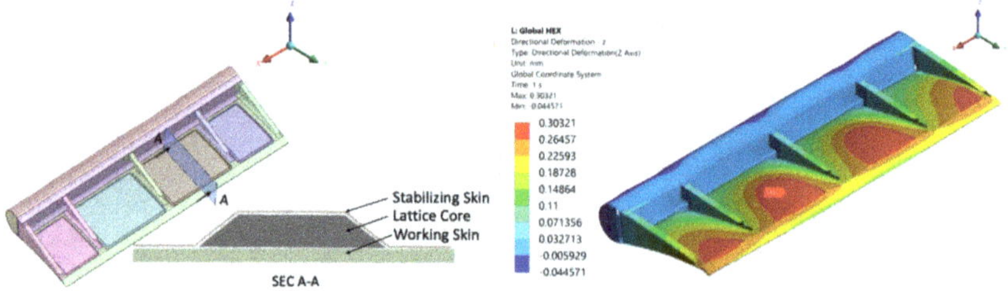

Figure 2 – a. Tiltrotor Control Surface: Schematic representation of asymmetric sandwich panel section view (Upper skin hidden). b. Hex-Core structure directional deformation [mm], z-direction (Upper skin hidden).

As shown in Table 1, the mechanical properties of the BCC cell are equivalent along the three cartesian directions; on the other hand, the WBCC, presents a preferred load direction along the three cartesian directions. For this reason, a second configuration WBCC2 is considered rotating the WBCC cell so that the material 3rd direction is parallel with global y-direction.

The z-direction displacements are investigated. Indeed, a global analysis of the homogenized core loses information on stress distribution in the lattice cell struts and the contact areas with the stabilizing and working skins. As a representative result in terms of stress, since no relevant difference appears in the four cases, Figure 3 presents the maximum stress values (σ_1) in the x-direction of the upper skin's 1st ply for the Hex configuration. No significant influence is appreciable on the σ_1 stress values at the observed layer. Figure 4 and Figure 5 show the z-direction deformation for upper and lower skins along the control surface span and chord. Overall, the Hex topology appears to be the more rigid solution, while WBCC1 is the least.

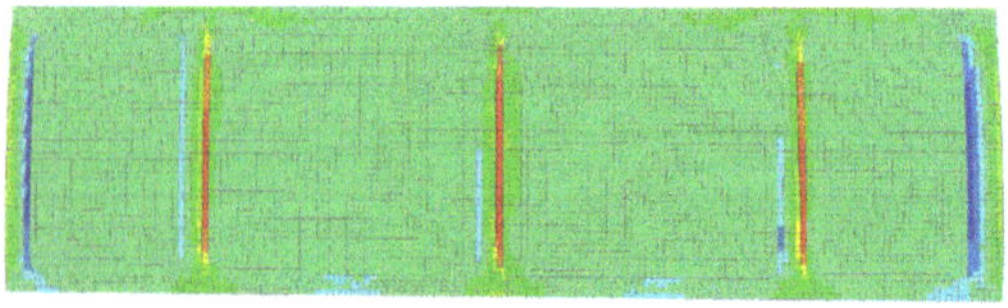

	σ_1 [MPa]	
	Max	Min
Hex	10.622	-7.967
BCC	10.775	-7.917
WBCC1	10.496	-7.970
WBCC2	11.045	-8.084

Figure 3 – Top skin 1st ply σ_1 stress contour map.

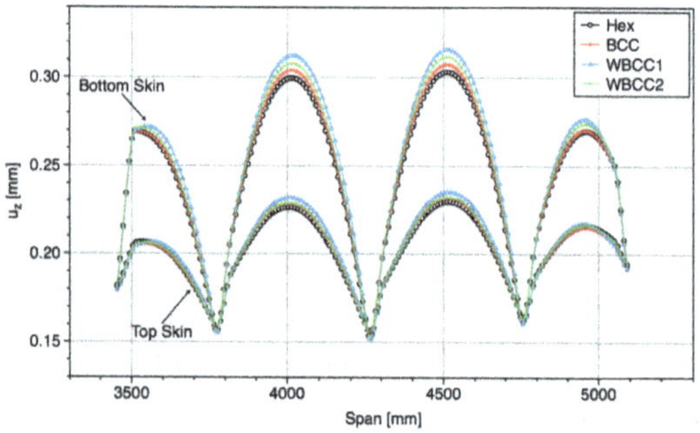

Figure 4 – Directional deformation along z-direction as a function of the control surface span. Top and Bottom skins.

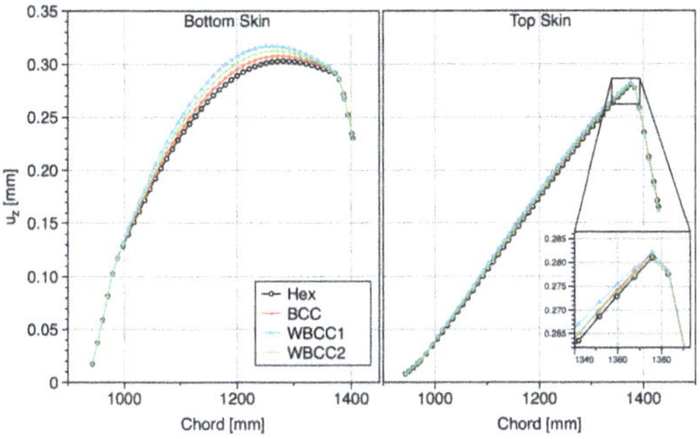

Figure 5 - Directional deformation along z-direction as a function of the control surface chord. Top and Bottom skins.

Conclusions

The DAVYD project sought novel structural configurations for the control surfaces of a new-generation tiltrotor. Various core options are proposed to reinforce the asymmetric sandwich panels. Displacements and stress contour maps indicate the potential use of topology-optimised lattice structures to enhance the overall response. Further studies will be carried out on the subject such as tailoring and optimising the orientation of the lattice cells.

Acknowledgements

The study was financially supported by M.U.R. for the DAVYD project (Grant ARS01_00940)

References

[1] B. Castanie, C. Bouvet, and M. Ginot, "Review of composite sandwich structure in aeronautic applications," *Composites Part C: Open Access*, vol. 1, p. 100004, Feb. 2020. https://doi.org/10.1016/j.jcomc.2020.100004.

[2] J. Bühring, M. Nuño, and K.-U. Schröder, "Additive manufactured sandwich structures: Mechanical characterization and usage potential in small aircraft," *Aerosp Sci Technol*, vol. 111, p. 106548, Feb. 2021. https://doi.org/10.1016/j.ast.2021.106548.

[3] N. Wicks and J. W. Hutchinson, "Optimal truss plates," *Int J Solids Struct*, vol. 38, no. 30–31, 2001. https://doi.org/10.1016/S0020-7683(00)00315-2.

[4] C. G. Ferro, S. Varetti, G. De Pasquale, and P. Maggiore, "Lattice structured impact absorber with embedded anti-icing system for aircraft wings fabricated with additive SLM process," *Mater Today Commun*, vol. 15, 2018. https://doi.org/10.1016/j.mtcomm.2018.03.007.

[5] K. Kohsaka, K. Ushijima, and W. J. Cantwell, "Study on vibration characteristics of sandwich beam with BCC lattice core," *Mater Sci Eng B Solid State Mater Adv Technol*, vol. 264, 2021. https://doi.org/10.1016/j.mseb.2020.114986.

[6] D. Tumino, A. Alaimo, G. Mantegna, C. Orlando, and S. Valvano, "Mechanical properties of BCC lattice cells with waved struts," *International Journal on Interactive Design and Manufacturing (IJIDeM)*, 2023. https://doi.org/10.1007/s12008-023-01359-9.

[7] A. Kumar, N. Muthu, and R. G. Narayanan, "Equivalent orthotropic properties of periodic honeycomb structure: strain-energy approach and homogenization," *International Journal of Mechanics and Materials in Design*, 2022. https://doi.org/10.1007/s10999-022-09620-x.

Thermal buckling analysis and optimization of VAT structures via layer-wise models

A. Pagani[1,a,*], E. Zappino[1,b], R. Masia[1,c], F. Bracaglia[1,d], and E. Carrera[1,e]

[1]Politecnico di Torino, DIMEAS, Corso Duca degli Abruzzi 24, Torino

[a]alfonso.pagani@polito.it, [b]enrico.zappino@polito.it, [c]rebecca.masia@polito.it, [d]francesca.bracaglia@studenti.polito.it, [e]erasmo.carrera@polito.it

Keywords: Thermal Bucklin, Variable Angle Tow (VAT), Layer Wise (LW), High-Order Structural Theories

Abstract. The present study investigates the combination of different manufacturing parameters, such as the curvature radius of the single fiber along its path on a symmetric stacking sequence of the Variable Angle Tows composite (VAT). Moreover, the study objective is to individuate the VAT configuration to maximize the critical thermal buckling load of a thermal-loaded composite square plate. Numerical simulations are performed in the Carrera Unified Formulation (CUF) framework, which allows for a high-order two-dimensional (2D) theory based on the Finite Element Method (FEM), enabling the Layer Wise (LW) discretization of the model. The linearized buckling problem is involved in the formulation, and the resolution of the eigenvalue problem leads to finding the thermal buckling critical temperature.

Introduction

Composite materials have superior strength and stiffness properties compared to traditional materials but are often influenced by the environment. Those involving an increase in material temperature induce expansion strains, and consequently, the buckling critical load can be dramatically affected by an over-temperature. The thermal environment is particularly relevant for space applications with a consistent radiation heating. Instead, for high-speed aeronautical applications, the heating is usually imposed on the structures by the drag. Due to the applied overtemperature, buckling deflection may occur suddenly under specific load conditions, and the deformation of the plate can significantly influence the structure behavior. If a constant along the thickness temperature profile is applied, deflection occurs only at the unique critical temperature.
Variable Angle Tows (VAT) materials exploit a new manufacturing technology introducing an additional degree of freedom in the fiber deposition that can follow curved shapes [1]. It is well known that VAT laminates can improve buckling performance compared to classical composites [2]. Furthermore, thermal buckling has been deeply studied for all the materials that can be subjected to thermal environments [3]. The present work focuses on the fiber deposition optimization of a thermally-loaded square plate to retard the thermal buckling phenomenon. The plate theory is applied within the Carrera Unified Formulation (CUF) framework combined with the Finite Element Method (FEM) approximation [4]. Different kinematic theories are employed, and the results are compared and discussed.

Model description

The decoupled approach is employed to describe the thermal problem. The primary variables of the problem are the displacements, denoted by $\boldsymbol{u}^T$:

$$\boldsymbol{u}^T(x, y, z) = \left(u_x, u_y, u_z\right) \quad (1)$$

where the symbol T denotes transposition. As reported in Eq. (2), the strains $\boldsymbol{\varepsilon}$ can be obtained from the displacement by the geometrical relations using the non-linear differential operator b. The stresses $\boldsymbol{\sigma}$ are calculated from the strains through the Hooke's law, by applying the material properties matrix C as in Eq. (3). Note that in the VAT case, the matrix C depends on the local fiber orientation and is not globally defined as in the case of classical configuration.

$$\boldsymbol{\varepsilon} = \left(\varepsilon_{xx}, \varepsilon_{yy}, \varepsilon_{zz}, \varepsilon_{xz}, \varepsilon_{yz}, \varepsilon_{xy}\right)^T = \boldsymbol{b}\,\boldsymbol{u} \tag{2}$$

$$\boldsymbol{\sigma} = \left(\sigma_{xx}, \sigma_{yy}, \sigma_{zz}, \sigma_{xz}, \sigma_{yz}, \sigma_{xy}\right)^T = \boldsymbol{C}\,\boldsymbol{\varepsilon} \tag{3}$$

The CUF introduces an indicial notation where the three-dimensional displacements are divided into two components, the first $\boldsymbol{F}_\tau(z)$ denotes the kinematic expansion function, and the second $\boldsymbol{u}_\tau(x,y)$ is the in-plane unknown vector. Furthermore, the FEM is employed, and $\boldsymbol{u}_\tau(x,y)$ can be expressed by combining the shape functions $\boldsymbol{N}_i(x,y)$ and the nodal displacement vector $\boldsymbol{q}_{i\tau}$ as represented in Eq. (4).

$$\boldsymbol{u}(x,y,z) = \boldsymbol{F}_\tau(z)\,\boldsymbol{u}_\tau(x,y) = \boldsymbol{F}_\tau(z)\boldsymbol{N}_i(x,y)\,\boldsymbol{q}_{i\tau} \quad \tau = 1,2,\dots,M \quad i = 1,2,\dots,N_n \tag{4}$$

where the double index means sum, M is the number of expansion terms, and N_n is the number of nodes. Via the CUF, high-order theories are employed to describe the in-thickness behavior of the analyzed plate, and among the various theories, Taylor Expansion (TE) and Lagrange Expansion (LE) functions are selected as $\boldsymbol{F}_\tau(z)$ for the present investigation. Furthermore, the laminate properties are described with the Equivalent Single Layer (ESL) approach in the TE case and using the Layer Wise (LW) approach in LE models.

The CUF allows an invariant formulation of the problem using the Fundamental Nuclei (FN) that are not formally mutated with the employed expansion or on the number of nodes [4]. The indicial notation lets the problem governing equation building. In the case of buckling, using the PVD, as reported in Eq. (5), to obtain a linearized formulation of the governing equation based on the tangent stiffness matrix $\boldsymbol{K}_T$ is possible. Where $\boldsymbol{K}_T$ is expressed in terms of FN, the complete procedure can be found in [5].

$$\delta^2(L_{int}) = \int_V \delta(\delta\boldsymbol{\varepsilon}^T\boldsymbol{\sigma})dV = \delta\boldsymbol{q}_{sj}^T\,\boldsymbol{K}_T^{ij\tau s}\delta\boldsymbol{q}_{\tau i} = \delta\boldsymbol{q}_{sj}^T(\,\boldsymbol{K}_0^{ij\tau s} + \boldsymbol{K}_\sigma^{ij\tau s})\,\delta\boldsymbol{q}_{\tau i} \tag{5}$$

The buckling critical temperature is considered coincident with the bifurcation point, which is the load point where two equilibrium configurations exist. As a result, an eigenvalue problem must be solved to obtain the critical load:

$$|\boldsymbol{K}_0 + \boldsymbol{\lambda}_{cr}\,\boldsymbol{K}_\sigma| = \boldsymbol{0} \tag{6}$$

Due to the linear approximation, the matrix $\boldsymbol{K}_\sigma$ is supposed to be proportional to λ_{cr}, which is the buckling critical load factor. $\boldsymbol{K}_\sigma$ is the geometric stiffness matrix strictly dependent on the internal pre-stress state due to the thermal over-temperature load.

Numerical results

In the present section, some numerical results about thermal buckling are presented. The VAT deposition angles are changed to obtain the best configuration retarding the buckling critical thermal load. Eq. (7) describes the fiber orientation θ. It depends on the three angles Φ, T_0, and T_1, which are, the reference system rotation angle, the fiber orientation in the center of the plate and,

Materials Research Proceedings 37 (2023) 300-303 https://doi.org/10.21741/9781644902813-65

the angle at the edge, respectively. The symbol d denotes a characteristic distance, and x' is the new reference direction [1].

$$\theta = \Phi + T_0 + \frac{T_1 - T_0}{d} x' \tag{7}$$

For the sake of simplicity, only T_0 and T_1 are changed during the optimization with Φ equal to zero, the staking sequence is fixed at $[\theta/-\theta]_s$, and the plate is simply supported. The square plate has 150 mm edges and 1.016 mm thickness, and a one-degree constant over-temperature is applied on the whole plate. Two plates are analyzed, the first composed of Carbon/Epoxy whose properties are E_1 = 147 GPa, E_2 =10.3 GPa, ν_{12}=0.27, G_{12}= 7.0 GPa, α_1= -0.9 x 10^{-6} 1/K, α_2= 27.0 x 10^{-6} 1/K. The second plate is E-Glass/Epoxy with E_1 = 41 GPa, E_2 =10.04 GPa, ν=0.28 G_{12}= 4.3 GPa, α_1= 7.0 x 10^{-6} 1/K, α_2= 26.0 x 10^{-6} 1/K. Each plate presents a discretization of 20x20 Q9 in-plane elements, 2nd order TE and 3rd order LE are employed as expansion functions.

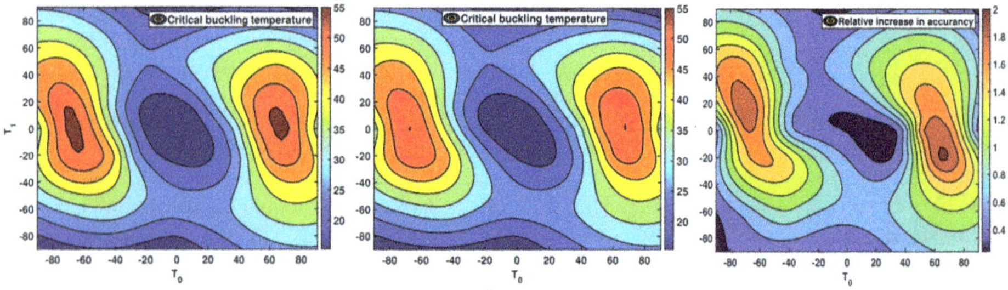

(a) TE2 *(b) LE3* *(c) Perc. difference*

Figure 1: (a), (b) Carbon/Epoxy critical buckling temperature varying the path angles, 20 x 20 Q9 in plane mesh. (a): Present method with ESL and TE2, (b) Present method with LW and LE3. (c) Percentage relative increasing accuracy from LE to TE2 ($\frac{\theta_{TE}-\theta_{LE}}{\theta_{TE}}\mathbf{x100}$)*. Carbon/Epoxy.*

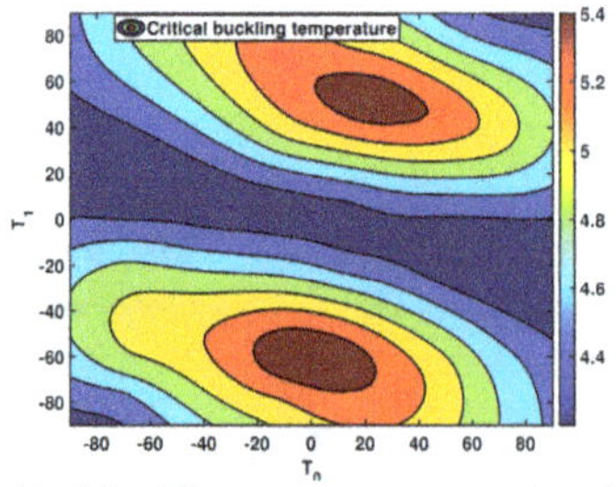

Figure 2: E-Glass/Epoxy critical buckling temperature varying the path angles, 20 x 20 Q9 in plane mesh, Present method with LW and LE3.

Table 1:Maximum critical buckling temperature for VAT and straight configuration with $\boldsymbol{\theta} = \mathbf{45°}$.

	T_0	T_1	ΔT_{cr}	$\Delta T_{straight}$
		Carbon/Epoxy		
20x20 Q9 LE3	-68.29°	0.45°	55.11°	41.32°
20x20 Q9 TE2	-67.39°	0.45°	56.27°	41.87°
Ref. [6]	69.00°	-5.71°	57.79°	42.21°
		E-Glass/Epoxy		
20x20 Q9 LE3	-4.97°	-60.15°	5.55°	5.38°
Ref. [6]	6.71°	58.04°	5.58°	5.43°

Conclusions

- As expected, the thermal buckling critical temperature graphs are symmetric to the line $T_0 = -T_1$ and present two maximum value zones for both configurations (Fig.1 and 2).
- Considering the adopted expansion theory, the 2nd-order TE model is more similar to the reference. Still, the LE model is more accurate even if it presents a greater computational cost. As depicted in Fig.1(c), the main difference is not in the critical temperature but mainly in the extension of the optimal zones. As a result, the zones with maximum temperature are abruptly reduced in the LE model reported in Fig.1(b) than in the less accurate model in Fig.1(a).
- Figure 1(c) clarifies that the increase in accuracy due to the LE model is not constant comparing different orientations angle but rises with the increase in the critical temperature.
- For the Carbon/Epoxy laminate, as collected in Table 1, the better path configuration allows reaching a critical over-temperature of 55.1° with a consistent gain to the straight deposition critical temperature of 41.3°.
- From Table 1 it is clear that E-Glass/Epoxy configuration does not present significant advantages using VAT deposition with a gain of 0.03% passing from a critical temperature of 5.384° to 5.55° in the VAT case.
- In the end, the choice of VAT deposition may be helpful to retard the buckling critical load, and, in Carbon/Epoxy case, the gain is more evident than in the Glass/Epoxy plate.

References

[1]. *Accurate stress analysis of variable angle tow shells by high-order equivalent-single-layer and layer-wise finite element models. A. R. Sánchez-Majano, R. Azzara, A. Pagani, E. Carrera.* S. l.: Materials, 14(21), 2021. https://doi.org/10.3390/ma14216486

[2]. *Prebuckling and buckling analysis of variable angle tow plates with general boundary conditions.* G. Raju, Z. Wu, B. C. Kim, P. M. Weaver. s.l: Composite Structures.

[3]. *Mechanical and thermal buckling loads of rectangular FG plates by using higher-order unified formulation.* M. Farrokh, M. Afzali. E. Carrera. s.l.: E. Mechanical of Advanced Materials and Structures, 2019.

[4]. *Finite Element Analysis of Structures through Unified Formulation.* E. Carrera, M. Cinefra, M. Petrolo, E. Zappino. s.l.: Wiley & Sons, 2014.

[5]. Geometrically nonlinear refined shell theories by Carrera Unified Formulation. B. Wu, A. Pagani, W. Q. Chen, E. Carrera. s.l.: Mechanics of Advanced Materials and Structures, 2021.

[6]. *Thermal buckling of composite plates with spatial varying fiber orientation.* A. V. Duran, N. A. Fasanella, V. Sundararaghavan, A. M. Waas. s.l.: Composite Structures, 2015.

Materials Research Proceedings 37 (2023) 304-307 https://doi.org/10.21741/9781644902813-66

Surface node method for the peridynamic simulation of elastodynamic problems with Neumann boundary conditions

Francesco Scabbia[1,a,*], Mirco Zaccariotto[1,2,b] and Ugo Galvanetto[1,2,c]

[1]Centro di Ateneo di Studi e Attività Spaziali "Giuseppe Colombo" (CISAS), Università degli Studi di Padova, Padova, 35131, Italy

[2]Industrial Engineering Department (DII), Università degli Studi di Padova, Padova, 35131, Italy

[a]francesco.scabbia@phd.unipd.it, [b]mirco.zaccariotto@unipd.it, [c]ugo.galvanetto@unipd.it

Keywords: Peridynamics, Surface Node Method, Elastic Wave Propagation, Neumann Boundary Conditions

Abstract. Peridynamics is a nonlocal theory that can effectively handle discontinuities, including crack initiation and propagation. However, near the boundaries, the incomplete nonlocal regions are the cause of the peridynamic surface effect, resulting in unphysical stiffness variation. Additionally, imposing local boundary conditions in a peridynamic (nonlocal) model is often necessary. To address these issues, the surface node method has been proposed for improving accuracy near the boundaries of the body. Although this method has been verified for a variety of problems, it has not been applied for elastodynamic problems involving Neumann boundary conditions. In this work we show a numerical example of this case, comparing the results with the corresponding peridynamic analytical solution. The numerical results exhibit no stiffness variations near the boundaries throughout the entire simulation timespan. Therefore, we conclude that the surface node method allows to effectively solve elastodynamic peridynamic problems involving Neumann boundary conditions, with improved accuracy near the boundaries.

Introduction to peridynamics

Peridynamics (PD) is a nonlocal continuum theory based on integro-differential equations in which discontinuities, such as cracks, in the displacement field can arise and evolve without mathematical inconsistencies [1,2]. In a PD body B, two points interact through a so-called *bond* if their distance is smaller than δ, named *horizon size*. The PD equation of motion for a generic point x at a time instant t in a 1D, homogeneous, linear elastic body [3] is given as

$$\ddot{u}(x,t) = v^2 \int_{H_x} \frac{u(x',t)-u(x,t)}{\delta(x'-x)^2} \mathrm{d}x', \tag{1}$$

where $H_x = \{x' \in B: |x'-x| \leq \delta\}$ is the set of points x' interacting with point x, u is the displacement, $\ddot{u}$ is the acceleration, and v is the wave speed.

By using the meshfree method with a uniform grid spacing Δx [4], in which every node represents a cell of length Δx (see Fig. 1), Eq. 1 is discretized in space as

$$\ddot{u}(x_i,t) = \frac{v^2}{\delta} \sum_{j \in H_i} \frac{u(x_j,t)-u(x_i,t)}{\left(x_j-x_i\right)^2} \beta_{ij} \Delta x \,, \tag{2}$$

where x_i and x_j are respectively the coordinates of node i and any node j within the neighborhood H_i of node i, and β_{ij} is the quadrature coefficient, namely the fraction of cell of node j which lies within the neighborhood H_i [5]. The explicit central difference method is used for time integration [4]:

$$u(x_i, t_{n+1}) = 2u(x_i, t_n) - u(x_i, t_{n-1}) + \frac{(v\Delta t)^2}{\delta} \sum_{j \in H_i} \frac{u(x_j,t) - u(x_i,t)}{(x_j - x_i)^2} \beta_{ij} \Delta x \, , \tag{3}$$

where Δt is the time step size and n stands for the index of the current time step.

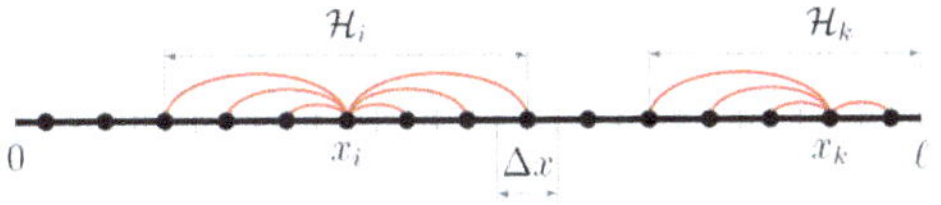

Figure 1: Each node (black dots) in the peridynamic body represents a cell of length Δx and interacts with all the nodes within its neighborhood through bonds (red lines).

As shown in Fig. 1, the nodes near the boundary of the body have an incomplete neighborhood. Due to this fact, the stiffness properties of the nodes close to the boundary are different from those of the nodes in the bulk. This undesired phenomenon is called *PD surface effect* [6-9]. Furthermore, boundary conditions in PD models should be imposed over a finite-thickness layer, in contrast with experimental measurements which are performed only at the boundary. Therefore, we use the Surface Node Method (SNM) to solve these issues [7-9].

Overview of the Surface Node Method

As shown in Fig. 2, the *fictitious nodes* are added in the peridynamic model to complete the neighborhoods of the interior nodes near the boundaries of the body. Furthermore, the *surface nodes* are introduced at the boundaries of the body to impose the boundary conditions.

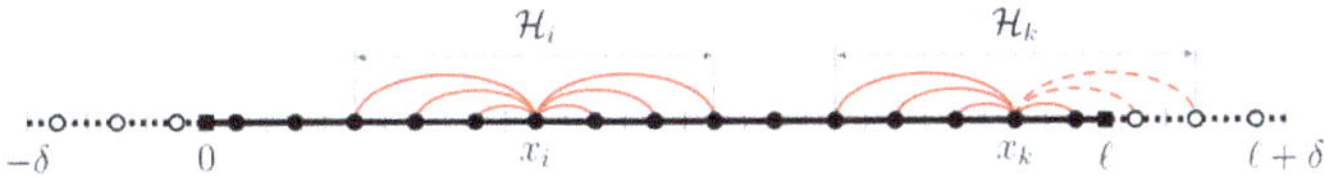

Figure 2: The fictitious nodes (empty dots) are introduced to reduce the PD surface effect by completing the neighborhoods of the interior nodes (solid dots). The surface nodes (solid squares) are introduced to impose the peridynamic boundary conditions through the fictitious bonds (red dashed lines).

The displacements of the fictitious nodes are determined by extrapolation from the displacements of the interior nodes. By using, for example, the linear Taylor-based method [7-9], the displacement of any fictitious node f is given as

$$u(x_f, t_n) = u(x_s, t_n) + (x_f - x_s) \frac{u(x_s,t_n) - u(x_p,t_n)}{x_s - x_p} \, , \tag{4}$$

where s is the index of the closest surface node and p is the index of the interior node closest to that surface node.

The surface nodes do not have interactions (bonds) with other nodes, the state of the fictitious bonds crossing them is governed by the equation of the force flux [7-9]:

$$\tau(x_s, t_n) = \frac{v^2}{\delta} \sum_{ij \in \mathrm{I}} \frac{u(x_j,t_n) - u(x_i,t_n)}{(x_j - x_i)^2} \beta_{ij} \Delta x^2 , \tag{5}$$

where $\tau(x_s, t_n)$ is the force flux at the surface node s and I is the set of all the fictitious bonds with positive direction crossing the boundary at x_s. Thanks to Eq. 5, constraints and loads can be applied to the surface nodes as one would do in a local model.

Numerical example

For the first time, we present a numerical example of a peridynamic model making use of the SNM to solve an elastodynamic problem involving Neumann boundary conditions. The initial boundary value problem is given as

$$\begin{cases} \ddot{u}(x,t) = \frac{v^2}{\delta}\int_{H_x} \frac{u(x',t)-u(x,t)}{(x'-x)^2}\,\mathrm{d}x' & \text{for } 0 < x < \ell,\ t > 0, \\ u(0,t) = 0,\ \tau(\ell,t) = 0 & \text{for } t > 0, \\ u(x,0) = 0.02e^{-100\left(\frac{x-0.5}{\ell}\right)^2},\ \dot{u}(x,0) = 0 & \text{for } 0 < x < \ell, \end{cases} \tag{6}$$

where $\dot{u}$ is the velocity and ℓ is the length of the peridynamic body. The analytical solution is computed by means of the method of separation of variables [3]:

$$u(x,t) = \sum_{m=1,3,5,\dots}^{\infty} \frac{0.004\sqrt{\pi}}{\ell} \sin\left(\frac{k_m \ell}{2}\right) e^{\frac{-k_m^2}{400}} \sin(k_m x) \cos\left(vt\sqrt{\frac{2}{\delta^2}\left[k_m\delta\,\mathrm{Si}(k_m\delta) + \cos(k_m\delta) - 1\right]}\right), \tag{7}$$

where $k_m = \frac{m\pi}{2\ell}$ and $\mathrm{Si}(\cdot)$ is the sine integral function. The analytical solution in Eq. 7, truncated at $m = 80$, will be used as a comparison for the numerical results in Fig. 3.

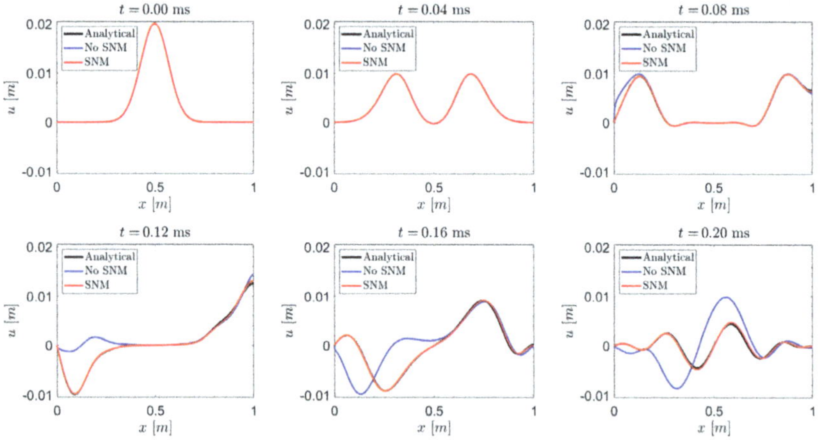

Figure 3: Plots of the propagating wave at different instants of time t for the analytical solution and the peridynamic models with and without the use of the Surface Node Method (SNM).

Fig. 3 shows the numerical results of the peridynamic model with and without the use of the SNM. If the SNM is not employed, the loads and the constraints are applied directly to the interior nodes closest to the boundary. The model parameters used to obtain those results are the following: $v = 5000$ m/s, $\ell = 1$ m, $\delta = 0.1$ m, $\Delta x = 0.001$ m, and $\Delta t = 0.2$ μs. As shown in Fig. 3, it is evident that the model with the SNM provides results closer to the analytical peridynamic solution. The major differences are noticeable in the region near the end of the bar, where Dirichlet boundary conditions are imposed. However, without employing any boundary correction, non-negligible errors are also present at the other end of the bar, where Neumann boundary conditions are applied. Thus, the SNM allows to considerably reduce the numerical errors due to the PD surface effect and the imposition of the boundary conditions.

Conclusions

In this work, we numerically solved a 1D peridynamic elastodynamic problem involving Neumann boundary conditions by using the Surface Node Method (SNM) to mitigate the PD surface effect and impose the boundary conditions. The analytical solution to this problem has been derived thanks to the method of separation of variables. The numerical results show that the use of the SNM significantly reduces the errors near the boundaries of the body when compared to the corresponding model without boundary corrections, both where Dirichlet and Neumann boundary conditions are applied.

Acknowledgements. The authors would like to acknowledge the support they received from MIUR under the research project PRIN2017-DEVISU and from the University of Padova under the research project BIRD2020 NR.202824/20.

References

[1] S.A. Silling, Reformulation of elasticity theory for discontinuities and long-range forces, J. Mech. Phys. Solids 48 (2000) 175-209. https://doi.org/10.1016/S0022-5096(99)00029-0

[2] S.A. Silling, M. Epton, O. Weckner, J. Xu, E. Askari, Peridynamic states and constitutive modelling, J. Elast. 88 (2007) 151-184. https://doi.org/10.1007/s10659-007-9125-1

[3] Z. Chen, X. Peng, S. Jafarzadeh, F. Bobaru, Analytical solutions of peridynamic equations. Part II: elastic wave propagation, Int. J. Eng. Sci. 188 (2023). https://doi.org/10.1016/j.ijengsci.2023.103866

[4] S.A. Silling, E. Askari, A meshfree method based on the peridynamic model of solid mechanics, Comput. Struct. 83 (2005) 1526-1535. https://doi.org/10.1016/j.compstruc.2004.11.026

[5] F. Scabbia, M. Zaccariotto, U. Galvanetto, Accurate computation of partial volumes in 3D peridynamics, Eng. Comput. 39 (2023) 959-991. https://doi.org/10.1007/s00366-022-01725-3

[6] Q.V. Le, F. Bobaru, Surface corrections for peridynamic models in elasticity and fracture, Comput. Mech. 61 (2018) 499-518. https://doi.org/10.1007/s00466-017-1469-1

[7] F. Scabbia, M. Zaccariotto, U. Galvanetto, A novel and effective way to impose boundary conditions and to mitigate the surface effect in state-based Peridynamics, Int. J. Numer. Methods. Eng. 122 (2021) 5773-5811. https://doi.org/10.1002/nme.6773

[8] F. Scabbia, M. Zaccariotto, U. Galvanetto, A new method based on Taylor expansion and nearest-node strategy to impose Dirichlet and Neumann boundary conditions in ordinary state-based Peridynamics, Comp. Mech. 70 (2022) 1-27. https://doi.org/10.1007/s00466-022-02153-2

[9] F. Scabbia, M. Zaccariotto, U. Galvanetto, A new surface node method to accurately model the mechanical behavior of the boundary in 3D state-based Peridynamics, J. Peridyn. Nonlocal Model (2023) 1-35. https://doi.org/10.1007/s42102-022-00094-1

Crack localization on a statically deflected beam by high-resolution photos

Andrea Vincenzo De Nunzio[1,a*], Giada Faraco[1,b*]
Nicola Ivan Giannoccaro[1,c] and Arcangelo Messina[1,d]

[1]Dipartimento di Ingegneria dell'Innovazione, Università del Salento, 73100 Lecce (LE), Italia

[a]andreavincenzo.denunzio@unisalento.it, [b]giada.faraco@unisalento.it,
[c]ivan.giannoccaro@unisalento.it, [d]arcangelo.messina@unisalento.it

Keywords: Structural Health Monitoring, Crack Detection, Image Processing, Optical Measurements

Abstract. In a context where the complexity of systems and their interconnection is increasing exponentially, the possibility of being able to monitor the structural integrity of crucial parts of structures is of considerable importance. In addition, the availability of modern and advanced tools opens the door to the advent of new diagnostic techniques. In this regard, the authors here deeply investigate and test a modern technique that allows to analyze a structure starting from a photo in order to identify and locate damage present in the structure in a rapid and non-destructive way. This allows to obtain an accurate location of the damage and consequently a quick evaluation of its state of integrity. Moreover, a further advantage lies in the possibility of carrying out the analysis in a non-invasive way without any physical interaction with the analyzed structure. The suitability of the technique is tested on a statically deformed beam in epoxy glass laminate. It has a notch, which represents the defect, and the goal is to determine the notch position, which is not visible in the photo. The basis of the proposed method is the correlation between the curvature that the beam presents under load conditions and its flexural stiffness. The damage on the beam, in fact, leads to a punctual alteration of its flexibility which is identified by sudden changes in the second derivative of the transversal deflection. The proposed methodology consists in taking a photo of the inflected beam; subsequently, the acquired photo is manipulated with specifically designed image processing tools, first to segment the beam shape and then to extract its axis. Finally, the second derivative is extracted using two different numerical differentiator filters (Lanczos filters and Gaussian wavelets) along with suitable processing to reduce the border distortions. The tests conducted demonstrate that it is possible to accurately detect the position of the notch. Although the authors realize that the technique can generally need sensibly large displacements, the results seem promising. Such a need is probably due to the resolution of the camera, which can sometimes represent a technological limit. It is believed that higher resolution would allow damage to be detected even for smaller displacements. A fundamental advantage is the speed of the methodology illustrated since it takes just a few moments from taking the photo to evaluating the results. This is accompanied by the ease of acquiring the measurement, which involves the use of the camera and its support without additional equipment.

Introduction

Structural health monitoring (SHM) is a crucial aspect of ensuring the safety and reliability of various engineering mechanical structures such as buildings, bridges etc. The ability to detect, assess, and manage structural degradation or damage is essential for preventing catastrophic failures and minimizing risks. SHM techniques generally consist in collecting data related to structure behavior, load distribution, vibrations, and material properties to analyze using sophisticated algorithms and models to detect any abnormalities or signs of deterioration [1].

Advantages of computer vision for SHM are several [2,3]: they are generally inexpensive, contactless, non-destructive and allow for the assessment of structural integrity without the installation of sensors directly on the structure, unlike traditional techniques that rely on data collection through accelerometers or strain gauges. Given the advantages, it is not surprising that vision-based SHM techniques are not limited to civil engineering structures but can also be extended to mechanical components (such as beam-type structures) characterized by operational conditions significantly different from those in civil engineering, such as small displacements and high vibration frequencies.

A computer-vision technique is adopted in [4–6] where a computer vision method for measuring the in-plane displacement field of cantilever beams is presented. The purpose of these articles is to carry out a structural health analysis starting from the deformed shape of the beam (acquired by the vision-based algorithm) by exploiting different damage detection techniques: second derivative algorithm, line segment algorithm and voting algorithm in [4], fractal dimension, wavelet transform and roughness methods in [5,6].

The authors' intent in this study is to develop and evaluate the potential of a vision-based method for SHM and damage detection. To achieve this goal, an algorithm was first developed to extract the deformation axis of a deflected beam starting from a photo of it. Then the acquired information is used as input for the damage detection procedure. Its basis is the correlation between the curvature that the beam exhibits under load conditions, and its flexural stiffness (e.g. [7,8]). The damage on the beam results in a localized change in its flexibility, manifesting as abrupt variations in the second derivative of the transverse deflection. Therefore, the proposed methodology can be summarized in two parts: first the acquisition and processing of the photo to extract the axis of the deformed beam, and second, the damage detection using the second derivative method. To conclude, the effectiveness of the proposed method was tested in the laboratory on a beam of which the structural integrity had already been evaluated in a previous study [9]. Furthermore, this allowed for a comparison between a traditional data acquisition methodology (PSD-triangular laser sensor) and a vision-based data acquisition technique.

Materials and Methods

The proposed methodology was tested on an epoxy glass laminate beam with dimensions equal to 500 x 10 x 25 mm; a notch, 2 x 5 x 25 mm, is made at 265 mm from the beam clamped end to simulate the damage.

With the purpose of evaluating suitability of the vision-based data acquisition methodology compared to a classical one (results are available in [9]), the beam was statically deformed to emulate its second mode shape.

The experimental setup (Fig. 1a) was composed by a frame, in which the beam was clamped at one end, two operating screws to impose the deflection shape, and a consumer-grade camera (Canon EOS 1200D) fixed on a tripod. Black cardboard was used as the background to make the beam more visible and to improve the segmentation processing.

The beam was deformed by setting its maximum displacements equal to those measured in [9]; subsequently, such an initial deflection was amplified by 2, 4, 6 and 10 times to test the method sensitivity.

In order to take non-distorted photos, camera calibration was carried out by the Matlab Camera Calibrator Toolbox.

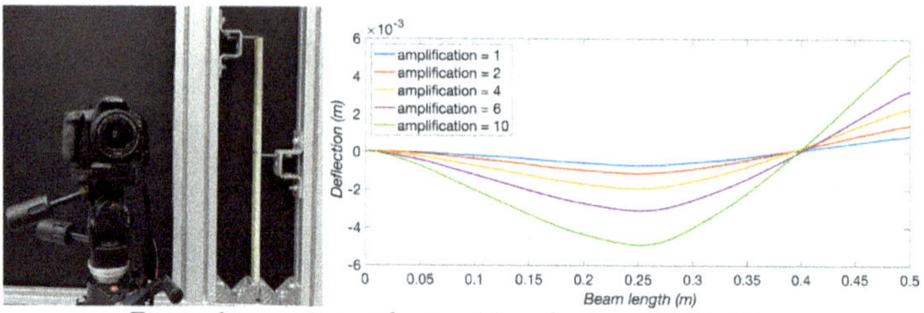

Figure 1: experimental setup (a) and extracted axis (b)

The segmentation algorithm was implemented in Matlab. Its output is a mask in which the beam is the foreground. The beam axis extraction algorithm is applied to this mask; such a procedure employs a $2\Delta_m$ x $2\Delta_n$ pixel window moving along the beam profile by one pixel per step. At every iteration the axis point coordinates (in pixels) are calculated by (1):

$$(x_P, y_P) = \left(\frac{m_{max}+m_{min}}{2} ; \frac{\sum_{i=1}^{4\cdot\Delta m\cdot\Delta n} I_i . n_i}{\sum_{i=1}^{4\cdot\Delta m\cdot\Delta n} I_i} \right) \quad (1)$$

where m_{max} and m_{min} are the coordinates, along the beam axis direction, of the extreme pixels of the window, n_i is the coordinate of the i-th pixel in the transversal direction and I_i its intensity. Afterwards, a scale factor S [10] transformed pixel-coordinates into physical units [m]; in particular, photos having a resolution equal to 5184x3456 pixels produced $S \sim 0.1$ mm/pixels.

This procedure allowed the extraction of the bent beam axis $y(x)$ (Fig. 1) which was subsequently processed to estimate its second derivative (Fig. 2).

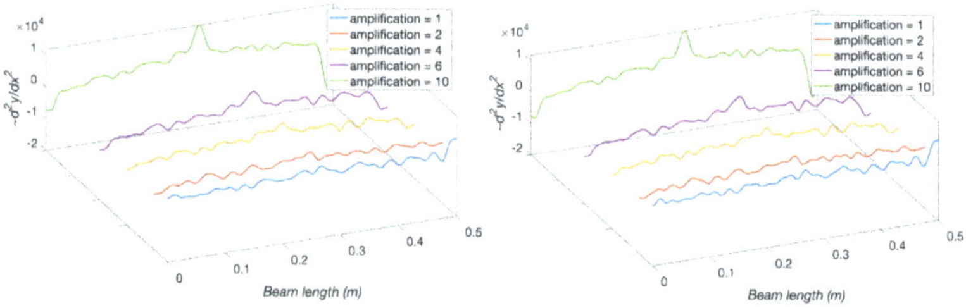

Figure 2: 2nd order derivative calculated by Lanczo's filter (a) and Continuous Wavelet Transform (b)

In particular, Figure 2 shows the second derivatives estimated with two different differentiator filters, i.e. Figure 2(a) represents the estimate through the Lanczos filter [11,12] while the estimate in Figure 2(b) is obtained by means of the continuous wavelet transform [12]. In both cases the second derivative was obtained by performing the numerical convolution (*) of the bent beam $y(x)$ twice through the respective filters of first derivative (h_1 in (2)). This to extend the physical signal through the *Rotation* option for each convolution step for reducing distortions at the edges [13]. Therefore, the following formula was applied for both filters.

$$\frac{d^2y}{dx^2} \cong h_1 * (h_1 * y_{ext})_{ext} \quad (2)$$

The length of the filters was calibrated through dilation parameters which provided similar performance to reject the experimental/numerical noise.

Results

It was observed that the displacements of the reference deformation [9] did not initially allow the damage to be identified with the proposed method. At least, this has been the case with the experimental setup herein used. Sensitivity analyses, based on different amplifications, suggested that a minimum deflection of 3 mm over a 500 mm long beam was needed to have a clear location of the damage.

Conclusions

The purpose of this work was to evaluate the suitability of a vision-based data acquisition method for damage detection compared to previous analyses. The results proved that the proposed vision-based method needs larger displacement than a PSD-triangular laser sensor to acquire valuable data. However, we found this technique is rather encouraging because it allowed to acquire a dense full-field displacement with a minimum effort in comparison to [9] which involved the acquisition of a large number of discrete points in a time consuming way.

Future research will be conducted by considering dynamic shapes of the beam excited by a shaker or, even, by environmental forces.

References

[1]H.-N. Li, L. Ren, Z.-G. Jia, T.-H. Yi, D.-S. Li, State-of-the-art in structural health monitoring of large and complex civil infrastructures, J Civil Struct Health Monit. 6 (2016) 3–16. https://doi.org/10.1007/s13349-015-0108-9

[2]C.-Z. Dong, F.N. Catbas, A review of computer vision–based structural health monitoring at local and global levels, Structural Health Monitoring. 20 (2021) 692–743. https://doi.org/10.1177/1475921720935585

[3]D. Feng, M.Q. Feng, Computer vision for SHM of civil infrastructure: From dynamic response measurement to damage detection – A review, Engineering Structures. 156 (2018) 105–117. https://doi.org/10.1016/j.engstruct.2017.11.018

[4]Z. Dworakowski, P. Kohut, A. Gallina, K. Holak, T. Uhl, Vision-based algorithms for damage detection and localization in structural health monitoring: Vision-based Algorithms for Damage Detection and Localization, Struct. Control Health Monit. 23 (2016) 35–50. https://doi.org/10.1002/stc.1755

[5]J. Shi, X. Xu, J. Wang, G. Li, Beam damage detection using computer vision technology, Nondestructive Testing and Evaluation. 25 (2010) 189–204. https://doi.org/10.1080/10589750903242525

[6]R. Kumar, S.K. Singh, Crack detection near the ends of a beam using wavelet transform and high resolution beam deflection measurement, European Journal of Mechanics - A/Solids. 88 (2021) 104259. https://doi.org/10.1016/j.euromechsol.2021.104259

[7]A.K. Pandey, M. Biswas, M.M. Samman, Damage detection from changes in curvature mode shapes, Journal of Sound and Vibration. 145 (1991) 321–332. https://doi.org/10.1016/0022-460X(91)90595-B

[8]U. Andreaus, P. Casini, Identification of multiple open and fatigue cracks in beam-like structures using wavelets on deflection signals, Continuum Mech. Thermodyn. 28 (2016) 361–378. https://doi.org/10.1007/s00161-015-0435-4

[9]B. Trentadue, A. Messina, N.I. Giannoccaro, Detecting damage through the processing of dynamic shapes measured by a PSD-triangular laser sensor, International Journal of Solids and

Structures. 44 (2007) 5554–5575. https://doi.org/10.1016/j.ijsolstr.2007.01.018

[10] A. Khaloo, D. Lattanzi, Pixel-wise structural motion tracking from rectified repurposed videos, Struct Control Health Monit. 24 (2017). https://doi.org/10.1002/stc.2009

[11] C. Lanczos, Applied Analysis (1956), Dover Publications, New York

[12] A. Messina, Detecting damage in beams through digital differentiator filters and continuous wavelet transforms, J. of Sound and Vib. 272(2004) 385-412. https://doi.org/10.1016/j.jsv.2003.03.009

[13] A. Messina, Refinements of damage detection methods based on wavelet analysis of dynamical shapes, International Journal of Solids and Structures. 45 (2008) 4068–4097. https://doi.org/10.1016/j.ijsolstr.2008.02.015

Materials Research Proceedings 37 (2023) 313-316 https://doi.org/10.21741/9781644902813-68

Adaptive finite elements based on Carrera unified formulation for meshes with arbitrary polygons

Maria Cinefra[1,a]* and Andrea Rubino[1,b]

[1]Politecnico di Bari, via Edoardo Orabona 4, Bari, Italy

[a]maria.cinefra@poliba.it, [b]andrea.rubino@poliba.it

Keywords: Carrera Unified Formulation, Adaptive Finite Elements, 3D Elements, Node-Dependent Kinematics, Arbitrary Polygons

Abstract. The new Adaptive Finite Elements presented are based on Carrera Unified Formulation (CUF) that permits to implement 1D and 2D elements with 3D capabilities. In particular, by exploiting the node-dependent kinematic approach recently introduced and incorporating the FEM shape functions with the CUF kinematic assumptions in unique 3D approximating functions, it is demonstrated that new mesh capabilities can be obtained with the use of presented elements by easy implementation. A classical patch test is performed to investigate the mesh distortion sensitivity.

Introduction

To alleviate meshing issues due to FEM, more advanced techniques can be used, resorting to numerical methods that are designed from the very beginning to provide arbitrary order of accuracy on more generally shaped elements. These techniques are based, for instance, on the Virtual Element Method (VEM) [1]. In fact, in contrast to FEM in which elements are typically triangular and quadrilateral in 2D or tetrahedral and hexahedral in 3D, the VEM permits arbitrary two-dimensional polygonal and three-dimensional polyhedral elements [2]. This allows the problem domain to be discretized by elements represented by arbitrary polygons, which can be concave and convex. Moreover, different polynomial consistency is allowed within the method and non-conforming discretizations can be handled, mainly for local refinement and so on, representing the key aspect of this method.

The newly proposed Adaptive Elements, which are Finite Elements based on Carrera Unified Formulation (CUF), represent more agile and manageable elements, based on a well-known technique (FEM), easy to implement and can produce the same benefits of using VEM. Recently, in the framework of CUF, a novel approach called Node-Dependent Kinematics (NDK) has been proposed to further increase the numerical efficiency of the models [3]. The Node-Dependent Kinematics approach, along with the 3D modelling of non-orthogonal geometries [4], opens the possibility to create elements with a non-conventional number of nodes, complex geometrical shapes, and different polynomial approximations along different spatial directions, leading to a certain freedom in modelling the problem considered. In this way, 3D elements can be recovered with a non-regular shape that can adapt to the edges of the domain considered, when the geometrical constraints or the kinematic behavior of the structure requires that.

Adaptive Finite Elements

In traditional beam and plate theories, it is difficult to model beams with variable sections or sections that are not orthogonal to the axis, as well as plates with varying thickness or edges not perpendicular to the mid-surface. However, with the Node-Dependent Kinematic approach [3], it is possible to extend the Carrera Unified Formulation models to incorporate more complex geometries. This is achieved by incorporating the CUF kinematic assumption with the FEM discretization to create a unique 3D approximation, as follows:

$$\mathrm{u} = (F_\tau^i N_i)\mathrm{q}_{\tau i} = L_{\tau i}(\xi, \eta, \zeta)\mathrm{q}_{\tau i}$$

Where where ξ, η, ζ are the natural coordinates corresponding to x, y, z and $\mathbf{u} = \{u_x, u_y, u_z\}$ is the 3D displacement field. A double summation over the indeces $i=1,\ldots, N_n$ and $\tau=1,\ldots,M$ is implied, where N_n is the number of the nodes in the element and M is the number of approximating functions adopted for the kinematic theory. F_τ and N_i are defined according to the type of element (beam or plate): N_i are the FEM shape functions that are used along the axis for beam elements or in the plane for the plate elements; F_τ are the approximating functions over the cross-section for the beam or along the thickness for the plate. Lagrange polynomials are chosen in this work for generating both F_τ and N_i, according to the CUF [4]. $q_{\tau i}$ are the generalized nodal displacements. Note that the function F_τ depends on the node 'i'.

The function $L_{\tau i} = (F^i_\tau N_i)$ represents a non-conventional 3D shape function in which the order of expansion can be different along one of the spatial directions. As a result, this approach overcomes the limitations of classical 3D finite element, related to its aspect ratio, with a consequent saving in degrees of freedom and the possibility to create polygonal elements.

In the following, the acronyms B2 (linear), B3 (quadratic) and B4 (cubic) are used both for the 1D discretization of the axis in beam models or for the approximation through the thickness in plate models; as well as the acronyms Q4 (linear), Q9 (quadratic) and Q16 (cubic) are used both for the 2D approximation of the cross-section in beam models and for the discretization of the in-plane domain in plate models.

Patch test

This standard test is classically used in order to investigate the mesh distortion sensitivity in plate bending problems. The in-plane dimensions of the square plate considered are 100×100 m^2 and the thickness is 5 m. The boundary conditions are clamped (CC). A transverse concentrated force is applied at the center of the top surface F = 4 N. The material is isotropic with the following properties: E = 10.92 Pa, $\nu = 0.30$ and $\rho = 1$ kg/m^3. The meshes initially used are a 4x4 Q4 elements and a 4x4 Q9 elements in the plane, with B3 approximation along the thickness in both cases.

Due to the symmetry of the problem, a quarter of the plate is considered. In particular, the bottom-left quarter of the plate is chosen and symmetry boundary conditions are applied along cut edges. F = 1 N is applied on the top-right corner. The resulting mesh of the plate quarter is 2×2 based on Q4 or Q9 plate elements.

Some beam elements are introduced in the mesh in order to obtain distorted elements with different number of nodes along the edges. Then, three different configurations are analyzed:

- 2xQ4 1xB2: in this configuration, the previous two elements at the top are substituted by a single 1D element with 2 nodes, having a Q9 approximation on the cross-section;
- 2xQ4 1xB3: as in the previous case, the two elements at the top are substituted by a single 1D element but having 3 nodes along the axis, where the cross-section is expressed as a Q9 element;
- 2xQ4 1xQ9: the number of nodes is identical with respect to the previous case, but the discretization of the two elements at the top is different, accounting for a single Q9 element having a B3 discretization along the thickness.

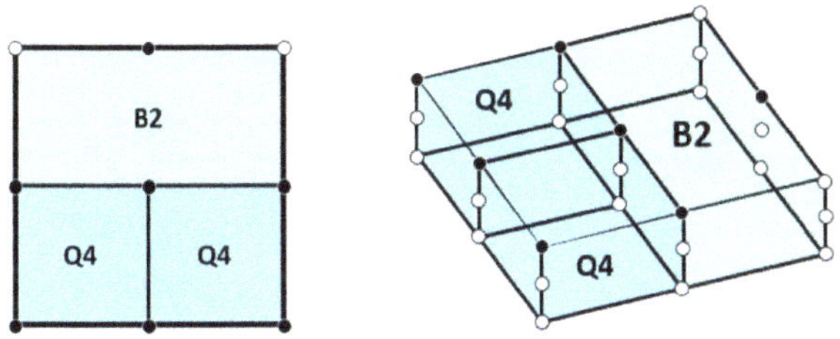

(a) 2D view (b) 3D view

Figure 1: Quarter plate: 2xQ4 1xB2 mesh

In Fig. 1 it is possible to visualize one of the described configurations: the black dots are the FEM nodes, while the white dots represent the CUF nodes added considering the thickness and the cross-section of the plate and the beam elements, respectively.

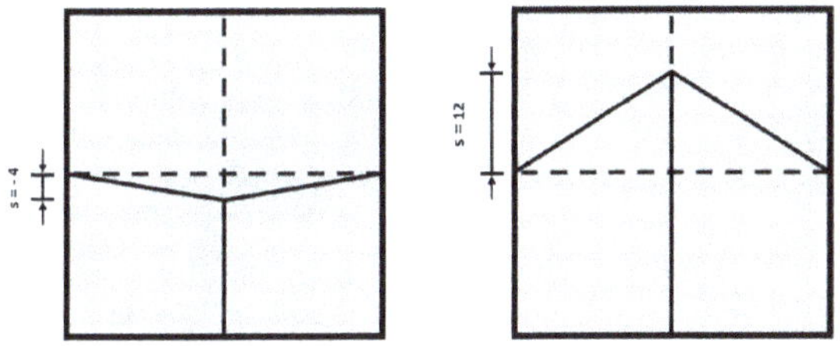

(a) Quarter plate - Parameter $s < 0$ (b) Quarter plate - Parameter $s > 0$

Figure 2: Quarter plate - mesh distortion defined by parameter 's'

The mesh distortion is characterized by the parameter s ∈ {−12,−10,−6, 0, 6, 10, 12}, which defines the coordinates of the central node of the plate quarter. Note that the parameter 's' can assume negative and positive values, as shown in Fig. 2: if the parameter s is negative, the considered element is convex; if the parameter s is positive, the element becomes non-convex.

A static analysis is performed and the results are computed in terms of transverse displacement $U_z = u_z(a, a, 0)$ (the center point on the midsurface of the entire plate) for the distorted mesh and it is normalized with respect to the reference value U_{z0} calculated for the regular mesh (s = 0). The results computed by varying the parameter s are reported in Fig. 3.

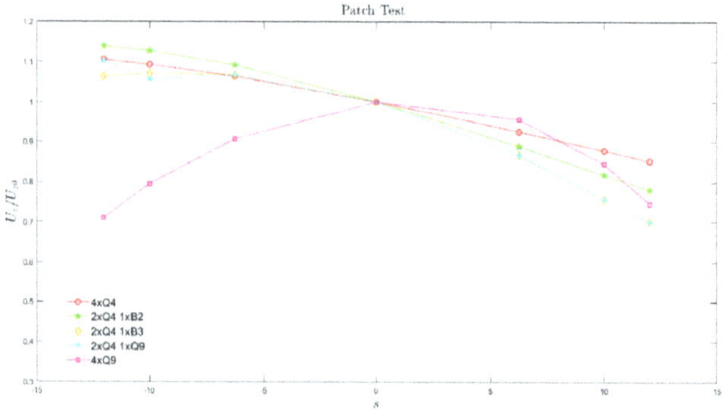

Figure 3: Variation of the ratio U_z/U_{z0} for CC quarter plate

One has to take into account that the 2x2Q9 solution is the most accurate with the highest number of degrees of freedom (DOFs=225), the 2x2Q4 (DOFs=81) is the least accurate and the other solutions are in the middle according to their related degree of approximation. The meshes 2xQ4 1xQ9 (DOFs=108) and 2xQ4 1xB3 (DOFs=108) are very similar to each other and they are compared to highlight that their solution are exactly the same, although the first uses a plate element (Q9) and the other one a beam element (B3).

Conclusions

The results demonstrate that the sensitivity to the distortion of the mixed meshes with beam elements is comparable to the regular meshes (2x2Q4 and 2x2Q9). In particular, the use of B2 beam element in the mesh 2xQ4 1xB2, with a number of DOFs equal to the mesh 2x2Q4 (DOFs=81), doesn't compromise the trend of the results when the mesh is distorted, although it permits to differentiate the number of nodes along the edges of the element (see Figure 3(a)) by reducing also the total number of the nodes in the element with respect to the other mixed meshes.

References

[1] H. Chi, L. Beirão Da Veiga, G.H. Paulino, Some basic formulations of the Virtual Element Method (VEM) for finite deformations, Comput. Methods Appl. Mech. Eng. 318 (2017) 148-192. https://doi.org/10.1016/j.cma.2016.12.020

[2] L. Beirão Da Veiga, F. Brezzi, L.D. Marini, A. Russo, The hitchhiker's guide to the virtual element method, Math. Models Methods Appl. Sci. 24:08 (2014) 1541-1573. https://doi.org/10.1142/S021820251440003X

[3] G. Li, E. Carrera, M. Cinefra, A.G. De Miguel, A. Pagani, E. Zappino, An adaptable refinement approach for shell finite element models based on node-dependent kinematics, Compos. B. Eng. 210 (2019) 1-19. https://doi.org/10.1016/j.compstruct.2018.10.111

[4] M. Cinefra, Non-conventional 1D and 2D finite elements based on CUF for the analysis of non-orthogonal geometries, Eur. J. Mech. A Solids 88 (2021) 104273. https://doi.org/10.1016/j.euromechsol.2021.104273

Analysis of the manufacturing signature on AFP-manufactures variable stiffness composite panels

Alfonso Pagani[1,a *] and Alberto Racionero Sánchez-Majano[2,b]

[1]Mul2 Lab, Department of Mechanical and Aerospace Engineering, Politecnico di Torino, Corso Duca degli Abruzzi 24, Turin, Italy

[a]alfonso.pagani@polito.it, [b]alberto.racionero@polito.it

Keywords: Variable Stiffness Composites, Automated Fiber Placement, Unified Formulation, Defect Modeling

Abstract. Variable stiffness composites broaden the design space, in comparison with straight-fiber composites, to meet fixed mechanical performance. Nevertheless, the manufacturing of these advanced composites incurs into the presence of undesired fabrication defects such as gaps and overlaps, which alter the mechanical behavior of the laminated parts. In this work, the authors couple the Defect Layer Method, utilized to model defects, with the Carrera Unified Formulation in order to study how the manufacturing signature affects the fundamental frequency of variable stiffness laminates.

Introduction

Novel manufacturing techniques, such as Automated Fiber Placement (AFP), have permitted to conceive new families of laminated structures, namely Variable Stiffness Composites (VSC) or Variable Angle Tow (VAT), in which fiber tows are steered conforming curvilinear paths. Olmedo and Gürdal [1] investigated the buckling of VAT plates for different boundary conditions and rotations of the fiber path. Nevertheless, in the cases that Olmedo and Gürdal considered, the minimum turning radius, which determines whether a laminate is manufacturable or not, was not taken into account. Therefore, not all the solutions were feasible from a fabrication point of view. Besides, due to the manufacturing features inherent to AFP, imperfections are prone to arise during the fabrication process, namely gaps and/or overlaps, and affect the structural performance [2].

Many authors have proposed numerical methods to investigate the effect of gaps and/or overlaps on the mechanical properties of variable stiffness composites. Blom et al. [2] investigated how gaps affect the strength and stiffness of VAT using the Finite Element Method (FEM). They concluded that increasing the laminate's total gap area deteriorates strength and stiffness. Fayazbakhsh et al. [3] proposed the Defect Layer Method (DLM), which permits to capture the gap and/or overlap areas that appear in the laminate without incurring into an excessive computational burden as in [2].

In this manuscript, we tackle the influence of gaps and overlaps on VAT laminates by coupling DLM with the Carrera Unified Formulation (CUF) [4]. CUF permits obtaining the governing equations of any structural theory without making *ad hoc* assumptions. So far, CUF has proven to predict accurate stress states [5], as well as capture the influence of multiscale uncertainty defects on the mechanical performance of VAT plates [6].

Variable stiffness composite plates and defect modeling

VAT laminated components are fabricated by steering fiber bands along curvilinear paths. Throughout the years, different variation laws have been investigated. This work focuses on the linear variation, in which the local fiber orientation, $\boldsymbol{\theta}$, varies along the $\boldsymbol{x'}$ direction as follows:

$$\theta(x') = \phi + T_0 + \frac{T_1 - T_0}{d}|x'|. \quad (1)$$

The physical meaning the parameters involved above is depicted in [1], and are illustrated in Figure 1.

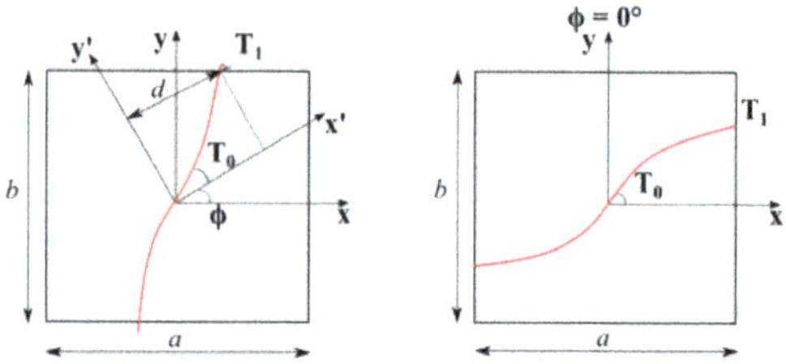

Figure 1. Representation of the fiber parameters involved in the definition of a VAT laminate.

Despite the steering capability of VAT plates, the AFP machines that manufacture them present some limitations, being one of them the curvature of the laid fiber path. A manufacturing feature inherent to AFP is the presence of gaps and overlaps in the final part, thereby affecting the structural performance. DLM [3] is considered to model these imperfections. As mentioned in [3], the defect area percentage is the only parameter to modify the elastic properties or the thickness associated to the finite element. Note that the elastic properties vary if gaps are considered, whereas an increase in the thickness is provided in the case of overlap.

Unified finite elements

2D FE are implemented within the CUF formalism. According to [4], the 3D field of displacement can be expressed in terms of arbitrary through-the-thickness expansion functions, $\boldsymbol{F_\tau(z)}$, of the 2D generalized unknowns laying over the $\boldsymbol{x-y}$ plane. That is,

$$\boldsymbol{u}(x,y,z) = F_\tau(z)\boldsymbol{u}_\tau(x,y).\ \tau = 1,\dots,M \tag{2}$$

Therein, M denotes the number of expansion terms, and $\boldsymbol{u}_\tau(x,y)$ is the vector containing the generalized displacements. The analysis of multi-layered structures is commonly conducted by following an Equivalent-Single-Layer (ESL) and Layer-Wise (LW) approach. In this manuscript, ESL models are built using Taylor polynomials as F_τ in the thickness direction. On the other hand, LW utilizes Lagrange polynomials over the single layers and then imposes the continuity of displacements at the layer interfaces, as in [7]. In this context, TEn denotes a TE of the n-th order, whilst LEn indicates the usage of an LE with n-th order polynomials. Moreover, XLEn means that X Lagrange polynomials of n-th order are used to describe each layer of the laminate.

Utilizing the FE and shape functions $N_i(x,y)$, the displacement field becomes:

$$\boldsymbol{u}(x,y,z) = N_i(x,y)F_\tau(z)\boldsymbol{q}_{\tau i}(x,y).\ i = 1,\dots,N_n \tag{3}$$

In Eq. (3), $\boldsymbol{q}_{\tau i}$ denotes the unknown nodal variables, and N_n indicates the number of nodes per element. In this work, 2D nine-node quadratic elements, referred to as Q9, are employed as N_i for the $x-y$ plane discretization. For the sake of brevity, the governing equations that calculate the fundamental frequency are not reported but can be found in [4].

Machine simulation: identification of defected regions

The steering of fiber bands along a fixed direction, and shifting the AFP head in its perpendicular direction to generate the subsequent fiber course, leads to the presence of gaps and/or overlaps. The location in which they appear depends not only on T_0 and T_1 but also on the steering strategy. Figure **2** illustrates the case of a [⟨0,45⟩] plate in which the fiber courses touch each other at the edge (Figure 2 left) and at the center of the plate (Figure 2 right). The yellow area indicates a gap area, whereas the green area highlights an overlap area.

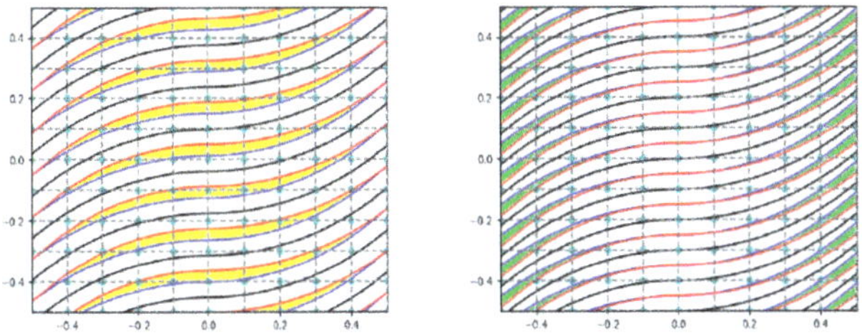

Figure 2. Example of a plate with $[\langle \mathbf{0}, \mathbf{45} \rangle]$ *stacking sequence with full gap (left) and full overlap (right) manufacturing strategy.*

The previous imperfections affect such large areas because the course width is kept constant throughout the steering process. In order to reduce the defect area, the course width has to decrease or increase whenever a course intersects the successive one or it does not reach the precedent course's edge, respectively. The increase or decrease of the course width is achieved by cutting an individual tow and restarting its deposition. This comports the generation of small triangular defected regions, as evidenced in Figure 3.

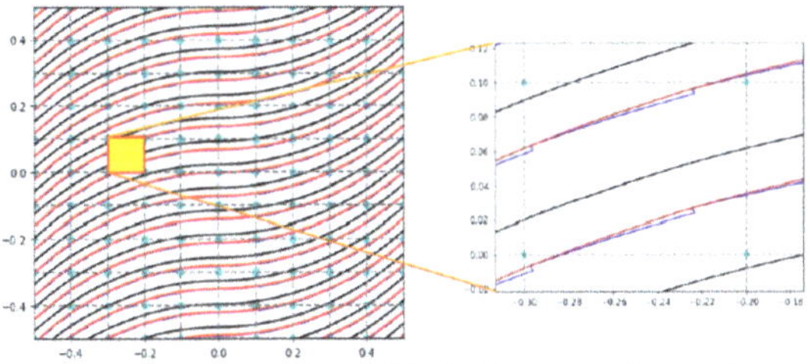

Figure 3. Gap defect correction over a $[\langle \mathbf{0}, \mathbf{45} \rangle]$ *ply The zoomed area shows the triangular gaps that are generated.*

Effect of gap defects on fundamental frequency

In this section, the effect of manufacturing defects is addressed. As depicted in the previous section, two manufacturing strategies are considered: without and with defect correction. These two strategies are denoted as Type 1 and Type 2, respectively. For the following analyses, the tow paths are conformed by sixteen tows, each with a width of $t_w = 3.125$ mm. The FE mesh comprises 10x10 Q9 elements, based on a convergence analysis performed beforehand.

Table 1 presents the effect of gaps on the first five fundamental frequencies of the $[\langle 0{,}45 \rangle, \langle -45, -60 \rangle, \langle 0{,}45 \rangle]$ plate from [8]. An LW-1 LE2 model is employed since, when gaps are considered, the thickness of the laminate remains unaltered, and no additional computational effort is required when compared to the ideal plate. It is observed that gaps lead to a decrease in the fundamental frequency because resin-rich areas are present within the layers. Type 1 defects present a much lower frequency than the ideal case, whereas Type 2 limits the fundamental frequency reduction.

Table 1. Effect of manufacturing gaps on the $[\langle \mathbf{0}, \mathbf{45} \rangle, \langle \mathbf{-45}, \mathbf{-60} \rangle, \langle \mathbf{0}, \mathbf{45} \rangle]$ *plate.*

Model	f_1 [Hz]	f_2 [Hz]	f_3 [Hz]	f_4 [Hz]	f_5 [Hz]
No defects	609.91	903.93	1216.18	1328.88	1469.58
Gap Type 1	547.82	797.10	1083.41	1159.71	1295.57
Gap Type 2	599.10	888.99	1193.21	1306.87	1442.75

Conclusions

This manuscript has discussed the effect of manufacture-induced gaps on the fundamental frequency of VSC laminates. The position where gaps and overlaps will appear during manufacturing has been predicted. These defects were incorporated into the FE model by means of the Defect Layer Method. As it was expected, the presence of gaps incurred a decrease in the fundamental frequency. Future works will be related to the optimization of VAT plates in which the aforementioned manufacturing imperfections are considered.

References

[1] R. Olmedo, Z. Gürdal. Buckling response of laminates with spatially varying fiber orientations. In 34th Structures, Structural Dynamics and Materials Conferences (1993), 1567. https://doi.org/10.2514/6.1993-1567

[2] A. W. Blom, C. S. Lopes, P. J. Kromwijk, Z. Gürdal, P. P. Camanho. A theoretical model to study the influence of tow-drop areas on the stiffness and strength of variable-stiffness laminates. Journal of Composite Materials 43(5) (2009), 403-425. https://doi.org/10.1177/0021998308097675

[3] K. Fayazbakhsh, M. A. Nik, D. Pasini, L. Lessard. Defect layer method to capture effect of gaps and overlaps in variable stiffness laminates made by automated fiber placement. Composite Structure 97 (2013), 245-251. https://doi.org/10.1016/j.compstruct.2012.10.031

[4] E. Carrera, M. Cinefra, M. Petrolo, E. Zappino. Finite Element Analysis of Structures through Unified Formulation. Wiley & Sons, Hoboken, New Jersey. 2014.

[5] A. R. Sánchez-Majano, R. Azzara, A. Pagani, E. Carrera. Accurate stress analysis of variable angle tow shells by high-order equivalent-single-layer and layer-wise finite element models. Materials 14(21) (2021), 6486. https://doi.org/10.3390/ma14216486

[6] A. Pagani, M. Petrolo, A.R. Sánchez-Majano. Stochastic characterization of multiscale material uncertainties on the fibre-matrix interface stress state composite variable stiffness plates. International Journal of Engineering Science 183 (2023), 103787. https://doi.org/10.1016/j.ijengsci.2022.103787

[7] E. Carrera. Theories and finite elements for multi-layered, anisotropic, composite plates and shells. Archives of Computational Methods in Engineering 9(2) (2002), 87-140.

[8] H. Akhavan, P. Ribeiro. Natural modes of vibration of variable stiffness composite laminates with curvilinear fibers. Composite Structures 93(11) (2011), 3040-3047. https://doi.org/10.1016/j.compstruct.2011.04.027

An analytical tool for studying the impact of process parameters on the mechanical response of composites

E. Zappino[1,a*], M. Petrolo[1,b], R. Masia[1,c], M. Santori[1,d] and N. Zobeiry[2,e]

[1]MUL2 Lab, Department of Mechanical and Aerospace Engineering, Politecnico di Torino, 10129 Turin, Italy

[2]Department of Materials Science and Engineering, University of Washington, Seattle, WA, 98195, USA

[a]enrico.zappino@polito.it, [b]marco.petrolo@polito.it, [c]rebecca.masia@polito.it, [d]martina.santori@polito.it, [e]navidz@uw.edu

Keywords: Virtual Manufacturing, Carrera Unified Formulation, Process-Induced Deformations, Residual Stresses

Abstract. The present work presents a numerical framework able to predict the impact of the manufacturing process on the mechanical performance of the composite component. A simple one-dimensional thermochemical model has been used to predict the evolution of the degree of cure of the resin for a given thermal cycle. The homogenized properties at the lamina level have been obtained through a classical mixtures law and employed to predict the process-induced deformations. A refined one-dimensional model, derived in the framework of the Carrera Unified Formulation, has been used to provide accurate results with reduced computational costs. The virtual manufacturing framework has been used to investigate the impact of the process parameters on process-induced defects of a simple composite part. Different curing cycles have been considered and their outcomes discussed. The results demonstrate the capability of the present numerical tool to correlate the manufacturing process parameters with the mechanical performances of the final component.

Introduction

Composite materials are increasingly being used in various applications in industry due to their superior mechanical properties [1]. The manufacturing of composite structures for aerospace applications must meet strict requirements in terms of process-induced defects. The use of in-autoclave processes ensures higher mechanical performances of the final component by reducing the presence of voids but, on the other hand, involves the use of high temperature and pressure that can lead to residual deformations and stresses. [2]. The process-induced defects may lead to an early failure of the component or to geometrical inaccuracies that make the structural assembly complicated. Numerical tools based on the finite element method (FEM) can be used to predict process-induced defects but, the three-dimensional nature of the problem requires solid models to be used with a consequent high computational cost. In this work, a refined one-dimensional model, developed within the field of Carrera Unified Formulation (CUF) [3], is used to obtain residual deformations due to the curing process accurately with low computational cost [4]. A one-dimensional thermochemical model is used to predict the evolution of temperature and degree of cure during the process of the composite structure [5]. The mechanical properties of the composite laminate are obtained from a micromechanical model based on the law of mixtures [6].

Equations and model

The cure model is based on the heat transfer governing equation through thickness, i.e., the one-dimensional Fourier thermal conduction equation:

$$\dot{Q} + k\frac{\partial^2 T}{\partial z^2} = \rho H_r V_r \frac{d\alpha}{dt} + k\frac{\partial^2 T}{\partial z^2} = \rho c_p \frac{\partial T}{\partial t} \qquad \text{for } T(z,t) \text{ in } (0 < z < l) \qquad (1)$$

where k, ρ, and c_p are the thermal conductivity, density, and specific heat of the composite, respectively, and are assumed constant during the process. α is the degree of cure, H_r is the total heat released from the resin reaction, V_r is the volume fraction of the resin. The thickness of the laminate is equal to l. T and t are temperature and time. The normal to the inplane dimension of the composite is the z-direction. The term $\dot{Q}$ represents the internal heat generated by the exothermic chemical reaction of the resin.

The cure rate for graphite/epoxy material follows this expression, where k_1, k_2 and k_3 are defined by the Arrhenius equation [5]:

$$\begin{aligned} &\frac{d\alpha}{dt} = (k_1 + k_2\alpha)(1-\alpha)(0.47-\alpha) && \text{for } (\alpha \le 0.3) \\ &\frac{d\alpha}{dt} = k_3(1-\alpha) && \text{for } (\alpha > 0.3) \\ &k_i = A_i e^{\frac{-\Delta E_i}{RT}} && \text{for } i = 1, 2, 3 \end{aligned} \qquad (2)$$

R is the universal gas constant, A_i are the pre-exponential coefficients, and ΔE_i are the activation energies.

The mechanical properties of the resin are strongly dependent on the curing process. The instantaneous modulus of the resin can be expressed as a function of the degree of cure according to the α mixing rule [6]:

$$\begin{aligned} &E_m = (1-\alpha_{mod})E_m^{\circ} + \alpha_{mod}E_m^{\circ} + \gamma\alpha_{mod}(1-\alpha_{mod})(E_m^{\infty} - E_m^{\circ}) \\ &\alpha_{mod} = \frac{\alpha - \alpha_{gel}^{mod}}{\alpha_{diff}^{mod} - \alpha_{gel}^{mod}} \end{aligned} \qquad (3)$$

The parameters E_m° and E_m^{∞} are the moduli of the fully uncured and fully cured resin, respectively. α_{gel}^{mod} and α_{diff}^{mod} are the extreme values of the degree of cure considered during the process and are assumed to be $\alpha_{gel}^{mod} = 0$ and $\alpha_{diff}^{mod} = 1$. The parameter γ allows quantification of the contrasting mechanisms of chemical hardening and stress relaxation. It is assumed null in this study. The instantaneous shear modulus of the resin is calculated by the relation of isotropic materials. During the process, the resin has a chemical shrinkage that induces a uniform strain for all principal directions. The mechanical properties of the fiber do not depend on the curing process and are assumed constant. The micromechanics model [6] evaluates homogeneous mechanical properties, chemical shrinkage, and thermal expansion coefficient from fiber and matrix properties.

Results

The considered laminate is 2.54 cm thick and is made of AS4/3501-6 material (graphite/epoxy). The thermophysical characteristics of the material are: $\rho = 1.52 \times 10^3$ kg/m^3, $c_p = 942$ J/(W °C), $k = 0.4457$ W/(m °C), $A_1 = 3.502 \times 10^7$ s^{-1}, $A_2 = -3.357 \times 10^7$ s^{-1}, $A_3 = 3.267 \times 10^3$ s^{-1}, $\Delta E_1 = 8.07 \times 10^4$ J/mol, $\Delta E_2 = 7.78 \times 10^4$ J/mol, $\Delta E_1 = 5.66 \times 10^4$ J/mol, $H_r = 198.9$ kJ/kg [6]. The one-dimensional model consists of ten elements. Convection boundary conditions are applied to the top and bottom of the laminate. The first study allows the present thermochemical model to be verified by comparing the trends in the degree of polymerization and temperature at the centerline of the laminate with those predicted by Bogetti [7] for a cure cycle. The curing process consists of two hold-on periods at temperatures of 389.25 K and 450.55 K for

70 min and 127 min, respectively. As can be seen in Fig.1(a), there is a good match between the two models. The small initial discrepancy in the degree of cure may be due to a different choice of the initial degree of cure not being specified in the reference. The temperature spikes when the air temperature is kept constant are due to the exothermic nature of the resin reaction.

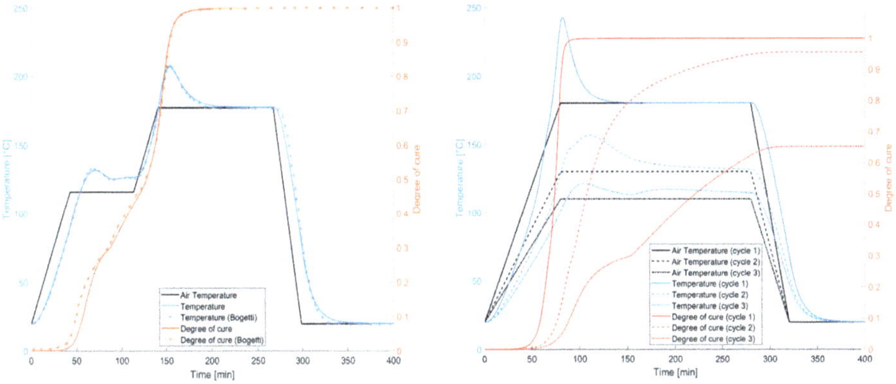

(a) *(b)*

Figure 1: (a) Comparison of the prediction of degree of cure and temperature for a cure cycle with Bogetti's prediction at the centreline of the laminate, (b) Degree of cure and temperature for three different cure cycles

After evaluating the model's accuracy, three cure cycles are applied to the same laminate as in the previous case. The first cycle reaches the maximum temperature of $T_1 = 453.15$ K, the second one the temperature $T_2 = 403.15$ K, and the third $T_3 = 383.15$ K and they are kept constant for 200 min. Fig. 1(b) shows the time trends of temperature and cure degree at the center of the laminate for the three cure cycles. The resin characteristics vary with the degree of cure, except for the Poisson and thermal expansion coefficients, which are assumed constant. The instantaneous resin modulus is calculated using Eq. (3) for each time step, knowing that $E_m^{\circ} = 3.447$ MPa and $E_m^{\infty} = 3.447 \times 10^3$ MPa. The fiber properties are kept constant. Through the micromechanical model, the homogeneous properties of the composite are obtained. These parameters are used as input to the analysis to calculate the deformation of an L-shaped component with an angle of 93° between the two flanges. Since the structure is symmetrical, only half is considered. The length of the single flange is 0.1 m. Refined one-dimensional kinematic model [4] can predict the structure's sping-in angle for the three cure cycles along the curvilinear coordinate x that follows the curvature of the flange. The greatest deviation of spring-in angle is obtained in the curved part of the L-shaped structure. For the three cycles the maximum values of spring-in angle are shown in Table 1.

Table 1: Maximum spring-in angle for the three cycles

	Cycle 1	*Cycle 2*	*Cycle 3*
Spring-in angle [deg]	*1.02*	*0.965*	*0.939*

Summary

A simple one-dimensional thermochemical model made it possible to accurately predict the trends in the degree of cure and temperature of the composite given a cure cycle. If the maximum temperature of the cure cycle is low, the resin is not fully cured at the end of the process. Through the micromechanical model, it was possible to obtain the homogeneous properties of the composite as the degree of cure changes and thus obtain the inputs to derive the deformations induced by the cure cycle. A refined one-dimensional kinematic model allowed the spring-in angle to be estimated. It was shown that the cure cycle influences the spring-in angle. The deformation decreases as the maximum temperature of the cycle decreases.

Further developments may include process optimization, the development of defect mitigation methodologies, or in the development of active process monitoring strategies.

Acknowledgments

This research work has been carried out within the project "Sviluppo di modelli per la manifattura virtuale basati sull'intelligenza artificiale per ridurre i difetti causati dal processo di cura di strutture in composito" funded by Ministero degli Affari Esteri e della Cooperazione Internazionale.

References

[1] M. Hojjati, SV. Hoa. Some Observations in Curing of Thick Thermosetting Laminated Composites. Science and Engineering of Composite Materials, vol. 4, no. 2, 1995, pp. 89-108. https://doi.org/10.1515/SECM.1995.4.2.89

[2] G. Fernlund, C. Mobuchon, N. Zobeiry. 2.3 autoclave processing. Comprehensive Composite Materials II, Elsevier; 2018, pp. 42–62 [chapter 2]. https://doi.org/10.1016/B978-0-12-803581-8.09899-4

[3] E. Carrera, M. Cinefra, M. Petrolo, E. Zappino. Finite element analysis of structures through unified formulation. John Wiley & Sons; 2014. https://doi.org/10.1002/9781118536643

[4] E. Zappino, N. Zobeiry, M. Petrolo, R. Vaziri, E. Carrera, A. Poursartip. Analysis of process-induced deformations and residual stresses in curved composite parts considering transverse shear stress and thickness stretching. Composite Structures, Volume 241, 2020. https://doi.org/10.1016/j.compstruct.2020.112057

[5] A. C. Loos, G. S. Springer. Curing of Graphite/epoxy Composites. Technical Report AFWAL-TR-83-4040, Air Force Wright Aeronautical Laboratories, Wright Patterson AFB, OH, 1983.

[6] T.A. Bogetti, J.W. Gillespie Jr. Process-Induced Stress and Deformation in Thick-Section Thermoset Composite Laminates. Journal of Composite Materials 26. (5), 1992. https://doi.org/10.1177/002199839202600502

[7] A. Johnston. An integrated model of the development of process-induced deformation in autoclave processing of composite structures. PhD thesis. University of British Columbia, 1997.

Acoustic characteristics evaluation of an innovative metamaterial obtained through 3D printing technique

L.M. Cardone[1*a], S. De Rosa[1b], G. Petrone[1c], G. Catapane[1d], A. Squillace[2e], L. Landolfi[2f], A.L.H.S. Detry[2g]

[1] PASTA-Lab (Laboratory for Promoting experiences in Aeronautical STructures and Acoustics), Department of Industrial Engineering - Aerospace Section, Università degli Studi di Napoli "Federico II", Via Claudio 21, Napoli, 80125, Italy

[2] Department of Chemical, Materials and Industrial Production Engineering, University of Naples Federico II, Naples, Italy

[a]luigimaria.cardone@unina.it, [b]sergio.derosa@unina.it, [c]giuseppe.petrone@unina.it, [d]giuseppe.catapane@unina.it, [e]antonino.squillace@unina.it, [f]luca.landolfi@unibg.it, [g]andrealorenzohenrisergio.detry@unina.it

Keywords: Noise Reduction, Sustainability, 3D Printing, Metamaterial

Abstract. The reduction of interior noise level in the transportation sector is a big problem to cope with in view to increase the comfort of passengers. For this reason a great emphasis from the research community is devoted to develop new technology which are able to satisfy the mechanical requirements with concrete benefits from the acoustic point of view. Currently, it does not exist a solution for wideband range of frequency. Indeed, porous materials are characterized by outstanding dissipation in the high frequency range but they exhibit poor performance in the low and medium frequency range, where instead resonant cavities systems have the best performances but with narrow-band sound absorption. For this reason, the design and development of new materials which offers a good acoustic absorption over a wide range of frequencies is requested. In this paper, a hybrid metamaterial is designed, by coupling resonant cavities with micro-porous material and obtained through additive manufacturing technique which enables to model complex geometries that could not be feasible with classical manufacturing. Numerical and experimental studies have been conducted on the manufactured samples of PLA, with an interesting focus on the effect of each parameter which affects the absorption properties.

Introduction

Materials have been used to control wave propagation for centuries. The evolution of these are the metamaterials: artificial structures, typically periodic, composed of small meta-atoms that, in the bulk, behave like a continuous material with unconventional effective properties. Acoustics Metamaterial are classified by the mechanism of sound absorption, i.e. how they can manipulate and control sound waves in ways that are not possible in conventional materials. Metamaterials with zero, or even negative, refractive index for sound offer new possibilities for acoustic imaging and for the control of sound at subwavelength scales [1].

In literature it is possible to find three macro-classes: porous metamaterials, membrane resonators and cavity resonators. The first acoustic meta-atoms were spherical metal cores coated with a soft rubber shell packed to a simple-cubic lattice in a host material, which could exhibit a Mie-type resonance frequency far below the wavelength-scale Bragg resonance frequency of the lattice [2]. Other architectures for acoustic metamaterials involve segments of pipes and resonators in the form of open and closed cavities. In 2006, these configurations composed a metamaterial characterized by a waveguide loaded with an array of coupled Helmholtz resonators [3]. At their collective resonance frequency, a low-frequency stopband is formed. Several other metamaterial-based approaches for realizing unusual acoustic refraction have been demonstrated, as example by coiling up space with labyrinthine structures [4], the sound propagation phase is delayed such that band folding with negative dispersion. Most of the acoustic metamaterial designs described above make use of periodic structures, but given that the concept of acoustic metamaterials is based on the local, internal mechanical response of the structure, there is no reason why metamaterials cannot be made from aperiodic architectures. This idea is beginning to be explored using metamaterials composed of a soft matrix containing an unstructured array of bubbles of a second material. This is the main idea of the concept that has been realized in the laboratory. The subject of this work is a hybrid metamaterial composed by resonant cavities and porous material obtained through 3D printing technique. The 3D printing technique has given to the researcher the possibility to invent different geometry that could be impossible to realize with classical manufacturing. The work will firstly illustrate the technical evolution that has led to the construction of the different type of metamaterials that nowadays are subject of study. In the middle it will be introduced the technique of 3D printing used to create the sample, and in the end the numerical simulation and the experimental results will be reported.

3D printing of porous materials

The samples in additive manufacturing used in this research have been obtained using a printer based on Fused Deposition Modeling (FDM) technology. FDM technology consists in depositing on a printing surface several layers of a material that, layer after layer, form a three-dimensional object. The raw material is usually a filament of a certain diameter, which is found on the market in the form of a coil. The filament is pushed by an extruder and passes into the heated nozzle at a temperature above the glass transition temperature of the material in use. The technology used during the realization phase consists in solubilizing the blowing agent in the print blanks, so that the material is expanded during printing. Unlike foaming the piece by solubilizing an expanding agent inside in a post-printing phase, this methodology, obtaining the desired morphology by controlling a rapid jump in pressure and temperature, allows you not to face problems related to the control of the geometry of the workpiece, because the high residual stresses due to printing deform the workpiece during foaming [5]. The production of polymer foams consists of several stages listed: (a) selection of granules and fibres; (b) extrusion of filament; (c) solubilization of the blowing agent in the filament; (d) AM foam printing. By controlling parameters such as the solubilization time, speed and temperature of the extrusion, it is possible to block filaments and then the foams at different densities (Fig. 1) [6]. In the case of study, in the realization of the specimens it was chosen to proceed with medium density foams. The versatility of this process has allowed the realization of the metamaterial of which it has been chosen to characterize the absorption.

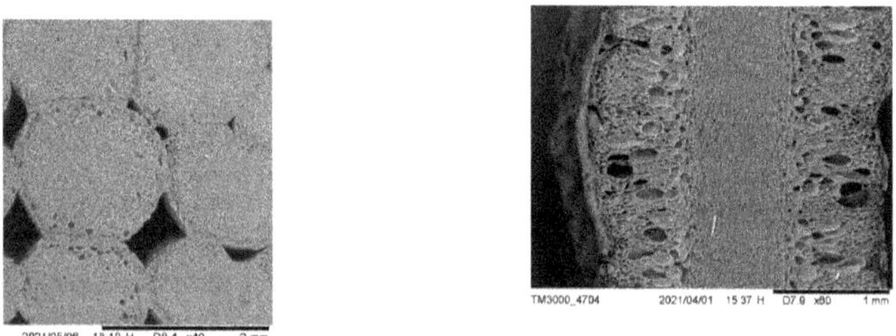

Figure 1 SEM Images of foamed filament (Tammaro et al., 2021)

Materials Research Proceedings 37 (2023) 325-328 https://doi.org/10.21741/9781644902813-71

Analyzed configuration

The tested specimens have the following geometrical data: height of 30mm and 19 holes with a diameter of 3mm, passed, arranged radially. A numerical analysis was immediately carried out using Comsol software to evaluate the acoustic performance of the identified geometry. Subsequently, the specimen were manufactured. In addition to the foaming parameter, it was also chosen to evaluate two possible printing logics: the first consists of printing the full cylinder and then drilling the holes using different techniques; the second, on the other hand, involves creating the holes during printing. This dual mode of realization showed how the production process clearly affects the behavior from the point of view of absorption.

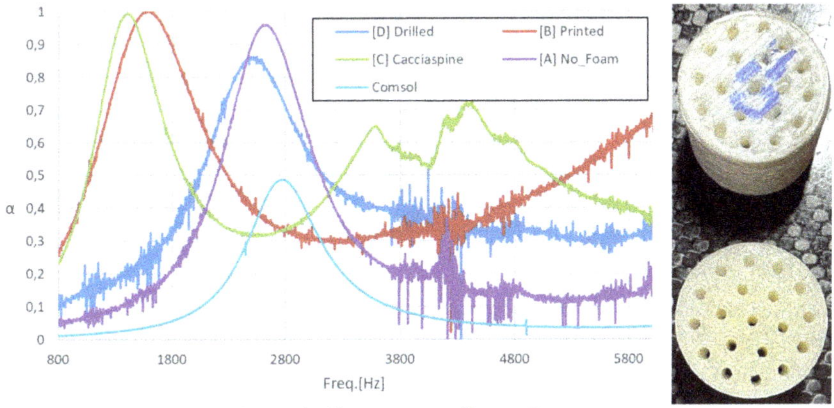

Figure 2: Absorption coefficient diagram

Several aspects are evident from the diagram in Fig. 2: 1) Validation of the numerical model: in order to obtain this comparison, the specimen was first printed using both modes mentioned above, then it was then tested in the impedance tube, complying with the regulations of test's standard, and finally after extracting the data, the curve of the numerical model (petrol) was compared with the experimental (violet). It is observed that the material (unfoamed PLA) is found to have a damping effect that amplifies the absorption curve at the frequency identified by the numerical model reaching almost unity. 2) Drilling method: it is interesting to note how the methodology used to make the holes alters the absorption behavior of the specimen. Three drilling methodologies were chosen to be analyzed. The first one is to drill the specimen in a cold running water bath by using a drill press. Despite the cold water immersion, the temperature reached between the drill bit and the material was such that the hole channel melted locally, thus occluding the internal porous structure of the filament, in fact obtaining the same absorption behavior as the specimen printed with unfoamed filament (blue curve). The second methodology consists of cold drilling using a drill bit and a hand vise to exert compression force. The effect of this second methodology is reported by the green curve. There is a copious shift present due to possible delaminating of the various layers of the specimen and imperfect flatness of the channels. Also from this curve, it is possible to see the presence of a new, higher bell, attributable to the behavior of the foams. Third and last methodology examined consists in directly printing the specimen with the hole locations. First fundamental difference from the previous methodologies lies in the definition of the channel of the hole. The mold in fact overlaps many circles thus obtaining the outer channel of the hole. This overlap is not always perfect and this kind of internal losses in the specimen, thus altering the path that the wave can follow. The result of this condition is the curve in orange. Further tests are in progress to obtain a drilling method that allows full use of the porous structure inside the filament.

Conclusion
Metamaterials are the subject of interest in the world's leading research centers, and the spread of 3D printing has facilitated the creation of prototypes characterized by geometries that cannot be obtained by classical mechanical machining processes. The aim of this work was to create a metamaterial capable of encapsulating all the advantages of the various sound absorption systems. Numerical and experimental analysis were conducted on one of the simplest configurations that could be realized. The results shows that the main difficult aspect is the realization of the holes. Those holes are the inlet for the acoustics wave to the porous foam of the inner part of the filament. If the channel of the hole is melted the only advantage of this technology is the lightness of the sample respect of an unfoamed one. Further analysis would be conducted to determine the best way of perforation and to implement new geometry with an consistent airgap inside the sample.

Reference
[1] L. Chang, A. Jiang, M. Rao, F. Ma, H. Huang, Z. Zhu, Y. Zhang, Y. Wu, B. Li, Y. Hu, Progress of low-frequency sound absorption research utilizing intelligent materials and acoustic metamaterials, RSC Adv. 11 (2021), 37784–37800. https://doi.org/10.1039/D1RA06493B
[2] L. Z., Zhang, X., Mao, Y., Zhu, Y.Y., Yang, Z., Chan, C.T., Sheng, P., Locally Resonant Sonic Materials. Science 289 (2000), 1734–1736. https://doi.org/10.1126/science.289.5485.1734

[3] N. Fang,, D. Xi, J. Xu, M. Ambati, W. Srituravanich, C. Sun, X. Zhang, Ultrasonic metamaterials with negative modulus, Nat. Mater. 5 (2006), 452–456. https://doi.org/10.1038/nmat1644

[4] M. Yang, P. Sheng, Sound Absorption Structures: From Porous Media to Acoustic Metamaterials. Annu. Rev. Mater. Res. 47 (2017), 83–114. https://doi.org/10.1146/annurev-matsci-070616-124032

[5] M.G.M. Marascio, J. Antons, D.P. Pioletti, P.E. Bourban, 3D Printing of Polymers with Hierarchical Continuous Porosity, Adv. Mater. Technol. 2 (2017), 1700145. https://doi.org/10.1002/admt.201700145

[6] D. Tammaro, A.L. Henry Detry, L. Landonfi, F. Napolitano, M.M. Villone, P.L. Maffettone, A. Squillace , Bio-Lightweight Structures by 3D Foam Printing, in: 2021 IEEE 6th International Forum on Research and Technology for Society and Industry (RTSI), IEEE, (2021) Naples, Italy, pp. 47–51. https://doi.org/10.1109/RTSI50628.2021.9597272

Hygrothermal effects in aeronautical composite materials subjected to freeze-thaw cycling

Christian Bianchi[1], Pietro Aceti[1*] and Giuseppe Sala[1]

[1] Department of Aerospace Science and Technology, Politecnico di Milano, Milan, Italy

*pietro.aceti@polimi.it

Keywords: Fibre Reinforced Composite, Humidity Absorption, Freeze-Thaw Cycle, Hygrothermal Effects

Abstract. Fiber-reinforced composites (FRC) are becoming increasingly popular in aerospace, automotive and energy sectors. Despite the advantages owed to their strength and lightweight, understanding their behavior in different environments poses challenges. Particularly, humidity, temperature, and freeze-thaw cycles can significantly affect the durability of FRC components. This study investigates the impact of humidity, temperature, and freeze-thaw cycles on FRC inter-laminar areas and the matrix/fiber interface. Experimental methods, including heat analysis, X-Ray tomography and mechanical testing will assess the material's response to changing environmental conditions. This research enhances our understanding of FRC behavior, crucial for designing and maintaining FRC components.

Introduction

Fiber-reinforced composites (FRC) are increasingly utilized in various industries, including aerospace, automotive and energy. However, the extensive application of these advanced materials introduces new challenges in understanding their component behavior throughout their operational lifespan. Environmental factors, such as humidity and temperature, can adversely affect the properties of the matrix and weaken the fiber/matrix interface. Research indicates that the interlaminar regions and the matrix/fiber interface are particularly susceptible to the influence of humidity and temperature. Moreover, the mechanical properties of the composite are significantly impacted by the combined presence of moisture and the freeze-thaw cycle, surpassing the effects of individual environmental factors. Microcracks are induced by freeze-thaw cycles, which increases moisture penetration into composite structures [1,2].

Case studies

A/c structures, designed to withstand challenging conditions while prioritizing strength, weight and safety, are highly specialized and technologically advanced. Keeping this in mind, two case studies are conducted to establish accurate and meaningful thermal cycles. One of them considers the environmental conditions of commercial flights, typically cruising at altitudes between FL320 and FL400. At these altitudes, the International Standard Atmosphere (ISA) indicates temperatures ranging from -48°C to -55°C. The atmospheric pressure varies between 27450Pa and 18750Pa. Due to the compression of air by the aircraft's body, the skin temperature of the aircraft is higher than the static air temperature of -55°C as described in the following equation (1):

$$T_s = T_\infty - T_\infty \cdot \frac{(k-1)\cdot Ma^2}{2} \tag{1}$$

Where T_s is the stagnation point temperature of an ideal gas at temperature T_∞ impacting an object with a Mach number Ma. The temperature cycle is performed considering the ASTM-D7792 each one constituted of a 3-hour static phase at 30°C and 90% relative humidity, followed by a 3-hour thermal ramp to -22°C. General aviation planes typically cruise between 4500 and

9500 feet in altitude (data collected considering the data-sheet of a Cessna-172). For the ISA at this altitude, temperature values span between 6°C and -3°C, far less in respect of a commercial flight. Each cycle consists of 3-hour static phase at 30°C and 90% relative humidity, followed by 3-hour thermal ramp to -22°C.

Interlaminar shear strength

The relevant literature shows how composite materials behave when conditioned and frozen-thawed [3,4]. To understand the phenomena, one must investigate the adhesion between reinforcing fibre and epoxy matrix. Internal microcracks, voids and water inclusion can reduce the material's interlaminar characteristics, lowering its mechanical properties and operational life. The interlaminar shear strength is a matrix-dominated feature that plays a significant role in a variety of applications, as interlaminar shear is frequently the mechanism for load transmission across various composite components. The ASTM D2344 is adopted for manufacturing and test.

Test specimens

Two types of aeronautical composite materials are investigated using 0° ASTM specimens. One consists of carbon-reinforced fibre, while the other consists of glass-reinforced fibers. In both material a differential scanning calorimetry test is performed to assess the glass transitional glass temperature, T_g values span between 140°C and 145°C.

Conditioning process

Temperature and relative humidity influence composite materials degradation. Moisture affects resin and fiber-resin interface under humid conditions. Temperature affects moisture absorption as well, expanding it in freezing conditions and contracting it at room temperature. Moisture absorption is a diffusion process, where water molecules move from high to low concentration until equilibrium. In moisture uptake conditioning, ASTM D5229's "B" procedure is followed. Specimens are immersed in distilled water at 80°C and weighed at defined intervals until saturation. Another parameter variation was investigated to better define composite material humidity absorption. To study carbon fiber-reinforced composites' humidity absorption, two types of specimens were created. The first type, according to ASTM D5229, is made of 0° fibers, while the second is made of 90° fibers.

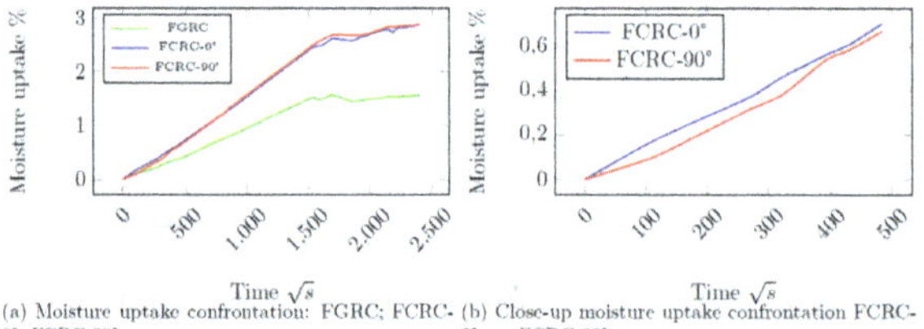

(a) Moisture uptake confrontation: FGRC; FCRC-0°; FCRC-90°
(b) Close-up moisture uptake confrontation FCRC-0° vs. FCRC-90°

Figure 1: Conditioning results in different experimental conditions

Fig.1 data shows that both materials reached saturation in approximately 30 days. Fig.1.b demonstrates an opposite initial humidity absorption trend than expected, considering the higher concentration of fiber ends on the long side of the specimen. Specifically, FCRC-0° exhibits a higher absorption coefficient compared to FCRC-90°. However, after an initial transient, both specimen types (FCRC-0° and FCRC-90°) show similar behaviors regarding the rate and nature of moisture absorption. Ultimately, both reach saturation simultaneously, indicating that the fiber orientation in the plies has minimal impact on the absorption process.

Computed Tomography

A North Star Imaging X25 CT X-Ray Inspection is employed to inspect intralaminar damage due to conditioning and thermal cycling. The techniques works well also if material to be analyzed is not radioopaque (as for example carbon fibers). In this case it previously moistened with a Zinc Iodide solution, which penetrates defects by capillarity, making them visible to X-rays. For non-radio-opaque composite materials an opaque enhanced dye penetrant was been used

Experimental results

Thermal and hygrothermal behavior is studied by separating the effects of water inclusion from those due to environmental temperature variation. The effects of thermal cycles are investigated as a final step.

Thermal effects. Temperature significantly affects the mechanical properties of composites, causing microstructural and property transformations including thermal expansion, degradation, and softening. To establish accurate testing conditions, dry specimens are referenced at a standard temperature of 23°C as a baseline. Shear strengths of carbon-reinforced and glass-reinforced composites are determined as 91.77±0.37MPa and 72.52±1.72MPa, respectively, exhibiting a linear elastic phase followed by sudden load decrease until fracture, in accordance with ASTM D2344. Temperature-dependent behavior is assessed from tests conducted at 23°C, 60°C, 80°C, 110°C and 125°C; the results are presented in Fig.2. Samples undergo controlled heating in a convection oven aided by infrared lights to maintain desired temperatures. Interpolation curves generate mean shear stress summary graphs. Macroscopic examination and X-Ray Tomography confirm the expected fracture behavior of composites across different temperatures.

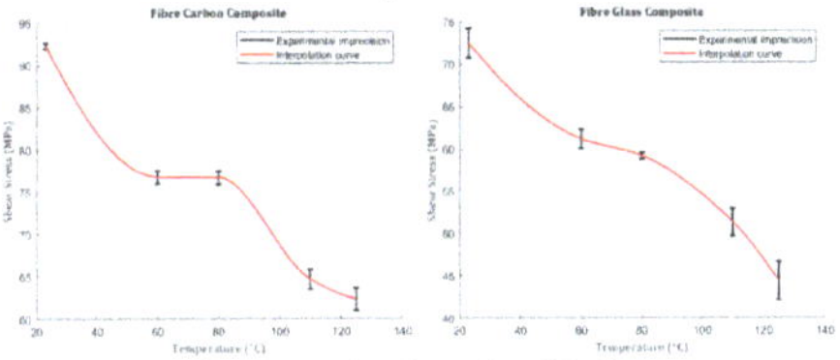

Figure 2: Interpolation ILSS results for FCRC and FGRC specimen at different temperatures.

Hygrothermal effects. Plasticization occurs when water molecules are absorbed into a material, transitioning it from a glassy to a rubbery state. This results in enhanced chain mobility and weaken intermolecular bonds. Consequently, there is a decrease in strength modulus and stiffness, while toughness and strain capacity increase. The ILSS test demonstrates that the carbon-reinforced composite exhibits plastic behavior with a less pronounced fracture compared to the dry material. Fracture occurs due to shear stress at mid-plane, confirmed by macroscopic examination, tomographic analysis (Fig. 3.a), and slow-motion videos. In contrast, the glass fiber reinforced composite initially shows an elasto-brittle behavior and the expected failure mode according to ASTM D2344. However, further investigation reveals an incorrect failure mode during the ILSS test. The wet glass-reinforced specimen collapses at the center until buckling occurs, as observed through tomography (Fig. 3.b).

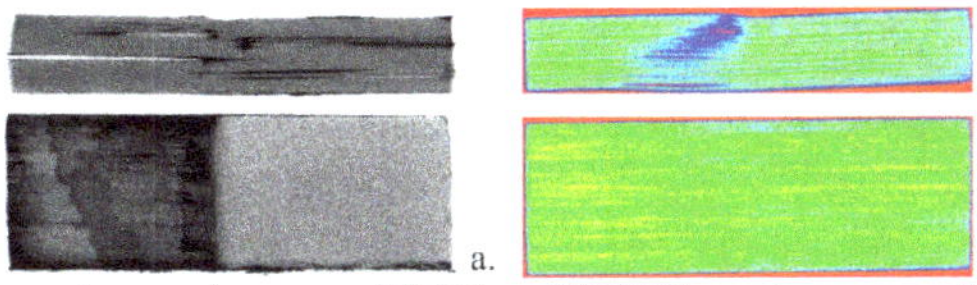

a. b.

Figure 3: Tomography results on an FCRC and FGRC specimen tested in "wet" condition.

Thermal cycling effects. A drying cycle is included in both types of cycles to restore the material to a dry condition once the cycle is completed. By removing the internal water in the composite, the impacts of cycles can be separated from the plasticization effects of the matrix. Fig.4.a and Fig.4.b compare the "Freeze-Thaw" and "Hot-Hot" cycling outcomes. According to previous hypotheses, the results should demonstrate a distinction between the two curves indicating the severity of degradation caused by the "F-T", due to the nature of water expansion within the material. As a matter of fact, the results obtained cannot demonstrate this phenomenon.

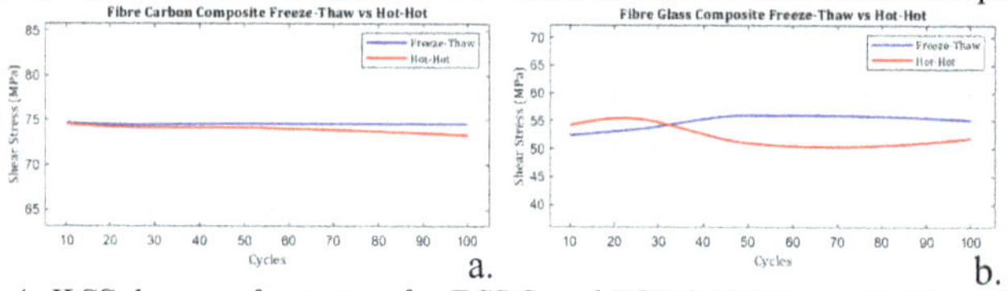

Figure 4: ILSS data confrontation for FCRC and FGRC "F-T" vs. "H-H" cycling.

Conclusions

The study is aimed to establish a representative coefficient indicating mechanical deterioration of materials under harsh environmental conditions, for timesaving in aeronautical industry destructive tests and standardized strength loss parameters in project planning. Two materials commonly used, carbon-reinforced and glass-reinforced composites, were selected. ILSS test investigated the phenomenon. Findings revealed significant shear strength reduction in both materials compared to the dry condition, with the "Hot-Wet" condition showing the most pronounced reduction. Further research is needed to understand failure mechanisms. Existing literature suggests a -25% shear strength reduction [1]. Research on multiple cycles and humidity absorption in composite materials coated with aeronautical paint is of interest, as it provides a more realistic understanding of aeronautical component behavior under hygrothermal effects.

References

[1] Mohammad Abedi, S. Ebrahim Moussavi Torshizi, and Roohollah Sarfaraz. Damage mechanisms in glass/epoxy composites subjected to simultaneous humidity and freezethaw cycles. Engineering Failure Analysis, 120:105041, 2 2021. https://doi.org/10.1016/j.engfailanal.2020.105041

[2] Laurent Cormier and Simon Joncas. Effects of cold temperature, moisture and freeze thaw cycles on the mechanical properties of unidirectional glass fiber-epoxy composites. American Institute of Aeronautics and Astronautics, 4 2010. https://doi.org/10.2514/6.2010-2823

[3] Pietro Aceti, Luca Carminati, Paolo Bettini and Giuseppe Sala. Hygrothermal ageing of composite structures. Part 1: technical review. Composite Structures, 117076, 2023. https://doi.org/10.1016/j.compstruct.2023.117076

[4] Pietro Aceti, Luca Carminati, Paolo Bettini and Giuseppe Sala. Hygrothermal ageing of composite structures. Part 2: mitigation technique, detection and removal. Composite Structures, 117076, 2023. https://doi.org/10.1016/j.compstruct.2023.117076

Polymer matrices for composite materials: monitoring of manufacturing process, mechanical properties and ageing using fiber-optic sensors

Davide Airoldi[1], Pietro Aceti[1*] and Giuseppe Sala[1]

[1] Department of Aerospace Science and Technology, Politecnico di Milano, Milan, Italy;

pietro.aceti@polimi.it

Keywords: Polymer Matrices, Humidity Absorption, Fiber Optic Sensors, Epoxy Resin, Hygrothermal Aging

Abstract. Composite materials have gained significant prominence in the field of aerospace engineering owing to their exceptional strength-to-weight ratio, making them well-suited for structural applications. However, these materials are susceptible to degradation due to exposure to environmental factors, such as humidity and temperature changes. Detecting and quantifying such damage presents considerable challenges, particularly in the case of cyclically loaded components. Fiber Bragg Grating (FBG) sensors provide a non-destructive means of monitoring composite material degradation by leveraging optical reflection to measure changes in strain and temperature. This research aims to assess and validate a methodology for employing FBG sensors to effectively monitor the degradation of composite material matrices. The investigation mainly consists in characterizing the correlation between FBG sensor wavelength shifts and the strains incurred due to the manufacturing process, moisture absorption, and thermal effects. The anticipated outcomes hold the potential to enhance the reliability and safety of composite structures employed within the aeronautical domain.

Introduction

The studies reported in this paper investigated the production process, conducted mechanical tests and evaluated the mechanical properties changes under different conditions in epoxy resin specimens. Understanding the behavior of epoxy resins is crucial due to their widespread usage in aircraft structures and their sensitivity to ageing caused by environmental conditions. Real-time monitoring is essential in aeronautic structures and two FBG sensors were used to monitor deformations and residual stresses during production and to monitor the strains developed during immersion in a hot-water environment. Tensile tests were performed to assess the mechanical properties, including the impact of moisture on resin performance and the recovery after drying. This work lays the foundation for further exploration of fiber optic sensors in moisture monitoring and highlights their potential for cost savings and enhanced aircraft safety [1].

FBG as humidity sensor

Several reference studies have demonstrated the feasibility of utilizing an FBG sensor as a humidity sensor by coating it with a moisture-sensitive coating. In this manner, when the sensor is exposed to a humid environment, the fiber detects a wavelength shift resulting from the strain induced by moisture absorption. The equation that relates the wavelength variations to the different contributions of relative humidity, mechanical deformations and temperature changes is reported here below:

$$\frac{\Delta\lambda_B}{\lambda_B} = (1-p_e)\varepsilon_M + (1-p_e)\frac{E_cA_c}{E_cA_c+E_fA_f}\beta_c\Delta RH + \left\{(1-p_e)\frac{E_cA_c\alpha_c+E_fA_f\alpha_f}{E_cA_c+E_fA_f}+\xi\right\}\Delta T$$

Where ε_M is the axial mechanical strain, ΔT is the temperature change, p_e is the photo-elastic constant, and ξ is the thermo-optical coefficient. Where E is the Young's modulus, A the cross-sectional areas, α the coefficient of thermal expansion, and β_c the constant coefficient of hygroscopic expansion of the polymer. The subscript c stands for the coating, while f stands for the bare fiber. Once decoupled the mechanical, thermal and humidity effects it becomes possible to determine each contribution and monitor the manufacturing and conditioning processes [2], [3], [4], [5].

Production of test specimens

Before facing the testing phase, it was necessary to produce the specimens in compliance with ASTM D638 Standard. Epoxy resin was adopted, since it is a commonly used matrix material in composites due to its excellent adhesive properties, low viscosity, and exceptional mechanical properties. It finds applications in aerospace, automotive, sports, and marine industries. The production process consists of different phases:

- Heat the resin in the oven to make it liquid.
- Place the resin in the vacuum oven performing a degassing process.
- Heat the resin again to reduce its viscosity.
- Cast the resin in the mold using a syringe.
- Cure the resin in the oven.
- Allow the specimen to cool down.

During this process, the strains developed in the specimen were monitored by the FBG sensor, obtaining the plot in Figure 1:

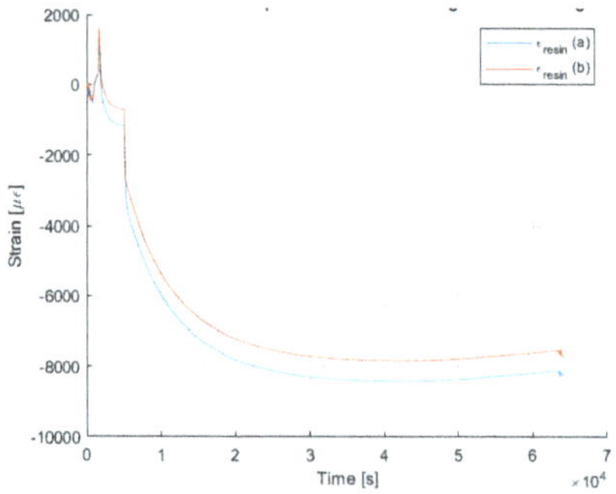

Figure 1: Total strain developed on the resin during manufacturing.

The entire production process caused a compression state-of-stress, leading to a total value of strain close to 8000 micro-strains and a residual internal stress of 25 MPa.

Conditioning process

To reproduce a hot-wet environment and accelerate the moisture absorption process, the test specimens were immersed in an 80°C water bath. The test lasted approximately 1500 hours, during which real-time monitoring of wavelength variations was carried out as shown in Figure 2. By extrapolating the collected data, it was possible to calculate the deformations of the specimens. The strains due to relative humidity shows a rapid increase in the first 100 hours, followed by a sharp decline until reaching a plateau. The non-monotonous behavior could be attributed to various factors, such as the slowdown of the hydrolysis process [6], [7]. The strains obtained in this phase are showed in the figure below:

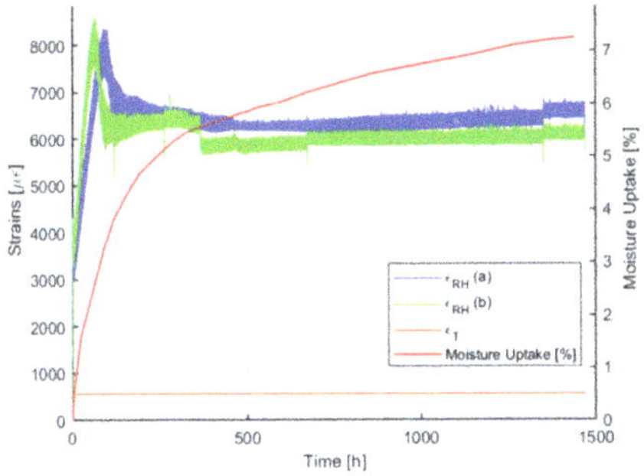

Figure 2: Strain comparison during conditioning.

Mechanical properties

Once having conditioned a series of specimens, their mechanical properties were compared to those characterizing unconditioned specimens and specimens dried after absorption. This was done to observe whether moisture absorption would degrade the material's performance and, if so, whether it could be recovered through a drying process. The figure 3 shows that the moisture absorption causes a degradation of the mechanical properties in terms of Young's modulus and resistance. After te drying process, the material is unable to regain its original properties. In fact, its performance worsens, exhibiting brittle behavior and being prone to premature failure. These tests were carried out using an electro-mechanical tensile machine in compliance with ASTM International Standards [8].

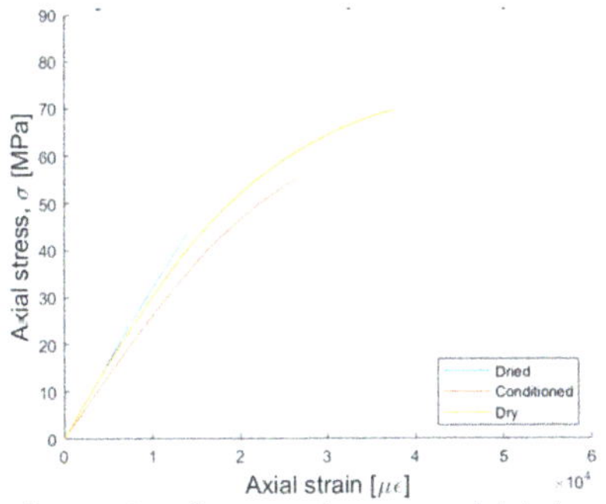

Figure 3: Comparison between dry, wet and dried specimens.

Conclusions

The research describes the technological challenges faced in studying the mechanical properties of epoxy resin. After developing production process to obtain defect-free specimens, tests were conducted to evauate the material's mechanical properties under dry, conditioned, and dried conditions. The specimens immersed in hot water showed a reduction in mechanical properties: -14.3% in Young's modulus, -24% in breakage stress, -31.5% in strain at break. The dried specimens were unable to fully recover their initial characteristics, and they show a further reduction of resistance properties: -37.7% in stress at break, -64.5% in breakage strain. Fiber optic sensors were used to monitor the strains during production and conditioning, showing a good response in detecting the humidity variations. After being immersed for 1500 hours, the fibers were not able anymore to detect deformations during a tensile test because they were damaged.

Further research should investigate the causes of sensor malfunction and study the effect of humid environments and temperature variations on complete composite materials used in the aerospace industry.

References

[1] V. Giurgiuntiu, Structural Health Monitoring of Aerospace Composite, Elsevier, 2016.

[2] P. Giaccari, H. G. Limberger e P. Kronenberg, «*Influence of humidity and temperature on polyimide-coated*» in Bragg Gratings, Photosensitivity, and Poling in Glass fibers and Waveguides: Applications and Fundamentals, Optical Society of America, p. BFB2, (2001). https://doi.org/10.1364/BGPP.2001.BFB2

[3] P. Aceti, «Sensing principles of FBG in humid environments,» Politecnico di Milano, 2023.

[4] Teck L. Yeo, Tong Sun, Kenneth T. V. Grattan, David Parry, Rob Lade, and Brian D. Powell, «Polymer-Coated Fiber Bragg Grating for Relative Humidity Sensing,» *IEEE Sensor Journal,* 2005.

[5] Pascal Kronenberg and Pramod K. Rastogi, Philippe Giaccari and Hans G. Limberger, «Relative humidity sensor with optical fiber Bragg gratings», *Optics Letters,* 2002. https://doi.org/10.1364/OL.27.001385

[6] Pietro Aceti, Luca Carminati, Paolo Bettini and Giuseppe Sala. Hygrothermal ageing of composite structures. Part 1: technical review. Composite Structures, 117076, 2023. https://doi.org/10.1016/j.compstruct.2023.117076

[7] Pietro Aceti, Luca Carminati, Paolo Bettini and Giuseppe Sala. Hygrothermal ageing of composite structures. Part 2: mitigation technique, detection and removal. Composite Structures, 117076, 2023. https://doi.org/10.1016/j.compstruct.2023.117076

[8] ASTM International. Standard Test Method tensile properties of plastics. D638

Materials Research Proceedings 37 (2023) 337-340 https://doi.org/10.21741/9781644902813-74

An energy-based design approach in the aero-structural optimization of a morphing aileron

Alessandro De Gaspari[1,a *], Vittorio Cavalieri[1,b] and Nicola Fonzi[1,c]

[1]Department of Aerospace Science and Technology, Politecnico di Milano, Via La Masa 34, 20156 Milano, Italy

[a]alessandro.degaspari@polimi.it, [b]vittorio.cavalieri@polimi.it, [c]nicola.fonzi@polimi.it

Keywords: Multi-Objective Optimization, Morphing, Aerodynamic Efficiency, Energy Efficiency

Abstract. This paper describes the application of an energy-based optimization procedure for the design of a morphing aileron as an alternative to replace a conventional hinged aileron. The design procedure starts with an aerodynamic shape optimization embedding skin structural constraints and energetic information. Different candidate morphing shapes able to provide reduced drag are obtained, and they differ for the required actuation level. The structural design is then performed through a dedicated multi-objective topology and sizing optimization, aimed at obtaining a structural configuration that achieves the target shape with minimum error and minimum actuation force. The energetic comparison between the designed solution and the hinged solution shows that morphing is convenient also from the energy viewpoint. Finally, a fluid-structure interaction simulation assesses the performances of the designed solution.

Introduction

Nowadays, there is a growing interest in mitigating the environmental impact of air transportation [1]. The morphing concept is one of the research topics that have the potential to improve aircraft efficiency and reduce pollutant emissions and noise [2].

The aim of this work is to design a morphing aileron and to compare its performances with a conventional hinged aileron. From the aerodynamic viewpoint, the smooth curvature change of morphing enables improved efficiency [3]. However, the morphing concept represents a valid alternative only if an overall benefit is achieved [4]. Therefore, the actuation requirement of the morphing solution is used as design objective and as performance index for validation.

The supercritical NASA SC(2)-0412 airfoil, modified to consider its shape change in the wing aileron region, is used as test case in a transonic flight condition (Reynolds=6000000, Mach=0.74). The design procedure is split in two levels. First, shape optimization is conducted, including aerodynamic analyses, skin structural constraints and energetic estimates. Second, structural optimization provides solutions according to the requirements of the previously defined target shapes. Finally, energetic and aero-structural assessments are performed.

Energy-based Shape Optimization

Parameterization technique. The Class-Shape Transformation (CST) method, specialized for morphing devices [5], is used for the airfoil identification and to introduce the morphing shape changes. The adopted approach allows the structural behavior of the morphing skin to be considered, with the analytical computation of the skin stresses from the geometrical description. The morphing shapes are parameterized using two design variables selected among the CST parameters, namely the trailing-edge equivalent rotation and the airfoil boat-tail angle variation.

Actuation energy estimate. The energy requirement for the morphing device is estimated as sum of strain energy and aerodynamic work [6]. The strain energy stored in the structure is due to the morphing deformation process and is computed analytically from the curvature variation of the skin. The aerodynamic work corresponds to the energy required to counteract the application of the aerodynamic forces during the morphing process. It can be estimated from the pressure coefficient distribution and the morphing shape change. The actuation system is assumed to be a linear sliding actuator connected to the lower skin. Assuming a linear variation of the force with the stroke, the maximum actuation force is estimated from the actuation energy.

Problem statement. The shape optimization is formulated as a multi-objective problem. Drag coefficient and actuation force are objectives to minimize. The lift coefficient is constrained to be higher than 1.15. Other constraints prevent axial stresses in the upper skin and limit the maximum curvature variation. The uncertainty associated to the structural design, namely the thickness distribution of the skin, is also considered in the optimization by parametrically solving the optimization problem for different thickness values. Response surface models of the objective functions and constraints are built to perform the optimization.

Response surfaces. Lift and drag coefficients are computed with Reynolds-averaged Navier-Stokes (RANS) equations, solved in SU2 [7]. CST is used to compute strain energy, aerodynamic work, actuation energy and force. All these outputs are computed for a Latin hypercube sample in the space of the design variables. These simulated values are used to construct response surface models through Radial Basis Function (RBF) interpolation.

Shape optimization results. The multi-objective optimization based on response surfaces is performed for four thicknesses values, resulting in several Pareto fronts, which are reported in Fig. 1. Different morphing solutions can provide reduced drag, but they differ for the actuation level required to achieve the morphing shape. Three candidate shapes are selected for the subsequent steps of analysis and design.

Aerodynamic comparison with the hinged aileron. The aerodynamic performances of the selected optimal shapes are compared with the hinged aileron rotated of 9.7 deg, which corresponds to the target lift coefficient requirement. The rigid solution is characterized by higher drag, resulting in lower aerodynamic efficiency at each angle of attack.

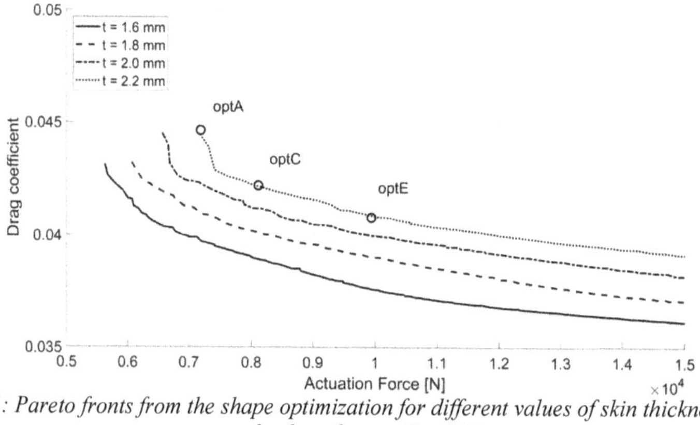

Figure 1: Pareto fronts from the shape optimization for different values of skin thickness t and equal value of target C_L=1.15.

Structural Topology and Sizing Optimization

The structural design of the morphing trailing-edge is based on a medium-fidelity FEM model consisting of a skin section and some internal beams connected to upper and lower skin.

Problem statement. A dedicated multi-objective genetic algorithm is used for the topology and sizing optimization. The design variables describe both the topology (beam attachment points, beam existence) and the sizing (in-plane dimensions of skin and beams). Different objectives are minimized: i) the least-square error (LSE) between the target shape and the morphing shape of the actuated device subject to the aerodynamic loads of the target shape; ii) the actuation force to achieve the morphing shape; iii) the LSE between the undeformed configuration and the undeployed deformed shape under the baseline aerodynamic loads.

Structural design results. The adopted optimization enables to achieve the target shape with minimum error and minimum force. The process is repeated for the three candidate shapes.

Aerodynamic validation. Aerodynamic analyses of the achieved morphing shapes are performed to evaluate if the target lift coefficient is achieved. When this is not guaranteed (*optC*, *optE*), a slightly increase of the actuator stroke allows the target lift coefficient to be met. The three deformed morphing shapes are compared with the hinged shape in Fig. 2.

Structural validation. Strain in the structure is below 0.5%, as depicted in Fig. 3 in case of solution *optE*.

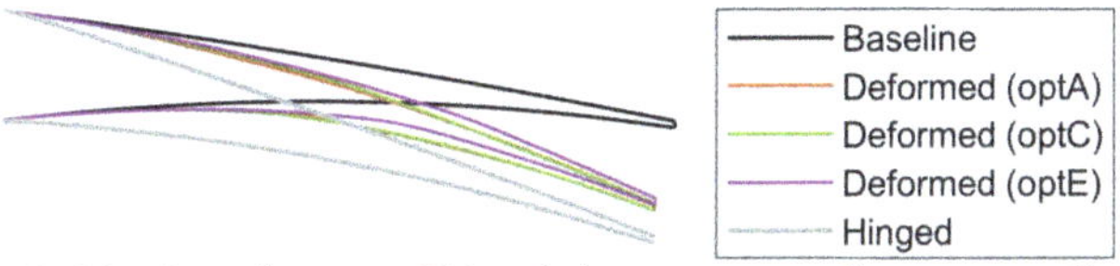

Figure 2: Morphing shapes and hinged shape corresponding to target C_L=1.15.

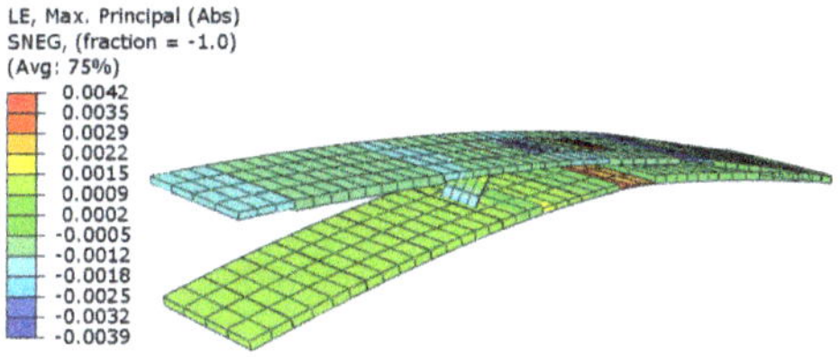

Figure 3: Strain distribution in the structure for solution optE.

Actuation Energy Evaluation

The designed structural solutions that guarantee the target lift coefficient are compared with the hinged solution from the actuation energy viewpoint. The actuation energy for the morphing solutions can be divided in a structural contribution (due to the strain energy) and an aerodynamic contribution (due to the aerodynamic loads), as reported in Fig. 4. The morphing results are compared with the hinged energy result, computed from hinge moment and aileron rotation, and totally due to aerodynamic loads. Although there is a strain energy contribution associated with the morphing process, the actuation energy for morphing solutions is lower than the energy required by hinged solution. This is possible because morphing solutions can achieve the target lift coefficient with smaller trailing-edge equivalent rotation with respect to the rigid rotation of the conventional aileron. Consequently, reduced drag is also achieved.

Performance Assessment

As final validation, a fluid-structure interaction (FSI) analysis, coupling RANS analyses and nonlinear structural analyses, is performed to evaluate how the aerodynamic performances are affected by the structural compliance. This FSI analysis shows that the morphing shape can be achieved as expected from FEM analyses, with negligible differences in the deployed configuration, that do not compromise the requested lift coefficient increment. However, the structural compliance has a small impact on the baseline configuration of the airfoil.

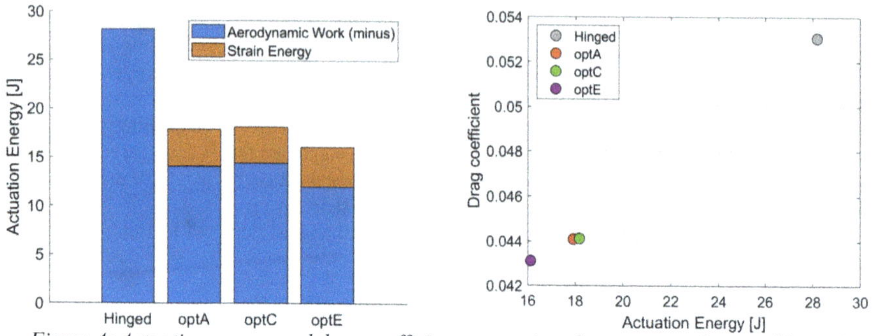

Figure 4: Actuation energy and drag coefficient comparison between morphing and hinged aileron.

Conclusions

This paper has presented the optimum design of a morphing aileron and the comparison of its performances with a corresponding hinged aileron. The proposed energy-based approach has proved successful in providing morphing solutions characterized by enhanced aerodynamic and energetic efficiency with respect to the conventional hinged solution.

References

[1] European Commission, Flightpath. 2050-Europe's vision for aviation: maintaining global leadership and serving society's needs. Publications Office of the EU, 2012.

[2] H. Monner, M. Kintscher, T. Lorkowski and S. Storm. Design of a smart droop nose as leading edge high lift system for transportation aircrafts. In 50th AIAA/ASME/ASCE/AHS/ASC Structures, Structural Dynamics, and Materials Conference, 2009. https://doi.org/10.2514/6.2009-2128

[3] B.K.S. Woods, O. Bilgen and M.I. Friswell. Wind tunnel testing of the fish bone active camber morphing concept. Journal of Intelligent Material Systems and Structures, 25(7):772–785, 2014. https://doi.org/10.1177/1045389X14521700

[4] T.A. Weisshaar. Morphing aircraft technology - new shapes for aircraft design, 2006.

[5] A. De Gaspari and S. Ricci. Knowledge-based shape optimization of morphing wing for more efficient aircraft. International Journal of Aerospace Engineering, 2015. https://doi.org/10.1155/2015/325724

[6] A. De Gaspari. Study on the actuation aspects for a morphing aileron using an energy–based design approach. In Actuators, volume 11, page 185. MDPI, 2022. https://doi.org/10.3390/act11070185

[7] R. Sanchez, R. Palacios, T.D. Economon, H.L. Kline, J.J. Alonso and F. Palacios. Towards a Fluid-Structure Interaction Solver for Problems with Large Deformations Within the Open-Source SU2 Suite. In 57th AIAA/ASCE/AHS/ASC Structures, Structural Dynamics, and Materials Conference, 2016. https://doi.org/10.2514/6.2016-0205

A boundary element method for thermo-elastic homogenization of polycrystals

Dario Campagna[1,a], Vincenzo Gulizzi[1,b], Alberto Milazzo[1,c] and Ivano Benedetti[1,d*]

[1]Department of Engineering, University of Palermo, Viale delle Scienze, Edificio 8, 90128, Palermo, Italy

[a] dario.campagna@unipa.it, [b] vincenzo.gulizzi@unipa.it, [c] alberto.milazzo@unipa.it, [c] ivano.benedetti@unipa.it

Keywords: Polycrystalline Materials, Steady-State Thermo-Elasticity, Computational Homogenization, Computational Micro-Mechanics, Multiscale Materials Modelling, Boundary Element Method

Abstract. A computational framework for thermo-elastic homogenization of polycrystalline materials is proposed. The formulation is developed at the crystal level and it is based on the explicit Voronoi representation of the micro-morphology. The crystal thermo-elastic equations are formulated in an integral form and numerically treated through the boundary element method. The presence of volume integrals, induced by the inherent physics of the thermo-elastic coupling, is addressed through a Dual Reciprocity Method (DRM), which allows recasting the formulation in terms of boundary integrals only. The developed methodology is applied for estimating the homogenized thermo-elastic constants of two widely employed ceramic materials. The method may find applications in multiscale analysis of polycrystalline structural component.

Introduction

Multiscale materials modelling, which focuses on understanding how mechanisms at different length/time scales interact and contribute to the emerging of materials properties at larger scales, is assuming increasing importance in engineering, thanks to developments in experimental materials nano/micro-characterization, which provide a wealth of detailed information about the materials constituents, and to the larger availability of high performance computing, which provides the means to process all the available information in complex modelling frameworks. This favours the understanding of existing materials and boosts the design of new ones, with desired properties at a given scale, consolidating the *materials-by-design* paradigm [1].

An essential item of multiscale modelling is materials homogenization, which generally focuses on inferring the materials aggregate properties at a given scale from the knowledge of the morphological and constitutive features of material constituents at lower scales. Examples are provided by the techniques focused on predicting the properties of composite materials at the laminate level from the characterization of the individual plies or even, at a lower scale, from the properties of fibers and matrices and their mutual arrangement. An important concept in materials homogenization is that of *representative volume element* that may be defined as a material sample small enough to be considered as a material point at the component scale, but large enough to contain a number of elementary micro-constituents sufficient to characterize the aggregate properties in an average sense, so that no meaningful fluctuations in average materials properties are induced by small variations of the specimen size [2].

In this work, an original framework for computational homogenization of polycrystalline materials, is discussed. Polycrystalline materials, which include metals, alloys, or ceramics, are widely employed in engineering and their properties at the component level emerge from the properties of individual crystals and their interactions. The framework is based on the employment of a multi-region boundary integral formulation for representing the thermo-mechanics of the aggregate. Differently from finite element approaches, it allows examining the thermo-elastic problem considering as primary variables only crystal boundary displacements, temperature, tractions and thermal fluxes, promoting remarkable savings in terms of overall number of degrees of freedom and computational storage memory and resolution time, which is of paramount importance for effective multiscale analysis.

This work summarizes the main aspects of the developed framework and reports some homogenization results. The interested readers may access Refs.[3-13] for further details.

Formulation overview

The developed framework is based on: *i*) algorithms for the generation of sets of Voronoi polycrystalline specimens [3,6]; *ii*) robust meshing algorithms able to deal with the statistical features of polycrystalline aggregates [6]; *iii*) a boundary integral representation of the thermo-elastic problem obtained through the *dual reciprocity method* (DRM) [13]; *iv*) algorithms for the discretization and robust numerical integration of the thermo-mechanical boundary integral equations [13]; *iv*) algorithms for the computation of volume averages of the micro-fields, needed for retrieving the *apparent* materials properties and estimating the *effective* ones [13].

Morphology generation/meshing. Voronoi-Laguerre tessellations are employed to represent polycrystalline morphologies retaining the main statistical features of real materials. Open-source packages, e.g. VORO++ (https://math.lbl.gov/voro++/) or NEPER (https://neper.info), are available to generate general 3D tessellations. Specialized algorithms for robust boundary elements meshing are presented in Ref.[6]. *Non-prismatic periodic realizations* are employed in this work, as they remove boundary walls distortions and enhance homogenization convergence.

Boundary integral formulation. Differently from finite elements, the starting point for the formulation is the single-crystal thermo-elastic boundary integral representation

$$c_{ij}(\boldsymbol{x})\, U_j(\boldsymbol{x}) + \int_\Gamma \hat{T}^*_{ij}(\boldsymbol{x},\boldsymbol{y}) U_j(\boldsymbol{y}) d\Gamma = \int_\Gamma \mathrm{U}^*_{ij}(\boldsymbol{x},\boldsymbol{y}) T_j(\boldsymbol{y}) d\Gamma + \int_\Omega \mathrm{U}^*_{ij}(\boldsymbol{x},\boldsymbol{y}) F_j(\boldsymbol{y}) d\Gamma \qquad (1)$$

where: $\boldsymbol{i},\boldsymbol{j}$ =1,...,4; $\boldsymbol{\Gamma}$ and $\boldsymbol{\Omega}$ denote the crystal boundary and domain; x and y are respectively the *collocation* and *integration* point; $\boldsymbol{U_j}$ and $\boldsymbol{T_j}$ are components of generalized thermo-elastic displacements and tractions respectively, which collect, respectively, components of displacements and the temperature jump and components of mechanical tractions and the thermal flux; $\mathbf{U}^*_{ij}$ and $\hat{\boldsymbol{T}}^*_{ij}$ are contain combinations of components of the elastic and thermal fundamental solutions, such as to introduce the thermo-elastic coupling in the integral representation, together with the volume terms $\boldsymbol{F_j}$, which contains components of the thermal gradient. In Eq.(1) the first integral on the left-hand side must be intended as Cauchy principal value.

The presence of the volume integral in Eq.(1) requires special consideration: its presence would call for volume discretization, thus reducing the attractiveness of the integral formulation. To retrieve the benefits of a pure *boundary* representation, this integral can be transformed into a sum of boundary integrals employing the dual reciprocity technique described in Ref.[13] and references therein. Once this transformation is performed, the pure boundary integral equations can be employed for modelling the thermo-mechanics of the aggregate: for each grain, they are collocated at the nodes of the boundary mesh and are numerically integrated using the boundary element method. After such operations, discrete systems expressed in terms of generalized thermo-elastic boundary nodal displacements and tractions are associated with each grain; such equations are then coupled with suitable interface continuity/equilibrium equations and with consistent boundary conditions enforced on the overall aggregate producing a system of the form

$$\begin{bmatrix} \mathbf{A} \\ I \end{bmatrix} \mathbf{X} = \begin{bmatrix} \mathbf{BY} \\ \mathbf{0} \end{bmatrix} \tag{2}$$

where the blocks A and B contain coefficients stemming from the boundary element integration, X and Y collect, respectively, unknown and known nodal components of displacements, temperature, tractions and thermal flux, and I implements the intergranular continuity equations. Eq.(2) is to be solved with sparse-matrix specialized solvers, due to its numerical structure.

Once the numerical solution of the system in Eq.(2) is available, the micro-fields can be solved. The homogenization is performed enforcing periodic thermo-elastic boundary conditions and computing volume averages of stresses and thermal fluxes, which allows retrieving the apparent elastic, conductivity, and thermo-elastic apparent constants. In this work a *statistical computational homogenization* is implemented, which employs both ensemble and volume averages for estimating apparent and effective properties. Assuming ergodicity, ensemble averages of volume averages computed over sets of polycrystals containing a selected number of gains are computed to associate apparent properties with that number of grains; the operation is repeated at increasing number of grains until convergence of the apparent properties is recorded, thus providing an estimate of the material effective properties, see Ref.[13].

Some numerical results

Statistical computational homogenization results about polycrystalline silicon carbide and alumina are presented here. Single crystal properties at room temperature are taken as in Ref.[13] and Fig.(1) shows the convergence of the apparent thermo-elastic constants at increasing number of grains. Ensemble averages are computed over sets of ten realizations and up to 100 grains per realization are considered. The estimated properties always fall within Reuss and Voigt bounds.

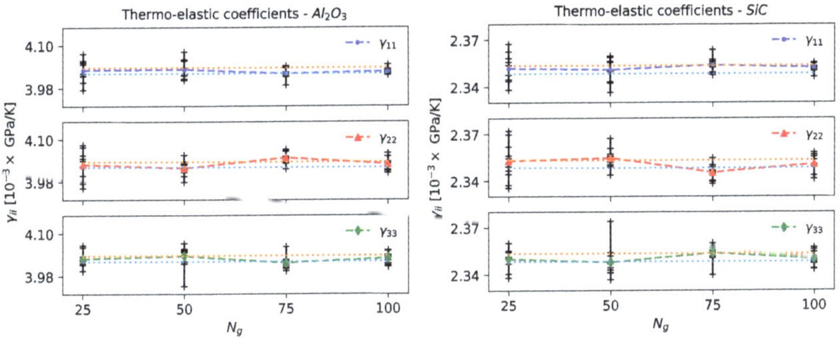

Fig.1: Computational homogenisation results for selected components of thermo-elastic coefficients for polycrystalline alumina and silicon carbide. The + markers identify volume averages over single realisations, the dashed curves correspond to ensemble averages, while the shaded area lies between the Reuss' and Voigt's bounds.

Summary

The development of an original multi-region dual reciprocity boundary elements framework for thermo-elastic homogenization of polycrystalline materials has been discussed, highlighting the benefits offered with respect to other popular approaches. The presented results highlight the effectiveness in estimating materials effective properties. The proposed tool may be employed as component of multiscale analysis tools.

References

[1] M. J. Buehler, Materials by design? A perspective from atoms to structures, MRS Bulletin 38 (2) (2013) 169--176. doi:10.1557/mrs.2013.26. https://doi.org/10.1557/mrs.2013.26

[2] S. Nemat-Nasser, M. Hori, Micromechanics: overall properties of heterogeneous materials, in: North-Holland Series in Applied Mathematics and Mechanics, Vol. 37, 1993.

[3] I. Benedetti, M. Aliabadi, A three-dimensional grain boundary formulation for microstructural modeling of polycrystalline materials, Comput. Mater. Sci. 67 (2013) 249–260. https://doi.org/10.1016/j.commatsci.2012.08.006

[4] I. Benedetti, M. Aliabadi, A three-dimensional cohesive-frictional grain-boundary micromechanical model for intergranular degradation and failure in polycrystalline materials, Comput. Methods Appl. Mech. Engrg. 265 (2013) 36–62. https://doi.org/10.1016/j.cma.2013.05.023

[5] I. Benedetti, M. Aliabadi, Multiscale modeling of polycrystalline materials: A boundary element approach to material degradation and fracture, Comput. Methods Appl. Mech. Engrg. 289 (2015) 429–453. https://doi.org/10.1016/j.cma.2015.02.018

[6] V. Gulizzi, A. Milazzo, I. Benedetti, An enhanced grain-boundary framework for computational homogenization and micro-cracking simulations of polycrystalline materials, Comput. Mech. 56 (4) (2015) 631–651. https://doi.org/10.1007/s00466-015-1192-8

[7] I. Benedetti, V. Gulizzi, V. Mallardo, A grain boundary formulation for crystal plasticity, Int. J. Plast. 83 (2016) 202–224. https://doi.org/10.1016/j.ijplas.2016.04.010

[8] V. Gulizzi, C. Rycroft, I. Benedetti, Modelling intergranular and transgranular micro-cracking in polycrystalline materials, Comput. Methods Appl. Mech. Engrg. 329 (2018) 168–194. https://doi.org/10.1016/j.cma.2017.10.005

[9] I. Benedetti, V. Gulizzi, A. Milazzo, Grain-boundary modelling of hydrogen assisted intergranular stress corrosion cracking, Mech. Mater. 117 (2018) 137–151. https://doi.org/10.1016/j.mechmat.2017.11.001

[10] I. Benedetti, V. Gulizzi, A grain-scale model for high-cycle fatigue degradation in polycrystalline materials, Int. J. Fatigue 116 (2018) 90–105. https://doi.org/10.1016/j.ijfatigue.2018.06.010

[11] I. Benedetti, V. Gulizzi, A. Milazzo, A microstructural model for homogenisation and cracking of piezoelectric polycrystals, Comput. Methods Appl. Mech. Engrg. 357 (2019) 112595. https://doi.org/10.1016/j.cma.2019.112595

[12] F. Parrinello, V. Gulizzi, I. Benedetti, A computational framework for low-cycle fatigue in polycrystalline materials, Comput. Methods Appl. Mech. Engrg. 383 (2021) 113898. https://doi.org/10.1016/j.cma.2021.113898

[13] Benedetti, I. An integral framework for computational thermo-elastic homogenization of polycrystalline materials, Comput. Methods in Appl. Mech. Engrg. 407 (2023): 115927. https://doi.org/10.1016/j.cma.2023.115927

Flutter instability in elastic structures

Davide Bigoni[1,a*], Francesco Dal Corso[1,b], Andrea Piccolroaz[1,c], Diego Misseroni[1,d] and Giovanni Noselli[2,e]

[1]DICAM, University of Trento, Italy

[2]SISSA – International School for Advanced Studies, Trieste, Italy

[a]davide.bigoni@unitn.it, [b]francesco.dalcorso@unitn.it, [c]andrea.piccolroaz@unitn.it, [d]diego.misseroni@unitn.it, [e]giovanni.noselli@sissa.it

Keywords: Flutter, Hopf Bifurcation, Non-Holonomic Systems

Abstract. Flutter instability caused by follower loads has become a reality after the invention of the "freely-rotating wheel device" by Bigoni and Noselli, of the "flutter machine", and of the device to generate Reut-type loads. Further research has proven that flutter instability, Hopf bifurcation, dissipation instabilities, and the Ziegler paradox are all possible in conservative systems, thus disproving an erroneous belief continuing since at least 50 years. Finally, a new type of flutter instability has been addressed, generated by the "fusion" of two structures which are separately stable, but become unstable when joined together. The analysis of instability involves here the treatment of a discontinuity in the curvature of a constraint.

Introduction

Flutter instability is a dynamic behaviour consisting in a blowing-up oscillatory motion, a phenomenon discovered in structural mechanics almost a century ago because of the application of non-conservative (follower forces) to structures. Beside aeroelastic flutter (which will not be considered here), research in this field embraces mechanobiology, growing of plant shoots, motility of cells through eukaryotic cilia, vertebral segmentation in embryos, control problems, the design of soft robotic actuators, graphene peeling, wire drawing, the deployment and retrieval of space tether systems, and solar sails. In the following, new findings are presented in this research topic, including experimental methodologies to follower forces [1-4], non-holonomic constraint [5], effects related to non-smoothness of the equations of motion [6].

The flutter machine

The concept of a follower force is controversial, because the application of these kind of forces to elastic structures leads to several unexpected and counter-intuitive effects: (i.) the lack of Eulerian instability; (ii.) the presence of a Hopf bifurcation; (iii.) the destabilizing effect of dissipation; (iv.) the so-called 'Ziegler paradox' [7]. In this context, Koiter [8] pointed out that follower forces should have been considered as mathematical abstractions, not reproducible in laboratories. As a consequence of that, several attempts to generate these forces have been criticized. A new approach was proposed by Bigoni and Noselli [1] and later perfected with the so-called 'flutter machine' by Bigoni et al. [2, 3], in which a tangentially follower force is generated through the sliding with friction of a freely rotating wheel against a moving support, Fig. 1.

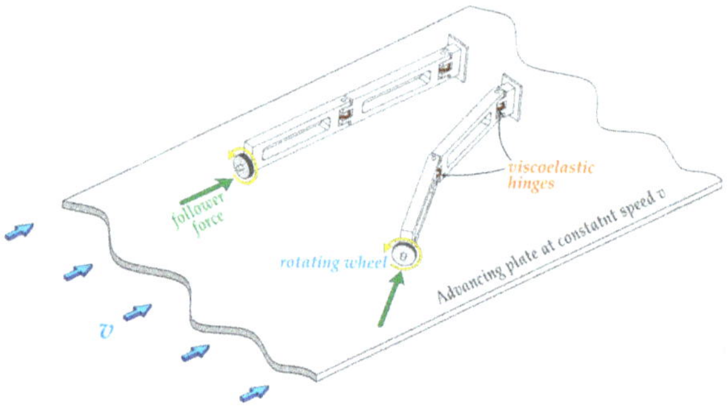

Fig. 1: The way to generate a follower force through the sliding of a freely-rotating wheel against a movable constraint. The wheel is mounted at the free end of a Ziegler double pendulum.

The new experimental set-up has permitted for the first time the systematic validation of the model of following forces and has led to a confirmation of the above listed features (i.)-(iv.). The concepts developed for follower forces also suggested the design of a device to produce forces of the Reut type [4].

Non-holonomic constraints

The concept of the sliding wheel has prompted the idea of substituting the wheel with a similar but purely non-holonomic constraint, Fig. 2. These constraints provide a prescription on the velocity of the end of structure, but not on its position.

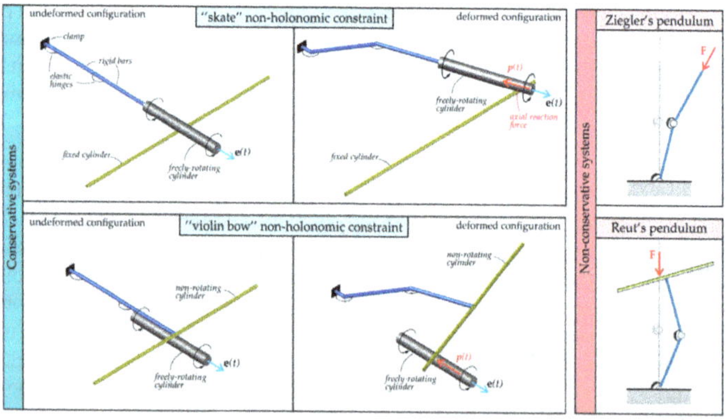

Fig. 2: Application of non-holonomic constraints to generate flutter instability in a double pendulum through the generation of forces similar to those pertinent to the Ziegler's and the Reut's structures.

The remarkable feature emerging from the analysis of non-holonomic constraint is that the latter preserves conservation of energy. Therefore, it is found that flutter instability, Hopf bifurcation, the destabilizing effects of viscosity, and the Ziegler paradox are all possible even within a conservative framework.

Non-smoothness of the equations of motion
As a continuation of the above-presented studies, an elastic structure subject to a follower force was considered in which one end can slide against a linear spring along a smooth constraint presenting a discontinuity in the curvature, as shown in Fig. 3.

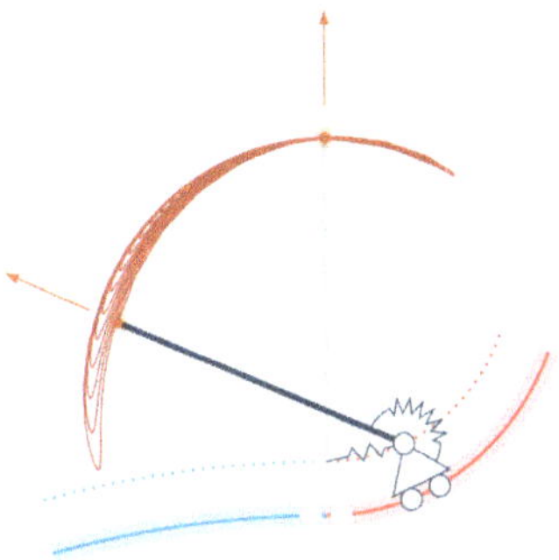

Fig. 3: A structure governed by a non-smooth system of differential equations showing flutter instability as a consequence of the application of a follower force together with a constraint with a jump in the curvature.

The dynamics of the structure is governed by a non-smooth system of differential equations leading to a flutter instability. The most relevant point is that such system can be viewed as the 'fusion' of two different structures, each with a circular sliding profile. Interestingly, the two separate structures are stable under the same load for which the compound structure is unstable. Therefore, the mechanical instability is the effect of both the follower nature of the load and of the discontinuity of the curvature of the sliding profile.

Acknowledgements
The authors gratefully acknowledge financial support from the European Research Council (ERC) under the European Union's Horizon 2020 research and innovation programme (Grant agreement No. ERC-ADG-2021-101052956-BEYOND).

References

[1] D. Bigoni and G. Noselli. Experimental evidence of flutter and divergence instabilities induced by dry friction. J. Mech. Phys. Solids, 59 (2011), 2208-2226. https://doi.org/10.1016/j.jmps.2011.05.007

[2] M. Tommasini, O. Kirillov, D. Misseroni, D. Bigoni. The destabilizing effect of external damping: Singular flutter boundary for the Pfluger column with vanishing external dissipation, J. Mech. Phys. Solids 91 (2016), 204-215. https://doi.org/10.1016/j.jmps.2016.03.011

[3] D. Bigoni, O. Kirillov, D. Misseroni, G. Noselli, M. Tommasini. Flutter and divergence instability in the Pfluger column: Experimental evidence of the Ziegler destabilization paradox, J. Mech. Phys. Solids 116 (2018), 99-116. https://doi.org/10.1016/j.jmps.2018.03.024

[4] D. Bigoni, D. Misseroni. Structures loaded with a force acting along a fixed straight line, or the "Reut's column" problem, J. Mech. Phys. Solids 134 (2020), 103741. https://doi.org/10.1016/j.jmps.2019.103741

[5] A. Cazzolli, F. Dal Corso, D. Bigoni. Non-holonomic constraints inducing flutter instability in structures under conservative loadings, J. Mech. Phys. Solids 138 (2020), 103919. https://doi.org/10.1016/j.jmps.2020.103919

[6] M. Rossi, A. Piccolroaz, D. Bigoni. Fusion of two stable elastic structures resulting in an unstable system, J. Mech. Phys. Solids 173 (2023), 105201. https://doi.org/10.1016/j.jmps.2023.105201

[7] H. Ziegler. Principles of structural stability, Birkhäuser, Basel und Stuttgart (1977). https://doi.org/10.1007/978-3-0348-5912-7

[8] W.T. Koiter.Unrealistic follower forces, J. Sound Vib. 194 (1996), 636–638. https://doi.org/10.1006/jsvi.1996.0383

Can we use buckling to design adaptive composite wings?

Chiara Bisagni[1,a *]

[1]Politecnico di Milano, Department of Aerospace Science and Technology, via la Masa 34, 20156, Milano, Italy

[a]chiara.bisagni@polimi.it

Keywords: Buckling, Composite Structures, Adaptive Wing

Abstract. In aeronautics, buckling has long been considered as a structural phenomenon to be avoided, because characterized by large out-of-plane displacements and therefore by losing the ability to sustain the designed loads. Several recent studies show the possibility to allow composite stiffened panels of primary aeronautical components to work in the post-buckling field so to potentially reduce the structural weight. The present study aims to control buckling behavior of composite structural components for future adaptive wings using novel tailorable and effective mechanisms. Instead of the traditional design against buckling, the idea is to use the nonlinear post-buckling response to control stiffness changes which redistribute the load in the wing structure. Numerical studies are at first conducted on a composite plate and then implemented in a simplified thin-walled composite wing box, where stiffness changes is controlled using buckling.

Introduction

The development of aviation has always been characterized by the search for maximum efficiency in a holistic and multidisciplinary way, through optimal structural, aerodynamic and propulsive efficiency characteristics. Two contrasting challenges have recently become evident: on one side the growth of air traffic, which implies the need to increase transport capacity and reduce travel times while maintaining or improving the level of safety, and on the other the search for a reduction of the environmental impacts of air transport. The Advisory Council for Aviation Research and Innovation in Europe (ACARE) provided a Strategic Research and Innovation Agenda (SRIA), that contains the roadmap for aviation research, development and innovation for reaching the so-called "Clean Aviation" [1], that is part of the European Green Deal.

It is evident that a simple evolutionary improvement of aircraft technologies will not be sufficient to fulfil the challenging targets requested in terms of environmental impact reduction.

In the last years, a few papers appeared in literature showing how buckling could be used for innovative applications, but in aeronautics they are mainly devoted to use buckling for the design of energy harvesters or for bistable structures with limited morphing applications [2-3].

The project NABUCCO (New Adaptive and BUCkling-driven COmposite aerospace structures), funded by the European Union through an ERC Advanced Grant, is proposing to design new adaptive buckling-driven composite structural concepts for the next generation of aircraft configurations, that will impact on two of the biggest levels for reduction of fuel burn needed for future clean aviation: reduced weight and increased efficiency.

NABUCCO Project

The NABUCCO project aims to design new adaptive buckling-driven composite structures for next generation of aircraft configurations. The project is proposing a paradigm shift in aerospace design concepts, considering buckling no more as a phenomenon to be avoided, but as a favorable behavior to be actively exploited.

Let's suppose for a moment to change radically the traditional design approach, and to analyze what happens if from the beginning a structure is designed to exhibit buckling, and not to avoid it.

The main drawbacks that are commonly seen towards buckling in the design of composite aerospace structures are summarized in Table 1, together with the way they are seen positively in NABUCCO.

Table 1. From drawbacks to advantages of buckling.

Buckling must be avoided because...	Buckling can be exploited because...
Buckling produces stiffness reduction.	The stiffness reduction can be used for shape variation, load redistribution and dynamic response change.
Buckling generates large nonlinear deformations.	The large nonlinear deformations can be exploited to significantly change the structural shape with a minimum amount of provided energy.
The transition from pre- to post-buckling can be instantaneous (snap-through).	The fast structural response can be adopted for passive control of peak loads, such as those due to heavy gust excitation or maneuvers.
Buckling strongly depends on geometrical imperfections, material variability, external loads, boundary conditions.	The potential design space can become very large, allowing for many and unusual combinations of structural configurations to obtain pre-defined shapes.

The composite structures have the capability to work safely in the deep post-buckling field with large out-of-plane displacement, as shown for example in [4]. Besides, if appropriately designed, these structures can undergo repeated loading-unloading-loading cycles remaining in the elastic field for the different post-buckling configurations.

The idea of the NABUCCO project is to modify and adapt the aircraft wing shape during the flight mission by the direct use of the buckling phenomena, taking advantage of their typical large nonlinear displacements and stiffness redistribution (Fig. 1).

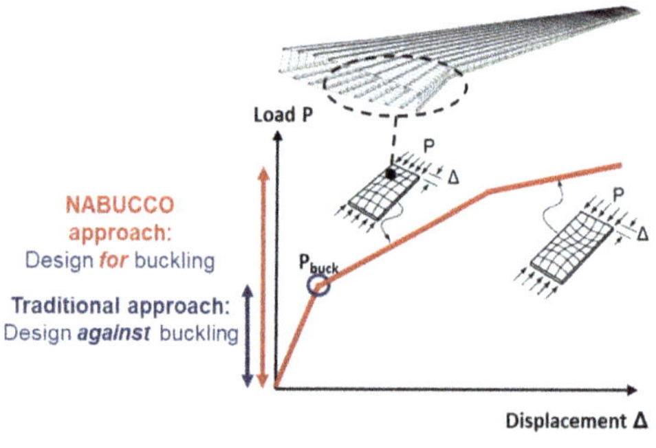

Fig. 1. Approach of NABUCCO project:
Design for buckling of composite aeronautical components.

Buckling-driven Mechanisms for Composite Adaptive Structures

All the potentialities offered by composite materials are used, thanks also to novel manufacturing processes, and the boundary conditions are modified to govern when buckling occurs so to tune multiple non-traditional post-buckling stable configurations.

Buckling-driven mechanisms are developed to obtain different stable configurations changing the boundary conditions, that require only small forces to be reconfigured. In this way it is possible to induce controlled buckling, to obtain local stiffness increase or reduction, infinite number of shapes variation, and load redistribution.

The first steps of the buckling-driven methodology under development consisted in the numerical investigation of a composite plate and then of a simplified thin-walled composite wing box, where stiffness changes were controlled using three buckling-driven mechanisms by restraining the out-of-plane buckling deformation using point, area and maximum displacement constraints [5]. In particular, in the investigation of the composite wing box, the post-buckling behavior of the spar web was controlled by out-of-plane deflection constraints, so that the wing twisting performances were tailored by the relative stiffness of the spar web compared to the rest of the structure. In this way, the nonlinear post-buckling response is used to control stiffness changes, as shown in Fig. 2.

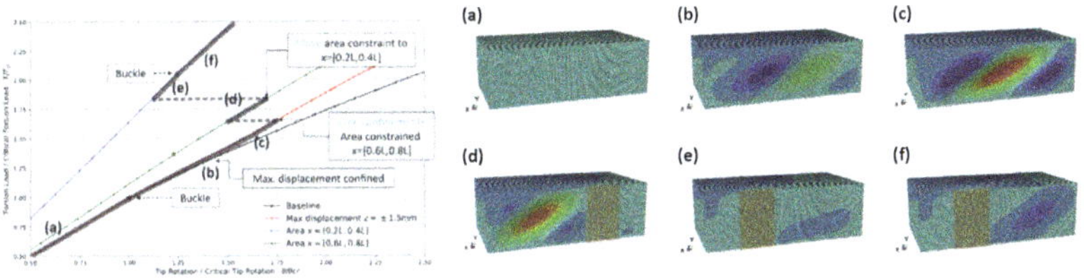

Fig. 2. Adaptive multi-stable composite wingbox:
controlled torque-rotation load path and corresponding wingbox deformation.

In this way, the load in the wing structure is redistributed controlling the stiffness changes, and the methodology allows to gain the real-time controllability of the post-buckling behavior and to enhance the adaptivity of the wing structure to meet multi-stable tailorable situations, with a limited amount of load and actuation energy.

The investigated structures still requires an extensive research of feasibility as they need to be validated experimentally. For this reason NABUCCO is developing a strongly coupled computational-experimental framework, to demonstrate the feasibility of designing aircraft components with controllable buckling, starting from simple composite panel, and then increasing complexity considering a wingbox, and later an adaptive wing for morphing application, as shown in Fig. 3.

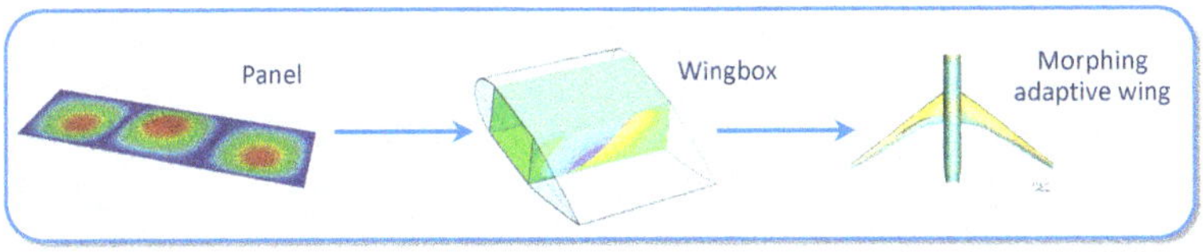

Fig. 3. Increasing complexity of the composite structures investigated in the NABUCCO project.

The first steps of the buckling-driven methodology under development for composite structures show that the design space is significantly enlarged and the designer can identify, manage and control the buckling phenomena.

Summary

The next generation aircraft will require to be lighter, more flexible and sustainable, with the same or increased level of safety. The relaxation of some of the established design constraints, together with the use of new lightweight and recyclable materials, can contribute to the development of new breakthrough technologies and design philosophies.

Structures designed to work in the post-buckling field and able to adapt their shape during different flight conditions are under development. The first steps of the methodology consisted in the numerical investigation of a composite plate and then of a simplified thin-walled composite wing box, where the post-buckling behavior of the structural component is controlled by out-of-plane deflection constraints.

Even if the investigated configurations still require an extensive research of feasibility as they need to be validated experimentally, they show the capability of controlling the stiffness changes through the nonlinear post-buckling response and to enhance the adaptivity of the wing structure to meet multi-stable tailorable situations, with a limited amount of load and actuation energy. These new composite structures will act on two of the biggest levers for the future of clean aviation: reduced weight and increased efficiency.

Acknowledgment

Funded by the European Union (ERC Advanced Grant, NABUCCO, project number 101053309). Views and opinions expressed are however those of the author only and do not necessarily reflect those of the European Union or the European Research Council Executive Agency. Neither the European Union nor the granting authority can be held responsible for them.

Funded by the

European Research Council

References

[1] Clean Aviation, Strategic Research and Innovation Agenda, December 2021. Information on https://clean-aviation.eu/sites/default/files/2022-01/CAJU-GB-2021-12-16-SRIA_en.pdf

[2] A. F. Arrieta, I. K. Kuder, M. Rist, T. Waeber, P. Ermanni, Passive load alleviation aerofoil concept with variable stiffness multi-stable composites, Composite Structures. 116 (2014) 235-242. https://doi.org/10.1016/j.compstruct.2014.05.016

[3] Y. Li, S. Pellegrino, A theory for the design of multi-stable morphing structures, Journal of the Mechanics and Physics of Solids. 136 (2020) 103772. https://doi.org/10.1016/j.jmps.2019.103772

[4] K. S. van Dooren, B.H.A.H. Tijs, J.E.A. Waleson, C. Bisagni, Skin-stringer separation in post-buckling of butt-joint stiffened thermoplastic composite panels, Composite Structures, 304 (2023) 116294304. https://doi.org/10.1016/j.compstruct.2022.116294

[5] J. Zhang, C. Bisagni, Buckling-driven mechanisms for twisting control in adaptive composite wings, Aerospace Science and Technology, 118 (2021) 107006. https://doi.org/10.1016/j.ast.2021.107006

Aeronautics and Astronautics - AIDAA XXVII International Congress Materials Research Forum LLC
Materials Research Proceedings 37 (2023) 353-356 https://doi.org/10.21741/9781644902813-78

Immersed boundary-conformal coupling of cylindrical IGA patches

Giuliano Guarino[1,2,a*], Pablo Antolin[2,b], Alberto Milazzo[1,c] and Annalisa Buffa[2,d]

[1]Department of Engineering, Università degli Studi di Palermo, V.le delle Scienze, Bld. 8, 90128, Palermo, Italy

[2] Institute of Mathematics, École Polytechnique Fédérale de Lausanne, CH-1015, Lausanne, Switzerland

[a]giuliano.guarino@unipa.it, [b]pablo.antolin@epfl.ch, [c]alberto.milazzo@unipa.it, [d]annalisa.buffa@epfl.ch

Keywords: Kirchhoff-Love Shells, Isogeometric Analysis, Interior Penalty Coupling

Abstract. In this work an Immersed-Boundary-Conformal coupling method for coupling shells is presented. The linear elastic static analysis is carried out using the Kirchhoff-Love shell model. The variational statement is discretized with an Isogeometric Analysis approach. The method employs auxiliary shell patches conformal to the interfaces which are coupled to the main ones using an Interior Penalty formulation. Results showing the potential of such approach to study multi-component shell structures are provided.

Introduction

Cylindrical beams are extensively used in many engineering sectors due to their high stiffness-to-weight ratios, with typical examples that can be found in civil, marine, and automotive sectors. Truss structure relying on cylindrical beams are usually investigated using one-dimensional elements that allow to characterize the general mechanical response of the assembly but do not describe accurately the displacement and stress fields near the interfaces between elements. Here, a two or three-dimensional analysis is typically required.

The Isogeometric Analysis (IGA) method [1] employs spline basis for both the description of the geometry and the approximation of the solution, allowing for a straightforward incrementation of the polynomial order while at the same time limiting the increase in the number of degrees of freedom. However, describing the geometry obtained from the intersection of two cylinders with splines is not an easy task. A possible approach consists in using an Immersed-Boundary description for each cylindrical patch. In details, an additional curve is used to embed each of the two parametric domains to describe the boundary corresponding to the interface between the intersecting cylinders. This approach results in a subdivision of each parametric domain in an active and a non-active region, and in a trimming of the elements in between the two. However, such approach requires a weakly imposed continuity between the patches and an ad-hoc refinement strategies for the region near the interface if a higher resolution is required.

In this work, the proposed coupling strategy relies on an auxiliary patch for each cylinder. These are constructed conformal to the interface on one side. As such, while the coupling of the rotational degrees of freedom is still performed in a weak sense, the coupling of the displacements is obtained in a strong sense by merging corresponding degrees of freedom. The other side of each patch is used to define the boundary of the corresponding parametric domain, and to enforce the continuity using an Interior Penalty approach. The main advantage of such strategy is the arbitrary of the auxiliary patches, that allows for a straightforward local refinement in the region close to the intersection.

To demonstrate the potential of the proposed approach, preliminary numerical results are provided regarding the creation of a boundary conformal patch for a Kirchhoff plate and the coupling of cylindrical shell through the Interior Penalty approach.

Kirchhoff-Love shells

The starting point of the proposed method is the Kirchhoff-Love shell equation. The weak form of the equation for a single patch shell is stated as finding the displacement field $\boldsymbol{u}$ such as $a(\boldsymbol{v},\boldsymbol{u}) = f(\boldsymbol{v}) \quad \forall \boldsymbol{v}$. Where the definition of the spaces for $\boldsymbol{u}$ and $\boldsymbol{v}$ depends on the essential boundary condition for the problem at hand (see [3]). The bilinear and the linear form are defined as:

$$a(\boldsymbol{v},\boldsymbol{u}) = \int_\Omega \boldsymbol{\varepsilon}(\boldsymbol{v}) : \boldsymbol{N}(\boldsymbol{u}) \mathrm{d}\Omega + \int_\Omega \boldsymbol{\kappa}(\boldsymbol{v}) : \boldsymbol{M}(\boldsymbol{u}) \mathrm{d}\Omega \tag{1a}$$

$$f(\boldsymbol{v}) = \int_\Omega \boldsymbol{v} \cdot \widetilde{\boldsymbol{F}} \mathrm{d}\Omega + \int_{\Gamma_{N_1}} \boldsymbol{v} \cdot \widetilde{\boldsymbol{T}} \mathrm{d}\Gamma + \int_{\Gamma_{N_2}} \theta_n(\boldsymbol{v}) \widetilde{M}_{nn} \mathrm{d}\Gamma + \textstyle\sum_{C \in \chi} \left(v_3 \tilde{S}\right)|_C \, , \tag{1b}$$

where $\boldsymbol{\varepsilon}$ and $\boldsymbol{\kappa}$ are the membrane and bending strain respectively, $\boldsymbol{N}$ and $\boldsymbol{M}$ are the generalized force and moment, $\widetilde{\boldsymbol{F}}, \widetilde{\boldsymbol{T}}, \widetilde{M}_{nn}$ and $\tilde{S}$ are the applied domain force, boundary force, bending moment and corner force, θ_n is the bending rotation, and Ω, Γ_{N_1}, Γ_{N_2}, and C are the surface, the boundary where force and moment boundary conditions are applied and the corners, respectively. The interested reader is referred to [2] for the details regarding the formulation.

Interior penalty coupling

When dealing with multiple patch shell structures, a coupling strategy is necessary. In some cases, when the patches are conformal to the interface, the coupling condition can be imposed strongly by appropriately selecting the spaces for $\boldsymbol{u}$ and $\boldsymbol{v}$. However, in general, the coupling condition needs to be imposed weakly through integrals along the interface. This is particularly true when the patches are trimmed in an immersed boundary approach. In such cases, the problem becomes finding $\boldsymbol{u}_h$ such as

$$\textstyle\sum_{p=1}^{N_p} a_p(\boldsymbol{v}_h, \boldsymbol{u}_h) + \sum_{i=1}^{N_i} b_i(\boldsymbol{v}_h, \boldsymbol{u}_h) = \sum_{p=1}^{N_p} f_p(\boldsymbol{v}_h) \quad \forall \boldsymbol{v}_h \tag{2}$$

where $\boldsymbol{u}_h$ and $\boldsymbol{v}_h$ are the discretized displacements and test function, respectively. $a_p(\boldsymbol{v}_h, \boldsymbol{u}_h)$ and $f_p(\boldsymbol{v}_h)$ represent the discretized version of Eq. (1) for the p-th of the N_p patches. $b_i(\boldsymbol{v}_h, \boldsymbol{u}_h)$ is the bilinear form associated to the i-th of the N_i interfaces and is defined as

$$\begin{aligned} b_i(\boldsymbol{v}_h, \boldsymbol{u}_h) = \int_{\Gamma_h^I} (\{\boldsymbol{T}(\boldsymbol{v}_h)\} \cdot [\![\boldsymbol{u}_h]\!] + [\![\boldsymbol{v}_h]\!] \cdot \{\boldsymbol{T}(\boldsymbol{h}_h)\}) d\Gamma + \\ + \int_{\Gamma_h^I} (\{M_{nn}(\boldsymbol{v}_h)\} [\![\theta_n(\boldsymbol{u}_h)]\!] + [\![\theta_n(\boldsymbol{v}_h)]\!] \{M_{nn}(\boldsymbol{v}_h)\}) d\Gamma + \\ + \mu_D \int_{\Gamma_h^I} [\![\boldsymbol{v}_h]\!] \cdot [\![\boldsymbol{u}_h]\!] d\Gamma + \mu_R \int_{\Gamma_h^I} [\![\theta_n(\boldsymbol{v}_h)]\!] [\![\theta_n(\boldsymbol{u}_h)]\!] d\Gamma \end{aligned} \tag{4}$$

where Γ_h^I is the i-th interface, $\boldsymbol{T}$ and M_{nn} are the fluxes associated with the Ersatz force and bending moment (see [3]), and $[\![a]\!]$ and $\{a\}$ are the standard jump and average operator. μ_D and μ_R are the penalty parameters associated with the displacements and the rotation, that are chosen here as $\mu_d = \beta \frac{Et}{h}$ and $\mu_r = \beta \frac{Et^3}{h}$, where E, t, and h are the maximum young modulus of the material, the shell thickness, and mesh size, and β is a problem independent constant chosen here as 10^3.

The Immersed Boundary-Conformal approach

In the immersed boundary paradigm, the shell surface is embedded within a trimming curve, resulting in the presence of entire, partial, and empty elements. On the partial elements boundary and coupling condition can only be imposed in a weak sense. In the proposed method, an auxiliary boundary conforming patch is generated along the trimming curve. Although coupling between the auxiliary patch and the internal domain becomes necessary, it occurs in a more internal region of the domain where local phenomena are less likely to occur. Furthermore, these auxiliary patches

Materials Research Proceedings 37 (2023) 353-356 https://doi.org/10.21741/9781644902813-78

can be refined as needed to capture these local phenomena more accurately. One of the difficulties associated with the proposed method lies in the integration in the trimmed elements. Here, the algorithm described in [4] is adopted, which is based on a high-order reparameterization of the trimmed elements.

(a) (b)

Figure 1 Same shell structure discretized using a trimmed single patch (a) and a combination of a main patch and two auxiliary boundary conformal ones (b).

In Fig. (1), it is shown an example of a shell structure with boundary conformal patches aligned with the trimmed domain. In a similar fashion, we believe that coupling two intersecting surfaces creating an interface conformal patch for each surface would be beneficial to the accuracy of the analysis and is currently under investigation.

Numerical Results

In this session some preliminary results leading to the immersed boundary conformal coupling are presented.

Kirchhoff plate with pseudo cut-outs: The plate shown in Fig. (2), characterized by edge size $L = 10$ m, Poisson ratio $\nu = 0.3$, thickness $t = 0.1$ m, Young modulus $E = 12(1-\nu^2)/t^3$Pa, and having simply supported boundary condition, is subjected to a constructed force that produces the displacement field described by $u_3 = \sin(2\pi x/L)\sin(2\pi y/L)$. In Fig. (2a) it is shown the parametric space with two pseudo cut-outs surrounded by boundary conformal patches. In this test the holes have been filled to easily retrieve the analytical solution. Fig(2b) shows the contour of the solution with superimposed mesh of the main and the auxiliary patches, it is pointed out the smoothness of the solution. Fig. (2c) shows the L_2 convergence for this test.

Intersecting cylindrical patches: Fig.(3a) shows a structure constituted of five intersecting cylindrical shells characterized by Young modulus $\nu = 0.3$, Poisson ratio $\nu = 0.3$, and thickness 10 mm. On the vertical cylinder, simply supported boundary conditions are applied at the opposite ends. A uniform domain force $f = [1,1,1]$ is applied on each patch. In Fig. (3b) it can be noted how the interior penalty method guarantees coupling conditions of the complex structure under investigation.

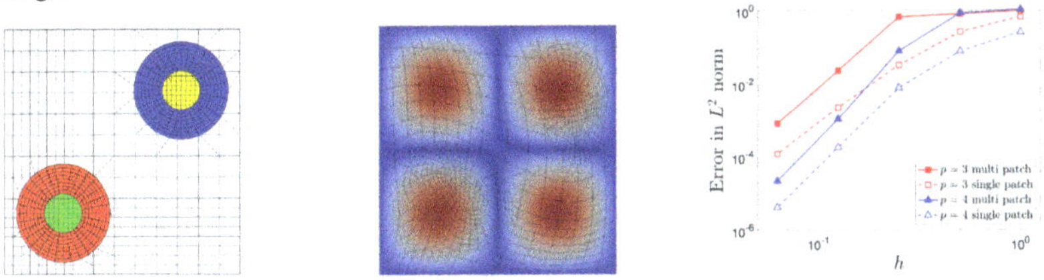

(a) (b) (c)

Figure 2 Kirchhoff plate with two pseudo cut-outs. (a) Parametric domain (b) physical domain with superimposed displacements contour (c) $\mathbf{L^2}$ *convergence of the displacements field.*

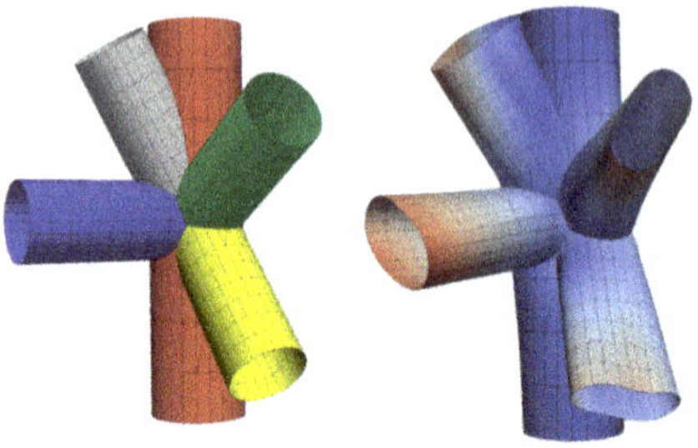

Figure 3 Shell structure with multiple cylindrical patch (a). Contour of the magnitude of the displacement (b).

Conclusions

This contribute introduces the boundary conformal method for the coupling of shell patches and provides initial results towards the implementation of a complete approach. The proposed formulation is based on the interior penalty method for coupling shell patches, the utilization of auxiliary interface conformal patches, and a high-order integration scheme for trimmed elements. The presented results suggest the potential of the method, which is currently being developed.

References

[1] T.J.R. Hughes, J.A. Cottrell, Y. Bazilevs, Isogeometric analysis: CAD, finite elements, NURBS, exact geometry and mesh refinement, Computer Methods in Applied Mechanics and Engineering, 194 (2005), 4135-4195. https://doi.org/10.1016/j.cma.2004.10.008

[2] J. Kiendl, K.-U. Bletzinger, J. Linhard, R. Wüchner, Isogeometric shell analysis with Kirchhoff–Love elements, Computer Methods in Applied Mechanics and Engineering, 198 (2009) 49-52. https://doi.org/10.1016/j.cma.2009.08.013

[3] J. Benzaken, J. A. Evans, S. F. McCormick, R. Tamstorf, Nitsche's method for linear Kirchhoff–Love shells: Formulation, error analysis, and verification, Computer Methods in Applied Mechanics and Engineering, 374 (2021), 113544. https://doi.org/10.1016/j.cma.2020.113544

[4] Wei, X., Marussig, B., Antolin, P. et al, Immersed boundary-conformal isogeometric method for linear elliptic problems. Computational Mechanics 68 (2021), 1385–1405. https://doi.org/10.1007/s00466-021-02074-6

Materials Research Proceedings 37 (2023) 357-362 https://doi.org/10.21741/9781644902813-79

Folding simulation of TRAC longerons via unified one-dimensional finite elements

Riccardo Augello[1,a*], Erasmo Carrera[2,b], Alfonso Pagani[2,c] and Sergio Pellegrino[2,d]

[1]California Institute of Technology, Pasadena, CA, 91125

[2]Department of Mechanical and Aerospace Engineering, Mul2 Lab, Politecnico di Torino, Torino, Italy

[a]raugello@caltech.edu, [b]erasmo.carrera@polito.it, [c]alfonso.pagani@polito.it, [d]sergiop@caltech.edu

Keywords: Deployable Booms, TRAC longerons, Carrera Unified Formulation, Contact Mechanics, Nonlinear Analysis

Abstract. This paper proposes a simulation of the folding phase of TRAC deployable booms using refined one-dimensional finite elements in the framework of the Carrera Unified Formulation. The mathematical model involves standard beam finite elements placed along the length of the longeron, and Lagrange polynomials as expansion functions for the cross-sectional domain. The nonlinear governing equations are written recalling the principle of virtual work, and they are linearized using the Newton-Raphson scheme. The contact between the two flanges is simulated with linear spring which activate when pre-defined node pairs approach under a fixed tolerance. Two simulations are carried out, including or not the contact behavior, respectively. The results highlight the capability of the proposed model to deal with large displacements and contact between the ultra-thin flanges of the structure.

Introduction

Over the years, deployable structures have become increasingly popular in space engineering, since they make it possible to obtain a consistent reduction of mass, volume, and consequently, the cost of a satellite. Deployable booms found a large number of applications such as for telescopes [1] and antennas [2].

The most common type of deployable booms is tape springs, which can be described as thin-walled elastic strips. A combination of those can be used to generate other types of booms, such as the Triangular Rollable And Collapsible (TRAC). Originally invented by Murphey and Banik [3] and developed by the Air Force Research Laboratory, it consists of two circular domains (flanges) made of tape-spring that are attached along the common edge (web). The resultant cross-section gives TRAC longerons larger bending stiffness than other popular options (such as the Collapsible Tubular Mast (CTM) [4], the Storable Tubular Extendable Member (STEM) [5] and bi-STEM [6]).

The present work deals with the simulation of the folding phase of TRAC booms. By employing the Carrera Unified Formulation [7], the ultra-thin boom can be modeled using refined one-dimensional beam finite elements. The nonlinear governing equations are solved using a Newton-Raphson linearization scheme along with a displacement control.

One dimensional model of the TRAC longeron

The deployable mechanism analyzed in this work is shown in Fig. 1, and it consists of a foldable TRAC longeron. The geometric properties are $L = 400$ mm, $h = 8$ mm, $r = 12.8$ mm, $\vartheta = 105°$ and $t = 80$ μm.

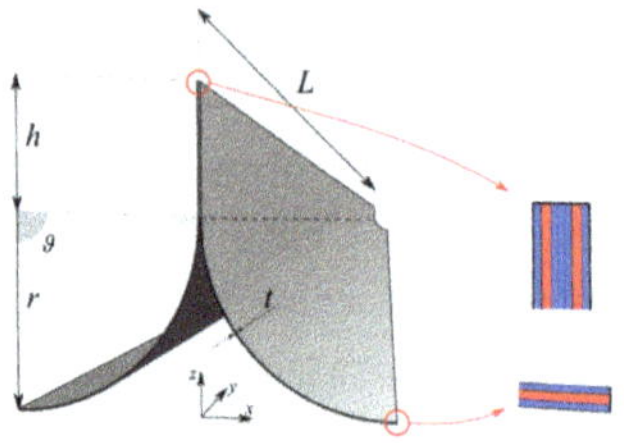

Figure 1 - Geometry of the considered TRAC longeron

As far as the material properties are concerned, the composite layup in the flanges of the longeron is [±45 GFPW/ 0 CF/ ±45 GFPW], and that in the web region is [±45 GFPW/ 0 CF/ $\pm45_3$ GFPW/ 0 CF/ ±45 GFPW], where CF represents a thin ply with unidirectional carbon fibers. GFPW represents plain weave scrim glass. Both plies are impregnated with resin, as described in [8].

For the numerical model, a Cartesian reference system is used, so that the y direction is placed in the length direction, and the x, z identify the cross-sectional domain. According to the Carrera Unified Formulation (CUF), the three-dimensional displacement field can be written in the following unified manner:

$$\boldsymbol{u}(x, y, z) = F_\tau(x.z)N_i(y)\boldsymbol{u}_{\tau i} \qquad \tau = 1,2,\dots,M \qquad i = 1,2,\dots,N_n \tag{1}$$

where $\boldsymbol{u}(\boldsymbol{x}, \boldsymbol{y}, \boldsymbol{z})$ is the displacement vector, whose components are expressed in the general reference system $(\boldsymbol{x}, \boldsymbol{y}, \boldsymbol{z})$ of Fig. 1, $\boldsymbol{F}_\tau$ represent the cross-sectional functions depending on the x, z coordinate, $\boldsymbol{\tau}$ is the sum index and M is the number of terms of the expansion in the cross-section plane assumed for the displacements. $\boldsymbol{N}_i$ stands for the $\boldsymbol{i}$th one-dimensional shape function, $\boldsymbol{u}_{\tau i}$ is the vector of the FE nodal parameters, $\boldsymbol{i}$ indicates summation and Nn is the number of the FEs nodes per element. In this work, a cubic interpolation for the axis direction is assumed. Figure 2 shows the resultant one-dimensional model of the TRAC longeron.

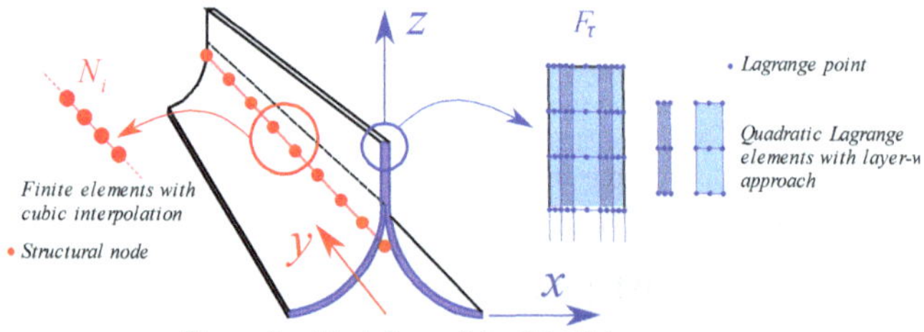

Figure 2 - Modeling of the TRAC longeron

Boundary conditions

The considered boundary conditions are reported in Fig. 3. Basically, one side of the TRAC can rotate around the z axis, whereas the other can rotate around the z axis and move along the y axis.

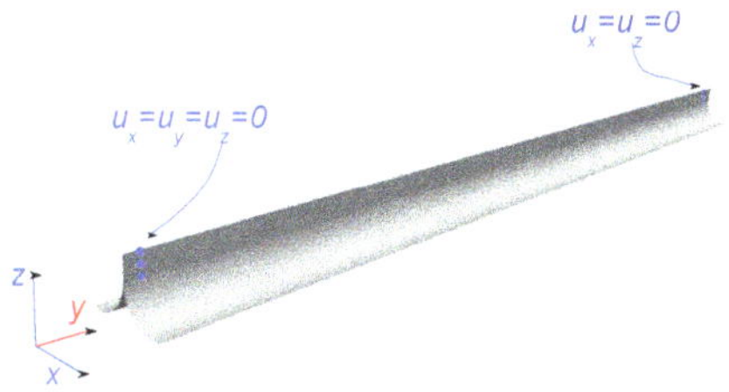

Figure 3 - Boundary conditions

Subsequently, the structure is subjected to a pinching in the middle, by imposed displacements, as depicted in Fig. 4(a). Figure 4(b) reports the folding process, where the rotations are ensured by pairs of displacements at the sides of the structure.

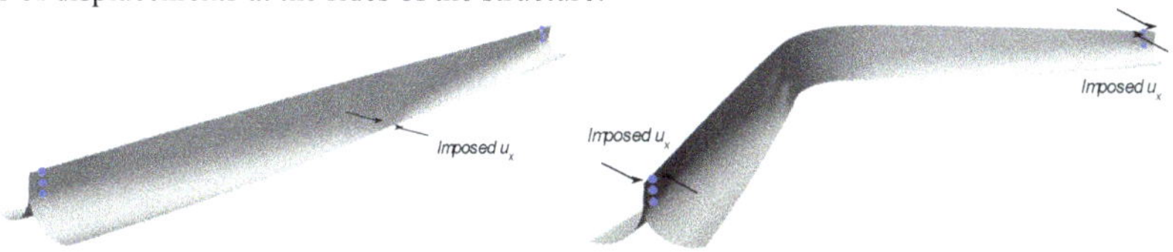

(a) (b)

Figure 4 - Pinching and folding of the TRAC longeron

Nonlinear governing equations

The set of the nonlinear governing equations, which full derivation is described in [10], is expressed in the following:

$$\boldsymbol{K_S q} - \boldsymbol{p} = \boldsymbol{0} \quad (2)$$

where K_S, q and p are the global, assembled FE stiffness, displacement and external force arrays of the final structure. Equation (2) represents a nonlinear algebraic system of equation for which an iterative method is needed. We employ here the same procedure detailed in [11], where a Newton–Raphson scheme is used by making use of a path following constraint. This procedure demands for the linearization of the nonlinear governing equations. As a result, we need to introduce the so-called tangent stiffness matrix $K_T = \frac{d(K_S q - p)}{dq}$: The explicit form of K_T is not given here, but it is derived in a unified form in [12]. Finally, in this work, a displacement control algorithm is introduced.

Contact mechanics

In the present work, contact between the flanges of the TRAC longeron is simulated. A node-to-node contact approach is employed, recalling the penalty technique. Basically, a linear spring is simulated between the nodes of the two flanges, as depicted in Fig. 5.

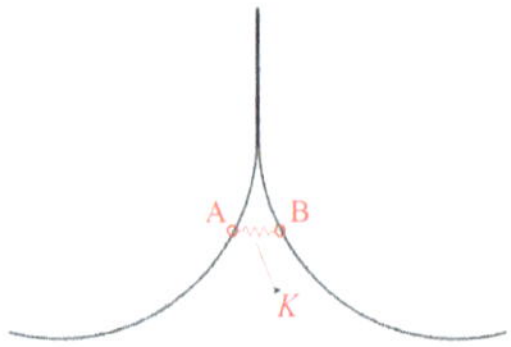

Figure 5 - Node-to-note contact simulation

The nodes are paired as input of the analysis, so the mesh coincidence between the inner surfaces of the two flanges must be ensured. The springs activate under a particular tolerance, which is set as 1 μm, and produces two opposite forces on the nodes. Thus, the force vector $\boldsymbol{p}$ and the stiffness matrix K_T are updated at each iteration step of the previously described Newton-Raphson algorithm.

Numerical results

The numerical results regard two simulation, including or not the contact between the two flanges. The developed mathematical model involves 20 cubic FEs for the axis discretization and 64 quadratic Lagrange polynomials for the cross-sectional domain (30 for each flange and 4 for the web). The total number of degrees of freedom is 74871.
As far as the simulation without contact is concerned, the pinching phase is reported in Fig. 6.

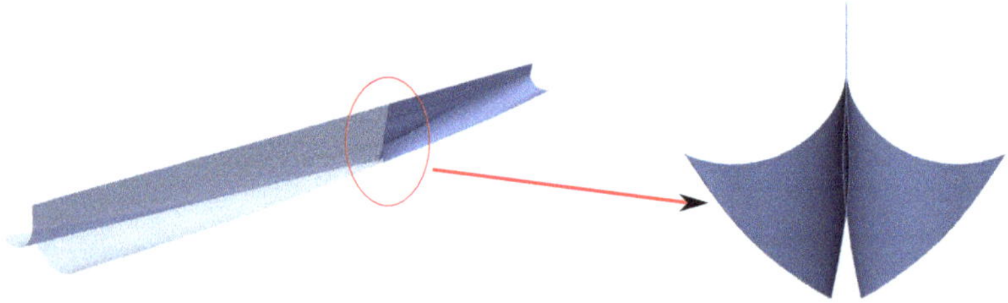

Figure 6 - Pinching simulation of the TRAC longeron without contact mechanics

As can be seen from Fig. 6, the penetration between the two flanges arises in a large domain of the boom. The complete folding simulation is reported in Fig. 7.

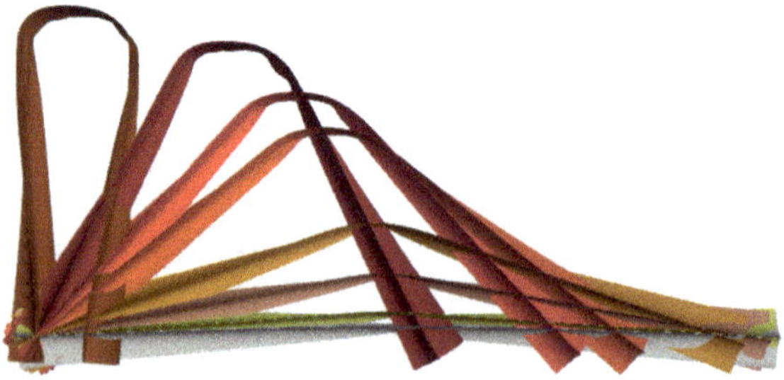

Figure 7 - Folding simulation of the TRAC longeron without contact mechanics

The folding of the TRAC boom can be simulated, although, the penetration between the two flanges persists during the overall simulation..

The same simulation is conducted including the contact between the two flanges. The results for the pinching simulation are reported in Fig. 8.

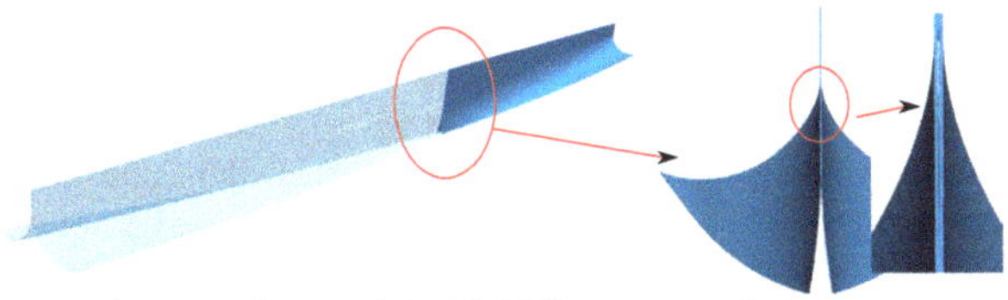

Figure 8 - Pinching simulation of the TRAC longeron including contact mechanics

In contrast to the case without contact, no penetration is observed between the two flanges. Moreover, since the web has one additional layer that the two flanges (see Fig. 1), the two flanges do not contact each other in the proximity of the web (see the enlargement in Fig. 8). The full folding simulation is reported in Fig. 10.

Figure 10 - Folding simulation of the TRAC longeron including contact mechanics

Finally, Fig. 10 reports the contact between the two flanges in one of the equilibrium state.

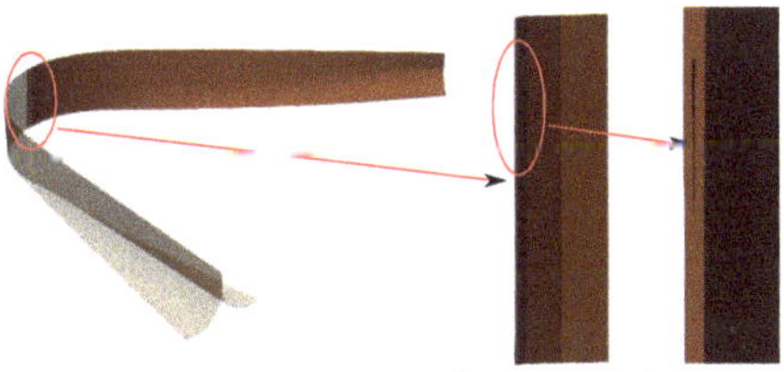

Figure 10 - Particular of the folding simulation of the TRAC longeron including contact mechanics

Conclusions

The present work focussed on the simulation of the folding phase of a deployable boom, namely the TRAC longeron. This composite ultra-thin structure was modelled with the Carrera unified formulation, thanks to which refined one-dimensional finite elements can be built, accounting for any cross-sectional deformation. The nonlinear governing equations were derived in terms of the fundamental nuclei, which represent the basic building block of the overall stiffness matrix. Then, a Newton-Raphson scheme was employed, and a consistent linearization of the governing equation was performed. The contact between the two flanges of the TRAC longeron is simulated with node-to-node linear springs, which are updated at every iteration of the nonlinear procedure. Two simulations with and without the contact were carried out. The results show the robustness of the present approach to deal with far nonlinear regimes and contact nonlinearities. Future works will deal with the introduction of nonlinear springs to improve the convergence rate of the nonlinear solution and for the dynamic nonlinear simulation of the deployment of the TRAC longerons.

Acknowledgment
This work is part of the project NOVITAS, funded by the European Union's Horizon Europe research and innovation program under the Marie Sklodowska-Curie grant agreement No 101059825.

References

[1] L. Blanchard. A tape-spring hexapod for deployable telescopes: Dynamics. In ESA Special Publication, pp. 1-5, 2006.

[2] F. Royer and S. Pellegrino. Ultralight ladder-type coilable space structures. In 2018 AIAA Spacecraft Structures Conference, pp. 1-14, 2018. https://doi.org/10.2514/6.2018-1200

[3] T.W. Murphey and J. Banik, Triangular rollable and collapsible boom, US Patent 7,895,795, 2011.

[4] D.S. Crouch, Mars viking surface sampler subsystem, 25th Conference on Remote Systems Technology, pp. 142–151, 1977.

[5] F. Roybal, J. Banik, and T.W. Murphey, Development of an elastically deployable boom for tensioned planar structures, 48th AIAA/ASME/ASCE/AHS/ASC Structures, Structural Dynamics, and Materials Conference, Honolulu, Hawaii, USA, pp. 1–14, 2007. https://doi.org/10.2514/6.2007-1838

[6] M.W. Thomson, Deployable and retractable telescoping tubular structure development, The 28th Aerospace Mechanisms Symposium, Cleveland, Ohio, USA, 1994.

[7] E. Carrera, M. Cinefra, M. Petrolo, and E. Zappino, Finite Element Analysis of Structures through Unified Formulation, Wiley, Chichester, West Sussex, UK, 2014. https://doi.org/10.1002/9781118536643

[9] N.H. Reddy, and S. Pellegrino, 2023. Dynamics of the Caltech SSPP deployable structures: structure–mechanism interaction and deployment envelope. https://doi.org/10.2514/6.2023-2065

[10] A. Pagani, and E. Carrera, 2018. Unified formulation of geometrically nonlinear refined beam theories. Mechanics of Advanced Materials and Structures, 25(1), pp.15-31. https://doi.org/10.1080/15376494.2016.1232458

[11] E. Carrera, A. Pagani, and R. Augello, 2020. Evaluation of geometrically nonlinear effects due to large cross-sectional deformations of compact and shell-like structures. Mechanics of Advanced Materials and Structures, 27(14), pp.1269-1277. https://doi.org/10.1080/15376494.2018.1507063

[12] A. Pagani, and E. Carrera, 2017. Large-deflection and post-buckling analyses of laminated composite beams by Carrera Unified Formulation. Composite Structures, 170, pp.40-52. https://doi.org/10.1016/j.compstruct.2017.03.008

Numerical-analytical evaluation about the impact in water of an elastic wedge using the SPH method

D. Guagliardo[1,a], E. Cestino[1,b*] and G. Nicolosi[1,c]

[1] Politecnico di Torino (DIMEAS), Corso Duca degli Abruzzi, 24 10129 Turin, Italy

[a]s272087@studenti.polito.it, [b]enrico.cestino@polito.it, [c]gabriele.nicolosi93@gmail.com

Keywords: Fluid-Structure Interaction, LS-DYNA, Smoothed Particle Hydrodynamics, Elastic Wedge

Abstract. In a preliminary study about the structural behaviour of a body, the material can be supposed infinitely rigid. This choice is useful to simplify the problem and to obtain results that well approximate reality. In this specific case under study, it proves how the effect of the material elasticity has a fundamental role on the pressures developed when a wedge impacts water. The analysis is made using ANSYS LS-DYNA software modelling the wedge through the FEM method, characterized by the material defined by the MAT_001-ELASTIC keyword, and the water defined by the SPH (Smoothed-Particle Hydrodynamics) method. The elastic body behaviour will be evaluated through the presence of a displacement in the middle point, on the face impacting water, and the comparison between the obtained pressure and the one predicted by the analytical theories of von Karman[1] and Wagner[2], which study the impact of a rigid wedge.

Introduction

The student team "TEAM S55" of Polytechnic of Turin was born in 2017 to rebuild the SIAI-Marchetti S55 seaplane on a 1:8 scale [1-3]. The following paper is created by the FSI section which studies the interaction between the aircraft and the water at ditching.

In a preliminary study of the phenomena [4, 5], the rigid wedge approximation can be used to predict the maximum slamming pressure.

The purpose of this paper is to evaluate the pressure trend and the displacement of the wedge's impacting surface during the penetration of a wedge into the water. The numerical analysis was done using the software ANSYS LS-DYNA, in which due to the high computational requirements inherent in the SPH method, the influence of air is neglected to reduce the computational time since it doesn't influence the accuracy of stress prediction [6].

Starting from analytical theory developed by von Karman [7] and Wagner [8] based on a rigid wedge, subsequent studies [3, 9] have been considered that go beyond rigid wedge considerations and move towards different aspects of elasticity. The effect of the structure elasticity on the pressure peak and distribution will be considered. Furthermore, along the impacting surface, the results of displacements will be considered as the speed varies.

Von Karman and Wagner theories

Wagner [8, 10] and Von Karman [7] theories are the most indicated theories in literature to estimate the ditching of a wedge in the water.

Von Karman theory, Eq. 1, is based on the application of the momentum theorem on a body whose weight increases at the impact with the water. The increase is caused by the water contained in a cylinder of diameter equal to the wedge width. Von Karman [7] aimed to find out how the dihedral angle affected the result of the pressure on landing.

$$P_{VK}(x) = \frac{\rho V_0^2}{2} \frac{\pi}{\left(1+\frac{\gamma \pi x^2}{2W}\right)^3} \cot(\alpha) \qquad (1)$$

$$P_{VK_{Max}} = \frac{\rho V_0^2}{2} \pi \cot(\alpha) \quad (2)$$

According to his study, the maximum pressure, Eq. 2, is found at the keel of the wedge and the pressure increases as the dihedral angle decreases.

Wagner [10] studied the impact to estimate the pressure considering the superposition principle. The first studied phenomenon is the wedge penetration in the water studied like a perfectly inelastic collision. The second one is the water molecules behavior. In fact, when the wedge penetrates the water, the stationary molecules start to move, reducing the internal pressure. This decrease is function of the wedge penetration: during the phenomena both the wedge and water molecules velocity have a decreasing trend that causes a minor reduction of pressure, Eq. 3 and Eq. 4.

$$P_W(x) = \frac{1}{2}\rho V^2 \left[\frac{\pi}{tan(\alpha)\sqrt{1-\frac{x^2}{L^2}}} - \frac{\frac{x^2}{L^2}}{1-\frac{x^2}{L^2}} + 2\,\frac{\ddot{y}}{V^2}\sqrt{L^2 - x^2} \right] \quad (3)$$

$$P_{W_{Max}} = \frac{1}{2}\rho V^2 \left[1 + \frac{\pi^2}{4\,tan^2(\alpha)}\right] \quad (4)$$

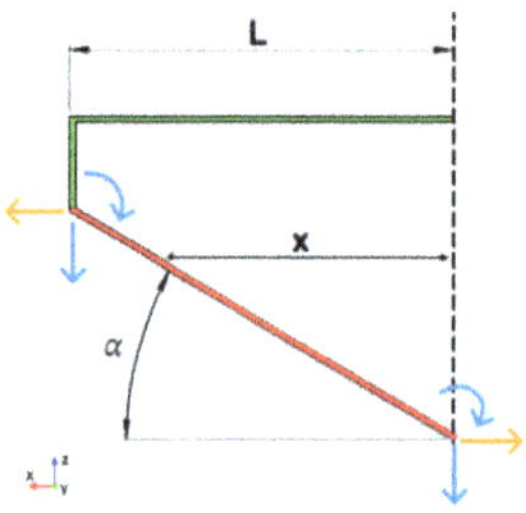

Figure 1: Model constraints

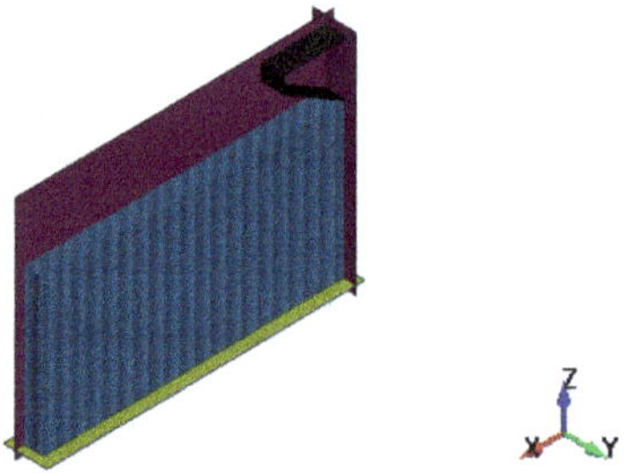

Figure 2: Half wedge model in ANSYS LS-DYNA

Model

The elastic model shown in Fig. 1 is a wedge modelled in ANSYS LS-DYNA PrePost. The wedge has a width of 127.4 mm, a height of 105 mm, a thickness of 2 mm and a dihedral angle α of 30°. It is 0.43 kg heavy, and the material is considered as isotropic. The surface impacting water, the oblique one, and the others are modelled respectively through 001 MAT_ELASTIC and 020 MAT_RIGID, both considering a steel with a Young's modulus of 190GPa. The decision to use MAT_RIGID in non-impact water surfaces is in according to reduce the computational cost and to model the boundary as desired.

The boundary conditions, shown in Fig. 1 and Fig. 2 are:

- displacements allow in x and z directions, and rotation allow along y direction for elastic surface;
- displacements constrain in x and y directions at the edge of elastic surface;
- displacements allowed just in z directions and all rotations blocked for the rigid surfaces;
- control volume is composed of a **RIGIDWALL_PLANAR,* in the bottom (yellow plane), a **BOUNDARY_SPH_NON_REFLECTING,* perpendicular to x-direction and on the side

Materials Research Proceedings 37 (2023) 363-367 https://doi.org/10.21741/9781644902813-80

opposite the wedge, and three **BOUNDARY_SPH_SYMMETRY-PLANE* (purple planes), two perpendicular to the y-direction and one perpendicular to the x direction and close to the wedge.

The material used for the water is 009 MAT_NULL and the Equation Of State is **EOS_GRUNEISEN*.

The total number of elements is 901200, which 1200 are shell elements and 900000 are SPH (Smoothed Particle Hydrodynamics). The computational cost for each simulation is about 15 h using 48 CPUs simultaneously.

Pressure and displacement results

The study results plotted in Fig. 3 and Fig. 4 show the pressure trend as function of the wedge surface and velocity respectively. The Fig. 3 represents the Von Karman and Wagner theories applied to our wedge and the results of the numerical analysis, when the impact velocity is 5.8 m/s. The Fig. 4 considers the pressure detection sensor, in all analyses, is positioned at 13.93 mm along the x direction. The maximum pressure expected is reached at the wedge's keel for Von Karman and Wagner theory both. It is interesting to notice that Von Karman predicts higher pressure value than Wagner along the surface. This Wagner trend is due to fluid dynamic effect that becomes increasingly negligible along the surface.

The numerical model results are included between the two theories values until they reach the middle of the impact surface. Over that position the numerical results estimate higher pressure values than the two theories.

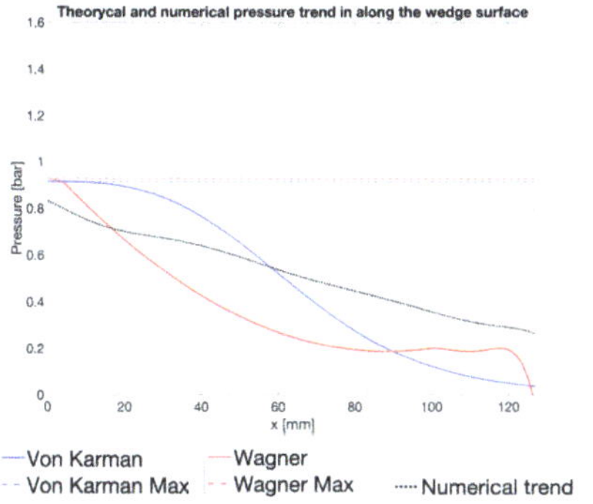

Figure 3: Pressure as function of the surface

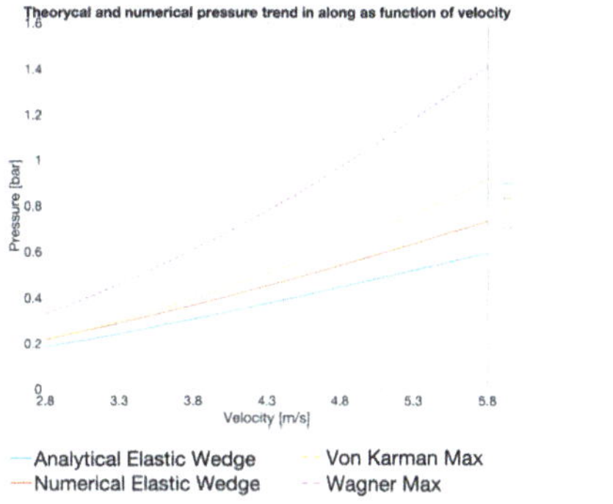

Figure 4: Pressure as function of the velocity

The phenomena studied correspond to a perfectly inelastic collision, so it is important to observe how the pressure changes as function of velocity. Fig. 4 compares the wedge behaviour assuming a rigid, as descripted by Von Karman and Wagner theories, and an elastic material, as reported in [3]. At the impact, the rigid material undergoes a higher-pressure magnitude than elastic material, because part of the impact energy is the source of the deformation of the impact surface and the remaining part is absorbed by the body. Moreover, the analytical and numerical results are different due to the boundary conditions considered; in [3] the panel is considered simply supported and greater freedom to deform have consequently a reduction of pressure.

In Fig. 5 is showed the maximum displacement for each node that makes up the mesh, not to be confused with the maximum deformation. As expected, the numerical results of maximum displacement depend on the vertical speed; as higher as speed, the higher the maximum displacement will be.

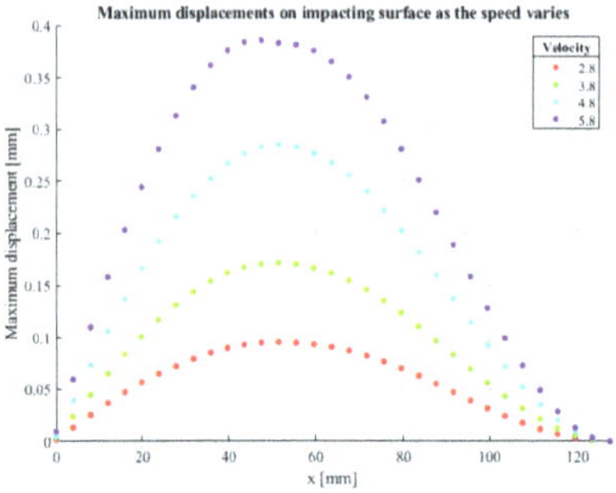

Figure 5: Displacement as function of the velocity

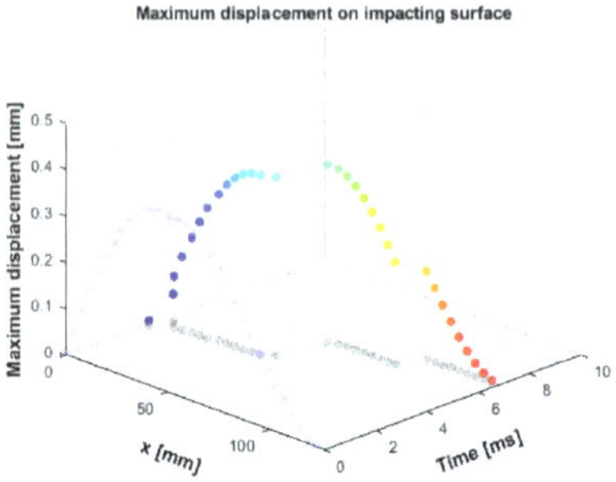

Figure 6: Displacement as function of the velocity

In Fig. 6 the vertical velocity considered is 5.8 m/s; being the maximum displacement in function of surface position, the blue shadow is the same curve plotted in Fig. 5. In Fig. 6, the added value is given by the providing of how the maximum displacement evolves over time. Being the keel to impact the water first, the maximum displacement moves with the increase of the wetted part.

Conclusions

A preliminary study on a rigid wedge impacting water be made neglecting the elasticity behaviour to predict the maximum slamming pressure. On the other hand, results of real impacting model can be conflicting and for pressure can be lower than expecting. An elastic model may be introduced trying to predict a better behaviour of the real nature of things.

The elastic model introduces the possibility of wedge deformation, this allows a local increase of dihedral angle with a decrease of pressure. In conclusion, the effect of flexibility on pressure distribution is more important as speed increases due to the higher displacement measured.

Acknowledgement

The authors thank the Politecnico di Torino and the DIMEAS department for the provision of resources and facilities. We thank DAUIN for computational resources provided through High Performance Computing Polito (HPC@POLITO), a project of Academic Computing within the Department of Control and Computer Engineering at the Politecnico di Torino (http://www.hpc.polito.it). Special thanks to ANSYS and EnginSoft SpA for the support and for providing us with the software licenses.

References

[1] CESTINO, E.; FRULLA, G.; SAPIENZA, V.; PINTO, P.; RIZZI, F.; ZARAMELLA, F.; BANFI, D. (2018) Replica 55 Project: A Wood Seaplane In The Era Of Composite Materials, In: Proc of 31st ICAS 2018 Congress, 9-14 September 2018, Belo Horizonte (Brasil)

[2] NICOLOSI G., VALPIANI F., GRILLI G., SAPONARO PIACENTE A., DI IANNI L., CESTINO E., SAPIENZA V., POLLA A., PIANA P. Design Of A Vertical Ditching Test. Proc. 32nd ICAS Congress 6-10 September 2021 - Shanghai, China

[3] Cestino, E., Frulla, G., Polla, A., Nicolosi, G. (2023). Equivalent Material Identification in Composite Scaled Hulls Vertical Impact Tests. In: Lopresto, V., Papa, I., Langella, A. (eds) Dynamic Response and Failure of Composite Materials. DRAF 2022. Lecture Notes in Mechanical Engineering. Springer, Cham. https://doi.org/10.1007/978-3-031-28547-9_6

[5] Valpiani F, Polla A, Cicolini P, Grilli G, Cestino E, Sapienza V. Early Numerical Evaluation of Fluid-Structure Interaction of a Simply Wedge Geometry with Different Deadrise Angle. AIDAA XXVI International Congress, Pisa, 2021.

[6] Fragassa C, Topalovic M, Pavlovic A, Vulovic S. Dealing with the Effect of Air in Fluid Structure Interaction by Coupled SPH-FEM Methods. *Materials*. 2019; 12(7):1162. https://doi.org/10.3390/ma12071162

[7] von Kármán T. The impact on seaplane floats during landing. NACA Technical Notes N.321, 1929.

[8] Wagner H. Über Stoß- und Gleitvorgänge an der Oberfläche von Flüssigkeiten. Zeitschrift Für Angewandte Mathematik Und Mechanik, Vol. 12, No. 4, 1932. https://doi.org/10.1002/zamm.19320120402

[9] I. Stenius, A. Rosén, and J. Kuttenkeuler, 'Explicit FE-modelling of hydroelasticity in panel-water impacts', International Shipbuilding Progress, vol. 54, no. 2–3, pp. 111–127, 2007.

[10] Shah S.A. Water Impact Investigations for Aircraft Ditching Analysis thesis, School of Aerospace, RMIT University, 2010.

Data-driven deep neural network for structural damage detection in composite solar arrays on flexible spacecraft

Federica Angeletti[1,a *], Paolo Gasbarri[1,b] and Marco Sabatini[1,c]

[1]Scuola di Ingegneria Aerospaziale, Sapienza University of Rome, Via Salaria 851, Rome, Italy

[a]federica.angeletti@uniroma1.it, [b]paolo.gasbarri@uniroma1.it, [c]marco.sabatini@uniroma1.it

Keywords: Structural Health Monitoring, Deep Learning, Flexible Spacecraft

Abstract. A data-driven approach based on Deep Neural Network (DNN) techniques is here proposed for Structural Health Monitoring of large in-orbit flexible systems. Damage scenarios are generated via a Finite Element commercial code to train and test the machine learning model, by considering equivalent properties of the composite material of the solar panels. The fully coupled 3D equations for the flexible spacecraft are integrated to test typical profiles of attitude manoeuvres in case of different damages. The DNN model is trained using sensor-measured time series responses, with each response associated with the label of the corresponding damage scenario, and tested via k-folding approach. This methodology offers a promising approach to detect structural damage in solar arrays on spacecraft using machine learning techniques.

Introduction

With the increasing use of composite materials in solar arrays on modern spacecraft, structural damages during the operational life have become a significant concern. Such events often lead to modifications of the control/structure interaction dynamics, thus posing an issue for the implemented system controller. Detecting failures in flexible structures can however be challenging: local damages may not cause significant changes in the global dynamics of the satellites to be detected by on-board sensors. Therefore, a set of sensors at structural level is beneficial to identify promptly the location and the entity of the damages.

As far as current state-of-the-art solutions for damage identification are concerned, there are mostly two adopted philosophies: physics-based and data-driven methods [1-3]. The purpose of this study is to propose a Deep Learning methodology with multi-classification damage capabilities, with respect to the research proposed by the authors' previous work [4-5]. Indeed, the present study aims at proposing an architecture and guidelines for performing SHM of space structured based on LSTM-NNs. A challenging study case in terms of impact of failures on the global spacecraft dynamics is selected, and a more complex problem in terms of higher dimensionality of the multi-class identification problem is addressed, giving information not only about the presence, but also the location of the damage. The structure and damage entity is implemented taking into account the equivalent properties and effects on a traditional composite space structure, in particular, an aluminum honeycomb.

The approach is carried out as follows. Firstly, the 3D mathematical model of a flexible spacecraft is implemented in a simulator for carrying out a wide set of in-orbit attitude maneuvers. Then, the spacecraft test case model is described, including the network of distributed sensors for the SHM and the damage configurations addressed herein. The implemented deep neural architecture is described, based on an LSTM variant. Finally, the main results about the performance of the trained classification network are discussed.

Bidirectional Long-Short Term Memory Network

The efficacy of Long Short-Term Memory (LSTM) models has been validated in various domains, notably within the area of time series prediction [6][7], as well as for both single-variable and multi-variable time sequence classification tasks [8]. LSTMs are purposefully crafted to exploit long-range dependencies, enabling them to effectively address scenarios where the present time step is distant from correspondent information. By incorporating the capability to effectively process historical data within a single cell, Deep Neural Network (DNN) architectures can benefit by establishing connections across multiple LSTM layers.

Details about the structure of a LSTM network can be found in [8]. In this study, we use a multivariate deep classification model composed of an input layer, two stacked Bi-LSTM layers, including a dropout layer to address overfitting issues, a Softmax layer, and a final classification layer.

Spacecraft Dynamics and Damages

This section briefly introduces a representative case of a spacecraft equipped with two solar panels of 3 x 1 m (composed of two sub-panels each), designed using MSC Nastran FEM tool based on information available in literature about dimensions, mass and shape of the panels. The first three modes of the assembled spacecraft are illustrated in Fig.1. Since the size and mass constraints at launch require solar panels to be lightweight, while strong and stiff, a composite material – an aluminium honeycomb here specifically - is selected for each sub-panel. Moreover, to reduce the complexity of FE model, an equivalent model of the multi-layer composite structure is here considered as a single-layer panel. The equivalent thickness t_{eq}, and stiffness moduli E_{eq} and G_{eq}, obtained by solving the equations in available literature [9], are

$$t_{eq} = \sqrt{3h_c^2 + 6h_c t_f + 4t_f^2} \qquad E_{eq} = (2t_f E_f)/t_{eq} \qquad G_{eq} = (2t_f G_f)/t_{eq} \tag{1}$$

where t_f is the skins thickness, h_c is the height of honeycomb core, E_f and G_f skins moduli. The equivalent data for a 10mm sandwich panel are $t_{eq} = 0.0156m$, $E_{eq} = 90GPa$, $G_{eq} = 3.31GPa$.

At the same time, the Modal Strain Energy (MSE) - defined as the amount of elastic energy stored in a finite element - associated to the flexible appendages was computed. The related MSE map (see Fig. 2) is used to identify the locations of the elements whose change in mechanical properties could be more problematic for the global dynamics of the system. The objective is to avoid building a heavy set of data including damages all over the structure (also damages associated with low risk, i.e. inducing a negligible change in the modal properties of the satellite), potentially leading to an excessively high dimension multivariate classification problem. Instead, the approach proposed here is to discriminate a set of potential critical damages, to be identified via the deep learning architecture, based on MSE concentration.

In this research, damages are considered as resulting from space debris hits, causing a perforation in the structure. The dimension of damage is assumed as not exceeding 5cm x 5cm, which is a representative size for high velocity impacts for aluminium honeycomb [10]. Damages are simulated only on one solar panel. Hence, a set of three-axis accelerometers sensors are installed on one side. In particular, the position of the sensors is depicted in Fig. 3.

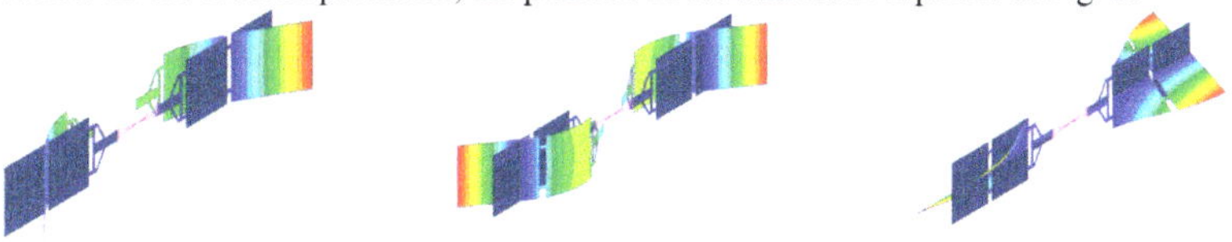

Fig. 1: Modal shapes. From left to right, 1st mode: 0.97 Hz, 2nd mode: 1.58 Hz, 3rd mode: 3.27 Hz

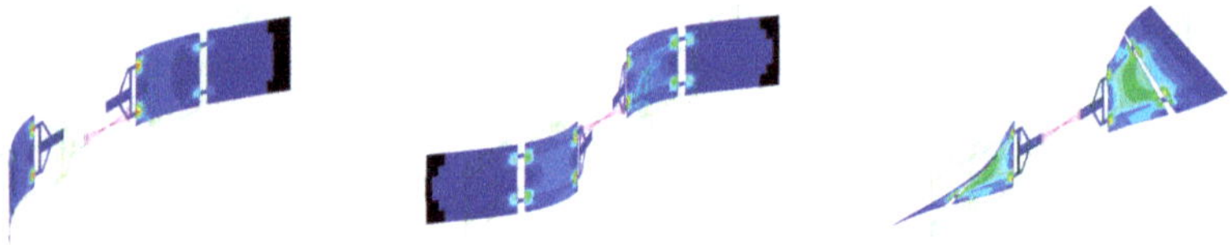

Fig. 2: Modal Strain Energy (MSE) map. From left to right, 1st mode, 2nd mode, 3rd mode

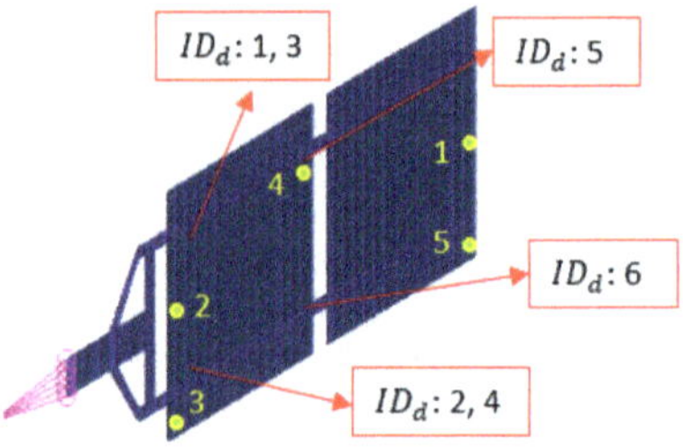

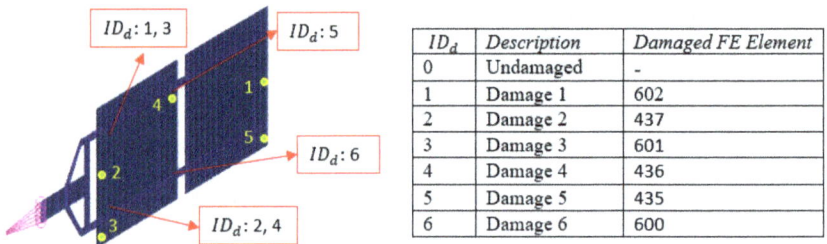

ID_d	*Description*	*Damaged FE Element*
0	Undamaged	-
1	Damage 1	602
2	Damage 2	437
3	Damage 3	601
4	Damage 4	436
5	Damage 5	435
6	Damage 6	600

Fig. 3: Overview of damages (indicated as ID_d) and sensors (indicated with numbers from 1 to 5).

Training and Validation

To generate the training set, a 3D simulator of in-orbit flexible spacecraft dynamics – developed in house [11-12] – is used to reproduce satellite attitude maneuvers. A finite element model is created for each damage scenario. Once the damaged structural sub-models, deriving from the original undamaged one, and the network of sensors are defined, the dataset generation for the training of the DNN can be set up. In detail, the followed steps are:

- The finite element structural models are imported in Matlab to perform the non-linear simulation of attitude maneuvers for the flexible spacecraft.
- Several one-, two- and three-axis attitude maneuvers are simulated by varying not only the target orientations, but also the gains of the quaternion-based PD control law to further diversify the dataset.
- Time histories from installed accelerometers sensors are recorded and saved to create the training and testing data, not before being pre-processed and normalized according to their mean and standard deviation.

Results

Several parameters are considered and optimized to improve the performance of the network: the DNN hyperparameters, the data pre-processing and the number and location of sensors. Finally, Table 1 shows the results in terms of accuracy (mean value and standard deviation) of the adopted classification network. In detail, four damages in the area of the structure associated with both highest MSE (red areas in Fig. 2) and lower MSE (green areas in Fig. 2) are considered. Despite inducing a reduced change in the system frequencies and modal shape (about 1% relative difference), the DNN shows a good classification accuracy even in this challenging case. The confusion matrix is illustrated in Fig. 4.

Table 1: Classification accuracy

Case	*Description*	*Rationale*	*Accuracy*
1	ID_d: 0, 1, 2, 5, 6 Sensors: 1, 2, 3, 4, 5	Identify the location of different damages in different MSE concentration areas (highest and second highest)	85.09% ± 4.87%

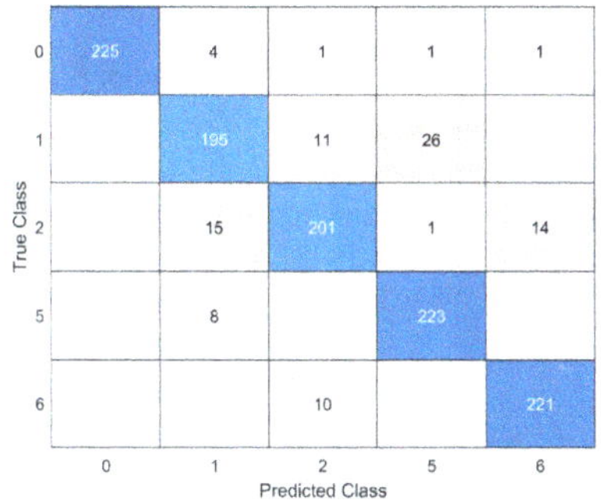

Fig. 4: Confusion matrix

Conclusions

This work has showcased the potential of LSTM networks in identifying damages in large space structures by analyzing time responses measured by sensors. The presented results are not only preparatory to carry out a laboratory experimental validation phase on a flexible spacecraft test-rig, but also propaedeutic to apply the system to several innovative composite materials. Indeed, future research endeavors could explore the possibility of training deep neural networks using real-world measured data, enabling them to operate in practical conditions and accurately predict actual damages.

Acknowledgments

We would like to thank Prof. Massimo Panella and Prof. Antonello Rosato for their expertise and insights, which were instrumental in shaping the direction and focus of this research.

Funding

The work was financed by the *European Union - NextGenerationEU* (National Sustainable Mobility Center CN00000023, Italian Ministry of University and Research Decree n. 1033 - 17/06/2022, Spoke 11 - Innovative Materials & Lightweighting). The opinions expressed are those of the authors only and should not be considered as representative of the European Union or the European Commission's official position. Neither the European Union nor the European Commission can be held responsible for them.

References

[1] E. Figueiredo, et al., Machine learning algorithm for damage detection, computational and experimental methods in structure, Vibration-Based Techniques for Damage Detection and Localization in Engineering Structures, 10 (2018) 1-39. https://doi.org/10.1142/9781786344977_0001

[2] J. Vitola, et al., Data-Driven Methodologies for Structural Damage Detection Based on Machine Learning Applications, Pattern Recognition - Analysis and Applications, (2016). https://doi.org/10.5772/65867

[3] G. B. Palmerini, F. Angeletti, P. Iannelli, Multiple Model Filtering for Failure Identification in Large Space Structures, Lecture Notes in Civil Engineering, 128 (2021) 171-181. https://doi.org/10.1007/978-3-030-64908-1_16

[4] P. Iannelli, F. Angeletti, P. Gasbarri, M. Panella, A. Rosato, Deep learning-based Structural Health Monitoring for damage detection on a large space antenna, Acta Astronautica, 193 (2022) 635-643, https://doi.org/10.1016/j.actaastro.2021.08.003

[5] F. Angeletti, P. Iannelli, P. Gasbarri, M. Panella, A. Rosato, A Study on Structural Health Monitoring of a Large Space Antenna via Distributed Sensors and Deep Learning, Sensors, 23(1) (2023) 368, https://doi.org/10.3390/s23010368

[6] F. Succetti, A. Rosato, R. Araneo, M. Panella, Deep neural networks for multivariate prediction of photovoltaic power time series, IEEE Access, 8 (2020) 211490-211505. https://doi.org/10.1109/ACCESS.2020.3039733

[7] Z. Ma, M. Yao, T. Hong and B. Li, Aircraft Surface Trajectory Prediction Method Based on LSTM with Attenuated Memory Window, Journal of Physics: Conference Series, 1215 (2019). https://doi.org/10.1088/1742-6596/1215/1/012003

[8] F. Karim, S. Majumdar, H. Darabi, S. Harford, Multivariate LSTM-FCNs for Time Series Classification, Neural network, 116 (2019) 237-245. https://doi.org/10.1016/j.neunet.2019.04.014

[9] J. K. Paik, A. K. Thayamballi and G. S. Kim, The Strength Characteristics of Aluminum Honeycomb Sandwich Panels, Thin-Walled Struct., 35 (1999) 205–31. https://doi.org/10.1016/S0263-8231(99)00026-9

[10] X. Kunbo, Z. Jiandong, G. Zizheng, C. Yan, Z. Pinliang et al., Investigation on solar array damage characteristic under millimetre size orbital debris hypervelocity impact, Proc. 7th European Conference on Space Debris, Darmstadt, Germany, 18–21 April 2017, published by the ESA Space Debris Office.

[11] F. Angeletti, P. Iannelli, P. Gasbarri, M. Sabatini, End-to-end design of a robust attitude control and vibration suppression system for large space smart structures, Acta Astronautica, 187 (2021) 416–428. https://doi.org/10.1016/j.actaastro.2021.04.007

[12] F. Angeletti, P. Iannelli, P. Gasbarri, Automated nested co-design framework for structural/control dynamics in flexible space systems, Acta Astronautica, 198 (2022) 445–453. https://doi.org/10.1016/j.actaastro.2022.05.016

A numerical parametric study on delamination influence on the fatigue behaviour of stiffened composite components

Angela Russo[1,a*], Andrea Sellitto[1,b], Concetta Palumbo[1,c], Rossana Castaldo[1,d] and Aniello Riccio[1,e]

[1]University of Campania "Luigi Vanvitelli", Department of Engineering, Via Roma 29, 81031 Aversa (CE), Italy

[a]angela.russo@unicampania.it, [b]andrea.sellitto@unicampania.it, [c]concetta.palumbo@unicampania.it, [d]rossana.castaldo@unicampania.it, [e]aniello.riccio@unicampania.it

Keywords: CFRP, Fatigue, SMXB, Delamination Growth

Abstract. This paper investigates the fatigue phenomenon in Carbon Fibre Reinforced Plastic (CFRP) composite materials. Fatigue is a major problem in composite materials, due to their complex microstructures and inhomogeneous properties. In composite materials, fatigue is caused by cyclic loading, which leads to the accumulation of damage and eventually failure. This is related to several factors such as material properties, geometry, loading conditions, and environmental conditions. The fatigue life of composite materials is usually much lower than that of metals, and it is often catastrophic and unpredictable. Therefore, it is mandatory to understand behaviour of composite materials subjected to cyclic loading condition and to develop strategies to improve their fatigue performance. To this end, a Paris Law-based module has been implemented in the well-establish SMart-time XB (SMXB) procedure, being able to accurately numerically simulate the delamination growth caused by cyclic loads in complex composite structures. This process, which takes advantages of the mesh and load independency of the SMXB method in the evaluation of the delamination growth, has been implemented in the Ansys Parametric Design Language to create a highly versatile and parametric procedure. A numerical parametric study has been carried out to investigate the behaviour of a pre-existing circular delamination under cyclic loading, to assess the influence of delamination radius and thickness on the delamination growth. The results of this study will provide important insights into how delamination radius and thickness affect the delamination growth and the durability of composite structures. This study will help to inform the design of composite structures for various applications.

Introduction

Damage mechanisms in composite materials are the main reason behind the limited use of these materials for aircraft component construction. Delamination, in particular, is one of the most serious failure events that composites may experience, as it is undetectable and can therefore propagate, causing a rapid structural collapse. Fatigue phenomenon amplifies the weaknesses in terms of damage propagation in composite materials [1].

The study of interlaminar damage propagation has been extensively addressed in the literature, both experimentally and numerically. In particular, the Virtual Crack Closure Technique (VCCT) [2] and Cohesive Zone Models (CZM) are the most widely used computational methodologies for the investigation of delamination propagation in finite element (FE) environment. Among the others, the numerical procedure SMart time XB (SMXB) [2,3] is a FE tool based on the VCCT, which, however, has the unique feature to simulate delamination without dependence on the size of the elements used for the discretization of the model and independent from the load step size used in the numerical analyses. These characteristics are the basis for a further development of the SMXB, which is the possibility of simulating the evolution of delamination due to fatigue, i.e. a load acting on the structure in a cyclic fashion. The fatigue delamination simulation method integrated in SMXB has been implemented using the Ansys Parametric Design Language and is based on the Paris law relation. In [4], the mesh independence of the results obtained with the FaTigue SMXB (FT-SMXB) tool has been demonstrated in the case of a Double Cantilever Beam specimen.

In this work, a parametric study has been performed to investigate the behaviour of a CFRP typical stiffened panel with a pre-existing circular delamination, which aims to simulate an impact damage. The panel has been subjected to cyclic loads in order to assess the influence of the delamination radius and depth on the global compressive behaviour of the structure. The principal added value of this work is the use of the FT-SMXB, due to its inherent characteristics. Standard VCCT and CZM probably would affect the results due to their dependence on the mesh and time step of the finite element analysis.

Methodology

The SMart Time XB procedure has been introduced for the first time in [2] by Pietropaoli and Riccio in 2010. The tool has been improved over the years with different capabilities, for example the introduction of double delamination front in the case of skin-stringer debonding [3] or also the possibility to consider the R-curve. It is still being studied and enhanced today, in particular, by introducing the capability to simulate the fatigue driven delamination [4].

Fatigue in composite materials can be simulated considering the Paris Law equation, which has been developed for metals and then extended for composite materials. Equation (1) reports the Paris Law equation for composites, where a is the crack length, N is the number of cycles, G is the Energy Release Rate and n, c are experimentally derived constant parameters. This equation expresses the crack growth rate as a function of the number of fatigue cycles.

$$\frac{da}{dN} = cf(G)^n \tag{1}$$

The function of ERR, $f(G)$, may assume different expressions. In the FT-SMXB, $f(G)$ can be selected among more than 4 expressions [], depending on the user needs.

The implemented tool considers the accumulation of damage on the crack front, so that the number of cycles at which the delamination propagates and the position of propagation on the crack front can be correctly assessed. Moreover, the procedure allows to take into account the load ratio and permits to evaluate the residual strength of the structure after a certain number of fatigue cycles.

Numerical application

The studied structure, shown in Figure 1, is a typical aeronautical component with two T-shaped stringers tied on a skin panel. It is characterized by a circular delamination, located in the middle of the bay, representing an example of typical impact damage. Two different configurations have been examined, both under cyclic loading conditions, which share all the geometrical properties except the radius of the delamination and its depth. Two radii have been considered, 30 mm and 40 mm, for each of them the delamination has been placed under 2 and 3 layers.

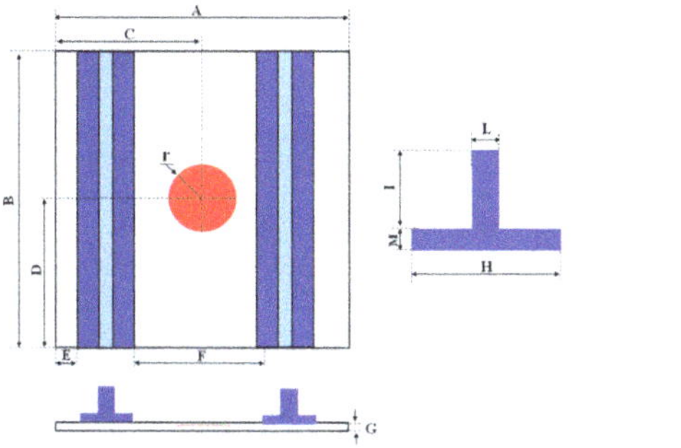

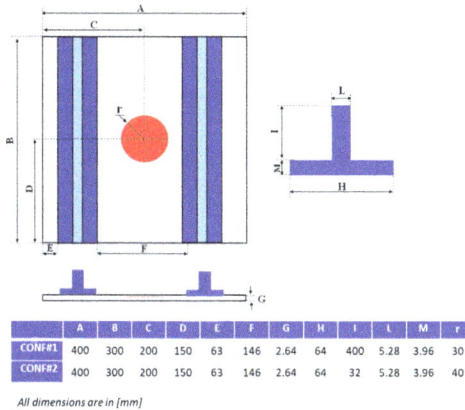

	A	B	C	D	E	F	G	H	I	L	M	r
CONF#1	400	300	200	150	63	146	2.64	64	400	5.28	3.96	30
CONF#2	400	300	200	150	63	146	2.64	64	32	5.28	3.96	40

All dimensions are in [mm]

Figure 1. Geometry of the analysed panel

The stacking sequence of the skin panel contains 16 layers with a layup of $(45°,0°,90°,-45°]_{2s}$, the foot stringer has 24 layers with a sequence of $(45°,0°,90°,-45°)_{3s}$, finally the web stringer has 32 layers with a layup of $(45°,0°,90°,-45°)_{4s}$.

The finite element model has been built in the Ansys Mechanical environment and a global-local approach has been used, whereby the structure has been discretised with a sparse mesh in the area not involved in the crack propagation phenomenon and a local area, characterised by the initial circular delamination and the propagation region, discretised on the contrary with a fine mesh. Both solid and shell parts have been considered, in order to reduce the computational cost of the analyses. The panel has been clamped on one side, while compressive cyclic load has been applied on a pilot node, linked to the other side nodes of the structure by means of rigid link elements.

Results

The panel has been analysed under cyclic loading conditions: 90% and 80% of the static onset load has been considered for the analyses. For each configuration static analysis has been performed in order to obtain the onset load on which the applied fatigue load is based. However, for the sake of brevity, only the results of fatigue simulations are reported in this section.

The results in terms of delaminated area as a function of the number of cycles are shown in Figure 2 for the configuration with 30 mm delamination radius. For simplicity, the configurations have been named CONF#1 for radius of 30 mm and CONF#2 for radius of 40 mm, in particular CONF#1-2 is the configuration with delamination radius of 30 mm under 2 plies.

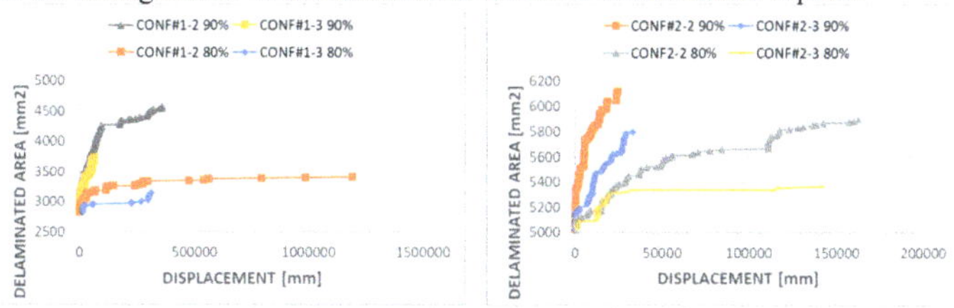

Figure 2. Delaminated area vs. Number of cycles

The analyses stop when propagation reaches the limit of propagation depending on the propagation region defined by the user (this is a numerical issue) or when the number of cycles overcome 1E6. According to Figure 2, in the selected configurations, only CONF#1-2 at 80% load exceeds one million cycles. The behaviour of the two configurations is similar: increasing the delamination depth reduces the delaminated area at the same number of cycles. Moreover, as expected, for a higher load the delamination is considerably unstable. Table 1 summarize the results in terms of delamination onset, buckling load and number of cycles to failure.

Table 1. Result summary

Configuration	Delamination buckling	Global panel buckling	Delamination onset load (static)	Number of cycles onset – 90%	Number of cycles failure – 90%	Number of cycle onset – 80%	Number of cycles failure – 80%
CONF#1-2	8264 N	207590 N	146117 N	148	358469	3174	>1E6
CONF#1-3	27325 N	207241 N	140209 N	385	64597	18000	309403
CONF#2-2	4696 N	206758 N	145539 N	92	24443	1208	162791
CONF#2-3	15610 N	205516 N	135655 N	69	33433	1650	144271

The local buckling load of delamination is different for all the configurations, it decreases by increasing the radius, while, at the same radius, it increases by increasing the delamination depth. The global buckling remains almost the same, between 205 and 207 kN.

Conclusions

In this paper the fatigue driven delamination has been studied in a composite stiffened panel with circular delamination. It has been found that the fatigue response of the panel depends on the delamination depth. Increasing the depth reduce the number of cycle to failure of the panel. the propagation trend remains similar as the delamination radius changes.

References

[1] C. B. Carlos G. Dávila, «Fatigue life and damage tolerance of postbuckled composite stiffened structures with initial delamination,» Composite Structures, pp. 73-84, 2017. https://doi.org/10.1016/j.compstruct.2016.11.033

[2] E. Pietropaoli e A. Riccio, « On the robustness of finite element procedures based on Virtual Crack Closure Technique and fail release approach for delamination growth phenomena. Definition and assessment of a novel methodology.,» Compos. Sci. Technol, p. 1288–1300, 2010. https://doi.org/10.1016/j.compscitech.2010.04.006

[3] A. Riccio, A. Raimondo e F. Scaramuzzino, « A robust numerical approach for the simulation of skin–stringer debonding growth in stiffened composite panels under compression.» Compos. Eng., pp. 131-142, 2015. https://doi.org/10.1016/j.compositesb.2014.11.007

[4] Russo, A., Riccio, A., & Sellitto, A. (2022). A robust cumulative damage approach for the simulation of delamination under cyclic loading conditions. Composite Structures, 281. https://doi.org/10.1016/j.compstruct.2021.114998

Materials Research Proceedings 37 (2023) 377-380 https://doi.org/10.21741/9781644902813-83

On the use of double-double design philosophy in the redesign of composite fuselage barrel components

Antonio Garofano[1,a *], Andrea Sellitto[1,b] and Aniello Riccio[1,c]

[1]University of Campania "Luigi Vanvitelli", Department of Engineering, Via Roma 29, 81031, Caserta, Italy

[a]antonio.garofano@unicampania.it, [b]andrea.sellitto@unicampania.it, [c]aniello.riccio@unicampania.it

Keywords: Composite Structures, Double-Double, Laminates Optimization, Mass Saving

Abstract. Mass minimization and mechanical performance maximization constitute the basic aspects of the structural optimization processes. In particular, the laminate redesign in terms of thickness and lay-up grants the main approach for the optimization of composite components. The innovative Double-Double laminate concept provides an effective approach to design composite components for weight and strength requirements, overcoming the use of the conventional 0°, 90° and ±45° ply orientations. In Double-Double designed components, 4-plies building blocks are stacked one upon the other to constitute a laminate. Each building block is made up of four [±Φ, ±Ψ] oriented plies. In the present work, the Double-Double approach has been adopted in the redesign of the composite lay-up and thickness profile of frames in a composite fuselage barrel. The DD optimized frames achieved a total mass reduction by up to 35% while ensuring mechanical performances comparable to the starting configuration.

Introduction

Laminate lay-up optimization represents the most feasible field in which structural engineers can work as compared to material changing and components redesign to design more efficient, lightweight and performing composite load-bearing structures in the aviation field. Lay-ups typically used in composite aviation components only involve 0°, 90° and ±45° plies orientations. Moreover, symmetry and balancing requirements must be accounted for manufacturing and performance needs, limiting the optimization processes for mass and mechanical performance requirements to sub-optimal solutions. One step forward is proposed by the newly Double-Double approach introduced by Professor S.W. Tsai [1]. Double-Double laminates are made up by stacking 4-plies building blocks without symmetry and balancing requirements. Each building block, or sub-laminate, consists of four plies based on double bi-axial angles [±Φ, ±Ψ]. Φ and Ψ angles can be tuned assuming each value between 0° and 90° to manage mechanical performances of laminates. DD angles and number of the required building blocks are directly optimized for the application and its load conditions. A new composite lay-up scheme able to optimize the components for weight and strength requirements is given, enabling to overcome the conventional 0°, 90° and ±45° plies orientations [2]. The Excel-based Lam-Search optimizer tool automatically determines the best [±Φ, ±Ψ] angles among all allowed combinations of DD angles to minimize the safety margin by computing the strength ratio R with respect to a selected failure criterion. Laminate thickness is updated according to the computed strength ratio value [3]. In this framework, the present paper is aimed to perform an optimization process on the frames of a composite fuselage barrel subjected to static loads through the Double-Double design approach for mass reduction purposes.

FE model description

The FE model of a regional aircraft's fuselage barrel has been developed in the Abaqus environment, as showed in Fig. 1. Main structural sub-components have been modelled as solid extruded parts while frames as 3D shell parts allowing to run sequential analyses with variable thickness during the optimization process. The optimized [±37.5, ±45] Double-Double lay-up has been employed for the skin [4]. A uniform $[90,45,0,45,-45,90,45,-45,0,-45,45,-45]_s$ stacking sequence characterized the frames in the starting configuration. The IMS/977-2 CFRP [4], a Woven Fabric [4] and the T700 CPLY64 [4] have been adopted for the composite components while Aluminum 2024-T42 for the metal parts. The FE model has been discretized by means of 8-node continuum shell elements (SC8R) and 8-node linear brick elements (C3D8R) for composite and aluminum components, respectively, while 4-node general-purpose shell elements (S4R) have been used for frames.

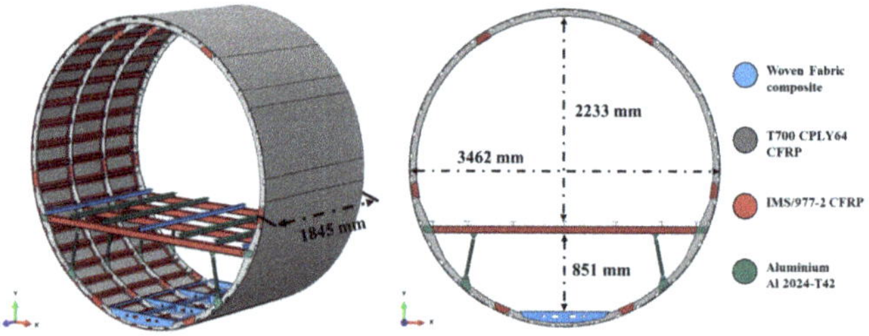

Figure 1. Fuselage barrel FE model and dimensions: a) isometric view; b) frontal view

According to the Lam-Search optimizer tool needs to perform the optimization process, frames have been subdivided in zones and cells. A zone has been defined for each frame, enabling one best [±Φ; ±Ψ] DD lay-up for each frame. Seven cells have been defined in each zone by considering only the half of each frame being the structure symmetry, as showed in Fig. 2a. Thereafter, results have been extended to the other half. The thickness profile in cells is computed according to the strength ratio R value with respect to the controlling cell.

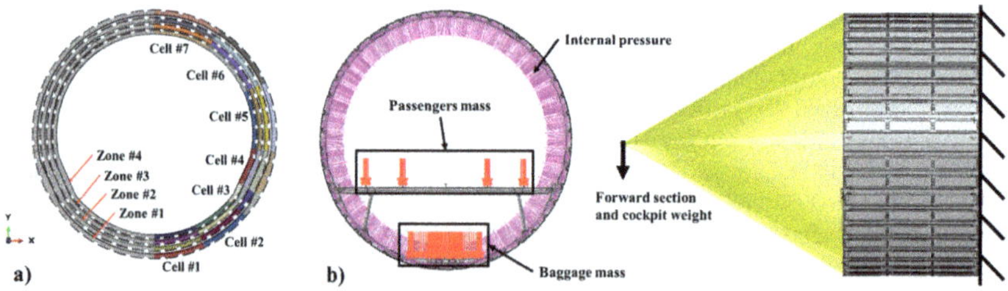

Figure 2. a) Zone and cells subdivision of the frames; b) applied loading and boundary conditions

An iterative process has been carried out to reach the strength optimized laminate design for frames through thickness and layup variation in cells of each zone. Linear static analyses considering the operating loads presented in Fig. 2b have been performed in each iteration. In detail, a pressure has been applied to the internal surfaces to simulate pressurization while a force representing the cockpit and forward section weight has been applied to a reference point and transferred to the structure. Mass loads of passengers and baggage have been considered.

Optimization process

A preliminary linear static FE analysis has been performed on the fuselage barrel in the starting configuration with quad laminates frames and assumed as Iteration 0. The average stress components in all cells of each zone have been used as input data to the Lam-Search optimizer tool, resulting in the best DD angles and cells thickness profiles for each zone as output. The same process has been individually applied to each zone. Results in Iteration 0 suggested the best DD angles and thickness profiles for cells in zones 1 to 4, respectively, leading to a total mass reduction of frames up to 85%. However, R values went above the unity. Consequently, the starting FE model has been updated according to Iteration 0 results and a new FE analysis has been carried out to update the stress components in the cells, generating new input data for the Lam-Search optimizer tool. This process has been repeated until identical DD angles in each zone between two iterations were achieved and any cell exhibited a strength ratio below unity.

At the end of Iteration 3, all cells in all zones achieved a strength ratio R greater than unity. Furthermore, the optimal DD angles in each zone resulted [±37.5; ±60], [±30; ±52.5], [±30; ±60], and [±45; ±45] for zones 1 to 4, respectively, and identical to those obtained in the previous iteration. The thickness optimization led to a total mass reduction of frames up to 71%. The time-history variation of the average value of strength ratio R for each zone across iterations is shown in Fig. 3. In Iteration 3, the thickness optimization led the R-values above and close the unity, offering an effective design solution in terms of mass saving.

ITERATION 0 — Tsai-Wu profile of strength ratio R

Cells	Zone 1	Zone 2	Zone 3	Zone 4
1	3.22	3.35	3.37	3.35
2	5.11	3.36	4.05	5.36
3	3.65	3.53	3.55	4.83
4	4.17	3.49	3.70	5.64
5	6.79	4.61	5.01	7.11
6	6.68	4.06	4.47	7.35
7	7.55	4.15	4.61	7.92

ITERATION 1 — Tsai-Wu profile of strength ratio R

Cells	Zone 1	Zone 2	Zone 3	Zone 4
1	0.73	0.63	0.67	0.93
2	0.76	0.61	0.64	0.90
3	0.80	0.61	0.68	0.91
4	0.81	0.63	0.68	0.87
5	0.75	0.71	0.69	0.84
6	0.79	0.69	0.70	0.86
7	0.74	0.65	0.69	0.89

ITERATION 2 — Tsai-Wu profile of strength ratio R

Cells	Zone 1	Zone 2	Zone 3	Zone 4
1	1.04	1.06	1.01	0.93
2	1.09	1.00	1.03	0.98
3	1.03	1.05	1.10	0.89
4	0.94	1.09	1.03	0.88
5	1.01	1.10	1.07	0.89
6	0.94	1.11	1.08	0.97
7	0.90	1.12	1.09	0.94

ITERATION 3 — Tsai-Wu profile of strength ratio R

Cells	Zone 1	Zone 2	Zone 3	Zone 4
1	1.03	1.05	1.00	1.04
2	1.07	1.01	1.02	1.08
3	1.02	1.04	1.09	1.01
4	1.02	1.08	1.02	1.06
5	1.00	1.09	1.06	1.05
6	1.01	1.10	1.07	1.06
7	1.02	1.11	1.08	1.01

Figure 3. Time-history variation of the strength Ratio R in cells in all zones across iterations (Note: R-values below the unity in red, R-values above the unity in green)

The optimized design for frames provided by the optimization process is characterized by a continuously scaled thickness across cells, not applicable to composite materials, as laminates have a discrete thickness. Hence, digitization is required to conservatively convert the cells thickness to discrete values according to the building block thickness. In addition, cells thickness within the same frame sub-components have been equalized to the thickest one to promote an easy manufacturing and uniformity of the sub-component, performing a tapering operation. Digitizing and tapering operations, showed in Fig. 4, reduced the mass saving given by Double-Double designed frames to 65% while providing an increase in the easy of manufacturing.

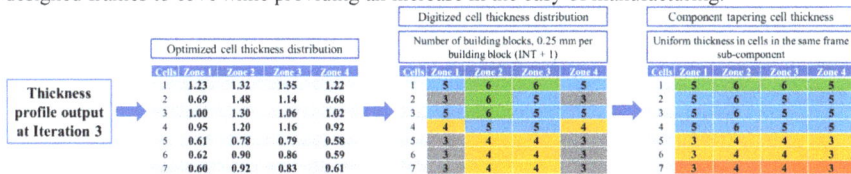

Thickness profile output at Iteration 3

Optimized cell thickness distribution

Cells	Zone 1	Zone 2	Zone 3	Zone 4
1	1.23	1.32	1.35	1.22
2	0.69	1.48	1.14	0.68
3	1.00	1.30	1.06	1.02
4	0.95	1.20	1.16	0.92
5	0.61	0.78	0.79	0.58
6	0.62	0.90	0.86	0.59
7	0.60	0.92	0.83	0.61

Digitized cell thickness distribution — Number of building blocks, 0.25 mm per building block (INT + 1)

Cells	Zone 1	Zone 2	Zone 3	Zone 4
1	5	6	6	5
2	3	6	5	3
3	5	6	5	5
4	4	5	5	4
5	3	4	4	3
6	3	4	4	3
7	3	4	4	3

Component tapering cell thickness — Uniform thickness in cells in the same frame sub-component

Cells	Zone 1	Zone 2	Zone 3	Zone 4
1	5	6	6	5
2	5	6	5	5
3	5	6	5	5
4	5	6	5	5
5	3	4	4	3
6	3	4	4	3
7	3	4	4	3

Figure 4. Thickness evolution in digitizing and tapering operations starting from the optimized design

Thickness variation led to a change in the frames total mass and consequently in the fuselage barrel total mass. The time history of the mass variation and mass saving in the frames and in the whole fuselage barrel, compared to the initial configuration is shown in Fig. 5. The frames optimization performed through the Lam-Search optimizer tool resulted in a frames design that reduced their total mass to 35% of the initial mass after tapering. Consequently, the mass variation in frames resulted in the mass reduction of the overall fuselage barrel by up to 12%.

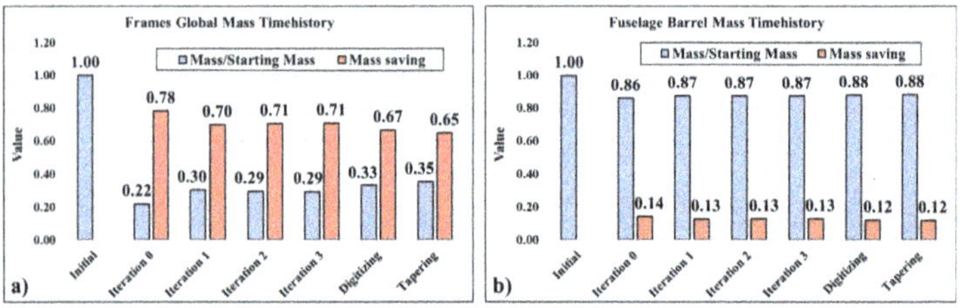

Figure 5. Mass and mass saving variation in frames and in the whole fuselage barrel through optimization

Conclusions

In the present work, the Double-Double lay-up concept has been applied to the frames of a composite fuselage barrel, considering the operating loads. Throughout iterations, the procedure identified the best DD composite lay-up and thickness profiles, providing mechanical performances comparable to the initial configuration while significantly reducing the components total mass. At the end of the optimization process, the initial configuration of frames with quad laminates has been replaced by an optimized configuration with the best DD lay-up for each of the frames, corresponding to [±37.5; ±60], [±30; ±52.5], [±30; ±60] and [±45; ±45] for zones 1 to 4, respectively. The thickness profile has been tailored to the acting loading condition ensuring a total mass reduction of frames by up to 35%. Thus, the use of the Double-Double laminates concept and the Lam-Search optimizer tool proved to be an effective method for optimizing the design of composite components for mass reduction purposes while ensuring a mechanical behavior comparable to the starting configuration.

References

[1] Shrivastava, S., Sharma, N., Tsai, S. W., & Mohite, P. M. (2020). D and DD-drop layup optimization of aircraft wing panels under multi-load case design environment. Composite Structures, 248, 112518. https://doi.org/10.1016/j.compstruct.2020.112518

[2] Tsai, S. W. (2021). Double–double: new family of composite laminates. AIAA Journal, 59(11), 4293-4305. https://doi.org/10.2514/1.J060659

[3] Vermes, B., Tsai, S. W., Riccio, A., Di Caprio, F., & Roy, S. (2021). Application of the Tsai's modulus and double-double concepts to the definition of a new affordable design approach for composite laminates. Composite Structures, 259, 113246. https://doi.org/10.1016/j.compstruct.2020.113246

[4] Garofano, A., Sellitto, A., Acanfora, V., Di Caprio, F., & Riccio, A. (2023). On the effectiveness of double-double design on crashworthiness of fuselage barrel. Aerospace Science and Technology, 108479. https://doi.org/10.1016/j.ast.2023.108479

Multifunctional composites as Solid-Polymer-Electrolytes (SPE) for Lithium Ion Battery (LIB)

Salvatore Mallardo[1,a *], Gennaro di Mauro[2,b], Michele Guida[2,c], Pietro Russo[1,d], Gabriella Santagata[1,e], Rosa Turco[3,f]

[1]Institute for Polymers, Composites and Biomaterials, National Research Council, Pozzuoli, Italy

[2]Department of Industrial Engineering, University of Naples Federico II, Naples, Italy

[3]Department of Chemical Sciences, University of Naples Federico II, Naples, Italy

[a]salvatore.mallardo@ipcb.cnr.it, [b]gennaro.dimauro@unina.it, [c]michele.guida@unina.it, [d]pietro.russo@ipcb.cnr.it, [e]gabriella.santagata@ipcb.cnr.it, [f]rosa.turco@unina.it

Keywords: Multifunctional Materials, Polymer Composites, Solid Polymer Electrolyte (SPE), Energy Generation, Aircraft Structures

Abstract. Novel solid-polymer-electrolytes (SPE) have been formulated as key components of structural multifunctional materials to develop Lithium Ion Battery (LIB). To this aim, SPE blends based on polyethylene oxide (PEO), different molecular weights polyethylene glycole (PEG), PEG-modified sepiolite (SEP) and lithium triflate have been prepared by one pot melt mixing. The films were obtained by compression moulding following a method easily scalable to industrial level. The different films have been characterized by structural (FTIR-ATR), thermal (DSC, TGA), morphological (SEM) and mechanical (tensile tests) analysis. The different properties could be mainly addressed to the diverse PEG both amounts and molecular weight and to the specific physical interaction occurring between PEO, PEG sepiolite and lithium ions strongly influencing crystallinity, thermal stability and mechanical response. Thus, SPE2 sample evidenced the highest both crystallinity and mechanical stiffness and toughness, whereas SPE1 and SPE3 film showed the best compromise between molecular crystallinity and mechanical performances, mostly as strain at break are concerned. Finally, SPE4 film, including the highest amount of PEG showed a peculiar increasing of mechanical rigidity in opposition to molecular plasticization effect exploited by PEG. The many features of SPE systems requires special attention and further research when it comes time to design structural multifunctional materials for LIB based batteries of Type-III.

Introduction

The emerging technologies of structural batteries (SBs), i.e., multi-component-based systems able to combine the functions of energy storage and the load-bearing ability in the attempt to improve such performances, are gaining even more attention to solve the environmental sustainability at decarbonizing the transportation sector [1]. SBs are based on solid polymer electrolytes (SPEs), i.e., solid polymeric materials with ability to transport lithium ions. Usually, SPEs consist of polymer matrices and lithium ions obtained after dissolution of polymer with lithium salts and following drying process. The result is the formation of a thin, inflammable, and flexible film of a coordination complex between the polymer and the lithium ions [2]. When a potential difference is applied across the solid electrolyte, the lithium ions move from the anode to the cathode through the polymer matrix, they react with the cathode material to produce a flow of electrons. Among the polymers, polyethylene oxide (PEO) is the most attractive since it shows excellent solubility for lithium salts, is a low cost, non-toxic, biocompatible, water soluble, high conductive material [4] with high energy density, high electrochemical stability, and excellent compatibility with inorganic salts. It is widely accepted that lithium-ion transport occurs at the expense of the flexible amorphous phase [5, 6]. In this paper, PEO as organic electrolyte, lithium triflate (lithium

Materials Research Proceedings 37 (2023) 381-385 https://doi.org/10.21741/9781644902813-84

triflouromethanesulfonate) $LiCF_3SO_3$ as electrolyte, sepiolite $(Si_{12}O_{30}Mg_8(OH)_4(H_2O)_4{\times}_8H_2O)$ (SEP) as a natural, low cost and abundant fibrous clay mineral enhancing the dissociation of lithium salts, have been used to develop SPEs [7, 8, 9]. To enhance SEP dispersion in PEO thus avoiding fiber aggregation, organic modification of SEP surface has been performed by blending it with water solutions of polyethylene glycol (PEG), a plasticizer able to reduce PEO crystallinity promoting the ion conductivity [10]. All components were melted, composed, and compressed into films in a "one pot solution" avoiding solvents, thus following an environmentally sustainable approach. The SPEs were characterized by thermal, mechanical, and morphological analysis to evaluate their chemical-physical performance.

Experimental: Materials and methods

PEG1000 and PEG1550 by Fluka, polyethylene oxide (PEO) Mw 5.000.000, Sepiolite and Lithium triflate by Sigma Aldrich. Mixtures of PEG and sepiolite (SEP) were prepared by SEP water dispersion and subsequently adding 5% and 10 % w/w of PEG1000 and PEG1550, as detailed in Table 1. The suspension was washed, centrifuged, and dried thus obtaining organo-modified sepiolite (SEPmod). The raw physical blends of SEPmod, Polyethylenoxide (PEO) and Lithium triflate (LiTfr), obtained by their physical mixing, were melt blended, using a Brabender Plasti-Corder mixer kept at 140°C and 80 rpm of screw rate. The resulting SPEs were recovered, cooled at room temperature, roughly powdered, dried and compression molded at 130°C by using a table hydraulic hot-press. FTIR-ATR, DSC, TGA, SEM and tensile tests have been performed on samples. DSC analysis were carried out using a Mettler DSC/822 calorimeter at a nitrogen gas flow of 50 ml/min. Samples were firstly heated from 25 to 150 °C at 10 °C min^{-1}, then cooled up to −80 °C at a rate of 50 °C min^{-1} and newly heated up 250 °C at 10 °C min^{-1}.

TGA analysis were performed using a thermogravimetric analyzer TGA/DTG Perkin-Elmer Pyris Diamond. The samples were heated from 25°C to 600°C at 10°C min^{-1}, under nitrogen at 30 mL/min. SEM analysis were carried out using a Quanta 200 FEG, 338 FEI scanning electron device. Prior to the observation, surfaces were coated with a homogeneous layer (18 ± 0.2 nm) of Au and Pd alloy. The micrographs were performed at room temperature and in high vacuum mode. Tensile tests were carried out by using a dynamometer model 4301, Instron equipped with a load cell of 1 kN. The tests were performed on dumbbell-shaped films. Young's modulus, stress and strain at break values were determined and reported data are the average values of six determinations. All the tests were carried out at room temperature and at a crosshead rate of 2mm/min.

***Table 1** Film percentage composition and identification codes.*

IC	***Percentage composition (wt%)***				
Sample	PEO	LiCF_3 SO_3	SEP	PEG1000	PEG1550
SPE_Li	91.5	8.5	-	-	-
SPE_0	87.5	8.13	4.36	-	-
SPE_1	87.5	8.13	4.15	-	0.21
SPE_2	87.5	8.13	4.15	0.21	-
SPE_3	87.5	8.13	3.96	-	0.40
SPE_4	87.5	8.13	3.96	0.40	-

Results and Discussion

The thermal parameters of the SPEs samples are showed in Table 2.

***Table 2** Thermal parameters of SPE based films and neat polymers measured by DSC and TGA analysis. The values of crystallization and melting enthalpy in parentheses (ΔH) and the crystallinity (χ) are normalized with respect to the weight fraction of the polymer in the blend.*

	DSC					TGA		Crystallinity
	$\Delta H_c[J/g]$ ± 2%	$T_{c,onset}$[°C] ± 2%	T_c[°C] ± 2%	$\Delta H_m[J/g]$ ± 1%	T_m[°C] ± 1%	T_{onset}[°C] ± 2%	T_{peak}[°C] ± 2%	χ_c(%)
PEO	146.9	-31.7	-40.6	121.8	66.4	366.7	400.2	59.4
SPE_0	118.2	-51.7	-64.0	117.4	62.4	391.9	419.7	65.4
SPE_1	117.4	-21.7	-36.4	118.4	61.3	388.7	419.0	66.0
SPE_2	140.7	-18.3	-30.5	133.5	60.5	383.8	421.3	74.4
SPE_3	119.2	-22.5	-33.8	136.1	61.2	386.	419.8	75.9
SPE_4	134.3	-20.8	-34.4	123.8	60.9	376.1	409.0	69.0

It is worthy to highlight that all the samples began to crystallize at very low temperatures. This result is due to the fast-cooling rate (50°C/min), which decreases the crystallization process as the polymer chains, entangled in the melt, do not have enough time to separate enough to form crystals, so the amorphous nature of the melt is "frozen into" the solid. In the investigated system, all the samples reported in table 1, started to crystallize before than neat PEO and before SPE0 composite and show higher crystallinity percentage (χc). SPE2 systems crystallize before the other systems due to the balanced effects of nucleating effect and plasticizing action of PEG1000. No substantial differences have been detected among SPE1, SPE3, SPE4 samples, except for a slight decrease in Tc compared to SPE2, overall ΔHc and an increase crystallinity. Neat PEO starts to crystallize at lower temperatures, as expected. The absence of PEG in SPE0 sample drastically decreased onset and peak crystallization temperatures. During the second heating run, except for PEO showing the highest melting temperature, all the samples evidenced similar Tm likely associated with the melting of quite regular crystalline pattern developed during the cooling from the melt.

TGA results are reported in table 1 as the onset of temperature degradation, (T_{onset}), taken as the temperature at which 5% weight loss (WL) of sample occurred, and the peak of DTG thermograms, corresponding to the temperature of maximum degradation rate of the polymer (T_{peak}). From the analysis of these data, it emerged that all samples degrade around 419 °C, while at higher temperatures, around 470 °C, lithium triflate begins to degrade.

In addition, all SPE based systems showed higher thermal stability than the neat polymer. The neat PEO thermal degradation process is ranged in a shorter space of time than other SPE based samples, as expected, since no reinforcing and stabilizing agents were included inside the polymer matrix. This experimental outcome suggests the formation of tight intermolecular entanglements between the components likely due to their physical compatibility leading to mutual stabilization except for SPE4 sample, strongly influenced by the prevailing effect of high concentrations of PEG1000, significantly affecting PEO thermal stability.

In Figure 1, morphological analysis (SEM) of SPE based films are reported. As for brevity of discussion, only SPE0, SPE1 and SPE2 micrographs have been investigated and discussed. SPE0 sample (Figure 1a), evidenced a quite uneven and highly structured polymer surface with a fine and homogenous distribution of discrete microdomains of rod-like sepiolite filler, visible as white dots on the polymer surface. Moreover, the morphology changed in presence of PEG plasticizers. In SPE1 sample (Figure 1b), a homogeneous, quite smoothed and continuous polymer surface can be observed with the very fine distribution of sepiolite particles between the polymer macromolecular chains. Low concentrations of the higher Mw PEG allow suitable polymer plasticization, improving the physical interactions among the polar residues and the ionic charges of all the components of the biocomposite.

As concerning SPE2 sample, a very different surface topography is found (Figure 1c). A continuous but coarsened polymer surface can be observed characterized by a good interfacial adhesion between the polymer matrix and the plasticizer, as expected since PEO and PEG show the same polymer network but very different molecular weights. Profuse well dispersed and well embedded sepiolite particles occur confirming that the very close-fitting polymer-based network can deeply entrap these particles to the point of being disguised and no longer visible.

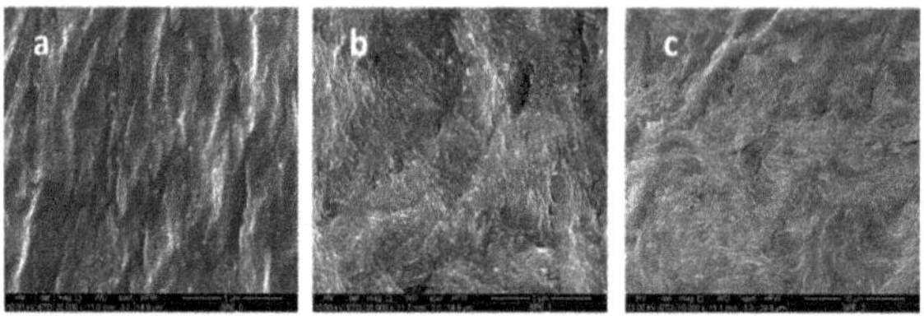

Figure 1 SEM micrographs of SPE0 (a), SPE1 (b) and SPE2 (c) surfaces.

The averaged values of elastic modulus (YM), stress at break (σ_b), strain at break (ε_b) are detailed in Table 3.

Table 3. Tensile test parameters for each adopted formulation

Sample	*Thickness (mm)*	*Young Modulus (YM)* [MPa]	*Strain at Break (εb)* [%]	*Stress at Break (σb)* [MPa]
SPE_Li	0.092	99.0±13.27	3.5±0.4	1.1±0.2
SPE_0	0.130	331.7 ± 30.0	4.5 ± 07	1.2 ± 0.1
SPE_1	0.145	305.6 ± 13.1	5.9 ± 2.6	1.1 ± 0.4
SPE_2	0.130	343.8 ± 10.8	4.8 ± 0.8	1.2 ± 0.2
SPE_3	0.148	397.9 ± 46.9	6.4 ± 1.5	0.8 ± 0.5
SPE_4	0.103	426.4 ± 65.3	3.8 ± 0.4	1.9 ± 0.1

The addition of lithium salt into the PEO matrix caused a general reduction of mechanical properties with respect to the other PEO based compositions including SEP. Indeed, from the analysis of the results, the system SPE0 evidenced the lowest YM. This finding is not surprising since the lack of inorganic filler is responsible for macromolecular chain mobility restriction and increasing stiffness and rigidity. In addition, also strain at the break of this system evidenced detrimental values. This last outcome could be due to the peculiar PEO Lithium salt interaction occurring between the polar ether groups of PEO and the positive charge of lithium hindering the regular flow of PEO macromolecular chains under tensile test. As concerning the other SPE based composites, the results indicate that usually, except for the SPE1 formulation, the addition of the plasticizer causes an increase in the tensile stiffness of the tested film. This effect is in line with the higher degree of crystallinity of the formulations found in DSC analysis. The crystallization behaviour of polymers the crystallinity, strongly influenced the macroscopic behaviour of polymers, such as the mechanical performances [11]. Both plasticizers induce an increase in this parameter with more marked effects in the case of PEG1000 compared to the PEG1550. This outcome is likely due to different behaviour induced by two PEG molecular weights. The most significant increase in stress for PEG 1000 samples is in line with the greater nucleating effect witnessed by Young's modulus values. Finally, about the strain undergone by the specimens, it seems that all the formulations are characterized by a value usually higher than that of the reference sample (SPE0) without plasticizers. In any case, for the system containing PEG1550, the strain at the peak of the curve seems to increase for a plasticizer content equal to 0.21% by weight and then decrease slightly for higher contents. The results indicate an exactly reversed trend for systems with PEG1000 as the content of this additive increases.

References

[1] European Commission. Clean Sky Benefits.

[2] Zhang, Y., Feng, W., Zhen, Y., Zhao, P., Wang, X, Effects of lithium salts on PEO-based solid polymer electrolytes and their all-solid-state lithium-ion batteries, Ionics, 28, (2022) 2751–2758. https://doi.org/10.1007/s11581-022-04525-3

[3] Long, L., Wang, S., Xiao, M., Meng, Y. (2016). J. Mater. Chem. A, 4: 10038–10069. https://doi.org/10.1039/C6TA02621D

[4] Christie, A.M., Lilley, S.J., Staunton, E. et al. Nature 433, (2005) 50. https://doi.org/10.1038/nature03186

[5] Meyer, W.H. (1998). Adv. Mater. 10: 439–448. https://doi.org/10.1002/(SICI)1521-4095(199804)10:6%3C439::AID-ADMA439%3E3.0.CO;2-I

[6] Zhu, Q., Ye, C., Mao, D. (2022). Solid-State Electrolytes for Lithium– Sulfur Batteries: Challenges, Progress, and Strategies, Nanomaterials, 12, 3612. https://doi.org/10.3390/nano12203612

[7] Zhao, Q., Stalin, S., Zhao, C.Z., Archer, L. A. (2020). Designing solidstate electrolytes for safe, energy-dense batteries, Nat. Rev. Mater. 5, 229–252. https://doi.org/10.1038/s41578-019-0165-5

[8] Han, L., Wang, J., Mu, X., Wu, T., Liao C., Wu, N., Xing, W., Song, L., Kan, Y., Hu, Y. (2021). Controllable magnetic field aligned sepiolite nanowires for high ionic conductivity and high safety PEO solid polymer electrolytes, Journal of Colloid and Interface Science 585, 596–604. https://doi.org/10.1016/j.jcis.2020.10.039

[9] Wright, P.V. Electrochim. Acta 43: 1998 1137–1998 1143. https://doi.org/10.1016/S0013-4686(97)10011-1

[10] Dhatarwal, P., Sengwa R. J. (2017). Indian J. Pure Appl. Phys., 55, 7.

[11] Kong, Y., Hay, J.N. (2002). The measurement of the crystallinity of polymer by DSC. Polymer, 43, 3873–3878. https://doi.org/10.1016/S0032-3861(02)00235-5

Virtual element method for damage modelling of two-dimensional metallic lattice materials

Marco Lo Cascio[1,a *], Ivano Benedetti[1,b] and Alberto Milazzo[1,c]

[1]Dipartimento di Ingegneria, Università degli Studi di Palermo, Viale delle Scienze, Ed. 8, Palermo, 90128, Italy

[a]marco.locascio01@unipa.it, [b]ivano.benedetti@unipa.it, [c]alberto.milazzo@unipa.it

Keywords: Virtual Element Method, Metallic Lattice Materials, Damage

Abstract. Additively-manufactured metallic lattice materials are a class of architectured solids that is becoming increasingly popular due to their unique cellular structure, which can be engineered to meet specific design requirements. Understanding and modelling the damage in these innovative materials is a significant challenge that must be addressed for their effective use in aerospace applications. The Virtual Element Method (VEM) is a numerical technique recently introduced as a generalisation of the FEM capable of handling meshes comprising an assemblage of generic polytopes. This advantage in creating domain discretisation has already been used to model the behaviour of materials with complex microstructures. This work employs a numerical framework based on a nonlinear VEM formulation combined with a continuum damage model to study the fracture behaviour of two-dimensional metallic lattice material under static loading. VEM's effectiveness in modelling lattice failure behaviour is assessed through several numerical tests. The influence of micro-architecture on the material's failure behaviour and macroscopic mechanical performance is discussed.

Introduction

The computational modelling of the behaviour of lattice materials is an active field of research aimed at complementing the experimental activity in the quest for a better understanding of the potential of these materials in engineering applications. The Virtual Element Method (VEM) [1] is a recent generalisation of the Finite Element Method (FEM) for the treatment of general polygonal/polyhedral mesh elements that has been already used for several problems in structural mechanics [2,3,4,5,6] applications.

Formulation

For the lowest-order VEM formulation herein adopted, for a general polygonal virtual element E, the element degrees of freedom are the values of the components of the displacement at each of its n vertex, collected into the vector $\boldsymbol{u}_E$. The displacements field is expressed as $\boldsymbol{u} = \boldsymbol{N}(x,y)\,\boldsymbol{u}_E$, where $\boldsymbol{N}(x,y)$ is the matrix containing the virtual shape functions $\boldsymbol{N}_v(x,y)$ associated with each vertex $\boldsymbol{v}$. Shape functions are known only on the element edges of $\boldsymbol{E}$, where they are globally continuous linear polynomials. An explicit expression for the strains is unavailable because the shape functions $\boldsymbol{N}_v$ are not explicitly known within the polygonal element. An approximated constant strain field $\boldsymbol{\varepsilon}_\Pi$ is assumed within each element, which can be computed from the degrees of freedom $\boldsymbol{u}_E$ as $\boldsymbol{\varepsilon}_\Pi = \boldsymbol{\Pi}_E\,\boldsymbol{u}_E$, where $\boldsymbol{\Pi}_E \in R^{3\times 2n}$ is the matrix representation of a projection operator defined as

$$\boldsymbol{\Pi}_E = \frac{1}{A_E}\sum_{v=1}^{n}\int_{e_v}\boldsymbol{N}_v^E\,\boldsymbol{N}(x,y)\,ds \quad (1)$$

where A_E is the area of the polygonal element E, bounded by its n edges e_v and $\boldsymbol{N}_v^E$ is the matrix containing the components n_x and n_y of the outward unit normal vector over each edge. Since the virtual shape functions $\boldsymbol{N}_v$ on the element edges are known polynomials, the integrals appearing at the right-hand side of previous equation are exactly computable.

The tangent stiffness matrix $\boldsymbol{K}_E$ for a general virtual element E is given by the sum of two terms. The first term, named the consistency term, is given by

$$\boldsymbol{K}_E^c = A_E\, \boldsymbol{\Pi}_E^T\, \boldsymbol{C}\, \boldsymbol{\Pi}_E \tag{2}$$

where $\boldsymbol{C}$ is the material tangent stiffness tensor in Voigt notation. $\boldsymbol{K}_E^s$ is a stabilization term whose presence is motivated by the need to avoid zero-energy modes not associated with rigid body motions. The loss of material integrity is governed by the internal damage variable ω, $0 \leq \omega \leq 1$. The constitutive equations for an isotropic damage model is

$$\boldsymbol{\sigma} = (1-\omega)\, \boldsymbol{C}^0\, \boldsymbol{\varepsilon}_\Pi \tag{3}$$

where $\boldsymbol{\sigma}$ and $\boldsymbol{\varepsilon}_\Pi$ collect the Voigt components of the stress and strain respectively, and $\boldsymbol{C}^0$ is the elasticity matrix for the pristine elastic material. The evolution of damage is governed by the linear softening law

$$\omega(\kappa) = \frac{\kappa_f}{\kappa_f - \kappa_0}\left(-\frac{\kappa_0}{\kappa}\right) \tag{4}$$

and loading-unloading conditions

$$\mathrm{f}(\boldsymbol{\varepsilon},\kappa) = \varepsilon_{eq}(\boldsymbol{\varepsilon}) - \kappa \leq 0, \qquad \dot{\kappa} \geq 0, \qquad \mathrm{f}(\varepsilon,\kappa)\,\dot{\kappa} = 0 \tag{5}$$

in which f is the damage loading function, ε_{eq} is the modified Von Mises equivalent strain [8], and κ is an internal variable that corresponds to the maximum level of equivalent strain ever reached in the previous history of the material. The stress at a generic point $\boldsymbol{x}$ and at a generic loading increment λ is given by $\boldsymbol{\sigma} = \boldsymbol{\sigma}(\lambda, \boldsymbol{x}, \boldsymbol{\varepsilon}_\Pi, \mathcal{H})$, where $\mathcal{H}$ contains the history variables of the damage model. The tangent material stiffness matrix $\boldsymbol{C}$ at a certain time t is consistently computed from the constitutive law as

$$\boldsymbol{C}(t, \boldsymbol{x}, \boldsymbol{\varepsilon}_\Pi, \mathcal{H}) = \frac{\partial\sigma}{\partial\varepsilon_\Pi} \tag{6}$$

To avoid damage localisation and mesh dependency of the solution, an integral-type nonlocal damage model has been employed. The adopted weight function is the truncated quadratic polynomial function [4].

Numerical Tests

The specimen design used for the following numerical tests is based on the Extended Compact Tension, EC(T), specimen, shown in Fig. 1(a). The EC(T) has been developed for fatigue and fracture testing of solid (dense) materials but has been adapted and used for characterizing lattice structures. The rectangular specimen consists of repeated unit cells of side length l. Each specimen is eleven unit cells wide (W), forty unit cells high (H, and two unit cells thick (t). Only half of the EC(T) specimens were modelled by applying appropriate symmetry boundary conditions as shown in Fig. 1(b). The unit cell of the lattice structure studied in this work is shown in Fig. 2(a) and is based on a two-dimensional representation of a body-centred cubic (GBCC) unit cell (Fig. 1(b)). Each unit cell has external dimensions of $l \times l$, with $l = 3.5$ mm. Each lattice unit cell is discretised with 176 polygonal virtual elements (Fig. 2(c), and the numerical model has 157882 degrees of freedom. Simulations are carried out under displacement control and plane-strain assumption. The material selected as the constituent material is an additive manufactured Ti-6Al-4V alloy whose mechanical and damage properties are [7]: Young's Modulus $E = 123$ GPa, Poisson's ratio $\nu = 0.3$, yield strength $\sigma_y = 932$ MPa and fracture strain $\kappa_f = 0.1105$. The interaction radius has been set to $R = 2$ mm. Two different unit cell configurations have been analysed, with truss diameter $d = 0.5$ mm and $d = 0.75$ mm. For each configuration, numerical tests have been performed with crack length $a = 2l$ and $a = 3l$.

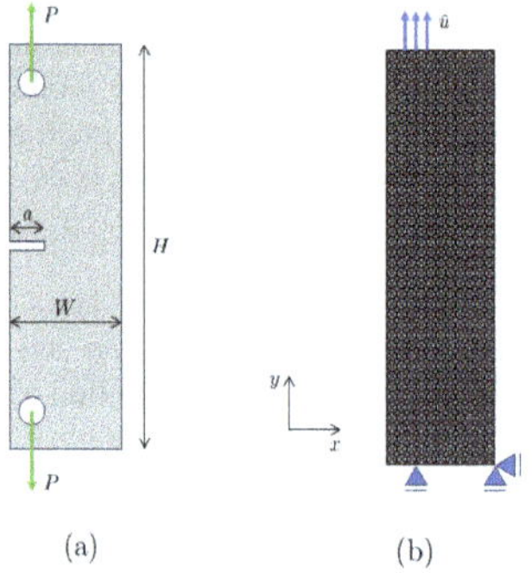

Figure 1: (a) Extended Compact Tension, EC(T), specimen; (b) computational model.

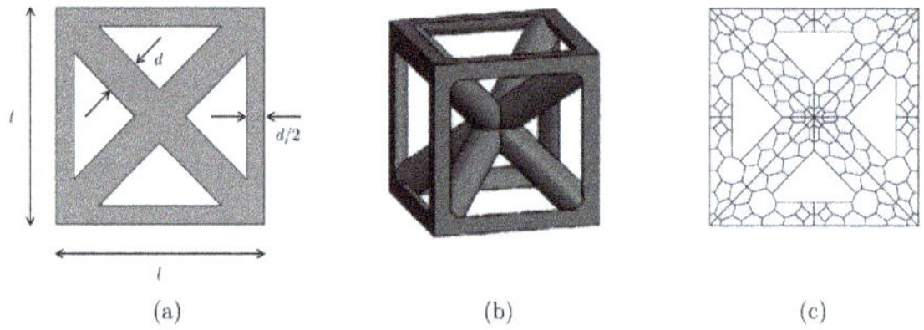

Figure 2: (a) 2D representation of the GBCC unit cell geometry; (b) actual 3D geometry.

Conclusions

A nonlinear VEM formulation combined with a continuum damage model has been employed to model the fracture behaviour of two-dimensional metallic lattice material under static loading. VEM's effectiveness in modelling complex morphologies such as lattice structures has been verified. The load as a function of the load-line displacement remains relatively linear for most specimens until close to reaching the critical load and a reduction of the critical load with initial crack extension can be observed for both unit cell configuration.

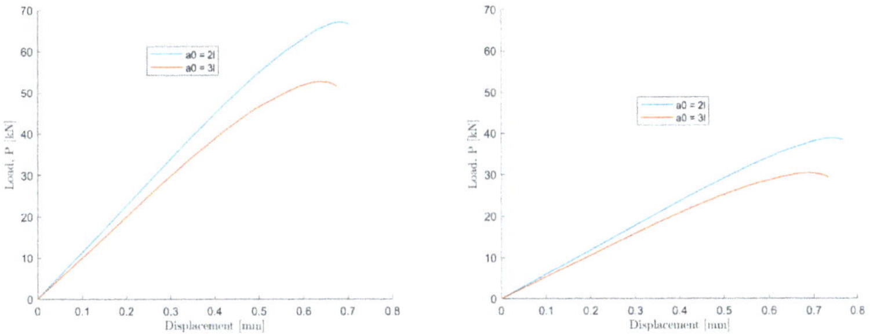

(a) (b)

Figure 3: Load-displacement responses of the (a) 0.75 mm and (b) 0.50 mm uniform diameter EC(T) specimens showing the applied load as a function of load-line displacement for different initial crack lengths.

Acknowledgments
The authors would like to acknowledge the support from Azione 1.2.1.03 – Specializzazione tecnologica territori – Accordo di Programma tra MIUR e Regione Siciliana - Codice Progetto PONPE03_00206_1 – "Advanced framework for Manufacturing Engineering and product Lifecycle Enhancement " - Acronimo "AMELIE" CUP G76I20000060007.

References

[1] Beirão da Veiga, L., Brezzi, F., Cangiani, A., Manzini, G., Marini, L. D., Russo, A., Basic principles of virtual element methods, Mathematical Models and Methods in Applied Sciences 23(01) (2013) 199-214. https://doi.org/10.1142/S0218202512500492

[2] Wriggers, P., Reddy, B. D., Rust, W., & Hudobivnik, B., Efficient virtual element formulations for compressible and incompressible finite deformations, Computational Mechanics, 60 (2017) 253-268. https://doi.org/10.1007/s00466-017-1405-4

[3] M. Lo Cascio, Marco, A. Milazzo, I. Benedetti, Virtual element method for computational homogenization of composite and heterogeneous materials. Composite Structures 232 (2020) 111523. https://doi.org/10.1016/j.compstruct.2019.111523

[4] M. Lo Cascio, Marco, A. Milazzo, I. Benedetti, A hybrid virtual–boundary element formulation for heterogeneous materials, International Journal of Mechanical Sciences 199 (2021) 106404. https://doi.org/10.1016/j.ijmecsci.2021.106404

[5] M. Lo Cascio, I. Benedetti, Coupling BEM and VEM for the Analysis of Composite Materials with Damage, Journal of Multiscale Modelling 13.01 (2022) 2144001. https://doi.org/10.1142/S1756973721440017

[6] M. Lo Cascio, Marco, A. Milazzo, I. Benedetti, Virtual element method: Micro-mechanics applications. In Key Engineering Materials 827 (2020) 128-133, Trans Tech Publications Ltd. https://doi.org/10.4028/www.scientific.net/KEM.827.128

[7] L. Dong, V. Deshpande, H. Wadley, Mechanical response of Ti–6Al–4V octet-truss lattice structures, International Journal of Solids and Structures 60 (2015) 107-124.

[8] J. H. P. De Vree, W.A.M. Brekelmans, M.A.J. van Gils, Comparison of nonlocal approaches in continuum damage mechanics, Computers and Structures 55.4 (1995) 581-588.

Materials Research Proceedings 37 (2023) 390-393 https://doi.org/10.21741/9781644902813-86

A meso-scale model of progressive damage and failure in LSI-produced ceramic matrix composites for aerospace applications

A. Airoldi[1,a *], M. Riva[1,b], E. Novembre[1,c], A.M. Caporale[1,d], G. Sala[1,e], M. De Stefano Fumo[2,f] and L. Cavalli[3,g]

[1]Dept. of Aerospace Science and Technology, Politecnico di Milano, Via La Masa 34, 20156, Milano, Italy

[2]CIRA, Italian Aerospace Research Centre, Via Maiorise, 81043 Capua (CE), Italy

[3]Petroceramics S.p.A., Viale Europa, 2, 24040 Stezzano (BG), Italy

[a]alessandro.airoldi@polimi.it, [b]marco.riva@polimi.it, [c]edoardo.novembre@polimi.it, [d]antoniomaria.caporale@polimi.it, [e]giuseppe.sala@polimi.it, [f]m.destefano@cira.it, [g]cavalli@petroceramics.com

Keywords: Ceramic Matrix Composites, Continuum Damage Mechanics, Binary Models, Stochastic Properties

Abstract. The paper is focused on the development of a modelling approach for Ceramic Matrix Composites (CMC) laminates, produced through a cost-affordable Liquid Silicon Infiltration (LSI) technique. The objective is the development of a tool capable of evaluating the design values for the material in the presence of technological defects and complex geometrical features, which could be used at the level of structural elements of details of reusable space vehicles. The model exploits a bi-phasic decomposition to capture four important aspect of the material: the non-linear behaviour occurring when load is not applied in the fibre direction, the significant bending to tensile strength ratio, the role of matrix fractures in the failure process and the role of delamination phenomena in the response. The correlation with tensile and bending tests performed with different lay-ups indicates that the developed approach can fulfil such objectives and may be used in the definition of structural details and of damage tolerance of innovative space vehicles.

Introduction

Ceramic Matrix Composites represent one of the most promising solutions for the development of structures capable of performing structural roles at temperatures beyond 1000 °C, such as the ones occurring in re-entry or single-stage-to-orbit vehicles in space missions, or in hypersonic vehicles and in propulsive systems in the aerospace field [1].

The increasing demand for truly reusable space and hypersonic transport systems introduces significant issues regarding the structural integrity and the damage tolerance of such hot structures. The material cost is another critical aspect for the development of the next generation of space vehicles. The LSI technique, can significantly reduce the cost of the CMC production, with respect to more traditional techniques, like chemical vapour infiltration or polymer infiltration and pyrolysis [2], but the porosity levels and the possible occurrence of technological defects can increase, thus emphasizing the need for controlling damage development in the material [4].

The activity presented in this paper was performed within the project AM^3aC^2A, funded by the Italian Space Agency (ASI), aimed at developing develop multi-scale approaches for the structural integrity of CMC in reusable aerospace components. The objective was the formulation of a non-linear numerical approach to be used in models of CMC laminates at the level of structural details or element, with a ply-wise (meso-scale) approach. The prediction of the structural response in the presence of geometrical features such as holes, cutouts, highly curved parts, and macro-porosity is one of the final goal of model, which in this paper is proved capable to idealize the most important failure mechanisms and to capture the quantitative response of multi-directional CMC laminates in tension and bending, thus providing a proof of its potential.

Experimental characterization

A test campaign was conducted with the aim of obtaining the basic properties of the orthotropic C/SiC fabric plies produced through the LSI technique and, at the same time, of providing validation experiments for the model to be developed. Figure 1-A reports the stress vs. strain responses recorded in four types of quasi-static tensile tests on laminates with lamination sequence of $[0]_{20}$ ("*T0*"), $[45/-45]_{10s}$ ("*T45*"), $[30/-30]_{10s}$ ("*T30*"), and $[0/45/90/-45]_{5s}$ (*"TQI"*).

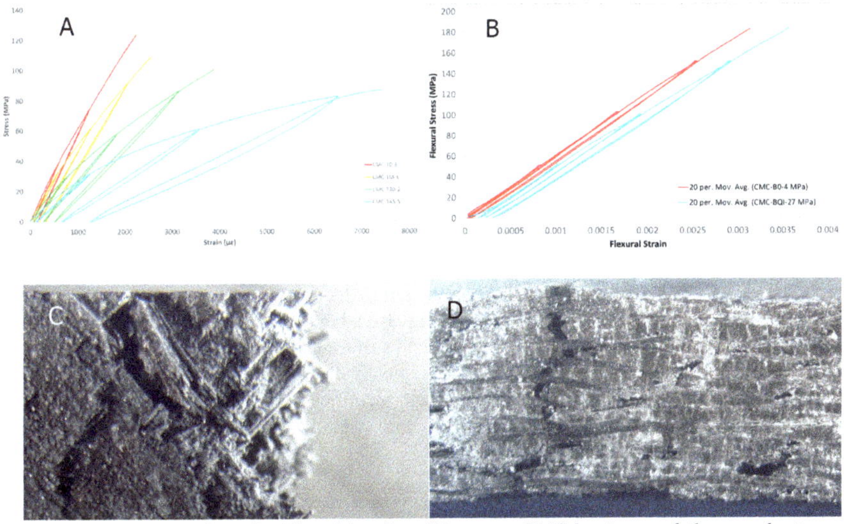

Fig. 1 – Response of tensile (A) and bending (B) test on CMC laminates, failure modes in "T45"(C) and "B0" tests (D)

The standard ASTM C1275-15 was taken as a guideline to design the specimens and perform the tests. All the specimens were dog-bone shaped, with (a total length of 200 mm, a gauge length of 45 mm, and an average thickness of 4.2 mm. All the tensile specimens were equipped with glass-fiber reinforced epoxy tabs to mitigate the risk of failures out of the grip. The stress vs. strain response reported in Fig. 1-B were obtained in two three-point bending tests performed on $[0]_{20}$ laminates (*"B0"*) and $[0/45/90/-45]_{5s}$ laminates ("BQI").

The response in the tensile tests was characterized by increasing non-linear behavior and ultimate strain at failure, as the percentage of fibers not aligned with the load application direction is increased. The maximum stress carried by the external plies of the laminate *B0* was significantly higher than the value recorded in the tensile "*T0*" stress. The failure mode in "*T45*" specimens indicated a diffused damage state in the matrix, as shown in Fig. 1-C, while the failure in "*B0*" test was characterized by a neat fracture in the 90°-oriented fiber yarns, while the 0°-oriented fibers do not show a clear fracture line (Fig. 1-D).

Overview of the numerical approach

The numerical approach adopted moved from the technique developed in [3,4] for polymer matrix composites. The CMC homogenized material model was decomposed into two idealized phases: fiber and matrix. The fiber were modelled through a layer of membrane element, which carried stress only in the fabric reinforcement directions, while the matrix was represented by solid elements. In general, the technique makes possible the representation of delamination without the use of zero-thickness cohesive elements, the development of different constitutive laws for matrix- and fiber-dominated responses, the representation of the interactions between matrix damage inside the plies and the delamination phenomena. The in-plane damage in the matrix was modelled by using a single scalar damage variable, as represented in Eq. 1. Such variable evolved with the distance from a threshold damage function, shaped as a Tsai-Wu criterion, expressed in Eq. 2. For the sake of brevity, the aspects related to the integration of delamination damage in the modelling technique are not reported (see [3,4] for further details).

$$\begin{bmatrix}\varepsilon_{11}^m\\ \varepsilon_{22}^m\\ \gamma_{12}^m\end{bmatrix}=\begin{pmatrix}\frac{1}{(1-d_m)E_{11}^m} & \frac{\nu_{21}^m}{E_{22}^m} & 0\\ \frac{\nu_{12}^m}{E_{11}^m} & \frac{1}{(1-d_m)E_{22}^m} & 0\\ 0 & 0 & \frac{1}{(1-d_m)G_{12}^m}\end{pmatrix}\begin{bmatrix}\sigma_{11}^m\\ \sigma_{22}^m\\ \tau_{12}^m\end{bmatrix} \tag{1}$$

$$f(\tilde{\sigma}^m)=\sqrt{F_1\tilde{\sigma}_{11}^m+F_2\tilde{\sigma}_{22}^m+F_{11}(\tilde{\sigma}_{11}^m)^2+F_{22}(\tilde{\sigma}_{22}^m)^2+2F_{12}\tilde{\sigma}_{11}^m\tilde{\sigma}_{22}^m+F_{66}(\tilde{\tau}_{12}^m)^2} \tag{2}$$

Results and Conclusions

A couple of iso-damage surfaces are shown in black and red colors in Fig. 3-A. Figures 3-B and C present the finite element models of the tensile and bending tests, respectively. The need of modelling the influence of matrix damage on the failure led to calibrate a new surface for the peak stress carried by the matrix phase (in green in Fig. 3-B), beyond which an exponentially decaying strain softening regime was modelled. Such second surface and the limit stress carried by the fiber phase were not fixed in the models, but were statistically distributed in the elements of the models. The parameters of such distribution have been identified through a Monte-Carlo approach, considering the correlation with the ultimate strength in all the tests as main performance index.

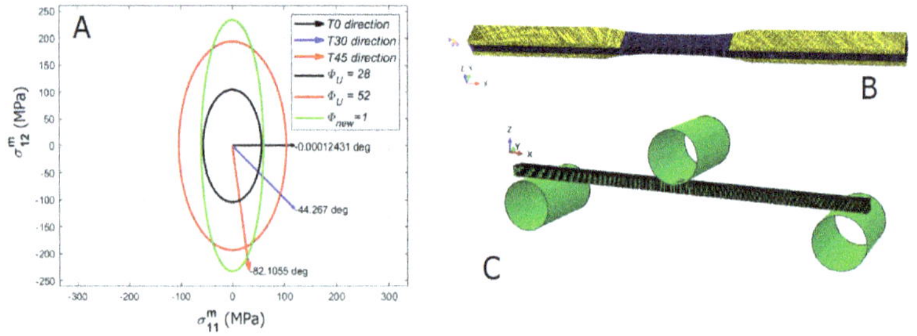

Fig. 3 – Tsai-Wu shaped surface for matrix damage evolution (A) and FE models of tensile (B) and bending (C) tests

The bi-phasic nature of the approach, the choice and the calibration of the damage evolution laws, the statistical distribution of the ultimate strength of the idealized phases led to obtain the results presented in Fig. 4, which indicate the all the quantitative and qualitative aspects of the responses are captured. In particular both the ultimate tensile strength in the "*T0*" test (Fig. 4-A) and the force vs. displacement response in the bending "*B0*" tests is obtained, thus indicating that the statistical distribution of properties can represent the bending/tensile strength ratio (Fig 4_D). Moreover, the localization of fracture can be represented, as shown in Fig. 4-E. For such localization, the integration of delamination damage in the analyses was found to play a fundamental role. Hence. the modelling approach was able of representing, without meshes at the sub-ply or microscopic levels, the fundamental quantitative and qualitative aspects of non-linear response and failure of the CMC material.

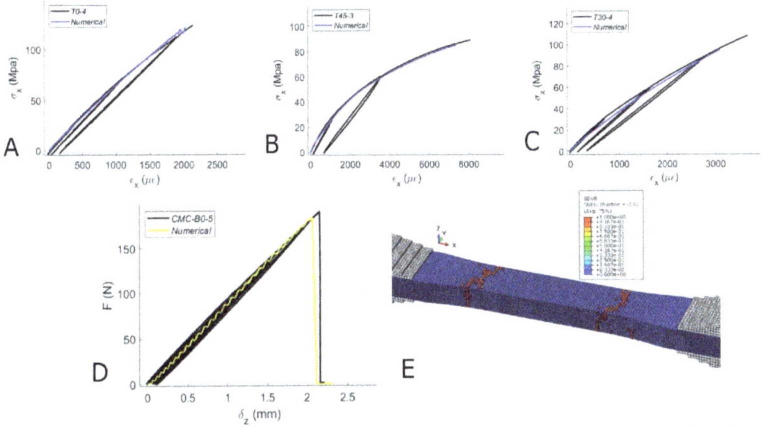

Fig. 4 – Numerical-experimental correlation in test "T0" (A), "T45" (B), "T30" (C), "B0" (D), and numerical failure mode (fiber damage) in "T30" analysis

Acknowledgements

The activities were funded by the Italian Space Agency (ASI), within the AM^{3}aC^{2}A project

References

[1] H. Hald, Operational limits for reusable space transportation systems due to physical boundaries of C/SiC materials. Aerospace Science and Technology 7(2003) 551–9, https://doi.org/10.1016/S1270-9638(03)00054-3

[2] W. Krenkel, F. Berndt, C/C–SiC composites for space applications and advanced friction systems, Materials Science and Engineering: A 412 (2005) 177–81. https://doi.org/10.1016/j.msea.2005.08.204

[3] A. Airoldi, C. Mirani, L. Principito, A bi-phasic modelling approach for interlaminar and intralaminar damage in the matrix of composite laminates, Composite Structures, 234 (2020), 111747. https://doi.org/10.1016/j.compstruct.2019.111747

[4] A. Airoldi, E. Novembre, C. Mirani, G. Gianotti , R. Passoni, C. Cantoni, A model for damage and failure of carbon-carbon composites: development and identification through Gaussian process regression, Materials Today Communication, 35 (2023) 106059. https://doi.org/10.1016/j.mtcomm.2023.106059

Aeronautics and Astronautics - AIDAA XXVII International Congress Materials Research Forum LLC
Materials Research Proceedings 37 (2023) 394-398 https://doi.org/10.21741/9781644902813-87

A peridynamics elastoplastic model with isotropic and kinematic hardening for static problems

Mirco Zaccariotto[1,a *], Atefeh Pirzadeh [1,b], Federico Dalla Barba[1,c], Lorenzo Sanavia[2,d], Florin Bobaru[3,e] and Ugo Galvanetto[1,f]

[1] Dep. of Industrial Engineering, University of Padova, Padova, Italy

[2] Dep. of Civil, Environmental and Architectural Eng., University of Padova, Padova, Italy

[3] Dep. of Mechanical and Materials Eng., University of Nebraska-Lincoln, Lincoln, USA

[a]mirco.zaccariotto@unipd.it, [b] atefeh.pirzadeh@phd.unipd.it, [c] federico.dallabarba@unipd.it, [d]lorenzo.sanavia@unipd.it, [e]fbobaru2@unl.edu, [f]ugo.galvanetto@unipd.it

Keywords: Ordinary State-Based Peridynamics, Elastoplastic Materials, Isotropic Hardening, Kinematic Hardening

Abstract. This study proposes a formulation equivalent to J2 plasticity with the associated flow rule to simulate the elastoplastic behavior of materials with isotropic or kinematic hardening in a peridynamic framework. The capabilities of the developed formulation are analysed through 2D and 3D case studies whose results (displacement and stress field) are compared with those obtained from the corresponding FEM models.

Introduction

The evaluation of the structural residual life of aerospace structures requires the ability to predict damage propagation. In recent years, Peridynamics (PD) [1], a new non-local continuum theory, attracted the attention of many researchers for its capability to simulate crack initiation, propagation and interaction. The theory has been widely used to model crack propagation in brittle materials, while the analysis of elastoplastic materials [2] has been mainly limited to the perfectly elastoplastic behavior [3,4]. Unfortunately, in the case of metals, experimental observations reveal a complex plastic behavior, which requires models with isotropic and kinematic hardening [2]. A PD constitutive model for 2D elastoplasticity with isotropic hardening is presented in [6] and extended to the 3D case in [7]. In this paper, an elastoplastic formulation equivalent to J2 plasticity is presented with which it is possible to simulate the elastoplastic behavior in the case of both isotropic and kinematic hardening.

Formulation

The PD equation of motion, for the static case, is [1]:

$$\int_{\mathcal{H}} (\underline{T}[\boldsymbol{x}]\langle \boldsymbol{x}' - \boldsymbol{x}\rangle - \underline{T}[\boldsymbol{x}']\langle \boldsymbol{x} - \boldsymbol{x}'\rangle) dV_{\boldsymbol{x}'} + \boldsymbol{b}(\boldsymbol{x}) = 0 \qquad (1)$$

where $\boldsymbol{b}(\boldsymbol{x})$ is the body force on the material point $\boldsymbol{x}$, $\underline{T}[\boldsymbol{x}]$ is the force state on the material point $\boldsymbol{x}$ corresponding to the bond vector $\boldsymbol{x}' - \boldsymbol{x}$ and $\mathcal{H}$ represents the neighborhood of $\boldsymbol{x}$ whose radius δ is the *horizon,* see Fig.1. The force vector state is a vector of modulus $\underline{t}$ (the scalar force state) and direction coincident with the deformed configuration of the bond (ordinary state-based PD).

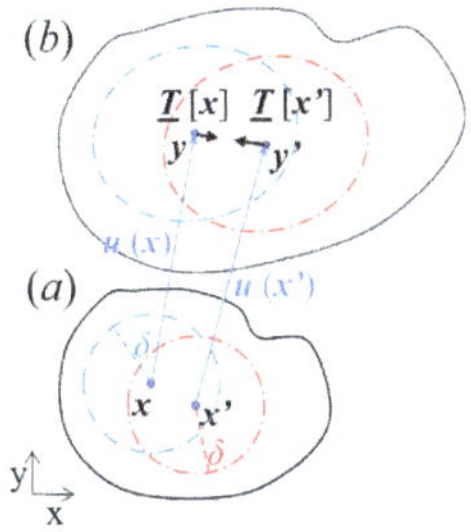

Figure 1: Positions of two interacting material points $\mathbf{x}$ *and* $\mathbf{x}'$*; (a) initial and (b) deformed configuration.*

The extension of a bond (a scalar quantity) is $\underline{e} = \underline{y} - \underline{x}$, where $\underline{y} = |\boldsymbol{y}' - \boldsymbol{y}|$ and $\underline{x} = |\boldsymbol{x}' - \boldsymbol{x}|$. $\boldsymbol{x}$ is the position of the material point in the initial configuration while $\boldsymbol{x}'$ denotes the generic point belonging to the neighborhood of $\boldsymbol{x}$. While $\boldsymbol{y}$ and $\boldsymbol{y}'$ are, respectively, the positions of $\boldsymbol{x}$ and $\boldsymbol{x}'$ in the deformed configuration.

In order to study problems involving elastoplastic behavior, it is necessary to distinguish between the isotropic and deviatoric component of the bond extension and of the scalar force state [1]. The extension of a bond is the sum of two components [1]: the isotropic and the deviatoric. Therefore, $\underline{e} = \underline{e}^{iso} + \underline{e}^{d}$, similarly for the scalar force state we have: $\underline{t} = \underline{t}^{iso} + \underline{t}^{d}$. In [3,4] it is emphasized that $\underline{t}^{iso}$ does not depend on $\underline{e}^{d}$ and the extension component $\underline{e}^{iso}$ is only elastic. Whereas $\underline{e}^{d}$ is itself the sum of two components, elastic and plastic respectively $\underline{e}^{d} = \underline{e}^{de} + \underline{e}^{dp}$. Finally, the deviatoric component of the scalar force state, in the case of materials with elastoplastic behavior, results in the 3D case, [4]:

$$\underline{t}^{d} = -5\mu\theta\frac{\underline{\underline{\omega x}}}{m} + \frac{15\mu}{m}\underline{\omega}(\underline{e} - \underline{e}^{dp}) \tag{3}$$

Where μ is the shear modulus, θ is the dilatation, m is the weighted volume and $\underline{\omega}$ is the influence function (see [1,4] for further details).

Furthermore, on the basis of classical plasticity theory, the load-unload conditions in the Kuhn-Tucker form and the consistency condition [2] must be fulfilled when solving elastoplastic problems. In the case of the peridynamic formulation [3] these conditions become:

$$\begin{cases} \lambda \geq 0, \ f(\underline{t}^{d}) \leq 0, \ \ \lambda f(\underline{t}^{d}) = 0 \\ \qquad\quad \lambda \dot{f}(\underline{t}^{d}) = 0 \end{cases} \tag{4}$$

where f is the yield function and λ is the continuum consistency parameter; while the plastic flow rule is [3]:

$$\underline{\dot{e}}^{dp} = \lambda \nabla^{d}\psi(\underline{t}^{d}) \tag{5}$$

in which $\nabla^{d}\psi(\underline{t}^{d})$ is the constrained Fréchet derivative of $\psi(\underline{t}^{d})$ defined hereafter.

In [3,4] based on the formulation introduced in [1] the following equation for the yield function for materials with perfectly plastic behavior is proposed:

$$f(\underline{t}^d) = \psi(\underline{t}^d) - \psi_0 = \frac{\|\underline{t}^d\|^2}{2} - \psi_0 \tag{6}$$

where $\|\underline{t}^d\|^2 = \int_{\mathcal{H}} (\underline{t}^d)^2 dV_{x'}$ and $\psi_0 = 25\sigma_Y^2/8\pi\delta^5$ in which σ_Y is the material's yield stress. Eq.6 is equivalent to the yield function $f = \sigma_{\mathrm{vM}} - \sigma_Y$ used in classical mechanics [2] where σ_{vM} is the von Mises stress.

The proposed formulation to study the behavior of materials with isotropic and kinematic hardening, was inspired by the corresponding yield function used in classical mechanics [2]

$$f = |\sigma_{\mathrm{vM}} - q| - (\sigma_Y + K\alpha) \tag{7}$$

In this equation, K is the isotropic hardening modulus, q is the back stress resulting from the kinematic hardening, and α is the internal hardening variable. q and α, initially zero, vary according to the following equations [2]:

$$\dot{q} = \dot{\varepsilon}_p H \quad \text{and} \quad \dot{\alpha} = \mathrm{sign}(\sigma_{VM} - q)\dot{\varepsilon}_p \tag{8}$$

Where H is the kinematic hardening modulus and ε_p is the equivalent plastic strain. Rewriting Eq. 7 in the context of the peridynamic formulation (for details see [5]) one obtains an equation analogous to Eq.6 in which, however, ψ_0 is no longer a constant and depends on the load increment at the material point considered. Therefore ψ_0 becomes:

$$\psi_0(\boldsymbol{x},t) = \frac{25[\sigma_Y + K\alpha + \mathrm{sign}(\sigma_{\mathrm{vM}} - q)q]^2}{8\pi\delta^5} \tag{9}$$

In Eq.9, q and α are found using Eq.8 in which the equivalent plastic strain should be replaced with the corresponding quantity expressed in the peridynamic formulation, which is a function of the deviatoric plastic extension [6,7]. It is worth noting that Eq.9 is obtained by considering small displacements.

The numerical implementation strategy of the proposed formulation involves discretizing the domain with a uniform grid of nodes. Then the non-linear static problem is solved using an incremental approach. Therefore, an iterative procedure using a return mapping algorithm was implemented for the determination of the deviatoric plastic extension and the various dependent quantities [5].

Numerical examples

In the following examples E=200 (GPa), ν=0.3, ρ=8000 (kg/m^3) and σ_{y0}=600 (MPa). The isotropic hardening modulus is K=20 (GPa), and the kinematic hardening modulus is H=20 (GPa). All cases were studied using the proposed PD formulation and classical FE simulations.

The first case is a thin plate with a central hole (Fig.2a) in plane stress conditions, subjected to an imposed displacement (Fig.2b) applied in increments of 0.0125 (mm).

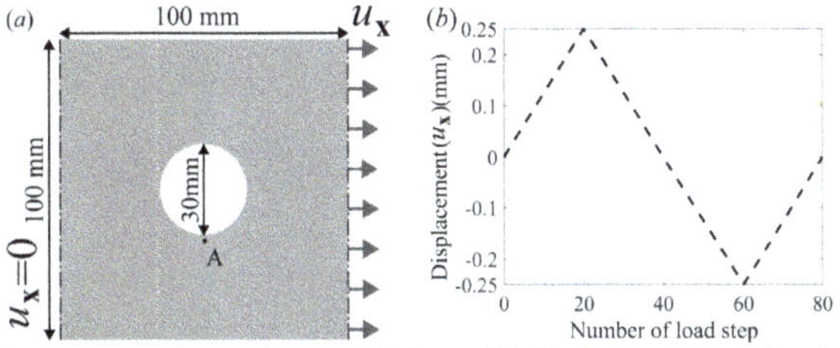

Figure 2: (a) Geometry and boundary conditions; (b) displacement loading in the x direction.

In Fig. 3, the displacements in the *x*-direction calculated by PD and FE at the 20th load step (u_x=0.25 mm) are compared. Good agreement can be observed between the results obtained from the two models.

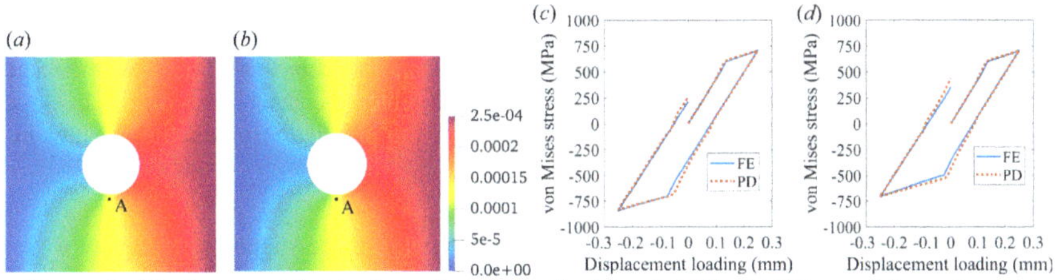

Figure 3: Displacement (m) in the x direction solved by: (a) FE, (b) PD. von Mises stress at point A vs loading displacement, (c) isotropic hardening case; (d) kinematic hardening case.

Fig.3c-d compares the von Mises stress at point A (Fig.2a) for the entire load history for both the isotropic (Fig.3c) and kinematic (Fig.3d) hardening cases. Point A is located in the region of the body where plastic deformation develops. The agreement between the PD and FE results is good: in particular, the PD model correctly estimates the maximum expected stress values obtained with the FE model. Note that in Fig.3c-d positive von Mises stresses are associated with a tensile load and negative von Mises stresses with a compressive load.

The second example studies a 3D structure whose dimensions, load and constraint conditions are shown in Fig.4. The material behavior is elastoplastic with isotropic hardening. The load cycle (imposed displacement) is similar to that shown in Fig.2b with a maximum displacement value of 0.35 (mm). Fig.4 compares the results obtained with the PD and FE models in terms of both displacement (u_y component) and von Mises stress in the *xy* plane containing the longitudinal axis of the specimen at the 20th load increment (at which the maximum displacement is applied). Good agreement is observed between the PD and FEM results despite the fact that no strategies were adopted to mitigate the surface effect in the PD model.

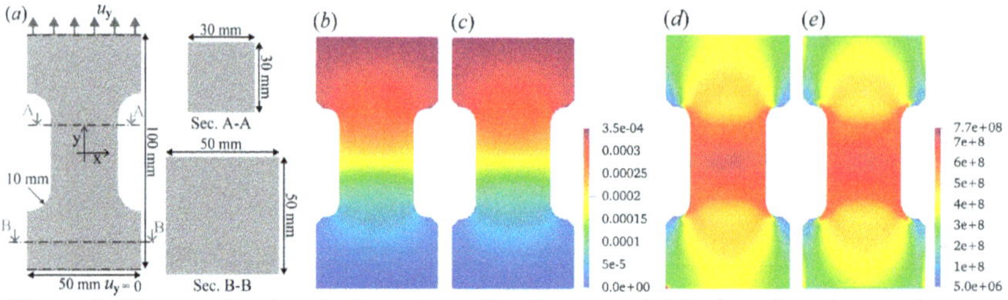

Figure 4: 3D example, a) main dimensions. Displacement (m) in the y direction on plane xy computed by: (b) FE, (c) PD ($u_y = 3.5 \cdot 10^{-4}$ m). Distribution of von Mises stress (Pa) on plane xy obtained by: (d) FE, (e) PD ($u_y = 3.5 \cdot 10^{-4}$ m).

Conclusions

This study presents an extension of the elastoplastic model in Peridynamics capable of describing isotropic and kinematic hardening behavior. The proposed formulation is equivalent to J2 plasticity with associated flow rule. 2D and 3D cases were studied, and the comparison between the results of the PD models and the corresponding FE models highlighted the capabilities of the developed approach, which represents the necessary first step for the simulation of ductile fracture in a peridynamic framework.

Acknowledgements

U. Galvanetto and M. Zaccariotto would like to acknowledge the support they received from University of Padua under the research project BIRD2020 NR202824/20.

References

[1] S.A. Silling, M. Epton, O. Weckner, J. Xu, E. Askari, Peridynamic states and constitutive modeling, J. Elasticity 88 (2007) 151-184. https://doi.org/10.1007/s10659-007-9125-1

[2] Simo, J.C., Hughes, T.J.R., Computational inelasticity. volume 7. Springer-Verlag, New York, 1998. https://doi.org/10.1007/b98904

[3] Mitchell, J.A., A nonlocal, ordinary, state-based plasticity model for peridynamics, Technical Report. Sandia National Laboratories, Albuquerque, NM, and Livermore, CA, (2011).

[4] Mousavi, F., Jafarzadeh, S., Bobaru, F., An ordinary state-based peridynamic elastoplastic 2D model consistent with J2 plasticity. Int. J. of Solids and Structures 229 (2021). https://doi.org/10.1016/j.ijsolstr.2021.111146

[5] A. Pirzadeh, F. Dalla Barba, F. Bobaru, L. Sanavia, M. Zaccariotto, U. Galvanetto, Elastoplastic Peridynamic formulation for materials with isotropic and kinematic hardening, submitted for publication (2023).

[6] Madenci, E., Oterkus, S., Ordinary state-based peridynamics for plastic deformation according to von mises yield criteria with isotropic hardening. J. of the Mech. and Ph. of Solids 86 (2016) 192–219. https://doi.org/10.1016/j.jmps.2015.09.016

[7] Liu, Z., Bie, Y., Cui, Z., Cui, X., Ordinary state-based peridynamics for nonlinear hardening plastic materials' deformation and its fracture process. Eng. Fract. Mech. 223 (2020). https://doi.org/10.1016/j.engfracmech.2019.106782

Structural batteries challenges for emerging technologies in aviation

Gennaro Di Mauro[1,a], Michele Guida[1,b,*], Gerardo Olivares[2,c], Luis Manuel Gomez[2,d]

[1]Department of Industrial Engineering, Università degli Studi di Napoli Federico II

[2]National Institute for Aviation Research (NIAR), Wichita State University, Kansas, USA

[a]gennaro.dimauro@unina.it, [b]michele.guida@unina.it; [c]gerardo.olivares@wichita.edu, [d]luismanuel@niar.wichita.edu

Keywords: Structural Batteries, Multifunctional Materials, Airworthiness Requirements, Certification by Analysis

Abstract. In a global context where modern societies need to move towards greater environmental sustainability, ambitious targets to limit pollutant emissions and combat climate change have been set out. Concerning the aviation sector, research centers and industries are carrying out new aircraft designs with increased use of electrical energy onboard aircraft both for non-propulsive and propulsive purposes, leading to the concepts of More Electric Aircraft (MEA), Hybrid Electric Aircraft (HEA) and All-Electric Aircraft (AEA). Despite the expected flight emissions reduction, new potential air transportation missions, safer flights, and enhanced design flexibility, there are some drawbacks hindering the trend to HEA solutions, strictly bounded to the limited performance of traditional battery systems. The reference is to low energy and power densities, which impact on aircraft weight and flight performances. A new technology, namely structural battery, combining energy storage and load-bearing capacity in multifunctional material structures, is now under investigation since capable to mitigate or even eliminate barriers to the electrification of air transport sector. Although, the deployment of this technology raises relevant questions regarding airworthiness requirements, which need to be applied when considering such multifunctional materials. The purpose of the presented activity is to take a step towards the definition of aircraft certification requirements when dealing with structural batteries, considering them both as a structure and as a battery, to maintain unchanged or even improve the level of safety in all normal and emergency conditions.

Introduction

Aviation is a continuously growing sector that has seen an increase of 6% per year in passengers and goods transported since 1950 and it is estimated that this number will increase by a further 44% by 2050 [1]. The need for reduce the environmental impact of the sector with the help of eco-friendly initiatives, to achieve the objectives, set by the Paris agreements, which provide for the reduction of total CO2 emissions. The importance of such a change is evident considering the eventuality of continuing to use fossil fuels as a source of power, there would be a 17% increase in CO_2 compared to the current situation. Instead, in a different scenario, in which the electric alternative was able to acquire a greater application space, a 45% reduction in CO2 would be achieved compared to the levels present in 2005. Among the solutions identified aimed at reducing the environmental impact of aviation the hybrid electric, electric and solar energy aircraft are applied, [2], there are already certified aircraft projects and others are in the experimental stage, and it is interesting how these projects exclusively concern aircraft of categories glider, very light aircraft and unmanned.

The main reason for such a circumscription and, consequently, for the non-inclusion of larger and heavier aircraft, lies in two parameters: specific energy density and specific energy, [3]. The first concerns the amount of energy available per unit mass; while the second is the amount of energy that can be contained in each volume. The batteries currently used are the lithium ones as they guarantee a reduced volume and mass compared to the competitors. By comparing the lithium batteries available on the market with the fossil fuels currently in use, such as kerosene and diesel, the parameters suggest that, with the same specific energy density, the batteries would weigh about 50 times more than their fossil alternatives. At the same time, for the same specific energy, batteries would occupy about 18 times the volume of their fossil alternatives. At this point, one wonders how to overcome these physical limitations. A potential solution, object of this thesis work, is represented by multifunctional materials, [6]. The latter are materials capable of fulfilling several functions simultaneously: in our case reference will be made to the structural batteries, which, as mentioned above, perform both a structural and energy storing function at the same time. This work deals with the most interesting points concerning the applications of Structural Batteries in the aeronautical field, [4]. The main environmental issues and the objectives that push aviation towards a 'green' perspective were discussed. Subsequently, a summary of the most relevant literature on the state of the art available regarding structural batteries was discussed. A focus on the mechanical point of view was made by discussing the main issues related to their manufacturing and certification requirements. The main current regulations issued by the FAA and EASA were discussed and a 'tailoring' between lithium-ion battery standards and composite material standards was discussed to find a certification process for this type of materials.

Material and Process Control

It is more important than ever to ensure that batteries are safe, strong, and dependable. However, the safety, performance, and longevity of a battery are only as excellent as the materials used to make them. *Electrodes, separators, current collectors, and electrolytes* must be extensively characterized and monitored from the moment they enter the production plant until they are incorporated into the finished product.

Solid electrode materials, it is well known that the parameters of the electrochemically active material have a substantial influence on the performance of battery electrodes. Following the formulation of the electrode, the components must be combined to produce a slurry or a dry blend. Relevant influencing variables include particle size, aggregate or agglomeration size, and particle size distributions of the different components. Furthermore, determining features such as powder rheological qualities or particle density and porosity can aid in tailoring the dry-mixing process. The characteristics of the powders are critical for functional and process-relevant factors. These include particle size distribution and mean particle size, particle shape, porosity, and specific surface area. They affect the flowability, dispersibility, and viscosity of the particles, as well as the sedimentation (stability) of the final slurry. The latter influences structural recovery after application, drying and calendering behaviour of the produced electrode layer, mechanical characteristics, and, lastly, the electrochemical parameters of the battery cell.

Electrode slurries, proper design and development of the mixing and coating procedures of the anode and cathode slurry is critical to battery performance. Knowing a slurry's rheological behaviour can also aid in the dimensioning of slurry pumps and anticipating slurry storage behaviour. The uniform thickness and density of the slurry layer are critical to ensuring the battery's lifetime, charge-discharge performance, and ion transfer rate - regardless of battery size. Finally, the proportion of solid particles in a slurry determines the quality and uniformity of the final coating.

Separators: A lithium-ion battery's separator is a thin, porous membrane that plays an important role in battery performance by avoiding a short circuit between the anode and cathode while permitting ion movement between them. Separators must be mechanically strong, stable in active battery conditions, and inert to other cell components while remaining porous enough to allow ion passage. The separator's through-pore size is an important characteristic for guaranteeing optimal battery performance since the holes must be tiny enough to prevent dendrites from growing across the separator while yet being large enough to allow ion movement between the cathode and anode. Larger pores or pin holes must also be searched for and avoided since they might contribute to short-circuit generation. Mechanical strength and structural qualities are another important criterion for separators. Measuring the amount of pre-tension necessary for the separator is critical for avoiding rupture or tearing during assembly as well as drooping after assembly.

Liquid electrolyte, lithium-ion electrolytes are essential in batteries because they allow charge transfer between the anode and cathode. Lithium-containing salts dissolved in an organic solvent are employed for this purpose. The salt lithium hexafluorophosphate (LiPF6) is most often utilized. Because of the reactivity of lithium in water, organic solvents serve as matrix in which lithium salts are embedded.

Raw material quality, adequate salt dissolving, and ion mobility are all critical elements to consider.

Performance-based regulations

Performance-based regulation is commonly regarded as a preferable way to regulation [7]. Rather than outlining the actions that regulated companies must take, performance-based regulation mandates the achievement of results and provides flexibility in how to reach them. Thus, assuming this approach is valid, it is possible, e.g., to treat a composite laminate in which a structural battery is integrated as a laminate in which a delamination has occurred evolving the current regulatory activities, and which meets the structural and electrical requirements considering that:

- regarding the structural aspect, the concept of damage tolerance and fatigue assessment of the structure, a strength assessment, detail design and fabrication shall demonstrate that catastrophic failure due to fatigue, manufacturing defects, deterioration, Environmental (ED) or Accidental Damage (AD) will be avoided for the operational life of the aircraft systems [13].
- regarding the aspect of the interaction between systems and structure, for aircraft equipped with systems that affect structural performance, either directly or because of a failure or malfunction, the influence of these systems and their failure conditions shall be considered when demonstrating compliance with airworthiness requirements, to evaluate the structural performance of airplanes equipped with these systems [13].

In addition to FAA requirements [11] and [12], the AGATE [9] and NCAMP [10] programs are two databases that provide a list of composite materials certified according to FAR part 25 regulations, reporting, in addition, the entire material certification procedure.

Certification Requirements for Composite Materials

One of the problems associated with this new technology concerns the certification process. This research activity seeks to carry out a 'tailoring' between the existing certification regulations of composite materials and lithium batteries, since the structural batteries, simultaneously fulfilling the functions of structural element and energy storing, must simultaneously comply with the two regulations.

For the structural airworthiness requirements, reference is made to the Advisory Circular 20-107 of the Federal Aviation Administration, which refers to the Code of Federal Regulation 14 parts 23, 25, 27 and 29.

In particular, the latter impose that the structure, regardless of the material it is made of, must be subjected to static tests, performed at 150% of the Design Limit Load, fatigue tests where the structure is subjected to cyclic loads with frequencies between 5 and 10 Hz, and impact tests and Damage Tolerance with impactors of different sizes.

At the first level (coupon level) we are interested in defining the mechanical properties of foil and laminate, thus defining the admissible ones.

The second level of analysis is dedicated to establishing the eligibility for critical structural details present in the project. The elements are still relatively simple structures, such as: glued or bolted joints, skin-stringer combinations, panels or laminates which in the design are loaded more in tension, compression and shear.

These elements are tested both at room temperature and in extreme environmental conditions, depending on the fatigue loads that could be encountered in real service conditions of the structure. Then the sub-component level analysis takes place in which are tested for example: wing panels reinforced by currents, simplified mobile surfaces, spars, ribs.

It is a generally unnecessary test that is only required when new materials are used in the project.

The last level of analysis is the full scale, which concerns the entire structure, which must be tested both statically and for durability and damage tolerance in different environmental conditions.

Focusing on the coupon level test, the legislation imposes the test execution methods and the selection of the samples to be tested (number, type and size). A-Basis type samples are distinguished with mechanical properties higher than 99% of the population with a confidence interval of at least 95% (for their determination, at least 55 different specimens are tested, obtained starting from 5 banks of material); B-Basis which have mechanical properties greater than 90% of the population with a confidence interval of 95% (for their determination at least 18 specimens obtained from 3 banks of material are tested).

The samples are tested according to regulations in various conditions and the tests are performed in compliance with the standards imposed by the American Society for Testing and Materials, which set out the test conditions, the methods of execution, the dimensions of the sample, the number of tests necessary and the conditions of acceptability of the test results.

From an energy storing point of view, the structural batteries must be certified according to Advisory Circular 20-174 which establishes the rules relating to safety, health monitoring, continuing airworthiness and maintenance of these devices.

Conclusions

In this report general considerations about certification requirements on composite material structures and the current trend of certification agencies to move towards a performance-based regulations have been addressed.

We are still a long way from certifying an aircraft of this type, the solutions currently available do indeed allow for 'full-electric' travel but at the expense of the aircraft's distance and duration ranges. In the meantime, other solutions are also being evaluated, such as integrating the batteries into the floor or within other constituent elements of the aircraft. In other fields of application such as the automotive one, research seems to have obtained better results, think of Tesla which has managed to integrate the battery pack into the powertrain or other solutions that provide for the integration of structural batteries into the vehicle body, [5]; clearly this is possible because on the ground it is not necessary to submit to the weight limits physically imposed by the aeronautical field.

Although the application of this technology, to date, is easier to apply for a field such as the automotive one, not subject to the physical and certification limitations imposed, however, in the aeronautical field, it is always necessary to continue to look to the future by pushing the research towards new frontiers and solutions.

References

[1] European Commission. Communication from the Commission—The European Green Deal.

[2] Yildiz, M. (2022). Initial airworthiness requirements for aircraft electric propulsion. Aircraft Engineering and Aerospace Technology, 94(8), 1357-1365. https://doi.org/10.1108/AEAT-08-2021-0238

[3] Sziroczak, D., Jankovics, I., Gal, I., Rohacs, D. (2020). Conceptual design of small aircraft with hybrid-electric propulsion systems. Energy, 204, 117937. https://doi.org/10.1016/j.energy.2020.117937

[4] Adam, T.J., Liao, G., Petersen, J., Geier, S., Finke, B., Wierach, P., Kwade, A., Wiedemann, M. (2018). Multifunctional Composites for Future Energy Storage in Aerospace Structures. Energies, 11, 335. https://doi.org/10.3390/en11020335

[5] Scholz, A.E., Hermanutz, A., Hornung, M. (2018). Feasibility Analysis and Comparative Assessment of Structural Power Technology in All-Electric Composite Aircraft. In Proceedings of the Deutscher Luftund Raumfahrtkongress, Friedrichshafen, Germany.

[6] Nguyen, S.N., Millereux, A., Pouyat, A., Greenhalgh, E.S., Shaffer, M.S.P., Kucernak, A.R.J., Linde, P. (2021). Conceptual Multifunctional Design, Feasibility and Requirements for Structural Power in Aircraft Cabins. Journal of Aircraft, 58, 677–687. https://doi.org/10.2514/1.C036205

[7] https://www.easa.europa.eu/sites/default/files/dfu/Report%20A%20Harmonised%20European%20Approach%20to%20a%20Performance%20Based%20Environment.pdf

[8] https://www.acquisition.gov/far/part-25

[9] https://agate.niar.wichita.edu/

[10] https://www.wichita.edu/industry_and_defense/NIAR/Research/ncamp.php

[11] FAA. AC20-107B: *Composite Aircraft Structure*. 2009.

[12] FAA. AC20-184: *Guidance on Testing and Installation of Rechargeable Lithium Battery and Battery Systems on Aircraft*. 2015.

[13] EASA. CS-25: Certification Specifications for Large Aeroplanes.

Dynamic buckling structural test of a CFRP passenger floor stanchion

Gennaro Di Mauro[1,a], Michele Guida[1,b,*], Fabrizio Ricci[1,c], Leandro Maio[1,d]

[1]Department of Industrial Engineering, University of Naples Federico II, Naples, Italy

[a]gennaro.dimauro@unina.it, [b]michele.guida@unina.it, [c]fabrizio.ricci@unina.it
[d]leandro.maio@unina.it

Keywords: Composite Material, Transient Analysis, Test Validation, Dynamic Buckling

Abstract. The work focuses on the study of the structural behavior of a composite floor beam in the cargo area of a commercial aircraft subjected to static and dynamic loads. Experimental tests have been performed in the laboratories of the Dept. of Industrial Engineering (UniNa) jointly with the development of numerical models suitable to correctly simulate the phenomenon through the LS-DYNA software. The definition of a robust numerical model allowed to evaluate the possibility of buckling triggering. The test article was equipped with potting supports on both ends of the tested beam, filling the pots with epoxy resin toughened with glass fiber nanoparticles. This allowed to uniformly load the beam ends in compression and to carry out the tests loading the specimen statically and dynamically, to observe the differences in the behavior of the beam in correspondence with the two different types of applied load. The result obtained through the comparison between the numerical model and the experimental test is that the dynamic buckling is triggered by a quantitatively smaller load than in the static case. Furthermore, it has been observed that the experimental compressive displacement to trigger the dynamic buckling instability is greater than the displacement observed in the static case.

Introduction

Buckling is an instability phenomenon that is common in "thin" structures. Buckling was formerly thought to be a totally static occurrence. The classic example is the Euler column buckling, in which a beam properly restrained and statically loaded in compression at the ends experiences the equilibrium instability usually in the linear elastic material behavior range.

Depending on the applied load, the beam can return to the initial equilibrium configuration (stable equilibrium), move to a new equilibrium condition different from the initial one (indifferent equilibrium), or move away from the initial equilibrium configuration indefinitely (unstable equilibrium). The buckling load is the smallest load for which equilibrium is indifferent.

Buckling can, however, be produced by varying loads over time. Many writers [1-3] have explored the application of a time-dependent axial stress to a beam, which causes lateral vibrations and can eventually lead to instability.

Dynamic buckling is a relatively new phenomenon. One of the first researchers to investigate dynamic buckling was [4], who proposed a theoretical solution for the situation of a simply supported rectangular plate exposed to variable floor loads over time. A criterion that connected dynamic buckling to load duration was developed in [5] and [6] where it was investigated the effects of a high intensity, short duration load. According to the findings, long-term critical dynamic buckling stresses may be less intense than matching static buckling loads.

According to the findings, long-lasting critical dynamic buckling stresses may be less intense than matching static buckling loads.

In this study, the structural behaviour of a composite floor beam subjected to low-frequency cyclic load conditions has been explored. Three distinct loads—below, near to, and over the static critical buckling load—have been taken into account to determine the structural response.

The aim is to conduct an experimental numerical investigation of the dynamic buckling phenomenology on a composite material beam. It was important to modify the test object to make sure the experimental test was carried out correctly. In specifically, two pottings were attached to the ends of the bar using epoxy resin castings toughened with glass fibre nanoparticles to guarantee that the compressive force was applied symmetrically.

The department laboratory's test equipment was used to conduct the experimental experiments. LS-DYNA software was used to mathematically recreate the treated phenomena. The Matlab working environment was used to process the results.

Case study description

The geometrical model and the numerical model, discretized in the Finite Element (FE) environment, are reported in Figure 1, while the mechanical properties of the composite lamina are reported in Table 1. The stacking sequence of the beam is [-45; 45; 90; 45; -45; 0; 0; 0; 0; 0; -45; 45; 90; 45; -45].

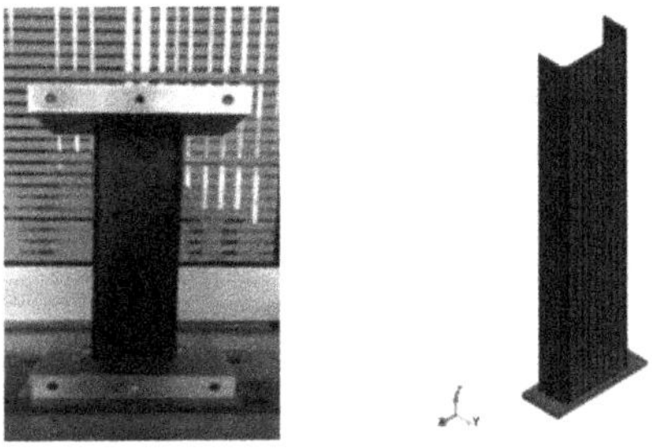

Figure 1. Left: geometrical model; right: numerical model.

Table 1. Mechanical properties of the lamina.

ρ [g/cm^3]	th [mm]	E_{11} [MPa]	E_{22} [MPa]	G_{12} [MPa]	G_{13} [MPa]	G_{23} [MPa]	ν_{12}	Xt [MPa]	Xc [MPa]	Yt [MPa]	Yc [MPa]	Sc [MPa]
1.6	0.186	135000	8430	4160	4160	3328	0.26	2257	800	75	171	85

To find the optimal balance between computational costs and the correctness of the findings in terms of expected stiffness, a preliminary mesh convergence study has been performed. As a result, several static linear studies with varied in-plane and through-the-thickness element sizes have been carried out. Particularly, three distinct mesh element sizes—coarser (8 mm), moderate (4 mm), and finer (2 mm) were taken into consideration. Nine mesh configurations have been examined, and analyzed, three through-the-thickness mesh configurations having been researched for each in-plane element size.

Results

The numerical model has been verified by comparison with the stiffness and failure findings of an experimental test programme, the compressive experimental test up to ultimate failure and reported in the last work [8] underlined the results:

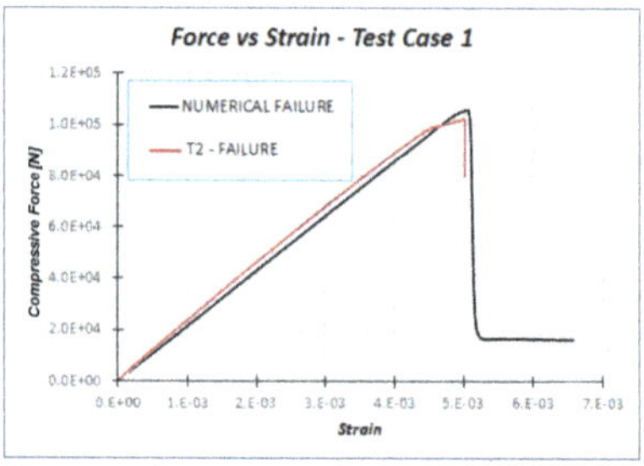

Figure 2. Numerical-experimental comparisons: Load vs. strain.

According to these experimental testing, the breaking load was 103KN, and up to the failure load, there are no buckling events, both experimentally.

The table 2 reports the failure displacements and loads, there is good agreement between the solutions of both formulations.

Table 2. Numerical-experimental comparisons: Failure displacement and failure load.

	Failure Displacement	Failure Load
Numerical	1.59 mm	107.3 kN
Experimental	1.60 mm	104.0 kN
Error	0.06%	3.1%

Then, a compressive experimental test designed to measure the structure's stiffness was repeated without taking the failure into account to confirm the linear deformation on the 315mm-long stanchion made of a composite material that combines carbon fibres with a highly toughened epoxy matrix and a 15-ply lamination sequence.

Figure 3 shows the good agreement in stiffness and failure load between the numerically predicted solutions and the outcomes of the experimental campaign testing. Additionally, the Hashin's failure criteria were used to compute the test article's failure mode, which was then predicted with a high degree of accuracy.

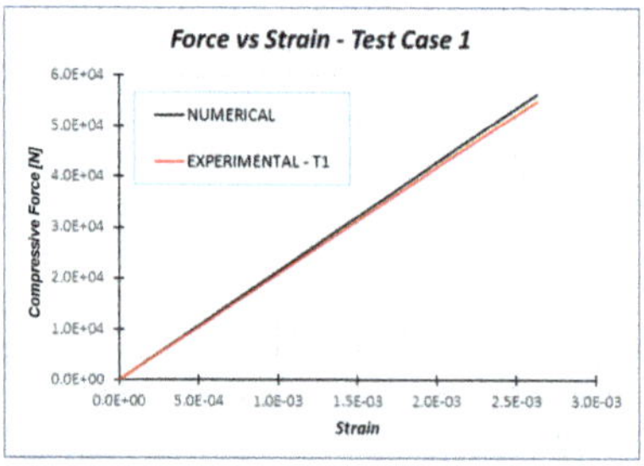

Figure 3. Numerical-experimental comparisons: Load vs. strain.

Once the numerical model is validated by static test, the dynamic experimental test is executed applying a load increase and constant speed of 100 mm/s. The explicit numerical investigations have been carried out to investigate how the dynamic buckling phenomena came to be. Every explicit analysis has been run with a structural dampening of 2%.

In figure 4 is plotted he F(t), from the analysis of the following graphs, the phenomenon of dynamic buckling instability can be observed, which occurs after 0.18 ms. Close to this time value, a buckling load equal to 89 kN and a displacement value equal to 0.063 mm are noted.

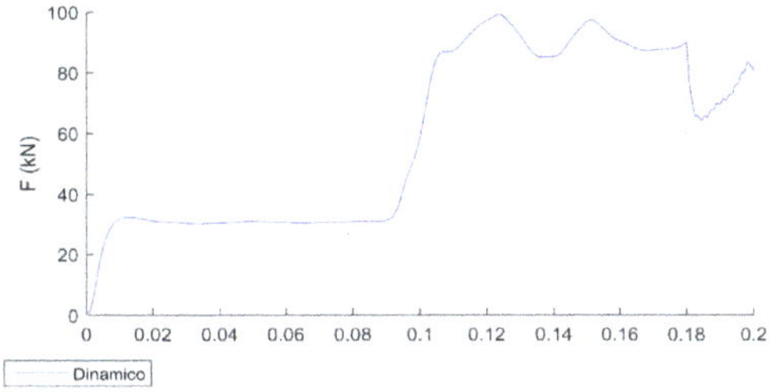

Figure 4. Time history of the force in the dynamic experimental test.

Conclusions

It is clear from an examination of the findings that: for composite materials, the buckling load in a dynamic compression test assumes a lower value than the buckling load in a quasi-static compression test. Additionally, in a dynamic compression test, a larger displacement than that required to cause buckling in a quasi-static compression test is required to cause the buckling phenomena. Finally, it can be shown that during the dynamic compression test, after buckling is attained, the structure does not collapse, but rather the bar tack may continue to operate in post buckling.

The figure 5 reports the differences between the numerical simulation about the post buckling static and dynamic, when the dynamic buckling value is exceeded in the left case, the post-buckling displacements are transmitted by the lower plate throughout the beam, whereas in the quasi-static compression test, the structure collapses and the displacements are not as effectively transmitted. In the case of the dynamic compression test, it is observed that the curve F(t) exhibits a plateau close to the buckling load. This is because some of the mechanical energy resulting from the application of the load is lost due to the deformation that took place after the beam protruded from its axis.

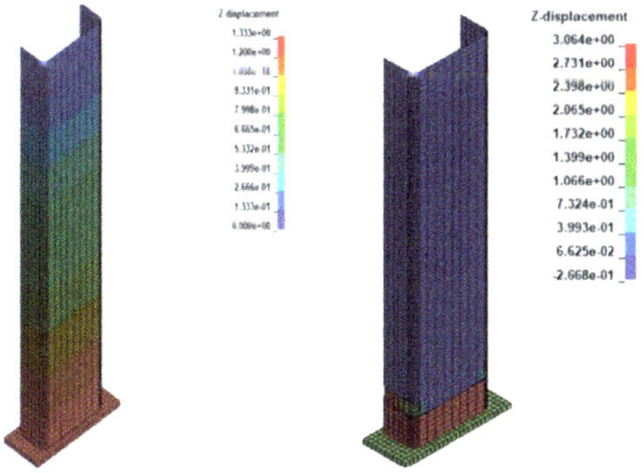

Figure 5. Post buckling in static and dynamic simulations.

References

[1] M. Amabili, M.P. Païdoussis. Review of studies on geometrically nonlinear vibrations and dynamics of circular cylindrical shells and panels, with and without fluid-structure interaction. Applied Mechanics Reviews. 56(4):349-356, 2003. https://doi.org/10.1115/1.1565084

[2] F. Alijani, M. Amabili. Non-linear vibrations of shells: A literature review from 2003 to 2013. International Journal of Non-Linear Mechanics. 58:233-257, 2014. https://doi.org/10.1016/j.ijnonlinmec.2013.09.012

[3] T. Kubiak. Static and dynamic buckling of thin-walled plate structures. In. Static and Dynamic Buckling of Thin-Walled Plate Structures, 2013. https://doi.org/10.1007/978-3-319-00654-3

[4] G.A. Zizicas. Dynamic buckling of thin elastic plates. Transactions of the ASME. 74(7):1257, 1952. https://doi.org/10.1115/1.4016090

[5] B. Budiansky, R.S. Roth. Axisymmetric dynamic buckling of clamped shallow spherical shells. NASA TN D-1510: 597-606, 1962.

[6] H.E. Lindberg, A.L. Florence. Dynamic Pulse Buckling, 1987. https://doi.org/10.1007/978-94-009-3657-7

[7] A. Sellitto, F. Di Caprio, M. Guida, S. Saputo A. Riccio. "Dynamic pulse buckling of composite stanchions in the sub-cargo floor area of a civil regional aircraft". Materials, 2020, vol. 13 issue 16, 3594. https://doi.org/10.3390/ma13163594

Materials Research Proceedings 37 (2023) 409-412 https://doi.org/10.21741/9781644902813-90

Deep learning algorithms for delamination identification on composites panels by wave propagation signals analysis

Ernesto Monaco[1,a *], Fabrizio Ricci[1,b]

[1] Department of Industrial Engineering – Aerospace Section – Università degli Studi di Napoli "Federico II", Italy

[a]ermonaco@unina.it, [b]fabricci@unina.it

Keywords: Structural Health Monitoring, Composites Structures, Deep Neural Networks

Abstract. Performances are a key concern in aerospace vehicles, requiring safer structures with as little consumption as possible. Composite materials replaced aluminum alloys even in primary structures to achieve higher performances with lighter components. However, random events such as low-velocity impacts may induce damages that are typically more dangerous and mostly not visible than in metals. Structural Health Monitoring deals mainly with sensorised structures providing signals related to their "health status" aiming at lower maintenance costs and weights of aircrafts. Much effort has been spent during last years on analysis techniques for evaluating metrics correlated to damages' existence, location and extensions from signals provided by the sensors networks. Deep learning techniques can be a very powerful instrument for signals patterns reconstruction and selection but require the availability of consistent amount of both healthy and damaged structural configuration experimental data sets, with high materials and testing costs, or data reproduced by validated numerical simulations. Within this work will be presented a supervised deep neural networks trained by experimental measurements as well as numerically generated strain propagation signals. The final scope is the detection of delamination into composites plates for aerospace employ. The approach is based on the production of images trough signal processing techniques and on employ of an image recognition convolutional network. The network is trained and tested on combinations of experimental and numerical data.

Introduction

Last developments in the modern aerospace industry push towards an improvement in flight efficiency and autonomy leading to a great increment in the usage of composite materials. They allow to lower the weight and obtain easily more complex shapes, but, due to their peculiar composition and fabrication methods, they are affected by delamination and defects. So, every aircraft's component is subject to time-scheduled maintenance sessions even when there is no clear evidence that it is required. This is a very expensive and time consuming process.

In this field, the Structural Health Monitoring technology (SHM system) [1,2,3], based on networks of distributed sensors embedded, or secondary bonded, throughout the whole structure under investigation, could be conveniently used for real-time health monitoring and/or as a data acquisition tool. Structural data, however, may constitute an enormous amount of information that in most cases is difficult to classify. Furthermore, since time is an important factor, the automation of the analysis process could be a significant advantage in this field. From this point of view, intelligent algorithms that can take decision in an autonomous manner reducing the human participation, like Deep Neural Networks (DNNs), may be useful to overcome this impasse.

Structural data may be adequately filtered with the aid of specific Deep Neural Networks designed and trained for the structural context and aimed to the classification and identification of significant parameters [4,5,6]. The DNNs, based on strategic engineering criteria, may represent an effective and efficient analysis tool to promote faster data analysis and classification. In the field of aircraft maintenance, this approach may lead, for example, to a faster awareness of a component health situation or predict failures. Neural Networks typically requires a relevant amount of data in order to be trained and to acquire the necessary reliability in classifying and recognizing the occurrence of the selected event but, once trained, they can be extremely effective and low-time consuming in analysing each single scenario to be classified. In this study the potentialities of deep learning with high frequency Lamb waves propagation based SHM methodologies are investigated employing explicit finite element simulations to collect propagation signals due to impact damages on a composite plate; this approach will also be employed in the next future to populate experimental data sets necessary for deep learning algorithm. Previous experimental signals acquired on the real impacted carbon fiber panel have been used to validate a numerical equivalent to allow the expansion of the dataset available [5].

Numerical time history signals have been collected for both the healthy and un-healthy state of the structure and transformed into RGB images. A well-known convolutional algorithm trained on healthy and damaged signals is used to identify anomalies in the form of delamination in the structure. The paper presents the pre-liminary results achieved by the authors..

SHM algorithm implementation

The goal of the work is to develop a neural network capable of detecting damage and its position in a composite panel exploiting Lamb waves. For this purpose, a numerical model was used to simulate the propagation of Lamb waves in a flat composite panel. Then, a Matlab algorithm has been created to transform the detected signals into images that the adopted neural network classifies as damaged or intact.

Numerical model, experimental set-up and signal analysis

The composite panel considered consists of 12 plies of three different pre-preg types oriented according to multiple directions [7,8]. The PZT sensors, utilized for lamb waves generation and acquisition, are applied on the panel according to the geometry in Fig. 1b.

The "pitch-catch" technique has been adopted for signals acquisition and damage detection, that is, a transducer behaves as an actuator, while the remaining sensors detect the signal that has been released inside the panel. Four different positions were considered for impact damage simulation.

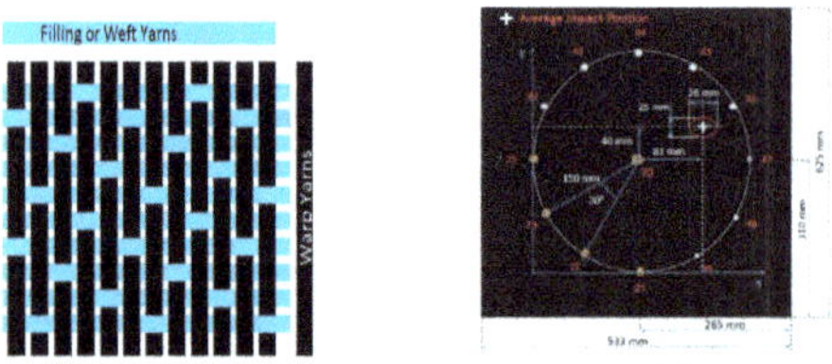

(a) (b)

Fig. 1. a) 5-Harness Satin Weave Layer; b) composite panel sensors configuration

A key point is the data analysis from rough signals to get a proper identification system. Features (time of flight, group velocity, signal transmission factor, wave energy) are extracted with an appropriate signal processing technique obtaining a diagnosis that presents location and/or severity of the damage.

Materials Research Proceedings 37 (2023) 409-412 https://doi.org/10.21741/9781644902813-90

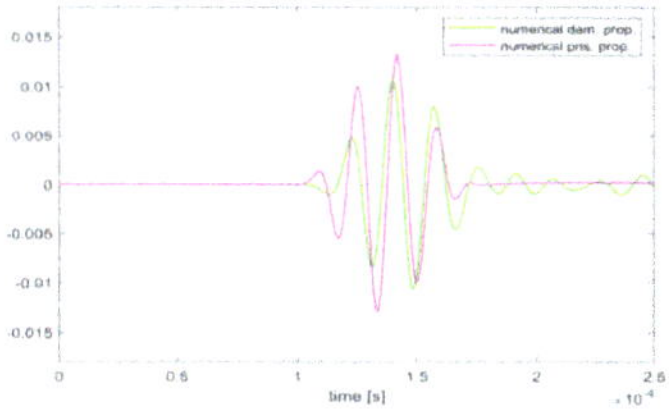

Fig. 2. Signals comparison between pristine and damaged model (path 20_46)

So, if "f" is the particular propagation feature considered, it is possible to define a damage index according to the following formulation:

$$\mathrm{DI} = \frac{(f_d - f_b) \cdot (f_d - f_b)}{f_b \cdot f_b}$$

Where f_d is the value obtained by the signal of the panel as it is at the analysis time, while f_b is the one extracted by the baseline propagation. Then, a damage index (DI) close to zero suggests a healthy-like propagation, while a value over a certain threshold warns for a failure.

RGB images generation ,analysis and results

To implement the convolutional approach, the acquired signals are transformed into RGB images exploiting a MATLAB code. RGB (Red Green Blue) is an additive color model, that is, an abstract mathematical model that allows to represent colors in numerical form, using the red, green, and blue color components. Each image obtained concerns a specific actuator-sensor path and consists of the overlap between the signal detected in the intact panel and that detected in the damaged panel. Going from top to bottom, the signals overlap consists of 10 intact signals, 10 signals in which a damaged and an intact alternate, and finally 10 damaged signals. As shown in Fig. 3, it is clear that the damaged images are characterized by a series of horizontal knurls (that are missing in typical healthy images). The latter are induced by disturbances in signal reception, attributable to damage.

Fig. 3: Generic experimental image damaged (above) and intact (below) obtained by RGB conversion

For images analysis the Google Net neural network, already present in MATLAB, was used. It is a deep convolutional neural network architecture codenamed Inception that achieves the new state of the art for classification and detection in the ImageNet Large-Scale Visual Recognition Challenge 2014 (ILSVRC14); is a convolutional network consisting of 144 layers, which requires an image as input and returns its classification in output. The network training phase was carried out using MATLAB's Deep Network Designer, a tool that allows to graphically train a neural network, entering the appro-priate training parameters such as learning speed, validation frequency and number of epochs. The network was trained in three different ways to evaluate the recognition efficiency of the experimental images and the relative reliability: numerical, experimental and hybrid training. The most interesting results related to the hybrid training, for which 30 intact images and 30 damaged images were selected, equally divided between numerical and experimental. In this case we were unable to obtain an accuracy higher than 83.3% in the training validation phase; thus, we have not been able to predict the behavior of this network. It was decided to test it anyway and evaluate the results. This condition was also analyzed because we wanted to simulate a situation in which few experimental data were obtained and, therefore, it was necessary to thicken the latter with data obtained numerically. Surprising results were provided by the hybrid network, trained with mixed signals. In fact, overall, it is the network that has shown a higher recognition rate. These results confirmed the high potential that characterizes the hybrid training obtained by combining validated experimental and numerical data.

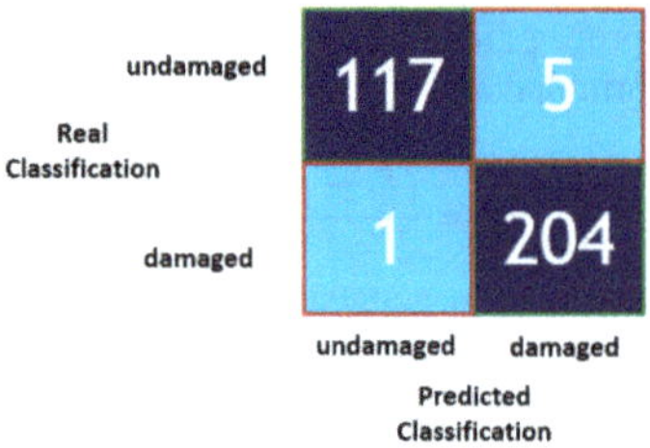

Fig. 4. Results obtained by the Net trained on "hybrid" results

References

[1] Ranasinghe K., Sabatini R., Gardi A., Bijjahalli S., Kapoor R., Fahey T., Thangavel K. Advances in Integrated System Health Management for mission-essential and safety-critical aerospace applications. Progress in Aerospace Sciences. Volume 128. 2022. https://doi.org/10.1016/j.paerosci.2021.100758

[2] Memmolo, V.; Maio, L.; Boffa, N.D.; Monaco, E.; Ricci, F. Damage detection tomography based on guided waves in composite structures using a distributed sensor network. Opt. Eng. 2015, 55, 011007. https://doi.org/10.1117/1.OE.55.1.011007

[3] E. Monaco, N.D. Boffa, F. Ricci, L. Maio, V. Memmolo, Guided waves for structural health monitoring in composites: a review and implementation strategies, Progress in Aerospace Sciences, Volume 129, 2022. https://doi.org/10.1016/j.paerosci.2021.100790

[4] Mahindra Rautela, J. Senthilnath, Jochen Moll, Srinivasan Gopalakrishnan, Combined two-level damage identification strategy using ultrasonic guided waves and physical knowledge assisted machine learning, Ultrasonics, Volume 115, 2021. https://doi.org/10.1016/j.ultras.2021.106451

[5] E. Monaco, N. D. Boffa, F. Ricci, M. Rautela, M. Cinque, Simulation of waves propagation into composites thin shells by FEM methodologies for training of deep neural networks aimed at damage reconstruction, Spie Smart Structures/NDE - Health Monitoring of Structural and Biological Systems XV Conference - March 2021. https://doi.org/10.1117/12.2583572

[6] Rautela M., Gopalakrishnan S., Monaco E. "Unsupervised deep learning-based delamination detection in aerospace composite panels" – Spie Smart Structures/NDE - Health Monitoring of Structural and Biological Systems XV Conference - March 2021. https://doi.org/10.1117/12.2582993

Deployment of a CubeSat radiative surface through an autonomous torsional SMA actuator

Filippo Carnier[1,a *], Alberto Riccardo Donati[1,b], Elena Villa[2,c], Daniela Rigamonti[1,d] Paolo Bettini[1,e]

[1] Dept. of Aerospace Science and Technology, Politecnico di Milano, Via La Masa 34, 20156 Milano, Italy

[2] Consiglio Nazionale delle Ricerche Istituto per l'Energetica e le Interfasi (CNR IENI), Unità Operativa di Supporto di Lecco, Corso Promessi Sposi, 29, 23900 Lecco, Italy

[a]filippo.carnier@mail.polimi.it, [b]albertoriccardo.donati@mail.polimi.it, [c]elena.villa@cnr.it, [d]daniela.rigamonti@polimi.it , [e]paolo.bettini@polimi.it

Keywords: Torsional Shape Memory Alloys, CubeSats, Deployable Radiator, Thermal Control

Abstract. This study aims to provide a proof of concept concerning the integration of an "S" shaped SMA tube into the thermal circuit of a 12U CubeSat. The torsion-based actuator utilizes the heat from the circulating fluid accumulated inside the satellite to enable the deployment of a radiator panel through the manifestation of the shape memory effect in the material, facilitating heat dissipation via radiation.

Introduction

CubeSats have proven over time to be a viable alternative to conventional systems, performing the same scientific operations in a considerably smaller volume. Despite their many innovative aspects, the miniaturization of this class of satellites still presents several challenges to overcome. In particular, integrating hardware components in such limited space restricts design flexibility and poses specific issues in developing adequate thermal control systems due to constrained power supplies.

A commonly adopted strategy for thermal management involves deploying radiator panels to dissipate heat generated by the system's internal components via radiation in the space environment. Among the various deployment mechanism solutions, using shape memory alloy actuators could represent a revolutionary choice.

SMA can lead to very convenient devices with a significant reduction in mechanical complexity and size and better reliability of the actuation system, providing an excellent technological opportunity to replace conventional electric, pneumatic or hydraulic actuators across all sectors, especially in the space segment [1].

The following work aims to develop a torsional SMA tubular actuator to be integrated on a 12U CubeSat's thermal fluid loop circuit in order to deploy a radiator panel, thus maintaining the satellite's internal environment within the appropriate temperature ranges. The torsional behavior of SMA actuators is not widely discussed in the literature and presents critical aspects that still require further investigation, such as cycling stability, a crucial property for optimal integration into a space system.

Concept description

In the proposed solution, the actuator exhibits an S-shaped tubular morphology [2] which enables the integration of the SMA into a closed-loop liquid circuit, allowing for thermal control operations and panel actuation to coexist within the same element, significantly reducing system complexity. In this approach, the mechanism governing the SMA activation is the same fluid flowing inside the circuit, which experiences localized heating within the satellite due to heat dissipation from internal components and external thermal loads. Through convective heat exchange, the tube, initially deformed in torsion in the martensitic phase with the panel fully closed, generates adequate torque as the SME unfolds, ensuring a 90-degree opening of the radiator element. As the deformation imposed on the tube's central section recovers, the end embedded within the panel is compelled to rotate rigidly, subsequently facilitating the panel's movement (Figure 1).

After the first recovery of the memorized form, neither the return to the low temperature (even below M_f) nor subsequent heating can induce further variations in the shape, until a deformation provided by an external element is set again. Since the actuator is required to operate cyclically depending on the satellite's thermal demands, a rearming strategy must be implemented (for instance, the rearming element can be represented by a torsional spring).

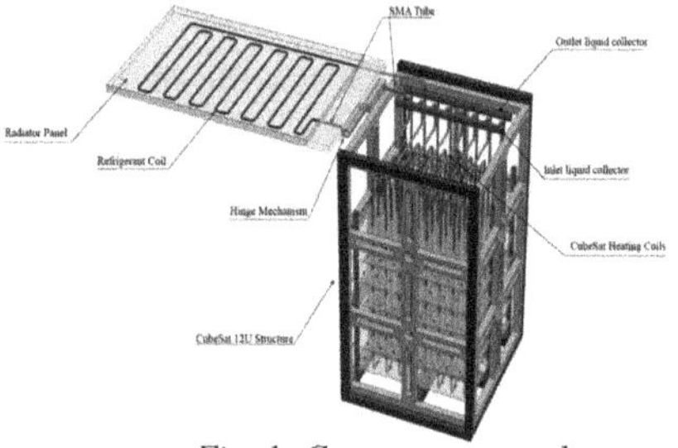

Fig. 1: Concept proposed

Manufacturing and thermomechanical characterization of the SMA actuator

The prototype design process began with the production of various actuator samples using unprocessed tubes made of NiTi alloy. Starting from the unprocessed material purchased, which has an outer diameter of 3 mm, a thickness of 0.24 mm, and pseudoelastic properties, it was necessary to implement heat treatments in order to obtain the desired shape and modify the characteristic temperatures set to achieve shape memory properties. Thermal treatments result in the generation of numerous precipitates inside the material, compromising the maximum performance that the actuator can provide. Therefore, for future developments, it will be necessary to employ tubes that already exhibit characteristic temperatures suitable for the final application. The tube is firstly inserted into a mold, designed in accordance with the geometry to be imposed on the material, followed by a two-phase furnace heating: a) preheating the tube to a temperature of 565°C and b) maintaining a constant temperature at that level for 45 minutes.

To assess transformation temperatures and behavior, DSC tests have been conducted on a single sample. The results reveal that M_f, M_s, A_s, and A_f are -11.99°C, 20.98°C, 20.43°C, and 43.35°C, respectively.

Rotary recovery tests have also been performed to gain a clear understanding of the actuator's performance, particularly in terms of the material's deformation state recovery capacity. The residual rotation detected at the end of each cycle is related to the vertical distance between the initial and final points of the hysteresis curve, indicating a deformation that the material will not be able to recover. This distance increases as the torsional load applied to the tube increases. For torque values greater than 0.07 Nm, the formation of a non-negligible residual deformation was detected.

Cycling tests have been performed to assess the number of cycles after which a complete recovery of the imposed deformation is no longer guaranteed, due to a permanent modification of the crystalline microstructure of the alloy. The tests were conducted with an applied load of T=0.0655 Nm, as the rotary recovery data indicate that this value represents the minimum load required to impose a 90° rotation on the material in the martensitic phase.

From the cycling tests, it was concluded that after 70 cycles, the material starts to exhibit a destabilization of performance. These results are extremely promising when compared to those of linear actuators with a high degree of precipitates within the matrix, in which the destabilization of shape memory properties emerges after a few cycles.

Materials Research Proceedings 37 (2023) 413-416 https://doi.org/10.21741/9781644902813-91

Prototype design and fabrication

A conceptual mockup has been designed (Figure 2) and constructed to evaluate the feasibility of the proposed solution through experimental tests conducted in a terrestrial environment, simulating only the internal heating within the CubeSat. Consequently, no rearming mechanism has been implemented. As a result, after each opening process, the system must be cooled inside refrigerators and manually rearmed.

The prototype features a liquid fluid loop integrated into a fixed frame, with dimensions identical to those of a 12U CubeSat structure, and a 3D printed frame free to rotate, representing the radiator panel. The actuator is housed inside a hinge mechanism necessary to ensure the alignment of the tube to the desired axis of rotation for the panel deployment and its connection to the fixed frame.

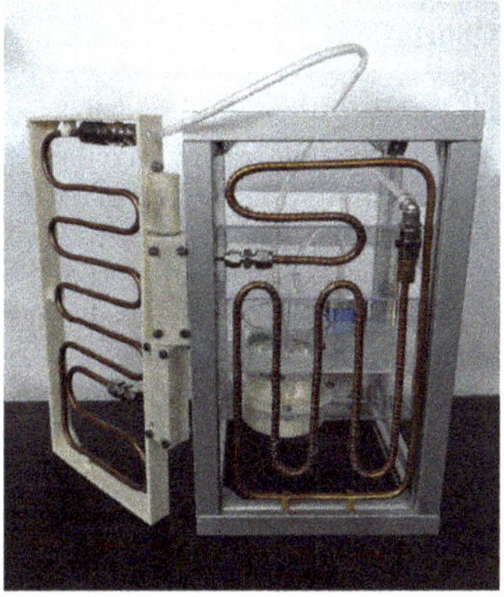

Fig. 2: Experimental mockup

The hinge mechanism is composed of two elements, one intended to be attached to the fixed frame and the other to the mobile frame, both capable of rotating with respect to each other.

The two components are then forced to rotate relative to each other by 90° thereby generating a torsional stress state in the central section of the tube and thus preloading it in order to mount the panel in a closed configuration. Subsequently, each element is connected to the corresponding frame, and finally, the support element for the panel is inserted and mounted onto the panel itself. The liquid fluid loop consists of two copper serpentines, each one connected to an end of the actuator, positioned respectively inside the CubeSat structure and the panel. The circulation of the liquid is mediated by a micropump, in turn, connected to the serpentines via PTFE flexible hoses that close the loop.

Test results

A test of the prototype was conducted to demonstrate the functionality of the design. The test started with the prototype at room temperature with the SMA tube already in the armed configuration (i.e., panel closed). The circulating liquid was heated at the internal coil within the fixed frame using an electric resistance wire, wrapped upstream of the actuator's inlet section, powered to dissipate 400W due to the Joule effect.

The entire process was monitored by a FLIR infrared thermal camera and two thermocouples, positioned at the inlet and outlet sections of the tube. The angles reached were measured using a graduated scale located beneath the panel.

The heating proved to be adequate, allowing the tube to reach temperatures suitable for complete austenitization of the alloy. As a result, the panel achieved a rotation of 85° in 155 seconds from the time the current supply was turned on, as shown in figure 4. The inability to reach a fully open position (i.e, 90° of rotation) is related to the thermal treatments the material underwent. These treatments lead to the formation of precipitates, which compromise the macroscopic recovery of the imposed deformation state

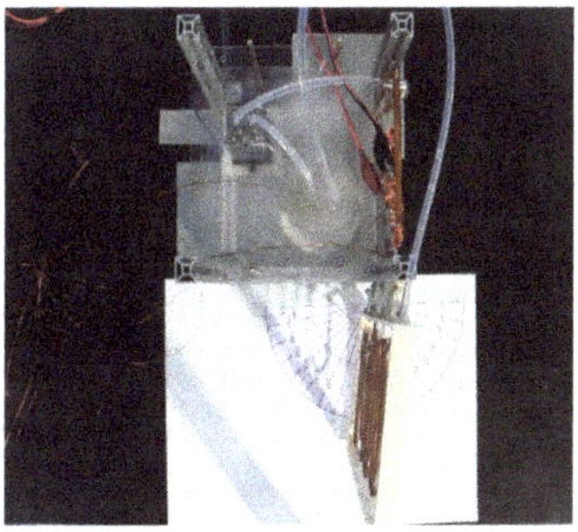

Fig. 3: Final angle reached

during the development of the shape memory effect at the material level.

Conclusions

The presented work aims to provide a proof of concept on the feasibility of developing a torsional tubular SMA actuator, which is activated by the internal circulation of a fluid heated to an appropriate temperature. The goal is to integrate this actuator into a 12U CubeSat's thermal fluid loop circuit for the purpose of deploying a radiator panel to 90° angle of rotation.
The fabrication process started with the production of various actuator samples using precursor tubes for stents, made of NiTi alloy, due to the purchasing easiness. The outer diameter of these tubes is 3 mm and the wall thickness is 0.24 mm. Since these precursor tubes exhibited pseudoelastic behavior, it was deemed necessary to implement thermal treatments in order to obtain the desired morphology and shape memory features. Torsion tests demonstrated that significant rotations could be achieved at low strain/stress levels, highlighting the suitability of this approach for the 90-degree deployment of a radiating panel on a small satellite. Moreover, cycling tests revealed that, despite the high degree of precipitates within the matrix, the material's stability is ensured for approximately 70 cycles. This result is particularly noteworthy, as it is well known that linear SMA actuators with high precipitate content tend to become unstable much earlier. Consequently, it has been shown that the choice to implement a torsion-based actuation system can be considered highly valid. Subsequently, a prototype was designed and developed to assess the feasibility of employing the actuator in a real satellite operational context. The system was then tested in a terrestrial environment, yielding highly interesting results, as the panel reached an angle of 85° within a relatively short time. The incomplete achievement of the desired rotation can be attributed to the high concentration of precipitates in the actuator, which compromises its maximum attainable performance. Nevertheless, these results are extremely promising, as, even with this non-optimal material, a comprehensive feasibility study of the system to be developed was provided, demonstrating a solid foundation for reliability.

Acknowledgments Authors would like to acknowledge gratefully ASI for the support under the collaboration agreement with Politecnico di Milano n.2018-5-HH.0 of the framework agreement n.2016-27-H.0

References

[1] Lagoudas D.C and Dimitris C. Shape Memory Alloy: modeling and engineering applications. Springer, 2008.
[2] D Rigamonti, P Bettini, L Di Landro, and G Sala. Development of a smart hinge based on shape memory alloys for space applications. In 25th Conference of the Italian Association of Aeronautics and Astronautics (AIDAA 2019), pages 1719–1742. AIDAA, 2019.

Development of an FBG-based hinge moment measuring system for wind tunnel testing

A. Taraborrelli[1,a], A. Gurioli[2,b], P. De Fidelibus[2,c], E. Casciaro[1,d], M. Boffadossi[1,e], P. Bettini[1,f]

[1]Department of Aerospace Science and Technologies, Politecnico di Milano, via La Masa 34, 20156, Milano (MI), Italy

[2]Air Vehicle Technologies, Wind Tunnel Testing, Leonardo Aircraft Division, via ing. Paolo Foresio 1, 21040, Venegono Superiore (VA), Italy

[a]alessandro.taraborrelli@mail.polimi.it, [b]alessandro.gurioli@leonardocompany.com, [c]paride.defidelibus@leonardocompany.com, [d]emanuele.casciaro@polimi.it, [e]maurizio.boffadossi@polimi.it, [f]paolo.bettini@polimi.it

Keywords: Additive Manufacturing, Fiber Bragg Gratings, Wind Tunnel Testing, Hinge Moment

Abstract. This paper presents the development and implementation of a hinge moment measuring system for wind tunnel tests based on Fiber Bragg Grating (FBG) sensors. These sensors, which are drawn directly into optical fibers, are capable of measuring strain and temperature variations and represent a precious addition to the aeronautical industry thanks to their peculiar characteristics, including high accuracy, low invasivity, embeddability and electromagnetic immunity. In detail, the development of the system exploits a combination of Fused Deposition Modeling technology and FemtoSecond® Gratings to design and create an independent, deformable structure in which a set of FBGs could be embedded within internal curved channels obtained during the 3D-printing process. This involved a complete re-design of the interface between the stabilizer and the elevator of a horizontal tail model. The material used for producing the structure is ULTEM 9085™, which made the development of the system particularly cost-effective and efficient. The paper also describes the installation of the FBGs, including the design of the channels, the selection of a glue, its injection technique and the following calibration procedure. Finally, the component is tested in the wind tunnel facility of Leonardo Aircraft Division in Venegono (VA, Italy), and the obtained results for some elevator's deflections are presented.

Introduction

Wind tunnel testing is a well-established discipline in Engineering, which, side-by-side with modern computational fluid dynamics, is used for the analysis of the behaviour and performance of aircrafts, cars, buildings and more during their design phase. In wind tunnel activities, the main focus is on measuring forces and moments, and for static measurements, this is typically done by relying on strain gauges. However, strain gauges still show many disadvantages, including inertial effects, high invasivity and susceptibility to electromagnetic interference among the others [1]. Optical fiber sensors represent a promising alternative to strain gauges, thank to their peculiar characteristics such as high sensitivity, faster response time, low invasivity and immunity to electromagnetic interference [2]. Furthermore, these sensors can be directly integrated into structures both during and after their manufacturing process, as demonstrated by S. Pinto [3], who successfully installed them in a component through internal, straight channels directly created during a 3D printing sequence. This paper describes the design and testing of an optical fiber-based hinge moment measuring system for the elevator of a wind tunnel aircraft model's horizontal tail. Since these measurements are highly critical and challenging to obtain by using traditional strain gauges sensors, especially because of the size constraints imposed by the component, the activity presented in this work focused on developing a system based on a minimally invasive technology. To do this, a complete re-design of the interface between the stabilizer and the elevator of the model was carried out, in order to integrate between them an independent, deformable structure capable of hosting several Fiber Bragg Gratings (FBGs) within internal curved passages obtained through the Fused Deposition Modeling (FDM) technique. The making of this work was made possible thank to Leonardo Company's Aircraft Division in Venegono (VA, Italy), with special recognition to the Wind Tunnel Department. The Company's printer, specifically the Stratasys Fortus® 400mc, was used to build the component with the ULTEM™ 9085 material, which guarantees excellent physical and mechanical properties for high-demand and special applications.

Design

The design of the intermediate deformable structure, shown in Fig. 1 and 2, is the result of an accurate trade-off between the overall structural stiffness of the assembly and the deformability of the sensitive sections. This was achieved by performing several Finite Element Analysis on the component, which were carried out by imposing a set of hinge moments corresponding to the expected loads for the experimental conditions under exam. The final design presents two deformable trusses per each measuring station (4 in total), since the idea was to keep a back-up channel which could have been useful in case of problems related to the adjacent one.

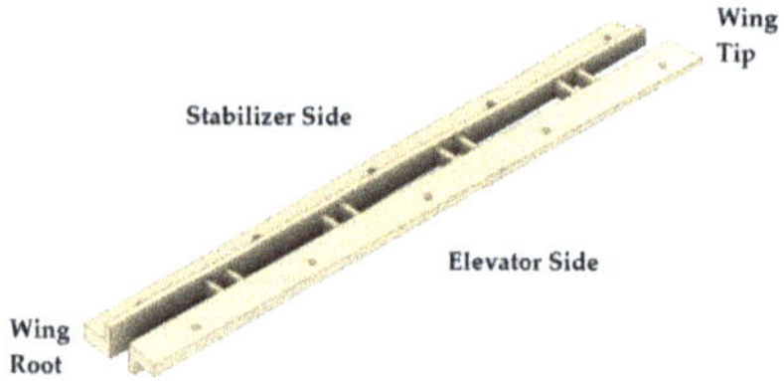

Figure 1: 3D-view of the deformable structure's design (without channels)

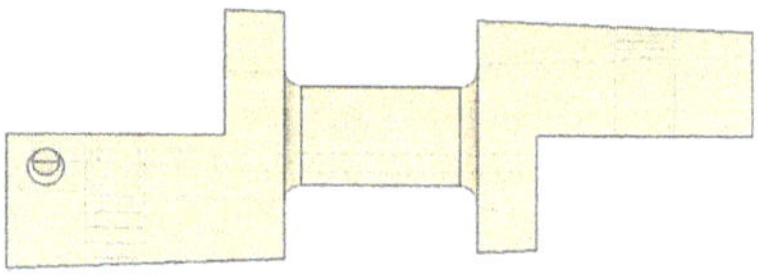

Figure 2: View from the wing root with detail of the system of channels

Manufacturing of the Internal Channels and Sensors' Installation

The design of the internal channels in the structure had to meet several requirements, including the presence of turns, conjunction points and areas with varying radius, as well as the compatibility with the size constraints of the structure and the printing resolution of the Fortus® 400mc. Several tests were conducted until finding out an optimum pattern and diameter for the passages. Small sinks were included in the design at the conjunction areas between the channels, to prevent the printer from depositing excessive material that could potentially obstruct the passages. Fig. 3 shows the intended position of five polyimide-coated Femto-Second gratings (in blue) inside the channels. To fix and secure the sensors in each measuring station, M-Bond 600 Adhesive by Micro Measurements was chosen and used, since it resulted compatible with ULTEM and guaranteed an optimal viscosity for the application, as well as a high performance for stress analyses. A compatible blue dye was added to the adhesive to make the procedure visible through the material. The mixture was inserted into the channels through the sinks with a 0.6 mm rounded-head nozzle, halting the injection as soon as it was observed to come out from the end of the channel on the interface side with the elevator. The component was then left for curing under heat lamps for 5 hours at 55°C.

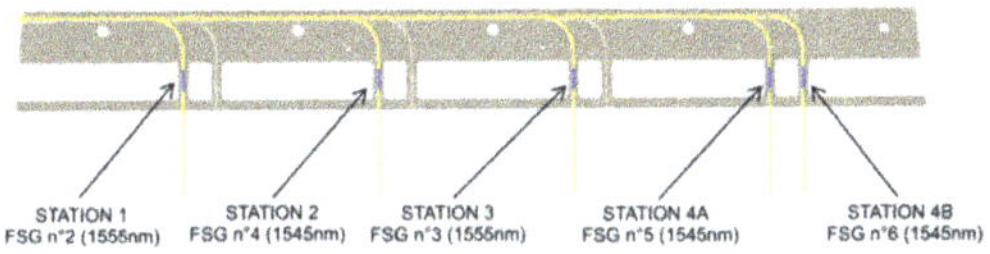

Figure 3: Designated position for each embedded sensor

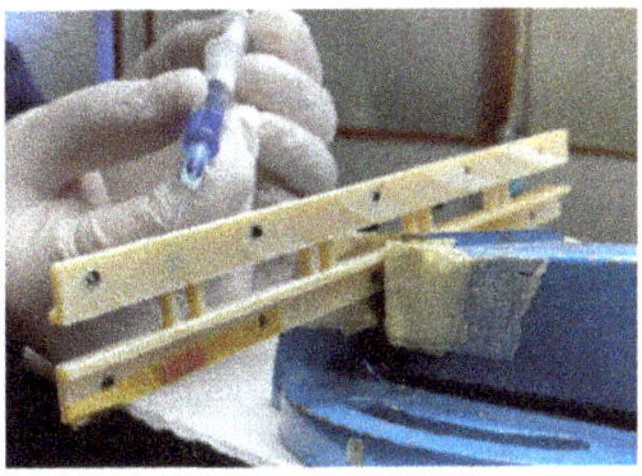

Figure 4: Injection of the glue

Calibration

The calibration of the system of sensors was performed by using the specialized equipment of the company. The deformable structure was secured to the other components with five screws per side, tightened with a torque wrench to guarantee the repeatability of the constraint after each installation. Five holes were drilled on a dummy elevator at 10 mm behind the hinge axis. Each hole was loaded singularly and then in combination with others to accurately reproduce a set of calibration hinge moments on the elevator, chosen in accordance with the expected loads during testing. A first-order calibration vector, which links the signals to the loads, was calculated by performing a regression between the vector of all applied hinge moments and the matrix containing the measured wavelength variations from all the sensors. The sensor embedded in station 4B (Fig.3) was excluded from this calculation as it was damaged during the installation and the output signals resulted altered with respect to the expected one.

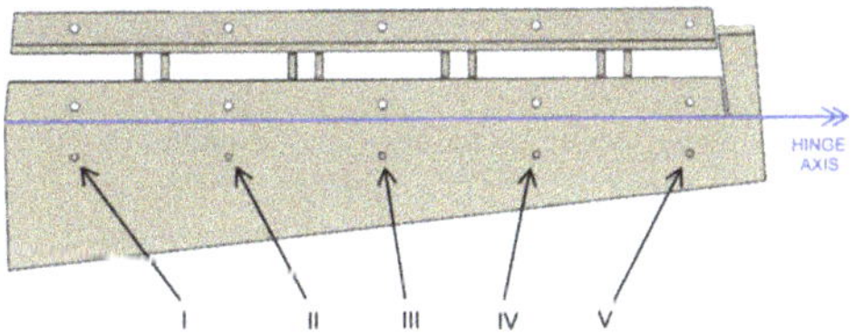

Figure 5: Calibration points on the elevator

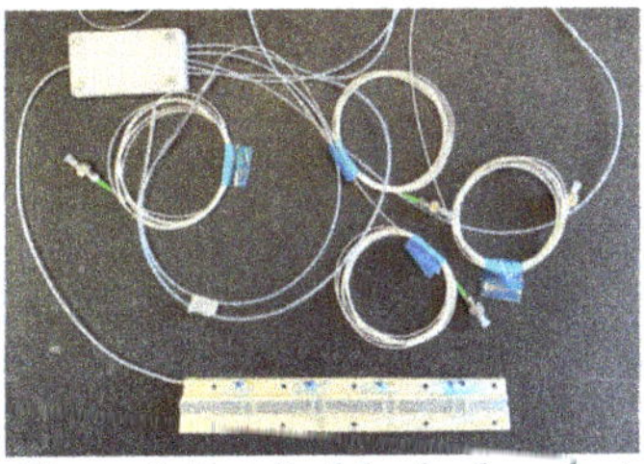

Figure 6: Detail of the final product

Results

Since FBGs are sensible to both mechanical and thermal deformation, the model was accurately designed in a way that the temperature-related wavelength shifts during testing resulted to be significantly smaller than the mechanical ones. Despite that, the effect of temperature oscillations during testing was not considered negligible, and it was decided to perform a thermal calibration of the sensors directly in the wind tunnel environment, by investigating their response during a heating cycle in relation to the data provided by a thermocouple installed inside the fuselage of the model. Temperature variations were also reduced by limiting the duration of each test, conducting them in a continuous sweep mode in pitch, with an α-sweep rate of 0.5°/s. The campaign involved installing the full horizontal tail assembly on the corresponding aircraft model (Fig. 7) and testing three different elevator deflections δ of 0°, +12.5° and -12.5° at wind speeds of 40 and 50 m/s. The tests were repeated several times to assess the level of repeatability of the measurements.
The averaged results for the examined conditions, normalized with respect to the maximum tested α and the maximum measured hinge moment, are presented in Fig. 8, 9 and 10.

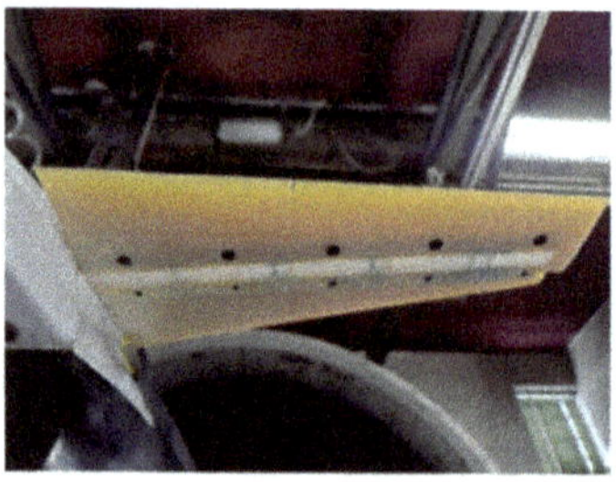

Figure 7: Full assembly mounted on the model

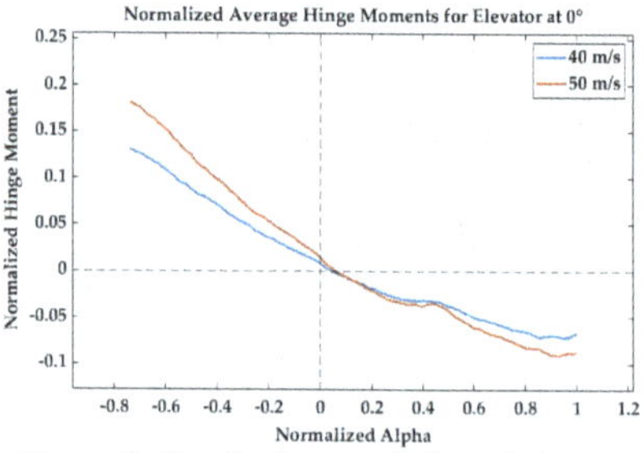

Figure 8: Results for non-deflected elevator

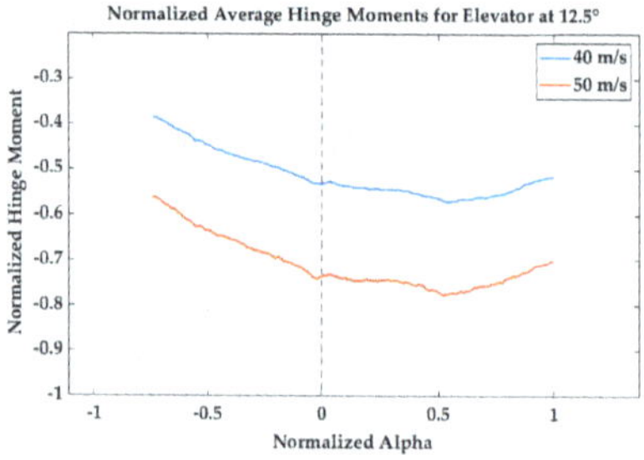

Figure 9: Results for +12.5° elevator deflection

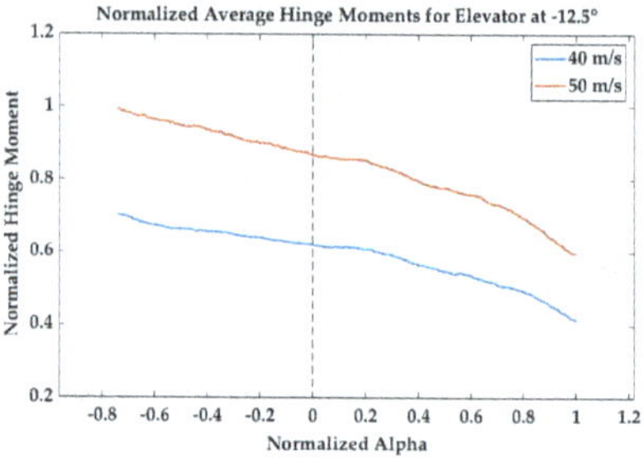

Figure 10: Results for -12.5° elevator deflection

Conclusion and future improvements

This work proves how the production of a wind tunnel model's component with embedded optical fiber sensors is possible with a low-invasive and re-usable solution and can produce successful results which otherwise could not be achieved with conventional measurement systems. In fact, the collected measurements resulted compatible with the expected ones, indicating that the experimental setup and procedures were appropriately designed and executed. However, to extend the application range of the component and allow for testing at higher wind speeds, it should be stiffened, for example by increasing the size of the deformable stations. Future developments should also explore other additive manufacturing techniques, including metallic 3D-printing, to improve the durability of the component and to overcome the structural limitations related to the use of polymeric materials. Finally, the back-up configuration proposed in this work could be exploited to include, for each station, an unconstrained sensor for compensation of temperature effects. All these upgrades clearly have the potential to improve the performance of the system.

References

[1] C. Tropea, A. Yarin, J. Foss, ''Springer Handbook of Experimental Fluid Mechanics'' (2007). DOI: https://doi.org/10.1007/978-3-540-30299-5

[2] A. Kersey, M.A Davis, H.J. Patrick, M. Leblanc, K. Koo, C.G. Askins, M. Putnam, E. Friebele, ''Fiber Grating Sensors'', Journal of Lightwave Technology (1997) vol. 15, issue 8, pp. 1442 – 1463. DOI: https://doi.org/10.1109/50.618377

[3] S. Pinto, ''Development and testing of optical fiber based monitoring systems for a wind tunnel application'', Politecnico Di Milano, URL: http://hdl.handle.net/10589/148372

Buckling and post-buckling response of curved, composite, stiffened panels under combined loads including pressurization

Luisa Boni[1,a*], Daniele Fanteria[1,b*], Tommaso Lucchesi[1]

[1]Department of Civil and Industrial Engineering – Aerospace division, University of Pisa, via G. Caruso, 8, Pisa, Italy

[a]luisa.boni@unipi.it, [b]daniele.fanteria@unipi.it

Keywords: Post-Buckling, Curved Composite Stiffened Panel, Finite Element Analysis, Combined Loads

Abstract. In recent years, metal stiffened shells for aerospace applications have been gradually replaced by composite shells, which are widely used in fuselage, tail, and wing structures due to their advantageous properties. Under operating conditions, stiffened panels are subjected to different types of loads, combined in various ways, which can lead to instability. Like their metallic counterparts, allowing post-buckling within the operational envelope could lead to significant weight reductions for composite structures, but unlike the metal case, their response in this state is not fully understood and the potential of composites is not fully exploited. In this context, the main objective of the present work is to investigate the buckling and post-buckling behavior of composite curved panels subjected to combined loads. The buckling behavior of a representative stiffened curved panel has been simulated by non-linear finite element analyses, from the simplest pure compression and pure shear cases to the final analysis of the panel subjected to pressurization, shear, and compression simultaneously. The results of this study quantify the reduction of the critical compression and shear loads due to their simultaneous action, as well as the effect of the pressurization load, which was generally beneficial, but remarkably so in the case of pure shear.

Introduction

Over the years, the use of composite materials has gradually increased, reaching levels of up to 50% of the structural weight of new generation aircraft such as the Airbus A350 or Boeing B787. In this context, metal stiffened shells, either flat or curved, have been replaced by their composite counterparts. Regardless of their constituent materials, stiffened panels must withstand a variety of complex loading conditions, any of which could cause the panel to buckle. Therefore, it is of paramount importance to establish methods that can effectively predict the structural behavior of composite panels beyond the first occurrence of instability in order to exploit their post-buckling capabilities. In the present work, the Finite Element Method (FEM) has been chosen as the main analysis tool. Indeed, the FEM has proven to be a valuable tool for investigating the structural response of stiffened panels [1].

Particular attention has been paid to the realistic modelling of geometry, loading and boundary conditions, avoiding the oversimplifications commonly found in the literature. The commercial software ABAQUS 2022 [2] was used to perform all the FE analyses. The modelling and simulation strategies are preliminarily validated on a metal panel whose data are available in the literature, as well as experimental and numerical results detailing its critical and post-critical behavior.

For this the lower fuselage panel studied by Rouse et alii in [3] was selected. Considerable effort has been made to apply realistic boundary conditions to the panel under analysis to avoid the D - BOX modelling used in [3] but not described in detail. The reference metal panel model was loaded with pure compression only. A composite version of the panel was then developed to investigate its stability under combined loading.

Reference panel and development of the composite version

The reference panel is made of aluminium alloy and consists of a curved skin stiffened by fifteen hat-section stringers and four zee-section frames. An aluminium tear strip is bonded to the skin underneath all the stiffening elements. While the stringers are attached directly to the skin, the frames are attached to the skin by shear clips and to the stringers by tension ties.

Quadrilateral, 4-node, stress/displacement shell elements with reduced integration and a large strain formulation (S4R) [2] were used to model all panel components. The tear strip was modelled implicitly by increasing the thickness of the skin under the stiffening elements and the connections between adjacent members were modelled using TIE constraints.

The eigenvectors provided by a preliminary linear buckling analysis were assigned as initial imperfections to a non-linear analysis to determine the pure compression buckling load of the structure. The loading and Boundary Conditions (BC) were assigned according to a typical experimental setup: one of the curved edges was fixed and the other was compressed by a concentrated axial force applied to the reference node, which shares its d.o.f. with all edge nodes. The analysis yielded a buckling load substantially in agreement with that reported in [3], thus qualifying the metal model as a reference to develop of the composite version.

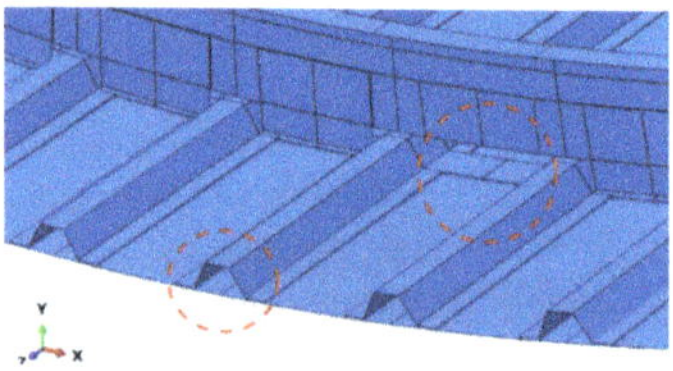

	Ply n.	CPT [mm]	Stacking sequence
Skin	14	1.75	$[+45/-45/90/+45/-45/0/0]_S$
Stringer	16	2	$[+45/-45/90/+45/-45/0/90/0]_S$
Frame	48	6	$[+45/-45/90/+45/-45/0/90/0]_{3S}$
Shear Clip	32	4	$[+45/-45/90/+45/-45/0/90/0]_{2S}$

Fig. 1 – Composite panel architecture and layup.

The composite panel has "omega" section stringers (15 as a reference, with the same spacing) as shown in Fig. 1. The attached flanges of adjacent stringers are extended to form a pad-up under the shear clips (see Fig. 1). The width of the shear clips has been increased while maintaining sufficient clearance for the stringers to pass through ('mouse holes'); the tension straps have been eliminated. All components are thin laminates of carbon-epoxy prepreg and have a symmetrical and balanced stacking sequence to avoid couplings (see Fig. 1).

Tab. 1 – Stiffness properties of the composite panel compared to the reference.

	Skin		
Material	AA Alloy	CFRP	
Thick. [mm]	1.6	1.75	
Esten. Stiff.	$Et/(1-v^2)$	A_{11}	A_{22}
[kN/mm]	131.2	109.4 (-16.6%)	77.6 (-40.9%)
Flex. Stiff.	$Et^3/12(1-v^2)$	D_{11}	D_{22}
[kN*mm]	28.0	16.4 (-41.4%)	25.2 (-10.0%)

	Stringer	
Material	AA Alloy	CFRP
Cross section	Hat	"Omega"
Cross Sect. Area [mm^2]	249.5	248.8 (-0.3%)
Axial Stiff. (EA) [KN]	17895	12560 (-29.8%)
Bend. Stiff. (EI_{yy}) [KNm^2]	2.61	1.34 (-48.7%)

The composite stringers are designed to have the same cross-sectional area as their metallic counterparts, and the composite skin has a thickness comparable to the metallic reference.

The frames and shear clips were dimensioned to carry the combined loads without causing instability problems. The stiffness properties of the composite panel are shown in Tab. 1. Overall, the reference panel has stiffer elements than the composite panel.

Materials Research Proceedings 37 (2023) 421-424 https://doi.org/10.21741/9781644902813-93

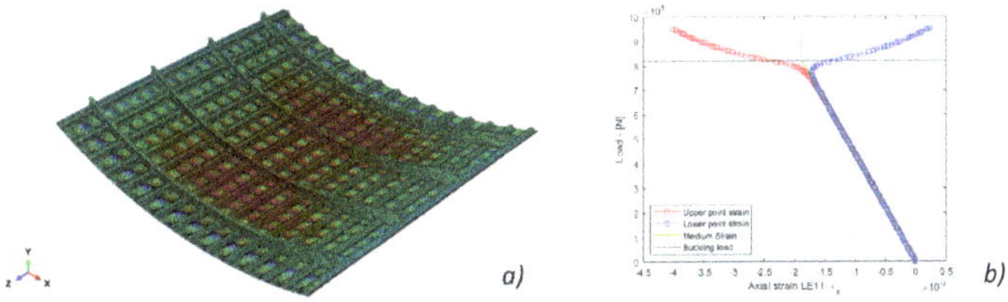

Fig. 2 – Compression buckling: radial displacement a) and load-strain diagram b)

The analyses on the composite panel were carried out using a mean element size close to that used for the reference panel, which was selected through a sensitivity analysis. Despite the lower stiffness, the compression buckling load of the composite panel is 820 KN (deformed shape in Fig. 2), significantly higher than that of the metal panel (570 KN).

The composite design was then frozen and used for all subsequent analyses.

Compression and Shear Buckling

The same procedure as for compression was used to calculate shear buckling: the eigenmodes of the linear analyses were used to perturb the geometry in the non-linear analyses. Again, using an experimental setup as a reference to enforce BC, one curved edge was fixed and the other was loaded by a pure torque (acting around the axis of the cylinder defined by the skin).

The strain analysis confirmed that the structure had been subjected to pure shear, as the longitudinal and transverse strains were zero prior to the instability, which manifested itself with skin buckles following patterns like those of metal panels. The skin between the stiffeners develops diagonal buckles at an angle of about 30°, as shown in Fig. 3a.

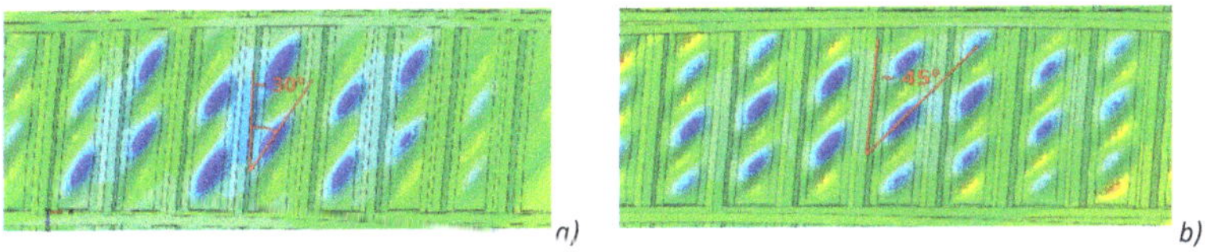

Fig.3 - Waves orientation for a) pure shear and b) shear and compression

The post-buckling configuration and stress state meet the *Incomplete Diagonal Tension Theory* formulated by Khun [4], widely accepted as the reference theory about shear buckling. The effect of restrained warping was also investigated and a strong reduction in buckling torque, quantifiable to about 50%, was found when warping is allowed. Eventually, an analysis was then carried out under the simultaneous action of compression and shear, with restrained warping. The effect of the compression is to increase the effectiveness of the shear loads in inducing the diagonal tension field; this synergy reduces the combined buckling load by about 40% with respect to the compression alone. Furthermore, in accordance with the reference theory, the angle of inclination of the buckles increased up to 45°, as shown in Fig. 3b.

The Effects of Pressurisation

The pressure was applied in advance to obtain realistic conditions: the radial displacement was left free, the symmetry is imposed along straight edges and the longitudinal stress that occurs in the real fuselage is introduced as an imposed displacement, evaluated through an ad hoc analysis.

In a second step, an increasing external compressive or shear load was applied to reach the buckling condition and to study the behavior of the panel in the post-buckling regime. The results of these analyses show that pressurization increases the buckling load significantly - 1540 KN vs. 821 KN for compression only (more than 85% increase) - or even remarkably: 3110 KNm vs. 1430 KNm for shear only (more than 115% increase).

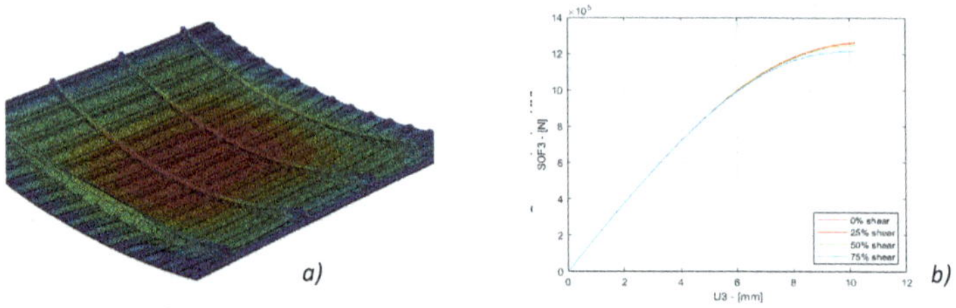

Fig.4 – Simultaneous application of pressurization, compression, and shear: a) Radial displacement, b) Longitudinal load-displacement curve.

Finally, some compression analyses were carried out at a given level of pressurisation with a constant shear load applied (increasing percentages of the pure shear buckling load were considered). For each value of the shear load, the deformed shape shown in Figure 4a is very different from those relevant to individual load cases (see Figures 2 and 3). The influence of the different shear levels on the longitudinal load-displacement curve is shown in Figure 4b. The buckling load is insensitive to the shear that slightly affects the post-buckling phase.

Conclusions

The study allowed the development of buckling and post-buckling modelling and analysis strategies under single and combined loading. Results relevant representative stiffened composite curved panel show that the buckling and post-buckling behavior in shear is consistent with the incomplete diagonal tension theory developed for metal structures.

The reduction in buckling load under simultaneous compression and shear loads is quantified and the effect of pressurisation is evaluated. Pressurisation is found to be remarkably beneficial in the case of pure compression; when shear is also present, it slightly affects the post-buckling phase, while the buckling load remains the same.

References

[1] L. Boni, D. Fanteria, and A. Lanciotti, Post-buckling behaviour of flat stiffened composite panels: Experiments vs. analysis, Composite Structures. (2012), pp. 3421 – 3433. https://doi.org/10.1016/j.compstruct.2012.06.005

[2] ABAQUS 2022 User's Manual, Dassault Systèmes Simulia Corp., Providence, USA (2022).

[3] M. Rouse, R. Young, and R. Gehrki, Structural stability of a stiffened aluminum fuselage panel subjected to combined mechanical and internal pressure loads, 44th AIAA/ASME/ ASCE/AHS/ASC Structures, Structural Dynamics, and Materials Conference. (2003). https://doi.org/10.2514/6.2003-1423

[4] P. Kuhn, J.P. Peterson, and L.R. Levin, A summary of diagonal tension Part I: methods of analysis, NACA TN 2661. (1952).

MOST Project

Insights on state of the art and perspectives of XR for human machine interfaces in advanced air mobility and urban air mobility

Sandhya Santhosh[1,a]*, Francesca DeCrescenzio[1], Millene Gomes Araujo[1], Marzia Corsi[1], Sara Bagassi[1], Fabrizio Lamberti[2], Filippo Gabriele Pratticò[2], Domenico Accardo[3], Claudia Conte[3], Francesco De Nola[4], Marco Bazzani[4], Joyce Adriano Losi[5]

[1]Department of Industrial Engineering, University of Bologna, Italy

[2]Department of Control and Computer Engineering, Polytechnic of Turin, Italy

[3]Department of Industrial Engineering, University of Naples Federico II, Italy

[4]Teoresi Group, Italy

[5]Accenture

* sandhya.santhosh2@unibo.it

Keywords: Urban Air Mobility, Advanced Air Mobility, Unmanned Aerial Systems, Immersive Technologies, Extended Reality, Human Machine Interfaces, U-Space

Abstract. With technological innovation and advancements, especially in autonomy, battery and digitization, the future of air transport and mobility is transiting towards a broader spectrum of Advanced Air Mobility (AAM) and Urban Air Mobility (UAM). UAM envisions safer, faster, and more sustainable air mobility for smarter cities and urban environments including passenger transport and goods delivery. Nevertheless, this concept is still considered extremely breakthrough and several technological and operational aspects are mostly undefined. In this context, a comprehensive approach to AAM/UAM may be to adapt cutting-edge technologies in developing sustainable framework and Human-Machine Interfaces (HMIs) in order to realize the challenges, benefits, and conditions of such transport system in advance for future safer, more reliable and globally approved operations. One of the technologies that can contribute to accelerate advancements through human centred simulating UAM processes and operations is XR (eXtended Reality). This paper presents the early steps of a multidisciplinary study performed under the framework of PNRR (Piano Nazionale di Ripresa e Resilienza) and MOST (Centro Nazionale Mobilità Sostenibile) project in analyzing the perspectives of XR based HMIs for UAM paradigm and potential AAM/UAM use case scenarios that can be simulated with XR in view of attaining efficient and effective future solutions. Furthermore, the work introduces the state-of-the-art overview on XR facilitated UAM applications and considers prospective potential use cases that can be developed through PNRR research study in demonstrating XR as an enabling technology in promising areas of the UAM framework.

Introduction

Air mobility, also referred to as AAM or UAM, has emerged as a transformative concept in the realm of transportation, offering new possibilities for efficient and sustainable movement of people and goods. According to the studies performed in the framework of the Italian AAM Strategic Plan, the global AAM market is expected to grow at a 20-25% rate from 2021 to 2030, reaching an estimated value of around USD 38-55 billion per year [1]. A significant interest is paid to its implementation in the context of urban environments where UAM represents a promising vision for the future of transportation of goods and passengers, aiming at providing efficient and sustainable aerial transportation solutions within urban areas [1][2]. According to Tojal et al.,

UAM is a mobility concept for urban areas that makes use of any kind of mainly Unmanned Aerial Systems (UASs) to perform any type of mission that is operated in the Very Low Level (VLL) airspace aiming at improving the welfare of individuals and organizations [3]. Thanks to technological advancements in UAM in conjunction with advanced materials, aircraft architecture, enhanced battery capacity, digitalisation of air traffic management etc., the commercial exploitations of such mobility system is expected to become a reality in Europe within 3 to 5 years [4]. However, the actual implementation of UAM comes with numerous challenges. The safe integration of UAM vehicles into urban airspace, the development of infrastructures such as vertiports and changing stations, regulatory frameworks, public acceptance, and efficient operations are among key considerations (as highlighted in Fig.1)[5][6]. These challenges necessitate a multidisciplinary approach that involves collaboration between industry stakeholders, policymakers, urban planners, aviation authorities, and technology innovators [7]. Therefore, attention has been increasing towards contemplating innovative technologies in simulating and developing advanced human-machine interfaces (HMIs) through human and user-centred approaches for future UAM scenarios and foreseeing the challenges in order to find efficient and effective solutions and support regulatory processes.

Immersive media comprising Virtual Reality (VR), Augmented Reality (AR), and Mixed Reality (MR) are amongst the currently fastest growing and promising tools for such innovative HMIs. These, also commonly referred to with the umbrella term XR, enable the users to experience immersive and interactive environments, and have been proven to enhance design validations, reduce training costs, enhance user engagement, improve communication and collaboration with seamless data access etc.[8], [9],[10],[11]. Through a comprehensive analysis of existing literature, case studies, and industry developments, the present work aims to provide insights into the current state of UAM scenarios and explores the potential role of XR-based HMIs and simulations. By understanding the complexity of UAM, we can better appreciate the significant impact it may have on urban transportation and facilitate its successful integration into our cities.

Fig. 1. Overview of XR simulation themes for UAM: (1) Types of UAM (2) Top concerns highlighted by EASA (3) Potential UAM themes for XR applications.

Related work on XR-based HMIs and Simulations for UAM

It is recognized that in the realm of the digital transformation of processes and the 4.0 industrial revolution, XR technologies have paved the way to advanced HMIs acting as a bridge connecting the gap between humans and machines [13]. Revenue in AR and VR market worldwide is expected to show an annual growth rate (CAGR 2023-2027) of 40.12%, resulting in a projected market volume of US$9.10bn by 2027 [14]. It is evident that XR and UAM are together rapidly growing markets. Besides this aspect it must be considered that the integration of XR-based HMIs and simulations for UAM presents numerous benefits. It facilitates the design and evaluation of user-centric interfaces that consider human factors, ergonomics, and cognitive workload in highly automated environments. Furthermore, XR-based simulations enable stakeholders to assess the feasibility and performance of UAM systems, optimize operational procedures and identify potential safety risks.

To this regard, we have performed a preliminary study on collecting a selection of the existing works in the field of XR-based UAM and categorised them into 3 different perspectives: Market, Scientific, and Industrial research (see Table 1).

Table 1. Selection of previous works relevant for this study

<table>
<tr><th></th><th>Reference</th><th colspan="3">Forecasts</th></tr>
<tr><td rowspan="2">Market Perspective</td><td>[1]</td><td colspan="3">Global AAM/UAM market research forecasts a growth of CAGR at 20/25% from 2021 to 2030</td></tr>
<tr><td>[14]</td><td colspan="3">AR and VR market worldwide is expected to show an annual growth rate (CAGR 2023-2027) of 40.12%</td></tr>
<tr><td colspan="2"></td><td>UAM Mission Scenarios</td><td>Description</td><td>XR technology</td></tr>
<tr><td rowspan="5">Scientific Perspective</td><td>[15]</td><td>Collaborative Decision Making</td><td>3D map rendering with planes, runways and waypoints demonstrating air traffic scenarios</td><td>AR</td></tr>
<tr><td>[16]</td><td>Simulation of Workspace</td><td>Taking off and landing a quadcopter</td><td>VR, CAVE</td></tr>
<tr><td>[17]</td><td>System integration and testing</td><td>Urban Traffic Management, UAS operations</td><td>HMI, AR CAVE</td></tr>
<tr><td>[18][19][20]</td><td>Public/ Social Acceptance</td><td>Auditory and Visual perception of drones, acceptance of Air taxis</td><td>AR, VR</td></tr>
<tr><td>[21][22]</td><td>Virtual Prototyping and Design</td><td>Urban Airport Infrastructure design, Air taxi cabin</td><td>VR, MR</td></tr>
<tr><td rowspan="3">Industry Perspective</td><td>[24][25]</td><td>Visualization of Airspace data</td><td>CLARITY: HMD for Air traffic control
-Drone Control with intuitive gestures</td><td>MR, VR
AR</td></tr>
<tr><td>[23][26][27]</td><td>Training and Simulation</td><td>-Real-time Tower and Apron Control Research Simulator (NARSIM)
-Pilot training program for eVTOL
-eVTOL Flight Simulator</td><td>Simulator, AR
AI, VR, MR
MR</td></tr>
<tr><td>[28][29]</td><td>Simulation</td><td>Drone Simulator</td><td>AR-to-gamepad interface</td></tr>
</table>

Conclusion

UAM is an emerging transport system with dedicated services that integrate aerial unmanned platforms for passengers and goods transport in urban environment. As UAM progresses, there is a growing need for advanced HMIs and simulations to enhance the design, operation, and ensure safety of these complex systems. With advent growth towards automation, technologies such as XR offers innovative solutions for creating immersive environments and interactive experiences for future UAM scenarios. In this context, this paper highlights a literature study on XR-based HMIs and simulations to support UAM services. We classified the information into three perspectives of scientific, industrial and market in view of highlighting the main areas of XR-based HMIs and simulations for future UAM scenarios. It has been observed that the literature identifies the key aspects relating to the fields of virtual prototyping, design, training, simulation, human factors evaluation, airspace visualization, collaborative decision making, system integration and testing, public engagement, and education/marketing.

Acknowledgments
This study was carried out within the MOST – Sustainable Mobility National Research Center and received funding from the European Union Next-Generation EU (PIANO NAZIONALE DI RIPRESA E RESILIENZA (PNRR) – MISSIONE 4 COMPONENTE 2, INVESTIMENTO 1.4 – D.D. 1033 17/06/2022, CN00000023). This manuscript reflects only the authors' views and opinions, neither the European Union nor the European Commission can be considered responsible for them.

References

[1] AAM National Strategic Plan (2021-2030) for the development of Advanced Air Mobility in Italy, www.enac.gov.it
[2] White Paper on Urban Air Mobility and Sustainable development, https://www.asd-europe.org (2023).
[3] M. Tojal, H. Hesselink, A. Fransoy, E. Ventas, V. Gordo, Y. Xu, Analysis of the definition of Urban Air Mobility –how its attributes impact on the development of the concept, Transportation Research Procedia, Volume 59, 2021, Pages 3-13, ISSN 2352-1465. https://doi.org/10.1016/j.trpro.2021.11.091
[4] EASA Urban Air Mobility https://www.easa.europa.eu/en/domains/urban-air-mobility-uam
[5] Bauranov, A., & Rakas, J. (2021). Designing airspace for urban air mobility: A review of concepts and approaches. Progress in Aerospace Sciences, 125, p.100726. https://doi.org/10.1016/j.paerosci.2021.100726
[6] Schweiger, K. and Preis, L., 2022. Urban Air Mobility: Systematic Review of Scientific Publications and Regulations for Vertiport Design and Operations. *Drones*, *6*(7), p.179. https://doi.org/10.3390/drones6070179
[7] Full report https://www.easa.europa.eu/sites/default/files/dfu/uam-full-report.pdf
[8] Santhosh, S., De Crescenzio, F. and Vitolo, B., 2022. Defining the potential of extended reality tools for implementing co-creation of user oriented products and systems. In Design Tools and Methods in Industrial Engineering II: ADM 2021, September 9–10, 2021, Rome, Italy (pp. 165-174). Springer International Publishing. https://doi.org/10.1007/978-3-030-91234-5_17
[9] Bagassi, S., De Crescenzio, F., Piastra, S., Persiani, C. A., Ellejmi, M., Groskreutz, A. R., & Higuera, J. (2020). Human-in-the-loop evaluation of an augmented reality based interface for the airport control tower. Computers in Industry, 123, 103291. https://doi.org/10.1016/j.compind.2020.103291
[10] Pratticò, F. G., & Lamberti, F. (2021). Towards the adoption of virtual reality training systems for the self-tuition of industrial robot operators: A case study at KUKA. Computers in Industry, 129, 103446. https://doi.org/10.1016/j.compind.2021.103446
[11] Sikorski, B., Leoncini, P., & Luongo, C. (2020). A glasses-based holographic tabletop for collaborative monitoring of aerial missions. In Augmented Reality, Virtual Reality, and Computer Graphics: 7th Int.Conf., AVR 2020, Lecce, Italy, September 7–10, 2020, Proceedings, Part I 7 (pp. 343-360). Springer International Publishing. https://doi.org/10.1007/978-3-030-58465-8_26
[12] https://www.easa.Europa.eu/en/light/topics/vertiports-urban-environment
[13] https://www.agendadigitale.eu/industry-4-0/hmi-cose-ladvanced-human-machine-interfaces-e-perche-e-utile-per-lindustria-4-0/, Accessed on 22/05/2023.
[14] https://www.statista.com/

[15] Malich T., Hanakova L., Socha V., Van den Bergh S., Serlova M., Socha L., Stojic S., Kraus J.Use of virtual and Augmented Reality in design of software for airspace (2019) ICMT 2019 - 7th International Conference on Military Technologies, Proceedings, art. no. 8870030. https://doi.org/10.1109/MILTECHS.2019.8870030
[16] Marayong, P., Shankar, P., Wei, J., Nguyen, H., Strybel, T. Z., & Battiste, V. (2020, March). Urban Air Mobility System Testbed using CAVE Virtual Reality Environment. In *2020 IEEE Aerospace Conference* (pp. 1-7). IEEE. https://doi.org/10.1109/AERO47225.2020.9172534
[17] Dao, Q. V., Homola, J., Cencetti, M., Mercer, J., & Martin, L. (2019, August). A Research Platform for Urban Air Mobility (UAM) and UAS Traffic Management (UTM) Concepts and Application. In International Conference on Human Interaction & Emerging Technologies (IHIET 2019) (No. ARC-E-DAA-TN68588).
[18] Aalmoes, R., & Sieben, N. (2021, March). Noise and visual perception of Urban Air Mobility vehicles. In Delft International Conference on Urban Air Mobility (DICUAM), virtual.
[19] Stolz, Maria and Tim Laudien. "Assessing Social Acceptance of Urban Air Mobility using Virtual Reality." 2022 IEEE/AIAA 41st Digital Avionics Systems Conference (DASC) (2022): 1-9. https://doi.org/10.1109/DASC55683.2022.9925775
[20] Janotta, F. & Hogreve, J (2021). Acceptance of AirTaxis – Empirical insights following a flight in virtual reality. https://doi.org/10.31219/osf.io/m62yd
[21] Stewart Birrell, William Payre, Katie Zdanowicz, Paul Herriotts, Urban air mobility infrastructure design: Using virtual reality to capture user experience within the world's first urban airport, Applied Ergonomics, Volume 105, 2022, 103843, ISSN 0003-6870. https://doi.org/10.1016/j.apergo.2022.103843
[22] T. Laudien, J. M. Ernst and B. Isabella Schuchardt, "Implementing a Customizable Air Taxi Simulator with a Video-See-Through Head-Mounted Display – A Comparison of Different Mixed reality Approaches," 2022 IEEE/AIAA 41st Digital Avionics Systems Conference (DASC), Portsmouth, VA, USA, 2022, pp. 1-10. https://doi.org/10.1109/DASC55683.2022.9925870
[23] https://aurora-uam.eu/#summary
[24] https://360.world/clairity/#:~:text=The%20CLAIRITY%20system%20takes%20the,unique%20HMD%20and%20camera%20solutions.
[25] Konstantoudakis K, Christaki K, Tsiakmakis D, Sainidis D, Albanis G, Dimou A, Daras P. Drone Control in AR: An Intuitive System for Single-Handed Gesture Control, Drone Tracking, and Contextualized Camera Feed Visualization in Augmented Reality. Drones. 2022 Feb 10;6(2):43. https://doi.org/10.3390/drones6020043
[26] https://www.volocopter.com/newsroom/cae-and-volocopter-partner-to-create-global-air-taxi-pilot-workforce/
[27] https://flyelite.com/evtol/
[28] https://www.geospatialworld.net/news/worlds-first-augmented-reality-drone-flight-simulator-app-launched-epson/
[29] https://www.dji.com/it/simulator/info

Aeronautics and Astronautics - AIDAA XXVII International Congress Materials Research Forum LLC
Materials Research Proceedings 37 (2023) 431-434 https://doi.org/10.21741/9781644902813-95

Electric conversion of a general aviation aircraft: a case study

Sergio Bagarello[1,2,a], Ivano Benedetti[1,2,b *]

[1] Department of Engineering, Università degli Studi di Palermo, Viale delle Scienze, Edificio 8, 90128 Palermo, Italy

[2] Sustainable Mobility Center - Centro Nazionale per la Mobilità Sostenibile – MOST – Italy

[a] sergio.bagarello@unipa.it, [b] ivano.benedetti@unipa.it

Keywords: Aircraft Electric Conversion, Electric Propulsion, Electric General Aviation, Aircraft Conceptual Design

Abstract. This study analyses the process required to convert a conventional, air-breathing, piston-powered, General Aviation (GA) airplane to fully electric propulsion. The work is configured as a feasibility study for such modifications with the intent of setting a path for similar electric conversion programs on GA airplanes. The motivation behind industries' interest in alternative propulsion is examined and a full comprehension of the characteristics of the plane in question is achieved through the acquisition of transversal knowledge, examining the aircraft both from the engineering and real-world user points of view. Electric motor, batteries, auxiliary systems and implementation considerations were all made in accordance with regulatory authorities' requirements, with the purpose of making the project to comply with EASA CS Part 23. The present work analyses the performances expected from the electric plane and compares them with the standard aircraft evaluating the project's pros and cons. Considerations regarding typical mission profiles show how the electric powerplant will allow the airplane to outperform his conventional counterpart in terms of rate of climb, pollutant emission reduction, noise levels and operating costs. Such gains are however counterbalanced by the detriment of range and endurance performances, which might be deemed acceptable considering the specific plane's intended use. The study shows how, even though close integration in electric GA aircrafts is desirable since the first stages of conceptual design, piston-to-electric conversions are possible and may indeed contribute to mitigate aviation climate impact.

Introduction

In recent years, the themes of sustainability and low/zero-emissions propulsion have assumed great relevance in aviation. Aviation represents a crucial asset and the only way to transport people and goods across the world within a day. In 2016, the sector drove $2.7 trillion in economic activity and supported 65.5 million jobs, which made up 3.6% of the global gross domestic product (GDP) [1]. However, the destructive environmental consequences of aviation are undeniable. Although flying makes up only 3.5% of total human-induced carbon emissions, it is one of the most challenging to decarbonize [1]. In the interest of sustaining policies like the MEA (More Electric Aircrafts) [2], a feasible path might consist in converting established, well-known, conventional airplanes to electric power via a process that makes the new project an interesting alternative with respect to the gas-powered counterpart, cutting the costs of designing a new plane from scratch. This study therefore analyzes the process needed to replace the propulsive system of a conventional internal combustion engine airplane with a state-of-the-art electric motor and the associated batteries for the new aircraft to be used *as a trainer and semi-acrobatic General Aviation plane*.

Conversion Process

Requirements. Considering the current state of the technology, flight time represents one of the major limitations in electrical aircraft's capability. Therefore, the logical target market for electric aircraft would be identified by applications for which flight time and range are not as decisive as they are in commercial aviation. Student pilots normally train in flights of about one hour. Anything more is generally considered counterproductive according to most flight instructors, so flight training represents a perfect target application [3]. The aircraft chosen as case study is the *SOCATA Tampico TB-9*, a French-made, all-metal, low-wing, four-seater, single engine aircraft in use in numerous flight schools [4]. It is well-known for its straightforward approach, forgiving aerodynamics characteristics and robust build.

Batteries mass. The sizing process needs to evaluate the batteries total mass m_b [kg]. Given the required run-time endurance E [s], the mass of the batteries can be computed as

$$m_b = \frac{P_{sh}\, E}{E_{sb}\ \eta_{b2s}} \tag{1}$$

where P_{sh} is the required shaft power during run-time [kW], E_{sb} is the battery specific energy [J/kg] and η_{b2s} represents the total system efficiency from battery to motor output shaft.

A safer and more conservative approach considers the power used equal to 120 kW, the maximum rated power that the selected motor can deliver, and the required batteries mass is found to be 387 kg. However, a more detailed analysis of the mission in hand, whose features are retrieved from the aircraft flight manual [5], requires the knowledge of the expected power levels during all flight operations, which resulted in $m_b = 346.3$ kg. For the sake of the study, the more conservative value has been considered.

Powerplant installation. The conventional aircraft is equipped with a four-cylinders, direct drive air-cooled Lycoming O-320-D2A engine capable of producing 160 hp at 2700 rpm. It is connected to the frontal bulkhead via a steel engine support structure, easily modifiable to account for the different form factors of the two powerplants [6]. After a careful market survey, the more suitable engine for the conversion is found to be the Safran ENGINeUS™ in the 120 kW configuration.

Energy is provided by state-of-the-art lithium-sulfur batteries produced by Sion Power with a specific energy of 500 Wh/kg. The cathode is made of sulfur and a conductive material while the anode is made of lithium or a lithium-alloy. The reaction between lithium and sulfur is highly energetic, which leads to a high energy density in the battery.

In the conversion process, the main constraint is the need to the comply with the Maximum Take-Off Weight (MTOW) of the standard model. EASA's Supplemental Type Certificate specifications allow a 5% deviation from the initial MTOW of the plane after the modifications, which will therefore retained as a requirement. The weight check – Table 1 – shows how the removal of the internal combustion engine, of the tanks and fuel system, of fuel and of the rear bench and passengers leaves enough mass available for the installation of the motor, batteries, and auxiliary systems even in the most demanding configuration of 120 kW continuos $\boldsymbol{P_{sh}}$. The payload reduction is clearly another cost of the conversion, but it is deemed acceptable considering the prospective use of the converted airplane as a GA trainer.

Centre of gravity preservation. Trim, stability, and structural considerations require that the new powerplant is integrated preserving the allowed CG envelope. The main components placement for the considered aircraft is schematized in Fig.1. The necessary checks are performed using the aircraft standard Mass and Balance modulus [7] as a reference – Table 2 – and verifying that in even in the most demanding configuration (two people plus baggage) the Center of Gravity stays inside its nominal envelope, Fig.1.

***Table 1**: Weight analysis for the converted aircraft.*

<table>
<tr><td colspan="2">TB-9 Maximum Take-Off Weight</td><td>1060 Kg</td><td>Total of Removed items</td></tr>
<tr><td rowspan="4">Removed items</td><td>Engine</td><td>120 Kg</td><td rowspan="4">- 448.76 Kg</td></tr>
<tr><td>Tanks and fuel lines</td><td>45 Kg</td></tr>
<tr><td>AVGAS100LL</td><td>113.76 Kg</td></tr>
<tr><td>Rear passengers</td><td>170 Kg</td></tr>
<tr><td rowspan="3">Added items</td><td>Electric motor</td><td>45 Kg</td><td rowspan="3">+442 Kg</td></tr>
<tr><td>Batteries</td><td>387 Kg</td></tr>
<tr><td>Cables and Insulation</td><td>10 Kg</td></tr>
<tr><td colspan="2">Electric TB-9 Take-Off Weight</td><td>1053 Kg</td><td>Total of Added items</td></tr>
</table>

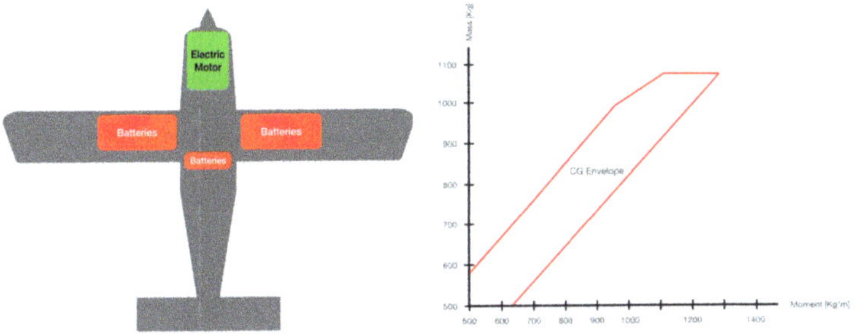

***Figure 1**: Components placement (left) and center of gravity envelope (right) for the converted aircraft.*

***Table 2**: mass and balance check.*

	Weight [kg]	**Arm** [m]	**Moment** [Kgm]
Aircraft	808	0.965	779.72
Pilot and co-pilot	170	1.155	196.35
Baggage	65	2.035	132.275
TOTAL	1043		1108.345

Performance evaluation. In conventional engines, the volumetric efficiency decreases significantly with altitude: electric engines, on the contrary, are not susceptible to altitude variations in terms of power output. The limiting factor in terms of ceiling of the electrified airplane is related to air density decreases through propeller and lifting surfaces. With the implementation of electric technology, at 80 KIAS, the rate of climb achievable, and more importantly, sustainable for a longer time, will be given by

$$V_V = \frac{\eta_p P_{sh}}{W} - \frac{V}{(L/D)} \tag{2}$$

where V is the flight speed, η_p is the propeller efficiency, W the aircraft weight [N] and $\frac{L}{D}$ is the lift-to-drag ratio. For the case in hand, Eq.(2) provides $V_V = 1212$ ft/min.

Range is strongly affected by the limitations of the employed technology: in the case in hand the usable run-time endurance will be limited to 60 min for normal operations plus sufficient energy supply to sustain a holding pattern of 30 min. The theoretical range can be estimated as

$$R = \frac{E_{sb}\,\eta_{b2s}\,\eta_p}{g}\left(\frac{m_b}{m}\right)\left(\frac{L}{D}\right) \tag{3}$$

where m_b/m is the battery mass fraction. For the considered case Eq.(3) provides $R = 353$ km. Operating costs significantly benefit from the electric conversion and, considering the price of electricity at the time of the development of the study of €$0.35/kW$h, 20% savings per flight-hour are estimated with respect to AVGAS operations, see Table 4.

***Table 4**: Performance analysis for the converted aircraft.*

Theoretical Range	−55.7%
Endurance	−68.75%
Payload	−43%
Climb rate	+61.6%
Cost per flight hour	−20%
Emissions	−100%

Summary

The electric conversion of a GA trainer aircraft has been assessed. Although range and endurance are heavily affected by the limitations of the current battery technology, as expected, the other performances remain acceptable or are even enhanced and the overall platform may be profitably employed in applications where range and endurance are not the main concern, as in the considered case of training aircraft. While the platform emissions are certainly reduced, overall emission reductions depend on the *green quality* of the energy sources employed for battery recharging.

Acknowledgements

The authors acknowledge the support of the Sustainable Mobility Center (Centro Nazionale per la Mobilità Sostenibile - CNMS) under Grant CN00000023 CUP B73C22000760001.

References

[1] Kousoulidou M, Lonza L, European Aviation Environmental Report 2016. EASA, EEA, EUROCONTROL; JRC99523, 2016

[2] Brelje, BJ, Martins, JRRA, Electric, hybrid, and turboelectric fixed-wing aircraft: A review of concepts, models, and design approaches. *Progress in Aerospace Sciences*, 104, 1-19, 2019. https://doi.org/10.1016/j.paerosci.2018.06.004

[3] Raymer, D., *Aircraft design: a conceptual approach*, American Institute of Aeronautics and Astronautics, 2012. https://doi.org/10.2514/4.869112

[4] GlobalAir.com, "Socata Tampico TB-9", GlobalAir, https://www.globalair.com/aircraft-for-sale/specifications?specid=523, 2023

[5] AeroClub Palermo DTO Flying School, *Manuale di impiego – Velivolo Socata TB-9,* Palermo (PA), 90137 Italy, 2019

[6] Lycoming Engines, *O-320 Series Operator's Manual,* Williamsport, PA 17701 USA, 2006

[7] AeroClub Palermo DTO Flying School, Modulo di caricamento e centraggio – Velivolo Socata TB-9, Modello CO5, Ed.1, Palermo (PA), 90137 Italy, 2019

Aeronautics and Astronautics - AIDAA XXVII International Congress Materials Research Forum LLC
Materials Research Proceedings 37 (2023) 435-439 https://doi.org/10.21741/9781644902813-96

Aerodynamic design of advanced rear end for large passenger aircraft

Salvatore Corcione[1,a*], Vincenzo Cusati[1,b] and Fabrizio Nicolosi[1,c]

[1] Industrial Engineering Department, University of Naples Federico II, Via Claudio 21, 80125, Naples, Italy

[a]salvatore.corcione@unina.it, [b]vincenzo.cusati@unina.it, [c]fabrnico@unina.it

Keywords: Aircraft Design, Forward Swept Tailplane, Leading Edge eXtension, Aerodynamics

Abstract. This paper focuses on the aerodynamic design of an advanced rear end concept for a large passenger aircraft, such as the Airbus A320. The aim was to reduce the size of the horizontal tailplane to minimize the aerodynamic drawbacks related to longitudinal stability and control requirements. This reduction would lead to improved aircraft performance by reducing fuel burn and rear-end weight. Assuming the same position of the aerodynamic center of the horizontal tailplane of a conventional aircraft, the results of this investigation showed that the required stabilizing performance of the tail could be achieved with a smaller tail surface. A reduction of 6% in tail planform area was achieved by leveraging the unique aerodynamic characteristics of a forward-swept tail, combined with the implementation of a leading-edge extension device. The reduced wetted area and the lower weight of the horizontal empennage could result in fuel savings of 100 to 120 kg of fuel per 1,000 km. This is equivalent to approximately 1.0 to 1.2% for the specific aircraft category being considered.

Introduction

Advancements in design and improvements in empennage efficiency and effectiveness have the potential to enhance aircraft performance by reducing fuel burn and weight through reductions in tail-plane size. The penalties associated with meeting both longitudinal and directional stability and control requirements constitute a significant portion of the total aircraft drag. Loads acting on aircraft tails contribute to the overall induced drag, compressibility, profile drag, structural weight, and maximum lift capability of the aircraft. The empennage of a typical Large Passenger Aircraft accounts for one-fifth to one-fourth of the total lifting surface and 3% up to 6% of the maximum take-off weight. It contributes 5% to 8% to the total trimmed drag in cruise conditions [1].

The simplest unconventional solution is represented by the Vee-tail [2,3]. This solution is sometimes used in remotely piloted aircraft and has also been implemented in mass-produced manned aircraft, such as the Beechcraft Bonanza M35. However, the results of the NEFA [4] project concluded that although a Vee-tail configuration offered performance improvements due to its reduced wetted area, the added complexity and additional system did not result in any weight or cost benefits over a conventional empennage. A comprehensive study on advanced rear-end configurations was recently conducted in the EU-funded project NACRE [5] demonstrated that these configurations could offer advantages in terms of reducing empennage drag, but not in terms of weight.

To further advance the implementation of rear-end concepts that effectively reduce drag and weight, the utilization of a forward-swept horizontal tailplane could represent a viable way.

The adoption of a forward-swept tailplane enables a structural configuration in which the connection of the horizontal tail to the rear end does not require a structural opening in a region of the fuselage that is heavily affected by structural loads [6]. By removing the structural opening at the rear end, the weight of the fuselage can be reduced. This solution also reduces fuselage deformations, resulting in a more efficient horizontal stabilizer surface [6].

Transonic aircraft wings typically have a positive sweepback. The main reason is linked to the aircraft encountering a vertical gust during its flight. In the case of positive sweepback, the bending deformation decreases the local angle of attack, resulting in a natural reduction of aerodynamic loads. In the case of a wing with a negative sweep angle, the effect is reversed. As a result, static divergence may occur, leading to structural failure. Forward-swept wings are capable of withstanding significantly higher gust loads compared to wings with positive sweepback, making them heavier. Despite this drawback, several studies have explored the potential of utilizing the aerodynamic advantages of a forward-swept wing [7,8] propose a solution to mitigate the coupling between flexional and torsional deformation by using aeroelastic tailoring techniques. In terms of structural sizing, aeroelasticity is less demanding for wings with a relatively low aspect ratio. Thus, in the case of horizontal tails, introducing negative sweep angles could be a viable solution to improve the performance of the rear-end and empennage.

Forward-swept lifting surfaces offer several aerodynamic advantages over conventional sweepback designs. For a given leading edge sweep angle, forward-swept wings exhibit a shock-sweep angle that is five degrees higher than that of aft-swept wings [9]. Therefore, the implementation of a forward-swept design requires a smaller leading edge sweep angle compared to a positively swept-back configuration with an equivalent sweep angle at the quarter chord line. Moreover, in a forward-swept wing, the airflow moves from the root to the tip, resulting in higher stall angles [10], increasing the maximum aerodynamic forces or reducing the tailplane area can yield the same maximum force, potentially leading to a decrease in drag and weight.

This paper deals with the aerodynamic design of an advanced rear-end configuration carried out within the EU-funded project named IMPACT [11]. The aim is to optimize the rear-end of the fuselage and empennage of large passenger aircraft to reduce drag, weight, and fuel burn. The investigation focuses on minimizing the surface area of the horizontal tailplane by utilizing the unique characteristics of a forward-swept lifting surface, which is further enhanced by passive leading edge extension devices.

Advanced Rear End Aerodynamic design

To fully catch the peculiar aerodynamic features, high-fidelity CFD RANS calculation were required. The high-fidelity analysis was performed using the commercial software STARCCM+. Details of the numerical model setup are reported Table 1, whereas Figure 1 shows a comprehensive overview of the fluid domain and the application of boundary conditions. The investigation started by comparing the reference isolated tailplane geometries: a conventional tail (HTP) and a reference forward-swept arrangement. Table 2 summarizes the main design parameters. Results clearly indicate that the isolated forward-swept tail arrangement effectively wash in the aerodynamic loads. Thanks to this peculiar behavior, the FSHTP exhibits better aerodynamic performance in terms of lift curve slope and maximum achievable negative lift coefficient, as indicated by the chart of Figure 2. This means that an aerodynamically equivalent forward-swept tail would require a smaller area to reach the same aircraft stability and control characteristics.

Since an additional component must be added, this element can be designed to enhance the lifting capacity of the tail. In this respect, the additional element would be a leading-edge extension device (LEX).

Table 1 Numerical setup and mesh characteristics for high-fidelity CFD analysis

Parameter	Value
Domain span	$16\ b_H/2$
Domain height	$30\ b_H/2$
Domain length	$100\ b_H/2$
Mesh type	unstructured polyhedral
On body minimum surface size	0.0042 (m)
On body target surface size	0.035 (m)
Number of volume cells	11 183 156
Number of prism layers	25
First cell wall distance	$1e^{-6}$ (m)
Turbulence model	SST $\kappa - \omega$
Flow model	Compressible
Inflow boundary condition	Free stream
Outflow boundary condition	Pressure outlet
Number of iterations	5000
Mach number	0.2
Flight altitude	sea level

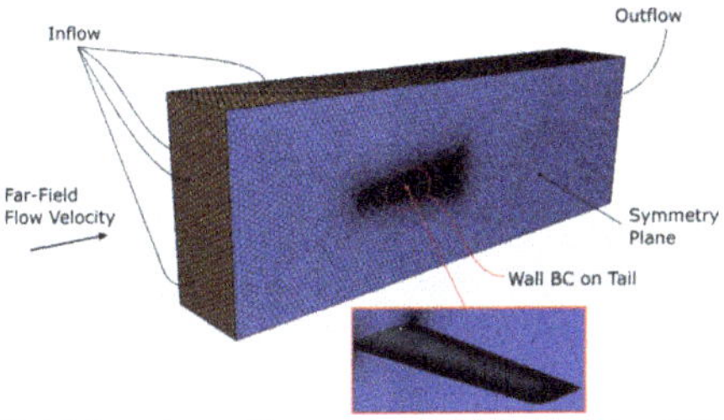

Figure 1 Fluid domain and boundary conditions using STAR-CCM+.

Table 2 Reference geometries, HTP and FSHTP.

	HTP	FSHTP
Sweep angle	32 deg	-15 deg
S_H	31.36 m^2	31.36 m^2
$b_H/2$	6.723 m	6.723 m
Taper Ratio	0.36	0.715
Aspect Ratio	5.765	5.765

Unfortunately, these advantages are lost when considering the fuselage-tail configuration. The forward-swept tail arrangement exhibits significant separation at the junction with the fuselage. This reflects on the lifting capabilities, as shown in Figure 3.

To design an effective advanced rear end that incorporates a forward-swept tailplane, an additional component must be introduced to prevent significant flow separation at the junction of the fuselage.

Leading Edge eXtensions (LEX) are aerodynamic features found on some aircraft, typically fighter jets. LEX refers to the forward extensions of the wing root area, usually in a triangular or trapezoidal shape. They are located at the junction between the wing and the fuselage. The primary purpose of LEX is to improve the aircraft's high angle-of-attack performance and enhance its manoeuvrability.

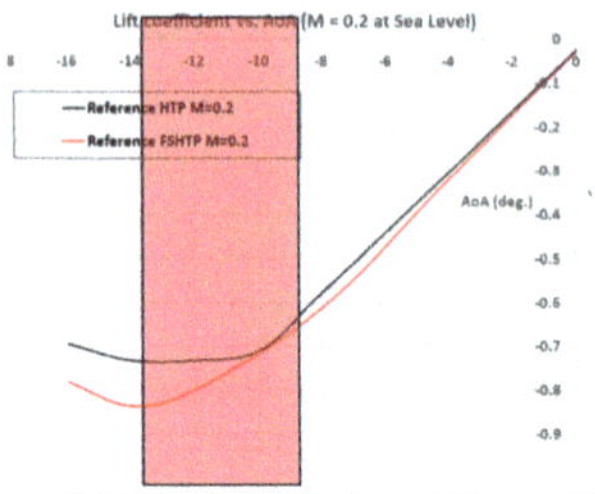

Figure 2 Comparison of the lift coefficient curves for conventional HTP and FSHTP configurations.

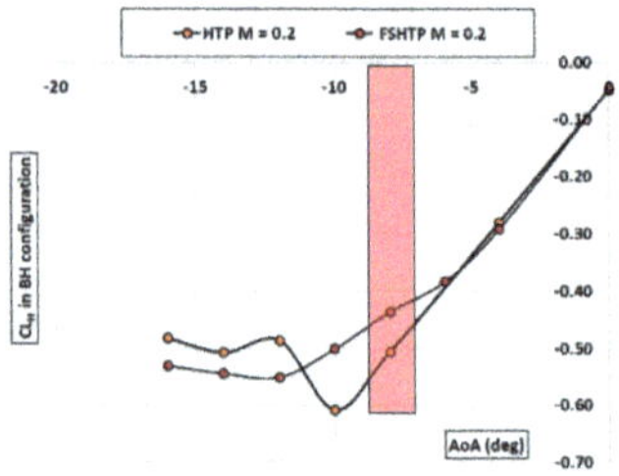

Figure 3 Comparison of the lift coefficient curves for HTP and FSHTP (body and horizontal tailplane).

By performing a Design of Experiment about several forward-swept tailplanes and LEX designs, see Figure 4, the best solution has been identified. As shown by the results of Figure 5, the maximum lift capabilities of the horizontal empennage can significantly be improved by introducing a forward-swept tailplane enhanced by a LEX device. The optimum solution (see the solid grey line in Figure 5), provides for a tailplane area which is 6% lower than the reference tails (see Table 3) giving a maximum negative lift coefficient which is approximatively 20% higher than the reference conventional tailplane.

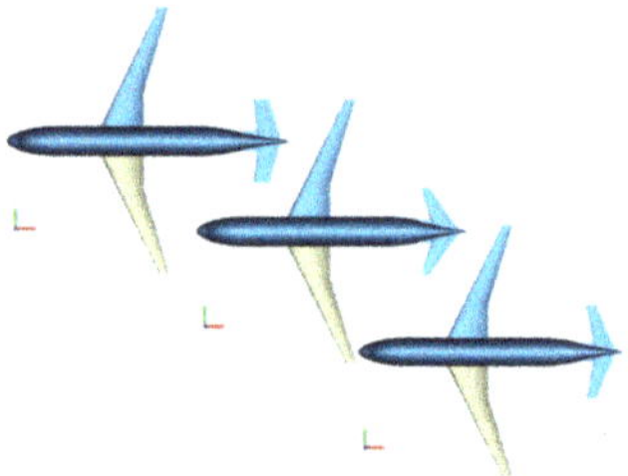

Figure 4 Some configurations investigated in the DoE execution.

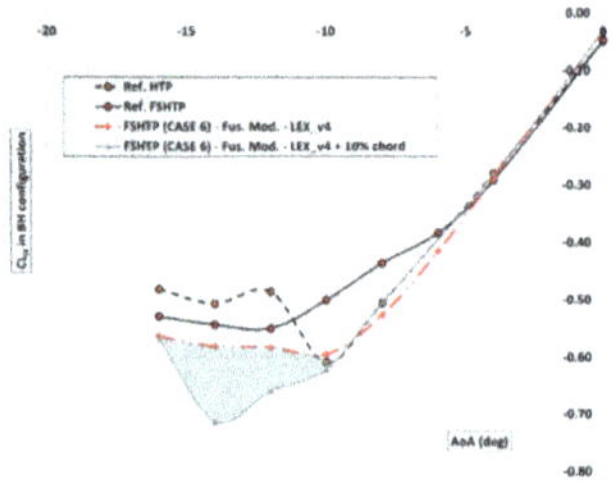

Figure 5 Lift coefficient curves for reference HTP, FSHTP and best FSHTP enhanced by a LEX device (in body and horizontal tailplane arrangement).

Table 3 Geometric parameters, reference HTP, FSHTP and Optimised FSHTP and LEX.

	HTP	FSHTP	Optimised FSHTP+LEX
Sweep angle	32 deg	-15 deg	-10 deg
S_H	31.36 m^2	31.36 m^2	29.45 m^2
$b_H/2$	6.723 m	6.723 m	6.723 m
Taper Ratio	0.36	0.715	0.620
Aspect Ratio	5.765	5.765	6.103
$(C_{LEX}+C_{rootH})/C_{rootH}$	---	---	1.804
b_{LEX}/b_H	---	---	0.217

References

[1] Anderson, J.D., Aircraft Performance and Design, WCB McGraw-Hill, Boston, MA, USA, 1999

[2] Sánchez-Carmona, A. and Cuerno-Rejado, C. and García-Hernández, L., Unconventional tail configurations for transport aircraft, Progress in Flight Physics 9 (2017) 127-148. https://doi.org/10.1051/eucass/2016090127

[3] Sánchez-Carmona, A. and Cuerno-Rejado, C., Vee-tail conceptual design criteria for commercial transport aeroplanes, Chinese Journal of Aeronautics 32 (2019) 595-610. Chinese Journal of Aeronautics. https://doi.org/10.1016/j.cja.2018.06.012

[4] CleanSky, New empennage for aircraft (NEFA), https://cordis.europa.eu/project/id/G4RD-CT-2002-00864

[5] Frota, J. and Nicholls, K. and Müller, M. and Gall, P.E. and Loerke, J. and Macgregor, K. and Schmollgruber, P. and Russell, J. and Hepperle, M. and Dron, S. and Plötner, K. and Gallant, G., Final Activity Report. New Aircraft Concept Research (NACRE), NACRE Consortium: Blagnac, France (2010). https://cordis.europa.eu/project/id/516068/reporting

[6] Llamas Sandin, R.C. and Luque Buzo, M., Aircraft Horizontal Stabilizer Surface, Google Patents (2010) US 2010/0148000 A1. https://patents.google.com/patent/US20100148000A1/en

[7] Seitz, A. and Kruse, M. and Wunderlich, T. and Bold, J. and Heinrich, L., The DLR Project LamAiR: Design of a NLF Forward Swept Wing for Short and Medium Range Transport Application, 29th AIAA Applied Aerodynamics Conference (2012). https://doi.org/10.2514/6.2011-3526

[8] Seitz, A. and Hübner, A. and Risse, K., The DLR TuLam project: design of a short and medium range transport aircraft with forward swept NLF wing, CEAS Aeronautical Journal 11 (2020) 449-459. https://doi.org/10.2514/6.2011-3526

[9] Roskam, J. and Lan, C.T.E., Airplane Aerodynamics and Performance, DARcorporation, Lawrence, Kansas 66044, USA, 1999

[10] Bertin, J.J. and Cummings, R.M., Aerodynamics for Engineers, Pearson Education International, Upper Saddle River, NJ, 2009

[11] AIcraft advanced rear end and eMpennage oPtimisaAtion enhanced by anti-iCe coaTings and devices (IMPACT), GA no. 885052. https://www.impact-cleansky-project.eu/

Towards multidisciplinary design optimization of next-generation green aircraft

Luca Pustina[1,a *], Matteo Blandino[1,b], Pietro Paolo Ciottoli[1,c] and Franco Mastroddi[1,d]

[1]Department of Mechanical and Aerospace Engineering, Sapienza University of Rome, Via Eudossiana 18, 00184 Rome, Italy

[a]luca.pustina@uniroma1.it, [b]matteo.blandino@uniroma1.it, [c]pietropaolo.ciottoli@uniroma1.it, [d]franco.mastroddi@uniroma1.it

Keywords: Green Aircraft, Multi-Disciplinary Optimization, Reduced Order Model

Abstract. Reducing greenhouse gas emissions is one of the most important challenges of the next future. The aviation industry faces increasing pressure to reduce its environmental footprint and improve its sustainability. This work is framed within the Italian national project "MOST- Spoke 1 - AIR MOBILITY - WP5," which studies innovative solutions for next-generation green aircraft. This paper proposes a multidisciplinary design optimization (MDO) framework for the design of new-generation green aircraft. Several propulsion solutions are analyzed, including fully electric and hydrogen fuel cells. The Multidisciplinary Design Optimization (MDO) framework considers several disciplines, including aerodynamics, structures, flight dynamics, propulsion, cost analysis, and life-cycle analysis for facing at the best the design challenge of next-generation green aircraft.

Introduction

For developing new-generation green aircraft, it is important to consider the interactions between the system's disciplines. Through early resolution of the multidisciplinary optimization (MDO) problem using state-of-the-art computational analysis tools, it is possible to enhance the design while concurrently minimizing the time and cost associated with the design cycle. In the realm of developing next-generation environmentally friendly aircraft, while the Maximum Take Off Mass (MTOM) remains a conventional optimization objective due to its strong correlation with the overall lifecycle cost of the aircraft, it is crucial to conduct a thorough-life cycle analysis and establish a metric for assessing the aircraft's overall environmental sustainability. Accurately modeling new-generation environmentally friendly propulsion systems and establishing a merit function for comparing the environmental friendliness of aircraft, such as the total equivalent CO2, is crucial. In addition to traditional disciplines like aerodynamics, structures, and flight dynamics, green propulsion system modeling and life cycle analysis are essential, even in early design stages. This study provides an overview of physical-based models for structural analysis, aerodynamics, green propulsion systems, and life cycle analysis for preliminary design. Moreover, a Multi-Disciplinary Optimization (MDO) architecture is presented, including considerations of Design Variables (DVs), constraints, and objective functions.

Structural modeling

Structural models are essential for optimizing aircraft performance, efficiency, and safety. They must withstand different loads and conditions while being lightweight to maximize range, payload capacity, and minimize operating costs. Various low fidelity structural models are used for aircraft optimization, such as analytical beam models. Beam models efficiently represent elongated components like wings, fuselages, and tails, considering bending, shear, torsion and axial loads. They provide insights into stress distribution, deflections, and dynamic response, assuming linear elasticity, small deflection theory, and homogeneous materials. Beam models are valuable for preliminary design, concept exploration, and rapid evaluation of structural configurations. However, with enhancements in computational capabilities, Finite Element Models (FEM) have gained popularity in aircraft optimization due to their versatility and accuracy. Moreover, FEM allows for a more detailed analysis of complex geometries. It is particularly beneficial for innovative aircraft configurations such as, for example, the blended wing body, and truss braced wings. In this work, a FEM model of the entire aircraft is generated using an in-house code called FUROR (Framework for aUtomatic geneRatiOn of aeRoelastic models) based on the input of assigned set of design variables. FUROR utilizes the open-source geometric library OpenCasCade to automatically generate the wing boxes and fuselage geometries, and the open-source code GMSH for automatic FEM grid generation. The generation of the aircraft FE model involves three steps. First, the aircraft geometry is defined based on main standard geometrical characteristics such as wingspan, dihedral, sweep, chord, and fuselage length (see Fig. 1). Additionally, the geometry of the main structural components, including wing spars, ribs, and stringers, is also defined. In the second step, a hybrid structured-unstructured FEM mesh is generated using GMSH. Finally, the complete FE aircraft model is generated, where the beam sections and shell thickness of the FEM model are defined in a standard Nastran input file. Furthermore, during this phase, the connections between the wings and the fuselage are established using a simplified approach that is compatible with the early design stage. Additionally, FUROR has the capability to generate the aerodynamic wing model, which will be discussed in detail in the following section.

Fig. 1: Generation of FE aircraft models with different main standard geometrical characteristics. In grey a standard regional aircraft, in blue an increased wingspan design and in red an increased wing sweep design.

Aerodynamic modeling

Aerodynamic modeling is crucial in the preliminary design for evaluating forces and mission performance. Various models exist with different accuracy and computational cost. Simplified physical-based aerodynamics models such as strip theory, doublet lattice, or vortex lattice methods are crucial in the field. Strip theory divides the wing into sections, evaluating the local angle of attack for each. Lookup tables determine each segment's lift and drag coefficients and the corresponding aerodynamic forces. A correction factor is applied for three-dimensional effects, and the Theodorsen method can be used for unsteady aeroelastic dynamic analyses. Strip theory enables the estimation of aerodynamic efficiency with drag estimation, also approximately incorporating viscous effects. However, strip theory lacks accuracy in estimating three-dimensional effects and wing interactions because simple analytical models are typically used in such cases. In preliminary design, the doublet lattice method is commonly employed as a three-dimensional model. It offers enhanced accuracy in lift evaluation, making it suitable for aeroelastic analysis and accounting for finite wake effects. In the FUROR software, both the strip theory approach and the doublet lattice method are available.

Propulsion Modeling

The proposed model for the propulsive system is based upon the work of De Vries et al. [1]. Authors aimed to propose a simple method for evaluating the benefits and technological needs of integrating a secondary energy source (batteries or fuel cells) with the primary thermal source (fuel) in the redesign of an existing aircraft. In this method, the various configurations are characterized by the supplied power ratio, which represents the ratio between the power output from the batteries and the total power output, taking into account the electric subsystem and the fuel. The proposed method combines constraint analysis, mission analysis, and subsequent aircraft mass estimation. The first step involves a classical constraint analysis that aims to meet all mission requirements by utilizing an aircraft propulsive power constraint diagram.

From this diagram [1], a specific design point is selected based on the power-to-weight ratio (P_p/W_{TO}) and wing loading (W_{TO}/S). Then, the next step consists of quasi-stationary mission analysis to size the aircraft in energy terms, having chosen a first attempt MTOM. A simplified mission profile is selected, consisting of different flight phases. Various authors [2,3] have derived analytical formulas similar to the Breguet equation or modified it for hybrid aircraft, but these formulas depend on the aircraft configuration, and there is no universally optimal choice. The presented design method explores different approaches for using battery and fuel in specific flight phases. The energy requirements from both fuel and battery sources are integrated over the entire mission duration. In the final phase, the aircraft's masses are estimated, including energy sources and propulsion system components. From these mass estimates and the mass outcomes of the structure subsystem, MTOM is derived, which corresponds to the energy required by the two energy sources to complete the mission. The last two steps of the method are iterated until the MTOM value of two consecutive iterations converges below a certain chosen tolerance. Lastly, a basic life cycle analysis should be conducted to estimate the total CO2 emissions. This analysis considers not only fuel emissions from gas turbine engines but also battery discharge for hybrid/full-electric aircraft, as well as the production and transportation of aircraft components. This approach ensures a fair comparison among the proposed configurations.

Multidisciplinary Design Optimization

The initial stage of aircraft design involves establishing the Top-Level Aircraft Requirements, e.g., desired overall performance, capabilities, and main characteristics of the aircraft. Traditionally, the optimization objective has focused on the Maximum Take Off Mass (MTOM) due to its close relationship with the overall lifecycle cost of the aircraft. A thorough life cycle analysis is crucial in developing eco-friendly aircraft to evaluate overall environmental sustainability. After defining the objective functions, the next step involves selecting the DVs and determining their ranges. Both structural and shape variables must be employed. Structural variables primarily focus on minimizing weight, while shape variables contribute to the optimization of aerodynamics which is of the utmost importance for the propulsion design to the evaluation of the energy requirements. Finally, to fully address the MDO problem, constraints need to be defined. This includes considering mission constraints such as range and altitude, conducting a stress assessment for extreme maneuver load cases to ensure aircraft safety, and incorporating constraints directly related to the propulsion system such as take-off distance and the estimated specific energy for the batteries and fuel cells that will be introduced in the weight estimation process. The design space can be systematically investigated by an optimizer (a possible choice is the Multi Objective Genetic Algorithm) to uncover the Pareto front of the selected objective functions (MTOM and total equivalent CO2). When the complete multidisciplinary analysis (MDA) is executed for every set of DVs, the MDO framework is referred to as Multi-Disciplinary Feasible (MDF) architecture [4]. This ensures the feasibility of each discipline for every set of DVs. The MDA involves conducting a trim and static structural analysis for various extreme maneuver load cases, followed by an aeroelastic analysis to ensure sufficient aeroelastic damping during maximum aircraft speed conditions (see Fig. 2). Additionally, the constraint and mission analysis are carried out to estimate the required battery pack (or fuel cells) mass necessary to accomplish the TLARs. Finally, the life cycle analysis is performed. To handle interdependencies among disciplines, an iterative approach is used, ensuring convergence of the MDA by matching the disciplines' output copy with the final output.

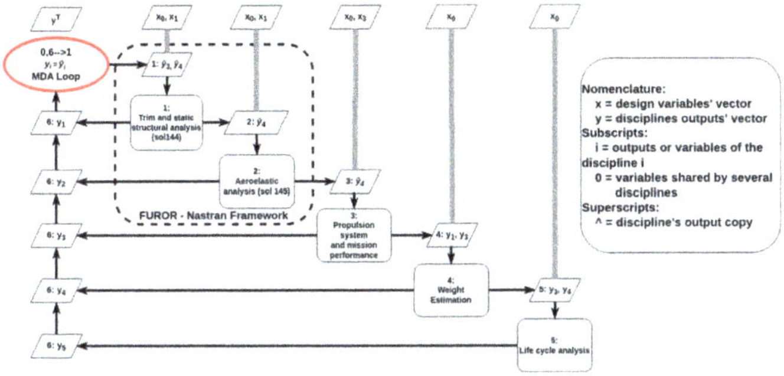

Fig. 2: Multi-Disciplinary Analysis Extended Design Structure Matrix [4]

Conclusions

This work presents an MDO architecture for advanced green aircraft, outlining the objective function, constraints, and design variables. Additionally, it describes the reduced order models (ROMs) for each discipline with appropriate fidelity for early design. The subsequent phases involve implementing the proposed MDO framework and conducting optimization on a reference regional aircraft, exploring various propulsion systems and innovative configurations (e.g., blended wing body, truss braced wing).

Acknowledgements

This work is part of the research activity developed by the authors within the framework of the "PNRR" CN4 MOST (Mobilità sostenibile): SPOKE 1 (Air Mobility), WP5: "Multidisciplinary design optimization and innovative solutions for next generation green aircraft with demonstrator".

References

[1] De Vries R., Brown M., and Vos R. Preliminary sizing method for hybrid-electric distributed propulsion aircraft. Journal of Aircraft, pages 2172–2188, 2019. https://doi.org/10.2514/1.C035388

[2] Voskuijl M., van Bogaert J., and Rao A. G. Analysis and design of hybrid electric regional turboprop aircraft. CEAS Aeronautical Journal, Vol. 9, pages 15–25, 2018. https://doi.org/10.1007/s13272-017-0272-1

[3] Wroblewski G. E. and Ansell P. J. A bréguet range equation for hybrid-electric jet aircraft sizing and analysis. AIAA Propulsion and Energy 2020 Forum, pages 1–24, 2020.

[4] Martins, Joaquim RRA, and Andrew B. Lambe. "Multidisciplinary design optimization: a survey of architectures." AIAA journal 51.9 (2013): 2049-2075. https://doi.org/10.2514/1.J051895

Improvements in on-board systems design for advanced sustainable air mobility

Claudia Conte[1,a*] and Domenico Accardo[1,b]

[1]Department of Industrial Engineering, University of Naples Federico II, Italy

[a]claudia.conte2@unina.it, [b]domenico.accardo@unina.it

Keywords: Model Based Systems Engineering, Embedded Systems, Unmanned Aerial Systems, U-Space

Abstract. This paper describes the activity proposed in the context of National Center for Sustainable Mobility (CN MOST) for designing an advanced core Guidance, Navigation, and Control system together with an effective on-board systems configuration for sustainable air mobility. A Model Based Systems Engineering strategy is adopted to support the design and development phases. The introduction of new sustainability objectives and the U-Space services to support the integration of unmanned air vehicles in the traditional Air Traffic Management drives the need of a full redesign of on-board systems that must be interfaced with different air platform categories. High performance processing units are considered for embedded systems, including but not limited to machine learning based, image processing and data fusion algorithms for advanced navigation. Three use-cases are presented as reference platform and mission types for validating the proposed systems configuration, specifically unmanned electric Vertical Take Off and Landing aircraft, fully electric general aviation aircraft, and hybrid-electric regional aircraft.

Introduction

The worldwide effort to increase technological improvement towards more sustainable air mobility is focusing on advanced propulsion systems, effective route planning, efficient airframe design and high-performance on-board Guidance, Navigation, and Control (GN&C) solutions [1-4]. Moreover, an innovative redesign of on-board equipment and systems is needed to meet new requirements and guarantee high safety levels according to the rapid changes of Air Traffic Management (ATM) that will include manned and unmanned vehicles with different specifications [5].

The need to integrate the so-called Unmanned Aerial Systems (UASs) with the existing traditional aviation is the focus of several programs [6]. The European response to this need has been implemented thanks to the definition of U-Space [7] for operations at Very Low Level (VLL) airspace in the context of the UAS Traffic Management (UTM). The definition of standard services - such as contingency management, traffic information, identification, collision avoidance, tracking - that allow to safely manage several vehicles with different tasks drives the design and development of specific on-board equipment with an increasing level of autonomy. Modular on-board systems configurations [8] can support the development of core enabling modules that can be designed for a wide range of vehicles also thanks to the use of low-cost Commercial Off The Shelf (COTS) processing units [9].

This paper aims at presenting an overview of the design activity focused on on-board systems configuration in the context of integrated ATM/UTM environment. Modular configurations are investigated to adapt innovative features of core GN&C systems to different air vehicles also assessing specific risk analysis evaluations for strategical and tactical mission management.

Overview of On-board Systems Design

The rapid evolution of air traffic requirements carries out the design and development of advanced on-board systems that must meet new mission profiles. Data fusion algorithms, Machine Learning based methods and modular configurations are the main drivers for innovative enabling technologies that support new generation air vehicles in complex environments.

A groundbreaking approach that follows all the design and development phases by using an advanced implementation method for systems modeling will support the adoption of advanced hardware and software components as well as effective real-time processing algorithms, such as the Model Based Systems Engineering (MBSE). The International Council On Systems Engineering (INCOSE) is developing several initiatives in the field of MBSE [10]. The National Aeronautics and Space Administration (NASA) started the MBSE Pathfinder in space topics [11]. The European Space Agency (ESA) founded the MB4SE and OSMOSE groups to introduce the MBSE concept [12].

The main aim of the proposed activity is the design of an advanced core GN&C system that can be adapted to different types of air vehicles and interfaced with the under development services of the integrated ATM/UTM. The general workflow description is reported in Fig.1 by following an MBSE approach [13] that helps to reduce the risk of unexpected redesign issues.

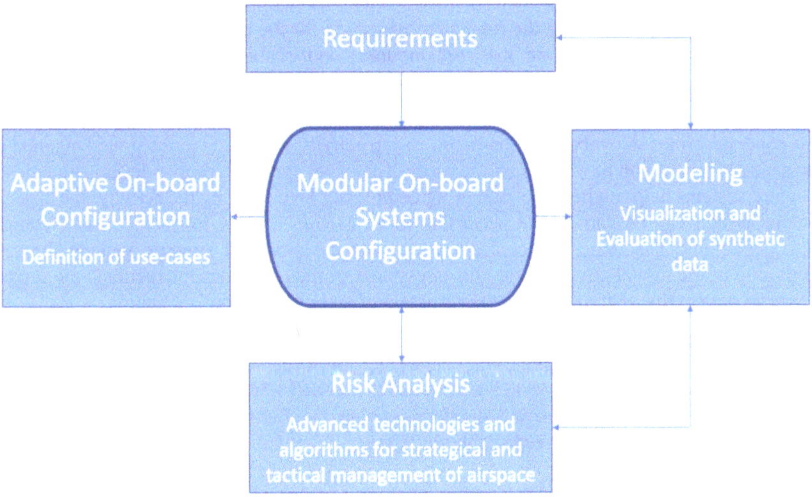

Fig. 1. Schematic description of the workflow for the proposed on-board systems design.

The requirements definition involves the selection of possible stakeholders and the regulatory framework analysis. Specific user needs must be highlighted, such as business goals, safety and security issues, and typical mission profiles must be identified. A model-based design strategy helps the development of a modular configuration and thanks to the evaluation and visualization in dedicated Human Machine Interfaces (HMIs) of synthetic output data, a risk analysis of the designed aerial system can be carried out to safely manage the mission at both strategical and tactical levels. The development of specific systems models allows to refine the initial requirements and associate each requirement to the related system module. Ad hoc use cases must be designed to validate the identified requirements and test the developed modular configuration in different scenarios.

Definition of Use Cases

The preliminary analysis of requirements for the proposed design solution involves the identification of reference platforms and missions. The aerial vehicles that are included in the analysis are: i) unmanned electric Vertical Take Off and Landing (eVTOL) aircraft, such as [14], ii) fully electric general aviation aircraft, such as [15], and iii) hybrid electric regional aircraft, such as [16].

The unmanned eVTOL platform is involved in the context of Smart Cities [17] for Urban Air Mobility (UAM) purposes. Typical mission profiles include package delivery and air-taxi operations.

The fully electric general aviation aircraft is one of the reference platforms for fully electric propulsion as well as electric on-board systems implementation. Personal transport and dedicated professional applications, such as surveillance and mapping are some relevant missions.

The hybrid electric regional aircraft involves the challenge of developing more electric on-board systems configuration for medium haul routes. The main designed task is related to commercial aviation transport with less than 100 seats also considering high efficiency sustainability features.

The mentioned platforms are the reference case study for the design of an adaptive on-board system configuration, identifying the core modules and enabling technologies to meet the developing requirements of air mobility.

Conclusion

This paper describes an overview of on-board design activity carried out in integrated manned and unmanned aviation with new needs and services. The development of proper design strategies for on-board systems helps address the increasing complexity level. During the activity, the on-board systems design can be carried out according to an MBSE strategy to manage more complex implementations. Tha main expected advantages of the proposed strategy include the rapid integration in the digital environment, the assessment of inter-modules interactions, the increase of overall system reliability level and the reduction of commercial implementation.

Acknowledgments

This study was carried out within the National Center for Sustainable Mobility (Centro Nazionale per la Mobilità Sostenibile CN MOST CN00000023 - PNRR - M4C2 Inv. 1.4). This manuscript reflects only the authors' views and opinions, neither the European Union nor the European Commission can be considered responsible for them.

References

[1] A. H. Epstein, S. M. O'Flarity, Considerations for Reducing Aviation's CO2 with Aircraft Electric Propulsion, Journal of Propulsion and Power, AIAA, 2019, vol 35:3, pp. 572-582, https://doi.org/10.2514/1.B37015

[2] V. Grewe, A. Gangoli Rao, T. Grönstedt, et al., Evaluating the climate impact of aviation emission scenarios towards the Paris agreement including COVID-19 effects, Nature Communications, 2021, 12, 3841, https://doi.org/10.1038/s41467-021-24091-y

[3] D. Eisenhut, N. Moebs, E. Windels, D. Bergmann, I. Geiß, R. Reis, A. Strohmayer, Aircraft Requirements for Sustainable Regional Aviation, Aerospace, 2021, 8, 61. https://doi.org/10.3390/aerospace8030061

[4] N. Avogadro, M. Cattaneo, S. Paleari, R. Redondi, Replacing short-medium haul intra-European flights with high-speed rail: Impact on CO2 emissions and regional accessibility, Transport Policy, 2021, vol. 14, pp. 25-39, https://doi.org/10.1016/j.tranpol.2021.08.014

[5] Information on https://www.sesarju.eu/newsroom/brochures-publications/state-harmonisation

[6] Information on https://www.faa.gov/uas/research_development/traffic_management.

[7] C. Barrado, M. Boyero, L. Brucculeri, G. Ferrara, A. Hately, P. Hullah, D. Martin-Marrero, E. Pastor, A.P. Rushton, A. Volkert, U-Space Concept of Operations: A Key Enabler for Opening Airspace to Emerging Low-Altitude Operations, Aerospace, 2020, 7, 24. https://doi.org/10.3390/aerospace7030024

[8] R. Nouacer, M. Hussein, H. Espinoza, Y. Ouhammou, M. Ladeira, R. Castiñeira, Towards a framework of key technologies for drones, Microprocessors and Microsystems, 2020, vol. 77, https://doi.org/10.1016/j.micpro.2020.103142.

[9] C. Conte, G. de Alteriis, G. Rufino and D. Accardo, An Innovative Process-Based Mission Management System for Unmanned Vehicles, 2020, IEEE 7th International Workshop on Metrology for AeroSpace (MetroAeroSpace), 2020, pp. 377-381, https://doi.org/10.1109/MetroAeroSpace48742.2020.9160121

[10] Information on https://www.incose.org/incose-member-resources/working-groups/transformational/mbse-initiative.

[11] Information on https://ntrs.nasa.gov/citations/20170009110

[12] Information on https://technology.esa.int/page/MBSE

[13] Information on https://www.mathworks.com/solutions/model-based-systems-engineering.html

[14] Information on https://flyingbasket.com/fb3

[15] Information on https://www.nasa.gov/specials/X57/

[16] Information on https://www.clean-aviation.eu/hybrid-electric-regional-aircraft

[17] Information on https://www.easa.europa.eu/en/domains/urban-air-mobility-uam

Recent developments about hybrid propelled aircraft: a short review

L.M. Cardone[1*a], S. De Rosa[1b], G. Petrone[1c], F. Franco[1d], C.S. Greco[1e]

[1] Department of Industrial Engineering, Università di Napoli Federico II, Via Claudio 21, 80125, Napoli, Italia

[a]luigimaria.cardone@unina.it, [b]sergio.derosa@unina.it, [c]giuseppe.petrone@unina.it, [d]francesco.franco@unina.it, [e]carlosalvatore.greco@unina.it

Keywords: Hybrid Propulsion Systems, Fuel Cell, Electric Aircraft

Abstract. Over the last decades, the rapid growth of the consumption of fossil fuel has generated an increased need for energy sustainability. The negative impact on the environment and now, as a global society, the dependence on fossil fuel have been questioned. For contrasting these adverse effects, NASA calls on the aeronautic industry to reduce aircraft fuel burn by 70% by 2025 in their N+3 concepts. Following high-profile government and industry studies, electric aircraft propulsion has emerged as an important research topic, this includes all-electric, hybrid electric, and turbo-electric architectures. The paper overviews the recent state-of-art about the innovative propulsion systems exploring the operating logic, their technological requirements, the ongoing research, and development in all the components necessary to make this technological change a feasible option for the future of passenger flight. It will be also reported the existing commercial products, prototypes, demonstrators for having a precise picture of the situation.

Introduction

Over the last decades, the economization of transportation has generated a sharp increase in fossil fuel consumption, causing a considerable increase in pollutant emissions. This increase in CO_2 emissions has prompted major industry companies and research institutions to develop new propulsion technologies for the aviation sector that are more efficient and have a reduced environmental impact. The new strong requirements about the emission pushed the academic and industrial researchers to investigates a new and sustainable type of engines for aerospace sector. Following the outcomes of these research Electric Propulsion Systems (EPS) and Hybrid Electric Propulsion System (HEPS) received a great emphasis and they have emerged as an important research topic and possible solution to satisfy the requirements imposed by NASA. In the next sections, the various innovative propulsion systems, identified in literature, will be discussed, labeling their advantages, disadvantages, and relative fields of application in relation to the type of aircraft that can accommodate them. Then the demonstrators will be reported that have been realize. In the end conclusion will be drawn.

Innovative Propulsion Systems

This section reports the main innovative propulsion systems identified in the literature. All the systems will briefly be introduced highlighting its advantages and disadvantages.

Sustainable Aviation Fuel SAF

Sustainable Aviation Fuel (SAF) can be defined as alternative fuel to conventional fossil-based jet fuel that is produced from either biological or non-biological sources. These fuels belong to the drop-in fuels class, characterized by having the same chemical and physical characteristics of the fuels used in aviation. This is a key aspect of its possible success because it allows the manufacturers do not make substantial modifications to the aircraft. The International Civil Aviation Organization (ICAO) reported that it would be physically possible to meet 100% of demand by 2050, which corresponds to a 63% reduction in emissions. So, it seems that SAF is a simple way to reduce CO_2 emission but there are also some disadvantages as the highly cost of production, that could be reduced only with a combination of policy incentives, capital investments, and time [1].

All Electric Propulsion Systems

All-electric aircraft configuration is the simplest of all the electric architectures in terms of layout, the power source, in this category is the rechargeable battery, connected directly to an electric motor through a power management system that drives a propeller. Globally this system is characterized by an higher efficiency compares with the classical propulsion system (83% vs 25% of the classical propulsion system). This category of aircraft has many advantages, including zero-local-emission, low noise, and operating cost reduction. The most critical disadvantage of this category of aircraft is the battery storage technology that has not achieved sufficient maturity to cover the same distances as jet fuel-driven aircraft [2].

An evolution of the classical battery storage system are the Fuel Cells propulsion system. This system is composed by three main components: the Fuel Cell, the hydrogen storage tank (gaseous or liquid), and a buffer battery. This category of aircraft is characterized by a much lower Maximum Take Off Weight (MTOW), as battery weight is almost eliminated, compared to an analogue battery storage aircraft and a similar refueling times and weights, comparable to the classical propulsion systems present on the market today. These systems are characterized by a great versatility since they do not require pre-storage of energy in batteries, but they only need to store a quantity of hydrogen required to accomplish the proposed mission. There are different types of F-Cells, a clever way to classified them is by the type of electrolyte used, the most common one used in the transport industry are the Proton Exchange Membrane (PEM) characterized by a low temperature of exercise, simplicity of manufacturing, reliability, and the competitive cost of production [3].

Hybrid-Electric Propulsion Systems HEPs

The implementation of full-electric propulsion systems, for classes of aircraft other than simple demonstrators, is critical because the technology of energy storage in batteries, as well as the realization of the various components associated with the distribution of energy on the aircraft, is still unsuitable for their realization, especially considering the deadlines imposed by the European Union. Instead, a more viable route is to go and complement the classical turbojet/turboprop systems with electric machines thus evaluating an additional propulsive methodology known as Hybrid Systems. This category includes 3 possible configurations examined: Hybrid Series configuration, Hybrid Parallel configuration, and Hybrid Series/Parallel configuration [4].

In a Series Hybrid configuration, the Electric Motor (EM) is the only component which drives the propeller, without any gearbox, this implies that the Internal Combustion Engine (ICE) can run always at its optimal regime. Due to this condition, fuel efficiency can remain high, and its lifespan can be lengthened. Another advantage is the flexibility of choosing the ICE location due to its the mechanical decoupling. The series configuration suffers of many disadvantages, such as massive power loss due the conversion system and an high cost of the components that have to be sizing to maximum power require during the mission.

In the Parallel Hybrid configuration type of system, the propeller is driven by both ICE and EM. From this configuration, multiply advantages are achieved. The most important advantage is that small ICE and small EM can be selected because all the main components of the system do not need to be sized to cope with maximum power requirements. Also, this configuration has some disadvantages, the main one is that it needs an efficiency energy management strategy, because it must optimize the power contribution of the ICE and the EM, to make them always work under the most efficient conditions.

The last analyzed configuration is the Hybrid Series–Parallel system. It is a mixture of two architectures show before, the main part of this configuration is the planetary gearbox at which the propulsion components, such as propellers, engine, EM, and generator, are connected to it. All the advantages and disadvantages of this system belong to the power-split part. The main advantage of using this device is that all the engine, ICE and EM, work in their maximum efficiency region. Despite this important advantage this type of system suffers of a constate dissipation rate due to the permanent connection between the part. These configurations are the most used architectures in aerospace and automotive sectors. Among them, the series configuration enables the engine to operate at its ideal operating condition, but as cited before, its system efficiency is relatively low since large power losses exist in the energy conversion. The series–parallel is the most functional, but the most complicated configuration out of the three architectures. It is the least popular configuration concerning aircraft application due to high complexity. it can therefore be inferred that, given current levels of technology, the parallel hybrid configuration appears to be the most versatile.

Flight Demonstrators

The attention is now focused on the attempts made by universities and research institutions in moving from purely theoretical models to prototypes able to fly, all the models have been classified according to their geometrical dimensions. With this parameter three categories are identified: small, medium, and large scale. The small scale is convenient for demonstrating the feasibility of the hybrid electric technology, several attempts were made by industries and research group, the most interested one is the Drone develop by Harris Aerial which use a fuel cell system with a small high-pressure tank of Hydrogen [5].

The medium scale is interesting because there is the opportunity to ensure the transportation of people. The most innovative flying configuration is the conversion kit installed on an ATR 72 proposed by ZeroAvia [6]. The last class of aircraft is the most critical one due to the technological limit of the component available. One interesting possible configuration is the one proposed by ZEROe program developed by Airbus which use two hybrid-hydrogen turboprop engines installed on an A350 [7].

So, the small/mid-scale sector, both academic and industrial institutes presented the flying or flying-capable demos. Furthermore, the studies of large hybrid aircrafts are still at the stage of concept designs and analysis due to the limitation of electrical and other technologies.

In Figure 1, it is interesting to observe that there is an area, indicated with the red square, are confined most of the different configurations before cited. It is evident that the maximum distance that can be covered is about 350 nm (648 km) while as far as the transport of passengers is concerned from the datasheets it has been found that these aircraft carry at most 4 people.

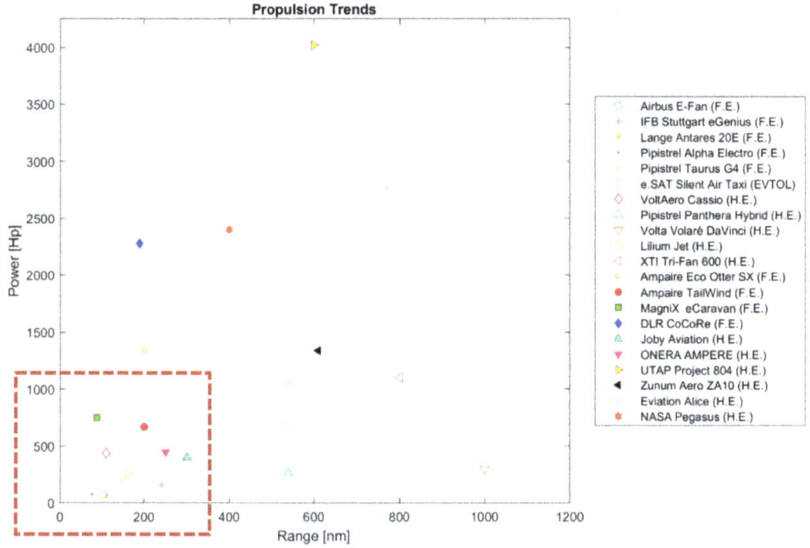

Figure 1 Propulsion Trends (F.E.: Full Electric; H.E.: Hybrid Electric)

Conclusion

The main purpose of this work is to quickly present the actual state of the art of new propulsion systems potentially adaptable to current aircraft configurations to achieve the CO_2 emission reduction targets set in 2050.

Without further technological development of batteries, it can be concluded that:

- Full-electric aircraft are implementable at most to the General Aviation category.
- Series/Parallel hybrid systems and fuel cells, characterized by a greater flexibility can be applied on a large scale of aircraft guaranteeing a partial abatement of CO_2.
- The main route remains SAF as they are solutions that can ideally be applied immediately since they do not require any modifications in the aircraft.

All the data reported in this short paper comes from a full paper under review [8].

Acknowledgments

This study was carried out within the MOST – Sustainable Mobility National Research Center and received funding from the European Union Next-Generation EU (PIANO NAZIONALE DI RIPRESA E RESILIENZA (PNRR) – MISSIONE 4 COMPONENTE 2, INVESTIMENTO 1.4 – D.D. 1033 17/06/2022, CN00000023). This manuscript reflects only the authors' views and opinions, neither the European Union nor the European Commission can be considered responsible for them.

References

[1] Ng, K.S., Farooq, D., Yang, A., 2021. Global biorenewable development strategies for sustainable aviation fuel production. Renew. Sustain. Energy Rev. 150, 111502. https://doi.org/10.1016/j.rser.2021.111502

[2] Brelje, B.J., Martins, J.R.R.A., 2019. Electric, hybrid, and turboelectric fixed-wing aircraft: A review of concepts, models, and design approaches. Prog. Aerosp. Sci. 104, 1–19. https://doi.org/10.1016/j.paerosci.2018.06.004

[3] Olabi, A.G., et alii, 2021. Fuel cell application in the automotive industry and future perspective. Energy 214, 118955. https://doi.org/10.1016/j.energy.2020.118955

[4] Singh, K.V., Bansal, H.O., Singh, D., 2019. A comprehensive review on hybrid electric vehicles: architectures and components. J. Mod. Transp. 27, 77–107. https://doi.org/10.1007/s40534-019-0184-3

[5] Harris Aerial, 2023. Carrier H6 Hybrid. https://www.harrisaerial.com/carrier-h6-hybrid-drone/

[6] ZeroAvia, 2022. ZA2000 2-5MW modular hydrogen-electric powertrain for 40-80 seat regional turboprops by 2027. https://zeroavia.com

[7] AIRBUS, 2022. ZEROe :Towards the world's first hydrogen-powered commercial aircraft, https://www.airbus.com/en/innovation/low-carbon-aviation/hydrogen/zeroe#:~:text=In%202022%2C%20we%20launched%20our,combustion%20propulsion%20system%20by%202025

[8] Cardone L. M., et alii, 2023. Review of the Recent Developments about the Hybrid Propelled Aircraft, submitted to Aerotecnica Missili e Spazio, under review.

Refined structural theories for dynamic and fatigue analyses of structure subjected to random excitations

Matteo Filippi[1,a *], Elisa Tortorelli[1,b], Marco Petrolo[1,c] and Erasmo Carrera[1,d]

[1]Department of Mechanical and Aerospace Engineering, Politecnico di Torino, Turin, Italy, 10129

[a]matteo.filippi@polito.it, [b]elisa.tortorelli@polito.it, [c]marco.petrolo@polito.it, [d]erasmo.carrera@polito.it

Keywords: Random Excitation, Refined Structural Model

Abstract. This paper presents the application of low- and high-fidelity finite beam elements to analyze the dynamic response of aerospace structures subjected to random excitations. The refined structural models are developed with the Carrera Unified Formulation (CUF), enabling arbitrary finite element solutions to be easily generated. The solution scheme uses power spectral densities and the modal reduction strategy to reduce the computational burden. The response of an aluminum box beam is studied and compared with a solution obtained by a commercial code. Considering the root-mean-square value of the axial stress, an estimation of the fatigue life of the structure is obtained.

Introduction

Fatigue is one of the prevalent causes of failure in structural and mechanical components. To correctly estimate the fatigue life of an element, it is essential to evaluate the stress distribution as accurately as possible. Time and frequency domain analyses can be employed to characterize the fatigue performance of a structure. In particular, the frequency domain approach is preferred thanks to its lower computational cost than direct integrations of the governing equation in the time domain. [1-2]. In particular, the Power Spectral Density (PSD) method is commonly used in structural dynamics and random vibration analysis [3]. The Finite Element (FE) method is a flexible and powerful tool for determining displacement and stress spectra. Previous works mostly adopted finite elements based on classical and first-order shear deformation theories [4-5]. While these kinematics expansions are suitable for various structural problems, they may not hold valid assumptions for other applications, such as laminated and thin-walled structures. Two- and three-dimensional (3D) FE formulations can be employed to address their limitations, but they often lead to a significant increase in computational costs. This study proposes an alternative approach by utilizing high-order finite beam elements. These elements provide an accurate and computationally efficient solution for predicting structural responses to random excitations. The Carrera Unified Formulation (CUF) enables the automatic implementation of various kinematic models through a recursive notation. Specifically, the response of a clamped-free box beam is given in terms of PSD and root-mean-square (RMS) of displacements and stresses, and a fatigue life prediction is provided given the value of RMS of axial stress.

One-dimensional finite elements

The one-dimensional (1D) model adopted in this work is based on the Carrera Unified Formulation (CUF). According to the CUF, the 3D displacement field of a solid beam with main dimension along the y-axis, can be expressed as a generic expansion of the generalized displacements $u_\tau(y,t)$:

$$\mathbf{u}(x,y,z,t) = F_\tau(x,z)N_i(y)\mathbf{u}_\tau(y,t), \qquad \tau = 1,2,\dots,M \text{ and } i = 1,\dots,nsn \tag{1}$$

In this equation, M represents the number of terms used in the expansion, while nsn represents the number of structural nodes of a single Finite Element (FE). Repeated subscripts indicate summation, $N_i(y)$ refers to the 1D FE shape functions, and $F_\tau(x, z)$ represents arbitrary functions on the cross-section. In this study, we adopt the Taylor (TE) and Lagrange (LE) expansion classes as polynomial bases. By introducing the Principle of Virtual Displacement (PVD) $\delta L_{int} = \delta L_{ext}$, it is possible to derive finite element matrices and vectors by assembling the so-called *Fundamental Nuclei*. These nuclei are the invariant of the methodology.

Theory of random response and fatigue life prediction

Using the Fourier transform of the equation of motion, it is possible to obtain the equation in frequency domain:

$$\mathbf{q_k}(\boldsymbol{\omega}) = [-\boldsymbol{\omega}^2\mathbf{M} + \mathbf{i}\boldsymbol{\omega}\mathbf{C} + \mathbf{K}]^{-1}\mathbf{F}_\mathbf{k}^* \qquad \mathbf{i} = \sqrt{-1} \tag{3}$$

Where $\mathbf{q_k}$ is the column vector that collects the degrees of freedom (DOF) of the FE model, k is an arbitrary non-null generalized coordinate, $\mathbf{F}_\mathbf{k}^*$ is the generalized force vector in frequency domain and it has only one nun-null term (equal to 1).

To reduce the computational cost, it is a common practice to employ a modal reduction strategy. The Power Spectral Density (PSD) function of a signal gives an indication of the average power contained in particular frequencies and the root mean square (RMS) represents the square root of the area below PSD curve. Given the input PSD function of the load $\mathbf{S_F}$, the output PSD of the three-dimensional displacement $\mathbf{S_u}$ and the stress $\mathbf{S_\sigma}$ components at various frequencies $(\boldsymbol{\omega})$ are obtained:

$$\mathbf{S}_{\mathbf{u_i}}(\boldsymbol{\omega}) = \overline{\mathbf{H}}_{\mathbf{u_i}}(\boldsymbol{\omega})\mathbf{S_F}(\boldsymbol{\omega})\mathbf{H}_{\mathbf{u_i}}^{\mathbf{T}}(\boldsymbol{\omega}) \qquad \mathbf{i} = \mathbf{1, 2, 3}$$

$$\mathbf{S}_{\sigma_\mathbf{j}}(\boldsymbol{\omega}) = \overline{\mathbf{H}}_{\sigma_\mathbf{j}}(\boldsymbol{\omega})\mathbf{S_F}(\boldsymbol{\omega})\mathbf{H}_{\sigma_\mathbf{j}}{}^{\mathbf{T}}(\boldsymbol{\omega}) \qquad \mathbf{j} = \mathbf{1}, \dots, \mathbf{6} \tag{4}$$

where $\overline{\mathbf{H}}(\boldsymbol{\omega})$ and $\mathbf{H}^\mathbf{T}(\boldsymbol{\omega})$ are the complex conjugate and the transpose of the transfer function and it can be computed with the FE method by performing as many frequencies response analysis as of the non-null terms (**nnz**) in the generalized force vector **F**:

$$\mathbf{H}_{\mathbf{q_k}}(\boldsymbol{\omega}) = \left[\mathbf{q}_{\mathbf{k}_1}\ \mathbf{q}_{\mathbf{k}_2}\ \cdots\ \mathbf{q}_{\mathbf{k}_L}\right] \qquad \mathbf{k} = \mathbf{1}, \dots, \mathbf{nnz} \qquad \mathbf{L} = \mathbf{1}, \dots, \mathbf{fs} \tag{5}$$

where **q** is derived from Eq. (3) and fs is the number of frequency steps. In this work, the structure is subjected to white noise excitations, thus the PSD of this type of noise is constant.

Numerical results

The numerical example refers to a clamped-free box beam made of aluminum alloy with E = 71.7 GPa, G= 27.6 GPa, ν= 0.3, ρ = 2700 kgm^{-3} and dimensions L=2.00 m, h = 0.05 m, b = 0.25 m, t=0.01 m. The beam is subjected to three-point loads (1 N) as shown in Figure 1.

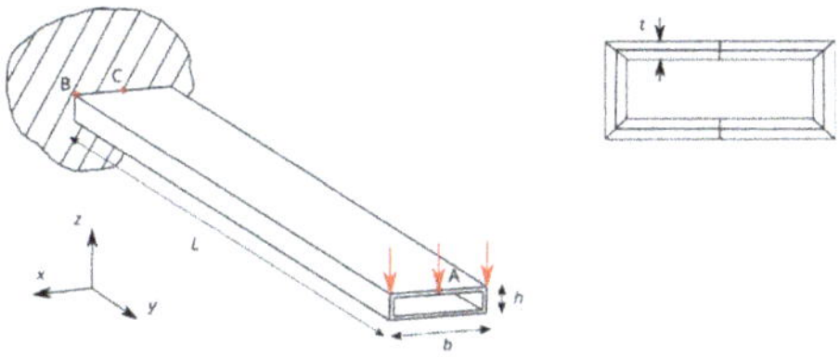

Figure 1. Boundary conditions and geometry of the clamped-free box beam subjected to clipped white noise. Scheme of the cross-section.

The structure was discretized using ten cubic beam elements. The first ten natural frequencies of the structure (Table 2) and the mass participation (Figure 2) were evaluated by conducting a normal mode analysis. The results have been obtained with the 12-LE9 model consisting of two Lagrange-type bi-quadratic elements for each lateral edge and four for the top and bottom surfaces. In the following analyses, forty modes were employed.

Table 1. First six natural frequencies obtained with NASTAN and CUF-FEM approach.

12-LE9	NASTRAN
13.90	13.88
56.13	56.17
84.03	83.09
178.68	170.8
223.17	216.08
324.78	324.67
406.34	378.13
461.23	424.19
613.37	539.52
629.52	563.01

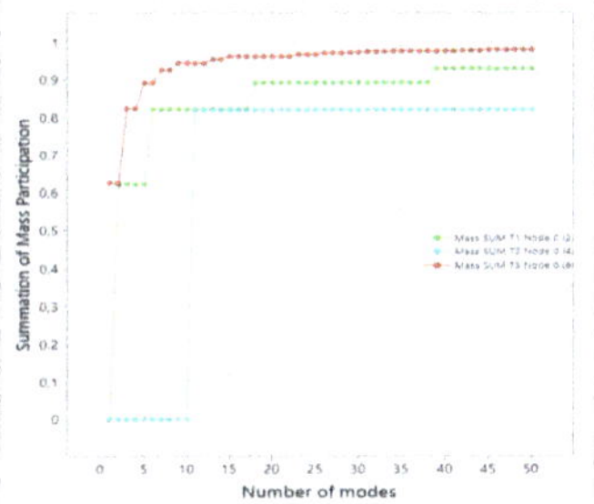

Figure 2. Mass participation versus number of modes of the box beam response.

Figure 3 shows point A's vertical displacement PSD and point C's axial stress PSD. In Figure 4, the distribution along the thickness corresponding to Point C of the root mean square of the axial stress is shown. Statistically speaking, the RMS stress value represents the 1σ value and will be experienced 68.3% of the time. A 2σ will be experienced 27.1% of the time, and a 3σ will be experienced 4.33%. These values represent 99.73% of the stresses the beam will experience at point C. Using Miner's cumulative damage $R_n = \frac{n_1}{N_1} + \frac{n_2}{N_2} + \frac{n_3}{N_3}$, it possible to obtained n, which is the number of cycles to fail: $1 = \frac{0.6831n}{N_1} + \frac{0.271n}{N_2} + \frac{0.0433n}{N_3}$. If the beam is vibrating at a frequency of 13.9 Hz (first natural frequency), then it will take approximately 2769 hours to fail. The results show that the 1D CUF approach is a valuable alternative to the common the 3D approach used by commercial codes.

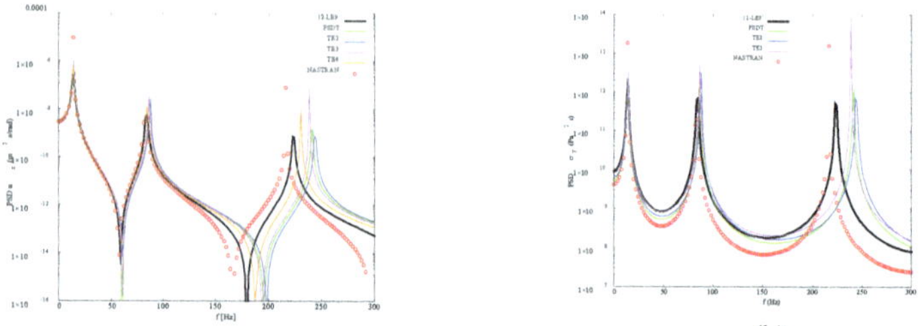

(a) (b)

Figure 3. PSD of: (a) vertical displacement of point A, (b) axial stress of point B.

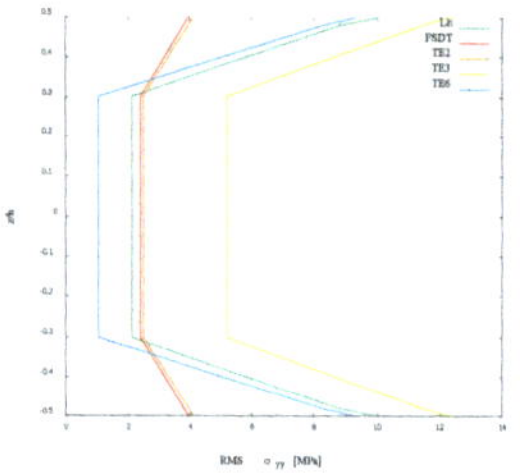

	1 RMS	2 RMS	3 RMS
S	10.06	20.12	30.18
N	4.3E10	1.8E8	7.6E6

Figure 4. RMS distributions of the axial stress along the thickness. On the right, a table with values taken from a fatigue curve of aluminium. For a given stress in [MPa], the number of cycles needed to cause failure is given.

Acknowledgments
This study was carried out within the MOST – Sustainable Mobility National Research Center and received funding from the European Union Next-Generation EU (PIANO NAZIONALE DI RIPRESA E RESILIENZA (PNRR) – MISSIONE 4 COMPONENTE 2, INVESTIMENTO 1.4 – D.D. 1033 17/06/2022, CN00000023). This manuscript reflects only the authors' views and opinions, neither the European Union nor the European Commission can be considered responsible for them.

References

[1] Muñiz-Calvente M. A comparative review of time- and frequency-domain methods for fatigue damage assessment. International Journal of Fatigue, 163, 107069 (2022). https://doi.org/10.1016/j.ijfatigue.2022.107069

[2] Halfpenny, A. A frequency domain approach for fatigue life estimation from finite element analysis. Key Engineering Materials, 167, pp 401-410 (1999). https://doi.org/10.4028/www.scientific.net/KEM.167-168.401

[3] Rafiee R., Sharifi P. Stochastic failure analysis of composite pipes subjected to random excitation. Construction and Building Materials, 224, pp 950-961 (2019). https://doi.org/10.1016/j.conbuildmat.2019.07.107

[4] Olson M. D., A consistent finite element method for random response problems. Computers & Structures, 2, pp 163-180 (1972). https://doi.org/10.1016/0045-7949(72)90026-0

[5] Dey, S. S., Finite element method for random response of structures due to stochastic excitation, Computer Methods in Applied Mechanics and Engineering, Vol. 20, No. 2, pp. 173–194 (1979). https://doi.org/10.1016/0045-7825(79)90016-1

Satellite and Space Systems

Facility for validating technologies for the autonomous space rendezvous and docking to uncooperative targets

M. Sabatini[1,a*], G.B. Palmerini[1,b], P. Gasbarri[1,c], F. Angeletti[1,d]

[1]Scuola di Ingegneria Aerospaziale, Sapienza University of Rome, Via Salaria 851 Rome

[a]marco.sabatini@uniroma1.it, [b]giovanni.palmerini@uniroma1.it, [c]paolo.gasbarri@uniroma1.it, [d]federica.angeletti@uniroma1.it

Keywords: Autonomous Proximity Operations; Relative Navigation; Free-Floating Platforms

Abstract. We present the latest advancements in the air-bearing facility installed at La Sapienza's GN Lab in the School of Aerospace Engineering. This facility has been utilized in recent times to validate robust control laws for simultaneous attitude control and vibration active damping. The instrumentation and testbed have been restructured and enhanced to enable simulations of close proximity operations. Relative pose determination, accomplished through visual navigation as either an auxiliary or standalone system, is the first building block. Leveraging the acquired knowledge, optimal guidance and control algorithms can be tested for contactless operations (e.g. on-orbit inspection), as well as berthing and docking tasks.

Introduction

Rendezvous and docking maneuvers have been successfully performed since the early days of spaceflight, but still pose challenges due to the increasing complexity of the mission scenario. From the early missions that required a human pilot, the goal of autonomous operations has already been achieved with notable examples in the ISS cargo resupply.

The current state of the art in proximity operations still involves some form of cooperation between the maneuvering satellite (the Chaser) and the Target satellite in the form of aids to relative navigation or in the form of mechanical interfaces to facilitate the docking phase. In either case, the Target satellite is usually controlled and fully operational. Much research is being done to extend these operations to a larger class of objects that are noncooperative and uncontrolled. Applications range from debris removal (see [1]) to orbital maintenance operations to repair or extend the operational life of a satellite.

Several missions have been developed to demonstrate the technology required for rendezvous and proximity operations (see the AVANTI demonstration [2] or the ELSA -D mission [3]).

A fundamental step in increasing the TRL of the required technologies is to conduct ground experiments to validate the components in a laboratory environment (typically TRL 4/5). For example, the work in[4] focuses on experimental verification by implementing and comparing real-time guidance algorithms on the Floating Spacecraft Simulator (FSS) testbed; a similar air-bearing testbed is used in [5] to validate the effectiveness of relative pose measurement systems. Various experimental approaches (e.g., using drones) can be used to test guidance and navigation subsystems (see [6]).

One of the most critical technologies to consolidate for successful close-proximity operations is the ability of an active spacecraft to accurately estimate its relative position and attitude (pose) with respect to an active/inactive Target (see [7]).

At the Guidance and Navigation Laboratory (GN Lab) of the School of Aerospace Engineering at La Sapienza, University of Rome, we have developed over the last decade a test bed consisting of a free-floating platform that can maneuver in a frictionless environment thanks to ON-OFF thrusters. The test bed, now in its second stage of development, has been used in the recent past to validate the active robust control system for attitude control of large flexible satellites as part of an ESA ITT study.

This paper shows the current equipment of the facility, consisting of two free-floating platforms (with a fleet to be expanded in the near future), a very accurate external measurement system (providing position and attitude measurements of the platforms with an accuracy of 0.1 mm and 0.01 degrees, respectively), and large screens for the projection of simulated orbital views that are changed in real time according to the motion of the platform.

With this configuration, it is now possible to validate critical short-range subsystems, such as visual navigation to determine the position of an uncooperative target in difficult lighting conditions and with limited computing power. Different algorithms can be tested for long-range (in which case a simple angle algorithm can be used) and short-range scenarios (feature-based algorithms or AI techniques are viable options). Maneuvers based on this information can eventually lead to docking, a phase where experimental validation of the behavior of the mechanisms and platforms during contact dynamics in a reactionless environment is critical. Successful validation of the critical aspects of GNC in a laboratory environment will pave the way for the realization of space-based enablers using microsatellites.

The first platform: early development of PINOCCHIO

The GN Lab has been developing a free-floating platform since 2012 (see [8]). In its first version, the platform, named PINOCCHIO (Platform Integrating Navigation and Orbital Control Capabilities Hosting Intelligence Onboard), was a 10-kg class platform, that utilized low-pressure air (10 atm) stored in an onboard tank for generating a thin film of air removing the friction between the air-bearings and the working table. It incorporated a second onboard tank supplying eight nozzles, serving as cold gas thrusters to enable horizontal movement and rotation around the vertical axis (yaw). Originally, a glass table (Fig. 1 - left) was used, but it was later upgraded to a black granite table due to its improved dimensional stability and operational workspace. PINOCCHIO operated autonomously, with its onboard computer managing all functions. The modular architecture allowed for the integration of various avionics components, including accelerometers, gyros, cameras, and lab star sensors, to instrument the guidance and control systems for both position and attitude. The platform's bus, which closely resembled the size and mass properties of a real microsatellite, proved instrumental in numerous studies conducted by the Sapienza team in recent years, as referenced in [9][10][11]. These studies encompassed areas such as rendezvous and docking guidance, optical navigation, and combined control of the platform's attitude during robotic arm motion (as depicted in Fig. 1). Additionally, the team explored the effects of flexibility in attitude control resulting from elastic appendages.

The need to further investigate the research field of attitude control of very large and flexible satellites using a multi-input multi-output control with attitude and elastic sensors and actuators, in the framework of the ESA ITT EXPRO-PLUS project "ACACLAS" (Advanced Collocated Active Control of Large Antenna Structures), called for the design, realization and characterization of a new platform to increase the performance and the capabilities of the test rig.

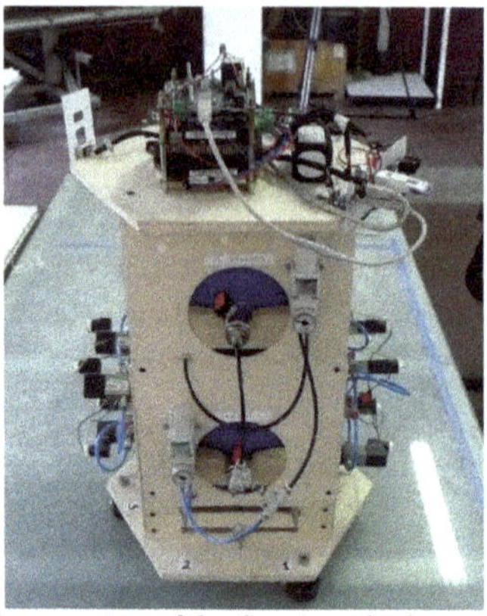

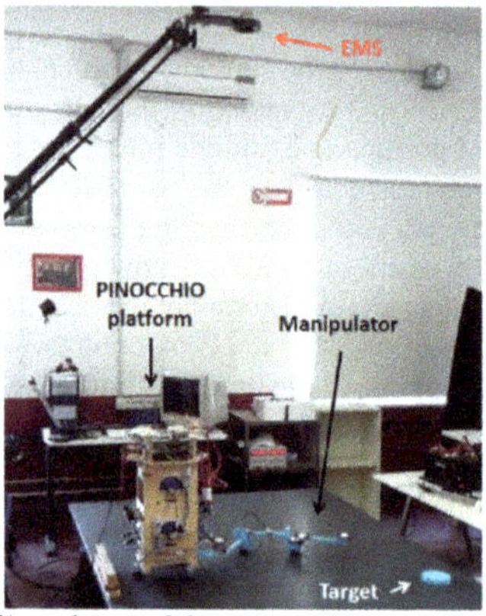

Fig. 1 The first version of the floating platform (left) and one of the experiments focused on the use of space robotic manipulators (right)

The second platform: a special focus on attitude control of highly flexible satellites

A new platform has been developed to enhance the performance of the initial platform in several aspects:

(a) Increasing the bus rigidity
(b) Extending the duration of experiments
(c) Including a dedicated electronics module for the payload
(d) Equipping the test rig with a high-accuracy metrology system.

The floating platform's structure has been completely redesigned using an Aluminum alloy (UNI 6060). It has overall dimensions of 300 mm × 350 mm × 480 mm (height) and consists of three compartments, which can be expanded due to its modular design, catering to specific purposes. The lower compartment houses the pneumatic system, while the second floor (300 mm × 350 mm × 100 mm) accommodates the electronics, including sensors, microcontrollers, power supplies, and wiring (see Fig. 2 - left). To achieve the required long experiment duration, a high-pressure tank with an operational pressure of 200 atm is used (see Fig. 2 - right). The third floor provides additional modules for specific missions. Currently, it houses the electronics required to control flexible appendages, such as solar panels and antennas (see Fig. 3 - left). These structures can be easily attached when experiments necessitate active control of flexible appendages. In such cases, a smaller, more precise working table is utilized since attitude maneuvers are typically the main focus (see Fig. 3 - right).

For accurate measurements of the structure's attitude and elastic displacement, a VICON system is employed. This technique utilizes retroreflective markers that are tracked by six infrared cameras. A detailed characterization of the platform can be found in reference [13], demonstrating navigation accuracy below 100 μm for position and 0.003 deg for attitude.

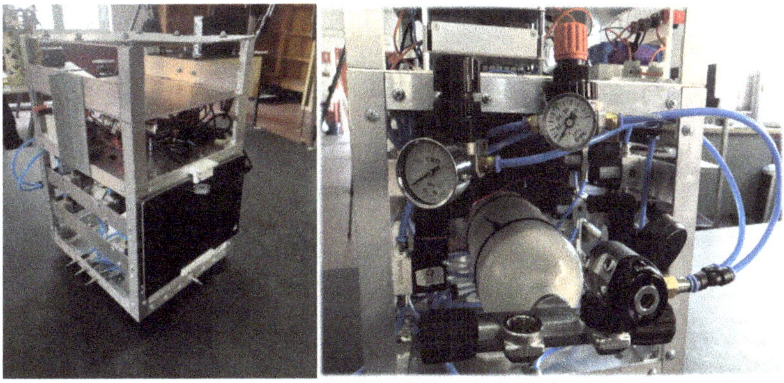

Fig. 2 The new version of the floating platform (left) with the high-pressure systems for air-bearings and thrusters (right).

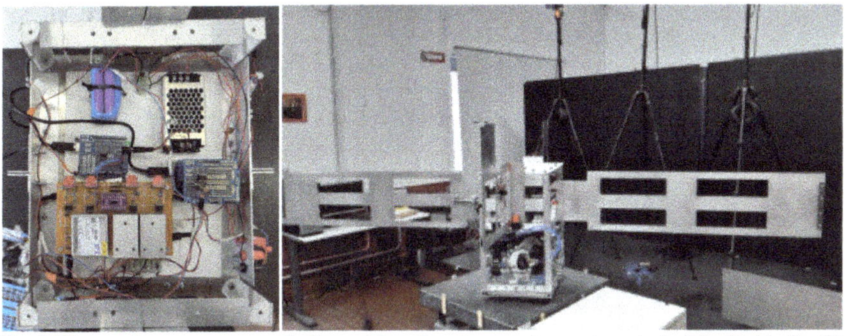

Fig. 3 Electronics compartment (left) used for vibration sensing and control of the highly flexible appendages (right).

A testbed for autonomous rendezvous and docking

The two platforms and the external metrology system have been now repurposed at the scope of testing guidance, control and, more specifically, optical navigation for the determination of the relative pose of a Chaser satellite with respect to a Target satellites. Active electro-optics systems can be used and tested, showing a promising performance (see [12]); these systems can be aided or partially substituted by vision systems (monocular or stereovision). For a hardware-in-the-loop test, the orbital environment should be reconstructed also from a visual point of view. To achieve this, an array of screens has been arranged to project realistic potential background images (Fig. 4 - left depicts the currently installed first screen). This allows the Chaser satellite's acquired image to include not only relevant Target satellite features but also various potential sources of visual disturbance, such as Earth landscapes, clouds, or stars (see Fig. 4 - right). Additionally, a lamp simulating the sun's position can be used to account for the constraint of avoiding blinding of the Chaser's cameras.

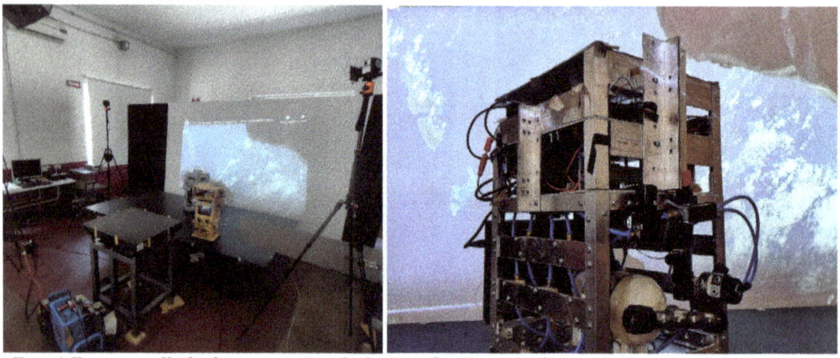

Fig. 4 First installed white screen with the two floating platforms (left) and a snapshot acquired by the navigation systems on board the Chaser satellite (right).

The onboard navigation systems provide relative pose measurements that must be validated thanks to benchmark metrology systems. The VICON system already mentioned in the previous section is a powerful mean in this sense. The inertial position of both platforms can be computed in real-time with a maximum frequency of 300 Hz, and with an accuracy which is much better than what achievable with the onboard systems. Fig. 5 reports an example of the visualization and the stream of data available with such a system, where each platform is identified by a different pattern of reflective markers.

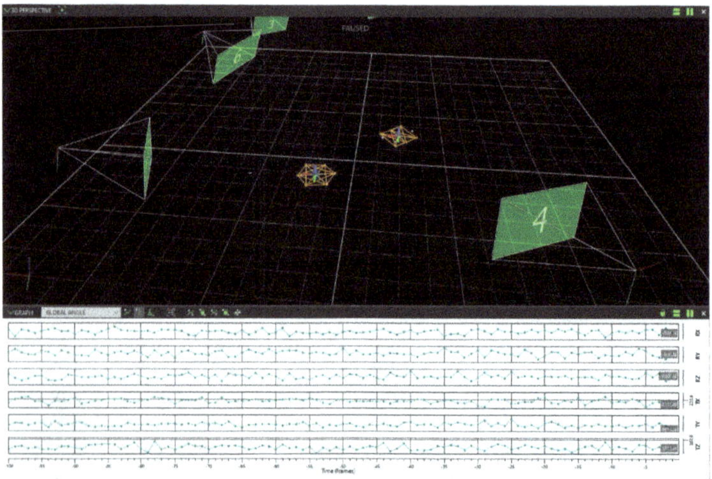

Fig. 5 Chaser and Target platforms acquired by the external metrology system

This set up can be used not only for visual based navigation, but also for testing range finders (LIDAR or radio frequency) and systems emulating differential GPS measurements. Relative pose determination is one of the building blocks of the autonomous rendezvous and docking operations. While it can be also tested using robotic arms (see [14]), with the air-bearing approach also the guidance and control algorithms can be tested in closed loop, including the contact forces and the post contact dynamics which are among the most difficult phases to simulate. The position and attitude control of the single platforms are accurate enough (see Fig. 6 for an example of an attitude slew maneuvers of 30 deg, with stationary error of 0.1 deg) to allow for a complete docking, including the validation of the performance of mechanical hardware.

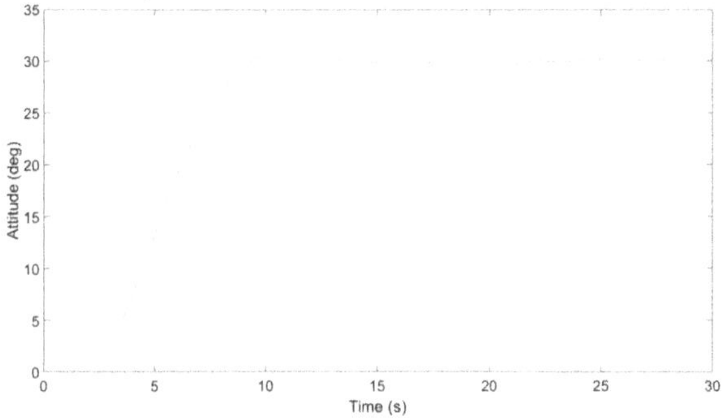

Fig. 6 Example of an attitude slew maneuver

Conclusions

We have showcased the key features and performance of an experimental testbed that has played a crucial role in validating attitude control algorithms for highly flexible spacecraft. Currently, we are expanding its capabilities to facilitate testing of space proximity operations, specifically focusing on control and relative navigation. We encourage the academic community to explore this platform for validating and testing various building blocks of the autonomous formation flying guidance, navigation, and control architecture, in the framework of future fruitful collaborations.

References

[1] C. Bonnal, J.-M. Ruault, M.-C. Desjean, "Active debris removal: Recent progress and current trends" Acta Astronautica 85 (2013) pp. 51–60. https://doi.org/10.1016/j.actaastro.2012.11.009

[2] G. Gaias and J.-S. Ardaens, "Flight Demonstration of Autonomous Noncooperative Rendezvous in Low Earth Orbit", Journal Of Guidance, Control, And Dynamics Vol. 41, No. 6, June 2018. https://doi.org/10.2514/1.G003239

[3] J. Forshaw, S. Iizuka, C. Blackerby, and N. Okada, " ELSA-d – A novel end-of-life debris removal mission: mission overview, CONOPS, and launch preparations", First Int'l. Orbital Debris Conf. (2019)

[4] R. Zappulla II, H. Park, J. Virgili-Llop and M. Romano, "Experiments on Autonomous Spacecraft Rendezvous and Docking Using an Adaptive Artificial Potential Field Approach", AAS 16-459

[5] Z. Wei, H. Wen, H. Hu, D. Jin, "Ground experiment on rendezvous and docking with aspinning target using multistage control strategy" Aerospace Science and Technology 104 (2020) 105967. https://doi.org/10.1016/j.ast.2020.105967

[6] T. Mahendrakar et al., "Autonomous Rendezvous with Non-Cooperative Target Objects with Swarm Chasers And Observers", AAS 23-423

[7] R. Opromolla, G.Fasano, G. Rufino, M. Grassi, "A review of cooperative and uncooperative spacecraft pose determination techniques for close-proximity operations", Progress in Aerospace Sciences 93 (2017) 53–72. https://doi.org/10.1016/j.paerosci.2017.07.001

[8] M. Sabatini, M. Farnocchia, G. Palmerini, "Design and Tests of a Frictionless 2D Platform for Studying Space Navigation and Control Subsystems", paper7.1406, IEEE Aerospace Conference 2012, BigSky, Montana, USA. https://doi.org/10.1109/AERO.2012.6187259

[9] M. Sabatini, G. B. Palmerini, P. Gasbarri, "Synergetic approach in attitude control of very flexible satellites by means of thrusters and PZT devices", Aerospace Science and Technology, Volume 96, 2020. https://doi.org/10.1016/j.ast.2019.105541

[10] M. Sabatini, P. Gasbarri, G. B. Palmerini, Coordinated control of a space manipulator tested by means of an air bearing free floating platform, Acta Astronautica, Volume 139, October 2017, Pages 296-305. https://doi.org/10.1016/j.actaastro.2017.07.015

[11] M. Sabatini, G. B. Palmerini, P. Gasbarri, "A Testbed For Visual Based Navigation And Control During Space Rendezvous Operations", Acta Astronautica, Volume 117, 1 December 2015, Pages 184-196. https://doi.org/10.1016/j.actaastro.2015.07.026

[12] A. Nocerino et al., "Experimental validation of inertia parameters and attitude estimation of uncooperative space targets using solid state LIDAR", Article in press, ttps://doi.org/10.1016/j.actaastro.2023.02.010

[13] M. Sabatini, P. Gasbarri, G. B. Palmerini, "Design, realization and characterization of a free-floating platform for flexible satellite control experiments", Acta Astronautica, 2023, in press. https://doi.org/10.1016/j.actaastro.2023.05.007

[14] R. Volpe et al., "Testing and Validation of an Image-Based, Pose and Shape Reconstruction Algorithm for Didymos Mission", Aerotec. Missili Spaz. 99, 17–32 (2020). https://doi.org/10.1007/s42496-020-00034-6

Materials Research Proceedings 37 (2023) 465-468 https://doi.org/10.21741/9781644902813-102

Simulation of in-space fragmentation events

Lorenzo Olivieri[1,a*], Cinzia Giacomuzzo[1,b], Stefano Lopresti[1,c] and Alessandro Francesconi[2,d]

[1] CISAS "G. Colombo", University of Padova, Via Venezia 15, 35131 Padova, Italy

[2] DII/CISAS, University of Padova, Via Venezia 1, 35131 Padova, Italy

[a]lorenzo.olivieri@unipd.it, [b]cinzia.giacomuzzo@unipd.it, [c]stefano.lopresti@unipd.it, [d]alessandro.francesconi@unipd.it

Keywords: Space Debris, Fragmentation, Break-Up, Numerical Simulations

Abstract. In the next years the space debris population is expected to progressively grow due to in-space collisions and break-up events; in addition, anti-satellite tests can further affect the debris environment by generating large clouds of fragments. The simulation of these events allows identifying the main parameters affecting fragmentation and generating statistically accurate populations of generated debris, both above and below detection thresholds for ground-based observatories. Such information can be employed to improve current fragmentation models and to reproduce historical events to better understand their influence on the non-detectable space debris population. In addition, numerical simulation can also be employed to identify the most critical object to be removed to reduce the risk of irreversible orbit pollution. In this paper, the simulation of historical in-orbit fragmentation events is discussed and the generated debris populations are presented. The presented case-studies include the COSMOS-IRIDIUM collision, the COSMOS 1408 anti-satellite test, the 2022-151B CZ-6A in-orbit break-up, and a potential collision of ENVISAT with a spent rocket stage; for these events, results are presented in terms of cumulative fragments distributions and debris orbital distributions.

Introduction

The increasing number of objects resident in Earth orbits is leading the debris environment dangerously close to the Kessler Syndrome, i.e. to a condition of self-sustained cascade impacts and break-ups that would strongly reduce the access and exploitation of near-Earth space [1]. Mitigation techniques and strategies to reduce the hazard of space debris are under evaluation by the scientific community and the main stakeholders [2]; however, it is still crucial to understand the physical processes involved in spacecraft collisions and fragmentations. Data on spacecraft breakup can be acquired by the observation of in-space fragmentation events [3-4], the execution of ground tests [5-6], and the performing of numerical simulations [7-8].

In this context, the University of Padova has developed the Collision Simulation Tool Solver (CSTS) to numerically evaluate in-space fragmentation events [9-10]. In the tool (see Fig. 1), the colliding bodies are modelled with a mesh of Macroscopic Elements (MEs) that represent the main parts of the satellite; structural links connect them forming a system-level net. In case of collision, the involved MEs are subjected to fragmentation, while structural damage can be transmitted through the links; this approach can be propagated through a cascade effect representative of the object fragmentation, allowing the simulation of complex collision scenarios and producing statistically accurate results.

In this work, the CSTS is employed to replicate three fragmentation events observed in orbit and the potential breakup of ENVISAT due to the collision with a spent rocket stage. For each case, a brief description of the model and the main simulation results are presented.

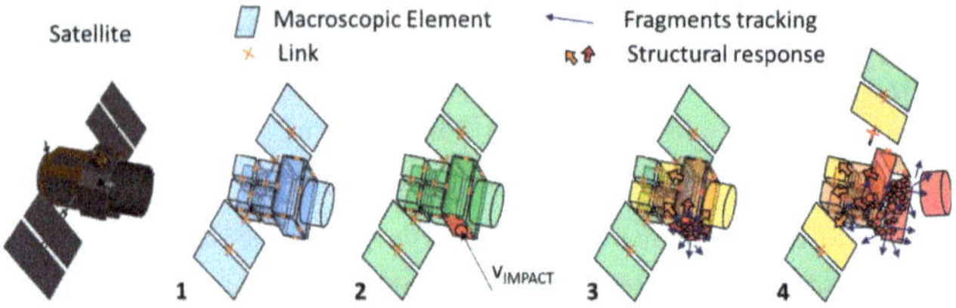

Fig. 1: CSTS modelling with MEs and links and simulation logic with cascade effect

In-space fragmentation case studies

1. COSMOS-IRIDIUM collision

This event, dating back to 2009, was the first collision between two intact spacecraft, the active IRIDIUM 33 and the defunct COSMOS 2251. In CSTS, two simulations replicating a central and a glancing impact have been performed. Fig. 2 shows the geometrical models for both cases and the obtained results in terms of cumulative characteristic length distribution; the glancing impact data (yellow) is clearly in accordance with the NASA SBM model.

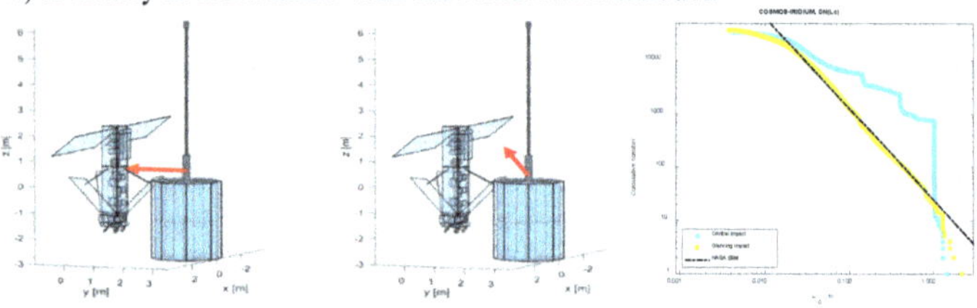

Fig. 2: COSMOS-IRIDIUM geometrical models for central (left) and glancing (centre) impacts and generated fragments cumulative distributions (right)

The Gabbard diagram in Fig. 3 compare CSTS data with the observed fragments for COSMOS 2251. Again, it is possible to notice an accordance between numerical data and observations.

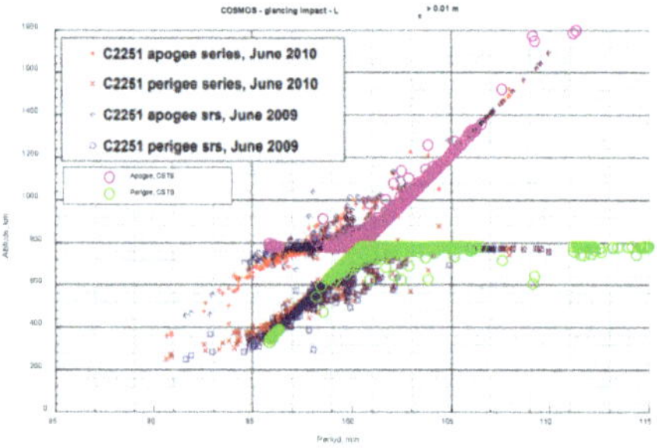

Fig. 3: Comparison of observed and simulated fragments (glancing impact) on the Gabbard diagram for COSMOS 2251 debris cloud

2. COSMOS 1408 anti-satellite test

In November 2021 a Russian anti-satellite test led to the break-up of the defunct COSMOS 1408 satellite. For this case, only partial information on the spacecraft and the kinetic impactor were available; the accuracy of CSTS model (see Fig. 4, left) is therefore limited, leading to an underestimation of the fragments cumulative number (in red in Fig. 4, center) with respect to observations (blue line) and NASA SBM model (black lines). However, as visible in the Gabbard diagram (Fig. 4, right), the orbital distribution of generated fragments is still in accordance with observations.

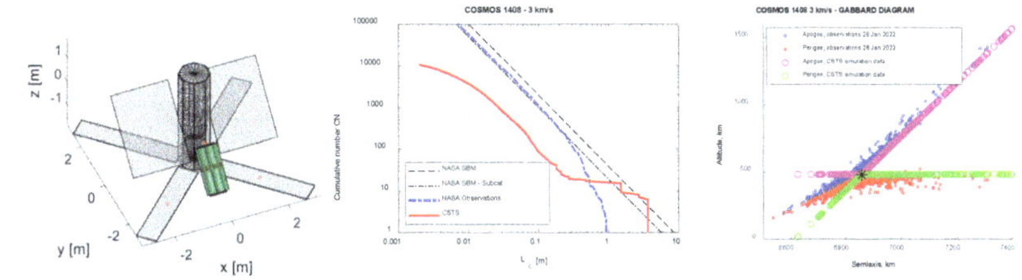

Fig. 4: geometrical model of COSMOS 1408, in gray, and the kinetic impactor, green (left); generated fragments cumulative distributions (center) and Gabbard diagram (left)

3. 2022-151B CZ-6A in-orbit break-up

In November 2022, the second stage of the CZ-6A fragmented after releasing its payload. This event was replicated with a dedicated CSTS simulation (Fig. 5), estimating the explosion of a tank. A total of more than 500 fragments were obtained by the simulation; numerical data are still compatible with the orbital distribution of observed fragments (Fig. 5).

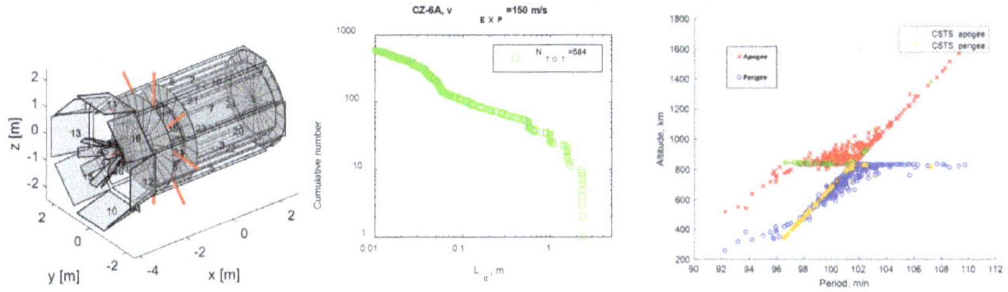

Fig. 5: CZ-6A geometrical model (left), generated fragments cumulative distributions (centre), and comparison of observed and simulated fragments on the Gabbard diagram (right)

4. Potential collision of ENVISAT with a spent rocket stage

Last, this potential collision scenario evaluated an impact of ENVISAT with a spent rocket stage (Fig. 6, left) at two different velocities, respectively of 1 km/s and 10 km/s. With CSTS it is possible to obtain the cumulative distributions reported in Fig. 6, right. As expected, the 10 km/s scenario generates more fragments due to the higher energy of the event, with about 100,000 fragments larger than 5 mm. The obtained distribution is below the estimation of this event performed by the NASA SBM; however, this breakup would strongly affect the already crowded 800 km sun-synchronous orbit currently occupied by ENVISAT.

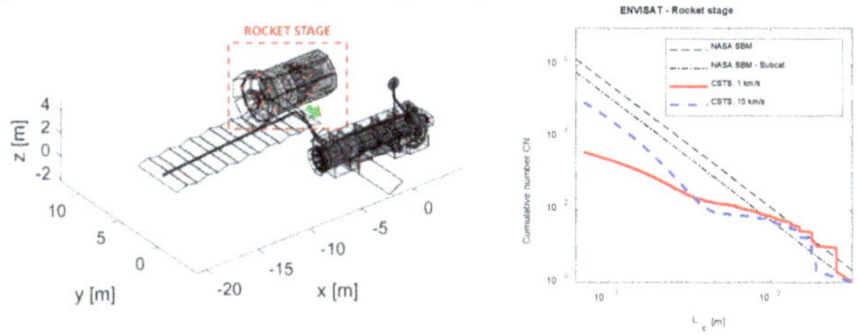

Fig. 6: ENVISAT Vs. rocket stage impact geometrical models (left) and generated fragments cumulative distributions (right)

Conclusions

This paper presented four simulation cases for in-space break-up events. It is shown that CSTS is capable to replicate complex fragmentation scenarios, providing statistically accurate results. These data will be employed to evaluate the effect of break-ups in the evolution of the non-detectable debris population and to assess the correlated risks.

Acknowledgements

The CSTS software was developed in the framework of ESA contracts No. 4000119143/16/NL/BJ/zk "Numerical simulations for spacecraft catastrophic disruption analysis" and No. 4000133656/20/D/SR, "Exploiting numerical modelling for the characterisation of collision break-ups; the COSMOS-IRIDIUM collision was modelled in the framework of the aforementioned ESA contract 4000133656/20/D/SR. The other simulations were performed in the framework of ASI-INAF Agreement "Supporto alle attivita` IADC e validazione pre-operativa per SST (N. 2020-6-HH.0)".

References

[1] Kessler, D. J., and Cour-Palais, B. G. "Collision frequency of artificial satellites: The creation of a debris belt." Journal of Geophysical Research: Space Physics 83.A6 (1978): 2637-2646. https://doi.org/10.1029/JA083iA06p02637
[2] Yakovlev, M. (2005, August). The "IADC Space Debris Mitigation Guidelines" and supporting documents. In 4th European Conference on Space Debris (Vol. 587, pp. 591-597).
[3] Krag, H., et al. "A 1 cm space debris impact onto the sentinel-1a solar array." Acta Astronautica 137 (2017): 434-443. https://doi.org/10.1016/j.actaastro.2017.05.010
[4] Kelso, T. S. "Analysis of the iridium 33 cosmos 2251 collision." (2009).
[5] Hanada, T., Liou, J. C., Nakajima, T., & Stansbery, E. (2009). Outcome of recent satellite impact experiments. Advances in Space Research, 44(5), 558-567. https://doi.org/10.1016/j.asr.2009.04.016
[6] Olivieri, L., Smocovich, P. A., Giacomuzzo, C., & Francesconi, A. (2022). Characterization of the fragments generated by a Picosatellite impact experiment. *International Journal of Impact Engineering*, *168*, 104313. https://doi.org/10.1016/j.ijimpeng.2022.104313
[7] McKnight, D., Maher, R., and Nagl, L.. "Fragmentation Algorithms for Strategic and Theater Targets (FASTT) Empirical Breakup Model, Ver 3.0." DNA-TR-94-104, December (1994).
[8] Sorge, M. E. "Satellite fragmentation modeling with IMPACT." AIAA/AAS Astrodynamics Specialist Conference and Exhibit. 2008. https://doi.org/10.2514/6.2008-6265
[9] Francesconi, A., et al. "CST: A new semi-empirical tool for simulating spacecraft collisions in orbit." Acta Astronautica 160 (2019): 195-205. https://doi.org/10.1016/j.actaastro.2019.04.035
[10] Olivieri L., et al. "Simulations of satellites mock-up fragmentation." Acta Astronautica, accepted on February 25th, 2023. https://doi.org/10.1016/j.actaastro.2023.02.036

Aeronautics and Astronautics - AIDAA XXVII International Congress Materials Research Forum LLC
Materials Research Proceedings 37 (2023) 469-473 https://doi.org/10.21741/9781644902813-103

Scientific activity of Sapienza University of Rome aerospace systems laboratory on the study of lunar regolith simulants, focusing on their effect on the microwave fields propagation

Andrea Delfini[1,a], Roberto Pastore[1,b *], Fabio Santoni[1,c], Michele Lustrino[2,d] and Mario Marchetti[3,e]

[1]Department of Astronautical, Electric and Energy Engineering, Sapienza University of Rome, Via Eudossiana 18, 00184 Rome, Italy

[2]Department of Earth Science, Sapienza University of Rome, P.le Aldo Moro 5, 00185 Rome, Italy

[3]AIDAA, Via Salaria 851, 00100 Rome, Italy

[a]andrea.delfini@uniroma1.it, [b]roberto.pastore@uniroma1.it, [c]fabio.santoni@uniroma1.it, [d]michele.lustrino@uniroma1.it, [e]marchettimario335@gmail.com

Keywords: Lunar Regolith, Moon Surface Absorption, Electromagnetic Microwave Propagation

Abstract. The forthcoming space missions aiming at developing new habits on the Moon and into deep space are opening new challenges for materials scientists in enabling in-situ efficient systems and subsystems. During the last decades, Space Agencies programs of long-term missions addressed to the future Moon colonization moved the aerospace research interest toward the knowledge of how the lunar conditions could represent scientific and technological tasks to be tackled, to deal with such a big challenge. Among very many matters, a still open question is to understand how proper the lunar environment would be for TLC systems daily used on Earth, or whether it should be necessary to establish different stable systems on the Moon by finding alternative solutions with respect to the Earth conventional technologies. This paper introduces the scientific activity developed during recent years at the Aerospace Systems Laboratory of Sapienza University of Rome, concerning the study of lunar regolith with focus on its effect on the microwave fields propagation. The research addresses such task by simulating several representative Moon environmental conditions, reproducing well defined chemical/physical background in terms of atmospheric parameters and soil compositions, as from the available literature data, and analyzing the microwave propagation characteristics to design efficiently mobile TLC systems operating on the Moon. With the further objective of considering regolith as main routine resource for drawing up systems and facilities constituting lunar living structures, the analysis of regolith-microwave interaction is thus focused on two specific paths, such as building airtight structures by means of ISRU methodologies and the EM compatibility (EMC) analysis of simulated lunar environment & TLC systems design. This work can be thus considered as linked to the forthcoming projects aimed at enhancing the research community knowledge about the Moon environment, by assessing scientific background and establishing technological processes for lunar TLC systems development.

Introduction

The required degree of technical innovation will allow to achieve scientific outcomes of huge impact and industrial fallout, taking into account the utmost interest for the Moon occurring nowadays. It has to be stressed that Long Term Evolution (LTE)/4G technology promises to revolutionize lunar surface communications by delivering reliable, high data rates while containing power, size and cost. Communications will be a crucial component for NASA's Artemis program, which will establish a sustainable presence on the Moon by the end of the decade. Deploying the first LTE/4G communications system in space will pave the way towards sustainable human presence on the lunar surface: the network, in fact, will provide critical communication capabilities for many different data transmission applications, including vital command and control functions, remote control of lunar rovers, real-time navigation and streaming of high definition video. These communication applications are all vital to long-term mission on the lunar surface, since reliable, resilient and high-capacity communications networks will be key for supporting a sustainable human presence on the Moon. The same LTE technologies that have met the world's mobile data and voice needs for the last decade are well suited to provide mission critical and state-of-the-art connectivity and communications capabilities for any future space expedition. Commercial off-the-shelf communications technologies, particularly the standards-based fourth generation cellular technology, are mature, proven reliable and robust, easily deployable, and scalable. NASA plans to leverage these innovations for its Artemis program, which will establish sustainable operations on the Moon by the end of the decade in preparation for an expedition to Mars. Thus, the presented research aims at lay down useful guidelines to deploy advanced TLC technologies in the most extreme environments, in order to validate the solution's performance and technology readiness level, and further optimize it for future terrestrial and space applications.

Work Summary

The scientific activity developed during recent years at the Aerospace Systems Laboratory (LSA – "Sapienza" University of Rome) concerned the study of lunar regolith, focusing on its effect on the microwave fields propagation. The research addresses such task by carrying out a full experimental characterization performed by simulating several representative Moon environmental conditions – i.e., by reproducing well defined chemical/physical background in terms of atmospheric parameters and soil compositions, as from the available literature data – and analyzing the microwave propagation characteristics in order to design efficiently mobile telecommunications systems operating on the Moon. With the further objective of considering regolith as main routine resource for drawing up systems and facilities constituting lunar living structures, the analysis of regolith-microwave interaction is thus focused on two specific paths:

- Building airtight structures by means of ISRU methodologies: the study is framed within the field of EM characterization of hybrid materials for space application currently carried out at LSA laboratories, focusing on the properties of several typologies of regolith-based bulk materials in terms of microwave reflection, absorption and transmission properties, in order to assess the basic conditions for remote operations.
- EM compatibility (EMC) analysis of simulated lunar environment & TLC systems design: the open question of understanding how the Moon environment would be fit for communications systems daily used on Earth, or whether it should be necessary to establish alternative solutions, is carried out by means of advanced facilities available at LSA laboratories, by reproducing fully constrained chemical/physical background conditions in terms of atmospheric and soil compositions, and analyzing the free space EM field propagation characteristics within lunar simulated environment.

The proposed study considers various types of lunar regolith reproduced on the basis of literature findings (see data in [1-3]). The base material used is dark powder of volcanic origin (Black Pyroclastite), which is enriched with the inclusion of various weight percentages of Ilmenite, in order to make it more similar to the regolith found on the surface of the basaltic lunar seas (main components: Ilmenite, Iron and Titanium Oxide). In order to obtain samples as much as possible responsive to the reality, the percentage of Titanium Oxide present in the Ilmenite sand is set as a parameter, and the correspondent percentage in weight of Ilmenite is added to the pyroclastic sand by mechanical mixing.

The electromagnetic experimental characterization is performed by means of a reverberation chamber (RC), due to the intrinsic capability provided by such experimental equipment to perform electromagnetic compatibility (EMC) tests with high level of accuracy. The facility adopted is the 'Space Environment Simulator' (SAS) of the Aerospace Systems Laboratory (LSA) available @ DIAEE – Sapienza University of Rome, which is a cylindrical vacuum chamber (volume around 5m^3) adapted to perform microwave characterization in conditions of chaotic EM propagation (see Fig.1, [4-6]). Basically, by using a reverberation chamber the knowledge of the absorbed power allows to retrieve the as-called absorption cross section (ACS) of the object under investigation. The inner SAS atmosphere is controlled in terms of vacuum level, temperature and percentage of humidity. In particular, the latter environmental parameter influence on materials ACS is investigated by exploring conditions of ultra-high vacuum, medium-low pressure, standard and increasingly moistened air, in order to discriminate the microwave absorption effects not due to lunar simulants intrinsic characteristics [7]. A representative result of the experimental characterization of the reverberation environment adopted is reported in Fig.2, where the measurements of the quality factor of the RC for different atmosphere conditions – i.e., max vacuum 'blank' level (~10^{-5} mbar), low pressure (~10^{-3} mbar) CO_2 filled and standard air – and the corresponding absorbing cross section (evaluated as discrepancy against the reference) are plotted over the microwave broadband considered. The influence of the partial presence of humidity in air is appreciable, especially at increasing frequency, giving evidence of the power absorbing effectiveness due to dipolar rotation resonances of the water vapor molecules in the EM range under investigation [8,9].

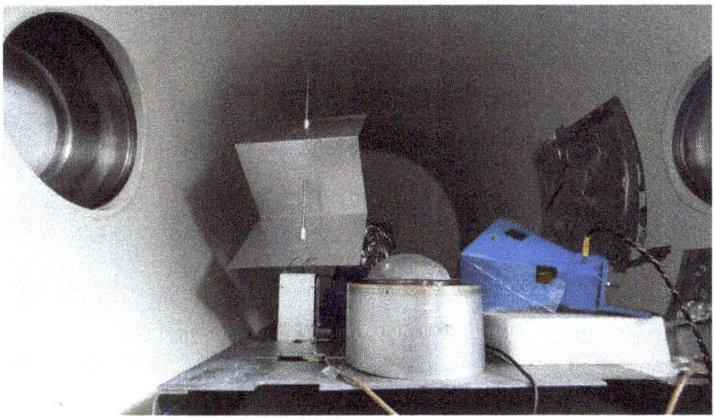

Fig.1. Inner view of the reverberation chamber for microwave characterization arranged by adapting the Space Environment Simulator of the Aerospace Systems Laboratory @ DIAEE – Sapienza Univ. of Rome.

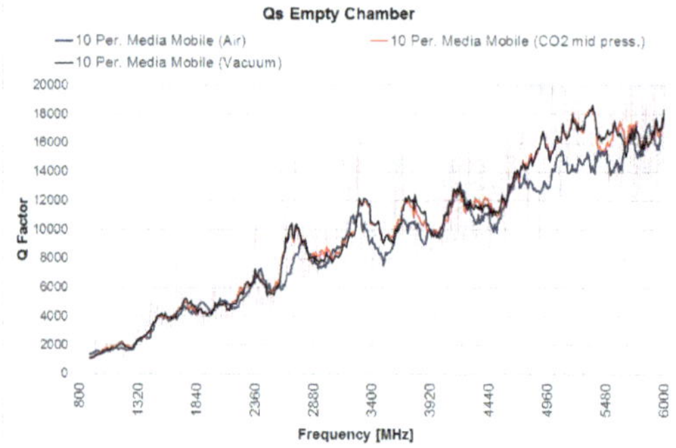

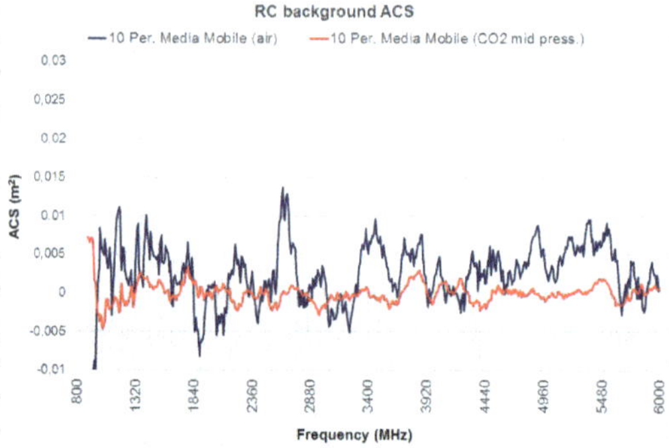

Fig.2. SAS chamber's quality factors and environments ACS measurements.

References

[1] D. S. McKay et al., Lunar Regolith in Lunar Sourcebook A User's Guide to the Moon, Cambridge University Press, 1991, 346.

[2] Y. Qian et al., Young lunar mare basalts in the Chang'e-5 sample return region, northern Oceanus Procellarum, Earth and Planetary Science Letters 555 (2021) 116702. https://doi.org/10.1016/j.epsl.2020.116702

[3] D. P. Moriarty III et al., The search for lunar mantle rocks exposed on the surface of the Moon, Nature Communications 12 (2021), 4659. https://doi.org/10.1038/s41467-021-24626-3

[4] D. Micheli, F. Santoni, A. Giusti, A. Delfini, R. Pastore, A. Vricella, M. Albano, L. Arena, F. Piergentili, M. Marchetti, Electromagnetic absorption properties of spacecraft and space debris, Acta Astronaut. 133 (2017) 128-135. https://doi.org/10.1016/j.actaastro.2017.01.015

[5] D. Micheli, A. Delfini, R. Pastore, M. Marchetti, R. Diana, G. Gradoni, Absorption cross section of building materials at mm wavelength in a reverberation chamber, Meas. Sci. Technol. 28 (2017) 024001. https://doi.org/10.1088/1361-6501/aa53a1

[6] D. Micheli, A. Delfini, F. Piergentili, et al., Measurement of Martian Soil Electromagnetic Absorption Cross Section from 800 MHz to 6 GHz for future Mars Cellular Telecommunication systems, 2022 IEEE 9th International Workshop on Metrology for AeroSpace, MetroAeroSpace

2022 - Proceedings, Pages 219-224.
https://doi.org/10.1109/MetroAeroSpace54187.2022.9855901

[7] Z. Wei, Moisture in tunnel influenced to transmission features of electromagnetic wave, Coal Science and technology 31 (2003) 39-41.

[8] T. Meissner, F. J. Wentz, The complex dielectric constant of pure and sea water from microwave satellite observations, IEEE T. Geosci. Remote 42 (2004) 1836-1849. https://doi.org/10.1109/TGRS.2004.831888

[9] M. Zribi, A. Le Morvan, N. Baghdadi N, Dielectric constant modelling with soil–air composition and its effect on SAR radar signal backscattered over soil surface, Sensors 8 (2008) 6810-6824. https://doi.org/10.3390/s8116810

The new transmitting antenna for BIRALES

G. Bianchi[1,a], S. Mariotti[1,b], M.F. Montaruli[2,c*], P. Di Lizia[2,d], M. Massari[2,e], M.A. De Luca[2,f], R. Demuru[3,g], G. Sangaletti[3,h], L. Mesiano[3,i], I. Boreanaz[3,l]

[1] Istituto Nazionale di Astrofisica, Via P. Gobetti 101, 40129, Bologna, Italy

[2] Politecnico di Milano, Department of Aerospace Science and Technology, Via G. La Masa 34, 20156, Milan, Italy

[3] SELT Aerospace & Defense, Viale delle Industrie 13/22, 20020, Arese (MI), Italy

[a]germano.bianchi@inaf.it, [b]sergio.mariotti@inaf.it, [c]marcofelice.montaruli@polimi.it, [d]pierluigi.dilizia@polimi.it, [e]mauro.massari@polimi.it, [f]mariaalessandra.deluca@polimi.it, [g]riccardo.demuru@selt-sistemi.com, [h]giovanni.sangaletti@selt-sistemi.com, [i]luca.mesiano@selt-sistemi.com, [l]ivan.boreanaz@selt-sistemi.com

Keywords: Space Debris, EUSST, Radar, Emitter

Abstract. In the last decades the increasing Resident Space Objects (RSOs) population is fostering many Space Surveillance and Tracking (SST) initiatives, which are currently based on the use of ground sensors. These can be distinguished in optical, laser and radar and categorized in tracking and survey sensors. In particular, the survey radars allow to determine the orbit of both catalogued and uncatalogued objects. Italy contributes to the European SST (EUSST) activities with the BIstatic RAdar for LEo Survey (BIRALES), whose transmitter is the Radio Frequency Transmitter (TRF), located at the Italian Joint Test Range of Salto di Quirra in Sardinia, and whose receiver is a portion of the Northern Cross Radio Telescope, located at the Medicina Radio Astronomical Station, near Bologna. The current sensor configuration is undergoing an upgrade process, including the receiver field of view extension and a new transmitter station. The purpose of the work is to present the new transmitting antenna of BIRALES, showing its technological progress and the potential for the monitoring of space debris. The final objective is to produce a high technological radar to improve the performance of the EUSST sensors network. In particular, the aim is to increase both the number of detectable objects and the sensitivity to detect fragments of a few centimetres up to an altitude of 2,000 km, with a remarkable improvement of orbit determination procedures and quality. The transmitting antenna has been designed to be very flexible for any type of observations, modifying its parameters depending on the observation needs and according to the service to offer (monitoring of fragments, re-entry or for collision avoidance). The work presents the system architecture and the transmitting antenna structure, and the performance are assessed through numerical simulations.

Introduction

Due to the growing complexity of the orbital environment, space-based assets are increasingly at risk of collision with other operational spacecraft or debris [1] [2]. At the same time, objects may re-enter the Earth's atmosphere and cause damage on the ground [3]. To address these concerns, the Space Surveillance and Tracking (SST) Consortium was established by the European Parliament and the Council in 2015. The EUSST system comprises a network of ground-based sensors aimed at surveying and tracking space objects, together with processing capabilities aiming at providing data, information and services [4].

BIRALES belongs to EUSST network, and this article focuses on the description of the transmitter and the final architecture of BIRALES, showing its capabilities.

BIRALES architecture

BIRALES is a bi-static radar [5] [6] [7] which uses the North-South branch of the Northern Cross radio telescope (Fig. 1) located in Medicina (BO) as receiving antenna (RX).

Fig. 1 BIRALES receiving antenna.

Fig. 2 BIRALES current (black) and future baseline (red).

The transmitting one is currently a parabolic dish located in Sardinia with a baseline of about 580 km. A new transmitter (TX), which will guarantee further performance increases, is under construction and will be installed closer to the receiver, forming a 250 km baseline. The new TX has been designed to work perfectly coupled with the RX and aligned as much as possible at the same terrestrial meridian of the RX, in order to maximize the overlap of the two antenna beams. The location identified for the transmitting antenna is 250 km southerly, as reported in the Fig. 2. The reduced baseline will also grant an increase in sensitivity. The receiving antenna has a collective area of about 11,000 m^2, and it is composed by 64 parallel parabolic cylindrical reflectors with 256 receivers installed on the focal lines (4 receivers per focal line).

Two types of observations are foreseen [8]: survey mode, for catalogue updating, observation of uncontrolled re-entry of large objects and monitoring to avoid collisions, and high-sensitivity mode, to observe fragmentations events and detect small objects (down to a few centimetres). The new TX is designed to adjust the Field of View (FoV) and gain according to the specific observation requested. In survey mode, the 64 cylindrical reflectors of the Northern Cross are collected in 8 groups of 8 cylinders, regularly spaced in elevation in order to obtain a coverage of the sky of about 45 deg in the N-S direction (each individual group of 8 antennas has a FoV of about 7x7 deg). Simultaneously, the TX is switched to the survey configuration mode, in order to have the same irradiated sky area of 7x45 deg (Fig. 3).

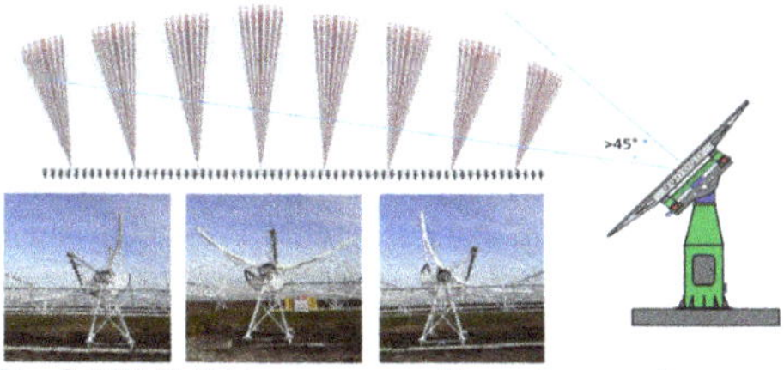

Fig. 3 BIRALES in survey configuration mode.

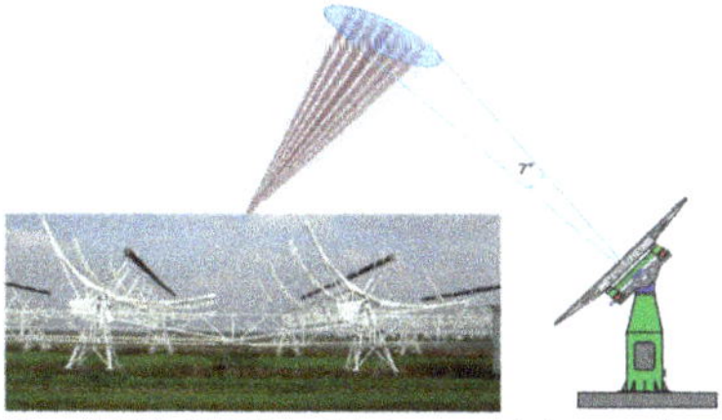

Fig. 4 BIRALES in high sensitivity mode.

In the high-sensitivity mode, the 64 cylindrical parabolic reflectors of the Northern Cross are pointed at the same declination to increase the antenna gain. At the same time, the transmitting antenna is switched to the high-sensitivity mode, increasing the gain and reducing the beam width at 7 deg (Fig. 4).

The selected architecture for the transmitting antenna is a patch array architecture (Fig. 5), composed of two sub-arrays which can be controlled to different beam apertures based on search and tracking operations (Fig. 6). The narrow beam sub-array is composed by a matrix of 8x4

Materials Research Proceedings 37 (2023) 474-477 https://doi.org/10.21741/9781644902813-104

antenna elements; each antenna element of the narrow beam array can sustain more than 312.5W in continuous wave (CW) mode. The wide beam sub-array is composed of a matrix of 1x8 antenna elements; each antenna element of the wide beam array can sustain more than 1.25KW of CW power. Considering the very demanding functional requirement for BIRALES, the antenna arrays and elements have been designed to optimize power factors, reflections, lobes, beam directivity and performance. The Antenna Control Unit (ACU) is installed on a two-axes pedestal able to tilt and rotate the antenna on azimuth and elevation axes. The implementation of a two-axes pedestal allows to relocate antenna beams and redirect monitoring and tracking waveforms to different sky areas (Fig. 7).

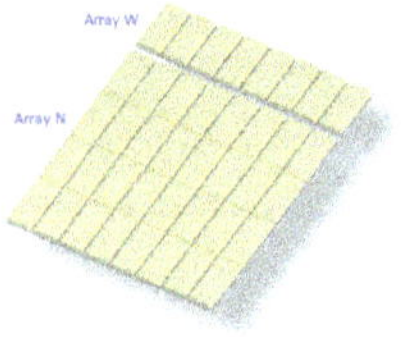

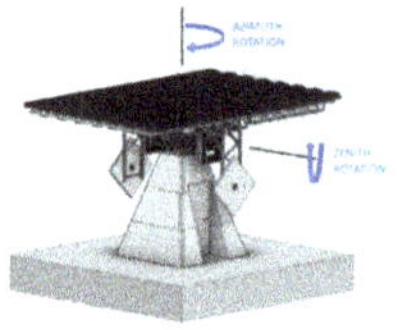

Fig. 5 Transmitting patch array antenna. Fig. 6 narrow (left) and wide (right) beams. Fig. 7 Transmitting antenna.

The new transmitting antenna will be fed by a power amplifier, able to provide an RF power up to 10 kW. Since the power amplifier is intended to operate continuously (24/7) in automatic and remote mode, the design needs to take into account the robustness and some redundancy criteria. The power amplifier will operate with two different modulations: continuous wave (CW) and pulse compression chirp modulation.

Radar performance

A simulation is carried out to evaluate the performance of the new configuration. In particular, it is fundamental to assess the importance of setting the transmitter station along the same meridian as the receiver one. To this purpose, the catalogue maintenance capability is examined in two configurations: the first configuration involves no misalignment (0 deg) between the transmitter and receiver meridians, while the second configuration introduces a misalignment of 0.2 deg between them. For both cases, a baseline of 250 km is considered. The study is conducted through the SpaceCraft and Objects Observation Planning (SCOOP) software, which, given a space objects catalogue, an analysis time window and the stations composing a survey sensor (like the ones involved in a bistatic radar), computes the observable transit. The space object catalogue is generally provided through the Two-Line Elements (TLEs), automatically downloaded from Spacetrack website, of the targets required by the user.

The simulation regards an analysis time window ranging from 00:00:00 of March 20th to 23:59:59 of March 26th, 2023 (according to a UTC time coordinate). The receiver station is considered pointed towards the zenith direction, with a FoV of 7x45 deg (7 deg in azimuth and 45 deg in elevation), that is in the survey operational mode. Both the transmitting stations are evaluated considering a 7x45 deg FoV, and three North pointing elevations are investigated: 45 deg, 60 deg and 75 deg. Results are shown in Table I and Table II, for the 0 deg and 0.2 deg offset from the receiver station meridian respectively. In particular, the number of detected objects, the number of transit and the median detection duration are reported.

Focusing on Table I it is possible to observe that the higher the pointing elevation, the larger the number of both detected objects and transits. Comparing Table I and Table II results, it is possible to observe that a 0.2 deg longitude offset with respect to the receiver meridian deteriorates the detection rate, that is the contribution to space objects catalogue maintenance.

Tab.I 0 deg offset case.

Pointing elevation	Number of objects	Number of transits	Median duration
45 deg	738	1147	7.1 s
60 deg	968	1889	5.5 s
75 deg	985	2040	4.8 s

Tab.II 0.2 deg offset case.

Pointing elevation	Number of objects	Number of transits	Median duration
45 deg	722	1104	7.4 s
60 deg	788	1321	6.6 s
75 deg	739	1221	6.4 s

Conclusions

This article describes the transmitter architecture of BIRALES. Numerical simulations highlighted that placing the transmitter station along the same meridian as the receiver one would represent a remarkable plus in terms of contribution to the building-up and maintenance of the space objects catalogue.

Acknowledgement

The research activities described in this paper were performed within the European Commission Framework Programme H2020 "SST Space Surveillance and Tracking" contracts N. 952852 (2-3SST2018-20).

References

[1] A. De Vittori et al., " Low-Thrust Collision Avoidance Maneuver Optimization", Journal of Guidance, Control, and Dynamics, 45(10), 1815-1829, 2022. https://doi.org/10.2514/1.G006630

[2] A. Muciaccia, et al., "Observation and Analysis of Cosmos 1408 Fragmentation." INTERNATIONAL ASTRONAUTICAL CONGRESS: IAC PROCEEDINGS. 2022.

[3] R. Cipollone, et al., "A Re-Entry Analysis Software Module for Space Surveillance and Tracking Operations." INTERNATIONAL ASTRONAUTICAL CONGRESS: IAC PROCEEDINGS. 2022.

[4] M.F. Montaruli, et al., "A software suite for orbit determination in Space Surveillance and Tracking applications." 9th European Conference for Aerospace Sciences (EUCASS 2022). 2022.

[5] G. Bianchi et al., "A new concept of bi-static radar for space debris detection and monitoring", 1st International Conference on Electrical, Computer, Communications and Mechatronics Engineering (ICECCME), 2021, pp. 1-6. https://doi.org/10.1109/ICECCME52200.2021.9590991

[6] G. Bianchi, et al. "Exploration of an innovative ranging method for bistatic radar, applied in LEO Space Debris surveying and tracking." Proceedings of the International Astronautical Congress, IAC. 2020.

[7] M.F. Montaruli, L. Facchini, P. Di Lizia, M. Massari, G. Pupillo, G. Bianchi, G. Naldi, "Adaptive track estimation on a radar array system for space surveillance", Acta Astronautica, 2022, ISSN 0094-5765. https://doi.org/10.1016/j.actaastro.2022.05.051

[8] G. Bianchi et al., "A New Concept of Transmitting Antenna on Bi-Static Radar for Space Debris Monitoring", 2nd nternational Conference on Electrical, Computer, Communications and Mechatronics Engineering (ICECCME), 2022. https://doi.org/10.1109/ICECCME55909.2022.9988566

Continuous empowering with laser power transmission technologies for ISRU moon assets: CIRA approach

Maria Chiara Noviello[1,a*], Nunzia Favaloro[1,b]

[1]Italian Aerospace Research Centre (CIRA), Via Maiorise, 81043, Capua, CE (Italy)

Space Exploration Technologies Laboratory

[a]m.noviello@cira.it, [b]n.favaloro@cira.it

Keywords: ISRU, LPT, Moon, Exploration

Abstract. Due to the potential possibility of changing the dynamics of the New Space Economy, In-Situ Resource Utilization (ISRU) is acquiring more and more importance within the Space Exploration scenario. Indeed, the closest space missions will return humans to the Moon, while planning the long-term stay. This aspect opens the way to the need for employment and processing of local resources, with the aim of reducing the dependence on Earth-based resources, also ensuring the financial sustainability of the space exploration programs. ISRU technologies will demand for energy values likely to be in the Megawatt range and, eventually, at Gigawatt levels, to be ensured in the harsh hazardous environmental conditions of the celestial bodies (e.g. Moon, Mars, Near Earth Asteroids). This work, performed by the CIRA TEES Laboratory, provides the CIRA approach to the feasibility study concerning the Laser Power Transmission (LPT) technologies for Moon assets empowering. The aim is to evaluate whether LPT can be a potentially efficient solution for continuous power delivery from an orbiting source device, considering long-distance wireless employments and severe environmental conditions, to drive ISRU Moon assets (habitats, rovers, local industrial plants, conveyance facilities, et cetera). For the purpose of this study, starting from the space mission identification, an increasing complexity multi-step approach was properly conceived by CIRA to design the dedicated LPT system responding to the evaluated mission requirements.

ISRU: CIRA LPT Model & Approach

The term ISRU (In-Situ Resource Utilization) refers to the use and processing of local resources directly found on the Moon, or other planetary bodies, to obtain raw materials supporting robotic and/or human space exploration missions. ISRU technologies aim at creating products or services fundamental for lunar and Mars long-term stays, also reducing the need for continuous resupply from the Earth, [1], with an unavoidable huge impact on the dynamics of the New Space Economy.

In general terms, the ISRU domain relies on three key concepts: identification (prospecting for recoverability), processing (mining, extraction, beneficiation) and use of local (natural and artificial) resources. Each concept will imply dedicated technologies, systems and capability development involving various technical disciplines, [2]. More in detail, local resources account for the ones recoverable on extra-terrestrial bodies.

This paper deals with the problem of the Moon ISRU assets' continuous power delivery. Indeed, for the lunar case, the severe environmental conditions of the surface, combined with the long periods of darkness due to its day-night cycle, make the energy supply a pivotal issue. Several approaches have recently been considered to store and provide energy to the surface of the Moon by means of ISRU (In-Situ Resource Utilization) technology, [3]. Among the various potential solutions identified in literature, the CIRA TEES laboratory is deepening the one based on the solar power caught by dedicated satellites (Solar Power Satellites, SPS) in proper orbits, equipped with a laser system, capable of generating a power laser beam driven for long distances to activate

Materials Research Proceedings 37 (2023) 478-482 https://doi.org/10.21741/9781644902813-105

Moon assets (rovers, habitats, infrastructures, etc.). From now on, the overall system this work refers to will be designated as LPT (Laser Power Transmission) system, portrayed in Figure 1.

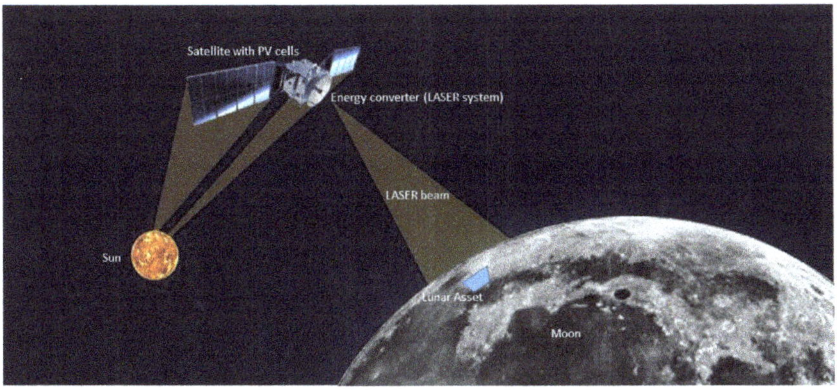

Figure 1 - Space Laser Power Transmission System Overview for Moon Assets Empowering

The space system, this work refers to, has been intended as promising in order to empower Moon assets in a continuous manner. Figure 2 reports a block schematization of the previous Figure 1, helpful to understand the here described work.

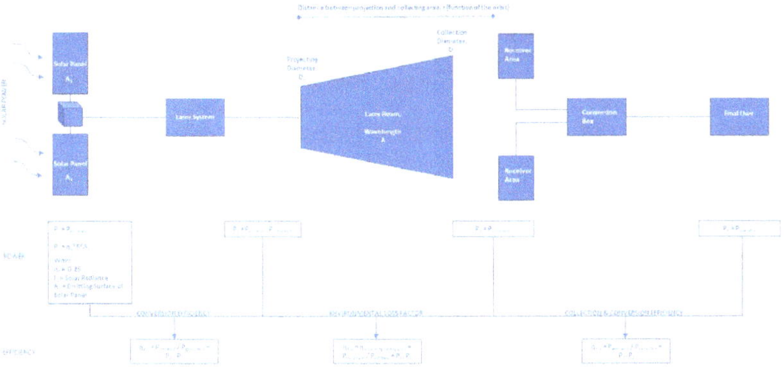

Figure 2 - CIRA LPT Preliminary Model Block Schematization

In particular, the overall system is composed by:

- One/more satellite/s, equipped with photovoltaic solar panels to catch the Sun power (the choice of the orbit is not the actual objective of this study);
- A laser system, using solar energy to generate a laser power beam;
- A laser power collecting station, with a proper receiving area;
- A conversion box, capable of converting the laser power beam into electrical energy to activate:
- A final user, which functioning needs to be guaranteed by the overall system.

Precisely, CIRA is now performing an accurate study to assess such a complex topic in terms of feasibility of the conceived system process. Thus, a multi-step dedicated approach has been developed: the main phases are depicted in Figure 3, with a focus on the Step 2.

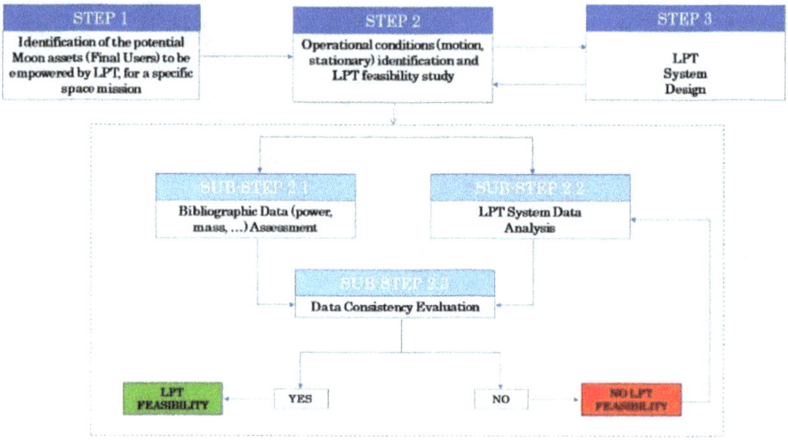

Figure 3 - LPT feasibility study for IRSU applications: the CIRA Approach

A first brief description of the conceived three steps is herein provided, while the Section 2 contains a more detailed characterization of the STEP 2, core of this work:

- STEP 1: dealing with the identification of the potential ISRU Moon assets to be empowered by means of an LPT system, for a given space mission. The Global Exploration Roadmap (GER, [6]) has been used in support of this stage to screen the possible final users of the LPT system, mentioning rovers, (for mobility, transportation, image acquisition, data measurement), landers (crew, cargo, robotic), habitats (pressurized, mono- or multi-module), industrial plants (for production and processing of materials), extraction instruments/equipment (bucket-wheel, auger/drill, scoop, pneumatic excavator), comminution instruments/equipment (size crushing, sorting), et cetera.
- STEP 2: among the final users, deriving from the Step 1, the rover has been chosen as the model to develop the preliminary LPT feasibility study in this stage. In particular, the Step 2 has been further exploded into three sub-steps:
  - ✓ SUB-STEP 2.1: after schematizing the possible operative conditions for the chosen systems (rovers), this sub-step collects literature power and mass data of (already dismissed/currently working) rovers for Moon and Mars exploration, also accounting instruments (magnetometers, spectrometers, etc.), cameras and actuators. An example of some collected data is reported in Table 1, for the Intrepid lunar rover, [4].

Table 1 - Collected mass and power data for Intrepid rover at system and sub-system level

		Operational Condition	Mass [Kg]	Power [W]	Reference
System	Intrepid Rover (Moon)	Driving Mode	371,00	239,00	Table 3-3, [4]
Sub-System	Instruments	Stationary Mode	11,50	16,00	Table 1-3, [4]
	Cameras		8,50	14,00	
	Actuators		48,31	60,00	Mass, Appendix D, [4]; Power, Table 3-2, [4]

- ✓ SUB-STEP 2.2: LPT system data analysis, by means of the implementation of analysis loops to process data resulting from the sub-step 2.1, [5], to be confronted with into the:
- ✓ SUB-STEP 2.3: dealing with the consistency assessment of the values deriving from the previous sub-steps.
- STEP 3: not object of the present work, regarding the LPT system design and optimization of design parameters, [7].

Operational Conditions Identification and LPT Feasibility Study: Focus on The Step 2
As before mentioned, the real core of this work is the Step 2, herein described more in detail.
At first, a collection of data from literature was indispensable to have an overview of the power and mass values related to still working/already dismissed Moon rover systems. Those values have been then used as key elements to drive the following two approaches, namely the direct and the reverse ones. In particular, referring to the preliminary model block scheme reported in Figure 2, the Table 2 summarizes the input and output parameters considered in both cases.

Table 2 - Direct and Reverse Approaches Input and Output Parameters for preliminary LPT Data System Analysis

Direct Approach		Reverse Approach	
Input Parameters	**Ordered Output Parameters**	**Input Parameters**	**Ordered Output Parameters**
η_0, Solar Panel Efficiency	P_1, Generated Power	P_4, Final User Available Power	$\eta_{3,2}$, Collecting Conversion Efficiency
I_R, Solar Radiance	P_2, Emitted Power	D_p, Projecting Beam Diameter	P_3, Collected Power
A_1, Emitting Surface of Solar Panels	$\eta_{3,2}$, Collecting Conversion Efficiency	D_c, Collected Beam Diameter	P_2, Emitted Power
$\eta_{2,1}$, Laser system Conversion Efficiency	P_3, Collected Power	r, Relative Distance Emitter - Receiver	P_1, Generated Power
D_p, Projecting Beam Diameter	P_4, Final User Available Power	λ, Power Beam Wavelength	A_1, Emitting Surface of Solar Panels
D_c, Collected Beam Diameter		η_0, Solar Panel Efficiency	
r, Relative Distance Emitter - Receiver		I_R, Solar Radiance	
λ, Power Beam Wavelength		$\eta_{2,1}$, Laser system Conversion Efficiency	
$\eta_{4,3}$, Final User Conversion Efficiency		$\eta_{4,3}$, Final User Conversion Efficiency	

Thanks to the direct approach, a sensitivity analysis of P_4 (Power Available for systems) as a function of A_1 (Solar Panel Area) has been carried out. Figure 4 reports the linear trend (based on the preliminary LPT system model) of $P_4 = P_4(A_1)$, by using Nd:YAG laser for Moon applications ($\eta_{3,2}$ resulted equal to 1, according with the lack of a consistent atmosphere).

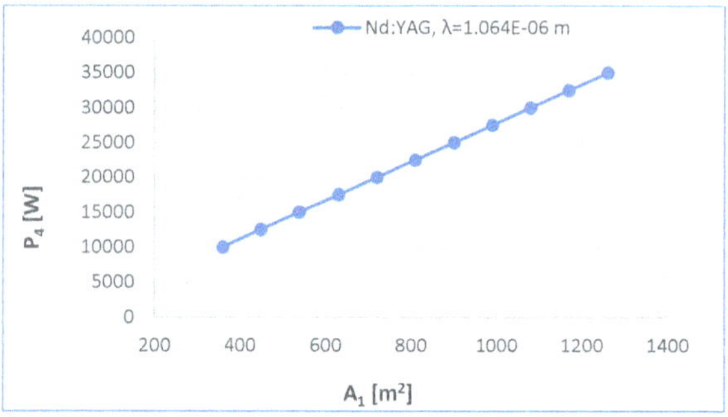

Figure 4 - Lunar Final User Power as a function of Satellite Solar Panel Area, P_4=$P_4(A_1)$

Conclusions

Referring to the challenge of the continuous empowering of Moon ISRU assets, the CIRA TEES laboratory developed a multi-step approach to conduct a preliminary feasibility study of a Solar Laser Power Transmission System for the power delivery across long distances, starting from an orbiting satellite.

At the current state, interesting results have been achieved in terms of the trend of the identified final user (rover system) available power as a function of the solar photovoltaic panels emitting area (/s, if more satellites will be considered in a more advanced system). The next step of this work will account the optimization of design parameters, such as the satellite orbit selection.

References

[1] M. Baldry, N. Gurieff, D. F. Keogh, "Imagining Sustainable Human Ecosystems with Power-to-x in-situ Resource Utilisation Technology", 4 November 2022. https://doi.org/10.20944/preprints202111.0508.v1

[2] In-Situ Resource Utilization Gap Assessment Report, International Space Exploration Coordination Group (ISECG), NASA, 2021

[3] M. F. Palos, P. Serra, S. Fereres, K. Stephenson, R. Gonzales-Cinca, "Lunar ISRU energy storage and electricity generation", Acta Astronautica Journal, Volume 170, May 2020, Pages 412-420. https://doi.org/10.1016/j.actaastro.2020.02.005

[4] NASA Intrepid Planetary Mission Concept Study Report, 2020

[5] C. Cougnet, E. Sein, A. Celeste, L. Summerer, "Solar Power Satellites For Space Exploration And Applications". November 2004

[6] Global Exploration Roadmap, GER, ISECG (globalspaceexploration.org), Posted on 23 May, 2023

[7] N. Favaloro et al., "Enabling Technologies for Space Exploration Missions: the CIRA-TEDS Program Roadmap Perspective", Aerotecnica Missili & Spazio, Springer Nature, June 2023. https://doi.org/10.1007/s42496-023-00159-4

Preliminary design of a CubeSat in loose formation with ICEYE-X16 for plastic litter detection

Francesca Pelliccia[1,a *], Raffaele Minichini[1], Maria Salvato[1], Salvatore Barone[1], Salvatore Dario dell'Aquila[1], Vincenzo Esposito[1], Marco Madonna[1], Andrea Mazzeo[1], Ilaria Salerno[1], Antimo Verde[1], Marco Grasso[1], Antonio Gigantino[1], Alfredo Renga[1]

[1]Università degli Studi di Napoli Federico II, Dipartimento di Ingegneria Industriale, Piazzale Tecchio 80, 80125, Napoli, Italy

[a]francesca.pelliccia2@studenti.unina.it

Keywords: Plastic Detection from Space, Multispectral/SAR Database, CubeSat Mission

Abstract. Every year, more than 14 million metric tons of plastic are estimated to enter rivers, lakes, and seas [1], becoming one of the main sources of pollution with significant economic and ecological impact on sensitive habitats, welfare, and vulnerable, endangered species. In this context, keeping track of plastic litter hot-spots and their evolution in time - both in open sea and coastal areas - becomes of fundamental importance. Plastic detection from space is at an early stage and, although some interesting capabilities have been demonstrated by multi-spectral imagery, hyperspectral sensing, and GNSS reflectometry, such technologies do not yet allow for the operational detection and monitoring of plastic from space on a global scale, with sufficient temporal and spatial coverage. The characteristics of Synthetic Aperture Radar (SAR) imagery would represent a keystone to realize almost continuous global monitoring of plastic litter at sea, but robust and reliable approaches for SAR-based plastic detection at sea are not available. The main problem is the lack of a large and assessed dataset to train and test new procedures and methods (e.g., deep learning) on large scales. Starting from this point, CROSSEYE (Combined in pendulum Remote Observation cubeSat System for icEYE) mission is presented with the objective to generate a wide dataset of multi-spectral and SAR images collected at the same time over the same areas. Plastic detection in multi-spectral images is mature enough to be used as a ground truth to cue SAR-based algorithms that autonomously perform the same task. CROSSEYE exploits a pre-existing SAR satellite belonging to ICEYE constellation - ICEYE-X16 - and completes it with an additional multi spectral camera equipped on a 6U CubeSat. The results coming from the preliminary design of CROSSEYE demonstrate the feasibility of a mission capable of detecting plastic debris from space by using state-of-the-art technologies.

Introduction

The primary objective of this mission is to validate an innovative measurement principle of combining acquisitions from different sensors, specifically electro-optical (EO) and radar systems. By leveraging the respective strengths of these technologies, namely the all-weather and all-time capabilities of radars and the spectral analysis capabilities of EO sensors, the aim is to detect plastic pollution floating in open sea (20 km or more from the coast) [2]. To achieve this, the mission will gather a comprehensive dataset of multi-spectral and synthetic aperture radar (SAR) images collected simultaneously (within four hours) over the same areas. The detection of plastic in the multi-spectral images will serve as a cue for SAR-based algorithms to perform the same task [3]. Furthermore, CROSSEYE will demonstrate how the aforementioned measurement principle can be enabled by means of a simple low-cost platform, both exploiting and enhancing the capabilities of other missions already in orbit. The mission will raise awareness of marine plastic litter among the general public, governments, and related organizations, stimulating interest in the topic and attracting investments towards finding solutions. These objectives are aligned with the United Nations Sustainable Development Goals (SDGs) 6 [4] and 14 [5] addressing clean water, sanitation, conservation and sustainable use of marine resources.

Payload

Electro-optical pushbroom scanners have a long history of successful implementation in space applications, often relying on readily available Commercial Off-The-Shelf (COTS) components. However, the requirements of the CROSSEYE mission pose a challenge in finding off-the-shelf sensors that can meet the mission criteria while being compatible with a 6U CubeSat standard platform in terms of mass and sizes. For plastic debris detection, a 20 m ground resolution is needed [6] as well as a wide spectral range from VNIR to SWIR bands. In particular, [6] and [7] report how the computation of certain indexes might aid the detection of floating debris: the Floating Debris Index (FDI), the Normalized Difference Vegetation Index (NDVI), and the Floating Algae Index (FAI). To calculate these indexes, the payload will need to detect several bands inspired from the Sentinel-2 Multi-Spectral Instrument (MSI) - B2, B3, B4, B6, B8, B11 [8]. Consequently, a custom multispectral pushbroom scanner (Table 1) is specifically designed to meet the needs of the mission. To accommodate the different wavelength ranges, two separate detectors are necessary as the technology employed for Visible and NIR wavelengths (VNIR) differs from that required for SWIR (Short-Wave InfraRed) wavelengths. To address CubeSat size limitations, the camera utilizes a single focal plane, where the two mentioned detectors share the same optic scheme rather than having separate pushbroom designs. Special band-pass filters are necessary for the detectors to effectively sense the specific bands listed before.

Table 1 - Pushbroom scanner main specifications

		VNIR	SWIR
Telescope parameters	Focal length [mm]	400	400
	Aperture [mm]	96.0	96.0
	F/#	4.17	4.17
	FOV across track [deg]	1.76	2.20
	Swath width [km]	16.0	20.0
	Ground sampling distance [m]	15.6	9.80
	MTF @ 32.5 lin/mm	0.66	0.66
Focal plane parameters	Number of pixel (H x V)	1024 x 25	2048 x 1
	Pixel size [μm]	12 x 12	7.5 x 7.5
	Number of bands	5	1
	Pixel depth [bits per pixel]	8	8
Mission parameters	Altitude [km]	520	
	Ground resolution [m]	20.0	
	Target frequency [lin/mm]	32.5	
Sizing parameters	Mass [kg]	4.09	
	Volume [mm^3]	1.59e+06 (< 2U)	
	Data rate [Mbps]	473	
	Power consumption (Standby-On) [W]	1.00-35.0 W	

Orbit design - pendulum formation

CROSSEYE mission concept is developed around the notion of pendulum configuration, a particular type of loose formation flying, also known as parallel orbits formation. CROSSEYE CubeSat flies at a safe distance from ICEYE-X16, without affecting, limiting, or altering its functionalities. The goal is to have CROSSEYE S/C as deputy and to have the chief role covered by ICEYE-X16 S/C, an X-band SAR satellite launched in January 2022. The orbital parameters of the chief satellite are reported in Table 2.

Table 2 - ICEYE X-16 orbital parameters

Altitude [km]	Inclination [deg]	LTAN	Eccentricity	AOP [deg]
525	97.485	10:00 PM	0.0011925	67.3522

According to the chosen configuration, CROSSEYE S/C orbit is defined in relation to the orbital parameters of ICEYE-X16 S/C: the formation is achieved by separating the orbits of the two spacecrafts in terms of Right Ascension of the Ascending Node (RAAN) Ω and true anomaly ν. $\Delta\Omega$ and $\Delta\nu$ are determined from geometrical considerations (Fig. 1), where R_B is the Slant Range at boresight, a is the angle between CROSSEYE S/C and ICEYE-X16 S/C nadiral direction of observation, q is the aperture angle of ICEYE-X16 SAR. Given the risk of collision, especially when flying where the orbits cross - at poles -, CROSSEYE S/C and ICEYE-X16 S/C are separated in eccentricity, too [9]. The pendulum formation parameters are listed in Table 3; the CROSSEYE S/C orbital parameters are derived (Table 4). The orbit is defined as SSO, and the Argument of Perigee (AOP) is chosen to make the orbit of CROSSEYE S/C frozen [10].

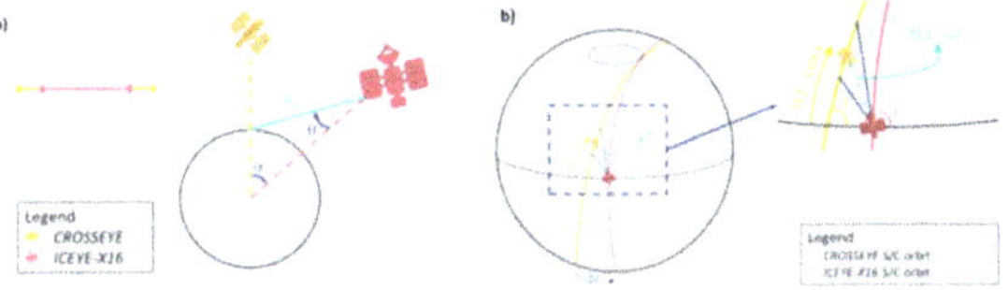

Fig. 1 - a) CROSSEYE and ICEYE-X16 relative observation geometry; b) Pendulum configuration: geometry

Table 3 - Pendulum formation parameters

$\Delta\Omega$ [deg]	$\Delta\nu$ [deg]	Δe
1.7263	-0.22488	-1.2149 e-04

Table 4 - CROSSEYE S/C orbital parameters

Altitude [km]	Inclination [deg]	LTAN	Eccentricity	AOP [deg]
525	97.485	10:00 PM	0.0010710	90

Space Segment

Other than the payload, CROSSEYE S/C consists of an attitude determination and control subsystem (ADCS), an on-board data-handling subsystem, an on-board software, a communication subsystem, an electric power subsystem (EPS), a passive thermal control subsystem (TCS), a chemical propulsion subsystem. A GNSS receiver is mounted on the platform to obtain accurate position and velocity measurements. ADCS guarantees Nadir pointing during observations with a minimum accuracy of 1% of the FOV of the equipped EO payload. As attitude control, a three-axis stabilized strategy is adopted by pairing reaction wheels with magnetorquers. Concerning the EPS, triple junction GaAs solar cells are selected for the solar arrays and Lithium-Polymer are chosen for the batteries. TCS consists of a thermal coating formed by a mixture of aluminium and white paint. Considering the significant data rate of the payload, TT&C subsystem is composed of an X-band antenna and a diplexer interfacing with an X-band transceiver. As for to the propulsion subsystem, thrusters use an ammonium dinitramide-based *green* monopropellant, in line with the sustainable nature of the mission. Except for the payload, all the components of the subsystems are COTS and easily integrable with each other, resulting in a standard 6U CubeSat architecture compliant with the CubeSat Design Specifications [11]. CROSSEYE space segment communicates with its ground segment. This relies on 11 of the 17 active Leaf Space network ground stations placed at middle latitudes all around the world.

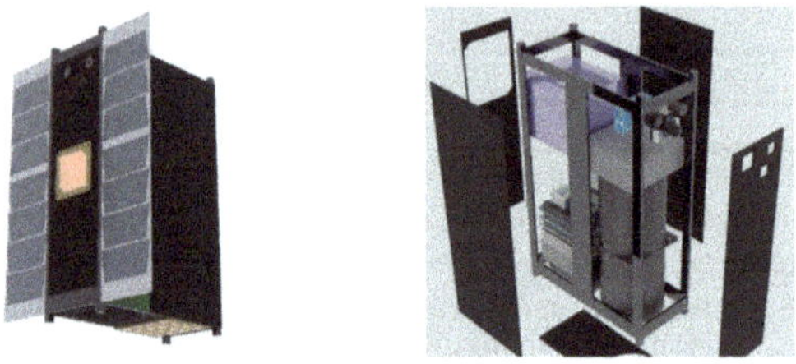

Fig. 2 - CROSSEYE platform CAD model and exploded view

Conclusions and further development

An incremental strategy can be implemented for CROSSEYE mission: by limiting the initial expenditure to a single CubeSat mission, this can be replaced or augmented with future missions of the same type to widen the quantity and quality of collected data. The design here proposed is easily adaptable to other SAR-equipped platforms that, mutatis mutandis, will have the capability to build a database that is tailored to fit the customer's demands (i.e., wildfire, coastal erosion, plastics). In general, the capabilities of CubeSat-related technologies are expected to improve in coming years: the combination of miniaturization challenges, and the potential for future technological advancements create an exciting prospect for the CROSSEYE mission. Else than technicalities, the mission will set the stage for significant contribution to plastic litter detection from space, promoting sustainability and furthering the understanding of Earth's ecosystem.

References

[1] Information on https://www.iucn.org/resources/issues-brief/marine-plastic-pollution

[2] Information on https://www.un.org/Depts/los/convention_agreements/texts/unclos/part2.htm

[3] Savastano S. et al., A First Approach to the Automatic Detection of Marine Litter in SAR Images Using Artificial Intelligence, IEEE International Geoscience and Remote Sensing Symposium IGARSS (2021) 8704-8707. https://doi.org/10.1109/IGARSS47720.2021.9737038

[4] Information on https://sdgs.un.org/goals/goal6

[5] Information on https://sdgs.un.org/goals/goal14

[6] Biermann L. et al., Finding Plastic Patches in Coastal Waters using Optical Satellite Data, Sci Rep 10, 5364 (2020). https://doi.org/10.1038/s41598-020-62298-z

[7] Hu C., A novel ocean color index to detect floating algae in the global oceans", Remote Sensing of Environment 113 (2009) 2118-2129. https://doi.org/10.1016/j.rse.2009.05.012

[8] Morales - Caselles C. et al., An inshore-offshore sorting system revealed from global classification of ocean litter, Nature Sustainability 4 (2021) 484-493. https://doi.org/10.1038/s41893-021-00720-8

[9] D'Errico M. et al., Relative Trajectory Design, Distributed Space Missions for Earth System Monitoring. Space Technology Library 31 (2019).

[10] Fasano G., Space Flight Dynamics lectures University of Naples Federico II (2022).

[11] The CubeSat Program, CubeSat Design Specification (1U - 12U) - Revision 14.1 (2022).

A revisited and general Kane's formulation applied to very flexible multibody spacecraft

D.P. Madonna[1,a], P. Gasbarri[2,b], M. Pontani[3,c], F. Gennari[4,d], L. Scialanga[4,e] and A. Marchetti[4,f]

[1]Department of Mechanical and Aerospace Engineering, Sapienza University of Rome, via Eudossiana 18, 00184 Rome, Italy

[2]School of Aerospace Engineering, Sapienza University of Rome, via Salaria 851, 00138 Rome, Italy

[3]Department of Astronautical, Electrical, and Energy Engineering, Sapienza University of Rome, via Salaria 851, 00138 Rome, Italy

[4]Thales Alenia Space Italia, via Saccomuro 24, 00131 Rome, Italy

[a]davidpaolo.madonna@uniroma1.it, [b]paolo.gasbarri@uniroma1.it, [c]mauro.pontani@uniroma1.it, [d]fabrizio.gennari@thalesaleniaspace.com, [e]luigi.scialanga@thalesaleniaspace.com, [f]andrea.marchetti@thalesaleniaspace.com

Keywords: Multibody Spacecraft Dynamics, Kane's Method, Flexible Spacecraft

Abstract. Current space missions require predicting the spacecraft dynamics with considerable reliability. Among the various components of a spacecraft, subsystems like payload, structures, and power depend heavily on the dynamic behavior of the satellite during its operational life. Therefore, to ensure that the results obtained through numerical simulations correspond to the actual behavior, an accurate dynamical model must be developed. In this context, an implementation of Kane's method is presented to derive the dynamical equations of a spacecraft composed of both rigid and flexible bodies connected via joints in tree topology. Starting from the kinematics of two generic interconnected bodies, a systematic approach is derived and the recursive structure of the equations is investigated. The Kane's formulation allows a relatively simple derivation of the equation of motion while obtaining the minimum set of differential equations, which implies lower computational time. On the other hand, this formulation excludes reaction forces and torques from the dynamical equations. Nevertheless, in this work a strategy to compute them a posteriori without further numerical integrations is presented. Flexibility is introduced through the standard modal decomposition technique, so that modal shapes obtained by FEA software can be directly utilized to characterize the elastic motion of the flexible bodies. A spacecraft composed of a rigid bus and several flexible appendages is modeled and numerical simulations point out that this systematic method is very effective for this illustrative example.

Introduction

The advanced level of technology in space missions and the substantial economic investment they require necessitate a high level of predictability in all aspects of the mission. The performance of critical subsystems, such as the payload, structures, and power subsystem, is directly influenced by the dynamic behavior of the satellite throughout its operational lifetime. Therefore, it is crucial to develop an accurate dynamical model that ensures the correspondence between numerical simulations and the actual behavior of the satellite. While it may be acceptable in some cases to model the spacecraft as a single rigid body, typically it is necessary to consider the satellite as a multibody system comprising both rigid and flexible elements. Various approaches exist to derive the dynamical equations of a multibody system. However, in this work, Kane's formulation is exclusively adopted due to its distinct advantages in terms of both algebraic and computational

Materials Research Proceedings 37 (2023) 487-494 https://doi.org/10.21741/9781644902813-107

aspects [1]. A practical version of Kane's equation, as provided in [2], has been extended in this research to encompass spacecraft consisting of both flexible and rigid bodies. The outcome is a concise matrix formulation that is also compatible with the results of modal analysis obtained through a finite element code such as NASTRAN.

Kinematics

Before providing the general form of Kane's equation, it is necessary to outline the kinematic quantities of interest.

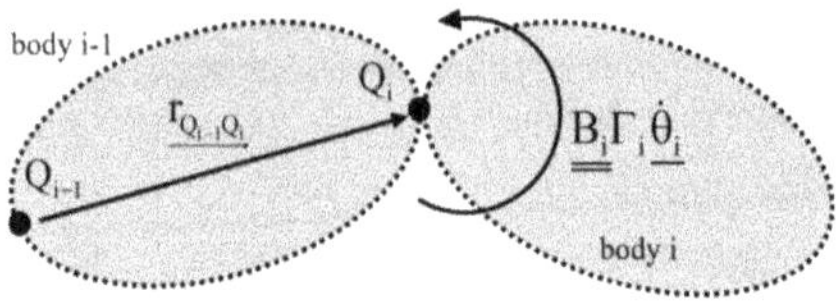

Fig. 1: two bodies connected via rotary joint

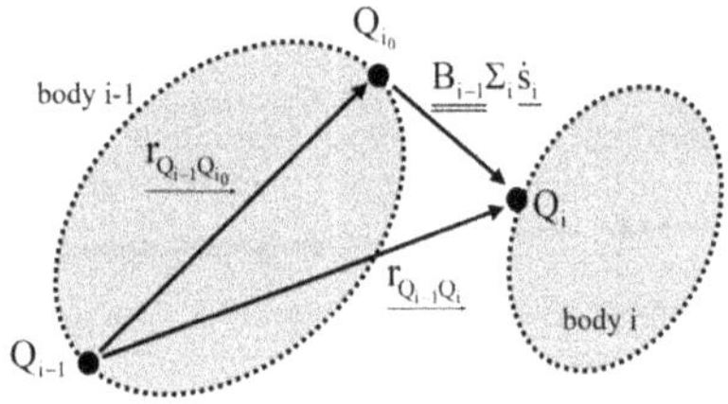

Fig. 2: two bodies connected via prismatic joint

Considering Figs. 1-2, the linear velocity of the connection point Q_i between two flexible bodies and the angular velocity of body i, both evaluated with respect to the inertial frame N, are

$$\text{rotary joint}: \begin{cases} {}^{N}\underline{v}^{Q_i} = {}^{N}\underline{v}^{Q_{i-1}} - \underline{\underline{B}}_{i-1}\widetilde{\underline{r}^{i-1}_{Q_{i-1}Q_i}}\,\underline{\omega}_{i-1} + \underline{\underline{B}}_{i-1}\sum_{k=1}^{n_F}\underline{\Phi_k^{(i-1)}(Q_i)}\dot{q}_k \\ {}^{N}\underline{\omega}^{B_i} = {}^{N}\underline{\omega}^{B_{i-1}} + \underline{\underline{B}}_i\Gamma_i\underline{\dot{\theta}_i} \end{cases} \tag{1}$$

$$\text{prismatic joint}: \begin{cases} {}^{N}\underline{v}^{Q_i} = {}^{N}\underline{v}^{Q_{i-1}} - \underline{\underline{B}}_{i-1}\widetilde{\underline{r}^{i-1}_{Q_{i-1}Q_i}}\,\underline{\omega}_{i-1} + \underline{\underline{B}}_{i-1}\sum_{k=1}^{n_F}\underline{\Phi_k^{(i-1)}(Q_{i_0})}\dot{q}_k + \underline{\underline{B}}_{i-1}\Sigma_i\underline{\dot{s}_i} \\ {}^{N}\underline{\omega}^{B_i} = {}^{N}\underline{\omega}^{B_{i-1}} \end{cases} \tag{2}$$

where $\underset{\rightarrow}{\square}$ denotes an invariant physical vector, $\underline{\square}$ denotes the components of a vector, $\underline{\underline{B}}_i = \begin{bmatrix} \vec{b}_1^i & \vec{b}_2^i & \vec{b}_3^i \end{bmatrix}$ is the vectrix associated with the i-th body frame [3], superscript "$\sqcup$" denotes the skew matrix associated with a vector, Γ_i and Σ_i are the i-th rotary and prismatic "joint partial" respectively, i.e. the $3\times n_j$ matrices (n_j is the number of degrees of freedom allowed by the joint) that, if post multiplied by the i-th joint velocity vector (angular $\underline{\dot{\theta}_i}$ or linear $\underline{\dot{s}_i}$), provide the relative velocity of the i-th body with respect to the (i-1)-th body [2]. Moreover, following the standard modal decomposition approach [4], $\underline{\Phi_k^{(i)}(P_i)}$ is the k-th modal shape associated with the i-th body and evaluated in the generic point P_i of body i, while q_k is the k-th modal amplitude and n_F is the total number of elastic modes. For the sake of clarity, it is important to notice that $\underline{q} = \begin{bmatrix} q_1 & \cdots & q_k & \cdots & q_{n_F} \end{bmatrix}$ contains the concatenation of the elastic modes of all the flexible bodies that compose the structure, so $\underline{\Phi_k^{(i)}(P_i)}$ is a zero vector when k corresponds to the elastic mode of a body different from the i-th one. The "Eulerian" velocities of Eqs. 1-2 are a function of the generalized velocities, i.e. the minimum-dimension set of velocities that completely describe the system dynamics. Considering a typical spacecraft topology and calling the bus "body 1", the vector of generalized velocities shows the following structure:

$$\underline{u} = \begin{bmatrix} \underline{v_{Q_1}}^T & \underline{\omega_1}^T & \left\{\underline{\dot{\theta}_i}^T\right\}_{N_{RJ}} & \left\{\underline{\dot{s}_i}^T\right\}_{N_{PJ}} & \dot{q}_1 & \cdots & \dot{q}_{n_F} \end{bmatrix} \tag{3}$$

where $\underline{v_1}$ and $\underline{\omega_1}$ are the components of linear and angular velocity of the bus respectively (being the root body, Q_1 can be any point of body 1), written with respect to $\underline{\underline{B_1}}$; the terms in parentheses refer to N_{RJ} revolute joints and N_{PJ} prismatic joints, respectively. To pass from generalized velocities to the Eulerian ones, it is necessary to introduce the partial velocity matrices, which play a crucial role in the Kane's formulation [5]. Specifically, the $3N_B \times 1$ vector (N_B is the number of bodies) containing the velocities of all points Q_i written with respect to $\underline{\underline{B_i}}$ is obtained by pre-multiplying $\underline{u}$ by the matrix of linear partial velocities V, while the angular velocities are provided by the use of the matrix of angular partial velocities Ω. Both V and Ω have dimensions $3N_B \times n$, where n is the total number of degrees of freedom of the structure. Each $3 \times n$ block is associated with a body, while each column is associated with a single degree of freedom of the system. Thanks to recursion, in Eqs. 1-2, it is possible to identify a repeating structure even in the partial velocities. Specifically, the i-th $3 \times n_{DOF}$ block shows the following structure:

$$V_i = \begin{bmatrix} \underset{i\leftarrow 1}{R} & -\underset{i\leftarrow 1}{R}\,\underline{\tilde{r}_{Q_1Q_i}^{(1)}} & \left\{-\underset{i\leftarrow j}{R}\,\underline{\tilde{r}_{Q_jQ_i}^{(j)}}\,\Gamma_j\right\}_{N_{RJ}} & \left\{\underset{i\leftarrow j}{R}\,\Sigma_{j+1}\right\}_{N_{PJ}} & \left\{\sum_{m=1}^{i-1}\underset{i\leftarrow m}{R}\,\underline{\Phi_k^{(m)}(Q_{m+1})}\right\}_{n_F} \end{bmatrix} \tag{4}$$

$$\Omega_i = \begin{bmatrix} 0_{3\times 3} & \underset{i\leftarrow 1}{R} & \left\{\underset{i\leftarrow j}{R}\,\Gamma_j\right\}_{N_{RJ}} & 0_{3\times n_{PJ}} & 0_{3\times n_F} \end{bmatrix} \tag{5}$$

where $\underset{i\leftarrow j}{R}$ is the rotation matrix from frame $\underline{\underline{B_j}}$ to frame $\underline{\underline{B_i}}$, superscript (i) in vectors specifies the frame with respect the components are written to, j refers to the body downstream of the joint whose degrees of freedom are being considered, n_{PJ} is the total number of degrees of freedom associated with prismatic joints. Moreover, the last component in Eq. 4 needs the introduction of the concept of "kinematic chain" to be explained. The kinematic chain can be seen as a branch of the tree topology of the multibody spacecraft. Every kinematic chain starts from the root body (body 1) and branches out to one of the terminal bodies: the number of kinematic chains of a structure corresponds to the number of end bodies. Hence, the index m in the last term of Eq. 4 proceeds only along bodies belonging to the same kinematic chain. Furthermore, all the terms in parentheses of Eqs. 4-5 must be replaced by blocks of zeros (with consistent dimensions) if the two considered bodies do not belong to the same kinematic chain.

To complete the kinematic description, accelerations must be derived. In Kane's formulation, it is important to identify the terms of the accelerations that do not depend on the time derivative of the generalized velocities. These terms are called "remainder accelerations" and, with reference to the building blocks in Figs. 1-2, show the following structure:

$$RJ: \begin{cases} \underline{a_i^{(R)}} = \underset{i\leftarrow i-1}{R}\,\underline{a_{i-1}^{(R)}}\Big|_{Q_i} \\ \underline{a_i^{(R)}}\Big|_{Q_{i+1}} = \underset{i\leftarrow i-1}{R}\,\underline{a_{i-1}^{(R)}}\Big|_{Q_i} - \underline{\tilde{r}_{Q_iQ_{i+1}}}\,\underline{\omega_i^*} - \underline{\tilde{\omega}_i}\,\underline{\tilde{r}_{Q_iQ_{i+1}}}\,\underline{\omega_i} + 2\underline{\tilde{\omega}_i}\sum_{k=1}^{n_F}\underline{\Phi_k^{(i)}(Q_{i+1})}\,\dot{q}_k \\ \underline{\alpha_i^{(R)}} = \underset{i\leftarrow i-1}{R}\,\underline{\alpha_{i-1}^{(R)}} + \dot{\Gamma}_i\dot{\theta}_i + \underline{\tilde{\omega}_i}\,\Gamma_i\dot{\theta}_i \end{cases} \tag{6}$$

$$
PJ: \begin{cases}
\underline{a_i^{(R)}} = \underset{i\leftarrow i-1}{R}\,\underline{a_{i-1}^{(R)}}\Big|_{Q_i} \\
\underline{a_i^{(R)}}\Big|_{Q_{i+1}} = \underset{i\leftarrow i-1}{R}\,\underline{a_{i-1}^{(R)}}\Big|_{Q_i} - \widetilde{\underline{r}_{Q_iQ_{i+1}}}\,\underline{\omega_i^*} - \left(\dot{\Sigma}_{i+1}s_{i+1} + 2\Sigma_{i+1}\dot{s}_{i+1}\right)^{\square}\underline{\omega_i} \\
\qquad - \widetilde{\underline{\omega}_i}\,\widetilde{\underline{r}_{Q_iQ_{i+1}}}\,\underline{\omega_i} + 2\widetilde{\underline{\omega}_i}\sum_{k=1}^{n_F}\underline{\Phi_k^{(i)}(Q_{i+1})}\,\dot{q}_k + \dot{\Sigma}_{i+1}\dot{s}_{i+1} \\
\underline{\alpha_i^{(R)}} = \underset{i\leftarrow i-1}{R}\,\underline{\alpha_{i-1}^{(R)}}
\end{cases} \tag{7}
$$

$$
\underline{\omega_i^*} = \underset{i\leftarrow 1}{\dot{R}}\,\underline{\omega_1} + \sum_{m=2}^{i-1}\left(\underset{i\leftarrow m}{\dot{R}}\,\Gamma_m\dot{\theta}_m + \underset{i\leftarrow m}{R}\,\dot{\Gamma}_m\dot{\theta}_m\right) + \dot{\Gamma}_i\dot{\theta}_i \tag{8}
$$

Actually, even in this case the structure must follow the scheme of kinematic chains: the passage from body i-1 to body i must be intended as a passage between two consecutive bodies on the same kinematic chain, not as a passage between two bodies with consecutive numeration. As for the last term of Eq. 4, the index m in Eq. 8 proceeds along the kinematic chains, not following the consecutive numeration.

Kane's equations

Applying the Kane's procedure to derive the dynamics of a multibody structure, the following expression is obtained:

$$
\begin{aligned}
&\left\langle V^T\{MV - S\Omega + B\Delta\} + \Omega^T\{SV - J\Omega + C\Delta\} + \Delta^T\left\{B^TV_F - C^T\Omega_F + Y\Delta\right\}\right\rangle \underline{\dot{u}} = \\
&= V^T\left\{-M\underline{a}^{(R)} + S\underline{\alpha}^{(R)} + [\underline{\tilde{\omega}}S\underline{\omega}] - 2[\underline{\tilde{\omega}}B]\right\} + \Omega^T\left\{-S\underline{a}^{(R)} - J\underline{\alpha}^{(R)} - [\underline{\tilde{\omega}}J\underline{\omega}] - 2[N\underline{\omega}]\right\} + \\
&\quad + \Delta^T\left\{-B^T\underline{a}_F^{(R)} - C^T\underline{\alpha}_F^{(R)} + \left[\underline{\omega}^TL\underline{\omega}\right] + 2\left[\underline{\omega}^T\underline{d}\right]\right\} - K\underline{q} - Z\underline{\dot{q}} + \underline{\hat{f}}
\end{aligned} \tag{9}
$$

where $\Delta = \begin{bmatrix} 0_{n_F\times n_R} & I_{n_F\times n_F}\end{bmatrix}$, being n_R and n_F the rigid and flexible degrees of freedom respectively of the whole system, M, S and J are matrices containing masses, static moments and inertia moments respectively, B and C are matrices containing translation and rotation modal participation factors respectively of all the flexible bodies of the structure; Y is the modal mass matrix, $\underline{a}^{(R)}$ and $\underline{\alpha}^{(R)}$ are $3N_B \times 1$ vectors containing respectively linear and angular remainder accelerations of all the bodies, subscript F in vectors and matrices indicates that only rows associated with flexible bodies must be retained, $\underline{\omega}$ is the $3N_B \times 1$ vector containing the angular velocities of all the bodies, N, L and D are three other modal integrals (in addition to B, C and G), K is the stiffness matrix, Z is the damping matrix, $\hat{f}$ is the vector containing the generalized active forces, i.e. the projection of external and interface forces and torques along the directions of partial velocities [5]. The structure of the terms appearing in Eq. 9 are reported in the Appendix.

Extraction of constraint reactions

The unavailability of constraint reactions is a significant limitation in Kane's formulation. However, for spacecraft with a tree topology configuration, it is possible to reconstruct the time histories of constraint reactions quite easily through a post-processing approach that relies on Newton/Euler equations, following the numerical integration of Kane's equations. In fact, unlike Kane's method, the Newton/Euler formulation for multibody structures incorporates constraint reactions in the state vector (however, this inclusion leads to longer computational times) [1, 6]. The procedure follows the subsequent steps: after the numerical solution of Kane's equations, one obtains $\underline{u}(t)$ and $\underline{\dot{u}}(t)$. Then, by utilizing the partial velocities and remainder accelerations, it is possible to reconstruct the temporal profiles of velocities and accelerations for all bodies within the structure. As a result, the constraint reactions become the only unknowns in the Newton-Euler equations that can be resolved through a post-processing module. During this operation, a top-down approach is necessary, starting from the bodies at the end of the kinematic chains. This is because each of these bodies has only a single joint, and one can then proceed backward along the kinematic chain toward the root body. In the example of Fig. 4,

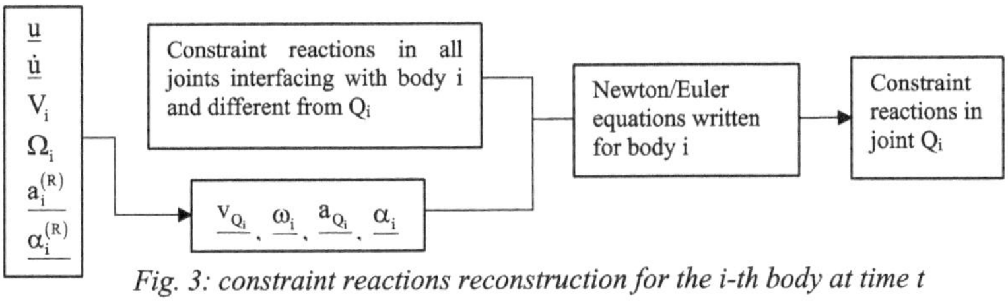

Fig. 3: constraint reactions reconstruction for the i-th body at time t

the constraint reactions must be computed first in joints Q_3, Q_4 and Q_6, and then in joints Q_2 and Q_5. The order of computing reactions for joints with the same subordination ranking can be any.

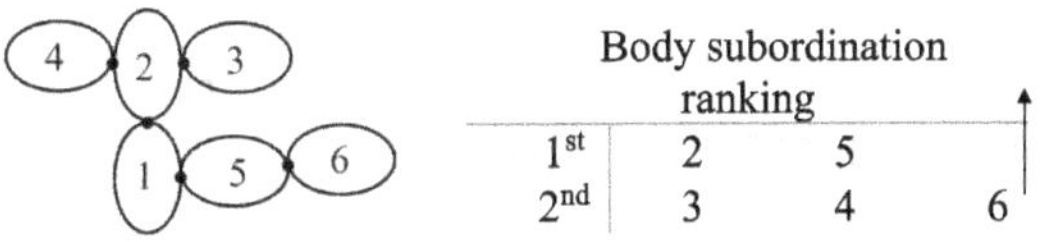

Fig. 4: example of top-down logic in deriving the constraint

Illustratuive simulation results

The presented formulation has been implemented in a numerical code to simulate the dynamic behavior of Explorer I, which is the same case studied in Reference [2]. The spacecraft consists of a cylindrical rigid bus and four appendages connected to the bus, as depicted in Figure 5. Similarly to the study in Reference [2], this investigation focuses on the spontaneous transition from a minor-axis to a major-axis spin caused by damping effects in the structure. However, there is a difference in the approach: while the analysis reported in [2] considered the appendages as rigid and introduced flexibility by incorporating a torsional spring-damper system at the interfaces between the appendages and the bus, in this work, the appendages are directly treated as flexible beams attached to the central the body of the spacecraft. Figure 6 illustrates the expected behavior of the bus angular velocity components, which exhibit the previously mentioned transition of the rotational behavior.

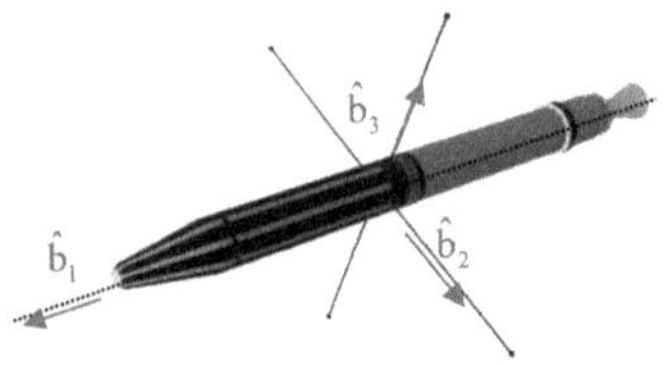

Fig. 5: sketch of the Explorer I [7]

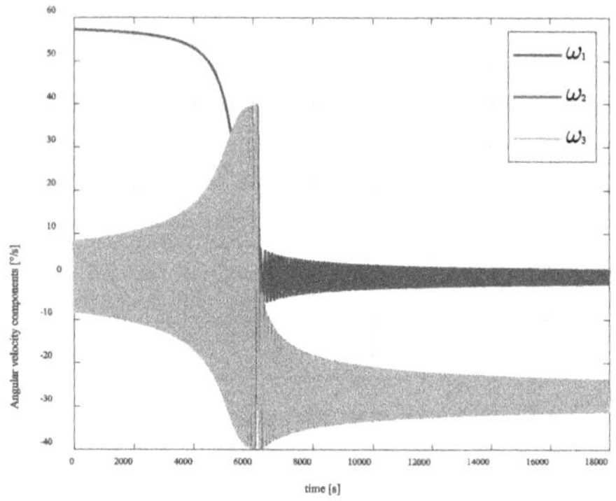

Fig. 6: time histories of the angular velocity components

Concluding remarks

A revised formulation of Kane's dynamical equations for a flexible multibody spacecraft is presented. By conducting a kinematic analysis, the expressions for partial velocities and remainder accelerations are derived, while emphasizing their recursive nature. The complete matrix formulation is provided, along with a proposed strategy for determining constraint reactions. Additionally, a numerical implementation of the formulation at hand is presented using the case of the Explorer I mission.

References

[1] A. Pisculli, L. Felicetti, M. Sabatini, P. Gasbarri and G. B. Palmerini, "A Hybrid Formulation for Modelling Multibody Spacecraft", Aerotecnica Missili & Spazio, no. 94, pp. 91-101, 2015. https://doi.org/10.1007/BF03404692

[2] E. T. Stoneking, "Implementation of Kane's Method for a Spacecraft", in AIAA Guidance, Navigation, and Control (GNC) Conference, Boston, 2013. https://doi.org/10.2514/6.2013-4649

[3] P. C. Hughes, "Spacecraft Attitude Dynamics", Wiley, 1986. https://doi.org/10.1017/S0001924000015578

[4] S. S. Rao, "Vibration of Continuous Systems", Wiley, 2007. https://doi.org/10.1002/9780470117866

[5] C. M. Roithmayr and D. H. Hodges , "Dynamics: Theory and Application of Kane's Method", Cambridge, 2016. https://doi.org/10.1115/1.4034731

[6] P. Santini and P. Gasbarri, "Dynamics of multibody systems in space environment; Lagrangian vs. Eulerian approach". Acta Astronautica, vol. 54, no. 1, pp. 1-24, 2004. https://doi.org/10.1016/S0094-5765(02)00277-1

[7] "Explorer 1 - Overview", NASA JPL, [Online]. Available: https://explorer1.jpl.nasa.gov/about/.

APPENDIX

A1. mass distribution and modal integrals

In the following expressions, the index "j" indicates the body, while the indices "k" and "l" identify the elastic mode.

- $m_j = \int_{B_j} dm$
- $\underline{b}^j_k = \int_{B_j} \underline{\Phi}^j_k \, dm$
- $\underline{s}_j = \int_{B_j} \underline{r}_{Q_jP_j} \, dm + \sum_{k=1}^{n_F} \underline{b}^j_k q_k$
- $\underline{c}^j_k = \int_{B_j} \widetilde{\underline{r}^j_{Q_jP_j}} \, \underline{\Phi}^j_k \, dm$
- $\underline{d}^j_{kl} = \int_{B_j} \widetilde{\underline{\Phi}^j_k} \, \underline{\Phi}^j_l \, dm$
- $\underline{g}^j_k = \underline{c}^j_k - \sum_{l=1}^{n_F} \underline{d}^j_{kl} q_k$
- $y^j_k = \int_{B_j} \left(\underline{\Phi}^j_k\right)^2 dm$
- $\underline{N}^j_k = -\int_{B_j} \widetilde{\underline{\Phi}^j_k} \, \widetilde{\underline{r}^j_{Q_jP_j}} \, dm$
- $J_j = -\int_{B_j} \widetilde{\underline{r}^j_{Q_jP_j}} \, \widetilde{\underline{r}^j_{Q_jP_j}} \, dm + \sum_{k=1}^{n_F} \left(\underline{N}^j_k + \underline{N}^{j\,T}_k\right) q_k$
- $\underline{L}^j_k = \underline{N}^j_k - \sum_{l=1}^{n_F} q_l \int_{B_j} \widetilde{\underline{\Phi}^j_l} \, \widetilde{\underline{\Phi}^j_k} \, dm$

A2. description of the terms of eq. 9

$$M = \begin{bmatrix} m_1 I_{3x3} & & & 0 \\ & m_2 I_{3x3} & & \\ & & \ddots & \\ 0 & & & m_{N_B} I_{3x3} \end{bmatrix} \quad S = \begin{bmatrix} \widetilde{\underline{s}_1} & & & 0 \\ & \widetilde{\underline{s}_2} & & \\ & & \ddots & \\ 0 & & & \widetilde{\underline{s}_{N_B}} \end{bmatrix} \quad J = \begin{bmatrix} J_1 & & & 0 \\ & J_2 & & \\ & & \ddots & \\ 0 & & & J_{N_B} \end{bmatrix}$$

$$B = \begin{bmatrix} \underline{b}^1_1 & \underline{b}^1_2 & \cdots & \underline{b}^1_{n_F} \\ \underline{b}^2_1 & \underline{b}^2_2 & \cdots & \underline{b}^2_{n_F} \\ \vdots & \vdots & \vdots & \vdots \\ \underline{b}^{N_{B,F}}_1 & \underline{b}^{N_{B,F}}_2 & \cdots & \underline{b}^{N_{B,F}}_{n_F} \end{bmatrix} \quad C = \begin{bmatrix} \underline{c}^1_1 & \underline{c}^1_2 & \cdots & \underline{c}^1_{n_F} \\ \underline{c}^2_1 & \underline{c}^2_2 & \cdots & \underline{c}^2_{n_F} \\ \vdots & \vdots & \vdots & \vdots \\ \underline{c}^{N_{B,F}}_1 & \underline{c}^{N_{B,F}}_2 & \cdots & \underline{c}^{N_{B,F}}_{n_F} \end{bmatrix} \quad G = \begin{bmatrix} \underline{g}^1_1 & \underline{g}^1_2 & \cdots & \underline{g}^1_{n_F} \\ \underline{g}^2_1 & \underline{g}^2_2 & \cdots & \underline{g}^2_{n_F} \\ \vdots & \vdots & \vdots & \vdots \\ \underline{g}^{N_{B,F}}_1 & \underline{g}^{N_{B,F}}_2 & \cdots & \underline{g}^{N_{B,F}}_{n_F} \end{bmatrix}$$

Materials Research Proceedings 37 (2023) 487-494 | https://doi.org/10.21741/9781644902813-107

$$Y=\begin{bmatrix} Y^1 & & & 0 \\ & Y^2 & & \\ & & \ddots & \\ 0 & & & Y^{N_{B,F}} \end{bmatrix} \quad \text{where} \quad Y^j=\begin{bmatrix} y_1^j & & \\ & \ddots & \\ & & y_{n_{F,j}}^j \end{bmatrix}$$

$N_{B,F}$ is the number of flexible bodies in the multibody spacecraft

$n_{F,j}$ is the number of elastic modes associated with the j-th body

$$[\underline{\tilde{\omega}}S\underline{\omega}]=\begin{bmatrix} \tilde{\omega}_1 \tilde{s}_1 \omega_1 \\ \tilde{\omega}_2 \tilde{s}_2 \omega_2 \\ \vdots \\ \tilde{\omega}_{N_B} \tilde{s}_{N_B} \omega_{N_B} \end{bmatrix} \qquad [\underline{\tilde{\omega}}B]=\begin{bmatrix} \tilde{\omega}_1 \sum_{k=1}^{n_F} b_k^1 \dot{q}_k \\ \tilde{\omega}_2 \sum_{k=1}^{n_F} b_k^2 \dot{q}_k \\ \vdots \\ \tilde{\omega}_{N_B} \sum_{k=1}^{n_F} b_k^{N_{B,F}} \dot{q}_k \end{bmatrix}$$

$$[\underline{\tilde{\omega}}J\underline{\omega}]=\begin{bmatrix} \tilde{\omega}_1 J_1 \omega_1 \\ \tilde{\omega}_2 J_2 \omega_2 \\ \vdots \\ \tilde{\omega}_{N_B} J_{N_B} \omega_{N_B} \end{bmatrix} \qquad [N^T\underline{\omega}]=\begin{bmatrix} \sum_{k=1}^{n_F} N_k^{1\,T} \dot{q}_k \omega_1 \\ \sum_{k=1}^{n_F} N_k^{2\,T} \dot{q}_k \omega_2 \\ \vdots \\ \sum_{k=1}^{n_F} N_k^{N_{B,F}\,T} \dot{q}_k \omega_{N_B} \end{bmatrix}$$

$$[\underline{\omega}^T L\underline{\omega}]=\begin{bmatrix} \sum_{j=1}^{N_{B,F}} \omega_j^T L_1^j \omega_j \\ \sum_{j=1}^{N_{B,F}} \omega_j^T L_2^j \omega_j \\ \vdots \\ \sum_{j=1}^{N_{B,F}} \omega_j^T L_{n_F}^j \omega_j \end{bmatrix} \qquad [\underline{\omega}^T \underline{d}]=\begin{bmatrix} \sum_{j=1}^{N_{B,F}} \omega_j^T \left(\sum_{k=1}^{n_F} d_{1k}^j \dot{q}_k\right) \\ \sum_{j=1}^{N_{B,F}} \omega_j^T \left(\sum_{k=1}^{n_F} d_{2k}^j \dot{q}_k\right) \\ \vdots \\ \sum_{j=1}^{N_{B,F}} \omega_j^T \left(\sum_{k=1}^{n_F} d_{n_F k}^j \dot{q}_k\right) \end{bmatrix}$$

$$K=\begin{bmatrix} 0_{n_R \times n_R} & & & \\ & (\lambda_1)^2 & 0 & 0 \\ & 0 & \ddots & 0 \\ & 0 & 0 & (\lambda_{n_F})^2 \end{bmatrix} \qquad Z=\begin{bmatrix} 0_{n_R \times n_R} & & & \\ & 2\zeta_1\lambda_1 & 0 & 0 \\ & 0 & \ddots & 0 \\ & 0 & 0 & 2\zeta_{n_F}\lambda_{n_F} \end{bmatrix}.$$

λ_k is the natural frequency of the k-th elastic mode

ζ_k is the damping factor of the k-th mode

Concept and feasibility analysis of the Alba CubeSat mission

M. Mozzato[1*], S. Enzo[1], R. Lazzaro[1], M. Minato[1], G. Bemporad[1], D. Visentin[1], F. Filippini[1], A. Dalla Via[1], A. Farina[1], E. Pilone[1], F. Basana[2], L. Olivieri[2], G. Colombatti[3], A. Francesconi[3]

[1] University of Padova, Via 8 Febbraio, 2 - 35122 Padova, Italy

[2] CISAS G. Colombo, Via Venezia, 15 - 35131 Padova, Italy

[3] DII/CISAS, Via Venezia, 1 - 35131 Padova, Italy

*monica.mozzato@studenti.unipd.it

Keywords: CubeSat; Feasibility Analysis, Debris, Fly Your Satellite! Design Booster

Abstract. Alba CubeSat is a 2U CubeSat which is being developed by a student team at the University of Padova. The Alba project aims to design, build, test, launch, and operate University of Padova's first student CubeSat, featuring four different payloads that aim to satisfy four independent objectives. The first goal is to collect data regarding the debris environment in LEO, the second goal is the study of the satellite vibrations, the third one is about CubeSat attitude determination through laser ranging technology and the fourth goal concerns satellite laser and quantum communication. The Alba CubeSat mission has been selected by ESA to join the Fly Your Satellite! Design Booster programme in December 2022. This paper presents the feasibility study of the Alba CubeSat mission reproduced in the framework of the "Space Systems Laboratory" class of M.Sc. in Aerospace Engineering at the University of Padova. In the beginning, a mission requirements definition was conducted. After that, the mission feasibility was considered, with preliminary requirements verification to assess the ability of the spacecraft to survive the space environment, including compliance with Debris Mitigation Guidelines, ground station visibility and minimum operative lifetime evaluation. The Alba mission sets a base for a better understanding of the space environment and its interaction with nanosatellites, and an improvement of the accuracy of debris models. Furthermore, this paper, describing the educational experience and the results achieved, will provide a useful example for future students' studies on CubeSat mission design.

Keywords: CubeSat; Feasibility Analysis, Debris, Fly Your Satellite! Design Booster

Introduction

One of the most common trends in the space sector is the evolution of CubeSats, micro satellites measuring just a few tens of centimeters in size. Their strength is not just the small dimensions and weight that guarantee a reduction of power consumption and costs, a CubeSat is also the perfect chance to sharpen the students' abilities and knowledge of the space industry. For that purpose, an accurate and comprehensive research of a CubeSat mission has been done by students, with a special focus on the requirements definition, starting with mission objectives. This activity has given the opportunity to face and address the issues and challenges of a space mission design. In the present work an alternative design of the 2U CubeSat mission of the student team Alba CubeSat of the University of Padova [1] is presented.

Students' work was aimed to define the requirements, based on which the commercial off-the-shelf (COTS) components have been selected for a preliminary design, while maintaining the original design of the four payloads. In addition, the students' team has identified and evaluated the risks and the success criteria, and they have carried out a wide variety of simulations in order to perform a complete feasibility analysis.

Mission overview

Alba CubeSat is a project that aims to design, build, test, launch, and operate University of Padova's first student-built 2U CubeSat, which features four distinct payloads that seek to achieve four independent objectives. In particular, the derived mission requirements are:

1. to collect in-situ measurements of the sub-mm space debris environment in LEO;
2. to study the micro-vibration environment on the satellite throughout different mission phases;
3. to do orbit and attitude determination through laser ranging;
4. to investigate alternative systems for possible Satellite Quantum Communication applications on nanosatellites using active retro-reflectors.

Starting from these, the system requirements were defined according to the process shown in Figure 1. The identification of requirements is a milestone that is the basis of any design activity. In order to define the requirements, assumptions were made, such as the altitude and type of operative orbit, the launch vehicle that will carry the CubeSat and the launch date. The study of the micro-vibration sensor and active retro-reflectors is beyond the scope of this paper.

For every requirement identified, one or more of these verification methods were assigned: analysis, test, review of design and inspection. Throughout the analysis, each requirement was subjected to review, update and tailoring as the mission development progressed and different needs or constraints emerged.

Another critical activity was the identification of risks and success criteria. The students compiled a risk register, in which the level of risk was evaluated and mitigation actions were proposed. Since this is a student-designed CubeSat project, the majority of the success criteria were linked to an educational purpose.

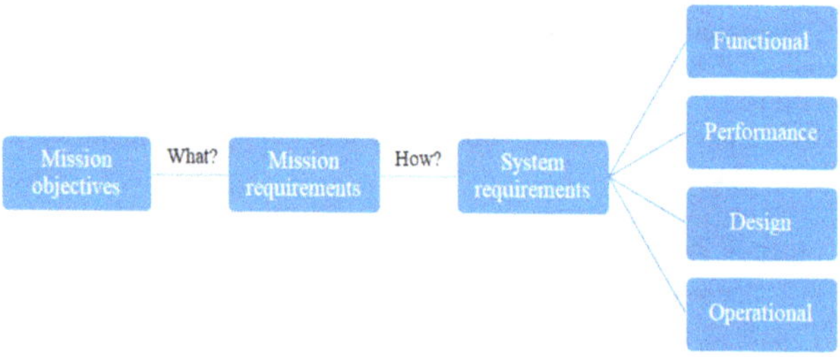

Figure 1: Logical scheme followed to identify requirements

Preliminary Analyses

An analysis of the possible target orbits has been performed considering the European Code of Conduct for space debris mitigation and the orbits commonly reached by other missions. A 500 km Sun-Synchronous Orbit (SSO) has been selected as the baseline for the mission. In order to be compatible with as many launches as possible, the LTAN has not been fixed. Therefore, the two extreme cases have been considered in the analyses, namely a worst hot case (WHC) scenario with an LTAN of 6AM (Dawn/Dusk), and a worst cold case (WCC) with an LTAN of 12AM (Noon/Midnight). The launch date has been assumed to be 30/03/2027 and the eccentricity of the orbit is 0.001.

With the chosen design (shown in the following section) mass budget and atmospheric reentry comply with ESA guidelines [2][3]. The thermal and power budgets have been calculated considering two extreme cases for on-board activities. At this stage, four operational modes have been defined: safe, idle, communication and payload. The two worst cases scenarios are representative of a safe mode for the minimum power consumption, and a communication mode for the maximum. Ground station visibility has been taken into account for link and data budget.

Environmental analyses have been performed to ensure component compatibility with thermal ranges, radiation and atomic oxygen interactions. Systema, an Airbus software, has been used for thermal and radiation analyses. The radiation analysis shows the accumulated radiation dose over the one-year operational lifespan (Figure 2). The thermal analyses show that the internal components reached a maximum temperature of 45 °C and a minimum of 39 °C in the WHC (Figure 3). In the WCC scenario, internal temperature ranges from -10 °C to 20 °C.

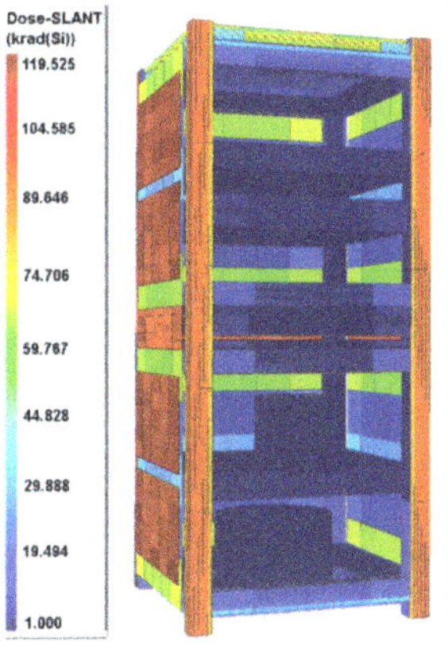

Figure 2: One-year cumulative radiation

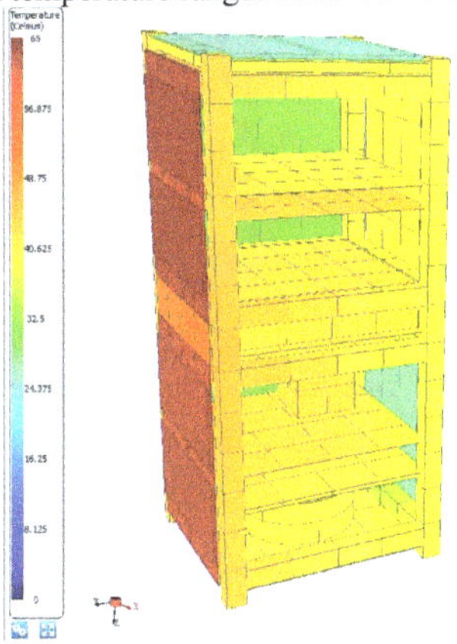

Figure 3: Component's temperature WHC

Subsystems and components selection

Component selection comes from the necessity to meet the requirements and system-level compatibility. After preliminary analyses, the following design choices have been made (Figure 4). Except the impact sensor, micro-vibration sensor and active retro-reflectors which are in-house developed, the other components are all COTS.

1. The four payloads are: the impact sensor which is a new system based on the technology demonstrator DRAGONS [4]; the laser ranging payload is composed by 6 CCRs with a 12.7 mm diameter; the micro-vibration sensor and the active retro-reflectors which are considered as black boxes with known specifications.
2. The 2U structure is made of an aluminum alloy (Al 6061) and is qualified according to ECSS-E-ST-10-03.
3. The ADCS is able to meet the three-axis stabilization pointing accuracy needed by the payloads (± 20° for each axis).
4. The power system includes: a 43 Wh battery pack and seven 1U solar panels.
5. Thermal management is based on passive conduction and radiation except for the battery pack which is equipped with a heater.
6. The OBC has been designed for space applications with limited resources. It fulfills the processing power, memory capacity, radiation tolerance and system-compatibility requirements.

7. The communication system comprehends a transceiver and an antenna. It is the most power-consuming subsystem during transmission, with a power consumption up to 3.3 W and an output power of 1 W.

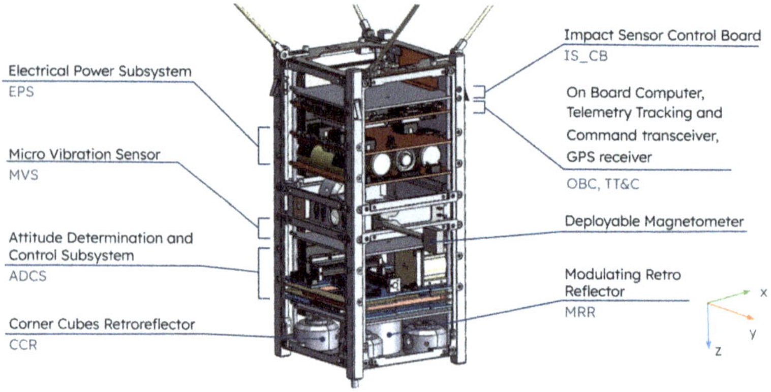

Figure 4: Internal components, solar panels have been removed

Conclusions

The successful development of this work involved several key tasks, including identifying requirements and their corresponding verification methods, identifying risks and find mitigation actions, designing and studying the functionality of two payloads (laser ranging and impact sensor), selecting appropriate commercial off-the-shelf (COTS) components for the subsystems, and conducting analyses to verify the specified requirements.

One of the critical points identified was the enclosure of all the components in a 2U, in particular the CCRs. Therefore, it is to be considered the development of a homemade structure for the CCRs to address this issue. Moreover, the power budget analysis revealed a potential insufficiency in power generation by the solar panels during the worst-case scenario (noon-midnight orbit). However, it is important to note that power consumption was likely overestimated.

The present work has contributed to enhancing students' understanding of how to conduct a feasibility study for a space mission. It also can serve as a useful reference, assisting anyone who is embarking on their first mission feasibility study.

References

[1] Development of a multi-payload 2U CubeSat: the Alba project - F.Basana, A.A. Avram, F.Fontanot, L.Lion, A.Francesconi - 4th Symposium on Space Educational Activities - Barcelona, April 2022

[2] European Space Agency - Margin Philosophy for Science Assessment Studies - Issue 1 - Revision 3 - 15/06/2012

[3] ESSB-HB-U-002 - ESA Space Debris Mitigation WG. - ESA Space Debris Mitigation Compliance, Verification Guidelines - Issue 1 - 19/02/2015

[4] LIOU, J.-C., et al. dragons-a micrometeoroid and orbital debris impact sensor. In: Nano-Satellite Symposium (NSAT). 2015

Space object identification and correlation through AI-aided light curve feature extraction

Chiara Bertolini[1,a], Riccardo Cipollone[1,b*], Andrea De Vittori[1,c], Pierluigi Di Lizia[1,d], Mauro Massari[1,e]

[1] Department of Aerospace Science and Technology, Politecnico Di Milano, Via Giuseppe La Masa 54 20156, Milano

[a]chiara.bertolini@polimi.it, [b]riccardo.cipollone@polimi.it, [c]andrea.devittori@polimi.it, [d]pierluigi.dilizia@polimi.it, [e]mauro.massari@polimi.it

Keywords: Light Curves, Neural Networks, Correlation

Abstract. With the constant growth of objects in orbit, the monitoring and cataloging of space population is essential. Light curves obtained from ground stations support this point, providing valuable information about the observed objects. The idea of using them to identify an object through correlation with a catalogued reference takes hold from their wide availability. This article focuses on the development of a tool for the analysis and correlation of two light curves, ARIEL. This tool is built through neural networks and declined in three strategies, each with its own goal: ROGUE, LINDEN and SIERRA. The light curves were retrieved via the database managed by the Mini-MegaTORTORA observatory and filtered using the Savitzky-Golay filter.

Introduction

The near-Earth environment is getting more populated, as commercial applications become a substantial part of the space economy, increasing the risk of collisions and fragmentations [1]. To keep track of this expanding population and to assess the risk of in-orbit collision and fragmentation, space agencies deploy Space Surveillance and Tracking (SST) systems [2]. Ground-based stations allow to retrieve orbital data of human-made objects [3]. When dealing with optical telescopes, photometry analysis can be performed, and light curves are generated as a consequence. Light curves, which represent object brightness variations, provide information on orbit regime, tumbling motion, and spacecraft geometry, enabling characterization of observed objects.

In general, traditional estimation-based methods, like the so-called Light Curve Inversion, have been extensively used for the identification of space objects [4]. However, complex models have to be considered and the resulting analysis is computationally time-consuming. Consequently, the state of the art is now drifting to the use of machine learning with bespoke Convolutional Neural Networks (CNN) or Recursive Neural Networks (RNN) ensuring up to 90% prediction accuracy [5][6].

This project proposes a novel approach to light curve characterization through the Machine Learning based Light curve Analysis (ARIEL) tool. Raw light curves are recovered from the database managed by the Mini-MegaTORTORA (MMT-9) observatory [7] and then pre-processed, before being fed into three different neural networks (NN): ROGUE, LINDEN, and SIERRA.

Performance for these networks is then assessed using different datasets obtained by varying the number of spacecraft platforms.

Theoretical background

As mentioned above, light curves are recovered from shots acquired with optical telescopes. An example of the observatory is represented by MMT-9 system [6], which predisposes a constantly updated database for human-made space objects. From that database, the main information recovered are the space object characteristics and corresponding light curves retrieved, each associated with a track ID and time-tag. The data are summarized in the following Table.

Table 1 – Data recovery from MMT-9 database

Data	Number of objects
Total number of objects recovered	6.314
Total number of light curves available	Over 150.000
Objects per category	
Type of orbit	LEO: 5.206; GEO: 174; Other: 934
Attitude regime	Periodic: 985; Aperiodic: 1550; Non variable: 3779
Type of object	Rocket bodies: 827; Debris: 839; Satellites: 4648

Before entering the NN, however, these raw light curves are pre-processed using the Savitzky-Golay filter [8], with smoothing properties particularly indicated for reducing high frequency noise. The outcome can be seen in Figure 1, where the grey signal is the raw light curve, while the red is the filtered one.

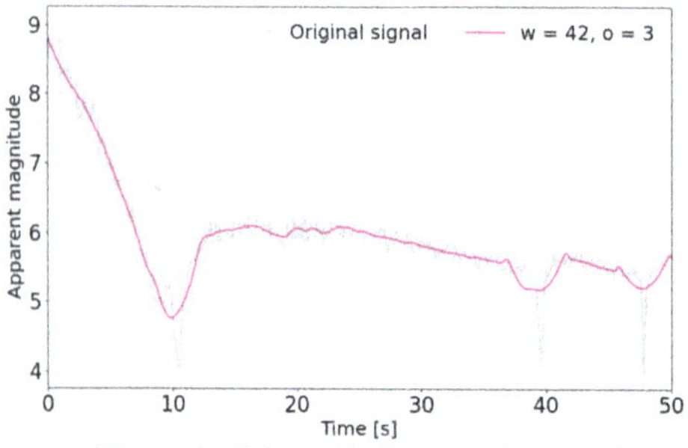

Figure 1 - Filtered light curve (cropped)

To assess the performances of the ARIEL networks, the focus was mainly on objects belonging to LEO or Low-Medium Orbit (LMO) regions, with periodic or aperiodic tumbling motion. The corresponding light curves have been filtered and stored in datasets, accompanied by the name and the type of object considered, i.e. Rocket body, Debris or Satellite. To avoid any bias towards a specific spacecraft, different platforms for each type are considered. For example, a dataset considers light curves belonging to Iridium and NOAA objects, but both labeled as "Satellites" – as stated in the MMT-9 database. Two different sets have therefore been considered. First, the Nominal dataset represent nominal conditions of operation of ARIEL, meaning a limited number of platforms a first version featuring periodic objects only, and a following one including aperiodic too. Then, a Variability test assesses the extent of ARIEL capabilities: different datasets are built considering an increasing number of platforms for each dataset, taking care that the three types data distribution is balanced out. All the objects considered have periodic or aperiodic attitude regime.

Deep learning networks are a subset of Machine learning models. Different NN structures can also be employed such as CNN and RNN: CNNs are particularly indicated to retain the general features of the input, while, RNN, such as the Long-Short Term Memory (LSTM) cells, take into account the input's time-dependence. After the NN setup, it needs to be trained and its performance assessed – mainly in terms of predictions' Accuracy. Particular attention has to be given in the model structure and dataset provided to avoid over- or underfitting of the network.

Siamese networks have a slightly different architecture [8]: the overall dataset is divided in Anchor, the reference, Positive and Negative, the closest and the farthest prediction from the reference. Then, an embedding model extracts features from the inputs and the network evaluates the distance Anchor-Positive and Anchor-Negative in order to bring the former closer and the latter farther. Thus a dedicated metrics, Similarity, is employed.

Architectures

Three different architectures are developed inside the ARIEL framework: ROGUE, LINDEN and SIERRA.

The Rocket bodies Light curves Identification (ROGUE) network aims at recognizing Rocket bodies among light curves of different types. The structure is a combination of CNN and LSTM cells. This test is conceived to verify the capability of the NN to identify a defined category of spacecraft.

The Light curve Identification and Correlation (LINDEN) NN compares two light curves and determines the correlation degree among the twos. Two models have to be therefore developed as shown in Figure 2:

- the Feature extraction part analyses the light curves and predicts the objects' type
- the Correlation evaluation block which, given the above-mentioned predictions, evaluates the distance between them.

The Feature extraction model is an improved version of the ROGUE model and the output gives a prediction vector over the class labels.

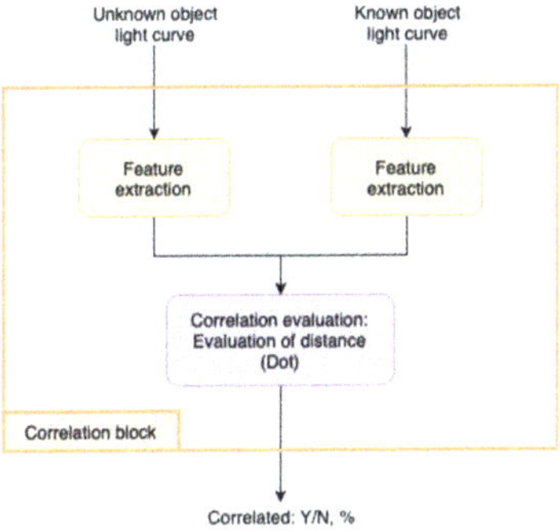

Figure 2 - LINDEN Structure

After having trained the Feature extraction part, it is inserted in the overall LINDEN Correlation block where the correlation between the prediction vectors is performed, thanks to a normalized dot product. As the Feature extraction model is frozen within the Correlation block, this allows to compute a correlation degree without being influenced by uncertainties in the model.

Siamese Network for Light curves Correlation (SIERRA) is a Siamese Network, which encompasses the above-mentioned Feature extraction block as embedding model.

Results

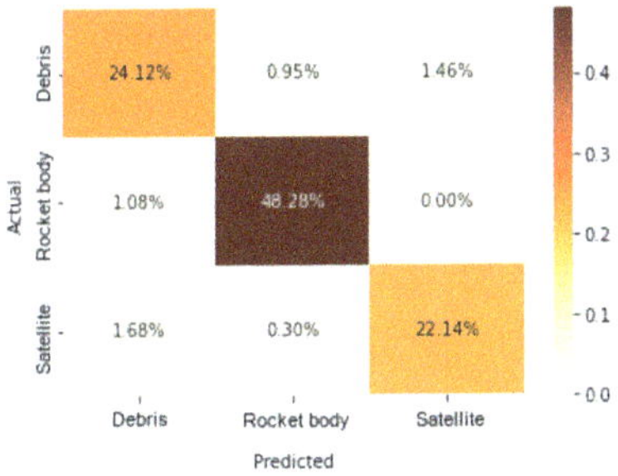

Figure 3 - Confusion matrix for LINDEN Feature extraction

Hereafter the results for ARIEL networks are summarized, obtained considering the above-mentioned datasets. The training has been performed using Google Colaboratory, where due to the limited GPU time availability, it has been divided in sessions from 100 to 200 epochs. The results are analyzed through confusion matrices, which compare predicted with actual labels. The more intense the color of the cell, the higher the prediction accuracy. An example can be shown in Figure 3 – where the results for LINDEN Feature Extraction for the second Nominal dataset are shown.

ROGUE: The results show around 97% accuracy for Nominal datasets while a drop can be observed for Variability sets – ranging from 95% to around 70% accuracy for increasing number of platforms. All in all, ROGUE can best differentiate the Rocket body type among up to 20 different platforms. However Nominal datasets do not present overfitting as Variability sets do.

LINDEN: As previously mentioned, the Feature extraction block is trained separately and then inserted in the overall Correlation block.

Feature extraction block: The results display over 95% accuracy in differentiating the type of objects inputted, for Nominal sets. Variability datasets still reach over 90% accuracy for lower numbers of satellite platforms, while severe overfitting can be noticed with increasing variety, with accuracy dropping down to 50% in the best-performing model. Moreover, the Debris type is completely missing among output predictions, which enforces the idea that they are more difficult to categorize due to their nature.

Correlation block: The performances observed prove that, as the levels reached in the Feature extraction block leave room for uncertainty, the results obtained in the Correlation block are quite scarce, in particular for the Variability datasets. Overfitting is observed also in Nominal conditions becoming even more relevant for the Variability datasets. However, the confusion observed is still below the 10% bound.

All in all, LINDEN proves its capabilities by granting an accurate type recognition, which allows correlation between the two inputs to be established properly. However only up to 20 platforms can be considered at the same time in order to obtain accurate results.

SIERRA: As expected, the overfitting present in the Feature extraction block propagates to the NN. The Similarity obtained in the different datasets is roughly giving a 10 % gap, therefore the Positive and Negative outcomes are properly distinguished. While using the Variability dataset with the lowest number of platforms – around 20 –, a remarkable confusion was observed. This was maybe due to Anchor and Negative having common characteristics not considered during the Feature extraction block.

Conclusions

ARIEL provides a strategy to identify objects according to their type and to establish a degree of correlation between the unknown object and a catalogued one. This is done by the analysis of light curves through a deep learning model combining CNN and LSTM layers that grasp general and time-dependent features at the same time. Three architectures are thus proposed, each focusing on a different aspect of the problem at hand: ROGUE, LINDEN and SIERRA.

The light curves are obtained from the MMT-9 database and have been pre-processed, in particular filtered with the Savitzky-Golay smoothing filter.

After extensive training using different datasets, the performances have been assessed, showcasing a resulting accuracy of around 90% in most test cases. The significant gap observed for the similarity in SIERRA proves these networks predict the type of object with little confusion. However these NN are limited by datasets including diverse platforms, where accurate type recognition is hampered, thus preventing the correlation to be performed. Moreover, overfitting is omnipresent: in some cases it becomes substantial, therefore impacting the accuracy of the predictions done.

Some options can hence be proposed to improve ARIEL, e.g. consider a smaller number of different platforms or restrict the problem to the recognition of platforms among a same type, or a same attitude regime (i.e. periodic, aperiodic or non variable), or even focus on the problem of the Debris type recognition. In fact it is the most mistaken type, as some of these objects are unused satellites or intact rocket body parts. Therefore, a dedicated analysis among Debris may be needed if those objects are involved.

References

[1] ESA Space Debris Office, "ESA annual space environment report," Tech. Rep. 6, Euro- pean Space Agency, Darmstadt, Germany, April 2022.

[2] EU SST, "European space surveillance and tracking," 2016. https://www.eusst.eu.

[3] Montaruli, Marco Felice & Facchini, Luca & Di Lizia, Pierluigi & Massari, Mauro & Pupillo, Giuseppe & Bianchi, Germano & Naldi, Giovanni. (2022). Adaptive track estimation on a radar array system for space surveillance. Acta Astronautica. 198. https://doi.org/10.1016/j.actaastro.2022.05.051

[4] B. Bradley and P. Axelrad, "Lightcurve in- version for shape estimation of GEO objects from space-based sensors," ISSFD, 2014.

[5] R. Linares and R. Furfaro, "Space object classification using deep convolutional neural networks," IEEE, 2016.

[6] E. Kerr, G. Falco, N. Maric, D. Petit, P. Talon, E. Geistere Petersen, C. Dorn, S. Eves, N. Sánchez-Ortiz, R. Dominguez Gonzalez, and J. Nomen-Torres, "Light curves for GEO object characterisation," (Darmstadt, Germany), ESA, ESA Space Debris Office, 4 2021. Proc. 8th European Conference on Space Debris.

[7] S. Karpov, E. Katkova, G. Beskin, A. Biryukov, S. Bondar, E. Davydov, E. Ivanov, A. Perkov, and V. Sasyuk, "Massive photometry of low altitude artificial satellites on Mini-Mega Tortora," 2015. Database: http://mmt.favor2.info/satellites

[8] N. Gallagher, "Savitzky-Golay smoothing and differentiation filter," 2020. https://eigenvector.com/wp-content/ uploads/2020/01/SavitzkyGolay.pdf

[9] H. Essam and S. Valdarrama, "Image similarity estimation using a Siamese Network with a Triplet loss," 2021. https://keras. io/examples/vision/siamese_network/

Development of a smart docking system for small satellites

Alex Caon[1,a*], Luca Lion[1,b], Lorenzo Olivieri[1,c], Francesco Branz[2,d], Alessandro Francesconi[2,e]

[1]C.I.S.A.S. - Centre of Studies and Activities for Space "G. Colombo", Via Venezia 15, Padova (Italy)

[2]Department of Industrial Engineering, University of Padova, via Venezia, 1, Padova (Italy)

[a]alex.caon@unipd.it, [b]luca.lion.1@phd.unipd.it, [c]lorenzo.olivieri@unipd.it, [d]francesco.branz@unipd.it, [e]alessandro.francesceoni@unipd.it

Keywords: Docking system, Autonomous system, Space Rider, Space Rider Observer Cube

Abstract. DOCKS is a smart docking system for space vehicles developed by the Department of Industrial Engineering, University of Padova, within the framework of the Space Rider Observer Cube (SROC) mission. The design and development of SROC is being conducted by a consortium of Italian entities under contract with the European Space Agency (ESA). The SROC mission is designed to be a payload on the ESA Space Rider (SR) spaceship. The main objective of the mission is to demonstrate the critical capabilities and technologies required to execute a rendezvous and docking mission in a safety-sensitive context. The space system is composed by a nanosatellite (approximately 12U CubeSat) and a deployment/retrieval mechanism mounted inside the payload bay of SR. During the mission, SROC will be released by SR, will perform inspection manoeuvres on SR and, at the end of the mission, will dock back inside the bay of SR, before re-entering Earth with the mothership. The docking functionality is provided by DOCKS. DOCKS is suitable for use onboard micro- and nanosatellites and merges a classical probe drogue configuration with a gripper–like design, to manage the connection between the parts. The system is equipped with a suite of sensors to estimate the relative pose of the target and with a dedicated computer, making it a smart standalone system. A laboratory prototype has been assembled and functionally tested, aiming at the validation of the capability to passively manage misalignments during the docking manoeuvre.

Introduction

The docking system (DOCKS) has been developed in the framework of the Space Rider Observer Cube (SROC) mission. The purpose of the mission is to demonstrate the capabilities and technologies required for rendezvous and docking with a target vehicle [1]. A brief description of the operations performed by SROC is:

1. Launch and early operations. SROC is stored inside the bay of Space Rider (SR). Once in orbit, SROC is pushed away from SR in order to begin its operative phases.
2. Proximity Operations. SROC is in a relative orbit with respect to SR in order to perform its observations.
3. Docking and Retrieval Phase. SROC approaches SR and docks with it in order to be re-stored inside the bay and return to Earth safely.

DOCKS overview

The DOCKing System (DOCKS) has been developed to be a standalone docking mechanism with an integrated set of sensors and a computer. In the following, all the parts of DOCKS will be described.

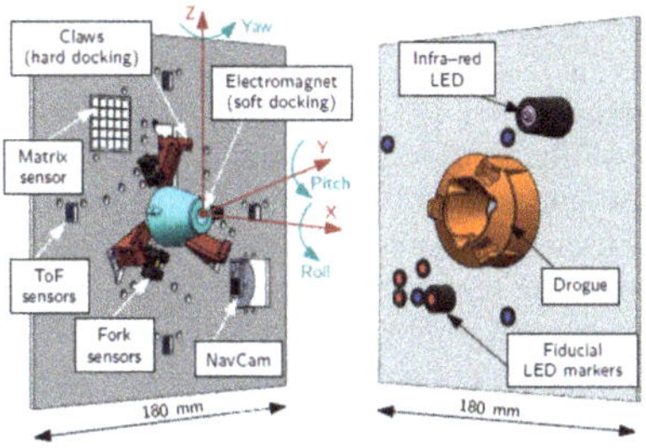

Figure 1: DOCKS-A and DOCKS-B. They are mounted on SROC and on SR respectively. In DOCKS-A it is also represented it frame of reference.

Docking mechanism

The mechanical connection between SROC and SR is provided by a docking mechanism that is composed by two main parts (Fig. 1). The first (DOCKS-A on SROC) is the active part with the mechanism, and the second (DOCKS-B on SR) is the counter part of the docking mechanism (the drogue) and the LEDs that allows the relative navigation.

The active part of docking mechanism is composed by two parts (shown in Fig. 2): a centring cone and three claws that provide the rigid mechanical connection.

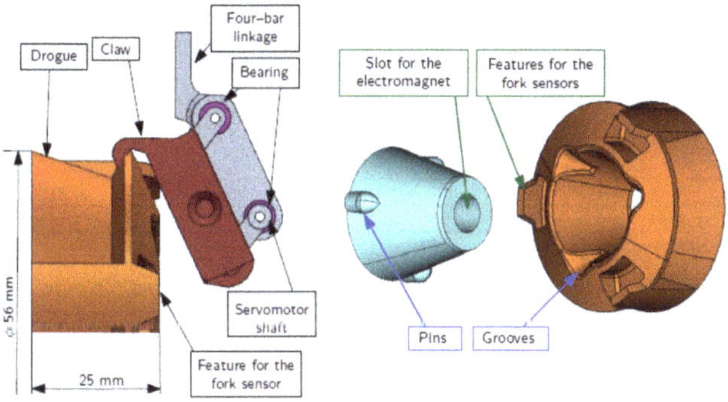

Figure 2: The centring cone and claws of DOCKS.

The centering cone has the purpose to force the alignment between SROC and SR when coupling with the drogue. In fact, the shape of the probe allows to tolerate 10 mm of lateral misalignment and 10 deg of pitch (and yaw) misalignment. In addition, the pins on the rim of the probe force the roll alignment, when they couple with the grooves on the drogue.

The hard docking is achieved with the three claws that ensure the rigid connection, by closing around the rim of the drogue. The claws are activated by a four-bar linkage in order to prevent linear actuations that could produce friction or jam the mechanism.

Sensor suite and estimation performances

As illustrated in Fig. 1, DOCKS-A is provided with three different sensors to measure the relative pose of DOCKS-B.

1. *A navigation camera*. Its purpose is to measure the entire relative pose of DOCKS-B. The NavCam with its computer, recognizes the pattern of LEDs on DOCKS-B and reconstruct its pose [2]. However, at distances lower than 50 mm, the camera is out of focus making the measurement unreliable.
2. *A set of four Time-of-Flight sensors*. They are employed to measure the distance and the relative pitch and yaw angles from 100 mm up to contact (as explained in Fig. 3) [3].
3. *A matrix sensor*. It is used to measure the relative position along the y and z axes (which is not measurable with the ToF sensors). The sensor employs a matrix of 5x5 phototransistors on DOCKS-A and an infrared LED on DOCKS-B. Depending on the relative position, the LED activates different pattern of phototransistors [4].

The ToF sensors are affected by a noise of approximately 4 mm at distances below 30 mm, causing an uncertainty on the measure of the relative angles of more than 5 deg. To improve the estimation, a Kalman filter has been applied to the measurements of the ToF sensors [5]. The tests performed on the sensor suite provided the estimation error reported in Tab.1.

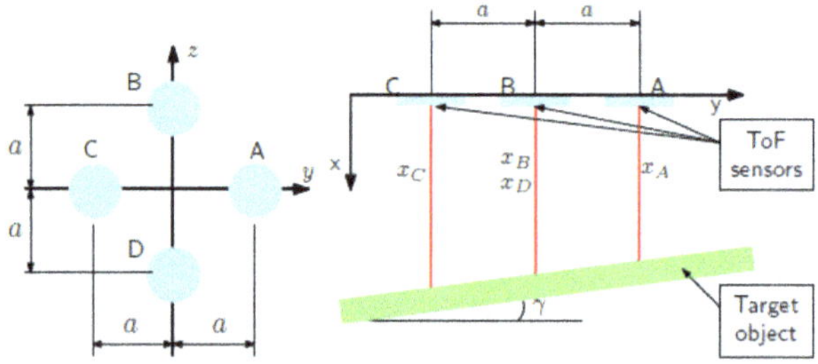

Figure 3: Measurement of the Time-of-Flight sensors.

Table 1: Estimation errors

	Error	X [mm]	Y [mm]	Z [mm]	Pitch [deg]	Yaw [yaw]
NavCam	Avg.	2.0	2.0	2.0	1.5	1.5
	Std. dev.	0.5	0.5	0.5	1.0	1.0
ToF + matrix	Avg.	0.14	1.16	1.41	0.52	0.11
	Std. dev.	0.43	0.13	2.13	0.64	0.68

Tests on DOCKS

DOCKS has undergone to a series of tests in order to verify its capabilities of DOCKS to tolerate relative misalignments and to self-align DOCKS-A to DOCKS-B. To this purpose, an ad-hoc experimental setup has been developed. DOCKS-A is mounted on the end-effector of a robotic arm, which has the purpose of moving DOCKS-A with an accuracy of 2 mm [6]. DOCKS-B is mounted on a frame DOCKS-B on a frame that blocks all the movements, except for the degree of freedom under tests for the self-alignment, as illustrated in Fig. 4.

At the beginning of the tests, the zero position has been defined as the perfect alignment between DOCKS-A and B. The tests have been conducted as follows: (1) a displacement has been imposed on DOCKS-B, (2) DOCKS-A has been moved along a linear trajectory to the zero position forcing the alignment between the parts. The test is considered successful if the claws close properly on the rim of the drogue without any residual displacement.

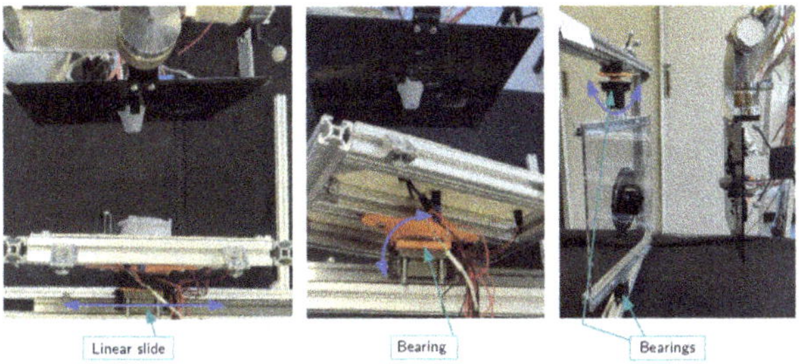

Figure 4: the experimental setups for the misalignment along the y axis, around the roll axis and around the yaw axis.

The results of the tests proved the capability of DOCKS to manage misalignment of 8.0 mm along the y and z axes, 9.0 deg around the yaw and pitch axes, and 10 deg around the roll axis.

Conclusions

This paper presents a brief description of an autonomous docking system, since it is equipped with (1) a set of sensors that, whith a Kalman filter, are able to estimate the relative pose of the target; (2) three actuators and an electromagnet to execute the soft and the hard docking; and (3) an integrated computer that applies the required algorithms to manage the sensors and actuators.

In addition, DOCKS has been designed to manage misalignment between DOCKS-A and DOCKS-B. To this aim, its centring cone with three features matches its counterpart on the target and forces the alignment between the parts. The test executed on DOCKS, proved that it is able to tolerate misalignment that are five to eight times the estimation errors of the sensors.

References

[1] S. Corpino, G. Ammirante, G. Daddi, F. Stesina, F. Corradino, A. Basler, A. Francesconi, F. Branz and J. Van den Eynde, "Space Rider Observer Cube - SROC: a CubeSat mission for proximity operations demonstration," in *73 International Astronautical Congress (IAC)*, Paris, France, 18-22 September 2022.

[2] F. Sansone, F. Banz and A. Francesconi, "A relative sensor for CubeSat based on LED fiducial markers," *Acta Astronautica,* 2018. https://doi.org/10.1016/j.actaastro.2018.02.028

[3] A. Caon, M. Peruffo, F. Branz and A. Francesconi, "Consensus sensor fusion to estimate the relative attitude during space capture operations," in *IEEE 9 International Workshop onMetrology for AeroSpace (MetroAeroSpace)*, 2022. https://doi.org/10.1109/MetroAeroSpace54187.2022.9856096

[4] A. Caon, F. Branz and A. Francesconi, "Characterization of a new positioning senosr for space capture," in *IEEE 8 International Workshop on Metrology for AeroSpace (MetroAeroSpace)*, 2021. https://doi.org/10.1109/MetroAeroSpace51421.2021.9511704

[5] A. Caon, F. Branz and A. Francesconi, "Smart capture tool for space robots," *Acta Astronautica,* 2023. https://doi.org/10.1016/j.actaastro.2023.05.014

[6] A. Caon, F. Branz and A. Francesconi, "Development and test of a robotic arm for experiments on close proximity operations," *Acta Astronautica, 195,* pp. 287-294, 2022. https://doi.org/10.1016/j.actaastro.2022.03.006

Materials Research Proceedings 37 (2023) 508-512 — https://doi.org/10.21741/9781644902813-111

AUTOMA project: technologies for autonomous in orbit assembly operations

Alex Caon[1,a*], Martina Imperatrice[2,b], Mattia Peruffo[1,c], Francesco Branz[2,d], Alessandro Francesconi[2,e]

[1]C.I.S.A.S. - Centre of Studies and Activities for Space "G. Colombo", Via Venezia 15, Padova (Italy)

[2]Department of Industrial Engineering, University of Padova, via Venezia, 1, Padova (Italy)

[a]alex.caon@unipd.it, [b]martina.imperatrice@studenti.unipd.it, [c]mattia.peruffo@unipd.it, [d]francesco.branz@unipd.it, [e]alessandro.francesceoni@unipd.it

Keywords: In-Orbit Servicing, Space Robots, Capture Tool, Autonomous System

Abstract. The possibility of manipulating objects in space is at the basis of the In-Orbit Servicing missions with the purpose to extend or improve the life of existing satellites. This can be obtained by equipping a target satellite with additional modules capable of providing additional basic functions, like power, thrust or communication. One of the most promising technologies to accomplish to these purposes is presented by space robots (satellites with one or more robotic manipulators) equipped with dedicated tool. The manipulators have the dual purposes to capture the additional module and to manipulate and attach it to the target satellite. In order to advance in IOS technologies, the Department of Industrial Engineer has funded the AUTOMA (AUtonomous Technologies for Orbital servicing and Modular Assembly) project[1]. The project aims to (1) upgrade an autonomous capture tool, (2) develop the additional module (EAU), and (3) execute tests in relevant laboratory scenarios. The autonomous tool is represented by SMACK (SMArt Capture Kit). SMACK is a capture system equipped with (1) different types of sensors to measure the relative pose during the entire approach for the capture and for the assembly; (2) a set of actuators to capture the module and keep a rigid connection during the manipulation; (3) a computer to execute locally the required software like guidance and navigation algorithms. The external module (Elementary Assembly Unit, EAU) is equipped with three features to be captured and manipulated by SMACK and a docking system to allow the assembly on the target structure. In order to test the assembly phase, SMACK has been mounted on the end-effector of a 6 degrees of freedom robotic arm in laboratory environment, while the target has been fixed on a frame. These tests proved the ability of SMACK to manage assembly tasks such as the control of a robotic arm with sufficient accuracy.

Introduction

In-Orbit Assembly missions have the purpose of assembling large structures in space such as telescopes or antennas [1]. Another fast-growing area is the possibility to install small building blocks onto existing satellites in order to provide additional functionalities such as communication, propulsion, power, etc. [2]. Both the objectives are achieved by the employment of small building blocks that are assembled by means of connecting ports. The assembly phase can be either performed by space robots that handle the blocks or executed autonomously by the blocks themselves.

The AUTOMA project has been funded in order to develop and test IOS technologies. In particular, the project focuses on the possibility to upgrade a capture tool, equipped with a set of

[1] This work was partially funded by Università degli Studi di Padova in the framework of the BIRD 2021 programme (BRAN_BIRD2121_01).

sensors, actuators and algorithms, able to autonomously perform different assembly tasks. To this aim, two main systems have been developed (see Fig. 1): the SMArt Capture Kit (SMACK) [3] and the Elementary Assembly Unit (EAU): the first has the purpose to catch and handle the second and assemble it on a target structure through a docking port.

The two systems have undergone a series of tests in order to evaluate their capabilities in terms of holding force of the gripper and the docking port, of the measurement error of each sensor, of the state estimation error and of the tolerated misalignment of both the gripper and of the docking port. A final functional system test consisted in letting SMACK to control the movements of the robot to perform the capture and the assembly of the EAU on the target structure. The use of the robotic arm ensures a positioning precision of 2 mm, which is lower than the errors of the sensors and algorithms of SMACK [4].

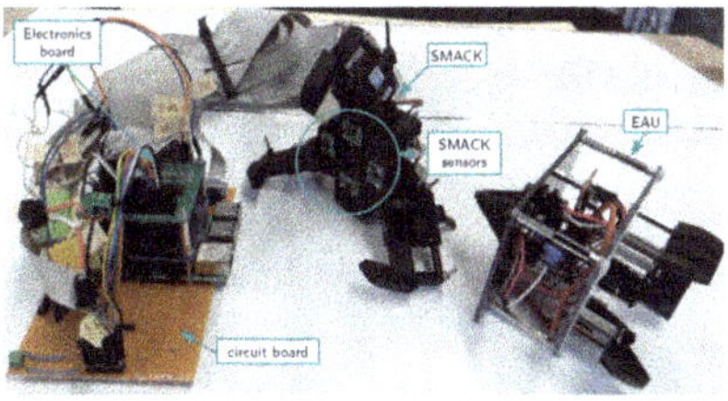

Figure 1: SMACK with its electronics and computer and the EAU.

SMArt Capture Kit (SMACK)

The SMArt Capture Kit (SMACK) has the purpose to identify, move and manage the assembly of the EAU on the target structure (or satellite). SMACK is equipped with a set of sensors, actuators and an integrated computer that renders it independent from the rest of the robotic arm on which it is mounted. The mechanical connection with the EAU is established by a gripper with three fingers. Each finger is individually actuated so that SMACK is able to capture the EAU even in case of relative misalignments by elongating differently each finger.

The sensors, coupled with the estimation algorithm, are employed to estimate the pose and the relative rates of the target. To this purpose, there are three types of sensors:

1. A NavCam that reconstructs the pose of a pattern of fiducial LED mounted on the target [5]. Measurement errors: 2.0 mm for the position and 1.5 deg for the attitude.
2. A set of four Time-of-Flight sensors that are employed to retrieve the distance and the relative yaw and pitch angles. In fact, if the target is tilted, the ToF sensors measure different distances, and allowing to indirectly measure the relative orientation, as illustrated in Fig. 2 [5]. They are coupled with a Kalman filter in order to improve their estimation capabilities. Estimation errors: 1.5 mm along the x axis and 1.5 deg for the yaw and pitch.
3. An in-plane matrix sensor that measures the position of the EAU along the x, y and z axes. The sensor is composed by a matrix of phototransistors activated by an infrared LED mounted on the EAU. The relative measure is computed based on the number of active phototransistors (second picture of Fig. 2) [6]. Measurement errors: 3.5 mm along the y and z axes.
4. A roll matrix to measure the relative roll angle. The sensors share the same working principle of the in-plane matrix (third picture of Fig. 2), but the active phototransistors are employed to measure the roll angle. Measurement error: 3.1 deg around the roll axis.

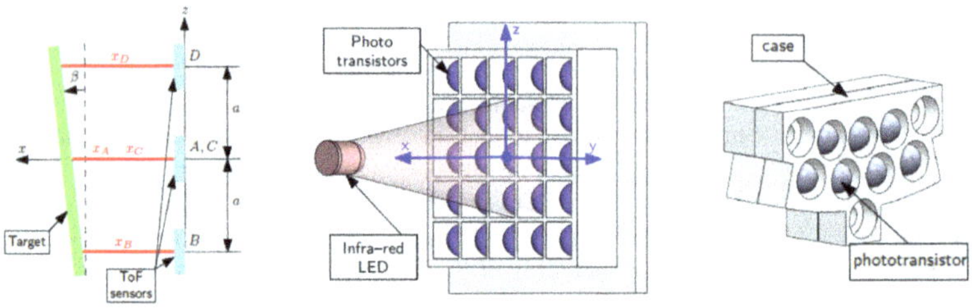

Figure 2: The measurement of a tilted object with the ToF sensors he in-plane matrix sensor, the roll matrix sensor and.

Elementary Assembly Unit (EAU)

The main purpose of the Elementary Assembly Unit (EAU) is to provide basic functionalities to the target interface, for example, additional power, communication, and thruster capabilities, etc. The EAU developed for this first stage is a box of size 100 mm by 100 mm by 50 mm, whose capabilities are limited to the mechanical connection by the means of a probe-drogue docking mechanism [8]. The tip of the probe can rotate in order to provide a rigid connection between the parts. The EAU is also equipped with three features to allow SMACK to catch it, a pattern of LEDs employed by the NavCam and two infrared LED beacons employed by the matrix sensors.

Assembly experiments

Both SMACK and the EAU have passed a series of tests to validate their capabilities in terms of (1) measurement and estimation capabilities; (2) holding force for both the gripper and the docking port; and (3) misalignment tolerance for the gripper and self-alignment management for the docking port. Then, a functional system test on the assembly procedure is required, to validate the involved mechanisms and the assembly procedure.

To execute the test, SMACK has been mounted on the end-effector of the robotic arm, the target frame has been mounted on a fixed frame, while the EAU has been placed in a known position. The procedure from the capture of the EAU to its assembly on the target frame has been divided into four main waypoints, which are transmitted to the robotic arm by SMACK, they are (referring to Fig. 3):

1. *Pre-capture*: in front of the EAU and aligned with it, but at a distance of 100 mm.
2. *Capture*: aligned with the EAU, in order to close the fingers and capture it.
3. *Post-capture*: after the capture, SMACK and the EAU are placed in front of the docking port at a distance of 100 mm.
4. *Docking*: the position where the docking mechanism can be activated.

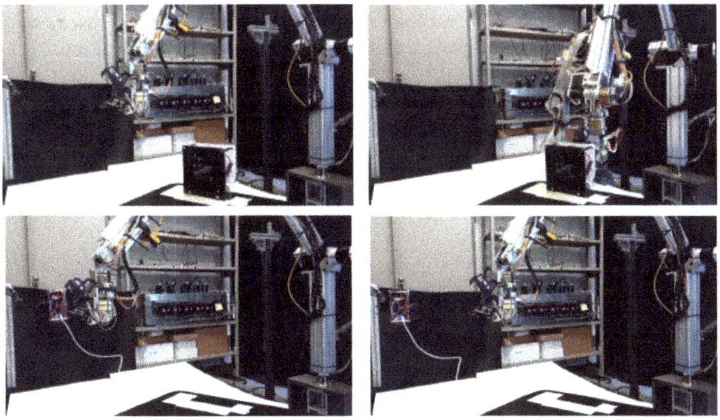

Figure 3: The waypoints of the phase.

Conclusions

This paper presents a brief overview of the AUTOMA project, that has the purpose to develop technologies in order to perform experiments on In-Orbit Assembly. Both the systems proved their capabilities through dedicated tests campaign. In particular, the relative state is estimated with an error lower than the tolerated misalignments, allowing SMACK to capture, handle and assemble the EAU to the target structure, as proved by the assembly experiment. The assembly phase and the involved mechanisms have been tested with a dedicated test which proved the abilities of the mechanisms to provide a rigid mechanical connection between the parts.

The next phase will focus on a closed-loop test in which SMACK has to perform an assembly with the use of its sensors and algorithms.

References

[1] W. R. Oegerle, L. R. Purves, J. G. Budinoff, R. V. Moe, T. M. Carnahan, D. Evans and C. K. Kim, "Concept for a large scalable space telescope: in-space assembly," *SPIE,* 2006. https://doi.org/10.1117/12.672244

[2] Northrop Grumman, "Space Logistics," 16 06 2023. [Online]. Available: https://www.northropgrumman.com/space/space-logistics-services/.

[3] A. Caon, F. Branz and A. Francesconi, "Smart capture tool for space robots," *Acta Astronautica,* 2023. https://doi.org/10.1016/j.actaastro.2023.05.014

[4] A. Caon, F. Branz and A. Francesconi, "Development and test of a robotic arm for experiments on close proximity operations," *Acta Astronautica, 195,* pp. 287-294, 2022. https://doi.org/10.1016/j.actaastro.2022.03.006

[5] A. Caon, M. Peruffo, F. Branz and A. Francesconi, "Consensus sensor fusion to estimate the relative attitude during space capture operations," in *IEEE 9 International Workshop onMetrology for AeroSpace (MetroAeroSpace)*, 2022. https://doi.org/10.1109/MetroAeroSpace54187.2022.9856096

[6] A. Caon, F. Branz and A. Francesconi, "Characterization of a new positioning senosr for space capture," in *IEEE 8 International Workshop on Metrology for AeroSpace (MetroAeroSpace)*, 2021. https://doi.org/10.1109/MetroAeroSpace51421.2021.9511704

[7] F. Sansone, F. Banz and A. Francesconi, "A relative sensor for CubeSat based on LED fiducial markers," *Acta Astronautica,* 2018. https://doi.org/10.1016/j.actaastro.2018.02.028

[8] F. Branz, L. Olivieri, F. Sansone and A. Francesconi, "Miniature docking mechanism for CubeSats," *Acta Astronautica,* pp. 510-519, 2020. https://doi.org/10.1016/j.actaastro.2020.06.042

Overview of spacecraft fragmentation testing

Stefano Lopresti[1,a]*, Federico Basana[1,b], Lorenzo Olivieri[1,c], Cinzia Giacomuzzo[1,d], Alessandro Francesconi[2,e]

[1] CISAS "G. Colombo", University of Padova, Via Venezia 15, 35131 Padova, Pd, Italy

[2] CISAS "G. Colombo" - DII, University of Padova, Via Venezia 15, 35131 Padova, Pd, Italy

[a]stefano.lopresti@unipd.it, [b]federico.basana@phd.unipd.it, [c]lorenzo.olivieri@unipd.it, [d]cinzia.giacomuzzo@unipd.it, [e]alessandro.francesconi@unipd.it

Keywords: Space Debris, Fragmentation Testing, Cumulative Distribution, Satellites Break Up

Abstract. Spacecraft fragmentation due to collisions with space debris is a major concern for space agencies and commercial entities, since the production of collisional fragments is one of the major sources of space debris. It is in fact believed that, in certain circumstances, the increase of fragmentation events could trigger collisional cascade that makes the future debris environmental not sustainable. Experimental studies have shown that the fragmentation process is highly complex and influenced by various factors, such as the material properties, the velocity and angle of the debris impact and the point of collision (e.g. central, glancing, on spacecraft appendages). In recent years, numerous impact tests have been performed, varying one or more of these parameters to better understand the physics behind these phenomena. In this context some tests have been also performed at the hypervelocity impact facility of the university of Padova. This paper provides an overview of the main experiments performed, the most critical issues observed and proposes some future directions for further research. Moreover, it summarizes the current state of research in spacecraft fragmentation, including the methods and techniques used to simulate debris impacts, the characterization of fragment properties and the analysis of the resulting debris cloud.

Introduction

The increasing presence of space debris poses a significant and escalating threat to the safety of space activities. Collisions with such debris are the primary sources of spacecraft fragmentation, leading to the generation of additional space debris and contributing to an increasingly congested orbital environment [1]. As a result, mitigating space debris has become a top priority for the international space community, necessitating the implementation of effective strategies to reduce the accumulation of space debris and ensure the safety of space operations [2]. Mathematical modelling of this phenomenon is very challenging due to the high velocities involved and the large energies generated during impact. It is therefore essential to perform impact tests that accurately represent the conditions that occur in orbit [3]. In addition, to properly calibrate the models, it is essential to perform parametric tests that allow the influence of the impact geometry (impact angle, velocity, point of impact, etc.) on debris generation to be isolated and studied individually.

To achieve these outcomes, it is crucial to have access to hyper-velocity facilities capable of executing impacts with excellent velocity control.

Hypervelocity impact facility

There are different types of Hypervelocity laboratories that perform impact tests, the most common are the powder-gas guns [4], it is possible to manufacture them in a two-stage light gas configuration that does not involve the use of explosive dust [5].

The conceptual process for both is similar: a piston is accelerated to compress a light gas (usually hydrogen) adiabatically inside a cylinder. The very quick compression leads to a sudden increase in the gas temperature and pressure (about 5000K and 4000Bar peak). When the pressure reaches its peak a valve is opened (usually is a rupture disk that breaks due to high pressure) and the gas is discharged onto a projectile that is fired at high velocity into the target in a vacuum chamber.

The difference in functioning between the two guns is in the way the cylinder is accelerated in the first stage: the light gas gun uses high-pressure gas (e.g., Helium at 120bar) while the second uses gunpowder as a propellant. Both methods are very efficient and it is difficult to compare them, however, from tests, it seems that gunpowder accelerators are able to reach higher velocities while the process with the light-gas guns has higher repeatability, this is because the combustion process is more unstable than unloading a pressurized tank. Moreover, the light gas gun requires significantly less maintenance.

In recent years, research is being directed toward the possibility of manufacturing three-stage accelerators [6]. This achievement would be very interesting because it involves adding a powder stage upstream to existing light gas guns. This could improve the performance of the guns and achieve peak velocities of more than 10 km/s.

Figure 1: Two stage light gas gun in the hypervelocity facility of the university of Padova.

Fragmentation test

The use of a target with a realistic material distribution is essential to obtain representative results for the physical parameters of the debris. In recent years, there has been increased interest in satellite fragmentation tests to study the response of complex geometries to hypervelocity impacts. The first such impact study was the SOCIT (Satellite Orbital debris Characterization Impact Test) tests series, which was made up of four hypervelocity impacts on representative satellite in space. In particular, in the fourth test, Socit4, the target was a flight ready Navy Transit 1960 era satellite. [7]

An aluminium projectile with a diameter of 4.7cm (150 grammes) fired at a speed of 7km/s was used for the test. The shot was performed on a model of an old generation satellite, therefore the materials are different from those mounted on new-generation spacecraft, for this reason it was decided to carry out a further experimental campaign with more modern targets.

Two hypervelocity tests were conducted in the DebriSat campaign. The first target was a representative upper stage model of a launch vehicle (DebriLV) and the second a 56kg satellite (DebriSat). For both tests, an aluminium cylinder measuring 8.6cm x 9cm was used as the projectile, which was fired at speeds of 6.8 and 6.9km/s, respectively.

DebriLV was composed of 2 pressurised gas tanks of different material and size, the rest of the structure was composed of other materials used in space such as aluminium 6061 and stainless steel [8].

Debrisat is a representative model of a LEO satellite, with a diameter of 60 cm and a height of 50 cm. In addition to using more advanced materials, it was decided to make the satellite 45% more massive than the Socit test. In order to better study the fragmentation of the satellite, each sector of DebriSat was built with different coloured material to better identify its origin [9].

Materials Research Proceedings 37 (2023) 513-517 | https://doi.org/10.21741/9781644902813-112

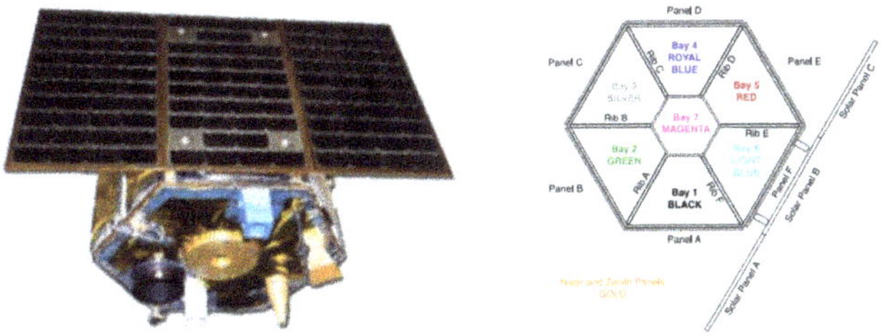

Figure 2: DebriSat photo (left) and schematic representation of the colour subdivision (right).

A rigorous procedure was developed to collect the fragments; the foam panels placed inside the firing chamber were first scanned by X-ray. Once scanned, fragments larger than 2mm were identified, extracted, a characteristic length is measured (the average of the three largest orthogonal dimensions of the fragment) and a unique identification number is assigned. The fragments were then sorted by material, shape and colour and scanned in 2D or 3D [10]. The data were collected and cumulative graphs of mass, shape and size distribution were produced, as well as characteristic length plots with area to mass ratios. This data will be used to improve the modeling of the relationships of these physical characteristics to each other and to calculate more accurate distributions of such parameters.

A team of researchers from CARDC (China Aerodynamics Research and Development Center) carried out a fragmentation test on three cubic aluminium mock-ups of 40x40x40 cm^3 with increasing weights of 7.3, 8.2, and 13.1 kg respectively. Inside these mockups there were a cylindrical central body and representative electronic boxes also made of aluminium, some parts of a printed circuit board were also included. The impact occurred at a speed of approximately 3.5km/s. The fragments were collected and cumulative debris distributions were made in terms of area to mass ratio and cross-sectional area. [11].

A further study was carried out at THIOT Ingénierie and included the fragmentation of a nanosatellite measuring 15x10x10 cm^3. On the satellite were mounted components representative of those used in space such as a 4-cells battery pack, electronic boards, inertia wheels and a solar panel (although it should be noted that they were non-flight acceptable). The projectile used was a 9mm-diameter polycarbonate equilateral cylinder incorporating a second 4mm-diameter aluminium equilateral cylinder fired at approximately 6.7km/s. The size and weight of the fragments were then collected by a six-axis robotic arm that also performed a 3D scan of each analysed fragment [12].

In this context, CISAS also decided to start a test campaign on complex structures. Two tests were performed on a mockup of a Picosatellite of size 50x50x50 mm^3. The first shot was central, with the impact face perpendicular to the projectile (a), while the second test was a glancing impact, performed with the picosatellite inclined at 45° with respect to the projectile direction (b). For these tests, fragments were manually collected, divided by size, weighed and measured to obtain cumulative distributions, characteristic lengths and shape diagrams [13]. The results obtained from the tests were then compared with those predicted by models in the literature (c).

a) *b)* *c)*

Figure 3: Experimental setups and characteristic length cumulative distribution.

The glancing impact produced less fragments compared to the other test and also compared to those predicted by the NASA SBM model. However, the inclination of the curves of the experimental and model data have a similar inclination.

Conclusion

Several tests have been performed in recent years to better understand the fragmentation dynamics of a hypervelocity impact. For the development of new models, it is of critical importance to have an increasingly rich and parameterized impact database available for the scientific community. Building new facilities or upgrading existing ones with the objective of reaching higher speeds and find fast and accurate fragment analysis procedures is a key target to achieve this goal.

References

[1] A. Rossi et al, "Modelling the evolution of the space debris population," *Planet Space Sci,* 1988.

[2] J.-R. Ribeiro et al, "Evolution of Policies and Technologies for Space Debris Mitigation Based on Bibliometric and Patent Analyses," *Space Policy,* pp. 40-56, 2018. https://doi.org/10.1016/j.spacepol.2018.03.005

[3] D. McKnight, R. Maher and L. Nag, "Refined algorithms for structural breakup due to hypervelocity impact," *International Journal of Impact Engineering,* pp. 547-558, 1995. https://doi.org/10.1016/0734-743X(95)99879-V

[4] T. J. Ringrose, H. W. Doyle, P. S. Foster and al, "A hypervelocity impact facility optimised for the dynamic study of high pressure shock compression," *Procedia Engineering,* 2017. https://doi.org/10.1016/j.proeng.2017.09.756

[5] A. Angrilli, D. Pavarin, M. De Cecco and A. Francesconi, "Impact facility based upon high frequency two-stage," *Acta Astronautica,* pp. 185 - 189, 2002. https://doi.org/10.1016/S0094-5765(02)00207-2

[6] A. J. Piekutowski and K. L. Poormon, "Development of a three-stage, light-gas gun at the University of Dayton Research Institute," *International Journal of Impact Engineering,* 2006. https://doi.org/10.1016/j.ijimpeng.2006.09.018

[7] J. Liou, J. Opiela, H. Cowardin, T. Huynh, M. Sorge, C. Griffice, P. Sheaffer, N. Fitz-Coy and M. Wilson, "Successful Hypervelocity Impacts of," *NASA Orbital Debris Quarterly News,* pp. 3-5, 2014.

[8] H. Cowardin, J.-C. Liou, P. Anz-Meador, M. Sorge, J. Opiela, N. Fitz-Coy, T. Huynh and P. Krisko, "Characterization of orbital Debris via Hiper-velocity laboratory-based Tests," in *European Conference on Space Debris*, 2017.

[9] M. Rivero, B. Shiotani, M. Carrasquilla, N. Fitz-Coy, J.-C. Liou, M. Sorge, T. Huynh, J. Opiela, P. Krisko and H. Cowardin, "DebriSat fragment Characterization System and Processung Status," in *IAC*, Bremen, Germany, 2018.

[10] M. Rivero, J. Kleespies and K. Patankar, "Characterization of Debris from the DebriSat Hypervelocity Test," in *IAC*, 15.

[11] S.-W. Lan, S. Liu, Y. Li, F.-W. Ke and j. Huang, "Debris area distribution of spacecraft under hypervelocity impact," *Acta Astronautica,* vol. 105, pp. 75-81, 2014. https://doi.org/10.1016/j.actaastro.2014.08.011

[12] H. Abdulhamid, D. Bouat, A. Collè and Al, "On-ground HVI on a nanosatellite. Impact test, fragments recovery and characterization, impact simulations.," in *8th European Conference in Space Debris*, 2021.

[13] L. Olivieri, P. A. Smocovich, C. Giacomuzzo and A. Francesconi, "Characterization of the fragments generated by a Picosatellite impact experiment," *International Journal of Impact Engineering,* vol. 168, 2022. https://doi.org/10.1016/j.ijimpeng.2022.104313

Feasibility analysis of a CubeSat mission for space rider observation and docking

Chilin Laura[1*], Bedendo Martina[1], Banzi Davide[1], Casara Riccardo[1], Costa Giovanni[1], Dolejsi Elisabetta[1], Quitadamo Vincenzo[1], Trabacchin Nicolò[1], Visconi Delia[1], Visentin Alessia[1], Basana Federico[2], Olivieri Lorenzo[2], Colombatti Giacomo[3], Francesconi Alessandro[3]

[1]University of Padova, Via 8 Febbraio 2, 35122 Padova

[2]CISAS G.Colombo, Via Venezia 15, 35131 Padova

[3]DII/CISAS, Via Venezia 1, 35131 Padova

*laura.chilin.1@studenti.unipd.it

Keywords: CubeSat, Manoeuvres, Inspection, Docking

Abstract. In the last few years the number of orbiting satellites has increased exponentially, in particular due to the development of the New Space Economy. Even if this phenomenon makes the space more accessible, bringing a great contribution to the scientific, economic and technological fields, on the other hand it contributes to the overpopulation of the space background. Therefore, it is necessary to develop new techniques to manage the space environment, such as in orbit servicing, which is a procedure that aims to refuel and repair satellites to extend their operational life. A first step to reach this goal is to inspect closely the object of interest to study its features. In this framework, the Space Rider Observer Cube (SROC) mission is being developed. SROC is a payload that will be deployed by Space Rider, an uncrewed and reusable robotic spacecraft designed by ESA. SROC is a 12U CubeSat, whose goal is to carry out inspection manoeuvres around the mothership, then re-enter on board using a safe docking system to come back to Earth. The feasibility of a mission similar to SROC has been simulated during a university class, starting from the definition of the system requirements with particular focus on the analysis of the payloads and subsystems, to ensure the achievement of the mission goals. In particular, the CubeSat is equipped with an optical instrument to capture high resolution images of Space Rider surface and a docking mechanism. Then the design of the orbit and the simulation of the effects of the space environment on the CubeSat have been studied using GMAT, SYSTEMA, MATLAB and other numerical tools. The results of the study are useful for future missions, aiming to inspect orbiting objects, such as operative satellites for in orbit servicing, space debris and dead satellites to study their geometries and plan their removal.

Introduction

As the number of orbiting objects around Earth is constantly rising, it is necessary to develop new strategies to manage the space environment. For this purpose, it's crucial to study the space objects in-situ with a close observation for future in orbit-servicing missions that will allow to extend the operative life of functional satellites. Examples of these missions are: Seeker, a 3U CubeSat used to complete autonomous mocking inspections [1] and AeroCube-10, a 1.5U CubeSat created to demonstrate precision satellite-to-satellite pointing [2]. However the number of space missions involving CubeSats meant to inspect other satellites remains low. The aim of this work is to contribute to this field studying a CubeSat inspired by SROC (Space Rider Observer Cube), a future mission designed by ESA in conjunction with Politecnico di Torino and Tyvak International, that aims to carry on inspections and docking manoeuvres with its mothership Space

Rider (SR), a reusable robotic spacecraft. In fact, SROC will reach its operative orbit inside Space Rider cargo bay and then will be deployed by it [3,4].

This project has been developed during a university class, and it started from the definition of the system requirements, followed by the selection of components and sizing of the subsystems, leading to the preliminary design of the system (Fig. 1).

Requirements and preliminary definition

Starting from the mission objectives of performing safe inspection and docking manoeuvres and transposing them into mission requirements, all the subsystems with their performance level have been defined. Afterwards, the design and operational requirements have been taken into account to reach a first iteration of requirements. Since the CubeSat is inspired by SROC, it has some similarities that have been included in the requirements such as its 12U structure, an imaging payload and a docking payload. It was established that Space Rider model orbit should be a 600 km dawn dusk 6 a.m. Sun Synchronous Orbit.

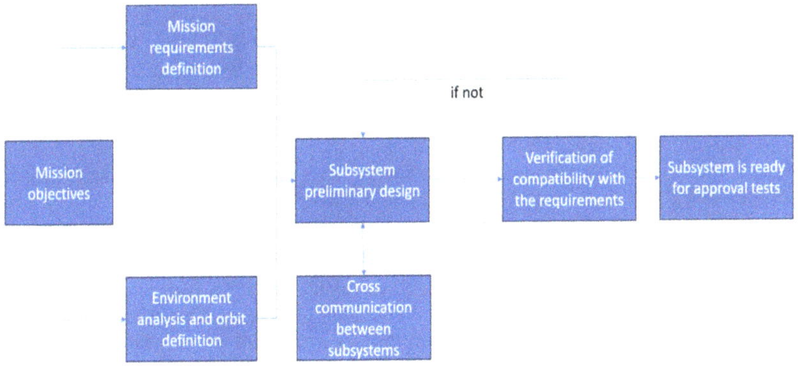

Fig. 1: Work procedure

Mission phases

The initial date of departure of the CubeSat from the mothership is set to be June 21, 2024, chosen to reduce the umbra periods and increase the electric power collected by the solar panels. Three reference frames have been designed for the simulations implemented in GMAT and MATLAB: the Earth inertial frame MJ2000Eq to define the mothership motion around the planet (Fig. 2a) [5]; radial in-track cross-track (RIC) centred in the mothership (Fig. 2a) and the CubeSat body frame to describe its relative orbit (Fig. 2b).

The motion of the CubeSat around the mothership is described by a Walking Safety Ellipse (WSE), chosen to be a safe orbit for both satellites. [6]

Materials Research Proceedings 37 (2023) 518-521 https://doi.org/10.21741/9781644902813-113

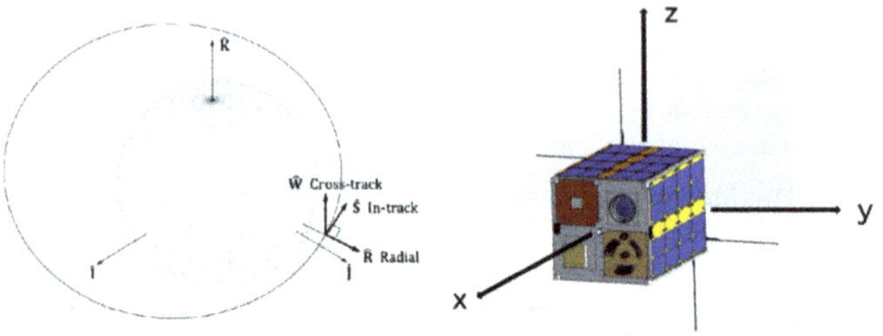

Fig. 2: a. MJ2000EQ and RIC frame system; b. Body system

The mission is designed to last 25 days, and its operative life can be divided in six phases:
1. Departure from Space Rider model (five hours): the CubeSat separates from the mothership and prepares for the first manoeuvre;
2. WSE entering phase (five hours): the CubeSat achieves the RIC coordinates to enter the WSE;
3. Inspection on a WSE orbital keeping: the CubeSat moves forward in the WSE for two days, then performs the orbital keeping manoeuvres to return to the initial conditions to repeat the inspection. This cycle is performed eleven times, to take pictures of the entire surface of the satellite in different conditions;
4. WSE departure: the CubeSat is brought to a stationary point, 100 m far from the mothership;
5. Hold Point approach (five hours): the CubeSat reaches a hold point at 50 m from SR model;
6. Rendezvous phase: the CubeSat reaches a 2 m distance from the mothership through a bang-bang technique.

CubeSat design

The CubeSat has the standard dimensions of a typical 12U CubeSat with a mass of 16kg. The skeleton of the external structure is made of Aluminium 6061, closed by graphene panels. Two other panels are added to the internal structure to create different zones inside the satellite and facilitate the storing of the components.

The satellite has two payloads: a camera (CMOS sensor and optical lens) that fits 1U and a partially external docking mechanism, whose dimensions are 1U. The camera is protected by a "Zerodur" layer. The images are taken and then transferred to the On-Board Computer (OBC) to be processed and then sent to Earth by the Telemetry and Tracking Control subsystem (TTC) antennas. The TTC communicates with ground stations using UHF band, through which the satellite also transmits telemetry data and receives commands.

The CubeSat is equipped with an Attitude Determination and Control Subsystem (ADCS) based on an Inertial Platform (Inertial Measurement Unit, IMU), navigation system GNSS, a magnetometer and multiple sun sensors to determine the configuration of the CubeSat around the Space Rider model. In addition, three reaction wheels are used for disturbances control and one for redundancy, together with three magnetorquers for their desaturation, while a momentum wheel is necessary to maintain the pointing towards SR during the inspection on the WSE.

The electric power, essential to all subsystems, is guaranteed by body mounted solar panels, positioned on the 6U faces and on the 4U face opposed to the camera lens, and a Li-Ion battery that comes into play when the energy provided by the panels is not sufficient. These components form the Electric Power Subsystem (EPS). The surface of the solar panels is covered with 125μm "Kapton". The thickness has been chosen to resist erosion due to the atomic oxygen.

The environment analysis (external and internal) has been realised with MATLAB, SPENVIS and Systema, and it suggested that a passive Thermal Control Subsystem (TCS) is sufficient to guarantee the maintenance of the operative temperature for individual components: for that purpose, the ADCS components have been covered with a Multi-Layer Insulation (MLI) material. Furthermore, it is not necessary to equip the satellite with a device for dissipating the internal thermal energy, i.e. a radiator.

The propulsive system consists in three cold gas B1 thrusters, fuelled by N2O propellant (nitrogen peroxide). This propellant is self-pressurising and has a good thermal control in space. This technology has been chosen because it presents a low power consumption, low mass and can perform every requested manoeuvre during the mission.

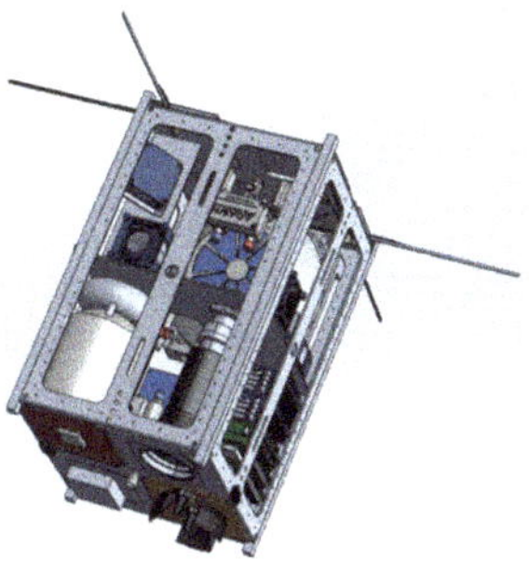

Fig. 3: CubeSat CAD without solar panels

In the end, the CubeSat is provided with a drag sail that will be used in case of docking failure to guarantee the compliance with the Space Debris Mitigation Guidelines.

Discussion and conclusions

During this study, some issues related to the CubeSat design arose. The resolution of the camera has been one of the most challenging aspects, because it is not only related to the pictures quality level, but it is also fundamental for the choice of a technology that can fit in the CubeSat. Moreover, the communication frequency band has been discussed, starting from a S-band communication to send images and a UHF band for the telemetry and command data. Due to power consumption issues, the UHF band has been chosen. In fact, the body mounted configuration is not able to provide the electric power necessary to guarantee an S-band communication. This led to the definition of two configurations for the EPS: the first one consisting of Sun tracking deployable solar panels, the second using the same structure without the Sun tracking mechanism. Both alternatives proved to be unsuitable as one of the mission objectives is to provide a way to dock safely with the mothership.

This paper showed the feasibility of the mission, giving results that will be useful for future missions developed with the goal of making in orbit servicing a reality. Further details on this work will be provided in a future extended paper.

References

[1] Brian Banker, Scott Askew "Seeker 1.0: Prototype Robotic Free Flying Inspector Mission Overview" NASA Johnson Space Center 2019

[2] Gangestad, Joseph W., Catherine C. Venturini, David Hinkley and Garrett Kinum. "A Sat-to-Sat Inspection Demonstration with the AeroCube-10 1.5U CubeSats." (2021).

[3]https://www.asi.it/en/technologies-and-engineering/micro-and-nanosatellites/esa-gstp-fly-program/sroc/ (accessed: June 2023)

[4]https://www.esa.int/Enabling_Support/Space_Transportation/Space_Rider_overview (accessed: June 2023)

[5] Kim, Eunhyouek & Han, Seungyeop & Sayegh, Amer. (2019). Sensitivity of the Gravity Model and Orbital Frame for On-board Real-Time Orbit Determination: Operational Results of GPS-12 GPS Receiver. Remote Sensing. 11. 1542. https://doi.org/10.3390/rs11131542

[6] Gaylor, David & Barbee, Brent. (2007). Algorithms for safe spacecraft proximity operations. Advances in the Astronautical Sciences. 127. 133-152.

Analysis of small spacecraft Mars aerocapture through a single-event drag modulation

Tobia Armando La Marca[1,a *], Giorgio Isoletta[2,b] and Michele Grassi[2,c]

[1]Scuola Superiore Meridionale, Largo San Marcellino 10, 80138 Naples, Italy

[2]Department of Industrial Engineering, University of Naples "Federico II", Piazzale Tecchio 80, 80125 Naples, Italy

[a] tobiaarmando.lamarca-ssm@unina.it, [b] giorgio.isoletta@unina.it, [c] michele.grassi@unina.it

Keywords: Aerocapture, Drag Modulation, Mars Exploration, Small Spacecraft

Abstract. In the last years, the scientific interest in Mars exploration has become more and more relevant, driving the development of technologies aimed at improving the current capabilities to land scientific payloads or to insert probes into stable orbits around the planet. In this framework, the use of low-cost small satellites could represent an advantageous solution for both the mission scenarios. In planetary exploration, the aerocapture manoeuvre is considered a promising technique to overcome the limits imposed by specific volume and mass ratio constraints on the design of the propulsion system. Based on these premises, this work focuses on the 2D aerocapture manoeuvre of a small spacecraft equipped with a Deployable Heat Shield (DHS). Specifically, the analysis aims at assessing the aerocapture manoeuvre feasibility exploiting a single shield surface variation.

Introduction

The aerocapture is an aero-assisted manoeuvre to transfer a vehicle from a hyperbolic orbit to a closed one at lower energy, by exploiting the aerodynamic drag force through a single atmospheric passage with properly designed decelerators, such as DHS, drag skirt or inflatable drag devices. Once out of the atmosphere, the spacecraft performs a subsequent Pericenter Raise Manoeuvre (PRM) to avoid repeated atmospheric passages and stabilize the spacecraft on a scientific orbit or on a parking orbit, ready for suppletive Post-Aerocapture Manoeuvres (PAM). If compared to a purely propulsive orbit injection (OI), the aerocapture manoeuvre allows to drastically increase the delivered mass payload thanks to the propellant savings and the smaller weight of the aerodynamic decelerators compared to the propellant needed for propulsive OI. The reduction of the propellant mass decreases the costs per kg of payload, thus enabling or enhancing many potential planetary mission profiles [1]. Moreover, the aerocapture benefits of inherent reduction of the manoeuvring time with respect to the aerobraking manoeuvre, which instead exploits multiple atmospheric passages for depleting the right amount of energy to reach the final orbit. Although the consensus of recent studies about the possibility to use aerocapture for science mission at Titan, Mars and possibly Venus [2], it has never been implemented to date because of environmental and object related uncertainties, e.g., the limited knowledge of the local atmospheric density and/or the lack of real-time navigation data. However, both the growing scientific interest in Mars exploration and the technological readiness acquired in atmospheric flights during the last decades motivate further investigations on Mars aerocapture. This contribution specifically focuses on the aerocapture technique for a small satellite equipped with a DHS, exploiting a single-event drag modulation. The aerocapture has been studied from a purely dynamical point of view, and a multiparametric analysis has been carried out to identify suitable aerocapture corridors. The results of the single-event drag modulation strategy are compared with the outcomes of fixed shield aperture strategy to assess the benefits of this technique in terms of number and characteristics of possible solutions.

Finally, the conductive thermal heat at the pericenter has been estimated to evaluate the thermodynamical loads the spacecraft will encounter during the atmospheric crossing.

Methodology

The present work analyses the aerocapture manoeuvre of a spacecraft initially moving on a hyperbolic approaching trajectory, resulting from a patched conics approximation of the Earth-to-Mars interplanetary transfer. According to this construction, the spacecraft dynamics can be modelled as a two-body problem. The trajectory is then propagated up to the Mars Atmospheric Interface (AI), usually set to 150 km. Once the spacecraft crosses the atmosphere, the drag perturbs the motion as according to the following equation:

$$\ddot{\vec{r}} + \frac{\mu_{Mars}}{r^3}\vec{r} = \overrightarrow{a_d} \quad (1)$$

where $\vec{r}$ is the spacecraft position vector in a 2D reference frame centred in Mars, μ_{Mars} is the planetary standard gravitational parameter, while $\overrightarrow{a_d}$ the atmospheric drag perturbing acceleration modelled as:

$$\overrightarrow{a_d} = -\frac{1}{2}\frac{\rho}{\beta} v\vec{v} \quad (2)$$

In Eq. 2, ρ is the atmospheric density, $\vec{v}$ is the spacecraft velocity vector relative to the atmosphere, v is its module and β is the ballistic coefficient defined as $\beta = \frac{m}{C_D S}$, being C_D the drag coefficient, S the shield cross-section and m the spacecraft mass. The velocity variation produced by the aerodynamic deceleration, Δv_{drag}, has been computed as the difference of the velocity at the pericentres of the arrival hyperbolic trajectory and the elliptical one obtained after the atmospheric crossing. Moreover, the impulsive burn Δv_{PRM} to circularize the elliptical exit orbit has been computed as a Hohmannian manoeuvre executed at the ellipse apoapsis as:

$$\Delta v_{PRM} = \sqrt{\frac{\mu_{Mars}}{r_a}} - \sqrt{\mu_{Mars}\left(\frac{2}{r_a} - \frac{1}{a_{exit}}\right)} \quad (3)$$

where r_a and a_{exit} are respectively the apocenter and the semi-major axis of the elliptical orbit. Finally, the Sutton and Graves [3] semi-empirical relation for stagnation-point convective heat rate has been employed to quantify the thermal conductive heat rate in W/m^2:

$$\dot{q}_c = K_m \left(\frac{\rho}{R_N}\right)^{0.5} v^3 \quad (4)$$

in which K_m is the Mars atmospheric conductive constant (1.898×10^{-4} $kg^{0.5}/m$) and R_N is the shield nosecone radius. Results are then converted into W/cm^2 to compare them with literature ones.

Results

Results here provided refer to spacecraft characteristics in [4] (and references therein). The spacecraft of *m = 150 kg* is equipped with a DHS providing a maximum surface extension of S_{max} = 7.065 m^2 and a R_N of 0.6 m. All the simulations have been conducted assuming the possibility of changing the shield surface after the spacecraft transit at the pericenter of the arrival trajectory. Several dynamics conditions have been evaluated, resulting from the combination of arrival velocities (v_∞ = 2, 3, 5 km/s), Keplerian arrival pericenters (h_p = 70,75,80,85,90 km) and trigger altitudes for the ballistic coefficient variation (h_{tr}), supposing two different possible values of β respectively equal to 2β and 3β, and analyzing all the trigger altitudes from the Keplerian pericenter up to the aerodynamic interface, with step of 5 km. Moreover, results here reported refers to the density nominal condition for September 1st, 2031, of the Mars Global Reference Atmospheric Model (Mars-GRAM) [4], and to a $C_D = 1$. Figure 1 shows the apocentric altitude h_a and eccentricity e_{exit} of the orbit obtained after the atmospheric passage as function of h_{tr} and β for $v_\infty = 3$km/s, while Table 1 lists the values of $\dot{q}_c$ at the pericenter, Δv_{drag} and Δv_{PRM} for some of the most relevant settings. In all the figures, dashed lines represent the solution achievable with the nominal (constant) ballistic coefficient.

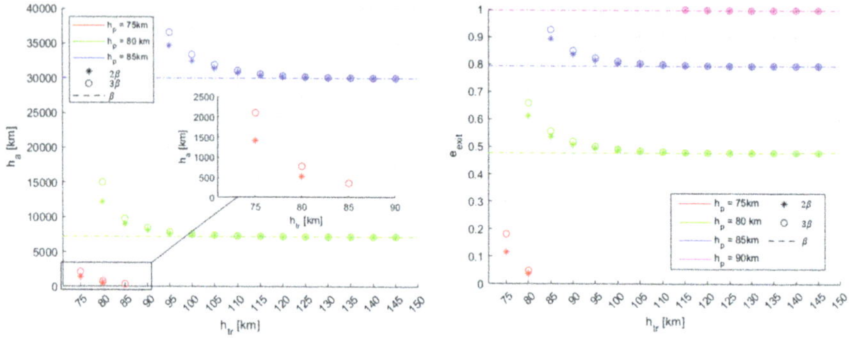

Figure 1 – exit orbit h_a and e_{exit} as function of h_{tr} for different h_p, β and v_∞ = 3 km/s

Table 1 - Most relevant results for v_∞ = 3 km/s

h_p [km]	h_{tr} [km]	$\Delta v_{drag,2\beta}$ [km/s]	$\Delta v_{drag,3\beta}$[km/s]	$\Delta v_{PRM,2\beta}$ [km/s]	$\Delta v_{PRM,3\beta}$ [km/s]	$\dot{q}_c$ $[W/cm^2]$
75	75	-1.607	-1.508	0.253	0.336	10.21
75	85	//	-1.780	//	0.0802	10.21
80	80	-0.906	-0.843	0.657	0.665	10.15
80	110	-1.086	-1.084	0.598	0.599	10.15
85	95	-0.645	-0.631	0.629	0.620	9.035
85	110	-0.669	-0.668	0.638	0.637	9.035

As expected, the largest variation of the exit orbit parameters is obtained when the ballistic coefficient modulation is triggered soon after the transit of the spacecraft at the pericenter. In particular, because of the larger density values encountered during the atmospheric crossing, the smaller is the h_p, the more circular is the exit orbit (i.e. the smaller are both e_{exit} and h_a). In turn, up to h_p=80km, Δv_{PRM} reduces with increasing h_{tr}, because of the larger amount of time spent with S_{max} at low altitudes, On the other hand, $\dot{q}_c$ values at the pericenters are in line with those expected from the literature [3], showing a small variation with h_p. Finally, an important result is that, thanks to the variation of β, it is possible to have solutions also at pericenter altitude where we have no solutions with the nominal β (i.e. 75 km), that are of high interest due to the possibility to reduce the eccentricity of the exit orbits. However, no solutions have been obtained for h_p smaller than 75 km or larger than 85 km. To better understand the range of aerocapture solutions, in Fig. 2 the results for v_∞ = 2 and 5 km/s are depicted, while Tab.2 focuses on most relevant results obtained.

When the hyperbolic speed is of 5 km/s, it is possible to have solution only for h_p = 75 km. On the contrary, when the satellite arrival speed is smaller, solutions have been obtained only for h_p higher than 75km, since the atmospheric drag at those altitudes is considerably smaller. However, as previously described, the final h_a is again severely affected by the h_p and v_∞ values.

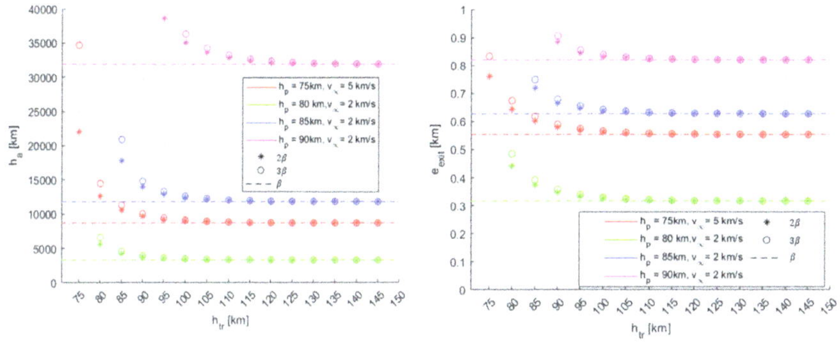

Figure 2 - exit orbit h_a and e_{exit} as function of h_{tr} for different h_p, β and v_∞ = 2 and 5 km/s

Table 2 - Most relevant results for v_∞ = 2 and 5 km/s

h_p [km]	h_{tr} [km]	$\Delta v_{drag,2\beta}$ [km/s]	$\Delta v_{drag,3\beta}$ [km/s]	$\Delta v_{PRM,2\beta}$ [km/s]	$\Delta v_{PRM,3\beta}$ [km/s]	$\dot{q}_c$ $[W/cm^2]$
75	75	-1.43	-1.34	0.661	0.627	15.72
75	110	-1.71	-1.71	0.626	0.627	15.72
80	80	-0.94	-0.88	0.553	0.583	8.74
80	110	-1.12	-1.12	0.442	0.443	8.74
90	95	-0.40	-0.38	0.614	0.606	6.72
90	120	-0.42	-0.42	0.633	0.633	6.72

Conclusions

This work presents an analysis of a single-event aerocapture manoeuvre for a small spacecraft equipped with a DHS. Results confirm the increasing number of solutions achievable with a ballistic coefficient modulation, although a strong dependency on the arrival conditions emerges. Additionally, aerocapture enables some otherwise unattainable exit orbits. For relatively high arrival speed, the atmospheric aerocapture corridor shortens in such a way to suggest further investigation of possible control techniques for a finer modulation of the deployable shield aperture. Thus, future development will focus on 3D aerocapture analysis with the definition of a deployable shield modulation logic.

References

[1] J. L. Hall, M. A. Noca, and R. W. Bailey, "Cost-Benefit Analysis of the Aerocapture Mission Set," *J Spacecr Rockets*, vol. 42, no. 2, pp. 309–320, Mar. 2005. https://doi.org/10.2514/1.4118

[2] T. R. Spilker *et al.*, "Qualitative Assessment of Aerocapture and Applications to Future Missions," *J Spacecr Rockets*, vol. 56, no. 2, pp. 536–545, Nov. 2018. https://doi.org/10.2514/1.A34056

[3] Z. R. Putnam and R. D. Braun, "Drag-Modulation Flight-Control System Options for Planetary Aerocapture," *J Spacecr Rockets*, vol. 51, no. 1, pp. 139–150, Aug. 2013. https://doi.org/10.2514/1.A32589

[4] G. Isoletta, M. Grassi, E. Fantino, D. de la Torre Sangrà, and J. Peláez, "Feasibility Study of Aerocapture at Mars with an Innovative Deployable Heat Shield," *J Spacecr Rockets*, vol. 58, no. 6, pp. 1752–1761, Jun. 2021. https://doi.org/10.2514/1.A35016

Onboard autonomous conjunction analysis with optical sensor

Luca Capocchiano[1,a], Michele Maestrini[1,b], Mauro Massari[1,c], Pierluigi Di Lizia[1,d]

[1]Department of Aerospace Science and Technology, Politecnico di Milano, Via Giuseppe La Masa 54 20156, Milano

[a]luca.capocchiano@mail.polimi.it, [b]michele.maestrini@polimi.it, [c]mauro.massari@polimi.it, [d]pierluigi.dilizia@polimi.it

Keywords: Relative Orbit Determination, Optical Sensor, Batch Filter, Conjunction Analysis

Abstract. The increasingly high number of spacecrafts orbiting our planet requires continuous observation to predict hazardous conjunctions. Direct onboard analysis would allow to ease the burden on ground infrastructure and increase the catalogued debris. A spaceborne optical sensor is used to assess the performance in terms of different targets visibility. A fast relative orbit determination algorithm is then proposed to compute the probability of collision for a particular case study and compared to a more accurate ground analysis.

Introduction

The growing number of satellite launches increases the risk of in-orbit collision, with potential cascade effects, further worsening the situation. Given the high number of close encounters every day, estimation processes must be frequently updated to avoid the occurrence of catastrophic collisions such as the Iridium-33/Cosmos-2251 event. The possibility to autonomously analyze any conjunction directly onboard would allow to significantly reduce the burden on ground infrastructure, leading to a faster update rate and lower risk of unexpected collisions. Therefore, in this research, a satellite is equipped with an optical sensor to determine the visibility performance with respect to a catalogue of potentially hazardous objects, with different parameters considered relevant for a significant statistical analysis. The closest encounters are then identified and a novel approach is presented for an accurate and computational efficient onboard orbit determination algorithm, providing results directly at the Time of Closest Approach (TCA), where the conjunctions are analyzed onto the B-plane.

Simulation design

The simulation consists of an asset spacecraft, based on the real satellite COSMO-SkyMed 4, operating in a Sun-Synchronous Low Earth Orbit, and a catalogue of 425 possible threats. The sensitivity analysis is carried out in the first eight days of September 2022. The chosen optical sensor is characterized by a limiting magnitude of 15, 30° field of view and 30° minimum Sun separation, operating in tracking mode; the camera is inertially pointed at the expected target direction, at the beginning of each visibility window, without any attitude information. The chosen sensor allows to see objects as small as 10 cm up to 6000 km [1]. The simulation is carried out though *SOPAC* (Space Object PAss Calculator), a Python library developed by *Politecnico di Milano* to compute all the possible observation opportunities for a given sensor network [2]. First, a sensitivity analysis is carried out, with the sensor performance evaluated both in terms of total visibility and revisit times for all the computed windows and compared to the asset observation uniformity along its orbit. Of the six main keplerian elements, only right ascension of the ascending node (or RAAN) shows a remarkable trend; higher semi-major axis may grant higher visibility, yet the considered catalogue is too limited for a significant inference. The total visibility time is shown on the left of Fig. 1, expressed in hours for three different sensor limits: pure geometry, only illumination and full limitations, as defined before. As expected, the presence of the Earth shadow strongly reduces the total time to values lower than 25 hours, while limiting magnitude and Sun separation have a relevant influence only for particularly small objects. The second relation highlights the total time as function of RAAN and uniformity index, computed dividing the asset orbit in 36 sections covering 10° in true anomaly and counting how many contains at least one potential observation. It is thus an important indication of the quality of measurements taken for the specific target.

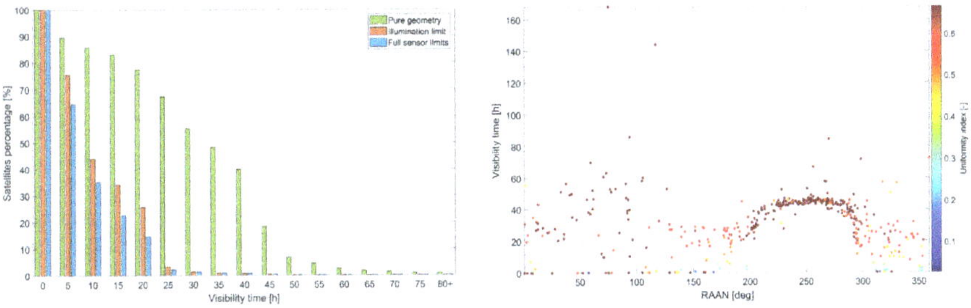

Figure 1 Total visibility time expressed in hours for three different sensor limits (left) and function of RAAN and uniformity index (right).

Objects with RAAN between 200 and 300 degrees are characterized by a higher time and higher uniformity index with no significant influence from initial target conditions. This is an expected result as COSMO-SkyMed 4 has a RAAN of about 70°, approximately 180° apart, resulting in encounters being mostly "head-on". On the contrary, objects with RAAN similar to the asset may reach higher values, though they are strongly dependent on the initial relative position.

An important parameter when scheduling observations is the revisit time, the time between two consecutive passages, reported on the left of Fig. 2. The maximum and minimum revisits are highlighted as function of RAAN, with an upper limit of 3000 seconds. At around 250 degrees, the two values are almost coincident at approximately 2000 seconds.

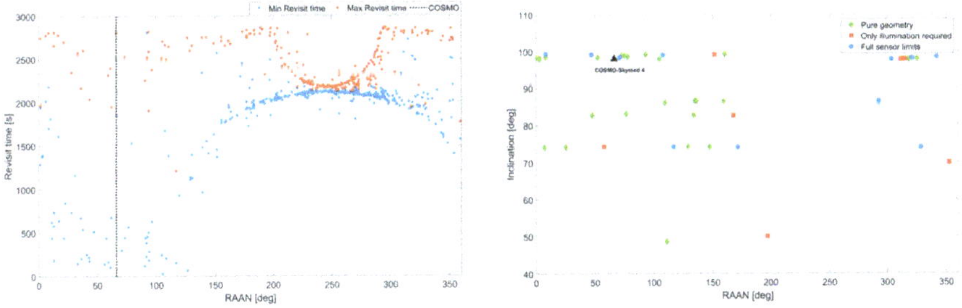

Figure 2 Maximum and minimum revisit (left), non-visible objects as function of inclination and RAAN (right)

Concerning non-visible objects, on the right of Fig. 2, out of the 425 objects, 25 are never visible due to geometric limitations, 7 due to lack of Sun illumination and 11 due to the specific sensor adopted. Once again, satellites with RAAN between 200° and 300° are always visible, regardless of the considered limit.

Materials Research Proceedings 37 (2023) 526-529 https://doi.org/10.21741/9781644902813-115

Methodology

The conjunction events are searched in the 8 days after the statistical analysis. The optical sensor accuracy is set at 0.01° for both right ascension and declination, to account for lower performance of an onboard system, as well as possible mounting errors. Since the proposed filtering method is based on the linear, minimum variance Least Squares, the propagation should be close to the real motion to ensure accurate results even with a single iteration. The best compromise is found by propagating directly from the TLEs using SGP4, allowing to keep a low computational time and high accuracy. On the contrary, the contribution of the State Transition Matrix (STM) is much less sensitive to inaccurate orbit modeling and a simple J2 effect can be implemented without excessively diminishing the estimation accuracy. The proposed method is based on the property of the STM to map each observation to the relative state directly at the TCA, without the need to further propagate the solution, which would result in a higher computational time and lower accuracy. If the primary spacecraft motion is assumed to be perfectly known, then the relative state deviation is equal to the target one, allowing to work in terms of absolute quantities, as reported in Algorithm 1. Although the asset spacecraft is tracked by ground stations providing accurate orbit determination, its trajectory is still affected by uncertainties, included in the filter post-processing through the Consider Covariance Analysis (CCA) [3]. Given the solution of the filtering process, relative position and covariance at the TCA are projected onto the B-plane, reducing the state dimension from six to only two positional coordinates, greatly simplifying the conjunction analysis and a potential CAM design [4]. The probability of collision is thus computed according to the Chan formulation, truncated to the third order [5]. The results provided by the onboard algorithm are compared to the ones obtained with a classic Unscented Kalman Filter (UKF), with observations coming from a set of three stations part of the EU SST sensor network (EU Space Surveillance and Tracking): S3TSR, MFDR and Cassini.

Algorithm 1: *OnBoard conjunction analysis*

```
Identify conjunction events and visibility windows
Simulate real measurements, according to the defined camera frame
Propagate reference trajectory from TLEs
For each measurement
    | Compute reference observations
    | Compute measurements deviation
    | Integrate STM to linearly map each observation to TCA
end
Compute target state and covariance at TCA
Project onto the b-plane and compute probability of collision
```

Conjunction result

The results of the onboard relative orbit determination and conjunction analysis are here presented for the closest event, involving Falcon 1 rocket body, orbiting in a quasi-equatorial orbit of 9° inclination at approximately 650 km of altitude. Despite the low visibility time of only 16 hours, 284 measurements are processed. As highlighted in Fig. 3, the estimated error is 23 m, with a covariance of 76 m computed from the Least Squares, and 120 m added from the CCA. Considering the miss distance of 871 meters, the probability of collision is effectively equal to zero. The ground-based analysis is performed for all the involved objects, computing error and covariance of the orbit determination process alone, in order to directly compare the two methodologies. The square root of the covariance trace and the positional error highlight similar trends for both the asset and the target spacecrafts. The final values are 33 and 23 meters respectively for COSMO, 110 m and 120 m for Falcon 1. The latter lies in correspondence of a spike, though the steady-state values are comparable to the asset. In general, the UKF results are slightly better, especially in terms of covariance, with the advantage of a real-time estimation process; however, the difference in computational time is considerable, with the Least Squares providing results in few seconds. Furthermore, the space-based platform is capable of seeing objects with very different orbits, still with a sufficiently high number of measurements, thus showing greater flexibility especially for low-inclination targets.

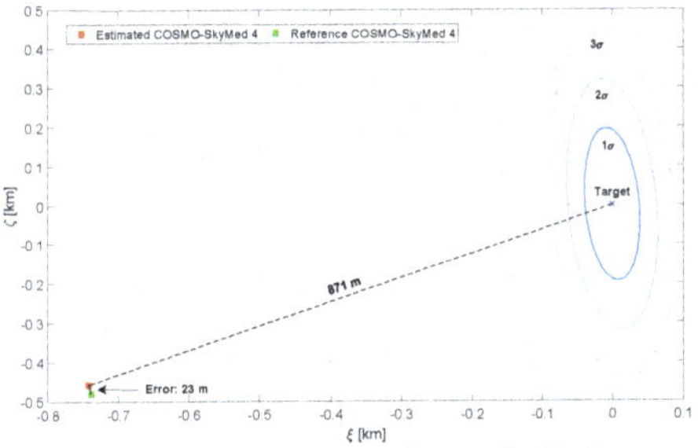

Figure 3 COSMO-SkyMed 4 and Falcon 1 projected onto the B-plane at the conjunction epoch. The combined covariance is centered at the target.

Conclusions

The proposed filtering method allows to estimate target position and covariance directly at the conjunction epoch in a fraction of the time required by typical ground-based algorithms. The sensitivity analysis performed for the onboard camera showed good performance in terms of target visibility for all the objects characterized by RAAN between 200 and 300 degrees. In future, the method could be expanded to include fully autonomous initial orbit determination and threat detection, allowing to reduce the burden on ground infrastructure and improve the sustainability of human activities in space.

References

[1] A. Morselli. High order methods for space situational awareness. 2014.

[2] N. Faraco, G. Purpura, P. Di Lizia, M. Massari, M. Peroni, A. Panico, A. Cecchini, F. Del Prete. SNOS: automatic optical observation scheduling for sensor networks. 9th European Conference for Aerospace Sciences (EUCASS 2022), pages 1–11, 2022.

[3] B. Schutz, B. Tapley, and G. H. Born. Statistical orbit determination. Elsevier, 2004. https://doi.org/10.1016/B978-012683630-1/50020-5

[4] A. De Vittori, M. F. Palermo, P. Di Lizia, and R. Armellin. Low-thrust collision avoidance maneuver optimization. Journal of Guidance, Control, and Dynamics, 45(10):1815–1829, 2022. https://doi.org/10.2514/1.G006630

[5] F. K. Chan et al. Spacecraft collision probability. Aerospace Press El Segundo, CA, 2008. https://doi.org/10.2514/4.989186

Mini-IRENE, a successful re-entry flight of a deployable heatshield capsule

Stefano Mungiguerra[1,a*], Raffaele Savino[1,b], Paolo Vernillo[2,c], Luca Ferracina[7,d], Francesco Punzo[3,e], Roberto Gardi[2,f], Maurizio Ruggiero[4,g], Renato Aurigemma[4,h], Pasquale Dell'Aversana[6,i], Luciano Gramiccia[5,j], Samantha Ianelli[8,k], Giovanni D'Aniello[5,l], Marta Albano[8,m]

[1]University of Naples Federico II, 80 Piazzale Tecchio, Naples, 80125, Italy

[2]CIRA "Italian Aerospace Research Centre", 81043 Capua, Italy

[3]ALI S.c.a r.l. Naples, Italy

[4]Euro.Soft, Naples, Italy

[5]SRS-ED, Naples, Italy

[6]Lead Tech, Naples, Italy

[7]ESA, European Space Agency, Noordwijk, The Netherlands

[8]ASI, Italian Space Agency, Rome, Italy

[a]stefano.mungiguerra@unina.it, [b]rasavino@unina.it, [c]p.vernillo@cira.it, [d]Luca.Ferracina@esa.int, [e]francesco.punzo@aliscarl.it, [f]R.Gardi@cira.it, [g]m.ruggiero@eurosoftsrl.eu, [h]r.aurigemma@eurosoftsrl.eu, [i]pasquale.dellaversana@leadtech.it, [j]luciano.gramiccia@srsed.it, [k]samantha.ianelli@asi.it, [l]giovanni.daniello@srsed.it, [m]marta.albano@asi.it

Keywords: Deployable Capsule, Sub-Orbital Flight, Flight Data, Aero-Thermo-Dynamic Loads

Abstract. This paper presents some of the results of the suborbital flight of the Mini-IRENE Flight Experiment (MIFE), flight demonstrator of the IRENE technology. A capsule equipped with a deployable heat shield was successfully launched with a VSB-30 suborbital rocket, achieving an apogee of 260 km, a peak deceleration of 12g and surviving the landing with successful retrieval. A huge set of telemetry data was acquired, including GPS, attitude, temperature, acceleration measurements. The capsule showed aerodynamic stability at all flight regimes. The derived drag coefficient was different from the predicted one, possibly due to flexible aero-brake deformation.

Introduction

The paper describes the main outcomes of the qualification flight of Mini-IRENE, a capsule launched with a Maser sounding rocket on 23rd November 2022 during the SSC S1X3-M15 campaign. The flight has represented the clou of the Mini-IRENE Flight Experiment (MIFE) project [1], funded by the Italian Space Agency (ASI) and managed by the European Space Agency (ESA) in the framework of a GSTP (General Support Technology Program). The project aimed at increasing the ripeness of an innovative technology for atmospheric (re-)entry up to TRL 6, developed by the companies of the ALI consortium, CIRA and University of Naples Federico II, as part of the wider IRENE program [2], and featured by an innovative deployable heat shield, resulting in a very low ballistic coefficient, allowing the exploitation of off-the-shelf materials for the thermal protection system, because of the acceptable heat fluxes, mechanical loads and final descent velocity [3]. The IRENE program aimed at developing a low-cost re-entry capsule, able to return payloads to Earth from the ISS and/or short-duration, scientific missions in Low Earth

Orbit (LEO). MIFE was the latest phase of the IRENE program, with the objectives to design and test a Ground Demonstrator (GD) for the thermal qualification in a Plasma Wind Tunnel, and realize a Flight Demonstrator (FD) to be qualified in a sub-orbital flight with a Sounding Rocket.

All the qualification tests have been performed successfully. The GD was qualified in the CIRA Scirocco Plasma Wind Tunnel for the thermal loads of a specific re-entry mission; the FD has successfully achieved the two main objectives of the sub-orbital flight mentioned above after ejection from the sounding rocket, namely the verification of the stability in every flight regime and the resistance of the heat shield under the thermal and mechanical loads due to the impact with the atmosphere.

This paper is focused on the analysis of the re-entry flight based both on recorded data and telemetry data. The recorded data have been retrieved from an Inertial Unit and a sensors suite while the telemetry data were transmitted to ground even in the supersonic regime.

The suborbital flight

A CAD model of the FD is shown in Fig. 1a, in the deployed configuration. For the launch, it was stowed inside an external shell composed of three panels, as shown in Fig. 1b.

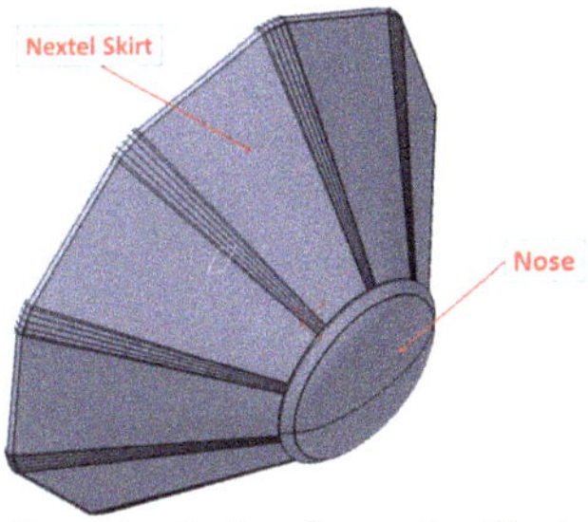

Fig. 1 (a) CAD of the deployed capsule, (b) picture of the stowed capsule before launch.

The capsule was launched from the Esrange base, in Sweden, on 23rd November 2022, onboard a VSB-30 rocket, in the Maser 15 mission (Fig. 2a). The capsule successfully completed its suborbital flight, with a correct aero-brake deployment, and was retrieved few minutes after landing thanks to GPS coordinates and telemetry data provided even after ground impact (Fig. 2b). The capsule trajectory was monitored by GPS and is compared with the nominal predicted trajectory in Fig. 3. A slightly lower apogee than expected was achieved (257 km instead of 261 km). The trajectory showed a sharp deviation in east direction, which was attributed to wind (thanks to balloon measurements) and not to any asymmetry in the capsule aerodynamics. Video recordings, together with magnetometers and accelerometers measurements, demonstrated that MIFE did not lose stability nor tumbled in any flight regime, even in the critical transonic phase. A significant spin rate was measured at the beginning of the continuum regime, possibly due to minor geometrical asymmetries in the TPS or center-of-mass unbalance.

Materials Research Proceedings 37 (2023) 530-533
https://doi.org/10.21741/9781644902813-116

Fig. 2 (a) A picture of the launch, (b) the capsule retrieved after landing

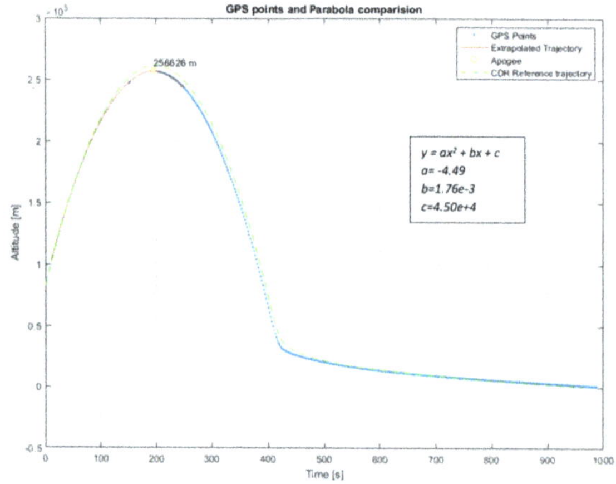

Fig. 3 Capsule parabolic trajectory, measured and predicted

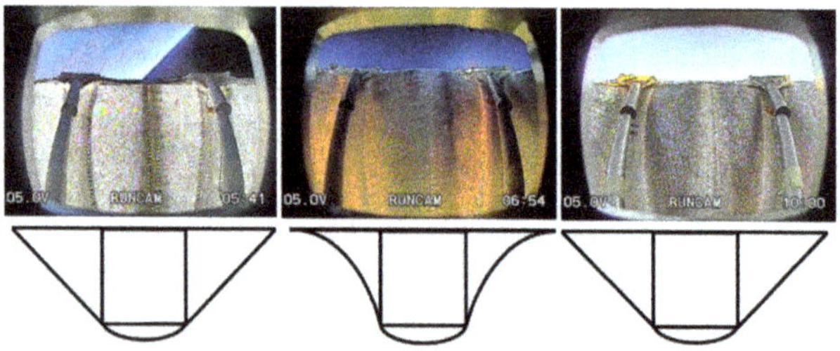

Fig. 4 Three images of the flexible TPS in free fall, during the peak of dynamic pressure, and during the final part of the flight; and a schematic representation

The capsule deceleration was higher than expected (12g versus 10.5g), and the peak of dynamic pressure occurred at a lower altitude. This allowed testing the system in a even harsher aerodynamic environment, which caused a deformation of the flexible TPS (Fig. 4), that had not been taken into account in the derivation of the aerodynamic database. This may be one of the reasons for the differences between the expected and rebuilt drag coefficient, whose trend with Mach number from supersonic to subsonic regime is shown in Fig. 5. The "flight" drag coefficient, computed on the basis of velocity and acceleration measurements, is lower than nominal in supersonic and higher in subsonic, even when considering a ±10% error on the standard atmospheric density used for the calculation. Further analyses, including CFD simulations of the aerodynamics of the deformed TPS, will be carried out for a deeper understanding of this discrepancy.

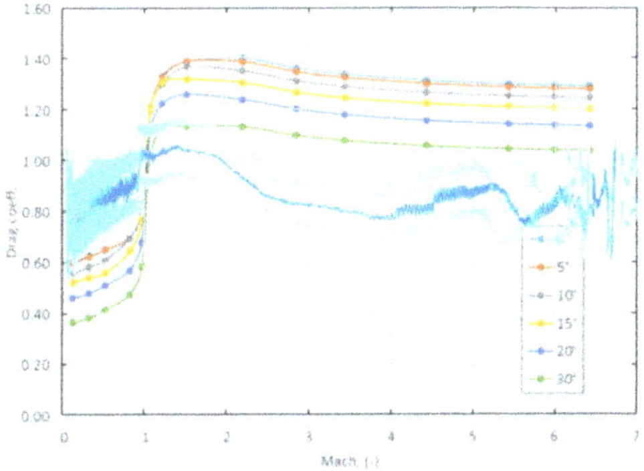

Fig. 5 Drag coefficient estimated based on flight data (blue line), with ±10% uncertainty (light blue lines), compared with numerical aerodynamic database, versus Mach number

Conclusions

The Mini-IRENE Flight Experiment (MIFE) project was successfully concluded with the suborbital launch of the flight demonstrator, which achieved all the mission objectives, including effective separation from the launcher, complete mechanism deployment, aerodynamic stability at all regimes, capsule survival to the aerothermodynamic loads, telemetry data acquisition and capsule retrieval after flight. Flight data analysis is still ongoing for a full comprehension of the capsule behavior and to learn important lessons for the future design of an orbital system based on the IRENE technology.

References

[1] A. Fedele, S. Mungiguerra, Aerodynamics and flight mechanics activities for a suborbital flight test of a deployable heat shield capsule, Acta Astronaut. 151 (2018) 324-333, https://doi.org/10.1016/j.actaastro.2018.05.044

[2] E. Bassano, et al., IRENE - Italian Re-Entry NacellE for Microgravity Experiments, 62nd International Astronautical Congress Cape Town, South Africa, 2011

[3] A. Alunni, et al., Pterodactyl: Trade Study for an Integrated Control System Design of a Mechanically Deployed Entry Vehicle, AIAA SciTech 2020 Forum, Orlando, FL, 2020. https://doi.org/10.2514/6.2020-1014

Electro-thermal dynamic simulations and results of a deorbiting tethered system

G. Anese[1,*], G. Colombatti[1], A. Brunello[1], A. Valmorbida[1], G. Polato[1], S. Chiodini[1], E.C. Lorenzini[1]

[1] University of Padova, Department of Industrial Engineering, Via Venezia 1, Padova

giovanni.anese@unipd.it

Keywords: Space Tethers, Electrodynamics, Numerical Simulations, Satellites Deorbiting

Abstract. Deorbiting techniques with small or better no propellant consumption are an important and critical field of space studies for the mitigation of orbital debris. Electrodynamic tethers (EDTs) are of particular interest because they make possible to deorbit space debris by exploiting the Lorentz force that is provided by the current flowing in the tether thanks to the interaction of the system with the Earth's magnetosphere and the ionosphere. This paper focuses on the differences between two software packages built at the University of Padova (FLEX and FLEXSIM) and their results in simulating various deorbiting scenarios. Both FLEXSIM and FLEX simulate the electro-thermal behaviour and the dynamics of an EDT. However, while the first one has the simplifying assumption that the tether is always aligned with the local vertical, the second one considers also the overall system attitude with respect to the radial direction and the tether flexibility. The computational times of these S/W are very different and it is important to understand the scenarios that are more appropriate for their use. Results aim to show the impact of different solar activity (simulations are done at different epochs) and lengths of conductive and non conductive segments of tether, in the range of a few hundreds of meters, on the total re-entry time. As expected, deorbiting is faster for high solar activity and conductive tether length but the performance must be balanced against the dynamics stability. The issue of stability over the deorbiting time is evaluated numerically for specific cases by using FLEX.

Introduction

The interest in the space exploration is increasing day by day because of a steady growth of orbital debris around the Earth. Thus, it's important to find ways to mitigate this growth and electrodynamic tethers (EDTs) represent a possible green and effective solution. Indeed, they don't use the traditional propulsive systems (chemical and electrical), because they can provide good drag forces (i.e., Lorentz forces) by operating in a passive way, that is, exploiting only environmental factors: the ionosphere and the Earth magnetic field.

This work presents briefly how these systems work and then focuses on some analysis of their performances in different deorbiting scenarios. Solar activity influences space environmental characteristics, like the atmospheric and plasma densities, and consequently also the deorbiting time. Consequently, specific simulations are performed to understand how the reentry time and the dynamics of a deorbiting kit based on an EDT change with respect to the solar activity.

Moreover, the effects of the tether length is also analyzed, because it impacts directly the current produced by the tether itself and consequently the Lorentz force generated by the system.

FLEX and FLEXSIM software packages (built by the University of Padova) are used to simulate the different mission scenarios, understand the differences between them and analyze the performance of EDT systems for deorbiting.

Electro-dynamic tether systems

Electrodynamic tether systems are one typology of space tethers [1]; they are based on a conductive tether (or tape) that links two satellites at its ends. If they operate in a passive way, they exploit the current that flows in the tether from the collection of ionospheric electrons on the bare tether anode, that are then re-emitted at the cathode at the opposite end of the tether. The current produces a Lorentz force through the interaction with the geomagnetic field, as follows:

$$\boldsymbol{F}_L = \int_0^L \boldsymbol{I} \times \boldsymbol{B}\, dx \approx I_{av} L\, (\boldsymbol{u}_t \times \boldsymbol{B}) \quad (1)$$

In the case of the passive mode, this force is opposite to the orbital velocity of the system, that is, a drag force that progressively decreases the altitude of the spacecraft.

However, electrodynamic tethers can be used in Low Earth Orbits (LEO) to have sizeable performances, because at higher altitude the electron density is too low to generate usable currents.

Software packages

FLEXSIM and FLEX are the chosen software packages to perform simulations. They were both built at the University of Padova for the E.T.PACK initiative [2] to study the electro-thermal dynamic of an EDT deorbiting system. They are different because in FLEXSIM a simplification is assumed: the tether is straight and always aligned with the local vertical and does neither oscillate nor flex, as in a real case. This assumption allows to use FLEXSIM as a first iteration step in the mission evaluation, especially for what concerns the deorbiting time and the computation of the average current in the tether. On the other hand, FLEX simulates the tether dynamics allowing the evaluation of dynamics stability, which is discussed later.

From the modelling point of view, while FLEXSIM considers the tether as a single element, FLEX divides the tether in different segments and computes the dynamics of each one. In FLEXSIM an average current of all the conductive length is considered and used in Eq. 1. In FLEX, instead, an average current and thus a Lorentz force is computed for each segment, leading to a more detailed dynamics of the tether, as stated above.

These two simplifications reduce drastically the computation time of FLEXSIM with respect to FLEX. The codes use a set of subroutines that allows to evaluate the environmental characteristics [3,4,5]. A critical element is the current subroutine, based on [6], that computes the current $I(x)$ as a function of the electron density, the motional-electric field and the voltage drop at the cathode.

Numerical Simulations

Several simulations were done to evaluate the performances of the deorbiting tethered system. As just mentioned, the environmental characteristics that determine the current on the tether are influenced by solar activity. Therefore, the same mission is simulated with both FLEX and FLEXSIM at three different epochs, with low, medium and high solar activity (based on the F10.7 index), in order to understand the solar activity impact on the reentry time.

All these simulations were conducted on an EDT with a tape of 500 m (450 m of Aluminum and 50 m of PEEK, the inert segment) and two tip masses of 12 kg. This is the baseline configuration of the In-Orbit Demonstration flight, presently planned for 2025 for E.T.PACK-F. The system is initially on a circular orbit at 600 km height inclined of 51.5°; the mission is considered completed when the system reaches an altitude of 250 km where the atmospheric density is strong enough to reenter the system in a few orbits. As mentioned earlier, the tether length is also investigated as a driving parameter. The investigated scenarios is the same of the previous simulations, but the tether length changes from 400 to 700 m, with a constant percentage of conductive (90%) and non-conductive (10%) tether for all cases. These simulations are done with FLEX, so that, not only the deorbiting time is evaluated, but also the system dynamics, because the variable length leads to a different behavior, allowing for the study of dynamics stability.

Results

Simulations results are now presented, starting from the effects of solar activity. As shown in Table 1 for the case of E.T.PACK-F with 500-m tether, the deorbiting time decreases at higher solar activities, since the electron density increases [7], leading to an improvement of electron collection on the tether and consequently higher currents (Fig. 1) and Lorentz forces.

Table 1 Deorbiting time [days] for different solar activities with FLEXSIM and FLEX

	Low solar activity	Medium solar activity	High solar activity
FLEXSIM [days]	94.152	25.161	19.706
FLEX [days]	88.182	23.267	18.075

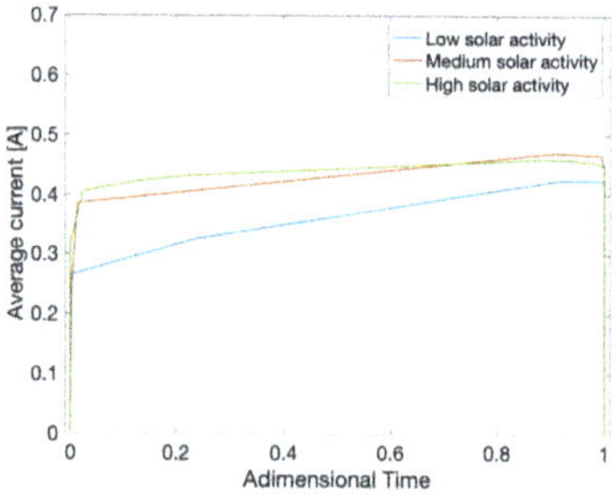

Figure 1 Envelope of current peaks vs time scaled (0 start of mission, 1 end of mission) for low, mean and high solar activity (FLEXSIM)

The differences between FLEXSIM and FLEX in deorbiting times is due to the tether dynamics, which helps the electromotive force (e.m.f.) thanks to the out-of-plane tether dynamics..

Varying the tether lengths at fixed environmental conditions (high solar activity), it is possible to see in Table 2 and Fig. 2, that increasing the length the time of reentry is lower. This is due to the the increase in the e.m.f. that generates higher currents and the increased overall Lorentz force that acts on a longer tether length. However, with a 700 m tether (or longer) instability occurred, the system begins to tumble and could not be controlled. This is due to the increased Lorentz force, and indeed longer tethers require heavier tip masses to be stabilised.

Table 2 Deorbiting time as function of tether length (with high solar flux)

Tether length [m]	400 (360 Al-40 PEEK)	450 (395 Al-45 PEEK)	500 (450 Al-50 PEEK)	600 (540 Al-60 PEEK)	650 (585 Al-65 PEEK)	700 (630 Al-70 PEEK)
Deorbiting time [days]	25.226	21.369	18.075	15.815	13.374	instability

Fig. 3 shows the in and out of plane angles for the configurations with 400 m and 650 m of tether (both stable since the angles never reach the 90 deg). It's possible to see that the blue lines have more peaks than the orange ones, indicating oscillations of longer tether are more persistent, hence these systems are less stable.

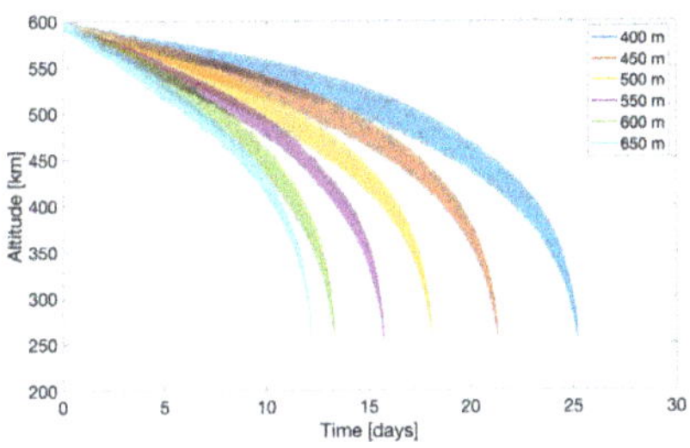

Figure 2 Center of mass altitude vs time as function of tether lengths (FLEX)

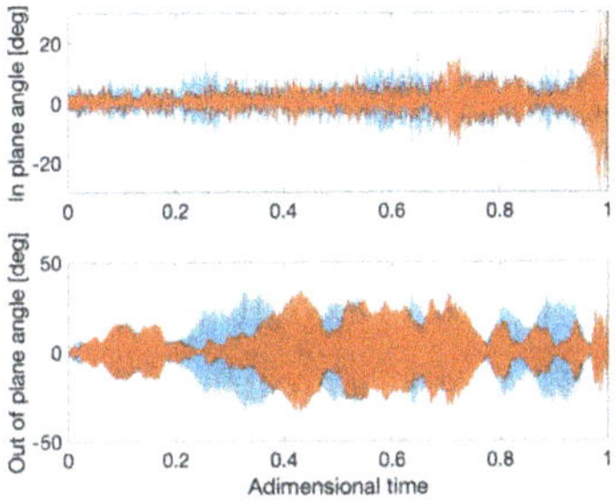

Figure 3 In and out of plane angles for tether length of 400 m (orange) and 650 m (blue)

Conclusions

The focus of this paper is the use of EDT systems for the deorbiting of space debris and the investigation of software packages (FLEX and FLEXSIM) that allow to study EDT performances and their dynamics. From the simulations done for this work, it is possible to confirm that EDTs have better performances at higher solar activities (where they can generate higher Lorentz forces) and with longer tethers. This increased performance must be balanced against the dynamics stability of the system that must be evaluated to assess the feasibility of the mission.

FLEX and FLEXSIM are valid software packages to investigate the performance of EDTs. They are also the starting point to evaluate other applications of tethered system, like EDT operating in active mode for generating thrust to allow a wider spectrum of orbital maneuvers.

Acknowledgments

This work was supported by Horizon Europe EIC Transition Programme under Grant Agreement No. 101058166 (E.T.PACK-F)

References

[1] M.L. Cosmo, E.C. Lorenzini, Tethers In Space Handbook, Third ed., Smithsonian Astrophysical Observatory, Cambridge (Massachussets), 1997.

[2] Information on https://etpack.eu/e-t-pack-f/.

[3] J. Picone et al., NRLMSISE-00 empirical model of the atmosphere: Statistical comparisons and scientific issues, J. Geophysical Research: Space Physsics, 107 (2002). https://doi.org/10.1029/2002JA009430

[4] D. Bilitza, B.W. Reinisch, International reference ionosphere 2007: Improvements and new parameters, Advances in space research, 42 (2008). https://doi.org/10.1016/j.asr.2007.07.048

[5] M. Mandea et al., International geomagnetic reference field—2000, Physics of the Earth and planetary interiors, 120 (2000). https://doi.org/10.1016/S0031-9201(00)00153-9

[6] M. Sanjurjo-Rivo et al., Efficient Computation of Current Collection in Bare Electrodynamic Tethers in and beyond OML Regime, J. Aerospace Engineering, 28 (2015), https://doi.org/10.1061/(ASCE)AS.1943-5525.0000479

[7] Liu, L. et al., The dependence of plasma density in the topside ionosphere on the solar activity level, Ann. Geophys., 25 (2007), 1337–1343. https://doi.org/10.5194/angeo-25-1337-2007

Mechanical and pneumatic design and testing of a floating module for zero-gravity motion simulation

Simone Galleani[1,a*], Thomas Berthod[1,b], Alex Caon[2,c], Luca Lion[2,d], Federico Basana[2,e], Lorenzo Olivieri[2,f], Francesco Branz[3,g], Alessandro Francesconi[3,h]

[1]University of Padova, Padova (Italy)

[2]C.I.S.A.S – Centre of Studies and Activities for Space "G. Colombo", Via Venezia 15, Padova (Italy)

[3]Department of Industrial Engineering, University of Padova, Via Venezia 1, Padova (Italy)

[a]simone.galleani@studenti.unipd.it, [b]thomas.berthod@studenti.unipd.it, [c]alex.caon@unipd.it, [d]luca.lion.1@phd.unipd.it, [e]federico.basana@phd.unipd.it, [f]lorenzo.olivieri@unipd.it, [g]francesco.branz@unipd.it, [h]alessandro.francesconi@unipd.it

Keywords: Floating Module, Low Friction Table, Pneumatic Tests

Abstract Close proximity operations demand an accurate control in a micro-gravity environment, hence they must be reproduced and simulated systematically. Consequently, laboratory tests are a crucial aspect to validate the performances of space systems. This paper presents the development of a floating pneumatic module, whose dimensions and mass are representative of a 12U CubeSat. The vehicle has been designed to perform planar low friction motion over a levelled table for docking experiments. The paper focuses on the pneumatic and mechanical designs and on the laboratory tests of the module. The pneumatic design regards the air-compressed pneumatic system. The major specifics have been determined by the requirement of performing a docking procedure by starting from a distance of 500 mm. The mechanical design has been guided by two main requirements. The first is the possibility to accommodate different docking systems (e.g.: docking port). The second is the possibility to control the position of the centre of mass of the module. Several tests have been performed to verify the capabilities of the vehicle, such as: (1) pneumatic tests to evaluate the thrust of the propulsion system through the execution of linear motions and (2) mechanical measurements with dedicated setups to improve the estimation of the position of the centre of mass from the CAD model of the system.

Introduction

A Close Proximity Operation (CPO) of an on-orbit spacecraft can be defined as a manoeuvre of one spacecraft (chaser) in a relative orbit with respect to another spacecraft (target) [1]. These operations are performed in micro-gravity conditions and include docking manoeuvres which require a systematic characterization of the forces and torques arising from the interaction of the chaser and the target. Therefore, laboratory tests and the realization of dedicated facilities are a critical aspect to validate the performances of docking mechanisms.

Among various microgravity simulation methods, such as parabolic flights, drop towers or robotic manipulators, an achievable solution for a laboratory environment is the use of planar Air-Bearings (ABs), which allows floating of tested devices [2] with the creation of a thin film of pressured gas between an internal porous structure and a surface. Thus, a planar 3 DoF motion can be achieved in a quasi-frictionless condition. Although a reduced number of DoF is obtained compared to the 6 DoF of an on-orbit motion, planar ABs are usually used as a support for dedicated vehicles which are equipped with thrusters and/or reaction wheels to simulate a satellite for CPOs and, specifically, docking manoeuvres experiments [2].

This paper presents the development of a floating pneumatic module, which has been designed to perform 3 DoF low-friction planar motion with three ABs over a levelled table. The module has a volume of 330x224x224 mm^3 with a mass of approximately 12 kg, so that it represents the mass properties of a 12U CubeSat.

With a dedicated propulsion system, the main goal of the vehicle is to simulate docking manoeuvres starting from a distance of 500 mm. The vehicle can operate both as a chaser, active mode, and as a target, passive mode, and it can accommodate different docking systems.

Pneumatic design

The pneumatic system has been designed to allow the vehicle to float with three round ABs and perform translational and rotational manoeuvres over a levelled table with a compressed air propulsion system.

Furthermore, the pneumatic system has been realized to satisfy the following requirements: (1) motion is provided by 8 thrusters (2 thrusts for each corner of the vehicle); (2) each thruster is activated by one Electro-Valve (EV); (3) the pneumatic circuit ensures a total floating time of 3 min and performs an acceleration of 50 mm/s^2.

Requirements (2) and (3) have led to the following specifications: (a) the three ABs with a diameter of 40 mm are able to lift a total weight of approximately 68 kg at an input pressure of 3.9 bar; (b) each thrust is composed by two nozzles with a throat diameter of 1.3 mm to improve the produced force and reduce the working pressure; (c) the total air volume should be at least 2 L at 10 bar. Therefore, the pneumatic circuit has been realized with a 2.5 L tank and a single pressure regulator to control ABs and thrusters at the same pressure.

Mechanical design

The mechanical design has revolved around the positioning and sizing of components, so that they would fit inside the total volume of the module. The mechanical design has been guided by three main requirements: (1) the module accommodates different docking systems with a dedicated volume of 100x224x224 mm^3 in the front part; (2) the three ABs are positioned in an equilateral triangular configuration and the Centre of Mass (CoM) of the module is controlled to be coincident (with an error of 1 mm) with the centroid of the ABs to guarantee uniform floating of the vehicle; (3) the centroid of the thrusters is aligned with the CoM of the system to allow pure rotational motions. Additionally, the centroid of the ABs coincides with the Geometrical Centre (GC) of the 330x224x224 mm^3 volume of the system.

Figure 1 shows (a) the CAD model of the vehicle with the reference frame at the GC used to refer the position of the CoM and (b) the assembled module.

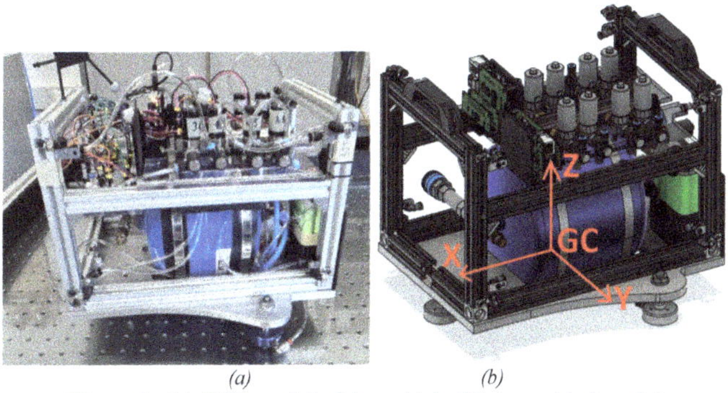

(a) (b)

Figure 1: (a) CAD model of the vehicle (b) Assembled module

An estimation of the position of the CoM has been obtained from the complete CAD model. With a total estimated mass of 8.2 kg with no payload, the CoM has been placed at -27.4 mm along the X axis and -0.9 mm along the Y axis with respect to the GC (Figure 1a). Moreover, the fully assembled module of Figure 2b has a greater total mass of 8.4 kg, because it includes electrical components and wiring which have not been considered in the CAD model.

A possible solution to control the position of the CoM of a system is the realization of custom masses which can be moved either automatically with motors [3] or manually [4]. For this reason and to get closer to the goal of a total mass of 12 kg, a group of manually movable steel masses has been designed and their masses have been determined from the CAD estimation.

In particular, three sets of masses have been designed: (1) a set of fixed masses (total mass of 620 g) to be mounted on the front part of the system to bring the CoM closer to the GC; (2) a couple of movable masses of 687 g each to control the X coordinate of the CoM; (3) a movable mass of 240 g to control the Y coordinate of the CoM.

Furthermore, considering a mass of 1 kg to represent a generic payload, by acting on the moving masses, it is possible to shift the X and Y coordinates of the CoM in a ±5 mm and ±2 mm ranges which contain the centroid of the ABs.

Tests on the pneumatic system

The tests on the pneumatic system have involved the execution of linear motions over the levelled table to estimate the provided thrust. The position of the module has been measured by a motion capture system (OptiTrack Prime^x 13 with an accuracy of ±0.2 mm). Figure 2 shows the setup for the tests on the pneumatic system.

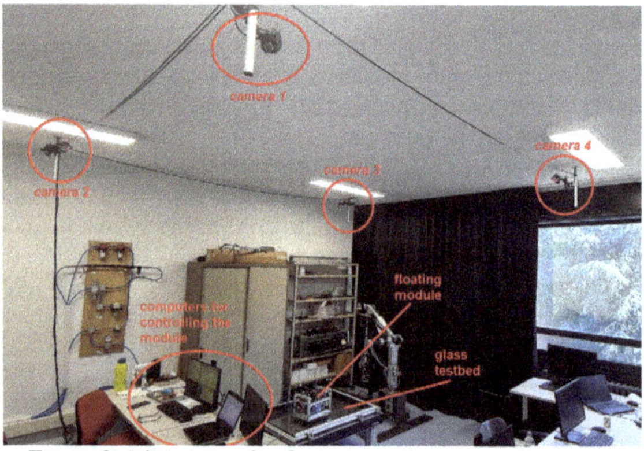

Figure 2: Main setup for the tests on the pneumatic systems

By commanding an impulse of 1 s to the EVs, two linear trajectories along the X and Y axes have been performed. By analysing the data of the measured position, the resulting thrusts have been estimated to be 1.541 N along the X axis and 1.531 N along the Y axis. The expected thrust has been calculated to be 2.225 N: the discrepancy could be related to the localized pressure losses in the pneumatic system.

Measurement of the position of the centre of mass

Mechanical measurements have been performed to estimate and balance the position of the CoM of the module. The position of the CoM has been measured with a setup of three load cells (rated load of 10 kg with an output of 2 ± 0.2 mV/V) placed under the supports of the three ABs. Figure

3 presents (a) the CAD model of the setup of the load cells and (b) the assembled setup with the module placed on top.

A Matlab algorithm has taken the three outputs of the load cells as inputs and converted them into mass values to compute the position of the CoM. With no payload and no fixed or movable masses (total mass of 8.4 kg), ten measurements have been acquired to account for the noise of the load cells. With an uncertainty of ±1 mm (~0.4% of the dimensions of the module), the mean value of the ten positions has placed the CoM at -25.6 mm and -3.1 mm along the X and Y axes with respect to the GC. The discrepancy from the CAD estimation is approximately 0.7%.
By mounting and acting on the moving masses, the CoM can be aligned with the GC within a ±1 mm range.

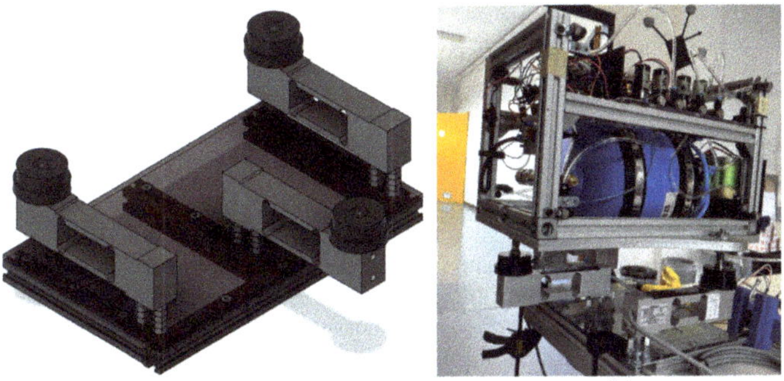

(a) *(b)*

Figure 3: (a) CAD model of the load cells setup – (b) Assembled load cells setup with module

Conclusions

This paper presents an overview of the pneumatic and mechanical designs and the performed tests and measurements of a floating pneumatic module which has been designed to execute 3 DoF planar low friction motion.

The thrust provided by the propulsion system has been quantified and the capability of the module to perform simple linear trajectories proven.

The position of the CoM has been measured with an uncertainty of ±1 mm, through a dedicated measuring setup.

The next steps of the development will involve the execution of rotational motions around the main axis over the levelled table to estimate the inertia of the module.

References

[1] Markus Wilde et al., Historical survey of kinematic and dynamic spacecraft simulators for laboratory experimentation of on-orbit proximity manoeuvres, Progress in Aerospace Sciences, https://doi.org/10.1016/j.paerosci.2019.100552

[2] Tomasz Rybus et al., Planar air-bearing microgravity simulators: Review of applications, existing solutions and design parameters, Acta Astronautica, https://doi.org/10.1016/j.actaastro.2015.12.018

[3] Chesi et al (2014), Automatic Mass Balancing of a Spacecraft Three-Axis Simulator: Analysis and Experimentation. Journal of Guidance, Control, and Dynamics, https://doi.org/10.2514/1.60380

[4] Marcello Romano et al., Acquisition, tracking and pointing control of the Bifocal Relay Mirror spacecraft, Acta Astronautica, https://doi.org/10.1016/S0094-5765(03)80011-5

Simulations for in-flight stellar calibration aimed at monitoring space instruments optical performance

Casini Chiara[1,2,5,a*], P. Chioetto[1,4], A. Comisso[1], A. Corso[1], F. Frassetto[1,5], P. Zuppella[1,4], V. Da Deppo[1,3]

[1] CNR-IFN, Via Trasea 7, 35131 Padova, Italy

[2] Centre of Studies and Activities for Space “Giuseppe Colombo”, Via Venezia 15, 35131 Padova, Italy

[3] INAF-OAPd, Vicolo dell'Osservatorio 5, 35122, Padova, Italy

[4] INAF-OAA, Largo Enrico Fermi 5, 50125 Firenze, Italy

[5] INAF-OATo, Via Osservatorio 20, 10025 Pino Torinese, Torino, Italy

[a]chiara.casini@pd.ifn.cnr.it

Keywords: In-Flight Calibration, Space Instruments, Metis, STC

Abstract. Stellar in-flight calibrations have a relevant impact on the ability of space optical instruments, such as telescopes or cameras, to provide reliable scientific products, i.e. accurate calibrated data. Indeed, by using the in-flight star images, the instrument optical performance can be checked and compared with the on-ground measurements. The results of the analysis of star images, throughout the whole instrument lifetime in space, will allow tracking the changes in instrument performance and sensitivity due to optical components degradation or misalignment. In this paper we present the concept, the necessary input and the available outputs of the simulations performed to predict the stars visible in the FoV of a specific space instrument. As an example of the method, its application to two specific cases, i.e, Metis coronagraph on-board Solar Orbiter and the stereo camera STC on-board BepiColombo, will be given. Indeed, due to their operation in proximity to the Sun, and also to Mercury for STC, both instruments operate in a hostile environment, are subjected to high temperatures and experience high temperature variations. Performance optical monitoring is thus extremely important.

Introduction

The proper calibration of a space instrument allows its optimal performance throughout the entire mission duration.

Space is a hostile environment: e.g. a mission going near the Sun experience hot temperatures, likely to induce component degradation, even if the mission is carefully planned and built.

The instrument response to a well-known source, e.g. star acquisition, is a valuable mean for monitoring the optical performance of space instrument and, if necessary, to update and correct the image calibration.

In this paper, a description of the possible in-flight stellar calibrations, and their related simulations, are given and then applied to two space instruments: on board of Solar Orbiter [1] on board of Solar Orbiter; and of SIMBIO-SYS have an original optical design [2] on board of BepiColombo. Both instrument have an original optical design and are working in very hostile environments, near the Sun, and at distances rarely reached so far by other space missions. Comparing on-ground calibration with simulations and in-flight calibration is a key element for assuring the correct performance of these space instruments.

In the next sections the emphasis will be put on: the importance of the calibration, a brief description of the instrumentation where the simulations are applied, how the simulations have been performed and the results obtained are then presented.

Calibrations

Calibration is a key and fundamental step in achieving accurate and reliable scientific measurements from space instruments. To this end, on-ground calibration is crucial. Both are important to validate instrument design, mitigate systematic errors, ensures data consistency. For the in-flight calibration, by comparing the instrument response to known stellar characteristics, any deviations or discrepancies in the optical performance can be detected and corrected. This process enhances the quality and reliability of the acquired scientific data, making calibration an essential component of space missions.

In-flight calibration basically consists in following a star moving across the detector during several minutes. Repeated stellar observations, at different times during the lifetime of the instrument, are important to track the changes in instrument sensitivity due to optical elements or detector degradation and other causes. Systematic observations of several stars will track sensitivity changes, and comparison between the in-flight and on-ground results can give important information on the status and performance of the instrument.

Following such stars requires achieving high accuracy acquisitions over high dynamic range, which is a fundamental challenge. Yet, it is necessary to determine an accurate absolute calibration for astronomical photometry. Direct comparisons of stellar outputs with calibrated flux sources can generally only be made for very bright stars.

Concretely, our goal is to find the stars on a Metis, and on a STC image. Our approach is based on-defining an upper limit on a grayscale image in intensity (mean value $+3\sigma$): every pixel with a higher intensity corresponds to a star. Then around the maximum value we define a box, usually is 10 pixels x 10 pixels, and we plot a 2D gaussian, from which we extrapolate information like PSF. Knowing the PSF all over the detector gives us the information on some defocus, vignetting and so on.

Besides, every detector has a linearity curve. On ground we can measure this curve with a uniform light source like an integration sphere, it can give very useful information about the efficiency of each pixel. In flight our light source is the light from the stars. We can use them as calibration sources, acquiring the light of the same star for different Integration Times (IT), and analyzing the response of the detector pixels.

And finally, through the passage of stars all over the detector we can identity defects such as shadows and bad pixels. It is important to know if such defects are stable over time or are changing, which would have an impact on imaging (the Sun or Mercury in our context). To do so we have to analyze the stars in any parts of the detector to know if it responds differently.

Instrumentations: Metis and STC

The Metis coronagraph and STC have innovative optical design due to their respective missions. Metis makes linearly polarized measurements of the solar corona in the visible spectral range, and simultaneously acquires images in the ultraviolet Ly-α neutral hydrogen line 121.6 nm, with an annular field of view from 1.5° to 2.9° and with an unprecedented temporal resolution among other space coronagraphs (10s). Metis will observe the Sun as close as 0.28 AU, so it is important to reduce the extremely high thermal load, therefore an Inverted Externally Occulted configuration is used to block the light of the solar disk. Indeed, to reduce the thermal load, the light of the photosphere enters in Metis through the Inverted External Occulter (IEO) and is then rejected towards the entrance aperture by the mirror M_0 that is acting as an occulter as shown in ***Figure 1*** (a). Coronal light is reflected by mirror M_1 towards mirror M_2, which also induces diffused light from real images of the edges of the IEO and the M_0 mirror. Therefore, an internal occulter (IO) and a Lyot Stop (LS) are respectively introduced, the rest of the diffused light being blocked by a Field Stop (FS). Then, the coronal light is reflected by mirror M_2 in the direction of the dichroic beam-splitter, the Interferential Filter (IF). The IF is optimized for narrowband spectral transmission in the ultraviolet (UV, 121.6 nm, H I Lyman-α), and broadband spectral reflection in the visible (VL, 580–640 nm). The visible light reflected by IF enters in a polarimetric unit and arrives on the detector. Both channels have a CMOS sensor, a 1024 x 1024 pixel matrix for the ultraviolet, and a 2048 x 2048 pixel matrix for the visible [1].

STC is a double wide-angle camera which main scientific aim is the mapping of the entire surface of Mercury in 3D. As shown in *Figure 1* (b), the STC camera consists of two sub-channels named High (H) and Low (L) with respect to the mounting interface on the spacecraft. There are also two different fore optics for each sub-channel plus a common modified Schmidt telescope. The light scattered by Mercury passes through the external baffle, is reflected inside a rhomboid prism, passes through a correcting doublet, the aperture stop (AS), and arrives on the spherical mirror M_1. It reflects on a telescope mirror, positioned off axis, which in turn reflects into a two-lens field corrector and arrives on the focal plane assembly (FPA). The STC detector can read a maximum of six specific windows. Nominally, these windows correspond to the areas of the 6 filter: two panchromatic (PAN) with FoV 5,3°×2,4° and four colored filter with FoV 5,3°×0,4° [2]. The tridimensional modeling of the instrument is available on [5].

Knowing the accurate location of the spacecraft and the optical path going through the instruments enables to perform simulations on what may be acquired by the instruments. This is of major importance to enhance *in fine* the performances of the in-flight calibrations.

Simulations

Simulations have emerged as powerful tools for addressing the challenges associated with in-flight stellar calibration. They enable scientists to create virtual environments that accurately mimic the behaviour of space instruments and their interaction with celestial objects. These virtual experiments allow for the optimization of calibration strategies, evaluation of instrument design choices, and testing of data analysis techniques. Simulations offer a cost-effective and efficient means of exploring a wide range of scenarios that may not be feasible in real-world settings.

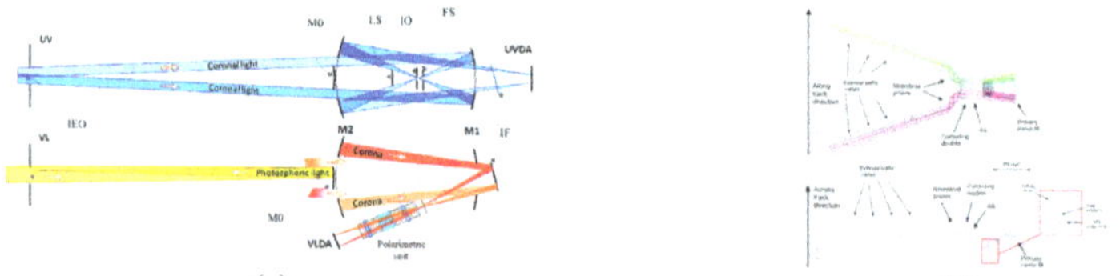

(a) (b)

Figure 1 Optical path inside the Metis coronagraph (a), and STC (b) [4].

The keystone of these simulations, performed in Python, is the SPICE kernel (Spacecraft, Planet, Instrument, Camera pointing, and Events). For each space instrument, the SPICE kernel gives, among other things, the information of the location of the spacecraft in the past and in the future, the boresight of each specific channel and the Field of View (FoV).

This kernel is used alongside the SIMBAD catalogue, which exhaustive for our purposes because it provides extensive stellar data. The Metis team and Slemer et al. for STC [3] performed the analysis for determining the maximum apparent magnitude of a star detectable by the instruments. Combining the SPICE kernel with the SIMBAD catalogue allowed us to obtain the results described in the next paragraph.

Results

The simulations reported for both instruments have been performed on the same day: 10-03-2026 at the 8:58:05, in order to highlight the fact that they are looking at different places. For this reason, the results of the stars seen by each channel is different.

The stars seen by the Metis coronagraph are presented in the *Figure 2*, respectively for the Visible channel (VL, (a)) and for the Ultraviolet channel (UV, (b)). Inside the Field of View there are 18 stars. Metis acquires in two channels but is looking at the same objects. The results appear mirrored because of the reflection inside the elements of the coronagraph.

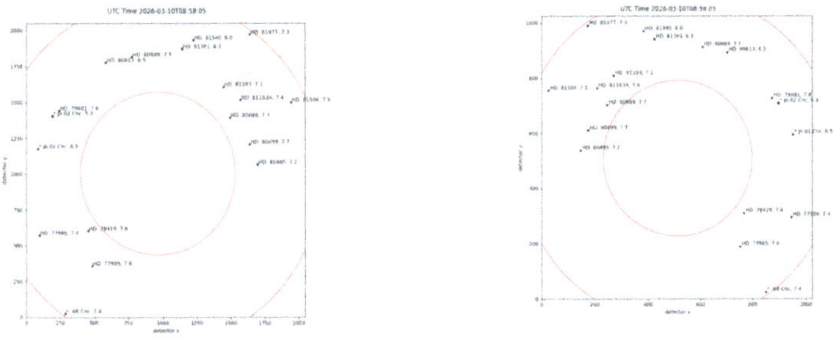

(a) (b)

Figure 2 Simulations of the stars seen by the Metis the day 10-03-2026 at the 8:58:05 in the visible (a) and UV (b).

For STC, in the FoV of the two panchromatic on March 10th 2026 at 8:58:05 are shown in *Figure 3* (a) and (b). The stars seen by the Pan L and Pan H are respectively, 24 and 29 stars. Because they are looking at 20° of difference respect to the Nadir.

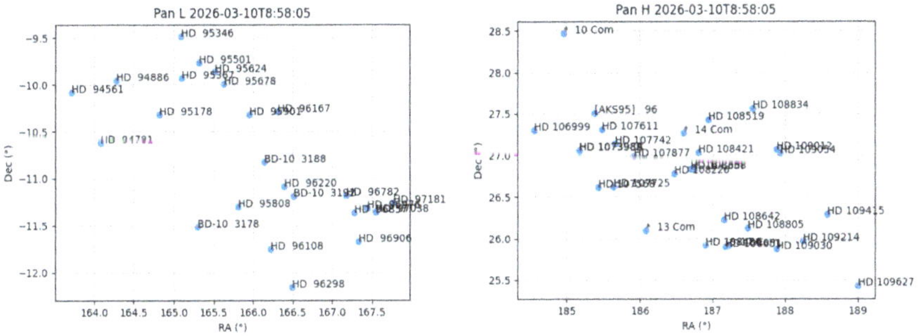

(a) (b)

Figure 3 Simulations of the stars seen by the panchromatic L (a) and H (b) the day 10-03-2026 at the 8:58:05.

The difference between the imaged simulated for Metis and STC are the display of the results. Knowing the stars can be seen from the instruments give us the opportunity to select the best target for the in-flight calibration.

Conclusion

In this paper, the problematics of stellar in-flight calibration have been presented, in particular the importance of the calibrations (on-ground and in-flight), and the crucial role of the stars as in-flight

Materials Research Proceedings 37 (2023) 542-546 https://doi.org/10.21741/9781644902813-119

calibration light sources have been highlighted. We report the simulations to determine the stars observable by two instruments: the Metis coronagraph on board of Solar Orbiter, and STC on board of BepiColombo. This simulation activity is part of an on-going work, associated with on-ground and in-flight, acquired and simulated images.

Fundings
This activity has been carried on in the framework of the ASI-INAF Contracts Agreement N. 2018-30-HH.0 and 2017-47-H.0.

References

[1] Fineschi, S. et al., Optical design of the multi-wavelength imaging coronagraph Metis for the solar orbiter mission. Exp Astron, 49, 239-263 (2020). https://doi.org/10.1007/s10686-020-09662-z

[2] G. Cremonese, et al., SIMBIO-SYS: Scientific Cameras and Spectrometer for the BepiColombo Mission, Space Sci Rev 216, 75, 2020. https://doi.org/10.1007/s11214-020-00704-8

[3] A.Slemer et al., Setting the parameters for the stellar calibration of the SIMBIO-SYS STC camera on-board the ESA BepiColombo Mission. Proc. of SPIE Vol. 11443 1144374-1. https://doi.org/10.1117/12.2560648

[4] V. Da Deppo et al., Optical design of the single-detector planetary stereo camera for the BepiColombo European Space Agency mission to Mercury. App. Opt. xx. https://doi.org/10.1364/AO.49.002910

[5] Information on https://www.dei.unipd.it/~dadeppo/STC.html

Reduced-order modelling of the deployment of a modified flasher origami for aerospace applications

A. Troise[1,a*], P. Celli[2,b], M. Cinefra[1,c], V. Netti[1,d], A. Buscicchio[1,e]

[1] Department of Mathematics Mechanics and Management, Polytechnic University of Bari, Via Edoardo Orabona, 4, 70125 Bari, Italy

[2] Department of Civil Engineering, Stony Brook University, Stony Brook, NY 11794, USA

[a]a.troise@phd.poliba.it, [b]paolo.celli@stonybrook.edu, [c]maria.cinefra@poliba.it, [d]vittorio.netti@poliba.it, [e]alessandro.buscicchio@poliba.it

Keywords: Origami Structures, Deployable Structures, Bar-and-Hinge, Mechanics

Abstract. In this paper, we simulate the nonlinear deployment mechanics of a modified flasher origami structure designed to be a deployable solar panel. We compare reduced-order bar-and-hinge simulations, where panels are modelled as bar assemblies connected by joints and torsional springs, with results obtained from commercial finite element software. Through this comparison, we demonstrate the ability of the bar-and-hinge approach to capture key features of the origami behaviour at a fraction of the time needed to perform regular finite-element simulations. We also provide details on how to properly tune the bar properties to simulate panels made bonding printed circuit boards to textile, and the joint properties to mimic folds that are made of fabric and flexible circuit interconnects.

Introduction

In the past few decades, origami structures have attracted significant attention in the field of science and engineering, due to their unique mechanical properties and reconfigurable and tuneable attributes. These properties make origami designs suitable for applications in fields such as robotics, medicine, and especially aerospace.

The task of modelling origami structures for space applications presents several challenges. First, modelling origami requires accounting for significant geometric nonlinearity due to the large rotations that the panels undergo during deployment and stowage. Additionally, real-life origami structures do not deploy following rigid body motions and are instead characterized by panel bending. Additional challenges appear when the origami systems to be modelled are made of multi-layer materials such as rigid-flex printed circuit boards (PCB). These materials are typically used in CubeSat applications, in which origami techniques are applied to deployable solar panels, communication devices and solar sails. Finally, origami simulations must yield information on the forces exerted by the deployment on the spacecraft.

Research investigations on the deployment of origami structures have been conducted utilizing a variety of finite element software and techniques, including ABAQUS 5, formulations based on Hamilton's equations to capture the dynamics of deployment with validation using ADAMS multibody dynamics [2], and quasi-static bar and hinge methods [3]. The fundamental principle of the bar-and-hinge approach, elucidated by Schenk and Guest [4] as well as Filipov et al. [5], centers on the simplification of the mechanics in origami, by replacing panels with assemblies of bars, hinges and torsional springs that limit out-of-plane rotations. This approach leverages the inherent limitations of permissible deformations within origami structures: in-plane stretching, out-of-plane folding along creases, and out-of-plane bending of panels. Bars are strategically positioned along straight fold lines and across panels to ensure in-plane stiffness. Rotational hinges are incorporated along the bars connecting panels to simulate crease folding, as well as along the bars traversing panels to replicate panel bending. The method solves the equilibrium equations iteratively, using a displacement-controlled algorithm. Despite having a limited number of degrees of freedom, the reduced-order bar-and-hinge model accurately predicts the overall mechanical behaviour of origami structures [6].

Here, we use the bar-and-hinge method to simulate the deployment of a modified flasher origami for space applications and compare the results to shell FEM results from ANSYS Motion.

Model generation workflow

The selected folding pattern is a modified Flasher origami, which is renowned for its radial deployment mechanism and has been notably utilized in the design of the NASA Starshade prototype, for which the modification amounts to an octagonal variation of the folding technique. The geometry is initially designed in 2D and saved as .svg. A specific color convention is used to distinguish between mountain folds and valley folds, as illustrated in Figure 1 (left): mountain folds are red, valleys are blue and boundary edges are black.

To simulate the deployment process, we need a closed version of this origami structure. This closed configuration is obtained using the interactive origami software (*Origami simulator*) by Ghassaei et al. [7]. The structure before and at a stage of partial folding are shown in Figure 1 (center and right).

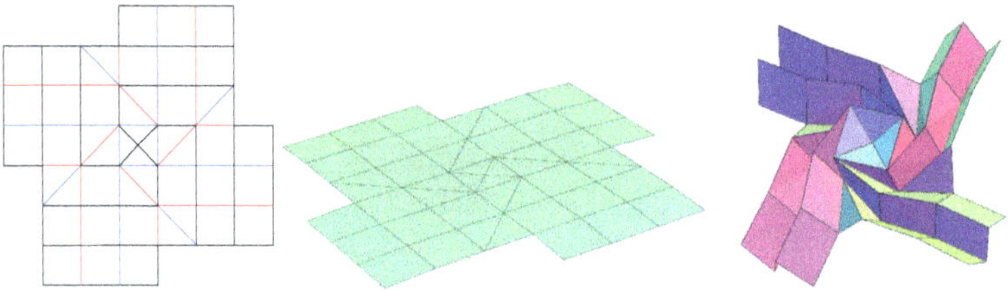

Figure 1 – Origami pattern in .svg file (left), 3D open configuration (center) and 3D partially closed (right). The image in the center and the one on the right have been rendered from the code in [7].

After obtaining the closed version of the origami structure, we export it as .obj file and import it in Merlin 2 (written in MATLAB) [2]. Prior to using it, the nodal coordinates from the .obj file are modified to better fit the desired geometry. The geometry of the origami structure is a 10x10x10 centimeters cube.

After importing the geometry, we set boundary conditions and loads. The only boundary condition imposed on the model is a complete translational block along the three axes to the central node of the horizontal panel. The loads are represented by four displacement constraints of 385 millimeters imposed on the top four vertices of the outermost panels.

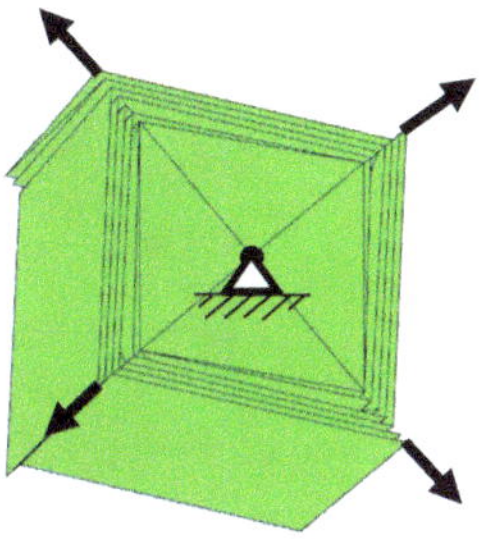

Figure 2 – Closed geometry of the flasher with graphical representation of boundary conditions and loads

In Merlin 2, the most important modelling parameters are the type of discretization of each panel, the material properties and the thickness of the geometry elements. The discretization alternatives for the analysis are called N4B5, which includes four nodes and five bars for every square panel and N5B8, including five nodes and eight bars for every square panel; here, we choose the latter.

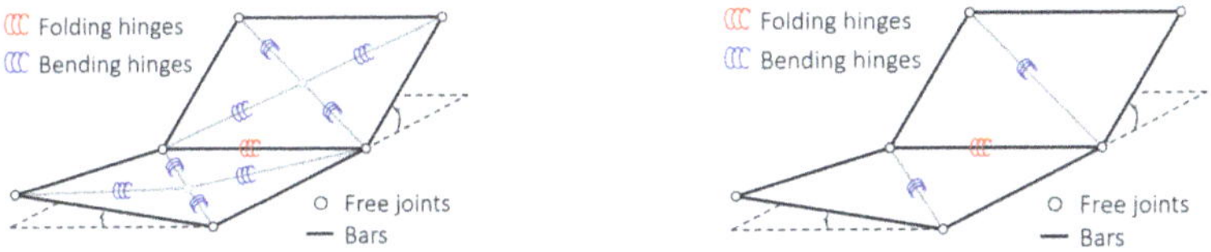

Figure 3 –N5B8 (left) and N4B5 (right) discretization, images from [2]

The parameters imposed for the material properties come from experimental data on a specific textile-based electronics substrate [8], used for the realization of the physical prototype that will be subjected to experimental tests to validate the data coming from both models.

The code allows to obtain a load-displacement curve for any node from the simulation, together with information concerning the stored energy of the bending and folding hinges in the geometry. The result of our analysis is shown in Figure 4. At large displacements, the load increases asymptotically since the deployment is complete at 360 millimeters and any further loading engages the high axial stiffness of the panels. The detail of the load-displacement curve, shown on the right, shows a gradual increase of the force from zero to 350 millimeters.

It can also be noticed that the first part of the plot shows a zero load, due to an initial free rotation given by the imperfect alignment of the forces in the model.

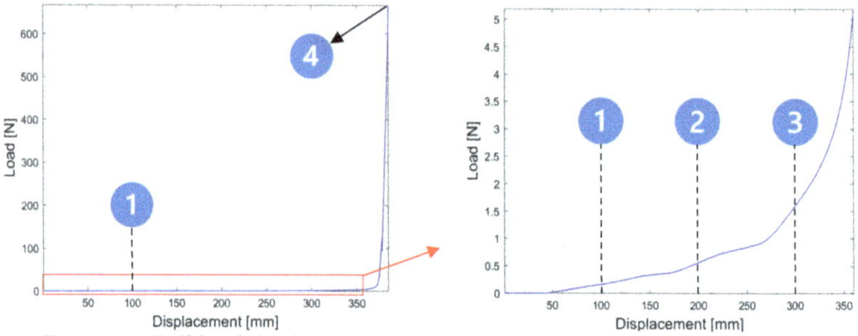

Figure 4 – Full load-displacement curve (left) and partial graph from 0 to 360 mm

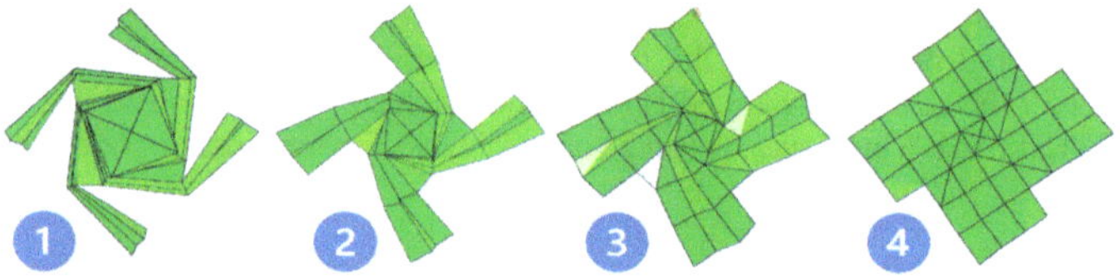

Figure 5 – Four configurations of flasher origami pattern during different phases of deployment, the numbers next to each configuration are related to the Load-Displacement curve in Figure 4

Ansys Motion

The same geometry is imported in the finite element software ANSYS Motion, with the objective of obtaining a comparison between the reduced order model and the finite-element one for validation purposes. To import a .obj geometry in ANSYS, we first import it in Solidworks, export it as a 2D geometry to the Ansys Workbench environment and successively modify it using SpaceClaim.

The material properties imposed to the ANSYS model are the same utilized for the reduced order model, with the approximation of elastic isotropic material, which is suitable for the expected large deformations and small strains. The contact constraint has been created for every panel to accurately model the interaction between the geometric elements.

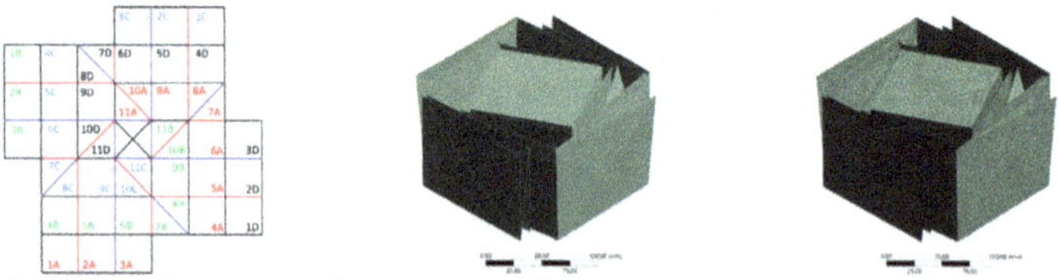

Figure 6 – From left to right: schematics used for constraints enumeration in ANSYS, geometry imported in ANSYS Workbench and geometry discretized using shell elements

The geometry in the closed configuration is renamed according to Figure 6 (left) to make the constraint-imposition process more efficient. The discretization is carried out using shell elements. All the boundary conditions and loads have been set up to create a simulation identical to the bar-and-hinge one. The main difference between the two models is the absence of folding springs in the ANSYS one, which causes zero resistance during deployment and therefore does not allow to validate load-displacement curves. As we can see, both models capture the same kinematics of deployment.

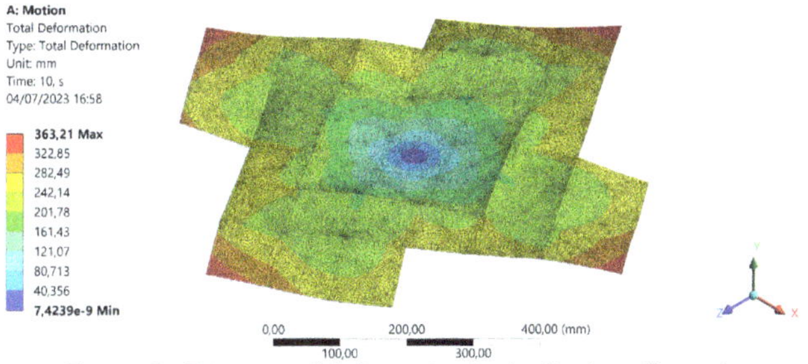

Figure 7 - Diagram of deformation in the final configuration

Conclusions and future developments

The finite element model and the bar and hinge one capture different aspects of the behaviour of the structure. FEM pursues this task through the utilization of higher-order elements, such as plates/shells or volumetric elements, with the same amount of information for the material properties under the approximation of linear isotropic behaviour. Bar-and-hinge models are an efficient tool for approximating the mechanical behaviour of origami structures. Despite their simplicity, these models can be used to capture out-of-plane bending and in-plane shearing, allowing to obtain a conspicuous amount of information with a reduced computational expense compared to the finite element models.

As a next step, we plan an in-depth analysis of the results of the two models (introducing torsional springs in the FEM as well), with a campaign of experimental tests on a physical prototype of the origami structure. This will allow a complete validation of the results, as well as the opportunity to refine the models and the material properties.

Up to now, deployment is simulated as outward radial applied displacements. To accurately capture the forces exerted on the spacecraft during deployment, we will implement a follower load in both models and change the boundary conditions by blocking rotations along the structure's axis; this should allow us to extract the moment produced on the structure during deployment – which will be useful to design the actuation device for deployment.

References

[1] Cai, Jianguo, et al. "Deployment simulation of foldable origami membrane structures." Aerospace Science and Technology 67 (2017): 343-353. https://doi.org/10.1016/j.ast.2017.04.002

[2] Jihui Li, Qingjun Li, Tongtong Sun, Zhiwei Zhu, Zichen Deng, A general formulation for simulating the dynamic deployment of thick origami, International Journal of Solids and Structures, Volume 274, 2023, 112279, ISSN 0020-7683. https://doi.org/10.1016/j.ijsolstr.2023.112279

[3] E.T. Filipov, K. Liu, T. Tachi, M. Schenk, G.H. Paulino, Bar and hinge models for scalable analysis of origami, International Journal of Solids and Structures, Volume 124, 2017, 26-45, ISSN 0020-7683. https://doi.org/10.1016/j.ijsolstr.2017.05.028

[4] Mark Schenk and Simon D Guest. "Origami folding: A structural engineering approach." In Origami 5, edited by Patsy Wang-Iverson, Robert J Lang, and Mark Yim, pp. 293–305. CRC Press, 2011.

[5] E. T. Filipov, K. Liu, T. Tachi, M. Schenk, and G. H. Paulino. "Bar and hinge models for scalable analysis of origami." International Journal of Solids and Structures 124 (2017), 26–45. https://doi.org/10.1016/j.ijsolstr.2017.05.028

[6] K. Liu and G. H. Paulino, "Highly efficient nonlinear structural analysis of origami assemblages using the MERLIN2 software." Origami, 2018, 7: 1167-1182.

[7] Amanda Ghassaei, Erik Demaine, and Neil Gershenfeld, "Fast, Interactive Origami Simulation using GPU Computation." (2018).

[8] Alessandro Buscicchio, et al. "FRET (Flexible Reinforced Electronics with Textile): A Novel Technology Enabler for Deployable Origami-Inspired Lightweight Aerospace Structures." In AIAA SCITECH 2023 Forum, p. 2081. 2023. https://doi.org/10.2514/6.2023-2081.

Deployment profile analysis for tethered deorbiting technologies

G. Polato[1,a*], A. Valmorbida[1,2], A. Brunello[2], G. Anese[1], S. Chiodini[1,2], G. Colombatti[1,2], E.C. Lorenzini[1,2]

[1]University of Padova, Department of Industrial Engineering, Via Venezia 1, Padova, Italy

[2]University of Padova, Center for Studies and Activities for Space (CISAS) "Giuseppe Colombo", Via Venezia 15, Padova, Italy

[a]giulio.polato@unipd.it

Keywords: Tether Deployment, Optimization, Deorbiting, Clean Space

Abstract. Over the past few decades, the man-made space debris has become an increasingly concerning problem for future space missions. Fortunately, some innovative "green" deorbiting technologies have been emerged. Among these strategies, electrodynamic tethers have demonstrated to be a promising option, thanks to their passive and fuel-free characteristics. By leveraging the Earth's ionosphere and the geomagnetic field, an electrodynamic tether generates a Lorentz drag force, that can significantly reduce the altitude of a satellite and ultimately cause it to re-enter the atmosphere. The goal of this research is to investigate a critical part of satellite tethered technology, namely the deployment phase. To accomplish this, we utilized a software tool developed by the University of Padova to simulate the dynamics of the deployment phase and optimize its trajectory, in order to meet the desired boundary conditions. This paper gives a description of the software and shows the results of a sensitivity analysis on the trajectory profile that examines the impact of variations in the release angle of the tether and the speed profile actuated by the motor that controls the deployment speed.

Introduction

Tethered satellites technology has been studied for almost 60 years (first mission, Gemini 11 in 1966) [1]. These satellites can be used for many different purposes such as generating artificial gravity, electrical power generation or electrodynamic thrust [2]. One of the most critical part of a tethered satellite mission is the deployment phase. Depending on the purpose of the mission, the deployment profile followed by the tether, in order to meet the desired final conditions, must be carefully controlled. There are two possible ways to control the tether deployment: by imposing either a tension profile or a length/velocity profile on the reeled-out tether [3]. For the purpose of our study, we chose the second option, where a velocity profile is given as input to a motor that controls the angular velocity of a spool, where the tether is winded. The goal is to reach a final condition where the two satellite are aligned along the local vertical. The advantage of the second strategy is that the reel-out velocity can be more easily measured than the tension on a moving tether and also it can be accurately controlled by a motor with a feedback velocity control.

Mathematical model

The dynamic of tethered satellite is strongly non-linear and is influenced by several parameters. In order to reduce the complexity of the problem, we use a classical Dumbbell model for the system

Materials Research Proceedings 37 (2023) 553-557 https://doi.org/10.21741/9781644902813-121

dynamics that assumes a rigid, straight, length-variable and mass-less tether [4]. The reference system and the parameters used are shown in Fig. 1.

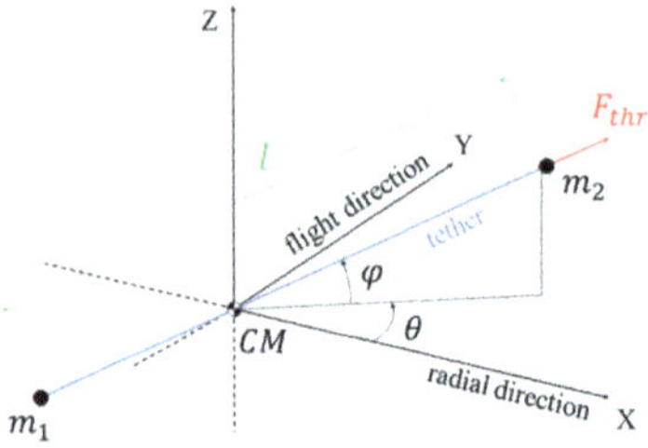

Fig. 1: Reference system frame for tethered satellite

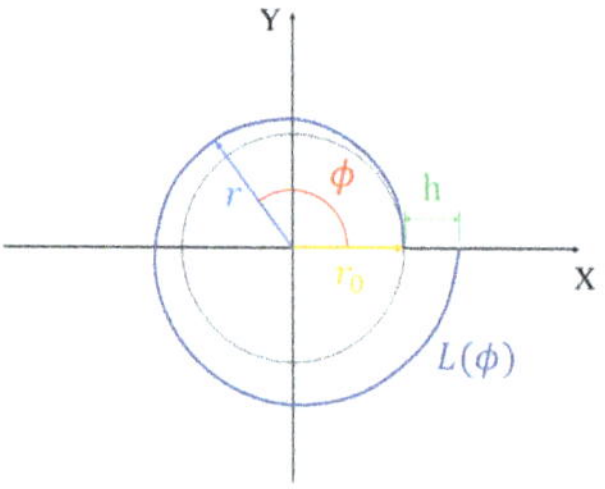

Fig. 2: Archimedean spiral parameters used to model the tape-shaped tether

The equations describing the deployment of a tethered system are the following:

$$\ddot{l} = l \cdot \left[\left(\dot{\theta} + \omega_O\right)^2(\varphi) - \omega_O^2 + \varphi^2 + 3\,\omega_O^2(\theta)\,(\varphi)\right] + \frac{F_{thr}}{m_2} - \frac{T_p}{m_R} \tag{1}$$

$$\ddot{\theta} = -2\left[\frac{l}{l} + \dot{\varphi}\; tan\, tan\,(\varphi)\right](\dot{\theta} + \omega_O) + 3\,\omega_O^2\; cos\, cos\,(\theta)\;\; sin\, sin\,(\theta) \tag{2}$$

$$\ddot{\varphi} = -2\, m_R \frac{l}{l}\dot{\varphi} - \left[\left(\dot{\theta} + \omega_O\right)^2 + 3\,\omega_O^2(\theta)\right] cos\, cos\,(\varphi)\;\; sin\, sin\,(\varphi) \tag{3}$$

where l is the tether length, θ the in-plane libration angle, φ the out-of-plane libration angle, F_{thr} the thrust applied on the second mass m_2, $m_R = \frac{m_1 m_2}{(m_1+m_2)}$ the reduced mass, T_p the tether tension, and ω_O the orbital angular velocity. For the reference deployment profile computation only the in-plane libration dynamics is considered ($\varphi, \dot{\varphi}, \ddot{\varphi} = 0$); the out of plane oscillation is used only for the sensibility analysis.

Finally, in order to derive the control profile of the motor that extracts the tether from the stationary spool, we use the Archimedean spiral to model the tape-shaped tether winded on the spool [5] (see Fig. 2):

$$r = r_0 + \frac{h}{2\pi}\phi \tag{4}$$

$$L = \frac{h}{4\pi}\left[\phi\sqrt{\phi^2+1} + ln\, ln\left(\phi + \sqrt{\phi^2+1}\right)\right] \tag{5}$$

where r_0 is the initial radius, h the tether width, ϕ the angular coordinate of the spiral, $r(\phi)$ the radial coordinate of the spiral, and L the tether length in the spool varying from r_0 to r.

Software implementation

The software utilized for implementing the dynamics of the deployment was developed by the E.T.PACK-F team of the University of Padova. It mainly consists of four phases: the first three phases are used to derive the reference profile trajectory [6], and the last one is used to derive the input motor profile.

First Phase. In order to simulate the separation phase at the beginning of the deployment, we impose an acceleration profile driven by an initial thrust provided by the thrusters on board the module with mass m_2. As boundary conditions we imposed the maximum tether velocity that we want to reach and a time of 50 s needed to reach it. These values are compatible with the performance of the thrusters and the deployment motor.

Second Phase. During this phase, starting from the final conditions of the first phase, the trajectory is optimized using the software BOCOP [7], that approximates the optimal control problem with a finite-dimension optimization problem (NLP) through a time discretization. The boundary conditions to solve this problem are: the initial and final state conditions, i.e., tether length, length rate, libration angle and libration rate, the total deployment time, and upper and lower bounds of the state variables during the optimization process.

Third Phase. The last step for the trajectory profile derivation is to smooth out the transition between the first and the second phase, and the final part of the trajectory. This phase is important because we want to avoid sudden change in the acceleration profile, hence a sudden change in the tension of the tether that can lead to its rupture. In order to obtain a continuity of the second derivative (the acceleration) and a final acceleration equal to zero, we use an 8th-order polynomial to approximate the length profile.

Fourth Phase: Finally, the velocity profile is converted to an angular velocity, using the Archimedean spiral model. The angular velocity is converted from deg/s to rpm which can be used as a reference profile for the motor that controls the spool mechanism.

Sensibility analysis

In order to study the stability of the reference trajectory we introduced errors in our simulations associated with the orientation misalignment of the modules at the beginning of the deployment ($\varepsilon_\theta = \pm 10\, deg, \varepsilon_\varphi = \pm 10\, deg$) on the in-plane and out-of-plane angles, respectively. In addition, we evaluated the influence of errors on the actual angular velocity of the deployment motor with respect to the reference profile ($\varepsilon_t = \pm 15\, deg/turns$), e.g., with an error of $\varepsilon_t = 9\, deg/turn$ the spool performs 1,025 turns instead of 1 turn, hence the motor is running faster than the reference velocity. The values considered for the ε_t error is exaggerated with respect to the actual performance of our motor, but we decided to use these values in order to have a better understanding of the general trend of the error.

The analysis was conducted by varying one driving variable at a time, in order to understand how each parameter affects the deployment dynamics and quantify its contribution to the error on the desired libration amplitude. For the demonstration flight of E.T.PACK-F, the requirement on the final libration amplitude is 10 deg [8].

Results

The following Fig. 3 show the results obtained while Table 1 summarizes the upper bounds of the limit cases. From the Fig. 3 it is possible to understand that the out-of-plane angle error has a smaller influence on the libration amplitude than the in-plane angle error.

Moreover from Fig. 4 it is shown the three trajectory obtained for the limit case of the three errors compared to the nominal trajectory. From this figure it is possible to visualize the effects of these three errors.

Table. 1 upper and lower bound error on the final libration angle

	ε_θ	ε_φ	ε_t
θ_{libr} lower bound error	0.9065 [deg]	$5.5247 \cdot 10^{-3}$ [deg]	-0.4385 [deg]
θ_{libr} upper bound error	-0.1651 [deg]	$0.1689 \cdot 10^{-3}$ [deg]	0.7656 [deg]

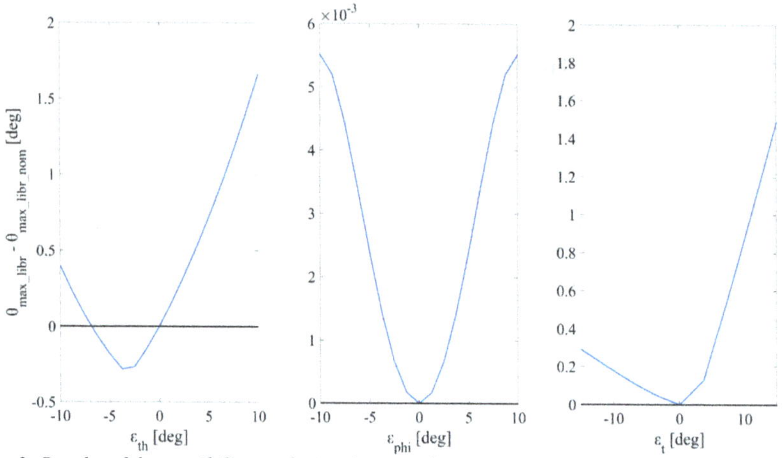

Fig. 3: Results of the sensibility analysis, where on the y-axis is indicated the difference between the free libration amplitude and the nominal libration amplitude

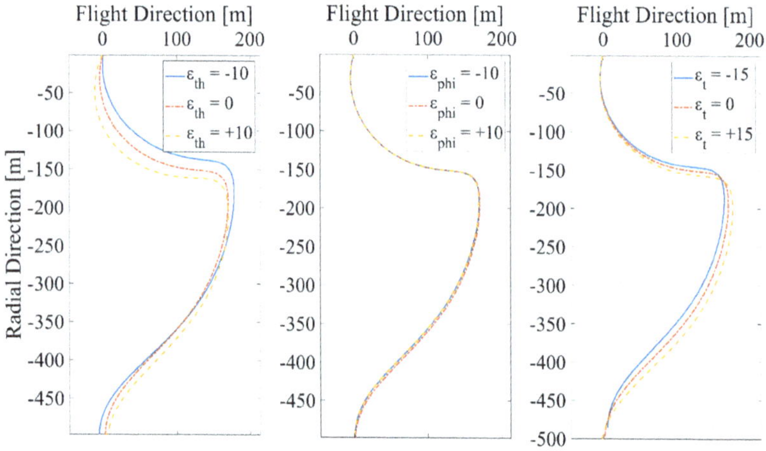

Fig. 4: Comparison of the trajectories for a the limit case of the three errors and the nominal trajectory

Finally, we can conclude also that having a higher velocity than the nominal brings to a higher error on the libration amplitude. Overall, the in-plane angle error has more influences on the libration amplitude, and it has the same order of error as the velocity error.

Conclusions

In this paper, we presented the software developed for the E.T.PACK-F project to compute the deployment reference trajectory and described the four phases of its operation. By considering initial and final conditions, we were able to determine an optimized trajectory that fulfills our deployment requirements. After that, we identified potential errors that could impact the post-deployment libration amplitude. To assess their influence, we conducted a sensitivity analysis evaluating the effects of each error individually. Based on the simulations conducted, we can conclude that the post-deployment libration amplitude is less than 2 deg, which satisfies the 10 deg requirement. This confirms that the computed reference trajectory for deployment shows a good robustness to the errors considered in this study.

Acknowledgments
This work was supported by Horizon Europe EIC Transition Programme under Grant Agreement No. 101058166 (E.T.PACK-F)

References

[1] Y. Chen, et al., History of the tether concept and tether missions: a review, International Scholarly Research Notices 2013 (2013). http://dx.doi.org/10.1155/2013/502973

[2] M. Kruijff, Tethers in Space: A propellantless propulsion in-orbit demonstration., Ph.D. Thesis, Technical Univ. of Delft, Delft, The Netherlands, (2011).

[3] E. C. Lorenzini, et al., Control and flight performance of tethered satellite small expendable deployment system-II, Journal of Guidance, Control, and Dynamics 19.5 (1996): 1148-1156. https://doi.org/10.2514/3.21757

[4] P. Mantri, Deployment Dynamic of Space Tether Systems, Ph.D. Thesis, Faculty of North Carolina State University (2007).

[5] from: https://en.wikipedia.org/wiki/Archimedean_spiral

[6] from: https://research.dii.unipd.it/mts/

[7] Team Commands, I. Saclay, BOCOP: an open-source toolbox for optimal control. (2017), See: http://bocop.org

[8] G. Sarego, et al., Deployment requirements for deorbiting electrodynamic tether technology, CEAS Space Journal 13.4 (2021): 567-581. https://doi.org/10.1007/s12567-021-00349-5

The Janus COM mechanism onboard the JUICE probe to the Jovian system

G. Colombatti*[,1,2], A. Aboudan[1], M. Bartolomei[1], S. Chiodini[1,2], A. Dattolo[3], G. Noci[3], F. Sarti[3], T. Bilotta[3], A. Colosimo[3]

[1]CISAS Giuseppe Colombo, University of Padova, via Venezia 15, Padova Italy

[2]Department of Industrial Engineering DII, University of Padova, via Venezia 1, Padova Italy

[3]Leonardo SpA, Firenze, Italy

*giacomo.colombatti@unipd.it

Keywords: Telescope, Jupiter, JUICE, Qualification

Abstract. After the successful launch of the JUICE (JUpiter ICy moons Explorer) on the 14th of April 2023 all the on board subsystems and instrument are testing their functionalities. The JANUS (Jovis, Amorum ac Natorum Undique. Scrutator) telescope is the imaging system on board the spacecraft and is an optical camera devoted to the study of global, regional and local morphology and processes on the Jovian moons, and to perform mapping of the clouds on Jupiter. Following the heritage of the successful design of the OSIRIS WAC camera, on board the Rosetta mission, the group of researchers at CISAS "Giuseppe Colombo"- Università degli studi di Padova, led by prof. S. Debei, in collaboration with colleagues of the Leonardo spa Company developed the mechanism responsible for the protection of the telescope during cruise phase. The COver Mechanism (COM) provides the external closure of the JANUS Optical Head Unit (OHU). It shields the optical parts from contamination, it is light and dust tight and works in the plane of the telescope entrance window avoiding the exposure of the inner surface of the cover itself and the core part of the telescope to the external dust and pollution. The lower part of the cover provides, also, a reference surface for the in-flight calibration of the telescopes. The main functional and environmental requirements of this mechanism can be identified and summarized as follows: the door provides optics and detector protection from sunlight and contamination; the subsystem, located at the main entrance of the JANUS OHU outer Baffle, provides the function of opening and closing of the cover. The opened Cover allows the JANUS OHU and detector to face the outer environment to perform planetary observations during science mission phases and to perform in-flight calibration observation of different targets (e.g., moons, stellar fields). This paper presents the design, the mechanical solutions adopted for a reliable system and the results of the test performed on ground in order to qualify the JANUS COM mechanism before flight.

Introduction

The astonishing JUICE mission, launched early this year, will perform a very detailed exploration of the ocean layers and analysis of subsurface water reservoirs; it will study the Ganymede's intrinsic magnetic field, its topographical, geological and compositional maps. The JUICE mission was designed to investigate the physical properties of the icy crusts and of the internal mass distribution of jovian moons [1]. It will arrive at Jupiter in January 2031 after 7.6-years using an Earth–Venus–Earth–Earth gravity assist sequence.

Among the JUICE payload, JANUS (Jovis Amorum ac Natorum Undique Scrutator) is the narrow-angle camera selected as the visible (from near UV to NIR) imager onboard JUICE. An overview of the scientific goals of JANUS, together with measurements needed to fulfil the specific goal, is given in the following. The detailed investigation of the Galilean icy satellites, which are believed to harbor subsurface water oceans, is central to elucidating the conditions for habitability of icy worlds. Visible wavelength imaging is needed to determine the formation and characteristics of magmatic, tectonic, and impact features, relate them to surface forming processes, constrain global and regional surface ages, and investigate the processes of erosion and deposition [2]. The JANUS instrument is composed by the following functional (and physically independent) subsystems:

- Optical Head Unit (OHU), mounted on the S/C Optical Bench
- Proximity Electronics Unit (PEU), located close to OHU on S/C Optical Bench
- Main Electronics Unit (MEU), located in the S/C vault
- Interconnecting harness

The COver Mechanism (COM) provides the external closure of the JANUS Optical Head Unit (OHU). It protects the optical parts from contamination; it is dust tight and works in the plane of the entrance window avoiding the exposure of the inner surface of the cover itself and the core part of the telescope to the external dust and pollution. The lower part of the cover provides, also, a reference surface for the in-flight calibration of the telescopes.

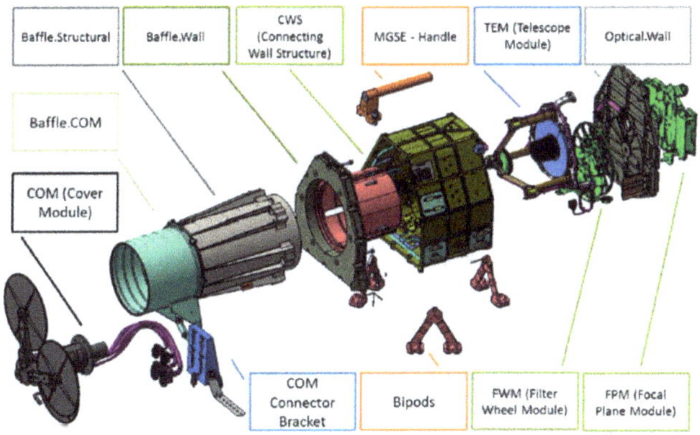

Fig. 1. OHU unit layout overview; on the left side of image the JANUS COM subsystem [1].

COM Subsystem Requirements and Baseline Design

The main functional and environmental requirements of this mechanism can be identified and summarized as follows:

- the cover provides optics and detector protection from sunlight and contamination;
- the subsystem, located at the main entrance of the JANUS OHU outer Baffle, provides the function of opening and closing the door;
- the subsystem mechanism implements the following needed functionalities in each flight condition:
    - Closed - Locked position: to protect of the optical elements and detector from contamination by the inner part of the door during launch (pre-loaded);
    - Closed - not Locked (Cruise) position: to protect the optical elements and detector from contamination during the cruise phase and to prevent long-term sticking of the COM door and the OHU Baffle I/F;

  - Open - Locked position: to allow the JANUS OHU and detector to face the outer environment to perform planetary observations during science mission phases and to perform in-flight calibration observation of different targets (e.g., moons, stellar fields).
- single-point failure tolerance requires redundancy and the ability to open the door permanently in the case if an irreversible system failure occurs (fail-safe device);
- requirement to validate open and closed positions;
- dynamic load during launch;
- non-operational temperature range (−40 to +35 °C) implies a design for high differential thermal loads within the mechanisms.

COM Subsystem
The COM subsystem is mainly composed by the following parts, which can be seen in Figure 2:

- the stepper motor (MT) for controlling the position of the cover;
- the interface flange (FL) for fixing the mechanism to the baffle of the telescope;
- the main body (BD) were the mechanical parts for allowing the movements are present;
- the main shaft (MS) governing the vertical movement;
- the bracket (BR) holding the cover shield;

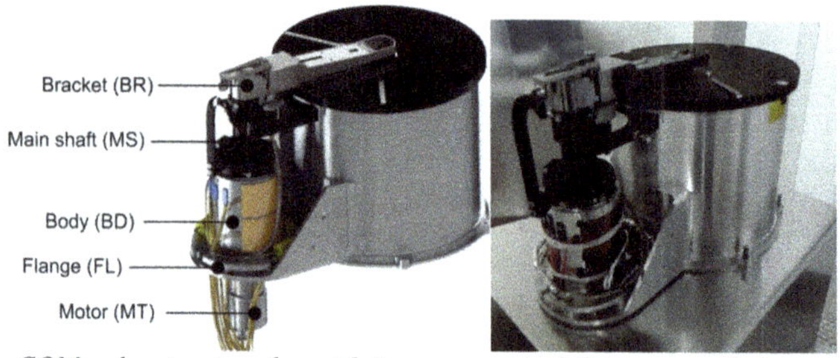

Figure 2: COM subsystem together with its components; left image: rendering; right image: flight model.

Mechanism Functional Test and Calibration
To demonstrate that mechanism functionality in the presence of environmental loads a life test has been executed during the thermal cycles. Due to waiting time between each activation, all the activations have been divided in 3 slots. The activations have been distributed along each cycle at different temperatures. The total activations are 5520 cycles divided as shown in Table 1.
The mechanism open and closed position has been calibrated with repeatability tests. Figure 3 shows repeatability test performed on EQM, data collected on 15 runs, both resistant torque and the switch status are reported, and highlights the repeatability of the mechanism performances.

Materials Research Proceedings 37 (2023) 558-562 https://doi.org/10.21741/9781644902813-122

Table 1 COM lifetime test activations

Phase	P [Pa]	T [°C]	Cycles
1	amb	amb	500
2	5e-3	Ambo	500
3	5e-3	-45…-20	2880
4	5e-3	-20…+10	720
5	5e-3	+10…+35	920

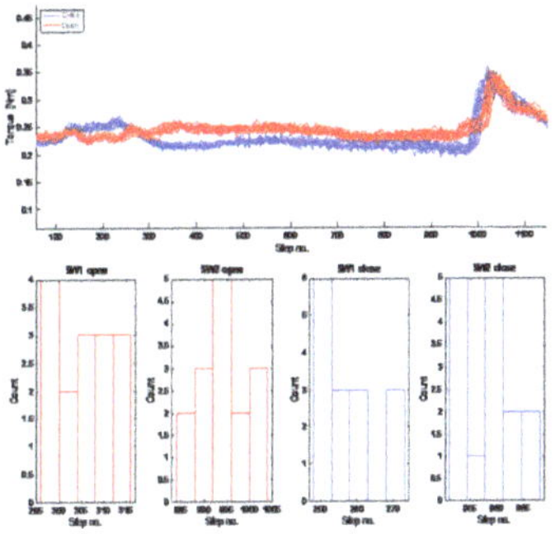

Figure 3: Repeatability test performed on EQM, data collected on 15 runs, both resistant torque and the switch status are reported, and highlights the repeatability of the mechanism performances.

Thermal Monitoring

COM temperature is measured by five Pt1000 thermistors, three are located on the external surface of the COM close to the heaters (see Figure 4), other two are mounted on the motor chassis.

COM temperature is controlled by means of two heaters (2 nominal and 2 redundant) located on the COM external surface. The survival thermal control subsystem is composed by the three Pt1000 plus two magnetically self-compensating heaters on the COM mantle.

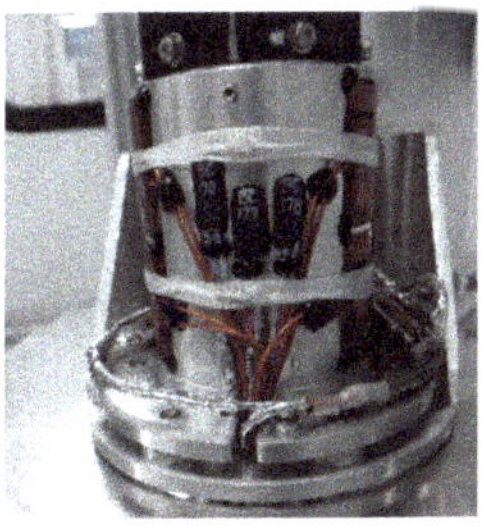

Figure 4: Monitoring PT1000: located in COM main body near the thermal regulation heater.

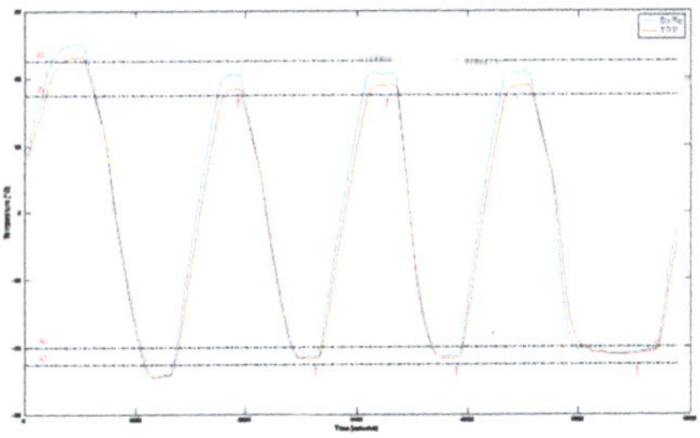

Figure 5: Thermal cycles (-45°C…+45°C); red arrows show the activation events; red cross shows the SMA activation event d highlights the repeatability of the mechanism performances.

Conclusion

The JANUS COM Team members are very grateful to prof. Stefano Debei for his inspiring scientific activity and for all the precious suggestions he gave us for the design, development and testing of the last space exploration equipment he lead at the CISAS "Giuseppe Colombo" space center of the University of Padova and for his valuable ideas collaborating with the members of Leonardo SPA- Florence company.

References

[1] Grasset, Olivier, et al. "JUpiter ICy moons Explorer (JUICE): An ESA mission to orbit Ganymede and to characterise the Jupiter system." Planetary and Space Science 78 (2013): 1-21. https://doi.org/10.1016/j.pss.2012.12.002

[2] Della Corte, Vincenzo, et al. "Scientific objectives of JANUS Instrument onboard JUICE mission and key technical solutions for its Optical Head." 2019 IEEE 5th International workshop on metrology for aerospace (MetroAeroSpace). IEEE, 2019. https://doi.org/10.1109/MetroAeroSpace.2019.8869584

Comparison of LARES 1 and LARES 2 missions - one year after the launch

Ignazio Ciufolini[1,a], Antonio Paolozzi[2,b], Emiliano Ortore[2,c], Claudio Paris[2,d*], Erricos C. Pavlis[3,e], John C. Ries[4,f] and Richard Matzner[5,g]

[1]Group of Astrodynamics for the Use of Space Systems (Gauss), Via Sambuca Pistoiese 70, 00138 Rome, Italy

[2]Scuola di Ingegneria Aerospaziale, Sapienza University, Via Salaria 851 – 00138 Rome, Italy

[3]GESTAR II - University of Maryland, Baltimore County (UMBC) & NASA Goddard, 61A, TRC #182, 1000 Hilltop Circle, Baltimore, Maryland, USA 21250

[4]Center for Space Research, University of Texas at Austin, Austin, USA

[5]Center for Gravitational Physics, Weinberg Center, University of Texas at Austin, Austin, USA

[a]ignazio.ciufolini@gmail.com, [b]antonio.paolozzi@uniroma1.it, [c]emiliano.ortore@uniroma1.it [d]claudio.paris@uniroma1.it, [e]epavlis@umbc.edu, [f]ries@csr.utexas.edu, [g]richard.matzner@sbcglobal.net

Keywords: LARES 2, Satellite Design, Laser Ranging, General Relativity, Frame-Dragging

Abstract. The LARES 1 and LARES 2 missions were designed to test an intriguing phenomenon predicted by the theory of general relativity: the *Lense-Thirring* (*frame-dragging*) effect. In particular, the LARES 2 mission was designed with the goal of reaching an accuracy 10 times better than that obtained with LARES 1, launched 10 years earlier. To reach this demanding goal a special orbit and a specific satellite design was required. Knowledge of the gravitational field of Earth of ever-increasing accuracy, thanks to the Follow-on GRACE space mission together with the spectacular orbital injection accuracy provided by the Avio-ASI-ESA launcher VEGA C, will make possible an even better accuracy after a few years of data analysis. In this paper the two missions are compared along with the results obtained from the LARES 1 mission and those expected from LARES 2.

Introduction

LARES 2, was successfully launched from the European spaceport in French Guyana on the inaugural flight of VEGA C (13 July, 2022). This launch occurred 10 years after the maiden flight of VEGA that carried as main payload the LARES 1 satellite. Both launch vehicles were developed, financed and managed by ASI, Avio and ESA. The two orbits are quite special. Particularly the LARES 2 orbit needed to be quite high compared to classical LEO orbits, and in comparison, to the case of LARES 1 there were very tight tolerances in the orbit parameters. The injection accuracy for LARES 1 was very high, but for LARES 2 the accuracy was spectacular; it matched the orbit to 10 times better accuracy than previously required, affording the prerequisites for even better results than originally designed [1]. This will allow an improvement in the accuracy of the frame-dragging (Lense-Thirring effect) measurement by one order of magnitude with respect to obtained using LARES 1 [2], allowing an accuracy of at least as good as a few parts per thousand. Frame-dragging is measured by observing how the node of a satellite orbit is shifted by the dragging of spacetime induced by the Earth rotation. In general relativity spacetime is deformed by mass-energy but also by currents of mass-energy, such as Earth's rotation. Laser ranging provides the most accurate ranging measurement achievable today in near-Earth space and is capable of providing the necessary data for the LARES missions. The main problem in measurement of frame dragging arises from classical gravitational and non-gravitational perturbations whose effects on the node are huge compared to frame-dragging. A combination of the data of the two LARES satellites and the two LAGEOS satellites is required, together with very accurate knowledge of the gravitational field of Earth is necessary to extract the frame dragging values during the analysis. (The gravitational fields from GRACE and GRACE Follow On missions are used.)

Frame-dragging of general relativity

General relativity (GR) is the best theory of gravitation interaction available today [3,4]. However, there are still open issues such as its reconciliation with quantum mechanics and the problem of spacetime singularities inside black holes where all known physical theories break down, and whose existence is a robust prediction of general relativity [5]. The accelerating expansion of the universe [6] is another mystery that increases the interest in experimental verification of GR. In this framework LARES 1 and LARES 2 missions find their natural environment, to measure the effect of Earth rotation on spacetime: the Lense-Thirring effect, named after the two Austrian physicists that derived it in 1918. In principle the measurement is relatively easy because it would be sufficient to measure the node shift of one satellite orbit and compare it with the prediction of GR. Unfortunately, the shift due to GR is only about 118.5 mas/y for LARES 1 and about 30.7 mas/y for LARES 2, translating to only about 4 m/y and 2 m/y respectively, while classical perturbations produce shifts about 7 orders of magnitude larger, translating to node shifts of many thousands of km per year. The original idea to circumvent this problem was proposed and published in references [7-12]; the key is to use two satellites orbiting at the same altitude but with supplementary inclinations or in other words to use a so-called butterfly configuration (Fig. 1). Originally, in the 80's, the mission was named LAGEOS 3 and was supposed to put a copy of LAGEOS 1 and 2 in the supplementary orbit now occupied by LARES 2. In 2012 ESA and ASI offered a launch opportunity with the inaugural flight of the VEGA launcher developed by Avio. The launch envelope of the inaugural flight was limited to 1500 km and so a different approach to eliminate the effect on the node of the even-zonal harmonics was devised [2]. In this case the combination of the three satellites LAGEOS 1, LAGEOS 2 and LARES 1 was necessary to eliminate major disturbances due to the uncertainties of the first two even-zonal harmonics: J_2 and J_4. Intermediate results were reported in the years that followed the launch (see for instance [13]), culminating 7 years after the launch in reaching the mission goal of 1-2% accuracy in the frame-dragging [2]. Frame-dragging around Earth is tiny and very difficult to measure, but observable with high accuracy with the LARES missions. In extreme astrophysical/cosmological phenomena such as the formation of accretion disks of rotating black holes and in explaining the fixed direction of jets in active galactic nuclei, frame dragging is very important, large, but difficult to determine to high accuracy. Also, black hole mergers which produce gravitational waves observed by the LIGO-Virgo-KAGRA laser interferometer detectors [14], are strongly influenced by frame-dragging.

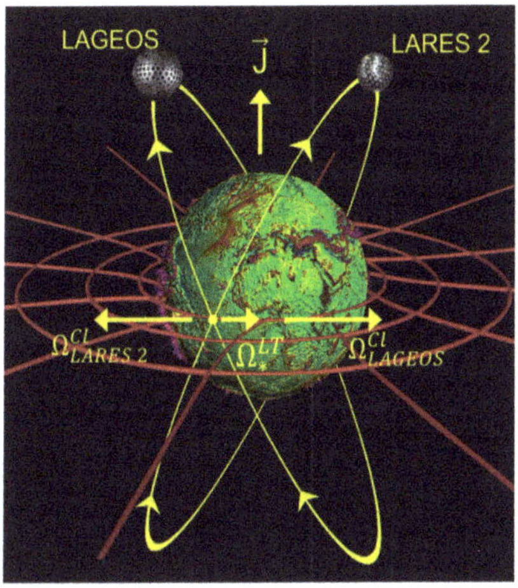

Fig. 1. The even zonal gravitational harmonics of the Earth produce the very large and well-known shift of the node Ω^{Cl} (the superscript Cl reminds us that the shift is due to classical perturbation) while $\Omega^{LT} = \Omega^{LT}_{LAGEOS} + \Omega^{LT}_{LARES\,2}$ (LT stands for: Lense-Thirring effect) is the tiny shift due to the dragging of inertial frames predicted by the theory of general relativity and **J** *is the Earth angular momentum. The small variations of the gravitational field of Earth are represented in false colors and report actual experimental values obtained for instance by the GRACE and GRACE FO space missions. The red lines are a pictorial representation on how spacetime is dragged by the Earth rotation.*

LARES 1 and LARES 2 missions

Both missions are based on the precise orbit determinations of the two satellites along with those of the two LAGEOS satellites. Laser ranging is used to measure the satellite orbits to a few millimeters' accuracy. Both LARES satellite bodies are made in one single piece, differently from all other geodetic passive satellites; LARES 2 has a novel CCR distribution which is not regular along the parallels. In Table 1 are compared other characteristics. In Fig. 2 is reported the last LT measurement obtained with LARES 1 mission that shows the mission goal was fulfilled.

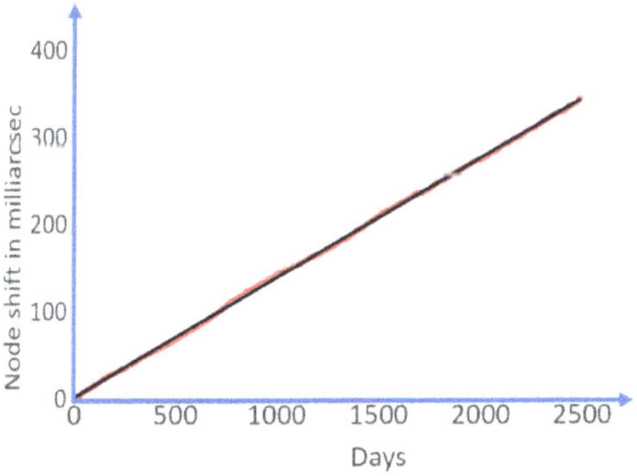

Fig. 2. Result of the LARES 1 mission [2]. The figure reports the LT effect measurement obtained by analyzing 7 years of orbital data of three satellites: the two LAGEOS and LARES 1. The horizontal axis reports the number of days. In red are the actual data and in black a linear fit. Taking 1 as the theoretical value of LT effect we have obtained 0.9910 ± 0.02, i.e., with 0.02 systematic error and with 0.001 formal error.

The orbital analysis of LARES 2 combined with that of LAGEOS 1 is in progress. The current availability of approximately one year of data is not sufficient to reduce the effects on the satellite nodes of the large perturbations due to some tides and, *so far*, to improve the accuracy of the measurement of LT in Fig. 2. By applying proper averaging and fits of such orbital perturbations, using data of a longer period of time, an improved measurement will be obtained in a few years.

Table 1. Comparison of the LAGEOS with the two LARES satellites. S and M are the surface and the mass of the satellites respectively. S/M is calculated relative to the value of LAGEOS 1.

	ORBITAL PARAMETERS			Metal alloy	D [m]	M [kg]	CCR No.	CCR dia [in]	S/M relative
	i [°]	a [km]	e						
LARES 1	69.44	7827.598	0.0009	W	0.364	386.8	92	1.5"	0.39
LARES 2	70.158	12266.198	0.00027	Ni	0.424	294.8	303	1.0"	0.69
LAGEOS 1	109.844	12269.988	0.004	Al-Cu	0.6	407	426	1.5"	1

Acknowledgments. The authors acknowledge the Italian Space Agency (ASI) for supporting the LARES 1 and LARES 2 missions and the International Laser Ranging Service (ILRS) for tracking the satellites. E. C. Pavlis acknowledges the support of NASA Grant 80NSSC22M0001.

References

[1] Ciufolini, I., Pavlis, E.C., Sindoni, G., ... Koenig, R., Paris, C. A new laser-ranged satellite for General Relativity and space geodesy: II. Monte Carlo simulations and covariance analyses of the LARES 2 experiment, European Physical Journal Plus, 2017, 132(8), 337. https://doi.org/10.1140/epjp/i2017-11636-0

[2] Ciufolini, I., Paolozzi, A., Pavlis, E.C., ...Gurzadyan, V., Penrose, R., An improved test of the general relativistic effect of frame-dragging using the LARES and LAGEOS satellites, European Physical Journal C, 2019, 79(10), 872. https://doi.org/10.1140/epjc/s10052-019-7386-z

[3] C.W. Misner, K.S. Thorne, J.A. Wheeler, Gravitation (Freeman, San Francisco, 1973).

[4] I. Ciufolini, J.A. Wheeler, Gravitation and Inertia (Princeton University Press, Princeton, New Jersey, 1995). https://doi.org/10.1515/9780691190198

[5] R. Penrose, Gravitational Collapse and Space-Time Singularities, Phys. Rev. Lett. 14, 57–59 (1965). https://doi.org/10.1103/PhysRevLett.14.57

[6] A. Riess et al., Astron. J., Observational evidence from supernovae for an accelerating universe and a cosmological constant, 116, 1009–1038 (1998). https://doi.org/10.1086/300499

[7] I. Ciufolini, Measurement of the Lense-Thirring drag on high-altitude, laser-ranged artificial satellites, Phys. Rev. Lett. 56, 278 (27 Jan 1986). https://doi.org/10.1103/PhysRevLett.56.278

[8] I. Ciufolini, A comprehensive introduction to the LAGEOS gravitomagnetic experiment: from the importance of the gravitomagnetic field in physics to preliminary error analysis and error budget, International Journal of Modern Physics A, 4, No. 13, pp. 3083-3145 (1989). https://doi.org/10.1142/S0217751X89001266

[9] B. Tapley, J.C. Ries, R.J. Eanes, M.M.Watkins, NASA-ASI Study on LAGEOS III, CSR-UT publication n. CSR-89-3, Austin, Texas (1989).

[10] I. Ciufolini et al., ASI-NASA Study on LAGEOS III (CNR, Rome, Italy, 1989).

[11] I. Ciufolini, Theory and experiments in general relativity and other metric theories. Ph.D dissertation, advisors: John A. Wheeler, Richard Matzner and Steven Weinberg.Univ. of Texas, Austin (Ann Arbor, Michigan, 1984).

[12] J.C. Ries, Simulation of an experiment to measure the LenseThirring precession using a second LAGEOS satellite. Ph.D Dissertation, Univ. of Texas, Austin, 1989.

[13] Ciufolini, I., Paolozzi, A., Pavlis, E.C., Sindoni, G., Paris, C., Preliminary orbital analysis of the LARES space experiment, European Physical Journal Plus, 2015, 130(7), 133. https://doi.org/10.1140/epjp/i2015-15133-2

[14] B.P. Abbott et al., Observation of Gravitational Waves from a Binary Black Hole Merger, Phys. Rev. Lett. 116, 061102 (2016). https://doi.org/10.1142/9789814699662_0011

Tracking particles ejected from active asteroid Bennu with event-based vision

Loïc James Azzalini[1,a *], Dario Izzo[1,b]

[1] Advanced Concepts Team, European Space Agency, European Space Research and Technology Centre (ESTEC), Keplerlaan 1, 2201 AZ Noordwijk, The Netherlands

[a]jazzalin@outlook.com, [b]dario.izzo@esa.int

Keywords: Neuromorphic Vision, Event-Based Sensing, Multi-Object Tracking

Abstract. Early detection and tracking of ejecta in the vicinity of small solar system bodies is crucial to guarantee spacecraft safety and support scientific observation. During the visit of active asteroid Bennu, the OSIRIS-REx spacecraft relied on the analysis of images captured by onboard navigation cameras to detect particle ejection events, which ultimately became one of the mission's scientific highlights. To increase the scientific return of similar time-constrained missions, this work proposes an event-based solution that is dedicated to the detection and tracking of centimetre-sized particles. Unlike a standard frame-based camera, the pixels of an event-based camera independently trigger events indicating whether the scene brightness has increased or decreased at that time and location in the sensor plane. As a result of the sparse and asynchronous spatiotemporal output, event cameras combine very high dynamic range and temporal resolution with low-power consumption, which could complement existing onboard imaging techniques. This paper motivates the use of a scientific event camera by reconstructing the particle ejection episodes reported by the OSIRIS-REx mission in a photorealistic scene generator and in turn, simulating event-based observations. The resulting streams of spatiotemporal data support future work on event-based multi-object tracking.

Introduction

An active asteroid shows evidence of mass loss caused by natural processes or as a result of human-planned activities, sharing characteristics more often associated with comets. Detecting and interpreting the dynamic properties of these small solar system bodies (SSSB) constitute important scientific objectives of sample-return and flyby missions as they may hold the key to understanding their past and future [1]. In the case of asteroid Bennu, such activity was only detected in situ as the navigation cameras of visiting spacecraft OSIRIS-REx captured centimetre-size rocks being ejected from the surface. Whether for situational awareness to guarantee spacecraft safety or scientific observation, missions to active asteroids or comets benefit from early detection and tracking of such events [2]. Currently, extensive offline image analysis is needed to balance the brightness of the body and the dimness of the particles (stray light reduction), to negate the static objects in the scene (image differencing) and to detect the relative motion of the particles (blinking) [1]. Given the time-constrained nature of these missions, techniques that automate the detection and tracking of particles, such as the frame-based multi-object tracking algorithm proposed in [3], could significantly contribute to the mission's scientific returns.

This work proposes a solution based on the principles of dynamic vision sensing and the event camera, a novel imaging sensor inspired by the neural pathways of the retina [4, 5]. Where a standard frame-based camera would capture redundant static information, an event camera would only report pixel-level brightness changes induced by the dynamics present in the scene. The sparse and asynchronous output of the sensor and its large dynamic range provide an effective low-power solution to detection and tracking problems in challenging lighting conditions [6]. Given particularly dynamic environments around SSSBs as a result of plumes, dust and/or particle ejecta, we theorise that event-based technology could augment onboard navigation and scientific cameras, streamlining current approaches based on offline image analysis.

The apparent advantages of these dynamic vision sensors for space applications have recently been reviewed in [7]. While the principles of dynamic vision have yet to be applied to the detection and tracking of particles around an active asteroid, similar work on event-based star tracking [8] and (event-based) space situational awareness [9, 10] provide valuable insights into the challenges posed by the problem at hand. Unlike the streaks produced by resident space objects (RSOs) crossing the field of view of a ground-based or in-orbit telescope, the particles of interest in this study undergo more complex dynamics locally, given the proximity of the SSSB, limiting the utility of classical approaches such as the Hough transforms. Moreover, the problem of associating observations to tracks is exacerbated by significant measurement noise that is inherent to the event camera. Popular multi-object tracking solutions such as the (probabilistic) multiple hypothesis tracking filters have been adapted to event-based representations of RSOs [9, 11], suggesting that these algorithms also lend themselves to the analysis of particle dynamics in the vicinity of an SSSB.

To support the evaluation of event-based multi-particle tracking, we use the navigation and ancillary information of the OSIRIS-REx mission in simulation as well as reports of notable particle ejection episodes [12] to highlight how an event-based sensor can augment visual data capture and contribute to the mission's scientific objectives. The dynamic scene, composed of asteroid Bennu and several centimetre-size particles ejected from its surface, is first reconstructed with photorealistic computer graphics tools from the point of view of the visiting spacecraft [13]. Dynamic vision sensing is subsequently simulated by emulating the sensitivity and noise characteristics of a real event camera [14, 15]. In turn, we demonstrate that, in the absence of frames, asynchronous streams of events capture sufficient information about the scene to enable fast and continuous tracking of multiple particles. This constitutes an essential step to support future research on event-based determination of the size and orbit of RSOs.

The following section introduces the principle of operation underlying the high temporal resolution and high dynamic range of the event camera. Section Simulation and Results describes the simulation environment in which the dynamic scene is reproduced and the preliminary results on event-based tracking of particle ejecta. Given the sparse and asynchronous output of the event camera, future work will focus on the formulation of a multi-object tracking problem that is compatible with the event-based representation of the particle tracks.

Theoretical Background

Event-Based Vision. The pixels of a dynamic vision sensor independently report changes in scene brightness, which is generally taken to be the log intensity (i.e., $log\, I$), resulting in a sparse and asynchronous stream of *events*. The pixel at location (x, y) in the sensor plane outputs a 1 (alternatively, 0) if the sensed brightness at time t increased (decreased) by the predefined magnitude Θ_{ON} (Θ_{OFF}) known as the contrast thresholds. Thus, an event can be represented by the 4-tuple (t, x, y, p), where p denotes the polarity of the brightness change. After triggering an event, the pixel resets its baseline from which to monitor subsequent changes in brightness to the current log intensity, as depicted in Figure 1 (a). Figure 1 (b) shows a synthetic example of a stream of events output by an event camera capturing a black dot spinning on a white disk. As only the motion of the black dot contributes changes in brightness in the scene, the spatiotemporal output of the sensor takes the form of a spiral.

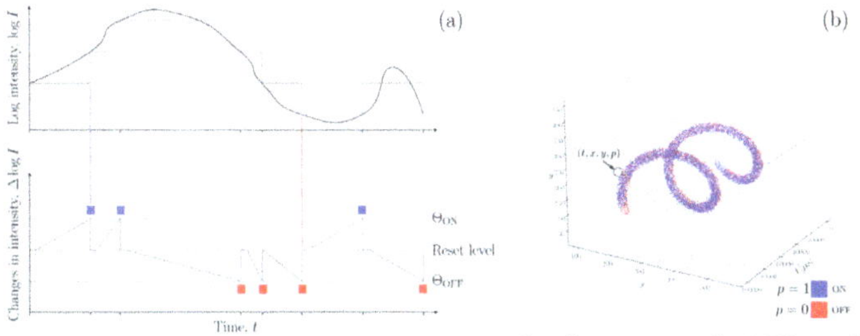

Figure 1 Dynamic vision sensing overview: (a) principle of operation of a DVS pixel (adapted from [4]) (b) example output from an event camera capturing a black dot spinning on a white disk (based on synthetic data from Prophesee's Metavision toolkit [16])

The sparsity of the output allows for high readout rates in the range of $2 - 1200\ MHz$, high temporal resolution (in the order of μs), low-power consumption and, owing to the logarithmic scale used by each pixel, high dynamic range [5]. While these are appealing characteristics for an onboard sensor, the event camera is inherently noisy due to the complex transistor circuits underlying its differential mode of operation. Figure 1 (b) captures some of these noisy contributions in the form of sporadic streaks of ON events at several pixel locations in the sensor plane. These artefacts are the result of nonidealities such as hot pixels and leak currents which produce events irrespective of the scene dynamics. In order to faithfully simulate the output of these sensors, event camera emulators [14, 15] model these sources of noise and other hardware-related nonidealities (e.g., shot noise).

Simulation and Results

The first step of the simulation of dynamic vision sensing around active asteroid Bennu consists in extracting the relative position of the OSIRIS-REx spacecraft, the Sun and the ejected particles. The Orbital C mission phase is of particular interest given that it was dedicated to particle monitoring, resulting in a higher cadence of imaging. Specifically, a time window centred on the particle ejection episode of September 13th, 2019, is considered in this work given that out of the 30 particles detected, 22 tracks led to successful orbit determination [12]. Figure 2 depicts the reconstructed particle ejection episode based on the interpolation of the relevant SPICE kernels.

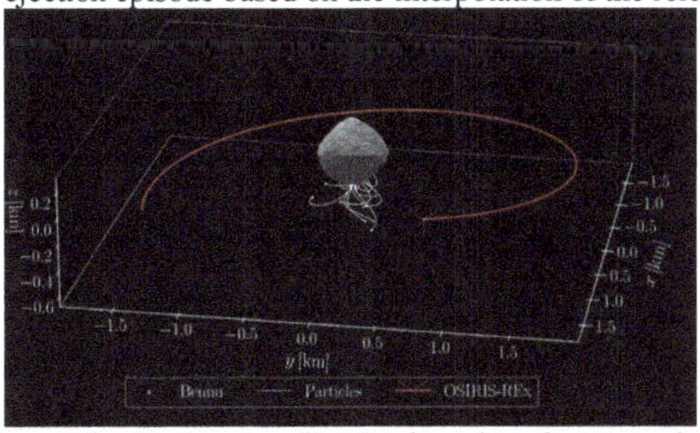

Figure 2 Bennu-fixed particle ejection visualization based on the interpolation of SPICE kernels from 2019-09-13T21:00:00 to 2019-09-14T00:00:00 [12]

Given the dynamic nature of the event camera, the ejection scene depicted in Figure 2 is then animated in Blender, using photorealistic models of the asteroid and the particles. Different textures and reflectance properties are used to reproduce the rubble pile appearance of Bennu [13] based on the position of the Sun during the Orbital C phase. The simulated ejecta consists of 14 particles with sizes ranging from $1 - 11\ cm$ according to the average diameters reported in [1]. We emulate the large field of view and image sensor size of the onboard navigation camera (NavCam 1) by configuring the Blender perspective camera with $W = 2592$ px, $H = 1944$ px and a horizontal field of view of $\vartheta = 44°$. Then, the positions of the particles and the emulated camera along the Orbital C trajectory are updated to capture frames of the ejection episode, as shown in Figure 3 (a) (cropped near the limb where the ejecta emanate).

Overall, 40 additional renders are generated and combined into a video at 30 FPS tracing the path of the particles from the surface of Bennu to the lower-left borders of the field of view. To study an event-based representation of the scene dynamics, the video is then passed to an event camera emulator [14, 15], which converts the video to spatiotemporal streams of events based on the operation principle described in Section Event-Based Vision. In the interest of qualitative comparison with the original renders, synthetic events are binned into event-frames over short time windows such that the original frame rate is recovered.

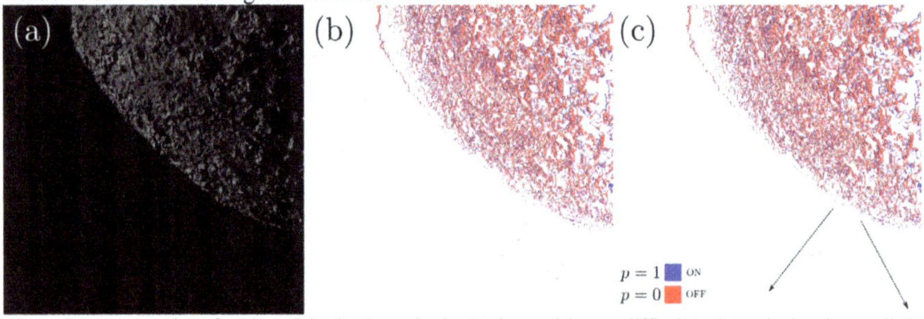

Figure 3 Reconstruction of a 14-particle ejection episode: (a) the particles are difficult to detect in the photorealistic render, (b) accumulation of synthetic events in a single frame with are more easily identifiable against a dynamic noisy background (b) and clearly visible in the absence of noise (c), where the arrows indicate the ejecta direction.

While the low-light conditions depicted in Figure 3 (a) significantly challenge the detection of any particles, the accumulation of synthetic events depicted in (b) allows the particles to be visually distinguished from the asteroid, despite significant noise contributions from the sensor emulation (to aid in locating the particles, frame (c) depicts a noiseless emulation of the same event camera).

Conclusion

This work showcases preliminary results on the scientific use of event cameras onboard a spacecraft in dynamic environments around small solar system bodies. After reconstructing the scene of a particle ejection episode from the SPICE kernels of the OSIRIS-REx mission, we emulate the viewpoint of an event camera to gain familiarity with event-based representations of particle tracks. In follow-up work, we will evaluate the formulation of a multi-object tracking problem, given event-based observations of an unknown (and potentially time-varying) number of particles. Multi-object tracking algorithms may be applied directly to the event-frames shown in Figure 3. However, this defeats the purpose of emulating an event camera in the first place, as by so doing, solutions do not take full advantage of the sparsity and temporal resolution of the data. Instead, we will consider approaches that process events in an online manner.

References

[1] Lauretta, D. S. et al. (2019). Episodes of particle ejection from the surface of the active asteroid (101955) Bennu. Science, 366 (6470), https://doi.org/10.1126/science.aay3544

[2] Chesley, S. R. et al. (2020). Trajectory estimation for particles observed in the vicinity of (101955) Bennu. Journal of Geophysical Research: Planets, 125, e2019JE006363. https://doi.org/10.1029/2019JE006363

[3] Liounis, A. J. et al. (2020). Autonomous detection of particles and tracks in optical images. Earth and Space Science, 7, e2019EA000843. https://doi.org/10.1029/2019EA000843

[4] Liu, S.-C., & Delbruck, T. (2010). Neuromorphic sensory systems. Current Opinion in Neuro- biology, 20 (3), 288-295. https://doi.org/10.1016/j.conb.2010.03.007

[5] Gallego, G. et al. (2022). Event-based vision: A survey. IEEE Transactions on Pattern Analysis and Machine Intelligence, 44 (1), 154-180. https://doi.org/10.1109/TPAMI.2020.3008413

[6] Lagorce, X. et al. (2015). Asynchronous Event-Based Multikernel Algorithm for High-Speed Visual Features Tracking. IEEE Transactions on Neural Networks and Learning Systems, vol. 26, no. 8, pp. 1710-1720. https://doi.org/10.1109/TNNLS.2014.2352401

[7] Izzo, D. et al. (2022). Neuromorphic computing and sensing in space. arXiv preprint arXiv:2212.05236. https://doi.org/10.48550/arXiv.2212.05236

[8] Chin, T.-J., et al. (2019). Star tracking using an event camera. In 2019 IEEE/CVF Conference on Computer Vision and Pattern Recognition Workshops (CVPRW) (p. 1646-1655). https://doi.org/10.1109/CVPRW.2019.00208

[9] Cheung, B. et al. (2018). Probabilistic multi hypothesis tracker for an event based sensor. In 2018 21st International Conference on Information Fusion (fusion) (p. 1-8). https://doi.org/10.23919/ICIF.2018.8455718

[10] Afshar, S. et al. (2020). Event-Based Object Detection and Tracking for Space Situational Awareness. IEEE Sensors Journal, vol. 20, no. 24, pp. 15117-15132. https://doi.org/10.1109/JSEN.2020.3009687

[11] Oliver, R. et al. (2022). Event-based sensor multiple hypothesis tracker for space domain awareness. In AMOS Conference 2022. https://doi.org/10.5167/uzh-231276

[12] Hergenrother, C. W. et al. (2020). Photometry of particles ejected from active asteroid (101955) Bennu. Journal of Geophysical Research: Planets, 125, e2020JE006381. https://doi.org/10.1029/2020JE006381

[13] Pajusalu M. et al. (2022) SISPO: Space Imaging Simulator for Proximity Operations. PLOS ONE 17(3): e0263882. https://doi.org/10.1371/journal.pone.0263882

[14] Gehrig, D. et al. (2020). Video to events: Recycling video datasets for event cameras. In 2020 IEEE/CVF Conference on Computer Vision and Pattern Recognition (CVPR) (p. 3583-3592). https://doi.org/10.1109/CVPR42600.2020 .00364

[15] Hu, Y. et al. (2021). v2e: From Video Frames to Realistic DVS Events. 2021 IEEE/CVF Conference on Computer Vision and Pattern Recognition Workshops (CVPRW), Nashville, TN, USA, 2021, pp. 1312-1321, https://doi.org/10.1109/CVPRW53098.2021.00144

[16] Prophesee.ai. Metavision SDK Docs – Recordings and Datasets, https://docs.prophesee.ai/

Totimorphic structures for space application

Amy Thomas[1,a]*, Jai Grover[1,a], Dario Izzo[1,a] and Dominik Dold[1,a]

[1]European Space Agency, Advanced Concepts Team, European Space Research and Technology Centre, Keplerlaan 1, 2201 AZ Noordwijk, The Netherlands

[a]{forename.surname}@esa.int

Keywords: Morphing Structure, Deployable Structure, Multi-Functional Metamaterial, Totimorphic, Programmable Material

Abstract. We propose to use a recently introduced *Totimorphic* metamaterial for constructing morphable space structures. As a first step to investigate the feasibility of this concept, we present a method for morphing such structures autonomously between different shapes using physically plausible actuations. The presented method is part of a currently developed Python library bundling non-rigid morphing and finite element analysis code, which will be made publicly available in the future. With this work, we aim to lay a foundation for exploring a promising and novel class of multi-functional, reconfigurable space structures.

Introduction

The last decade has seen rapid expansion and interest in the field of advanced materials and structures, especially in the development of programmable, multi-functional, and morphable structures. Such structures are particularly suited for space environments, where payload mass and volume are tightly constrained, making light structures capable of performing multiple functions highly desirable. For instance, Origami principles have already been used in the development of deployable solar panels and antennae, and applications for other advanced structures, such as NASA's starshade [1], have been proposed [2,3]. However, most deployable space structures are currently limited between strict configurations (typically stowed and deployed), often prohibiting any reconfiguration thereafter.

Here, we explore a recently proposed metamaterial called a *Totimorphic structure* [4] whose characteristic properties might enable designs capable of redeployment and reconfiguration into many different shapes after initial deployment. This is especially intriguing for constructing adaptive structures, as many recent papers [5,6] have demonstrated the feasibility of changing a structure's mechanical properties through geometric alterations alone. Such systems might enable mission designs capable of complex and efficient post-launch reconfiguration, adjusting to changing mission goals or conditions in situ and providing space missions with greater flexibility, thus fundamentally changing the types of missions possible in the future.

In the following, we first explain the concept of totimorphic structures before introducing a computational method for obtaining the actuations needed to morph them into different shapes.

Methodology

Totimorphic structures are composed of neutrally-stable unit cells (also called Totimorphic Unit Cells, TUCs) [4] shown in Fig. 1a,b. A TUC consists of a beam with a ball joint in its middle (A-P-B in Fig. 1a), a lever connecting to the joint (P-C) and two springs connecting the ends of the beam with the end of the lever (A-C and B-C). The neutrally stable behaviour of TUCs arises from the lever-spring motive: if the two springs are zero-length springs with identical spring stiffness, any position of the lever results in zero moment acting on the lever. The result of this property is propagated across larger structures built from TUCs (e.g. Fig. 1c), such that in the absence of external forces (e.g. gravity), the totimorphic structure will retain its shape while remaining completely compliant to any external force or displacement. By selective locking and unlocking of the structure's joints, we predict that the totimorphic property will allow the structure to be smoothly morphed between different shapes via beam/lever rotations alone – which we exploit in our method presented in the next section. Once the desired target shape is reached, the structure can be made rigid by locking all joints.

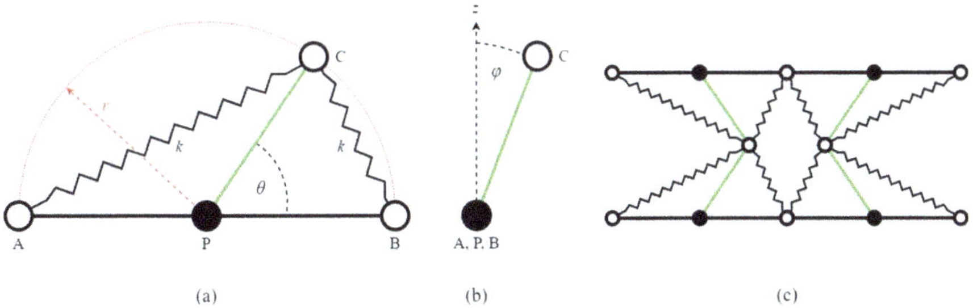

Figure 1: (a) Illustration of a TUC with lever rotation θ. (b) The TUCs beam rotation is given by φ, i.e., A-P-B is seen from the side here. (c) A simple auxetic structure built from four TUCs.

Before presenting our approach for morphing such structures, we first introduce the used mathematical description. Each TUC is geometrically fully defined by its position vector P, the beam vector AB, the lever length r, the angle between beam and lever θ, and the roll angle of the lever from the vertical φ (Fig. 1b). TUCs may be connected to each other at the A, B, C nodes, however for this paper's analysis, only A-B and C-C connections were considered. Additionally, r was set to be equal for all TUCs. If the coordinates of A, B or C are already known (i.e. the unit cell is attached to another unit cell), the unit cell can be described just by AB, θ, and φ – which is useful for implementing the morphing method.

Results

Our morphing method works as follows. We first define the initial and target geometries, set a maximally allowed change in AB, θ, and φ per iteration (same for all TUCs), and a subset of nodes are set as 'pinned' (i.e. their position is fixed, but they are allowed to rotate). We also 'activate' the pinned cells, (i.e. allow them to morph). Then we run the following iteration until the structure's geometry is within some tolerance of the desired target geometry: each activated TUC changes AB, θ and φ as much as possible to reach its target configuration while still maintaining structural cohesion (i.e. no beams breaking/prolonging/squeezing or TUCs separating). Then they become 'fixed' (cannot be activated anymore in this iteration, so the TUCs are frozen in place) and inform their neighbouring cells that they should move next (they activate their neighbours). This process is repeated until all cells are fixed. Consequently, all TUCs are unfixed and the next iteration begins. Intuitively, a wave of 'cell activation' moves through the structure, where activated cells are moved closer to their target and become immovable (fixed) afterwards.

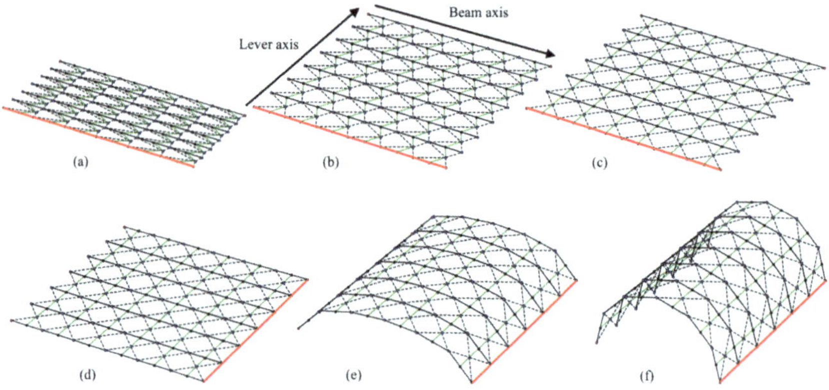

Figure 2: A totimorphic structure morphing continuously from a stowed plane to a square plane (a) θ=30 (b) θ=60 (c) θ=90, and then into a semi-cylinder (d-f). Red lines indicate pinned nodes with fixed coordinates.

The morphing method is illustrated in Fig. 2 for continuously morphing a structure to different target shapes: (a-c) turning a stowed geometry into a large flat surface, and (d-f) turning the flat surface into a half-cylinder. Although we halt at the shape in Fig. 2f here, in principle the structure can be mapped to any developable surface of an equal or smaller surface area [4], so long as there is a valid target geometry for the method.

In comparison with an Origami structure, which must necessarily unfold and change its thickness during deployment, we can see in (a-c) that the stowed totimorphic structure can continuously increase its surface area without any increase in the stowing thickness, resulting in a large stowed-to-deployed surface area ratio. Further, the stowing of the totimorphic structure does not constrain its final structural geometry, since it is possible for the structure to morph into intermediary states that enable better deployment before morphing into the final configuration.

Discussion

For the presented method to work, one requires knowledge about the exact initial and target configuration of the totimorphic structure (including angles) instead of just the shapes. Moreover, the choice of pinned nodes is crucial to ensure convergence to the target shape. Thus, we plan to develop an improved non-rigid morphing methodology that is both more robust and general, based on a model that predicts the whole structure's response to small perturbations. Such a model might further allow us to use inverse design approaches from artificial intelligence [6] to autonomously find configurations that possess desired effective mechanical properties.

Although totimorphic structures are more flexible in their morphing capabilities than Origami-based structures, they come with the drawback of having a high number of degrees of freedom as well as many movable mechanical parts. We anticipate that this will lead to challenges in the production of a physical prototype, which have to be overcome first for totimorphic structures to compete with alternative approaches. For instance, the neutrally stable behaviour will most likely not be perfectly realised in a physical system due to effects such as bending of the beam in a TUC from radial forces applied through the lever – an effect that has to be considered during the design and testing stages of a real prototype.

We are confident that totimorphic structures are ideally suited for deployment in extremely-low gravity environments such as orbits and deep space, where no external loads due to gravity interfere with the morphing process and the neutrally stable behaviour can be utilised to its fullest. As noted in [7], the gravitational forces experienced on planetary bodies will impede the morphing of totimorphic structures, the extent of which has to be investigated in future work.

To lock and unlock the joints as well as induce beam/lever rotations, we envision a thin and foldable support layer glued to the totimorphic structure through which lock, unlock and rotation commands, e.g., from a microcontroller, can be relayed electronically. For instance, since only a small number of cells are unlocked at a time, electrical pulses could be used with a shared bus to realise an efficient and economic solution.

Conclusion

Totimorphic structures offer unique structural characteristics that make them ideal for space missions – providing a high degree of flexibility coupled with a low mass and volume. They are suitable for a multitude of potential space applications; such as deployable habitats and structures, tools (e.g. nets for grabbing something), or for building moving structures by creating locomotion from morphing. Hence, we believe that totimorphic structures are a promising candidate in the search for technologies that enable morphable, multi-functional space structures.

Acknowledgments

We would like to thank Derek Aranguren van Egmond and Michael Mallon for helpful discussions, and our colleagues at ESA's Advanced Concepts Team for their ongoing support. AT and DD acknowledge support through ESA's fellowship and young graduate trainee programs.

References

[1] Information on https://exoplanets.nasa.gov/exep/technology/starshade/. Accessed: 05/07/23.

[2] Ynchausti, C., Roubicek, C., Erickson, J., Sargent, B., Magleby, S. P., and Howell, L. L.Hexagonal Twist Origami Pattern for Deployable Space Arrays ASME Open J. Engineering ASME. January 2022 1 011041 doi: https://doi.org/10.1115/1.4055357

[3] Biswas, A., Zekios, C.L., Ynchausti, C. et al. An ultra-wideband origami microwave absorber. Sci Rep 12, 13449 (2022). https://doi.org/10.1038/s41598-022-17648-4

[4] Chaudhary, G., Ganga Prasath, S., Soucy, E. and Mahadevan, L., 2021. Totimorphic assemblies from neutrally stable units. Proceedings of the National Academy of Sciences, 118(42), p.e2107003118. https://doi.org/10.1073/pnas.2107003118

[5] Yazdani, H., Aranguren van, D., Esmail, I., Genest, M., Paquet, C., Ashrafi, B., Bioinspired Stochastic Design: Tough and Stiff Ceramic Systems. Adv. Funct. Mater. 2022, 32, 2108492. https://doi.org/10.1002/adfm.202108492

[6] Dominik Dold, Derek Aranguren van Egmond, Differentiable graph-structured models for inverse design of lattice materials, arXiv (2022), https://doi.org/10.48550/arXiv.2304.05422

[7] Schenk, M. and Guest, S.D., 2014. On zero stiffness. Proceedings of the Institution of Mechanical Engineers, Part C: Journal of Mechanical Engineering Science, 228(10), pp.1701-1714. https://doi.org/10.1177/0954406213511903

Pushing the limits of re-entry technology: an overview of the Efesto-2 project and the advancements in inflatable heat shields

Giuseppe Guidotti[1], Giuseppe Governale[2,*], Nicole Viola[2], Ingrid Dietleinc, Steffen Callsen[3], Kevin Bergmann[3], Junnai Zhai[4], Roberto Gardi[5], Barbara Tiseo[5], Ysolde Prevereaud[6], Yann Dauvois[6], Giovanni Gambacciani[7], Giada Dammacco[7]

[1]DEIMOS Space S.L.U, Tres Cantos 28760, Spain

[2]Department of Mechanical and Aerospace Engineering, Politecnico di Torino, Turin – Italy

[3]Deutsches Zentrum für Luft- Und Raumfahrt e.V. (DLR), Bremen 28359, Germany

[4]Deutsches Zentrum für Luft- Und Raumfahrt e.V. (DLR), Köln, Germany

[5]Centro Italiano Ricerche Aerospaziali (CIRA), Capua – Caserta, Italy

[6]Office National d'Etudes et de Recherches Aerospatiales (ONERA), Toulouse, France

[7]Pangaia Grado Zero SRL (PGZ), Firenze 50056, Italy

*giuseppe.guidotti@deimos-space.com; giuseppe.governale@polito.it

Keywords: Re-Entry Technology, Inflatable Heat Shields, Thermal Protection System, Horizon Europe

Abstract. As space exploration technology advances, the need for reliable re-entry systems becomes increasingly critical. The European Flexible Heat Shields: Advanced TPS Design and Tests for Future In-Orbit Demonstration – 2 (EFESTO-2) project is a Horizon Europe-funded initiative aimed at improving the Technology Readiness Level of Inflatable Heat Shields (IHS), an innovative thermal protection system that can be deployed during re-entry. The project seeks to further advance the work achieved in the EFESTO project, with a focus on expanding investigations into critical aspects of IHS and increasing the confidence level and robustness of the tools and models used in the field. The EFESTO-2 project is built on four pillars, including consolidating the use-case applicability through a business case analysis for a meaningful space application, extending the investigation spectrum of the father project EFESTO to other critical aspects of the IHS field, increasing the confidence level and robustness of tools/models, and consolidating the roadmap to guarantee continuity in presiding the IHS field in Europe among the scientific and industrial community. This paper provides an overview of the EFESTO-2 project's objectives, achievements, ongoing activities, and planned activities up to completion. The project's advancements in the fields of thermal protection systems, inflatable heat shields, and technology readiness level are described in detail, highlighting the project's contributions to the European re-entry technology roadmap. Through this project, the European Space Program aims to push the limits of re-entry technology and reinforce its position as a leader in innovative technology for space exploration. This project has received funding from the European Union's Horizon Europe research and innovation program under grant agreement No 1010811041.

Introduction
The EFESTO-2 project builds upon the success of the previous H2020 EFESTO project and aims to advance European expertise in the field of Inflatable Heat Shields (IHS). With the increasing demand for reusable space transportation systems, the development of innovative thermal protection solutions is crucial for safe and cost-effective space missions. Inflatable Heat Shields have shown great potential in enabling controlled re-entry and recovery of spacecraft, making them a promising technology also for future space exploration missions.

Project Objectives
The EFESTO-2 project, funded by the European Union's Horizon Europe program, aims to achieve the following objectives. Firstly, it seeks to consolidate the use-case applicability of Inflatable Heat Shields (IHSs) through a comprehensive business case analysis for a meaningful space application. Secondly, it aims to expand the investigation spectrum by conducting extensive tests focused on aerodynamics and mechanical aspects, complementing the previous EFESTO project. Thirdly, the project aims to enhance the confidence level and robustness of the tools and models developed in EFESTO by incorporating test data. Lastly, it aims to consolidate the definition of a roadmap towards the near-future development of IHS technology up to Technology Readiness Level 7 (TRL7). The study-logic applicable within the EFESTO-2 initiative for implementing the planned effort is represented in Figure 1.

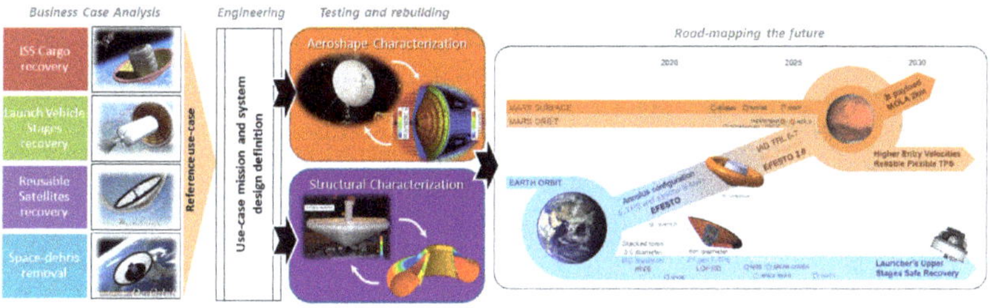

Figure 1 EFESTO-2 project study-logic

Business Case Analysis
The Business Case Analysis (BCA) was conducted as the initial task in the EFESTO-2 project to identify the most promising use-case application for inflatable heat shields (IHSs) and guide the subsequent design study for a reference mission/system. The BCA focused on exploring potential applications of IHSs in the re-entry and recovery of space systems meant for reuse. Examples of potential applications included the recovery of launch system stages, ISS cargo systems, and de-orbiting and recovery of reusable satellites.

The BCA workflow involved several stages. An overview of target markets for IHS technology was conducted, followed by the identification of the most promising commercial applications using a trade-off analysis. Qualitative evaluations were performed using frameworks such as SWOT and PESTEL, considering market trends, substitutes, competitors, and potential customers. Additionally, a cost-oriented assessment of a reference use-case for re-entry and recovery was carried out.

The evaluation of IHSs application scenarios considered different planetary re-entry scenarios, including Earth, Mars, and others. The trade-off criteria included market size, market timeline, complexity, and technological score. The outcomes indicated that stage reusability, small payload recovery, and space mining cargo recovery were promising applications for IHS adoption.

Materials Research Proceedings 37 (2023) 576-580 https://doi.org/10.21741/9781644902813-126

In the Earth re-entry scenario, LV stage recovery, small payload recovery, and space mining cargo recovery were identified as the most commercially interesting cases. These use-cases were further assessed using SWOT and PESTEL frameworks, considering factors such as political, economic, social, technological, environmental, and legal aspects. Based on the SWOT/PESTEL assessment, the best candidate use-case for Earth re-entry was identified as "LV stage recovery," which demonstrated relevant characteristics for both micro and macro IHSs, achievable marketability in a short time, and good profitability opportunities.

Therefore, the recovery of an LV stage will serve as the reference use-case for the subsequent work presented in the conference paper.

Reference Mission

A review of over 70 launch systems worldwide resulted in the identification of 20 potential candidates. Key parameters and indicators were considered to classify the launch systems into four clusters. Cluster II was selected as the most promising due to its compatibility with the IHS technology development and a significant technology development step already taken during the EFESTO project. The concept of operations (ConOps) for the Inflatable Heat Shield exploitation is based on the recovery of a launch vehicle upper stage. The ConOps includes two main phases: LEOP/ORBITAL and RECOVERY. The baseline strategy for the EFESTO-2 project is to execute the recovery via mid-air retrieval by helicopter. The engineering effort focused on the re-entry part of the recovery, and the descent and mid-air retrieval phases are out-of-scope.

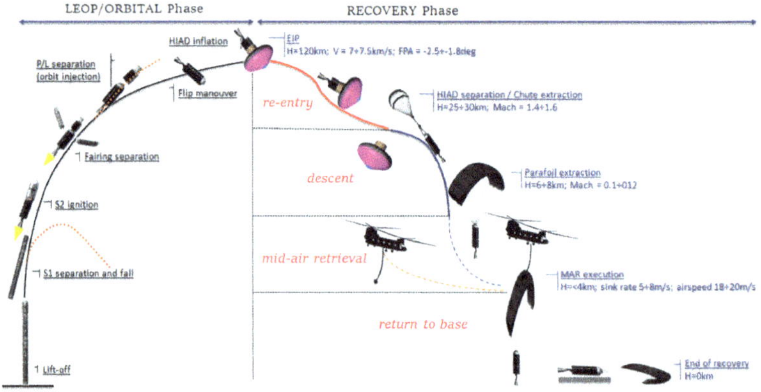

Figure 2 EFESTO-2 baseline ConOps

Mission Analysis / Aerodynamics and Aerothermodynamics

A parametric analysis was conducted to determine the combination of boundary conditions that offer a good initial flight path angle range and compliance with system constraints. Reference and sizing trajectories were calculated, and a Monte Carlo analysis confirmed compliance with all constraints.

Aero-shapes with varying parameters were investigated, and an aerodynamic database was developed for the selected flight points. CFD simulations were conducted to evaluate aerodynamic and aerothermodynamic behavior and obtain load distributions along the body.

System Design

A system design loop was performed to obtain a coherent layout for the IHS and its subsystems integrated with the Firefly Alpha upper stage. An option with a diameter of 5.32 m and a half cone angle of 60°, was selected based on qualitative assessment, aerodynamic performance, and mass estimation. Further efforts were made to reduce system mass, resulting in a minor reduction in the diameter of the inflated IHS.

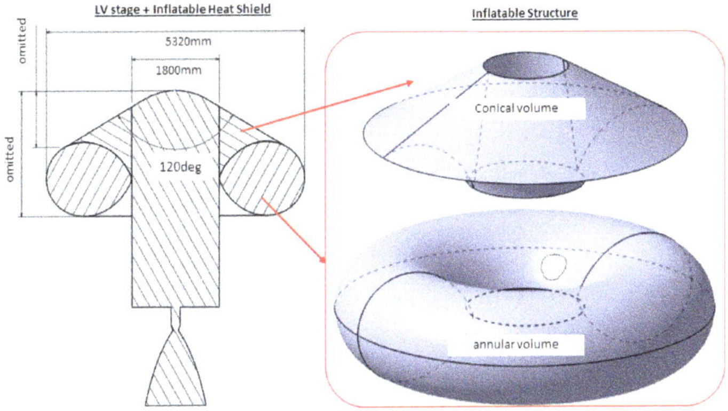

Figure 3 LV-stage and Inflatable Heat Shield integration (left), Inflatable Structure model (right)

Future Work

In the future, the project will focus on conducting ground tests for aerodynamics and flying qualities, as well as mechanical characterization of the Inflatable Structure. The aerodynamics testing will involve wind tunnel experiments with subscale models to study stability, while the mechanical characterization will use a ground demonstrator to evaluate structural behavior. The collected data will be used to improve numerical models and enhance the understanding of inflatable heat shield technology. The project aims to consolidate the technology up to TRL7 and has made progress in the initial stages, including a Business Case Analysis and reference system design. The project will conclude with test campaigns and the dissemination of findings through additional papers.

Acknowledgement

The EFESTO-2 project acknowledges the funding received from the European Union Horizon Europe research and innovation programme under grant agreement No 1010811041.

References

[1] Hughes, S. J. et al. 2013. Hypersonic Inflatable Aerodynamic Decelerator (HIAD) Technology Development Overview. In: 13rd International Planetary Probe Workshop.

[2] Marraffa, L., Boutamine, D. 2006. IRDT 2R Mission, First Results". In Proceedings of the 5th European Workshop of Thermal Protection Systems and Hot Structures. Edited by K. Fletcher. ESA SP-631.

[3] LOFTID flight: https://www.nasa.gov/feature/loftid-inflatable-heat-shield-test-a-success-early-results-show

[4] Dillman, R. A. et al. 2018. Planned Orbital Flight Test of a 6m HIAD. In: 15th International Planetary Probe Workshop.

[5] Guidotti, G. et al. 2022. The EFESTO project: Advanced European re-e ntry system based on inflatable heat shield. In: 2nd International Conference on Flight Vehicles, Aerothermodynamics and Re-entry Missions & Engineering, Heilbronn, Germany, 2022.

[6] Schleutker, T. et al. 2022. Flexible TPS design and testing for advanced European re-entry system based on inflatable heat shield for EFESTO project. In: 2nd International Conference on

Flight Vehicles, Aerothermodynamics and Re-entry Missions & Engineering, Heilbronn, Germany, 2022.

[7] Gardi, R. et al. 2022. Design development and testing of the Inflatable Structure and its Demonstrator for the EFESTO project. In: 2nd International Conference on Flight Vehicles, Aerothermodynamics and Re-entry Missions & Engineering, Heilbronn, Germany, 2022.

[8] DeRoy S.R., et al. 2016. Vulcan, Aces And Beyond: Providing Launch Services For Tomorrow's Spacecraft. AAS 16-052.

[9] Chandra, A., Thangavelautham, J. 2018. De-orbiting Small Satellites Using Inflatables. Space and Terrestrial Robotic Exploration Laboratory, Department of Aerospace and Mechanical Engineering, University of Arizona

[10] Andrews, J. et al. 2011. Nanosat Deorbit and Recovery System to Enable New Missions. In: 25th annual AIAA/USU Conference on Small Satellites. SSC11-X-3.

The ATEMO device: a compact solution for earth monitoring

Federico Toson[1,a*], Alessio Aboudan[1,b], Carlo Bettanini[1,2,c], Giacomo Colombatti[1,2,d], Irene Terlizzi[3,e], Sebastiano Chiodini[1,2,f], Lorenzo Olivieri[1,g]

[1] CISAS G. Colombo, University of Padova, Italy

[2] DII, University of Padova, Italy

[3] DAFNAE, University of Padova, Italy

[a]federico.toson@phd.unipd.it, [b]alessio.aboudan@unipd.it, [c]carlo.bettanini@unipd.it, [d]giacomo.colombatti@unipd.it, [e]irene.terlizzi@unipd.it, [f]sebastiano.chiodini@unipd.it, [g]lorenzo.olivieri@unipd.it

Keywords: Earth Monitoring, Light Pollution, Air Pollution, Vegetation Indices

Abstract. In today's context, where climate change is increasingly topical and of global interest, the ATEMO (Atmospheric Technologies for Earth Monitoring and Observation) project proposes a multi-purpose solution, which can be integrated on board drones or stratospheric balloons, to provide a framework for environmental assessment in areas of interest. Specifically, ATEMO, equipped with a set of cameras and sensors, aims to enrich data on air pollution, light pollution and vegetation health with high spatial resolution and rapid deployment.

Introduction

In environmental protection, monitoring activities are crucial for determining preventive strategies or large-scale solutions; particular interest is arisen regarding Earth observation from space [1]. However, due to low spatial and, mainly, temporal resolution of such instrumentation, information about an area or site of interest is often not sufficient and may need to be validated by other measurement and analysis methods.

The ATEMO project therefore aims to increase knowledge on environmental issues with the development of a versatile device that can be used in various contexts and environments, from the analysis of atmospheric composition, through the detection of substances and pollutants such as VOCs (Volatile Organic Compounds), NO_x (Oxides of Nitrogen), SO_x (Oxides of Sulphur), CO_2 (Carbon Dioxide), O_3 (Ozone), to the determination of ground-based light sources and their characterisation. Furthermore, in order to understand how these factors may influence ecosystems, ATEMO also aims to determine ground vegetation indices (e.g. the NDVI - Normalized Difference Vegetation Index). The ultimate goals are comparing these data with the ones obtained from satellites [2] and to look for a correlation between the various investigated parameters. In fact, while the influence of air pollutants and night lighting on ecosystems, plants and humans is already known [3], recent studies suggested that the combined effect of light and air pollution leads to the formation of harmful and worsening reaction by-products [4] of the already critical situation.

In the remainder of this paper, the ATEMO device will be introduced, and the main design solutions will be described: the selected solution allows compactness and integrability on different vehicles (drones, tethered balloons and stratospheric balloons). Last, the tests campaign planned to validate the device functional requirements will be briefly introduced.

Design of ATEMO experiment

Due to the desired versatility of ATEMO, the device is strictly constrained in both mass and size. The device has a weight of 2.5 kg, a base of 17 cm x 17 cm and a height of 25 cm; as shown in Fig. 1, the instrument consists of an external aluminium frame to which are attached 3D-printed components that guarantee both the safe integration of the various sensors and cameras and their easy attachment and removal in the event of a change of setup. In fact, one of ATEMO's main strengths is its modularity and scalability, which allows it to best adapt its physical and measurement characteristics to the application context.

Regarding the sensing part of the device, ATEMO is equipped with two cameras (one monochrome and one colour) Basler ace 2 both with Sony's IMX546, an 8MP square CMOS sensor with a wide relative response spectrum (from approximately 400 nm to 1000 nm). In addition, a FLIR typology Vue Pro R thermal imaging camera is mounted on board. There are also several sensors for temperature, pressure, humidity, detection of O_3, CO_2 and other substances, a GPS tracking system and an SQM-L (Sky Quality Meter - L type), a sensor that allows to measure the brightness of the night sky in magnitudes per square arc second.

The two cameras are used for two main purposes: the estimation of vegetation indices during the day and the analysis of light sources at night. For the two cases, the setup of the cameras obviously slightly changes in function of the required the light filtering. In the case of the vegetation analysis, the colour camera is equipped with a triple-band filter (475, 550 and 850 nm) and the monochrome camera with a band-pass filter (735 nm), both of which are designed to reproduce indices already calculated by means of Planet satellites [5]. In the night-time application, however, the two cameras are used with a filtering solution similar to that of the MINLU experiment [6], another project conducted by our research group. In order to easily switch filters, a filter wheel driven by a stepper motor has been chosen; its attachment points can be seen in Fig.1, at the centre of the camera compartment in the view from below.

The cameras field of view (FOV) is another additional key point to consider. In fact, the choice of optics is made according to the host vehicle; in the case of integration on drones or balloons tethered at a height of 50-100 m, optics with large apertures are used, while in the case of stratospheric balloons, optics with much narrower apertures are favoured.

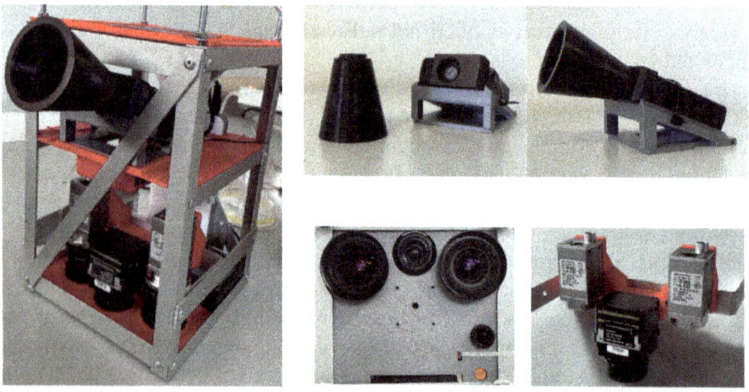

Figure 1: On the left: the complete assembled experiment. On the right the main components: the SQM with its 3D-printed baffled and the cameras at the bottom.

The system is controlled by a Raspberry Pi 3, programmed with Python, which, thanks to a 20000 mAh lithium battery, acquires data from on-board instrumentation and manages electrical and electronic systems for more than 6 hours. ATEMO's power consumption is in the range between 6 W and 14 W, depending on the sampling frequency and the subsystems in use.

For a correct interpretation of the data provided by ATEMO, a calibration using a commercial spectrometer (Black Comet C-SR-14) was necessary; the on-board cameras were characterised in the various filtering configurations since the input information is affected by the transmission of the optics, the filter, and the response of the CMOS sensor. Using a selection of known sources and the spectrometer, the response function of the system was thus defined, making the collected results consistent and comparable with those of other measurement systems (e.g., satellites).

It shall be underlined that the current design choices may be affected by the experimental activities planned in the next years and briefly introduced in the following section; thanks to its modularity, it is expected to easily adapt ATEMO with only minor modifications to the bus.

Test programme

Several test campaigns have been planned over the next three years to verify the functioning of ATEMO. These campaigns retrace previous experiments already conducted by the research group, but with a more compact configuration that encompasses them all.

This summer (2023), some experiments have already begun in collaboration with the Department of Agronomy, Food, Natural Resources, Animals, and the Environment (DAFNAE) of the University of Padova. The aim is to evaluate the effects of differential irrigation of soybean fields in order to understand whether it is possible to obtain abundant harvests with a considerable saving of water. In this situation, the experiment is integrated aboard a tethered balloon and stationed for a day at the same location at a height of 50 metres (Fig. 2).

Of course, this is only one of the possible configurations in which ATEMO will be involved in measurement campaigns. In the coming months, in fact, other collaborations will be launched, e.g. with the Chilean PUCV (Pontificia Universidad Católica de Valparaíso) University for the determination of light pollution.

Figure 2: In the first two photos, on the left, the release phase, on the right, the tethered balloon at an altitude of 50 metres.

Conclusions

In this paper, the ATEMO device for Earth monitoring has been presented. To date, the device is in the early stages of testing; the different application cases will be investigated in the next months. Functional and operational tests will be consequently performed by the research team, which has an excellent knowledge of drone flights [7], stratospheric balloons [8] and the assessment of air [9] and light pollution [10].

In conclusion, ATEMO is promising for its potential to provide a global environmental image of an area of interest, thus proving useful in a variety of fields, from atmospheric monitoring in populated areas to precision agriculture and in the verification of satellite data.

Acknowledgments

Design, manufacturing, testing and validation of ATEMO's acquisitions are possible thanks to the collaborations between CISAS G. Colombo with the Department of Industrial Engineering (DII) and the Department of Agronomy, Food, Natural Resources, Animals, and the Environment (DAFNAE) of the University of Padova. Soybean analyses are performed in the framework of the Prin (Project of Significant National Interest) 'Rewatering'; the light pollution activities are in collaboration with the Chilean PUCV University.

References

[1] "Europe's eyes on Earth," U.E., 2022. [Online]. Available: https://www.copernicus.eu/en.

[2] "Sentinel Online," ESA, [Online]. Available: https://sentinels.copernicus.eu.

[3] J. Schwartz, "The Distributed Lag between Air Pollution and Daily Deaths," *Epidemiology,* vol. 11, no. 3, pp. 320-326, May 2000. https://doi.org/10.1097/00001648-200005000-00016

[4] F. E. Blacet, "Potochemistry in the lower Atmosphere," in *XIIth International Congress of Pure and Applied Chemistry*, New York, 1951.

[5] Planet, "Daily Earth Data to See Change and Make Better Decisions," Planet, 2023. [Online]. Available: https://www.planet.com.

[6] C. Bettanini, M. Bartolomei, A. Aboudan, G. Colombatti and P. Fiorentin, "Evaluation of light pollution sources over Tuscany with an autonomous payload for sounding balloons," in *Aerospace Europe Conference* , Warsaw, 2021.

[7] G. M. Bolla, M. Casagrande, A. Comazzetto, R. Dal Moro, M. Destro, E. Fantin, G. Colombatti, A. Aboudan and E. C. Lorenzini, "ARIA: Air Pollutants Monitoring Using UAVs," in *5th IEEE International Workshop on Metrology for AeroSpace (MetroAeroSpace)*, Rome, 2018. https://doi.org/10.1109/MetroAeroSpace.2018.8453584

[8] M. Fulchignoni, A. Aboudan, F. Angrilli, M. Antonello, S. Bastianello, C. Bettanini, G. Bianchini, G. Colombatti, F. Ferri and e. al, "A stratospheric balloon experiment to test the Huygens atmospheric structure instrument (HASI)," *Planetary and Space Science,* vol. 52, pp. 867-880, 18 May 2004. https://doi.org/10.1016/j.pss.2004.02.009

[9] F. Toson, M. Pulice, M. Furiato, M. Pavan, S. Sandon, D. Sandu and R. Giovanni, "Launch of an Innovative Air Pollutant Sampler up to 27,000 Metres Using a Stratospheric Balloon," *Aerotecnica Missili & Spazio,* vol. 102, no. 2, pp. 127-138, June 2023. https://doi.org/10.1007/s42496-023-00151-y

[10] C. Bettanini, M. Bartolomei, P. Fiorentin, A. Aboudan and S. Cavazzani, "Evaluation of Sources of Artificial Light at Night With an Autonomous Payload in a Sounding Balloon Flight," *IEEE Journal of selected topics in applied Earth observations and remote sensing,* vol. 16, pp. 2318-2326, 2023. https://doi.org/10.1109/JSTARS.2023.3245190

The Hera Milani mission

M. Cardi[1,a,*], M. Pavoni[1,b], D. Calvi[1,c], F. Perez[2], P. Martino[2], I. Carnelli[2] and Milani consortium members

[1]Tyvak International srl, Via Orvieto 19, 10149 Torino (TO), Italy

[2]European Space Research & Technology Centre (ESTEC), ESA, Postbus 299, 2200 AG, Noordwijk, The Netherlands

[a]margherita@tyvak.eu, [b]marco.pavoni@tyvak.eu, [c]daniele.calvi@tyvak.eu

Keywords: Asteroid, Nanosatellite, Hera, Milani, Impact, Didymos

Abstract. Hera is the European part of the Asteroid Impact & Deflection Assessment (AIDA) international colaboration with NASA who is responsible for the DART (Double Asteroid Redirection Test) kinetic impactor spacecraft. Hera will be launched in October 2024 and will arrive at Didymos in January 2027. The Hera mothercraft will accommodate two 6U Nanosatellite, Milani and Juventas. The Milani Nanosatellite is developed by Tyvak International leading a consortium of European Universities, Research Centers and Firms from Italy, Czech Republic, Finland. During the cruise to the Asteroid (+2 years), Milani Nanosatellite will be hosted inside the Hera mothercraft, periodically checked for health and charged. At arrival it will be deployed and commissioned while HERA is performing the Didymos detailed characterization phase, at about 10 to 20 km distance from the asteroid. The Milani mission objectives are defined as to add scientific value to the overall Hera mission: i) Map the global composition of the Didymos asteroids, ii) Characterize the surface of the Didymos asteroids, iii) Evaluate DART impacts effects on Didymos asteroids and support gravity field determination, iv) Characterize dust clouds around the Didymos asteroid, enhancing the scientific return of the whole HERA mission. The scientific payloads supporting the achievement of these objectives are the main Payload "ASPECT" (developed by VTT, Finland), a SWIR, NIR and VIS imaging spectrometer and the secondary Payload "VISTA" (developed by INAF, Italy), a thermogravimeter aiming at collecting and characterizing volatiles and dust particles below 10μm. The Milani mission and the project team is facing challenges such as, among others, the use of COTS components in deep space environment, optical navigation implementation, interfaces management with the HERA mothercraft since the very beginning of the design up to the mission. Tyvak International work focuses on the development and integration of the Milani vehicle, including mission specifics development enabling the mission and vehicle models enabling early interface testing with Hera mothercraft.

Introduction

In 2027, the Hera spacecraft will rendezvous with the binary asteroid 65803 Didymos as the European contribution to the AIDA (Asteroid Impact and Deflection Assessment) international collaboration. NASA is responsible for the Double Asteroid Redirection Test (DART) kinetic impactor spacecraft. Hera and DART have been conceived to be mutually independent, however, their value is increased when combined. Indeed, Hera is a planetary defense mission aimed to investigate the effect of DART impact, with clear scientific objectives as a bonus. In proximity to the target, Hera will release two 6U Nanosatellites called Milani and Juventas. The two nanosatellites will be the first Nanosatellites to orbit in the close proximity of a small body and the first to perform scientific and technological operations around a binary asteroid.

Tyvak International is responsible for the Milani system development and is leading (as Prime Contractor) a large consortium made by 10+ entities from Italy, Czech Republic and Finland. Milani will contribute to the scientific value of the Hera planetary defense mission, mainly through the visual inspection of the asteroid (main payload: ASPECT) and dust detection (secondary payload: VISTA).

Didymos properties

Didymos is a binary Near-Earth Asteroid (NEA) of S-type discovered in 1996 formed by Didymos, or D1 (the primary) and Dimorphos, or D2 (the secondary). Up-to-date data about Didymos and Dimorphos are reported in the following tables:

Table 1. Binary system parameters (semi-major axis, eccentricity, inclination, revolution period)

System parameters			
a	e	i	T
1.66446 AU	0.3839	3.4083 deg	770 days

Table 2. Didymos and Dimorphos mass and spin periods properties

System parameters			
M1	M2	T1	T2
5.226×10^{11} kg	4.860×10^{9} kg	2.26h	11.92h

The orbital properties are retrieved from the up-to-date kernels of the Hera mission. In the up-to date reference model, Dimorphos and Didymos are assumed to share the same equatorial plane on which their relative motion occurs and Dimorphos is assumed to be in a tidally locked configuration with Didymos. In this work, two reference frames are used. "DidymosEclipJ2000" is a quasi-inertial reference frame, centered in the system barycenter with the axis directed as the inertial EclipJ2000 reference frame. This frame can be considered inertial for intervals of time negligible with respect to Didymos heliocentric motion. "DidymosEquatorialSunSouth" is a non-inertial reference frame in which the trajectories are shown. It is centered in the system barycenter and has the X-Y plane on the asteroid equatorial plane. The X axis is aligned with the projection of the Sun vector on the equatorial plane and the Z-axis is aligned to the south pole of Didymos.

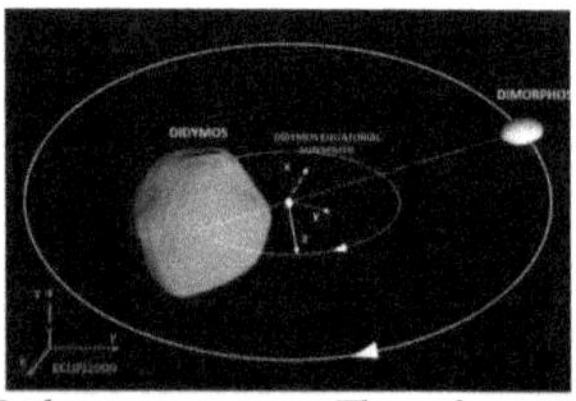

Figure 1. Didymos geometry. The reference frames are highlighted. The red frame is the inertial Eclip2000 which corresponds to the quasi-inertial DidymosEclipJ2000 when centred in the system barycentre. The yellow frame is the Didymos Equatorial Sun South (Courtesy: Politecnico di Milano)

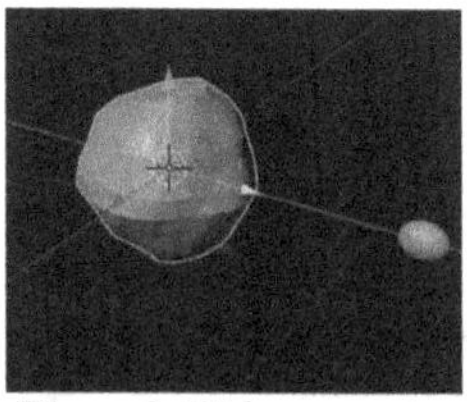

Figure 2. Didymos system geometry: polyhedral radar shape model of D1 and triaxial ellipsoidal model of D2 in D1_Body reference frame. (Courtesy: Politecnico di Milano)

Scientific goals and operational constraints

Milani scientific phases design has been mostly driven by its main payload, ASPECT. ASPECT is a passive payload, equipped with a four-channel visible to near-infrared hyperspectral imager and will be used on Milani to perform global mapping of the asteroids with detailed observation of the DART crater on Dimorphos. ASPECT main scientific goals can be summarized in three actions:

1. Imaging both the asteroids with a spatial resolution better than 2 m/pixel
2. Imaging the secondary asteroid with a spatial resolution better than 1 m/pixel
3. Imaging the DART crater with a spatial resolution better than 0.5 m/pixel at phase angle (Sun-Asteroid-Milani angle) in the range [0-10] deg and [30-60] deg.

In terms of trajectory design, spatial resolution requirements drive the maximum range at which scientific observations can be performed. From an operational point of view, Milani's communication with ground will be performed via Inter-Satellite Link (ISL) using Hera as data relay. For this reason, data downlink and uplink must be performed within the same communication windows used by Hera. Operations will be scheduled considering:

- Hera mission operations requirements
- Milani Nanosatellite mission operations requirements
- Mission Data downlink (Milani-to-Hera)
- Communication window (Hera-to-Earth)

In order to avoid open-loop manoeuvres, Milani needs to select the manoeuvring frequency to be as close as possible to Hera's pattern (4-3 days). This is not mandatory, however, it ensures the compatibility of the strategy with the requirement on the Turn-Around time (TAT)1 of 48 h.

Scientific goals and operational constraints are the results of an initial phase of requirements definition and consolidations, led by Politecnico di Torino team and have been the main driver for the detailed design of the main phases of Milani's mission: Far Range Phase (FRP) and Close Range Phase (CRP). The scientific goals that mostly drove the mission design of Milani have been derived from its main payload, ASPECT, presented in the following sections.

Milani Mission profile and Concept of Operations (ConOps)

The Milani Mission is designed by Politecnico di Milano (PoliMI). Milani trajectory design has been mainly driven by the main scientific goals of the mission, but it has also been influenced by both technical and operational constraints. Due to the low gravity environment around the asteroids, selecting Keplerian orbits as nominal trajectories would require a demanding station keeping strategy to counteract the SRP effect. For this reason, a patched-arc manoeuvring strategy that leverages the SRP acceleration to target pre-selected waypoints has been implemented. This strategy has flight heritage in small-body environment. It is the one currently envisaged by the Hera spacecraft during its operational phases and previously performed by the Rosetta spacecraft during its initial scientific phase, after rendezvous with comet 67-P/Churyumov-Gerasimenko. The waypoints selection has been mostly influenced by the passive nature of Milani's payload as well by the on-board navigation strategy, which forces the Nanosatellite to avoid the night-side. The resulting trajectories are loop orbits with manoeuvres points placed as far away from each other as possible to maximize the time spent in proximity to the system. Main Milani mission phases are hereafter presented:

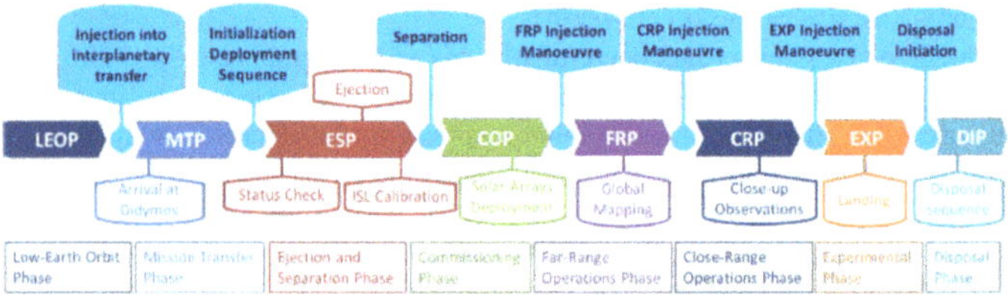

Figure 3. Milani mission phases

- **Low Earth Orbit Commissioning Phase (LEOP)**, will be done on Hera spacecraft upon launch; a specific list of checkout tests will be executed also on Milani Nanosatellite to verify the basic functionalities that can be verified in stowed and integrated Configurations
- **Mission Transfer Phase (MTP)**, or interplanetary cruise, will be characterized by regular checkout tests to be executed on Milani Nanosatellite to verify the basic functionalities
- **Ejection and separation phase (ESP)**, will start upon arrival to the asteroids and will be characterized by checkout test in stowed configuration, ejection of Milani Nanosatellite outside Hera, pre-deployment checkout in exposed configuration, Milani Nanosatellite separation from Hera
- **Commissioning Phase (COP)**, checkout, stabilization, and calibrations
- **Far Range Operations Phase (FRP)**, transfer to the operative orbits, first global mapping, and technologies demonstration
- **Close Range Operations Phase (CRP)**, transfer to the operative orbits closer to the asteroids, Close-up observation of Didymos bodies, additional technology demonstration, observation of the DART impact crater
- **Experimental Phase (EXP)**, foreseeing the landing on the asteroids or transfer on a heliocentric graveyard orbit, currently under evaluation
- **Disposal Phase (DIP)**, Passivation

System Overview

The Nanosatellite leverages on Tyvak Trestles platform architecture, avionics technology Mark II. This is a standard platform, however, some customizations were made specifically for Milani mission. In the following figure, the vehicle configuration is shown.

Figure 4. Milani nanosatellite – Deployed configuration

The system is composed of the following elements:

- Avionics (Tyvak Mark II technology), including Flight Computer, Electrical Power System, ADCS
- Primary Payload (ASPECT)
- Secondary Payload (VISTA)

- Cold-gas propulsion system, enabling technology
- External Inter-satellite link (ISL) radio + antennas
- Navigation Camera
- COTS components
- Mission Specific Interfaces (such as Payload Interface Board, PIB)
- Interfaces with the Hera mothercraft:
    - Milani is integrated into the Deep Space Deployer (DSD) developed by ISIS, providing also a specific CubeSat Interface Board to interface the Milani CubeSat with the DSD
    - The main interface with the assembly constituted by the DSD and Milani Nanosat with the Hera mothercraft is the Life Support Interface Board (LSIB), developed by KUVA Space and allowing the exchange of power and data between the two spacecraft and so the execution of the checkout tests during the stowed and exposed configuration.

A radiation-related analysis was executed to mitigate risks associated to the execution of the mission in deep space environment. The radiation analysis effort was led by Politecnico di Torino team and included both fault injection approaches and dedicated radiation testing on a subset of components identified as critical for the mission.

Conclusions

To date, the Milani project is in Assembly Integration and Test phase. Upon successful vehicle qualification, a System Validation Testing (SVT) Phase will be foreseen aiming at testing the end-to-end communication with the Hera mothercraft at ESTEC. A risk mitigation approach was implemented during the project through the reduced EM and Structural and Thermal Interface Model (STIM) development and delivery.

Acknowledgement

This work has been performed within the scope of ESA Contract No. 1222343567/62/NL/GLC. The authors would like to acknowledge the support received by the whole Milani Consortium and European Space Agency team following and supporting the Milani mission development.

References

[1] https://www.sciencedirect.com/science/article/pii/S0273117720309078

[2] https://meetingorganizer.copernicus.org/EPSC2021/EPSC2021-732.html

[3] Analysis of Asteroid 65803 Didymos for the European Space Agency Asteroid Impact Mission" by P. Michel et al. (2016)

[4] The Double Asteroid Redirection Test (DART): Deflection and Impact Monitoring (DIM) Mission" by Carnelli et al. (2020)

Aeronautics and Astronautics - AIDAA XXVII International Congress Materials Research Forum LLC
Materials Research Proceedings 37 (2023) 590-594 https://doi.org/10.21741/9781644902813-129

Small celestial body exploration with CubeSat Swarms

Emmanuel Blazquez[1,a *], Dario Izzo[1,b], Francesco Biscani[2], Roger Walker[3] and Franco Perez-Lissi[3]

[1]European Space and Research Technology Centre, Advanced Concepts and Studies Office, European Space Agency, Kepleerlan 1, 2201AZ Noordwijk (The Netherlands)

[2] European Space Operations Centre, , European Space Agency, Robert-Bosch-Straße 5, 64293 Darmstadt (Germany)

[3] European Space and Research Technology Centre, Cubesat Systems Unit, European Space Agency, Kepleerlan 1, 2201AZ Noordwijk (The Netherlands)

[a]emmanuel.blazquez@esa.int, [b]dario.izzo@esa.int

Keywords: CubeSat Swarms, Small Celestial Bodies, Mission Analysis

Abstract. This work presents a large-scale simulation study investigating the deployment and operation of distributed swarms of CubeSats for interplanetary missions to small celestial bodies. Utilizing Taylor numerical integration and advanced collision detection techniques, we explore the potential of large CubeSat swarms in capturing gravity signals and reconstructing the internal mass distribution of a small celestial body while minimizing risks and Delta V budget. Our results offer insight into the applicability of this approach for future deep space exploration missions.

Introduction

In the last decade CubeSats have emerged as an innovative and cost-effective platform for testing new satellite technologies, with applications ranging from Earth observation to deep space missions. For instance, the HERA interplanetary mission that will explore the Didymos binary system in 2025 will embark two CubeSats to perform a detailed exploration of the system. [1] In this context, distributed swarms of CubeSats are a promising strategy for future exploration missions to small celestial bodies, enabling increased scientific return while minimizing risks associated with operating in an unknown environment. [2]

In this work, we present a large-scale simulation study of the deployment and operation of many CubeSats around small celestial bodies, with the aim of assessing the applicability of large CubeSats swarms for distributed operations during interplanetary missions to asteroids and comets. Our approach leverages the *cascade* and *heyoka* C++/Python libraries to propagate the evolution of the swarm, reliably detecting close encounters and collisions using high-order Taylor numerical integration and collision detection. [3] We assume that the swarm is deployed sequentially by a "mother" spacecraft operating in proximity of the celestial body of interest, and that each CubeSat has the capability of performing 6-DoF orbital manoeuvres with limited ΔV budget. We use mascon mass models to have a representation of the mass distribution of the bodies around which the swarm is orbiting and consider CubeSat deployment uncertainty in position and velocity as well as a fully automated trajectory control strategy aimed at keeping the overall Delta V budget under control while maximizing scientific return and controlling the risk of conjunctions. Our simulations offer insights into the trade-offs involved in deploying and maintaining large CubeSat swarms for distributed operations in deep space.

Background and Models

In this study, we consider a prospective advanced mission concept inspired by the European Space Agency's HERA mission. [1] We assume that a mothership satellite is inserted into orbit around a small celestial body of irregular shape and will consequently deploy a swarm of CubeSats from a common orbit. The swarm's objective is to capture gravity signal around the body to assess its mass distribution at a later stage following the procedure described in the *geodesyNETs* project. [4] The swarm CubeSats will have to orbit around the body without colliding with each other, staying within a spherical region defined by a minimum safety radius to the body's center of mass and a maximum radius of operations. These operations will have to be performed with minimum ΔV budget requirements.

Small celestial bodies of interest for this study will have their internal mass distribution represented by a heterogeneous mascon model. In other words, they will be modelled as a set of point masses located in specific positions and with some given mass. The mascon models used for this study were developed in the *geodesyNETs* project. [4] The dynamics of the swarm around the spacecraft are simulated via cascade simulations making use of the *heyoka* high-performance Taylor integrator. The equations of motion are represented by a gravitational N-body problem where N is the number of mascons, plus the solar radiation pressure integrated in an inertial reference frame and consider a rotating celestial body. Terminal events are added to the equations of motion to enforce a maneuver when a spacecraft enters the safety sphere around the body or exits the operations sphere.

We assume that each CubeSat of the swarm has the capability of performing maneuvers with full degrees of freedom whenever it gets close to another element of the swarm, to the central body or flies too far away. A 2-phase collision detection algorithm is used to detect when two elements of the swarm get too close to each-other. It exploits the availability, at each step, of Taylor polynomial expansions of the system dynamics that are the product of the integration scheme offered by *heyoka* and therefore come at no additional cost. *Cascade* makes use of these expansions to compute Axis Aligned Bounding Boxes (AABB) encapsulating the positions of each orbiting object within a chosen collisional timestep. The algorithm operates in two phases: a broad phase with a Bounding Volume Hierarchy (BVH) over the 4-dimensional AABB defined in cartesian coordinates, followed by a narrow phase more computationally demanding making use of a polynomial root finding algorithm. This approach is well-suited to parallelization and particularly performant in non-collision rich environments such as the ones considered in this study.

Simulations and Results

We apply our framework to simulate swarm missions to Itokawa, Bennu and Ryugu. [5,6,7] For each body four swarm configurations are considered, with 5, 10, 25 and 50 spacecraft respectively. Itokawa is modelled using 24800 mascons, Bennu with 17400 mascons and Ryugu with 26800 mascons. The mothership is initially located on a circular orbit around the body of interest with a semi-major axis of 1.5km, an inclination of 90° and a mean anomaly of 90°. The solar radiation pressure acting on the swarm spacecraft is computed considering a wet area of 1 m^2, with the sun direction being selected arbitrarily in the ecliptic plane. The spacecraft must remain in an ellipsoidal region around the asteroid, with a no-entry safety ellipsoid whose dimensions are those of the celestial body with a factor 1.3 and a no-exit sphere of radius 3 km for Itokawa and 2 km for Bennu and Ryugu. The relative release velocities of the swarm spacecrafts with respect to the mothership are randomly selected in [1.5, 3.5] cm/s. 100 collisional timesteps are processed in parallel to account for CubeSat collisions, where one timestep is defined as $1/60^{th}$ of the orbital period of the mothership spacecraft and is also representative of the release frequency of the CubeSats in the swarm. The collisional radius, the minimum relative distance required between two CubeSats to avoid collisions, is set to 5 m. The dynamics and operations of the swarm are simulated in each case for a total duration of 4 days.

Table 1: Simulation results for swarms orbiting around Itokawa/Bennu/Ryugu,

	Itokawa				Bennu				Ryugu			
Swarm size	5	10	25	50	5	10	25	50	5	20	25	50
ΔV_{min} [m/s]	0.1	0.1	0.1	0.1	0	0	0	0.03	0	0	0	0
ΔV_{max} [m/s]	1.4	2.6	3	3.7	0,66	2.9	2.6	4.5	0.26	0.26	1.04	1.63
ΔV_{mean} [m/s]	0.82	0.85	0.97	1.4	0.39	0.68	0.7	0.92	0.08	0.05	0.12	0.34
Collision events	2	2	5	12	2	9	51	89	4	4	21	77
Safety events	7	9	39	191	34	15	40	78	4	4	12	36
Re-entry events	39	82	192	393	46	114	259	481	4	9	26	142

Table 1 showcases the Results obtained from the simulations. The number of events requiring maneuvers is in general dominated by re-entry requirements when the satellites are in danger of exiting their operational region around the asteroid. As expected, the size of the swarm has a direct impact on the ΔV budget for the mission. The increase observed in the maximum ΔV required for swarms with a higher number of spacecrafts is partially due to the increased number of collisional events recorded, especially for Bennu and Ryugu. However, re-entry events account for most of the budget and increase proportionally with the number of satellites in the swarm. This is because, on average, every CubeSat of the swarm will need to be actuated several times to stay within the operational region for the mission. Safety events are more predominant for Itokawa due to the elongated shape of the body that makes trajectory correction maneuvers more likely to cause future safety events.

Of particular interest is the fact that the trade-off for the number of satellites in the swarm, therefore gravity signal recovery, versus the ΔV budget of the mission is dependent on the shape and mass distribution of the body. Itokawa sees an 18% increase in mean ΔV going from 5 to 25 satellites in the swarm, for a drastic increase in recovered gravity signal and therefore internal shape reconstruction. But the increase from 25 to 50 satellites leads to a 44% increase in mean ΔV: the number of collision events increases enough to create a cascade of safety events that increase the overall budget significantly. The tradeoff for Bennu and Ryugu appears at different sizes for the swarm but exhibits a similar behavior: an increased number of collisional events leads to an increased number of safety and re-entry events and therefore hinders the mission budget. This could signify that the number of collision events dominates the trade-off analysis and could be a suitable preliminary indicator to size the swarm when gravity signal reconstruction is sufficient. Regularity in the shape of the body seems to be correlated with a higher number of collisional events but overall lower maintenance budgets, as the cascade phenomenon is not as pronounced. This analysis will be refined in future work with a more robust optimization setup, the purpose of this work being principally to propose and make available a framework for the efficient parallel simulation of CubeSat swarm operations around small celestial bodies.

Fig. 2 presents a top-view of a 50-spacecraft swarm simulation orbiting Itokawa as well as the asteroid shape reconstruction from the gravity signal gathered from the swarm after 4 days of operations and using the procedure described in the *geodesyNETs* project, along with the ΔV budget distribution for the swarm. [4]

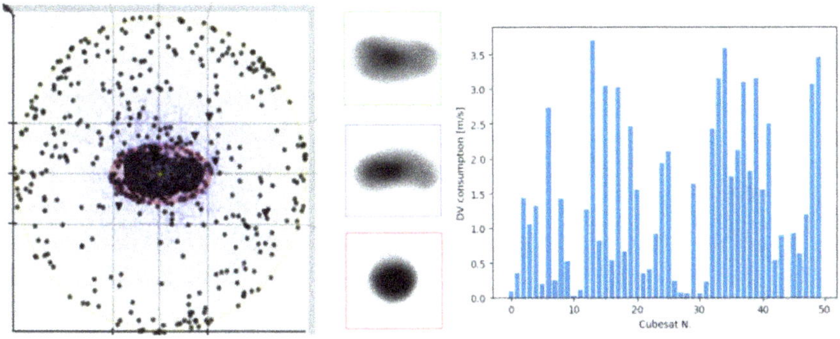

Figure 2: Itokawa study case, (left) simulation of a collision-free swarm of 50 elements. The position of the swarm elements during the gravity field measurements are shown in blue. Impulsive maneuvers to keep the swarm in safe proximity are shown in black (collision avoidance, triangles). (middle) reconstruction of the asteroid shape after 4 days of operation. (right) ΔV budget distribution across the CubeSats in the swarm.

Conclusion

In this work we have applied our framework to simulate CubeSat swarm missions to Itokawa, Bennu and Ryugu. We show how swarms of different sizes, ranging from a few CubeSats to several dozen, can efficiently capture gravity signal and reconstruct spherical harmonic expansions while minimizing the risk of conjunction and counteracting the effects of environmental disturbances. Our results showcase the potential of large CubeSat swarms for future exploration missions to small celestial bodies and propose an Open-source framework for preliminary mission analysis studies in such scenario. The code used for this study will be made publicly available at https://gitlab.com/EuropeanSpaceAgency/collision-free-swarm, and the *cascade* library used for this work is an open source python module available at https://github.com/esa/cascade.

References

[1] Michel, P., Kueppers, M., Sierks, H., Carnelli, I. et al. (2018). European component of the AIDA mission to a binary asteroid: Characterization and interpretation of the impact of the DART mission. In Advances in Space Research (Vol. 62, Issue 8, pp. 2261–2272). Elsevier BV. https://doi.org/10.1016/j.asr.2017.12.020

[2] Stacey, N., Dennison, K., and D'Amico, S. (2022). Autonomous Asteroid Characterization through Nanosatellite Swarming. In 2022 IEEE Aerospace Conference (AERO). 2022 IEEE Aerospace Conference (AERO). IEEE. https://doi.org/10.1109/aero53065.2022.9843328

[3] Biscani, F., and Izzo, D. (2021). Revisiting high-order Taylor methods for astrodynamics and celestial mechanics. In Monthly Notices of the Royal Astronomical Society (Vol. 504, Issue 2, pp. 2614–2628). Oxford University Press (OUP). https://doi.org/10.1093/mnras/stab1032

[4] Izzo, D., and Gómez, P. (2022). Geodesy of irregular small bodies via neural density fields. In Communications Engineering (Vol. 1, Issue 1). Springer Science and Business Media LLC. https://doi.org/10.1038/s44172-022-00050-3

[5] Fujiwara, A., Kawaguchi, J., Yeomans, D. K. et al. (2006). The Rubble-Pile Asteroid Itokawa as Observed by Hayabusa. In Science (Vol. 312, Issue 5778, pp. 1330–1334). American Association for the Advancement of Science (AAAS). https://doi.org/10.1126/science.1125841

[6] Saiki, T., Takei, Y., Takahashi, T. et al. (2021). Overview of Hayabusa2 Asteroid Proximity Operation Planning and Preliminary Results. In Transaction of the Japan Society for Aeronautical and Space Sciences, Aerospace Technology Japan (Vol. 19, Issue 1, pp. 52–60). Japan Society for Aeronautical and Space Sciences. https://doi.org/10.2322/tastj.19.52

[7] Lauretta, D. S., DellaGiustina, D. N., Bennett, C. A. et al. (2019). The unexpected surface of asteroid (101955) Bennu. In Nature (Vol. 568, Issue 7750, pp. 55–60). Springer Science and Business Media LLC. https://doi.org/10.1038/s41586-019-1033-6

Space Flight Mechanics

Low-energy earth-moon mission analysis using low-thrust optimal and feedback control

A. Almonte[1,a], I. Ziccardi[1,b], A. Adriani[1,c], A. Marchetti[1,d], M. Pontani[2,e]

[1]Thales Alenia Space Italia (TAS-I), Via Saccomuro 24, 00131 Rome, Italy

[2]Department of Astronautical, Electrical, and Energy Engineering, Sapienza University of Rome, via Salaria 851, 00138 Rome, Italy

[a]alessio.almonte@gmail.com, [b]irene.ziccardi@thalesaleniaspace.com, [c]andrea.adriani@thalesaleniaspace.com, [d]andrea.marchetti@thalesaleniaspace.com, [e]mauro.pontani@uniroma1.it

Keywords: Earth-Moon Transfers, Low Energy Transfers, Low-Thrust Spacecraft, Optimal Control, Feedback Guidance and Control, Particle Swarm Optimization

Abstract. This work is focused on designing a low-energy orbit transfer in the Earth-Moon system, aimed at reaching stable capture in a highly elliptical lunar orbit, with the use of low-thrust propulsion. The mission at hand includes three different phases: low-energy ballistic transfer starting from Earth, low-thrust minimum-fuel arc, and low-thrust lunar orbit insertion using variable-thrust nonlinear orbit control. First, a reference trajectory is generated in the framework of the Patched Planar Circular Restricted Three-Body Problem (PPCR3BP), leveraging invariant manifold dynamics. Trajectory propagation is performed using the Bicircular Restricted Four-Body Problem (BR4BP) model. Particle swarm optimization is applied for trajectory refinement and to detect the subsequent minimum-fuel low-thrust arc. Finally, the lunar orbit is entered thanks to the use of variable-thrust nonlinear orbit control.

Introduction

Low-energy Earth-Moon transfers have been studied extensively in the last decades. Some missions have already exploited the results from these studies, leading to considerable propellant savings and other advantages, such as flexibility in target orbit selection, extended launch windows, and more relaxed operational schedules. At the end of the 60s Conley [1] used elements of dynamical systems theory to identify temporary lunar capture conditions. Three decades ago, Belbruno and Miller [2] developed the Weak Stability Boundary (WSB) technique and applied it to lunar transfers, discovering a low-energy transfer through the equilibrium regions of the Sun-Earth-Moon system. Koon *et al* [3] obtained similar results by following the Conley methodology, taking advantage of invariant manifolds of planar Lyapunov orbits around the Earth-Moon L2 libration point. More recently Mingotti *et al* [4] designed low-energy low-thrust transfers, using PPCR3BP for the low-energy trajectory arc and optimal control with a direct method for the low-thrust lunar capture arc. An alternative approach to reach a stable capture orbit is represented by variable-thrust nonlinear orbit control, as described by Gurfil in [5] and Pontani *et al.* in [6]. In fact, using a feedback control law allows applying real-time control and compensate perturbations, with no need of a reference trajectory.

This work is focused on designing a low-energy Earth-Moon transfer starting from a GTO parking orbit and aimed at reaching a stable lunar capture orbit, with the use of low-thrust propulsion. Several design approaches are being employed: (a) use of the invariant manifold dynamics, to obtain a low-energy planar reference trajectory from Earth to Moon, (b) optimal control with the use of the particle swarm algorithm (PSO) to detect the subsequent minimum-fuel low-thrust arc, and (c) variable-thrust nonlinear orbit control to enter the desired lunar orbit. This work intends to show that the mission design techniques proposed in this study represent a convenient approach to preliminary Earth-Moon mission analysis.

MISSION Analysis and results

The mission is composed of two phases, analyzed using different reference frames, to simplify the design. Phase 1 covers the low-energy low-thrust Earth-Moon transfer. A planar low-energy exterior transfer is found using invariant manifold dynamics in the PPCR3BP framework. First, two convenient Jacobi constants $C_{SE} = 3.0075$ and $C_{EM} = 3.15$ are selected for the Sun-Earth and Earth-Moon three body system. Then, Sun-Earth L2 and Earth-Moon L2 planar Lyapunov orbits associated with those constants are found by exploiting their symmetry properties with the use of PSO. Invariant manifolds are propagated from Lyapunov orbits. Invariant manifolds in PCR3BP are a subset of the phase space, and separate bouncing trajectories from transit orbits. Taking advantage of this property, invariant manifolds are cut with Poincaré sections, reducing the phase space dimension to 2. Manifold cuts and Poincaré sections are shown in Figure 1, where section S_A cuts stable manifold $W_S^+(\gamma_2)$ and section S_B cuts unstable manifold $W_U^-(\gamma_2)$ of Sun-Earth L2 Lyapunov orbit γ_2; instead, section S_C cuts stable manifold $W_S^+(\delta_2)$ of Earth-Moon L2 Lyapunov orbit δ_2.

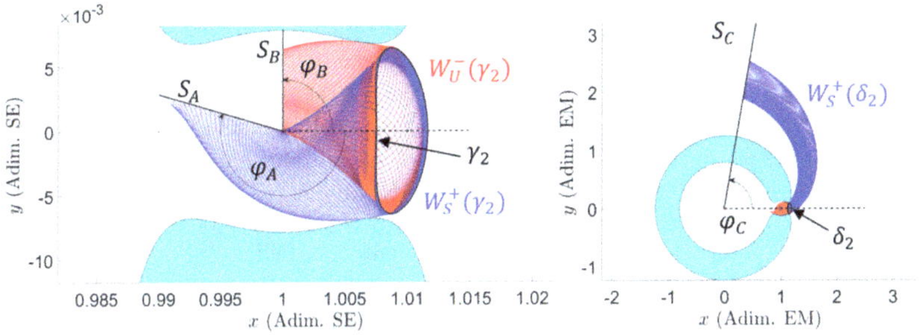

Figure 1: Manifolds cut and Poincaré sections

Intersections are evaluated in the $r_2 - \dot{r}_2$ plane, shown in Figure 2, where r_2 is the distance and $\dot{r}_2$ the radial velocity with respect to Earth, secondary body in the Sun-Earth three-body system.

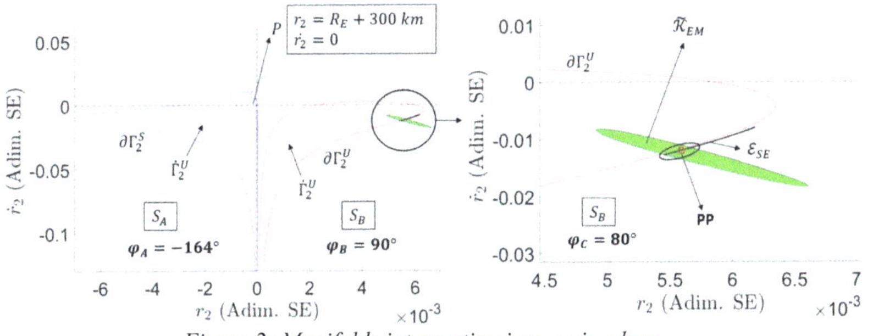

Figure 2: Manifolds intersection in $r_2 - \dot{r}_2$ plane

From figure 2 it is possible to see that the angle φ_A is tuned to move the $W_S^+(\gamma_2)$ stable manifold cut $\partial\Gamma_2^S$ close to the starting point P corresponding to GTO pericenter conditions. Trajectories starting from P in S_A end in the $\mathcal{E}_{SE}$ set on section S_B, at external points close to the $W_U^-(\gamma_2)$ unstable manifold cut $\partial\Gamma_2^U$. Tuning angle φ_C allows moving the set $\widetilde{\mathcal{K}}_{EM}$, until intersection with $\mathcal{E}_{SE}$ occurs. The set $\mathcal{E}_{SE} \cap \widetilde{\mathcal{K}}_{EM}$ represents temporary ballistic capture trajectories and here the Patch Point (PP) between Sun-Earth and Earth-Moon trajectory arcs is chosen. The PP conditions are then integrated backward using the Bicircular Restricted Four-Body Problem (BR4BP), to reach (through backward integration) the GTO orbit, with parameters $a = 24363.57$ km, $e = 0.7036$, $i = 23.45°$, $\Omega = 0°$, $\omega = 163.72°$, $\theta_* = 0°$. The velocity change to perform Translunar Injection (TLI) is $\Delta v^{(TLI)} = 671.8$ m/s and the time of flight is $\Delta t_1 = 142.47$ days. The small maneuver necessary at patch point to link the trajectories is substituted with a low-thrust arc obtained with minimum-fuel optimal control. The low thrust propulsion system is identified by $u_T^{(\max)} = \frac{T_{max}}{m_0} = 2 \cdot 10^{-4}$ m/s and $c = g_0 I_{sp} = 18.142$ km/s. The state vector is defined as $\boldsymbol{X} = \left[x, y, v_x, v_y, \frac{m}{m_0}\right]^T = [x_1, x_2, x_3, x_4, x_5]^T$ and the control vector is $\boldsymbol{u} = [u_T, \alpha]$, where $u_T = T/m_0$ with $0 \le u_T \le u_T^{(\max)}$ and α is the angle between the thrust direction and the line from the Sun and the Earth-Moon barycenter. The objective of minimum-fuel optimal control is to find the control $\boldsymbol{u_T}$ and the constant parameters vector $\boldsymbol{p}$ such that the cost function $J = -m_f$ is minimized, while satisfying the state equations $\dot{\boldsymbol{X}} = \boldsymbol{f}(\boldsymbol{X}, \boldsymbol{u}, t, \boldsymbol{p})$ and the boundary conditions $\boldsymbol{\Psi}(\boldsymbol{X_0}, \boldsymbol{X_f}, t_0, t_f, \boldsymbol{p}) = \boldsymbol{0}$. Additional constraints are added to the Lunar Orbit Insertion (LOI) condition, limited to orbits with $e \in (0.6, 0.7)$, to arrive at a capture orbit, and $r_P = R_M + 100$ km to avoid Moon impact. These constraints are written in terms of equality constraints using the parameter vector $\boldsymbol{p}$. Exploiting the necessary optimality conditions and the Pontryagin minimum principle it is possible to obtain the control law, depending on co-state vector $\boldsymbol{\lambda} = [\lambda_1, \lambda_2, \lambda_3, \lambda_4, \lambda_5]^T$. The minimum set of unknown parameters is $\boldsymbol{\chi} = \{\boldsymbol{\lambda_0}, t_f, \boldsymbol{p}\}$ and is found using PSO. The minimum time of flight is $t_f = 13.30$ days, with $\frac{m_f}{m_0} = 0.994$. The time histories of the thrust angle are shown in Figure 3.

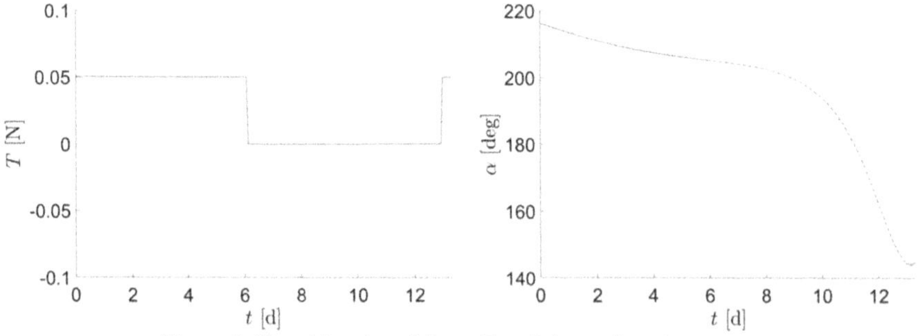

Figure 3: Time histories of thrust T and thrust direction α

In Phase 2 of the mission nonlinear control was employed to enter a stable lunar orbit. This control enjoys quasi-global stability properties and allows compensating perturbations [6]. The main objective of this phase is convergence to the target orbit, while compensating perturbations due to Earth and Sun. In this framework the state vector is given by the Modified Equinoctial

Elements (MEE) and the mass ratio, i.e. $\boldsymbol{X} = \left[p, l, m, n, s, q, \frac{m}{m_0}\right]^T = [x_1, x_2, x_3, x_4, x_5, x_6, x_7]^T = [\boldsymbol{z}, x_6, x_7]^T$, whereas the control vector is $\boldsymbol{u_T} = \boldsymbol{T}/m_0$ with $0 \leq u_T \leq u_T^{(\max)}$. The target set is defined in terms of MEE as

$$\boldsymbol{\Psi} = \left[x_1 - a_d(1 - e_d^2),\ x_2 - e_d cos(\Omega_d + \omega_d),\ x_3 - e_d sin(\Omega_d + \omega_d),\ x_4 - \tan\left(\frac{i_d}{2}\right)\cos(\Omega_d),\ x_5 - \tan\left(\frac{i_d}{2}\right)\sin(\Omega_d)\right]^T.$$

The dynamics is governed by the Lagrange Equations with MEE [6]. The control law is derived in [6] and yields $\boldsymbol{a_T}$, i.e. the thrust acceleration as a function of the state, the boundary condition violation, and the perturbing acceleration. The latter includes the effect of Earth and Sun as third bodies. Matrix $\boldsymbol{K}$ is a positive definite diagonal matrix of gains, selected after trial-and-attempt tuning. The target orbit is reached in a time of flight $t_f = 77.52$ days, with $\frac{m_f}{m_0} = 0.920$. After 100 days from Lunar Orbit Injection (LOI) the mass ratio reduces to $\frac{m_f}{m_0} = 0.909$, because propellant is used to compensate the perturbations. The orbit elements of the planar capture orbit reached at the end of Phase 1, together with target parameters and parameters after 100 days of propagation, are shown in Table 1. The trajectory in Phase 2 and the full trajectory are shown in Figure 4.

	a [km]	e	i [deg]	Ω [deg]	ω [deg]
LOI	108058	0.7000	0	-23.78	48.14
Target orbit	9751	0.6870	55.70	120.00	90.00
Final	9772	0.6871	55.71	120.01	89.99

Table 1: orbit elements at LOI and along the target orbit

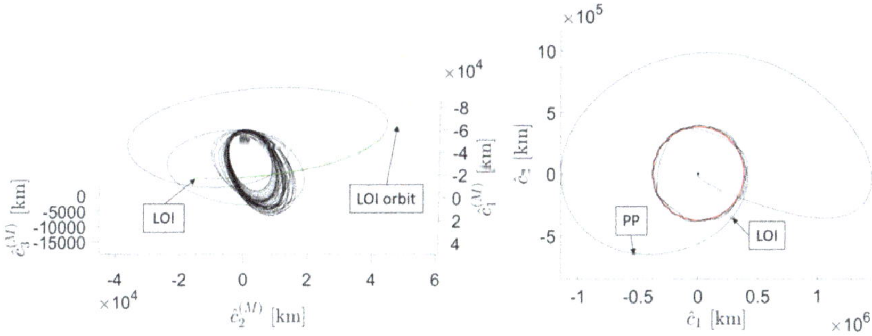

Figure 4: a) Trajectory in Phase 2, in MCI reference frame b) Full transfer in ECI reference frame

Concluding Remarks

This paper proposes a preliminary mission analysis for a low-energy low-thrust Earth-Moon transfer, starting from a GTO orbit and aimed at reaching a lunar highly elliptical orbit. In Phase 1 of the mission, regarding the Earth-Moon transfer, invariant manifold dynamics is used to obtain a low-energy planar reference trajectory, thus reducing the trajectory design to the research of a point in the phase space. The PSO algorithm allows further refininement of the trajectory in the framework of the BR4BP. Then, the same algorithm is employed to find the subsequent minimum-fuel low-thrust arc. In Phase 2 the final highly elliptic lunar orbit is reached using variable thrust nonlinear orbit control, with perturbations compensation and no need of a reference trajectory.

Assuming that the departure from GTO is demanded to the launch vehicle, as a part of its operations, the overall propellant consumption for the spacecraft equals 8% of its initial mass. In the end, the combination of the techniques described in this study allows defining an Earth-Moon mission profile with modest propellant consumption. In principle, the methodology at hand is also appliable to a variety of departure and target orbits in the Earth-Moon system.

References

[1] Conley, C., On the Ultimate Behavior of Orbits with Respect to an Unstable Critical Point I. Oscillating, Asymptotic, and Capture Orbits, Journal of Differential Equations, No. 5, 1969, pp 136–158. https://doi.org/10.1016/0022-0396(69)90108-9

[2] Belbruno, E. A., Miller, J. K., Sun-Perturbed Earth-to-Moon Transfers with Ballistic Capture, Journal of Guidance, Control and Dynamics, Vol. 16, No.4, 1993, pp 770-775. https://doi.org/10.2514/3.21079

[3] Koon, W. S., Lo, M. W., Marsden, J. E., Ross, S. D., Low Energy Transfer to the Moon, Celestial Mechanics and Dynamical Astronomy, Vol. 81, 2001, pp 63-73. https://doi.org/10.1007/978-94-017-1327-6_8

[4] Mingotti, G., Topputo, F., Bernelli-Zazzera, F., Efficient Invariant-Manifold, Low-Thrust Planar Trajectories to the Moon, Communications in Nonlinear Science and Numerical Simulation, Vol. 17, Issue 2, 2012, pp 817-831. https://doi.org/10.1016/j.cnsns.2011.06.033

[5] Gurfil, P., Nonlinear feedback control of low-thrust orbital transfer in a central gravitational field, Acta Astronautica, Vol. 60, 2007, pp 631-648. https://doi.org/10.1016/j.actaastro.2006.10.001

[6] Pontani, M., Pustorino, M., Nonlinear Earth orbit control using low-thrust propulsion, Acta Astronautica, Vol. 179, 2021, pp 296–310. https://doi.org/10.1016/j.actaastro.2020.10.037

The application of modal effective mass for PCB friction lock compliance against spacecraft launch random vibration spectrum

Mark Wylie[1*], Leonardo Barilaro[2]

[1]South East Technological University (SETU), Aerospace, Mechanical and Electronic Engineering Department, Carlow campus, Kilkenny Rd, Moanacurragh, Co. Carlow, Ireland

[2]The Malta College of Arts, Science & Technology, Aviation Department, Triq Kordin, Paola PLA 9032, Malta

* mark.wylie@setu.ie

Keywords: Modal Finite Element Analysis, Electronics, Spacecraft Structures, Friction-Locking Devices, Random Vibration Analysis

Abstract. Modern spacecraft design requires high density, low mass, modular electronic system architectures. This format often utilises a common backplane with Printed Circuit Boards (PCBs) interconnects. Adaptable electronic systems, such as modular Data Acquisition (DAq) systems, allow for configuration via insertion and removal of modules to meet the mission requirements. Common methods to mechanically fix the PCB to the chassis are by using stand-offs, with the primary function to minimise displacement through structural rigidity and to provide strain relief to the electronic connectors. Other methods, such as PCB friction lock allow for strain relief, improved thermal grounding of the PCB to the chassis but also allows for easy insertion and removal of the PCBs. One disadvantage of this system is that the retention force of the PCB is carried by a friction lock device and under acceleration loads, typically experienced in the launch environment, may cause failure. This paper presents a method to establish compliance of PCB friction lock devices using modal Finite Element Analysis (FEA) to predict the resonant frequencies and their Mass Participation Factor (MPF). Using this data, it is proposed that the use of an adaptation of the Miles Equation along with an equivalent g-RMS estimation can be used to determine the Random Vibration Load Factors (RVLF). A comparison of the RVLF with the retention force of the friction lock device can then give insight to the friction joint compliance.

Introduction

Quasi-Static Load (QSL), Shock Response Spectrum (SRS), sine and random vibration Acceleration Spectral Density (ASD) are typical acceleration loads required for qualification of space bound hardware. They represent loads during maneuvers (e.g., roll/tilt and orbital), pyrotechnic events (separation and fairing jettison) and motor induced vibrations. Payload flight equipment is designed against a defined set of these acceleration loads but are often equated to Load Factors (LF) which are equivalent accelerations (expressed in g's) and applied through the Centre of Gravity (CoG) of the structure [1]. This paper considers Random Vibration Load Factor (RVLF) and its equivalent g (force) for verification of PCBs friction lock mechanisms (see Fig 1). This efficient design allows for high density electronic architectures and modularization via insertion/extraction into a chassis by accessing front side only. This verification technique is an effort saving technique for the purposes of Proto Flight development.

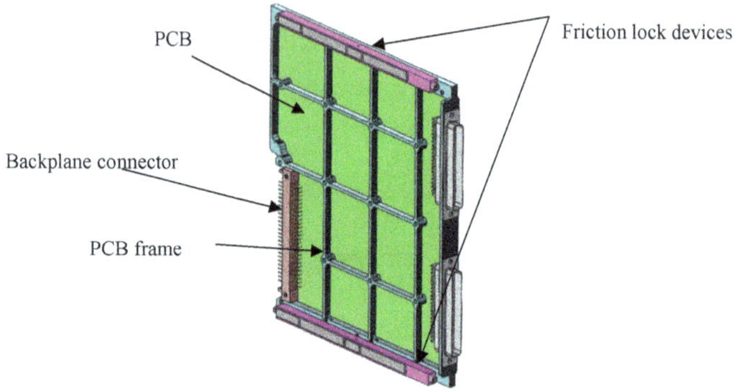

Fig 1: PCB on AA6082-T6 PCB frame with QTY 2 friction locking devices (NVENT).

In theory, structures may contain millions of Degrees of Freedom (D.O.F), each having a resonant frequency. Of these modes, high resonant frequencies (>2000 Hz or > 10,000 Hz) may be deemed to have low structural impact, one; because it is beyond the typical random vibration spectrum upper frequency limit (2000 Hz) and SRS (10,000 Hz) and two; because displacement is inversely proportional to modal frequency and often do not constitute a ductile failure mode (not withstanding that these modes should still be considered for load spectra for the purposes of fatigue compliance and failure modes for brittle ceramic components).The amount of mass moving in any direction is a function of the mode participation factor and the effective mass at that mode. This is sometimes called Mass Participation Factor (MPF) and it is common, using modal FEA, to evaluate MPF such that the summation accounts for > 90 % of the total structural mass in all orthogonal directions [2]. Significant modes of interest are extracted from this data (typically with MPF > 5 %). The resonant frequency of the structure in each orthogonal axis can also be identified as the first significant mode (i.e., lowest mode frequency typically with > 5 % MPF) in each orthogonal direction. On consideration for structural analysis of the PCB friction lock compliance, the following data is of importance; the mode frequency (Hz), the MPF and the orthogonal direction in which it acts. Mode frequency because it is inversely proportional to displacement and using Steinberg studies can determine electronic component survivability [3], MPF because of the inertia involved with this mode frequency and finally direction (mode shape), because in some cases, the direction in which it acts may coincide with friction locking mechanism or its deformation may cause collision with adjacent structures (i.e., adjacent PCBs components). This analysis considers the modes that are parallel to direction of the friction lock and the MPF at these modes. From this, the force response In-Plane (IP) with the PCB friction lock can be compared to the friction force holding the PCB in-situ. In summary, this method aims to determine if the PCB friction lock will be compromised by random vibration ASD load. Displacement and translation can cause a critical failure, especially for electronic systems that rely on a common backplane for interconnects. In the next section a proposed pass/fail criteria to determine if the friction lock joint will survive the random vibration acceleration load case is presented.

Method

The random vibration spectrum is non-deterministic but when analysed statistically over a period of time the g-RMS is constant and the likelihood of peaks (g) outside of the RMS are given by a Gaussian distribution (3σ). The random vibration spectrum is a base excitation to the hardware and typically ranges from 20-2000 Hz but allows for numerous frequencies to be excited at the same time. A conservative assumption for the analysis of PCBs within an electronic chassis is to assume no attenuation of the base excitation to the PCBs. However, it has been shown for shock inputs that structural discontinuities can significantly attenuate the shock pulse, this is known as the 3-joint rule [4]. For frequencies above 2000 Hz or for systems where the resonant frequencies are unknown, the RVLF can be approximated by multiplying the overall g-RMS by 3 (i.e. 3σ). Standard methods used to formulate the RVLF are based on the Miles Equation [1][5] where the base input amplitude is taken at the resonant frequency and the amplitude (g^2/Hz) is taken as a maximum value from the ASD plateau and considered constant across the entire frequency domain. This is over conservative unless the ASD is flat or within one octave either side of the resonant frequency [7] because typically the ASD plot has a ramp up (+3 dB) to the knee point and decline (-5 dB) after the plateau knee point (e.g. MIL-STL-1540C [6]) where levels are lower in these regions. A more accurate method would be to predict the resonant frequencies and MPF using modal FEA and, using Q (often estimated as Q = 20 or approximated using $Q = \sqrt{Fn_{PCB}}$) [3] calculate the g-RMS ($\ddot{X}_{g-RMS}$) using the following equation Eq 1.

$$\ddot{X}_{g-RMS} = 3\sigma.\sqrt{\sum_{i=1}^{N}\frac{\left[1+(2\xi(\frac{F_i}{Fn})^2\right]}{\left[1-(\frac{F_i}{Fn})^2\right]^2+\left[2\xi(\frac{F_i}{Fn})\right]^2}} \quad (1)$$

The derivation of this method, taken by others [5][7] and adapted for PCB friction lock compliance, is based on the system's response to a typical random vibration ASD. This method allows for the inclusion of the ASD amplitude to vary across the frequency range of interest, i.e. representing the ramp and declines typically found in random vibration ASDs. Once the g-RMS been calculated the total RVLF can be calculated using Eq 2:

$$RVLF = m.\left(MPF.\ddot{X}_{g-RMS,mode\ i} + ((1-MPF).\ddot{X}_{g-RMS,ASD}\right) \quad (2)$$

Where; m = mass of PCB and PCB frame, MPF = MPF at resonant frequency of the system in direction of the friction lock, $X_{g\text{-RMS, mode i}}$ = g-RMS at resonant frequency using Q = 20, ξ=1/2Q, 3σ and $X_{\text{RMS, spectrum}}$ = g-RMS of overall spectrum multiplied by 3σ. The RVLF is an estimation of the forces acting on the PCB friction lock device in the direction of slippage due to the random vibration load case. The retention force (F_{max}) from a friction lock device (typical values NVENT CardLok systems F_{lock} = 400-3000 N) opposing this is Fmax and is calculated using Eq 3:

$$F_{max} = \mu.F_{lock}.number\ of\ friction\ lock\ devices \quad (3)$$

Friction coefficient (μ) is taken as 0.3 for static aluminum-aluminum interfaces [8]. Failure of the locking device is established when the RVLF exceeds the Fmax. A sample calculation is provided for the PCB and PCB frame in Fig. 1 against the random vibration ASD in Tab. 1. with g-RMS of 14.7 g. A modal FEA study of the assembly predicts an IP resonant frequency of 1318 Hz with an MPF of 0.2 %. This was the only significant mode within an order of magnitude below 2000 Hz.

Tab 1: Random vibration ASD used in study.

Frequency	**g^2/Hz**
20	0.08
100	0.4
300	0.4
2000	0.017

Results

This paper presents a methodology for the analysis of PCB and PCB frame assemblies that are fixed using friction lock devices against random vibration acceleration load case that are experienced during spacecraft launch. A sample calculation based on the assembly in Fig.1 is presented:

$$RVLF = 0.415 \times [(0.002 \times 120.8) + (0.998 \times 44.1)] = 18.35\,N \quad (4)$$

Where:
$M_{Frame+PCB} = 0.415$ kg
$MPF_1 = 0.002$ @ 1317.8 Hz
$X_{RMS,\ mode\ i} = 120.8$ g (3 σ and Q = 20), see Fig. 2 and Eq. 1, implemented using MS Excel
$X_{RMS,\ spectrum} = =14.7$ g-rms *3 σ = 44.1 g

As per Eq. 3, assuming the use of NVENT Schroff Series 48-5 (1418 N retention force each)

$$F_{max} = 0.3 \times 1481 \times 2 = 888.6\,N \quad (5)$$

Therefore: $M.o.S = \frac{888.6}{18.35} - 1 = 47.4$ (6)

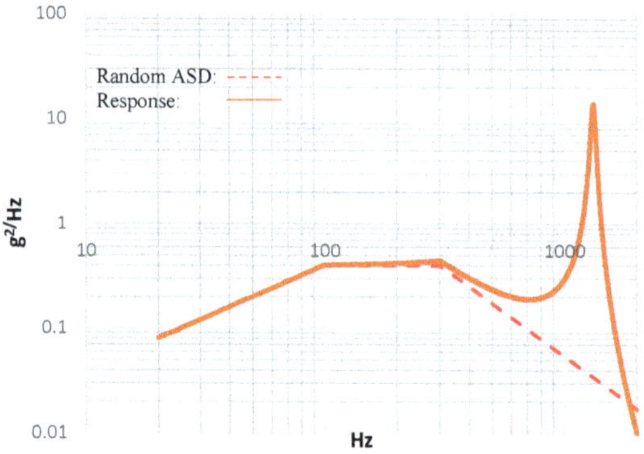

Fig 2: Random vibration ASD response based on Eq. 1 and resonant frequency of PCB assembly.

Conclusion

This research is based on flight hardware destined to be launched on the Ariane 6-2 in 2026 and is currently under development. The hardware has passed system qualification testing for acceleration load cases. The sample calculation shows very low force response acting on the PCB assembly IP (18.35 g) which is primarily due to the low MPF (0.2 %) within the ASD limit range (2000 Hz). It is also due to the vehicle IP random vibration requirements of 14.7 g-rms. The Out-of-Plane (OOP) is greater at 22.7 g-rms. Nevertheless, given the retention force of the NVENT product and excluding any Factors of Safety (FoS), Local Design Factors (LDF), Qualification Loads (QL), Design Loads (DL) etc., the MoS for slip is high at 47.4 and presents a low risk for movement or failure. This study would benefit from a more detailed validation of the PCB assembly modes by locally instrumenting low-mass triaxial accelerometers on PCB locations and spectra analysis of multiple accelerometers across the PCB which can allow for validation of the MPF and mode shapes. It is a laborious technique and one such example of this is presented by Sandia National Laboratories [9]. Furthermore, an experimental apparatus capable of measuring forces on such friction lock devices could be used to establish the random vibration ASD thresholds for slip and failure.

References

[1] L. Trittoni & M. Martini, Force-limited Acceleration Spectra Derivation by Random Vibration Analysis, Alenia Spazio, www.vibrationdata.com, (2004).

[2] Xie J.,Sun D.,Xu C. and Wu J. The Influence of Finite Element Meshing Accuracy on a Welding Machine for Offshore Platform'S Modal Analysis. Polish Maritime Research, Vol.25 (I3), pp. 147-153. (2018). https://doi.org/10.2478/pomr-2018-0124

[3] T. Irvine, Extending Steinberg's Fatigue Analysis of Electronics Equipment Methodology to a Full Relative Displacement vs. Cycles Curve. Rev C, www.vibrationdata.com, (2013).

[4] V. Babuska, S P. Gomez, S A. Smith, C Hammetter and D Murphy. "Spacecraft Pyroshock Attenuation in Three Parts," AIAA 2017-0633. 58th AIAA Structures, Structural Dynamics, & Materials Conference, (2017). https://doi.org/10.2514/6.2017-0633

[5] J.W. Miles, "On Structural Fatigue under Random Loading", Acoustical Society of America Journal, vol. 29, no. 1, p. 176, doi:10.1121/1.1918447, (1957). https://doi.org/10.1121/1.1918447

[6] MIL-STD-1540C, Test Requirements for Launch, Upper-Stage and Space Vehicles, (1994).

[7] T. Irvine, An Introduction to the Random Vibration Spectrum, Section 17. http://www.vibrationdata.com/tutorials2/Tom_book_12_1_19.pdf, (2019).

[8] D. Fuller. Excerpt from "Theory and Practice of Lubrication for Engineers". Coefficients of Friction. https://web.mit.edu/8.13/8.13c/references-fall/aip/aip-handbook-section2d.pdf, (1970).

[9] R. Mayes et al, Efficient Method of Measuring Effective Mass of a System, Experimental Mechanics, NDE and Model Validation Department, Sandia National Laboratories, (2014).

Near-optimal feedback guidance for low-thrust earth orbit transfers

D. Atmaca[1,a] and M. Pontani[2,b]

[1]M.S. in Space and Astronautical Engineering, Sapienza University of Rome, via Eudossiana 18, 00184 Rome, Italy

[2]Department of Astronautical, Electrical, and Energy Engineering, Sapienza University of Rome, via Salaria 851, 00138 Rome, Italy

[a]mauro.pontani@uniroma1.it, [b]direncatmaca@gmail.com

Abstract. This research proposes a near-optimal feedback guidance based on nonlinear control for low-thrust Earth orbit transfers. For the numerical simulations, two flight conditions are defined: (i) nominal conditions and (ii) nonnominal conditions that account for the orbit injection errors and the stochastic failures of the propulsion system. Condition (ii) is studied through an extensive Monte Carlo Analysis, to demonstrate the nonlinear feedback guidance's numerical stability and convergence properties. To illustrate the performance under both conditions, an orbit transfer from low Earth orbit to geostationary orbit is considered. Near-optimality of the feedback guidance comes from carefully selecting the nonlinear control gains. Comparison of the transfer with an existing study that uses optimal control reveals that orbit transfers based on feedback orbit control are very close to the optimal solution.

Keywords: Earth Orbit Transfers, Low-Thrust Spacecraft, Feedback Guidance and Control

Introduction

Orbit control is a critical part of spacecraft control design and was extensively studied over the last century. Most studies focus on impulsive transfers. However, near-optimal and nonlinear strategies for low-thrust transfers are becoming popular since they allow compensation of orbital perturbations.

The study of nonlinear and near-optimal feedback guidance for low-thrust spacecraft is a reasonably new topic, with significant publications taking place over the last three decades. An important contribution is due to Gurfil [1], who utilizes nonlinear control with modified equinoctial orbit elements for low-thrust orbit transfers. The study guarantees asymptotic convergence from an initial elliptical orbit to any final elliptical orbit using Gauss's variational equations. Pontani and Pustorino [2] have recently applied nonlinear control strategies to orbit injection and maintenance problems where the control scheme takes advantage of Lyapunov stability combined with LaSalle's invariance principle. Gao [3] presents a linear feedback guidance approach that exhibits near-optimality for low-thrust Earth orbit transfers using orbital averaging. Kluever [4] proposed a simple closed-loop feedback-driven scheme for low-thrust orbit transfers that allows calculating sub-optimal trajectories. Petropoulos [5] developed a simple strategy based on candidate Lyapunov functions for low-thrust orbit transfers while coining the term proximity quotient or Q-Law. There are several other studies based on Q-Law [6, 7], and they focus on mitigating the sub-optimality of this strategy.

This research proposes a near-optimal feedback guidance based on nonlinear control for low-thrust Earth orbit transfers. Both eclipse condition and orbit perturbations (i.e., several Earth gravitational harmonics, solar radiation pressure, aerodynamic drag, and gravitational attraction due to Sun and Moon) are modeled. Two flight conditions are defined: (i) nominal conditions and (ii) nonnominal conditions that account for orbit injection errors and stochastic failures of the propulsion system. An orbit transfer from low Earth orbit to geostationary orbit is considered. The initial and final orbit elements are taken from an existing study on optimal orbit control [8] for the purpose of comparing and demonstrating the near-optimality of the nonlinear feedback guidance.

Nonlinear orbit control using modified equinoctial elements

Orbit elements lead to singularities in the Gauss planetary equations for circular and equatorial orbits. To avoid similar issues, this study utilizes Modified Equinoctial Orbit Elements (MEE), defined as

$$p = a\left(1-e^2\right) \quad l = e\cos(\Omega+\omega) \quad m = e\sin(\Omega+\omega) \quad n = \tan\frac{i}{2}\cos\Omega \quad s = \tan\frac{i}{2}\sin\Omega \quad q = \Omega+\omega+\theta_* \tag{1}$$

where $a, e, \Omega, \omega, i, \theta_*$ are semimajor axis eccentricity, right ascension of the ascending node (RAAN), argument of periapsis, inclination, and true anomaly, respectively. Five of them are collected in z, defined as $z = [x_1 \quad x_2 \quad x_3 \quad x_4 \quad x_5]^T = [p \quad l \quad m \quad n \quad s]^T$, and subject to the governing equation

$$\dot{z} = \mathbf{G}(z, x_6)\boldsymbol{a} \tag{2}$$

The last element is $x_6 = q$. The $\boldsymbol{a}$ term in (2) includes the projections of both perturbing and thrust acceleration onto the LVLH-frame. The explicit expression of $\mathbf{G}$ and the governing equation for x_6 are reported in [2]. Two (constant) parameters identify the characteristics of the low-thrust propulsion system: $u_T^{(max)} = T_{max}/m_0$ and c, where T_{max}, m_0, and c denote respectively maximum thrust magnitude, initial mass, and effective exhaust velocity. As a result, letting $x_7 = m/m_0$ (where m is the instantaneous mass), one obtains $\dot{x}_7 = -u_T/c$, with $u_T = T/m_0$. Thus, $\boldsymbol{a}_T = \boldsymbol{u}_T/x_7$ defines the instantaneous thrust acceleration. The term $\boldsymbol{a}$ includes two contributions, $\boldsymbol{a} = \boldsymbol{a}_T + \boldsymbol{a}_P$, where the $\boldsymbol{a}_P$ term refers to the perturbing acceleration. For this study, four types of orbital perturbations are considered: (a) the Earth gravitational harmonics (with $|J_{l,m}| > 10^{-6}$), (b) solar radiation pressure, (c) third-body attraction due to the Sun and the Moon, and (d) aerodynamic drag. The drag is modeled by assuming a reference surface area of $23.569\ \text{m}^2$ and ballistic coefficient equal to $0.0576\ \text{m}^2/\text{kg}$. In addition, the solar radiation pressure is modeled using a fully reflective surface area where the reflective coefficient is equal to 2. In the end, $\boldsymbol{x} = \left[z^T \quad x_6 \quad x_7\right]^T = [x_1 \quad x_2 \quad x_3 \quad x_4 \quad x_5 \quad x_6 \quad x_7]^T$ identifies the complete state vector in compact form, whereas $\boldsymbol{u}_T$ is the control vector.

Nonlinear orbit control allows identifying a feedback law that can drive the spacecraft toward the desired orbit while ensuring global asymptotic stability. For the problem at hand, the target set, associated with the final orbit, is $\boldsymbol{\psi} = \left[x_1 - p_d \quad x_2^2 + x_3^2 - e_d^2 \quad x_4^2 + x_5^2 - \tan^2(i_d/2)\right]^T$, where subscript d denotes the desired value of the respective variable. The feedback law

$$\boldsymbol{u}_T = -u_T^{(max)} \frac{x_7(\boldsymbol{b}+\boldsymbol{a}_P)}{\max\left\{u_T^{(max)}, \left|x_7(\boldsymbol{b}+\boldsymbol{a}_P)\right|\right\}}, \text{ with } \boldsymbol{b} = \mathbf{G}^T\left(\frac{\partial \boldsymbol{\psi}}{\partial z}\right)^T \mathbf{K}\boldsymbol{\psi} \text{ and } \mathbf{K} = \text{diag}\{k_1, k_2, k_3\} \tag{3}$$

is proven to enjoy quasi global stability [2], using the Lyapunov direct method, in conjunction with the LaSalle's principle. However, the choice of the three gains (k_1, k_2, k_3) plays a crucial role for the purpose of speeding up convergence to the target set. This study proposes and applies a gain selection method composed of two sequential steps:

Step 1. Exhaustive table search that includes different gain combinations; each gain is changed with increment by $10^{0.1}$, in the interval $1 \le k_i \le 10^6$.

Step 2. Using the values found at step 1, the native "fminsearch" MATLAB routine, which employs a Nelder-Mead simplex algorithm, is used.

The preceding two steps are completed for different initial orbits, associated with identical values of semimajor axis, eccentricity, RAAN, and argument of perigee, and different initial inclinations. The propulsion parameters for the gain optimization process are assumed to be $c = 30$ km/s and $u_T^{(\max)} = 10^{-4} g_0$ with $g_0 = 9.8065$ m/s^2.

Numerical results

The near-optimal feedback guidance proposed in this study is tested under nominal and nonnominal conditions. For both cases, initial and final orbit elements and the propulsion parameters are taken from an existing study focusing on optimal orbit control [8]. The final orbit is geostationary, whereas the initial orbit is circular, with $a_0 = 6927$ km and $i_0 = 28.5°$. The propulsion parameters are $c = 32.361$ km/s and $u_T^{(\max)} = 3.348 \cdot 10^{-4}$ m/s^2, and they characterize a low-thrust propulsion system. The gain values are selected from the preceding systematic study and are $k_1 = 0.9722$, $k_2 = 1056$, and $k_3 = 967$. For the transfer time and final mass ratios reported in this section, the following criteria are used to indicate the end of the transfer:

$$|p - p_d| \leq 10 \text{ km} \qquad e \leq 0.005 \qquad i \leq 0.5° \tag{4}$$

Nominal Conditions

This subsection reports the numerical results under nominal conditions and compares the proposed nonlinear feedback guidance with the existing optimal solution. Figure 1 shows the near-optimal transfer path, with eclipse arcs (where propulsion is unavailable) highlighted in blue.

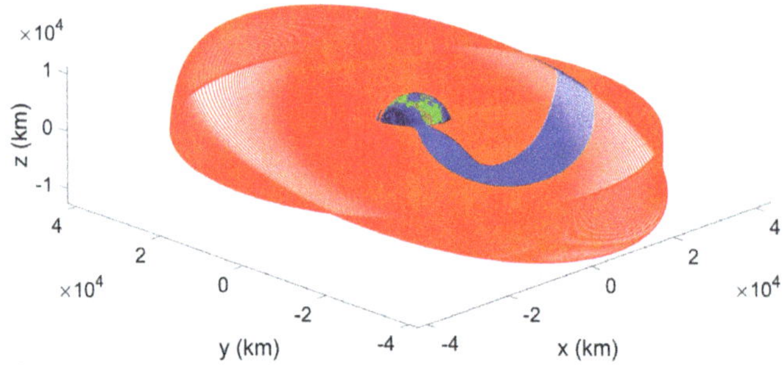

Figure 1: Cartesian motion of the spacecraft in the ECI frame (blue lines indicate eclipse)

At the beginning of the transfer, the perturbing acceleration is higher than the thrust acceleration. However, the perturbing term quickly decays to low values as the osculating radius increases. Using the criteria defined by (4), the transfer time is $t_f = 228.2$ days with a final mass ratio, $x_{7f} = 0.8245$. This result is compared to the optimal solution found in [8], which considers shadowing (but neglects orbit perturbations). The optimal path is completed in 215.9 days, with final mass ratio $x_{7f} = 0.8394$. Therefore, with the proposed nonlinear feedback approach, the transfer time is only 5.71% higher, and the final mass ratio is 1.78% lower than the optimal solution. Hence, this demonstrates that nonlinear feedback control can generate a transfer path very close to the optimal, minimum-time solution.

Monte Carlo Analysis

This subsection concentrates on nonnominal flight conditions, which account for the orbit injection errors and the stochastic failures of the propulsion system. The propulsion parameters, initial and final orbits, and gain values are the same as in the nominal case. Orbit injection errors are modeled by randomizing the initial orbit elements, using a uniform distribution. More specifically, the perigee and apogee radii (r_p and r_a) and inclination i have uniform distribution in the following ranges: $r_p = [350, 549]\ \text{km} + R_E$ and $r_a = [549, 750]\ \text{km} + R_E$ (where R_E is the Earth radius), and $i = [22.5, 34.5]\ \text{deg}$. Moreover, RAAN, argument of perigee, and true anomaly have uniform distribution in the entire range of definition. The stochastic failure of the propulsion system is modeled by specifying the starting point of failure and its duration. These two stoachastic variables, denoted respectively with t_{fail} and t_{dur} have uniform distribution as well, i.e. $t_{fail} = [1, 100]$ days and $t_{dur} = [5, 20]$ days.

In spite of initial errors at orbit injection and stochastic propulsion failures, feedback control successfully drives the spacecraft to the desired orbit. Table 1 reports the Monte Carlo Analysis's statistical results, based on 1000 simulations, and compares the proposed feedback guidance and existing optimal solutions. These results testify to the excellent stability properties of feedback control as well as to the effectiveness of the gain selection method.

Table 1: Statistical results of the Monte Carlo Analysis and comparison with the optimal solution

	Mean	Std. Dev.	Optimal Solution
t_f (days)	236.41	10.51	215.94
x_7	0.8234	0.0063	0.8394

Concluding Remarks

This paper proposes and applies a near-optimal feedback guidance strategy based on nonlinear orbit control to low-thrust Earth orbit transfers. Feedback guidance utilizes some fundamental principles of Lyapunov stability theory and LaSalle's invariance principle. A novel gain selection strategy that involves an exhaustive table search and a numerical optimization algorithm provides near-optimality of the optimal paths traveled through feedback guidance. Two different flight conditions are considered: *(i) nominal conditions and (ii) nonnominal conditions that account for the orbit injection errors and the stochastic failures of the propulsion system. The numerical results* testify to the excellent stability properties of feedback control, as well as to the effectiveness of the gain selection method, even in nonnominal flight conditions.

References

[1] P. Gurfil, "Nonlinear Feedback Control of Low-Thrust Orbital Transfer in a Central Gravitational Field," *Acta Astronautica,* vol. 60, no. 8 & 9, pp. 631-648, 2007. https://doi.org/10.1016/j.actaastro.2006.10.001

[2] M. Pontani and M. Pustorino, "Nonlinear Earth Orbit Control Using Low-Thrust Propulsion," *Acta Astronautica,* vol. 179, pp. 296-310, 2021. https://doi.org/10.1016/j.actaastro.2020.10.037

[3] Y. Gao, "Linear Feedback Guidance for Low-Thrust Many-Revolution Earth-Orbit Transfers," *Journal of Spacecraft and Rockets,* vol. 46, no. 6, pp. 1320-1325, 2009. https://doi.org/10.2514/1.43395

[4] C. A. Kluever, "Simple Guidance Scheme for Low-Thrust Orbit Transfers," *Journal of Guidance, Control, and Dynamics,* vol. 21, no. 6, pp. 1015-1017, 1998. https://doi.org/10.2514/2.4344

[5] A. E. Petropoulos, "Low-Thrust Orbit Transfers Using Candidate Lyapunov Functions with a Mechanism for Coasting," in *AIAA/AAS Astrodynamics Specialist Conference and Exhibit*, Rhode Island, 2004. https://doi.org/10.2514/6.2004-5089

[6] B. B. Jagannatha, J.-B. H. Bouvier and K. Ho, "Preliminary Design of Low-Energy, Low-Thrust Transfers to Halo Orbits Using Feedback Control," *Journal of Guidance, Control, and Dynamics,* vol. 42, no. 2, pp. 1-12, 2018. https://doi.org/10.2514/1.G003759

[7] H. Holt, R. Armellin, A. Scorsoglio and R. Furfaro, "Low-thrust Trajectory Design using Closed-loop Feedback-driven Control Laws and State-dependent Parameters," in *AIAA Scitech 2020 Forum*, Orlando, 2020. https://doi.org/10.2514/6.2020-1694

[8] M. Pontani and F. Corallo, "Optimal Low-Thrust Orbit Transfers with Shadowing Effect Using a Multiple-Arc Formulation," in *72nd International Astronautical Congress*, Dubai, 2021. https://doi.org/10.1016/j.actaastro.2022.06.034

Reduced-attitude stabilization for spacecraft boresight pointing using magnetorquers

Fabio Celani

School of Aerospace Engineering, Sapienza University of Rome, Via Salaria 851 00138 Roma Italy

fabio.celani@uniroma1.it mail

Keywords: Boresight Pointing, Magnetorquers, Reduced Attitude, Stability

Abstract. This paper presents a method for achieving a desired boresight pointing of an instrument on a spacecraft using only magnetorquers as torque actuators. The desired pointing direction is inertially fixed. The proposed method is of proportional-derivate type and stabilizes the boresight pointing. Numerical simulations illustrate the effectiveness of the method and show that convergence to the desired pointing direction occurs faster than employing a three-axis stabilization approach.

Introduction

Pointing of the boresight of an instrument is a common task of many spacecraft during their operational life. For example, it might be required to point a telescope at a star, to point a transmitting or receiving antenna at a ground station, or to point the Sun with solar panels. A common approach for boresight control is performing three-axis attitude control which requires knowledge of the full attitude of the spacecraft. However, in boresight pointing applications, only the pointing direction is relevant and the rotation about the boresight is not considered. In this sense, full attitude knowledge is not required for boresight control, and it suffices to measure the reduced-attitude vector defined on the two-dimensional sphere. Motivated by this practical consideration this research presents a method for reduced-attitude stabilization for a spacecraft that uses only magnetorquers as torque actuators.

Magnetorquers are planar current-driven coils rigidly placed on the spacecraft typically along three orthogonal axes. The interaction between the magnetic dipole moment generated by those coils and the Earth magnetic field creates a torque that attempts to align the magnetic dipole moment in the direction of the field. Magnetorquers present the following benefits: (i) they are simple, reliable, and low cost; (ii) they need only renewable electrical power to be operated; (iii) they save weight with respect to any other class of torque actuators. On the other hand, magnetorquers have the following important limitations: (i) the control torque generated by magnetorquers is constrained to belong to the plane orthogonal to the Earth magnetic field; (ii) the maximum torque they can generate is substantially smaller than for other types of torque actuators. Due to these limitations, using only magnetorquers for attitude stabilization leads to smaller pointing accuracy and slower convergence compared to other torque actuators. Thus, it is considered a feasible option especially for low-cost micro and nano satellites, and for satellites with a failure in the main torque actuators.

This work proposes a boresight stabilization law for an inertially pointing spacecraft equipped only with magnetorquers as torque actuators. Numerical simulations for a case study show the effectiveness of the proposed law.

Spacecraft Boresight Pointing

Consider the problem of aligning a body-fixed boresight axis of an instrument on a spacecraft with an inertially-fixed reference axis. In addition, once the alignment condition is achieved the spacecraft must not rotate about the reference axis to avoid blurred measurements from the instrument. The body-fixed boresight axis is expressed by the constant column matrix of its body coordinates $\boldsymbol{a} \in \mathbb{S}^2 = \{\boldsymbol{v} \in \mathbb{R}^3 : \boldsymbol{v}^T\boldsymbol{v} = \mathbf{1}\}$. The inertially-fixed reference axis is expressed by the constant column matrix $\boldsymbol{a}_r^i \in \mathbb{S}^2$ of its coordinates along the standard Earth-Centered Inertial (ECI) frame [1]. Let $\boldsymbol{R}$ be the rotation matrix that transforms a column matrix of body coordinates into the corresponding column matrix of ECI coordinates. Thus, the reference axis expressed in body-coordinates is given by $\boldsymbol{a}_r^b = \boldsymbol{R}^T\boldsymbol{a}_r^i$. Column matrix $\boldsymbol{a}_r^b$ describes a reduced attitude and obeys the following kinematics [2]

$$\dot{\boldsymbol{a}}_r^b = \boldsymbol{a}_r^b \times \boldsymbol{\omega} \tag{1}$$

where $\boldsymbol{\omega}$ is the column matrix of body coordinates of the angular velocity of the body frame with respect to the ECI frame. The spacecraft is modeled as a rigid body and its attitude dynamics are given by [2]

$$\boldsymbol{J}\dot{\boldsymbol{\omega}} + \boldsymbol{\omega} \times (\boldsymbol{J}\boldsymbol{\omega}) = \boldsymbol{T}_c + \boldsymbol{T}_d \tag{2}$$

where $\boldsymbol{J}$ is the spacecraft inertia matrix, $\boldsymbol{T}_c$ is the column matrix of body coordinates of the control torque, and $\boldsymbol{T}_d$ is column matrix of body coordinates of the sum of all disturbance torques acting on the spacecraft. The control torque is generated by three magnetorquers aligned with the body axes. Thus, it can be expressed as follows

$$\boldsymbol{T}_c = \boldsymbol{m}_c \times \boldsymbol{b}^b \tag{3}$$

where $\boldsymbol{m}_c$ is the column matrix of the control magnetic dipole moments generated by the three magnetorquers, and $\boldsymbol{b}^b$ is the column matrix of body coordinates of the geomagnetic induction.

The proposed boresight stabilization law is given by

$$\boldsymbol{m}_c = \boldsymbol{b}^b \times [k_p(\boldsymbol{a} \times \boldsymbol{a}_r^b) - k_d\,\boldsymbol{\omega}] \tag{4}$$

where k_p and k_d are positive scalar gains. The stabilization law is obtained by modifying the proportional-derivative like law for fully actuated spacecraft presented in [2]. Specifically, the cross product with $\boldsymbol{b}^b$ is introduced to enforce that $\boldsymbol{m}_c$ is perpendicular to $\boldsymbol{b}^b$. The latter property allows to save energy since Eq. 3 shows that a term in $\boldsymbol{m}_c$ parallel to $\boldsymbol{b}_b$ does not give any contribution to the control torque $\boldsymbol{T}_c$.

Stability Analysis

In this section a stability result of the proposed boresight stabilization law is presented. The stability analysis is carried out by adopting the inclined dipole model for the geomagnetic field [1]. Consider a circular orbit for the spacecraft and let $\boldsymbol{b}^i(t)$ denote the column matrix of ECI coordinates of the geomagnetic induction along the orbit. The adoption of the inclined dipole model leads to an almost periodic time behavior for $\boldsymbol{b}^i(t)$. Consider matrix

$$\boldsymbol{\Gamma}^i(t) = \boldsymbol{b}^i(t)^T\boldsymbol{b}^i(t)\boldsymbol{I}_{3\times3} - \boldsymbol{b}^i(t)\boldsymbol{b}^i(t)^T \tag{5}$$

where $\boldsymbol{I}_{3\times3}$ is the identity matrix. Note that $\boldsymbol{\Gamma}^i(\boldsymbol{t})$ is also almost periodic, and consider the following average value

$$\mathbf{\Gamma}_{av}^{i} = \lim_{T\to\infty} \frac{\mathbf{1}}{T} \int_{\mathbf{0}}^{T} \mathbf{\Gamma}^{i}(\tau)\, d\tau \qquad (6)$$

It can be shown that the following stability result holds true.

Proposition. If $\mathbf{det}(\mathbf{\Gamma}_{av}^{i}) \neq \mathbf{0}$ then it is possible to determine positive values for k_p and k_d so that the equilibrium $(\boldsymbol{a}_r^b, \boldsymbol{\omega}) = (\boldsymbol{a}, \mathbf{0})$ of the closed-loop in Eqs. 1-4 with $\boldsymbol{T}_d = \mathbf{0}$ is asymptotically stable.

It can be obtained that condition $\mathbf{det}(\mathbf{\Gamma}_{av}^{i}) \neq \mathbf{0}$ is fulfilled if the orbit inclination is not too low.

Numerical Simulations

The goal of this section is to validate by numerical simulations the boresight stabilization law in Eq. (4). Consider a spacecraft with inertia matrix equal to $\boldsymbol{J} = \mathrm{diag}[27\ 25\ 17]\ \mathrm{kg\ m}^2$ which follows a circular orbit with 450 km altitude, 87 deg inclination, and zero RAAN. The orbital period is about 5600 sec. The maximum value for each element of $\boldsymbol{m}_c$ is set to 10 A m^2. The body-fixed boresight axis and the inertially-fixed reference axes are set equal to $\boldsymbol{a} = \boldsymbol{a}_r^i = [0\ 0\ 1]^T$. Disturbance torques included in the simulations are gravity-gradient torque, residual magnetization torque, and aerodynamic drag torque modeled as in [3]. The gains of the stabilization law are set to $k_p = 4.9\ 10^4$ $k_d = 1.2\ 10^9$. Since $\boldsymbol{a} = \boldsymbol{a}_r^i$ the required boresight pointing can also be achieved through a three-axis attitude stabilization action that aligns the body frame with the ECI frame. Thus, the performances of the proposed boresight stabilization law are compared with those of the following proportional-derivative-like three-axis attitude stabilization law [2]

$$\boldsymbol{m}_c = \boldsymbol{b}^b \times \left[-k_p \sum_{i=1}^{3} (\boldsymbol{e}_i \times \boldsymbol{R}\boldsymbol{e}_i) - k_d\, \boldsymbol{\omega}\right] \qquad (7)$$

in which $\boldsymbol{e}_i$ is the i-th column of the identity matrix $\boldsymbol{I}_{3\times3}$. The numerical values of k_p and k_d are the same as for the boresight stabilization law.

A Monte Carlo campaign of 40 simulations is run by considering random initial attitude and random initial angular velocity with maximum amplitude of 3 deg/sec. The time behaviors of the pointing error and of the control magnetic dipole moment are reported in Fig. 1. In each simulation the pointing accuracy is evaluated through the magnitude $\Theta_{max\ ss}$ which denotes the maximum pointing error in steady-state. Table 1 reports the mean values $\overline{\Theta_{max\ ss}}$ and the standard deviations $\sigma(\Theta_{max\ ss})$ obtained with the two types of stabilization.

Table 1. Statistics of the Monte Carlo campaign

type of stabilization	$\overline{\Theta_{max\ ss}}$ [deg]	$\sigma(\Theta_{max\ ss})$ [deg]
boresight	17.68	3.22
three-axis	18.25	0.60

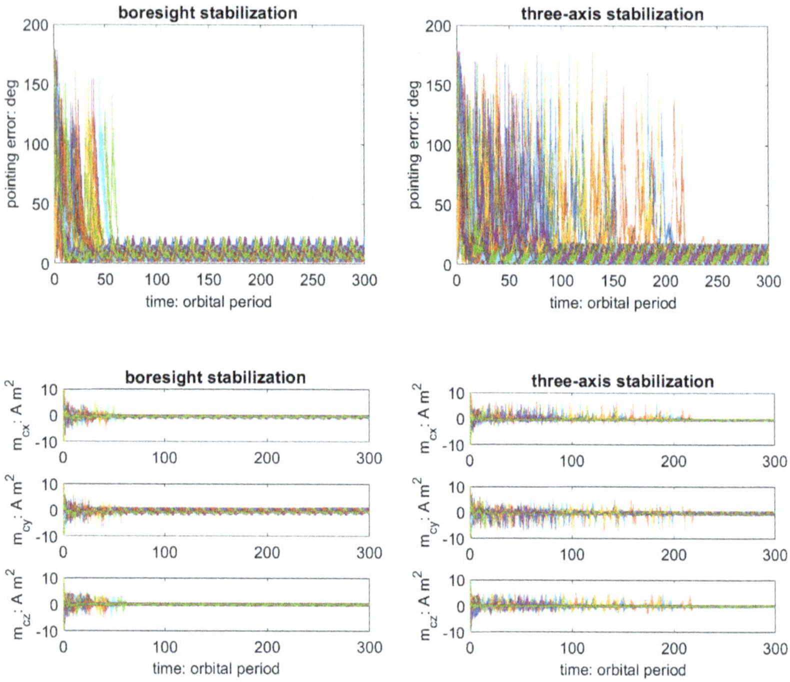

Figure 1. Time histories of the pointing error and of the control magnetic dipole moment.

The statistics show that the pointing accuracies of the two types of stabilization are similar. However, through visual inspection of the time behaviors of the pointing error obtain that boresight stabilization achieves a faster convergence to the steady-state behavior. Specifically, by employing boresight stabilization converge is obtained after approximately 60 orbital periods in the worst simulation run whereas with three-axis stabilization it is obtained only after about 250 orbital periods in the worst case. Faster convergence is probably due to the fact that the objective of boresight stabilization is achieving the desired alignment only for the boresight axis rather than for of all three body axis.

References

[1] J.R. Wertz (Ed.), Spacecraft Attitude Determination and Control, Kluwer Academic, Dordrecht, 1978.

[2] N.A. Chaturvedi, A.K. Sanyal, N.H. McClamroch, Rigid-body attitude control, IEEE Control Systems Magazine 31 (2011), 30-51. https://doi.org/10.1109/MCS.2011.940459

[3] F. Celani, Spacecraft attitude stabilization using magnetorquers with separation between measurement and actuation, Journal of Guidance, Control, and Dynamics 39 (2016), 2184-2191. https://doi.org/10.2514/1.G001804

Trajectory optimization and multiple-sliding-surface terminal guidance in the lifting atmospheric reentry

Edoardo Maria Leonardi[1,a *] and Mauro Pontani[2,c]

[1]Department of Astronautical, Electrical, and Energy Engineering, Sapienza University of Rome, via Salaria 851, 00138 Rome, Italy

[a]edoardomaria.leonardi@uniroma1.it, [b]mauro.pontani@uniroma1.it

Keywords: Lifting Reentry, Optimal Guidance, Sliding-Mode

Abstract. In this paper the problem of guiding a vehicle from the entry interface to the ground is addressed. The Space Shuttle Orbiter is assumed as the reference vehicle and its aerodynamics data are interpolated in order to properly simulate its dynamics. The transatmospheric guidance is based on an open-loop optimal strategy which minimizes the total heat input absorbed by the vehicle while satisfying all the constraints. Instead, the terminal phase guidance is achieved through a multiple-sliding-surface technique, able to drive the vehicle toward a specified landing point, with desired heading angle and vertical velocity at touchdown, even in the presence of nonnominal initial conditions. The time derivatives of lift coefficient and bank angle are used as control inputs, while the sliding surfaces are defined so that these two inputs are involved simultaneously in the lateral and vertical guidance. The terminal guidance strategy is successfully tested through a Monte Carlo campaign, in the presence of stochastic winds and wide dispersions on the initial conditions at the Terminal Area Energy Management, in more critical scenarios with respect to the orbiter safety criteria.

Introduction

The development of an effective guidance architecture for atmospheric reentry and precise landing represents a crucial issue for the design of reusable vehicles capable of performing safe planetary reentry. Unsurprisingly, the interest in guidance and control technologies for atmospheric reentry and landing of winged vehicles has increased [1-2], as the flexibility and controllability of the reentry trajectory can be increased through the employment of lifting bodies. However, this implies a greater sensitivity to the environmental conditions. Thus, the usefulness of a real-time guidance algorithm, able to generate online trajectories, is evident, for the purpose of guaranteeing safe descent and landing even in the presence of nonnominal conditions and dispersions caused by the preceding transatmospheric phase.

The guidance and control strategy of the Space Shuttle relied on the modulation of the bank angle to follow a pre-computed reference drag profile, and could only account for small deviations from the nominal conditions [3]. Mease and Kremer and Mease et al. [4] revisited the Shuttle reentry guidance, using nonlinear geometric methods. Later on, Benito and Mease [5] developed and applied a new controller based on model prediction, where the bank angle is modulated to minimize an effective cost function which accounts for the error in drag acceleration and downrange. Nonlinear predictive control was employed by Minwen and Dayi to generate skip entry trajectories for low lift-to-drag vehicles [6]. Most recently, Lu [7] considered a unified guidance methodology based on a predictor-corrector algorithm, for vehicles with different aerodynamic efficiency, while satisfying the boundaries on the thermic flux and load factor. Instead, a more limited number of papers addressed the terminal descent and landing, which is traveled after the Terminal Area Energy Management (TAEM) interface. Kluever [8] developed a guidance scheme for an unpowered vehicle with limited normal acceleration capabilities. Bollino et al. [9] employed a pseudospectral-based algorithm for optimal feedback guidance of reentry spacecraft, in the presence of large uncertainties and disturbances. Fahroo and Doman [10] used again a pseudospectral method in a mission scenario with actuation failures. Finally, reinforcement learning was used for autonomous guidance algorithms for precise landing [11]. Recently, sliding mode control was proposed as an effective nonlinear approach to yield real-time feedback control laws able to drive an unpowered space vehicle toward a specified landing site [3,12]. Depending on the instantaneous state and the desired final conditions, sliding mode control was already shown to be effective for generating feasible atmospheric paths leading to safe landing in finite time, even when several nonnominal flight conditions may occur that can significantly deviate the vehicle from the desired trajectory, e.g. winds or atmospheric density fluctuations [13].

In this work, an open-loop optimal guidance is developed for the transatmospheric arc, capable of minimizing the total heat input while driving the vehicle toward the TAEM. The Space Shuttle Orbiter is taken as the reference vehicle and an analytical method is employed to keep the maximum thermic flux below the safety limit, while accounting for the saturation on the control variables. Finally, the multiple-sliding-surface guidance is employed in order to drive the vehicle from the TAEM to the landing point, with accurate aerodynamic modelling, while including stochastic winds and large dispersions on the initial values of the state and control variables.

Reentry dynamics

The reentry vehicle is modelled as a 3-DOF lifting body and the position of the centre of mass is identified by a set of three spherical coordinates (r,λ_g,φ), representing respectively the instantaneous radius, the geographical longitude and the latitude. The additional variables are given by the relative velocity with respect to the Earth surface v_r, the heading angle ζ_r and the flight path angle γ_r. The trajectory equations describe the motion of the center of mass due to the effect of the forces acting on it [14].

The Space Shuttle Orbiter is taken as the reference vehicle for numerical simulations. It is assumed that the lift and drag coefficients (C_L and C_D) depend only on the angle of attack α and Mach number M, while the sideslip coefficient C_Q depends only of the sideslip angle β and Mach number M. The aerodynamics coefficients are obtained from wind tunnel tests [15] and are interpolated in order to derive their expressions as continuous functions of the aerodynamic angles and Mach number ($C_L = C_L(\alpha, M)$, $C_D = C_D(\alpha, M)$, $C_Q = C_Q(\beta, M)$).

Transatmospheric phase

The transatmospheric guidance drives the vehicle from the entry interface towards the TAEM, while keeping the thermic flux per unit area at the stagnation point q_s below the maximum value and minimizing the cost function

$$J = k_r\Delta_r + k_y\Delta_y + k_z\Delta_z + k_v\Delta_v + k_\zeta\Delta_\zeta + k_\gamma\Delta_\gamma + k_q\int_0^{t_f} q_s\,dt \tag{1}$$

where the coefficients k are chosen to balance the different contributions, while the terms Δ represent the deviations on the state variables at the final time, located at the TAEM. The reentry trajectory is sampled at equally-spaced time instants t_k from the entry interface to the TAEM and the guidance law is determined through parametric optimization of the following parameters:

- sampled values of the bank angle;
- sampled values of the angle of attack from $M = 6$ to the TAEM;
- the total time of flight t_f from the reentry interface to the TAEM;

- the Mach number M^* at the end of the costant-angle-of-attack flight profile;
- the argument of latitude u_0 at the initial time;

The boundary conditions reflect the typical discent profile of the Space Shuttle [16] ($r_0 = 122000$ m, $v_{r0} = 7300$ m/s, $\gamma_{r0} = -1.4°$, $r_f = 25000$ m, $v_{rf} = 762$ m/s, $\varphi_f = 29.41°$, $\lambda_g = -81.46°$, $\zeta_f = -60.24°$) and the algorithm must keep the thermic flux below the maximum allowable value, equal to 681.39 kW/m^2, even lower than the typical value reported in the scientific literature, i.e. 794.43 kW/m^2 [17]. The dynamic pressure must be less than 16.375 kPa.

Thermic flux saturation. The thermic flux at the leading edge can be computed as $q_s = q_a q_r$, where $q_r = a\sqrt{\rho}(bv_r)^n$ and $q_a = c_0 + c_1\alpha + c_2\alpha^2 + c_3\alpha^3$, with a = 17700, b = 0.0001 and n = 3.07 [17]. The derivative of the thermic flux can be easily computed as $\dot{q}_s = Fq_a + q_r G\dot{\alpha}$, where F and G are auxiliary functions that do not depend on the input variable ($\dot{\alpha}$). Therefore, the time derivative of the lift coefficient can be computed as

$$\dot{C}_l = \frac{dC_L}{d\alpha}\dot{\alpha} = -\frac{dC_L}{d\alpha}\frac{F}{G}\frac{q_a}{q_r} \quad (2)$$

Guidance strategy. The descent of the vehicle through the atmosphere is controlled through modulation of the angle of attack and bank angle. In particular, the variation of the angle of attack follows the succession of four distinct flight profiles:

- constant-angle-of-attack flight from the entry interface to $M = M^*$;
- variable-angle-of-attack flight as described by Eq. 10;
- variable-angle-of attack flight following a sinusoidal profile from $M = M^*$ to $\tilde{M} = 6$;
- variable-angle-of-attack flight optimized by the guidance algorithm.

Numerical results. Table 1 reports the results of the optimization. The guidance algorithm is able to drive the vehicle through the atmosphere, with limited dispersions on the final state at the TAEM (cf. Table **1**), along a descent path close to the actual trajectory of the Orbiter [16].

Table 1: displacements of the state variables from the boundary values at TAEM

Q [MJ/m^2]	Δr [m]	Δy [m]	Δz [m]	Δv_r [m/s]	$\Delta\zeta_r$ [°]	$\Delta\gamma_r$ [°]
325.87	2.05	0.32	53.95	9.50	$6.35 \cdot 10^{-4}$	$1.26 \cdot 10^{-5}$

Fig. *1* and 2 highlight the time history of the angle of attack, which keeps the thermic flux below the maximum value. Saturation of the thermal flux occurs after about 200 s, as shown

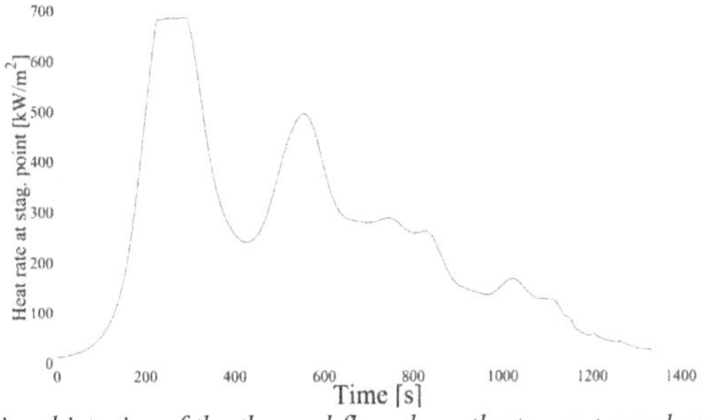

Fig. 1: time histories of the thermal flux along the transatmospheric arc

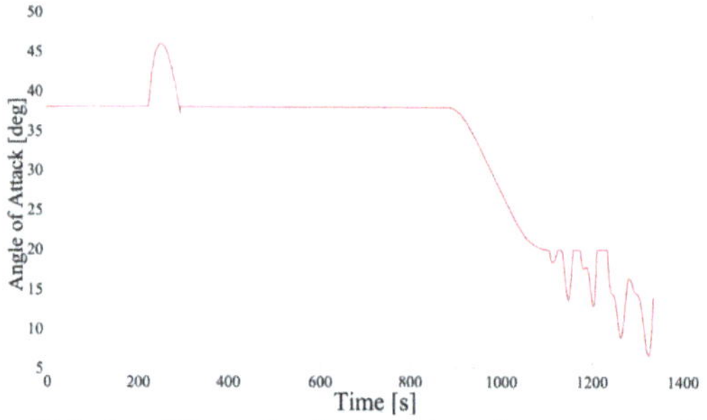

Fig. 2: time histories of the angle of attack along the transatmospheric arc

Terminal guidance

Along the transatmospheric arc, different factors may modify the reentry trajectory from the reference profile. Therefore, the terminal guidance must be able to drive the vehicle despite a wide range of initial conditions. In a previous work, sliding-mode control was already employed as a nonlinear approach to yield real-time feedback guidance laws in an accurate dynamic framework, including winds and large deviations from the initial trajectory variables [13]. In this study, significant improvements are developed with respect to the previous research:

- sliding-mode guidance is tested for a longer time period (i.e. from the TAEM to ground) and the aerodynamic modeling is based on real data rather than approximate analytical expressions;
- the saturation of the control variables is accounted inside the expression of the control input, so that only feasible trajectories are generated;
- the guidance gains are updated through an adaptive strategy, allowing further extension of the capability of the algorithm.

Numerical results. A total number of 500 simulations are run and the initial conditions are randomly generated with upper/lower bounds set to $\pm 2\sigma_s$ (where σ_s denotes the standard deviation of the variable of interest). Stochastic wind is also accounted for, whose intensity and direction is stronger than the safety limits prescribed for the Space Shuttle Orbiter landing [18]. Table **2** collects the initial conditions and associated standard deviations, which reflect the actual reference flight profile of the Space Shuttle [16]. Instead, Table **3** collects the results of the Monte Carlo campaign.

Table 2: initial conditions and standard deviations

Variable	$x(0)$ [m]	$y(0)$ [m]	$z(0)$ [m]	$v_r(0)$ [m/s]	$\zeta_r(0)$ [deg]	$\gamma_r(0)$ [deg]	$C_L(0)$ [-]	$\sigma(0)$ [deg]
Initial Conditions	24050	-54850	95932	762	ζ_f	-8	0.3969	0
Std. Dev.	0	2500	2500	15	5	1	0.01	5

Table 3: results of the Montecarlo campaign

Variable	r_{down} [m]	r_{cross} [m]	$\dot{x}$ [m/s]	v_r [m/s]	ζ_r [deg]	γ_r [deg]	α [deg]	σ [deg]
Mean	761.61	4.41	-1.02	138.69	-60.23	-0.43	6.26	-0.07
Std. Dev.	$-5.66 \cdot 10^{-3}$	$5.04 \cdot 10^{-3}$	0.10	16.63	0.02	0.09	1.74	1.17

From inspection of Table 3, it is evident that the algorithm is able to drive the vehicle to the prescribed landing point, which is located 762 m beyond the runway threshold, with limited crossrange component and vertical velocity at touchdown, and the proper alignment with the runway [16]. Figure 3 shows the stream of trajectories from the TAEM to the landing runway.

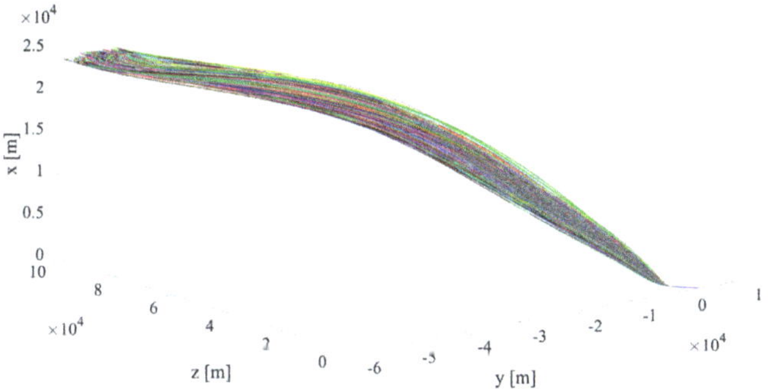

Fig. 3: stream of trajectories

Concluding remarks

This paper addresses the problem of driving a winged vehicle (i.e. the Space Shuttle Orbiter) from the entry interface to landing, while satisfying all the constraints. The transatmospheric guidance is based on an open-loop algorithm that minimizes the total heat input and saturates the maximum thermic flux. The terminal guidance is based on a multiple-sliding-surface strategy, which allows online generation of trajectories. The simulation setup includes a complete dynamic framework with an accurate aerodynamics modeling based on wind tunnel tests. *The numerical results* show the ability of the proposed guidance to modulate the angle of attack to avoid exceeding the maximum thermal flux, while compensating for winds and dispersions of position and velocity from the nominal trajectory during the terminal phase. The vehicle reaches the landing point with the proper alignment with the runway and a safe vertical velocity.

References

[1] R. Haya, L. T. Castellani and A. Ayuso, "Reentry GNC concept for a reusable orbital platform (space rider)", *69th International Astronautical Congress,* Bremen, Germany, *IAC-18 D*, Vol. 2, 2018, p. 5.

[2] Z. Krevor, R. Howard, T. Mosher and K. Scott, "Dream Chaser Commercial Crewed Spacecraft Overview", *17th AIAA International Space Planes and Hypersonic Systems and Technologies Conference,* AIAA Paper 2011-2245, 2011. https://doi.org/10.2514/6.2011-2245

[3] N. Harl and S. Balakrishnan, "Reentry terminal guidance through sliding mode control", *Journal of Guidance, Control and Dynamics,* Vol. 33, No. 1, pp. 186-199, 2010. https://doi.org/10.2514/1.42654

[4] K. D. Mease and J.-P. Kremer, "Shuttle Entry Guidance Revisited Using Nonlinear Geometric Methods", *Journal of Guidance, Control and Dynamics,* Vol. 17, No. 6, 1994, pp. 1350-1356. https://doi.org/10.2514/3.21355

[5] J. Benito and K. D. Mease, "Nonlinear Predictive Controller for Drag Tracking in Entry Guidance", *AIAA/AAS Astrodynamics Specialist Conference and Exhibit,* AIAA Paper 2008-7350, 2008. https://doi.org/10.2514/6.2008-7350

[6] G. Minwen and W. Dayi, "Guidance Law for Low-Lift Skip Reentry Subject to Control Saturation Based on Nonlinear Predictive Control", *Aerospace Science and Technoogy,* Vol. 37, No. 6, 2014, pp. 48-54. https://doi.org/10.1016/j.ast.2014.05.004

[7] P. Lu, "Entry Guidance: A Unified Method", *Journal of Guidance, Control and Dynamics,* Vol. 37, No. 3, 2014, pp. 713-728. https://doi.org/10.2514/1.62605

[8] C. A. Kluever, "Unpowered Approach and Landing Guidance Using Trajectory Planning", *Journal of Guidance, Control and Dynamics,* Vol. 27, No. 6, 2004, pp. 967-974. https://doi.org/10.2514/1.7877

[9] M. R. K. Bollino and D. Doman, "Optimal Nonlinear Feedback Guidance for Reentry Vehicles", *AIAA Guidance, Navigation and Control Conference and Exhibit,* AIAA Paper 2006-6074, 2006. https://doi.org/10.2514/6.2006-6074

[10] F. Fahroo and D. Doman, "A Direct Method for Approach and Landing Trajectory Reshaping with Failure Effect Estimation", *AIAA Guidance, Navigatin and Control Conference and Exhibit,* AIAA Paper 2004-4772, 2004. https://doi.org/10.2514/6.2004-4772

[11] B. Gaude, R. Linares and R. Furfaro, "Deep reinforcement learning for six degree-of-freedom planetary landing", *Advances in Space Research,* Vol. 65, No. 7, pp. 1723-1741, 2020. https://doi.org/10.1016/j.asr.2019.12.030

[12] X. Liu, F. Li, Y. Zhao, "Approach and Landing Guidance Design for Reusable Launch Vhicle Using Multiple Sliding Surfaces Techniques", *Chinese Journal of Aeronautics*, Vol. 30, No. 4, 2017, pp. 1582-1591. https://doi.org/10.1016/j.cja.2017.06.008

[13] A. Vitiello, E. M. Leonardi and M. Pontani, "Multiple-Sliding-Surface Guidance and Control for Terminal Atmospheric Reentry and Precise Landing", *Journal of Spacecraft and Rockets*, Vol. 60, N. 3, pp. 912-923, 2023. https://doi.org/10.2514/1.A35438

[14] M. Pontani, *Advanced Spacecraft Dynamics*, 1st ed., Edizioni Efesto, Rome, 2023, pp. 253-259.

[15] C. Weiland, *Aerodynamic Data of Space Vehicles,* 1st ed., Springer Berlin, Heidelberg, 2014, pp. 174-197. https://doi.org/10.1007/978-3-642-54168-1_1

[16] D. R. Jenkins, *Space Shuttle: The History of the National Space Transportation System: The First 100 Missions*, D. R. Jenkins, Cape Canaveral, 2008, pp. 260-261.

[17] J. T. Betts, *Practical methods for optimal control and estimation using nonlinear programming,* 2nd ed., SIAM, Philadelphia, 2010, pp. 247-256. https://doi.org/10.1137/1.9780898718577

[19] S. Siceloff, "Nasa-Space Shuttle Weather Launch Commit Criteria and KSC End of Mission Weather Landing Criteria", 2003.

Aeronautics and Astronautics - AIDAA XXVII International Congress Materials Research Forum LLC
Materials Research Proceedings 37 (2023) 621-624 https://doi.org/10.21741/9781644902813-135

Analytic formulation for J2 perturbed orbits

Silvano Sgubini[1,a *], Giovanni B. Palmerini[1,b]

[1] Scuola di ingegneria Aerospaziale, Sapienza Università di Roma via Salaria 851 – 00138 Roma, Italy

[a] silvano.sgubini@gmail.com, [b] giovanni.palmerini@uniroma1.it

Keywords: Orbit Propagation, Orbit Perturbations, J2 Effects, Formation Flying

Abstract. The paper deals with a technique developed along the years at the Scuola di Ingegneria Aerospaziale to provide an exact solution for J2 perturbed orbits, here applied to spacecraft formations. Analytic solutions are useful in the design phase and can help in operations to identify and to efficiently maintain a suitable configuration. The approach is based on the elaboration, conveniently performed by means of a symbolic software tool, of a set of equations analogous to the Lagrange planetary relations. Resulting parameters are expressed through Fourier series depending only on the initial conditions. Comparison with standard, longer to obtain and less accurate numerical propagation clarify the advantage of the technique, which is limited only by the number of terms taken into account in the expansion.

Introduction

Numerical propagation of orbits gained widespread acceptance due to the availability of large computation resources and to the possibility to include the effects of all perturbations. However, analytic formulations – when available - offer an exact and really fast solution and helps in the understanding of the problem, with obvious advantages in design. It is well known that Keplerian trajectories can be expressed as an expansion of terms, providing an analytical solution, even if practically limited by the number of terms taken into account. Taylor expansions in powers of the time or of the eccentricity and Fourier expansion in terms of the anomaly are possible, with a bound on eccentricity values in order to ensure convergence [1, 2]. It is interesting to similarly act for real orbits, where perturbations have to be considered. There is a large interval of orbital altitudes, between 600 and 900 km, where – for standard spacecraft, i.e. the ones missing extremely large appendages – the dominant perturbation is the one due to the aspherical gravitational potential of the Earth. Furthermore, the second harmonic of the Earth potential, the one representing the oblate or polar-flattened Earth and shortly indicated as J2, is definitely the most relevant term, so that the analysis can be conveniently limited to it. Interestingly, this interval of altitudes is highly important for Earth observation missions. In such a frame, an analytic solution – with obvious advantages with respect to numerical propagation in terms of time and accuracy – can be of significant interest. The present study is inspired to the original approach by Broglio [3], and has been step-by-step improved and applied to tracking and orbit determination and in previous works by present authors [4,5,6]. In this paper the focus is mostly on the computation of the distances among satellites, i.e. in the field of formation flying. The ultimate goal would be to obtain results similar to the ones provided by extensive numerical simulations aimed to identify J2 invariant formations [7] and to help in the relevant analysis [8], with the target to limit the effort required to control the configuration [9].

Materials Research Proceedings 37 (2023) 621-624 https://doi.org/10.21741/9781644902813-135

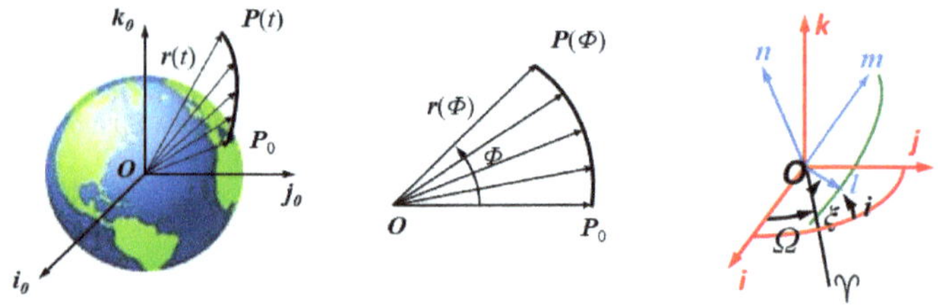

Fig. 1 – Sketch of a J2-perturbed orbit (not anymore laying on a plane) with the parameters adopted to describe the position of the satellite (adapted from [4]).

Approach

The position of the satellite along the orbit can be defined, according to the frame reported in Fig. 1, by the radius and the three angles Ω, i and ξ . The dynamics (Laplace planetary equations) can be written as

$$i' = a_n \frac{r^3}{K^2}\cos\xi \qquad \Omega' = a_n \frac{r^3}{K^2}\frac{\sin\xi}{\sin i} \qquad \xi' = 1 - a_n \frac{r^3}{K^2}\frac{\sin\xi}{\tan i} \tag{1}$$

$$\frac{K'}{K} = a_m \frac{r^3}{K^2} \qquad t' = \frac{r^2}{K} \qquad \left(\frac{1}{r}\right)'' + \frac{1}{r} = -\frac{r^2}{K^2}\left[a_l + r\left(\frac{1}{r}\right)' a_m\right]$$

where K is the angular momentum, t the time and derivatives, represented by the apex ($'$), refer to the angular variable and the a coefficients, if we limit to the case of the J2 effect, are simply given as

$$a_l = \frac{\mu}{r^2} - \frac{3}{2}\mu J_2 R_\oplus^2 \frac{1}{r^4}(1 - 3\,sin^2\,i\,sin^2\,\xi)$$

$$a_m = -\frac{3}{2}\mu J_2 R_\oplus^2 \frac{1}{r^4} sin^2\,i\,sin\,2\,\xi \qquad a_n = -\frac{3}{2}\mu J_2 R_\oplus^2 \frac{1}{r^4} sin\,2\,i\,sin\,\xi \tag{2}$$

After a significant mathematical elaboration (see [5]), the set of Eq.(1) leads to an expression for r and for the three angles Ω, i and ξ as in following Eq. 3. Coefficients depend on the initial conditions only, and can be evaluated until the desired order. Notice that nowadays such an elaboration has been helped by symbolic software (e.g. MATLAB [10] in the present case).

$$r = R'_0 + \sum_h \left[R'_h \cos\left(h\frac{2\pi}{T}(t - t_{iniz})\right) + R''_h \sin\left(h\frac{2\pi}{T}(t - t_{iniz})\right)\right]$$

$$\Omega = \Omega'_0 - \frac{3}{2}J_2 \cos i_{iniz}\left(\frac{R_\oplus \mu_\oplus}{K_{iniz}^2}\right)^2 \frac{2\pi(t - t_{iniz})}{T} + \sum_h \left[\Omega'_h \cos\left(h\frac{2\pi}{T}(t - t_{iniz})\right) + \Omega''_h \sin\left(h\frac{2\pi}{T}(t - t_{iniz})\right)\right] \tag{3}$$

$$i = I'_0 + \sum_h \left[I'_h \cos\left(h\frac{2\pi}{T}(t - t_{iniz})\right) + I''_h \sin\left(h\frac{2\pi}{T}(t - t_{iniz})\right)\right]$$

$$\xi = \xi'_0 + \frac{2\pi}{T}(t - t_{iniz}) - \frac{3}{2}J_2 \cos i_{iniz}\left(\frac{R_\oplus \mu_\oplus}{K_{iniz}^2}\right)^2 \frac{2\pi(t - t_{iniz})}{T} + \sum_h \left[\xi'_h \cos\left(h\frac{2\pi}{T}(t - t_{iniz})\right) + \xi''_h \sin\left(h\frac{2\pi}{T}(t - t_{iniz})\right)\right]$$

The correctness of the solution can be easily estimated by the comparison with a standard numerical propagation (see Fig. 2 for examples relevant to two parameters of interest).

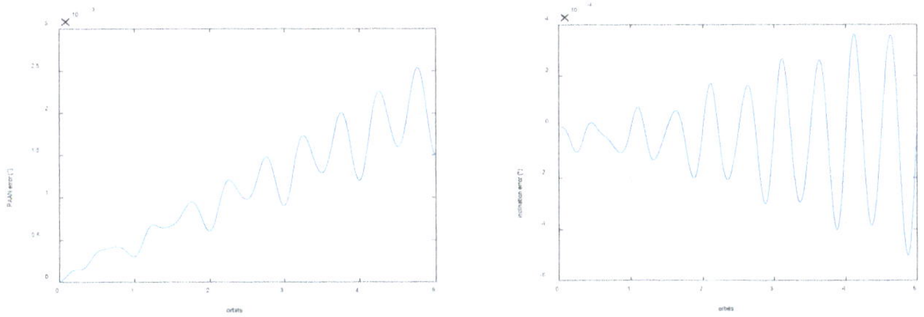

Fig. 2 – Differences between analytical approach and numerical integration.

Formations

The very same approach can be iterated for different spacecraft. However, it is extremely important to remark the relevance of the initial conditions to be imposed to the satellites belonging to the formation. A simple computation of the distance between generic, yet close initial conditions gives the results presented in Fig. 3 for two spacecraft. Notice that the distance is given by the difference between the two vectors representing the radii, with three backward rotations in the angles Ω_1, i_1 , ξ_1 for the first spacecraft and Ω_2, i_2 and ξ_2 for the second one to obtain the components along the inertial frame's axes.

Within the concept of formation flying, it is desired that a configuration with limited inter-satellite distances should last in time. So, additional constraints can be applied among the parameters referred to the two - or more – spacecraft belonging to the formation. A first preliminary indication can be given by imposing the same energy to the two satellites (Fig. 4).

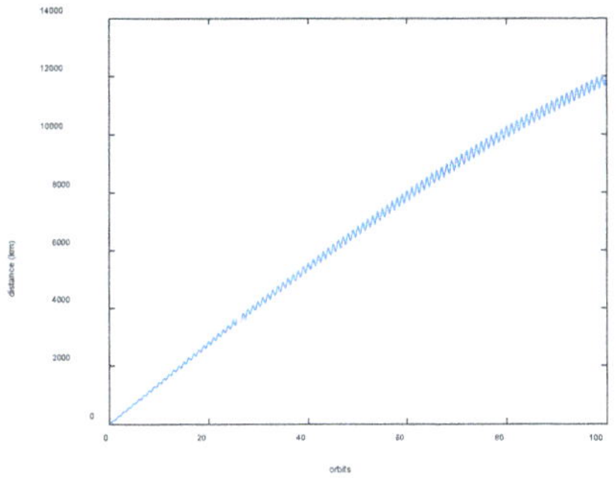

Fig. 3 – Distance between two close satellites.

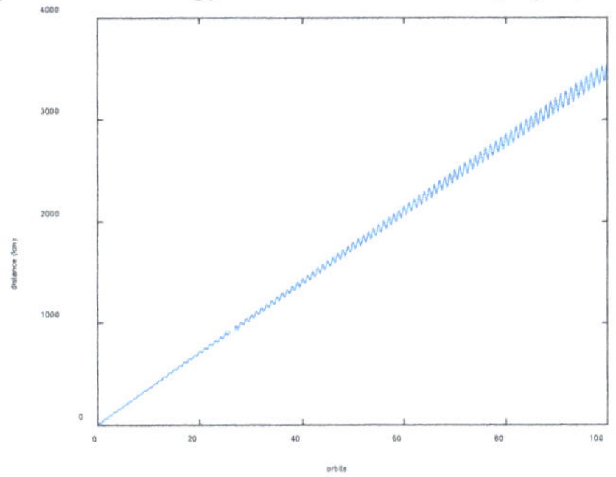

Fig. 4 – Distance imposing equal energy.

A first constraint is given by equal period, that is a requirement to avoid divergence. A second constraint is related to the inclination, that has to be assumed as quite close for the satellites to stay in formation: in fact, even 1 degree if difference would create a distance in the order of 120 km for orbits of 600 km altitude. Furthermore, larger differences end up in a different environment in terms of other perturbations, so an almost equal inclination can be reasonably assessed. A third constraint is given by the equal precession of the ascending nodes. Note that the analytical solution gives a secular term: this term vanishes for some critical inclinations. Once all of these constraints are imposed, the results plotted in Fig. 5-6 can be obtained.

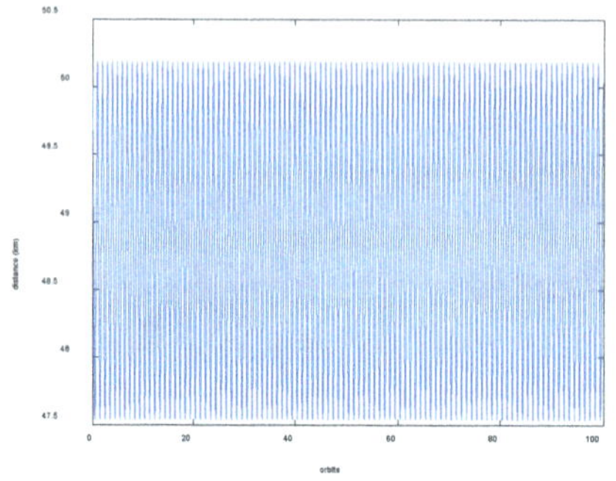

Fig. 5 – Distance imposing the condition of an equal location after a short time interval (8s).

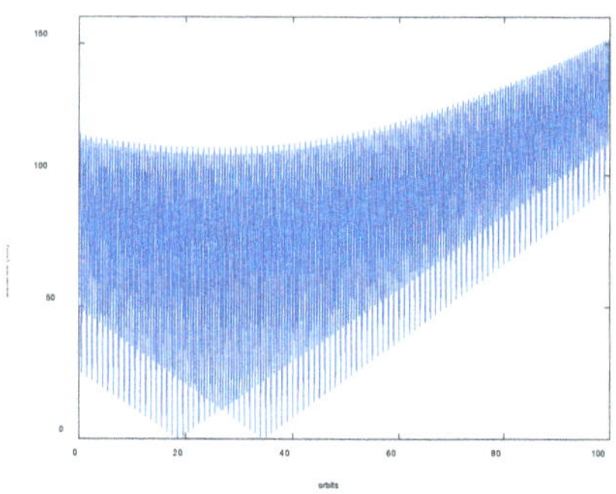

Fig. 6 Initial conditions as per Fig.4 adding constraints of equal period and secular drift.

Final Remarks

Design and operations phases of formation flying missions can be helped by the availability of analytic solutions taking into account the oblateness effects. The work, following the path pursued along the years by the authors, proposes analytic formulas for the distances between satellites considering the dominant effect of the J2 term, and prove their correctness and their appeal even if only a limited number of terms in the expansion should be used.

References

[1] R.F. Moulton. Introduction to Celestial Mechanics. McMillian, New York, 1914.

[2] S.A. Klioner. Lecture Notes on Basic Celestial Mechanics, Technische Universitaet Dresden, 2016. https://doi.org/10.48550/arXiv.1609.00915

[3] L. Broglio. Lezioni di Meccanica del Volo Spaziale (a cura di S. Sgubini), Scuola di Ingegneria Aerospaziale, Università di Roma "La Sapienza", 1973.

[4] S. Sgubini, G.B. Palmerini. A fast approach for relative orbital determination in spacecraft formations, in "IEEE Aerospace Conference Proceedings", Big Sky (MT, USA), 2012.

[5] S. Sgubini, G.B. Palmerini. Efficient Techniques for Relative Motion Analysis between Eccentric Orbits under J2 Effect, in "IEEE Aerospace Conference Proceedings", Big Sky (MT, USA), 2013.

[6] S. Sgubini, G.B. Palmerini. Fast Analytical Evaluation of Near-Keplerian Orbit Evolution in Multiple Launch Release. Congresso Nazionale AIDAA, Roma, 2019.

[7] M. Sabatini, D. Izzo, G.B. Palmerini. Analysis and control of convenient orbital configuration for formation flying missions. Adv. Astronaut. Sci. 124, (2006) 313–330.

[8] S.R. Vadali, P. Sengupta, H. Yan, K.T. Alfriend. On the fundamental frequencies of relative motion and the control of satellite formations. Adv. Astronaut. Sci. 129 (2008) 2747–2766.

[9] M. Sabatini, D. Izzo, G.B. Palmerini. Minimum control for spacecraft formations in a J2 perturbed environment, Celest. Mech. Dyn. Astron. 105 (2009) 141-157. https://doi.org/10.1007/s10569-009-9214-5

[10] Symbolic Math Toolbox™ User's Guide, The MathWorks Inc., 2023.

Low-thrust maneuver anomaly detection of a cooperative asset using publicly available orbital data

Riccardo Cipollone[1,a*], Pierluigi di Lizia[1,b]

[1]Department of Aerospace Science and Technology, Politecnico Di Milano, Via Giuseppe La Masa 54 20156, Milano, Italy

[a]riccardo.cipollone@polimi.it, [b]pierluigi.dilizia@polimi.it

Keywords: Low-Thrust Maneuver Estimation, Semi-analytic Method, Anomaly Detection

Abstract. This work presents a novel method to estimate perturbations with respect to nominal maneuver planning by exploiting Two-Line-Element (TLE) data as initial step, then moving on to Global Positioning System (GPS) processed data. The case study is a low-thrust engine validation mission in Low Earth Orbit. The first algorithm exploits a couple of TLEs as boundary conditions to set up a least-squares problem and find the tangential thrust magnitude and firing duration to best fit the bounding orbital states, making use of Taylor differential algebra and Picard iterations. The second one makes use of a sequence of GPS states to apply multistep finite differences and a root-finding algorithm to retrieve information about both thrust profile and firing bounding times.

Introduction

Over the past few decades, there has been a steady growth in the quantity of scientific and commercial Earth-bound space missions. This upward trend has led to a significant expansion of activities and programs focused on Space Situational Awareness. Up-to-date orbital data obtained by updating ephemeris data with new observations play a vital role in effectively tracking cooperative target assets, offering valuable information to ensure a mission's success [1]. The main product provided by the processing pipelines used to exploit measurements' information are regularly maintained Resident Space Object catalogs [2].

The kind of anomaly detection performed in this framework is usually linked with the orbital motion of the satellite and to its routine maneuvers. The high-level workflow starts by exploiting available intel about known objects to build predictions of a target's nominal behavior and compare it to the actual incoming acquisitions. If any properly defined distance from the nominal path trespasses some user defined threshold, the anomalous event is recorded, and further analysis can be carried out. This specific case of anomaly identification widely overlaps with maneuver detection and characterization of a tracked object. The reason for this is that according to how much it is known about the nominal trajectory and control policy of an object, any anomalous event involving the dynamics of the target can be modeled as a maneuver, steering it away from the nominal path, and characterized as such, in terms of an equivalent acceleration or expense. Examples in literature are diverse, most of them stemming from the theory of maneuver target tracking, dealing with observations that partially describe the state at observation epoch. These techniques are usually based on adaptive Kalman filtering modeling an input term as a stochastic process or a deterministic input to be estimated when the measurement innovation term fails a Gaussian test, meaning that the modeled uncertainty is no longer enough to explain deviations from the prediction [3]. As for a more SST-tailored application, the work in [4], shows how State Transition Matrix theory can be remodeled to linearly map small variations of control, modeled as an impulsive ΔV, to variations in the final orbital state. Following the usual optimal control assumption, the residual between the predicted final state and the actual one is minimized, linking the pre- and post-maneuver orbit with an impulsive magnitude and a firing epoch. This last method represents the basis for the one proposed in this work, allowing to connect two TLEs, by leveraging some assumptions based on the knowledge of the target features, by a high-order Taylor expansion of both thrust and firing epochs. As for the module exploiting GPS states, due to lack of rich literature, for the low-thrust case, a finite different method has been envisaged as first approach to test if an acceleration profile and accurate onset and termination times can be directly extracted from the states sequence.

Fundamentals and method

The theoretical tools used for the TLE-based maneuver anomaly detection are mainly Taylor Differential Algebra (DA) and Picard iterations, allowing for the condensation of iterative function evaluations in a single polynomial map, function of both thrust magnitude and time variations with respect to the reference onset and termination epochs.

DA provides a solution to analytical problems through an algebraic approach by means of the Taylor polynomial algebra. A Taylor expansion up to an arbitrary order k can be used to represent any deterministic function f of v variables that is C^{k+1} in the domain of interest $[-1,1]^v$ (scaled according to the needs), with limited computational effort:

$$f(x) = \sum_{k=0}^{N} \frac{f^{(k)}(x_0)}{k!}(x - x_0)^k \tag{1}$$

Thus, variables are represented as truncated power series around an expansion point x_0, instead of standard types [5]. The DA framework is implemented in a C + + computational environment through the DACE library. The key feature of DA leveraged for this technique is the flow expansion of an Ordinary Differential Equation (ODE): this feature relieves the processing burden due to iterative integrators embedding the whole integration scheme in a single function evaluation.

As for Picard iterations, they are implemented exploiting the same C++ computational environment following the scheme to obtain a k-order time expansion [6]:

$$\phi_0(t) = y_0 \tag{2}$$

$$\phi_{k+1}(t) = y_0 + \int_{t_0}^{t} f(s, \phi_k(\tau))dk \tag{3}$$

Where $\phi_{k+1}(t)$ is the order k expansion, y_0 is the function value at the center of the expansion and f is ϕ function's time derivative expression, used to iteratively integrate around the reference time t_0.

As regards the GPS data-based technique, multi-step, high-order finite differences are used to provide an estimate of the acceleration and jerk directly from the velocity provided with the GPS states. The general formulation used to compose the coefficients involved in the differentiation scheme are derived from the Fronberg method reported in [7] to compute a 1-D derivative of arbitrarily high order on a uniformly spaced grid for a given number of points. To detect the onset and termination epochs of each firing instead, a thresholding method is employed, based on the variance computed on the cumulative sequence of data to detect abrupt changes in it, such as peaks or steps [8]. A final refinement step is performed to resolve the above-mentioned epochs below the sampling frequency of the given GPS states by exploiting a non-linear system root-finding algorithm (MATLAB fsolve) to put to zero the residuals corresponding to the first sampled GPS state downstream the discontinuities as a function of their corresponding epoch.

Concerning the method itself, the TLE-based module is based on several assumptions to limit the complexity of the problem: constant and tangential thrust during the firing, constant burning rate and known onset time. The process starts by propagating the state coming from the first TLE file by augmenting the state with thrust and mass and by using the nominally controlled trajectory as reference for the expansion. The number of firing arcs to be included between the TLE couple

Materials Research Proceedings 37 (2023) 625-629 https://doi.org/10.21741/9781644902813-136

can vary, increasing the number of variables of the problem and, therefore, the number of possible solutions. The DA variables are initialized to order 4. They model the constant thrust norm perturbations of each firing arc together with its termination time, computed by 4 Picard iterations around the expected termination time. The process is repeated for every firing arc, composing TPSs mapping perturbations for each of them to the one on the orbital state. The process outputs a propagated final state used to build a residual with respect to the one derived from the second TLE file. This residual function is then fed to MATLAB fsolve so to find perturbations putting it to zero with limited computational effort. The firings' modeling process is summed up in Fig. 1.

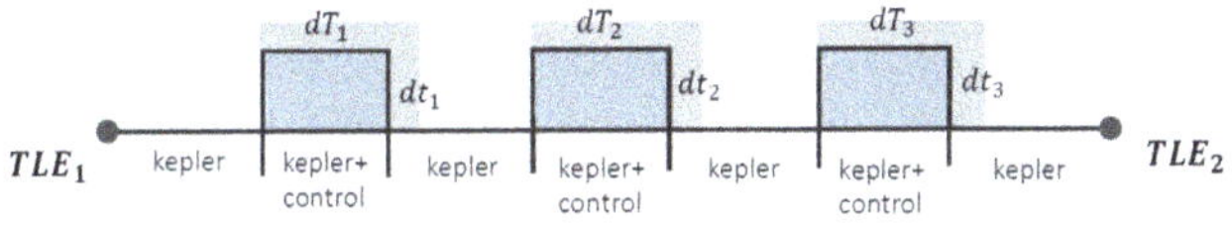

Figure 1. DA variables configuration for a 3-firing case, dT_n being thrust magnitude perturbations, dt_n termination time perturbations.

As for the GPS-based technique, the assumptions are two: the thrust is constant and tangential throughout the firing and so is the burning rate. It starts by simply computing the second time derivative of the velocity extracted from the sampled states. The norm of this quantity (jerk) is then employed to build residuals with respect to the second derivative norm of the expected trajectory velocity, computed by differencing only once, starting from the expression of the nominal dynamics. This sequence of residuals is scanned for abrupt changes by analyzing the variance computed on a progressively higher number of elements and checking whether it overcomes a predefined threshold.

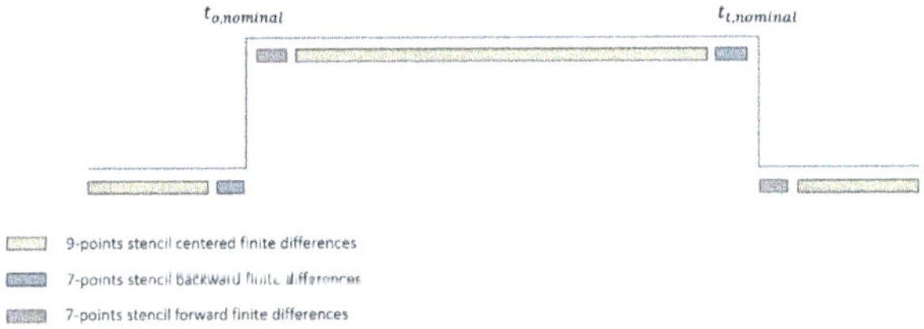

Figure 2. Segments of data used to accurately differentiate the GPS velocity time series.

The discontinuity points identified are used to perform a composite finite difference method, splitting the time series of velocities into 7 segments for every arc, divided by the onset and termination times (as shown in Fig. 2). The method is composed of a backward finite differencing scheme of order 6 for the samples immediately before the discontinuities, a forward scheme of order 6 for the values just after it and a centered scheme of order 8 for the rest. In this way, a first estimate of the thrust profile can be obtained to give reliable boundary conditions to the boundary time estimates refinement. This last step is performed with a propagation across each discontinuity that is only a function of the related epoch. Residuals on the downstream state are defined and put to zero by MATLAB fsolve. Once the actual onset and termination times are obtained, the composite finite differencing scheme is performed once again with updated epochs and mass profile, and the final thrust estimate is retrieved.

Uncertainty propagation is embedded in both modules by means of linear projection of available covariances to the estimated quantities space (via the Jacobian of the transform).

Results

The scenario used to test the TLE-based module is simulated starting from a LEO object with the following Keplerian elements:

$$\boldsymbol{e} = [7.5805e + 03\ km, 0.0760, 0.7151\ rad, 5.9935\ rad, 2.1723\ rad] \tag{4}$$

performing from 1 to 3 nominal firing arcs of 1100 s duration with a constant tangential thrust of 10 mN. The nominal trajectory is then perturbed with a 10 mN thrust parasitic magnitude and a 7 s firing termination time delay for the single firing case to generate the post-maneuver TLE. The same is done for the 2-firing (10 mN, 7 s for the first one, -7 mN, -3 s for the second one) and 3-firing cases (10 mN, 7 s, -7 mN, -3 s, -3 mN, 5 s). The tests show errors in below 1e-4 mN in thrust perturbation magnitude and 1e-5 s in termination time perturbation both in the 1 and the 2-firing cases, while the 3-firing one results in convergence to wrong local minima, due to solution multiplicity of the non-linear system.

As for the GPS-based module test scenario, the same target is involved and the thrust magnitude, onset and termination times perturbations are (4 mN thrust magnitude perturbation, -17 s onset perturbation, 35 s termination perturbation), used to sample the GPS states with a 5 s time step. In this case results show fair performance with 1e-1 mN as average error order of magnitude on thrust profile and errors of 1e-4 s for both onset and termination time perturbations.

Conclusions

This work presents a method to effectively exploit publicly available orbital data to perform maneuver anomaly detection on a cooperative asset. A crucial detail to understand the current performance of the technique resides in the fact that, due to the kind of preliminary study conducted on them to be then further elaborated, these first tests have been conducted without adding any noise to the states used as measurements. The first further step to take is in facts to study how sensitive these techniques are to measurement noise and whether to integrate the pipeline with filtering techniques to take this aspect into account and even it out.

References

[1] Montaruli, Marco Felice, Purpura, Giovanni, Cipollone, Riccardo, De Vittori, Andrea, Di Lizia, Pierluigi, Massari, Mauro, Peroni, Moreno, Panico, Alessandro, Cecchini, Andrea, and Rigamonti Marco, 'A Software Suite for Orbit Determination in Space Surveillance and Tracking Applications', EUCASS-3AF 2022. https://doi.org/10.13009/EUCASS2022-7338

[2] ESA Space Debris Office, 'Esa's Annual Space Environment Report', tech. rep., European Space Agency, April 2022

[3] Rong Li X. , Jilkov Vesselin P. , A Survey of Maneuvering Target Tracking—Part IV: Decision-Based Methods, Proceedings of SPIE Conference on Signal and Data Processing of Small Targets, Orlando, FL, USA, April 2002

[4] Pastor, G. Escribano, and D. Escobar, "Satellite maneuver detection with optical survey observations," Advanced Maui Optical and Space Surveillance Technologies Conference, 2020.

[5] Wittig A., Di Lizia P., Armellin R., Makino K., Bernelli-Zazzera F., and Berz M., Propagation of large uncertainty sets in orbital dynamics by automatic domain splitting. Celestial Mechanics and Dynamical Astronomy, 122(3):239–261, 2015. https://doi.org/10.1007/s10569-015-9618-3

[6] Vitolo M., Maestrini M., Di Lizia P., Sampling-Based Strategy for On-Orbit Satellite Inspection, 25th Conference of the Italian Association of Aeronautics and Astronautics (AIDAA 2019)

[7] Fornberg, Bengt. 'Generation of Finite Difference Formulas on Arbitrarily Spaced Grids'. Mathematics of Computation 51, no. 184 (1988): 699–706. https://doi.org/10.1090/S0025-5718-1988-0935077-0

[8] Killick R., P. Fearnhead, and I.A. Eckley. "Optimal detection of changepoints with a linear computational cost." Journal of the American Statistical Association. Vol. 107, Number 500, 2012, pp.1590-1598. https://doi.org/10.1080/01621459.2012.737745

Efficient models for low thrust collision avoidance in space

Juan Luis Gonzalo[1,a *], Camilla Colombo[1], Pierluigi Di Lizia[1], Andrea De Vittori[1], Michele Maestrini[1], Pau Gago Padreny[2], Marc Torras Ribell[2], Diego Escobar Antón[2]

[1]Politecnico di Milano, Department of Aerospace Science and Technology, Via la Masa 34, 20158 Milan (Italy)

[2]GMV, Calle de Isaac Newton 11, 28760 Tres Cantos, Madrid (Spain)

[a]juanluis.gonzalo@polimi.it

Keywords: Space Collision Avoidance, Low Thrust Propulsion, Spaceflight Mechanics, Analytical Models, Space Traffic Management

Abstract. A family of analytical and semi-analytical models for the characterization and design of low thrust collision avoidance manoeuvres (CAMs) in space is presented. The orbit modification due to the CAM is quantified through the change in Keplerian elements, and their evolution in time is described by analytical expressions separating secular and oscillatory components. Furthermore, quasi-optimal, piecewise-constant control profiles are derived from impulsive CAM models. The development of these models is part of an ESA-funded project to advance existing tools for collision avoidance activities.

Introduction

The number of objects in orbit around the Earth is growing at an accelerating pace, due to both the increasing number of public entities and private companies leveraging space assets for diverse applications and the accumulation of space debris. Regarding active satellites, their numbers are soaring owing to the deployment of large constellations like Starlink, and the democratization of access to space enabled by lower launch costs and smaller, cost-effective platforms like CubeSats. The increasing congestion in space presents multiple challenges for Space Traffic Management and Space Situational Awareness, and demands the implementation of space debris mitigation actions like end-of-life deorbiting and collision avoidance (COLA) to curb the increase of space debris. However, an increasing portion of satellites are equipping low thrust propulsion systems, which, although more efficient, present a smaller control authority and complicate the design of disposal and collision avoidance manoeuvres (CAMs), particularly for last-time scenarios. As a result, new models and tools are needed to support low thrust COLA activities in congested orbital regions like low Earth orbit (LEO) and geostationary orbit (GEO).

In this context, GMV, UC3M and Politecnico di Milano are developing the ELECTROCAM project, funded by the European Space Agency to advance their models and tools for the analysis of low thrust COLA activities. The project covers several aspects of COLA activities, including the assessment of current capabilities of low thrust satellites [1], propagation of uncertainties [2,3], efficient analytical and semi-analytical models for CAMs [4], and update of ESA's operational software tool ARES.

This work deals with some of the analytical and semi-analytical CAM models developed within the ELECTROCAM project. The models are focused on computational efficiency, to serve as initial guesses for more accurate, higher-cost numerical algorithms, and to perform sensitivity analyses over large sets of data. To this end, a single-averaging of the thrust-perturbed equations of motion is performed to express the evolution of the Keplerian elements as the combination of linear and oscillatory contributions, both expressed through exact or approximated analytical expressions. A simplified control law is also proposed, leveraging analogous models for the impulse propulsion case. The performance of the models has been assessed through sensitivity analyses in relevant COLA scenarios, and some selected results are presented here.

Single-averaged low-thrust CAM model

Let us consider a predicted close approach (CA) at a given time of closest approach (TCA) between a low-thrust-capable spacecraft and a debris (in general, a non-collaborative object). The CAM is modelled following the scheme proposed in [5], composed of three steps. First, the orbit modification due to the CAM is quantified through the change $\delta\boldsymbol{\alpha}$ of its vector of Keplerian elements $\boldsymbol{\alpha}=[a,e,i,\Omega,\omega,M]$, whose components are, respectively, semimajor axis, eccentricity, inclination, right ascension of the ascending node, argument of pericentre, and mean anomaly. Second, the change in orbital elements is mapped to changes in position and velocity at the TCA using a linearized relative motion model with the nominal trajectory as reference orbit. Because the displacements associated to CAMs in practical scenarios are typically small, the loss of accuracy due to the use of a linearized model is limited. Finally, the outcome from the CAM is projected and characterized in the nominal encounter plane at TCA, and the change in collision probability quantified.

The most complex step is the derivation of (semi-)analytical expressions for $\delta\boldsymbol{\alpha}$ under a continuous thrust acceleration $\mathbf{a}_t$. The models considered here are based on the single-averaging of Gauss's planetary equations over one revolution, under the assumption of small thrust acceleration. This assumption allows to linearize the equations of motion in thrust acceleration and study the tangential and normal components separately (denoted by superscripts t and n):

$$\dot{\boldsymbol{\alpha}} = \mathbf{G}(\boldsymbol{\alpha}, t; \boldsymbol{a}_t) \approx \mathbf{G}^t(\boldsymbol{\alpha}_0 + \delta\boldsymbol{\alpha}^t, t; a_t^t) + \mathbf{G}^n(\boldsymbol{\alpha}_0 + \delta\boldsymbol{\alpha}^n, t; a_t^n) + \dot{\boldsymbol{\alpha}}_0 = \delta\dot{\boldsymbol{\alpha}}^t + \delta\dot{\boldsymbol{\alpha}}^n + \dot{\boldsymbol{\alpha}}_0. \quad (1)$$

Where $\mathbf{G}$ represents Gauss's planetary equations, and $\dot{\boldsymbol{\alpha}}_0$ is the evolution of mean anomaly for the unperturbed orbit. The solutions for $\delta\boldsymbol{\alpha}^t$ and $\delta\boldsymbol{\alpha}^n$ from Eq. 1 fall into three categories. First, some Keplerian elements are unaffected by the corresponding thrust component (Ω and i for tangential, a, i and Ω for normal). Then, some elements have only oscillatory behaviours, and their expressions can be integrated directly (ω for tangential, e for normal). Finally, the rest of elements combine secular behaviours with time scale proportional to the thrust magnitude, and oscillatory components with period linked to the orbital one. This is the case for a and e under tangential thrust, and ω under normal thrust. The secular expressions are obtained changing the independent variable in the equations of motion from time to eccentric anomaly E, and averaging the ODEs over 1 revolution. The resulting secular terms are linear, with slopes function of complete elliptic integrals of the first and second kind of the reference eccentricity e_{ref}. The oscillatory terms are obtained as a series expansion in e_{ref} and involve harmonics of the orbital period in E. The detailed derivation of the single-averaged analytical models and resulting expressions can be found in [5,6,7,4].

The proposed expressions for $\delta\boldsymbol{\alpha}$ have E as independent variable. A time law for $E(t)$ can be obtained from the differential equation for E, introducing the analytical approximation obtained for $\boldsymbol{\alpha}(E)= \boldsymbol{\alpha}_0+ \delta\boldsymbol{\alpha}(E)$. The resulting time law is explicit to compute time as function of anomaly, $t(E)$, but implicit to compute $E(t)$. This situation is analogous to that of the Kepler's equation, to which it reduces for $\mathbf{a}_t$=0, and prevents the solution from being fully analytical

Quasi-optimal control law

The previous models allow to evaluate the outcome of a CAM without a numerical integration, but to optimize the CAM they should be used within an iterative numerical optimizer. A more computationally efficient approach is proposed based on analogous models developed for impulsive CAMs [8]. The optimization problem for impulsive CAMs can be reduced to an eigenproblem, where the eigenvector associated to the largest eigenvalue determines the optimal direction of thrust. Then, a piecewise-constant low-thrust CAM can be defined by dividing the thrust arc in segments and assigning to each of them the orientation of the impulsive CAM at its middle point. Furthermore, given that optimal fuel manoeuvres follow bang-bang structures, the magnitude of thrust at each segment will be either maximum or 0. To define in which segments is more convenient to thrust, it can be proven that the eigenvalue of the impulsive CAM at each segment serves as proxy for the local efficiency of the CAM compared to the other segments.

Test case

Multiple scenarios for low-thrust CAMs have been studied within the ELECTROCAM project. For brevity, a single case is presented here. The Keplerian elements at TCA of the spacecraft are $\boldsymbol{\alpha}_{sc}$ = [7552.1 km, 0.0012, 87.93 deg, 1.95 deg, 127.53 deg, 5.10 deg], while for the debris $\boldsymbol{\alpha}_{deb}$ = [7575.5 km, 0.0100, 89.39 deg, 179.03 deg, 112.64 deg, 295.72 deg]. The elements of the combined covariance matrix in the B-plane (defined as in [5]) are σ_ξ=0.0179 km, σ_ζ=0.0214 km, $\rho_{\xi\zeta}$=-0.0524. Fig. 1 shows the results in terms of probability of collision PoC and displacement inside the B-plane δb, for a single-thrust-arc CAM with fixed thrust magnitude (10^{-8} km/s or 10^{-9} km/s) and different values of the thrust arc duration, Δt_{CAM}, and the coast arc between thrust end and TCA, Δt_{coast}. Times are expressed as fractions of the nominal orbital period T.

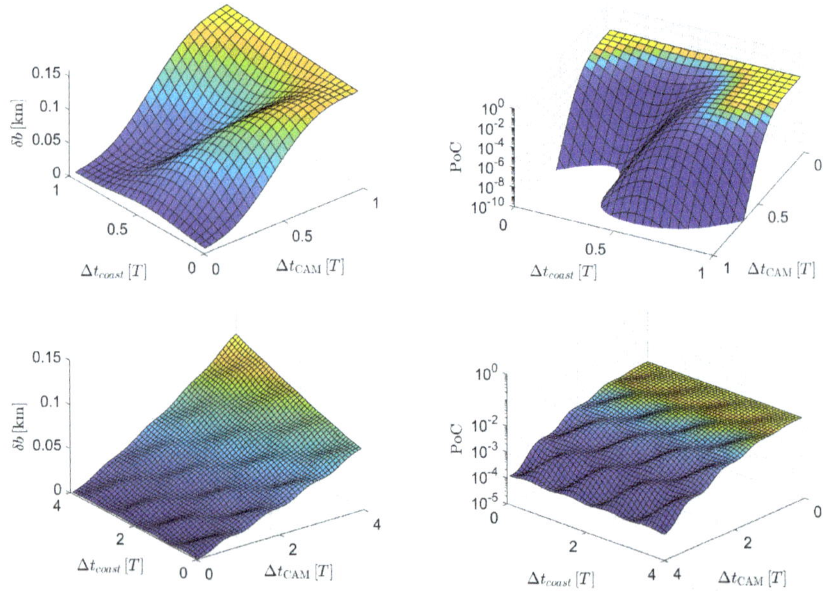

Figure 1. Displacement and PoC for test case in LEO, with $a_t=10^{-8}$ km/s (top) and $a_t=10^{-9}$ km/s (bottom)

Conclusions

The latest developments in a novel family of analytical and semi-analytical models for low-thrust CAM computation and design have been presented. These models rely on the single-averaging of the equations of motion in Keplerian elements to derive approximate analytical solutions separating the secular and oscillatory components for the orbit evolution induced by the CAM. For

CAM design, instead, quasi-optimal piecewise-constant thrust profiles are derived from the impulsive counterpart of the models. Part of these developments have been performed within the ESA-funded ELECTROCAM project.

Acknowledgements
This work has received funding from the European Space Agency through the project "ELECTROCAM: Assessment of collision avoidance manoeuvre planning for low-thrust missions" (call AO/1-10666/21/D/SR).

Juan Luis Gonzalo also thanks the funding of his research position by the Italian Ministero dell'Università e della Rierca, Programma Operativo Nazionale (PON) "Ricerca e Innovazione" 2014-2020, contract RTDA – DM 1062 (REACT-EU).

References

[1] C. Colombo, A. De Vittori, M. Omodei, J.L. Gonzalo, M. Maestrini, P. Di Lizia, P. Gago Padreny, M. Torras Ribell, A. Gallego Torrego, D. Escobar Antón, Roberto Armellin, Sensitivity analysis of collision avoidance manoeuvre with low thrust propulsion, Aerospace Europe Conference 2023 (10th EUCASS - 9th CEAS), Lausanne, Switzerland, 9-13 July 2023

[2] M. Maestrini, A. De Vittori, P. Di Lizia, C. Colombo, Dynamics-Based Uncertainty Propagation with Low-Thrust, 2022 AAS/AIAA Astrodynamics Specialist Conference, Charlotte, NC, USA, 7-11 Aug. 2022

[3] M. Maestrini, A. De Vittori, J.L. Gonzalo, C. Colombo, P. Di Lizia, J. Míguez Arenas, M. Sanjurjo Rivo, A. Diez Martín, P. Gago Padreny, D. Escobar Antón, ELECTROCAM: Assessing the Effect of Low-Thrust Uncertainties on Orbit Propagation, 2nd ESA NEO and debris detection conference, Darmstadt, Germany, 24-26 Jan. 2023

[4] J.L. Gonzalo, C. Colombo, P. Di Lizia, Single-averaged models for low-thrust collision avoidance under uncertainties, 73rd IAC, Paris, France, 18-22 Sept. 2022

[5] J.L. Gonzalo, C. Colombo, P. Di Lizia, Introducing MISS, a new tool for collision avoidance analysis and design, Journal of Space Safety Engineering, 7:3 (2020), 282-289. https://doi.org/10.1016/j.jsse.2020.07.010

[6] J.L. Gonzalo, C. Colombo, P. Di Lizia, A semianalytical approach to low-thrust collision avoidance manoeuvre design, 70th IAC, Washington, D.C., USA, 21-25 Oct. 2019. Paper no. IAC-19.A6.2.3

[7] J.L. Gonzalo, C. Colombo, P. Di Lizia, Computationally efficient approaches for low-thrust collision avoidance activities, 72nd IAC, Dubai, UAE, 25-29 Oct. 2021. Paper no. IAC-21-A6.10-B6.5.5

[8] J.L. Gonzalo, C. Colombo, P. Di Lizia, Analytical framework for space debris collision avoidance maneuver design, J. Guid. Control Dyn., 44:3 (2021), 469-487. https://doi.org/10.2514/1.G005398

An overview of the ArgoMoon and LICIAcube flight dynamics operations

Marco Lombardo[1,a*], Luis Gomez Casajus[2], Marco Zannoni[1,2,b], Igor Gai[2], Edoardo Gramigna[1], Paolo Tortora[1,2], Elisabetta Dotto[3], Marilena Amoroso[4], Simone Pirrotta[4], Valerio Di Tana[5], Biagio Cotugno[5], Silvio Patruno[5], Francesco Cavallo[5], and the *LICIACube* Team[1,2,3,4,5]

[1]Dipartimento di Ingegneria Industriale (DIN), Alma Mater Studiorum – Università di Bologna, Via Fontanelle 40, 47121, Forlì, Italy

[2]Centro Interdipartimentale di Ricerca Industriale Aerospaziale (CIRI AERO), Alma Mater Studiorum— Università di Bologna, Via Baldassarre Carnaccini 12, 47121 Forlì, Italy

[3]INAF Osservatorio Astronomico di Roma, Monte Porzio Catone, Italy

[4]Agenzia Spaziale Italiana, Via del Politecnico, 00133 Roma, Italy

[5]Argotec S.r.l., Via Cervino 52, 10155 Torino, Italy

[a]marco.lombardo14@unibo.it, [b]m.zannoni@unibo.it

Keywords: ArgoMoon, LICIACube, Deep Space, Cis-Lunar, Navigation, Orbit Determination, Flight Path Control, IRIS, Moon, Dimorphos, Didymos, Artemis-1, DART

Abstract CubeSats are becoming a reliable alternative for low-cost space applications in deep space, as mission companions or as standalone missions. The use of CubeSats in deep space requires to address many operational challenges, particularly those related to navigation. LICIACube and ArgoMoon are the first two 6U CubeSat missions to the outer space funded and managed by the Italian Space Agency, whose spacecrafts have been developed and operated by Argotec. The flight dynamics operations of both missions were performed by the flight dynamics team of the University of Bologna using NASA/JPL's navigation software MONTE. This paper gives a brief presentation of the flight dynamics operations of ArgoMoon and LICIACube and presents the obtained results highlighting the challenges of cis-lunar and deep space CubeSat navigation as well as the achieved successes.

Introduction

The CubeSat standard identifies a category of small satellites whose design is based on an elementary form factor of 1 U, that corresponds to a cube of 10 cm of latus [1]. The use of compact and lightweight spacecraft (S/C) enables a reduction of costs (design, assembly, integration, and launch) without sacrificing research objectives. Many CubeSats have been launched in Low Earth Orbit (LEO) [2], demonstrating that this small satellite technology is also trustworthy for complex missions other than educational or technological demonstrations. These characteristics made CubeSats appealing for deep space and cis-lunar exploration [3], either as stand-alone missions or as companions to conventional, larger spacecrafts. Two recent and relevant cis-lunar and deep space CubeSat missions, that successfully proved the capabilities of the platform and ground teams to operate in the outer space, are LICIACube and ArgoMoon.

Light Italian CubeSat for Imaging of Asteroids (LICIACube)

LICIACube is a 6U CubeSat mission of the Italian Space Agency (ASI) [4] that participated to the Double Asteroid Redirection Test (DART) mission of NASA [5]. The DART mission aimed to perform a technology demonstration to examine asteroid redirection by performing a controlled high velocity impact [6]. The objective of LICIACube was to take thorough and relevant photos of the effects of DART impact on Dimorphos. LICIACube was equipped with an active attitude determination and control system, a cold-gas orbital Propulsion System (PS), and a star tracker, while the core of the scientific payload was composed of two optical cameras [7]. LICIACube was placed into DART as a secondary payload and, after a cruise of 10 months, it was released into space on September 11, 2022, 15 days before the planned impact [8].

LICIACube ground-based navigation has been performed by the Radio Science and Planetary Exploration Laboratory (RSLab) of the University of Bologna (UNIBO), and independently also by the NASA's Jet Propulsion Laboratory (JPL), both using the NASA/JPL's orbit determination software MONTE [9]. The navigation strategy was based on two-way radiometric observables, Doppler and range, acquired by the Deep Space Network (DSN) [10]. The navigation aimed to fly LICIACube through a specific region of the Dimorphos B-plane defined from the mission high-level scientific requirements. The primary navigation requirements were to maintain the nominal trajectory and fly-by conditions to prevent collisions with the impact ejecta debris, maintain the impact scene within the cameras field of view, and prevent the saturation of the reaction wheels during the high-rate rotation phase to point at Dimorphos near the closest approach. The mission timeline included a calibration maneuver (CAL1) to check the thrusters, a targeting maneuver (Orbital Maneuver 1 - OM1) to address the aimpoint, and two clean-up maneuvers (OM2, OM3) to clear potential trajectory deviations during the operations.

On the first hours after the LICIACube deployment from DART, the acquired Doppler data showed a signature compatible with a tumbling motion caused by the S/C, which entered in safe mode because of the release event. However, the tumbling motion was then successfully damped by the CubeSat before the end of the first tracking pass. To properly fit the data of the first two tracking passes, a Doppler bias of ~1.86 Hz had to be estimated. Then, a commanded re-configuration of the LICIACube transponder during the second pass removed the latter bias, leaving only a small residual bias of ~0.025 Hz to be estimated. The Doppler biases were caused by a quantization error on the on-board digital IRIS [11] transponder. Stochastic accelerations were implemented and estimated to fit the data to the noise level. A detailed inspection of the stochastic accelerations shown signatures currently attributed to larger than expected non-gravitation accelerations. Despite the challenges, the team managed to design all the necessary maneuvers to reach the target point and to reconstruct the trajectory of LICIACube, satisfying all the navigation requirements. Figure 1 reports the Orbit Determination (OD) solutions of OM1, OM2, and OM3 deliveries, on the Dimorphos B-plane. As can be seen from the presented results, after OM2, the predicted trajectory and its 3-σ state uncertainty were widely contained in the requirement region and, as consequence, OM3 was scrubbed.

On September 26, 2022, LICIACube flew by Dimorphos at a distance of ~58 km and at a relative velocity of ~6.1 km/s [8], successfully acquiring the images before and after the DART impact on the asteroid including the ejection plume.

ArgoMoon

NASA selected 10 CubeSats as secondary payloads of the Space Launch System (SLS) for the Artemis-1 mission. Among them, ASI's ArgoMoon was chosen as an essential technological demonstrator [12]. The ArgoMoon S/C is based on the same 6U CubeSat platform of LICIACube, and the main goals of the mission were to autonomously fly around the Interim Cryogenic Propulsion Stage (ICPS), to capture images of the stage, and to confirm that the other CubeSats

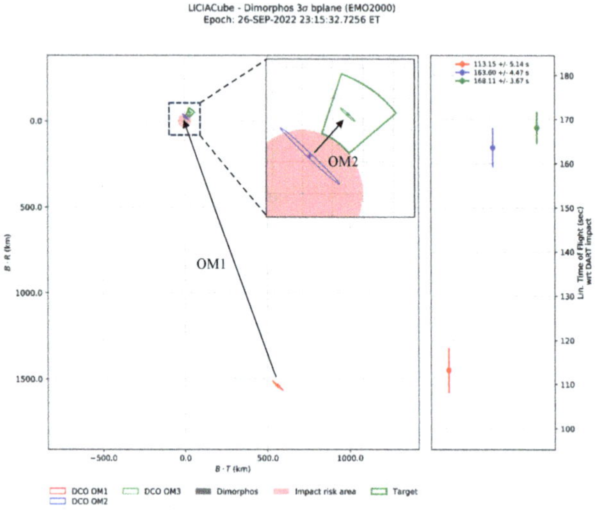

Figure 1: LICIACube predicted trajectory and 3-σ uncertainties mapped to the Dimorphos B-plane for the OD solutions of OM1, OM2 and OM3 deliveries.

were deployed during the first six hours of the mission. After the deployment, the mission foresaw a highly elliptical geocentric orbit for 180 days, with multiple encounters with the Moon.

As for LICIACube, the cis-lunar radiometric navigation of ArgoMoon was performed by the RSLab of UNIBO [13] using MONTE and exploited two-way radiometric data, Doppler and range, acquired by the DSN and the European Space Tracking (ESTRACK). The flight path control aimed to follow the reference trajectory through a dedicated optimal control strategy [13]. The ArgoMoon navigation requirements were designed to guarantee the correct pointing of the S/C from the DSN stations, prevent impacts with Earth and Moon, and allow the S/C disposal in a heliocentric orbit at the end of the mission.

ArgoMoon was successfully launched on November 16, 2022, at 06:47:44 UTC by the SLS. The S/C was released from ICPS after 3 hours and 49 minutes and the first signal acquisition successfully occurred at 10:37 UTC. During its mission, ArgoMoon performed 4 orbital maneuvers (Orbit Trim Maneuver 1B - OTM1B, Statistical Trim Maneuver 0 - STM0, STM1, STM2). However, no maneuver reached the commanded ΔV, showing a significantly underperforming thruster. Moreover, after STM0, a signature on the stochastic accelerations raised the doubt that the S/C thruster could have been leaking, but this assumption is still under investigation. A successful fly-by of the Moon was accomplished on November 21 at 16:07 UTC, about 48 minutes earlier than scheduled. Due to the differences between the commanded ΔV and that actually produced by thruster, the error on the B-plane with respect to the reference trajectory was of 5000 km, as can be seen from UBO007-10 of Figure 2. After the Moon's fly-by, ArgoMoon flew by the Earth at ~166000 km on November 24, 2022, at 18:38 UTC. The S/C was unable to follow a geocentric orbit due to the altered geometry of the Moon's fly-by, effectively entering into a heliocentric orbit.

During the operations, the navigation team delivered a total of 10 OD solutions (UBO001 to UBO010) and 5 orbital maneuvers (where 4 of them have been commanded and executed, and one used as backup) [14]. The quality of the ArgoMoon radiometric data were mostly affected by the

S/C's rotational dynamics, the IRIS radio design and configuration, and the on-board activates (for example safe mode, reboot, desaturation maneuvers).

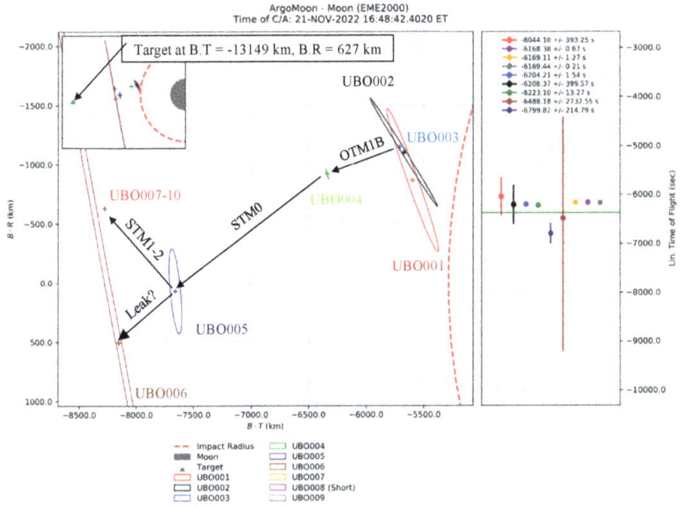

Figure 2: Summary of the delivered ArgoMoon OD solutions mapped on the Moon B-plane.

Conclusions

Using CubeSats in deep space requires addressing many operational challenges, particularly those related to the navigation, given the platform limitations related to the off-the-shelf components that are employed in the CubeSat philosophy and the strong requirements, similar to classical large deep space missions.

The navigation of LICIACube and ArgoMoon has proven the capability of CubeSats platforms to operate in deep space and achieve complex objectives. The navigation results show a performance on the residuals as good as any typical deep space mission. The Doppler and range residuals of LICIACube show a Root Mean Square (RMS) of 0.05 mm/s and 80 cm, respectively, while ArgoMoon had a RMS of 0.1 mm/s on Doppler and 30 cm on range. For both missions, the UNIBO navigation team was able to fulfill the navigation requirements, even with a very stringent contingency timeline, proving the reliability of the designed navigation procedures. Thanks to the obtained success, the pioneering flights carried on by LICIACube and ArgoMoon will surely provide a relevant heritage for the upcoming deep space and cis-lunar CubeSats missions.

References

[1] Johnstone, Alicia. CubeSat design specification Rev. 14.1 the CubeSat program. Cal Poly SLO (2022).

[2] Kulu, Erik. Nanosatellite Launch Forecasts-Track Record and Latest Prediction. (2022).

[3] Walker, Roger, et al. Deep-space CubeSats: thinking inside the box. Astronomy & Geophysics 59.5 (2018). https://doi.org/10.1093/astrogeo/aty232

[4] Dotto, Elisabetta, et al. LICIACube - The Light Italian Cubesat for Imaging of Asteroids In support of the NASA DART mission towards asteroid (65803) Didymos. Planet. Space Sci. 199, 105185 (2021). https://doi.org/10.1016/j.pss.2021.105185

[5] Rivkin, Andrew S., et al. The Double Asteroid Redirection Test (DART): Planetary Defense Investigations and Requirements, The Planetary Science Journal, 3(7), 153 (2022).

[6] Cheng, Andrew F. et al. Momentum transfer from the DART mission kinetic impact on asteroid Dimorphos, Nature (2023).

[7] Tortora, Paolo and Valerio, Di Tana. LICIACube, the Italian Witness of DART Impact on Didymos, 2019 IEEE 5th International Workshop on Metrology for AeroSpace (MetroAeroSpace), Turin, Italy, pp. 314-317 (2019). https://doi.org/10.1109/MetroAeroSpace.2019.8869672

[8] Dotto, Elisabetta and Zinzi, Angelo. Impact observations of asteroid Dimorphos via Light Italian CubeSat for imaging of asteroids (LICIACube). Nat Commun 14, 3055 (2023). https://doi.org/10.1038/s41467-023-38705-0

[9] Evans, Scott, et al. MONTE: The next generation of mission design and navigation software, CEAS Space Journal 10.1, pp. 79-86 (2018). https://doi.org/10.1007/s12567-017-0171-7

[10] Gai, Igor, et al. Orbit Determination of LICIACube: Expected Performance and Attainable Accuracy. Geophysical Research Abstracts. Vol. 21 (2019).

[11] Duncan, Courtney, and Amy Smith. IRIS Deep Space CubeSat Transponder. Proceedings of the 11th CubeSat Workshop, San Luis Obispo, CA (2014).

[12] Pirrotta, Simone, et al. ArgoMoon: the Italian cubesat for Artemis1 mission. European Planetary Science Congress (2021). https://doi.org/10.5194/epsc2021-879

[13] Lombardo, Marco, et al. Design and Analysis of the Cis-Lunar Navigation for the Argo-Moon CubeSat Mission, Aerospace, MDPI (2022). https://doi.org/10.3390/aerospace9110659

[14] Lombardo, Marco (2023). Navigation design and flight dynamics operations of the ArgoMoon Cis-Lunar CubeSat (Doctoral Thesis). Available on AMSDottorato, Alma Mater Studiorum - University of Bologna, Bologna, Italy.

Re-entry predictions of space objects and impact on air traffic

Franco Bernelli-Zazzera[1,a*], Camilla Colombo[1,b], Mattia Recchia[1,c]

[1] Department of Aerospace Science and Technology, Politecnico di Milano, Via La Masa 34, 20156 Milano, Italy

[a] franco.bernelli@polimi.it, [b] camilla.colombo@polimi.it, [c] mattia.recchia@mail.polimi.it

Keywords: Spacecraft Re-Entry, Breakup, Air Traffic

Abstract. This work focuses on predicting the re-entry of an uncontrolled re-entry vehicle (RV) and how this affects air traffic. It includes the propagation of the nominal trajectory and that of the fragments resulting from the breakup of the object. The breakup does not occur at a fixed altitude but is a consequence of the thermal and dynamic loads acting on the RV as it re-enters the atmosphere. The purpose of the analysis is to identify a dangerous area at specific heights (flight levels) to be evacuated in time for air traffic. The hazard area is defined as that which includes all the impact points of the fragments at this altitude taking into account the additional safety margins. The study also considers the presence of uncertainties affecting the initial state of the vehicle. Accordingly, a Monte Carlo analysis is performed to predict the worst-case scenario and to better estimate the hazard area. Once the area has been defined, an evacuation algorithm calculates, for each aircraft, the trajectory changes necessary to clear or avoid the zone over time.

Introduction

When a spacecraft, usually at the end of its life, leaves its nominal operating orbit, either due to some planned maneuver or natural decay caused by disturbances, and begins to approach increasingly dense atmospheric layers, it is said to be re-entering the Earth. Some reentry vehicles are designed to survive in Earth's atmosphere and be recovered, so they may have additional capabilities, such as the ability to develop lift forces to perform a soft landing [1]. Other vehicles, on the other hand, are not designed to withstand aerodynamic and thermal loads during the final phase of the trajectory and can suffer partial or total fragmentation. This case is called destructive reentry [2].

In this research, starting from the state of the vehicle at an altitude of 120 km, in which the reentry is supposed to start, the trajectory is propagated until the breakup conditions are met, thus defining the breakup point. At this altitude, the reentry vehicle (RV) is assumed to experience complete fragmentation caused by the high dynamic and thermal loads to which it is subjected. Therefore, a debris cloud is generated at the breakup point, composed of fragments each characterized by different parameters. Once the cloud of debris is generated, all fragments fall until they reach the level of interest (Flight Level) at which their dispersion is evaluated.

Since reentries are subject to many uncertainties, a Monte Carlo (MC) analysis is performed to account for some errors that may be present in the initial state of the vehicle and to evaluate how they affect the expected impact location. All fragment trajectories resulting from the MC analysis are then used to define the Hazard Area (HA) which includes all predicted debris locations and represents the area posing a risk to local air traffic. Once the danger zone has been defined, the affected aircraft must be redirected to evacuate or avoid the zone.

The procedures explained are applied to a simulated re-entry event in which real-traffic data are used to simulate a realistic scenario.

Methodology

The analysis starts by assigning the RV initial conditions, typically referring to the beginning of the reentry, arbitrary set at an altitude of 120 km [2]. The nominal trajectory, corresponding to the initial re-entry vehicle state, is propagated in time. At each propagation step, it is verified if the breakup conditions are satisfied. When the conditions are met, the RV breakups and the debris cloud is formed. For the cloud generation, it is assumed that the object experiences a complete fragmentation at a single altitude and all the fragments are generated at the same time. Once the debris are generated, for each one of them the new initial conditions are computed taking into account both the pre-breakup state and the velocity increment with the proper direction. At this point, all the fragments are considered as single independent entities and their trajectory is evaluated until they reach the altitude of interest, that could be ground level or an altitude corresponding to a particular Flight Level. Finally, the fragments dispersion is evaluated, in terms of longitude and latitude, at the altitude of interest. This debris distribution will be useful for the following study analyzing the effects on the air traffic.

Breakup Models. Two fragmentation methods are implemented in this work, both providing very similar results.

The first implemented model is based on the NASA Standard Breakup Model [3] and for this reason is called the NASA-Based Breakup Model (NBBM). The NASA Standard Breakup Model derives from the analyses of the fragmentation, due to both explosions and collisions, of spacecraft and rocket bodies in Low Earth Orbit (LEO) and it aims at defining each fragment with three different parameters: the characteristic length (L_c), the Area-to-Mass ratio (AM) and the velocity variation imparted (ΔV). All these features are described in terms of probability distributions. The models used in this research assume that the RV breakup occurs as a consequence of a fictitious collision with air. For this reason, the implemented power law distribution that provides the number of fragments of a given size and larger (N_{Lc}) is the one used in the NASA Breakup Model for collision events.

The second model proposed in this work is called Independent-Based Breakup Model (IBBM). This method tries to merge some features of the NASA Standard Breakup Model [3] with others implemented in the Independent discrete fragmentation model [4], which is applied mainly for asteroid entry analyses. Specifically, the IBBM implements the same distributions of the NASA Breakup Model for the computation of the fragment's characteristic length and Area-to-Mass ratio. The main difference between the IBBM and the NBBM is in the computation of the velocity variation.

In this work, both the dynamic and thermal loads are supposed to be able to cause the RV's breakup. In particular, the complete fragmentation is triggered whether the dynamic pressure acting on the vehicle exceed its ultimate tensile strength or if the temperature reaches the melting point of the material composing the RV. For this purpose, the RV is assumed to be made entirely of aluminum and a relation linking the aluminum ultimate tensile strength to the material temperature is implemented.

Dynamics. The nominal trajectory and the post-breakup fragments trajectories are evaluated adopting the following assumptions. The RV is a non-lifting object, not capable of generating any lift force ($L = 0$). The motion is over a spherical, non-rotating Earth ($\omega_E = 0$). A ballistic entry is assumed, with no thrust force ($T = 0$) and no propellant mass flow ($\dot{\boldsymbol{m}} = \boldsymbol{0}$). The mass ablation of both the RV and the related fragments is neglected.

With the assumptions just mentioned, the equations of motion become the following set of six first order ordinary differential equations (ODEs) [1]:

$$\begin{cases} \dot{h} = v \sin \gamma \\ \dot{\lambda} = \frac{v \cos \gamma \sin \psi}{r \cos \varphi} \\ \dot{\varphi} = \frac{v}{r} \cos \gamma \cos \psi \\ \dot{v} = -\frac{D}{m} - g \sin \gamma \\ \dot{\gamma} = -\left(g - \frac{v^2}{r}\right) \frac{\cos \gamma}{v} \\ \dot{\psi} = \frac{v}{r} \cos \gamma \sin \psi \tan \varphi \end{cases} \tag{1}$$

where D is the drag force and the state $x = \{h; \lambda; \varphi; v; \gamma; \psi\}$, is composed, respectively, by the altitude, longitude, latitude, velocity, climb angle and heading angle.

Monte Carlo Analysis. Earth re-entries are affected by lots of uncertainties that could have non negligible effects on the prediction of the hazard area that poses risk to the air traffic. To statistically predict that area, a Monte Carlo (MC) analysis is performed. Each sample generated for the MC analysis represents a set of new initial conditions at the nominal altitude of 120 km. The re-entry simulation is therefore repeated for each sample and the resulting debris dispersion at Flight Level 400 (FL400) are recorded. Once the MC simulation is completed, an area enclosing all the fragments footprints at FL400 can be defined. The final Hazard Area is then retrieved by adding some additional safety margin. Figure 1 shows both the pre and post-breakup trajectories, while Figure 2 reports both the fragments dispersion at the altitude of interest and the computed Hazard Area.

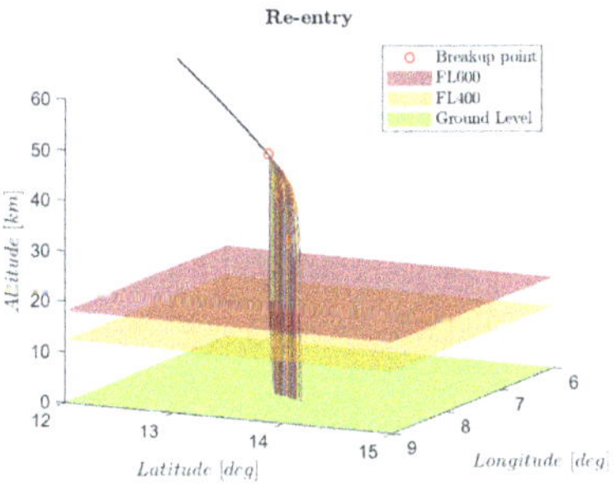

Figure 1: Re-entry trajectories.

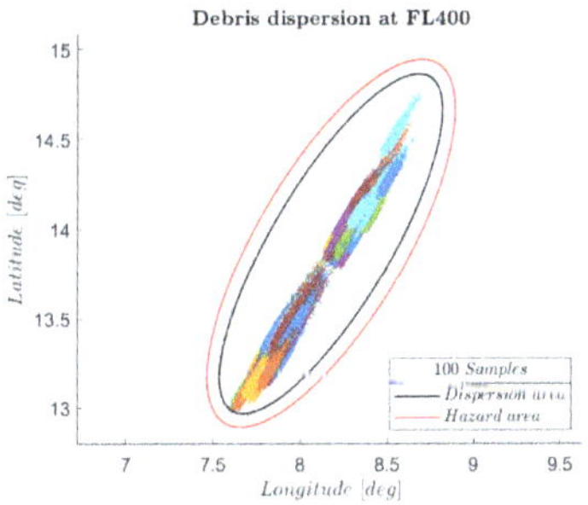

Figure 2: Hazard area and debris dispersion.

Air Traffic management. Two different actions, inferred from [5], are proposed for the management of the air traffic in presence of an hazard area.

The first algorithm is used to evacuate aircraft that are within the hazard area at the time it is computed. The algorithm's logic is to find the required changes in the aircraft heading which allow the shortest evacuation time. The following assumptions are made: 1) the aircraft are assumed to move with constant velocity during all the operations; 2) after completing the required turns, the aircraft move on a straight trajectory; 3) it is assumed that aircraft can only perform horizontal maneuvers.

The second algorithm is applied to an aircraft which is outside the hazard area at the moment it is computed but it is expected to enter it. The following assumptions are made: 1) the aircraft are assumed to move with constant velocity throughout the path; 2) the nominal path is assumed to be aligned in the same direction of the initial velocity vector; 3) turn maneuvers are not considered in the computation of the alternative path and aircraft are supposed to be able to turn instantaneously; 4) only horizontal maneuvers are taken into account. The algorithm then computes an alternative flight path that allows the aircraft to avoid the hazard area.

Figures 3 and 4 show examples of the evacuation and avoidance procedures.

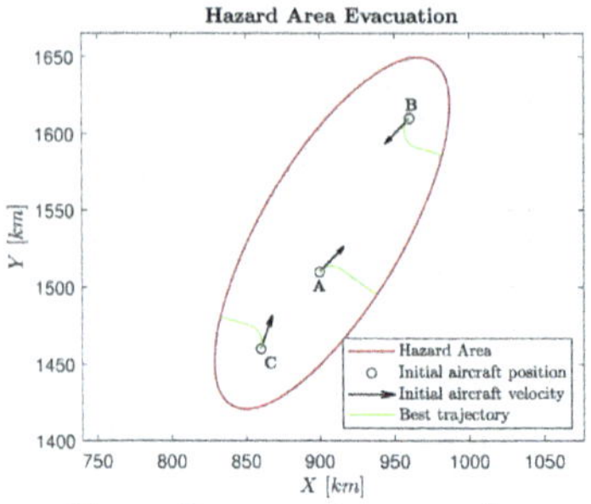

Figure 3: evacuation procedures.

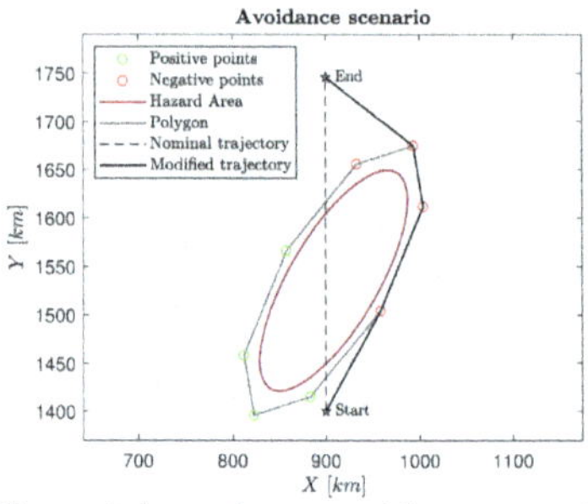

Figure 4: hazard area avoidance procedure.

Real Scenario

A generic reentry event is analyzed, leading to the definition of a Hazard Area. Then, to assess the re-entry impact on the local air traffic, real traffic data are retrieved from Flightradar24 [6] filtering only the flights at the altitude corresponding to the FL400 (40000 ft) and at a particular time instant. Finally, the algorithms discussed are used to manage the air traffic according to the aircraft positions. The simulated path are shown in Figure 5 in which stars represent the aircraft initial positions while dots the final ones.

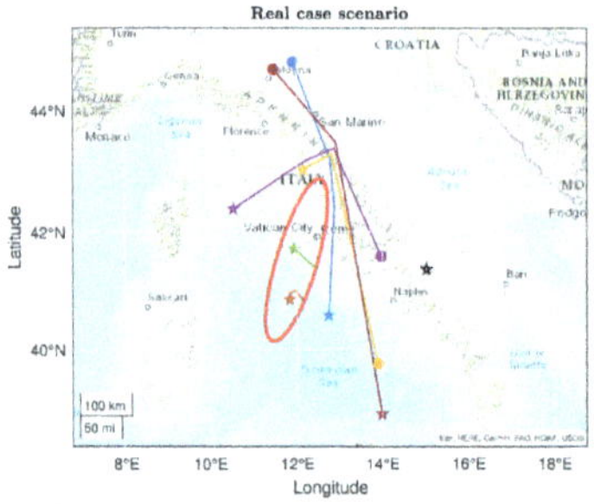

Figure 5: Evacuation and avoidance simulated paths.

Conclusions

This work has shown a preliminary assessment of the impacts Earth's re-entries have on the air traffic. Some future developments can be easily integrated while keeping the overall structure intact. Particularly, improvements in the breakup models with the introduction of new distribution functions that better describe the atmosphere's fragmentation can be integrated. Also the air traffic management algorithms could be upgraded with new procedures that ensure minimization of the impact on the routes while still avoiding collisions.

References

[1] F. J. Regan and S. M. Anandakrishnan. Dynamics of Atmospheric Re-Entry. AIAA Education Series. American Institute of Aeronautics and Astronautics, 1993. ISBN 9781600860461. https://doi.org/10.2514/4.861741

[2] F. Sanson. On-ground risk estimation of reentering human-made space objects. PhD thesis, 09 2019.

[3] N.L. Johnson, P. Krisko, J.-C Liou, and P. Anz-Meador. Nasa's new breakup model of evolve 4.0. Advances in Space Research, 28:1377–1384, 12 2001.. https://doi.org/10.1016/S0273-1177(01)00423-9

[4] P. Mehta, E. Minisci, and M. Vasile. Breakup modelling and trajectory simulation under uncertainty for asteroids. 04 2015.

[5] Ganghuai W., Zheng T., T. Masek, and J. Schwartz. A monte carlo simulation tool for evaluating space launch and re-entry operations. In 2016 Integrated Communications Navigation and Surveillance (ICNS), pages 9A3–1–9A3–15, 2016. dhttps://doi.org/10.1109/ICNSURV.2016.7486392

[6] Flightradar24. flightradar24:live air traffic, Accessed on 25 May 2023.

Space Propulsion

Update on green chemical propulsion activities and achievements by the University of Padua and its spin-off T4I

F. Barato[1]*, A. Ruffin[2], M. Santi[2], M. Fagherazzi[2], N. Bellomo[2] and D. Pavarin[1,2]

[1]Department of Industrial Engineering, University of Padua, Via Venezia 1, 35131, Padua, Italy

[2]Technology for Propulsion and Innovation S.p.a, Via Emilia 15, 35043, Monselice, Italy

* francesco.barato@unipd.it

Keywords: Hydrogen Peroxide, Green Chemical Propulsion, Hybrid Rockets, Liquid Thrusters, Sounding Rocket, Throttling, Regenerative Cooling, Additive Manufacturing

Abstract. In recent years, there has been a great research interest on green propulsion, both for environmental, cost and ease-of-use considerations, further accelerated by the needs of the NewSpace Economy. Hydrogen peroxide is a green and versatile propellant that is suitable for a lot of different uses in space applications. Following a previous AIDAA publication of 2019, this paper updates the research performed on hydrogen peroxide-based propulsion by the University of Padua and its spin-off T4i with the latest achievements. Starting from the simplest propulsion systems, several monopropellant thrusters have been successfully designed and tested, ranging from a propulsion module of 1 N, to a 10 N and 200 N flight-weight items. The thrusters can operate in blowdown or pressure-regulated mode, and they have been tested for hundreds of seconds of continuous operation and for thousands of pulses. A 450 N liquid bipropellant motor that burns the monopropellant exhausts with diesel fuel has also been developed and tested. The motor uses an unconventional internal vortex flow field to achieve stability, efficiency, and self-cooling of the chamber. The nozzle throat region temperature is kept under control by regenerative cooling channels fed by the peroxide. All thrusters make extensive use of additive manufacturing. The hydrogen peroxide technology has also been applied on hybrid propulsion, which was the initial main expertise of the Padua University propulsion group. Hundreds of tests have been performed at lab-scale, mainly with paraffin wax and polyethylene as fuels, with burning time up to 80 seconds. The motors are able to start, stop and restart multiple times. A cavitating pintle valve has been developed in house in order to control the oxidizer mass flow. With this valve, the hybrid motors are able to throttle the thrust in a range of 1:12.6. A similar valve has been also employed in the integrated monopropellant propulsion system of a lunar drone, composed by a 400 N throttleable engine together with 4 small 14 N on-off attitude control thrusters. Moreover, several dozens of hybrid tests have been performed at 5-10 kN scale up to 50 seconds. Finally, a composite sounding rocket powered by a pressure-regulated 5 kN hybrid rocket has been fully designed and successfully flight tested.

Introduction

The chemical propulsion group at University of Padua was established around 2006, working on hybrid propulsion with green oxidizer (N_2O, GOX) and plastic fuels, mainly HDPE and paraffin wax. For the story before the shift to hydrogen peroxide, the reader is referred to a previous AIDAA paper [1]. Since 2014, the propulsion team and its spin-off Technology for Propulsion and Innovation (T4i) have been focused their effort on the development of hydrogen peroxide-based propulsion systems.

The choice of hydrogen peroxide is due to the fact that is a very versatile green propellant because it can be decomposed relatively easily in liquid phase and can be used in restartable and throttleable liquid monopropellants, bipropellants, hybrids and gas generators. Moreover, it can be

Materials Research Proceedings 37 (2023) 645-653 https://doi.org/10.21741/9781644902813-140

stored at room temperatures at any pressure and can feed the engines with very repeatable performance.

The primary feature of this research effort is the use of stabilized hydrogen peroxide concentrated in-situ from commercial feedstock. A distillation plant capable of concentrating 1 kg/hour of hydrogen peroxide from 60% to 92% has been operated for years with little maintenance. The plant runs autonomously 24/7 and has concentrated several tons of propellant up to now. In little less than a decade the group has performed nearly a thousand monopropellant, bipropellant and hybrid rocket tests with hydrogen peroxide. In the following paragraphs the different types of motors will be described.

Monopropellant propulsion

Several monopropellant systems, ranging from 1N to 200 N, have been developed in order to operate as main engines and/or attitude control thrusters for space vehicles [2-3].

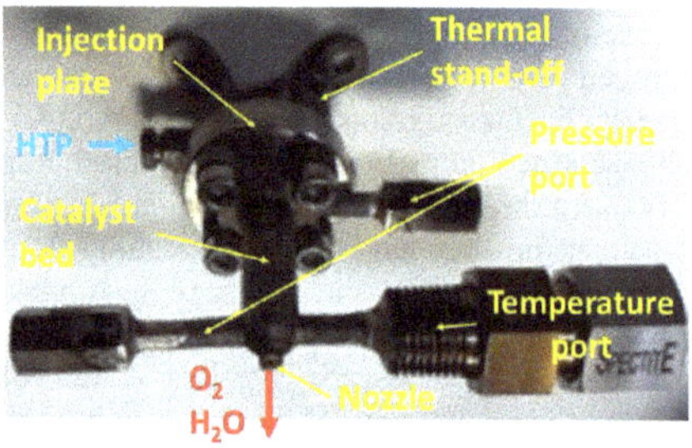

Fig 1. 1 N monopropellant thruster engineering model

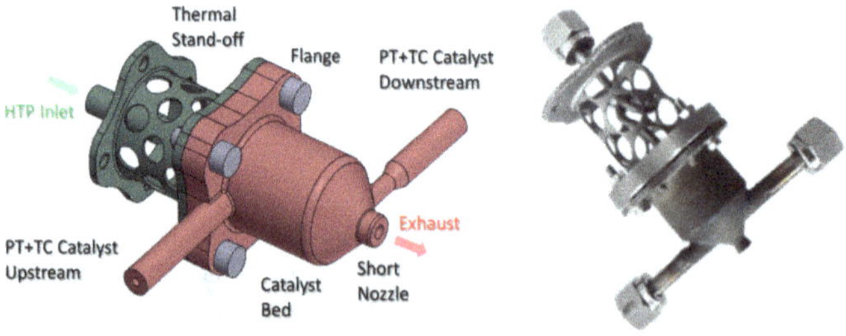

Fig 2. 10 N monopropellant thruster engineering model

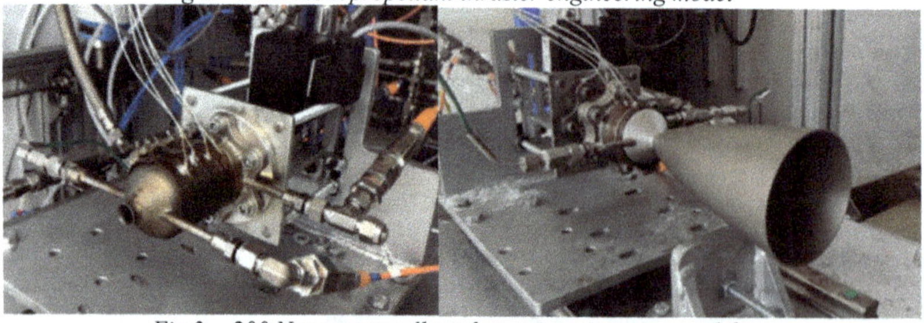

Fig 3. 200 N monopropellant thruster engineering models

Materials Research Proceedings 37 (2023) 645-653 https://doi.org/10.21741/9781644902813-140

The thrusters can be used both in blowdown and pressure-regulated mode from 5 bar to more than 40 bar. The thrusters are actuated with on-off solenoid valve and operated in continuous and bang-bang mode. Efficiencies above 95% have been achieved with continuous firing times up to above 1000 s. More than 4000 pulses have also been demonstrated. Depending on thruster size, valve timing can go from 30 to 100 ms. The catalyst is able of cold starting without pre-heating. The thrusters are manufactured with 3D printing Selective Laser Melting (SLM) technology. The engineering models have flanges to disassemble the thrusters and pressure/temperature sensor ports while the flight weight units have a minimum number of interfaces/components.

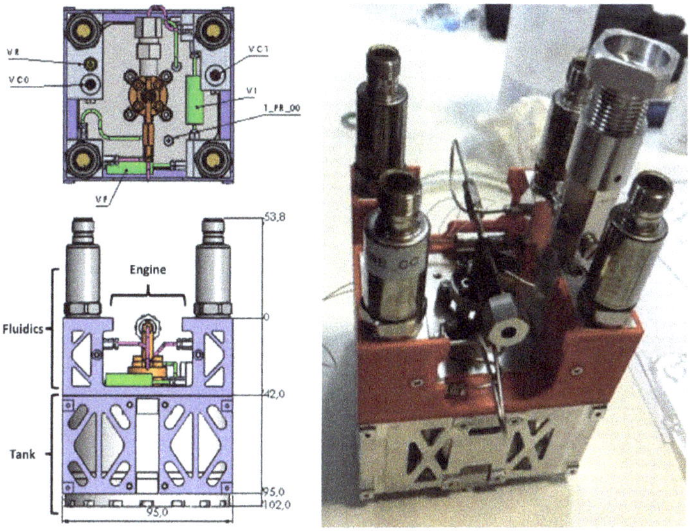

Fig 4. 1U monopropellant propulsion module engineering model

An entire monopropellant propulsion module has been developed in the frame of the PM^3 project, a modular multi-mission platform founded by the Italian Ministry of Education, Universities and Research [4]. The fundamental objective of the project was the study of a 50 kg class satellite platform characterized by the ability to accommodate multiple interoperable payloads. The propulsion system architecture is based on a simple unregulated blowdown discharge starting from a MEOP of 50 bars. The engine initial thrust in vacuum is 1N and then slowly decreases to 0.5 N at EOL as the tank pressure decreases. The fluidic line is composed by a custom piston-separated tank and few COTS components: an isolation valve, the fill and drain valves, and the firing valve. The tank is designed to be easily extended to increase propellant volume and meet the additional total impulse that may be required for other missions. Despite the tested system being an engineering model, the overall design has been flight-oriented, including the required amount of propellant in the 1U envelope. Only pressure sensors used only for test monitoring have been accommodated outside this envelope.

Fig 5. Moon Drone propulsion system (laid upside down): 400 N throttleable main engine (top) and four bang-bang 14 N attitude thrusters (bottom)

Fig 6. Moon Drone 400 N Main Engine Flow Control Valve

Another program involving monopropellant hydrogen peroxide is Moon drone, a small platform scouting the surrounding environment that has been proposed in support of a rover mission on the Moon [5]. The design of the Moon Drone has been performed through an ESA TRP study lead by Thales Alenia Space with partners GMV, Brno University of Technology and T4i. An Earth-related flight prototype will be tested within the program and T4i is in charge of all the thrusters' development with the support of the University of Padova. After a trade-off between several possibilities performed by TAS with T4i/UNIPD support, the propulsion design proceeded with a configuration composed by a single throttleable main engine used for the main displacements aided by 4 small thrusters operated in bang-bang for attitude control. All the thrusters were fed by the same tank, which is pressurized by nitrogen. The main engine has 400 N maximum thrust; it has a continuous regulating cavitating pintle flow control valve driven by a stepper motor in feedback, a development from a previous one already developed in-house for a hybrid rocket. The rearranged flow control valve differs from the older version for the valve body that has been optimized and reduced consistently in weight. The attitude control thrusters have a maximum thrust of 14 N, they are operated with a commercial on-off solenoid valve controlled at 10 Hz. Both types of thrusters have been designed, manufactured and thoroughly tested, demonstrating the proper fulfillment of the defined specifications and they are ready for the flight campaign of the Moon Drone prototype.

Bipropellant propulsion

A 450 N hydrogen peroxide-based bipropellant liquid engine has been designed and tested. The motor is a staged combustion engine that features a vortex-cooled combustion chamber based on a swirled oxidizer injection and uses standard automotive diesel as fuel, which is injected on the catalytic decomposed peroxide stream [6]. The cooling solution for the thrust chamber is characterized by a double co-spinning counter-flowing vortex flow. It is well known that a swirled flow improves mixing and residence time thus enhancing the combustion efficiency. Moreover, this particular flowfield allows the flame to be trapped in the inner vortex while the outer one composed only by the oxidizer act as a shield that extracts heat from the chamber walls. The motor has been successfully tested, achieving smooth ignition and shut down, stable steady combustion and efficiencies above 96%.

Afterwards a regenerative cooling for the nozzle throat region with H_2O_2 has been designed through a numerical steady 1-D code [7-8]. The nozzle with its internal channels has been produced by additive manufacturing in Inconel® 718. The cooling has been tested successfully, demonstrating the capability to keep the metal parts at reasonable temperatures in steady state and showing only moderate heating of the liquid H_2O_2.

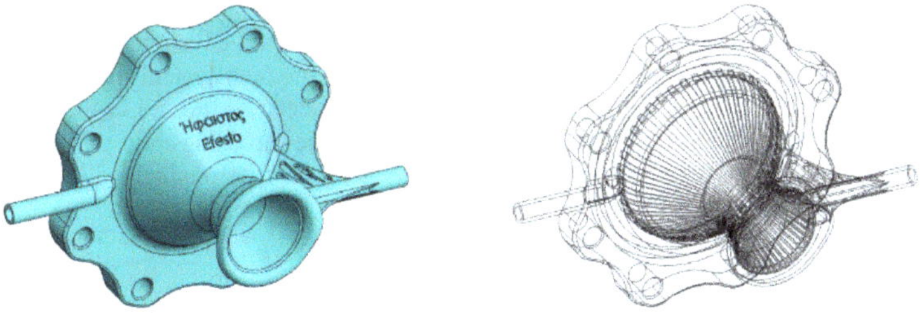

Fig 7. 450 N liquid thruster regenerative nozzle: external view (left), lattice view (right)

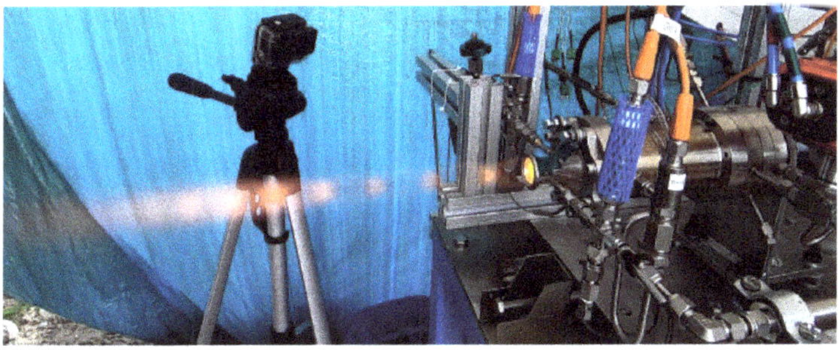

Fig 8. 450 N liquid bipropellant fire test with regenerative cooling

Hybrid propulsion

Starting in 2014, hundreds of hybrid rocket tests with hydrogen peroxide have been performed at lab-scale (100-1000 N). Hybrid firing times up to 80 s have been achieved [9]. More than 50 scale up tests with thrust above 5 kN (sea level) have also been performed up to date.

Thanks to the catalyst decomposition of the hydrogen peroxide, the hybrid motors have the capability to cold start, to run stable and efficiently, to stop and restart multiple times and to be throttled. The motors can be adapted to different missions in terms of thrust and burning times, tailoring the regression rate level of the fuel by varying the intensity of the swirled injection [10].

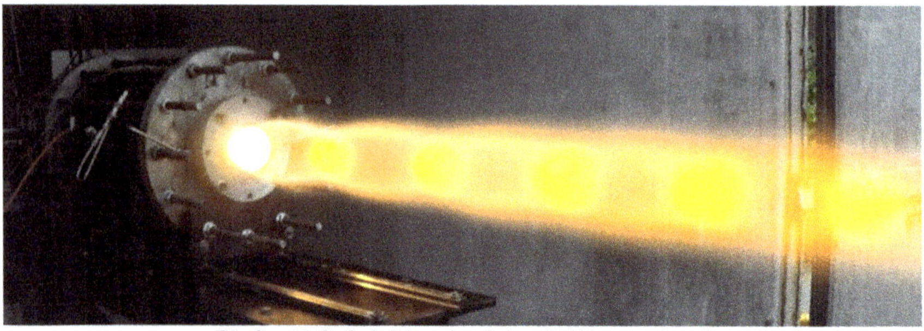

Fig 9. H_2O_2 hybrid rocket firing (5 kN at sea level)

A cavitating venturi variable pintle flow control valve has been developed in house [11]. The valve is able to choke the oxidizer mass flow and decouple the feed system from the combustion chamber dynamic. Afterwards a stepper electric motor has been connected to the movable flow control valve [12]. With this set-up an outstanding real time throttling ratio of 12.6:1 has been achieved showing the possibility to perform different thrust profiles on demand [13]. A remotely controlled human manual throttling test has been also performed.

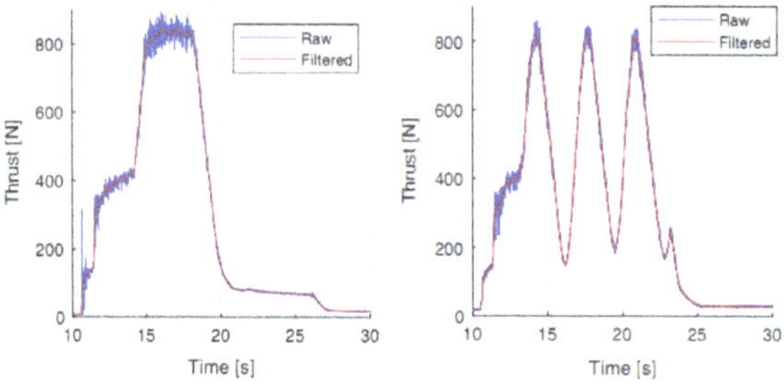

Fig 10. Throttling of a H_2O_2 hybrid rocket: step command (left), sinusoidal command (right)

Fig 11. Throttling of a H_2O_2 hybrid rocket: max thrust (left), min thrust (right)

Afterwards, the team started to develop a 200 mm diameter, 6 m long sounding rocket propelled by a 5 kN thrust hydrogen peroxide-paraffin hybrid rocket [14]. The aim of the passive, aerodynamically stabilized, sounding rocket was to serve as a flight test bed for new technologies in the structures and propulsion system. The sounding rocket was finally launched succesfully on February 24, 2022, from the Poligono Interforze of Salto di Quirra (PISQ) in Sardinia, within the project Aviolancio (Air-launch), coordinated by the Italian Research Center (Consiglio Nazionale delle Ricerche, CNR) and the Italian Air Force (Aeronautica Militare Italiana, AMI).

Fig 12. 5 kN H_2O_2 sounding rocket: on the ramp (left), at launch (right)

Conclusions

The University of Padua and its spin-off company T4i have been conducting research on green propulsion using hydrogen peroxide as a propellant since 2014. Hydrogen peroxide is a very versatile chemical that can be used in multiple propulsive applications.

The team has successfully designed and tested various monopropellant thrusters, ranging from 1 N to 200 N. These thrusters can operate in blowdown or pressure-regulated mode and have been tested for continuous and pulsed operations. Efficiencies above 95% have been achieved with continuous firing times up to above 1000 s. More than 4000 pulses have also been demonstrated. Depending on thruster size, valve timing can go from 30 to 100 ms. An entire 1U-1N unregulated blowdown pressure-fed propulsion unit has also been developed and tested.

The integrated propulsion system for a lunar drone, which includes a 400 N throttleable monoprop main engine and four 14 N bang-bang attitude control thrusters have been also developed and successfully tested. The main engine is actuated by a cavitating venturi pintle flow control valve developed in-house.

The team has also developed and tested a 450 N liquid bipropellant motor that burns the monopropellant exhausts with diesel fuel. This motor utilizes an unconventional internal vortex flow field for stability, efficiency, and self-cooling of the combustion chamber. The nozzle throat region is regeneratively cooled with the H_2O_2. The motor has been successfully tested, achieving smooth ignition and shut down, efficiencies above 96%, stable steady combustion and proper thermomechanical behavior.

Both the monopropellant thrusters and the liquid motor extensively employ additive manufacturing techniques to reduce the number of parts and allow complex design features.

Finally, the hydrogen peroxide technology has also been applied to hybrid propulsion, initially the main expertise of the Padua University propulsion group. Numerous lab-scale tests have been carried out, mainly with paraffin wax and polyethylene as fuels, achieving burning times of up to 80 seconds. The hybrid motors can start, stop, and restart multiple times, again utilizing a cavitating pintle valve to control oxidizer mass flow and enable deep thrust throttling. Additionally, several dozen hybrid tests have been performed at 5-10 kN scale for up to 50 seconds. Lastly, a composite sounding rocket powered by a pressure-regulated 5 kN hybrid rocket has been designed and successfully flight tested.

References

[1] F. Barato, N. Bellomo, A. Ruffin, E. Paccagnella, M. Santi, M. Franco and D. Pavarin, Status and Achievements of the Hydrogen Peroxide Chemical Propulsion Research at Padua University, Italian Association of Aeronautics and Astronautics (AIDAA) XXV International Congress, 9-12 September 2019, Rome, Italy.

[2] M. Santi, I. Dorgnach, F. Barato, D. Pavarin. Design and Testing of a 3D Printed 10 N Hydrogen Peroxide Monopropellant Thruster, AIAA Propulsion and Energy Forum, Indianapolis, IN, USA (2019). https://doi.org/10.2514/6.2019-4277

[3] D. Nissan, M. Santi, F. Barato, D. Pavarin, Testing of a Small HTP Monopropellant Thruster for Space Applications, International Astronautical Congress, Paris, France, 18-22 September 2022.

[4] M. Santi, L. Gerolin, D. Antelo, B. Montanari, M. Fagherazzi, F. Barato, D. Pavarin, Development and testing of an engineering model of a hydrogen peroxide based 1N propulsion unit, International Astronautical Congress, Paris, France, 18-22 September 2022.

[5] A. Ruffin, M. Fagherazzi, N. Bellomo, F. Barato, D. Pavarin, M. Pessana, Development of the Propulsion System for a Moon Drone Vehicle Demonstrator, AIAA 2021-3564, AIAA Propulsion and Energy 2021 Forum, 9-11 August, Virtual Event. https://arc.aiaa.org/doi/abs/10.2514/6.2021-3564

[6] M. Santi, M. Fagherazzi, F. Barato, D. Pavarin, Design and Testing of a Hydrogen Peroxide Bipropellant Thruster, AIAA 2020-3827, AIAA Propulsion and Energy 2020 Forum, 24-28 August, Virtual Event. https://doi.org/10.2514/6.2020-3827.

[7] M. Fagherazzi, M. Santi, F. Barato, D. Pavarin, Design and Testing of a 3D Printed Regenerative Cooled Nozzle for a Hydrogen Peroxide based Bi-Propellant Thruster, AIAA 2021-3235, AIAA Propulsion and Energy 2021 Forum, 9-11 August, Virtual Event. https://arc.aiaa.org/doi/10.2514/6.2021-3235.c1

[8] M. Fagherazzi, M. Santi, F. Barato, M. Pizzarelli, Simplified Thermal Analysis Model for Regeneratively Cooled Rocket Engine Thrust Chambers and Its Calibration with Experimental Data, Aerospace 2023, 10(5), 403; https://doi.org/10.3390/aerospace10050403

[9] E. Paccagnella, M. Santi, A. Ruffin, F. Barato, D. Pavarin, G. Misté, G. Venturelli, N. Bellomo. Testing of a Long-Burning-Time Paraffin-based Hybrid Rocket Motor. Journal of Propulsion and Power, Vol. 35, No. 2, pp. 432-442 (2019). https://doi.org/10.2514/1.B37144

[10] M. Franco, F. Barato, E. Paccagnella, M.Santi, A. Battiston, A. Comazzetto and D. Pavarin, Regression Rate Design Tailoring Through Vortex Injection in Hybrid Rocket Motors, Journal of Spacecraft and Rockets, 7 November 2019. https://doi.org/10.2514/1.A34539

[11] A. Ruffin, F. Barato, M. Santi, E. Paccagnella, N. Bellomo, G. Misté, G. Venturelli, D. Pavarin. Development of a Cavitating Pintle for a Throttleable Hybrid Rocket Motor, 7th European Conference for Aeronautics and Space Sciences (EUCASS), Milan, Italy (2017).

[12] A. Ruffin, M. Santi, E. Paccagnella, F. Barato, N. Bellomo G. Miste, G. Venturelli, D. Pavarin. Development of a Flow Control Valve for a Throttleable Hybrid Rocket Motor and Throttling Fire Tests, 54th AIAA/SAE/ASEE Joint Propulsion Conference, Cincinnati, OH, USA (2018). https://doi.org/10.2514/6.2018-4664

[13] A. Ruffin, E. Paccagnella, M. Santi, F. Barato, D. Pavarin. Real Time Deep Throttling Tests of a Hydrogen Peroxide Hybrid Rocket Motor, Journal of Propulsion and Power, Published Online: 4 April 2022. https://doi.org/10.2514/1.B38504

[14] F. Barato, D. Pavarin, Advanced Low-Cost Hypersonic Flight Test Platforms, HiSST: 2nd International Conference on High-Speed Vehicle Science Technology, 11–15 September 2022, Bruges, Belgium.

1D numerical simulations aimed to reproduce the operative conditions of a LOx/LCH$_4$ engine demonstrator

Angelo Romano[1,a*], Daniele Ricci[1,b], Francesco Battista[1,c]

[1]CIRA – Italian Aerospace Research Center, Italy

[a]a.romano@cira.it, [b]d.ricci@cira.it, [c]f.battista@cira.it

Keywords: Liquid Rocket Engine, LOx/LCH$_4$, EcosimPro Simulations, Regenerative Cooling

Abstract. The present paper describes the results of the numerical simulations performed by means of the *"EcosimPro"* software, aimed at reproducing the operative conditions of the regenerative thrust chamber "DEMO-0A" designed by the Italian Aerospace Research Center. The operative conditions simulated are both cold flow and firing conditions. A validation of the numerical cold flow results has been performed by comparing them with the experimental data gathered during a cold flow campaign. Once validated the cold flow numerical model, various hot test conditions of the demonstrator have been simulated by considering different heat wall exchange coefficient correlations, in order to obtain information about the thermal power released during the combustion process and to assess the simulation capabilities of the *"EcosimPro"* software in predicting the behaviour of the demonstrator in firing conditions by modelling it with a 1-D approach.

Introduction

The utilization of liquid oxygen/liquid methane couple (LOx/LCH$_4$) as a potential candidate to substitute hypergolic propellants in the next future propulsion systems has arisen an increasing interest due to the advantages offered in terms of high specific impulse, cooling capabilities, re-usability and low environmental impact [1]. In this perspective, the Italian Aerospace Research Center manages the "HYPROB" research program, which includes also the realization of a LOx/LCH$_4$ demonstrator engine named "DEMO-0A".

The thermal exchange in a liquid rocket engine regeneratively cooled represents a coupled heat transfer problem between the hot gases, the chamber wall and the coolant in the channels. Various approaches have been used to solve this problem: the possibility to use a 3-D modelling for the heat conduction through the wall and a 2-D approach, based on semiempirical correlations for the coolant and the hot gas flows, has been evaluated in [2], [3]. To overcome the complexity and the computational cost introduced by 3-D approaches, simplified quasi-2-D models have been extensively used to solve the heat transfer problem ([4], [5], [6]).

The aim of the present paper is to model the "DEMO-0A" engine designed and realized by the Italian Aerospace Research Center (CIRA) by means of 1-D components offered by the *EcosimPro* software. In particular, a series of numerical simulations have been performed to simulate cold flow and firing conditions, in which the coupled heat transfer problem between the hot gases, the chamber liner and the coolant has been solved by a 1-D approach. The scopes of these simulations, performed with a 1-D approach, consisted in: 1) validating the numerical results of the cooling system by a comparison with the experimental cold flow results; 2) investigating the effects on the thermal power released by the combustion chamber by considering different wall heat exchange semiempirical correlations; 3) assessing the capabilities of the 1-D model implemented in *EcosimPro* to predict the behaviour of the demonstrator in firing conditions.

Materials Research Proceedings 37 (2023) 654-659 https://doi.org/10.21741/9781644902813-141

The "DEMO-0A" Thrust chamber assembly

The "DEMO-0A" is a 30 kN thrust class demonstrator, fed with LOx/LCH_4, technologically representative of a thrust chamber assembly of an expander engine, regeneratively cooled by a counter-flow cooling jacket made up of 96 axial channels. The cooling channels of the "DEMO-0A" are obtained by joining an inner liner made of a cooper alloy (CuCrZr) with two outer layers (the first one made of cooper and the second one of nickel). The outer layers are deposited on the inner one by means of the electrodeposition technology. In a former version of the demonstrator the cooling channels were obtained by brazing the inner layer with an Inconel outer layer. Both liquid methane and water can be used to cool the engine ([7], [8]).

Figure 1 shows a model of the demonstrator that includes the igniter, the injector head with 18 coaxial recessed injectors and the thrust chamber with inlet/outlet manifolds for LCH_4 (or water).

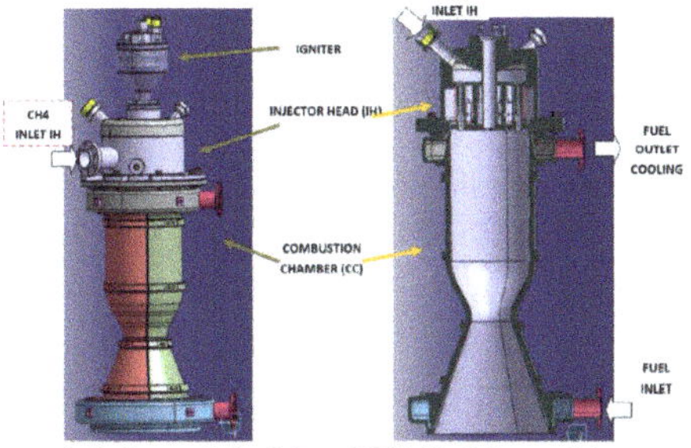

Figure 1:3-D model of the «DEMO-0A» architecture

The main geometrical and performance parameters are reported in Table 1.

Table 1

"DEMO-0A" performance and geometrical parameters			
Performance		Geometry	
T [kN] @ sea level	23.7	L [mm]	440
I_{sp} [s]	286	$D_{chamber}$ [mm]	119.6
$\dot{m}$ LCH_4 [kg/s]	1.92	D_{throat} [mm]	59.8

Methodology

The numerical simulations have been performed by means of the *EcosimPro* software, a simulation tool that allows to model continuous and discrete systems.

The equations solved in the cold flow condition simulation are the mass conservation and the momentum conservation equations written under steady state conditions hypothesis ([9], [10]).

On the contrary, the system of governing equations of the combustor and cooling jacket coupling consists in the continuity equation written for the vaporized propellants, the momentum equation and the energy equation, all written under the hypothesis of steady state conditions [9]. The vapours are assumed to be fully released at the first node of the grid discretizing the combustor component assuming a characteristic vaporization time τ_{vap} set to 0.1 ms. The energy equation contains the term modelling the heat exchange between the hot gases, the chamber liner and the cooling jacket. Both convection and radiation are considered and *EcosimPro* offers three empirical correlations to calculate the wall heat exchange coefficient h_c: the Bartz, the modified Bartz and the Pavli correlations. The heat conduction in the cooling jacket liner is modelled by the Fourier equation. Regarding the combustion modelling a delayed equilibrium model has been chosen, in so doing a non-equilibrium combustor is simulated by introducing a time delay between the equilibrium condition and the actual burnt gases composition.

The discretization scheme considered is the AUSM with a 2nd order accuracy and the integration method is the «CVODE_BDF_SPARSE» [9] with transient and steady tolerances set to $1 * 10^{-6}$ as recommended in [9] in order to reduce the simulation time. The final time of the simulation has been set to 20 s.

Results

The cold flow experimental campaign was carried out at the AVIO/ASI FAST2 facility in Colleferro (Rome) and was devoted to perform tests in order to measure the pressure drops along the cooling channels of the regenerative cooling system and to define a characteristic law to predict the pressure drops of the "DEMO-0A" cooling system.

A mesh sensitivity analysis has been performed in order to individuate the discretizing mesh that allows to obtain a good quality solution with the minimum number of nodes and in the end the mesh independence turned out to be a discretizing grid of 35 nodes.

The results of three out of the six cold flow tests simulations performed in *EcosimPro* in terms of coolant pressure drop are reported in Table 2. The numerical results slightly underestimate the experimental ones and it is possible to see that the estimation error is between 2% and 3%.

The water pressure profile along the cooling jacket is reported in Figure 2(a) (red line) and allows to appreciate that the larger pressure gradients are concentrated in the throat region, as expected.

Table 2

		Cold flow tests results		
		EcosimPro	Experimental	$\varepsilon = \frac{\Delta X_{Eco} - \Delta X_{Exp}}{\Delta X_{Exp}}$
Test01	Δ_P_cool. [bar]	60.73	62.37	-2.63%
	Water MFR [kg/s]	5.12	5.12	0%
Test02	Δ_P_cool. [bar]	46.96	48.03	-2.23%
	Water MFR [kg/s]	4.47	4.49	-0.4%
Test03	Δ_P_cool. [bar]	35.83	36.68	-2.32%
	Water MFR [kg/s]	3.91	3.91	0%

By relating the pressure drops obtained during the six experimental tests with the squared water mass flow rates it is possible to note that their relation can be modelled as linear (Figure 2(b)).

The firing test campaign has been carried out at the AVIO/ASI FAST2 facility in Colleferro (Rome) and consisted in three tests. The mesh independence analysis, considered for the coupling of the cooling jacket with the combustor, differs from the one used in the cold flow simulations and is a discretizing grid made of 25 nodes.

In order to predict the temperature rise of the water inside the cooling channels, the three empirical correlations offered by *EcosimPro* to compute the wall heat exchange coefficient have been considered and the results in terms of water temperature rise and heat flux have been compared. These comparisons are here presented and they refer to the first firing test, labelled as «FT01».

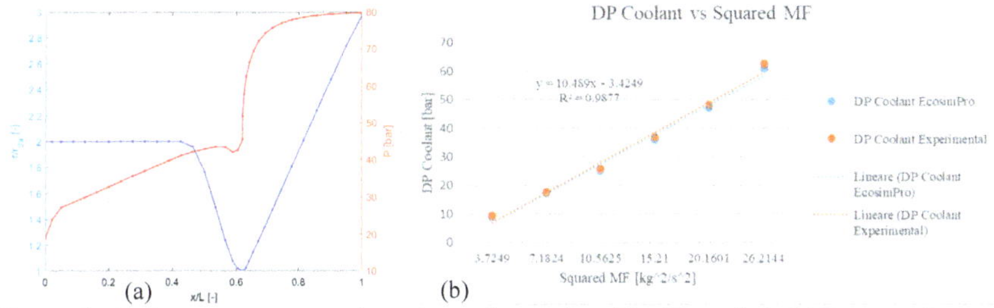

Figure 2: (a) Pressure variation along the cooling jacket. (Test01 results). (b) Pressure drop vs squared water MFR linear relation.

Figure 3(a) compares the water temperature profiles obtained with the Bartz, modified Bartz and Pavli correlations. The Bartz correlation underestimates the temperature increase while the modified Bartz and the Pavli correlations overestimate it.

The heat fluxes coming from the combustor are compared in Figure 3(b) and show how a peak is reached in the throat section, phenomenon due to the fact that in this region the exchange area (i.e. the lateral surface through which the thermal power coming from the combustor is exchanged) is the minimum. The heat flux obtained with the Pavli wall heat exchange correlation is characterized by the highest peak, on the contrary the Bartz correlation provides the lowest peak. Since the modified Bartz correlation provides the lowest error in estimating the water temperature rise, it has been considered as the correlation to use to predict the firing condition behaviour.

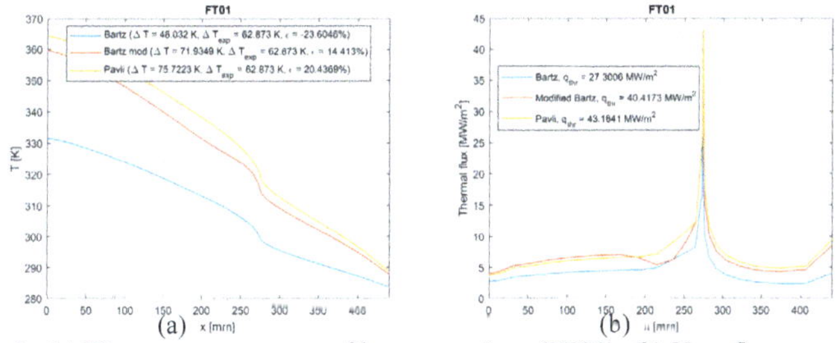

Figure 3: (a) Water temperature profiles comparison (FT01). (b) Heat fluxes comparison (FT01).

Table 3 reports a summary of the numerical and experimental results of the three firing tests. The numerical results well predict the pressure chamber, the coolant pressure drop and the thrust, on the contrary the coolant temperature rises are overpredicted. As expected, the higher thrusts occur during the tests where the chamber pressure is higher. Furthermore, the combustion efficiencies for the three tests are 93%, 89% and 93 %, respectively.

Table 3: Firing tests results

	FT01		FT02		FT03	
	Experimental	EcosimPro	Experimental	EcosimPro	Experimental	EcosimPro
P_CC [bar]	37.99	38	36.61	35.98	45.66	45.76
Water MFR [kg/s]	5.55	5.55	5.46	5.43	5.52	5.52
Water ΔP [bar]	67.86	69.19	65.72	66.99	66.85	68.46
Water ΔT [K]	62.87	71.93	61	73.16	71.92	83.23

Conclusions

The simulations performed by means of the *EcosimPro* software on the «DEMO-0A» to reproduce the cold flow operative conditions turned out to be very accurate for the prediction of the pressure drops experienced by the water in the cooling jacket that are slightly underestimated with respect to the experimental ones. This underestimation can be ascribed to the fact that the schematic used does not include the inlet and outlet manifolds that add further pressure drops. The simulations reproducing the firing conditions of the demonstrator had the aim to investigate the effects of the different wall heat exchange coefficient correlations on the thermal behaviour of the engine and the outcome of this study shows that the results obtained by considering the Bartz correlation underestimates the coolant temperature rise and thermal flux, on the contrary the modified Bartz and Pavli correlations overestimate them and the empirical correlation that provides results comparable to the experimental ones is the modified Bartz correlation. The results in terms of chamber pressure and temperature, pressure drop of the cooling jacket and thrust delivered are better predicted. In the end, the outcomes of the numerical simulations aimed at reproducing the firing conditions suggest that the *EcosimPro* software can be used to perform preliminary simulations able to provide accurate results in terms of chamber pressure, temperature and thrust, but with thermal results characterized by a lower accuracy due to the fact that the model developed is 1-D and does not take into account some phenomena that can occur in an engine and that are typically 2-D or 3-D , such as the thermal stratification of the coolant inside the cooling channels that changes the coolant thermophysical properties and consequently the thermal exchange occurring between the combustion chamber and the cooling jacket.

References

[1] C. D. Brown, "Conceptual Investigations for a Methane-Fueled Expander Rocket Engine.," *40th AIAA/ASME/SAE/ASEE Joint Propuls. Conf. & Exhib.*, 2004. https://doi.org/10.2514/6.2004-4210

[2] J. Jokhakar and M. Naraghi, "A CFD-RTE model for thermal analysis of regeneratively cooled rocket engines.," *44th AIAA/ASME/SAE/ASEE Joint Propulsion Conference & Exhibit. 2008.*, p. 4557, 2008. https://doi.org/10.2514/6.2008-4557

[3] H. Kawashima, H. Negishi, T. Tomita, K. Obase and T. Kaneko, "Verification of Prediction methods for Methane Heat Transfer Characteristics.," *48th AIAA/ASME/SAE/ASEE Joint Propulsion Conference & Exhibit.*, p. 4120, 2012. https://doi.org/10.2514/6.2012-4120

[4] M. Pizzarelli, S. Carapellese and F. Nasuti, "A quasi-2-D model for the prediction of the wall temperature of rocket engine cooling channels.," *Numerical Heat Transfer, Part A: Applications,* pp. 1-24, 2011. https://doi.org/10.1080/10407782.2011.578011

[5] M. Leonardi, F. Di Matteo, J. Steelant, B. Betti, M. Pizzarelli, F. Nasuti and M. Onofri, "A zooming approach to investigate heat transfer in liquid rocket engines with ESPSS propulsion simulation tool.," *8th Aerothermodynamics Symposium,* 2015.

[6] C. H. Marchi, F. Laroca, A. F. C. D. Silva and J. N. Hinckel, "Numerical solutions of flows in rocket engines with regenerative cooling.," *Numerical Heat Transfer, Part A: Applications,* pp. 699-717, 2004. https://doi.org/10.1080/10407780490424307

[7] F. Battista, D. Ricci, P. Natale and et al., "The HYPROB demonstrator line: status of the LOX/LCH4 propulsion activities.," *8th European Conference for Aeronautics and Space Sciences, EUCASS2019-FP0621,* 2019.

[8] Empresarios Agrupados, "EcosimPro ESPSS User Manual," 2020.

[9] S. Omori, W. G. Klaus and A. Krebsbach, "Wall temperature distribution calculation for a rocket nozzle contour.," *NASA-TN-D-6825,* 1972.

[10] B. Betti, M. Pizzarelli and F. Nasuti, "Coupled Heat Transfer Analysis in Regeneratively Cooled Thrust Chambers.," *Journal of Propulsion and Power,* pp. 360-367, 2014. https://doi.org/10.2514/1.B34855

[11] D. Ricci, F. Battista, M. Ferraiuolo and et al., "Development of a Liquid Rocket Ground Demonstrator through thermal analyses," *Heat Transf. Eng.,* pp. 1100-1116, 2020. https://doi.org/10.1080/01457632.2019.1600879

Fast reconfiguration maneuvers of a micro-satellite constellation based on a hybrid rocket engine

Antonio Sannino[1,a*], Stefano Mungiguerra[1,b], Sergio Cassese[1,c], Raffaele Savino[1,d], Alberto Fedele[2,e], Silvia Natalucci[2,f]

[1]Departement of Industrial Engineering, University of Naples Federico II, 80 Piazzale Tecchio, Naples, 80125, Italy

[2]Italian Space Agency, Via Del Politecnico, Rome, 00133, Italy

[a]antonio.sannino2@unina.it*, [b]stefano.mungiguerra@unina.it, [c]sergio.cassese@unina.it, [d]raffaele.savino@unina.it, [e]alberto.fedele@asi.it, [f]silvia.natalucci@asi.it

Keywords: Hybrid Rocket Engine, Hydrogen Peroxide, Cubesat Formation Flying, Orbital Maneuvers

Abstract. In this work, the formation flight of the CubeSat cluster RODiO (Radar for Earth Observation by synthetic aperture DIstributed on a cluster of cubesats equipped with high-technology micro-propellers for new Operative services [1]) with respect to a small satellite in LEO (Low Earth Orbit) has been analyzed. RODiO is an innovative mission concept funded by the Italian Space Agency (ASI) in the context of the Alcor program [2]. The small satellite is equipped with an antenna that allows it to function as a transmitter, whereas RODiO functions as a receiver. The extension of the virtual *SAR (Synthetic Aperture Radar) antenna* can be achieved by establishing an along-track baseline performing an orbital coplanar maneuver (a phasing maneuver). Another interesting scenario is the possibility to create a cross-track baseline performing an inclination change maneuver, useful for stereoradargrammetric applications. Such formation reconfiguration maneuvers can be achieved in relatively short times only by use of a high-thrust propulsion system, i.e. based on conventional chemical technologies. From the study of maneuvers, it is possible to identify the required ΔV, which represents an input parameter for the design of propulsion system. Among the different kinds of propulsion systems, a *Hybrid Rocket Engine* was chosen for its safety, compactness and re-ignition and throttle capabilities.

Introduction

In recent years, the use of CubeSats has become increasingly popular due to their simplicity of construction, cost and reduced production time compared to conventional satellites. These miniaturized satellites are well suited to formation flight for telecommunication and imaging purposes. In this study, the formation flight of a 16U CubeSat constellation (RODiO, consisting of four micro-satellites) was analyzed with respect to a LEO-satellite. The objective is to perform maneuvers to extend the virtual *SAR antenna*. The LEO-satellite is moving on a quasi-circular Sun-Synchronous Orbit (eccentricity $\approx 10^{-3}$, inclination $\approx 97°$) at a mean altitude of $\approx$ 400 km (World Geodetic System-84), and RODiO cluster follows it on this orbit.

In the following sections the *orbital maneuvers* considered were described. Using *GMAT (General Mission Analysis Tool)* an estimation of the maneuvers ΔV budget was obtained. Identified the maneuvers costs, a preliminary design of the *Hybrid Rocket Engine* for the CubeSat was carried out, complying with the requirements for propulsion unit volume (<1.5U) and mass (<2kg).

Phasing maneuver

In this maneuver, the objective is to bring one of the satellites of the RODiO cluster, which follows the LEO-satellite in its orbit, from a distance in the range [-90 km, -50 km], to a distance in the range [+50 km, +90 km]. In this way a *multistatic SAR data collection* with a triplet of acquisitions over the same area at three different observation angles is possible.

To this aim, a ΔV in the opposite direction of the motion must be applied. In this way the satellite RODiO reaches an elliptical orbit with an orbital period smaller than the period of the initial orbit and, after one orbit, the RODiO satellite reduces the along-track distance. After few orbits, the RODiO satellite reaches a new position beyond LEO-satellite and, at this point, a ΔV in the motion direction must be applied to establish a constant *along-track baseline*. The challenging point of this mission is the need to apply two ΔV but in opposite direction and evaluating the re-ignition capability of the propulsion system.

To study the relative motion between two satellites, the Hill reference frame defined in [3] was used, assuming that the LEO-satellite is the chief, while RODiO is the deputy. Combining the equations for the Hohmann transfer reported in [4] (under the assumptions of Keplerian mechanics), it is possible to write Eq. 1, which provides an initial estimate of the along-track baseline variation per orbit (ΔY_{orbit}):

$$\Delta Y_{orbit} = (\tau_{ci} - \tau_e)\, V_{ci} \quad (1)$$

where "τ_{ci}" is the orbital period of initial circular orbit, "τ_e" is the period of elliptical orbit, and "V_{ci}" is the velocity on the initial circular orbit. In Table 1 different cases are presented, and a simulation of relative motion using *GMAT* considering the presence of the atmosphere (*Jacchia-Roberts* model), the non-sphericity of the earth (*Earth Geopotential Model 96*), and solar radiation pressure, has been performed (results in Fig.1).

Table 1: Possible phasing maneuvers in the along-track distance ranges considered for different ΔV. (Y_i is the initial distance between a RODiO satellite and the LEO-satellite, Y_f is the final distance).

ΔV of single burn [m/s]	Y_i [km]	ΔY_{orbit} [km]	Number of orbits	Y_f [km]
2.5	-62.25	41.5	3	+62.25
3	-75	50	3	+75

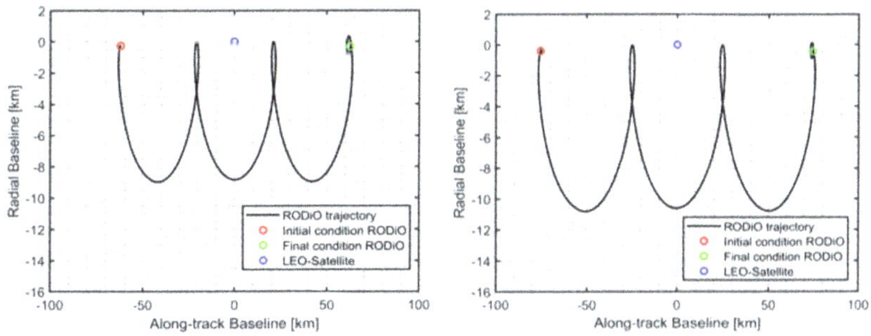

Fig. 1: RODiO trajectory in Hill reference frame with respect to LEO-Satellite (on the left, first case indicated in Table 1, on the right, second case indicated in Table 1).

Inclination change maneuver

The purpose of this maneuver is to change the inclination of the orbit of a satellite of the RODiO constellation by applying a normal ΔV to the orbital plane when the satellite arrives in the ascending node. In this way a relative drift of the nodes starts, and, after a certain time, a *cross-*

track baseline is established. Applying an impulsive burn in the same direction but where the satellite reaches the descending node, the *cross-track baseline* remains constant. An advantage of this maneuver is that the direction of ΔV does not change.

In *GMAT* the study of relative motion between two satellites of RODiO cluster has been analyzed (RODiO-1 is the chief, RODiO-2 is the deputy). Under the assumptions of a quasi-circular orbit, considering J2 effect, it is possible to simplify the equations of relative motion reported in [3] and calculate the *cross-track baseline* from Eq. 2:

$$z \approx a\delta\Omega \sin i \tag{2}$$

where "z" is the *cross-track baseline*, "a" is the semi-major axis, "δΩ" is the relative drift of ascending node consequent to the inclination change, and "i" is the inclination. In the Fig.2 the results obtained by *GMAT* for a ΔV = 3 m/s (that yields a Δi of 0.02 deg [4]) for each burn and a waiting time of 15 days are shown.

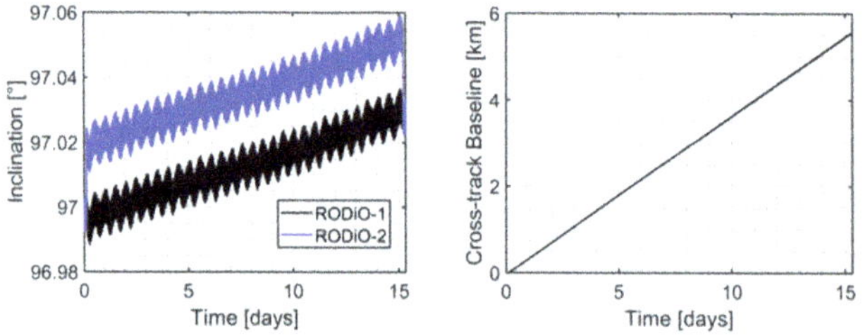

Fig. 2: Trend of inclination (on the left) and trend of cross-track baseline (on the right) with respect to mission time for ΔV = 3 m/s.

Hybrid Rocket Engine preliminary design

A mixture with *Hydrogen Peroxide* as oxidant (*91wt%*) and *ABS (Acrylonitrile Butadiene Styrene)* as fuel grain is selected to evaluate the performance of the *Hybrid Rocket Engine*. In first approximation, the regression rate for a hybrid rocket motor is related to oxidizer mass flux (Eq.3):

$$\dot{r} = a\,(G_{ox})^n \tag{3}$$

where "a" and "n" are experimental coefficients which change for each couple of propellants. Using the procedure described in [5] and considering experimental value of "a" and "n" obtained from test conducted on this engine scale, an estimate of the performance of the thrust chamber was performed. Considering an oxidant flow rate of 3.5 g/s, a circular port fuel grain with an initial port diameter of 10 mm, a nozzle throat diameter of 2 mm and an Area Ratio of 15, the performance of the propulsion system in terms of thrust and specific impulse can be evaluated. The mass of the CubeSat is 22 kg, but for an initial estimation of the performance and sizing of the rocket, a 10 kg margin on the mass budget (total mass of 32 kg) and a 100% margin on the ΔV (total ΔV required for maneuvers 12 m/s instead of 6 m/s) are considered. Table 2 shows the performance of the *Hybrid Rocket Engine*. The *Hydrogen Peroxide* total mass required is 106.75 g. From the performance analysis, it is possible to preliminarily size the propulsion system, in particular the thrust chamber, which will include a case containing the fuel grain and nozzle. Fig. 3 shows a sectional view of the thrust chamber with all dimensions of interest indicated from which it is possible to observe a pre-combustion chamber (upstream to fuel grain) of 10 mm length, a post-

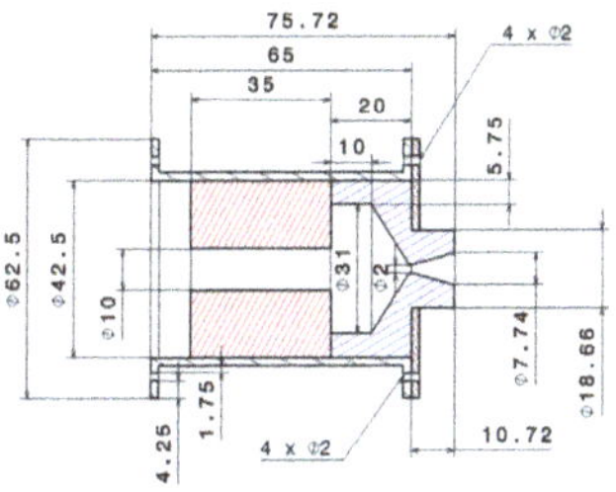

Fig. 3: CAD model section of thrust chamber (dimensions in mm). In blue the conical nozzle, in red the fuel grain, in black the external case, in brown a closing flange (wall thickness flange and case: 2 mm).

combustion chamber (downstream of the fuel grain) consisting of a cylindrical section (10 mm long) and the converging nozzle section (10 mm long). The pre- and post-chamber lengths shall be better defined by a more accurate study of the rocket's internal thermo-fluid dynamics. Considering graphite as material, the conical nozzle has a mass of 55 g. To reduce the length of the diverging section, a bell-shaped nozzle can be considered. In first approximation, the external case made up of steel, and its mass is 179 g. The mass of ABS fuel grain is 46 g. Since the dimensions of the thrust chamber are considerably smaller than the imposed limits (1.5 U, 10 cm x 10 cm x 15), it is reasonable to assume that there is sufficient space for the feed line, tanks, and catalytic chamber. With an appropriate choice of materials, the total mass requirement (<2kg) is also satisfied.

Table 2: Summary of mixture performance H_2O_2 (%wt 91) - ABS. Average values of regression rate, OF, vacuum thrust, specific vacuum impulse, fuel flow rate, chamber pressure, chamber temperature, and final grain diameter were reported (combustion efficiency and nozzle efficiency about 95%. Overall efficiency 90.25%).

t_b [s]	$\dot{r}$[mm/s]	OF	T_{vac}[N]	Isp_{vac}[s]	$\dot{m}_f$[g/s]	P_c [bar]	T_c [K]	D_{final} [mm]
30.5	0.53	2.31	12.6	255	1.53	22.6	1930	42.5

Concluding remarks

From numerical *GMAT* simulations it is possible to conclude that the *along-track baseline* variations are possible to move one RODiO satellite from the original position to a final position in few orbits, whereas a *cross-track baseline* variations up to few kilometers are possible with maneuvers duration of the orders of few days. Future developments could include a detailed design of the other main subsystems, numerical analysis of fluid dynamic, thermal and structural aspects, and the possibility of developing breadboards for ground testing of the propulsion unit.

References

[1] A. Renga, et al., Design considerations and performance analysis for RODiO distributed SAR mission, Acta Astronautica, Volume 210, September 2023, Pages 474-482. https://doi.org/10.1016/j.actaastro.2023.04.001

[2] G. Leccese, et al., Overview and Roadmap of Italian Space Agency Activities in the Micro- and Nano-Satellite Domain, 73rd IAC, Paris, France, 2022, 18 – 22 September.

[3] M. D'Errico (ed.), Distributed Space Missions for Earth System Monitoring, Space Technology Library, Springer Science Business Media, New York 2013, pp. 125-162. https://doi.org/10.1007/978-1-4614-4541-8_3

[4] D.A. Vallado, Fundamentals of Astrodynamics and Applications, Space Technology Library, Springer Dordrecht, edition 2, pp. 317-422.

[5] S. Mungiguerra, et al., Characterization of novel ceramic composites for rocket nozzles in high-temperature harsh environments, International Journal of Heat and Mass Transfer, 163 (2020) 120492. https://doi.org/10.1016/j.ijheatmasstransfer.2020.120492

Tests and simulations on 200N paraffin-oxygen hybrid rocket engines with different fuel grain lengths

Stefano Mungiguerra[1,a *], Daniele Cardillo[2,b], Giuseppe Gallo[3,c], Raffaele Savino[1,d], Francesco Battista[2,e]

[1]Department of Industrial Engineering, University of Naples Federico II, 80 Piazzale Tecchio, Naples, 80125, Italy

[2]CIRA "Italian Aerospace Research Centre", 81043 Capua, Italy

[3]Department of Mechanical and Space Engineering, Hokkaido University, Hokkaido 060-0808, Japan

[a]stefano.mungiguerra@unina.it, [b]d.cardillo@cira.it, [c]gallo@eng.hokudai.ac.jp, [d]rasavino@unina.it, [e]f.battista@cira.it

Keywords: Hybrid Rockets, Paraffin-Based Fuels, Characteristic Velocity, Nozzle Heat Transfer

Abstract. An experimental campaign, in the framework of the HYPROB-NEW hybrid rocket studies, was carried out on a 200N-thrust class hybrid rocket engine, using gaseous oxygen as the oxidizer and paraffin wax-based fuel, to investigate the effect of fuel grain length on motor performance and internal ballistics. Numerical analysis have been also performed to support the experimental findings. It was observed that, for given oxidizer flow rate, fuel grain length directly affects the characteristic velocity, because of its influence on residence time and mixing efficiency, so that the shortest grain configuration displayed the lowest performance. Moreover, CFD simulations provided an estimation of the regression rate profile along the grain length, providing a possible interpretation for the measured space-time-averaged fuel regression rate. Finally, a method for the rebuilding of the convective heat-transfer coefficient in the nozzle was used, based on a combination of numerical simulations and experimental acquisitions.

Introduction

The application field of hybrid rockets is currently still limited much probably for the low fuel regression rate compared to solid rockets, especially when the use of conventional polymeric fuels is foreseen, because of the diffusion-limited phenomena affecting grain regression [1]. One of the most investigated solutions to overcome limitation in fuel regression rate and thus rocket thrust is the use of liquefying fuels, characterized by a relevant liquid droplet entrainment component, which can substantially enhance fuel mass flow rate [2].

Within this framework, the HYPROB-NEW project, funded by Italian Ministry of Research and managed by the Italian Aerospace Research Centre (CIRA), envisaged a collaboration between CIRA and University of Naples Federico II, and was focused on the study of paraffin as a potential high-performance fuel for hybrid rockets. Among the various activities, firing test campaigns were carried out on a 200 N thrust-class hybrid rocket engine, using axially injected gaseous oxygen as oxidizer and a paraffin wax-based fuel. Different cylindrical fuel grain lengths were adopted to extend fuel characterization under different operating conditions, and to evaluate rocket performances and internal ballistics in the different configurations. In addition to data collected with 220 mm propellant grain length [[3]] (labeled as L), two further test campaigns were carried out considering 130 mm (labeled as M) and 70 mm (labeled as S) grain lengths. Full details on the experimental setup and results are reported in [4]. In this work, based on measurements of pressures, temperatures, thrust and mass flow rate, and with the support of Computational Fluid

Materials Research Proceedings 37 (2023) 664-667 https://doi.org/10.21741/9781644902813-143

Dynamic (CFD) simulations, some considerations on performance, regression rate and graphite nozzle heat transfer.

Characteristic velocity analysis
In this subsection, the motor performances of the three configurations are discussed. Fig. 1a represents the characteristic velocity obtained in the firing tests and compared with the ideal one computed by CEA software [5]. It can be seen that the motor length affects the mixing of the oxygen with the fuel. The motor shows a combustion efficiency close to 1 in the L configuration. This value decreases from 1 to about 0.9 and 0.75 in the M and S configurations, respectively.

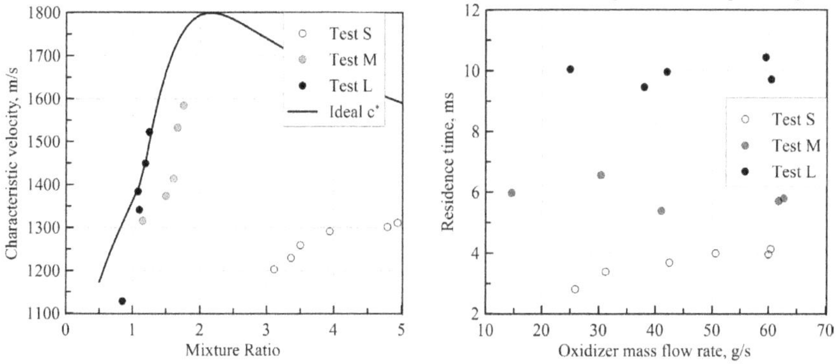

(a) (b)

Fig. 1 (a) Ideal and experimental characteristic velocity of Test S, M and L versus O/F, (b) estimated gas residence time versus oxidizer mass flow rate in the different configurations.

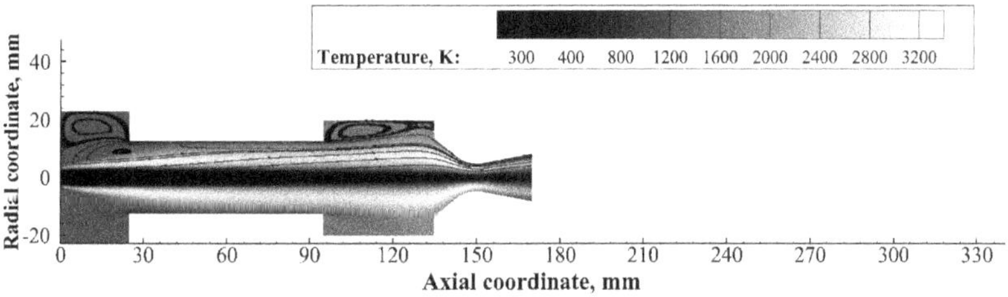

Fig. 2 Temperature contour plot with overlapped streamlines (top half) and mixture fraction iso-lines (bottom half) of test S at $\dot{m}_{ox} \approx 40\ g/s$ and at the average grain port diameter.

Therefore, it can be inferred that the grain length increases the motor combustion efficiency, because it increases the residence time of the gas mixture, which can be computed (Fig. 1b) as the ratio between the total motor length and the mean flow velocity in the chamber (weighted average among the flow velocities in pre-combustion chamber, grain port and post-combustion chamber, calculated by mass conservation using CEA software). The highest residence time (which is quite insensitive to mass flow rate) is shown by configuration L, which shows the highest combustion efficiency.

When the residence time is extremely low as in the case of Test S, the experimental characteristic velocity is also little affected by the change in the overall mixture ratio. This is likely due to the fact that, when the residence time is too low, the fuel released from the grain does not have enough time to reach the axis and therefore mixing efficiency is lower, as highlighted by Fig.

2, showing the results of CFD simulations carried out, by means of the model presented in [6], for test S at $\dot{m}_{ox} \approx 40\ g/s$ with the corresponding average grain port diameter.

Fuel regression rate analysis

Another major experimental finding was that the space-time-averaged fuel regression rate appeared to be affected by fuel grain length. First, regression rate in tests L was roughly 10-15% higher than for tests S. Moreover, axial recession was observed for tests M. To provide a possible explanation for these experimental observations, a CFD simulation was performed by the same model used above for the configuration L at $\dot{m}_{ox} \approx 40\ g/s$, computing the wall heat flux along the fuel grain wall, whose normalized trend is shown in Fig. 3. The wall heat flux behavior can be outlined as the product of the heat transfer coefficient, h_c, and ΔT. Indeed, $\dot{q}$ increases with the temperature, reaching a maximum at the axial coordinate between 125 mm and 150 mm, where the stoichiometric conditions are achieved; then, it decreases because a fuel-rich mixture is obtained. Three stations are highlighted in the picture, which correspond to the lengths of the configuration S, M and L. It can be seen that the space-averaging process leads to an apparent underestimation of the fuel regression rate for the shortest configuration. Moreover, as shown, the peak of wall heat flux is approximately achieved at the end of configuration M, which explains the reason why this configuration was affected by the axial recession.

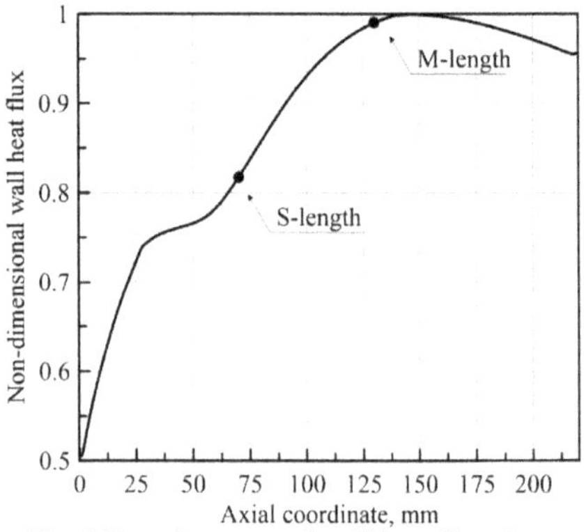

Fig. 3 Non-dimensional wall heat flux along the motor axis

Nozzle heat transfer rebuilding

Finally, measurement of temperature inside the graphite nozzle allowed for a rebuilding of wall heat flux in the S configuration (the most oxidizing) [4]. An iterative procedure was used to determine a profile of the convective heat transfer coefficient h_c matching the experimental temperature measurement, by solving the unsteady energy equation inside the nozzle with CFD simulations. The obtained coefficient was compared to that provided by empirical correlations. Fig. 4 shows this comparison for a test at $\dot{m}_{ox} \approx 40\ g/s$ and O/F ≈ 3.5, and a test at $\dot{m}_{ox} \approx 60\ g/s$ and O/F ≈ 4.79. It can be observed that, in the less oxidizing condition, the experimental trend of h_c deviates more from the empirical correlations, likely because in that condition the wall temperature remains almost always below the gasification temperature of the fuel, therefore creating a liquid paraffin layer at wall acting as insulator and heat sink.

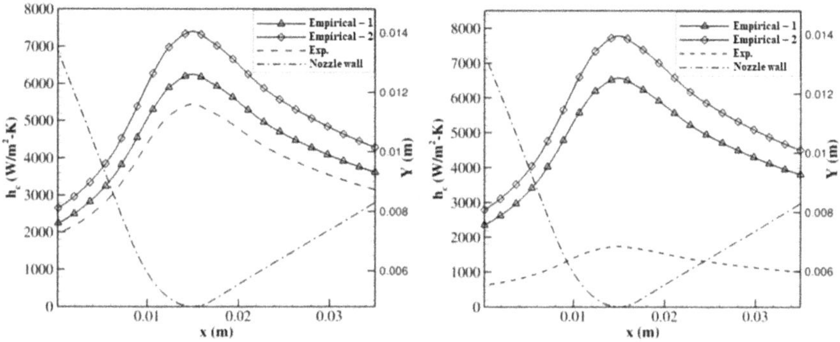

(a) (b)

Fig. 4 Comparison of h_c predictions with different methods: tests at $\dot{m}_{ox} \approx 60\ g/s$ (a) and $\dot{m}_{ox} \approx 40\ g/s$ (b)

Conclusions

The following main conclusions can be drawn:

- The characteristic velocity is affected by the grain length, with increasing efficiency for increasing length, because of the corresponding increase of gas mixture residence time, enhancing propellant mixing.
- The space-averaging process in the regression rate calculation is affected by the axially increasing grain consumption. Longer grains exhibit a higher space-time-averaged regression rate for a given Gox, but the regression rate trend in time is similar for all the configurations in the upstream region
- At low O/F, a significant part of the convective heat transfer to nozzle wall is absorbed by fuel gasification, leading to an overestimation of wall heat transfer by empirical correlation laws, while above a certain O/F threshold, the nozzle wall temperature is higher than fuel gasification temperature and the empirical correlations work properly.

References

[1] D. Altman, et al., Hybrid Rocket Propulsion Systems, in: R.W. Humble, et al. (Eds.), Space Propulsion Analysis and Design, 1st ed., McGraw–Hill, New York, 1995, 365–370.

[2] A. Mazzetti, et al., Paraffin-based hybrid rocket engines applications: a review and a market perspective, Acta Astronaut. 126 (2016) 286–297

[3] G. D. Di Martino, et al., Two-Hundred-Newton Laboratory-Scale Hybrid Rocket Testing for Paraffin Fuel-Performance Characterization, J. Propuls. Power 35 (2019) 224-235, https://doi.org/10.2514/1.B37017

[4] D. Cardillo, et al., Experimental Firing Test Campaign and Nozzle Heat Transfer Reconstruction in a 200 N Hybrid Rocket Engine with Different Paraffin-Based Fuel Grain Lengths. Aerosp. **2023**, 10, 546. https:// doi.org/10.3390/aerospace10060546

[5] S. Gordon, et al., Computer program of complex chemical equilibrium compositions and applications, NASA Ref. Publ. 1311 (1994).

[6] G. Gallo, et al., New Entrainment Model for Modelling the Regression Rate in Hybrid Rocket Engines, J. Propuls. Power 37 (2021) 893-909, https://doi.org/10.2514/1.B38333

Thermite-for-demise (T4D): thermite characteristics heuristic optimization on object- and spacecraft-oriented re-entry models

Alessandro Finazzi[1,a *], Filippo Maggi[1,b] and Tobias Lips[2,c]

[1]Dipartimento di Scienze e Tecnologie Aerospaziali, Politecnico di Milano, Via La Masa 34, 20156 Milano – Italy

[2]HTG - Hyperschall Technologie Göttingen GmbH, Am Handweisergraben 13, 37120 Bovenden – Germany

[a]alessandro.finazzi@polimi.it, [b]filippo.maggi@polimi.it, [c]t.lips@htg-gmbh.com

Keywords: Atmospheric Re-Entry, Genetic Algorithm, Thermite, Spacecraft Demise

Abstract. The major hazard associated with uncontrolled atmospheric re-entry is the casualty risk on ground. An innovative concept to support spacecraft demise that is now under investigation is the use of exothermic reactions. Thermites are good candidates for this role, being capable of releasing a noticeable amount of heat upon ignition. An appropriate selection of the metal-metal oxide couple can grant a formulation that is compliant with the main space operation needs, e.g., that is relatively insensitive to external stimuli and non-toxic. To support the selection of the energetic material for the experimental tests in the ESA-founded project SPADEXO and to preliminarily size the charge to be placed on board, the object-oriented code TRANSIT has been developed. This software has been compared to ESA's spacecraft-oriented code SCARAB (developed by HTG), that is capable to predict spacecraft re-entry with the highest possible level of detail. Both the models were subjected to a genetic algorithm optimization process to identify the best thermite properties and the foreseen energetic material mass for simple geometries applications. In this paper, the SCARAB results obtained for one geometry will be presented and compared with the ones retrieved by TRANSIT.

Introduction

The threat posed by space debris to the access and use of space is becoming more and more urgent every year. Recently, both the European Space Agency (ESA) and the Inter-Agency Space Debris Coordination Committee (IADC) have published their reports on the status of the space environment [1,2]. Considering the protected Low Earth Orbits (LEO) region as defined by IADC [3], the space traffic is now 10 times the level observed in 2000. However, both the cited documents demonstrate that the adoption of the space debris mitigation measures is insufficient. The ESA report [1] reveals that the 93% of small satellites (<10 kg) are naturally compliant with the 25-year rule, but for larger payloads the compliant share is significantly lower. Only between the 40% and 70% of the total payload mass is estimated to reach its end-of-life (EOL) in an orbit compliant with the current mitigation rules. If naturally compliant objects are discarded, until 2017 only between 10% and 40% of spacecrafts respected the mitigation guidelines [2]. Even if the trend in the last years has been generally positive, it is evident that the compliance rate is significantly lower than the internationally declared objective (90% [1,3,4]). Therefore, post-mission disposal (PMD) is still a problematic topic and the adoption of the current, if not even more stringent, mitigation rules is of paramount importance to reach a sustainable exploitation of space.

However, the necessity of de-orbiting spacecrafts from LEO involves an implicit casualty risk on ground. The current international guidelines [3] impose a maximum threshold for this risk of 10^{-4}. A first strategy to meet this requirement is to perform a high-thrust manoeuvre to force the impact on an uninhabited area, with the subsequent impacts on costs (e.g., additional fuel) and design. A second option is to perform an uncontrolled re-entry, possibly after a low-thrust manoeuvre or the deployment of passive devices to limit the residence time in orbit coherently with the 25-years limit. In this case, the associated costs and impacts on the post-mission disposal reliability would be lower. The main limitation for the use of this second strategy is the casualty risk limit. Its value must be computed using re-entry software like ESA's Debris Risk Assessment and Mitigation Analysis (DRAMA) [5] or Spacecraft Atmospheric Re-entry and Aerothermal Breakup (SCARAB) [6]. The uncontrolled re-entry is allowed only if the casualty risk threshold is respected. As its value is strictly connected to the spacecraft mass and to the robustness of its components, a rather new field of research is now becoming more and more important. This approach, named Design-for-Demise (D4D), consists in the intentional design of the spacecraft to promote its demise during the atmospheric re-entry, to limit the number and the mass of the fragments reaching the ground. One of the D4D strategies involves the maximization of the available heat to aid the demise of the most robust equipment. The main proposed solutions to act in this sense are the modification of the ballistic coefficient of the spacecraft, the exploitation of particular shapes to increase the local heat fluxes, or the use of exothermic reactions to provide additional enthalpy. The latter approach is the focus of this paper.

The hereby named Thermite-for-Demise (T4D) technology consists in placing a pyrotechnic charge into the structural voids of some spacecraft components. The energetic material, once ignited, provides the additional enthalpy necessary to induce the demise of the equipment. Thermites are particularly interesting for this application, thanks to their high energetic density, high adiabatic flame temperature, tunability and intrinsic safety [7]. An appropriate selection of the starting metal-metal oxide couple allows to meet both performance and operational requirements [8]. Recent patents proposed the use of thermites to promote spacecraft demise [9,10], and wind tunnel tests proved the concept in relevant environment [11,12]. A systematic study on the topic is currently ongoing in the framework of ESA-TRP SPADEXO project [13].

One of the main aspects of this technique that needs to be defined are the minimum thermite mass to be used and the best ignition time. Even if these parameters are strictly connected to the particular application (e.g., re-entry path, equipment material and shape) a heuristic optimization is hereby proposed for a selected simple geometry. A genetic algorithm is used on both an object- and a spacecraft-oriented re-entry software to minimize the thermite mass. The impacts of the different level of detail of the numerical models will be assessed, as well as the ratio between the component and the thermite masses for the selected conditions.

Methodology

TRANSIT. The object-oriented re-entry software used for this analysis is the TRANsatmosferic SImulation Tool (TRANSIT), developed to support the research on the T4D technology. The objective of this Python novel numerical model is to provide simple and fast simulations for a preliminary assessment on the efficacy of a T4D strategy for a given application. The selected atmospheric model is the NRL-MSISE00 [14] and the non-spherical shape of the Earth is described through a fourth-degree zonal harmonic description. The dynamic model that represents the ballistic re-entry is lumped and considers three degrees of freedom [15]. The aerodynamic model can handle three different geometries (sphere, cylinder, and box). The correlations for the computation of the coefficients of drag are taken from [16]. Shape factors are used to relate the heat load on the three randomly tumbling geometries with the stagnation heat flux on a flat plate [17] (for free molecular regime) or on a sphere [18] (for continuum regime). The hot air after the shock is considered as a non-calorically perfect gas in chemical equilibrium [19]. The thermal model is lumped, and the surface is assumed to regress uniformly.

SCARAB. The spacecraft-oriented software used for this study is SCARAB. It was developed since 1995 under the lead of HTG in the frame of ESA/ESOC contracts. It has been used to model the re-entry of numerous European satellites and launcher stages, as well as for rebuilding test campaigns in hypersonic wind tunnels. It has been compared to other re-entry prediction tools and validated with in-flight measurements. The main characteristic that differentiates SCARAB with respect to the more common object-oriented codes is the panel-based description of the spacecraft. This discretization allows the use of the complete 6 degree-of-freedom equations for the trajectory computation and the more detailed description of the temperature field in the spacecraft and of its break-up process. This representation consents to abandon the common random tumbling assumption and to include conductive heat transfer in the space object. An arbitrarily complex geometry can be reconstructed and then studied.

Thermite model. The additional enthalpy provided by the thermite is modelled as an internal heat source, that is activated once the spacecraft reaches the ignition temperature. The effective heat transfer Q_{eff} from the thermite to its vessel is quantified as per Eq. 1, where η is the heat transfer efficiency (hereby considered equal to 0.6 [13]), m_{th} is the thermite mass on board, and Q_{react} is the theoretical reaction heat release (3958.20 kJ/kg).

$$Q_{eff} = \eta \cdot m_{th} Q_{react} \tag{1}$$

The additional enthalpy is released according to five different profiles (constant, gaussian, early triangular, late triangular, and centred triangular). The user can select the duration of the reaction. In SCARAB, the heat source is applied only on the internal panels of the geometry. The reaction is started once the mean temperature of the spacecraft (in TRANSIT) or the local temperature (in SCARAB) reaches the ignition threshold. The thermal inertia of the thermite charge is considered modifying the specific heat and the mass of the vessel.

Optimization approach. When it comes to determining the impact of a thermite charge on a re-entering spacecraft, the inherent complexity of the re-entry process must be considered. Trade-off effects can be difficult to estimate. For example, a higher thermite filling for a hollow object implies both a higher additional enthalpy release upon ignition and a higher thermal inertia for the system. A lower temperature of ignition could imply an early decrease in mass, with a change in the ballistic coefficient that can be beneficial or not. Moreover, an early ignition could anticipate so much the maximum temperature reached during the descent that could provoke a temperature increase not sufficient to reach the melting point of the spacecraft material. To consider the complexity of the process, the heuristic optimization adopted in this study involves the use of a genetic algorithm, based on the open-source Python package PyGAD [20]. An aluminium sphere, with radius of 0.5 m and thickness of 0.03 m, was selected for the optimization. The initial conditions and the boundaries of the optimization variables are respectively presented in Table 1 and Table 2. Table 3 shows the main genetic algorithm parameters for the optimization in TRANSIT and in SCARAB. Notice that the lower level of detail of the object-oriented code consented to perform an extensive number of simulations in a reasonable time, while the generation and population numbers for the SCARAB optimizations are more limited due to time constraints. The material properties used in both re-entry software were taken from ESA's ESTIMATE database [21]. No demise is predicted for the random tumbling cases without the additional enthalpy released by the thermite. In the SCARAB case in which the dynamic module was activated, the demise of around the 20% of the stating mass is registered for the case without thermite. The fitness function used for the optimization is shown in Eq. 2, where f is the fitness, m_f and m_{sp} are respectively the final and the initial mass of the spacecraft, and m_{th} is the thermite charge mass.

$$f = \frac{1}{m_f + \frac{m_{th}}{m_{sp}} + 0.0000001} \tag{2}$$

Table 1: Initial conditions of the test cases.

Variable	Value
Longitude [°]	0
Latitude [°]	0
Altitude [km]	120
Velocity [m/s]	7273
Flight path angle [°]	-2.612
Heading angle [°]	42.35
Temperature [K]	300

Table 2: Optimization variables and boundaries for the genetic algorithm. Notice that the melting temperature T_{melt} of the material of the test case is used as the upper boundary for the ignition temperature.

Optimization variables	Interval
Profile [-]	[1,5]
Burning time [s]	[1,100]
Thermite density [kg/m^3]	[781,1095]
Filling factor [-]	[0.1,1]
Ignition temperature [K]	[573, T_{melt}]

Table 3: Main genetic algorithm parameters used for the optimization processes in TRANSIT and in SCARAB.

Genetic algorithm parameters	Value, TRANSIT	Value, SCARAB
Number of generations	100	20
Population per generation	50	12
Number of parents mating	15	4

Results

Table 4 shows the results obtained for the genetic algorithm optimization in TRANSIT and SCARAB. Complete demise was reached in all cases, therefore the fitness function value is the ratio between the starting spacecraft mass and the thermite charge one. It can be seen how its value is rather similar between all the simulations, around the 20-25% of the initial sphere mass. The optimization performed in TRANSIT appears as the worst case, implying the highest pyrotechnic charge mass. In the object-oriented code the best case is represented by a brief Gaussian heat release, while in the spacecraft-oriented one the best result is given by a centred triangle profile, for both the analysed conditions. In the one performed with SCARAB considering the dynamics of the sphere as computed by the dynamic module a rather long duration is preferred. This could be due to the lower impact of the burning time in SCARAB numerical model. Notice that these profiles inherently imply a delay between the ignition time and the maximum thermite heat release equal to the half of the burning time. Moreover, it must be considered that both the cases in random tumbling condition foresee a maximum temperature in case of failed ignition around 700 K. On the contrary, the third case already experiences partial demise without the thermite action. This behaviour explicates why a significant difference in the best ignition temperature can be observed between the random tumbling cases and the one in which the dynamic module of SCARAB is activated. Summarizing, all the optimized results show a release of additional enthalpy that is concentrated on the last phase of the re-entry, when the aerodynamic heat is more pronounced. This behaviour suggests that a late ignition is beneficial.

Table 4: Results of the genetic algorithm optimization for TRANSIT and SCARAB.

Variable	**Variable value, TRANSIT (random tumbling)**	**Variable value, SCARAB (random tumbling)**	**Variable value, SCARAB (dynamic module)**
Profile [-]	Gaussian	Centred Triangle	Centred Triangle
Burning time [s]	10.16	20.34	83.18
Thermite density [kg/m^3]	861.10	784.70	781.25
Filling factor [-]	0.16	0.14	0.16
Ignition temperature [K]	639.44	650.11	767.66
Fitness [-]	4.11	5.17	4.51

Conclusions

A genetic algorithm was applied on an object- and on a spacecraft-oriented code for a simple geometry, aiming at quantifying the best thermite properties for a T4D application, in terms of burning time, temperature of ignition and heat release profile. For the selected application, full demise was obtained in all cases. The methodology hereby presented could be applied to an arbitrary re-entry application. The TRANSIT result was more conservative than the ones achieved with SCARAB. It is suggested that this behaviour is due to the variation of shape that is considered in SCARAB once the geometry has started the demise process. This could imply that an object-oriented code could be a proper tool for a first sizing of the pyrotechnic charge, later to be verified and further optimized using a more detailed software. An extension of the presented study to other geometries and materials could strengthen this insight.

References

[1] AA. VV., ESA's Annual Space Environment Report. Issue 6. European Space Agency Space Debris Office. Darmstadt, Germany. (2023).

[2] AA. VV., IADC Report on the Status of the Space Debris Environment. Issue 1, Revision 0. Inter-Agency Space Debris Coordination Committee. (2023).

[3] AA. VV., IADC Space Debris Mitigation Guidelines. Revision 2. Inter-Agency Space Debris Coordination Committee. (2020).

[4] AA. VV., Tri-Agency Reliability Engineering Guidance: Post Mission Disposal and Extension Assessment. ESA-TECQQD-TN-025375 / CAA-2021025 / NASA/SP-20210024973. ESA, JAXA, and NASA. (2022).

[5] AA.VV., Final Report - Upgrade of DRAMA's Spacecraft Entry Survival Analysis Codes. Contract No. 4000115057/15/D/SR. Issue 3, Revision 1.0.2. Hyperschall Technologie Göttingen GmbH. (2019).

[6] Koppenwallner, G., B. Fritsche, T. Lips and H. Klinkrad, SCARAB - A multi-disciplinary code for destruction analysis of space-craft during re-entry. In: 5th European Symposium on Aerothermodynamics for Space Vehicles. Cologne, Germany. (2005).

[7] Fischer, S.H. and N.C. Grubelich, Theoretical Energy Release of Thermites, Intermetallics, and Combustible Metals. 24th International Pyrotechnics Seminar. Monterey, CA. (1998). https://doi.org/10.2172/658208

[8] Finazzi, A., F. Maggi, L. Galfetti, C. Paravan, S. Dossi, A. Murgia, T. Lips, G. Smet, Thermite-for-Demise (T4D): Material selection for exothermic reaction-aided spacecraft demise during re-entry. In : 2nd International Conference on Flight Vehicles, Aerothermodynamics and Re-entry Missions & Engineering (FAR). Heilbronn, Germany. (2022).

[9] Dihlan, D., and P. Omaly, Élement de véhicule spatial a capacité d'autodestruction ameliorée et procedure de fabrication d'un tel elément. Patent FR 2975080B1. (2011).

[10] Seiler, R., and G. Smet, Exothermic reaction aided spacecraft demise during re-entry. Patent EP 3604143A1. (2018).

[11] Monogarov, K.A., A.N. Pivkina, L.I. Grishin, Yu.V. Frolov and D. Dilhan, Uncontrolled re-entry of satellite parts after finishing their mission in LEO: Titanium alloy degradation by thermite reaction energy. Acta Astronautica. 135:69-75. (2017). https://doi.org/10.1016/j.actaastro.2016.10.031

[12] Schleutker, T., A. Gülhan., B. Esser, and T. Lips., ERASD – Exothermic Reaction Aided Spacecraft Demise – Proof of Concept Testing. Test Report. DLR, Supersonic and Hypersonic Technologies Department. (2019).

[13] Maggi, F., A. Finazzi, P. Finocchi, C. Paravan, L. Galfetti, S. Dossi, A. Murgia, T. Lips, G. Smet, K. Bodjona, Thermite-for-Demise: Preliminary on-Ground Heat Transfer Experimental Testing. In: AIAA SCITECH 2023 Forum. National Harbor, MD & Online. (2023). https://doi.org/10.2514/6.2023-1778

[14] J.M. Picone, A.E. Hedin, D.P. Drob and A.C. Aikin, NRL-MSISE-00 Empirical Model of the Atmosphere: Statistical Comparisons and Scientific Issues, J. Geophys. Res., Vol. 107, Issue A12, SIA 15-1:15-16. (2003). https://doi.org/10.1029/2002JA009430

[15] A. Tewari, Atmospheric and Space Flight Dynamics - Modeling and Simulation with MATLAB and Simulink, Birkhäuser, Boston. (2007).

[16] M. Trisolini, Space System Design for Demise and Survival, PhD Thesis, University of Southampton, Faculty of Engineering and the Environment, Department of Astronautics. (2018).

[17] R.D. Klett, Drag coefficients and heating ratios for right circular cylinders in free-molecular and continuum flow from Mach 10 to 30, Report SC-RR-64-2141, Sandia Laboratory, Albuquerque. (1964). https://doi.org/10.2172/4630398

[18] J.A. Fay and F.R. Riddel, Theory of stagnation point heat transfer in dissociated air, Journal of the Aerospace Sciences, Vol. 25, No.2, 73:85. (1958). https://doi.org/10.2514/8.7517

[19] J.D. Anderson Jr, Hypersonic and high temperature gas dynamics, American Institute of Aeronautic and Astronautics, 2nd edition, Reston, Virginia, USA. (2006). https://doi.org/10.2514/4.861956

[20] A.F. Gad, PyGAD: An Intuitive Genetic Algorithm Python Library, arXiv 2106.06158. (2021).

[21] Anon., European Space maTerIal deMisability dATabasE [Online], European Space Agency. https://estimate.sdo.esoc.esa.int/ . Last access: 10/06/2023.

Aeronautics and Astronautics - AIDAA XXVII International Congress | Materials Research Forum LLC
Materials Research Proceedings 37 (2023) 675-678 | https://doi.org/10.21741/9781644902813-145

Numerical and experimental assessment of a linear aerospike

Emanuele Resta[1,a *], Gaetano Maria Di Cicca[1,b], Michele Ferlauto[1,c] and Roberto Marsilio[1,d]

[1]DIMEAS, Politecnico di Torino, Corso Duca degli Abruzzi 24, Turin, Italy

[a]emanuele.resta@polito.it, [b]gaetano.dicicca@polito.it, [c]michele.ferlauto@polito.it, [d]roberto.marsilio@polito.it

Keywords: Aerospike, Nozzle, Experimental, Numerical

Abstract. In the present work a linear aerospike nozzle model has been studied with cold flow experiments in various working conditions. A series of numerical 3D RANS simulations have been performed in order to directly compare numerical and experimental results. Mean pressure distributions have been measured on the nozzle model symmetry plane, in order to characterize the flow evolution along the walls of the plug. The presented results show a good agreement between numerical and experimental results.

Introduction

Key requirements of future space transportation systems are a drastic reduction of launch costs and an increased reliability. Single-Stage-to-Orbit (SSTO) and Two-Stage-to-Orbit (TSTO) configurations are being studied as possible architectures of future launchers. However, the performance of the rocket engines heavily influences the possibility of realization of these vehicles. The performances of existing rocket engines are always lower than the theoretical values because of the presence of several loss mechanisms. Some examples are: the imperfect mixing of oxidizer and fuel in the combustion chamber, losses due to the process of combustion, losses for divergence and non-uniformity of the exiting flow and a non-ideal expansion of the propellants [1]. The latter source of losses is the most important. In fact, as was for the Space Shuttle Main Engine (SSME), the non-adaptation of the exhaust gases can cause up to 15% decrease in performance during certain phases of the mission [2].

A possible solution for the design of the engine of SSTO vehicles may be found in aerospike nozzles (also called plug nozzles), which represent a valid alternative to conventional bell nozzles. In fact, aerospike nozzles provide, at least theoretically, continuous altitude adaptation up to their geometrical area ratio. For high area ratio nozzles with relatively short length, plug nozzles perform better than conventional bell nozzles [2,3]. For these reasons, experimental, analytical and numerical research on plug nozzles have been performed since the 50s worldwide. One notable example of an aerospike engine project and prototype is the linear aerospike engine XRS-2200, which was selected as candidate propulsion system for the Venture Star/X33 SSTO spaceplane in the 1990s. Moreover, the use of linear aerospike nozzles has also started to spark interest regarding the propulsion of high-speed aircraft [4]. Nozzle flow control, including thrust vectoring and external flow interactions, has become increasingly relevant for the controllability of aircrafts. Fluidic thrust vectoring techniques in particular, are gaining significant interest for their advantages over traditional methods, both for use with conventional and unconventional nozzles [5,6]. One example is, for instance, differential throttling, a simple control strategy which can be applied in clustered aerospike engines with multiple independent combustion chambers. Since the mass flow rate and the pressure can be controlled independently, it is possible to generate an asymmetry in thrust, creating a lateral force component [7]. Another possible solution could be the Shock Vector Control approach, which consists in injecting a secondary flow from the plug wall. This causes the flow to separate, generating a recirculation zone on the wall and a shock wave that deflects the incoming flow. As a result, the pressure distribution on the nozzle walls becomes asymmetric, generating a lateral thrust component [8].

The aim of the study is to compare the pressure measurements obtained with an experimental setup, which was previously designed at Politecnico di Torino [9,10], with numerical simulation results. This experimental system will be able to be fitted with different types of nozzle geometries. The mean pressure is measured along the nozzle walls on the symmetry plane, and is compared with 3D numerical results for analogous working conditions.

Test Rig Set-Up and Instrumentation

The test-rig is composed of two subsystems: the air-supply control system and the nozzle model. The first subsystem provides the prescribed inlet flow conditions and is able to manage interchangeable nozzle models, both axisymmetric and linear, for example bell/dual-bell nozzles, and aerospikes. An interfacing duct may be required to generate the correct inlet flow conditions in the axisymmetric or 2D/3D case. The test rig is positioned on a frame, as shown in Figure 1. A corrugated metal hose with an inner diameter of 25 mm is used to provide compressed air to the system. It is connected to a diffuser, followed by a flow straightener.

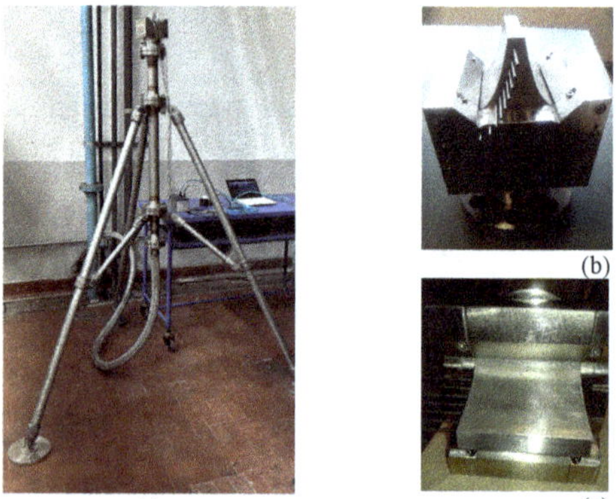

(a) (b) (c)

Fig. 1: Assembled Test-Rig (a). Two close-up views of the plug nozzle model (b-c)

For this work, a linear plug nozzle model was used, with a width to height throat ratio b/h_t equal to 30.41. The model is connected to the flange at the exit of the downstream duct. The exit Mach number M_e is equal to 1 and the design Nozzle Pressure Ratio NPR_d is equal to 200. Additional details on the dimensions of the nozzle plug are available in a previous paper [9].

The splines describing the plug surfaces of the nozzle have been designed using the method proposed by Angelino [11], with a tilt angle ϑ equal to 68.1° at the throat. The aerospike plug geometry is truncated at 40% of the ideal length.

Mean pressure distributions are measured along the nozzle plug using a Scanivalve® DSA5000 pressure scanner. This device is capable of taking up to 16 individual pressure measurements at different locations along the plug, with the use of 16 temperature compensated piezo-resistive pressure transducers. Moreover, individual 24-bit A/D converters are included for each pressure sensor. This feature allows fully synchronous data collection and data stream up to 5000Hz (samples/channel/second). One resistance temperature detector (RTD) per pressure sensor is integrated in the unit and each RTD utilizes its own 24-bit A/D converter. The system accuracy is ± 0.04 % FS for a pressure range from 0 up to 250 psi (from 0 up to 17.2 bar).

The static wall pressures are measured thanks to the orifices (with a diameter equal to 0.6 mm) drilled perpendicularly into the nozzle plug wall (see Figure 1-a). The distance between two adjacent pressure ports is equal to 7 mm. These ports are connected through small steel tubes and Teflon tubes to the Scanivalve® pressure scanner.

Numerical and Experimental Results

Numerical simulations have been performed for the same values of Nozzle Pressure Ratio (NPR) used in the experiments. The commercial solver Star-CCM+ has been utilized to perform the steady RANS simulations. The numerical scheme is second order accurate in space and first order accurate in pseudo-time. A grid of about 1.2 million cells has been utilized for these simulations.

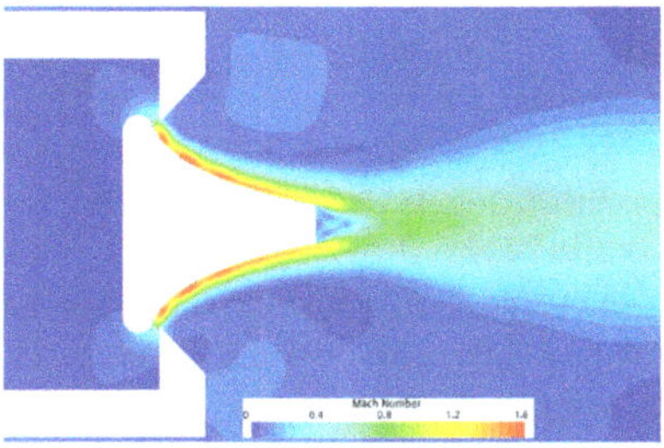

Fig. 2: Mach contour result of the flowfield in the symmetry plane of the nozzle

The Mach contour results from the numerical simulations are shown for the symmetry plane in Figure 2. The structure of the expanding flow is consistent with results from the literature regarding overexpanded flows in aerospike nozzles [2]. The presence of regions of compression and expansion can be clearly seen in this figure. A zone of recirculation on the base of the plug is also visible. In this region pressure is higher than on the walls or in the plume and that gives a positive contribution to thrust. Figure 3 shows a comparison in terms of wall static pressure between the measurements in the present experiment runs and the numerical results from the simulations. The measurements were performed for different NPRs, with the nozzle working in over-expanded conditions. The waviness of the pressure distributions highlights the presence of compression and expansion waves in the region close to the plug surface. The pressure distribution for NPR equal to 3 and 3.7 perfectly match the experimental data. What's more, there are some small differences between the numerical and experimental data for the remaining values of NPR. This could be due to numerical error of the simulation or imperfections in the manufacturing of the physical nozzle model. However, the results still match remarkably.

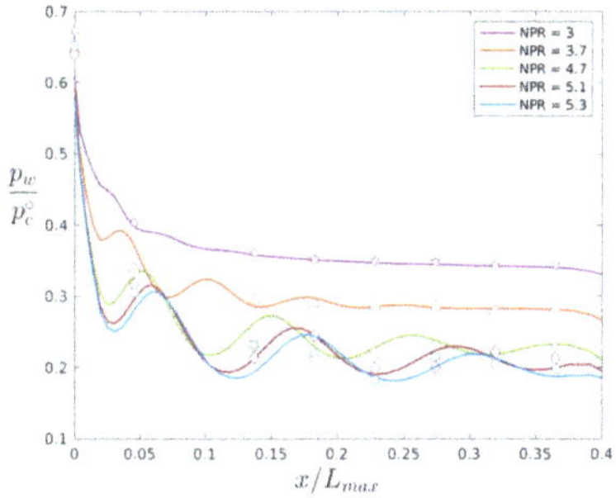

Fig. 3: Comparison between experimental and numerical wall mean pressure distributions.

Conclusions

An aerospike nozzle model has been tested experimentally to characterize the flow evolution along the walls of the plug. The experiments have been carried out in cold flow conditions at different values of NPR. The experimental results have been compared with 3D numerical simulation results, and are presented in terms of wall mean pressure distributions, and a very good agreement has been found.

References

[1] D. Manski, G. Hagemann, "Influence of rocket design parameters on engine nozzle efficiencies", AIAA 30th Joint Prop. Conference, 1994. https://doi.org/10.2514/6.1994-2756

[2] G. Hagemann, H. Immich, T. V. Nguyen, and G. E. Dumnov, "Advanced rocket nozzles," Journal of Propulsion and Power, vol. 14, no. 5, pp. 620–634, 1998. https://doi.org/10.2514/2.5354

[3] R. C. Parsley, K. J. van Stelle, "Altitude Compensating Nozzle Evaluation", AIAA Paper 92-3456, Jul. 1992. https://doi.org/10.2514/6.1992-3456

[4] T. Tomita, H. Tamura, and M. Takahashi, "An experimental evaluation of plug nozzle flow field," in 32nd AIAA Joint Prop. Conference, 1996. https://doi.org/10.2514/6.1996-2632

[5] M. Ferlauto, R. Marsilio "Influence of the External Flow Conditions to the Jet-Vectoring Performances of a SVC Nozzle", AIAA Joint Prop. Conference, Indianapolis, IN, 2019. https://doi.org/10.2514/6.2019-4343

[6] S. Eilers, M. Wilson, S. Whitmore, and Z. Peterson, "Side-force amplification on aerodynamically thrust-vectored aerospike nozzle," Journal of Propulsion and Power, 2012. https://doi.org/10.2514/6.2011-5531

[7] M. Ferlauto, A. Ferrero, M. Marsicovetere, and R. Marsilio, "A comparison of different technologies for thrust vectoring in a linear aerospike", ECCOMAS WCCM 2020, Paris, 2021. https://doi.org/10.23967/wccm-eccomas.2020.006

[8] E Resta, R Marsilio, M Ferlauto "Thrust Vectoring of a Fixed Axisymmetric Supersonic Nozzle Using the Shock-Vector Control Method", Fluids. 6(12):441, 2021. https://doi.org/10.3390/fluids6120441

[9] V. Bonnet, F. Ortone, G. M. Di Cicca, R. Marsilio, and M. Ferlauto, "Cold gas measurement system for linear aerospike nozzles," IEEE Metrology for Aerospace 2022, Pisa, Italy, 2022. https://doi.org/10.1109/MetroAeroSpace54187.2022.9855979

[10] G. M. Di Cicca, J. Hassan, E. Resta, R. Marsilio, M. Ferlauto, "Experimental Characterization of a Linear Aerospike Nozzle Flow", IEEE Metrology for Aerospace 2023, Milan, Italy, 2023. https://doi.org/10.1109/MetroAeroSpace57412.2023.10189981

[11] G. Angelino, "Approximate method for plug nozzle design", AIAA Journal, vol. 2, no. 10, pp. 1834–1835, 1964. https://doi.org/10.2514/3.2682

Numerical suite for the design, simulation and optimization of cathode-less plasma thrusters

Nabil Souhair[1,2,a *], Mirko Magarotto[3,b], Raoul Andriulli[1,c], Fabrizio Ponti[1,d],

[1] Alma Propulsion Laboratory, Department of Industrial Engineering, University of Bologna, Forlì, 47122, Italy

[2] LERMA Laboratory, Aerospace and Automotive Engineering School, International University of Rabat , Sala al Jadida, Rabat, 11100, Morocco

[3] Department of Electric Engineering, University of Padova, Padova, 35131, Italy

[a]nabil.souhair2@unibo.it, [b]mirko.magarotto@unipd.it, [c]raoul.andriulli@unibo.it, [d]fabrizio.ponti@unibo.it

Keywords: Space Propulsion, Electric Propulsion, Cathode-Less Plasma Thrusters, Plasma Thrusters, Numerical Simulations, Global Plasma Models, Multi Fluid Approach, Particle-In-Cell

Abstract. A numerical suite developed for the analysis and design of cathodeless plasma thrusters is presented. The suite includes a Global Model that estimates the thruster's propulsive performance by means of a balance of electron energy and population density, and a 3D numerical strategy to assess plasma behavior. The suite incorporates a FLUID and EM modules to solve plasma transport and electromagnetic wave propagation within the discharge chamber. The PLUME module, managed by the Starfish code, handles plasma dynamics in the magnetic nozzle using the electrostatic particle-in-cell approach. The suite has been validated against thrust measurements from a Helicon Plasma Thruster demonstrating the suite's potential for optimizing the design and operation of cathodeless plasma thrusters for space propulsion applications.

I Introduction

In the last decade, an active research has been conducted on the electric propulsion field [1,2]. Particular effort has been put in the development of cathodeless systems such as Helicon Plasma Thrusters (HPT) [3,4] and Electron Cyclotron Resonance Thrusters (ECRT) [5]. In these devices, plasma is produced within a dielectric tube where the neutral gas propellant is injected. An antenna, operated in the Radio Frequency (RF) or in the microwave frequency range, sustains the discharge coupling Electromagnetic (EM) power to the plasma [6]. Magnets produce a magnetostatic field that has three main functions: (i) increasing the efficiency of the source by enhancing the plasma confinement [7], (ii) driving the power coupling between the antenna and the plasma [8], (iii) improving the propulsive performance via the magnetic nozzle effect downstream the thruster outlet [9-12].

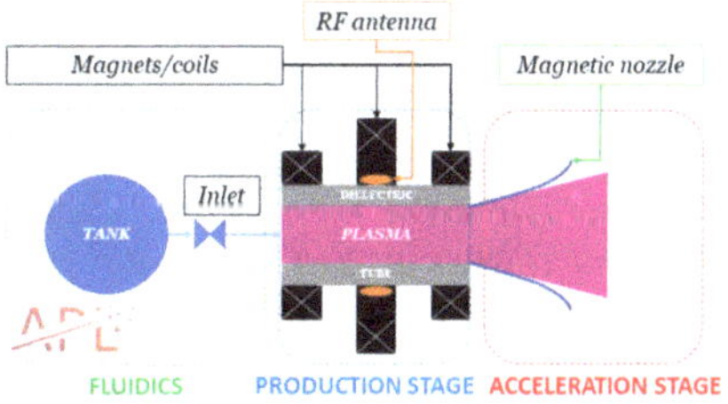

Figure 1. Schematic of a cathodeless plasma thruster. Magnetic field lines highlighted within the plasma source.

The key physical phenomena that govern the plasma dynamics in the production stage, with reference to Figure 1, are the EM wave propagation [13], the plasma transport [14], and their mutual coupling [15]. Instead, the acceleration and detachment phenomena [16] take place downstream the plasma source in the region identified as acceleration stage (see Figure 1). The latter is characterized by the formation of a plume where the plasma is more rarefied (density in the range 10^{16} - 10^{18} m^{-3}) than in the production stage [17]. Particle collisions and the geometry of the applied magnetostatic field govern the plasma behaviour in the region closer to the thruster [18]. Further downstream, the plasma expansion is driven by the thermal pressure, and the ambipolar diffusion [18]. Several analytical [19-21] and numerical approaches have been pursued in the literature for modelling both the production stage and the acceleration stage. The most relevant are: fluid [22], kinetic [23], Particle In Cell with Monte Carlo collisions (PIC-MCCs) [24], and hybrid [25].

In this work, a numerical suite for cathodeless plasma thrusters is presented and its exploitation in a low power case is discussed. A Global Model is devoted to the preliminary estimation of the propulsive performance [26]. More advanced tools have been developed to predict the plasma dynamics throughout the thruster. The 3D-VIRTUS code solves, with a fluid approach, the plasma transport within the production stage. The PIC tools Starfish [27] and have been customized to simulate the plasma dynamics within the magnetic nozzle. Starfish, which handles 2D axisymmetric domains, has been coupled to 3D-VIRTUS in order to estimate the propulsive performance.

II Methodology

II.A Global Model

The main assumptions associated to the Global Model [26] are: (i) cylindrical geometry of the plasma source, (ii) axisymmetric magnetostatic field, (iii) the presence of cusps in the source can be simulated, (iv) the paraxial approximation holds in the acceleration stage [19], (v) in the acceleration stage plasma is frozen to the field lines up to the detachment. The dynamics of the source is solved relying on the conservation of mass (Eq.1) and electron energy (Eq.2) equations

$$\frac{dn_I}{dt} = R^I{}_{chem} - R^I{}_{wall} - R^I{}_{ex} + R^I{}_{in} \tag{1}$$

$$\frac{d}{dt}\left(\frac{3}{2}n_e T_e\right) = Pw - P_{chem} - P_{wall} - P_{ex} \tag{2}$$

n_I is the number density of the I-th species, being this study focused on a xenon plasma I=e, i, *, g for electrons, ions (Xe^+), excited (Xe^*), and ground state (Xe) particles respectively. T_e is the electron temperature in eV. $R^I{}_{chem}$, $R^I{}_{wall}$, $R^I{}_{ex}$, $R^I{}_{in}$, are the I-th particle source/sink terms associated to plasma reactions, wall losses, particles outflow and inflow respectively. Pw, P_{chem}, P_{wall}, P_{ex} are the power coupled to the plasma, along with the source/sink terms associated to plasma reactions, wall losses, and particles outflow respectively. The plasma reactions considered are elastic scattering, ionization and excitation (see Table 1), therefore the $R^I{}_{chem}$ and P_{chem} terms read [29]

$$R^I{}_{chem} = \sum_J K_{JI} n_J n_e - \sum_J K_{IJ} n_I n_e \tag{3}$$

$$P_{chem} = \sum_I \sum_J K_{IJ} n_I n_e \, \Delta U_{IJ} + \sum_I K_{II} n_I n_e \frac{3m_e}{m_I} T_e \tag{4}$$

Where K_{IJ} is the rate constant for the inelastic transitions from species I to species J, K_{II} is the rate constant for elastic collisions between species I and electrons, ΔU_{IJ} is the energy difference (in eV) between species I and species J, along with m_e and m_I are the electron mass and I species mass respectively [30]. Assuming the Bohm sheath criterion at the source walls [26] and a sonic thruster outlet, similar expressions hold for R^I_{wall} and R^I_{ex}

$$R^I{}_{wall} = \frac{S^I{}_{wall}}{V} \Gamma^I{}_{wall} \tag{5a}$$

$$R^I{}_{ex} = \frac{S^I{}_{ex}}{V} \Gamma^I{}_{ex} \tag{5b}$$

Where V is the volume of the source, S^I is the equivalent surface of the source, and Γ^I is the particles flux. For ions and electrons $S^e = S^i$ and its expression, which accounts for the non-uniformity of the plasma within the source, can be found in [30]. Similarly $\Gamma^e = \Gamma^i = n_e u_B$ where u_B is the Bohm speed [31]. For neutrals $R^g{}_{wall} = -R^e{}_{wall}$ assuming total recombination at the walls [26]. Instead $S^g{}_{ex}$ is equal to the physical thruster outlet surface and, assuming the neutrals are in the molecular regime, $\Gamma^g = 1/4\ n_g u_{th}$ [26]; u_{th} is the neutrals thermal speed [31]. From the Bohm sheath criterion, the energy terms read [26]

$$P_{wall} = R^e{}_{wall} \left(2 + \log \sqrt{\frac{m_i}{2\pi m_e}} \right) T_e \tag{6a}$$

$$P_{ex} = R^e{}_{ex} \left(2 + \log \sqrt{\frac{m_i}{2\pi m_e}} \right) T_e \tag{6b}$$

Regarding the gas inflow, only neutral species are assumed to be injected into the source

$$R^g{}_{in} = \frac{\dot{m}_0}{V m_g} \tag{7}$$

where $\dot{m}_0$ is the mass flow rate.

Table 1: plasma reactions considered.

Reaction	Formula	Reference
Elastic scattering	e + Xe → e + Xe	[29]
Ionization	e + Xe → 2e + Xe^+	[29]
Excitation	e + Xe → e + Xe^* → e + Xe + hν	[29]

The thrust is computed according to the model presented in [19]. The contribution from the plasma is

$$F_p = \frac{M_{det}{}^2 + 1}{M_{det}} q n_e T_e S^e{}_{ex} \tag{8}$$

where q is the elementary charge, and M_{det} is the magnetic Mach number (v/u_B) at the detachment point. M_{det} is computed according to the detachment criterion prescribed in [19]. The contribution to the thrust due to neutral gas expansion is

$$F_g = p_g S^g{}_{ex} \tag{9}$$

where p_g is the neutral pressure. Total thrust and specific impulse, being g_0 the standard gravity, read

$$F = F_p + F_g \tag{10a}$$

$$I_{sp} = \frac{F}{g_0 \dot{m}_0} \tag{10b}$$

II.B Source Solver

The plasma source is handled with the 3D-VIRTUS code [15]. Plasma transport and EM wave propagation are solved self-consistently by means of two distinct modules, namely the Fluid module and the EM module, run iteratively. In the former, the plasma transport is solved in a 2D domain while the latter relies on a 3D domain [13]. The governing equations of the fluid module are continuity, energy and Poisson

$$\frac{\partial n_I}{\partial t} + \nabla \cdot \boldsymbol{\Gamma}_I = R^I{}_{chem} \tag{11a}$$

$$\frac{\partial n_\varepsilon}{\partial t} + \nabla \cdot \boldsymbol{\Gamma}_\varepsilon - \nabla\varphi \cdot \boldsymbol{\Gamma}_e = Pw - P_{chem} \tag{11b}$$

$$\nabla^2 \varphi = -q\left(\frac{n_i - n_e}{\varepsilon_0}\right) \tag{11c}$$

The species considered are electrons, ions (Xe^+) and neutrals (Xe). Where $n_\varepsilon = 3/2\ n_e\ T_e$ is the energy density, φ is the plasma potential and ε_0 is the permittivity of vacuum. Formally, the terms R^I_{chem} and P_{chem} are reported in Eq.3 and Eq.4, but in 3D-VIRTUS they are scalar fields which depend on the position. Reactions considered are listed in Table 1. $\boldsymbol{\Gamma}_I$ is the particle flux of the species I that, according to the drift diffusion approximation, reads

$$\boldsymbol{\Gamma}_I = \pm \bar{\bar{\mu}}_I n_I \nabla\varphi - \bar{\bar{D}}_I \nabla n_I + \boldsymbol{u}_0 n_I \tag{12}$$

Where μ_I and D_I are the mobility and the diffusion coefficients of the species I whose values are prescribed in [32]. $\boldsymbol{u}_0$ is the convection speed assumed aligned with the thruster axis and equal, in modulus, to $1/4\ v_{th}$ [7]. $\boldsymbol{\Gamma}_\varepsilon$ is the energy flux that reads

$$\boldsymbol{\Gamma}_\varepsilon = \bar{\bar{\mu}}_\varepsilon n_\varepsilon \nabla\varphi - \bar{\bar{D}}_\varepsilon \nabla n_\varepsilon + \boldsymbol{u}_0 n_\varepsilon \tag{13}$$

Where μ_ε and D_ε are derived according to the Einstein relations [15]. The power deposition profile is computed via the EM module and it reads [15]

$$Pw = \frac{1}{2q} Re(\boldsymbol{J}_{RF}{}^* \cdot \boldsymbol{E}_{RF}) \tag{14}$$

Where $\boldsymbol{J}_{RF}$ and $\boldsymbol{E}_{RF}$ are the complex values of the current density and electric field induced by the RF antenna onto the plasma.

A Robin type boundary condition is imposed to the electrons continuity and energy to enforce the Bohm sheath criterion [15]. A zero gradient Neumann type condition is imposed to the continuity of ions. At the walls, a Neumann condition is imposed to the continuity of the neutrals in order to enforce the recombination of the charged species [15]. At the thruster outlet, the neutral density gradient is assumed null [7]. Finally, a Neumann type boundary condition is imposed to Poisson's equation to enforce the equality between ions and electrons fluxes at the walls [26]. The thruster outlet is grounded [15].

II.C Plume Solver

The plasma in the plume has been simulated with a fully kinetic Particle In Cell (PIC) solver, namely Starfish [27-28]. This handles axisymmetric domains even though the particles' speed is solved in 3D. Starfish has been coupled to 3D-VIRTUS in order to solve the plasma dynamics in the overall thruster and, in turn, to estimate the propulsive performance. The plasma in the acceleration stage is assumed collisionless [33], nonetheless the dynamics of both neutral (Xe) and charged species, namely electrons and ions (Xe^+), are tracked. Particles' speed is solved from the discrete equation of motion

$$\frac{\boldsymbol{v}^{t+\frac{1}{2}}-\boldsymbol{v}^{t-\frac{1}{2}}}{\Delta t}=\frac{q_I}{m_I}\left(\boldsymbol{E}^t+\frac{\boldsymbol{v}^{t+\frac{1}{2}}+\boldsymbol{v}^{t-\frac{1}{2}}}{2}\times\boldsymbol{B}^t\right) \tag{15a}$$

$$\frac{\boldsymbol{r}^{t+1}-\boldsymbol{r}^t}{\Delta t}=\boldsymbol{v}^{t+\frac{1}{2}} \tag{15b}$$

Where $\boldsymbol{r}^t$ is the particle position at time step t, v is the particle speed, Δt is the time step, $\boldsymbol{E}$ is the electric field and $\boldsymbol{B}$ the magnetic field. A Boris scheme is used to advance particles [34]. Since the RF power deposition in the acceleration stage is assumed negligible [35] the EM fields in the plasma are calculated via the Poisson's equation. Therefore, $\boldsymbol{B}$ consists in the background magnetostatic field and $\boldsymbol{E}=-\nabla\varphi$ where

$$\varepsilon_0\nabla^2\varphi=-\rho \tag{16}$$

and ρ is the charge density computed from particles position via a linear deposition scheme [33].

Regarding boundary conditions, Eq.17 holds for the Poisson's equation

$$\varphi=0 \quad \text{Thruster outlet} \tag{17a}$$

$$\varphi=\varphi_\infty \quad \text{Thruster case} \tag{17b}$$

$$\frac{d\varphi}{dk}+\frac{1}{r}(\varphi-\varphi_\infty)=0 \quad \text{External boundary} \tag{17c}$$

Where k is the direction normal to the external boundary, r is the distance between the centre of the thruster and the boundary, along with φ_∞ is the potential at infinity. Eq.17c derives from the assumption that $\varphi\approx 1/r$ for $r\to\infty$ [36]. Regarding particles dynamics, ions and electrons that reach the thruster outlet and the thruster case are removed from the simulation domain [35]. The same condition holds for the ions at the external boundary [35] where an energy-based condition is defined to account for the electrons "trapped" by the potential drop across the plume [37]. The total energy of each electron that reaches the external boundary is computed according to Eq.18

$$E_{tot}=\frac{1}{2}m_I v^2-q_I\varphi \tag{18}$$

Materials Research Proceedings 37 (2023) 679-689 https://doi.org/10.21741/9781644902813-146

If $|E_{tot}|>|q\varphi_{\infty}|$ the electron is absorbed since its energy is high enough to escape at the infinity. If $|E_{tot}|\leq|q\varphi_{\infty}|$ the particle is considered "trapped" and it is subject to a specular reflection.
A control loop has been implemented in order to enforce the current free and the quasi-neutrality conditions at the thruster outlet. The value of the potential at infinity is updated according to Eq.19.

$$\varphi_{\infty}{}^{t+1} = \varphi_{\infty}{}^{t} + \frac{1}{C}\left(I_{iB}{}^{t} + I_{eB}{}^{t}\right)\Delta t \tag{19}$$

Where C is an equivalent capacitance used to tune φ_{∞} so that the current free condition holds. The ion flux at the thruster outlet is assumed constant (Eq.20a) [35] instead the electron flux is varied according to Eq.20b to ensure quasi-neutrality

$$I_{i0} = \frac{q_i \dot{m}_0}{M} \tag{20a}$$

$$I_{e*}{}^{t+1} = K\frac{n_{i0}{}^{t}}{n_{e0}{}^{t}}\, I_{e*}{}^{t} \tag{20b}$$

K is a positive constant and I_{e*} is the emitted electron flux. It is worth noting that the total electron flux (I_{e0}) depends also on the number of electrons absorbed at the
thruster outlet which is mainly driven by the value of φ_{∞} [35]. As a consequence the control strategy implemented relates φ_{∞} to the flux of particles injected into the domain and allows to compute self-consistently the total potential drop across the plume (namely $-\varphi_{\infty}$).
Finally, the thrust is computed according the Eq.21 [25]

$$F = \int \sum_{I} \left(m_I n_I v_{Iz} \boldsymbol{v}_z \cdot \hat{\boldsymbol{k}} + p_I \hat{\boldsymbol{z}} \cdot \hat{\boldsymbol{k}}\right) dS_B \tag{21}$$

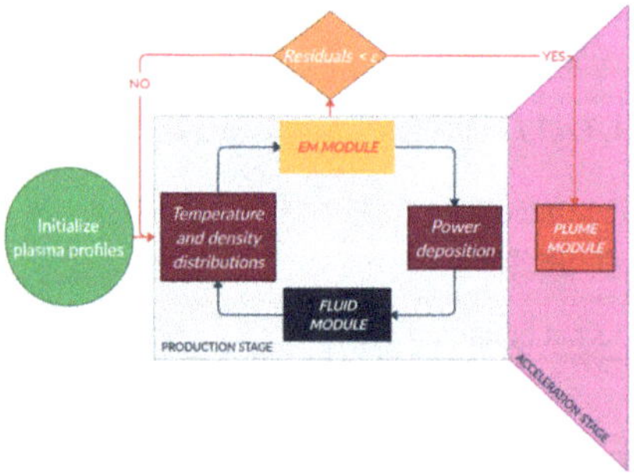

Figure 2. Schematic of the 2D thruster simulation tool. The Source solver (3D-VIRTUS) and the Plume solver (Starfish) run sequentially.

The coupling strategy between the production stage (i.e., 3D-VIRTUS) and the acceleration stage (i.e., Starfish) is schematically depicted in Figure 2. First, the source solver provides plasma profiles at the thruster outlet assuming a sonic condition for this boundary. Second the plume solver takes these profiles as an input and propagates the solution of the plasma expansion.

III Results

The thruster analysed in this work is a low power (50 W range) HPT. The plasma source is a cylinder whose dimensions are length L=0.060 m and diameter D=0.014 m. Magnetic field is generated by two rings of permanent magnets. At thruster outlet, the intensity of the magnetic field on the axis is B_0=600 G. The antenna is a five turns coil whose dimensions are length L_A=0.020 m, diameter D_A=0.020 m, wire width w_A=0.002 m. A mass flow rate of $\dot{m}_0$=0.15 mg/s has been assumed. The power provided to the antenna (Pw) varies in the range from 10 W up to 70 W. An antenna efficiency of η_A=0.8 has been assumed, so the power actually coupled to the plasma is $Pw_p = \eta_A\, Pw$.

A 2D simulation of the overall plasma thruster has been accomplished for an input power Pw=55 W. In Figure 3, the input power deposition profile adopted is shown, and is consistent with an inductive coupling mode [39,40].

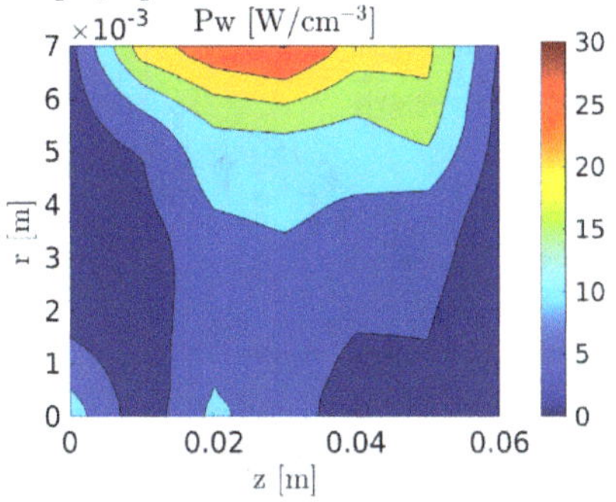

Figure 3. Deposited power map (Pw) function of the radial and axial positions (r-z) in the source.

The estimation of the propulsive performance obtained with both the GM and the 2D simulation of the HPT has been reported in Figure 4.

For what concerns the GM, due to the several assumptions considered, an uncertainty band of ±20% has been attributed to numerical results. The latter is mainly associated to the assumptions on plasma profiles (Eq.5) [26], interactions between the plasma and the walls of the source (Eq.6) [26], cross sections of the plasma reactions (Table 1) [26], and detachment criterion (Eq.8) [7]. The propulsive performance increases with the power coupled to the plasma. Trends predicted numerically and evaluated experimentally are in agreement. Moreover, numerical and experimental uncertainty bands overlap. This result can be considered sufficiently accurate for the scope of the GM which is meant for the preliminary characterization.

Regarding the 2D simulation, thrust has been computed coupling the solution of the production stage with both the semi-analytical formulation implemented in the Global Model (i.e., Eqs.8-10) and the results provided by Starfish (i.e., Eq.21). In both cases, the estimation of the propulsive performance matches better the experimental benchmark with respect to the Global Model. The most accurate result is provided coupling 3D-VIRTUS and Starfish being the difference between numerical and experimental results less than 20%.

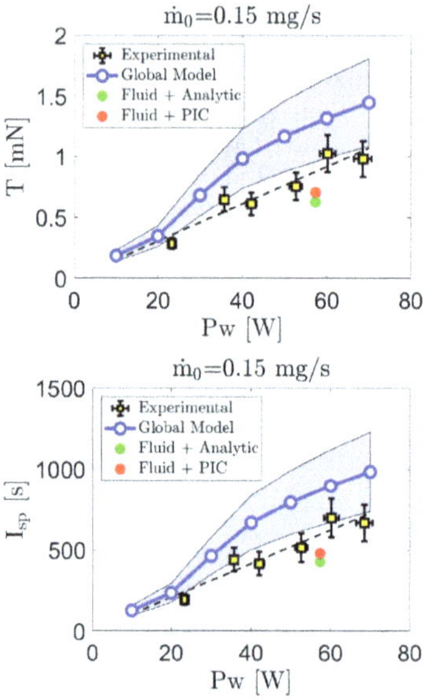

(a)

(b)

Figure 4. Performance predicted numerically compared against measures. (a) Thrust (T), and (b) specific impulse (I_{sp}) in function of the coupled power (Pw).

VI Conclusions

A numerical suite capable of simulating the propulsive performance and the plasma dynamics in a cathodeless plasma thruster is shown. It consists on a Global Model [26] for the preliminary simulation of the propulsive performance, the 3D-VIRTUS code [15] for the estimation of the plasma profile in the production stage, along with Starfish [27] for the solution of the acceleration stage. The results of the Global Model and of the coupling 3D-VIRTUS / Starfish have been benchmarked against measurements of the propulsive performance (i.e., thrust and specific impulse). The agreement between experiments and the Global Model is always better than 50%, whereas differences reduce to 20% with a multi-dimensional approach.

In future works the interfacing strategy between 3D-VIRTUS and Starfish will be improved adopting an iterative approach.

Finally, it is worth noting that the numerical suite presented in this work can simulate the plasma dynamics also in applications different from the electric propulsion as plasma antennas [43-47] and water treatment reactors [48].

References

[1] S. Mazouffre (2016). Electric propulsion for satellites and spacecraft: established technologies and novel approaches. Plasma Sources Science and Technology, 25(3), 033002. https://doi.org/10.1088/0963-0252/25/3/033002

[2] D. M. Goebel, I. Katz (2008). Fundamentals of electric propulsion: ion and Hall thrusters (Vol. 1). John Wiley & Sons. https://doi.org/10.1002/9780470436448

[3] K. Takahashi (2019). Helicon-type radiofrequency plasma thrusters and magnetic plasma nozzles. Reviews of Modern Plasma Physics, 3(1), 1-61. https://doi.org/10.1007/s41614-019-0024-2

[4] M. Manente, et al. (2019). REGULUS: A propulsion platform to boost small satellite missions. Acta Astronautica, 157, 241-249. https://doi.org/10.1016/j.actaastro.2018.12.022

[5] F. Cannat, T. Lafleur, J. Jarrige, P. Chabert, P. Q. Elias, D. Packan (2015). Optimization of a coaxial electron cyclotron resonance plasma thruster with an analytical model. Physics of Plasmas, 22(5), 053503. https://doi.org/10.1063/1.4920966

[6] D. Arnush, F. F. Chen (1998). Generalized theory of helicon waves. II. Excitation and absorption. Physics of Plasmas, 5(5), 1239-1254. https://doi.org/10.1063/1.872782

[7] M. Magarotto, M. Manente, F. Trezzolani, D. Pavarin (2020). Numerical model of a helicon plasma thruster. IEEE Transactions on Plasma Science, 48(4), 835-844. https://doi.org/10.1109/TPS.2020.2982541

[8] M. Magarotto, D. Melazzi, D. Pavarin (2019). Study on the influence of the magnetic field geometry on the power deposition in a helicon plasma source. Journal of Plasma Physics, 85(4). https://doi.org/10.1017/S0022377819000473

[9] E. Ahedo, M. Merino (2010). Two-dimensional supersonic plasma acceleration in a magnetic nozzle. Physics of Plasmas, 17(7), 073501. https://doi.org/10.1063/1.3442736

[10] N. Bellomo, et al. (2021). Design and In-orbit Demonstration of REGULUS, an Iodine electric propulsion system. CEAS Space Journal. https://doi.org/10.1007/s12567-021-00374-4

[11] A. V. Arefiev, B. N. Breizman (2004). Theoretical components of the VASIMR plasma propulsion concept. Physics of Plasmas, 11(5), 2942-2949. https://doi.org/10.1063/1.1666328

[12] K. Takahashi (2021). Magnetic nozzle radiofrequency plasma thruster approaching twenty percent thruster efficiency. Scientific reports, 11(1). https://doi.org/10.1038/s41598-021-82471-2

[13] D. Melazzi, V. Lancellotti (2014). ADAMANT: A surface and volume integral-equation solver for the analysis and design of helicon plasma sources. Computer Physics Communications, 185(7), 1914-1925. https://doi.org/10.1016/j.cpc.2014.03.019

[14] J. Zhou, D. Pérez-Grande, P. Fajardo, E. Ahedo (2019). Numerical treatment of a magnetized electron fluid model within an electromagnetic plasma thruster simulation code. Plasma Sources Science and Technology, 28(11), 115004. https://doi.org/10.1088/1361-6595/ab4bd3

[15] M. Magarotto, D. Melazzi, D. Pavarin (2020). 3D-VIRTUS: Equilibrium condition solver of radio-frequency magnetized plasma discharges for space applications. Computer Physics Communications, 247, 106953. https://doi.org/10.1016/j.cpc.2019.106953

[16] A. V. Arefiev, B. N. Breizman (2005). Magnetohydrodynamic scenario of plasma detachment in a magnetic nozzle. Physics of Plasmas, 12(4), 043504. https://doi.org/10.1063/1.1875632

[17] G. Sánchez-Arriaga, J. Zhou, E. Ahedo, M. Martínez-Sánchez, J. J. Ramos (2018). Kinetic features and non-stationary electron trapping in paraxial magnetic nozzles. Plasma Sources Science and Technology, 27(3), 035002. https://doi.org/10.1088/1361-6595/aaad7f

[18] F. Cichocki, A. Domínguez-Vázquez, M. Merino, E. Ahedo (2017). Hybrid 3D model for the interaction of plasma thruster plumes with nearby objects. Plasma Sources Science and Technology, 26(12), 125008. https://doi.org/10.1088/1361-6595/aa986e

[19] T. Lafleur (2014). Helicon plasma thruster discharge model. Physics of Plasmas, 21(4), 043507. https://doi.org/10.1063/1.4871727

[20] E. Ahedo, & J. Navarro-Cavallé (2013). Helicon thruster plasma modeling: Two-dimensional fluid-dynamics and propulsive performances. Physics of Plasmas, 20(4), 043512. https://doi.org/10.1063/1.4798409

[21] A. Fruchtman, K. Takahashi, C. Charles, R. W. Boswell (2012). A magnetic nozzle calculation of the force on a plasma. Physics of Plasmas, 19(3), 033507. https://doi.org/10.1063/1.3691650

[22] M. Magarotto, D. Pavarin (2020). Parametric study of a cathode-less radio frequency thruster. IEEE Transactions on Plasma Science, 48(8), 2723-2735. https://doi.org/10.1109/TPS.2020.3006257

[23] E. Ahedo, S. Correyero, J. Navarro-Cavallé, M. Merino (2020). Macroscopic and parametric study of a kinetic plasma expansion in a paraxial magnetic nozzle. Plasma Sources Science and Technology, 29(4), 045017. https://doi.org/10.1088/1361-6595/ab7855

[24] M. Li, M. Merino, E. Ahedo, H. Tang (2019). On electron boundary conditions in PIC plasma thruster plume simulations. Plasma Sources Science and Technology, 28(3), 034004. https://doi.org/10.1088/1361-6595/ab0949

[25] Á. Sánchez-Villar, J. Zhou, E. Ahedo, M. Merino (2021). Coupled plasma transport and electromagnetic wave simulation of an ECR thruster. Plasma Sources Science and Technology, 30(4), 045005. https://doi.org/10.1088/1361-6595/abde20

[26] N. Souhair, M. Magarotto, E. Majorana, F. Ponti, D. Pavarin (2021). Development of a lumping methodology for the analysis of the excited states in plasma discharges operated with argon, neon, krypton and xenon. Physics of Plasmas, 28(9). https://doi.org/10.1063/5.0057494

[27] L. Brieda, M. Keidar (2012). Development of the Starfish Plasma Simulation Code and Update on Multiscale Modeling of Hall Thrusters. 48th AIAA Joint Propulsion Conference, Atlanta, GA, AIAA-2012-4015. https://doi.org/10.2514/6.2012-4015

[28] J. F. Roussel, et al. (2008). SPIS open-source code: Methods, capabilities, achievements, and prospects. IEEE transactions on plasma science, 36(5), 2360-2368. https://doi.org/10.1109/TPS.2008.2002327

[29] P. Chabert, J. Arancibia Monreal, J. Bredin, L. Popelier, A. Aanesland (2012). Global model of a gridded-ion thruster powered by a radiofrequency inductive coil. Physics of Plasmas, 19(7), 073512. https://doi.org/10.1063/1.4737114

[30] M. A. Lieberman, A. J. Lichtenberg (2005). Principles of plasma discharges and materials processing. John Wiley & Sons. https://doi.org/10.1002/0471724254

[31] F. F. Chen (2012). Introduction to plasma physics. Springer Science & Business Media.

[32] N. Souhair, M. Magarotto, M. Manante, D. Pavarin, F. Ponti (2021). Improvement of a numerical tool for the simulation of a Helicon plasma thruster. Space propulsion 2020+1, On line conference, SP2020-00070.

[33] G. Gallina, M. Magarotto, M. Manente, D. Pavarin (2019). Enhanced biDimensional pIc: an electrostatic/magnetostatic particle-in-cell code for plasma based systems. Journal of Plasma Physics, 85(2), 905850205. https://doi.org/10.1017/S0022377819000205

[34] H. Qin, S. Zhang, J. Xiao, J. Liu, Y. Sun, W. M. Tang (2013). Why is Boris algorithm so good?. Physics of Plasmas, 20(8), 084503. https://doi.org/10.1063/1.4818428

[35] S. Di Fede, M. Magarotto, S. Andrews, D. Pavarin (2021). Simulation of the plume of a Magnetically Enhanced Plasma Thruster with SPIS. Journal of Plasma Physics. https://doi.org/10.1017/S0022377821001057

[36] D. J. Griffiths (2005). Introduction to electrodynamics. https://doi.org/10.1016/B978-1-85573-953-6.50026-X

[37] M. Martinez-Sanchez, J. Navarro-Cavallé, E. Ahedo (2015). Electron cooling and finite potential drop in a magnetized plasma expansion. Physics of Plasmas, 22(5), 053501. https://doi.org/10.1063/1.4919627

[38] F. Trezzolani, M. Magarotto, M. Manente, D. Pavarin (2018). Development of a counterbalanced pendulum thrust stand for electric propulsion. Measurement, 122, 494-501. https://doi.org/10.1016/j.measurement.2018.02.011

[39] F. Romano, et al. (2020). RF helicon-based inductive plasma thruster (IPT) design for an atmosphere-breathing electric propulsion system (ABEP). Acta Astronautica, 176, 476-483. https://doi.org/10.1016/j.actaastro.2020.07.008

[40] A. R. Ellingboe, R. W. Boswell (1996). Capacitive, inductive and helicon-wave modes of operation of a helicon plasma source. Physics of Plasmas, 3(7), 2797-2804. https://doi.org/10.1063/1.871713

[41] A. E. Vinci, S. Mazouffre (2021). Direct experimental comparison of krypton and xenon discharge properties in the magnetic nozzle of a helicon plasma source. Physics of Plasmas, 28(3), 033504. https://doi.org/10.1063/5.0037117

[42] P. Molmud (1960). Expansion of a rarefied gas cloud into a vacuum. The Physics of fluids, 3(3), 362-366. https://doi.org/10.1063/1.1706042

[43] G. Mansutti, et al. (2020). Modeling and design of a plasma-based transmit-array with beam scanning capabilities. Results in Physics, 16, 102923. https://doi.org/10.1016/j.rinp.2019.102923

[44] A. Daykin-Iliopoulos, et al. (2020). Characterisation of a thermionic plasma source apparatus for high-density gaseous plasma antenna applications. Plasma Sources Science and Technology, 29(11), 115002. https://doi.org/10.1088/1361-6595/abb21a

[45] M. Magarotto, P. De Carlo, G. Mansutti, F. J. Bosi, N. E. Buris, A. D. Capobianco, D. Pavarin (2020). Numerical suite for gaseous plasma antennas Simulation. IEEE Transactions on Plasma Science, 49(1), 285-297. https://doi.org/10.1109/TPS.2020.3040008

[46] P. De Carlo, M. Magarotto, G. Mansutti, A. Selmo, A. D. Capobianco, D. Pavarin (2021). Feasibility study of a novel class of plasma antennas for SatCom navigation systems. Acta Astronautica, 178, 846-853. https://doi.org/10.1016/j.actaastro.2020.10.015

[47] P. De Carlo, M. Magarotto, G. Mansutti, S. Boscolo, A. D. Capobianco, D. Pavarin (2021). Experimental Characterization of a Plasma Dipole in the UHF band. IEEE Antennas and Wireless Propagation Letters. https://doi.org/10.1109/LAWP.2021.3091739

[48] M. Saleem (2020). Comparative performance assessment of plasma reactors for the treatment of PFOA; reactor design, kinetics, mineralization and energy yield. Chemical Engineering Journal, 382, 123031. https://doi.org/10.1016/j.cej.2019.123031

Particle migration modeling in solid propellants

Raoul Andriulli[1,a *], Nabil Souhair[2,b], Luca Fadigati[1,c], Mattia Magnani[1,d] and Fabrizio Ponti[1,e]

[1] Alma Propulsion Lab, Alma Mater Studiorum-Università di Bologna, Via Fontanelle 40, 40121 Forlì, FC, Italy

[2] LERMA Laboratory Aerospace and Automotive Engineering School, International University of Rabat Sala al Jadida, Rabat, 11100, Morocco

[a]raoul.andriulli@unibo.it, [b]nabil.souhair2@unibo.it, [c]luca.fadigati2@unibo.it, [d]mattia.magnani8@studio.unibo.it, [e]fabrizio.ponti@unibo.it

Keywords: Solid Rocket Motors, Casting Process, Particle Migration, Numerical Simulation, OpenFOAM

Abstract. This work presents the development of an OpenFOAM solver to predict the migration of solid particles in concentrated suspensions under non-uniform shear flow. The solver modifies the *pimpleFoam* solver by implementing the conservation equation for particle volume fraction. It adapts the equation of motion for non-Newtonian flows and establishes a model for the viscous field using Krieger's correlation. The code is successfully validated by the experimental results from literature.

Introduction

Solid Rocket Motors (SRM) are commonly used in space propulsion. They provide additional thrust in the first phase of flight, either in conjunction with liquid engines (Ariane 5 and 6) or as the main thrust system (Vega C). While numerical modelling has considerably advanced [1, 2], there are still complex phenomena and uncertainties that make comprehensive performance descriptions challenging [3, 4]. Consequently, expensive testing is necessary to design and certify each SRM. The burning rate of the propellant plays a crucial role in predicting flight performance [2, 6, 7], affected by factors such as grain shape, particle orientation, air bubbles, and more [5, 8-14]. Manufacturing processes also impact ballistic performance [15-17], with macroscopic and microscopic phenomena potentially altering thrust profiles [18, 19]. One notable phenomenon is particle migration and segregation caused by the casting process, which gives rise to grain heterogeneity. All these aspects produce the so-called Hump effect, and its empirical evaluation is typically incorporated into internal ballistic simulations. However, a complete understanding and reliable prediction method for the Hump effect have not yet been achieved.

This study focuses on designing a CFD code using OpenFOAM [20] to simulate and describe particle migration. A bidimensional channel flow case is examined, and the simulation results are compared and validated against literature data. The validated approach can then be applied to simulate segregation phenomena in a motor. The paper is divided into two sections, one describing the developed model and the other presenting the results and validation using experimental data [21].

Theoretical model for particle migration

During the past few decades, it has become common knowledge that initially uniformly distributed particles will assume extremely nonuniform concentration distribution when subjected to a nonuniform shear flow. This non-homogeneous distribution is the result of a migration process which occurs at small values of particle Reynolds number ($\sim 10^{-4}$), so that the importance of inertial effects can be precluded. In general, solid particles will move from regions characterized by a higher shear rate to the neighbouring areas where lower values are measured.

The diffusive-flux model for the prediction of particle distribution ($\boldsymbol{\phi}$) resulting from shear-induced migration, has been developed by Phillips et al. [22]. In particular, two separate contributions to the segregation process are present, the first being related to the frequency of impacts between solid particles while the other depending mainly on the non-uniform viscosity field.

The conservation equation for solid particles in the Eulerian frame may be written as:

$$\frac{D\boldsymbol{\phi}}{Dt} = \boldsymbol{a}^2\boldsymbol{K}_c(\boldsymbol{\phi}^2\nabla\dot{\gamma} + \boldsymbol{\phi}\dot{\gamma}\nabla\boldsymbol{\phi}) + \boldsymbol{a}^2\boldsymbol{K}_\eta\nabla\cdot(\dot{\gamma}\boldsymbol{\phi}^2\nabla\ln\eta(\boldsymbol{\phi})) \tag{1}$$

where $\dot{\gamma} = (\mathbf{2D}:\mathbf{D})^{1/2}$ (D being the deformation rate tensor) is the local value of the shear rate, $\boldsymbol{K}_c$ and $\boldsymbol{K}_\eta$ are two empirical constants, $\boldsymbol{a}$ is the solid particle diameter, and $\boldsymbol{\eta}$ is the viscosity.

The mathematical model for the viscosity is set following the work performed by Krieger in [23] through empirical observations.

$$\eta(\boldsymbol{\phi}) = \eta_0\left(\mathbf{1} - \frac{\boldsymbol{\phi}}{\boldsymbol{\phi}_m}\right)^{-c} \tag{2}$$

where c is obtained by fitting the experimental data. The original fitting was performed for volume fractions in the range $\mathbf{0.01} < \boldsymbol{\phi} < \mathbf{0.5}$, however, for the sake of simplicity, many papers assume the model to be verified on a wider range, that is $\mathbf{0.01} < \boldsymbol{\phi} < \mathbf{0.68}$. The asymptote in correspondence of the volume fraction $\boldsymbol{\phi}_m$ is due to the fact that, beyond such limit, usually referred to as maximum packing fraction, the dispersed particles will create a rigid structure, and the fluid will cease to flow. The maximum packing fraction is evaluated as a function of the geometry of the solid particles and the microstructure that they form within the fluid, the value for a monomodal packing of rigid spheres is usually set to $\mathbf{0.63}$, while for poly-dispersed packing, values are usually higher (up to $\mathbf{0.75} \div \mathbf{0.80}$). In this paper the chosen value for $\boldsymbol{\phi}_m$ is $\mathbf{0.68}$ (in accordance with [24]).

Solver validation

The simulation of the bidimensional channel flow has been carried out for the duration of 12 seconds (time estimated for the reaching of steady conditions). The simulated channel is 35 mm and 1.7 mm wide. The time assumption shall be directly verified by considering the value of the variation of the volume fraction ($\Delta\phi$).

Considering the section of the channel at the position $L = 0.238$ m, i.e., where the data for the validation are going to be gathered, and the time t = 12 s, the maximum recorded value amounts to:

$$max\{|\Delta\phi|\}_L \cong max\left\{\left|\frac{\partial\phi}{\partial t}\right|\right\}_L \Delta t \cong 5.2\cdot 10^{-7} \tag{3}$$

It is evident how the variation has not achieved a zero value yet, however, since the order of magnitude is considerably narrow, and the profile peak is already close to the maximum value of packing, it is realistic to assume that the steady condition has been reached. It is hence possible to compare the results produced by OpenFOAM with the two references.

This comparison is performed in Fig. 1; with respect to the set of experimental data measured by Lyon & Leal [21] and the 1D and 2D models introduced by Ilyoung Kwon et al. in [24].

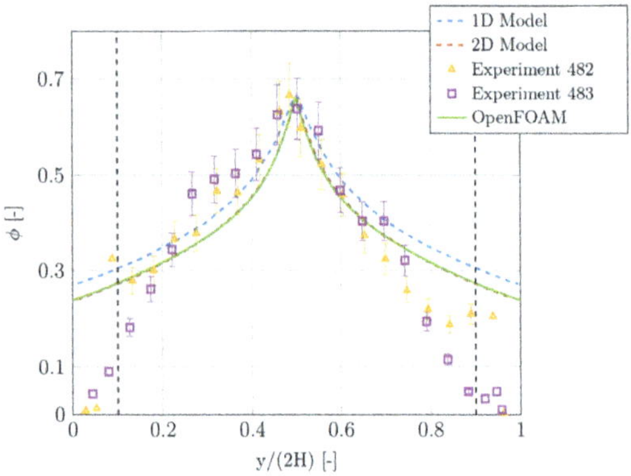

Figure 1: Comparison between OpenFOAM results and literature data of the particle distribution across the section of the channel (L = 0.238 m, t = 12 s). The 1D and 2D have been developed in [24], while the experimental data are measured in [21].

As it can be noticed, the OpenFOAM results emulate in a quite precise manner the behaviour of the 2D model. The global trend of the experimental data is respected, as the peak of the volume fraction is reached at the midpoint of the channel according to the theory and the observations. Furthermore, the right part of the curve, in the proximity of the centreline, appears to be quite close to the values of the two experiments. Adversely, on the left side, the situation is different, nonetheless, the reason may be also related to the precision of the measurements. Approaching the walls, the observed volume fraction diverges from the one predicted via CFD, the motivation might be due to the inaccuracy of the method adopted (i.e., LDV), which had the tendency to underestimate the particle concentration due to the drop of the signal-to-noise ratio caused by the solid boundaries.

Conclusion

The aim of this work was the development of an OpenFOAM solver aimed at correctly predicting the migration phenomenon that is experienced by the solid particles in concentrated suspensions when subjected to non-uniform shear flows. In order to simulate such behaviour, the conservation equation expressing the time variation of the particle volume fraction has been implemented in OpenFOAM. The chosen preexisting solver that has been modified is *pimpleFoam*, which discretizes the Navier-Stokes system of equation through the PIMPLE algorithm. As a first step, the formulation of the equation of motion has been adapted to correctly solve non-Newtonian flows. Successively, the model for the viscous field was established. The code implements the Krieger's correlation to include the viscosity variation over the domain due to the heterogeneity of the particle spatial distribution. Subsequently, the iterative cycle for the solution of the migration equation has been included within the time loop.

The above-mentioned code has been successfully validated by taking into account the measured data provided by the experiment of Lyon & Leal [21] and the results of the CFD code developed by Ilyoung Kwon et al. [24]. The flow that has been simulated in order to verify the capacity of the solver to approach the migration problem is a 2D channel flow. From the comparison between the results produced via OpenFOAM, with the finite volume discretization, and the data from literature it has been verified that the particles volume fraction ϕ is predicted in a quite satisfactory way. The various operations that compose the solver *migrationPimpleFoam* are suitable for a 3D flow; therefore, more complex scenarios might be simulated.

References

[1] REN P, WANG H, ZHOU G, et al (2021) Solid rocket motor propellant grain burnback simulation based on fast minimum distance function calculation and improved marching tetrahedron method. Chinese Journal of Aeronautics 34:208–224. https://doi.org/10.1016/j.cja.2020.08.052

[2] Wang X, Jackson TL, Massa L (2004) Numerical simulation of heterogeneous propellant combustion by a level set method. Combustion Theory and Modelling 8:227–254. https://doi.org/10.1088/1364-7830/8/2/003

[3] Sutton GP, Biblarz O (2016) Rocket Propulsion Elements, 9th ed. John Wiley & Sons

[4] Breton P Le, Ribereau D (2002) Casting Process Impact on Small-Scale Solid Rocket Motor Ballistic Performance. J Propuls Power 18:1211–1217. https://doi.org/10.2514/2.6055

[5] Ghose B, Kankane DK (2008) Estimation of location of defects in propellant grain by X-ray radiography. NDT & E International 41:125–128. https://doi.org/10.1016/j.ndteint.2007.08.005

[6] Wang X, Jackson TL, Buckmaster J (2007) Numerical simulation of the 3-dimensional combustion of aluminized heterogeneous propellants. Proceedings of the Combustion Institute

[7] Jackson TL (2012) Modeling of Heterogeneous Propellant Combustion: A Survey. AIAA Journal 50:993–1006. https://doi.org/10.2514/1.J051585

[8] Kochevets S, Buckmaster J, Jackson TL, Hegab A (2001) Random Packs and Their Use in Modeling Heterogeneous Solid Propellant Combustion. J Propuls Power 17:883–891. https://doi.org/10.2514/2.5820

[9] Plaud M, Gallier S, Morel M (2015) Simulations of heterogeneous propellant combustion: Effect of particle orientation and shape. Proceedings of the Combustion Institute 35:2447–2454. https://doi.org/10.1016/j.proci.2014.05.020

[10] Ponti F, Mini S, Annovazzi A (2020) Numerical Evaluation of the Effects of Inclusions on Solid Rocket Motor Performance. AIAA Journal 58:4028–4036. https://doi.org/10.2514/1.J058735

[11] Ponti F, Mini S, Fadigati L, et al (2021) Effects of inclusions on the performance of a solid rocket motor. Acta Astronaut 189:283–297. https://doi.org/10.1016/j.actaastro.2021.08.030

[12] Ponti F, Mini S, Fadigati L, et al (2021) INFLUENCE OF NOZZLE RADIATION ON SOLID ROCKET MOTORS TAIL-OFF THRUST. International Journal of Energetic Materials and Chemical Propulsion 20:45–64. https://doi.org/10.1615/IntJEnergeticMaterialsChemProp.2021038491

[13] Mini S, Ponti F, Brusa A, et al (2023) Prediction of Tail-Off Pressure Peak Anomaly on Small-Scale Rocket Motors. Aerospace 10:169. https://doi.org/10.3390/aerospace10020169

[14] Mini S, Ponti F, Annovazzi A, Gizzi E (2020) Impact of Thermal Protections Insulation Layer on Solid Rocket Motor Performance. In: AIAA Propulsion and Energy 2020 Forum. American Institute of Aeronautics and Astronautics, Reston, Virginia

[15] Le Breton P, Ribereau D, Marraud C, Lamarque P (2002) Experimental and Numerical Study of Casting Process Effects on Small Scale Solid Rocket Motor Ballistic Behavior. International Journal of Energetic Materials and Chemical Propulsion 5:132–145. https://doi.org/10.1615/IntJEnergeticMaterialsChemProp.v5.i1-6.160

[16] Mini S, Ponti F, Annovazzi A, et al (2022) A novel procedure to determine the effects of debonding on case exposure of solid rocket motors. Acta Astronaut 190:30–47. https://doi.org/10.1016/j.actaastro.2021.09.016

[17] Ponti F, Mini S, Fadigati L, et al (2022) Theoretical Study on The Influence of Debondings on Solid Rocket Motor Performance. International Journal of Energetic Materials and Chemical Propulsion 21:21–45. https://doi.org/10.1615/IntJEnergeticMaterialsChemProp.2021039436

[18] Ao W, Liu X, Rezaiguia H, et al (2017) Aluminum agglomeration involving the second mergence of agglomerates on the solid propellants burning surface: Experiments and modeling. Acta Astronaut 136:219–229. https://doi.org/10.1016/j.actaastro.2017.03.013

[19] Emelyanov VN, Teterina IV, Volkov KN (2020) Dynamics and combustion of single aluminium agglomerate in solid propellant environment. Acta Astronaut 176:682–694. https://doi.org/10.1016/j.actaastro.2020.03.046

[20] (2023) OpenFOAM foam-extend toolbox. https://sourceforge.net/projects/foam-extend/. Accessed 26 Apr 2023

[21] LYON MK, LEAL LG (1998) An experimental study of the motion of concentrated suspensions in two-dimensional channel flow. Part 1. Monodisperse systems. J Fluid Mech 363:25–56. https://doi.org/10.1017/S0022112098008817

[22] Phillips RJ, Armstrong RC, Brown RA, et al (1992) A constitutive equation for concentrated suspensions that accounts for shear-induced particle migration. Physics of Fluids A: Fluid Dynamics 4:30–40. https://doi.org/10.1063/1.858498

[23] Krieger IM (1972) Rheology of monodisperse latices. Adv Colloid Interface Sci 3:111–136. https://doi.org/10.1016/0001-8686(72)80001-0

[24] Kwon I, Jung HW, Hyun JC, et al (2018) Particle migration in planar Couette–Poiseuille flows of concentrated suspensions. J Rheol (N Y N Y) 62:419–435. https://doi.org/10.1122/1.4989416

Validation of a numerical strategy to simulate the expansion around a plug nozzle

Marco Daniel Gagliardi[1,a], Luca Fadigati[1,b], Nabil Souhair[1,c] and Fabrizio Ponti[1,d *]

[1]Alma Propulsion Lab, Alma Mater Studiorum-Università di Bologna, Via Fontanelle 40, 40121 Forlì, FC, Italy

[a]marco.gagliardi2@studio.unibo.it, [b]luca.fadigati2@unibo.it, [c]nabil.souhair@unibo.it, [d]fabrizio.ponti@unibo.it

Keywords: Aerospike Engine, Plug Nozzle, HLLC Scheme, Numerical Simulation, OpenFOAM

Abstract. Rocket engines currently use traditional bell-shaped nozzles that have a fixed area ratio and can only operate at maximum efficiency at a given altitude. Plug nozzles have been proposed as an alternative solution to achieve higher performance over a larger altitude range. Unlike bell nozzles, the flow is free to expand along the plug, as it is no longer surrounded by solid boundaries. Therefore, plug nozzles can adapt to different altitudes by expanding the flow to ambient pressure, resulting in continuous altitude adaptation. Due to the high surface area that needs to be cooled, one of the main challenges of plug nozzle design is thermal management. However, the introduction of aerospike geometry, which is essentially a truncated plug nozzle, has helped mitigate this issue. Simulating an aerospike engine is challenging due to the interaction between the plume and the external flow, which is necessary to accurately predict thrust. In this work, a numerical strategy for predicting the performance of an aerospike engine, during a static fire, was developed and validated.

Introduction

Rocket engines commonly employ traditional bell-shaped nozzles with fixed area ratios, limiting their maximum efficiency to a specific altitude corresponding to the design Nozzle Pressure Ratio (NPR). Therefore, bell shaped nozzles operate sub-optimally for a significant portion of a launcher's flight. Various solutions have been proposed to address these limitations, but none have proven suitable for practical flight operations. In the 1950s, plug nozzles and aerospikes were introduced to achieve higher performance across a broader altitude range. Unlike other nozzle concepts, they offer continuous altitude adaptation. The advent of additive manufacturing techniques has sparked new developments in aerospike technology worldwide, as economically viable processes can address the geometrical complexity of the engine. Several research groups are actively exploring aerospike technology. The Beijing University of Aeronautics and Astronautics has developed an optimization method to design aerospike contours that maximize total impulse from sea level to the design altitude [1]. Technische Universität Dresden has studied thrust vectoring control systems and ceramic additive manufacturing techniques [2, 3]. Accurate thrust prediction in aerospike engines requires considering the interaction between the plume and external flow, posing a significant simulation challenge. This paper focuses on the development and validation of a numerical strategy aimed at predicting the performance of an aerospike engine. By utilizing advanced numerical simulations, the aim is to obtain crucial information impractical to measure during physical tests, such as pressure distribution along the plug and in the plume. The paper is divided into two main sections: the first one describes the developed model, the second one the results obtained and the validation using experimental data coming from literature.

Materials Research Proceedings 37 (2023) 695-698 https://doi.org/10.21741/9781644902813-148

Model structure

The CFD software employed to carry out the simulation is OpenFOAM, the chosen solver is *dbnsTurbFoam* that is contained in foam-extend. It has been chosen because it is considered particularly suitable for simulating supersonic turbulent flows and results the only compressible solver exploiting the Harten-Lax-van Leer-Contact (HLLC) approximate Riemann solver, crucial to correctly capture shock waves without smearing them [4]. In [5] the fidelity of *dbnsTurbFoam* is assessed: taking as references the Sod's shock tube (analytical solution, [6]) and the Onera S8 Transonic Channel (experimental data, [7]) scenarios, it is shown how *dbnsTurbFoam* is suitable to study high speed compressible flows. Turbulence is taken into account by means of the k-ω SST model developed by Menter [8]. A key feature of the SST model is the implementation of the stress limiter parameter, a_1, in the definition of the eddy viscosity ν_t [9].

$$\nu_t = \frac{a_1 k}{max(a_1\omega;\ \Omega F_2)} \tag{1}$$

Such formulation of the eddy viscosity has been considered in the model in order to better account for the transport of the turbulent shear stress inside the boundary layer of compressible flows, largely improving the performance in case of adverse pressure gradients. Ω is the absolute value of vorticity, and F_2 is a blending function, equal to one for boundary layer flows and to zero for free shear layers, that, in this latter case, allows to return to the original definition $\nu_t = k/\omega$ [8]. The default value employed by Menter for a_1 is 0.31.

Results

The experimental work of S. B. Verma and M. Viji [10] consists in testing a linear plug nozzle. The nozzle is fed with air stored in a tank at ambient temperature and pressure, but both values are not explicitly provided in the reference paper; in this work, they have been assumed of 1 bar, as done also in [11], and 300 K, respectively. Fig. 1 shows the experimental setup and the nozzle geometry as given in [10]; the design Nozzle Pressure Ratio for the internal nozzle is 14, and the overall nozzle design NPR is 36, corresponding to a Mach 3±0.1 exit flow; all tests have been performed in over-expansion conditions. The nozzle is provided with nine pressure ports distributed along the plug axis, as shown in Fig. 1, that allow measurements for the reconstruction of the time-averaged pressure distribution along the spike wall.

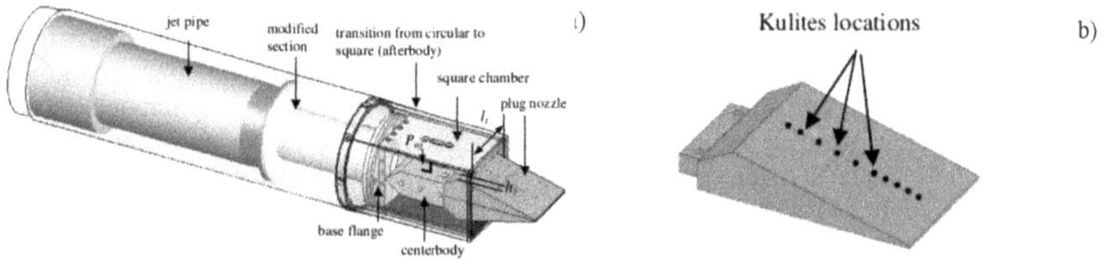

Fig. 1: Experimental setup for the analyses conducted in [10]: a) shows a section of the experimental setup, b) shows the position of the pressure measurements on the plug surface. Images from [10]

The simulations have been run until the jet achieved stationary conditions, and the obtained results have been compared with the data available in the two reference papers ([10] and [11]) in terms of flow topology and time-averaged pressure distribution along the spike wall. In Fig. 2, for example, the numerical results at NPR = 3.1, from [11] are compared with those obtained in this work. Fig. 3 shows the comparison between the experimental result and the ones from the simulations. On the horizontal axis, there is the coordinate along the spike normalized by L = 0.1m, that is the length of the spike itself from the throat section to the tip. The results are very close to the experimental.

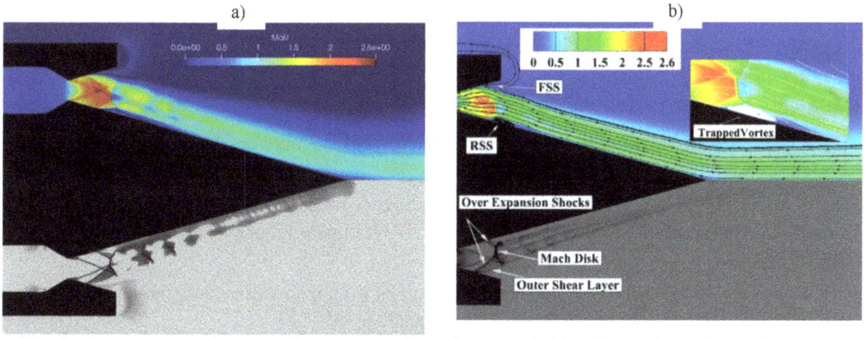

Fig. 1: A comparison between the numerical results provided by OpenFOAM (a) and ANSYS Fluent (b) at NPR = 3.1 [11].

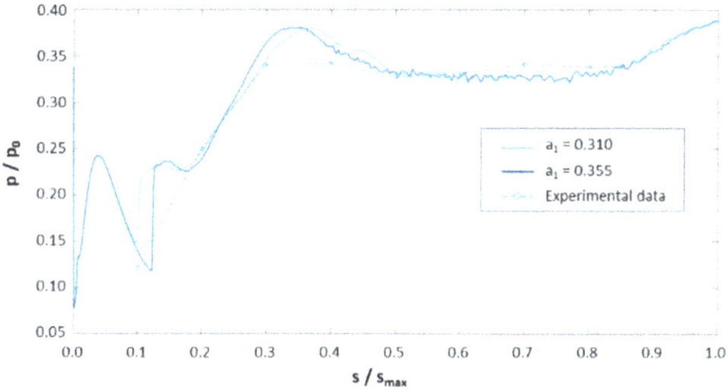

Fig. 3: Experimental spike wall pressure distribution (dots and linear interpolation) versus its numerical counterpart computed in OpenFOAM using two different values of the stress limiter in the turbulence model (NPR = 3.1).

Conclusion

The paper presents the definition of a model capable to simulate the supersonic flow at the exit of a plug nozzle. The simulation environment has been developed in OpenFOAM and based on the solver *dbnsTurbFoam*. The $\boldsymbol{k-\omega}$ SST model has been used to simulate flow turbulence and different parametrizations of the model have been discussed and simulated. The model has been validated using experimental data found in literature, collected on a linear plug nozzle, both in terms of pressure measurements on the plug wall, and in terms of Schlieren pictures of the flow. From the results that have been obtained, despite all the uncertainties associated to the reference experimental setup, the performance of the solver can be definitely considered appropriate.

References

[1] Wang C-H, Liu Y, Qin L-Z (2009) Aerospike nozzle contour design and its performance validation. Acta Astronaut 64:1264–1275. https://doi.org/10.1016/j.actaastro.2008.01.045

[2] Sieder-Katzmann J, Propst M, Tajmar M, Bach C (2021) INVESTIGATION OF AERODYNAMIC THRUST-VECTOR CONTROL FOR AERO- SPIKE NOZZLES IN COLD-GAS EXPERIMENTS

[3] Schwarzer-Fischer E, Abel J, Sieder-Katzmann J, et al (2022) Study on CerAMfacturing of Novel Alumina Aerospike Nozzles by Lithography-Based Ceramic Vat Photopolymerization (CerAM VPP). Materials 15:3279. https://doi.org/10.3390/ma15093279

[4] Toro EF (2019) The HLLC Riemann solver. Shock Waves 29:1065–1082. https://doi.org/10.1007/s00193-019-00912-4

[5] Chandramouli S, Gojon R, Fridh J, Mihaescu M (2017) Numerical Characterization of Entropy Noise With a Density Based Solver

[6] Sod GA (1978) A survey of several finite difference methods for systems of nonlinear hyperbolic conservation laws. J Comput Phys 27:1–31. https://doi.org/10.1016/0021-9991(78)90023-2

[7] Delery J (1978) Analyse du décollement résultant d'un interaction choc-chouche limite turbulent en transsonique. La Recherche Aérospatiale 6:305–320

[8] Menter FR (1994) Two-equation eddy-viscosity turbulence models for engineering applications. AIAA Journal 32:1598–1605. https://doi.org/10.2514/3.12149

[9] Huang PG, Bradshaw P, Coakley TJ (1992) Assessment of closure coefficients for compressible-flow turbulence models. Ames Research Center, Moffett Field,California 94035-1000

[10] Chutkey K, Viji M, Verma SB (2017) Effect of clustering on linear plug nozzle flow field for overexpanded internal jet. Shock Waves 27:623–633. https://doi.org/10.1007/s00193-017-0707-y

[11] Soman S, Suryan A, Nair PP, Dong Kim H (2021) Numerical Analysis of Flowfield in Linear Plug Nozzle with Base Bleed. J Spacecr Rockets 58:1786–1798. https://doi.org/10.2514/1.A34992

Special Session in Memory of Professor Debei

Solar simulator facility for the verification of space hardware performance

C. Bettanini[1,2] *, M. Bartolomei[2], S. Chiodini[1,2], L. Tasinato[2], P. Ramous[2], F. Dona[2] and S. Debei[1,2]

[1] Department of Industrial Engineering, University of Padova, via Venezia 1, Padova (Italy)

[2] CISAS - Center for Studies and Activities for Space "Giuseppe Colombo", University of Padova, via Venezia 15, Padova (Italy)

* carlo.bettanini@unipd.it

Keywords: High Flux Solar Simulator, Flux Homogeneity, Test of Satellite Hardware

Abstract. The paper presents the main characteristics of the high flux solar simulator facility designed and developed at University of Padova as key enabling technology to evaluate the effectiveness of satellite hardware for missions to the inner planets of the Solar System. The designed solar simulator can reproduce the intensity and spectral distribution of the Sun's radiation up to 8 Solar constants (around 10000 Watt/m^2) and the emitted flux can be directed to the viewport of a Thermal Vacuum Chamber in order to test the performance of space equipment under representative pressure and temperature conditions. Angles of incidence between 30° and 90° can be achieved using a motorised setup within the thermal chamber while different intensities of sunlight can be obtained by properly choosing the emitting lamp and regulating the electric power. After the verification of optical path alignment, a series of tests has been conducted to evaluate the flux homogeneity installing a commercial pyranometer on cartesian reference and moving the slide within the target area. A final Class A classification for the spatial non-uniformity of irradiance as for ASTM E927-19 has been achieved for the central target area. The facility has afterwards operated for validation campaign of satellite radiators in simulated orbital condition, verifying the repeatability of reproduced flux during continuous long-term operation.

Introduction

Solar simulators are used to mimic the light and heat conditions for materials, equipment and instruments subject to direct sunlight during operation; outdoor experiments can in fact be carried out but are strongly affected by variability of an uncontrollable environment. The Solar simulator design can be linked to two main categories of application: non-concentrating solar applications, employed for testing photovoltaic and solar water collectors and concentrating solar applications, for testing components and materials for high-temperature thermal and thermochemical applications.

The achieved flux density of simulators can so range from less than one tenth of the solar constant to tens of thousands of solar constants as in the so called High-Flux Solar Simulators (HFSS); examples are the SynLight built by DLR [1] or the solar simulator developed by ETH-Zurich [2] and [3] used for testing advanced high-temperature materials.

Several light sources can be considered for application in solar simulators to produce a radiation that approximates the natural light spectrum under controlled and repeatable conditions. A comparison of lamp wavelength spectrum, lamp intensity, cost, stability, durability, and hazards associated with use is provided in [4] .

In particular, for space applications the European Space Agency has developed several solar simulators able to provide illumination beams of different diameter and intensity under vacuum to allow qualification of satellites for Earth and planetary missions. The most powerful solar simulator (the Large Space Simulator) consists of an array of 19 xenon lamp modules, each with 25 kW power, capable of simulating 10 solar constants into an a target area of 2.7 m and has been used to simulate the BepiColombo satellite operational condition at Mercury.

Solar simulator design

The designed solar simulator can reproduce the intensity and spectral distribution of the Sun's radiation up to 8 Solar constants and has been installed near a Thermal Vacuum Chamber provided with a viewport in order to be able to test the performance of any space equipment under a range of pressure and temperature conditions, including different intensities of sunlight and different angles of incidence. A picture of the realised Solar simulator near the Thermal vacuum chamber is shown in figure 1.

The optical design of the simulator is aimed at guaranteeing repeatable values of flux and high levels of flux homogeneity across the target area; the light source is a Xenon arc lamp mounted vertically in the focus of a truncated ellipsoid reflector. The light path is afterward guided by multiple different reflecting surfaces mounted on a common optical bench; after the folding mirror, a primary mirror (spherical with 150 mm diameter), a fly-eye integrator mirror and finally a secondary mirror (spherical 500 with mm diameter) direct the flux onto the target area (Viewport with 200mm diameter). An overview of the optical path is provided in figure 2.

The fly-eye mirror has been realised to homogenize the uneven brightness of the light source by arranging multiple single lenses preserving the brightness of the beam and achieving a nearly-collimated beam with low divergence.

Figure 1 Solar simulator installed near the Optical aperture of Thermal Vacuum Chamber

Figure 2 3D reconstruction of the Solar Simulator optical path

The presence of high energy concentrated onto the small mirror areas can lead to overheating and potential damage so dedicated cooling systems have been implemented in the setup. The emitting Xenon lamp is cooled by three fans, while thermal control of the folding mirror and the fly-eye integrator is achieved by heat exchangers cooled by a mixture of ethylene glycol-water pumped within a cooled loop circuit. The heat dissipation in the secondary spherical mirror and the folding mirror is achieved by directing a high-velocity stream of air across the rear area (creating a "knife-like" airflow) using a compressed air circuit.

Evaluation of achieved Solar flux uniformity

Flux homogeneity is a critical characteristic of a solar simulator that refers to the uniformity of the irradiance across the target area. This uniformity is essential because it ensures that the devices being tested receive a consistent and repeatable level of illumination. Inaccuracies in the irradiance level can cause significant variations in the performance of the device under test, making it challenging to compare results and draw meaningful conclusions about its performance.

To evaluate flux homogeneity, the distribution of radiative flux across the target area of the simulator has been measured using direct mapping flux measurement system: a LP PYRA 02 AC4 pyranometer by Delta Ohm S.r.l. positioned on an equidistant grid of 4 cm. A Xenon arc emitting lamp with nominal flux of 1 Solar constant has been used for the test.

Figure 3 provides an illustration of the reconstructed distribution, depicting the mapping of radiative flux at the viewport of the thermal vacuum chamber with the lamp in operation. The flux distribution approaches an ellipsoidal Gaussian distribution, with the peak flux reaching about 1355 W/m2.

To assess the homogeneity of the flux, the measured flux's deviation from the interpolation of the measures with an ellipsoidal Gaussian distribution was evaluated; this analysis provides valuable insights into the uniformity of the irradiance across the target area of the solar simulator.

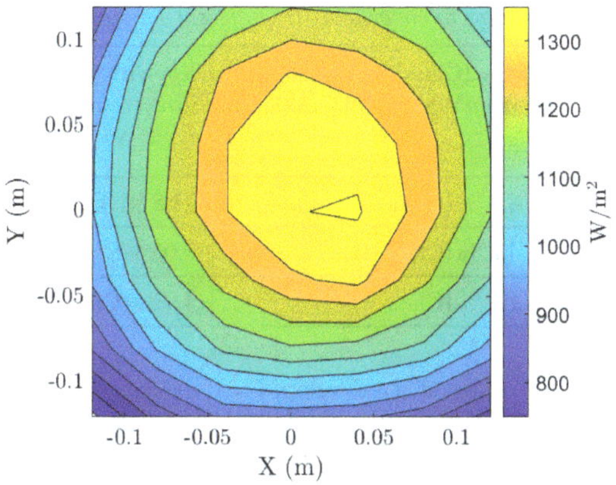

Figure 3 Mapping of radiative flux at the viewport of the thermal vacuum chamber as measured by the pyranometer.

Figure 4 depicts the differences between the measured and interpolated Solar fluxes. Panel (b) shows a histogram of the frequency of deviations from the interpolated flux, while panel (a) maps the inhomogeneities between the measured and interpolated fluxes.

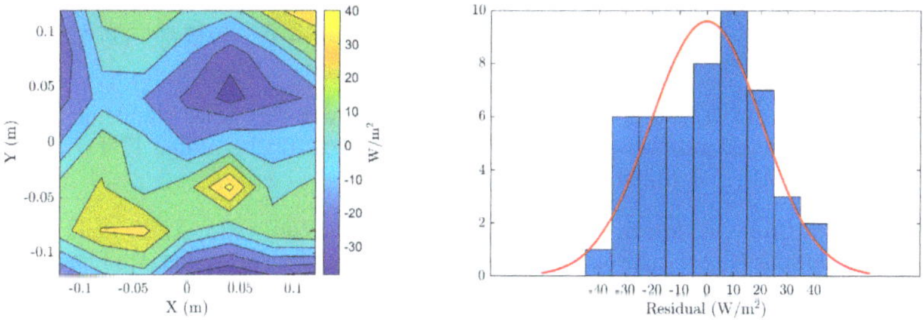

Figure 4 (a) Map of inhomogeneities between measured Solar flux and flux interpolated via Gaussian ellipsoid. (b) Distribution of inhomogeneities for measured points. The histogram shows the frequency of deviations from the flux interpolated via Gaussian ellipsoid."

The spatial non-uniformity of irradiance S_{NE} was evaluated according to ASTM E927-19 [5][5] from the interpolated solar flux W given by Table 1 parameters for different TV chamber viewport diameter. The S_{NE} is given by Equation (3).

$$S_{NE} = 100\frac{\max W - \min W}{\max W + \min W} \quad (3)$$

Table 1 shows the class corresponding to the S_{NE} calculated for the various diameters. Considering an entrance window with a diameter of 9 cm, the solar simulator is of Class A for the spatial non-uniformity of irradiance. The Class passes to B if we consider the standard deviation of the interpolation residuals.

Table 1 Spatial non-uniformity of irradiance calculated according to ASTM E927-19 without and considering the measurement uncertainty for different TV chamber viewport diameter.

TV chamber viewport diameter	**S_{NE} (Classification as from ASTM E927-19)**	**S_{NE} with 95% confidence uncertainties (Classification as from ASTM E927-19))**
20 cm	8.5 % (Class C)	11.8 % (Class U)
15 cm	4.8 % (Class B)	8.0 % (Class C)
9 cm	1.7 % (Class A)	4.9 % (Class B)

Conclusions

The Solar Simulator Facility designed and developed at University of Padova has been tested to evaluate the flux homogeneity and repeatability using a 1 Solar Constant rated Xenon arc lamp.

The results show achieved spatial non-uniformity of the light beam on the target area allows to reach Class A on a 9 cm diameter target according to ASTM E927-19.

The facility can so be successfully used to test space hardware in representative pressure, temperature and direct sun illumination conditions with the capability to reproduce intensity and spectral distribution of the Sun's radiation up to 8 Solar constants.

References

[1] K. Wieghardt, D. Laaber, V. Dohmen, P. Hilger, D. Korber, K.-H. Funken, B. Hoffschmidt, Synlight-A new facility for large-scale testing in CSP and solar chemistry, in: AIP Conference Proceedings, vol. 2033, AIP Publishing LLC, 2018, pp. 040042. https://doi.org/10.1063/1.5067078

[2] D. Hirsch, P. Zedtwitz and et al. 2003 A new 75 kW high-flux solar simulator for high-temperature thermal and thermochemical research J. Sol. Energy Eng. 125(1) 117-120. https://doi.org/10.1115/1.1528922

[3] J. Petrasch, P. Coray, A. Meier, M. Brack, P. Häberling, D. Wuillemin, A. Steinfeld A Novel 50kW 11,000 suns High-Flux Solar Simulator Based on an Array of Xenon Arc Lamps J. Sol. Energy Eng., 129 (2006), pp. 405-411D. Codd and A. e. a. Carlson 2010 A low cost high flux solar simulator Sol. Energy 84 (12) 2202-2212. https://doi.org/10.1115/1.2769701

[4] M. Tawfik, X.Tonnellier,C. Sansom , "Light source selection for a solar simulator for thermal applications: A review", Renewable and Sustainable Energy Reviews , Volume 90, July 2018, Pages 802-813. https://doi.org/10.1016/j.rser.2018.03.059

[5] ASTM E927-19 Standard Classification for Solar Simulators for Electrical Performance Testing of Photovoltaic Devices.

Trajectory reconstruction by means of an event-camera-based visual odometry method and machine learned features

S. Chiodini[1,a*], G. Trevisanuto[2,b], C. Bettanini[1,c], G. Colombatti[1,d] and M. Pertile[3,e]

[1]Department of Industrial Engineering, University of Padova, via Venezia 1, Padova (Italy)

[2]CISAS - Center for Studies and Activities for Space "Giuseppe Colombo", University of Padova, via Venezia 15, Padova (Italy)

[3]School of Engineering, University of Padova, Via Gradenigo 6/a, Padova, Italy

[a]sebastiano.chiodini@unipd.it, [b]giovanni.trevisanuto@studenti.unipd.it, [c]carlo.bettanini@unipd.it, [d]giacomo.colombatti@unipd.it, [e]marco.pertile@unipd.it

Keywords: Visual Odometry, Computer Vision, Machine Learning

Abstract. This paper presents a machine learned feature detector targeted to event-camera based visual odometry methods for unmanned aerial vehicles trajectory reconstruction. The proposed method uses machine-learned features to enhance the accuracy of the trajectory reconstruction. Traditional visual odometry methods suffer from poor performance in low light conditions and high-speed motion. The event-camera-based approach overcomes these limitations by detecting and processing only the changes in the visual scene. The machine-learned features are crafted to capture the unique characteristics of the event-camera data, enhancing the accuracy of the trajectory reconstruction. The inference pipeline is composed of a module repeated twice in sequence, formed by a Squeeze-and-Excite block and a ConvLSTM block with residual connection; it is followed by a final convolutional layer that provides the trajectories of the corners as a sequence of heatmaps. In the experimental part, a sequence of images was collected using an event-camera in outdoor environments for training and test.

Introduction

Bio-inspired systems are becoming increasingly widespread in the field of robotics. The advantages are related to the reduced use of resources, both in terms of power consumption and computational load. In terms of perception, Event-based vision sensors, such as Dynamic Vision Sensor (DVS) devices represent one of the intriguing advancements in image sensor technology. These devices incorporate in-pixel circuitry that can detect temporal changes in intensity and communicate these changes as binary "events" to the external world. Essentially, only the pixels that detect changes in light intensity transmit data, enabling data compression at the sensor level and facilitating low-latency operations. This is made possible because individual pixel changes can be transmitted without the need to read out full frame image frames [1]. Event-based cameras offer significant advantages over traditional cameras. Latency, which is the time delay in processing sensor data, is a critical factor, event-based cameras drastically reduce latency by transmitting data through events, which have microsecond-level latencies. Furthermore, event-based cameras possess a remarkably high dynamic range of 130 dB compared to the 60 dB range of standard cameras. This makes them well-suited for scenes with substantial illumination changes.

Certainly, one of the most promising applications for this type of camera is in the navigation of highly agile robots such as drones [2], as well as for the aspects of entry, descent, and landing of planetary probes [3]. To utilize these sensors for such purposes, it is necessary to adapt or invent new algorithms for Visual Odometry (VO). This ensures that the cameras can effectively support the navigation and mapping tasks required in these dynamic scenarios.

For VO, it is crucial to have keypoints that are repeatable and accurate across consecutive frames. Currently, there are handcrafted methods inspired by classical computer vision theory that allow the extraction of a series of features, such as the eHarris-based approach [3]. Inspired by the work of [4], we have chosen to utilize machine-learned features that exhibit a certain level of temporal stability. In this work, we present the method for training these machine-learned features, demonstrate how to integrate them into a visual odometry system, and showcase some preliminary results.

Method

The adopted event keypoint detection method is adapted from work of [4] and is based on receiving an event tensor (also called event cube) $E(x, y, t)$ of dimension $H \times W \times B$ as input and predicts a set of heatmaps as keypoint location. Regarding the event tensor input, H and W represent the height and width of the image sensor, respectively, and B indicates the number of temporal bins, which is 12 in our case. Generation of the event tensor involves several steps: as first the change in light $L_{xy,i}$ at pixel (x_i, y_i) crosses a threshold, a spike is generated; then the event camera outputs a spike stream with coordinates (x_i, y_i) and timestamp t_i, finally event stream is converted into an event tensor by considering an integration period Δt.

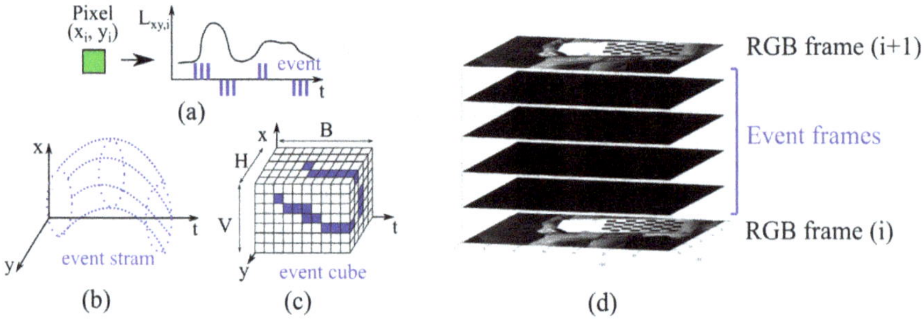

Figure 1 Event tensor generation: (a) when the change in light $\mathbf{L_{xy,i}}$ pixel $(\boldsymbol{x_i}, \boldsymbol{y_i})$ crosses a threshold a spike is generated, (b) the event camera outputs a spike stream in time $\mathbf{t_i}$ and space $(\boldsymbol{x_i}, \boldsymbol{y_i})$, (c) the event stream is converted in an event tensor considering and integration period $\boldsymbol{\Delta t}$. (d) Event tensor $\mathbf{E}(\boldsymbol{x}, \boldsymbol{y}, \boldsymbol{t})$ used for detector training, the training points are detected using Harris on grayscale frame and interpolating their position on the event frames.

The event tensor E(x, y, t) is utilized for detector training, where training points are detected using Harris on grayscale frames and then their positions are interpolated between two consecutive event frames and filtered based on epipolar constraint. The whole event tensor generation and training point selection is depicted in Figure 1.

The loss function is based on the Binary Cross Entropy (BCE) between the predicted heatmaps $\mathcal{H}_h(x, y)$ and the interpolated keypoints positions $\widehat{\mathcal{H}}_h(x, y)$:

$$\mathcal{L} = \sum_{h\in[1,n_h]} \sum_{(x,y)} \mathrm{BCE}(\mathcal{H}_h(x, y), \widehat{\mathcal{H}}_h(x, y)) \quad (1)$$

The first sum is over the n_h predicted heatmaps. The second sum is over the image locations (x, y). The neural network architecture used in this work is based on [5] and consists of a fully convolutional network with five layers, each utilizing 3x3 kernels. Each layer has 12 channels and incorporates residual connections. ConvLSTMs are employed in the second and fourth layers. The final layer, responsible for heatmap prediction, is a conventional convolutional layer. The remaining feed-forward layers utilize Squeeze-Excite (SE) connections. The training parameters are given in Table 1.

Table 1 Training parameters.

Num epochs	Learning rate	Num ev cubes	Num ev	Num keypoints	Num keypoint/ev frame (avg)
40	0.0001	6750	13.5*10^6	5442164	67.187

Results

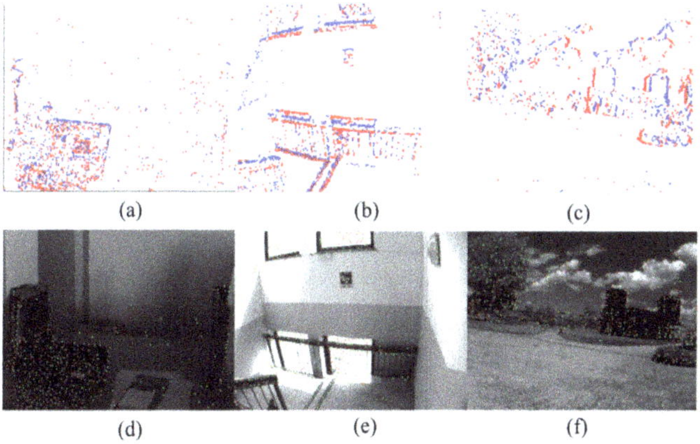

Figure 2 (a-c) Events frames obtained from the integration of 2000 events. (d.e) Corresponding RGB images of keypoints extracted from the respective event frames using the machine-learned detector (red) and a handcrafted feature detector such as eHarris (green).

To gather the frames and events required for training, sequences of images were collected in both outdoor and indoor environments. The DAVIS 346 camera from Inivation was employed for the acquisitions. The DAVIS 346 camera is a DVS event camera with a resolution of 346 x 260 pixels and includes an active pixel frame sensor. Figure 2 shows the event frames obtained from the integration of 2000 events and the corresponding RGB images with the detected keypoints. During the testing phase, the event cubes were provided as input to the machine-learned detector to extract the corresponding peak heatmaps. To verify the stability of the keypoints, a Nearest Neighbor (NN) algorithm was employed to track the keypoints in subsequent event frames. Figure 3 shows the graphs depicting the number of keypoints extracted, matched (between two consecutive frames using NN and filtered with RANSAC), and tracked (i.e., keypoints that, after being merged into tracks, belong to a track spanning at least 20 event frames).

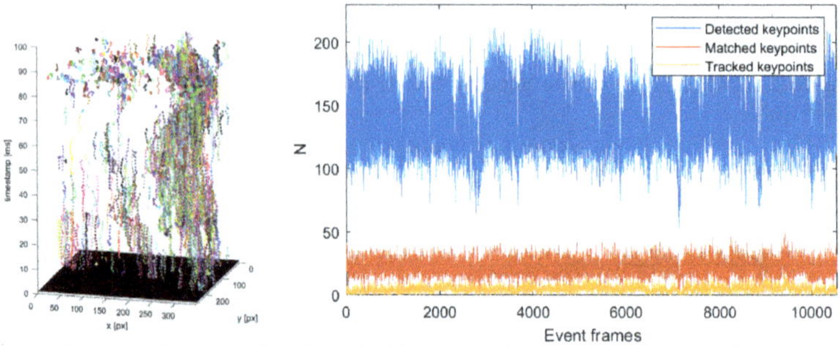

Figure 3 Detected, matched and tracked keypoints for the event stream collected outdoor.

Conclusions

In this work, the initial steps have been taken towards utilizing an event camera for the autonomous navigation of highly agile robots such as drones and entry descent and landing probes. The training of a stable keypoints detector across consecutive frames has been conducted. In future work, we will integrate this keypoint detector into a Visual Odometry pipeline and test the system on a tethered balloon.

Acknowledgements

The DVS/DAVIS technology was developed by the Sensors group of the Institute of Neuroinformatics (University of Zurich and ETH Zurich), which was funded by the EU FP7 SeeBetter project (grant 270324).

References

[1] Lichtsteiner, P., Posch, C., & Delbruck, T. (2008). A 128$\times $128 120 dB 15$\mu $ s latency asynchronous temporal contrast vision sensor. IEEE journal of solid-state circuits, 43(2), 566-576. https://doi.org/10.1109/JSSC.2007.914337

[2] Mueggler, E., Rebecq, H., Gallego, G., Delbruck, T., & Scaramuzza, D. (2017). The event-camera dataset and simulator: Event-based data for pose estimation, visual odometry, and SLAM. The International Journal of Robotics Research, 36(2), 142-149. https://doi.org/10.1177/0278364917691115

[3] Sikorski, O., Izzo, D., & Meoni, G. (2021). Event-based spacecraft landing using time-to-contact. In Proceedings of the IEEE/CVF Conference on Computer Vision and Pattern Recognition (pp. 1941-1950). https://doi.org/10.1109/CVPRW53098.2021.00222

[4] Vasco, V., Glover, A., Bartolozzi, C.: Fast Event-Based Harris Corner Detection Exploiting the Advantages of Event-Driven Cameras. In: IEEE/RSJ International Conference on Intelligent Robots and Systems (IROS). pp. 4144–4149 (2016). https://doi.org/10.1109/IROS.2016.7759610

[5] Chiberre, P., Perot, E., Sironi, A., & Lepetit, V. (2022). Long-Lived Accurate Keypoint in Event Streams. arXiv preprint arXiv:2209.10385.

The wide angle camera of rosetta

Giampiero Naletto

Department of Physics and Astronomy, University of Padova, Via Marzolo 8, Padova, Italy

giampiero.naletto@unipd.it

Keywords: Space Instrument, Wide Angle Camera, Rosetta

Abstract. Rosetta was the ESA cornerstone missions which investigated the comet 67P/Churuymov-Gerasimenko. One of the on board instruments was the OSIRIS imaging system, which included the Wide Angle Camera. This camera was designed, realized, integrated, aligned and tested at the Padova University. Several challenges had to be faced and solved, due to stringent requirements and the very peculiar mission profile which foresaw a long interplanetary travel with very different thermal environments. Rosetta has been a very successful mission which returned plenty of fundamental information about comets and more in general about the solar system origin. In this paper we summarize the main characteristics of the WAC and describe one of the many scientific results that it had returned.

Introduction

Rosetta was one of the ESA cornerstone missions, and it was dedicated to the investigation of the comet 67P/Churuymov-Gerasimenko. Launched on 2 March 2004, after a 10-year journey it arrived at the comet on 6 August 2014 and orbited around it until the controlled landing on the comet nucleus on 30 September 2016. One of the on-board instruments was the OSIRIS [1] imaging system, which consisted of two cameras, the Narrow Angle Camera and the Wide Angle Camera (WAC). The latter was realized under Italian responsibility, with contributions from other European partners. In particular, this instrument has been designed, realized, integrated, aligned and tested at the Padova University, with a coordinate effort of several departments.

The realization and testing of the WAC lasted several years during which also many young researchers have been involved, at their first work experience on a space instrument. The project started with the optical design and tolerance analysis, then the thermomechanical design followed. Different breadboard models were realized to check the subsystem's performance; then the camera qualification model was integrated and fully checked, and finally the flight model was prepared for launch after calibration. Several challenges had to be faced and solved during the instrument realization, due to the very peculiar mission profile which foresaw a long interplanetary travel with very different thermal environments. Rosetta has been a very successful mission which returned plenty of fundamental information about comets and more in general about the solar system origin.

The OSIRIS Wide Angle Camera

The two cameras of OSIRIS had complementary objectives. On one side, the Narrow Angle Camera had to be a system with high spatial resolution, to study the comet nucleus comet and in particular its structure and geological features. On the other, the WAC had a lower spatial resolution but a wider field of view and higher sensitivity, for observing the gas and dust flow around the nucleus [2]. The WAC optical configuration was an all-reflective two-mirror design, realized with an off-axis convex oblate ellipsoidal mirror followed by an almost on-axis concave oblate ellipsoidal mirror. The field of view was about 12°×12°, collected by a 2048×2048 CCD providing an average image scale of 21 arcsec/pixel. The optical performance was diffraction limited. The covered spectral range, 240-750 mm was selectable by 14 different filters, defined on the basis of the scientific requirements.

Materials Research Proceedings 37 (2023) 709-712 https://doi.org/10.21741/9781644902813-151

The WAC structure was based on a closed box made of aluminum alloy (see Fig. 1); it was lightweighted by electro-erosion machining. The structure was designed to prevent noise induced through vibration and to minimize vibration amplification at interfaces with mechanisms. Very challenging was the thermo-mechanical design [3], that had to satisfy the optical alignment tolerances. In fact, it was verified by ray-tracing simulation that the system could tolerate a relative shift of ±10 μm between the two mirrors, that is over a distance of 30 cm, during all mission lifetime: by thermal modelling, it was verified that to maintain this tolerance, the operational temperature of the WAC optical bench had to be maintained within the 12±5°C range. For an instrument that had to operate from 4 au to 1 au, that is with a great thermal excursion, having a square entrance aperture of about 15 cm side through with heat is exchanged with the space environment, this was a very critical requirement to be satisfied.

To guarantee the optimal optical performance notwithstanding the great thermal excursion over the mission, great care was dedicated to the thermo-structural design. In particular, the WAC interfaced with the payload spacecraft panel by three kinematic mounting feet, which minimized the heat exchange. A truss structure was designed to improve the thermal decoupling between the large external baffle (this was the most critical element for the thermal design, and at the end a glass reinforced epoxy structure with absorber coating, thermally insulated from the camera, was shown to be the best solution) and the optical bench and to minimize the temperature gradient. Finally, the telescope was covered by a thermal blanket. In addition, the optic supports were made of the same material as the optical bench to minimize distortion. Even if this allowed to limit the heat exchange of the WAC with the environment, it was necessary to introduce an active control system in the camera by suitable radiators (visible on the bottom of the optical bench in Fig. 1 left).

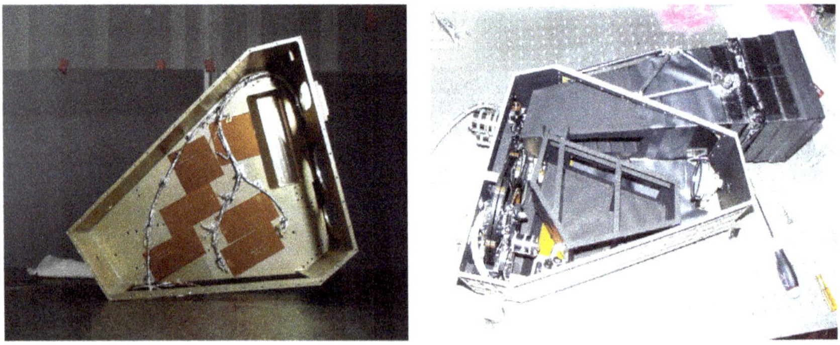

Fig. 1. On the left there is the WAC structure, obtained by excavating a single Al piece. On the right the WAC structure populated with all the optical elements.

As an example of the various mechanisms installed on the WAC, we describe here some of the characteristics of its shutter [4] (see Fig. 2), which has been the most operated one over the whole mission (more than 70k cycles). The shutter could expose a 28 × 28 mm^2 area (the sensor sensitive area was about 26 × 26 mm^2) with a uniformity better than 1/500. The shutter was realized by two blades travelling in front of the CCD, driven by a four-bar mechanism actioned by brushless dc motors. A customized encoder for each blade was mounted to the motor shaft and a position sensor at the final position verified that the first blade completed its travel, when reaching the lock device which kept it in open position. Then, when the exposure was completed, the second blade was released and unlocked the first, to back-travel together to the rest position by means of springs.

Fig. 2. WAC shutter mechanism; the blades are at the bottom, each joined to two moving bars.

WAC scientific results highlights

Rosetta, and in particular its imaging system OSIRIS, had a very large scientific return (e.g. [5],[6],[7]). Here we limit to recall one of the investigations realized with the WAC, and that took advantage of its extremely high contrast performance. In fact, this camera had as a target the observation of dust and gas in the comet coma and, because of the extremely low irradiance of these comet components, it was designed with an unobstructed optical configuration to minimize diffraction tails in the point spread function and nominally reach a contrast of more than 1000 at the comet nucleus-coma edge, and larger beyond. This optical design, coupled to the 16 bit detector dynamic range, allowed to detect extremely faint features close to the comet nucleus.

One of the key investigations of OSIRIS was monitoring the cometary jet activity and to look at the characteristics of the nucleus surface at the origin of these jets. In [8] the study of jets emitted in the period between Dec. 2014 and Oct. 2015, so including the perihelion, allowed to map the locations of jet sources on comet surface as a function of time. This confirmed the difference between the two comet "hemispheres", with different behaviors between North and South, following a seasonal trend. Thanks to the great quality images provided by the WAC, several extremely faint jets have been studied and their source located using a reverse propagation analysis (see, as an example, Fig. 3).

Acknowledgements

One of the main contributors to the WAC realization was Stefano Debei, who had been involved in the definition of the mechanical and thermal characteristics of the instrument since the beginning of his career. This paper is dedicated to his memory.

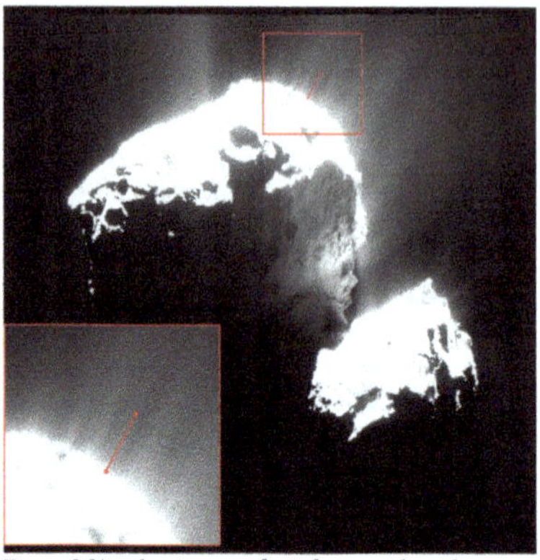

Fig. 3. WAC image (2014 Dec. 30): the image brightness has been adjusted to highlight the jets otherwise not visible. In the insert it is shown an example of jet identification. Image from [8].

References

[1] H.U. Keller, et al., "OSIRIS - The Scientific Camera System Onboard Rosetta", Space Science Reviews 128, Issue 1-4, pp. 433-506 (2007). http://doi.org/10.1007/s11214-006-9128-4

[2] G. Naletto, et al., "Optical design of the Wide Angle Camera for the Rosetta mission", Appl. Opt. 41(7), pp. 1446-1453 (2002). http://doi.org//10.1364/AO.41.001446.

[3] B. Saggin, et al., Thermal design of the Wide Angle Camera for Rosetta, presented at the International Astronautical Federation (IAF) Congress Bremen, Germany. Published Online (2012). http://doi.org/10.2514/6.IAC-03-Q.5.04

[4] De Cecco, M.,et al., "High performance shutter for space applications", SPIE Proc. 4771, pp. 186-197 (2002). HTTP://DOI.ORG/10.1117/12.482160

[5] H. Sierks, et al. "On the nucleus structure and activity of comet 67P/Churyumov-Gerasimenko", Science 347 Issue 6220, id. aaa0444, p. 6 (2015). http://doi.org/10.1126/science.aaa1044.

[6] A. Rotundi, et al. "Dust Measurements in the Coma of Comet 67P/Churyumov-Gerasimenko Inbound to the Sun Between 3.7 and 3.4 AU", Science 347 Issue 6220, id. aaa03905, p. 7 (2015). http://doi.org/10.1126/science.aaa3905

[7] N. Thomas, et al. "The Morphological Diversity of Comet 67P/Churyumov-Gerasimenko", Science 347 Issue 6220, id. aaa0440, p. 7 (2015). HTTP://DOI.ORG/10.1126/science.aaa0440.

[8] M.I. Schmitt, et al., "Long-term monitoring of comet 67P/Churyumov-Gerasimenko's jets with OSIRIS onboard Rosetta", MNRAS 469 (Suppl. 2), pp. S380–S385 (2017). http://doi.org/10.1093/mnras/stx1780

Development of a modular central electronic unit (CEU) for data handling and management in Martian atmosphere investigations

L. Marrocchi[1] *, M. Marchetti[2], F. Costa[1] and S. Debei[3,4]

[1] TEMIS Srl, via Donizetti 20, Corbetta (MI) (Italy)

[2] ART Spa, via Via Voc. Pischiello 20, Passignano sul Trasimeno (PG) (Italy)

[3] Department of Industrial Engineering, University of Padova, via Venezia 1, Padova (Italy)

[4] CISAS - Center for Studies and Activities for Space "Giuseppe Colombo", University of Padova, via Venezia 15, Padova (Italy)

* luca.marrocchi@temissrl.com

Keywords: Telemetry, Space Avionics, Sensor Acquisition, Data Merging

Abstract. This article describes the development of a highly modular Central Electronic Unit (CEU) for data handling and management in Martian atmosphere investigations as part of the DREAMS experiment in ExoMARS 2016 Mission. The CEU was developed by TEMIS and Centro Interdipartimentale di Studi e Attività Spaziali G. Colombo (CISAS) to handle, acquire and transmit medium and low-rate data from sensors dedicated to investigating the Martian atmosphere. The CEU subsystem was designed to communicate with an external host to receive commands and transfer telemetry, transform power from different sources, perform sensors' switch on and off, acquire scientific sensors and their housekeeping monitors, and provide CEU health status. The high modularity of the CEU architecture makes it easy to customize and interface with different types of equipment and instrumentation for various missions. The CEU subsystem is housed on Entry and Descent Module (EDM) and serves as the data handling platform of the DREAMS payload for the EXOMARS 2016 Mission. The CEU has proven to be an effective and flexible solution for data handling and management in Martian atmosphere investigations.

Introduction

The Dust characterization, Risk assessment, and Environment Analyser on the Martian Surface (DREAMS) [1] mission was part of the ExoMARS 2016 mission, which aimed to explore the Martian environment and gather data on meteorological conditions, hazards, and atmospheric electric phenomena. The Central Electronic Unit (CEU) subsystem was developed by TEMIS and CISAS to serve as the data handling platform [2] for the DREAMS payload and was installed inside the warm compartment of Entry and Descent Module (EDM) [3].

The CEU subsystem consisted of primary boards and expansion boards, which distributed functionalities among various modules. The primary boards included the CEU Backplane and the On-Board Data Handling (OBDH) Board, while the expansion boards included the Analog-to-Digital Converter (ADC) Board and the Central Processing Unit (CPU) Board.

Fig. 1 - CEU subsystem

The DREAMS payload included additional subsystems such as MarsTem (thermometer), DREAMS-P (pressure sensor), DREAMS-H (humidity sensor), MetWind (wind sensor),

MicroARES (electric probe), SIS (Solar Irradiance Sensor), and a primary battery. These subsystems worked in conjunction with the CEU to fulfil the mission objectives.

CEU Architecture

The CEU subsystem [4] was designed to fulfil its objectives through the implementation of two sets of boards: primary boards and expansion boards. The primary boards are responsible for supplying power, managing the entire subsystem (including sensor configuration), and processing and transmitting data from the sensors to the external host. On the other hand, the expansion boards are specifically designed to interface with various types of sensors (analog and/or digital) by adapting the electronics accordingly. To ensure modularity and efficient distribution of functionalities, the CEU was developed with a plug-in design and a common backplane, allowing for the integration of different boards and modules.

Primary boards

1) CEU Backplane:

The CEU was a modular equipment designed to grant power connections to supply all the boards and enable communication interfaces between the peripheral boards and the CEU Core (OBDH Board). The backplane of CEU (Fig. 2) was a fully passive PCB populated with one connector for each board. The passive backplane strategy was useful during integration as it allowed inspection of the assembled PCB to ensure compliance with requirements without specific electrical tests. This streamlined the AIT phase and enabled adding one board at a time using the backplane as a consolidated part.

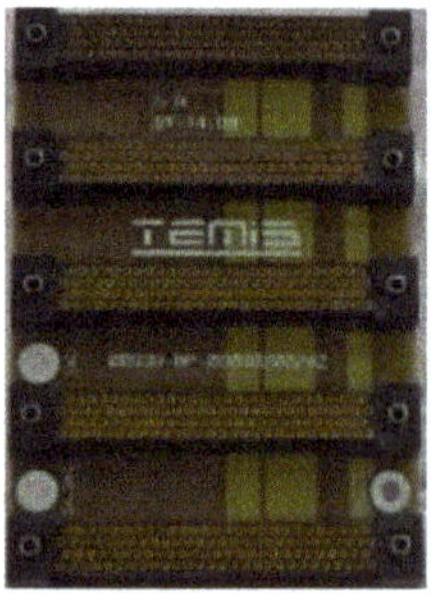

Fig. 2 - CEU Backplane

2) OBDH Board

The OBDH Board (Fig. 3) comprised several main blocks, namely a host interface for communication with an external unit, a microprocessor based on LEON2 (Atmel AT697F) along with its associated support memories, an FPGA (RT PROASIC3L) for expanding the microprocessor interfaces and handling low-level control of peripherals, and a non-volatile mass memory (1 Gbyte NAND FLASH). The interfaces used for communication with the expansion boards were implemented on the FPGA, offering customization options based on user requirements within the limitations of the RT PROASIC3L technology. The microprocessor had the capability to operate with a core clock of up to 100 MHz, providing a performance of 86 MIPS (Dhrystone 2.1). The communication interface between the microprocessor and the FPGA could be selected from the following choices: a PCI interface up to 33 MHz, an asynchronous 32-bit memory-mapped interface up to 20 Mbyte/s, or up to 2 UART interfaces up to 250 Kbit.

Fig. 3 - OBDH Board

3) DCDC Board

The DC/DC Board (Fig. 4) was responsible for managing the primary power supply source and distributing power to all the CEU boards. It generated the necessary secondary power rails, including +/-12V, 5V, 3.3V, and 1.8V.

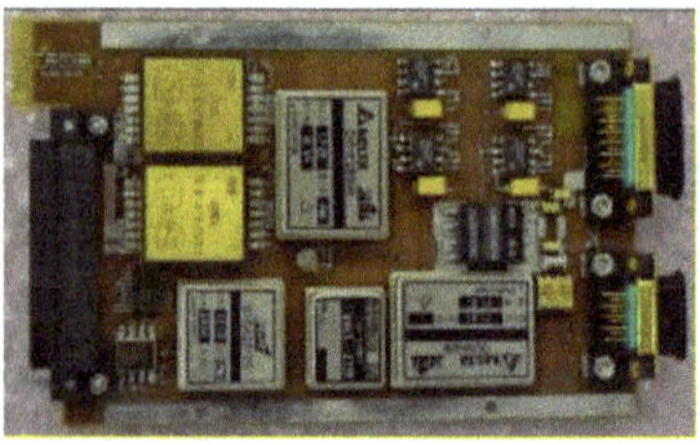

Fig. 4 - DCDC Board

Extension boards

1) ADC board:

The ADC board (Fig. 5) was specifically designed to facilitate the interface between the CEU and sensors that required analog-to-digital conversion and dedicated conditioning electronics. Its main components consisted of an ADC section utilizing the Aeroflex 14-bit Analog-to-Digital Converter (RHD5950) with 16 input channels, and an Acquisition Handler FPGA (RT PROASIC3L device) responsible for managing acquisitions and establishing communication with the OBDH board.

Fig. 5 - ADC Board

2) CPU board:

The CPU board (Fig. 6) was developed to establish connections with digital sensors. It included an Interfaces section that could be customized to meet the specific requirements of the sensors, an Acquisition Handler FPGA (RT PROASIC3L device) for managing acquisitions and interfacing with the OBDH board, and a 256 Mbyte SDRAM memory connected to the FPGA to support onboard data manipulation.

Fig. 6 - CPU Board

Both boards were equipped with onboard FPGAs, allowing for the manipulation of scientific data acquired from the sensors before transmitting them to the CEU's mass memory. Additionally, the FPGAs handled low-level control of peripherals and established communication with the OBDH board.

Power Consumption

The overall efficiency of the CEU was around 70%. To save power, the CEU was designed to have a "low power" mode where only the FPGA on the OBDH board was powered on while all other parts of the OBDH and other CEU boards were powered off. This configuration allowed the CEU system to consume less than 1 W of power.

The interfaces used to communicate with the expansion boards were implemented on FPGA and could be customized based on user requirements. The microprocessor worked with a core clock up to 100 MHz, resulting in a performance of 86 MIPS (Dhrystone 2.1). The communication interface between the microprocessor and the FPGA could be chosen from options such as PCI interface up to 33 MHz, Asynchronous 32-bit memory-mapped interface up to 20 Mbyte/s, or Up to 2 UART interfaces up to 250 Kbit.

Failures management

To automatically detect and correct failures, a detailed monitoring system was implemented to track critical parameters of the CEU boards and DREAMS sensors. This information was used by the CEU's Application Software (ASW) to make decisions to ensure safety and guarantee mission continuity even with reduced functionalities or performance degradations.

Conclusions

The CEU subsystem played a vital role in handling data, managing power, and controlling interfaces for the DREAMS payload during the EXOMARS 2016 mission. The modular design and flexibility of the CEU allowed customization and integration of different boards and sensors, making it suitable for various scientific missions. The successful deployment and operation of the CEU subsystem contributed to the scientific success of the EXOMARS 2016 mission by providing valuable data about the Martian environment during the dust storm season.

During the descent of EDM of the Exomars 2016 mission and its subsequent impact on the Martian surface, both the CEU subsystem and the DREAMS payload performed their designated tasks flawlessly. The CEU subsystem remained fully functional throughout the descent process, demonstrating its reliability in the challenging Martian environment. Likewise, the DREAMS payload successfully carried out its scientific measurements and data acquisition during the descent and impact phase. This successful performance of the CEU and DREAMS subsystems further solidified their capabilities and effectiveness in capturing vital scientific data in extreme conditions on Mars.

In addition to their successful performance during the descent and impact of the ExoMARS 2016 EDM, the CEU subsystem and the DREAMS payload also demonstrated their reliability and optimal health during the intermediate checkout phase while the spacecraft was en route to Mars. During this period, the CEU and DREAMS experiment effectively conducted the necessary checkouts and provided crucial health information about the experiments. Their flawless operation and optimal health state reassured the mission team about the robustness and functionality of the CEU and DREAMS systems, further enhancing confidence in their capabilities to carry out scientific investigations on Mars.

References

[1] F. Esposito, et al., DREAMS for the ExoMars 2016 mission: a suite of sensors for the characterization of Martian environment" (PDF). European Planetary Science Congress 2013, EPSC Abstracts Vol. 8, EPSC2013-815 (2013)

[2] L. Marrocchi et al., "High performance, high configurability modular telemetry system for launch vehicles," 2014 IEEE Metrology for Aerospace (MetroAeroSpace), Benevento, Italy, 2014, pp. 594-598. https://doi.org/10.1109/MetroAeroSpace.2014.6865994

[3] Ball, Andrew & Blancquaert, Thierry & Bayle, Olivier & Lorenzoni, Leila & Haldemann, A.. (2022). The ExoMars Schiaparelli Entry, Descent and Landing Demonstrator Module (EDM) System Design. Space Science Reviews. 218. https://doi.org/10.1007/s11214-022-00898-z

[4] M. Marchetti et al., "Data handling equipment for payload sub-systems," 2014 IEEE Metrology for Aerospace (MetroAeroSpace), Benevento, Italy, 2014, pp. 456-461. https://doi.org/10.1109/MetroAeroSpace.2014.6865968

SIMBIO-SYS, the remote sensing instruments on board the BepiColombo mission

G. Cremonese[1a*], C. Re[1b], and the SIMBIO-SYS team

[1]Osservatorio Astronomico di Padova, INAF, vicolo Osservatorio 5, Padova, Italy

[a]gabriele.cremonese@inaf.it, [b]cristina.re@inaf.it

Keywords: Remote Sensing Instruments, Space Missions, Planets: Mercury

Abstract. The SIMBIO-SYS (Spectrometer and Imaging for MPO BepiColombo Integrated Observatory SYStem) is a complex instrument suite part of the scientific payload of the Mercury Planetary Orbiter for the BepiColombo mission, the last of the cornerstone missions of the European Space Agency (ESA) Horizon + science program. It will explore Mercury, the closest planet to the Sun. The SIMBIO-SYS instrument will provide all the science imaging capability of the BepiColombo MPO spacecraft. It consists of three channels: the STereo imaging Channel (STC), with broad spectral band in the 410-930 nm range and medium spatial resolution (up to 60 m/px), that will provide Digital Terrain Model of the entire surface of the planet with an accuracy better than 80 m; the High Resolution Imaging Channel HRIC), with broad spectral bands in the 530-900 nm range and high spatial resolution (up to 6 m/px), that will provide high resolution images of about 10% of the surface, and the Visible and near-Infrared Hyperspectral Imaging channel (VIHI), with high spectral resolution (up to 6 nm) in the 400-2000 nm range and spatial resolution up to 120 m/px, it will provide the global coverage at 480 m/px with the spectral information, assuming the first orbit around Mercury with periherm at 480 km from the surface.
It has been funded by the two space agencies, ASI (Italian Space Agency) and CNES (French Space Agency) and it is the result of the collaboration between more than 100 scientists and engineers of 12 different countries all over the world, with the Italian prime contractor Leonardo spa. It is the first time that a planetary mission has three remote sensing instruments integrated in a system, sharing the Main Electronics, and under the responsibility of one team.

Introduction

The Spectrometer and Imagers for MPO Bepicolombo Integrated Observatory SYStem (SIMBIO-SYS) is a suite of three independent optical heads that will provide images and spectroscopic observations of Mercury's surface. The SIMBIO-SYS instrument on board the Mercury Planetary Orbiter (MPO), one of the two modules of the BepiColombo mission, is composed of HRIC (High Resolution Imaging Camera), STC (STereo Channel), and VIHI (Visible and Infrared Hyperspectral Imager). The scientific objectives at mission level are to obtain a global mapping of the surface with STC and VIHI in the first 6 months of the 1 year nominal mission. Both channels will provide data on the surface composition, the surface geology as well as Digital Terrain Models (DTMs) of the entire planet.

The observing strategy of SIMBIO-SYS is based on the global mapping requirement and includes high-resolution images of 10% of the surface (HRIC). In the second 6 months, STC and VIHI will fill in gaps possibly left in the global mapping. In this phase, VIHI will observe selected regions at a spatial resolution of a factor of four better than during the previous phase. STC will acquire 4 color images of selected regions.

The BepiColombo mission will arrive at Mercury in December 2025, the instrument commissioning will be in March 2026 with the starting nominal mission in April 2026.

It is the first time that a planetary mission has three remote sensing instruments integrated in a system sharing some hardware components. The Main Electronics (ME) and the onboard software

are the same, allowing for synergistic management of operations, data handling, and compression of all the acquired data. A single factory provided all the detectors that are similar for the three cameras. From an engineering point of view, the systemic approach to the design has had several advantages:

- the technical management structure allowed the management of the developing and testing phases with the main target of optimizing the overall performance and capabilities of SIMBIO-SYS;
- the integration of the common parts allowed a better control and the optimization of the resources as mass and power;
- mechanical, thermal, and electrical interfaces towards the spacecraft have been handled at a unified level.

From a scientific point of view, the systemic approach with the same management will provide many advantages:

- cross-calibration and co-alignment of the three channels is included in the commissioning and operations;
- co-registration and data fusion are easier and included in the scientific activities;
- common science planning is applied.

Scientific objectives

The SIMBIO-SYS integrated package [1](Figure 1) aims to provide answers to almost all the main scientific questions of the BepiColombo mission concerning the Mercury surface and composition, and to provide important contributions to the understanding of its interior and exosphere. The main scientific questions can be summarized in the following topics:

- shape and morphology;
- crustal mineralogy;
- geological mapping and stratigraphy;
- volatiles;
- interior;
- Hermean extreme environment;
- exosphere;
- surface changes;
- opportunity science

Technical description

To fulfil the scientific goals of the mission, in compliance with the limited resources available on the MPO and the harsh operative environment, several solutions and technologies have been implemented, namely:

- A very compact design for HRIC
- A wide spectral coverage wih a single channel instrument for VIHI
- A single detector dual channel design for STC
- A special coating (ITO) and baffle (Stavroudis design) for heat load rejection
- Diamond turned mirrors in RSA905 Aluminum alloy
- Spectral and radiometric in-flight calibration unit with no moving parts
- Large use of composite materials for structural parts
- Capabilities of stereo imaging

Materials Research Proceedings 37 (2023) 717-720 https://doi.org/10.21741/9781644902813-153

In addition, an instrument architecture with three dedicated Proximity Electronics (PE) and a common Main Electronics (ME) was designed, aimed at the best use of the available electrical resources thanks to the sharing of functions among the three channels.

Each channel of the SIMBIO-SYS Flight Model (FM) has been successfully tested and characterized at Leonardo SpA premises by means of two dedicated Optical Ground Support Equipment (OGSE) which have been developed and manufactured for this purpose.

Figure 1: The three channels together representing SIMBIO-SYS, including the Proximity Electronics (PE), on the left VIHI, in the middle STC and on the right HRIC

High Resolution Camera

HRIC is a camera operating in the 530 nm to 900 nm spectral range with 6 m resolution on ground from the orbital altitude of 480 km.

The camera is equipped with a main panchromatic channel and three broadband spectral channels centred at 550 nm, 750 nm and 880 nm with 40 nm bandwidth.

The HRIC optical system [2] is composed of a telescope with a Ritchey-Chretien configuration modified with three-lenses corrector aimed to guarantee the optical quality over the whole squared 1.47° FoV. The instrument has a focal length of 800 mm and is equipped with a dioptric image corrector adapting the FoV to a 2048 x 2048 pixels detector with a pixel size of 10 µm. The focal ratio is F#8.9, to be diffraction limited at 400 nm and to optimize radiometric flux and overall mechanical dimensions. Due to the large amount of heat flux coming from Mercury, the camera aperture is protected from the planet radiation by means of a baffle, in Stavroudis configuration, and a filter, named Thermal Infrared Rejection Device (TIRD), which can be considered as part of the HRIC thermo-optical system.

The selected Stavroudis configuration for the baffle is composed of six elliptical and five hyperbolic internally high-reflecting surfaces interconnected together. All the surfaces have two common foci and this provides the baffle with the geometrical property to reject externally, after a given number of multiple reflections, all the rays impacting on the conical surfaces from an angular direction within the two foci.

Stereo Camera

STC represents one of the first push-frame stereo cameras on board a planetary mission. Based on a new concept, STC integrates the compactness of a single-detector telescope with the photogrammetric capabilities of bidirectional cameras [3]. Two separate incoming optical path oriented at ±20° with respect to nadir allow the instrument to acquire images of the same surface region with a different viewing angle at two very close moments, taking advantage of the along-track movement of the S/C. The optical system is an advanced compact catadioptric layout concept, which consists of a front-end optical separation group of mirrors inclined in order to realize ± 20° stereo channels and a common telescope, based on an off-axis modified 'Schmidt corrector' design.

Visible Infrared Hyperspectral Imager

The VIHI channel has been designed to perform hyperspectral imaging observations of the whole Hermean surface in the VIS-NearIR range [1]. The channel concept is based on a collecting telescope and a diffraction-grating spectrometer ideally joined on the telescope focal plane, where the spectrometer entrance slit is located. The image of the slit is dispersed by the diffraction grating on a bi-dimensional detector. The instantaneous acquisition on the bi-dimensional detector consists of the slit image diffracted by the grating over the spectral range. The image itself is built in time by subsequent slit acquisitions, matching the S/C track speed with the slit size projected on ground (push broom mode). The final result is an hyperspectral cube, which associates a VIS-NearIR spectrum to each pixel on ground. VIHI has been designed to achieve an IFoV of 250 μrad (corresponding to 120m @ 480km altitude) and to cover a spectral range of 400-2000nm with a spectral sampling of 6.25nm.

References

[1] G.Cremonese, et al., SIMBIO-SYS: scientific cameras and spectrometer for the BepiColombo mission. 2020, Space Scie.Rev., 216, 75.

[2] M.Zusi, R.Paolinetti, V.Della Corte, G.Marra, M.Baroni, P.Palumbo, G.Cremonese, Optical design of the High Resolution Imaging Channel of SIMBIO-SYS. 2019, Applied Optics, 58, 4059. https://doi.org/10.1364/AO.58.004059

[3] V.Da Deppo, G.Naletto, G.Cremonese, L.Calamai, Optical design of the single-detector planetary stereo-camera for the BepiColombo ESA mission to Mercury, 2010, Applied Optics, 49, 2910. https://doi.org/10.1364/AO.49.002910

The ESA PANGAEA programme: training astronauts in field science

Matteo Massironi [1,2*], Francesco Sauro [3,4], Samuel J. Payler [3], Riccardo Pozzobon [1,2], Harald Hiesinger [5], Nicolas Mangold [6], Charles S. Cockell [7], Jesus Martínez Frias [8], Kåre Kullerud [9], Leonardo Turchi [3,10], Igor Drozdovskiy [3], Loredana Bessone [3]

[1] Dipartimento di Geoscienze, University of Padua, Italy

[2] CISAS University of Padova, Italy

[3] Directorate of Human and Robotics Exploration, European Space Agency, France

[4] Miles Beyond Srl, Italy

[5] Institut für Planetologie, Westf¨alische Wilhelms-Universit¨at Münster; Germany

[6] Laboratoire Planetologie et Geosciences, CNRS, Nantes, France

[7] School of Physics and Astronomy, University of Edinburgh, UK

[8] Instituto de Geociencias, IGEO (CSIC-UCM), Spain

[9] Norwegian Mining Museum, Kongsberg, Norway

[10] Spaceclick Srl, Italy

*matteo.massironi@unipd.it

Keywords: Planetary Geology, Human Space Exploration, Astronauts Training

Abstract. PANGAEA (Planetary ANalogue Geological and Astrobiological Exercise for Astronauts) is a field training course designed by the European Space Agency (ESA) that, since 2016, has imparted to ESA and NASA astronauts, and Roscosmos cosmonauts the basic theoretical and practical knowledge of geology and astrobiology and trained them in the field. Hence developing independent field skills, including working with a remotely located science team, is a key part of the training. For this reason, classroom and field lessons are tightly interwoven so that the concepts introduced in the classroom are shown in the field soon afterwards. The primary field sites selected for the course are the Permo-Triassic sedimentary sequences in the Italian Dolomites, analogue to the Martian alluvial plains ones, the impact geological environment of the Ries Crater, Germany, a comprehensive suite of volcanic emplacements and deposits in Lanzarote, Spain, and the anorthosite outcrops, analogue to lunar highlands rocks, in Lofoten, Norway. Each site is used as a base to deliver the main learning sessions, respectively: 1) Earth geology, rock recognition and sedimentology on Earth and Mars, 2) Lunar geology and impact cratering, 3) volcanism on Earth, Moon, and Mars, and astrobiology 4) intrusive rocks and lunar primordial crustal evolution. The four sessions are designed to increase the trainees' autonomy in the field up to autonomously executed geological traverses including sampling activities. Whilst PANGAEA's primary focus is astronaut training, where appropriate, technologies being developed for future missions are used to evaluate their performances in analogue field envi-ronments and to train the astronauts in using technologies that might support future missions.

Introduction

In 2016 some of the authors proposed to ESA the first field course for astronauts in geological and astrobiological planetary exploration named PANGAEA (Planetary ANalogue Geological and Astrobiological Exercise for Astronauts). This was possible thanks to the far-sightedness of prof.

Debei which immediately understood the potential of a course like PANGAEA for future human exploration of the Moon and Mars and made available all the needed facilities of the University of Padua to carry out the first edition which was attended by the ESA astronauts Luca Parmitano, Pedro Duque and Matthias Maurer. Other 4 editions of the training have been implemented since 2016, with a sixth one foreseen in 2023. In total, 10 astronauts from ESA, NASA and Roscosmos and additional 5 non-astronaut trainees including space engineers, EVA and operation specialists have attended the course. In this work we will summarise the course strategy, structure and lessons learned, but for a more detailed description the reader is referred to Sauro et al. [1] .

Goals and structure

The course forms part of the basic and pre-assignment training for European astronauts, and is open to trainees from other agencies. Significant focus is given to skills relevant to future field exploration, such as practical geological and geobiological field training, execution of self directed traverses in the field, ability to provide clear scientific descriptions of geological landscapes and features, and efficient documentation of sampling sites. For this reason, although minor portions of the course are taught in classrooms, most of the activities are in analogue geological environments, as done in the seventies during the preparation for the Apollo missions [1, 2]. PANGAEA course integrates both geology and astrobiology (including planetary protection) enabling overlapping concepts and ideas to be explored thoroughly. Trainees also have the opportunity to practice conducting field science under the additional constraints imposed by realistic spaceflight operational conditions.

Teaching in analogue sites

The primary field sites selected for the course are Permo-Triassic terrigenous sequences in the Italian Dolomites, impact lithologies in the Ries Crater, Germany, a comprehensive suite of volcanic deposits in Lanzarote, Spain and the Flakstadøy intrusive complex in the Lofoten archipelago, Norway. Each is used as a base to deliver the main learning sessions, respectively; 1) Earth geology, rock recognition and sedimentology on Earth and Mars, 2) Lunar geology and impact cratering, and 3) volcanism on Earth, Moon and Mars, execution of geological traverses, and sampling techniques and 4) Anorthosite and Mg-Suite intrusive complexes and lunar highlands. Classroom lessons are conducted at these field sites using local facilities often provided by Geoparks (e.g. Bletterbach Geopark in Italy, Ries Geopark in Germany and Lanzarote Geopark in Canary Islands). For the field work component, trainees are initially shown the basics of field geology and astrobiology during the first two sessions. In the third session they begin a process of becoming independent field scientists conducting geological traverses with realistic scientific goals, such as determine the contact relationship between geological units and the relative timing of events, recognise stratigraphic and tectonic structures, and sample rocks representative of the location or affected by different alteration processes (e.g. hydrothermal alteration). Trainees are initially accompanied by their instructors, but the coaching is progressively reduced until the field crew is supported by a remote science team, simulating a science back room in mission control.

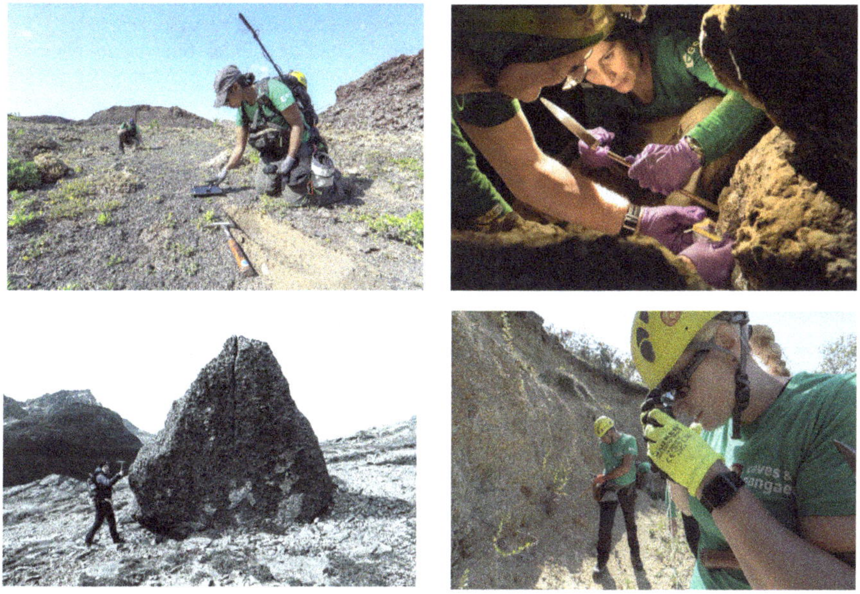

Fig. 1. *a) Self directed traverse in Lanzarote by ESA astronaut Alexander Gerst and NASA astronaut Stephanie Wilson. b) ESA astronaut Samantha Cristoforetti acquiring a microbilogical sample in a lava tube in Lanzarote. c) ESA astronaut Matthias Maurer sampling an anorthositic block in Lofoten.; d) NASA astronaut Kathleen Rubins observing at a sample in Ries.*

Lessons learned

Astronauts tend to be practically minded, and long theoretical lessons do not provide them with the best opportunity to learn. For this reason each theoretical lesson must be always coupled with field activities and firmly connected to examples on how the specific knowledge can be applied in the future missions. Dissociating lessons from these goals can lead trainees to lose interest and eventually question the point of the learning. For this reason the classroom has been conducted in facilities close to the field sites so that classroom lessons and field trips often occur on the same day.

Typically, field trips used for academic teaching involve visiting locations to learn more about a particular environment in context. For training astronauts, these types of field trips are still important, however bringing in some elements of planetary surface exploration and in field scientific research realism is pivotal to sustain trainees motivation and knowledge retention. PANGAEA achieves these objectives performing increasingly autonomous traverses through geological landscapes with predefined science goals and remotely located science teams. This contextualises the knowledge trainees gained through the regular field trips in a planetary exploration setting and promote the trainee's independence forcing a flexible execution in the field. Adding to the traverses real scientific objectives, defined with the local experts, with potential for further science advancements and publications, is also a very important motivational factor fostering the engagement of the trainees.

The course is also very enriching for the scientists involved who had to learn about operational constraints and simplify the communication to be the most effective as possible.

When available, traverses should include technological capabilities relevant to planetary exploration. Although the exact set of capabilities for EVAs on the lunar or Martian surface are yet to be designed and built, the core concepts of scientific data collection, documentation, and communication can still be effectively demonstrated by using Commercial off-the-shelf (COTS) hardware, prototypes and low TRL (technology readiness level) instruments to emulate the capabilities that should be available to astronauts conducting geological exploration on the Moon or Mars. This includes ways to collect images, notes, spectra, and other scientific data [e.g. 3]. Incorporating such systems into the training was found to be a highly effective way to allow astronauts to understand the types of information required for performing effective science during planetary surface exploration with remotely located science teams. Overall, it allowed them to appreciate the importance of quality data collection for enabling scientific interpretation.

Conclusion

The time constraints and areal extent of lunar surface exploration will be extended significantly during future Artemis missions thanks to the technological advancements made since the Apollo era. It is therefore particularly important to train astronauts to self-directed field work and efficient communication with ground based science teams. Astronauts that will be not specifically launched to the Moon or Mars, will be in any case likely involved in the planning, preparation and implementation of such missions. Training key personnel in geology and astrobiology in terrestrial field analogues as in the ESA-PANGAEA course will be essential to the success of such expensive and high-risk endeavours

References

[1] Sauro F. , Payler S. J., Massironi M., Pozzobon R., Hiesinger H., Mangold N., Cockell C. S. , Martínez Frias J., Kullerud K., Turchi L., Drozdovskiy I., Bessone L., (2023). Training astronauts for scientific exploration on planetary surfaces: the ESA PANGAEA programme. Acta Astronautica 204, 222-238. https://doi.org/10.1016/j.actaastro.2022.12.034

[2] Schmitt H. H, Snoke, A., Helper, M., Hurtado, J., Hodges, K. and Rice, J., (2011) Motives, methods, and essential preparation for planetary field geology on the Moon and Mars. Geological Society of America Special Papers, 483, 1-1. https://doi.org/10.1130/2011.2483(01)

[3] Turchi Payler, S.J., Sauro, F., Pozzobon, R., Massironi, M. and Bessone, L. (2021) The Electronic FieldBook: A system for supporting distributed field scienceoperations during astronaut training and human planetary exploration. Planetary and Space Science, 197, 105-164. https://doi.org/10.1016/j.pss.2021.105164

Materials Research Proceedings 37 (2023) 725-729 https://doi.org/10.21741/9781644902813-155

Front cover for space optical telescopes. a legacy from ROSETTA to JUICE

G. Parzianello[1,a *], M. Bartolomei[2,b], S. Chiodini[2,c], M. Zaccariotto[2,d], G. Colombatti[2,e], A. Aboudan[2,f], C. Bettanini[2,g], S. Debei[2,h]

[1] European Space Research and Technology Centre – European Space Agency - ESA-ESTEC, Kepleerlaan 1 - Noordwijk (NL)

[2] CISAS - Center for Studies and Activities for Space "Giuseppe Colombo", University of Padova, via Venezia 15, Padova (Italy)

[a] giorgio.parzianello@esa.int, [b]mirco.bartolomei@unipd.it, [c]sebastiano.chiodini@unipd.it, [d]mirco.zaccariotto@unipd.it, [e]giacomo.colombatti@unipd.it, [f]alessio.aboudan@unipd.it, [g]carlo.bettanini@unipd.it

Keywords: Front Door, Cover, Telescope, Space Mechanism, Rosetta, JUICE

Abstract. The reliability of a telescope's front cover for a space mission constitutes its primary demand, not to compromise the functionality of the entire instrument. Avoidance to expose possible reference surfaces to the external environment and contamination often drives the selection of the motion of the cover. The Front Door Mechanism (FDM) for the OSIRIS experiment of the ROSETTA mission has been developed and optimised to provide protection of the telescope, reliable functioning and being single point failure tolerant. Its combined translational and rotational motion allowed also the preservation of internal calibration surface for the entire duration of the mission. The ROSETTA mission was launched in 2004 and reached the comet in 2014 after several gravity assists, and deep space hibernation. It has been orbiting around the comet for roughly two years collecting enormous amount of data of the peculiar celestial body and ended in September 2016. As legacy of the successful design of the front cover for the ROSETTA mission, another cover mechanics (Cover Mechanism – COM) for the JANUS Optical Head Unit of the JUICE mission has been prepared with minor modifications to the initial design. All qualification has been performed and the JUICE mission has been launched successfully on 14 April 2023 and is now travelling towards Jupiter and its moons.

Introduction

Observation and collection of scientific data by means of telescopes is a very widely used approach for instrumentation for space missions. Such instruments allow coverage of very wide spectral ranges, with applications ranging from infrared to X-ray bands and with various techniques from imaging to spectroscopy.

To provide the quality needed to achieve the increasingly demanding scientific objectives, the instruments have to be protected from external contamination that would dramatically degrade the performance. Protection must take place both before and during the mission lifetime, especially with the increase of duration of planned space missions. Cleanliness measures are put in place during the entire assembly phases at different levels, from unit, to instrument, to satellite and also, after integration into the launcher, including purging, where needed.

After the fairing separation, still at low altitudes, the possible contamination comes from residual air density and from auto contamination from satellite outgassing. Finally, for missions with celestial bodies or planetary observation, the presence of an atmosphere or of emissions from the surface can constitute an important source of contamination.

Contamination protection of observation instruments such as telescopes can be increased by the usage of cover systems that close the telescope's baffle opening, thus avoiding direct exposure of the inner parts of the instrument.

By their own nature, covers constitute also one of the highest risks for the instrument itself. A failure of actuation of the cover would in fact result in the entire loss of the instrument.

Reliability of such systems plays therefore the main focus in their design and development, and redundant solutions to make them failure tolerant is one of the prime objectives of a proper cover design approach.

Several papers have been published describing the design and test results of the Front Door Mechanism (FDM) of the OSIRIS instrument [2, 3]. This paper deals mainly with the main concepts and the relevant aspects that enhanced the proven robustness and flexibility of the FDM and the heritage of its design for the JANUS instrument of the JUICE mission, The two missions are significantly different, but with similar telescope's protection needs.

FDM design

OSIRIS (Optical, Spectroscopic and Infrared Remote Imaging System) is one of the instruments of the ROSETTA mission, an ESA cornerstone science spacecraft launched in 2004 to study in close proximity a comet.

The OSIRIS instrument is composed by two telescopes, the Wide Angle Camera (WAC) and the Narrow Angle Camera (NAC) providing wide and detailed optical imaging of the comet. Some subsystems of both telescopes are identical, such as the Front Door Mechanism (FDM), the Shutter Mechanisms (SHM), the Shutter Electronics (SHE), the Filter Wheel Mechanism (FWM) [1].
Among the peculiarities of the ROSETTA mission, for the instruments protection concept, the long journey in hibernation to reach the comet and the long observation period around the comet during its rising activity while approaching the Sun are the two main drivers of the conceptual selection for the protection cover system for both telescopes. Rather than a significantly simpler one-off opening system, a fully reversible mechanism approach has been chosen to allow the repeated closure and protection of the telescope in case of adverse conditions.

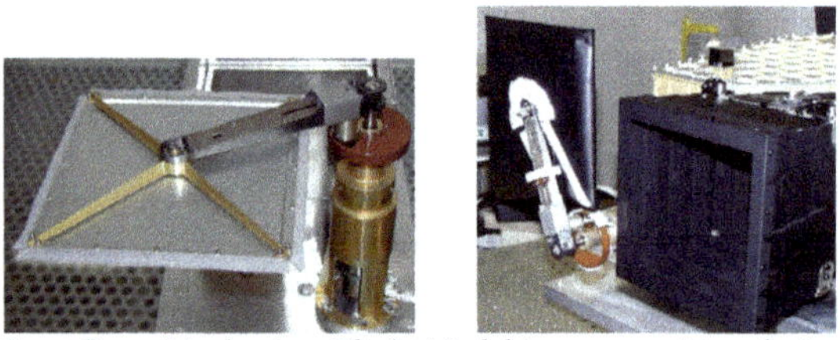

Figure 1: Front Door Mechanism Flight Model in open position during characterisation (left) and mounted on the Flight Model WAC telescope (right)

The possibility to select a variable lift of the cover from the baffle, appeared quickly as a significant advantage allowing the selection of tightness of the closure and avoiding any possible adhesion risk during the long hibernation phase. The inner surface of the cover is used as reference for the instrument internal calibration, and as such, avoiding its direct exposure to the external space, to avoid its contamination, became also a concept driver. A 3D motion achieved by the combination of a double cam system, linked by an internal pin has been finally selected for the actuation mechanism.

The Front Door Mechanism (FDM) during the opening motion performs four different phases:

1) Unlocking

2) Initial translational lift
3) Rototranslation to achieve final lift and rotation of 90 degrees
4) Locking

The cover motion is determined by the joint interaction of the internal pin, the fixed external cam and the moving internal cam, driven by a stepper motor. Fig. 2 explains the coupling concept. By the definition of the shape of the internal and external cam the desired motion can be achieved. This allows great flexibility in the motion to be performed by the cover.

One peculiarity of the cam design is its self-locking feature, both in closed and in open position. This allows the maintenance of the position without any power and is obtained by the kinematic design of the cams. A rotation of the inner cam of more than 45 degrees is needed to unlock the system from any of the two locked positions. No motion of the cover (except for compliance in the cams and couplings) is possible before the resting position, fully closed or fully open, is unlocked.

Various measures are also implemented to minimise any risk of jamming of the relative motion between the cams, pin and cover shaft, including main guiding bearings, various bushes and additional kinematical bearings. Moreover, on top of the standard redundancies included in the mechanism design, another additional fail-safe system, completely independent from the nominal motion and capable of opening irreversibly the field of view has been implemented inside the cover arm.

Light and dust tight closure is achieved by preloading the cover against the baffle using the stiffness of the arm, and also variable preload is possible.

On top of verification of the qualification requirements tests, including lifetime tests (> 10000 activations) on the QM model, an accurate characterisation of the performance and the repeatability of the parameters, especially those available in the telemetry, has been performed for each model.

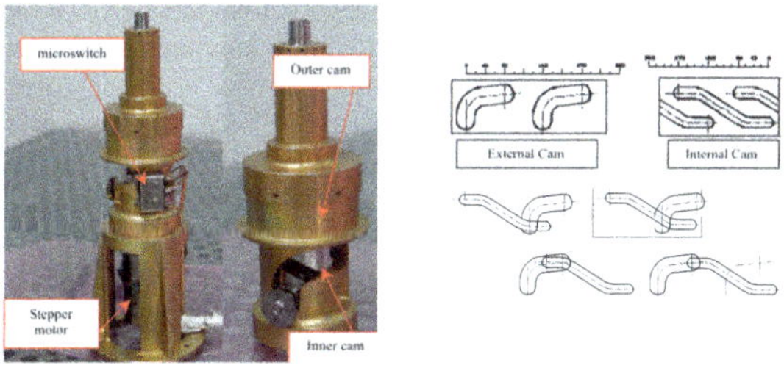

Figure 2: Front Door mechanism description (left) and cam design and conceptual interaction during the four phases(right)

In orbit

ROSETTA was launched on 2nd March 2004 for a long journey toward the encounter with the comet 67P/Churyumov-Gerasimenko that eventually happened in August 2014, after a series of gravity assist manoeuvres, Mars and asteroids fly-bys and after more than three years of hibernation and six months of orbits correction and approach to manage orbiting around the comet with a newly discovered and quite irregular shape [4].

During all the phases of the mission the Front Door Mechanism both of WAC and of NAC cameras operated flawlessly from the commissioning phase to the end of mission. They allowed the tight closure of the telescopes baffles openings during the initial phases, provided the telescopes' internal calibration reference surface and protected from contamination the optical elements during the spacecraft journey and when the instrument was not in use.

Its design proved its robustness, versatility and reliability allowing the extraordinary achievements of the OSIRIS experiment of the pioneering ROSETTA mission [5, 6, 7].

Legacy: JANUS instrument for the JUICE mission

On 14th April 2023, almost 20 years after the launch of ROSETTA, the JUICE satellite was launched, carrying onboard as one of the instruments JANUS, a narrow angle camera imager in the visible range. The JUICE mission aims to explore the Jovian moons (Ganymede, Europa and Callisto), including ocean layers and subsurface water reservoirs, magnetic field and geological features. It will reach Jupiter in January 2031 after a 7.6-year travel with various Earth–Venus–Earth–Earth gravity assist manoeuvres. JANUS instrument will enable visible wavelength imaging, crucial for understanding the formation and characteristics of various geological features, surface processes and erosion/deposition processes on icy satellites.

JANUS consists of independent subsystems, including the Optical Head Unit (OHU), Proximity Electronics Unit (PEU), Main Electronics Unit (MEU), and Interconnecting harness. The COver Mechanism (COM), used to protect JANUS and its delicate optical elements, is a direct derivation of the innovative design used for the FDM for the OSIRIS telescopes. It ensures protection of the OHU from contamination and provides a reference surface for telescope calibration during the mission [8].

Considering the very successful design of the Front Door Mechanism (FDM) for the OSIRIS experiment on ROSETTA, along with similar requirements and the nominal operation of the mechanism throughout an Interplanetary mission, it was decided to adapt the FDM to craft the COver Mechanism (COM) for JANUS.

The main functional and environmental requirements for JANUS' cover mechanism are in fact similar: protection of optics and detectors from sunlight and contamination, offer different positions for various flight conditions (closed-locked for launch, closed-not locked for cruise, open-locked for observation and calibration. Reliability and single-point failure tolerance is also obviously required, as it is the case for all instrument covers, as already discussed.

The main differences in the design of JANUS' COver Mechanism (COM) compared to ROSETTA's FDM are as follows.

Some geometric parameters needed adjustment such as the arm size and the body height and a partial redesign of the fail safe system. The geometry of the internal cams was also optimised for minimal resistant torque. Moreover, the material for the sealing under the door has been modified, while maintaining the same goals of dust and light tightness and vibration damping.

Fig. 3 shows the JANUS COM subsystem final design and flight model.

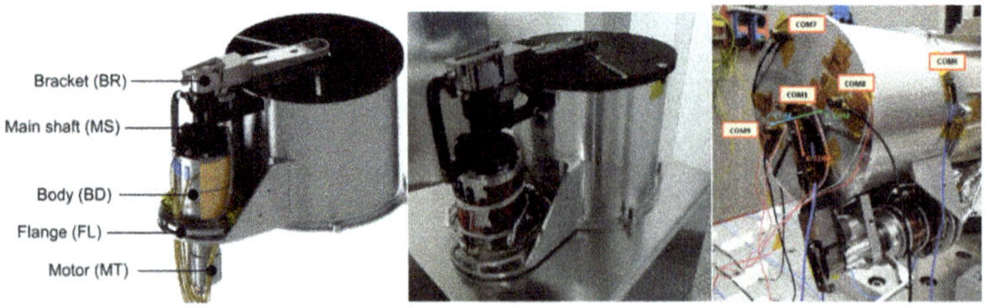

Figure 3 : COM subsystem and its components in rendering (left), flight model (centre), during vibration tests (right).

The JANUS COM successfully underwent the full qualification campaign maintaining its functional performance unaltered. It has been successfully integrated in JANUS instrument and onboard JUICE satellite and successfully launched in April 2023.

Conclusions

The innovative design of the cover system developed for the protection of the telescopes of the OSIRIS experiment of the ROSETTA mission, allowed a variety of options in its utilisation, from full dust and light tightness, to just partial detachment from the baffle surface and also provided the reference surface for internal calibration of the instrument.

Its reliability and robustness in wide range of operative conditions has been the main focus due to the extreme criticality of its potential failure, and it has been demonstrated during the entire mission

Acknowledgements:

This paper has been intended as a short overview and reflection on how a robust and adaptable design of a space unit could serve the needs of very different missions, decades away in time. Gratitude goes to all the many ones who contributed in several ways to realisation of the units from the very first prototype to the recently flown ones. A special mention and memory is deserved for our colleague and friend **Stefano**. He made, among many other things, also this legacy possible. We all owe him a lot!

References

[1] H. U. Keller, C. Barbieri, P. Lamy et al., "OSIRIS – The scientific camera system onboard Rosetta", Space Science Reviews 128(1): 433-506 – February 2007

[2] S. Debei, M. De Cecco, G. Parzianello, A. Francesconi, F Angrilli, "Design, Qualification and Acceptance of the Front Door Mechanism for the OSIRIS experiment of ROSETTA mission", International Symposium on Optical Science and Technology – 9th September 2002– Seattle, WA (USA) – Proceedings Volume 4771. https://doi.org/10.1117/12.482161

[3] G. Parzianello, S. Debei, F. Angrilli, "Reliability oriented design and testing of a telescope's front cover for long life space missions", 54th Congress of the International Astronautical Federation (IAF) – 29 September – 03 October 2003 – Bremen, Germany. https://doi.org/10.2514/6.IAC-03-I.P.10

[4] F. Preusker, F. Scholten et al, "The global meter-level shape model of comet 67P/Churyumov-Gerasimenko", Astronomy & Astrophysics, Vol 607, L1 (2017).

[5] H. Sierks et al., "E-type asteroid (2867) steins as imaged by OSIRIS on board Rosetta", Science, vol. 327, no. 5962, pp. 190-193, 2010. https://doi.org/10.1126/science.1179559

[6] H. Sierks et al., "On the nucleus structure and activity of comet 67P/Churyumov-Gerasimenko", Science, vol. 347, no. 6220, January 2015.

[7] A. Accomazzo, P. Ferri, S. Lodiot, J.L. Pellon-Bailon, A. Hubault, J. Urbanek, et al., "The final year of the Rosetta mission", Acta Astronautica, vol. 136, pp. 354-359, July 2017. https://doi.org/10.1016/j.actaastro.2017.03.027

[8] Della Corte, Vincenzo, et al. "Scientific objectives of JANUS Instrument onboard JUICE mission and key technical solutions for its Optical Head." 2019 IEEE 5th International workshop on metrology for aerospace (MetroAeroSpace). IEEE, 2019. https://doi.org/10.1109/MetroAeroSpace.2019.8869584

Vibroacoustics

In-vacuo structured fabrics for vibration control

Paolo Gardonio[1,a*], Sofia Baldini[1,b], Emiliano Rustighi[2,c], Ciro Malacarne[3,d], Matteo Perini[4,e]

[1]DPIA, Università degli Studi di Udine, Via Delle Scienze 206, 33100, Udine, Italy

[2]DIEA, Università degli Studi di Trento, Via Sommarive 9, 38123 Trento, Italy

[3]ProM Facility – Trentino Sviluppo S.p.A., Via Fortunato Zeni 8, 38068 Rovereto (TN), Italy

[a]paolo.gardonio@uniud.it, [b]baldini.sofia@spes.uniud.it, [c]emiliano.rustighi@unitn.it, [d]ciro.malacarne@trentinosviluppo.it, [e]matteo.perini@trentinosviluppo.it

Keywords: Tunable Structured Fabric, In Vacuo Material, Adaptive Material, Tuneable Vibration Absorber

Abstract. This paper presents a study on a new tuneable material for vibration control purposes. The material is formed by a structured fabric wrapped in a deflated bag. The fabric is made of an interwoven mesh of rigid truss-like particles. The vacuum inside the casing produces a jamming effect. Hence, the elasticity and damping of the packaged fabric can be tuned by changing the level of vacuum in the bag. This material can be conveniently employed to develop new vibration control treatments and devices. In this respect, the paper first presents the tests carried out with a six-point bending machine to characterise the static and dynamic stiffness of the material as well as its damping properties. Then it demonstrates an application, where the material is used as a tuneable vibration absorber.

Introduction

The material considered in this paper presents is formed by a structured fabric wrapped in a vacuum casing [1]. The fabric is made by an interwoven mesh of rigid truss-like particles, which forms a loose flexible construction, such as for example a chain mail armour. The fabric is packaged into a deflated plastic bag, whose level of vacuum is controlled online with a micro-compressor. The vacuum generated inside the casing produces a jamming effect, which results from both interlocking and friction between neighbouring particles [2]. In this way, the elasticity and damping of the packaged fabric can be conveniently tuned by changing the level of vacuum in the case. The result is thus a tuneable lightweight material, which can be effectively employed to develop new treatments, e.g., tuneable liners, and devices, e.g., Tuneable Vibration Absorbers (TVA), for passive and semi-active vibration and noise control [3,4].

Material layouts and test facilities

This paper presents two types of experiments on prototype beam-like in-vacuo structured fabrics, which, as shown in Fig. 1 and summarised in Table 1, encompass either single- or double-mails made with cubic, octahedral, spheric truss-like particles. As shown in Fig. 2a, In the first experiment, the beam-like in-vacuo fabrics are pinned at the two ends and excited in the middle by a shaker via a pinning jaw such that they work as pinned beams excited in bending. As shown in Fig. 2b, in the second experiment, the beam-like in-vacuo fabrics are mounted on a shaker such that they work as "flapping vibration absorbers". Table 1 summarises the principal properties of the structured fabrics analysed in this study.

Materials Research Proceedings 37 (2023) 731-735 https://doi.org/10.21741/9781644902813-156

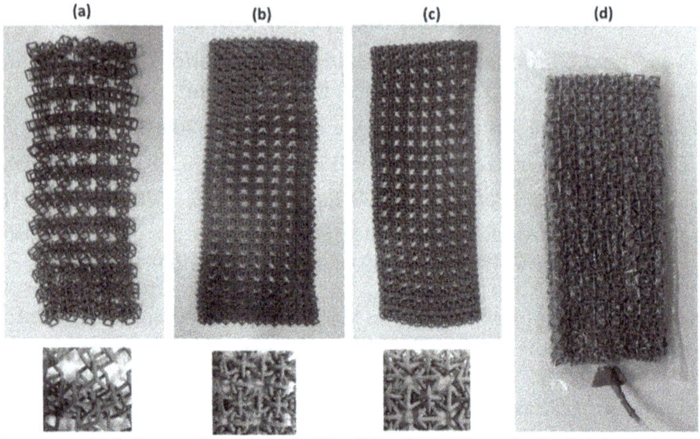

Fig. 1 structured fabrics made with (a) cubic, (b) spherical octahedral, (c) octahedral chain mails. (d) Beam specimen made by the fabric in a deflated bag.

Table 1: Prototyped structured fabrics studied in this paper with dimensions and weight

Name	Geometry	Width (mm)	Length (mm)	Thickness (mm)	Mass (g)
Spheres		100	210	10	49
Octahedra		110	240	15	62
Cubes		110	190	15	39

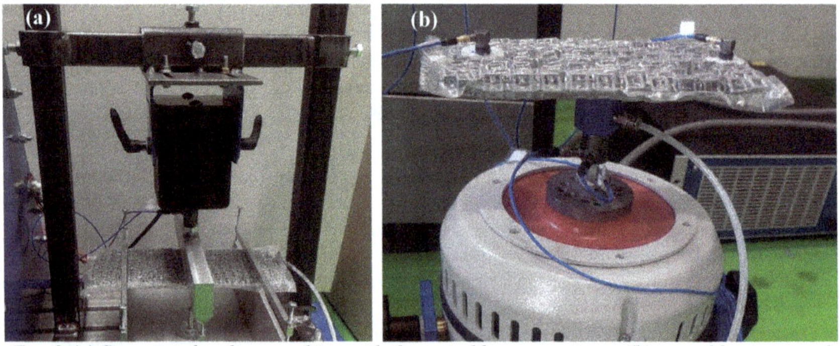

Fig. 2 (a) Six points bending machine with the tested beam specimen. (b) tuneable structured fabric vibration absorber mounted on a shaker vibration source.

Mechanical properties

Figure 3a shows the modulus and phase of the dynamic stiffness measured with the six-point bending machine with respect to the vacuum pressure in the deflated bag with the cubic structured fabric. The plot shows the typical spectrum of the middle point dynamic stiffness of a simply supported beam, which is characterised by a stiffness-like asymptotic response at low frequencies and a mass-like asymptotic response at high frequencies. At mid frequencies there is a sharp resonance through due indeed to the resonant response of the fundamental bending mode of the beam specimen. The graph shows that the vacuum shifts to higher frequencies the static bending stiffness of the material but has no effect on its apparent mass. As a result, it shifts to higher frequency the fundamental resonance too. Indeed, the graph in Figure 3b shows that, when pressure is increased from 5 kPa to 80 kPa, the resonance frequency of the specimens grows by about 20% to 25%. Also, the specimens with double layer fabrics have about 50 % to 100 % higher resonance frequency than their single layer counterparts. In general, the specimens encompassing the fabrics made with cubic grains show the highest bending stiffness and thus the highest, and widest, resonance frequency ranges. The graph in Figure 3c indicates that the loss factor of the specimens, derived with the half power bandwidth method at the first resonance frequency, are characterised by quite similar values for the three types of fabrics, which are confined between 5 % and 7 % for the single layer configuration and between 5.5 % and 9 % for the double-layer configuration.

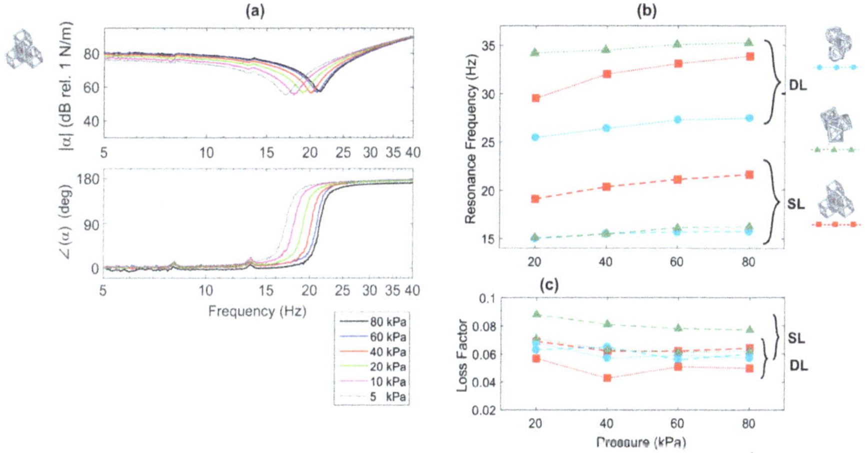

Fig. 3 Dynamic stiffness frequency response function (a) and resonance (b), loss factor (c) parameters with respect to the vacuum pressure.

Tunable Vibration Absorber

The beam specimen can be suitably used to construct a TVA. For instance, Fig. 2b shows a prototype device, which is mounted on the vibration table of a big shaker. In this case, the structured fabrics are insert in a fully sealed plastic bag, which is equipped in the middle span with a sealed inlet port built in plastic using 3D printing technology. In this way the in vacuo structured fabric acted as a two-arms beam clamped in the middle span to the inlet port, which acts as a post too. Indeed, the inlet port was designed in such a way as it served both as a connector for the vacuum tube and as a mechanical joint to fix the two-arms in-vacuo structured fabric beam to the hosting mechanical system. The vacuum was generated with an off the shelf pump and a simple circuit encompassing two valves and a vacuum gauge.

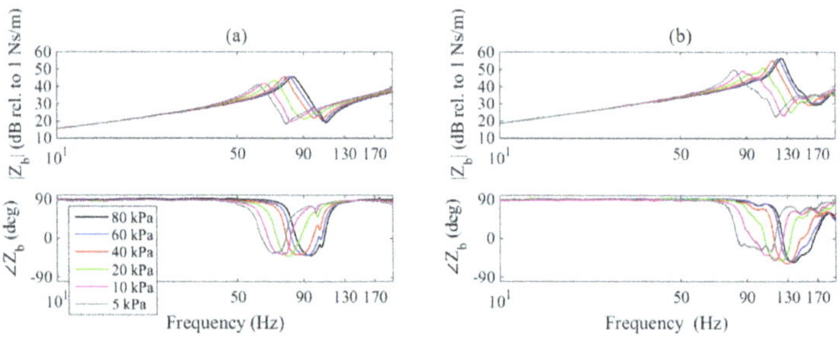

Figure 4: Impedance response for cubes assembled with one (a) or two (b) overlapped layers.

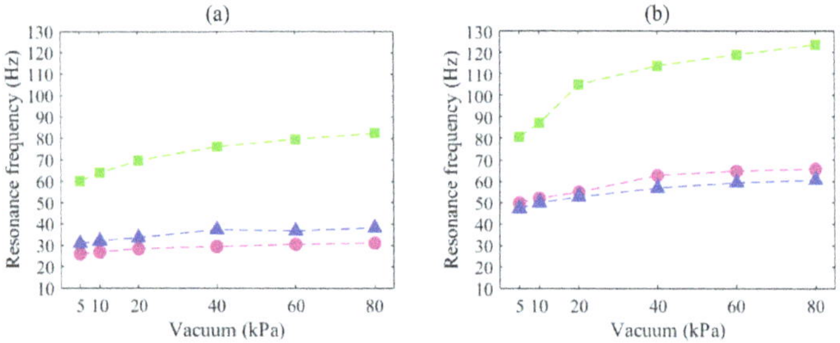

Figure 5: Measured resonance frequency of the tuneable structured fabrics assembled with one (a) or two (b) overlapped layers. Magenta for spheres, blue for octahedra and green for cubes.

Figure 4 shows the modulus and phase of the base impedance measured on the specimens encompassing either a single (left hand side plots) or a double (right hand side plots) cubic fabric vibration absorber with reference to vacuum pressure that grows from 5 kPa to 80 kPa. The spectra show the typical base impedance of a seismic mass connected to a base mass via a spring damper lumped element. Indeed, at low frequencies the spectrum is characterised by a modulus that rises proportionally to the circular frequency and a constant phase value of about +90°. There is then a resonance peak followed by an antiresonance through, with the phase that initially falls down to −90° and then recovers to +90°. At higher frequencies the spectrum shows again a modulus that rises proportionally to the circular frequency and a phase that maintains the +90° value. All this indicates that at low and high frequencies the device presents a mass effect. More precisely, measurements taken with the accelerometers have shown that at low frequencies the in-vacuo structured fabric beam behaves as a solid body together with the junction component, with negligible flapping effects. Therefore, the base impedance is given by the whole mass of the beam and base component. Alternatively, at higher frequencies, despite the base vibrations, the two ends of the in-vacuo structured fabric are characterised by little vibrations such that the base impedance is controlled by the mass of the base component. At resonance frequency the two ends of the in-vacuo structured fabric display large counter oscillations to the base oscillations, which generate the desired vibration absorption effect. The extent of these oscillations is controlled by the damping of the in-vacuo structured beam. Normally, to counteract harmonic vibrations, the vibration absorber is tuned in such a way as it resonates at the tonal disturbance and its damping is kept to the minimum possible value such that the hosting structure faces a large impedance load.

Alternatively, for broadband vibrations the vibration absorber is tuned in such a way it resonates at the resonance frequency of the hosting system, but in this case the damping is brought up such that the double resonant response of the combined hosting structure-TVA modes is optimally dampened.

Overall, the graphs shown in Fig. 5 indicates that the resonance frequency of the TVAs can be suitably shifted to higher/lower values by increasing/lowering the level of vacuum in the bags. For the 5 to 80 kPa pressure range, the single layer TVAs are characterised by resonance frequencies comprised between 60-85 Hz (cube grains), 25-30 Hz (sphere grains), 30-40 Hz (octahedra grins). Alternatively, the double layer TVAs are characterised by resonance frequencies comprised between 80-125 Hz (cube grains), 50-65 Hz (sphere grains), 45-60 Hz (octahedra grains). This confirms that the resonance frequency of the TVAs can be suitably tuned over significant ranges by varying the vacuum pressure. Moreover, these ranges can be further enlarged by adopting multiple layers with combinations of different grains.

Summary

The experiments presented in this paper have shown that the bending stiffness of the in vacuo structured fabric beams can be suitably varied by changing the vacuum pressure in the bag. This effect depends on the type of elementary truss grains of the fabric and on the number of layers wrapped in the bag. For instance, the cubic and octahedral fabrics offers the highest bending stiffness respectively for the single- and double-layer configurations. When the vacuum pressure is raised from 5 to 80 kPa, the stiffness of two-layers cubic fabric beam is doubled. Also, the double layer octahedral structured fabric shows 5 times higher stiffness than the single layer one. These properties can be suitably employed to shift the resonance frequency of the fundamental flapping flexural mode when the in vacuo structured fabric is operated as a vibration absorber. For instance, the study has shown that, for 5 to 80 kPa pressure range, the resonance frequency of either the single- or double-layer vibration absorbers can be increased by 30% to 40%. Also, the resonance frequency of the double layer vibration absorbers is about 50% higher than that of the single layer absorber.

References

[1] Y. Wang, L. Li, D. Hofmann, J. E. Andrade, and C. Daraio, Structured fabrics with tunable mechanical properties, Nature, 596 no. 7871, (2021) 238–243. https://doi.org/10.1038/s41586-021-03698-7

[2] R. P. Behringer and B. Chakraborty, The physics of jamming for granular materials: a review," Reports on Progress in Physics, 82 no. 1, (2018). https://doi.org/10.1088/1361-6633/aadc3c

[3] E. Rustighi, P. Gardonio, N. Cignolini, S. Baldini, C. Malacarne, M. Perini, Vibration response of tuneable structured fabrics, In proocedings of the International Conference on Noise and Vibration Engineering (ISMA 2022), Katholieke Universiteit Leuven, Belgium, 12–14 September 2022.

[4] S. Baldini, E. Rustighi, P. Gardonio, C. Malacarne, M. Perini, Invacuo structured fabric tuneable vibration absorber, In Proceedings of the X ECCOMAS Thematic Conference on Smart Structures and Materials (SMART 2023), Patras, Greece, 3 - 6 July 2023. https://doi.org/10.7712/150123.9779.444503

Experimental application of pseudo-equivalent deterministic excitation method for the reproduction of a structural response to a turbulent boundary layer excitation

Giulia Mazzeo[1,a *], Mohamed Ichchou[1,b], Giuseppe Petrone[2,c], Olivier Bareille[1,d], Francesco Franco[2,e], Sergio De Rosa[2,f]

[1]LTDS— Laboratoire de Tribologie et Dynamique des Systèmes, Ecole Centrale de Lyon Ecully 69130, France

[2]PASTA LAB — Laboratory for Promoting Experiences in Aeronautical Structures and Acoustics, University "Federico II" of Naples, Naples 80125, Italy

[a]giulia.mazzeo@ec-lyon.fr, [b]mohamed.ichchou@ec-lyon.fr, [c]giuseppe.petrone@unina.it, [d]olivier.bareille@ec-lyon.fr, [e]francesco.franco@unina.it, [f]sergio.derosa@unina.it

Keywords: Turbulent Boundary Layer (TBL), Wall-Pressure Fluctuations (WPFs), Structural Vibrational Response, Deterministic Forces

Abstract. The use of wind tunnels for studying the vibrational response of structures subjected to turbulent flows presents various challenges, such as background noise and complex setup requirements. This work introduces an alternative experimental method called X-PEDEm (eXperimental Pseudo-Equivalent Deterministic Excitation) that aims to reproduce an equivalent structural response to a Turbulent Boundary Layer (TBL) excitation without the need for a wind tunnel. X-PEDEm involves coupling the experimental acquisition of the structure's vibrational response with deterministic forces, such as an impulse force from a hammer, followed by post-processing. The method has been validated for different boundary conditions and flow speeds, offering versatility in recreating various types of TBL. While not an exact reproduction of turbulent flow-induced responses, X-PEDEm provides an optimal approximation with low time and resource requirements, making it easy to implement experimentally.

Introduction

In the field of transportation vehicle design and production, such as aircraft, ships, trains, and automobiles, there is ongoing research on the sound emissions caused by the interaction between fluids and structures. A key area of focus is predicting how structures respond to Wall-Pressure Fluctuations (WPFs) generated by a Turbulent Boundary Layer (TBL).

Currently, researchers rely on semi-empirical models, often represented as 2-points Cross-Spectral Density (CSD) functions like the Corcos model [1]. However, these models can be computationally intensive when applied in Finite Element Analysis (FEA) and have limitations in representing responses across a broad frequency range [2]. Consequently, there is a growing interest in alternative experimental methods for predicting the structural response to TBL excitation. Many researchers have used loudspeakers as a means to reproduce a TBL-like pressure field, but this approach can also present challenges [3, 4].

In this study, the authors aim to experimentally validate an alternative method based on the Pseudo-Equivalent Deterministic Excitation method (PEDEm) [5]. The potential of PEDEm and its numerical validation for experimental purposes have been previously presented [6]. Here, the focus is on experimentally validating the application of PEDEm, referred to as X-PEDEm, considering different sample panels under different boundary conditions and subjected to TBL excitation at various flow velocities.

Background theory and methodology

Leaving the reader free to explore how PEDEm was created and developed [5, 6], here the main formulations are presented in Eq. 1-3 and discussed.

$$[S_{FF}(\omega)] = \sum_{i=1}^{NG} d_i(\omega)\{\Theta^{(i)}\}\{\Theta^{(i)}\}^T \tag{1}$$

$$\{w(\omega, i)\} = [\Phi][H(\omega)][\Phi]^T\{\Theta^{(i)}\}\sqrt{d_i(\omega)} \tag{2}$$

$$[S_{WW}(\omega)] = \sum_{i=1}^{NG}\{w(\omega, i)\}^*\{w(\omega, i)\}^T \tag{3}$$

PEDEm is based on the reformulation of the CSD displacement matrix $[S_{WW}(\omega)]$ as shown in Eq. 2 and Eq. 3 by considering the modal decomposition of the CSD load matrix $[S_{FF}(\omega)]$ in eigenvectors $\{\Theta^{(i)}\}$ and eigenvalues $d_i(\omega)$ as expressed in Eq. 1. In particular, PEDEm considers two asymptotic behaviors of these eigensolutions:

- in a low frequency (LF) domain, the eigensolutions represent a spatial distribution totally correlated, for which the eigenvector matrix $[\Theta]$ is an all-1 matrix and only the first eigenvalue is non-null;
- in a high frequency (HF) domain, the eigensolutions represent a spatial distribution totally uncorrelated, for which the eigenvector matrix is an identity matrix, and all eigenvalues are equal and non-null.

X-PEDEm uses the same equations and the same asymptotic behaviors of PEDEm for the post-processing phase of experimental data that can be obtained with an easy experimental campaign as a hammer test. Indeed, X-PEDEm requires just the acquisition of the experimental Frequency Response Functions (FRFs) between acquisition points and excitation points. The acquisition points can be chosen randomly, and they should not be less than five, while the excitation points must respect the position configuration shown in Fig. 1a and they should not be less than ten.

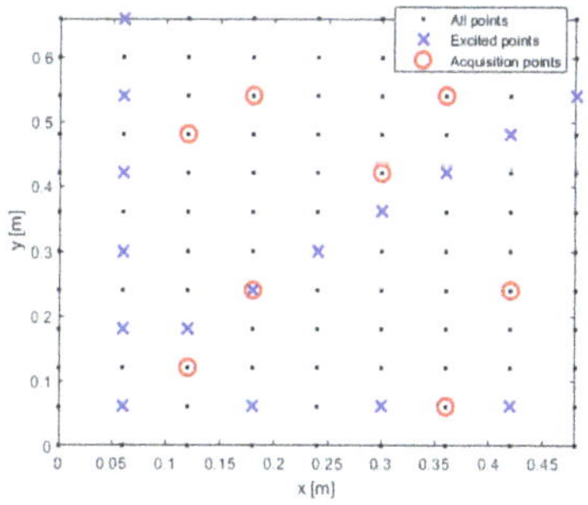

(a)

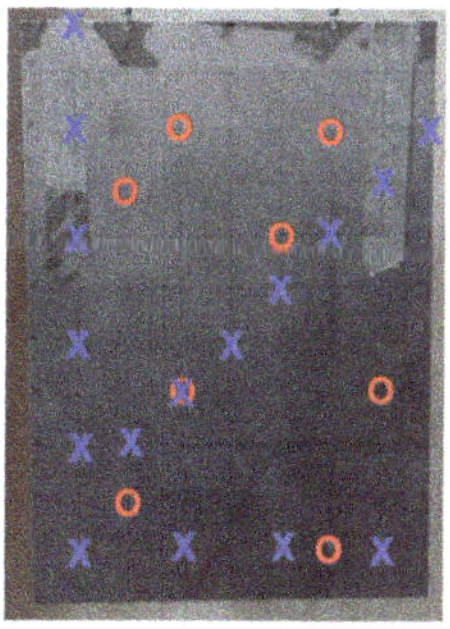

(b)

***Fig. 1** – Selection of acquisition points (red circles) and excitation points (blue crosses) over the structural mesh of sample panel "PAN_A". (a) Numerical mesh; (b) experimental mesh.*

Experimental method validation

X-PEDEm has been experimentally validated with a hammer test campaign performed over three different sample panels, for three different boundary conditions and for different flow velocities. For a matter of space, only the results for panel "PAN_A" (Fig. 1b), with totally free edges as boundary conditions, are here shown in Fig. 2. The numerical FSR is here considered as reference solution and it is calculated by using the Corcos model [1] with the following empirical coefficients values: $\alpha_x = 0.116$, $\alpha_y = 0.700$ and $U_c = 0.8U_0$. X-PEDEm, on the other hand, has been evaluated by using the experimental FRFs collected during the hammer test campain by considering the acquisition and excitation points shown in Fig. 1b. The numerical formulation of X-PEDEm, developed with a MATLAB© code, is shown too for further validation. It is possible to appreciate how X-PEDEm (numerical and experimental) is able to follow the numerical FSR solution for three different flow velocities U_0. With the increase of U_0, the convective coincidence frequency f_c increases too. f_c is used as approximated indicator to establish which asymptotic behavior one should refer to: below the f_c, the asymptotic behaviour for LF domain is considered, while above f_c, it is chosen the asymptotic behavior for HF domain.

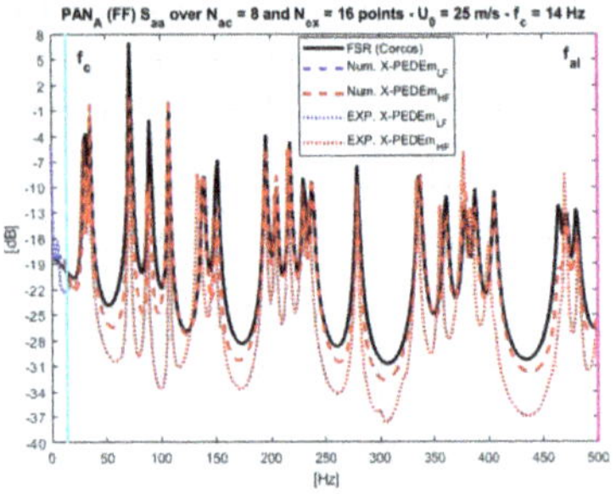

(a)

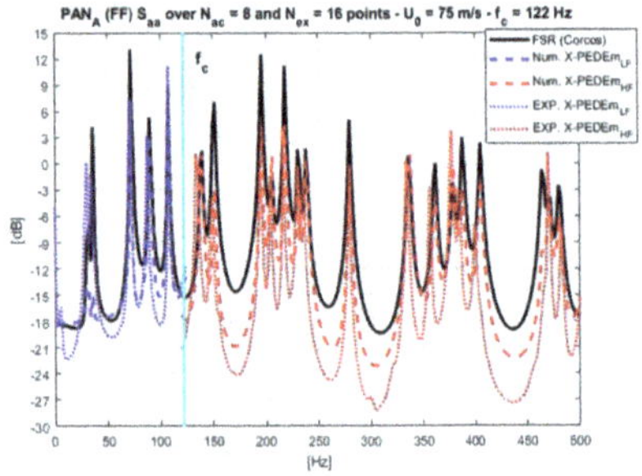

(b)

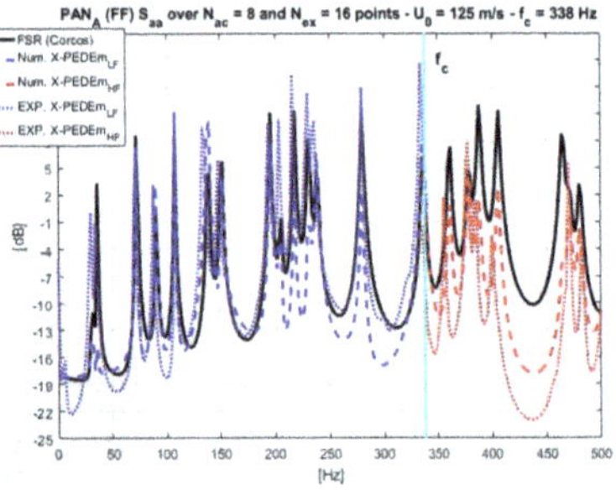

(c)

Fig. 2 *– Comparison between the FSR calculated with Corcos model (solid black line), numerical X-PEDEm solution (dashed line) and experimental X-PEDEm obtained with a hammer test (dotted line); blue color for X-PEDEm in the LF domain, red color for X-PEDEm in the HF domain. (a) Solution for* $U_0 = 25$ *m/s; (b) solution for* $U_0 = 75$ *m/s; (c) solution for* $U_0 = 125$ *m/s.*

Conclusions

X-PEDEm proves to be valid as alternative experimental method for the reproduction of the structural response to a TBL excitation. It can be performed with an easy experimental set-up and with fast post-processing of the experimental data. It ensures versatility for what concerning type of panels, boundary conditions and asymptotic flow velocities. It may be pointed out that a hammer test is not able to maintain reliable FRFs in a broadband frequency region, but X-PEDEm can be performed with different experimental tools, as for example a shaker.

Nevertheless, there are still some open issues:

- It is important to find an accurate indicator that establish the frequency limits for the two asymptotic behaviors.
- A direct comparison between the experimental results obtained in a wind tunnel and the ones obtained with X-PEDEm is required for a final validation of the method.

References

[1] G. M. Corcos, Resolution of Pressure in Turbulence, J. Acoust. Soc. Am. 35, 192–199 (1963). https://doi.org/10.1121/1.1918431

[2] S. De Rosa, F. Franco, Exact and numerical responses of a plate under a turbulent boundary layer excitation, J. Fluids Struct. 24, 212-230 (2008). https://doi.org/10.1016/j.jfluidstructs.2007.07.007

[3] M. Aucejo, L. Maxit, J.-L. Guyader, Experimental simulation of turbulent boundary layer induced vibrations by using a synthetic array, J. Sound Vib. 331, 3824-3843 (2012). https://doi.org/10.1016/j.jsv.2012.04.010

[4] O. Robin, A. Berry, S. Moreau, Experimental Synthesis of Spatially-Correlated Pressure Fields for the Vibroacoustic Testing of Panels. In: E. Ciappi, S. De Rosa, F. Franco, J.-L. Guyader, S. Hambric (Eds), Flinovia - Flow Induced Noise and Vibration Issues and Aspects, Springer, Cham. 2015. https://doi.org/10.1007/978-3-319-09713-8_8

[5] S. De Rosa, F. Franco, E. Ciappi, A simplified method for the analysis of the stochastic response in discrete coordinates, J. Sound Vib. 339, 359-375 (2015). https://doi.org/10.1016/j.jsv.2014.11.010

[6] G. Mazzeo, M. Ichchou, G. Petrone, O. Bareille, S. De Rosa, F. Franco, Pseudo-equivalent deterministic excitation method application for experimental reproduction of a structural response to a turbulent boundary layer excitation, J. Acoust. Soc. Am. 152, 1498–1514 (2022). https://doi.org/10.1121/10.0013424

Labyrinth quarter-wavelength tubes array for the reduction of machinery noise

G. Catapane[1,a *], G. Petrone[1,b], O. Robin[2,c], J.-C. Gauthier-Marquis[3,d], S. De Rosa[1,e]

[1] PASTA-Lab (Laboratory for Promoting experiences in Aeronautical STructures and Acoustics), Università degli Studi di Napoli "Federico II", Via Claudio 21, Napoli, 80125, Italy

[2] Centre de Recherche Acoustique-Signal-Humain, Université de Sherbrooke, 2500 boulevard de l'Université, Sherbrooke, J1K 2R1, Quebec, Canada

[3] Innovation Maritime, 53, rue Saint-Germain Ouest, Rimouski, 5L 4B4, Quebec, Canada

[a]giuseppe.catapane@unina.it, [b]giuseppe.petrone@unina.it, [c]olivier.robin@USherbrooke.ca, [d]jcgmarquis@imar.ca, [e]sergio.derosa@unina.it

Keywords: Underwater Noise, Acoustic Resonators, Noise Control

Abstract. Anthropogenic noise from navigation is a major contributor to the disturbance of the acoustic soundscape in underwater environments. The noise generated by ship's machinery exhibits energetic tonal harmonic peaks at multiples of the rotating and firing frequency, that occur in the 20-200 Hz frequency range and difficult to control with classical soundproofing materials. Quarter wavelength tubes (QWT) can be a concrete solution since their absorption peaks are harmonic odd integers of the first resonance frequency. The main issue of QWT is their tuning length, which equals 1.43 m for a 60 Hz resonator. The problem is solved by coiling the tube into a labyrinth. Three labyrinth quarter wavelength tubes are tuned respectively at 60, 90 and 120 Hz. Samples are printed with filament 3D additive manufacturing techniques using PLA and tested with a square impedance tube designed for low-frequency measurements. Measurement results are in good agreement with analytical and numerical predictions. An array including four 60 Hz, four 90 Hz and four 120 Hz labyrinths QWTs is finally tested.

Introduction

The most important source of noise of an internal combustion engine powered vehicle or equipment is usually the engine itself. The sound disturbance produced by a ship is dominated by machinery noise at low speed, and by cavitation and propeller noise at higher speeds.
The low frequency spectrum of the noise produced by an internal combustion engine is dominated by the engine rotating and firing frequencies and the associated harmonics [1]. Most of the internal combustion engines are four-stroke engines and those will be under consideration in this work. The explosion frequency f_{ex} of a four-stroke engine with N cylinders at a given speed rpm can be calculated following:

$$f_{ex} = \frac{N}{2}\frac{rpm}{60}, \qquad (1)$$

Let's take as an example a diesel engine that is a 6 cylinder - 4 stroke engine. With symmetric firings, 6 explosions occur on a complete cycle (two crankshaft rotations, and thus three ignition events occur per crankshaft revolution). Vibration measurements were conducted on a CAT C9 diesel engine at *Innovation Maritime*. With an average rotation speed of 1800 rpm, the engine exhibits an explosion frequency $f_{ex} = 90$ Hz. This frequency and its harmonics (multiples of f_{ex}) are generally the most energetic peaks in the noise and spectrum spectra. The two other important noise generation mechanisms are those associated with the rotation frequency of the crankshaft (in this case, $\frac{rpm}{60} = 30$ Hz) and those generated by the ignition imperfections or unbalance on a complete cycle (typically half of the rotation frequency, $\frac{1}{2} \cdot \frac{rpm}{60} = 15$ Hz). In Figure 1, significant peaks are identified each 15 Hz with a dominance at f_{ex} and its harmonics. This frequency spectrum is common to diesel engines used in the marine industry. Quarter wavelength tubes could be an efficient solution to limit this disturbance. Indeed, a quarter wavelength tube (QWT) is an open-closed tube (Figure 2a) that has resonant frequencies f_{qwt} when the tube length L is an odd-integer multiple of the quarter of the acoustic wavelength $\lambda = c_0/f$:

$$f_{qwt} = \frac{(2m-1)c_0}{4L}. \qquad \boldsymbol{m = 1, 2, 3 \ldots} \qquad (2)$$

At their successive harmonic resonances, large sound absorption can be achieved, above all for the first harmonics. This behavior is well adapted to the defined problem of reducing the harmonic noise generated by reciprocating engines. One issue is related to the required length for a QWT. To cope with this, the QWT is stretched into a labyrinth (Figure 2b) or following a spiral path, depending on the application, which guarantees similar sound absorption properties compared with a straight tube.

Theory and numerical implementation for labyrinth resonators

The acoustic impedance of a labyrinth resonator (LR) is studied according to the analytical approach proposed by Magnani et al. [2], where the labyrinth resonator is evaluated as a perforated plate followed by a QWT. The QWT of length $\boldsymbol{L}$ has an impedance equal to $\boldsymbol{Z_{QWT} = -jZ_{eff}\cot(k_{eff}L_{eff})}$, based on Low Reduced Frequency theory (LRF) [3], which studies the sound wave propagation with a lossy Helmholtz equation which takes into account viscous and thermal dissipation by modelling the effective density $\boldsymbol{\rho_{eff}}$ and speed of sound $\boldsymbol{c_{eff}}$, and consequently the effective impedance $\boldsymbol{Z_{eff} = \rho_{eff}c_{eff}}$ and the effective wavenumber $\boldsymbol{k_{eff} = \omega/c_{eff}}$. $\boldsymbol{L_{eff} = L - (4-\pi)\frac{d}{2}(n-1)}$ is the effective length to tune a labyrinth resonator, with $\boldsymbol{n}$ number of branches and $\boldsymbol{d}$ width of the channel [4].

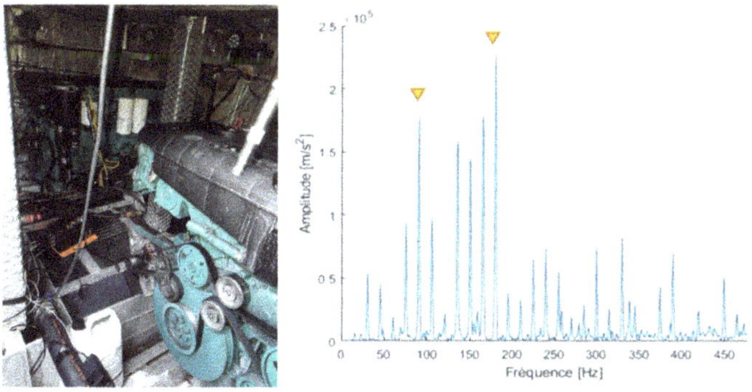

Figure 1: Result of a vibration measurement conducted on the crankshaft axis of a C9 diesel engine. The yellow triangles indicate the firing frequency of the pistons (90 Hz), as well as its first harmonic (180 Hz).

The plate of thickness $\boldsymbol{t_d}$ with the square inlet hole of side-length $\boldsymbol{d}$ is defined through Johnson-Champoux-Allard (JCA) approach [5]. The impedance of the labyrinth resonator $\boldsymbol{Z_{LR}}$ is:

$$Z_{LR} = \frac{1}{\phi_{inlet}}\left[Z_{d,JCA}\frac{-jZ_{QWT}\cot\left(k_{d,JCA}t_d\right) + Z_{d,JCA}}{Z_{QWT} - jZ_{d,JCA}\cot\left(k_{d,JCA}t_d\right)}\right], \tag{3}$$

where $Z_{d_{JCA}} = \rho_{JCA}c_{JCA}$ and $k_{d_{JCA}} = \omega/c_{JCA}$ are respectively the impedance and the complex wavenumber of the perforated plate, with $c_{JCA} = \sqrt{K_{JCA}/\rho_{JCA}}$, ρ_{JCA} and K_{JCA} effective speed of sound, density and bulk modulus. $\phi_{inlet} = A_{hole}/A_{plate}$ is the perforatio ratio between the hole area and the plate area. Labyrinth resonators are positioned inside a box to have several peaks at the desired frequencies. According to the electro-acoustic analogy, this implies that the impedance of the global system Z_{tot} and the sound absorption coefficient α_{theory} are respectively:

$$Z_{tot} = \left[\sum_{i=1}^{N}\frac{1}{Z_{LR,i}}\right]^{-1} \qquad \alpha_{theory} = \frac{4Re(Z_{tot}/Z_0)}{|Z_{tot}/Z_0|^2 + 2Re(Z_{tot}/Z_0) + 1} \tag{4}$$

The sound absorption of the box is experimentally evaluated with a 1 ft x 1 ft impedance tube: a speaker placed at one end of the tube excites it with a normal plane wave radiation, and the sample is placed at the opposite end, backed by a rigid wall. Two microphones separated by a distance s evaluate sound pressure in the tube, with P_2 at a distance x_2 respect to the sample, and P_1 at a distance $x_1 = x_2 + s$. The sound absorption coefficient under normal incidence is estimated according to the ISO 10534-2 1998 standard. Numerical simulations mimic this experimental measurement. Analyses are made with COMSOL Multiphysics, *Pressure Acoustics Module*, with the impedance tube and labyrinth walls considered as rigid.

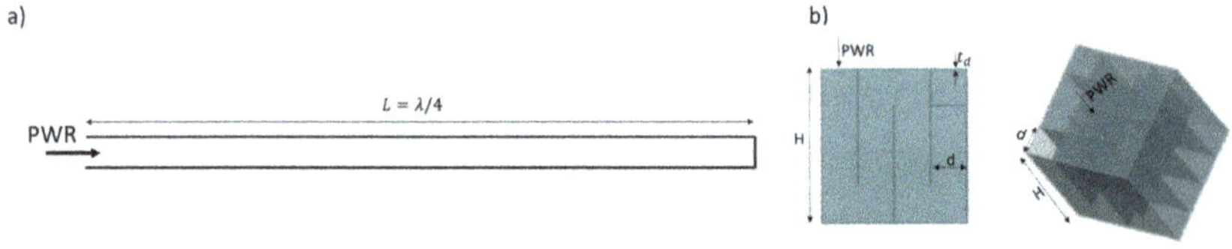

Figure 2: a) Quarter wavelength tube excited by a plane wave radiation (PWR); b) labyrinth resonator excited by a PWR.

Results

Three labyrinth resonators are modelled with their fundamental resonance peak at 60, 90, 120 Hz, and 3D printed to form a box with multiple resonators to have high sound absorption at 60, 90, 120 and 180 Hz. The latter is the second harmonic of the 60 Hz labyrinth. Their height (thickness of the sample) is fixed at 100 mm, to place them in a limited space. The 60 Hz labyrinth has lateral dimension of 97 mm times 97 mm, the 90 Hz has lateral dimension of 145 mm times 49 mm, and the 120 Hz has lateral dimension of 73 mm times 73 mm. A wooden box is designed to include all the resonators is a single unit of 1 foot by 1 foot area, that can be directly installed inside the impedance tube (of 1 foot by 1 foot square section), with four 60 Hz, four 90 Hz and four 120 Hz LRs (Figure 3a). In Figure 3b), the experimental sound absorption coefficient of the box with multiple resonators is plotted, comparing it with theory and simulation. Resonances are clearly distinguished even for experiments, with the sound absorption at each peak always bigger than 0.7

up to 500 Hz. Each tone is at its precise tuned frequency. In addition, their combination gives non-zero absorption in the entire frequency range studied.

Conclusion

The multi-resonator box shows convincing sound absorption properties and opens interesting perspectives. Experimental results are compared with analytical and numerical methods: the experimental peaks appear at the predicted frequencies, with an amplitude difference that can be attributed to the rigid wall hypothesis. Future developments will be pursued with an experimental campaign, in a water basin or ideally in the engine cabin of a ship.

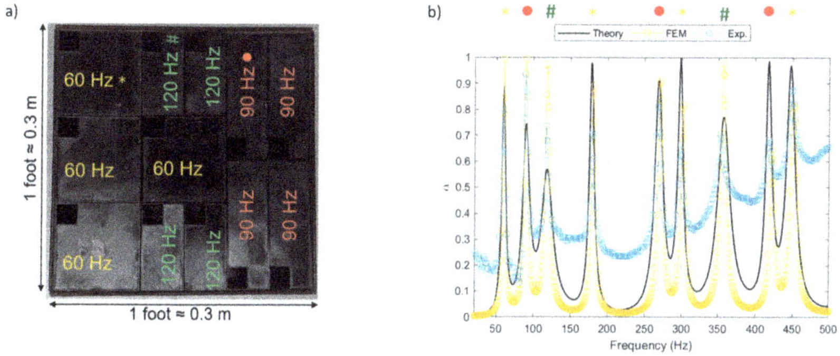

Figure 3: a) Picture of the multi-resonators box; b) Sound absorption of the labyrinth resonator box excited by a normally incident plane wave – comparison of analytical, numerical, and experimental results.

Acknowledgments

This research was supported by the Réseau Québec Maritime through the PLAINE research program (grant number PLAINE-2022PS04).

References

[1] B. M. Spessert, E.-A.-H. Jena, M. Fischer, and B. Kühn, "Combustion Engine Noise-a Historical Review, Internoise 2019, Madrid, Spain.

[2] A. Magnani, C. Marescotti, and F. Pompoli, "Acoustic absorption modeling of single and multiple coiled-up resonators," *Applied Acoustics*, vol. 186 (2022). https://doi.org/10.1016/j.apacoust.2021.108504

[3] C. Zwikker and C. W. Kosten, "Sound Absorbing Materials," *Elsevier Publishing company*, 1949.

[4] G. Catapane, D. Magliacano, G. Petrone, A. Casaburo, F. Franco, and S. De Rosa, "Labyrinth Resonator Design for Low-Frequency Acoustic Meta-Structures," in *Recent Trends in Wave Mechanics and Vibrations*, 2022, pp. 681–694. https://doi.org/10.1016/j.apacoust.2021.108504

[5] N. Atalla and F. Sgard, "Modeling of perforated plates and screens using rigid frame porous models," *J Sound Vib*, vol. 303 (1–2), 195–208 (2007). https://doi.org/10.1016/j.jsv.2007.01.012

Comparative study of shock response synthesis techniques for aerospace applications

Ada Ranieri[1,a *], Simone De Carolis[1,b], Giuseppe Carbone[1,c], Michele Dassisti[1,d], Andrea De Cesaris[2,e], Leonardo Soria[1,f]

[1] Polytechnic University of Bari, Bari, Italy

[2]Sitael, Mola di Bari (BA), Italy

[a]ada.ranieri@poliba.it, [b]simone.decarolis@poliba.it, [c]giuseppe.carbone@poliba.it, [d]michele.dassisti@poliba.it, [e]andrea.decesaris@sitael.com, [f]leonardo.soria@poliba.it

Keywords: Shock Response Spectrum, Synthesis, Wavelets, Damped Sinusoids, Enveloped Sinusoids

Abstract. The Shock Response Spectrum (SRS) is a widely used tool for analyzing and characterizing the response of mechanical systems to shock and transient events. In the aerospace industry, the SRS is used to compute the severity of the shock event on the electrical and optical equipment of a spacecraft. However, the SRS only provides magnitude information and does not retain temporal or phase information. Moving to the time domain is not a straightforward process because a time history has a unique SRS, but the converse is not true. Therefore, it is challenging to find the right time history that reproduces an SRS when simulating a given input profile using pyrotechnic devices or when computing the response to a shock input profile in the time-domain. For a given SRS an infinite combination of time pulses is possible. Synthesizing an SRS involves recovering a time-domain pulse that can accurately replicate the given SRS. There are many methods which are already widely utilized in the aerospace industry, including the use of damped sinusoids, enveloped sinusoids and wavelets. In this paper we compare different techniques, with the objective of identifying the most suitable method based on the considered frequency range and type of impulse. The case study under consideration is an SRS input profile corresponding to a real industrial case. Three artificial SRS accelerations have been generated to replicate the input, and the percentage errors of each method in comparison to the reference signal have been assessed. Further development will involve the use of optimization algorithms to generate the SRS profile with the smallest possible error.

Introduction

One of the most significant challenges in the space industry is the design and testing of aerospace structures and systems for reliable and safe operation in harsh environments, including the sudden and impulsive loads occurring during the launch phase. The most intense events are commonly caused by pyrotechnic devices actuating at the base of the spacecraft. The firing of these devices results in impulsive loads characterized by high peak acceleration, high-frequency content, and short duration. This poses a significant threat to the reliability and safety of electrical and optical components of the spacecraft, which are sensitive to high frequency loads. To demonstrate its compliance to shock requirements, the structure has to be tested by applying the shock load on the base interface. The accepted standard for implicit description of the pyroshock environment is the Shock Response Spectrum (SRS), which is a useful tool for estimating the damage potential of the shock pulse and for test level specification. The SRS finds its first applications in the 50's by the seismic and aerospace community. An SRS is generated by plotting in the frequency domain the peak response of a series of Single Degree of Freedom (SDoF) oscillating systems subjected to the same transient base acceleration input. The damping is usually assumed to be 5% (Q=10), while the natural frequency of each SDoF system is chosen to be different. The primary limitation of the SRS is its inability to provide temporal or phase information, as it only gives magnitude information. As a result, when subjecting a structure to electro-dynamic shaker testing for shock qualification, the SRS cannot be directly utilized [1]. Instead, it becomes necessary to synthesize an SRS-compatible acceleration time history. A similar challenge arises when analyzing nonlinear structures, where a modal approach is not feasible, and a modal transient analysis must be conducted to account for the phase among the peak responses of individual modes.

The aforementioned waveform can be obtained using a series of sinusoids [1,2] or wavelets [3], tailored to resemble an actual pyrotechnic shock pulse.

Shock Response Spectrum Synthesis

While a unique impulse in the time domain corresponds to a specific SRS, the opposite is not true. In fact, an SRS corresponds to an infinite number of possible pulses. As a result, there are several techniques available to obtain SRS-compatible acceleration time history. In this work we will investigate the accuracy of SRS synthesis throughout the summation of damped sines, enveloped sines and wavelets.

Wavelets. A wavelet is a discrete waveform of limited duration that is suited for approximating data with sharp discontinuities [4]. The original signal can be reconstructed as a summation of a set of wavelets with specified parameters. The equation of a single wavelet $W_m(t)$ is:

$$W_m(t) = \begin{cases} 0, \ \text{for } t < t_{dm} \\ A_m \sin\left[\frac{2\pi f_m}{N_m}(t - t_{dm})\right] \sin[2\pi f_m(t - t_{dm})] \ , \ \text{for } t_{dm} \le t \le \left[t_{dm} + \frac{N_m}{2f_m}\right] \\ 0, \ \text{for } t > \left[t_{dm} + \frac{N_m}{2f_m}\right] \end{cases}. \quad (1)$$

A discrete wavelet has a sinusoidal motion with a finite and odd number of half sine oscillations N_m with unique parameters for frequency f_m, amplitude A_m and time delay t_{dm}.

Damped sinusoids. The sinusoid approach shows a difference in the way the rise, peak and decay of the waveform is controlled, compared to the previously presented method. In this case the parameters to control are slightly different:

$$W_m(t) = \begin{cases} 0, \ \text{for } t < t_{dm} \\ A_m e^{-\xi_m 2\pi f_m (t - t_{dm})} \sin\left[\frac{2\pi f_m}{N_m}(t - t_{dm})\right] \sin[2\pi f_m(t - t_{dm})] \ , \ \text{for } t \ge t_{dm} \end{cases}. \quad (2)$$

It can be noted an extra term ξ_m, that is the damped sinusoid damping ratio.

Enveloped sinusoids. The enveloped sinusoids with random phase angles approach is similar to the one of damped sinusoids. The equation for enveloped sinusoids is given by:

$$W_m(t) = E(t) A_m \sin(2\pi f_m t + \varphi_m). \quad (3)$$

Where φ_m are random phase angles for each frequency n. The rise, plateau and decay of $W_m(t)$ is controlled by an envelope function $E(t)$ rather than damping.

For all the three methods, iterations for the parameters of a set of m waveforms a time t yield a synthesized acceleration that is expressed as:

$$\ddot{x}(t) = \sum_{m=1}^{N_m} W_m(t). \tag{4}$$

An example of a synthetized time history from the SRS input in Table 1 with a duration of $T = 0.06\ s$ can be seen in Fig.1.

Table 1. Shock load input

Frequency [Hz]	Amplitude [g]
100	56
1000	2820
10000	2820

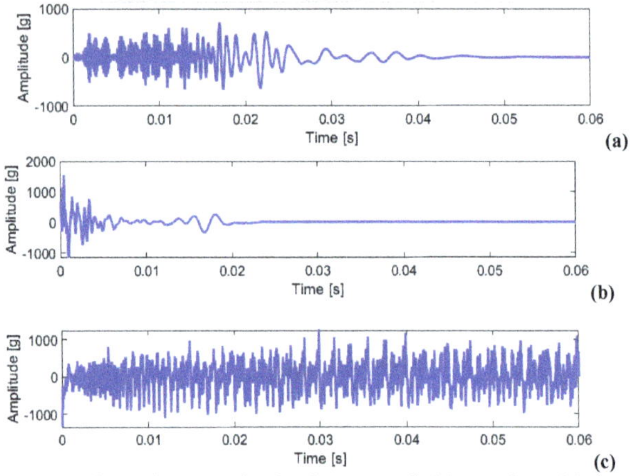

Figure 1. Reconstructed time history of a shock input with (a) wavelets, (b) damped sines, (c) enveloped sines

The synthetized accelerations have been converted to SRS and compared to the reference input as shown in Figure 2.

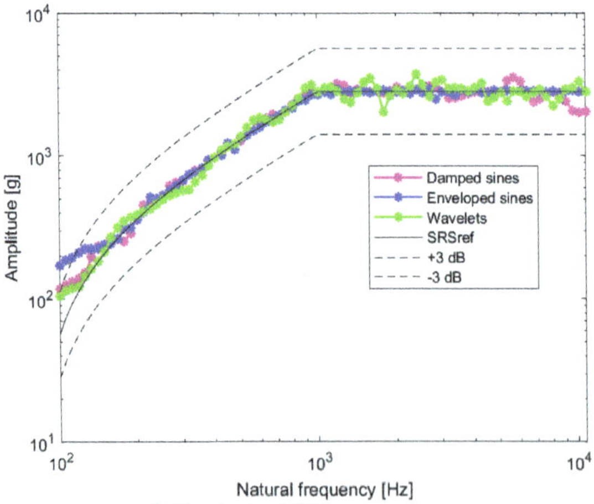

2. Synthetized SRS comparison

Table 2. Synthesis correlation coefficient

	Damped sines	Enveloped sines	Wavelets
100-200 Hz	0.98186	0.99566	0.99773
200-1000 Hz	0.97651	0.98699	0.9659
1000-10K Hz	0.98934	0.99421	0.98707

Furthermore, the Synthesis Correlation Coefficient (COR) [5] in Table 2 has been computed in low, middle and high frequency range to compare the efficiency.

$$COR = \frac{|\sum_{f_1}^{f_2} SRS_r(f_n) SRS_s(f_n)|^2}{\sum_{f_1}^{f_2} SRS_r(f_n)^2 \sum_{f_1}^{f_2} SRS_s(f_n)^2}. \quad (5)$$

Where SRS_r and SRS_s are respectively the reference and synthetized SRS. Globally, a good level of accuracy (near the unity) has been achieved. In particular, the enveloped sines method seems to be the most effective. It can be observed that the methods are less accurate in the middle frequency range (200-1000 Hz).

Conclusions and future developments

In conclusion, the investigated techniques, namely the summation of damped sines, enveloped sines, and wavelets, have shown good levels of accuracy in reproducing the desired SRS input. Further studies should be conducted by exploring different parameter settings and types of input profiles to enhance the understanding of these techniques. Additionally, the development of an optimization algorithm, such as the least square fitting method or genetic algorithm, should be pursued to combine the methods and synthesize a single SRS that minimizes the error and achieves a higher level of accuracy.

References

[1] T. Irvine, Shock Response Spectrum Synthesis Via Damped Sinusoids, Vibrationdata, 2012

[3] J. E. Alexander, A New Method to Synthesize an SRS Compatible Base Acceleration with Energy and Temporal Moments to Improve MDOF System Response, IMAC 36 Orlando, 7 Sound & Vibration, 2018.

[2] T. Irvine, Shock Response Spectrum Synthesis Via Wavelets, Vibrationdata, 2000.

[4] A. Grasps, An Introduction to Wavelets, IEEE Computational Science and Engineering vol. 2, num. 2, published by the IEEE Computer Society, Los Alamitos, 1995. https://doi.org/10.1109/99.388960

[5] R.G. Allemang, Vibrations: Experimental Modal Analysis, 1999. UC-SDRL-CN-20-263-662

Vibro-acoustic analysis of additively manufactured acoustic metamaterial via CUF adaptive finite elements

M. Rossi[1a], M.C. Moruzzi[1b*], S. Bagassi[1c], M. Corsi[1d], M. Cinefra[2e]

[1]Department of Industrial Engineering, University of Bologna, Italy

[2]Department of Mechanical, Mathematics and Management, Polytechnic of Bari, Italy

[a]mattia.rossi17@studio.unibo.it, [b]martinocarlo.moruzz2@unibo.it, [c]sara.bagassi@unibo.it, [d]marzia.corsi2@unibo.it, [e]maria.cinefra@poliba.it

Keywords: Aircraft Noise, Vibro-acoustics, Acoustic Metamaterials, Carrera's Unified Formulation

Abstract. In the field of noise and vibrations control inside the cabin, passive noise solutions coupled with the development of new unconventional materials, called Acoustic Metamaterials (AMMs) can be very promising to stop incoming noise and guarantee the passenger's comfort without an increase in aircraft weight. Within the framework of Carrera's Unified Formulation (CUF), we study the acoustic properties of double pierced AMM plate printed with Fused deposition modelling technique (FDM). The influence of several geometrical parameters is investigated, such as the size and location of the holes and the perforation ratio. The properties of this AMM are derived from vibro-acoustic analyses of the finite element software, Mul2, developed by *Politecnico di Torino*, that exploits the CUF. In order to study the AMM complex structure in the CUF framework, the Adaptive finite elements are exploited. This new class of 2D elements, recently developed, allows us to model with shell elements the AMM structure, which presents several discontinuities in the mid-surface due to the presence of corners and internal cavities.

Introduction

The problem of noise in commercial aircraft is usually related to the emission of sound waves outwards, resulting in noise pollution problems in areas near airports or landing and take-off paths [1]. However, noise is also an internal problem of the aircraft, where sound waves follow paths internal to the aircraft structures or are generated by sources that are already internal (e.g., the air conditioning system) [2]. Internal noise negatively affects the comfort of passengers and crew during flight, weakening the competitiveness of the aircraft compared to other methods of travel.

In order to decrease the amount of noise in the passenger cabin, there are several proposed solutions. The simplest solution in terms of both implementation and logic leads to shielding the passenger cabin by means of insulating panels in the cavity between the fuselage and the cabin panel [3]. In this work, we focus on the low and medium frequencies that are traditionally difficult to shield with conventional materials. For this reason, it was decided to use materials with special acoustic properties, acoustic metamaterials (AMMs). With AMMs, high absorption coefficients in a certain frequency range can be achieved with lightweight structures by using appropriate elementary cell geometries repeated periodically in the absorber panel. However, these new materials present problems related to the field of numerical analysis compared to the results of the current structure. The use of numerical models, the Finite Element Method (FEM) for the low and medium frequencies, is essential both in the preliminary stages to have extensive design flexibility, but even more in the later stages to study the AMM in the aircraft environment, where the analysis is scaled from the characterization of a simple plate to that of several panels in the aircraft fuselage.

The major problem is in the process of numerical homogenization of the AMM, which brings with it the hypothesis of a perfect repetition of the elementary cell. Small manufacturing errors can lead to significant differences between experimental and numerical results. To avoid this problem, the following work has chosen to use an AMM produced with additive printing based on literature [4], that can guarantee high precision in the production of the AMM and the study of a geometry difficult to obtain with conventional production methods.

A second problem is the numerical method used in the finite element field to solve the vibro-acoustic problem. Indeed, as will be further described, the proposed AMM has internal cavities pierced with the presence of double fluid-structure interfaces that would require the use of solid elements with conventional FEM methods. In order to avoid an excessive number of Degrees of Freedom (DoF) without losing accuracy, Carrera's Unified Formulation (CUF) has been chosen [5], which allows us to exploit different theories for plates, both Equivalent Single Layer (ESL) and Layer Wise (LW). The geometry of the material also required integration into the formulation of a new class of elements, the Adaptive finite elements [6, 7], to handle corners and intersections between the various elements.

Vibro-acoustic problem in the CUF-FEM framework

The vibro-acoustic problem that describes the AMM of this work is that it includes an elastic structure coupled to internal cavities. The following system must then be resolved before information on dynamic behaviour and Transmission Loss (TL) can be obtained:

$$\begin{bmatrix} -\omega^2\boldsymbol{M}+\boldsymbol{K} & \boldsymbol{S} \\ -\rho_f\omega^2\boldsymbol{S} & -\omega^2\boldsymbol{Q}+\boldsymbol{H} \end{bmatrix} \cdot \begin{Bmatrix} U \\ P \end{Bmatrix} = \begin{Bmatrix} \boldsymbol{F} \\ 0 \end{Bmatrix}$$

where we have the mass matrix **M**, **Q** and stiffness **K**, **H** of structure and fluid respectively, the coupling matrix **S** and the vector of the external loads on the structure **F**. The previous matrices and vectors are defined according to the fundamental nuclei formulation [8]. The vectors of the unknown are U and P, the three displacements and the pressure respectively. The properties of the fluid are defined by the density ρ_f and speed of the sound c_f, while those of the structure are defined by the elasticity matrix **C**. The problem is solved in the frequency domain, represented by the angular frequency ω. In the FEM approximation, the unknowns are the nodal values, the transition to the continuous field is guaranteed by the shape functions N.

According to the CUF, the three-dimensional field of a shell can be split into a two-dimensional field on the shell plane and an expansion on the thickness F_τ, called thickness function [5]. The choice of thickness functions leads to the use of different plate models. In this work, the Lagrange polynomials are used to have a Layer Wise approach. In addition, the integration of the Adaptive finite element leads to the use of a new three-dimensional shape function L, which combines the shape and thickness functions, in the calculation of the integral and the Jacobian. Then the displacement field is defined by the following equation:

$$u_\tau^k(x,y,z) = \left(N_i F_\tau^{ki}\right)\boldsymbol{\mathcal{U}}_{\tau i}^k = L_{\tau i}^k(x,y,z)\boldsymbol{\mathcal{U}}_{\tau i}^k$$

For the pressure field, being defined by three-dimensional elements, the conventional FEM formulation is applied.

Acoustic solution

The design of sandwich panels has been the subject of various studies during the past decades [9]. Since the panels' faces are not designed to absorb sound, the choice of the core turned out to be particularly crucial for the reduction of noise [10]. Instead, by using plates with small holes and an air cavity in the centre, the so-called micro-perforated panels (MPPs), the sandwich plate can have a positive impact on noise reduction, improving sound absorption. The research by Meng et al. [4] is chosen as a reference since it focuses on the influences of holes' diameter of an additively produced AMM.

It comes to light that the porosity, location, and diameter of the holes all affected the sound transmission loss. Following this direction, various AMM models have been developed to comprehend how the noise operates at low- to middle-frequency levels in various design parameter combinations. The dimension of the holes, the position of the holes, and the perforation ratio are the three main parameters of interest for the model definition in this study. The porosity or perforation ratio $PR_{\%}$, which is typically given in percentages, describes how much of a plate is made up of holes:

$$A_h = PR_{\%} \cdot A_e$$

where A_h is the circular area of the holes and A_e is the elementary cell area.

Then, a series of models are created, shown in Tab. 1 and the third one is reported in Fig. 1(a). The production method chosen is the fused deposition modelling technique (FDM) with a Fortus 250 mc and ABS Plus p430, as a material, on which preliminary studies have already been carried out [11-13]. Although it may not be one of the most precise printing techniques, it allows a sufficiently robust structure to be obtained and to direct the filament deposition path to increase the accuracy of the AMM's holes, especially on the oblique plates. All models were printed correctly except the fourth due to the excessively small size of the holes (i.e., 0.5 mm). Fig. 1(b) shows one of the printed models.

Table 1: AMM models' properties, the first model is not pierced.

Models	AMM density [kg/m^3]	$PR_{\%}$	Holes diameter [mm]	Holes location
1	1040	-	-	No holes
2	1039	0.349	1	Upper plate
3	1037	0.349	1	Upper plate & core
4	1037	0.349	0.5	Upper plate & core
5	1037	0.698	2	Upper plate & core
6	1035	0.175	1	Upper plate & core
7	1039	0.175	1	Upper plate & core

The numerical study of the models allows us to derive a priori the TL. In order to avoid errors, the model has been pre-validated by comparing its natural frequencies with those obtained from commercial software. Then, a direct frequency analysis is performed on a simply supported AMM plate, loaded with constant amplitude load (1 N). The results in terms of TL show that the presence of the holes, their size and position is decisive in defining the acoustic behaviour of the AMM. Among the various results reported in a complete way in Rossi's work [14], it is interesting to note that the third and fifth models (which have the same homogenized density) have significant TLs between 1000 and 1900 Hz in Fig. 2, although for the fifth with several peaks and discontinuities. While the fundamental frequency of the seventh model is the highest among the six models. The fourth model was not produced; therefore, it was not studied.

Conclusions

This work is intended to provide a preliminary basis for further study of AMMs produced with additive techniques for noise reduction in aircraft. The results demonstrate the excellent acoustic properties of these structures and the possibility to study them with innovative numerical methods.

In the future, it is necessary to validate the numerical results through experimental tests in the anechoic chamber or with an impedance tube and then scale the problem to study the AMM in the working environment. Finally, FDM is only one of the available additive techniques. The advantages or disadvantages of other technologies for AMM production need to be evaluated with appropriate research.

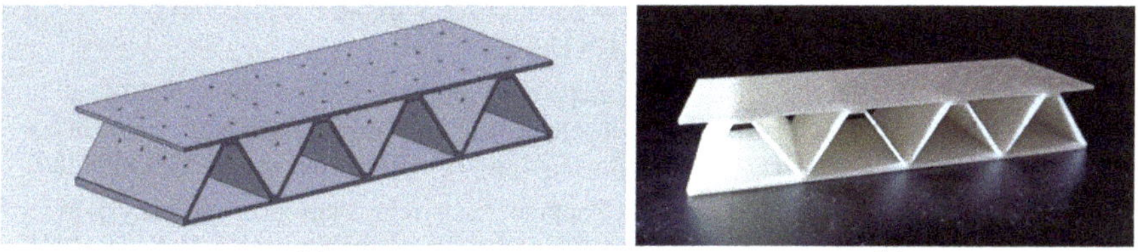

(a) (b)

Figure 2: The AMM designed in this work and based on the work by Meng et al [4]. (a) The third model isometric representation. (b) The first model after the production process.

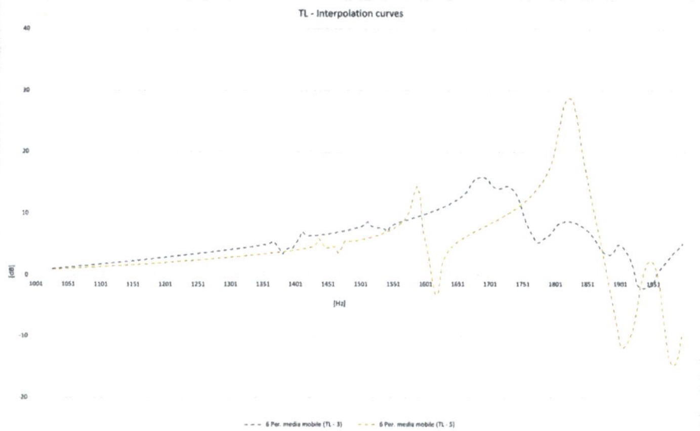

Figure 2: The average TL [dB] from 1000 Hz to 2000 Hz for the third model (grey line) and the fifth one (yellow line).

References

[1] LL. Beranek. The noisy dawn of the jet age. Sound and Vibration, 41:94–99, 2007.

[2] N. Hu, H. Buchholz, M. Herr, C. Spehr, and S. Haxter. Contributions of different aeroacoustic sources to aircraft cabin noise. In 19th AIAA/CEAS Aeroacoustics Conference, 2013. https://doi.org/10.2514/6.2013-2030

[3] MC. Moruzzi, M. Cinefra, and S. Bagassi. Vibroacoustic analysis of an innovative windowless cabin with metamaterial trim panels in regional turboprops. Mechanics of Advanced Materials and Structures, 28:1–13, 2019. https://doi.org/10.1080/15376494.2019.1682729

[4] H. Meng, MA. Galland, M. Ichchou, O. Bareille, FX. Xin, TJ. Lu. Small perforations in corrugated sandwich panel significantly enhance low frequency sound absorption and

transmission loss, Composite Structures, Volume 182: 1-11, 2017. https://doi.org/10.1016/j.compstruct.2017.08.103

[5] E. Carrera, M. Cinefra, M. Petrolo, and E. Zappino. Finite element analysis of structures through unified formulation. Wiley, 2014. https://doi.org/10.1002/9781118536643

[6] M. Cinefra. Non-conventional 1d and 2d finite elements based on cuf for the analysis of non-orthogonal geometries. European Journal of Mechanics/A Solids, 88:104273, 2021. https://doi.org/10.1016/j.euromechsol.2021.104273

[7] MC. Moruzzi, M. Cinefra, S. Bagassi, E. Zappino. Vibro-acoustic analysis of multi-layer cylindrical shell-cavity systems via CUF finite elements, 33rd Congress of the international council of the aeronautical sciences, 2022. https://doi.org/10.1016/j.compstruct.2020.113428

[8] M. Cinefra, MC. Moruzzi, S. Bagassi, E. Zappino, E. Carrera. Vibro-acoustic analysis of composite plate-cavity systems via CUF finite elements, Composite Structures, Vol. 259, 2021. https://doi.org/10.1016/j.compstruct.2020.113428

[9] V. Birman and GA. Kardomateas. Review of current trends in research and applications of sandwich structures. In: Composites Part B: Engineering 142.221–240, 2018. https://doi.org/10.1016/j.compositesb.2018.01.027

[10] MP. Arunkumar, J. Pitchaimani, KV. Gangadharan, MC. Lenin Babu. Influence of nature of core on vibro acoustic behavior of sandwich aerospace structures, Aerospace Science and Technology, Volume 56: 155-167, 2016. https://doi.org/10.1016/j.ast.2016.07.009

[11] M. Corsi, S. Bagassi, MC. Moruzzi, and L. Seccia. Correlation between production parameters and mechanical properties of ABS Plus p430 fused deposition material. Proceedings of the Institution of Mechanical Engineers, Part C: Journal of Mechanical Engineering Science, 236(5):2478–2487, Mar 2022. https://doi.org/10.1177/09544062211027205

[12] MC. Moruzzi, S. Bagassi, M. Cinefra, M. Corsi, M. Rossi. Design of additively manufactured metamaterial for cabin noise and vibrations reduction, XXVI AIDAA International Congress, 2021.

[13] H. Rezayat, W. Zhou, A. Siriruk, D. Penumadu, and SS. Babu. Structure–mechanical property relationship in fused deposition modelling. Materials Science and Technology, 31(8):895–903, 2015. https://doi.org/10.1179/1743284715Y.0000000010

[14] M. Rossi. Vibro-acoustic analysis of additively manufactured acoustic metamaterials. Ms Thesis. University of Bologna, 2022.

XR and Human Factors for Future Air Mobility

A brief review of pilots' workload assessment using flight simulators: subjective and objective metrics

Giuseppe Iacolino[1, a *], Antonio Esposito[1, b], Calogero Orlando[1, c] and Andrea Alaimo[1, d]

[1]Faculty of Engineering and Architecture, Kore University of Enna, 94100 Enna, EN, Italy

[a]giuseppe.iacolino@unikorestudent.it, [b]antonio.esposito@unikore.it, [c]calogero.orlando@unikore.it, [d]andrea.alaimo@unikore.it

Keywords: Human Factor, Subjective Methods, Objective Methods, Full Flight Simulator

Abstract. This study explores subjective and objective metrics to assess pilots WL, with a particular focus on the use of full flight simulators (FFS). The results show that FFS-based research demonstrates no significant differences compared to real flight experiences, highlighting the validity of FFS as a tool for studying pilots' performance. However, further research is needed to understand the impact of other parameters on pilots' performance, and to address human factor-related risks for enhanced aviation safety.

Introduction

Nowadays, workload (WL) can be considered one of the biggest challenges in the aeronautical field. The continuous growth of air traffic has led pilots to work under pressure and for extended periods of time. As reported in [1], 75% of aircraft accidents are linked to human errors and most of them are related to high levels of mental WL and fatigue. This is important in critical situations such as take-off, landing or emergency procedures. WL is associated with the onset of stress, a phenomenon with multiple facets ranging from neuroendocrine to psychological [2]. Pilots require training to effectively respond to both internal and external stimuli. Hence, using a flight simulator represents the most reliable, safe, time and cost savings way to study pilot's performances today [3]. Consequentially, it is necessary to employ some metrics that can better capture the pilots' stress level while performing the task. To do this, subjective and objective methods are widely exploited, and this work aims to describe such measures, with particular attention paid to the studies conducted using a full flight simulator (FFS).

Subjective and Objective Metrics

The concept of mental WL is described in [4] as a collection of mental, and composite brain states that influence human performance in various perceptual, cognitive, and sensorimotor abilities. Thus, it is crucial to choose the right metrics to evaluate it. Subjective and objective measures represent an important contribution on pilots' WL level evaluation. In detail, subjective WL measures comprise some tests which are administered while, before or after the performance. Being such subjective, they can be highly influenced by the pilot's psychological condition. That's why it is good practice to link them with the objective ones which represent the physical condition of the subject. The objective measures are mainly obtained via sensors which can detect some physiological parameters. In this work, subjective methods will be described first and objective ones after. The most reliable subjective methods are here mentioned.

The *NASA Task Load Index (TLX)* was first developed by Hart and Staveland in 1988. The TLX is widely recognized and extensively employed globally. The test has three stages to evaluate the pilot's WL. The first one obtains a global index across six dimensions: mental, physical, time demands, performance, effort, and frustration. Each dimension is divided into twenty intervals. In the second stage, pairs of dimensions are assessed, determining the ones with the greatest impact on WL. The third stage assigns scores to selected dimensions, with 0 for insignificant dimensions and 5 for the most important one [5]. On the other hand, the *Subjective Workload Assessment Technique (SWAT)* involves obtaining subjective ratings from participants regarding three dimensions of WL: time load, mental effort load, and psychological stress load. During the activity, participants are encouraged to assess their workload across three dimensions and rate it on a scale of low, medium, or high. A mental workload range scale is then calculated by combining the scores from all three dimensions. This assessment method relies on the self-report of participants [6]. Other subjective techniques are not mentioned for the sake of conciseness.

Regarding the objective metrics, some of them are reported here as well. As evidence, in [7] the eye movement provides indications on the pilot's visual perception, or [8] aimed to investigate the potential utility of α −Amy levels as a biomarker for stress in pilots operating within a high-stress environment. However, the heart rate variability (HRV) is one of the most employed parameters today. HRV is the spontaneous fluctuation in time between consecutive heartbeats, as measured by the distance between two successive R peaks on an electrocardiogram. This parameter can be studied in both frequency and time domain. In particular, the time domain measures also include the NN interval series standard deviation (SDNN). Based on the existing literature, a reduction in this parameter indicates an elevation in both mental WL and physical demands [9]. As previously mentioned, the study of HRV extends to the frequency domain. In this context, the fast Fourier transform is employed to estimate the Power Spectral Density associated with frequency bands, specifically focusing on high frequency and low frequency. These frequency bands provide valuable information about sympathetic and parasympathetic activities [10].

FFS for Human Factor Studies

Due to the rarity of FFS, there is a limited body of literature that specifically addresses their utilization. Most studies, in fact, primarily focus on simulators that provide visual cues to pilots through screens but lack the comprehensive flight sensation that an FFS can deliver. In 2018, the European Union Aviation Safety Agency (EASA) released the latest specifications for Airplane Flight Simulation Training devices (CSFSTD-A), which classify flight simulators into four levels of qualification: A, B, C, and D. The D-level FFS are the most advanced and reliable. For instance, to achieve a D-level qualification, the FFS must include a real-time feedback tool that allows the instructor to monitor the training envelope and prevent the airplane's operating limits from being exceeded. In [11] it is reported that acute effects on HRV and anxiety during a real-time flight are not significantly different from those experienced in a simulator, indicating that the simulated task planning and design closely approximate real-world conditions; in this case, an operational F-5M by Indra Company flight simulator was used. It is also reported that future flight simulators should incorporate immersive virtual reality technology simulating G forces and vibration. Nevertheless, in today's context, D-level FFS can be used to do reliable research and prevent in real-life risky decisions. In essence, pilots must be conscious on which kind of decisions they have to take while flying, both in high and normal WL condition. Additionally, the study conducted in [12] employed the CESSNA Citation C560 XLS FFS. The main findings of the study demonstrated that WL levels can be effectively differentiated by analyzing various domains of HRV, including time, frequency, and non-linear measures. These results highlight the importance of HRV indexes in assessing WL, and suggest the potential development of real-time, non-invasive instruments for evaluating it. In a separate study [13], the application of an FFS was investigated within typical flight scenarios. The research collected pilots' objective and subjective metrics to establish an evaluation index system. Research has revealed that pilot's errors have become a significant obstacle, impeding the progress towards enhancing flight safety within the aviation industry. The study also found some deficiencies in the design layout of the cockpit as in the flight crew operation process that cause pilots to make mistakes easily. The authors proposed that *designing aircrafts suitable* for pilots instead of *aircraft requiring pilots to adapt* can make the civil aircraft cockpit more humanized and highly automated. Also, the authors proposed a feasible method to analyze human factors (HF) in typical faults and incidents of transport category airplanes using comprehensive methods that combine subjective and objective measurements. In [14] it is investigated the correlation between subjective and objective indicators of fatigue, factors such as WL and work scheduling. It also examined whether the WL experienced by pilots during a simulator mission could serve as a moderator for increased fatigue after the mission. The study utilized a JAR STD 1A FFS. Results indicated that both subjective and objective measures of fatigue significantly rose during the three-hour experimental procedure. These findings suggest the presence of enduring effects from sleep deficit and propose a multifactorial model for assessing fatigue risks. In [15] two subjective WL measurements and three psychophysiological measurements were compared in both a simulator and a flight test. The comparisons were made across three flight scenarios using an ARJ21-700 FFS and a corresponding aircraft. Both flight scenarios and the flight environment significantly influenced NASA-TLX, eye blink rate, and HRV. Moreover, strong correlations were observed between the NASA-TLX and HRV, between the simulator and the flight test. These findings suggest that NASA-TLX and HRV can serve as consistent measures of WL in both FFS and real flight tests.

Conclusions

This short review aimed to highlight some subjective and objective metrics in order to understand the impact of stress on pilots' performance. The findings indicate that subjective and objective metrics play an important role for better understanding pilots' WL level. Furthermore, research involving FFS has shown promising results and benefits when compared to real flight experiences. Additionally, to emphasize the need for more studies is important. Despite current high safety standards, incidents related to HF still occur, highlighting the importance of ongoing investigation and improvement. By conducting more research and continuously improving understanding, pilot performance can be enhanced, risks can be mitigated, and safer aviation operations can be ensured for everyone involved.

Acknowledgment

This work was partially funded by the project "SAMOTHRACE" (Sicilian MicronanoTech Research And Innovation Center) – Funds: MUR, PNRR-M4C2, ECS_00000022, activity of Spoke 3 affiliated partner Kore University of Enna: WP3.4 SAfety MObility Sensors TEsting Platform - SA.MO.S.TE.P.

References

[1] Kharoufah, H., Murray, J., Baxter, G., & Wild, G. (2018). A review of human factors causations in commercial air transport accidents and incidents: From to 2000–2016. *Progress in Aerospace Sciences*, *99*, 1-13. https://doi.org/10.1016/j.paerosci.2018.03.002

[2] Masi, G., Amprimo, G., Ferraris, C., & Priano, L. (2023). Stress and Workload Assessment in Aviation—A Narrative Review. *Sensors*, *23*(7), 3556. https://doi.org/10.3390/s23073556

[3] Mohanavelu, K., Poonguzhali, S., Ravi, D., Singh, P. K., Mahajabin, M., Ramachandran, K., & Jayaraman, S. (2020). Cognitive Workload Analysis of Fighter Aircraft Pilots in Flight Simulator Environment. *Defence Science Journal*, *70*(2). https://doi.org/10.14429/dsj.70.14539

[4] Kramer, A. F., & Parasuraman, R. (2007). Neuroergonomics: Applications of neuroscience to human factors.

[5] Hart, S. G., & Staveland, L. E. (1988). Development of NASA-TLX (Task Load Index): Results of empirical and theoretical research. In *Advances in psychology* (Vol. 52, pp. 139-183). North-Holland. https://doi.org/10.1016/S0166-4115(08)62386-9

[6] Reid, G. B., & Nygren, T. E. (1988). The subjective workload assessment technique: A scaling procedure for measuring mental workload. In *Advances in psychology* (Vol. 52, pp. 185-218). North-Holland. https://doi.org/10.1016/S0166-4115(08)62387-0

[7] Oberhauser, M., & Dreyer, D. (2017). A virtual reality flight simulator for human factors engineering. *Cognition, Technology & Work*, *19*, 263-277. https://doi.org/10.1007/s10111-017-0421-7

[8] Iizuka, N., Awano, S., & Ansai, T. (2012). Salivary alpha-amylase activity and stress in Japan air self-defense force cargo pilots involved in Iraq reconstruction. *American Journal of Human Biology*, *24*(4), 468-472. https://doi.org/10.1002/ajhb.22247

[9] Watson, D. W. (2001, June). Physiological correlates of heart rate variability (HRV) and the subjective assessment of workload and fatigue in-flight crew: a practical study. In *2001 People in Control. The Second International Conference on Human Interfaces in Control Rooms, Cockpits and Command Centres* (pp. 159-163). IET. https://doi.org/10.1049/cp:20010453

[10] Bilan, A., Witczak, A., Palusiński, R., Myśliński, W., & Hanzlik, J. (2005). Circadian rhythm of spectral indices of heart rate variability in healthy subjects. *Journal of electrocardiology*, *38*(3), 239-243. https://doi.org/10.1016/j.jelectrocard.2005.01.012

[11] Fuentes-García, J. P., Clemente-Suárez, V. J., Marazuela-Martínez, M. Á., Tornero-Aguilera, J. F., & Villafaina, S. (2021). Impact of real and simulated flights on psychophysiological response of military pilots. *International Journal of Environmental Research and Public Health*, *18*(2), 787. https://doi.org/10.3390/ijerph18020787

[12] Alaimo, A., Esposito, A., Faraci, P., Orlando, C., & Valenti, G. D. (2022). Human Heart-Related Indexes Behavior Study for Aircraft Pilots Allowable Workload Level Assessment. *IEEE Access*, *10*, 16088-16100. https://doi.org/10.1109/ACCESS.2022.3145043

[13] Xing, G., Sun, Y., He, F., Wei, P., Wu, S., Ren, H., & Chen, Z. (2023). Analysis of Human Factors in Typical Accident Tests of Certain Type Flight Simulator. *Sustainability*, *15*(3), 2791. https://doi.org/10.3390/su15032791

[14] Hörmann, H. J., Gontar, P., & Haslbeck, A. (2015). Effects of workload on measures of sustained attention during a flight simulator night mission.

[15] Zheng, Y., Lu, Y., Jie, Y., & Fu, S. (2019). Predicting workload experienced in a flight test by measuring workload in a flight simulator. *Aerospace medicine and human performance*, *90*(7), 618-623. https://doi.org/10.3357/AMHP.5350.2019

Applying an interior VR co-design approach for the medical deployment vehicle of the future

J. Herzig[1,a], F. Reimer[1,b], M. Lindlar[2,c], P. Weiand[3,d], J. Biedermann[1,e], F. Meller[1,f], B. Nagel[1,g]

[1]Institute of System Architectures in Aeronautics, Department Cabin and Payload Systems, German Aerospace Center e.V. (DLR), Hein-Saß-Weg 22, 21129 Hamburg, Germany

[2]Institute of Aerospace Medicine, Department Sleep and Human Factors Research, German Aerospace Center e.V. (DLR), Linder Höhe, 51147 Köln, Germany

[3]Institute of Flight Systems, Department Rotorcraft, German Aerospace Center e.V. (DLR), Lilienthalpl. 7, 38108 Braunschweig, Germany

[a]jessica.herzig@dlr.de, [b]fabian.reimer@dlr.de, [c]markus.lindlar@dlr.de, [d]peter.weiand@dlr.de, [e]jörn.biedermann@dlr.de, [f]frank.meller@dlr.de, [g]björn.nagel@dlr.de

Keywords: VR, Co-Design, Cabin Design, User Centered

Introduction

The early arrival of qualified medical personnel at the scene of an accident is essential for a successful and effective first-aid treatment of emergency patients. Due to an increasing shortage of emergency medical personnel, as well as a decreasing hospital density in Germany, the continuity of medical care will be more challenging in the future. [1] [2]

According to a study by the Bertelsmann Foundation in the summer of 2019, the density of hospitals and clinics in Germany could be reduced from the current number of 1900 to 600. This immediately leads to the concentration of healthcare centers, making hospitals no longer equally accessible to the entire population. [3]

The resulting and extended time of arrival at the place of an accident immensely affects the adherence of the prehospital time and help time, which can seriously affect the health condition of the emergency patient. [1]

To ensure emergency medical care in the future, especially in structurally weak or densely populated regions, the DLR (German Aerospace Center) has set itself the goal of the "Rescue Helicopter 2030", to develop the aircraft of air rescue in Germany under the aspects of the future rescue service and to present new concepts.

Currently, research is being conducted on the concept of a medical deployment vehicle. This vehicle aims to transport medical professionals to the accident site in the shortest possible time to provide on-site initial care until the ambulance arrives for further transportation.

To maximize flight speed and range, this future rescue helicopter has been significantly reduced in size and weight.

The design of a tailored cabin concept for this purpose is characterized by high complexity in all aspects. In addition to the technical and mission-specific requirements of the helicopter system, new cabin designs must meet the individual needs to provide a high level of functionality and usability for all user groups. Furthermore, prototyping, planning, and conducting user tests are extremely time-consuming and costly, which increases the challenges in the development process [4]. To develop novel and highly complex cabin concepts that closely align with user requirements, the combination of user-centered design thinking methodology [5] and an immersive prototyping and feedback process has proven to be an effective approach [6] [7]

According to Burkett et. al., the concept of Co-design offers an approach that engages consumers and product users in the design process, aiming to foster enhancements and drive innovation. [8]. Furthermore, co-creation is widely acknowledged as "practices where a design practice and one or more communities of practice participate in creating new desired futures" (Lee, 2018) [9]. Co-design involves the empowerment of individuals, granting them the opportunity to exert substantial influence over designs. User groups are regarded as central experts, uniquely positioned to contribute their expertise and insights throughout the entire process. [10]

The focus of this approach is on implementing a VR Co-Design process through the execution and evaluation of user workshops, with a particular emphasis on medical personnel and pilots with experience in air rescue.

Method

In the initial development phase, information regarding general experiences and challenges in air rescue were gathered from practicing pilots and emergency physicians. In addition to a subsequent description of the medical deployment vehicle concept, participants had the opportunity to familiarize themselves with the project and list initial requirements for the concept based on their own experiences. [10]

Participation in an online questionnaire and an initial engagement with the concept, as well as raising awareness about the topic, were fundamental prerequisites for attending a collaborative online workshop.

The following section will describe the structure of the study and the key findings of the second phase.

PARTICIPANTS

The online survey was completed by 25 participants from the air rescue sector,[1] including four pilots, six emergency medical technicians (HEMS TC), and 15 doctors. The age of the participants ranged from 22 to 60 years, with a mean age of 39.4 years.

After 37.5% of the online questionnaire respondents expressed interest in participating in a Co-Design workshop, a total of two workshops were conducted, each with three participants.

Special attention was given to the composition of the participants, aiming to bring together different user groups to combine diverse perspectives in the collaborative development process. The composition of the groups was as follows:

- Group 1: Person A (emergency physician), Person B (pilot), Person C (medical technician)
- Group 2. Person A (emergency physician), Person B (pilot), Person C (engineer)

PROCEDURE

At the beginning of the Co- Design workshop, the participants were presented with the requirements and mission scenario for the medical deployment vehicle. The interactive online tool *Conceptboard* was used for this purpose. Subsequently, all participants were asked to conceptually design an initial cabin layout based on their own requirements within ten minutes.

[1] The online survey was internally circulated among the employees of ADAC (Allgemeine Deutsche Automobil-Club e.V)

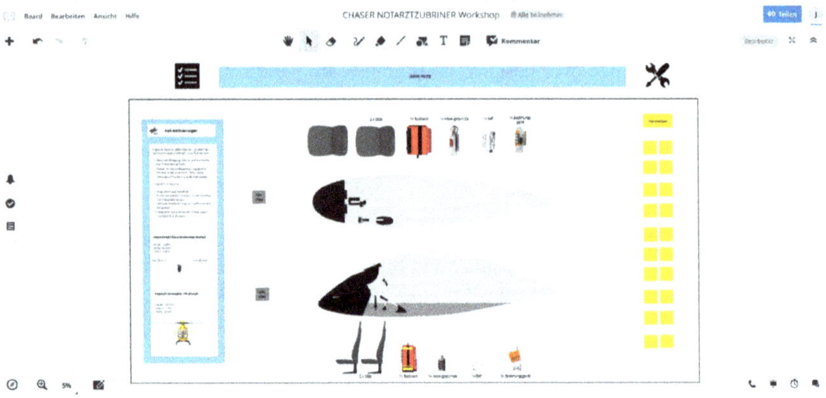

Figure 1: Example of the concept board work space for each participant including a description of the scenario, requirements and mission (left), the layout in two perspectives including interior parts and equipment (center) and notes for ideas (right)

Each participant received their own model where they could place objects on the helicopter floor plan. Additionally, there was an option to use post-it notes to add missing objects and include comments (see Figure 1). After presenting their concepts, the participants' designs were photographed and imported into the VR modeling program *Gravity Sketch*.

The 2D concepts created by the participants were placed in *Gravity Sketch* alongside the 3D model of the helicopter to provide an overview of the concepts and allow for a collaborative immersive incorporation of selected designs into the model.

A 3D cabin concept developed by the DLR cabin design team was positioned alongside as a comparative reference to the corresponding 2D solution approaches. Participation in the virtual reality session was facilitated through screen sharing of the collaboration app of *Gravity Sketch*.

After the participants were taken through the model via the shared screen, concept proposals could be expressed and implemented in real-time within the VR design model. With the aid of the newly gained spatial perspective, the concepts could be collectively reviewed, reevaluated, and optimized from the viewpoints of the three different professional groups regarding the usability of functions and positioning of modules (see Figure 2).

Figure 2: VR design concept as baseline for participants evaluation and further development process

Results

The workshop participants evaluated the Co-Design process of the medical deployment vehicle concept in virtual reality as an efficient and effective design method. The VR representation provided participants with an enhanced spatial awareness, making sizes, distances, and positions more comprehensible. This enabled immersive and rapid prototyping and efficient facilitated the evaluation of potential use cases.

The combination of the Co-Design process with immersive prototyping and optimization in virtual reality enables new and more effective design possibilities for user-centered and targeted cabin design.

This method allows for time and cost savings, as initial concepts can be developed in a short period. Additionally, it facilitates the direct implementation of experiences and requirements from different user groups into a virtual product, integrating the feedback and optimization process directly into the product. The findings of this paper provide a basis for adapting the method to other concepts with an expanded range of requirements. In addition to involving a higher number of participating stakeholder groups, direct integration into virtual reality is also conceivable.

References

[1] ADAC Luftrettung GmbH, "Multikopter im Rettungsdienst," München, 2010.

[2] S. M. A. u. H. D.Lauer, "Veränderungen und Entwicklungen in der präklinischen Notfallversorgung: Zentrale Herausforderungen für das Rettungsdienstmanagement," Bundedsgesundheitsblatt-Gesundheitsforschung-Gesundheitsschutz, 2022.

[3] J. Böken, "Neuordnung der Krankenhaus-Landschaft. Eine bessere Versorgung ist nur mit weniger Kliniken möglich," Bertelsmann Stiftung Nr.02, Gütersloh, 2019.

[4] J. O. S.Ahmed, "Assessment of Types of Prototyping in Human-Centered Product Design," Assessment of Types of Prototyping in Human-Centered Product Design. In: Duffy, V.

(eds) Digital Human Modeling. Applications in Health, Safety, Ergonomics, and Risk Management. DHM 2018. Lecture Notes in Computer Science(), vol 10917. Springer, 2018. https://doi.org/10.1007/978-3-319-91397-1_1

[5] Hasso Plattner Institute of Design, "An Introduction in Design Thinking Process Guide," Hasso Plattner Institute of Design, Stanford, 2010.

[6] I. Moerland-Masic, "Advanced Air Mobility: The Cabin of the Future Rescue Helicopters,," AHFE2022, 2022. https://doi.org/10.54941/ahfe1002492

[7] S.Cornelje, "Mixing Realities: Combining Extended and Physical Reality in Co-Creative Design for the Galley of the Flying-V," TU Delft, Delft, Netherlands; Hamburg, Germany, 2023.

[8] Dr.I.Burkett, "Dr.I.Burkett, Co-designing for Social Good: The Role of Citizens in Designing and Delivering Social Services, Part One," University of NSW: Knode.

[9] M. A. T. R. a. M. J.J.Lee, "Design Choices Framework for Co-creation Projects," International Journal of Design, Helsinki/Singepore, 2018.

[10] F. Reimer, "Closing The Loop: The Immersiv and User Centered Co-Design Process for Future Rescue Helicopter Cabin Concepts," DLRK, 2023.

[11] T. u. F.S.Visser, "Lost in Co-X Interpretations of Co-Design and Co-Creation," T.Mattelmäki und F.S.Visser, Lost in Co-X Interpretations of Co-Design and Co-Creation, Aalto University School of Art and Design/Delft University of Technology: Proceedings of IASDR'11, 4th World Conference on Design Research, 2011.

Aeronautics and Astronautics - AIDAA XXVII International Congress Materials Research Forum LLC
Materials Research Proceedings 37 (2023) 763-766 https://doi.org/10.21741/9781644902813-163

Innovative ideas for the use of augmented reality devices in aerodrome control towers

Jürgen Teutsch

Royal Netherlands Aerospace Centre (NLR), Anthony Fokkerweg 2, 1059 CM Amsterdam, Netherlands

juergen.teutsch@nlr.nl

Keywords: Augmented Reality, Air Traffic Control, Aerodrome Control Tower, Attention Guidance, Human Machine Interface, SESAR, NARSIM

Abstract. In recent years Augmented Reality (AR) has become one of the major focus points of user interface development. With the rapidly increasing computing power and developments in software and hardware applications during the last two decades, it has moved from theoretical approaches towards industry-wide application and mass production. The Royal Netherlands Aerospace Centre, NLR, tested several devices in the past, but only recent developments made it possible to effectively use them in an Air Traffic Control (ATC) working environment for Aerodrome Control Towers. In 2021 NLR carried out innovative technology experiments on their high-fidelity real-time air traffic control simulation and validation platform, NARSIM. These experiments were part of the SESAR 2020 project Digital Technologies for Tower (DTT) and focused on advanced HMI interaction modes for aerodrome tower controllers. A proposed Attention Capturing and Guidance concept with an AR device was evaluated inside an aerodrome control tower environment for Amsterdam Airport Schiphol. This paper reflects on the technology development activities that took place at NLR during the last decade and describes the different steps taken to apply the technology in a conventional control tower environment. It is shown that the recent technology developments must be seen as a big step forward in practical application of AR devices for ATC. Furthermore, an outlook into the expected future use of AR devices in conventional control tower environments will be given that goes beyond abovementioned concept elements. This outlook considers additional developments for standardization of digitized airport information and communication between different stakeholders and general performance improvements for AR devices.

Introduction

The technology used for AR combines virtual objects or information generated by a computer with the real world. These computer-generated overlays enhance user perception of the physical environment and, with added sensory input technology, lead to an interactive and immersive user experience. The history of AR devices already dates back several decades, with the technology evolving and advancing over time in the areas of power supplies, display and sound elements, and gyros for orientation. Through this, a major step from a simple display of data towards the inclusion of 3D-images in the real-world view could be achieved [1].

Royal NLR has been actively exploring applications for the use of HMD devices in the aerodrome tower ATC environment for over a decade. Initial research began in 2010 when NLR tested the nVisor ST™ HMD from NVIS in the NARSIM Tower environment, the in-house developed platform for highly realistic real-time simulations of ATC tower operations. The HMD device served as a demonstrator, displaying basic flight strip information that could either be static or change depending on the controller's line of sight.

In 2016, NLR integrated a Google Glass™ device (now known as Glass™) into the NARSIM environment to showcase additional capabilities with improved comfort to a selected group of air

Materials Research Proceedings 37 (2023) 763-766 https://doi.org/10.21741/9781644902813-163

traffic controllers from ATC The Netherlands (LVNL). The device streamed video feeds from remote cameras. These feeds were then displayed on Google Glass and switched automatically based on the user's direction of view, which was continuously tracked. This demonstration aimed to illustrate how tower controllers could gain visibility into apron areas by looking beyond physical obstructions in the line of sight, such as buildings.

In late 2019, with the arrival of the Microsoft HoloLens 2™ in the AR device market, NLR acquired two of these units. Although NLR had previously used HoloLens devices between 2016 and 2019 for various purposes like aircraft maintenance training, simulation debriefings, and projecting simulation results onto aircraft components, the potential of AR devices in the context of air traffic management and control had not yet been explored [2].

Evaluation of an AR Device for Attention Capturing and Guidance

With abovementioned development steps in mind, NLR continued the goal of investigating the possibilities of AR devices for enhancing the effectiveness and efficiency of tower ATC operations. In 2020 NLR joined a consortium for carrying out a project that focused on advanced HMI interaction modes for aerodrome tower controllers. That project was part of SESAR 2020, the second Single European Sky Advanced Research Programme, and was called Digital Technologies for Tower, DTT [3].

While the display of weather-adaptive static information (buildings and outlines) and flight phase-adaptive traffic labels as well as air gestures to interact with the labels and the system was carried out by other partners of the consortium, NLR focused this activity on the evaluation of an Attention Capturing & Guidance (AC&G) concept [4].

The AC&G concept was demonstrated in 2021 based on visual and auditory cues displayed in the AR device to alert and guide controllers in case of critical events. In order to find relevant events that would trigger that process, two existing Schiphol runway controller alerting systems were considered, the Runway Incursion Alerting System (RIAS) and the Go-around Detection System (GARDS). Both systems were previously prototyped on NARSIM and thus available in the simulation environment.

Different cues for each type of event were designed within the HoloLens application with different types of symbols for information display and user guidance. Various shapes and colours were tested, but also different information content. Aircraft labels available from the NARSIM A-SMGCS servers were also visualized inside the HoloLens and were used as attention getters and guidance elements, increasing the SA of the tower controller.

The test programme consisted of different events and combinations of events that happened while two experienced tower controllers carried out routine work in the NARSIM environment for Schiphol airport. Pseudo-pilots were in control of aircraft movements and communicated with the tower controllers. Similar traffic scenarios were used to compare working with and without the HoloLens. Results were gathered in different ways, using questionnaires after each test run and performing debriefings and interviews [5].

Fig. 1: Go-around Detection Alert as Seen through the AR Device

Based on the described evaluation experiment, it was determined that the AC&G operational concept for aerodrome tower controllers using an AR device is feasible and has potential. Although the experiment had a limited operational scope, and feedback was provided for improving certain aspects of the concept, the general outcome was positive. The feedback primarily focused on enhancing the symbology and timing of attention guidance cues. Overall, this result provided us with a solid foundation to further advance and refine the concept moving forward.

Additional Ideas about the Future Use of AR Devices in the Aerodrome Tower

Through discussions with the simulation participants, additional ideas emerged regarding potential future development steps and a vision for tower controller work with AR devices was developed. In the conventional context of improving existing visual operations within tower buildings, the inclusion of AR technology holds significant promise. AR presents the opportunity to enhance the outside view for controllers without the need to add further equipment to the working position and without forcing controllers to look down at the working position to acquire vital operational information. The latter means that head-up time is increased.

The augmentation of relevant information in the outside view including integrated 3D aural cues allows for several attention capturing and guidance capabilities that more effectively improve operations than non-AR technology. Intelligent virtual augmentation could include aspects, such as highlighted stop bars and their statuses, enhanced runway- and taxiway edges, indicated cleared routes for taxi operations, building contours and the outlines of other static obstacles or restricted areas. Such augmentation could be adapted automatically to the current visibility condition.

Traffic labels and their appearance and contents could be adapted to visibility conditions, the role of the controller, or the currently known flight status as well [6]. Taking it one step further, it could be investigated whether adaptation of what is shown or highlighted to support a controller could also take place in terms of the amount of traffic controlled or any signs of stress or high workload situations. This would mean that attention guidance would not only occur in case of a safety-relevant event already taking place, as in the case of safety net alerts being triggered, but much earlier in order to prevent such safety-critical situations in the first place.

Other technology additions may be considered as well, such as air gestures or automatic speech recognition. Air gestures generally would help in terms of system interaction, such as pulling up menus to change various settings, or system input operations, such as the selection of a route or a clearance limit as label information. Speech recognition, while generally meant to improve system input by itself, could be used together with AR to also highlight the labels of pilots calling in or, vice versa, callsigns being addressed by the controller, thus increasing SA and reducing workload once again.

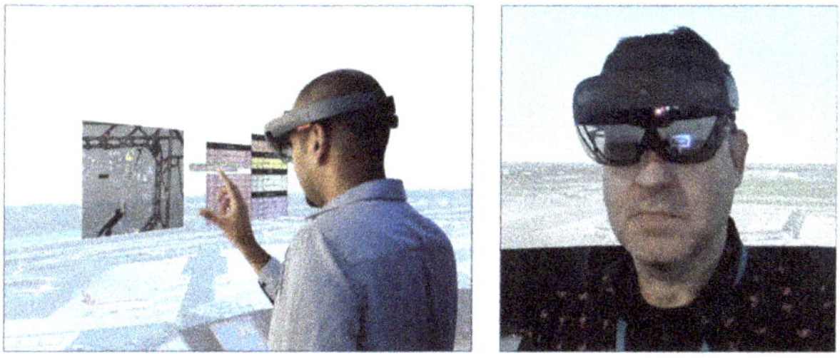

Fig. 2: Innovative Use of AR Devices in the Aerodrome Tower (e.g. EFS, Adaptive Labelling)

Additionally, AR has the potential to redefine the roles and responsibilities of controllers, with the AR system being informed of the sequence of operations and actions required by each individual in the tower and guiding human actions in accordance with the expected procedures.

Naturally, such innovative arrangements would necessitate a high level of automation and a well-defined delegation of authority, particularly during system failures or contingencies. Nonetheless, a potential future milestone could involve a complete redefinition of all existing working arrangements in the pursuit of an optimal operation.

Conclusions

In summary, the potential of using AR devices in the aerodrome tower operational context is very promising but has not fully been investigated yet. Many innovative ideas and functions have not been addressed yet or even been conceived.

For the future, research with an exploratory character should be carried out to show the full potential of the functionalities sketched above, perhaps even from scratch, i.e. without being constrained by current working concepts and organisational structures. The focus of such research should be directed at the main capabilities of AR devices in terms of improvements in safety and operational efficiency, namely to augment controller vision by highlighting operationally relevant elements in an adaptive fashion inserting the most important information directly into the field-of-view. Other functionality exploiting the availability of different sensors (e.g. surveillance and video streams) or adding enhanced prediction capabilities on the basis of Artificial Intelligence [7] could be added to also guide controllers in carrying out operations using different kinds of visual and auditory cues with situation-adaptive information. Last but not least combinations of AR and automated system interaction and input technologies, such as speech recognition with Artificial Intelligence support, should be investigated further [8].

References

[1] R.B. Wood, P.J. Howells, Head-up Display, in: C.R. Spitzer, U. Ferrell, T. Ferrell (Eds.), Digital Avionics Handbook, third edition, CRC Press, Taylor & Francis Group, Boca Raton (FL), 2015, pp. 17-1 to 17-27. https://doi.org/10.1201/b17545

[2] S. Park, S. Bokijonov, Y. Choi, Review of Microsoft HoloLens Applications over the Past Five Years, Applied Sciences, 2021; 11(16):7259. https://doi.org/10.3390/app11167259

[3] CORDIS EU Research Results, PJ05-W2 Digital Technologies for Tower, European Commission Website, Brussels, 2023. https://doi.org/10.3030/874470

[4] J. Teutsch, T.J.J. Bos, G.D.R. Zon, Appendix A - Technological Validation Exercise 001 Report, in: SESAR 2020 PJ.05-W2 Sol. 97.1 and Sol. 97.2 TVALR, TRL4 Data Pack, Deliverable D3.1.050, SESAR-JU, Brussels, 2023. https://doi.org/10.3030/874470

[5] J. Teutsch, T.J.J. Bos, M.C. van Apeldoorn and L. Camara, Attention Guidance for Tower ATC Using Augmented Reality Devices, 2022 Integrated Communication, Navigation and Surveillance Conference (ICNS), Dulles/Herndon, VA, USA, 2022, pp. 1-12. https://doi.org/10.1109/ICNS54818.2022.9771479

[6] R. Santarelli, S. Bagassi, J. Teutsch, R. Garcia Lasheras et al., Towards a digital control tower: the use of augmented reality tools to innovate interaction modes, SESAR Innovation Days (SID) 2022, SESAR-JU, Brussels, 2022

[7] P. Ortner, R. Steinhöfler, E. Leitgeb and H. Flühr, Augmented Air Traffic Control System - Artificial Intelligence as Digital Assistance System to Predict Air Traffic Conflicts, AI Journal, MDPI, 2022, 3(3), 623-644. https://doi.org/10.3390/ai3030036

[8] European Aviation Artificial Intelligence High Level Group, The Fly AI Report - Demystifying and Accelerating AI in Aviation/ATM, EUROCONTROL, Brussels, 2020.

ADS-B driven implementation of an augmented reality airport control tower platform

Tommaso Fadda[1,a*], Sara Bagassi[1,b] and Marzia Corsi[1,c]

[1]Department of Industrial Engineering, University of Bologna, Forlì, Italy

[a]tommaso.fadda2@unibo.it, [b]sara.bagassi@unibo.it, [c]marzia.corsi2@unibo.it

Keywords: Augmented Reality, Air Traffic Control, ADS-B, Human Factors

Abstract. This paper describes a real-world implementation of the solutions developed within the SESAR DTT Solution 97.1-EXE-002 project, which tested in a simulated scenario the use of Augmented Reality (AR) to assist the airport control tower operators (TWR). Following a user-centred design methodology, the requirements of a real-world live AR platform join with design concepts validated in previous projects, namely the Tracking Labels, the weather interface, and a low-visibility overlay, all used to increase the TWR situational awareness, performance and reactivity while reducing the workload. The designed AR platform performs the live tracking and visualization of real aircraft and surveillance information in the airport traffic zone. It bases on three key processes: the transmission of an ADS-B data flow to a Microsoft™ HoloLens2, the registration process of the AR platform, and the rendering of a real-time tracking system and other surveillance overlays. The concept has been first validated with the help of a TWR, preceding a technical validation to ensure the repeatability and reproducibility of the results. The results allow for defining new guidelines for the deployment in a control tower environment.

Introduction

Being an airport control tower operator (TWR) implies overseeing the aircraft in the airport manoeuvring area and departing and arriving traffic, relying on the out-of-the-window (OTW) tower's view to provide separation and clearances. In this high-risk, high-concentration and time-critical job, controllers are concerned with keeping a smooth traffic flow while ensuring the overall safety of airport operations, with the sum of these tasks resulting in a heavy workload.

While performances are a priority in a worldwide ever-growing traffic scenario, the main upgrades in TWRs' job were due to the addition of visual interfaces, which are better suited to ensure an increase in safety, with a collateral increase in cognitive workload due also to the continuous shift of focus between the outside view (head-up position) and the head-down interfaces at the work position providing surveillance and traffic information [1].

In this scenario, Augmented/Mixed Reality (AR) is suitable for helping the TWR [2], moving the surveillance information from the head-down interfaces to a collinear vision within the OTW airport traffic, promising to solve the chronic safety over performance compromise. At the University of Bologna, Solution 97.1-EXE-002 of the SESAR's funded "Digital Technologies for Tower" (DTT) project [3,4,5] tested this possibility, using a system of virtual tracking labels to pinpoint each aircraft in the user's view, providing context-related flight information, tailored to different control roles (ground vs runway) and environmental conditions. TWRs could then keep the focus on the live traffic, with increased situational awareness and reduced workload, while improving the overall safety and efficiency (temporal and thus economical) of airport operations in every condition, especially high-traffic and low-visibility.

These studies were conducted in a simulated real-time airport scenario, focusing on developing the overall AR concept in a safe, non-critical, and fully controllable environment. This paper describes the subsequent implementation of the developed concepts into the real world, resulting in a live application tracking operative aircraft and acquiring their surveillance information. The research addressed the tasks needed when dealing with a real-time physical world application. A new set of requirements have been defined, dealing with a global-coordinates registration process for AR and using live ADS-B data to retrieve the desired traffic information, as will be later described.

Methods

User-centred design (UCD) method is the obvious choice when dealing with a tricky task such as airport tower control. The design process is iterative, always considering the user needs before, during and after the development, ensuring that the final concept helps the TWR without relevant contraindications. This research aimed to increase the maturity of the solutions developed in a simulated environment, performing the critical steps to move up from a previous level 4 of the technology readiness scale, by transferring the concept into a relevant environment. For this purpose, some requirements were identified, joining outcomes of the lab validation campaign with new requirements for a real-world implementation. Table 1 summarizes all the requirements to implement and validate a real-time AR platform for the control tower.

Table 1. Summary of the design requirements

Solutions developed in previous projects	Requirements for real-world implementation
Use of Microsoft HoloLens2 as AR device	Real-time acquisition of accurate aircraft position
AR application developed with Unity3D	Accurate real-time tracking of aircraft in AR layer
Registration strategy using MRTK WLT	Management of multiple aircraft at the same time
Design of the tracking labels	Render tracking holograms minimizing latency
Definition of a runway overlay	Evaluate and counteract possible error sources
Design of the weather interface	Ensure reliability and precision of tracking system
Use of adaptive human-machine interface	Implement additional surveillance overlays
Ergonomic aspects	Validate

Design of the AR-based control tower platform

Starting from the requirements, a preliminary concept has been defined, using an ADS-B receiver to detect the aircraft in real-time and retransmitting the data stream to a HoloLens2 device using a user datagram protocol (UDP). The AR application in the device processes the data, identifying aircraft position and rendering the tracking labels. Figure 1 shows the architecture of the platform.

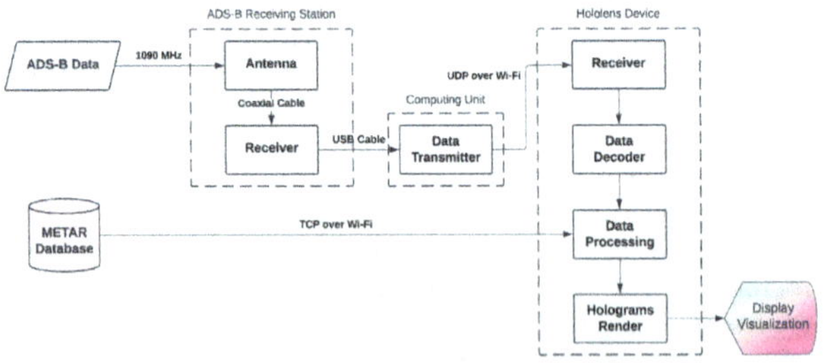

Figure 1. Architecture of the AR platform

ADS-B data flow. The vector state information is acquired from ADS-B data, which is constantly sent in real-time by every commercial aircraft. These data are read on a pc unit over serial communication and retransmitted over a UDP to the HoloLens, where they are used to constantly track the aircraft in the virtual world and provide surveillance information.

Calibration. A key aspect of AR applications is the registration process, meaning the continuous alignment of the virtual overlay over the intended elements in the physical world. Most AR technologies rely on a tracking system which can identify the user's position inside an environment and project the holographic overlay accordingly, with a real-time adaptation of the virtual elements' alignment over the physical world. This procedure is suited to position holograms with respect to a local frame of reference. Since aircraft positions are in global (geodetic) coordinates, a calibration procedure is required for the registration to occur. HoloLens2's applications can only fix the position of a virtual reference frame inside a room. Thus, a procedure was developed to identify the global position and orientation of this reference frame and then convert the relative position of aircraft with respect to the reference frame into cartesian coordinates using a conversion algorithm from geodetic coordinates to an East-North-Up (ENU) frame. Figure 2 shows the alignment of the reference frames and how the HoloLens 2 reference frame was positioned inside the environment. The virtual reference frame geodetic position is identified, and then the aircraft position is converted for an ENU frame with an origin coinciding with the virtual one. Finally, a rotation is applied to convert ENU coordinates into those of the virtual frame.

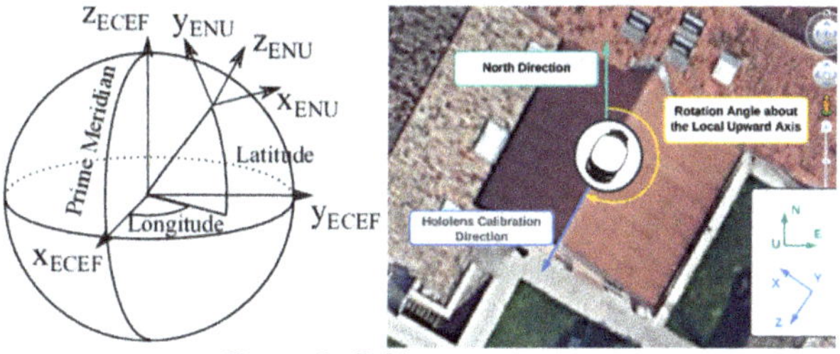

Figure 2. Calibration steps

Correction of position information. Some corrections had to be applied to the aircraft position data to obtain an adequate matching of the holograms. In particular, the altitude data comes from the pressure altimeter, with the uncorrected barometric altitude (H_{QNE}) - computed by the onboard barometer with respect to the standard sea-level value of 1013.25hPa - transmitted for collision avoidance purposes. The near-ground effect is also an issue at landings. Thus, the altitude was corrected considering the current sea-level pressure at the airport location (p_{QNH}), and a corrective term for landing gear height and ground effect (p_{LG}). The correction formula reports the height over ground (H_{QFE}) as the uncorrected altitude plus a corrective term for the different reference pressure, times a conversion term for the variation of pressure with altitude:

$$H_{QFE}(m) = H_{QNE}(m) + \left(p_{QNH} - (1013.25hPa - p_{LG})\right) x\, 8{,}23 m/hPa \quad (1)$$

Labels implementation. The tracking label design was retrieved from that already validated in the EXE-002 of Solution 97.1. The labels contain the aircraft height over ground (H_{QFE}), velocity, heading and callsign, with other information already computable or available, such as the distance from any point of the airport, the vertical rate, and H_{QNE}. The labels are positioned at a constant focal distance from the user, with a small sphere pinpointing the aircraft and a surmounting canvas containing the aircraft data. Colour coding is used for landing and departing traffic.

Real-time labels management. The whole purpose of the application is to track each detected aircraft keeping its position and information updated in real-time. The data processing chain starts with the ADS-B data being sent to the AR device as soon as they are detected. On the HoloLens2, the data are decoded and organized into a database containing the most updated information for each aircraft, which is then used to calculate the holograms' position, update the label content, and render the holograms in a seamless cyclic loop run at least at 60Hz, which is the rendering frame frequency. Being HoloLens2 capable of tracking the user's position relative to the virtual reference frame, the holograms stay in place while both aircraft and users move.

Additional overlays. The application was completed with additional overlays helping the controller, in particular a weather interface, using real data from METAR, and a runway overlay for low-visibility conditions, triggered by the visibility distance indicated by the METAR.

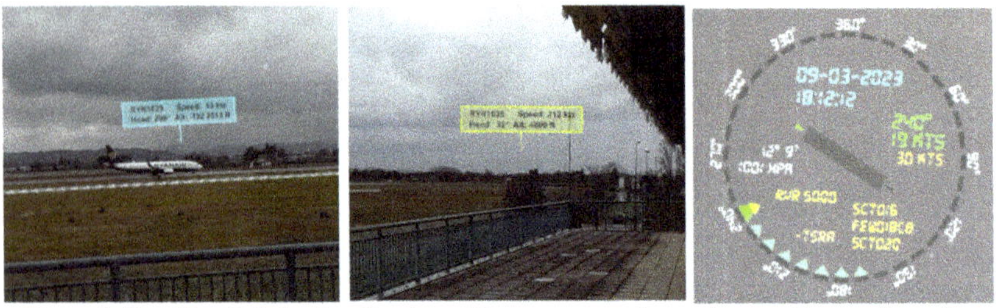

Figure 3. Tracking Labels visualization with runway overlay, colour coding, and weather interface

Technical validation. As part of the UCD, a technical validation was needed to ensure the goodness of the design. After that, the fulfilment of all the requirements was verified in different environments to verify the repeatability and reproducibility of the configuration.

Conclusion

This paper presented an AR control tower application working in a real-world scenario. Real-time data were used, and the matching of real and virtual worlds was addressed.

Future development. Starting from this concept, it will be possible to improve the application by testing it in shadow mode in a control tower, while integrating other surveillance information and solutions coming from the SESAR programme.

References

[1] Pinska, E., "An investigation of the head-up time at tower and ground control positions", Eurocontrol (2006).

[2] R. Reisman, D. Brown, "Design of Augmented Reality Tools for Air Traffic Control Towers", 6th AIAA Aviation Technology, Integration and Operations Conference – ATIO (2006). https://doi.org/10.2514/6.2006-7713

[3] S. Bagassi, F. De Crescenzio, S. Piastra, C.A. Persiani, M. Ellejmi, A.R. Groskreutz, J. Higuera, "Human-in-the-loop evaluation of an augmented reality based interface for the airport control tower", in *Computers in Industry*, Volume 123, article number 103291 (2020). https://doi.org/10.1016/j.compind.2020.103291

[4] R. Santarelli, S. Bagassi, M. Corsi, J. Teutsch, R. Garcia Lasheras, M.A. Amaro Carmona, A.R. Groskreutz, "Towards a digital control tower: the use of augmented reality tools to innovate interaction modes", 12th SESAR Innovation Days, Budapest, Hungary, 5-8 December 2022.

[5] Digital technologies for tower, https://www.sesarju.eu/projects/dtt. (2023).

Maturity-based taxonomy of extended reality technologies in aircraft lifecycle

Sara Bagassi[1,a], Marzia Corsi[1,b*], Francesca De Crescenzio[1,c], Martino Carlo Moruzzi[1,d], Sandhya Santhosh[1,e]

[1]Department of Industrial Engineering, University of Bologna, Forlì, Italy

[a]sara.bagassi@unibo.it, [b]marzia.corsi2@unibo.it, [c]francesca.decrescenzio@unibo.it, [d]martinocarlo.moruzz2@unibo.it, [e]sandhya.santhosh2@unibo.it

Keywords: Aircraft Lifecycle, Extended Reality, Aircraft Design, Aircraft Operations

Abstract. EXtended Reality (XR) is a fast growing and rapidly evolving technology. In the aeronautical sector, XR can be exploited for the entire aircraft lifecycle, however, different levels of maturity can be identified for applications in each one of the lifecycle's phases. This paper, by outlining the TRL of current XR applications over the aircraft lifecycle, aims to be a foundation to identify the possible future improvements and applications of immersive technologies in the aeronautical sector.

Introduction

EXtended Reality (XR) is a fast growing and rapidly evolving technology, a comprehensive term given to all computer-generated environments that either merge the physical and the virtual worlds or create an entirely immersive experience for the user. Even if XR technologies such as Virtual (VR) and Augmented Reality (AR) have already reached maturity as they are considered mainstream for a few domains of application, in the aeronautical sector this is not yet entirely true. XR can be exploited for the entire aircraft lifecycle, however, different levels of maturity can be identified for applications in each one of the lifecycle's phases (Figure 1).

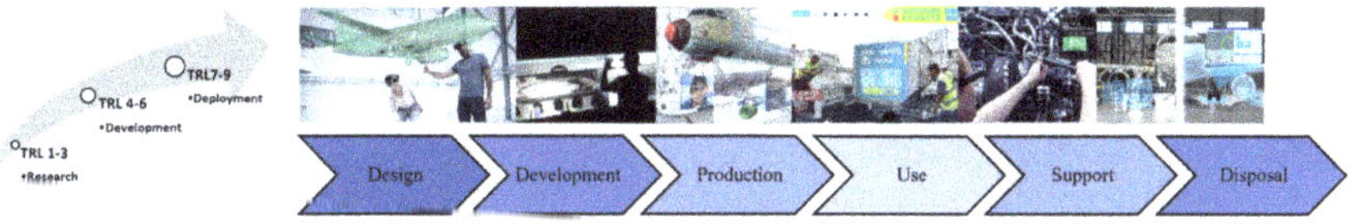

Figure 1: Aircraft lifecycle's phases and Technology Readiness Level (TRL)

This paper, by outlining the TRL of current XR applications over the aircraft lifecycle, aims to be a foundation to identify the possible future improvements and applications of immersive technologies in the aeronautical sector.

Extended Reality in Aircraft Lifecycle

Extended Reality technologies can find a number of applications in aeronautics. Synthetic visualisation and multimodal interaction tools target almost all aircraft lifecycle phases. From design to manufacturing and maintenance, passing through training and operations [1] and since the early stages of industrialisation, the main aeronautical companies are relying on those technologies [2].

Figure 2: Industry-first concept based on Mixed Reality technologies for Cabin Customisation and Design for the A320 Family (courtesy of Airbus[1]).

In the design and development phases, rather mature applications are established in the field of synthetic visualization which allow for a better understanding of the performance achieved by a certain design solution[2]. Furthermore, XR has proven to be beneficial for collaborative design, design reviews and user assessment throughout the product development process [3]. This application is still in the early stages of research, begin now developed as a proof of concept [4]. One of the biggest opportunities provided by XR is the one of including user feedback in the design process, as proved in a study conducted to virtually test the comfort of newly conceived aircraft cabin designs. Not only the project exploited V/AR technology to validate the proposed design approach and strategy in the design of aircraft cabins, but it demonstrated how XR can be a useful tool, starting from the very early stages, to the entire process which results to be more open and flexible, time efficient and cost-effective [5]. As a matter of fact, early stages of design, when prototypes are still in the digital phase, are ideal for leveraging eXtended Reality technologies. Recent research[3,4] [6], makes use of advanced VR technologies to enable the different stakeholders to experience and be immersed in the full scale design, allowing them to be involved in the process starting from the Inspiration and Ideation Phase [7-8].

In view of the production phase, a number of applications are developed within the framework of Industry 4.0 which aim at exploiting XR to make aircraft systems production and final assembly more efficient in terms of time, quality and dependability. In particular, if VR is mainly applied during the design phase, AR is mostly used in the production line. The usage of synthetic visualisation tools leading to early detection of non-conformities allows not only to avoid expensive corrections but also to improve the quality control process and reduce inspection time. Nevertheless, users' familiarity and training with the specific devices and technology are crucial aspects and must be considered to achieve the best possible results [9].

Airbus, among other companies such as Boeing and Embraer, foresaw the potential and applicability of digital technologies since the early 2000s, and in 2009 the European company collaborated to create an AR solution (*MiRA - Mixed Reality Application*) consisting of a tablet PC and related sensor pack to scan parts and detect errors. On the A380 MiRA allowed to check 80.000 brackets in the fuselage reducing the time needed from weeks to hours and the late discovery of mispositioned or damaged brackets by 40% [10]. Currently, Airbus is industrializing a first application to be deployed across various sites either to help with the drilling and fitting of brackets or to support quality control activities[1].

[1] https://www.airbus.com/en/newsroom/stories/2023-06-mixed-reality-to-meet-future-challenges
[2] https://www.holoforge.io/en/project/safran-marketing/
[3] https://www.dlr.de/as/en/desktopdefault.aspx/tabid-18135/28809_read-74794/
[4] https://www.dlr.de/en/images/2020/1/indicad-fuselage-designs-for-the-future

AR devices such as head-mounted displays, can be used not only to assist the assembly procedure, but also for communication purposes in human-autonomy system teams as proved by a recent research in which the hybrid team has to rivet stringers onto a fuselage [11]. Hazardous and physically demanding tasks were assigned to the mobile robotic system, while the human workers performed tasks requiring experience, knowledge and multi-sensory sensitivity.

The operational environment is benefiting a lot from the application of XR, both in training and real-life operations. One of the first applications of Augmented Reality is the cockpit head-up display (HUD), which enables the pilot to get synthetic information collimated with the real view (Figure 3). This is the only optical-based XR application that already reached complete maturity in the aeronautical sector.

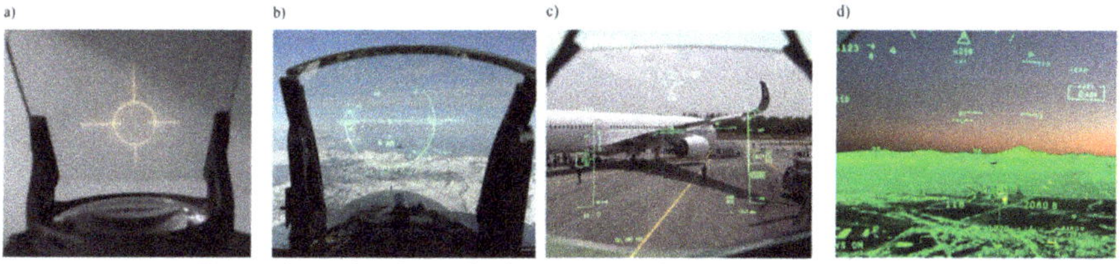

Figure 3: Technological advancement of the cockpit HUD [14].

More recently, different research projects explored a similar operational concept applied to airport control towers [12-13] reaching the development phase. The projects demonstrated how XR technologies in control towers are beneficial for the situational awareness of the controllers especially when the aerodrome is affected by low visibility conditions. The capability to make degraded visual information visible during low visibility and low-altitude operations given by XR can also enhance spatial awareness and safety during low visibility helicopters or ground operations [14-16]. Furthermore, ground operations are good candidates for technical solutions that make use of XR technologies.

Moving on to the support phase, a few quite mature applications of XR in maintenance, [17] inspection and repair operations are possible to identify. In 2015, Boeing adopted AR technology to reduce the impact of assembly harness errors. In 2017, General Electric used AR technology to communicate to mechanics whether they are properly tightening and sealing jet engine B-Nuts, and the working efficiency increased by 8-12% [18]. Maintenance is of paramount importance in the aviation industry therefore inspection and maintenance personnel have to be trained with appropriate tools. Training methods need to keep up with advanced technologies while ensuring safety and cost-effectiveness. Exploiting eXtended Reality as a training tool can overcome this issue providing a safe learning environment drastically reducing the risks that may occur during training with reduced costs, and improving labor efficiency [19].

The presented immersive technologies solutions are only a few of the numerous ones already available in the aeronautical sector. The proposed preliminary assessment of maturity-based taxonomy of XR technologies in aircraft lifecycle (Table 1) aims to serve as a base to identify the future possible improvements and applications of these technologies in the aircraft lifecycle. Furthermore, serves as a useful tool in understanding and addressing the number of challenges, ranging from technology to human factors, the aeronautical sector will face while maturing XR solutions in specific phases of aircraft lifecycle.

Table 1: Preliminary taxonomy assessment of XR technologies in aircraft lifecycle.

Lifecycle Phase		Taxonomy assessment		Reference
Design & Development	Scientific visualisation	Development	TRL 4-6	^2
	Collaborative design	Research	TRL 1-3	[3-4]
	Human-centered design	Research	TRL 1-3	[5-8] - ^3-4
Production	Assembly	Deployment Development	TRL 7-9 TRL 4-6	[9-10] ^1
	Human-autonomy teaming	Research	TRL 1-3	[11]
Operation	Cockpit HUD	Deployment	TRL 7-9	[14]
	Airport Control Tower	Research	TRL 3	[12-13]
	Ground Operations	Development	TRL 4-6	[14-16]
Support	Maintenance	Deployment	TRL 7-9	[17-19]
	Digital Twin interaction	Research	TRL 1-3	[20]

Conclusions

As a matter of fact, most of the presented solutions may reach maturity in the next 20 years. The vision is that, in the future, XR will be an extension of the human senses necessary to understand, interact, control, and command complex systems belonging to the digital era.

In the context of aeronautics, a more advanced level of development would regard XR as a means to visualize and interact with digital twins throughout the entire aircraft lifecycle [20]. EXtended Reality will be the technical enabler to interact with such systems that will be more autonomous and enhanced by artificial intelligence. Nevertheless, a number of challenges and barriers are still limiting the evolution of such technology in aeronautics, making its development slower than expected. XR still relies on cumbersome equipment, previous attempts to overcome this obstacle have proven unsuccessful. Incremental improvements are being made to enhance the wearability of hardware. Additionally, more automated solutions are needed to fully exploit the potential of such technology. Whilst a few Mixed Reality devices are already mainstream, non-intrusive solutions are not mature enough. The cost of equipment along with some big company trends are also acting as barriers to the deployment of business solutions based on XR technology. Although Industry 4.0 was expected to boost XR growth, it has also introduced some standardization issues in a number of application domains. Lastly, human factors and user acceptance are considered among the main risks for eXtended Reality massive use in safety-critical environments.

A maturity-based taxonomy of immersive technologies aims to be a tool to understand and overcome the number of challenges, ranging from technology to human factors, the aeronautical sector will face while maturing XR solutions in specific phases of aircraft lifecycle.

References

[1] Neretin, E. S., Kolokolnikov, P. A., & Mitrofanov, S. Y. (2021, June). Prospect for the application of augmented and virtual reality technologies in the design, production, operation of aircraft and training of aviation personnel. In *Journal of Physics: Conference Series* (Vol. 1958, No. 1, p. 012030). IOP Publishing. https://doi.org/10.1088/1742-6596/1958/1/012030

[2] Frigo, M. A., da Silva, E. C., & Barbosa, G. F. (2016). Augmented reality in aerospace manufacturing: A review. *Journal of Industrial and Intelligent Information*, *4*(2). https://doi.org/10.18178/jiii.4.2.125-130

[3] S. Santhosh, and F. De Crescenzio, 2022. A Mixed Reality Application for Collaborative and Interactive Design Review and Usability Studies. In Advances on Mechanics, Design Engineering and Manufacturing IV: Proceedings of the International Joint Conference on Mechanics, Design Engineering & Advanced Manufacturing, JCM 2022, June 1-3, 2022, Ischia, Italy (pp. 1505-1515). Cham: Springer International Publishing. https://doi.org/10.1007/978-3-031-15928-2_131

[4] M. Fuchs, F. Beckert, J. Biedermann and B. Nagel. Collaborative knowledge-based method for the interactive development of cabin systems in virtual reality. Computers in Industry, Vol. 136, 103590, ISSN 0166-3615, 2022. https://doi.org/10.1016/j.compind.2021.103590

[5] F. De Crescenzio, S. Bagassi, S. Asfaux, N. Lawson (2019). Human centred design and evaluation of cabin interiors for business jet aircraft in virtual reality. International Journal on Interactive Design and Manufacturing (IJIDeM), 13, 761-772. https://doi.org/10.1007/s12008-019-00565-8

[6] I. Moerland-Masic, F. Reimer, T. M. Bock, F. Meller, & B. Nagel, (2021). Application of VR technology in the aircraft cabin design process. *CEAS Aeronautical Journal*, 1-10. https://doi.org/10.1007/s13272-021-00559-x

[7] Schuchardt, B. I., Becker, D., Becker, R. G., End, A., Gerz, T., Meller, F., ... & Zhu, C. (2021). Urban air mobility research at the DLR German Aerospace Center–Getting the HorizonUAM project started. In *AIAA Aviation 2021 Forum* (p. 3197). https://doi.org/10.2514/6.2021-3197

[8] Schuchardt, B. I. (2021, September). HorizonUAM Project Overview. In *HorizonUAM Symposium 2021*.

[9] de Souza Cardoso, L. F., Mariano, F. C. M. Q., & Zorzal, E. R. (2020). Mobile augmented reality to support fuselage assembly. *Computers & Industrial Engineering*, *148*, 106712. https://doi.org/10.1016/j.cie.2020.106712

[10] Glockner, H., Jannek, K., Mahn, J., & Theis, B. (2014). Augmented reality in logistics: Changing the way we see logistics-a DHL perspective. DHL Customer Solutions & Innovation, 28.

[11] Luxenburger, A., Mohr, J., Spieldenner, T., Merkel, D., Espinosa, F., Schwartz, T., ... & Stoyke, M. (2019, December). Augmented reality for human-robot cooperation in aircraft assembly. In 2019 IEEE International Conference on Artificial Intelligence and Virtual Reality (AIVR) (pp. 263-2633). IEEE. https://doi.org/10.1109/AIVR46125.2019.00061

[12] S. Bagassi, F. De Crescenzio, S. Piastra, C. A. Persiani, M. Ellejmi, A. R. Groskreutz, J. Higuera, Human-in-the-loop evaluation of an augmented reality based interface for the airport control tower, Computers in Industry 123 (2020) 103291. https://doi.org/10.1016/j.compind.2020.103291

[13] R. Santarelli, S. Bagassi, M.Corsi, J. Teutsch, R. Garcia Lasheras, M. Angel Amaro Carmona, A. R. Groskreutz, Towards a digital control tower: the use of augmented reality tools to innovate interaction modes, in: Sesar Innovation Days 2022, 2022.

[14] Stanton, N. A., Plant, K. L., Roberts, A. P., & Allison, C. K. (2019). Use of Highways in the Sky and a virtual pad for landing Head Up Display symbology to enable improved helicopter pilots situation awareness and workload in degraded visual conditions. *Ergonomics*, *62*(2), 255-267. https://doi.org/10.1080/00140139.2017.1414301

[15] Blundell, J., Collins, C., Sears, R., Plioutsias, A., Huddlestone, J., Harris, D., ... & Lamb, P. (2022). Low-visibility commercial ground operations: An objective and subjective evaluation of a multimodal display. *The Aeronautical Journal*, In-press. https://doi.org/10.1017/aer.2022.81

[16] Blundell, J., & Harris, D. (2023). Designing augmented reality for future commercial aviation: a user-requirement analysis with commercial aviation pilots. *Virtual Reality*, 1-15. https://doi.org/10.1007/s10055-023-00798-9

[17] H. Eschen, T. Kötter, R. Rodeck, M. Harnisch, T. Schüppstuhl, (2018). Augmented and virtual reality for inspection and maintenance processes in the aviation industry. Procedia manufacturing, 19, 156-163. https://doi.org/10.1016/j.promfg.2018.01.022

[18] Hongli, S., Qingmiao, W., Weixuan, Y., Yuan, L., Yihui, C., & Hongchao, W. (2021). Application of AR technology in aircraft maintenance manual. In *Journal of Physics: Conference Series* (Vol. 1738, No. 1, p. 012133). IOP Publishing. https://doi.org/10.1088/1742-6596/1738/1/012133

[19] Wu, W. C., & Vu, V. H. (2022). Application of Virtual Reality Method in Aircraft Maintenance Service—Taking Dornier 228 as an Example. Applied Sciences, 12(14), 7283. https://doi.org/10.3390/app12147283

[20] Albuquerque, G., Fischer, P. M., Azeem, S. M., Bernstein, A. C., Utzig, S., & Gerndt, A. (2023). Digital Twins as Foundation for Augmented Reality Applications in Aerospace. In *Springer Handbook of Augmented Reality* (pp. 881-900). Cham: Springer International Publishing. https://doi.org/10.1007/978-3-030-67822-7_35

Keyword Index

j

www.ingramcontent.com/pod-product-compliance
Lightning Source LLC
Chambersburg PA
CBHW071312210326
41597CB00015B/1205